Modern Engineering Mathematics

Third Edition

We work with leading authors to develop the
strongest educational materials in mathematics,
bringing cutting-edge thinking and best learning
practice to a global market.

Under a range of well-known imprints, including
Prentice Hall, we craft high quality print and
electronic publications which help readers to
understand and apply their content,
whether studying or at work.

To find out more about the complete range of our
publishing please visit us on the World Wide Web at:
www.pearsoneduc.com

Modern
Engineering
Mathematics

Third Edition

Glyn James	*Coventry University*
and	
David Burley	*University of Sheffield*
Dick Clements	*University of Bristol*
Phil Dyke	*University of Plymouth*
John Searl	*University of Edinburgh*
Jerry Wright	*AT&T Shannon Laboratory*

An imprint of **Pearson Education**

Harlow, England · London · New York · Reading, Massachusetts · San Francisco · Toronto · Don Mills, Ontario · Sydney
Tokyo · Singapore · Hong Kong · Seoul · Taipei · Cape Town · Madrid · Mexico City · Amsterdam · Munich · Paris · Milan

Pearson Education Limited
Edinburgh Gate
Harlow
Essex CM20 2JE
England

and Associated Companies throughout the world

Visit us on the World Wide Web at:
www.pearsoneduc.com

First published 1992
Second edition 1996
Third edition 2001

ISBN 0 130 18319 9

British Library Cataloguing-in-Publication Data
A catalogue record for this book can be obtained from the British Library.

Library of Congress Cataloging-in-Publication Data

Modern engineering mathematics / Glyn James . . . [et al. – 3rd ed.
    p. cm.
  First-2nd eds. under James.
  Includes index.
  ISBN 0-13-018319-9 (pbk.)
    1. Engineering mathematics.   I. James, Glyn. II. James, Glyn. Modern engineering
mathematics.

  TA330 .J36 2001
  510′.2462–dc21

                                00-050135

10 9 8 7 6 5
06 05 04 03

Typeset by 35
Printed and bound in Italy by G.Canale & C. S.p.A.

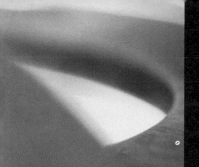

Contents

Chapter 12　Introduction to Fourier Series　811

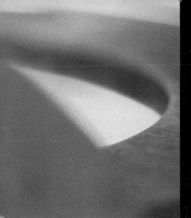

Preface

As with the previous editions, the range of material covered in this third edition is regarded as appropriate for a first level core studies course in mathematics for undergraduate courses in all engineering disciplines. Whilst designed primarily for use by engineering students it is believed that the book is also highly suitable for students of the physical sciences and applied mathematics. Additional material appropriate for second level undergraduate core studies, or possibly elective studies for some engineering disciplines, is contained in the companion text *Advanced Modern Engineering Mathematics*.

The objective of the authoring team remains that of achieving a balance between the development of understanding and the mastering of solution techniques with the emphasis being on the development of the student's ability to use mathematics with understanding to solve engineering problems. Consequently, the book is not a collection of recipes and techniques designed to teach students to solve routine exercises, nor is mathematical rigour introduced for its own sake. To achieve the desired objective the text contains:

- Worked examples
  Approximately 400 worked examples, many of which incorporate mathematical models and are designed both to provide relevance and to reinforce the role of mathematics in various branches of engineering. In response to feedback from users, additional worked examples have been incorporated within this revised edition.
- Applications
  To provide further exposure to the use of mathematical models in engineering practice, each chapter contains sections on engineering applications. These sections form an ideal framework for individual, or group, case study assignments leading to a written report and/or oral presentation; thereby helping to develop the skills of mathematical modelling necessary to prepare for the more open-ended modelling exercises at a later stage of the course.
- Exercises
  There are numerous exercise sections throughout the text and at the end of each chapter there is a comprehensive set of review exercises. While many of the exercise problems are designed to develop skills in mathematical techniques, others are designed to develop understanding and to encourage learning by doing, and some are of an open-ended nature. This book contains over 1000 exercises and answers to all the questions are given. It is hoped that this provision, together with the large number of worked examples and style of presentation,

also makes the book suitable for private or directed study. Again in response to feedback from users, the frequency of exercises sections has been revised. The numbering of questions within each section has also been changed so that questions are now ordered according to level of difficulty. To improve the range of exercises additional questions have been added to many of the sections.

● Numerical methods
Recognizing the increasing use of numerical methods in engineering practice, which often complement the use of analytical methods in analysis and design and are of ultimate relevance when solving complex engineering problems, there is wide agreement that they should be integrated within the mathematics curriculum. Consequently the treatment of numerical methods is integrated within the analytical work throughout the book. Algorithms are written in pseudocode and are, therefore, readily transferable to any specific programming language by the user.

Much of the feedback from users relates to the role and use of software packages, particularly symbolic algebra packages. Whilst fully recognizing both the importance and role of such packages in the teaching/learning process the authoring team continues to be of the opinion that it is inappropriate to restrict the text to a limited subset. Without making it an essential requirement the authors have attempted to highlight throughout the text situations where the user could make effective use of a software package. This also applies to exercises and, indeed, a limited number of exercises have been introduced for which the use of such a package is essential. Whilst any appropriate piece of software can be used, the authors recommend the use of MATLAB or MAPLE, and these are referenced to within the text. Throughout the text two icons are used:

● An open screen 🖥 indicates that use of a software package would be useful (e.g. for checking solutions) but not essential

● A closed screen 🖥 indicates that the use of a software package is essential or highly desirable.

Regarding content, the absence in the second edition of the section on eigenvalues/eigenvectors and the introductory chapter on Fourier series included in the first edition, was seen as regrettable by many users. Consequently these have been reintroduced into this third edition. Apart from these two topics, feedback indicated that content and level is viewed as satisfactory, so there has been no other significant change in subject content. Rather, emphasis has been placed on updating and improving the presentation of the material throughout the text, including the use of a second colour.

Available with this text is a CD-based testing and assessment package, with interactive multi-choice, multi-answer and hot-spot questions that allow students to test their understanding of the key topics. Ideal for reinforcing learning during the course or pre-examination revision, all questions will provide detailed student feedback on-screen, and also direct students back to the relevant section or page in the text for further study.

A comprehensive Solutions Manual is obtainable free of charge to lecturers using this textbook. It will also be available for download via the Web at www.booksites.net/james.

Acknowledgements

The authoring team is extremely grateful to all the reviewers and users of the text who have provided valuable comments on previous editions of this book. Most of this has been highly constructive and very much appreciated. The team has continued to enjoy the full support of a very enthusiastic production team at Pearson Education and wishes to thank all those concerned. Finally I would like to thank my wife, Dolan, for her full support throughout the preparation of this text and its previous editions.

Glyn James
Coventry
June 2000

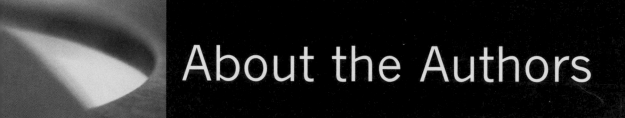

About the Authors

Glyn James has recently retired as Dean of the School of Mathematical and Information Sciences at Coventry University and is now Emeritus Professor in Mathematics at the University. He graduated from the University College of Wales, Cardiff in the late 1950s, obtaining first class honours degrees in both Mathematics and Chemistry. He obtained a PhD in Engineering Science in 1971 as an external student of the University of Warwick. He has been employed at Coventry since 1964 and held the position of the Head of Mathematics Department prior to his appointment as Dean in 1992. His research interests are in control theory and its applications to industrial problems. He also has a keen interest in mathematical education, particularly in relation to the teaching of engineering mathematics and mathematical modelling. He was co-chairman of the European Mathematics Working Group established by the European Society for Engineering Education (SEFI) in 1982, a past chairman of the Education Committee of the Institute of Mathematics and its Applications (IMA), and a member of the Royal Society Mathematics Education Subcommittee. In 1995 he was chairman of the Working Group that produced the report 'Mathematics Matters in Engineering' on behalf of the professional bodies in engineering and mathematics within the UK. He is also a member of the editorial/advisory board of three mathematical education journals. He has published numerous papers and is co-editor of five books on various aspects of mathematical modelling. He is a former member of the Council and past Vice-President of the IMA and has also served a period as Honorary Secretary of the Institute. He is a Chartered Mathematician and a Fellow of the IMA.

David Burley has recently retired from the University of Sheffield. He graduated in mathematics from King's College, University of London in 1955 and obtained his PhD in mathematical physics. After working in the University of Glasgow, he spent most of his academic career in the University of Sheffield, being Head of Department for six years. He has long experience of teaching engineering students and has been particularly interested in encouraging students to construct mathematical models in physical and biological contexts to enhance their learning. His research work has ranged through statistical mechanics, optimization and fluid mechanics. Current interests involve the flow of molten glass in a variety of situations and the application of results in the glass industry.

Dick Clements is Reader in the Department of Engineering Mathematics at Bristol University. He read for the Mathematical Tripos at Christ's College, Cambridge in the late 1960s. He went on to take a PGCE at Leicester University School of Education before returning to Cambridge to research a PhD in Aeronautical Engineering. In 1973 he was appointed Lecturer in Engineering Mathematics at Bristol University and has taught mathematics to engineering students ever since. He has undertaken research in a wide range of engineering topics but is particularly interested in mathematical modelling and the development of new ways of teaching mathematics to engineers. He has published numerous papers and one previous book, *Mathematical Modelling: A Case Study Approach*. He is a Chartered Engineer, a Member of the Royal Aeronautical Society, a Chartered Mathematician, a Fellow of the Institute of Mathematics and its Applications and a Member of the Royal Institute of Navigation.

Phil Dyke is Professor of Applied Mathematics and Head of School of Mathematics and Statistics at the University of Plymouth. After graduating with first class honours in Mathematics from the University of London, he gained a PhD in coastal sea modelling at Reading in 1972. Since then, Phil Dyke has been a full-time academic initially at Heriot-Watt University teaching engineers followed by a brief spell at Sunderland. He has been at Plymouth since 1984. He still engages in teaching and is actively involved in building mathematical models relevant to environmental issues.

John Searl is Director of the Edinburgh Centre for Mathematical Education at the University of Edinburgh. As well as lecturing on mathematical education, he teaches service courses for engineers and scientists. His current research concerns the development of learning environments that make for the effective learning of mathematics for 16–20 year olds. As an applied mathematician who has worked collaboratively with (among others) engineers, physicists, biologists and pharmacologists, he is keen to develop the problem-solving skills of his students and to encourage them to think for themselves.

Jerry Wright is a Principal Member of Technical Staff at the AT&T Shannon Laboratory, New Jersey, USA. He graduated in Engineering (BSc and PhD at the University of Southampton) and in Mathematics (MSc at the University of London) and worked at the National Physical Laboratory before moving to the University of Bristol in 1978. There he acquired wide experience in the teaching of mathematics to students of engineering, and became Senior Lecturer in Engineering Mathematics. He held a Royal Society Industrial Fellowship for 1994, and is a Fellow of the Institute of Mathematics and its Applications. In 1996 he moved to AT&T Labs (originally part of Bell Labs) to continue his research in spoken language understanding and human/computer dialogue systems.

1 Numbers, Algebra and Geometry

Chapter 1 Contents

1.1 # Introduction

Mathematics plays an important role in our lives. It is used in everyday activities from buying food to organizing maintenance schedules for aircraft. Through applications developed in various cultural and historical contexts, mathematics has been one of the decisive factors in shaping the modern world. It continues to grow and to find new uses, particularly in engineering and technology.

Mathematics provides a powerful, concise and unambiguous way of organizing and communicating information. It is a means by which aspects of the physical universe can be explained and predicted. It is a problem-solving activity supported by a body of knowledge. Mathematics consists of facts, concepts, skills and thinking processes – aspects that are closely interrelated. It is a hierarchical subject in that new ideas and skills are developed from existing ones. This sometimes makes it a difficult subject for learners who, at every stage of their mathematical development, need to have ready recall of material learned earlier.

In the first two chapters we shall summarize the concepts and techniques that most students will already understand and we shall extend them into further developments in mathematics. There are four key areas of which students will already have considerable knowledge.

- numbers
- algebra
- geometry
- functions

These areas are vital to making progress in engineering mathematics (indeed, they will solve many important problems in engineering). Here we shall aim to consolidate that knowledge, to make it more precise and to develop it. In this first chapter we will deal with the first three topics; functions are considered in Chapter 2.

1.2 Number and arithmetic

1.2.1 Number line

Mathematics has grown from primitive arithmetic and geometry into a vast body of knowledge. The most ancient mathematical skill is counting, using, in the first instance, the natural numbers and later the integers. The term **natural numbers** commonly refers to the set $\mathbb{N} = \{1, 2, 3, \ldots\}$, and the term **integers** to the set $\mathbb{Z} = \{0, 1, -1, 2, -2, 3, -3, \ldots\}$. The integers can be represented as equally spaced points on a line called the **number line** as shown in Figure 1.1. In a computer the integers can be stored exactly. The set of all points (not just those representing integers) on the number line represents the **real numbers** (so named to distinguish them from the complex numbers, which are

Figure 1.1
The number line.

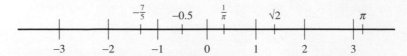

discussed in Chapter 3). The set of real numbers is denoted by $\mathbb{R}$. The general real number is usually denoted by the letter x and we write 'x in $\mathbb{R}$', meaning x is a real number. A real number that can be written as the ratio of two integers, like $\frac{3}{2}$ or $-\frac{7}{5}$, is called a **rational number**. Other numbers, like $\sqrt{2}$ and π, that cannot be expressed in that way are called **irrational numbers**. In a computer the real numbers can be stored only to a limited number of figures. This is a basic difference between the ways in which computers treat integers and real numbers, and is the reason why the computer languages commonly used by engineers distinguish between integer values and variables on the one hand and real number values and variables on the other.

1.2.2 Rules of arithmetic

The basic arithmetical operations of addition, subtraction, multiplication and division are performed subject to the **Fundamental Rules of Arithmetic**. For any three numbers a, b and c:

(a1) the commutative law of addition

$$a + b = b + a$$

(a2) the commutative law of multiplication

$$a \times b = b \times a$$

(b1) the associative law of addition

$$(a + b) + c = a + (b + c)$$

(b2) the associative law of multiplication

$$(a \times b) \times c = a \times (b \times c)$$

(c1) the distributive law of multiplication over addition and subtraction

$$(a + b) \times c = (a \times c) + (b \times c)$$
$$(a - b) \times c = (a \times c) - (b \times c)$$

(c2) the distributive law of division over addition and subtraction

$$(a + b) \div c = (a \div c) + (b \div c)$$
$$(a - b) \div c = (a \div c) - (b \div c)$$

These operations are called **binary** operations because they associate with every two members of the set of real numbers a unique third member; for example,

$$2 + 5 = 7 \quad \text{and} \quad 3 \times 6 = 18$$

A further operation used with real numbers is that of **powering**. For example, $a \times a$ is written as a^2, and $a \times a \times a$ is written as a^3. In general the product of n a's where n is a positive integer is written as a^n. (Here the n is called the **index** or **exponent**.) Operations with powering also obey simple rules:

$$a^n \times a^m = a^{n+m} \tag{1.1a}$$

$$a^n \div a^m = a^{n-m} \tag{1.1b}$$

$$(a^n)^m = a^{nm} \tag{1.1c}$$

From rule (1.1b) it follows, by setting $n = m$ and $a \neq 0$, that $a^0 = 1$. It is also convention to take $0^0 = 1$. The process of powering can be extended to include the fractional powers like $a^{1/2}$. Using rule (1.1c),

$$(a^{1/n})^n = a^{n/n} = a^1$$

and we see that

$$a^{1/n} = \sqrt[n]{a}$$

the nth root of a. Also, we can define a^{-m} using rule (1.1b) with $n = 0$, giving

$$1 \div a^m = a^{-m}, \qquad a \neq 0$$

Thus a^{-m} is the reciprocal of a^m. In contrast with the binary operations $+$, $\times$, $-$ and $\div$, which operate on two numbers, the powering operation $(\ )^r$ operates on just one element and is consequently called a **unary** operation. Notice that the fractional power

$$a^{m/n} = (\sqrt[n]{a})^m = \sqrt[n]{(a^m)}$$

is the nth root of a^m. If n is an even integer, then $a^{m/n}$ is not defined when a is negative. When $\sqrt[n]{a}$ is an irrational number then such a root is called a **surd**.

Example 1.1   Find the values of

(a) $27^{1/3}$ (b) $(-8)^{2/3}$ (c) $16^{-3/2}$

(d) $(-2)^{-2}$ (e) $(-1/8)^{-2/3}$ (f) $(9)^{-1/2}$

Solution   (a) $27^{1/3} = \sqrt[3]{27} = 3$

(b) $(-8)^{2/3} = (\sqrt[3]{(-8)})^2 = (-2)^2 = 4$

(c) $16^{-3/2} = (16^{1/2})^{-3} = (4)^{-3} = \frac{1}{64}$

(d) $(-2)^{-2} = \dfrac{1}{(-2)^2} = \frac{1}{4}$

(e) $(-1/8)^{-2/3} = [\sqrt[3]{(-1/8)}]^{-2} = [\sqrt[3]{(-1)}/\sqrt[3]{(8)}]^{-2} = [-1/2]^{-2} = 4$

(f) $(9)^{-1/2} = (3)^{-1} = \frac{1}{3}$

Example 1.2   Express (a) in terms of $\sqrt{2}$ and rationalize (b) to (d).

(a) $\sqrt{18} + \sqrt{32} - \sqrt{50}$ (b) $6/\sqrt{2}$

(c) $\dfrac{2}{1 - \sqrt{3}}$ (d) $\dfrac{1 - \sqrt{2}}{1 + \sqrt{6}}$

Solution   (a) $\sqrt{18} = \sqrt{(2 \times 9)} = \sqrt{2} \times \sqrt{9} = 3\sqrt{2}$

$$\sqrt{32} = \sqrt{(2 \times 16)} = \sqrt{2} \times \sqrt{16} = 4\sqrt{2}$$

$$\sqrt{50} = \sqrt{(2 \times 25)} = \sqrt{2} \times \sqrt{25} = 5\sqrt{2}$$

Thus $\sqrt{18} + \sqrt{32} - \sqrt{50} = 2\sqrt{2}$.

(b) $6/\sqrt{2} = 3 \times 2/\sqrt{2}$

Since $2 = \sqrt{2} \times \sqrt{2}$, we have $6/\sqrt{2} = 3\sqrt{2}$.

(c) $\dfrac{2}{1 - \sqrt{3}}$ can be simplified by multiplying 'top and bottom' by $1 + \sqrt{3}$ (notice the sign change in front of the $\sqrt{\ }$). Thus

$$\frac{2}{1 - \sqrt{3}} = \frac{2(1 + \sqrt{3})}{(1 - \sqrt{3})(1 + \sqrt{3})}$$

$$= \frac{2(1 + \sqrt{3})}{1 - 3}$$

$$= -1 - \sqrt{3}$$

(d)   Using the same technique as in part (c) we have

$$\frac{1 - \sqrt{2}}{1 + \sqrt{6}} = \frac{(1 - \sqrt{2})(1 - \sqrt{6})}{(1 + \sqrt{6})(1 - \sqrt{6})}$$

$$= \frac{1 - \sqrt{2} - \sqrt{6} + \sqrt{12}}{1 - 6}$$

$$= -(1 - \sqrt{2} - \sqrt{6} + 2\sqrt{3})/5$$

This process of expressing the irrational number so that all of the surds are in the numerator is called **rationalization**.

When evaluating arithmetical expressions the following rules of precedence are observed:

- the operation $(\ )^r$ is performed first
- then $\times$ and/or $\div$
- then $+$ and/or $-$

When two operators of equal precedence are adjacent in an expression the left-hand operation is performed first. For example

$$12 - 4 + 13 = 8 + 13 = 21$$

and

$$15 \div 3 \times 2 = 5 \times 2 = 10$$

The precedence rules are overridden by brackets; thus

$$12 - (4 + 13) = 12 - 17 = -5$$

and

$$15 \div (3 \times 2) = 15 \div 6 = 2.5$$

Example 1.3   Evaluate $7 - 5 \times 3 \div 2^2$.

Solution   Following the rules of precedence, we have

$$7 - 5 \times 3 \div 2^2 = 7 - 5 \times 3 \div 4 = 7 - 15 \div 4 = 7 - 3.75 = 3.25$$

1.2.3 Inequalities

The number line (Figure 1.1) makes explicit a further property of the real numbers – that of **ordering**. This enables us to make statements like 'seven is greater than two' and 'five is less than six'. We represent this using the comparison symbols

$>$, 'greater than'
$<$, 'less than'

It also makes obvious two other comparators:

$=$, 'equals'
$\neq$, 'does not equal'

These comparators obey simple rules when used in conjunction with the arithmetical operations. For any four numbers a, b, c and d:

$(a < b$ and $c < d)$	implies	$a + c < b + d$	**(1.2a)**
$(a < b$ and $c > d)$	implies	$a - c < b - d$	**(1.2b)**
$(a < b$ and $b < c)$	implies	$a < c$	**(1.2c)**
$a < b$	implies	$a + c < b + c$	**(1.2d)**
$(a < b$ and $c > 0)$	implies	$ac < bc$	**(1.2e)**
$(a < b$ and $c < 0)$	implies	$ac > bc$	**(1.2f)**
$(a < b$ and $ab > 0)$	implies	$\dfrac{1}{a} > \dfrac{1}{b}$	**(1.2g)**

Example 1.4   Show, without using a calculator or tables, that $\sqrt{2} + \sqrt{3} > 2^4(\sqrt{6})$.

Solution   By squaring we have that

$$(\sqrt{2} + \sqrt{3})^2 = 5 + 2\sqrt{2}\sqrt{3} = 5 + 2\sqrt{6}$$

Also

$$(2\sqrt{6})^2 = 24 < 25 = 5^2$$

implying that $5 > 2\sqrt{6}$. Thus

$$(\sqrt{2} + \sqrt{3})^2 > 2\sqrt{6} + 2\sqrt{6} = 4\sqrt{6}$$

and, since $\sqrt{2} + \sqrt{3}$ is a positive number, it follows that

$$\sqrt{2} + \sqrt{3} > \sqrt{(4\sqrt{6})} = 2^4(\sqrt{6})$$

1.2.4 Modulus and intervals

The size of a real number x is called its modulus and is denoted by $|x|$ (or sometimes by mod (x)). Thus

$$|x| = \begin{cases} x & (x \geqslant 0) \\ -x & (x < 0) \end{cases}$$
(1.3)

where the comparator $\geqslant$ indicates 'greater than or equal to'. (Likewise $\leqslant$ indicates 'less than or equal to'.)

Geometrically $|x|$ is the distance of the point representing x on the number line from the point representing zero. Similarly $|x - a|$ is the distance of the point representing x on the number line from that representing a.

The set of numbers between two numbers, a and b say, defines an **open interval** on the real line. This is the set $\{x : a < x < b, x \text{ in } \mathbb{R}\}$ and is usually denoted by (a, b). (Set notation will be introduced in Chapter 6; here $\{x : P\}$ denotes the set of all x that have property P.) Here the double-sided inequality means that x is greater than a and less than b; that is, the inequalities $a < x$ and $x < b$ apply simultaneously. An interval that includes the end points is called a **closed interval**, denoted by $[a, b]$, with

$$[a, b] = \{x : a \leqslant x \leqslant b, x \text{ in } \mathbb{R}\}$$

Note that the distance between two numbers a and b might either be $a - b$ or $b - a$ depending on which was the larger. An immediate consequence of this is that

$$|a - b| = |b - a|$$

since a is the same distance from b as b is from a.

Example 1.5   Find the values of x so that

$$|x - 4.3| = 5.8$$

Solution $|x - 4.3| = 5.8$ means that the distance between the real numbers x and 4.3 is 5.8 units, but does not tell us whether $x > 4.3$ or whether $x < 4.3$. The situation is illustrated in Figure 1.2, from which it is clear that the two possible values of x are -1.5 and 10.1.

Figure 1.2
Illustration of
$|x - 4.3| = 5.8$.

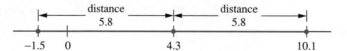

Example 1.6   Express the sets $\{x : |x - 3| < 5, x \text{ in } \mathbb{R}\}$ and $\{x : |x + 2| \leqslant 3, x \text{ in } \mathbb{R}\}$ as intervals.

Solution $|x - 3| < 5$ means that the distance of the point representing x on the number line from the point representing 3 is less than 5 units, as shown in Figure 1.3(a). This implies that

$$-5 < x - 3 < 5$$

Adding 3 to each member of this inequality, using rule (1.2d), gives

$$-2 < x < 8$$

and the set of numbers satisfying this inequality is the open interval $(-2, 8)$.

Figure 1.3
(a) The open interval
(−2, 8). (b) The closed
interval [−5, 1].

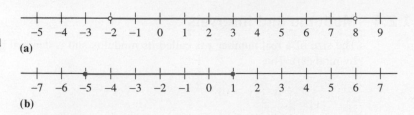

(a)

(b)

Similarly $|x + 2| \leqslant 3$, which may be rewritten as $|x - (-2)| \leqslant 3$, means that the distance of the point x on the number line from the point representing −2 is less than or equal to 3 units, as shown in Figure 1.3(b). This implies

$$-3 \leqslant x + 2 \leqslant 3$$

Subtracting 2 from each member of this inequality, using rule (1.2d), gives

$$-5 \leqslant x \leqslant 1$$

and the set of numbers satisfying this inequality is the closed interval [−5, 1].

We note in passing the following results. For any two real numbers x and y:

$|xy| = |x||y|$	**(1.4a)**
$|x| < a, a > 0,$ implies $-a < x < a$	**(1.4b)**
$|x + y| \leqslant |x| + |y|,$ known as the 'triangle inequality'	**(1.4c)**
$\frac{1}{2}(x + y) \geqslant \sqrt{(xy)},$ when $x \geqslant 0$ and $y \geqslant 0$	**(1.4d)**

Result (1.4d) is proved in Example 1.7 below and may be stated in words as

the arithmetic mean $\frac{1}{2}(x + y)$ of two positive numbers x and y is greater than or equal to the geometric mean $\sqrt{(xy)}$. Equality holds only when $y = x$.

Results (1.4a) to (1.4c) should be verified by the reader, who may find it helpful to try some particular values first.

Example 1.7

Prove that for any two positive numbers x and y, the arithmetic–geometric inequality

$$\tfrac{1}{2}(x + y) \geqslant \sqrt{(xy)}$$

holds.

Deduce that $x + \dfrac{1}{x} \geqslant 2$ for any positive number x.

Solution

The quantity xy can be interpreted as the area of a rectangle with sides x and y. The quantity $(x + y)^2$ can be interpreted as the area of a square of side $(x + y)$. Comparing areas in Figure 1.4, where the broken lines cut the square into 4 equal quarters of size A and it is assumed that $x > y$

Figure 1.4
Illustration of
$x^2 + y^2 \geqslant 2xy$.

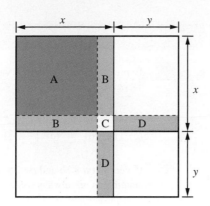

$$\left.\begin{array}{l} x^2 = A + 2B + C \\ y^2 = A - 2D - C \end{array}\right\} x^2 + y^2 = 2A + 2B - 2D$$

$$xy = A - B + D$$

Since B > D, (B = D + C), it follows that

$$x^2 + y^2 > 2xy$$

In the particular case when $x = y$ then B = D = 0 and

$$x^2 + y^2 = 2xy$$

so in general

$$x^2 + y^2 \geqslant 2xy$$

Also, from Figure 1.4, we see that

$$(x + y)^2 = x^2 + y^2 + 2xy$$

Hence we deduce

$$(x + y)^2 \geqslant 4xy$$

and since x and y are both positive we have

$$x + y \geqslant 2\sqrt{(xy)}$$

which is equivalent to

$$\tfrac{1}{2}(x + y) \geqslant \sqrt{(xy)}$$

In the special case when $y = \dfrac{1}{x}$ we have

$$x + \frac{1}{x} \geqslant 2\sqrt{\left(x\frac{1}{x}\right)}$$

that is,

$$x + \frac{1}{x} \geqslant 2$$

1.2.5 Exercises

1 Simplify the following expressions, giving the answers with positive indices and without brackets:

(a) $2^3 \times 2^{-4}$ (b) $2^3 \div 2^{-4}$ (c) $(2^3)^{-4}$

(d) $3^{1/3} \times 3^{5/3}$ (e) $(36)^{-1/2}$ (f) $16^{3/4}$

2 The expression $7 - 2 \times 3^2 + 8$ may be evaluated using the usual implicit rules of precedence. It could be rewritten as $((7 - (2 \times (3^2))) + 8)$ using brackets to make the precedence explicit. Similarly rewrite the following expressions in fully bracketed form:

(a) $21 + 4 \times 3 \div 2$

(b) $17 - 6^{2+3}$

(c) $4 \times 2^3 - 7 \div 6 \times 2$

(d) $2 \times 3 - 6 \div 4 + 3^{2-5}$

3 Express the following in the form $x + y\sqrt{2}$ with x and y rational numbers:

(a) $(7 + 5\sqrt{2})^3$ (b) $(2 + \sqrt{2})^4$

(c) $\sqrt[3]{(7 + 5\sqrt{2})}$ (d) $\sqrt{(\frac{11}{2} - 3\sqrt{2})}$

4 Show that

$$\frac{1}{a + b\sqrt{c}} = \frac{a - b\sqrt{c}}{a^2 - b^2 c}$$

Hence express the following numbers in the form $x + y\sqrt{n}$ where x and y are rational numbers and n is an integer:

(a) $\dfrac{1}{7 + 5\sqrt{2}}$ (b) $\dfrac{2 + 3\sqrt{2}}{9 - 7\sqrt{2}}$

(c) $\dfrac{4 - 2\sqrt{3}}{7 - 3\sqrt{3}}$ (d) $\dfrac{2 + 4\sqrt{5}}{4 - \sqrt{5}}$

5 Show that, if $\sqrt{a} + \sqrt{b} > \sqrt{c}$ then $a + 2\sqrt{(ab)} + b > c$, and hence that $4ab > (c - a - b)^2$ provided that $c - a - b > 0$. Use a similar method to determine, without using a calculator or tables, the larger of $\sqrt{5} + \sqrt{13}$ and $\sqrt{3} + \sqrt{19}$.

6 Express the following sets as intervals:

(a) $\{x : |x - 4| \leqslant 6\}$ (b) $\{x : |x + 3| < 2\}$

(c) $\{x : |2x - 1| \leqslant 7\}$ (b) $\{x : |\frac{1}{4}x + 3| < 3\}$

7 Express the following intervals as sets in the form $\{x : |ax + b| < c\}$ or $\{x : |ax + b| \leqslant c\}$:

(a) $(1, 7)$ (b) $[-4, -2]$

(c) $(17, 26)$ (d) $[-\frac{1}{2}, \frac{3}{4}]$

8 Given that $a < b$ and $c < d$, which of the following statements are always true?

(a) $a - c < b - d$ (b) $a - d < b - c$

(c) $ac < bd$ (d) $\dfrac{1}{b} < \dfrac{1}{a}$

In each case either prove that the statement is true or give a numerical example to show it can be false.

If, additionally, a, b, c and d are all greater than zero, how does that modify your answer?

9 The average speed for a journey is the distance covered divided by the time taken.

(a) A journey is completed by travelling for the first half of the *time* at speed v_1 and the second half at speed v_2. Find the average speed v_a for the journey in terms of v_1 and v_2.

(b) A journey is completed by travelling at speed v_1 for half the *distance* and at speed v_2 for the second half. Find the average speed v_b for the journey in terms of v_1 and v_2.

Deduce that a journey completed by travelling at two different speeds for equal distances will take longer than the same journey completed at the same two speeds for equal times.

10 Find the difference between 2 and the squares of

$$\frac{1}{1}, \frac{3}{2}, \frac{7}{5}, \frac{17}{12}, \frac{41}{29}, \frac{99}{70}$$

Verify that successive terms of the sequence stand in relation to each other as m/n does to $(m + 2n)/(m + n)$. Show that if m/n is a good approximation to $\sqrt{2}$ then $(m + 2n)/(m + n)$ is a better one, and that the errors in the two cases are in opposite directions. Find the next three terms of the above sequence.

Algebra

The idea, first introduced in the seventeenth century, of using letters to represent unspecified quantities led to the development of algebraic manipulation based on the elementary laws of arithmetic. This development greatly enhanced the problem-solving power of mathematics – so much so that it is difficult now to imagine doing mathematics without this resource.

1.3.1 Algebraic manipulation

Algebraic manipulation made possible concise statements of well-known results, such as

$$(a + b)^2 = a^2 + 2ab + b^2$$

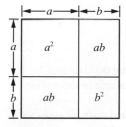

Figure 1.5
Illustration of
$(a + b)^2 = a^2 + 2ab + b^2$.

Previously these results had been obtained by a combination of verbal reasoning and elementary geometry as illustrated in Figure 1.5.

Other elementary results that are often used are

$$a^2 - b^2 = (a + b)(a - b) \qquad \textit{difference of squares}$$

$$a^2 - 2ab + b^2 = (a - b)^2$$

$$ax^2 + bx + c = a\left(x + \frac{b}{2a}\right)^2 + c - \frac{b^2}{4a} \qquad \textit{completing the square}$$

Example 1.8 A pipe has the form of a hollow cylinder as shown in Figure 1.6. Find its mass when

(a) its length is 1.5 m, its external diameter is 205 mm, its internal diameter is 160 mm and its density is 5500 kg m^{-3},

(b) its length is l m, its external diameter is D mm, its internal diameter is d mm and its density is ρ kg m^{-3}.

Solution (a) Standardizing the units of length, the internal and external diameters are 0.16 m and 0.205 m respectively. The area of cross-section of the pipe is

$$\pi(0.205^2 - 0.160^2)/4 \, \text{m}^2$$

and hence the volume of the material of the pipe is

$$\pi(0.205^2 - 0.160^2) \times 1.5/4 \, \text{m}^3$$

Hence the mass of the pipe is

$$5500 \times \pi(0.205^2 - 0.160^2) \times 1.5/4 \, \text{kg}$$

Evaluating this last expression by calculator gives the mass of the pipe as 106 to the nearest kilogram.

(b) The internal and external diameters of the pipe are $d/1000$ and $D/1000$ metres, respectively, so that the area of cross-section is

$$0.25\pi(D^2 - d^2)/1\,000\,000 \, \text{m}^2$$

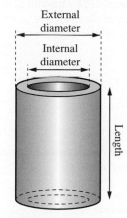

External diameter

Internal diameter

Length

Figure 1.6
Cylindrical pipe
of Example 1.8.

The volume of the pipe is

$$0.25\pi l(D^2 - d^2)/10^6 \, \text{m}^3$$

Hence the mass M kg of the pipe is given by the formulae

$$M = \pi\rho l(D^2 - d^2)/(4 \times 10^6) = 2.5\pi\rho l(D + d)(D - d) \times 10^{-5}$$

Example 1.9

Prove that

$$ab = \tfrac{1}{4}[(a + b)^2 - (a - b)^2]$$

Given $70^2 = 4900$ and $36^2 = 1296$, calculate 53×17.

Solution

Since

$$(a + b)^2 = a^2 + 2ab + b^2$$

and

$$(a - b)^2 = a^2 - 2ab + b^2$$

we deduce

$$(a + b)^2 - (a - b)^2 = 4ab$$

and

$$ab = \tfrac{1}{4}[(a + b)^2 - (a - b)^2]$$

The result is illustrated geometrically in Figure 1.7. Setting $a = 53$ and $b = 17$, we have

$$53 \times 17 = \tfrac{1}{4}[70^2 - 36^2] = 901$$

This method of calculating products was used by the Babylonians and is sometimes called the 'quarter-squares' algorithm. It is used in some analogue devices and simulators.

Figure 1.7
Illustration of $ab = \tfrac{1}{4}[(a + b)^2 - (a - b)^2]$.

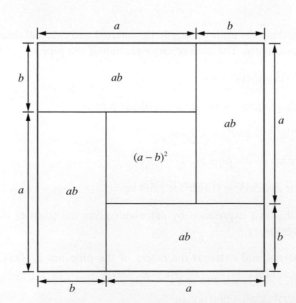

Example 1.10   Express as a single fraction

$$\frac{1}{(x + 1)(x + 2)} - \frac{2}{x + 1} + \frac{3}{x + 2}$$

Solution   The lowest common multiple of the denominators of these fractions is $(x + 1)(x + 2)$ so we may write

$$\frac{1}{(x + 1)(x + 2)} - \frac{2}{x + 1} + \frac{3}{x + 2} = \frac{1}{(x + 1)(x + 2)} - \frac{2(x + 2)}{(x + 1)(x + 2)} + \frac{3(x + 1)}{(x + 1)(x + 2)}$$

$$= \frac{1 - 2(x + 2) + 3(x + 1)}{(x + 1)(x + 2)}$$

$$= \frac{1 - 2x - 4 + 3x + 3}{(x + 1)(x + 2)}$$

$$= \frac{x}{(x + 1)(x + 2)}$$

Example 1.11   Use the method of completing the square to manipulate the following quadratic expressions into the form of a number + (or −) the square of a term involving x.

(a) $x^2 + 3x - 7$ (b) $5 - 4x - x^2$

(c) $3x^2 - 5x + 4$ (d) $1 + 2x - 2x^2$

Solution   Remember $(a + b)^2 = a^2 + 2ab + b^2$.

(a) To convert $x^2 + 3x$ into a perfect square we need to add $(\frac{3}{2})^2$. Thus we have

$$x^2 + 3x - 7 = [(x + \tfrac{3}{2})^2 - (\tfrac{3}{2})^2] - 7$$

$$= (x + \tfrac{3}{2})^2 - \tfrac{37}{4}$$

(b) $5 - 4x - x^2 = 5 - (4x + x^2)$

To convert $x^2 + 4x$ into a perfect square we need to add 2^2. Thus we have

$$x^2 + 4x = (x + 2)^2 - 2^2$$

and

$$5 - 4x - x^2 = 5 - [(x + 2)^2 - 2^2] = 9 - (x + 2)^2$$

(c) First we 'take outside' the coefficient of x^2:

$$3x^2 - 5x + 4 = 3(x^2 - \tfrac{5}{3}x + \tfrac{4}{3})$$

Then we rearrange

$$x^2 - \tfrac{5}{3}x = (x - \tfrac{5}{6})^2 - \tfrac{25}{36}$$

so that $3x^2 - 5x + 4 = 3[(x - \tfrac{5}{6})^2 - \tfrac{25}{36} + \tfrac{4}{3}] = 3[(x - \tfrac{5}{6})^2 + \tfrac{23}{36}]$.

(d) Similarly

$$1 + 2x - 2x^2 = 1 - 2(x^2 - x)$$

and

$$x^2 - x = (x - \tfrac{1}{2})^2 - \tfrac{1}{4}$$

so that

$$1 + 2x - 2x^2 = 1 - 2[(x - \tfrac{1}{2})^2 - \tfrac{1}{4}] = \tfrac{3}{2} - 2(x - \tfrac{1}{2})^2$$

The number 45 can be factorized as $3 \times 3 \times 5$. Any product from 3, 3 and 5 is also a factor of 45. Algebraic expressions can be factorized in a similar fashion. An algebraic expression with more than one term can be factorized if each term contains common factors (either numerical or algebraic). These factors are removed by division from each term and the non-common factors remaining are grouped into brackets.

Example 1.12   Factorize $xz + 2yz - 2y - x$.

Solution   There is no common factor to all four terms so we take them in pairs:

$$xz + 2yz - 2y - x = (x + 2y)z - (2y + x)$$
$$= (x + 2y)z - (x + 2y)$$
$$= (x + 2y)(z - 1)$$

Alternatively, we could have written:

$$xz + 2yz - 2y - x = (xz - x) + (2yz - 2y)$$
$$= x(z - 1) + 2y(z - 1)$$
$$= (x + 2y)(z - 1)$$

to obtain the same result.

The expansion of $(a + b)^2$ is a special case of a general result for $(a + b)^n$ known as the binomial expansion. This is discussed again in Sections 1.3.4 and 7.7.2. Here we shall look at the cases for $n = 0, 1, \ldots, 6$.

Writing these out, we have

$$(a + b)^0 = 1$$
$$(a + b)^1 = a + b$$
$$(a + b)^2 = a^2 + 2ab + b^2$$
$$(a + b)^3 = a^3 + 3a^2b + 3ab^2 + b^3$$
$$(a + b)^4 = a^4 + 4a^3b + 6a^2b^2 + 4ab^3 + b^4$$
$$(a + b)^5 = a^5 + 5a^4b + 10a^3b^2 + 10a^2b^3 + 5ab^4 + b^5$$
$$(a + b)^6 = a^6 + 6a^5b + 15a^4b^2 + 20a^3b^3 + 15a^2b^4 + 6ab^5 + b^6$$

This table can be extended indefinitely. Each line can easily be obtained from the previous one. Thus, for example,

$$(a + b)^4 = (a + b)(a + b)^3$$

$$= a(a^3 + 3a^2b + 3ab^2 + b^3) + b(a^3 + 3a^2b + 3ab^2 + b^3)$$

$$= a^4 + 3a^3b + 3a^2b^2 + ab^3 + a^3b + 3a^2b^2 + 3ab^3 + b^4$$

$$= a^4 + 4a^3b + 6a^2b^2 + 4ab^3 + b^4$$

The coefficients involved form a pattern of numbers called Pascal's triangle, shown in Figure 1.8. Each number in the interior of the triangle is obtained by summing the numbers to its right and left in the row above, as indicated by the arrows in Figure 1.8. This number pattern had been discovered prior to Pascal by the Chinese mathematician Chu Shih-chieh.

Figure 1.8
Pascal's triangle.

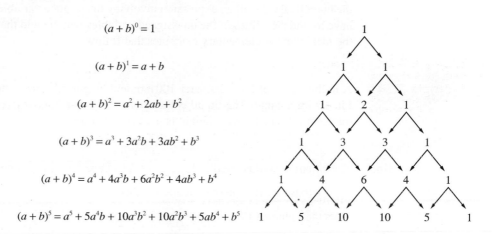

$$(a + b)^0 = 1$$

$$(a + b)^1 = a + b$$

$$(a + b)^2 = a^2 + 2ab + b^2$$

$$(a + b)^3 = a^3 + 3a^2b + 3ab^2 + b^3$$

$$(a + b)^4 = a^4 + 4a^3b + 6a^2b^2 + 4ab^3 + b^4$$

$$(a + b)^5 = a^5 + 5a^4b + 10a^3b^2 + 10a^2b^3 + 5ab^4 + b^5$$

Example 1.13 Expand

(a) $(2x + 3y)^2$ (b) $(2x - 3)^3$ (c) $\left(2x - \dfrac{1}{x}\right)^4$

Solution (a) Here we use the expansion

$$(a + b)^2 = a^2 + 2ab + b^2$$

with $a = 2x$ and $b = 3y$ to obtain

$$(2x + 3y)^2 = (2x)^2 + 2(2x)(3y) + (3y)^2$$

$$= 4x^2 + 12xy + 9y^2$$

(b) Here we use the expansion

$$(a + b)^3 = a^3 + 3a^2b + 3ab^2 + b^3$$

with $a = 2x$ and $b = -3$ to obtain

$$(2x - 3)^3 = 8x^3 - 36x^2 + 54x - 27$$

(c) Here we use the expansion

$$(a + b)^4 = a^4 + 4a^3b + 6a^2b^2 + 4ab^3 + b^4$$

with $a = 2x$ and $b = -1/x$ to obtain

$$\left(2x - \frac{1}{x}\right)^4 = (2x)^4 + 4(2x)^3(-1/x) + 6(2x)^2(-1/x)^2 + 4(2x)(-1/x)^3 + (-1/x)^4$$

$$= 16x^4 - 32x^2 + 24 - 8/x^2 + 1/x^4$$

1.3.2 Equations, inequalities and identities

It commonly occurs in the application of mathematics to practical problem-solving that the numerical value of an expression involving unassigned variables is specified and we have to find the values of the unassigned variables which yield that value. We illustrate the idea with the elementary examples that follow.

Example 1.14 A hollow cone of base diameter 100 mm and height 150 mm is held upside down and filled with a liquid. The liquid is then transferred to a hollow circular cylinder of base diameter 80 mm. To what height is the cylinder filled?

Solution The situation is illustrated in Figure 1.9. The capacity of the cone is

$$\tfrac{1}{3}(\text{base area}) \times (\text{perpendicular height})$$

Thus the volume of liquid contained in the cone is

$$\tfrac{1}{3}\pi(50^2)(150) = 125\,000\pi \text{ mm}^3$$

The volume of the liquid in the circular cylinder is

$$(\text{base area}) \times (\text{height}) = \pi(40^2)h \text{ mm}^3$$

Figure 1.9
The cone and cylinder
of Example 1.14.

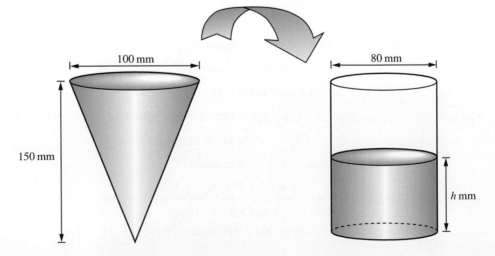

where h mm is the height of the liquid in the cylinder. Equating these quantities (assuming no liquid is lost in the transfer) we have

$$1600\pi h = 125\,000\pi$$

This **equation** enables us to find the value of the unassigned variable h:

$$h = 1250/16 = 78.125$$

Thus the height of the liquid in the cylinder is 78 to the nearest millimetre.

Example 1.15   A dealer bought a number of equally priced articles for a total cost of £120. He sold all but one of them, making a profit of £1.50 on each article with a total revenue of £135. How many articles did he buy?

Solution   Let n be the number of articles bought. Then the cost of each article was £$(120/n)$. The selling price of each article was £$(135/(n-1))$ so that the profit per item was

$$£\left\{\frac{135}{n-1} - \frac{120}{n}\right\}$$

which we are told is equal to £1.50. Thus

$$\frac{135}{n-1} - \frac{120}{n} = 1.50$$

This implies

$$135n - 120(n-1) = 1.50(n-1)n$$

Dividing both sides by 1.5 gives

$$90n - 80(n-1) = n^2 - n$$

Simplifying and collecting terms we obtain

$$n^2 - 11n - 80 = 0$$

This **equation** for n can be simplified further by factorizing the quadratic expression on the left-hand side

$$(n - 16)(n + 5) = 0$$

This implies either $n = 16$ or $n = -5$, so the dealer initially bought 16 articles (the solution $n = -5$ is not feasible).

Example 1.16   Using the method of completing the square, obtain the formula for finding the roots of the general quadratic equation

$$ax^2 + bx + c = 0 \quad (a \neq 0)$$

Solution   Dividing throughout by a gives

$$x^2 + \frac{b}{a}x + \frac{c}{a} = 0$$

Completing the square leads to

$$\left(x + \frac{b}{2a}\right)^2 + \frac{c}{a} = \left(\frac{b}{2a}\right)^2$$

giving

$$\left(x + \frac{b}{2a}\right)^2 = \frac{b^2}{4a^2} - \frac{c}{a} = \frac{b^2 - 4ac}{4a^2}$$

which on taking the square root gives

$$x + \frac{b}{2a} = \pm\frac{\sqrt{(b^2 - 4ac)}}{2a}$$

or

$$x = \frac{-b \pm \sqrt{(b^2 - 4ac)}}{2a} \tag{1.5}$$

Comment The formula given in (1.5) makes clear the three cases: where for $b^2 > 4ac$ we have two real roots to the equation, for $b^2 < 4ac$ we have no real roots, and for $b^2 = 4ac$ we have one real root (which is repeated). The quadratic equation has many important applications. One, which is of historical significance, concerned the electrical engineer Oliver Heaviside. In 1871 the telephone cable between England and Denmark developed a fault caused by a short circuit under the sea. His task was to locate that fault. The cable had a uniform resistance per unit length. His method of solution was brilliantly simple. The situation can be represented schematically as shown in Figure 1.10.

Figure 1.10
The circuit for the telephone line fault.

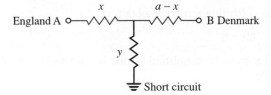

In the figure the total resistance of the line between A and B is a ohms and is known; x and y are unknown. If we can find x, we can locate the distance along the cable where the fault has occurred. Heaviside solved the problem by applying two tests. First he applied a battery, having voltage E, at A with the circuit open at B, and measured the resulting current I_1. Then he applied the same battery at A but with the cable earthed at B, and again measured the resulting current I_2. Using Ohm's law and the rules for combining resistances in parallel and in series, this yields the pair of equations

$$E = I_1(x + y)$$

$$E = I_2\left[x + \left(\frac{1}{y} + \frac{1}{a - x}\right)^{-1}\right]$$

Writing $b = E/I_1$ and $c = E/I_2$, we can eliminate y from these equations to obtain an equation for x:

$$x^2 - 2cx + c(a + b) - ab = 0$$

which, using (1.5), has solutions

$$x = c \pm \sqrt{[(a - c)(b - c)]}$$

From his experimental data Heaviside was able to predict accurately the location of the fault.

The condition for equality of the roots of a quadratic equation also occurs in practical applications, and we shall illustrate this in Chapter 2, Example 2.43 after considering the trigonometric functions.

In some problems we have to find the values of unassigned variables such that the value of an expression involving those variables satisfies an inequality condition (that is, it is either greater than, or alternatively less than, a specified value). Solving such inequalities requires careful observance of the rules for inequalities (1.2a–1.2g) set out in Section 1.2.3.

Example 1.17 Find the values of x for which

$$\frac{1}{3 - x} < 2 \tag{1.6}$$

Solution (a) When $3 - x > 0$, that is $x < 3$, we may, using (1.2e), multiply (1.6) throughout by $3 - x$ to give

$$1 < 2(3 - x)$$

which, using (1.2d, e), reduces to

$$x < \tfrac{5}{2}$$

so that (1.6) is satisfied when both $x < 3$ and $x < \tfrac{5}{2}$ are satisfied; that is, $x < \tfrac{5}{2}$.

(b) When $3 - x < 0$, that is $x > 3$, we may, using (1.2f), multiply (1.6) throughout by $3 - x$ to give

$$1 > 2(3 - x)$$

which reduces to $x > \tfrac{5}{2}$ so that (1.6) is also satisfied when both $x > 3$ and $x > \tfrac{5}{2}$; that is, $x > 3$.

Thus inequality (1.6) is satisfied by values of x in the ranges $x > 3$ and $x < \tfrac{5}{2}$.

Comment A common mistake made is simply to multiply (1.6) throughout by $3 - x$ to give the answer $x < \tfrac{5}{2}$, forgetting to consider both cases of $3 - x > 0$ and $3 - x < 0$. We shall return to consider this example from the graphical point of view in Example 2.34.

Example 1.18

A food manufacturer found that the sales figure for a certain item depended on its selling price. The company's market research department advised that the maximum number of items that could be sold weekly was 20 000 and that the number sold decreased by 100 for every 1p increase in its price. The total production cost consisted of a set-up cost of £200 **plus** 50p for every item manufactured. What price should the manufacturer adopt?

Solution

The data supplied by the market research department suggests that if the price of the item is p pence, then the number sold would be $20\,000 - 100p$. (So the company would sell none with $p = 200$, when the price is £2.) The production cost in pounds would be $200 + 0.5 \times$ (number sold), so that in terms of p we have the production cost £C given by

$$C = 200 + 0.5(20\,000 - 100p)$$

The revenue £R accrued by the manufacturer for the sales is (number sold) × (price), which gives

$$R = (20\,000 - 100p)p/100$$

(remember to express the amount in pounds). Thus, the profit £P is given by

$$P = R - C$$
$$= (20\,000 - 100p)p/100 - 200 - 0.5(20\,000 - 100p)$$
$$= -p^2 + 250p - 10\,200$$

Completing the square we have

$$P = 125^2 - (p - 125)^2 - 10\,200$$
$$= 5425 - (p - 125)^2$$

Since $(p - 125)^2 \geqslant 0$, we deduce that the maximum value of P is 5425 and to achieve this weekly profit, the manufacturer should adopt the price £1.25.

It is important to distinguish between those equalities that are valid for a restricted set of values of the unassigned variable x and those that are true for all values of x. For example

$$(x - 5)(x + 7) = 0$$

is true only if $x = 5$ or $x = -7$. In contrast

$$(x - 5)(x + 7) = x^2 + 2x - 35 \tag{1.7}$$

is true for all values of x. The word 'equals' here is being used in subtly different ways. In the first case '=' means 'is numerically equal to'; in the second case '=' means 'is algebraically equal to'. Sometimes we emphasize the different meaning by means of the special symbol ≡, meaning 'algebraically equal to'. (However, it is fairly common practice in engineering to use '=' in both cases.) Such equations are often called **identities**. Identities that involve an unassigned variable x as in (1.7) are valid for all values of x, and we can sometimes make use of this fact to simplify algebraic manipulations.

Example 1.19

Find the numbers A, B and C such that

$$x^2 + 2x - 35 \equiv A(x-1)^2 + B(x-1) + C$$

Solution

Method (a): Since $x^2 + 2x - 35 \equiv A(x-1)^2 + B(x-1) + C$ it will be true for any value we give to x. So we choose values that make finding A, B and C easy.

Choosing $x = 0$ gives $-35 = A - B + C$
Choosing $x = 1$ gives $-32 = C$
Choosing $x = 2$ gives $-27 = A + B + C$

So we obtain $C = -32$, with $A - B = -3$ and $A + B = 5$. Hence $A = 1$ and $B = 4$ to give the identity

$$x^2 + 2x - 35 \equiv (x-1)^2 + 4(x-1) - 32$$

Method (b): Expanding the terms on the right-hand side, we have

$$x^2 + 2x - 35 \equiv Ax^2 + (B - 2A)x + A - B + C$$

The expressions on either side of the equals sign are algebraically equal, which means that the coefficient of x^2 on the left-hand side must equal the coefficient of x^2 on the right-hand side and so on. Thus

$$1 = A$$

$$2 = B - 2A$$

$$-35 = A - B + C$$

Hence we find $A = 1$, $B = 4$ and $C = -32$, as before.

Note: Method (a) assumes that a valid A, B and C exist.

Example 1.20

Find numbers A, B and C such that

$$\frac{x^2}{x-1} \equiv Ax + B + \frac{C}{x-1}, \qquad x \neq 1$$

Solution

Expressing the right-hand side as a single term, we have

$$\frac{x^2}{x-1} \equiv \frac{(Ax + B)(x-1) + C}{x-1}$$

which, with $x \neq 1$, is equivalent to

$$x^2 \equiv (Ax + B)(x-1) + C$$

Choosing $x = 0$ gives $0 = -B + C$
Choosing $x = 1$ gives $1 = C$
Choosing $x = 2$ gives $4 = 2A + B + C$

Thus we obtain

$C = 1$, $B = 1$ and $A = 1$, yielding

$$\frac{x^2}{x-1} \equiv x + 1 + \frac{1}{x-1}$$

1.3.3   Suffix, sigma and pi notation

We have seen in previous sections how letters are used to denote general or unspecified values or numbers. This process has been extended in a variety of ways. In particular, the introduction of suffixes enables us to deal with problems that involve a high degree of generality or whose solutions have the flexibility to apply in a large number of situations. Consider for the moment an experiment involving measuring the temperature of an object (for example, a piece of machinery or a cooling fin in a heat exchanger) at intervals over a period of time. In giving a theoretical description of the experiment we would talk about the total period of time in general terms, say T minutes, and the time interval between measurements as h minutes, so that the total number of measurements would be given by $(T/h + 1)$ assuming that the initial and final temperatures are recorded. In practice we would obtain a set of experimental results, as illustrated partially in Figure 1.11.

Figure 1.11
Experimental results:
temperature against
lapsed time.

Lapsed time (*minutes*)	0	5	10	15	. . .	170	175	180
Temperature (°C)	97.51	96.57	93.18	91.53	. . .	26.43	24.91	23.57

Here we could talk about the twenty-first reading and look it up in the table. In the theoretical description we would need to talk about any one of the $(n + 1)$ temperature measurements. To facilitate this we introduce a suffix notation. We label the times at which the temperatures are recorded $t_0, t_1, t_2, \ldots, t_n$ where t_0 corresponds to the time when the initial measurement is taken, t_n to the time when the final measurement is taken, and

$$t_1 = t_0 + h, \; t_2 = t_0 + 2h, \ldots, t_n = t_0 + nh$$

so that $t_n = t_0 + T$. We label the corresponding temperatures by $\theta_0, \theta_1, \theta_2, \ldots, \theta_n$. We can then talk about the general result θ_k as measuring the temperature at time t_k.

In the analysis of the experimental results we may also wish to manipulate the data we have obtained. For example, we might wish to work out the average value of the temperature over the time period. With the specific experimental results given in Figure 1.11 it is possible to compute the average directly as

$$(97.51 + 96.57 + 93.18 + 91.53 + \ldots + 23.57)/37$$

In general, however, we have

$$(\theta_0 + \theta_1 + \theta_2 + \ldots + \theta_n)/(n + 1)$$

A compact way of writing this is to use the **sigma notation** for the extended summation $\theta_0 + \theta_1 + \ldots + \theta_n$. We write

$$\sum_{k=0}^{n} \theta_k \qquad (\Sigma \text{ is the upper-case Greek letter sigma.})$$

to denote

$$\theta_0 + \theta_1 + \theta_2 + \ldots + \theta_n$$

Thus

$$\sum_{k=0}^{3} \theta_k = \theta_0 + \theta_1 + \theta_2 + \theta_3$$

and

$$\sum_{k=5}^{10} \theta_k = \theta_5 + \theta_6 + \theta_7 + \theta_8 + \theta_9 + \theta_{10}$$

The suffix k appearing in the quantity to be summed and underneath the sigma symbol is the 'counting variable' or 'counter'. We may use any letter we please as a counter, provided that it is not being used at the same time for some other purpose. Thus

$$\sum_{i=0}^{3} \theta_i = \theta_0 + \theta_1 + \theta_2 + \theta_3 = \sum_{n=0}^{3} \theta_n = \sum_{j=0}^{3} \theta_j$$

Thus, in general, if $a_0, a_1, a_2, \ldots, a_n$ is a sequence of numbers or expressions, we write

$$\sum_{k=0}^{n} a_k = a_0 + a_1 + a_2 + \ldots + a_n$$

Another shorthand that is sometimes useful is for the extended product $a_0 a_1 a_2 \ldots a_n$, which we write as

$$\prod_{k=0}^{n} a_k = a_0 a_1 a_2 \ldots a_n \qquad \text{(Π is the upper-case Greek letter pi.)}$$

Thus

$$\prod_{k=0}^{3} a_k = a_0 a_1 a_2 a_3$$

and

$$\prod_{k=5}^{8} a_k = a_5 a_6 a_7 a_8$$

Example 1.21 Given $a_0 = 1$, $a_1 = 5$, $a_2 = 2$, $a_3 = 7$, $a_4 = -1$ and $b_0 = 0$, $b_1 = 2$, $b_2 = -2$, $b_3 = 11$, $b_4 = 3$, calculate

(a) $\sum_{k=0}^{4} a_k$ (b) $\sum_{i=2}^{3} a_i$ (c) $\sum_{k=1}^{3} a_k b_k$

(d) $\sum_{k=0}^{4} b_k^2$ (e) $\prod_{j=1}^{3} a_j$ (f) $\prod_{k=2}^{4} b_k$

Solution (a) $\displaystyle\sum_{k=0}^{4} a_k = a_0 + a_1 + a_2 + a_3 + a_4$

Substituting the given values for a_k $(k = 0, \ldots, 4)$ gives

$$\sum_{k=0}^{4} a_k = 1 + 5 + 2 + 7 + (-1) = 14$$

(b) $\displaystyle\sum_{i=2}^{3} a_i = a_2 + a_3 = 2 + 7 = 9$

(c) $\displaystyle\sum_{k=1}^{3} a_k b_k = a_1 b_1 + a_2 b_2 + a_3 b_3 = (5 \times 2) + (2 \times (-2)) + (7 \times 11) = 83$

(d) $\displaystyle\sum_{k=0}^{4} b_k^2 = b_0^2 + b_1^2 + b_2^2 + b_3^2 + b_4^2 = 0 + 4 + 4 + 121 + 9 = 138$

(e) $\displaystyle\prod_{j=1}^{3} a_j = a_1 a_2 a_3 = 5 \times 2 \times 7 = 70$

(f) $\displaystyle\prod_{k=2}^{4} b_k = b_2 b_3 b_4 = -2 \times 11 \times 3 = -66$

1.3.4 Factorial notation and the binomial expansion

The special extended product of integers

$$1 \times 2 \times 3 \times \ldots \times n = n \times (n-1) \times (n-2) \times \ldots \times 1$$

has a special notation and name. It is called **factorial n** and is denoted by $n!$. Thus with

$$n! = n(n-1)(n-2) \ldots (1)$$

as examples

$$5! = 5 \times 4 \times 3 \times 2 \times 1 \quad \text{and} \quad 8! = 8 \times 7 \times 6 \times 5 \times 4 \times 3 \times 2 \times 1$$

Notice that $5! = 5(4!)$ so that we can write in general

$$n! = (n-1)! \times n$$

This relationship enables us to define $0!$, since $1! = 1 \times 0!$ and $1!$ also equals 1. Thus $0!$ is defined by

$$0! = 1$$

Example 1.22   Evaluate

(a) 4!   (b) 3! × 2!   (c) 6!   (d) 7!/(2! × 5!)

Solution   (a) $4! = 4 \times 3 \times 2 \times 1 = 24$

(b) $3! \times 2! = (3 \times 2 \times 1) \times (2 \times 1) = 12$

(c) $6! = 6 \times 5 \times 4 \times 3 \times 2 \times 1 = 720$

Notice that $2! \times 3! \neq (2 \times 3)!$.

(d) $\dfrac{7!}{2! \times 5!} = \dfrac{7 \times 6 \times 5 \times 4 \times 3 \times 2 \times 1}{2 \times 1 \times 5 \times 4 \times 3 \times 2 \times 1} = \dfrac{7 \times 6}{2} = 21$

Notice that we could have simplified the last item by writing

$7! = 7 \times 6 \times (5!)$

then

$\dfrac{7!}{2! \times 5!} = \dfrac{7 \times 6 \times (5!)}{2! \times 5!} = 21$

An interpretation of $n!$ is the total number of different ways it is possible to arrange n different objects in a single line. For example, the word SEAT comprises four different letters, and we can arrange the letters in $4! = 24$ different ways.

SEAT	EATS	ATSE	TSEA
SETA	EAST	ATES	TSAE
SAET	ESAT	AETS	TESA
SATE	ESTA	AEST	TEAS
STAE	ETSA	ASET	TAES
STEA	ETAS	ASTE	TASE

This is because we can choose the first letter in four different ways (S, E, A or T). Once that choice is made, we can choose the second letter in three different ways, then we can choose the third letter in two different ways. Having chosen the first three letters, the last letter is automatically fixed. For each of the four possible first choices, we have three possible choices for the second letter, giving us twelve (4×3) possible choices of the first two letters. To each of these twelve possible choices we have two possible choices of the third letter, giving us twenty-four $(4 \times 3 \times 2)$ possible choices of the first three letters. Having chosen the first three letters, there is only one possible choice of last letter. So in all we have $4!$ possible choices.

Example 1.23   In how many ways can the letters of the word REGAL be arranged in a line, and in how many of those do the two letters A and E appear in adjacent positions?

Solution The word REGAL has five distinct letters so they can be arranged in a line in $5! = 120$ different ways. To find out in how many of those arrangements the A and E appear together, we consider how many arrangements can be made of RGL(AE) and RGL(EA), regarding the bracketed terms as a single symbol. There are $4!$ possible arrangements of both of these, so of the 120 different ways in which the letters of the word REGAL can be arranged, 48 contain the letters A and E in adjacent positions.

The introduction of the factorial notation facilitates the writing down of many complicated expressions. In particular it enables us to write down the general form of the binomial expansion discussed earlier in Section 1.3.1. There we wrote out longhand the expansion of $(a + b)^n$ for $n = 0, 1, 2, \ldots, 6$ and noted the relationship between the coefficients of $(a + b)^n$ and those of $(a + b)^{n-1}$, shown clearly in Pascal's triangle of Figure 1.8.

If

$$(a + b)^{n-1} = c_0 a^{n-1} + c_1 a^{n-2}b + c_2 a^{n-3}b^2 + c_3 a^{n-4}b^3 + \ldots + c_{n-1}b^{n-1}$$

and

$$(a + b)^n = d_0 a^n + d_1 a^{n-1}b + d_2 a^{n-2}b^2 + \ldots + d_{n-1}ab^{n-1} + d_n b^n$$

then, as described on p. 15 when developing Pascal's triangle,

$$c_0 = d_0 = 1, \quad d_1 = c_1 + c_0, \quad d_2 = c_2 + c_1, \quad d_3 = c_3 + c_2, \ldots$$

and in general

$$d_r = c_r + c_{r-1}$$

It is easy to verify that this relationship is satisfied by

$$d_r = \frac{n!}{r!(n-r)!}, \quad c_r = \frac{(n-1)!}{r!(n-1-r)!}, \quad c_{r-1} = \frac{(n-1)!}{(r-1)!(n-1-r+1)!}$$

and it can be shown that the coefficient of $a^{n-r}b^r$ in the expansion of $(a + b)^n$ is

$$\frac{n!}{r!(n-r)!} = \frac{n(n-1)(n-2)\ldots(n-r+1)}{r(r-1)(r-2)\ldots(1)} \tag{1.8}$$

This is a very important result, with many applications. Using it we can write down the general binomial expansion

$$(a + b)^n = \sum_{r=0}^{n} \frac{n!}{r!(n-r)!} a^{n-r}b^r \tag{1.9}$$

The coefficient $\dfrac{n!}{r!(n-r)!}$ is called the **binomial coefficient** and has the special notation

$$\binom{n}{r} = \frac{n!}{r!(n-r)!}$$

Thus we may write

$$(a + b)^n = \sum_{r=0}^{n} \binom{n}{r} a^{n-r} b^r \qquad (1.10)$$

which is referred to as the general **binomial expansion**.

Example 1.24   Expand the expression $(2 + x)^5$.

Solution   Setting $a = 2$ and $b = x$ in the general binomial expansion we have

$$(2 + x)^5 = \sum_{r=0}^{5} \binom{5}{r} 2^{5-r} x^r$$

$$= \binom{5}{0} 2^5 + \binom{5}{1} 2^4 x + \binom{5}{2} 2^3 x^2 + \binom{5}{3} 2^2 x^3 + \binom{5}{4} 2 x^4 + \binom{5}{5} x^5$$

$$= (1)(2^5) + (5)(2^4)x + (10)(2^3)x^2 + (10)(2^2)x^3 + (5)(2)x^4 + 1x^5$$

since $\binom{5}{0} = \dfrac{5!}{0!5!} = 1$, $\binom{5}{1} = \dfrac{5!}{1!4!} = 5$, $\binom{5}{2} = \dfrac{5!}{2!3!} = 10$ and so on. Thus

$$(2 + x)^5 = 32 + 80x + 80x^2 + 40x^3 + 10x^4 + x^5$$

1.3.5 Exercises

11 Simplify the following expressions

(a) $x^3 \times x^{-4}$ (b) $x^3 \div x^{-4}$ (c) $(x^3)^{-4}$

(d) $x^{1/3} \times x^{5/3}$ (e) $(4x^8)^{-1/2}$ (f) $\left(\dfrac{3}{2\sqrt{x}} \right)^{-2}$

(g) $\sqrt{x}\left(x^2 - \dfrac{2}{x} \right)$ (h) $\left(5x^{1/3} - \dfrac{1}{2x^{1/3}} \right)^2$

(i) $\dfrac{2x^{1/2} - x^{-1/2}}{x^{1/2}}$ (j) $\dfrac{(a^2 b)^{1/2}}{(ab^{-2})^2}$

(k) $(4ab^2)^{-3/2}$

12 Factorize

(a) $x^2 y - xy^2$

(b) $x^2 yz - xy^2 z + 2xyz^2$

(c) $ax - 2by - 2ay + bx$

(d) $x^2 + 3x - 10$

(e) $x^2 - \frac{1}{4}y^2$ (f) $81x^4 - y^4$

13 Simplify

(a) $\dfrac{x^2 - x - 12}{x^2 - 16}$ (b) $\dfrac{x - 1}{x^2 - 2x - 3} - \dfrac{2}{x + 1}$

(c) $\dfrac{1}{x^2 + 3x - 10} + \dfrac{1}{x^2 + 17x + 60}$

(d) $(3x + 2y)(x - 2y) + 4xy$

14 If

$$\frac{3c^2 + 3xc + x^2}{3c^2 + 3yc + y^2} = \frac{yV_1}{xV_2}$$

find the positive value of c when

$$x = 4, y = 6, V_1 = 120, V_2 = 315$$

15 Solve for p the equation

$$\frac{2p + 1}{p + 5} + \frac{p - 1}{p + 1} = 2$$

16 Rearrange the following formula to make s the subject

$$m = p\sqrt{\frac{s+t}{s-t}}$$

17 Given $u = \dfrac{x^2+t}{x^2-t}$, find t in terms of u and x.

18 Solve for t

$$\frac{1}{1-t} - \frac{1}{1+t} = 1$$

19 An isosceles trapezium has non-parallel sides of length 20 cm and the shorter parallel side is 30 cm. The perpendicular distance between the parallel sides is h cm. Show that the area of the trapezium is $h(30 + \sqrt{(400 - h^2)})$ cm^2.

20 An open container is made from a sheet of cardboard of size 200 mm × 300 mm using a simple fold, as shown in Figure 1.12. Show that the capacity C ml of the box is given by

$$C = x(150 - x)(100 - x)/250$$

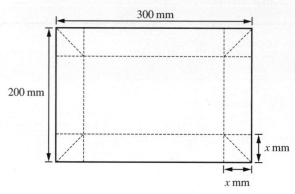

Figure 1.12 Sheet of cardboard of Question 20.

21 A rectangle has a perimeter of 30 m. If its length is twice its breadth, find the length.

22 (a) A4 paper is such that a half sheet has the same shape as the whole sheet. Find the ratio of the lengths of the sides of the paper.

(b) Foolscap paper is such that cutting off a square whose sides equal the shorter side of the paper leaves a rectangle which has the same shape as the original sheet. Find the ratio of the sides of the original page.

23 Find the values of x for which

(a) $\dfrac{5}{x} < 2$ (b) $\dfrac{1}{2-x} < 1$

(c) $\dfrac{3x-2}{x-1} > 2$ (d) $\dfrac{3}{3x-2} > \dfrac{1}{x+4}$

24 Find the values of x for which

$$x^2 < 2 + |x|$$

25 Rearrange the following quadratic expressions by completing the square.

(a) $x^2 + x - 12$ (b) $3 - 2x + x^2$

(c) $(x-1)^2 - (2x-3)^2$ (d) $1 + 4x - x^2$

26 Prove that

(a) $x^2 + 3x - 10 \geqslant -\left(\frac{7}{2}\right)^2$

(b) $18 + 4x - x^2 \leqslant 22$

(c) $x + \dfrac{4}{x} \geqslant 2$ where $x > 0$

27 Find the values of A and B such that

(a) $\dfrac{1}{(x+1)(x-2)} \equiv \dfrac{A}{x+1} + \dfrac{B}{x-2}$

(b) $3x + 2 \equiv A(x-1) + B(x-2)$

(c) $\dfrac{5x+1}{\sqrt{(x^2+x+1)}} \equiv \dfrac{A(2x+1) + B}{\sqrt{(x^2+x+1)}}$

28 Find the values of A, B and C such that

$$2x^2 - 5x + 12 \equiv A(x-1)^2 + B(x-1) + C$$

29 Given $a_0 = 2$, $a_1 = -1$, $a_2 = -4$, $a_3 = 5$, $a_4 = 3$ and $b_0 = 1$, $b_1 = 1$, $b_2 = 2$, $b_3 = -1$, $b_4 = 2$, calculate

(a) $\displaystyle\sum_{k=0}^{4} a_k$ (b) $\displaystyle\sum_{i=1}^{3} a_i$ (c) $\displaystyle\sum_{k=1}^{2} a_k b_k$

(d) $\displaystyle\sum_{j=0}^{4} b_j^2$ (e) $\displaystyle\prod_{k=1}^{3} a_k$ (f) $\displaystyle\prod_{k=1}^{4} b_k$

30 Evaluate

(a) $5!$ (b) $3!/4!$ (c) $7!/(3! \times 4!)$

(d) $\dbinom{5}{2}$ (e) $\dbinom{9}{3}$ (f) $\dbinom{8}{4}$

31 Using the general binomial expansion expand the following expressions

(a) $(x-3)^4$ (b) $(x + \frac{1}{2})^3$

(c) $(2x+3)^5$ (d) $(3x+2y)^4$

1.4 Geometry

1.4.1 Coordinates

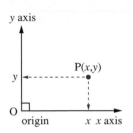

Figure 1.13

In addition to the introduction of algebraic manipulation another innovation made in the seventeenth century was the use of coordinates to represent the position of a point P on a plane as shown in Figure 1.13. Conventionally the point P is represented by an ordered pair of numbers contained in brackets thus: (x, y). This innovation was largely due to Descartes and consequently we often refer to (x, y) as the **cartesian coordinates** of P. This notation is the same as that for an open interval on the number line introduced in Section 1.2.4, but has an entirely separate meaning and the two should not be confused. Whether (x, y) denotes an open interval or a coordinate pair is usually clear from the context.

1.4.2 Straight lines

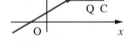

Figure 1.14

The introduction of coordinates made possible the algebraic description of the plane curves of classical geometry and the proof of standard results by algebraic methods.

Consider, for example, the point P lying on the line AB as shown in Figure 1.14. Let P divide AB in the ratio $\lambda : 1 - \lambda$. Then AP/AB $= \lambda$ and, by similar triangles,

$$\frac{AP}{AB} = \frac{PQ}{BC} = \frac{AQ}{AC}$$

Let A, B and P have coordinates (x_0, y_0), (x_1, y_1) and (x, y) respectively, then from the diagram

$$AQ = x - x_0, \ AC = x_1 - x_0, \ PQ = y - y_0, \ BC = y_1 - y_0$$

Thus

$$\frac{PQ}{BC} = \frac{AQ}{AC} \quad \text{implies} \quad \frac{y - y_0}{y_1 - y_0} = \frac{x - x_0}{x_1 - x_0}$$

from which we deduce, after some rearrangement,

$$y = \frac{y_1 - y_0}{x_1 - x_0}(x - x_0) + y_0 \tag{1.11}$$

which represents the equation of a straight line passing through two points (x_0, y_0) and (x_1, y_1).

More simply, the equation of a straight line passing through the two points having coordinates (x_0, y_0) and (x_1, y_1) may be written as

$$y = mx + c \tag{1.12}$$

where $m = \dfrac{y_1 - y_0}{x_1 - x_0}$ is the gradient (slope) of the line and $c = \dfrac{y_0 x_1 - y_1 x_0}{x_1 - x_0}$ is the intercept on the y axis.

Thus equations of the form

$$y = mx + c$$

represent straight lines on the plane and, consequently, are called **linear equations**.

Example 1.25   Find the equation of the straight line that passes through the points $(1, 2)$ and $(3, 3)$.

Solution   Taking $(x_0, y_0) = (1, 2)$ and $(x_1, y_1) = (3, 3)$

$$\text{slope of line} = \frac{y_1 - y_0}{x_1 - x_0} = \frac{3 - 2}{3 - 1} = \frac{1}{2}$$

so from formula (1.11) the equation of the straight line is

$$y = \tfrac{1}{2}(x - 1) + 2$$

which simplifies to

$$y = \tfrac{1}{2}x + \tfrac{3}{2}$$

Example 1.26   Find the equation of the straight line passing through the point $(3, 2)$ and parallel to the line $2y = 3x + 4$. Determine its x and y intercepts.

Solution   Writing $2y = 3x + 4$ as

$$y = \tfrac{3}{2}x + 2$$

we have from (1.12) that the slope of this line is $\tfrac{3}{2}$. Since the required line is parallel to this line, it will also have a slope of $\tfrac{3}{2}$. (The slope of the line perpendicular to it is $-\tfrac{2}{3}$.) Thus from (1.12) it has equation

$$y = \tfrac{3}{2}x + c$$

To determine the constant c, we use the fact that the line passes through the point $(3, 2)$, so that

$$2 = \tfrac{9}{2} + c \qquad \text{giving} \qquad c = -\tfrac{5}{2}$$

Thus the equation of the required line is

$$y = \tfrac{3}{2}x - \tfrac{5}{2} \qquad \text{or} \qquad 2y = 3x - 5$$

The y intercept is $c = -\tfrac{5}{2}$.

To obtain the x intercept we substitute $y = 0$, giving $x = \tfrac{5}{3}$, so that the x intercept is $\tfrac{5}{3}$.

The graph of the line is shown in Figure 1.15.

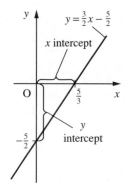

Figure 1.15
The straight line
$2y = 3x - 5$.

1.4.3 Circles

A circle is the planar curve whose points are all equidistant from a fixed point called the centre of the circle. The simplest case is a circle centred at the origin with

Figure 1.16
(a) A circle of centre origin, radius r. (b) A circle of centre (a, b), radius r.

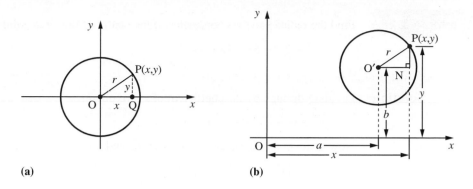

(a)   **(b)**

radius r, as shown in Figure 1.16(a). Applying Pythagoras' theorem to triangle OPQ we obtain

$$x^2 + y^2 = r^2$$

(Note that r is a constant.) When the centre of the circle is at the point (a, b), rather than the origin, the equation is

$$(x - a)^2 + (y - b)^2 = r^2 \qquad (1.13a)$$

obtained by applying Pythagoras' theorem in triangle O'PN of Figure 1.16(b). This expands to

$$x^2 + y^2 - 2ax - 2by + (a^2 + b^2 - r^2) = 0$$

so that the general equation

$$x^2 + y^2 + 2fx + 2gy + c = 0 \qquad (1.13b)$$

represents a circle having centre $(-f, -g)$ and radius $\sqrt{(f^2 + g^2 - c)}$. Notice that the general circle has three constants f, g and c in its equation. This implies that we need three points to specify a circle completely.

Example 1.27 Find the equation of the circle with centre $(1, 2)$ and radius 3.

Solution Using Pythagoras' theorem, if the point $P(x, y)$ lies on the circle then from (1.13a)

$$(x - 1)^2 + (y - 2)^2 = 3^2$$

Thus

$$x^2 - 2x + 1 + y^2 - 4y + 4 = 9$$

giving the equation as

$$x^2 + y^2 - 2x - 4y - 4 = 0$$

Example 1.28 Find the radius and the coordinates of the centre of the circle whose equation is

$$2x^2 + 2y^2 - 3x + 5y + 2 = 0$$

Solution Dividing through by the coefficient of x^2 we obtain

$$x^2 + y^2 - \tfrac{3}{2}x + \tfrac{5}{2}y + 1 = 0$$

Now completing the square on the x terms and the y terms separately gives

$$(x - \tfrac{3}{4})^2 + (y + \tfrac{5}{4})^2 = \tfrac{9}{16} + \tfrac{25}{16} - 1 = \tfrac{18}{16}$$

Hence, from 13(a), the circle has radius $(3\sqrt{2})/4$ and centre $(3/4, -5/4)$.

Example 1.29 Find the equation of the circle which passes through the points $(0, 0)$, $(0, 2)$, $(4, 0)$.

Solution Method (a): From (1.13b) the general equation of a circle is

$$x^2 + y^2 + 2fx + 2gy + c = 0$$

Substituting the three points into this equation gives three equations for the unknowns a, b and c.

Thus substituting $(0, 0)$ gives $c = 0$, substituting $(0, 2)$ gives $4 + 4g + c = 0$ and substituting $(4, 0)$ gives $16 + 8f + c = 0$. Solving these equations gives $g = -1$, $f = -2$ and $c = 0$ so that the required equation is

$$x^2 + y^2 - 4x - 2y = 0$$

Figure 1.17
The circle of
Example 1.29.

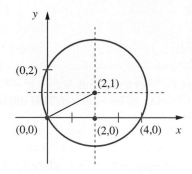

Method (b): From Figure 1.17 using the geometrical properties of the circle, we see that its centre lies at $(2, 1)$ and since it passes through the origin its radius is $\sqrt{5}$. Hence, from 1.13(a), its equation is

$$(x - 2)^2 + (y - 1)^2 = (\sqrt{5})^2$$

which simplifies to

$$x^2 + y^2 - 4x - 2y = 0$$

as before.

Figure 1.18
Standard equations
of the four conics.

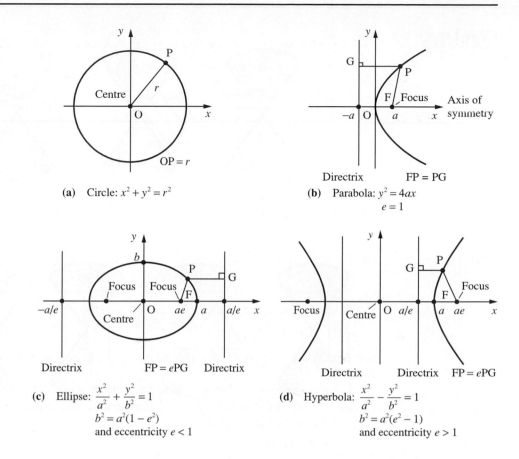

(a) Circle: $x^2 + y^2 = r^2$

(b) Parabola: $y^2 = 4ax$
$e = 1$

(c) Ellipse: $\dfrac{x^2}{a^2} + \dfrac{y^2}{b^2} = 1$
$b^2 = a^2(1 - e^2)$
and eccentricity $e < 1$

(d) Hyperbola: $\dfrac{x^2}{a^2} - \dfrac{y^2}{b^2} = 1$
$b^2 = a^2(e^2 - 1)$
and eccentricity $e > 1$

1.4.4 Conics

The circle is one of the conic sections (Figure 1.18) introduced around 200 BC by
Apollonius, who published an extensive study of their properties in a textbook that he
called *Conics*. He used this title because he visualized them as cuts made by a 'flat' or
plane surface when it intersects the surface of a cone in different directions, as illustrated in Figures 1.19(a–d). Note that the conic sections degenerate into a point and
straight lines at the extremities, as illustrated in Figures 1.19(e–g). Although at the time
of Apollonius his work on conics appeared to be of little value in terms of applications,
it has since turned out to have considerable importance. This is primarily due to the fact
that the conic sections are the paths followed by projectiles, artificial satellites, moons
and the Earth under the influence of gravity around planets or stars. The early Greek
astronomers thought that the planets moved in circular orbits, and it was not until 1609
that the German astronomer Johannes Kepler described their paths correctly as being
elliptic, with the Sun at one focus. It is quite possible for an orbit to be a curve other
than an ellipse. Imagine a meteorite or comet approaching the Sun from some distant
region in space. The path that the body will follow depends very much on the speed at
which it is moving. If the body is small compared with the Sun, say of planetary dimensions, and its speed relative to the Sun is not very high, it will never escape and will
describe an *elliptic* path about it. An example is the comet observed by Edward Halley
in 1682 and now known as Halley's comet. He computed its elliptic orbit, found that it

Figure 1.19

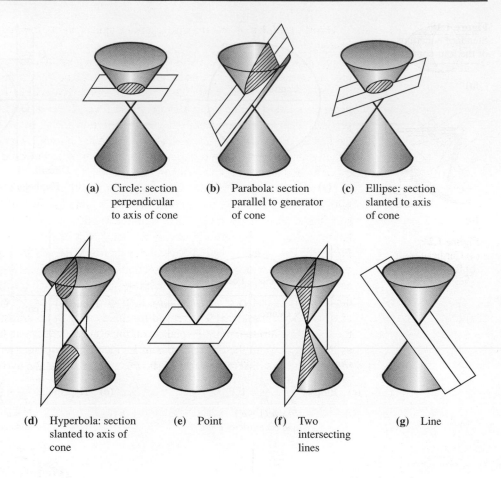

(a) Circle: section perpendicular to axis of cone

(b) Parabola: section parallel to generator of cone

(c) Ellipse: section slanted to axis of cone

(d) Hyperbola: section slanted to axis of cone

(e) Point

(f) Two intersecting lines

(g) Line

Figure 1.20

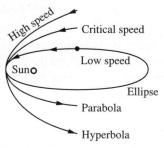

was the same comet that had been seen in 1066, 1456, 1531 and 1607, and correctly forecast its reappearance in 1758. It was most recently seen in 1986. If the speed of the body is very high, its path will be deviated by the Sun but it will not orbit forever around the Sun. Rather, it will bend around the Sun in a path in the form of a **hyperbola** and continue on its journey back to outer space. Somewhere between these two extremes there is a certain critical speed that is just too great to allow the body to orbit the Sun, but not great enough for the path to be a hyperbola. In this case the path is a **parabola**, and once again the body will bend around the Sun and continue on its journey into outer space. These possibilities are illustrated in Figure 1.20.

(a)

(b)

Figure 1.21
(a) Car headlamp.
(b) Radio telescope.

Other examples of where conic sections appear in engineering practice include the following.

(a) A parabolic surface, obtained by rotating a parabola about its axis of symmetry, has the important property that an energy source placed at the focus will cause rays to be reflected at the surface such that after reflection they will be parallel. Reversing the process, a parallel beam impinging on the surface will be reflected on to the focus. This property is involved in many engineering design projects: for example the design of a car headlamp or a radio telescope, as illustrated in Figures 1.21(a) and (b) respectively. Other examples involving a parabola are the path of a projectile and the shape of the cable on certain types of suspension bridge.

(b) A ray of light emitted from one focus of an elliptic mirror and reflected by the mirror will pass through the other focus as illustrated in Figure 1.22. This property is sometimes used in designing mirror combinations for a reflecting telescope. Ellipses have been used in other engineering designs, such as aircraft wings and stereo styli. Formerly, in order to avoid bursts due to freezing, water pipes were sometimes designed to have an elliptic cross-section. As described earlier, every planet orbits around the Sun in an elliptic path with the Sun at one of its foci. The planet's speed depends on its distance from the Sun; it speeds up as it nears the Sun and slows down as it moves further away. The reason for this is that for an ellipse the line drawn from the focus S (Sun) to a point P (planet) on the ellipse sweeps out areas at a constant rate as P moves around the ellipse. Thus in Figure 1.23 the planet will take the same time to travel the two different distances shown, assuming that the two shaded regions are of equal area.

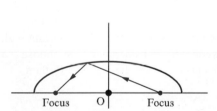

Figure 1.22 Reflection of a ray by an elliptic mirror.

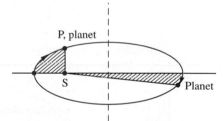

Figure 1.23

(c) Consider Concorde flying over land. As it breaks the sound barrier (that is, it travels faster than the speed of sound, which is about 750 mph ($331.4\,\mathrm{m\,s^{-1}}$)), it will create a shock wave, which we hear on the ground as a *sonic boom* – this being one of the major disadvantages of supersonic aircraft. This shock wave will trail behind the aircraft in the form of a cone with the aircraft as vertex. This cone will intersect the ground in a *hyperbolic curve* as illustrated in Figure 1.24. The sonic boom will hit every point on this curve at the same instant of time, so that people living on the curve will hear it simultaneously. No boom will be heard by people living outside this curve, but eventually it will be heard at every point inside it.

Figure 1.18 illustrates the conics in their standard positions, and the corresponding equations may be interpreted as the standard equations for the four curves. More generally the conic sections may be represented by the general second-order equation

Figure 1.24
Sonic boom.

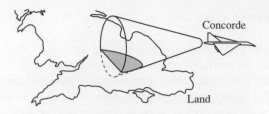

Concorde

Land

$$ax^2 + by^2 + 2fx + 2gy + 2hxy + c = 0 \qquad\qquad \textbf{(1.14)}$$

Provided its graph does not degenerate into a point or straight lines, equation (1.14) is representative of

- a circle if $a = b \neq 0$ and $h = 0$
- a parabola if $h^2 = ab$
- an ellipse if $h^2 < ab$
- a hyperbola if $h^2 > ab$

The conics can be defined mathematically in a number of (equivalent) ways as we will illustrate in the next examples.

Example 1.30

A point P moves in such a way that its total distance from two fixed points A and B is constant. Show that it describes an ellipse.

Solution

The definition of the curve implies that $AP + BP = \text{constant}$ with the origin O being the midpoint of AB. From symmetry considerations we choose x and y axes as shown in Figure 1.25. Suppose the curve crosses the x axis at P_0 then

$$AP_0 + BP_0 = AB + 2AP_0 = 2OP_0$$

so the constant in the definition is $2OP_0$ and for any point P on the curve

$$AP + BP = 2OP_0$$

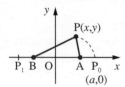

Figure 1.25
Path of Example 1.30.

Let $P = (x, y)$, $P_0 = (a, 0)$, $P_1 = (-a, 0)$, $A = (c, 0)$ and $B = (-c, 0)$. Then using Pythagoras' theorem we have

$$AP = \sqrt{[(x - c)^2 + y^2]}$$

$$BP = \sqrt{[(x + c)^2 + y^2]}$$

so that the defining equation of the curve becomes

$$\sqrt{[(x - c)^2 + y^2]} + \sqrt{[(x + c)^2 + y^2]} = 2a$$

To obtain the required equation we need to 'remove' the square root terms. This can only be done by squaring both sides of the equation. First we rewrite the equation as

$$\sqrt{[(x - c)^2 + y^2]} = 2a - \sqrt{[(x + c)^2 + y^2]}$$

and then square to give

$$(x - c)^2 + y^2 = 4a^2 - 4a\sqrt{[(x + c)^2 + y^2]} + (x + c)^2 + y^2$$

Expanding the squared terms we have

$$x^2 - 2cx + c^2 + y^2 = 4a^2 - 4a\sqrt{[(x + c)^2 + y^2]} + x^2 + 2cx + c^2 + y^2$$

Collecting together terms, we obtain

$$a\sqrt{[(x + c)^2 + y^2]} = a^2 + cx$$

Squaring both sides again gives

$$a^2[x^2 + 2cx + c^2 + y^2] = a^4 + 2a^2cx + c^2x^2$$

which simplifies to

$$(a^2 - c^2)x^2 + a^2y^2 = a^2(a^2 - c^2)$$

Noting that $a > c$ we write $a^2 - c^2 = b^2$, to obtain

$$b^2x^2 + a^2y^2 = a^2b^2$$

which yields the standard equation of the ellipse

$$\frac{x^2}{a^2} + \frac{y^2}{b^2} = 1$$

The points A and B are the foci of the ellipse, and the property that the sum of the focal distances is a constant is known as the **string property** of the ellipse since it enables us to draw an ellipse using a piece of string.

For a hyperbola, the *difference* of the focal distances is constant.

Example 1.31

A point moves in such a way that its distance from a fixed point F is equal to its perpendicular distance from a fixed line. Show that it describes a parabola.

Solution

Suppose the fixed line is LL′ shown in Figure 1.26, choosing the coordinate axes shown. Since PF = PN for points on the curve we deduce that the curve bisects FM, so that if F is $(a, 0)$, then M is $(-a, 0)$. Let the general point P on the curve have coordinates (x, y). Then by Pythagoras' theorem

$$PF = \sqrt{[(x - a)^2 + y^2]}$$

Also PN = $x + a$, so that PN = PF implies that

$$x + a = \sqrt{[(x - a)^2 + y^2]}$$

Squaring both sides gives

$$(x + a)^2 = (x - a)^2 + y^2$$

which simplifies to

$$y^2 = 4ax$$

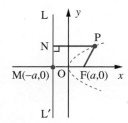

Figure 1.26
Path of point in
Example 1.31.

the standard equation of a parabola. The line LL′ is called the **directrix** of the parabola.

1.4.5 Parametric representation

In some practical situations the equation describing a curve in cartesian coordinates is very complicated and it is easier to specify the points in terms of a parameter. Sometimes this occurs in a very natural way. For example, in considering the trajectory of a projectile, we might specify its height and horizontal displacement separately in terms of the flight time. In the design of a safety guard for a moving part in a machine we might specify the position of the part in terms of an angle it has turned through. Such representation of curves is called **parametric representation** and we will illustrate the idea with an example. Later, in Section 2.6.6, we shall consider the polar form of specifying the equation of a curve.

The drawing of curves specified parametrically is easily performed using a graphical computer package or calculator.

Example 1.32 The horizontal and vertical displacements of a projectile at time t are x and y, respectively, as illustrated in Figure 1.27 where $x = ut$ and $y = vt - \frac{1}{2}gt^2$ where u and v are the initial horizontal and vertical velocities and g is the acceleration due to gravity. Show that its trajectory is a parabola.

Figure 1.27
Path of a projectile.

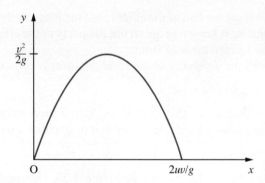

Solution Since $x = ut$ we may write $t = x/u$. Substituting this into the expression for y gives

$$y = \frac{vx}{u} - \frac{gx^2}{2u^2}$$

which is the equation of a parabola.

Completing the square we obtain

$$y = \frac{v^2}{2g} - \frac{g}{2u^2}\left(\frac{uv}{g} - x\right)^2$$

from which we can see that the projectile attains its maximum height, $\dfrac{v^2}{2g}$, at $x = uv/g$.

1.4.6 Exercises

32 Find the equation of the straight line

(a) with gradient $\frac{3}{2}$ passing through the point $(2, 1)$,

(b) with gradient -2 passing through the point $(-2, 3)$,

(c) passing through the points $(1, 2)$ and $(3, 7)$,

(d) passing through the points $(5, 0)$ and $(0, 3)$,

(e) parallel to the line $3y - x = 5$, passing through $(1, 1)$,

(f) perpendicular to the line $3y - x = 5$, passing through $(1, 1)$.

33 Write down the equation of the circle with centre $(1, 2)$ and radius 5.

34 Find the radius and the coordinates of the centre of the circle with equation

$$x^2 + y^2 + 4x - 6y = 3$$

35 Find the equation of the circle with centre $(-2, 3)$ that passes through $(1, -1)$.

36 Find the equation of the circle that passes through the points $(1, 0)$, $(3, 4)$ and $(5, 0)$.

37 A rod, 50 cm long, moves in a plane with its ends on two perpendicular wires. Find the equation of the curve followed by its midpoint.

38 Find the coordinates of the focus and the equation of the directrix of the parabola whose equation is

$$3y^2 = 8x$$

The chord which passes through the focus parallel to the directrix is called the **latus rectum** of the parabola. Show that the latus rectum of the above parabola has length 8/3.

39 For the ellipse $25x^2 + 16y^2 = 400$ find the coordinates of the foci, the eccentricity, the equations of the directrices and the lengths of the semi-major and semi-minor axes.

40 For the hyperbola $9x^2 - 16y^2 = 144$ find the coordinates of the foci and the vertices and the equations of its asymptotes.

41 Plot the curve whose parametric equations are $x = t(t + 4)$, $y = t + 1$. Show that it is a parabola.

42 Sketch the curve (the Cissoid of Diocles) given by

$$x = \frac{2t^2}{t^2 + 1}, \quad y = \frac{2t^3}{t^2 + 1}$$

Show that the cartesian form of the curve is

$$y^2 = x^3/(2 - x)$$

1.5 Numbers and accuracy

Arithmetic that only involves integers can be performed to obtain an exact answer (that is, one without rounding errors). In general this is not possible with real numbers and when solving practical problems such numbers are rounded to an appropriate number of digits. In this section we shall review the methods of recording numbers, obtain estimates for the effect of rounding errors in elementary calculations and discuss the implementation of arithmetic on computers.

1.5.1 Representation of numbers

For ordinary everyday purposes we use a system of representation based on ten **numerals**: 0, 1, 2, 3, 4, 5, 6, 7, 8, 9. These ten symbols are sufficient to represent all numbers if a **position notation** is adopted. For whole numbers this means that, starting from the right-hand end of the number, the least significant end, the figures represent the number of units, tens, hundreds, thousands, and so on. Thus one thousand, three hundred and sixty-five is represented by 1365, and two hundred and nine is represented by 209. Notice the role of the 0 in the latter example, acting as a position keeper. The use of a decimal point makes it possible to represent fractions as well as whole numbers. This system uses ten symbols. The number system is said to be 'to base ten' and is called the **decimal** system. Other bases are possible: for example, the Babylonians used a number system to base sixty, a fact that still influences our measurement of time. In some societies a number system evolved with more than one base, a survival of which can be seen in imperial measures (inches, feet, yards, . . .). For some applications it is more convenient to use a base other than ten. Early electronic computers used **binary** numbers (to base two); modern computers use **hexadecimal** numbers (to base sixteen). For elementary (pen-and-paper) arithmetic a representation to base twelve would be more convenient than the usual decimal notation because twelve has more integer divisors (2, 3, 4, 6) than ten (2, 5).

In a decimal number the positions to the left of the decimal point represent units (10^0), tens (10^1), hundreds (10^2) and so on, while those to the right of the decimal point represent tenths (10^{-1}), hundredths (10^{-2}) and so on. Thus, for example

$$
\begin{array}{ccccc}
2 & 1 & 4 & \cdot \quad 3 & 6 \\
\downarrow & \downarrow & \downarrow & \downarrow & \downarrow \\
10^2 & 10^1 & 10^0 & 10^{-1} & 10^{-2}
\end{array}
$$

so

$$214.36 = 2(10^2) + 1(10^1) + 4(10^0) + 3(\tfrac{1}{10}) + 6(\tfrac{1}{100})$$

$$= 200 + 10 + 4 + \tfrac{3}{10} + \tfrac{6}{100}$$

$$= \tfrac{21436}{100} = \tfrac{5359}{25}$$

In other number bases the pattern is the same: in base b the position values are $b^0, b^1, b^2, \ldots$ and $b^{-1}, b^{-2}, \ldots$. Thus in binary (base two) the position values are units, twos, fours, eights, sixteens and so on, and halves, quarters, eighths and so on. In hexadecimal (base sixteen) the position values are units, sixteens, two hundred and fifty-sixes, and so on, and sixteenths, two hundred and fifty-sixths, and so on.

Example 1.33   Write (a) the binary number 1011101_2 as a decimal number and (b) the decimal number 115_{10} as a binary number.

Solution   (a) $1011101_2 = 1(2^6) + 0(2^5) + 1(2^4) + 1(2^3) + 1(2^2) + 0(2^1) + 1(2^0)$

$$= 64_{10} + 0 + 16_{10} + 8_{10} + 4_{10} + 0 + 1_{10}$$

$$= 93_{10}$$

(b)   We achieve the conversion to binary by repeated division by 2. Thus

$$115 \div 2 = 57 \quad \text{remainder 1} \quad (2^0)$$

$$57 \div 2 = 28 \quad \text{remainder 1} \quad (2^1)$$

$$28 \div 2 = 14 \quad \text{remainder } 0 \qquad (2^2)$$
$$14 \div 2 = 7 \quad \text{remainder } 0 \qquad (2^3)$$
$$7 \div 2 = 3 \quad \text{remainder } 1 \qquad (2^4)$$
$$3 \div 2 = 1 \quad \text{remainder } 1 \qquad (2^5)$$
$$1 \div 2 = 0 \quad \text{remainder } 1 \qquad (2^6)$$

so that

$$115_{10} = 1110011_2$$

Example 1.34 Represent the numbers (a) two hundred and one, (b) two hundred and seventy-five, (c) five and three-quarters and (d) one-third in

(i) decimal form using the figures 0, 1, 2, 3, 4, 5, 6, 7, 8, 9;

(ii) binary form using the figures 0, 1;

(iii) duodecimal (base 12) form using the figures 0, 1, 2, 3, 4, 5, 6, 7, 8, 9, Δ, ε.

Solution (a) two hundred and one

(i) $= 2$ (hundreds) $+ 0$ (tens) and 1 (units) $= 201_{10}$

(ii) $= 1$ (one hundred and twenty-eight) $+ 1$ (sixty-four) $+ 1$ (eight) $+ 1$ (unit) $= 11001001_2$

(iii) $= 1$ (gross) $+ 4$ (dozens) $+ 9$ (units) $= 149_{12}$

Here the subscripts 10, 2, 12 indicate the number base.

(b) two hundred and seventy-five

(i) $= 2$ (hundreds) $+ 7$ (tens) $+ 5$ (units) $= 275_{10}$

(ii) $= 1$ (two hundred and fifty-six) $+ 1$ (sixteen) $+ 1$ (two) $+ 1$ (unit) $= 100010011_2$

(iii) $= 1$ (gross) $+ 10$ (dozens) $+$ eleven (units) $= 1\Delta\varepsilon_{12}$
(Δ represents ten and ε represents eleven)

(c) five and three-quarters

(i) $= 5$ (units) $+ 7$ (tenths) $+ 5$ (hundredths) $= 5.75_{10}$

(ii) $= 1$ (four) $+ 1$ (unit) $+ 1$ (half) $+ 1$ (quarter) $= 101.11_2$

(iii) $= 5$ (units) $+ 9$ (twelfths) $= 5.9_{12}$

(d) one-third

(i) $= 3$ (tenths) $+ 3$ (hundredths) $+ 3$ (thousandths) $+ \ldots = 0.333 \ldots_{10}$

(ii) $= 1$ (quarter) $+ 1$ (sixteenth) $+ 1$ (sixty-fourth) $+ \ldots = 0.010101 \ldots_2$

(iii) $= 4$ (twelfths) $= 0.4_{12}$

1.5.2 Rounding, decimal places and significant figures

The Fundamental Laws of Arithmetic are, of course, independent of the choice of representation of the numbers. Similarly, the representation of irrational numbers will always be incomplete. Because of these numbers and because some rational numbers have recurring representations (whether the representation of a particular rational number is recurring or not will of course depend on the number base used – see Example 1.34d), any arithmetical calculation will contain errors caused by truncation. In practical problems it is usually known how many figures are meaningful, and the numbers are 'rounded' accordingly. In the decimal representation, for example, the numbers are approximated by the closest decimal number with some prescribed number of figures after the decimal point. Thus, to two decimal places (dp),

$$\pi = 3.14 \quad \text{and} \quad \tfrac{5}{12} = 0.42$$

and to five decimal places

$$\pi = 3.141\,59 \quad \text{and} \quad \tfrac{5}{12} = 0.416\,67$$

Normally this is abbreviated to

$$\pi = 3.141\,59 \text{ (5dp)} \quad \text{and} \quad \tfrac{5}{12} = 0.416\,67 \text{ (5dp)}$$

Similarly

$$\sqrt{2} = 1.4142 \text{ (4dp)} \quad \text{and} \quad \tfrac{2}{3} = 0.667 \text{ (3dp)}$$

In hand computation, by convention, when shortening a number ending with a five we 'round to the even'. For example,

$$1.2345 \quad \text{and} \quad 1.2335$$

are both represented by 1.234 to three decimal places. In contrast, most calculators and computers would 'round up' in the ambiguous case, giving 1.2345 and 1.2335 as 1.235 and 1.234 respectively.

Any number occurring in practical computation will either be given an error bound or be correct to within half a unit in the least significant figure (sf). For example

$$\pi = 3.14 \pm 0.005 \quad \text{or} \quad \pi = 3.14$$

Any number given in scientific or mathematical tables observes this convention. Thus

$$g_0 = 9.806\,65$$

implies

$$g_0 = 9.806\,65 \pm 0.000\,005$$

that is,

$$9.806\,645 < g_0 < 9.806\,655$$

as illustrated in Figure 1.28.

Figure 1.28

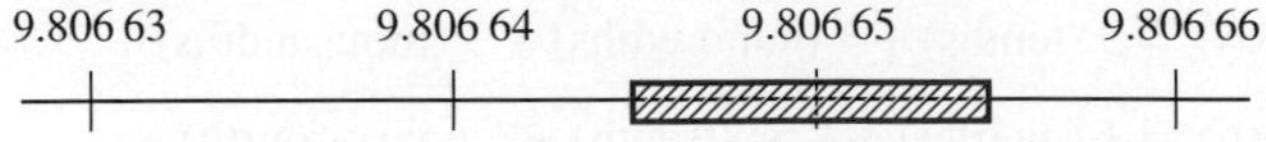

Sometimes the decimal notation may create a false impression of accuracy. When we write that the distance of the Earth from the Sun is ninety-three million miles, we mean that the distance is nearer to 93 000 000 than 94 000 000 or 92 000 000, not that it is nearer to 93 000 000 than 93 000 001 or 92 999 999. This possible misinterpretation of numerical data is avoided by either stating the number of significant figures, giving an error estimate or using scientific notation. In this example the distance d miles is given in the forms

$$d = 93\,000\,000 \quad (2\text{sf})$$

or

$$d = 93\,000\,000 \pm 500\,000$$

or

$$d = 9.3 \times 10^7$$

Notice how information about accuracy is discarded by the rounding-off process. The value ninety-three million miles is actually correct to within fifty thousand miles, while the convention about rounded numbers would imply an error bound of five hundred thousand.

The number of significant figures tells us about the relative accuracy of a number when it is related to a measurement. Thus a number given to 3sf is relatively ten times more accurate than one given to 2sf. The number of decimal places, dp, merely tells us the number of digits including leading zeros after the decimal point. Thus

$$2.321 \quad \text{and} \quad 0.000\,059\,71$$

both have 4sf, while the former has 3dp and the latter 8dp.

It is not clear how many significant figures a number like 3200 has. It might be 2, 3 or 4. To avoid this ambiguity it must be written in the form 3.2×10^3 (when it is correct to 2sf) or 3.20×10^3 (3sf) or 3.200×10^3 (4sf). This is usually called **scientific notation**. It is widely used to represent numbers that are very large or very small. Essentially, a number x is written in the form

$$x = a \times 10^n$$

where $1 \leqslant |a| < 10$ and n is an integer. Thus the mass of an electron at rest is 9.11×10^{-28} g, while the velocity of light in a vacuum is 2.9978×10^{10} cm s^{-1}.

Example 1.35

Express the number 150.4152

(a) correct to 1, 2 and 3 dp;   (b) correct to 1, 2 and 3 sf.

Solution

(a) $150.4152 = 150.4 \qquad (1\text{dp})$

$\qquad\qquad\quad = 150.42 \quad\;\, (2\text{dp})$

$\qquad\qquad\quad = 150.415 \quad (3\text{dp})$

(b) $150.4152 = 1.504\,152 \times 10^2$

$\qquad\qquad\quad = 2 \times 10^2 \qquad\;\, (1\text{sf})$

$\qquad\qquad\quad = 1.5 \times 10^2 \qquad (2\text{sf})$

$\qquad\qquad\quad = 1.50 \times 10^3 \quad\; (3\text{sf})$

1.5.3 Estimating the effect of rounding errors

Numerical data obtained experimentally will often contain rounding errors due to the limited accuracy of measuring instruments. Also, because irrational numbers and some rational numbers do not have a terminating decimal representation, arithmetical operations inevitably contain errors arising from rounding off. The effect of such errors can accumulate in an arithmetical procedure and good engineering computations will include an estimate for it. This process has become more important with the widespread use of computers. When users are isolated from the computational chore, they often fail to develop a sense of the limits of accuracy of an answer. In this section we shall develop the basic ideas for such sensitivity analyses of calculations.

Example 1.36   Compute

(a) $3.142 + 4.126$ (b) $5.164 - 2.341$ (c) 235.12×0.531

Calculate estimates for the effects of rounding errors in each answer and give the answer as a correctly rounded number.

Solution   (a) $3.142 + 4.126 = 7.268$

Because of the convention about rounded numbers, 3.142 represents all the numbers a between 3.1415 and 3.1425, and 4.126 represents all the numbers b between 4.1255 and 4.1265. Thus if a and b are correctly rounded numbers, their sum $a + b$ lies between $c_1 = 7.2670$ and $c_2 = 7.2690$. Rounding c_1 and c_2 to 3dp gives $c_1 = 7.267$ and $c_2 = 7.269$. Since these disagree, we cannot give an answer to 3dp. Rounding c_1 and c_2 to 2dp gives $c_1 = 7.27$ and $c_2 = 7.27$. Since these agree, we can give the answer to 2dp; thus $a + b = 7.27$, as shown in Figure 1.29.

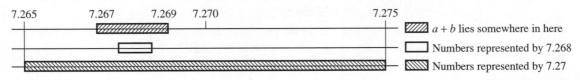

Figure 1.29

(b) $5.164 - 2.341 = 2.823$

Applying the same 'worst case' analysis to this implies that the difference lies between $5.1635 - 2.3415$ and $5.1645 - 2.3405$, that is between 2.8220 and 2.8240. Thus the answer should be written 2.823 ± 0.001 or, as a correctly rounded number, 2.82.

(c) $235.12 \times 0.531 = 124.84872$

Clearly, writing an answer with so many decimal places is unjustified if we are using rounded numbers, but how many decimal places are sensible? Using the 'worst case' analysis again, we deduce that the product lies between 235.115×0.5305 and 235.125×0.5315, that is between $c_1 = 124.7285075$ and $c_2 = 124.9689375$. Thus the answer should be written 124.85 ± 0.13. In this example, because of the place where the number occurs on the number line, c_1 and c_2 only agree when we round them to 3sf (0dp). Thus the product as a correctly rounded number is 125.

A competent computation will contain within it estimates of the effect of rounding errors. Analysing the effect of such errors for complicated expressions has to be approached systematically.

Definitions

(a) The **error** in a value is defined by

error = approximate value − true value

This is sometimes termed the dead error. Notice that the true value equals the approximate value minus the error.

(b) Similarly the **correction** is defined by

true value = approximate value + correction

so that

correction = −error

(c) The **error modulus** is the size of the error, $|\text{error}|$, and the **error bound** (or **absolute error bound**) is the maximum possible error modulus.

(d) The **relative error** is the ratio of the size of the error to the size of the true value:

$$\text{relative error} = \left| \frac{\text{error}}{\text{value}} \right|$$

The **relative error bound** is the maximum possible relative error.

(e) The **percent error** (or percentage error) is $100 \times$ relative error and the **percent error bound** is the maximum possible percent error.

In some contexts we think of the true value as an approximation and a remainder. In such cases the remainder is given by

remainder = −error

= correction

Example 1.37

Give the absolute and relative error bounds of the following correctly rounded numbers

(a) 29.92   (b) −0.015 23   (c) 3.9×10^{10}

Solution

(a) The number 29.92 is given to 2dp, which implies that it represents a number within the domain 29.92 ± 0.005. Thus its absolute error bound is 0.005, half a unit of the least significant figure, and its relative error bound is 0.005/29.92 or 0.000 17.

(b) The absolute error bound of −0.015 23 is half a unit of the least significant figure, that is, 0.000 005. Notice that it is a positive quantity. Its relative error bound is 0.000 005/0.015 23 or 0.000 33.

(c) The absolute error bound of 3.9×10^{10} is $0.05 \times 10^{10} = 5 \times 10^{8}$ and its relative error bound is 0.05/3.9 or 0.013.

Figure 1.30

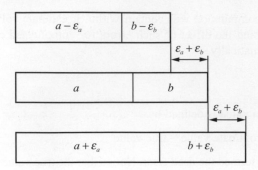

Usually, because we do not know the true values, we estimate the effects of error in a calculation in terms of the error bounds, the 'worst case' analysis illustrated in Example 1.36. The error bound of a value v is denoted by ε_v.

Consider, first, the sum $c = a + b$. When we add together the two rounded numbers a and b their sum will inherit a rounding error from both a and b. The true value of a lies between $a - \varepsilon_a$ and $a + \varepsilon_a$ and the true value of b lies between $b - \varepsilon_b$ and $b + \varepsilon_b$. Thus the smallest value that the true value of c can have is $a - \varepsilon_a + b - \varepsilon_b$, and its largest possible value is $a + \varepsilon_a + b + \varepsilon_b$. (Remember that ε_a and ε_b are positive.) Thus $c = a + b$ has an error bound

$$\varepsilon_c = \varepsilon_a + \varepsilon_b$$

as illustrated in Figure 1.30. A similar 'worst case' analysis shows that the difference $d = a - b$ has an error bound that is the sum of the error bounds of a and b:

$$d = a - b, \qquad \varepsilon_d = \varepsilon_a + \varepsilon_b$$

Thus for both addition and subtraction the error bound of the result is the sum of the individual error bounds.

Next consider the product $p = a \times b$, where a and b are positive numbers. The smallest possible value of p will be equal to the product of the least possible values of a and b; that is,

$$p > (a - \varepsilon_a) \times (b - \varepsilon_b)$$

Similarly

$$p < (a + \varepsilon_a) \times (b + \varepsilon_b)$$

Thus, on multiplying out the brackets, we obtain

$$ab - a\varepsilon_b - b\varepsilon_a + \varepsilon_a\varepsilon_b < p < ab + a\varepsilon_b + b\varepsilon_a + \varepsilon_a\varepsilon_b$$

Ignoring the very small term $\varepsilon_a\varepsilon_b$, we obtain an estimate for the error bound of the product:

$$\varepsilon_p = a\varepsilon_b + b\varepsilon_a, \qquad p = a \times b$$

Dividing both sides of the equation by p, we obtain

$$\frac{\varepsilon_p}{p} = \frac{\varepsilon_a}{a} + \frac{\varepsilon_b}{b}$$

Now the relative error of a is defined as the ratio of the error in a to the size of a. The above equation connects the relative error bounds for a, b and p:

$$r_p = r_a + r_b$$

Here $r_a = \varepsilon_a/|a|$ allowing for a to be negative, and so on.

A similar worst case analysis for the quotient $q = a/b$ leads to the estimate

$$r_q = r_a + r_b$$

Thus for both multiplication and division, the relative error bound of the result is the sum of the individual relative error bounds.

These elementary rules for estimating error bounds can be combined to obtain more general results. For example, consider $z = x^2$; then $r_z = 2r_x$. In general, if $z = x^y$, where x is a rounded number and y is exact, then

$$r_z = yr_x$$

Example 1.38

Evaluate 13.92×5.31 and $13.92 \div 5.31$.

Assuming that these values are correctly rounded numbers, calculate error bounds for each answer and write them as correctly rounded numbers which have the greatest possible number of significant digits.

Solution

$13.92 \times 5.31 = 73.9152$; $13.92 \div 5.31 = 2.621\,468\,927$

Let $a = 13.92$ and $b = 5.31$, then $r_a = 0.000\,36$ and $r_b = 0.000\,94$, so that $a \times b$ and $a \div b$ have relative error bounds $0.000\,36 + 0.000\,94 = 0.0013$. We obtain the absolute error bound of $a \times b$ by multiplying the relative error bound by $a \times b$. Thus the absolute error bound of $a \times b$ is $0.0013 \times 73.9152 = 0.0961$. Similarly, the absolute error bound of $a \div b$ is $0.0013 \times 2.6215 = 0.0034$. Hence the values of $a \times b$ and $a \div b$ lie in the error intervals

$$73.9152 - 0.0961 < a \times b < 73.9152 + 0.0961$$

and

$$2.6215 - 0.0034 < a \div b < 2.6215 + 0.0034$$

Thus $73.8191 < a \times b < 74.0113$ and $2.6181 < a \div b < 2.6249$.

From these inequalities we can deduce the correctly rounded values of $a \times b$ and $a \div b$:

$$a \times b = 74 \quad \text{and} \quad a \div b = 2.62$$

and we see how the rounding convention discards information. In a practical context, it would probably be more helpful to write:

$$73.81 < a \times b < 74.02$$

and

$$2.618 < a \div b < 2.625$$

Example 1.39

Evaluate

$$6.721 - \frac{4.931 \times 71.28}{89.45}$$

Assuming that all the values given are correctly rounded numbers, calculate an error bound for your answer and write it as a correctly rounded number.

Solution

Using a calculator, the answer obtained is

$$6.721 - \frac{4.931 \times 71.28}{89.45} = 2.791\,635\,216$$

To estimate the effect of the rounding error of the data, we first draw up a tree diagram representing the order in which the calculation is performed. Remember that $+$, $-$, $\times$ and $\div$ are binary operations, so only one operation can be performed at each step. Here we are evaluating

$$a - \frac{b \times c}{d} = e$$

We calculate this as $b \times c = p$, then $p \div d = q$ and then $a - q = e$, as shown in Figure 1.31(a). We set this calculation out in a table as shown in Figure 1.31(b), where the arrows show the flow of the error analysis calculation. Thus the value of e lies between $2.790\,235\ldots$ and $2.793\,035\ldots$, and the answer may be written as 2.7916 ± 0.0015 or as the correctly rounded number 2.79.

Figure 1.31

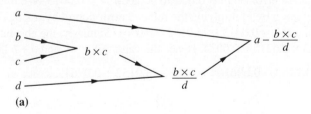

(a)

Label	Value	Absolute error bound	Relative error bound
b	4.931	0.0005	0.0005/4.931 = 0.0001
c	71.28	0.005	0.005/71.28 = 0.000 07
p	351.481 68		0.000 17
d	89.45	0.005	0.005/89.45 = 0.000 06
q	3.929 364 784	0.0009 = 0.000 23 × 3.9	0.000 23
a	6.721	0.0005	
e	2.791 635 216	0.0014	

(b)

1.5.4 Exercises

43 Find the decimal equivalent of 110110.101_2.

44 Find the binary and octal (base eight) equivalents of the decimal number $16\,321$. Obtain a simple rule that relates these two representations of the number, and hence write down the octal equivalent of 1011100101101_2.

45 Find the binary and octal equivalents of the decimal number 30.6. Does the rule obtained in Question 44 still apply?

46 Use binary arithmetic to evaluate

(a) $100011.011_2 + 1011.001_2$

(b) $111.10011_2 \times 10.111_2$

47 State the numbers of decimal places and significant figures of the following correctly rounded numbers:

(a) 980.665   (b) 9.11×10^{-28}

(c) 2.9978×10^{10} (d) 2.00×10^{33}

(e) 1.759×10^7 (f) 6.67×10^{-8}

48 In a right-angled triangle the height is measured as 1 m and the base as 2 m, both measurements being accurate to the nearest centimetre. Using Pythagoras' theorem, the hypotenuse is calculated as 2.236 07 m. Is this a sensible deduction? What other source of error will occur?

49 Determine the error bound and relative error bound for x, where

(a) $x = 35\,\text{min} \pm 5\,\text{s}$

(b) $x = 35\,\text{min} \pm 4\%$

(c) $x = 0.58$ and x is correctly rounded to 2dp.

50 A value is calculated to be 12.9576, with a relative error bound of 0.0003. Calculate its absolute error bound and give the value as a correctly rounded number with as many significant digits as possible.

51 Using exact arithmetic, compute the values of the expressions below. Assuming that all the numbers given are correctly rounded, find absolute and relative error bounds for each term in the

expressions and for your answers. Give the answers as correctly rounded numbers.

(a) $1.316 - 5.713 + 8.010$

(b) 2.51×1.01

(c) $19.61 + 21.53 - 18.67$

52 Evaluate $12.42 \times 5.675/15.63$, giving your answer as a correctly rounded number with the greatest number of significant figures.

53 Evaluate

$$a + b, \quad a - b, \quad a \times b, \quad a/b$$

for $a = 4.99$ and $b = 5.01$. Give absolute and relative error bounds for each answer.

54 Complete the table below for the computation

$$9.21 + (3.251 - 3.115)/0.112$$

and give the result as the correctly rounded answer with the greatest number of significant figures.

Label	Value	Absolute error bound	Relative error bound
a	3.251		
b	3.115		
$a - b$			
c	0.112		
$(a-b)/c$			
d	9.21		
$d + (a-b)/c$			

55 Evaluate $uv/(u + v)$ for $u = 1.135$ and $v = 2.332$, expressing your answer as a correctly rounded number.

56 Working to 4dp, evaluate

$$E = 1 - 1.65 + \tfrac{1}{2}(1.65)^2 - \tfrac{1}{6}(1.65)^3 + \tfrac{1}{24}(1.65)^4$$

(a) by evaluating each term and then summing,

(b) by 'nested multiplication'

$$E = 1 + 1.65(-1 + 1.65(\tfrac{1}{2} + 1.65(-\tfrac{1}{6} + \tfrac{1}{24}(1.65))))$$

Assuming that the number 1.65 is correctly rounded and that all other numbers are exact, obtain error bounds for both answers.

1.5.5 Computer arithmetic

The error estimate outlined in Example 1.39 is a 'worst case' analysis. The actual error will usually be considerably less than the error bound. For example, the maximum error in the sum of 100 numbers, each rounded to three decimal places, is 0.05. This would only occur in the unlikely event that each value has the greater possible rounding error. In contrast, the chance of the error being as large as one-tenth of this is only about 1 in 20.

When calculations are performed on a computer the situation is modified a little by the limited space available for number storage. Arithmetic is usually performed using floating-point notation. Each number x is stored in the **normal form**

$$x = (\text{sign})b^n(a)$$

where b is the number base, usually 2 or 16, n is an integer, and the **mantissa** a is a fraction with a fixed number of digits such that $1/b \leqslant a < 1$. As there are a limited number of digits available to represent the mantissa, calculations will involve intermediate rounding. As a consequence, the order in which a calculation is performed may affect the outcome – in other words the Fundamental Laws of Arithmetic may no longer hold! We shall illustrate this by means of an exaggerated example for a small computer using a decimal representation whose capacity for recording numbers is limited to four figures only. In large-scale calculations in engineering such considerations are sometimes important.

Consider a computer with storage capacity for real numbers limited to four figures; each number is recorded in the form $(\pm)10^n(a)$ where the exponent n is an integer, $0.1 \leqslant a < 1$ and a has four digits. For example,

$$\pi = +10^1(0.3142)$$

$$-\tfrac{1}{3} = -10^0(0.3333)$$

$$5764 = +10^4(0.5764)$$

$$-0.000\,971\,3 = -10^{-3}(0.9713)$$

$$5\,764\,213 = +10^7(0.5764)$$

Addition is performed by first adjusting the exponent of the smaller number to that of the larger, then adding the numbers, which now have the same multiplying power of 10, and lastly truncating the number to four digits. Thus $7.182 + 0.053\,81$ becomes

$$+10^1(0.7182) + 10^{-1}(0.5381) = 10^1(0.7182) + 10^1(0.005\,381)$$

$$= 10^1(0.723\,581)$$

$$= 10^1(0.7236)$$

With $a = 31.68$, $b = -31.54$ and $c = 83.21$, the two calculations $(a + b) + c$ and $(a + c) + b$ yield different results on this computer:

$$(a + b) + c = 83.35, \qquad (a + c) + b = 83.34$$

Notice how the symbol '=' is being used in the examples above. Sometimes it means 'equals to 4sf'. This computerized arithmetic is usually called **floating-point arithmetic**, and the number of digits used is normally specified.

1.5.6 Exercises

57 Two possible methods of adding five numbers are

$$(((a + b) + c) + d) + e$$

and

$$(((e + d) + c) + b) + a$$

Using 4dp floating-point arithmetic, evaluate the sum

$$10^1(0.1000) + 10^1(0.1000) - 10^0(0.5000)$$
$$+ 10^0(0.1667) + 10^{-1}(0.4167)$$

by both methods. Explain any discrepancy in the results.

58 Find $(10^{-2}(0.3251) \times 10^{-5}(0.2011))$ and $(10^{-1}(0.2168) \div 10^2(0.3211))$ using 4-digit floating-point arithmetic.

59 Find the relative error resulting when 4-digit floating-point arithmetic is used to evaluate

$$10^4(0.1000) + 10^2(0.1234) - 10^4(0.1013)$$

1.6 Engineering applications

In this section we illustrate through two examples how some of the results developed in this chapter may be used in an engineering application.

Example 1.40

A continuous belt of length L m passes over two wheels of radii r and R m with their centres a distance l m apart as illustrated in Figure 1.32. The belt is sufficiently tight for any sag to be negligible. Show that L is given approximately by

$$L \approx 2[l^2 - (R - r)^2]^{1/2} + \pi(R + r)$$

Find the error inherent in this approximation and obtain error bounds for L given the rounded data $R = 1.5$, $r = 0.5$ and $l = 3.5$.

Figure 1.32

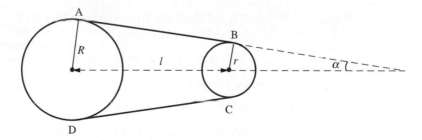

Solution

The length of the belt consists of the straight sections AB and CD and the wraps round the wheels $\overset{\frown}{BC}$ and $\overset{\frown}{DA}$. From Figure 1.33 it is clear that BP = OQ = l and $\angle$OAB is a right-angle. Also, AP = AO − OP and OP = QB so that AP = $R - r$. Applying Pythagoras' theorem to the triangle PAB gives

$$AB^2 = l^2 - (R - r)^2$$

Figure 1.33

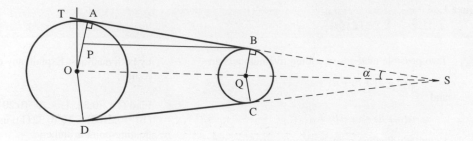

Since the length of an arc of a circle is the product of its radius and the angle (measured in radians) subtended at the centre (see equation 2.16), the length of wrap $\overset{\frown}{DA}$ is given by

$$\pi R + 2R(\angle AOT)$$

where the angle is measured in radians. By geometry, $\angle AOT = \angle OST = \alpha$, so that

$$\overset{\frown}{DA} = \pi R + 2R\alpha$$

Similarly, the arc $\overset{\frown}{BC} = \pi r - 2r\alpha$. Thus the total length of the belt is

$$L = 2[l^2 - (R - r)^2]^{1/2} + \pi(R + r) + 2(R - r)\alpha$$

Taking the length to be given approximately by

$$L \approx 2[l^2 - (R - r)^2]^{1/2} + \pi(R + r)$$

the error of the approximation is given by $-2(R - r)\alpha$, where the angle α is expressed in radians (remember that error = approximation − true value). The angle α is found by elementary trigonometry, since $\sin \alpha = (R - r)/l$. (Trigonometric functions will be reviewed in Section 2.6.)

For the (rounded) data given we deduce, following the procedures of Section 1.5.3, that for $R = 1.5$, $r = 0.5$ and $l = 3.5$ we have an error interval for α of

$$\left[\sin^{-1}\left(\frac{1.45 - 0.55}{3.55}\right),\ \sin^{-1}\left(\frac{1.55 - 0.45}{3.45}\right)\right] = [0.256, 0.325]$$

Thus $\alpha = 0.29 \pm 0.035$, and similarly $2(R - r)\alpha = 0.572 \pm 0.111$.

Evaluating the approximation for L gives

$$2[l^2 - (R - r)^2]^{1/2} + \pi(R + r) = 12.991 \pm 0.478$$

and the corresponding value for L is

$$L = 13.563 \pm 0.589$$

Thus, allowing for both the truncation error of the approximation and for the rounding errors in the data, the value 12.991 given by the approximation has an error interval [12.974, 14.152]. Its error bound is the larger of $|12.991 - 14.152|$ and $|12.991 - 12.974|$, that is, 1.16. Its relative error is 0.089 and its percent error is 8.9%, where the terminology follows the definitions given in Section 1.5.3.

Example 1.41

A cable company is to run an optical cable from a relay station, A, on the shore to an installation, B, on an island, as shown in Figure 1.34. The island is 6 km from the

Figure 1.34

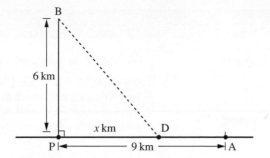

shore at its nearest point, P, and A is 9 km from P measured along the shore. It is proposed to run the cable from A along the shoreline and then underwater to the island. It costs 25% more to run the cable underwater than along the shoreline. At what point should the cable leave the shore in order to minimize the total cost?

Solution   Optimization problems frequently occur in engineering and technology and often their solution is found algebraically.

If the cable leaves the shore at D, a distance x km from P, then the underwater distance is $\sqrt{(x^2 + 36)}$ km and the overland distance is $(9 - x)$ km, assuming $0 < x < 9$. If the overland cost of laying the cable is £c per kilometre, then the total cost £C is given by

$$C(x) = [(9 - x) + 1.25\sqrt{(x^2 + 36)}]c$$

We wish to find the value of x, $0 \leqslant x \leqslant 9$, which minimizes C. To do this we first change the variable x by substituting

$$x = 3\left(t - \frac{1}{t}\right)$$

such that $x^2 + 36$ becomes a perfect square:

$$x^2 + 36 = 36 + 9(t^2 - 2 + 1/t^2)$$
$$= 9(t + 1/t)^2$$

Hence C(x) becomes

$$C(t) = [9 - 3(t - 1/t) + 3.75(t + 1/t)]c$$
$$= [9 + 0.75(t + 9/t)]c$$

Using the arithmetic–geometric inequality, we know that

$$t + \frac{9}{t} \geqslant 18$$

and that the equality occurs where $t = 9/t$, that is where $t = 3$.

Thus the minimum cost is achieved where $t = 3$ and $x = 3(3 - 1/3) = 8$.

Hence the cable should leave the shore after laying the cable 1 km from its starting point at A.

1.7 Review exercises (1–22)

1 (a) A formula in the theory of ventilation is

$$Q = \frac{\sqrt{H}}{K}\sqrt{\frac{A^2 D^2}{A^2 + D^2}}$$

Express A in terms of the other symbols.

(b) Solve the equation

$$\frac{1}{x+2} - \frac{2}{x} = \frac{3}{x-1}$$

2 Factorize the following

(a) $ax - 2x - a + 2$ (b) $a^2 - b^2 + 2bc - c^2$

(c) $4k^2 + 4kl + l^2 - 9m^2$ (d) $p^2 - 3pq + 2q^2$

(e) $l^2 + lm + ln + mn$

3 (a) Two small pegs are 8 cm apart on the same horizontal line. An inextensible string of length 16 cm has equal masses fastened at either end and is placed symmetrically over the pegs. The middle point of the string is pulled down vertically until it is in line with the masses. How far does each mass rise?

(b) Find an 'acceptable' value of x to three decimal places if the shaded area in Figure 1.35 is 10 square units.

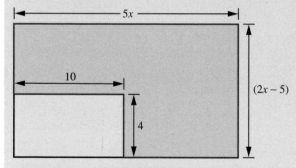

Figure 1.35 Shaded area of Question 3(b).

4 The impedance Z ohms of a circuit containing a resistance R ohms, inductance L henries and capacity C farads, when the frequency of the oscillation is n per second, is given by

$$Z = \sqrt{\left(R^2 + \left(2\pi nL - \frac{1}{2\pi nC}\right)^2\right)}$$

(a) Make L the subject of this formula.

(b) If $n = 50$, $R = 15$ and $C = 10^{-4}$ show that there are two values of L which make $Z = 20$ but only one value of L which will make $Z = 100$. Find the values of Z in each case to two decimal places.

5 Expand out (a) and (b) and rationalize (c) to (e).

(a) $(3\sqrt{2} - 2\sqrt{3})^2$

(b) $(\sqrt{5} + 7\sqrt{3})(2\sqrt{5} - 3\sqrt{3})$

(c) $\dfrac{4 + 3\sqrt{2}}{5 + \sqrt{2}}$

(d) $\dfrac{\sqrt{3} + \sqrt{2}}{2 - \sqrt{3}}$

(e) $\dfrac{1}{1 + \sqrt{2} - \sqrt{3}}$

6 Find integers m and n such that

$$\sqrt{(11 + 2\sqrt{30})} = \sqrt{m} + \sqrt{n}$$

7 Show that

$$\sqrt{(n + 1)} - \sqrt{n} = \frac{1}{\sqrt{(n+1)} + \sqrt{n}}$$

and deduce that

$$\sqrt{(n + 1)} - \sqrt{n} < \frac{1}{2\sqrt{n}} < \sqrt{n} - \sqrt{(n - 1)}$$

for any integer $n \geqslant 1$. Deduce that the sum

$$\frac{1}{\sqrt{1}} + \frac{1}{\sqrt{2}} + \frac{1}{\sqrt{3}} + \ldots + \frac{1}{\sqrt{(9999)}} + \frac{1}{\sqrt{(10\,000)}}$$

lies between 198 and 200.

8 Express each of the following subsets of $\mathbb{R}$ in terms of intervals:

(a) $\{x : 4x^2 - 3 < 4x,\ x \text{ in } \mathbb{R}\}$

(b) $\{x : 1/(x + 2) > 2/(x - 1),\ x \text{ in } \mathbb{R}\}$

(c) $\{x : |x + 1| < 2,\ x \text{ in } \mathbb{R}\}$

(d) $\{x : |x + 1| < 1 + \frac{1}{2}x,\ x \text{ in } \mathbb{R}\}$

9 It is known that of all plane curves that enclose a given area, the circle has the least perimeter. Show that if a plane curve of perimeter L encloses an area A then $4\pi A \leq L^2$. Verify this inequality for a square and a semicircle.

10 Show that if $a < b$, $b > 0$ and $c > 0$ then

$$\frac{a}{b} < \frac{a+c}{b+c} < 1$$

Obtain a similar inequality for the case $a > b$.

11 (a) If $n = n_1 + n_2 + n_3$ show that

$$\binom{n}{n_1}\binom{n_2 + n_3}{n_2} = \frac{n!}{n_1! n_2! n_3!}$$

(This represents the number of ways in which n objects may be divided into three groups containing respectively n_1, n_2 and n_3 objects.)

(b) Expand the following expressions

(i) $\left(1 - \frac{x}{2}\right)^5$ (ii) $(3 - 2x)^6$

12 (a) Evaluate $\sum_{n=-2}^{3} [n^{n+1} + 3(-1)^n]$

(b) A square grid of dots may be divided up into a set of L-shaped groups as illustrated in Figure 1.36.

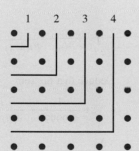

Figure 1.36

How many dots are inside the third L shape? How many extra dots are needed to extend the 3 by 3 square to one of side 4 by 4? How many dots are needed to extend an $(r-1)$ by $(r-1)$ square to one of size r by r? Denoting this number by P_r, use a geometric argument to obtain an expression for $\sum_{r=1}^{n} P_r$ and verify your conclusion by direct calculation in the case $n = 10$.

13 Find the equations of the straight line:

(a) which passes through the points $(-6, -11)$ and $(2, 5)$;

(b) which passes through the point $(4, -1)$ and has gradient $\frac{1}{3}$;

(c) which has the same intercept on the y axis as the line in (b) and is parallel to the line in (a).

14 Find the equation of the circle which touches the y axis at the point $(0, 3)$ and passes through the point $(1, 0)$.

15 Find the centres and radii of the following circles.

(a) $x^2 + y^2 + 2x - 4y + 1 = 0$

(b) $4x^2 - 4x + 4y^2 + 12y + 9 = 0$

(c) $9x^2 + 6x + 9y^2 - 6y = 25$

16 For each of the two parabolas

(i) $y^2 = 8x + 4y - 12$, and

(ii) $x^2 + 12y + 4x = 8$

determine

(a) the coordinates of the vertex;

(b) the coordinates of the focus;

(c) the equation of the directrix;

(d) the equation of the axis of symmetry.

Sketch each parabola.

17 Find the coordinates of the centre and foci of the ellipse with equation

$$25x^2 + 16y^2 - 100x - 256y + 724 = 0$$

What are the coordinates of its vertices and the equations of its directrices? Sketch the ellipse.

18 Find the duodecimal equivalent of the decimal number 10.386 23.

19 Show that if $y = x^{1/2}$ then the relative error bound of y is one-half that of x. Hence complete the table in Figure 1.37.

20 Assuming that all the numbers given are correctly rounded, calculate the positive root together with its error bound of the quadratic equation

$$1.4x^2 + 5.7x - 2.3 = 0$$

	Value	*Absolute error bound*	*Relative error bound*
a	7.01	0.005 $\longrightarrow$	0.0007
$\sqrt{a}$	2.6476	0.0009 $\longleftarrow$	0.000 35
b	52.13		
$\sqrt{b}$			
c	0.010 11		
$\sqrt{c}$			
d	5.631×10^{11}		
$\sqrt{d}$			
Correctly rounded values	$\sqrt{a}$ $\sqrt{b}$ $\sqrt{c}$ $\sqrt{d}$ 2.65		

Figure 1.37

Give your answer also as a correctly rounded number.

21 The quantities f, u and v are connected by

$$\frac{1}{f} = \frac{1}{u} + \frac{1}{v}$$

Find f when $u = 3.00$ and $v = 4.00$ are correctly rounded numbers. Compare the error bounds obtained for f when

(a) it is evaluated by taking the reciprocal of the sum of the reciprocals of u and v,

(b) it is evaluated using the formula

$$f = \frac{uv}{u + v}$$

22 If the number whose decimal representation is 14 732 has the representation $152\,112_b$ to base b, what is b?

2 Functions

Chapter 2 Contents

2.1 Introduction

As we have remarked in the introductory section of Chapter 1, mathematics provides a means of solving the practical problems that occur in engineering. To do this, it uses concepts and techniques that operate on and within the concepts. In this chapter we shall describe the concept of a function – a concept that is both fundamental to mathematics and intuitive. We shall make the intuitive idea mathematically precise by formal definitions and describe why such formalism is needed for practical problem-solving.

The function concept has taken many centuries to evolve. The intuitive basis for the concept is found in the analysis of cause and effect, which underpins developments in science, technology and commerce. As with many mathematical ideas, many people use the concept in their everyday activities without being aware that they are using mathematics, and many would be surprised if they were told that they were. The abstract manner in which the developed form of the concept is expressed by mathematicians often intimidates learners but the essential idea is very simple. A consequence of the long period of development is that the way in which the concept is described often makes an idiomatic use of words. Ordinary words which in common parlance have many different shades of meaning are used in mathematics with very specific meanings.

The key idea is that of the values of two variable quantities being related. For example, the amount of tax paid depends on the selling price of an item; the deflection of a beam depends on the applied load; the cost of an article varies with the number produced, and so on. Historically, this idea has been expressed in a number of ways. The oldest gave a verbal recipe for calculating the required value. Thus, in the early Middle Ages, a very elaborate verbal recipe was given for calculating the monthly interest payments on a loan which would now be expressed very compactly by a single formula. John Napier, when he developed the logarithm function at the beginning of the seventeenth century, expressed the functional relationship in terms of two particles moving along a straight line. One particle moved with constant velocity and the other with a velocity that depended on its distance from a fixed point on the line. The relationship between the distances travelled by the particles was used to define the logarithms of numbers. This would now be described by the solution of a differential equation. The introduction of algebraic notation led to the representation of functions by algebraic rather than verbal formulae. That produced many theoretical problems. For example, a considerable controversy was caused by Fourier when he used functions that did not have the same algebraic formula for all values of the independent variable. Similarly, the existence of functions that do not have a simple algebraic representation caused considerable difficulties for mathematicians in the early nineteenth century.

2.2 Basic definitions

2.2.1 Concept of a function

The essential idea that flows through all of the developments is that of two quantities whose values are related. One of these variables, the **independent** or **free variable**,

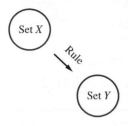

Figure 2.1
Schematic
representation
of a function.

may take any value in a set of values. The value it actually takes fixes uniquely the value of the second quantity, the **dependent variable**. Thus for each value of the independent variable there is one and only one value of the dependent variable. The way in which that value is calculated will vary between functions. Sometimes it will be by means of a formula, sometimes by means of a graph and sometimes by means of a table of values. Here the words 'value' and 'quantity' cover many very different contexts, but in each case what we have is two sets X and Y and a rule that assigns to each element x in the set X precisely one element y from the set Y. The elements of X and Y need not be numbers, but the essential idea is that to every x in the set X there corresponds exactly one y in the set Y. Whenever this situation arises we say that there is a **function** f that maps the set X to the set Y. Such a function may be illustrated schematically as in Figure 2.1.

We represent a functional relationship symbolically in two ways: either

$$f{:}x \to y \quad (x \text{ in } X)$$

or

$$y = f(x) \quad (x \text{ in } X)$$

The first emphasizes the fact that a function f associates each element x of X with exactly one element y of Y: it 'maps x to y'. The second method of notation emphasizes the dependence of the elements of Y on the elements of X under the function f. In this case the value or variable appearing within the brackets is known as the **argument** of the function; we might say 'the argument x of a function $f(x)$'. In engineering it is more common to use the second notation $y = f(x)$ and to refer to this as the function $f(x)$, while modern mathematics textbooks prefer the mapping notation, on the grounds that it is less ambiguous. The set X is called the **domain** of the function and the set Y is called its **codomain**. Knowing the domain and codomain is important in computing. We need to know the type of variables, whether they are integers or reals, and their size. When $y = f(x)$, y is said to be the **image** of x under f. The set of all images $y = f(x)$, x in X, is called the **image set** or **range** of f. It is not necessary for all elements y of the codomain set Y to be images under f. In the terminology of Chapter 6 the range is a subset of the codomain. We may regard x as being a variable that can be replaced by any element of the set X. The rule giving f is then completely determined if we know $f(x)$, and consequently in engineering it is common to refer to the function as being $f(x)$ rather than f. Likewise we can regard $y = f(x)$ as being a variable. However, while x can freely take any value from the set X, the variable $y = f(x)$ depends on the particular element chosen for x. We therefore refer to x as the **free** or **independent** variable and to y as the **dependent** variable. The function $f(x)$ is therefore specified completely by the set of ordered pairs (x, y) for all x in X. For real variables a graphical representation of the function may then be obtained by plotting a graph determined by this set of ordered pairs (x, y), with the independent variable x measured along the horizontal axis and the dependent variable y measured along the vertical axis. Obtaining a good graph by hand is not always easy but there are now available excellent graphics facilities on computers and calculators which assist in the task. Even so some practice is required to ensure that a good choice of 'drawing window' is selected to obtain a meaningful graph.

Example 2.1

The relationship between the temperature T_1 measured in degrees Celsius (°C) and the corresponding temperature T_2 measured in degrees Fahrenheit (°F) is

$$T_2 = \tfrac{9}{5}T_1 + 32$$

Interpreting this as a function with T_1 as the independent variable and T_2 as the dependent variable:

(a) What are the domain and codomain of the function?

(b) What is the function rule?

(c) Plot a graph of the function.

(d) What is the image set or range of the function?

(e) Use the function to convert the following into °F:

     (i) 60°C,   (ii) 0°C,   (iii) −50°C

Solution

(a) Since temperature can vary continuously, the domain is the set $T_1 \geqslant T_0 = -273.16$ (absolute zero). The codomain can be chosen as the set of real numbers $\mathbb{R}$.

(b) The function rule in words is

     multiply by $\tfrac{9}{5}$ and then add 32

or algebraically

$$f(T_1) = \tfrac{9}{5}T_1 + 32$$

(c) Since the domain is the set $T_1 \geqslant T_0$, there must be an image for every value of T_1 on the horizontal axis which is greater than −273.16. The graph of the function is that part of the line $T_2 = \tfrac{9}{5}T_1 + 32$ for which $T_1 > -273.16$.

(d) Since each value of T_2 is an image of some value T_1 in its domain, it follows that the range of $f(T_1)$ is the set of real numbers greater than −459.69.

(e) The conversion may be done graphically by reading values of the graph, as illustrated by the broken lines in Figure 2.2, or algebraically using the rule

Figure 2.2
Graph of
$T_2 = f(T_1) = \tfrac{9}{5}T_1 + 32$.

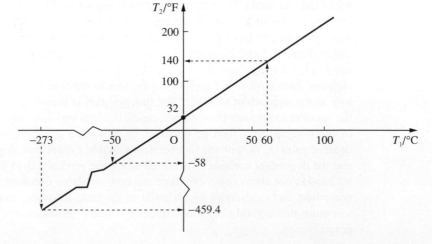

$$T_2 = \tfrac{9}{5}T_1 + 32$$

giving the values

(i) 140°F,   (ii) 32°F,   (iii) −58°F

A value of the independent variable for which the value of a function is zero, is called a **zero** of that function. Thus the function $f(x) = (x − 1)(x + 2)$ has two zeros, $x = 1$ and $x = −2$. These correspond to where the graph of the function crosses the x axis, as shown in Figure 2.3. We can see from the diagram that, for this function, its values decrease as the values of x increase from (say) −5 up to $−\tfrac{1}{2}$, and then its values increase with x. The function is said to be a **decreasing function** for $x < −\tfrac{1}{2}$ and an **increasing function** for $x > −\tfrac{1}{2}$. More formally, a function is said to be increasing on an interval if $f(x_2) > f(x_1)$ when $x_2 > x_1$ for all x_1 and x_2 both lying in the interval. Similarly for decreasing functions, we have $f(x_2) < f(x_1)$ when $x_2 > x_1$. The value of a function at the point where its behaviour changes from decreasing to increasing is a **minimum** (*plural* **minima**) of the function. Often this is denoted by an asterisk superscript f^* and the corresponding value of the independent variable by x^* so that $f(x^*) = f^*$. Similarly a **maximum** (*plural* **maxima**) occurs when a function changes from being increasing to being decreasing. Maxima and minima are jointly referred to as **optimal values** and as **extremal values** of the function. The point (x^*, f^*) of the graph of $f(x)$ is often called a turning point of the graph, whether it is a maximum or a minimum. These properties will be discussed in more detail in Chapter 8, Section 8.2.7 and Chapter 9, Section 9.2.

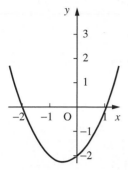

Figure 2.3
Graph of
$y = (x − 1)(x + 2)$.

Example 2.2

Draw graphs of the functions below, locating their zeros, intervals in which they are increasing, intervals in which they are decreasing and their optimal values.

(a) $y = 2x^3 + 3x^2 − 12x + 32$ (b) $y = (x − 1)^{2/3} − 1$

Solution

(a) The graph of the function is shown in Figure 2.4. From the graph we can see that the function has one zero at $x = −4$. It is an increasing function on the intervals

Figure 2.4
Graph of $y = 2x^3 + 3x^2$
$− 12x + 32$.

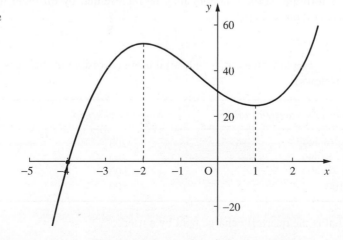

Figure 2.5
Graph of
$y = (x - 1)^{2/3} - 1$.

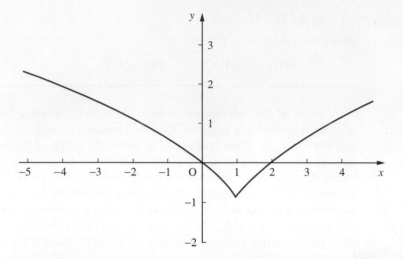

$-\infty < x < -2$ and $1 < x < \infty$ and a decreasing function on the interval $-2 < x < 1$. It achieves a maximum value of 52 at $x = -2$ and a minimum value of 25 at $x = 1$.

(b) The graph of the function is shown in Figure 2.5. (Note that to evaluate $(x - 1)^{2/3}$ on some calculators/computer packages it has to be expressed as $((x - 1)^2)^{1/3}$ for $x < 1$.)

From the graph, we see that the function has two zeros, one at $x = 0$ and the other at $x = 2$. It is a decreasing function for $x < 1$ and an increasing function for $x > 1$. This is obvious algebraically since $(x - 1)^{2/3}$ is greater than or equal to zero. This example also provides an illustration of the behaviour of some algebraic functions at a maximum or minimum value. In contrast to (a) where the function changes from decreasing to increasing at $x = 1$ quite smoothly, in this case the function changes from decreasing to increasing abruptly at $x = 1$. Such a minimum value is called a **cusp**.

It is important to appreciate the difference between a function and a formula. A function is a mapping that associates one and only one member of the codomain with every member of its domain. It may be possible to express this association, as in Example 2.1, by a formula. Some functions may be represented by different formulae on different parts of their domain.

Example 2.3

A gas company charges its industrial users according to their gas usage. Their tariff is as follows:

Quarterly usage/10^3 units	Standing charge/£	Charge per 10^3 units/£
0–19.999	200	60
20–49.999	400	50
50–99.999	600	46
$\geqslant 100$	800	44

What is the quarterly charge paid by a user?

Solution   The charge c paid by a user for a quarter's gas is a function, since for any number of units u used there is a unique charge. The function f: usage $\to$ cost must, however, be expressed in the form $c = f(u)$, where

$$f(u) = \begin{cases} 200 + 60u & (0 \leqslant u < 20) \\ 400 + 50u & (20 \leqslant u < 50) \\ 600 + 46u & (50 \leqslant u < 100) \\ 800 + 44u & (100 \leqslant u) \end{cases}$$

Functions that are represented by different formulae on different parts of their domains arise frequently in engineering and management applications.

2.2.2   Exercises

1   A straight horizontal road is to be constructed through rough terrain. The width of the road is to be 10 m, with the sides of the embankment sloping at 1 (vertical) in 2 (horizontal), as shown in Figure 2.6. Obtain a formula for the cross-sectional area of the road and its embankment, taken at right-angles to the road, where the rough ground lies at a depth x below the level of the proposed road. Use your formula to complete the table below, and draw a graph to represent this function.

x/m	0	1	2	3	4	5
Area/m^2	0		28			100

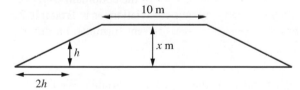

Figure 2.6

What is the value given by the formula when $x = -2$, and what is the meaning of that value?

2   A hot-water tank has the form of a circular cylinder of internal radius r, topped by a hemisphere as shown in Figure 2.7. Show that the internal surface area A is given by

$$A = 2\pi rh + 3\pi r^2$$

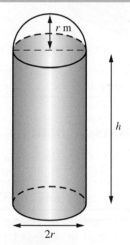

Figure 2.7

and the volume V enclosed is

$$V = \pi r^2 h + \tfrac{2}{3}\pi r^3$$

Find the formula relating the value of A to the value of r for tanks with capacity 0.15 m^3. Complete the table below for A in terms of r and draw a graph to represent the function.

r/m	0.10	0.15	0.20	0.25	0.30	0.35	0.40
A/m^2	3.05		1.71			1.50	

The cost of the tank is proportional to the amount of metal used in its manufacture. Estimate the value of r that will minimize that cost, carefully listing the assumptions you make in your analysis.

3   An oil storage tank has the form of a circular cylinder with its axis horizontal, as shown in Figure 2.8. The volume of oil in the tank when the depth is h is given in the table below.

h m	0.5	1.0	1.5	2.0	2.5	3.0	3.5	4.0
$V/1000l$	7.3	19.7	34.4	50.3	66.1	80.9	93.9	100.5

Draw a careful graph of V against h, and use it to design the graduation marks on a dipstick to be used to assess the volume of oil in the tank.

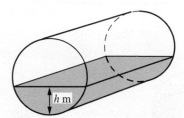

Figure 2.8

4   The initial cost of buying a car is £6000. Over the years, its value depreciates and its running costs increase, as shown in the table below.

t	1	2	3	4	5	6
Value after t years	4090	2880	2030	1430	1010	710
Running cost in year t	600	900	1200	1500	1800	2100

Draw up a table showing (a) the cumulative running cost after t years, (b) the total cost (that is, running cost plus depreciation) after t years and (c) the average cost per year over t years. Estimate the optimal time to replace the car.

5   Plot graphs of the functions below, locating their zeros, intervals in which they are increasing, intervals in which they are decreasing and their optimal values.

(a) $y = x^2(x^2 - 2)$ (b) $y = 1/[x(x - 2)]$

2.2.3   Inverse functions

In some situations we may need to use the functional dependence in the reverse sense. For example we may wish to use the function

$$T_2 = f(T_1) = \tfrac{9}{5}T_1 + 32 \tag{2.1}$$

of Example 2.1, relating T_2 in °F to the corresponding T_1 in °C to convert degrees Fahrenheit to degrees Celsius. In this simple case we can rearrange the relationship (2.1) algebraically

$$T_1 = \tfrac{5}{9}(T_2 - 32)$$

giving us the function

$$T_1 = g(T_2) = \tfrac{5}{9}(T_2 - 32) \tag{2.2}$$

having T_2 as the independent variable and T_1 as the dependent variable. We may then use this to convert degrees Fahrenheit into degrees Celsius.

Looking more closely at the two functions $f(T_1)$ and $g(T_2)$ associated with (2.1) and (2.2), we have the function rule for $f(T_1)$ as

multiply by $\tfrac{9}{5}$ and then add 32

If we reverse the process, we have the rule

take away 32 and then multiply by $\tfrac{5}{9}$

which is precisely the function rule for $g(T_2)$. Thus the function $T_1 = g(T_2)$ reverses the operations carried out by the function $T_2 = f(T_1)$, and for this reason is called the **inverse function** of $T_2 = f(T_1)$.

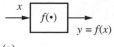

(a)

$x = f^{-1}(y)$ $f^{-1}(\bullet)$ y

(b)

Figure 2.9
Block diagram of
(a) function and
(b) inverse function.

In general, the inverse function of a function f is a function that reverses the operations carried out by f. It is denoted by f^{-1}. Writing $y = f(x)$, the function f may be represented by the block diagram of Figure 2.9(a), which indicates that the function operates on the input variable x to produce the output variable $y = f(x)$. The inverse function f^{-1} will reverse the process, and will take the value of y back to the original corresponding values of x. It can be represented by the block diagram of Figure 2.9(b).

We therefore have

$$x = f^{-1}(y), \quad \text{where } y = f(x) \tag{2.3}$$

that is, the independent variable x for f acts as the dependent variable for f^{-1}, and correspondingly the dependent variable y for f becomes the independent variable for f^{-1}.

Since it is usual to denote the independent variable of a function by x and the dependent variable by y, we interchange the variables x and y in (2.3) and define the inverse function by

$$\text{if } y = f^{-1}(x) \quad \text{then } x = f(y) \tag{2.4}$$

Again in engineering it is common to denote an inverse function by $f^{-1}(x)$ rather than f^{-1}. Writing x as the independent variable for both $f(x)$ and $f^{-1}(x)$ sometimes leads to confusion, so you need to be quite clear as to what is meant by an inverse function. It is also important not to confuse $f^{-1}(x)$ with $[f(x)]^{-1}$, which means $1/f(x)$.

Finding an explicit formula for $f^{-1}(x)$ is often impossible and its values are calculated by special numerical methods. Sometimes it is possible to find the formula for $f^{-1}(x)$ by algebraic methods. We illustrate the technique in the next two examples.

Example 2.4

Obtain the inverse function of the real function $y = f(x) = \frac{1}{5}(4x - 3)$.

Solution

Here the process can be done algebraically. First rearranging

$$y = f(x) = \tfrac{1}{5}(4x - 3)$$

to express x in terms of y gives

$$x = f^{-1}(y) = \tfrac{1}{4}(5y + 3)$$

Interchanging the variables x and y then gives

$$y = f^{-1}(x) = \tfrac{1}{4}(5x + 3)$$

as the inverse function of

$$y = f(x) = \tfrac{1}{5}(4x - 3)$$

As a check, we have

$$f(2) = \tfrac{1}{5}(4 \times 2 - 3) = 1$$

while

$$f^{-1}(1) = \tfrac{1}{4}(5 \times 1 + 3) = 2$$

Example 2.5   Obtain the inverse function of $y = f(x) = \dfrac{x + 2}{x + 1}$, $x \neq -1$.

Solution   We rearrange $y = \dfrac{x + 2}{x + 1}$ to obtain x in terms of y. Thus

$$y(x + 1) = x + 2 \quad \text{so that} \quad x(y - 1) = 2 - y$$

giving $x = \dfrac{2 - y}{y - 1}$, $y \neq 1$

Thus $f^{-1}(x) = \dfrac{2 - x}{x - 1}$, $x \neq 1$

If we are given the graph of $y = f(x)$ and wish to obtain the graph of the inverse function $y = f^{-1}(x)$ then what we really need to do is interchange the roles of x and y. Thus we need to manipulate the graph of $y = f(x)$ so that the x and y axes are interchanged. This can be achieved by taking the mirror image in the line $y = x$ and relabelling the axes as illustrated in Figures 2.10(a) and (b). It is important to recognize that the graphs of $y = f(x)$ and $y = f^{-1}(x)$ are symmetrical about the line $y = x$, since this property is frequently used in mathematical arguments. Notice that the x and y axes have the same scale.

Figure 2.10
The graph of
$y = f^{-1}(x)$.

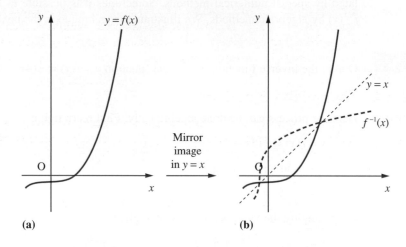

(a)   (b)

Example 2.6   Obtain the graph of $f^{-1}(x)$ when $f(x) = x^2$.

Solution   The graph of $y = x^2$ is shown in Figure 2.11(a). Its mirror image in the line $y = x$ gives the graph of Figure 2.11(b). We note that this graph is not representative of a function according to our definition, since for all values of $x > 0$ there are two images – one positive and one negative – as indicated by the broken line. This follows because $y = x^2$ corresponds to $x = +\sqrt{y}$ or $x = -\sqrt{y}$. In order to avoid this ambiguity, we define the

Figure 2.11
Graphs of $f(x) = x^2$
and its inverse.

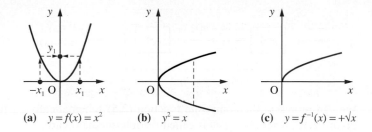

(a) $y = f(x) = x^2$ **(b)** $y^2 = x$ **(c)** $y = f^{-1}(x) = +\sqrt{x}$

inverse function of $f(x) = x^2$ to be $f^{-1}(x) = +\sqrt{x}$, which corresponds to the upper half of the graph as illustrated in Figure 2.11(c). $\sqrt{x}$ therefore denotes a positive number (cf. calculators), so the range of $\sqrt{x}$ is $x \geqslant 0$. Thus the inverse function of

$$y = f(x) = x^2 \quad (x \geqslant 0)$$

is

$$y = f^{-1}(x) = +\sqrt{x}$$

Note that the domain of $f(x)$ had to be restricted to $x \geqslant 0$ in order that an inverse could be defined.

We see from Example 2.6 that there is no immediate inverse function corresponding to $f(x) = x^2$. This arises because for the function $f(x) = x^2$ there is a codomain element that is the image of two domain elements x_1 and $-x_1$, as indicated by the broken arrowed lines in Figure 2.11(a). That is, $f(x_1) = f(-x_1) = y_1$. If a function $y = f(x)$ is to have an immediate inverse $f^{-1}(x)$, without any imposed conditions, then *every* element of its range must occur *precisely once* as an image under $f(x)$. Such a function is known as one-to-one (1–1) correspondence.

2.2.4 Composite functions

In many practical problems the mathematical model will involve several different functions. For example, the kinetic energy T of a moving particle is a function of its velocity v, so that

$$T = f(v)$$

Also, the velocity v itself is a function of time t, so that

$$v = g(t)$$

Clearly, by eliminating v, it is possible to express the kinetic energy as a function of time according to

$$T = f(g(t))$$

A function of the form $y = f(g(x))$ is called a **function of a function** or a **composite** of the functions $f(x)$ and $g(x)$. In modern mathematical texts it is common to denote the composite function by $f \circ g$ so that

$$y = f \circ g(x) = f(g(x)) \tag{2.5}$$

Figure 2.12
The composite
function $f(g(x))$.

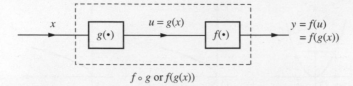

$$f \circ g \text{ or } f(g(x))$$

We can represent the composite function (2.5) schematically by the block diagram of Figure 2.12, where $u = g(x)$ is called the intermediate variable.

It is important to recognize that the composition of functions is not in general commutative. That is, for two general functions $f(x)$ and $g(x)$

$$f(g(x)) \neq g(f(x))$$

Algebraically, given two functions $y = f(x)$ and $y = g(x)$, the composite function $y = f(g(x))$ may be obtained by replacing x in the expression for $f(x)$ by $g(x)$. Likewise, the composite function $y = g(f(x))$ may be obtained by replacing x in the expression for $g(x)$ by $f(x)$.

Example 2.7   If $y = f(x) = x^2 + 2x$ and $y = g(x) = x - 1$, obtain the composite functions $f(g(x))$ and $g(f(x))$.

Solution   To obtain $f(g(x))$ replace x in the expression for $f(x)$ by $g(x)$, giving

$$y = f(g(x)) = (g(x))^2 + 2(g(x))$$

But $g(x) = x - 1$, so that

$$y = f(g(x)) = (x - 1)^2 + 2(x - 1)$$
$$= x^2 - 2x + 1 + 2x - 2$$

That is,

$$f(g(x)) = x^2 - 1$$

Similarly,

$$y = g(f(x)) = (f(x)) - 1$$
$$= (x^2 + 2x) - 1$$

That is,

$$g(f(x)) = x^2 + 2x - 1$$

Note that this example confirms the result that in general $f(g(x)) \neq g(f(x))$.

Given a function $y = f(x)$, two composite functions that occur frequently in engineering are

$$y = f(x + k) \quad \text{and} \quad y = f(x - k)$$

where k is a positive constant. As illustrated in Figures 2.13(b) and (c), the graphs of these two composite functions are readily obtained given the graph of $y = f(x)$ as in Figure 2.13(a). The graph of $y = f(x - k)$ is obtained by displacing the graph of $y = f(x)$

Figure 2.13
Graphs of $f(x)$,
$f(x - k)$ and $f(x + k)$,
with $k > 0$.

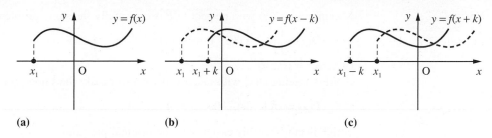

(a) (b) (c)

by k units to the right, while the graph of $y = f(x + k)$ is obtained by displacing the graph of $y = f(x)$ by k units to the left.

Viewing complicated functions as composites of simpler functions often enables us to 'get to the heart' of a practical problem, and to obtain and understand the solution. For example, recognizing that $y = x^2 + 2x - 3$ is the composite function $y = (x + 1)^2 - 4$, tells us that the function is essentially the squaring function. Its graph is a parabola with minimum point at $x = -1$, $y = -4$ (rather than at $x = 0$, $y = 0$).

Example 2.8

An open conical container is made from a sector of a circle of radius 10 cm as illustrated in Figure 2.14, with sectional angle θ (radians). The capacity $C\,\mathrm{cm}^3$ of the cone depends on θ. Find the algebraic formula for C in terms of θ and the simplest associated function that could be studied if we wish to maximize C with respect to θ.

Figure 2.14
Conical container
of Example 2.8.

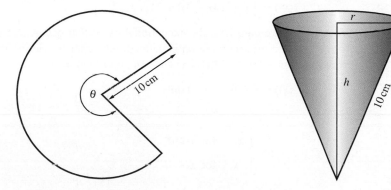

Solution Let the cone have base radius $r\,\mathrm{cm}$ and height $h\,\mathrm{cm}$. Then its capacity is given by $C = \frac{1}{3}\pi r^2 h$ with r and h dependent upon the sectorial angle θ (since the perimeter of the sector has to equal the circumference of the base of the cone). Thus, by Pythagoras' theorem,

$$10\theta = 2\pi r \quad \text{and} \quad h^2 = 10^2 - r^2$$

so that

$$C(\theta) = \frac{1}{3}\pi\left(\frac{10\theta}{2\pi}\right)^2\left[10^2 - \left(\frac{10\theta}{2\pi}\right)^2\right]^{1/2}$$

$$= \frac{1000}{3}\pi\left(\frac{\theta}{2\pi}\right)^2\left[1 - \left(\frac{\theta}{2\pi}\right)^2\right]^{1/2}, \quad 0 \leq \theta \leq 2\pi$$

Maximizing $C(\theta)$ with respect to θ is essentially the same problem as maximizing

$$D(x) = x(1-x)^{1/2}, \quad 0 \leq x \leq 1$$

(where $x = (\theta/2\pi)^2$).

Maximizing $D(x)$ with respect to x is essentially the same problem as maximizing

$$E(x) = x^2(1-x), \quad 0 \leq x \leq 1$$

which is considerably easier than the original problem.

(We will see later, in Chapter 8, that the optimal value of x is 2/3, which implies that the optimal value of $\theta = 2\pi\sqrt{(2/3)}$.)

When we compose a function with its inverse function, we usually obtain the identity function $y = x$. Thus from Example 2.4, we have

$$f(x) = \tfrac{1}{5}(4x - 3) \quad \text{and} \quad f^{-1}(x) = \tfrac{1}{4}(5x + 3)$$

and

$$f(f^{-1}(x)) = \tfrac{1}{5}\{4[\tfrac{1}{4}(5x+3)] - 3\} = x$$

and

$$f^{-1}(f(x)) = \tfrac{1}{4}\{5[\tfrac{1}{5}(4x-3)] + 3\} = x$$

We need to take care with the exceptional cases that occur, like the square root function, where the inverse function is defined only after restricting the domain of the original function. Thus for $f(x) = x^2$ ($x \geq 0$) and $f^{-1}(x) = \sqrt{x}$ ($x \geq 0$), we obtain

$$f(f^{-1}(x)) = x, \quad \text{for } x \geq 0 \text{ only}$$

and

$$f^{-1}(f(x)) = \begin{cases} x, & \text{for } x \geq 0 \\ -x, & \text{for } x \leq 0 \end{cases}$$

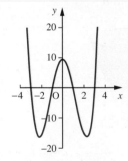

Figure 2.15 Graph of an even function.

2.2.5 Odd, even and periodic functions

Some commonly occurring functions in engineering contexts have the special properties of oddness or evenness or periodicity. These properties are best understood from the graphs of the functions.

An **even function** is one that satisfies the functional equation

$$f(-x) = f(x)$$

Thus the value of $f(-2)$ is the same as $f(2)$, and so on. The graph of such a function is symmetrical about the y axis, as shown in Figure 2.15.

In contrast, an **odd function** has a graph which is antisymmetrical about the origin, as shown in Figure 2.16, and satisfies the equation

$$f(-x) = -f(x)$$

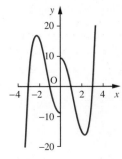

Figure 2.16 Graph of an odd function.

We notice that $f(0) = 0$ or is undefined.

Polynomial functions like $y = x^4 - x^2 - 1$, involving only even powers of x, are examples of even functions, while those like $y = x - x^5$, involving only odd powers of x, provide examples of odd functions. Of course, not all functions have the property of oddness or evenness.

Example 2.9 Which of the functions $y = f(x)$ whose graphs are shown in Figure 2.17 are odd, even or neither odd nor even?

Figure 2.17
Graphs of
Example 2.9.

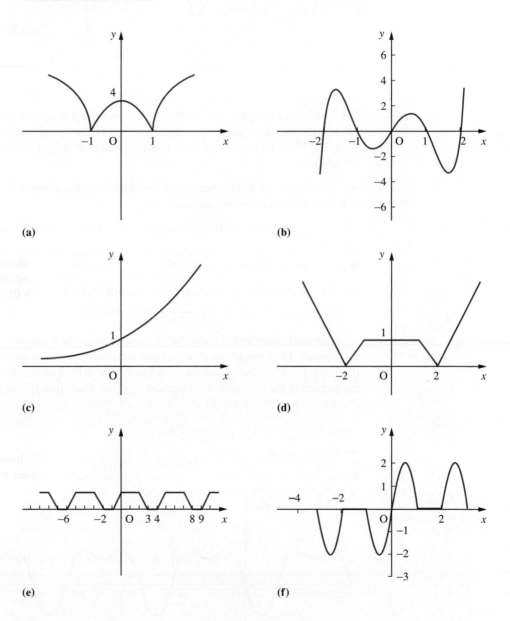

(a)

(b)

(c)

(d)

(e)

(f)

Solution (a) The graph for $x < 0$ is the mirror image of the graph for $x > 0$ when the mirror is placed on the y axis. Thus the graph represents an even function.

Figure 2.18

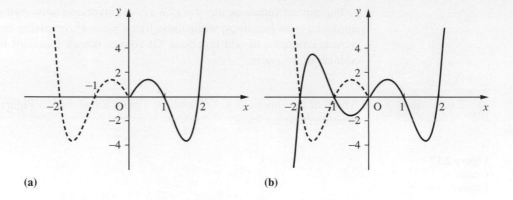

(a)

(b)

(b) The mirror image of the graph for $x > 0$ in the y axis is shown in Figure 2.18(a). Now reflecting that image in the x axis gives the graph shown in 2.18(b). Thus Figure 2.17(b) represents an odd function since its graph is antisymmetrical about the origin.

(c) The graph is neither symmetrical nor antisymmetrical about the origin, so the function it represents is neither odd nor even.

(d) The graph is symmetrical about the y axis so it is an even function.

(e) The graph is neither symmetrical nor antisymmetrical about the origin, so it is neither an even nor an odd function.

(f) The graph is antisymmetrical about the origin, so it represents an odd function.

A **periodic function** is such that its image values are repeated at regular intervals in its domain. Thus the graph of a periodic function can be divided into 'vertical strips' that are replicas of each other, as shown in Figure 2.19. The width of each strip is called the **period** of the function. We therefore say that a function $f(x)$ is periodic with period P if for all its domain values x

$$f(x + nP) = f(x)$$

for any integer n.

Figure 2.19
A periodic function
of period P.

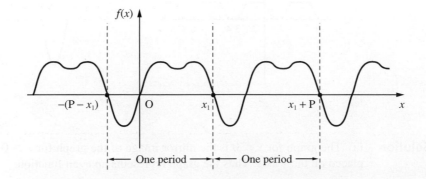

To provide a measure of the number of repetitions per unit of x, we define the **frequency** of a periodic function to be the reciprocal of its period, so that

$$frequency = \frac{1}{period}$$

The Greek letter v ('nu') is usually used to denote the frequency so that $v = 1/P$. The term **circular frequency** is also used in some engineering contexts. This is denoted by the Greek letter ω ('omega') and is defined by

$$\omega = 2\pi v = \frac{2\pi}{P}$$

It is measured in radians per unit of x. When the meaning is clear from the context the adjective 'circular' is commonly omitted.

Example 2.10

A function $f(x)$ has the graph on [0, 1] shown in Figure 2.20. Sketch its graph on [−3, 3] given that

(a) $f(x)$ is periodic with period 1

(b) $f(x)$ is periodic with period 2 and is even

(c) $f(x)$ is periodic with period 2 and is odd.

Figure 2.20
$f(x)$ of Example 2.10
defined on [0, 1].

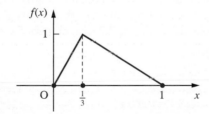

Solution

(a) Since $f(x)$ has period 1, strips of width 1 unit are simply replicas of the graph between 0 and 1. Hence we obtain the graph shown in Figure 2.21.

(b) Since $f(x)$ has period 2 we need to establish the graph over a complete period before we can replicate it along the domain of $f(x)$. Since it is an even function and we

Figure 2.21
$f(x)$ having period 1.

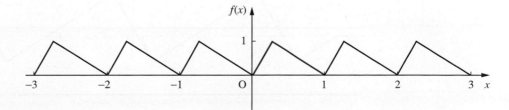

Figure 2.22
$f(x)$ periodic with
period 2 and is even.

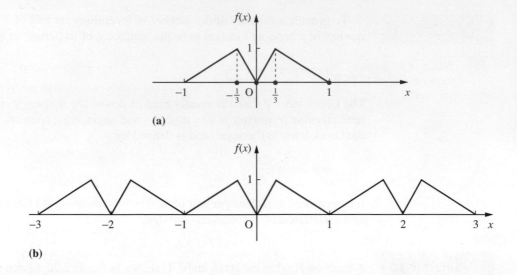

(a)

(b)

know its values between 0 and 1, we also know its values between -1 and 0. We can obtain the graph of $f(x)$ between -1 and 0 by reflecting in the y axis, as shown in Figure 2.22(a). Thus we have the graph over a complete period, from -1 to $+1$, and so we can replicate along the x axis, as shown in Figure 2.22(b).

(c) Similarly, if $f(x)$ is an odd function we can obtain the graph for the interval $[-1, 0]$ using antisymmetry and the graph for the interval $[0, 1]$. This gives us Figure 2.23(a) and we then obtain the whole graph, Figure 2.23(b), by periodic extension.

Figure 2.23
$f(x)$ periodic with
period 2 and is odd.

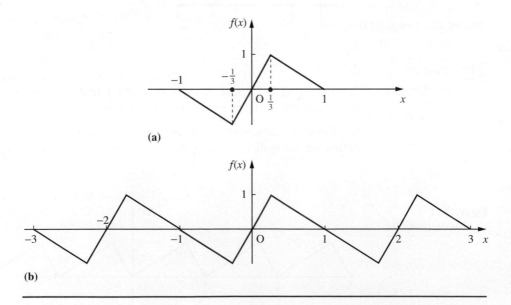

(a)

(b)

2.2.6 Exercises

6 Which of the functions $y = f(x)$ whose graphs are shown in Figure 2.24 are odd, even or neither odd nor even?

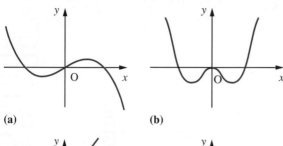

(a)   **(b)**

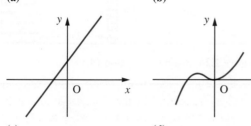

(c)   **(d)**

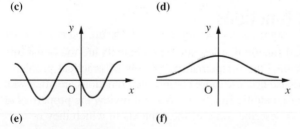

(e)   **(f)**

Figure 2.24 Graphs of Question 6.

7 Three different functions, $f(x)$, $g(x)$ and $h(x)$, have the same graph on $[0, 2]$ as shown in Figure 2.25. On separate diagrams, sketch their graphs for $[-4, 4]$ given that

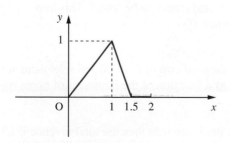

Figure 2.25 Graph of Question 7.

(a) $f(x)$ is periodic with period 2

(b) $g(x)$ is periodic with period 4 and is even

(c) $h(x)$ is periodic with period 4 and is odd.

8 A function $f(x)$ is defined by $f(x) = \frac{1}{2}(10^x + 10^{-x})$, for x in $\mathbb{R}$. Show that

(a) $2(f(x))^2 = f(2x) + 1$

(b) $2f(x)f(y) = f(x + y) + f(x - y)$

9 Draw separate graphs of the functions f and g where

$$f(x) = (x + 1)^2 \text{ and } g(x) = x - 2$$

The functions F and G are defined by

$$F(x) = f(g(x)) \text{ and } G(x) = g(f(x))$$

Find formulae for $F(x)$ and $G(x)$ and sketch their graphs. What relationships do the graphs of F and G bear to those of f and g?

10 A function f is defined by

$$f(x) = \begin{cases} 0 & (x < -1) \\ x + 1 & (-1 \leqslant x < 0) \\ 1 - x & (0 \leqslant x \leqslant 1) \\ 0 & (x > 1) \end{cases}$$

Sketch on separate diagrams the graphs of $f(x)$, $f(x + \frac{1}{2})$, $f(x + 1)$, $f(x + 2)$, $f(x - \frac{1}{2})$, $f(x - 1)$ and $f(x - 2)$.

11 Find the inverse function (if it is defined) of the following functions:

(a) $f(x) = 2x - 3$ $(x \text{ in } \mathbb{R})$

(b) $f(x) = \dfrac{2x - 3}{x + 4}$ $(x \text{ in } \mathbb{R}, x \neq -4)$

(c) $f(x) = x^2 + 1$ $(x \text{ in } \mathbb{R})$

If $f(x)$ does not have an inverse function, suggest a suitable restriction of the domain of $f(x)$ that will allow the definition of an inverse function.

12 Show that

$$f(x) = \dfrac{2x - 3}{x + 4}$$

may be expressed in the form

$$f(x) = g(h(l(x)))$$

where

$$l(x) = x + 4$$

$$h(x) = 1/x$$

$$g(x) = 2 - 11x$$

Interpret this result graphically.

13 The stiffness of a rectangular beam varies directly with the cube of its height and directly with its breadth. A beam of rectangular section is to be cut from a circular log of diameter d. Show that the optimal choice of height and breadth of the beam in terms of its stiffness is related to the value of x which maximizes the function

$$E(x) = x^3(d^2 - x), \quad 0 \leqslant x \leqslant d^2$$

14 A beam is used to support a building as shown in Figure 2.26. The beam has to pass over a 3 m brick wall which is 2 m from the building. Show that the minimum length of the beam is associated with the value of x which minimizes

$$E(x) = (x + 2)^2 + 9\left(1 + \frac{2}{x}\right)^2$$

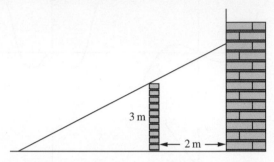

Figure 2.26 Beam of Question 14.

2.3 Linear and quadratic functions

Among the more commonly used functions in engineering contexts are the linear and quadratic functions. This is because the mathematical models of practical problems often involve linear functions and also because more complicated functions are often well approximated locally by linear or quadratic functions. We will review the properties of these functions and in the process describe some of the contexts in which they occur.

2.3.1 Linear functions

The **linear function** is the simplest function that occurs in practical problems. It has the formula $f(x) = mx + c$ where m and c are constant numbers and x is the unassigned or independent variable as usual. The graph of $f(x)$ is the set of points (x, y) where $y = mx + c$, which is the equation of a straight line on a cartesian coordinate plot. Hence, the function is called the linear function. To determine the formula for a particular linear function the two constants m and c have to be found. This implies that we need two pieces of information to determine $f(x)$.

Example 2.11 A manufacturer produces 5000 items at the total cost of £10 000 and sells them at £2.75 each. What is the manufacturer's profit as a function of the number x of items sold?

Solution Let the manufacturer's profit be £P. If x items are sold then the total revenue is £2.75x, so that the amount of profit $P(x)$ is given by

$$P(x) = \text{revenue} - \text{cost} = 2.75x - 10\,000$$

Here the domain of the function is [0, 5000] and the range is [−10000, 3750]. This function has a zero at $x = 3636\frac{4}{11}$. Thus to make a profit, the manufacturer has to sell more than 3636 items. (Note the modelling approximation in that, strictly, x is an integer variable, not a general real variable.)

If we know the values that the function $f(x)$ takes at two values, x_0 and x_1, of the independent variable x we can find the formula for $f(x)$. Let $f(x_0) = f_0$ and $f(x_1) = f_1$, then

$$f(x) = \frac{x - x_1}{x_0 - x_1}f_0 + \frac{x - x_0}{x_1 - x_0}f_1 \tag{2.6}$$

This formula is known as **Lagrange's formula**. It is obvious that the function is linear since we can arrange it as

$$f(x) = x\left[\frac{f_1 - f_0}{x_1 - x_0}\right] + \left[\frac{x_1 f_0 - x_0 f_1}{x_1 - x_0}\right]$$

The reader should verify from (2.6) that $f(x_0) = f_0$ and $f(x_1) = f_1$.

Example 2.12 Use Lagrange's formula to find the linear function $f(x)$ where $f(10) = 1241$ and $f(15) = 1556$.

Solution Taking $x_0 = 10$ and $x_1 = 15$ so that $f_0 = 1241$ and $f_1 = 1556$ we obtain

$$f(x) = \frac{x - 15}{10 - 15}(1241) + \frac{x - 10}{15 - 10}(1556)$$

$$= \frac{x}{5}(1556 - 1241) + 3(1241) - 2(1556)$$

$$= \frac{x}{5}(315) + (3723 - 3112) = 63x + 611$$

The **rate of change** of a function, between two values $x = x_0$ and $x = x_1$ in its domain, is defined by the ratio of the change in the values of the function to the change in the values of x. Thus

$$\text{rate of change} = \frac{f(x_1) - f(x_0)}{x_1 - x_0} \left(= \frac{\text{change in values of } f(x)}{\text{change in values of } x} \right)$$

For a linear function with formula $f(x) = mx + c$ we have

$$\text{rate of change} = \frac{(mx_1 + c) - (mx_0 + c)}{x_1 - x_0}$$

$$= \frac{m(x_1 - x_0)}{x_1 - x_0} = m$$

which is a constant. If we know the rate of change m of a linear function $f(x)$ and the value f_0 at a point $x = x_0$, then we can write the formula for $f(x)$ as

$$f(x) = mx + f_0 - mx_0$$

Example 2.13   The labour cost of producing a certain item is £21 per 10 000 items and the raw materials cost is £4 for 1000 items. Each time a new production run is begun, there is a set-up cost of £8. What is the cost, £$C(x)$, of a production run of x items?

Solution   Here the cost function has a rate of change comprising the labour cost per item (21/10 000) and the materials cost per item (4/1000). Thus the rate of change is 0.0061. We also know that if there is a production run with zero items, there is still a set-up cost of £8 so $f(0) = 8$. Thus the required function is

$$C(x) = 0.0061x + 8$$

2.3.2 Least squares fit of a linear function to experimental data

Because the linear function occurs in many mathematical models of practical problems, we often have to 'fit' linear functions to experimental data. That is, we have to find the values of m and c which yield the best overall description of the data. There are two distinct mathematical models that occur. These are given by the functions with formulae

(a) $y = ax$ and   (b) $y = mx + c$

For example, the extension of an ideal spring under load may be represented by a function of type (a), while the velocity of a projectile launched vertically may be represented by a function of type (b).

From experiments we obtain a set of data points (x_k, y_k), $k = 1, 2, \ldots, n$. We wish to find the value of the constant(s) of the linear function that best describes the phenomenon the data represents.

Case (a): the theoretical model has the form $y = ax$

The difference between theoretical value ax_k and the experimental value y_k at x_k is $(ax_k - y_k)$. This is the 'error' of the model at $x = x_k$. We define the value of a for which $y = ax$ best represents the data to be that value which minimizes the sum S of the squared errors:

$$S = \sum_{k=1}^{n} (ax_k - y_k)^2$$

(Hence the name 'least squares fit': the squares of the errors are chosen to avoid simple cancellation of two large errors of opposite sign.)

It is easy to find the minimizing value of a since S is essentially a quadratic expression in a. (All the x_k's and y_k's are numbers.) Rewriting, we have

$$S = \sum_{k=1}^{n} (a^2 x_k^2 - 2a x_k y_k + y_k^2)$$

$$= \sum_{k=1}^{n} (a^2 x_k^2) + \sum_{k=1}^{n} (-2a x_k y_k) + \sum_{k=1}^{n} y_k^2$$

$$= a^2 \sum_{k=1}^{n} x_k^2 - 2a \sum_{k=1}^{n} x_k y_k + \sum_{k=1}^{n} y_k^2$$

(Notice the 'taking out' of the common factors a^2 and $-2a$ in these sums.) Writing

$$P = \sum_{k=1}^{n} x_k^2, \quad Q = \sum_{k=1}^{n} x_k y_k \quad \text{and} \quad R = \sum_{k=1}^{n} y_k^2$$

we have

$$S = Pa^2 - 2aQ + R$$

On 'completing the square'

$$S = P\left(a - \frac{Q}{P}\right)^2 + \frac{RP - Q^2}{P}$$

and we see that the minimizing value of a is given by Q/P, when the first term is zero. Thus S is minimized when

$$a = \frac{\displaystyle\sum_{k=1}^{n} x_k y_k}{\displaystyle\sum_{k=1}^{n} x_k^2} \tag{2.7}$$

Example 2.14   Find the value of a which provides the least squares fit to the model $y = ax$ for the data given in Figure 2.27.

Figure 2.27
Data of Example 2.14.

k	1	2	3	4	5	6
x_k	50	100	150	200	250	300
y_k	5	8	9	11	12	15

Solution   From (2.7) the least squares fit is provided by

$$a = \left(\sum_{k=1}^{6} x_k y_k\right) \Big/ \left(\sum_{k=1}^{6} x_k^2\right)$$

Here

$$\sum_{k=1}^{6} x_k y_k = 250 + 800 + 1350 + 2200 + 3000 + 4500 = 12\,100$$

and

$$\sum_{k=1}^{6} x_k^2 = 50^2 + 100^2 + 150^2 + 200^2 + 250^2 + 300^2 = 227\,500$$

so that $a = 121/2275 = 0.053$.

Case (b): the theoretical model has the form $y = mx + c$

Analogous to case (a), this can be seen as minimizing the sum

$$S = \sum_{k=1}^{n} (mx_k + c - y_k)^2$$

The algebraic approach to this minimization uses completion of squares in two variables. The details are complicated but are given below. Working through the details provides useful practice and consolidation of the use of the sigma notation.

Multiplying out the terms gives

$$S = m^2 \sum_{k=1}^{n} x_k^2 - 2m \sum_{k=1}^{n} x_k y_k + 2mc \sum_{k=1}^{n} x_k - 2c \sum_{k=1}^{n} y_k + nc^2 + \sum_{k=1}^{n} y_k^2$$

Now $\sum_{k=1}^{n} x_k = n\bar{x}$ and $\sum_{k=1}^{n} y_k = n\bar{y}$, where $\bar{x}$ and $\bar{y}$ are the **mean values** of the x_k's and y_k's respectively, so S can be written

$$S = m^2 \sum_{k=1}^{n} x_k^2 - 2m \sum_{k=1}^{n} x_k y_k + 2mcn\bar{x} - 2cn\bar{y} + nc^2 + \sum_{k=1}^{n} y_k^2$$

Completing the square with terms involving n gives

$$S = n(c - \bar{y} + m\bar{x})^2 + m^2 \left\{ \sum_{k=1}^{n} x_k^2 - n\bar{x}^2 \right\} - 2m \left\{ \sum_{k=1}^{n} x_k y_k - n\bar{x}\bar{y} \right\} + \sum_{k=1}^{n} y_k^2 - n\bar{y}^2$$

Now completing the square with the remaining terms involving m we have

$$S = n(c - \bar{y} + m\bar{x})^2 + p(m - q/p)^2 + r - q^2/p$$

where

$$p = \sum_{k=1}^{n} x_k^2 - n\bar{x}^2 \quad \text{and} \quad q = \sum_{k=1}^{n} x_k y_k - n\bar{x}\bar{y} \quad \text{and} \quad r = \sum_{k=1}^{n} y_k^2 - n\bar{y}^2$$

Thus S is minimized where

$$m = \frac{\displaystyle\sum_{k=1}^{n} x_k y_k - n\bar{x}\bar{y}}{\displaystyle\sum_{k=1}^{n} x_k^2 - n\bar{x}^2} \quad \text{and} \quad c = \bar{y} - m\bar{x} \tag{2.8}$$

To avoid loss of significance, the formula for m is usually expressed in the form

$$m = \frac{\sum_{k=1}^{n}(x_k - \bar{x})(y_k - \bar{y})}{\sum_{k=1}^{n}(x_k - \bar{x})^2} \qquad (2.9)$$

We can observe that in this case the best straight line passes through the average data point $(\bar{x}, \bar{y})$, and the best straight line has the formula

$$y = mx + c$$

with $c = \bar{y} - m\bar{x}$.

Example 2.15 Find the values of m and c which provide the least squares fit to the linear model $y = mx + c$ for the data given in Figure 2.28.

Figure 2.28
Data of Example 2.15.

k	1	2	3	4	5
x_k	0	1	2	3	4
y_k	1	1	2	2	3

Solution From (2.9) the least squares fit is provided by

$$m = \frac{\sum_{k=1}^{n}(x_k - \bar{x})(y_k - \bar{y})}{\sum_{k=1}^{n}(x_k - \bar{x})^2}$$

Here $\bar{x} = \frac{1}{5}(10) = 2.0$, $\bar{y} = \frac{1}{5}(9) = 1.8$, $\sum_{k=1}^{n}(x_k - \bar{x})(y_k - \bar{y}) = 5.0$ and $\sum_{k=1}^{n}(x_k - \bar{x})^2 = 10$, so that

$$m = 0.5$$

and hence $c = 1.8 - 0.5(2) = 0.8$.

Thus the best straight line fit to the data is provided by $y = 0.5x + 0.8$.

The formula for case (b) is the one most commonly given on calculators and in computer packages (where it is called **linear regression**). It is important to have a theoretical justification to fitting data to a function, otherwise it is easy to produce nonsense. For example, the data in Example 2.14 actually related to the extension of a soft spring under a load so that it would be inappropriate to fit that data to $y = mx + c$. A non-zero value for c would imply an extension with zero load!

2.3.3 The quadratic function

The general quadratic function has the form

$$f(x) = ax^2 + bx + c$$

where a, b and c are constants and $a \neq 0$. By 'completing the square' we can show that

$$f(x) = a\left[\left(x + \frac{b}{2a}\right)^2 + \frac{4ac - b^2}{4a^2}\right]$$

which implies that the graph of $f(x)$ is either a 'cup' ($a > 0$) or a 'cap' ($a < 0$), as shown in Figure 2.29, and is a parabola.

We can see that, because the quadratic function has three constants, to determine a specific quadratic function requires three data points. The formula for the quadratic function $f(x)$ taking the values f_0, f_1, f_2 at the values x_0, x_1, x_2, of the independent variable x, may be written in Lagrange's form:

$$f(x) = \frac{(x - x_1)(x - x_2)}{(x_0 - x_1)(x_0 - x_2)}f_0 + \frac{(x - x_0)(x - x_2)}{(x_1 - x_0)(x_1 - x_2)}f_1 + \frac{(x - x_0)(x - x_1)}{(x_2 - x_0)(x_2 - x_1)}f_2$$

(2.10)

The right-hand side of this formula is clearly a quadratic function. The reader should spend a few minutes verifying that inserting the values $x = x_0, x_1$ and x_2 yields $f(x_0) = f_0$, $f(x_1) = f_1$ and $f(x_2) = f_2$.

Figure 2.29
(a) $a > 0$; (b) $a < 0$.

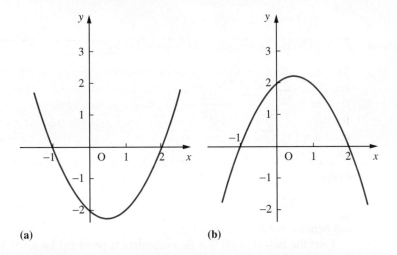

(a)

(b)

Example 2.16   Find the formula of the quadratic function which satisfies the data points (1, 2), (2, 4) and (3, 8).

Solution   Choose $x_0 = 1$, $x_1 = 2$ and $x_2 = 3$ so that $f_0 = 2$, $f_1 = 4$ and $f_2 = 8$. Then using Lagrange's formula (2.10) we have

$$f(x) = \frac{(x-2)(x-3)}{(1-2)(1-3)}(2) + \frac{(x-1)(x-3)}{(2-1)(2-3)}(4) + \frac{(x-1)(x-2)}{(3-1)(3-2)}(8)$$

$$= (x-2)(x-3) - 4(x-1)(x-3) + 4(x-1)(x-2)$$

$$= x^2 - x + 2$$

Lagrange's formula is not always the best way to obtain the formula of a quadratic function. Sometimes we wish to obtain the formula as an expansion about a specific point, as illustrated in Example 2.17.

Example 2.17   Find the quadratic function in the form

$$f(x) = A(x-2)^2 + B(x-2) + C$$

which satisfies $f(1) = 2, f(2) = 4, f(3) = 8$.

Solution   Setting $x = 1$, 2 and 3 into the formula for $f(x)$ we obtain

$$f(1): A - B + C = 2$$

$$f(2): \qquad\quad C = 4$$

$$f(3): A + B + C = 8$$

from which we quickly find $A = 1, B = 3$ and $C = 4$. Thus

$$f(x) = (x-2)^2 + 3(x-2) + 4$$

The way we express the quadratic function depends on the problem context. The form $f(x) = ax^2 + bx + c$ is convenient for values of x near $x = 0$, while the form $f(x) = A(x-x_0)^2 + B(x-x_0) + C$ is convenient for values of x near $x = x_0$. (The second form here is sometimes called the **Taylor expansion** of $f(x)$ about $x = x_0$.) This is discussed for the general function in Section 9.5, where we make use of the differential calculus to obtain the expansion.

Since we can rewrite

$$f(x) = ax^2 + bx + c$$

as

$$f(x) = a\left[\left(x + \frac{b}{2a}\right)^2 + \frac{4ac - b^2}{4a^2}\right]$$

we see that when $b^2 > 4ac$ we can factorize $f(x)$ into the product of two linear factors and $f(x)$ has two zeros given as in (1.5) by

$$x = \frac{-b \pm \sqrt{(b^2 - 4ac)}}{2a}$$

When $b^2 < 4ac$, $f(x)$ cannot be factorized and does not have a zero. In this case it is called an **irreducible quadratic function**.

Example 2.18

Complete the squares of the following quadratics and specify which are irreducible.

(a) $y = x^2 + x + 1$ (b) $y = 3x^2 - 2x - 1$

(c) $y = 4 + 3x - x^2$ (d) $y = 2x - 1 - 2x^2$

Solution

(a) Using the method of completing the square we have

$$y = x^2 + x + 1 = (x + \tfrac{1}{2})^2 + \tfrac{3}{4} = (x + \tfrac{1}{2})^2 + (\tfrac{\sqrt{3}}{2})^2$$

Since this is a sum of squares, like $A^2 + B^2$, it cannot, unlike a difference of squares, $A^2 - B^2 = (A - B)(A + B)$, be factorized. Thus this is an irreducible quadratic function.

(b) $y = 3x^2 - 2x - 1 = 3(x^2 - \tfrac{2}{3}x - \tfrac{1}{3})$

$$= 3[(x - \tfrac{1}{3})^2 - \tfrac{4}{9}] = 3[(x - \tfrac{1}{3}) - \tfrac{2}{3}][(x - \tfrac{1}{3}) + \tfrac{2}{3}]$$

$$= 3[x - 1][x + \tfrac{1}{3}] = (x - 1)(3x + 1)$$

Thus this is not an irreducible quadratic function.

(c) $y = 4 + 3x - x^2 = 4 + \tfrac{9}{4} - (x - \tfrac{3}{2})^2$

$$= \tfrac{25}{4} - (x - \tfrac{3}{2})^2 = [\tfrac{5}{2} - (x - \tfrac{3}{2})][\tfrac{5}{2} + (x - \tfrac{3}{2})]$$

$$= (4 - x)(1 + x)$$

Thus y is a product of two linear factors and $4 + 3x - x^2$ is not an irreducible quadratic function.

(d) $y = 2x - 1 - 2x^2 = -1 - 2(x^2 - x)$

$$= -1 + \tfrac{1}{2} - 2(x - \tfrac{1}{2})^2 = -\tfrac{1}{2} - 2(x - \tfrac{1}{2})^2$$

$$= -2[\tfrac{1}{4} + (x - \tfrac{1}{2})^2]$$

Since the term inside the square brackets is the sum of squares, we have an irreducible quadratic function.

The quadratic function

$$f(x) = ax^2 + bx + c$$

has a maximum when $a < 0$ and a minimum when $a > 0$, as illustrated earlier in Figure 2.29. The position and value of that extremal point (that is, of the maximum or the minimum) can be obtained from the completed square form of $f(x)$

$$f(x) = a\left(x + \frac{b}{2a}\right)^2 + \frac{4ac - b^2}{4a}$$

The extremal value occurs where

$$x + \frac{b}{2a} = 0$$

Thus, when $a > 0$, $f(x)$ has a minimum value $(4ac - b^2)/(4a)$ where $x = -b/(2a)$. When $a < 0$, $f(x)$ has a maximum value $(4ac - b^2)/(4a)$ at $x = -b/(2a)$. This result is important in engineering contexts when we are trying to optimize costs or profits or to produce an optimal design. (See Section 2.10.)

Example 2.19   Find the extremal values of the functions

(a) $y = x^2 + x + 1$ (b) $y = 3x^2 - 2x - 1$

(c) $y = 4 + 3x - x^2$ (d) $y = 2x - 1 - 2x^2$

Solution   This uses the completed squares of Example 2.18.

(a) $y = x^2 + x + 1 = (x + \frac{1}{2})^2 + \frac{3}{4}$

Clearly the smallest value y can take is $\frac{3}{4}$ and this occurs when $x + \frac{1}{2} = 0$; that is, when $x = -\frac{1}{2}$.

(b) $y = 3x^2 - 2x - 1 = 3(x - \frac{1}{3})^2 - \frac{4}{3}$

Clearly the smallest value of y occurs when $x = \frac{1}{3}$ and is equal to $-\frac{4}{3}$.

(c) $y = 4 + 3x - x^2 = \frac{25}{4} - (x - \frac{3}{2})^2$

Clearly the largest value y can take is $\frac{25}{4}$ and this occurs when $x = \frac{3}{2}$.

(d) $y = 2x - 1 - 2x^2 = -\frac{1}{2} - 2(x - \frac{1}{2})^2$

Thus the maximum value of y equals $-\frac{1}{2}$ and occurs where $x = \frac{1}{2}$.

2.3.4 Exercises

15   Obtain the formula for the linear functions $f(x)$ such that

(a) $f(0) - 3$ and $f(2) = -1$

(b) $f(-1) = 2$ and $f(3) = 4$

(c) $f(1.231) = 2.791$ and $f(2.492) = 3.112$

16   Calculate the rate of change of the linear functions given by

(a) $f(x) = 3x - 2$

(b) $f(x) = 2 - 3x$

(c) $f(-1) = 2$ and $f(3) = 4$

17   The total labour cost of producing a certain item is £43 per 100 items produced. The raw materials cost £25 per 1000 items. There is a set-up cost of £50 for each production run. Obtain the formula for the cost of a production run of x items.

The manufacturer decides to have a production run of 2000 items. What is its cost? If the items are sold at £1.20 each, write down a formula for the manufacturer's profit if x items are sold. What is the breakeven number of items sold?

18   Find the formulae of the quadratic functions $f(x)$ such that

(a) $f(1) = 3, f(2) = 7$ and $f(4) = 19$

(b) $f(-1) = 1, f(1) = -1$ and $f(4) = 2$

19 Find the numbers A, B and C such that

$$f(x) = x^2 - 8x + 10$$
$$= A(x - 2)^2 + B(x - 2) + C$$

20 Determine which of the following quadratic functions are irreducible.

(a) $f(x) = x^2 + 2x + 3$ (b) $f(x) = 4x^2 - 12x + 9$

(c) $f(x) = 6 - 4x - 3x^2$ (d) $f(x) = 3x - 1 - 5x^2$

21 Find the maximum or minimum values of the quadratic functions given in Question 20.

22 For what values of x are the values of the quadratic functions below greater than zero?

(a) $f(x) = x^2 - 6x + 8$ (b) $f(x) = 15 + x - 2x^2$

23 Find the least squares fit to the linear function $y = ax$ of the data given in Figure 2.30.

k	1	2	3	4	5
x_k	10.1	10.2	10.3	10.4	10.5
y_k	3.10	3.12	3.21	3.25	3.32

Figure 2.30 Table of Question 23.

24 Find the least squares fit to the linear function $y = mx + c$ for the experimental data given in Figure 2.31.

k	1	2	3	4	5
x_k	55	60	65	70	75
y_k	107	109	114	118	123

Figure 2.31 Table of Question 24.

2.4 Polynomial functions

A **polynomial function** has the general form

$$f(x) = a_n x^n + a_{n-1} x^{n-1} + \ldots + a_1 x + a_0, \quad x \text{ in } \mathbb{R} \qquad (2.11)$$

where n is a positive integer and a_r is a real number called the coefficient of x^r, $r = 0, 1, \ldots, n$. The index n of the highest power of x occurring is called the **degree of the polynomial**. For $n = 1$ we obtain the linear function

$$f(x) = a_1 x + a_0$$

and for $n = 2$ the quadratic function

$$f(x) = a_2 x^2 + a_1 x + a_0$$

and so on.

We obtained in Sections 2.3.1 and 2.3.3 Lagrange's formulae for linear and for quadratic functions. The basic idea of the formulae can be used to obtain a formula for a polynomial of degree n which is such that $f(x_0) = f_0, f(x_1) = f_1, f(x_2) = f_2, \ldots,$ $f(x_n) = f_n$. Notice we need $(n + 1)$ values to determine a polynomial of degree n. We can write Lagrange's formula in the form.

$$f(x) = L_0(x)f_0 + L_1(x)f_1 + L_2(x)f_2 + \ldots + L_n(x)f_n$$

where $L_0(x), L_1(x), \ldots, L_n(x)$ are polynomials of degree n such that

$$L_k(x_j) = 0, \quad x_j \neq x_k \text{ (or } j \neq k)$$

$$L_k(x_k) = 1$$

This implies that L_k has the form

$$L_k(x) = \frac{(x - x_0)(x - x_1)(x - x_2) \ldots (x - x_{k-1})(x - x_{k+1}) \ldots (x - x_n)}{(x_k - x_0)(x_k - x_1)(x_k - x_2) \ldots (x_k - x_{k-1})(x_k - x_{k+1}) \ldots (x_k - x_n)}$$

(It is easy to verify that L_k has degree n and that $L_k(x_j) = 0, j \neq k$ and $L_k(x_k) = 1$.)

Example 2.20

Find the cubic function such that $f(-3) = 528, f(0) = 1017, f(2) = 1433$ and $f(5) = 2312$.

Solution

Notice that we need four data points to determine a cubic function. We can write

$$f(x) = L_0(x)f_0 + L_1(x)f_1 + L_2(x)f_2 + L_3(x)f_3$$

where $x_0 = -3, f_0 = 528, x_1 = 0, f_1 = 1017, x_2 = 2, f_2 = 1433, x_3 = 5$ and $f_3 = 2312$. Thus

$$L_0(x) = \frac{(x - 0)(x - 2)(x - 5)}{(-3 - 0)(-3 - 2)(-3 - 5)} = -\tfrac{1}{120}(x^3 - 7x^2 + 10x)$$

$$L_1(x) = \frac{(x + 3)(x - 2)(x - 5)}{(0 + 3)(0 - 2)(0 - 5)} = \tfrac{1}{30}(x^3 - 4x^2 - 11x + 30)$$

$$L_2(x) = \frac{(x + 3)(x - 0)(x - 5)}{(2 + 3)(2 - 0)(2 - 5)} = -\tfrac{1}{30}(x^3 - 2x^2 - 15x)$$

$$L_3(x) = \frac{(x + 3)(x - 0)(x - 2)}{(5 + 3)(5 - 0)(5 - 2)} = \tfrac{1}{120}(x^3 + x^2 - 6x)$$

Notice that each of the L_k's is a cubic function, so that their sum will be a cubic function

$$f(x) = -\tfrac{1}{120}(x^3 - 7x^2 + 10x)(528) + \tfrac{1}{30}(x^3 - 4x^2 - 11x + 30)(1017)$$

$$- \tfrac{1}{30}(x^3 - 2x^2 - 15x)(1433) + \tfrac{1}{120}(x^3 + x^2 - 6x)(2312)$$

$$= x^3 + 10x^2 + 184x + 1017$$

2.4.1 Basic properties

Polynomials have two important mathematical properties.

Property (i)
If two polynomials are equal for all values of the independent variable then corresponding coefficients of the powers of the variable are equal. Thus if

$$f(x) = a_n x^n + a_{n-1} x^{n-1} + \ldots + a_1 x + a_0$$

$$g(x) = b_n x^n + b_{n-1} x^{n-1} + \ldots + b_1 x + b_0$$

and

$$f(x) = g(x) \quad \text{for all } x$$

then

$$a_i = b_i \quad \text{for } i = 0, 1, 2, \dots, n$$

This property forms the basis of a technique called **equating coefficients**, which will be used in determining partial fractions in Section 2.5.

Property (ii)

Any polynomial with real coefficients can be expressed as a product of linear and irreducible quadratic factors.

Example 2.21 Find the values of A, B and C that ensure that

$$x^2 + 1 = A(x - 1) + B(x + 2) + C(x^2 + 2)$$

for all values of x.

Solution Multiplying out the right-hand side, we have

$$x^2 + 0x + 1 = Cx^2 + (A + B)x + (-A + 2B + 2C)$$

Using Property (i), we compare, or equate, the coefficients of x^2, x and x^0 in turn to give

$$C = 1$$
$$A + B = 0$$
$$-A + 2B + 2C = 1$$

which we then solve to give

$$A = \tfrac{1}{3}, \quad B = -\tfrac{1}{3}, \quad C = 1$$

Checking, we have

$$Cx^2 + (A + B)x + (-A + 2B + 2C) = x^2 + (\tfrac{1}{3} - \tfrac{1}{3})x + (-\tfrac{1}{3} - \tfrac{2}{3} + 2) = x^2 + 1$$

2.4.2 Factorization

Although Property (ii) was known earlier, the first rigorous proof was published by Gauss in 1799. The result is an 'existence theorem'. It tells us that polynomials can be factored, but does not indicate how to find the factors!

Example 2.22 Factorize the polynomials

(a) $x^3 - 3x^2 + 6x - 4$ (b) $x^4 - 16$ (c) $x^4 + 16$

Solution (a) The function $f(x) = x^3 - 3x^2 + 6x - 4$ clearly has the value zero at $x = 1$. Thus $x - 1$ must be a factor of $f(x)$. We can now divide $x^3 - 3x^2 + 6x - 4$ by $x - 1$ using algebraic division, a process akin to long division of numbers. The process may be set out as follows.

Step 1

$$x - 1)x^3 - 3x^2 + 6x - 4($$

In order to produce the term x^3, $x - 1$ must be multiplied by x^2. Do this and subtract the result from $x^3 - 3x^2 + 6x - 4$.

$$x - 1)x^3 - 3x^2 + 6x - 4(x^2$$
$$\underline{x^3 - x^2}$$
$$-2x^2 + 6x - 4$$

Step 2

Now repeat the process on the polynomial $-2x^2 + 6x - 4$. In this case, in order to eliminate the term $-2x^2$, we must multiply $x - 1$ by $-2x$.

$$x - 1)x^3 - 3x^2 + 6x - 4(x^2 - 2x$$
$$\underline{x^3 - x^2}$$
$$-2x^2 + 6x - 4$$
$$\underline{-2x^2 + 2x}$$
$$4x - 4$$

Step 3

Finally we must multiply $x - 1$ by 4 to eliminate $4x - 4$ as follows.

$$x - 1)x^3 - 3x^2 + 6x - 4(x^2 - 2x + 4$$
$$\underline{x^3 - x^2}$$
$$-2x^2 + 6x - 4$$
$$\underline{-2x^2 + 2x}$$
$$4x - 4$$
$$\underline{4x - 4}$$

Thus

$$f(x) = (x - 1)(x^2 - 2x + 4)$$

The quadratic factor $x^2 - 2x + 4$ is an **irreducible factor**, as is shown by 'completing the square':

$$x^2 - 2x + 4 = (x - 1)^2 + 3$$

(b) The functions $f_1(x) = x^4$ and $f_2(x) = x^4 - 16$ have similar graphs, as shown in Figures 2.32(a) and (b). It is clear from these graphs that $f_2(x)$ has zeros at two values of x, where $x^4 = 16$: that is, at $x^2 = 4$ ($x^2 = -4$ is not allowed for real x). Thus the zeros of f_2 are at $x = 2$ and $x = -2$, and we can write

$$f_2(x) = x^4 - 16 = (x^2 - 4)(x^2 + 4)$$

$$= (x - 2)(x + 2)(x^2 + 4)$$

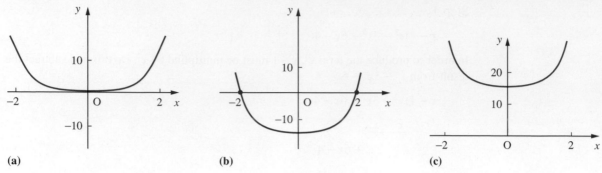

Figure 2.32 Graphs of (a) $y = f_1(x) = x^4$, (b) $y = f_2(x) = x^4 - 16$ and (c) $y = f_3(x) = x^4 + 16$.

(c) The functions $f_1(x) = x^4$ and $f_3(x) = x^4 + 16$ have similar graphs, as shown in Figures 2.32(a) and (c). It is clear from these graphs that $f_3(x)$ does not have any real zeros, so we expect it to be factored into two quadratic terms. We can write

$$x^4 + 16 = (x^2 + 4)^2 - 8x^2$$

which is a difference of squares and may be factored.

$$(x^2 + 4)^2 - 8x^2 = (x^2 + 4)^2 - (x\sqrt{8})^2 = [(x^2 + 4) - x\sqrt{8}][(x^2 + 4) + x\sqrt{8}]$$

Thus we obtain

$$f_3(x) = x^4 + 16 = (x^2 - 2x\sqrt{2} + 4)(x^2 + 2x\sqrt{2} + 4)$$

Since $x^2 \pm 2x\sqrt{2} + 4 = (x \pm \sqrt{2})^2 + 2$, we deduce that these are irreducible quadratics.

2.4.3 Nested multiplication and synthetic division

In Example 2.22(a) we found the image value of the polynomial at $x = 1$ by direct substitution. In general, however, the most efficient way to evaluate the image values of a polynomial function is to use **nested multiplication**. Consider the cubic function

$$f(x) = 4x^3 - 5x^2 + 2x + 3$$

This may be written as

$$f(x) = [(4x - 5)x + 2]x + 3$$

We evaluate this by evaluating each bracketed expression in turn, working from the innermost. Thus to find $f(6)$, the following steps are taken:

(1) Multiply 4 by x and subtract 5; in this case $4 \times 6 - 5 = 19$.
(2) Multiply the result of step 1 by x and add 2; in this case $19 \times 6 + 2 = 116$.
(3) Multiply the result of step 2 by x and add 3; in this case $116 \times 6 + 3 = 699$.

Thus $f(6) = 699$.

On a computer this is performed by means of a simple recurrence relation. To evaluate

$$f(x) = a_n x^n + a_{n-1} x^{n-1} + \ldots + a_0$$

at $x = t$, we use the formulae

$$b_{n-1} = a_n$$

$$b_{n-2} = tb_{n-1} + a_{n-1}$$

$$b_{n-3} = tb_{n-2} + a_{n-2}$$

$$\vdots$$

$$b_1 = tb_2 + a_2$$

$$b_0 = tb_1 + a_1$$

$$f(t) = tb_0 + a_0$$

which may be summarized as

$$\left. \begin{array}{l} b_{n-1} = a_n \\[2mm] b_{n-k} = tb_{n-k+1} + a_{n-k+1} \quad (k = 2, 3, \ldots, n) \\[2mm] f(t) = tb_0 + a_0 \end{array} \right\} \tag{2.12}$$

(The reason for storing the intermediate values b_k will become obvious below.)

Having evaluated $f(x)$ at $x = t$, it follows that for a given t

$$f(x) - f(t) = 0$$

at $x = t$; that is, $f(x) - f(t)$ has a factor $x - t$. Thus we can write

$$f(x) - f(t) = (x - t)(c_{n-1}x^{n-1} + c_{n-2}x^{n-2} + \ldots + c_1x + c_0)$$

Multiplying out the right-hand side, we have

$$f(x) - f(t) = c_{n-1}x^n + (c_{n-2} - tc_{n-1})x^{n-1} + (c_{n-3} - tc_{n-2})x^{n-2} + \ldots + (c_0 - tc_1)x + (-tc_0)$$

so that we may write

$$f(x) = c_{n-1}x^n + (c_{n-2} - tc_{n-1})x^{n-1} + (c_{n-3} - tc_{n-2})x^{n-2} + \ldots + (c_0 - tc_1)x + f(t) - tc_0$$

But

$$f(x) = a_n x^n + a_{n-1}x^{n-1} + a_{n-2}x^{n-2} + \ldots + a_1x + a_0$$

So, using Property (i) of Section 2.4.1 and comparing coefficients of like powers of x, we have

$$c_{n-1} = a_n$$

$$c_{n-2} - tc_{n-1} = a_{n-1} \quad \text{implying} \quad c_{n-2} = tc_{n-1} + a_{n-1}$$

$$c_{n-3} - tc_{n-2} = a_{n-2} \quad \text{implying} \quad c_{n-3} = tc_{n-2} + a_{n-2}$$

$$\vdots \qquad\qquad \vdots \qquad\qquad \vdots$$

$$c_0 - tc_1 = a_1 \quad \text{implying} \quad c_0 = tc_1 + a_1$$

$$f(t) - tc_0 = a_0 \quad \text{implying} \quad f(t) = tc_0 + a_0$$

Thus c_k satisfies exactly the same formula as b_k, so that the intermediate numbers generated by the method are the coefficients of the quotient polynomial. We can then write

$$f(x) = (b_{n-1}x^{n-1} + b_{n-2}x^{n-2} + \ldots + b_1x + b_0)(x - t) + f(t) \qquad \textbf{(2.13)}$$

or

$$\frac{f(x)}{x - t} = b_{n-1}x^{n-1} + b_{n-2}x^{n-2} + \ldots + b_1x + b_0 + \frac{f(t)}{x - t}$$

Result (2.13) tells us that if the polynomial $f(x)$ given in (2.11) is divided by $x - t$ then this results in a quotient polynomial $q(x)$ given by

$$q(x) = b_{n-1}x^{n-1} + \ldots + b_0$$

and a remainder $r = f(t)$ that is independent of x. Because of this property, the method of nested multiplication is sometimes called **synthetic division**.

The coefficients b_i, $i = 0, \ldots, n - 1$, of the quotient polynomial and remainder term $f(t)$ may be determined using the formulae (2.12). The process may be carried out in the following tabular form:

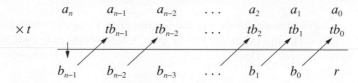

After the number below the line is calculated as the sum of the two numbers immediately above it, it is multiplied by t and placed in the next space above the line as indicated by the arrows. This procedure is repeated until all the terms are calculated.

The method of synthetic division could have been used as an alternative to algebraic division in Example 2.22.

Example 2.23   Show that $f(x) = x^3 - 3x^2 + 6x - 4$ is zero at $x = 1$, and hence factorize $f(x)$.

Solution   Using the nested multiplication procedure to divide $x^3 - 3x^2 + 6x - 4$ by $x - 1$ gives the tabular form

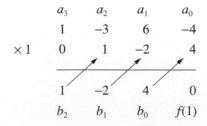

Since the remainder $f(1)$ is zero, it follows that $f(x)$ is zero at $x = 1$. Thus

$$f(x) = (x^2 - 2x + 4)(x - 1)$$

and we have extracted the factor $x - 1$. We may then examine the quadratic factor $x^2 - 2x + 4$ as we did in Example 2.22(a) and show that it is an irreducible quadratic factor.

Sometimes in problem-solving we need to rearrange the formula for the polynomial function as an expansion about a point, $x = a$, other than $x = 0$. That is, we need to find the numbers $A_0, A_1, \ldots, A_n$ such that

$$f(x) = a_n x^n + a_{n-1} x^{n-1} + \ldots + a_1 x + a_0$$

$$= A_n(x - a)^n + A_{n-1}(x - a)^{n-1} + \ldots + A_1(x - a) + A_0$$

This transformation can be achieved using the technique illustrated for the quadratic function in Example 2.17 which depends on the identity property of polynomials. It can be achieved more easily using **repeated synthetic division**, as is shown in Example 2.24.

Example 2.24 Obtain the expansion about $x = 2$ of the function $y = x^3 - 3x^2 + 6x - 4$.

Solution Using the numerical scheme as sct in Example 2.23 we have

$$
\begin{array}{r}
 1 \quad -3 \quad 6 \quad -4 \\
\times 2 \quad 0 \quad 2 \quad -2 \quad 8 \\
\hline
 1 \quad -1 \quad 4 \quad 4
\end{array}
$$

so that

$$x^3 - 3x^2 + 6x - 4 = (x - 2)(x^2 - x + 4) + 4$$

Now repeating the process with $y = x^2 - x + 4$, we have

$$
\begin{array}{r}
 1 \quad -1 \quad 4 \\
\times 2 \quad 0 \quad 2 \quad 2 \\
\hline
 1 \quad 1 \quad 6
\end{array}
$$

so that

$$x^2 - x + 4 = (x - 2)(x + 1) + 6$$

and

$$x^3 - 3x^2 + 6x - 4 = (x - 2)[(x - 2)(x + 1) + 6] + 4$$

Lastly,

$$x + 1 = (x - 2) + 3$$

so that

$$y = (x - 2)[(x - 2)^2 + 3(x - 2) + 6] + 4$$

$$= (x - 2)^3 + 3(x - 2)^2 + 6(x - 2) + 4$$

For hand computation the whole process can be set out as a single table:

$$
\begin{array}{rrrrr}
& 1 & -3 & 6 & -4 \\
\times 2 & 0 & 2 & -2 & 8 \\
\hline
& 1 & -1 & 4 & \vdots 4 \\
\times 2 & 0 & 2 & 2 & \\
\hline
& 1 & 1 & \vdots 6 & \\
\times 2 & 0 & 2 & & \\
\hline
& 1 & \vdots 3 & &
\end{array}
$$

Here, then, 1, 3, 6 and 4 provide the coefficients of $(x-2)^3$, $(x-2)^2$, $(x-2)^1$ and $(x-2)^0$ in the Taylor expansion.

2.4.4 Roots of polynomial equations

Polynomial equations occur frequently in engineering applications, from the identification of resonant frequencies when concerned with rotating machinery to the stability analysis of circuits. It is often useful to see the connections between the roots of a polynomial equation and its coefficients.

Example 2.25   Show that any real roots of the equation

$$x^3 - 3x^2 + 6x - 4 = 0$$

lie between $x = 0$ and $x = 2$.

Solution   From Example 2.24 we know that

$$x^3 - 3x^2 + 6x - 4 \equiv (x-2)^3 + 3(x-2)^2 + 6(x-2) + 4$$

Now if $x > 2$, $(x-2)^3$, $(x-2)^2$ and $(x-2)$ are all positive numbers so that for $x > 2$

$$(x-2)^3 + 3(x-2)^2 + 6(x-2) + 4 > 0$$

Thus $x^3 - 3x^2 + 6x - 4 = 0$ does not have a root that is greater than $x = 2$.

Similarly for $x < 0$, x^3 and x are both negative and $x^3 - 3x^2 + 6x - 4 < 0$ for $x < 0$. Thus $x^3 - 3x^2 + 6x - 4 = 0$ does not have a root that is less than $x = 0$. Hence all the real roots of

$$x^3 - 3x^2 + 6x - 4 = 0$$

lie between $x = 0$ and $x = 2$.

We can generalize the results of Example 2.25. Defining

$$f(x) = \sum_{k=0}^{n} A_n(x-a)^n$$

then the polynomial equation $f(x) = 0$ has no roots greater than $x = a$ if all of the A_k's have the same sign and has no roots less than $x = a$ if the A_k's alternate in sign.

The roots of a polynomial equation are related to its coefficients in more direct ways. Consider, for the moment, the quadratic equation with roots α and β. Then we can write the equation as

$$(x - \alpha)(x - \beta) = 0$$

which is equivalent to

$$x^2 - (\alpha + \beta)x + \alpha\beta = 0$$

Comparing this to the standard quadratic equation we have

$$a(x^2 - (\alpha + \beta)x + \alpha\beta) \equiv ax^2 + bx + c$$

Thus $-a(\alpha + \beta) = b$ and $a\alpha\beta = c$ so that

$$\alpha + \beta = -b/a \quad \text{and} \quad \alpha\beta = c/a$$

This gives us direct links between the sum of the roots of a quadratic equation and its coefficients and between the product of the roots and the coefficients. Similarly, we can show that if α, β and γ are the roots of the cubic equation

$$ax^3 + bx^2 + cx + d = 0$$

then

$$\alpha + \beta + \gamma = -b/a, \quad \alpha\beta + \beta\gamma + \gamma\alpha = c/a, \quad \alpha\beta\gamma = -d/a$$

In general, for the polynomial equation

$$a_n x^n + a_{n-1}x^{n-1} + a_{n-2}x^{n-2} + \ldots + a_1 x + a_0 = 0$$

the sum of the products of the roots, k at a time, is $(-1)^k a_{n-k}/a_n$.

Example 2.26

Show that the roots, α, β of the quadratic equation

$$ax^2 + bx + c = 0$$

may be written in the form

$$\frac{-b - \sqrt{(b^2 - 4ac)}}{2a} \quad \text{and} \quad \frac{2c}{-b - \sqrt{(b^2 - 4ac)}}$$

Obtain the roots of the equation

$$1.0x^2 + 17.8x + 1.5 = 0$$

Assuming the numbers given are correctly rounded, calculate error bounds for the roots.

Solution Using the formula for the roots of a quadratic equation we can select one root, α say, so that

$$\alpha = \frac{-b - \sqrt{(b^2 - 4ac)}}{2a}$$

Then, since $\alpha\beta = c/a$, we have

$$\beta = \frac{c}{a\alpha} = \frac{2c}{-b - \sqrt{(b^2 - 4ac)}}$$

Now consider the equation

$$1.0x^2 + 17.8x + 1.5 = 0$$

whose coefficients are correctly rounded numbers. Using the quadratic formula we obtain the roots

$$\alpha \approx -17.715\,327\,56$$

and

$$\beta \approx -0.084\,672\,44$$

Using the results of Section 1.5.3 we can estimate error bounds for these answers as shown in Figure 2.33. From that table we can see that using the form

$$\frac{-b - \sqrt{(b^2 - 4ac)}}{2a}$$

to estimate α we have an error bound of 0.943, while using

$$\frac{-b + \sqrt{(b^2 - 4ac)}}{2a}$$

to estimate β we have an error bound of 0.062. As this latter estimate of error is almost as big as the root itself we might be inclined to regard the answer as valueless. But calculating the error bound using the form

$$\beta = \frac{2c}{-b - \sqrt{(b^2 - 4ac)}}$$

Figure 2.33
Estimating error bounds for roots.

Label	Value	Absolute error bound	Relative error bound
a	1.0	0.05	0.05
b	17.8	0.05	0.0028
c	1.5	0.05	0.0333
b^2	316.84	1.77	0.0056
$4ac$	6.00	0.50	0.0833
$b^2 - 4ac$	310.84	2.27	0.0073
$d = \sqrt{(b^2 - 4ac)}$	17.630 66	0.065	0.0037
$-b - d$	−35.430 66	0.115	0.0032
$(-b - d)/(2a)$	−17.715 33	0.943	0.0532
$-b + d$	−0.169 34	0.115	0.6791
$(-b + d)/(2a)$	−0.084 67	0.062	0.7291
$2c/(-b - d)$	−0.084 67	0.003	0.0365

gives an estimate of 0.003. Thus we can write

$$\alpha = -17.7 \pm 5\% \quad \text{and} \quad \beta = -0.085 \pm 4\%$$

The reason for the discrepancy between the two error estimates for β lies in the fact that in the traditional form of the formula we are subtracting two nearly equal numbers, and consequently the error bounds dominate.

The numerical method most often used for evaluating the roots of a polynomial is the Newton–Raphson procedure. This will be described in Chapter 9, Section 9.5.8.

2.4.5 Curve sketching of polynomials

It is often useful to an engineer to be able to draw a quick sketch of a function that includes the main features of the latter but is not necessarily an accurately scaled and drawn graph of it. The ability to factorize a polynomial is an important stage in this process. Notice that near the origin the terms involving the lowest powers of x (i.e. $a_0 + a_1 x$) are the most significant and that for large values of x, the highest powers of x (i.e. $a_n x^n$) dominate.

Example 2.27 Sketch the graphs of the functions $f(x) = x^3 - 7x + 6$ and $g(x) = -2x^3 + 3x^2 - 2x + 1$.

Solution Factorization yields $x^3 - 7x + 6 = (x - 1)(x - 2)(x + 3)$. Hence $f(1) = 0$, $f(2) = 0$ and $f(-3) = 0$. We now know where the function crosses the x axis. Next we find out where the function crosses the y axis. This is at the point $(0, f(0))$, in this case at $(0, 6)$. Finally, we note that when $|x|$ is very large the function $f(x)$ is dominated by the term in x^3, which will be much larger than the term in x or the constant. Thus when x is large and positive so is $f(x)$. When x is large and negative so is $f(x)$. We also notice that the graph must have at least one maximum between $x = -3$ and $x = 1$, since at $x = -3$ it is an increasing function (going from negative to positive values) and at $x = 1$ it is a decreasing function. Similarly, there must be a minimum between $x = 1$ and $x = 2$. We can summarize the information we have about $f(x)$ so far in Figure 2.34. This suggests that the sketch of the function is as shown in Figure 2.35.

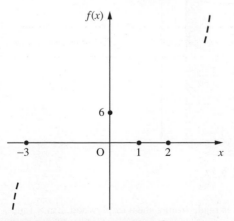

Figure 2.34 Essential features of the graph of $f(x) = x^3 - 7x + 6$.

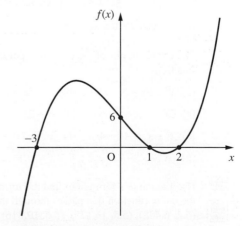

Figure 2.35 Graph of $f(x) = x^3 - 7x + 6$.

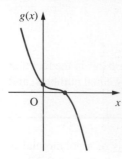

Figure 2.36
Graph of $g(x) = -2x^3 + 3x^2 - 2x + 1$.

Sketching the graph of $g(x)$ is very similar. We find that $g(x) = -(x-1)(2x^2 - x + 1)$. The factor $2x^2 - x + 1$ is irreducible, so the only zero of $g(x)$ is at $x = 1$. Since $g(0) = 1$, the function crosses the y axis at $y = 1$. When $|x|$ is large $g(x)$ is dominated by $-2x^3$. This means that when x is large and positive $g(x)$ is large and negative, and vice versa. Putting all this information together leads to a sketch like Figure 2.36.

Example 2.27 shows how basic sketching of polynomial functions can be achieved. We shall see in Chapter 8 how the methods of calculus can yield more information that can help in sketching functions.

2.4.6 Exercises

25 Factorize the following polynomial functions:

(a) $x^3 - 2x^2 - 11x + 12$

(b) $x^3 + 2x^2 - 5x - 6$

(c) $x^4 + x^2 - 2$

(d) $2x^4 + 5x^3 - x^2 - 6x$

(e) $2x^4 - 9x^3 + 14x^2 - 9x + 2$

(f) $x^4 + 5x^2 - 36$

26 Find the coefficients A, B, C, D and E such that
$$y = 2x^4 - 9x^3 + 145x^2 - 9x + 2$$
$$= A(x-2)^4 + B(x-2)^3 + C(x-2)^2$$
$$+ D(x-2) + E$$

27 Show that the zeros of
$$y = x^4 - 5x^3 + 5x^2 - 10x + 6$$
lie between $x = 0$ and $x = 5$.

28 Sketch the graphs of the following polynomial functions:

(a) $x^2 + 3x - 4$ (b) $x^2 - 5x + 6$

(c) $2x^2 - 3x + 2$ (d) $x^3 + 2x^2 - x - 2$

(e) $x^3 - x^2 - 4x + 4$ (f) $2x^3 - 4x^2 - 8x - 6$

(g) $6x^4 - x^3 - 7x^2 + x + 1$

29 Use Lagrange's formula to find the formula for the cubic function that passes through the points $(5.2, 6.408)$, $(5.5, 16.125)$, $(5.6, 19.816)$ and $(5.8, 27.912)$.

30 Find a formula for the quadratic function whose graph passes through the points $(1, 403)$, $(3, 471)$ and $(7, 679)$.

31 (a) Show that if the equation $ax^3 + bx + c = 0$ has a repeated root α then $3a\alpha^2 + b = 0$.

(b) A can is to be made in the form of a circular cylinder of radius r (in cm) and height h (in cm) as shown in Figure 2.37. Its capacity is to be 0.5 l. Show that the surface area A (in cm^2) of the can is

$$A = 2\pi r^2 + \frac{1000}{r}$$

Using the result of (a), deduce that A has a minimum value when $6\pi r^2 - A = 0$. Hence find the corresponding values of r and h.

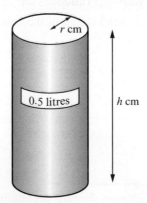

Figure 2.37

32 A box is made from a sheet of plywood, $2\,\text{m} \times 1\,\text{m}$, with the waste shown in Figure 2.38(a). Find the

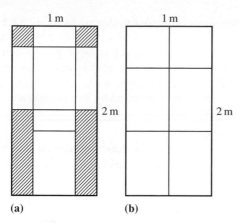

(a)   (b)

Figure 2.38

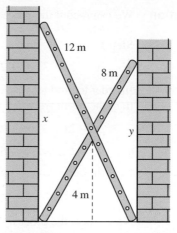

Figure 2.39

maximum capacity of such a box and compare it with the capacity of the box constructed without the wastage, as shown in Figure 2.38(b).

33 Two ladders, of lengths 12 m and 8 m, lean against buildings on opposite sides of an alley, as shown in Figure 2.39. Show that the heights x and y (in metres) reached by the tops of the ladders in the positions shown satisfy the equations

$$\frac{1}{x} + \frac{1}{y} = \frac{1}{4} \quad \text{and} \quad x^2 - y^2 = 80$$

Show that x satisfies the equation

$$x^4 - 8x^3 - 80x^2 + 640x - 1280 = 0$$

and that the width of the alley is given by $\sqrt{(12^2 - x_0^2)}$, where x_0 is the positive root of this equation. By first tabulating the polynomial over a suitable domain and then drawing its graph, estimate the value of x_0 and the width of the alley. Check your solution of the quartic (to 2dp) using a suitable software package.

2.5 Rational functions

Rational functions have the general form

$$f(x) = \frac{p(x)}{q(x)}$$

where $p(x)$ and $q(x)$ are polynomials. If the degree of p is less than the degree of q, $f(x)$ is said to be a **strictly proper rational function**. If p and q have the same degree then $f(x)$ is a **proper rational function**. It is said to be an **improper rational function** if the degree of p is greater than the degree of q.

An improper or proper rational function can always be expressed as a polynomial plus a strictly proper rational function, for example, by algebraic division.

Example 2.28   Express the improper rational function

$$f(x) = \frac{3x^4 + 2x^3 - 5x^2 + 6x - 7}{x^2 - 2x + 3}$$

as the sum of a polynomial function and a strictly proper rational function.

Solution We can record the process of division in a manner similar to that of Example 2.22.

Step 1

$$x^2 - 2x + 3)3x^4 + 2x^3 - 5x^2 + 6x - 7($$

In order to produce the term $3x^4$, $x^2 - 2x + 3$ must be multiplied by $3x^2$. Do this and subtract the result from $3x^4 + 2x^3 - 5x^2 + 6x - 7$.

$$x^2 - 2x + 3)3x^4 + 2x^3 - 5x^2 + 6x - 7(3x^2$$
$$\underline{3x^4 - 6x^3 + 9x^2}$$
$$8x^3 - 14x^2 + 6x - 7$$

Step 2

Now repeat the process on the polynomial $8x^3 - 14x^2 + 6x - 7$. In this case, in order to eliminate the term $8x^3$ we must multiply $x^2 - 2x + 3$ by $8x$.

$$x^2 - 2x + 3)3x^4 + 2x^3 - 5x^2 + 6x - 7(3x^2 + 8x$$
$$\underline{3x^4 - 6x^3 + 9x^2}$$
$$8x^3 - 14x^2 + 6x - 7$$
$$\underline{8x^3 - 16x^2 + 24x}$$
$$2x^2 - 18x - 7$$

Step 3

Finally, to eliminate the $2x^2$ term, we must multiply $x^2 - 2x + 3$ by 2.

$$x^2 - 2x + 3)3x^4 + 2x^3 - 5x^2 + 6x - 7(3x^2 + 8x + 2$$
$$\underline{3x^4 - 6x^3 + 9x^2}$$
$$8x^3 - 14x^2 + 6x - 7$$
$$\underline{8x^3 - 16x^2 + 24x}$$
$$2x^2 - 18x - 7$$
$$\underline{2x^2 - 4x + 6}$$
$$-14x - 13$$

We cannot eliminate the $-14x - 13$ terms, so we have

$$f(x) = 3x^2 + 8x + 2 - \frac{14x + 13}{x^2 - 2x + 3}$$

Any strictly proper rational function can be expressed as a sum of simpler functions whose denominators are linear or irreducible quadratic functions. For example:

$$\frac{x^2 + 1}{(1 + x)(1 - x)(2 + 2x + x^2)} = \frac{1}{1 + x} + \frac{1}{5(1 - x)} - \frac{4x + 7}{5(2 + 2x + x^2)}$$

These simpler functions are called the **partial fractions** of the rational function, and are often useful in the mathematical analysis and design of engineering systems.

The construction of the partial fraction form of a rational function is the inverse process to that of collecting together separate rational expressions into a single rational function. For example:

$$\frac{1}{1+x} + \frac{1}{5(1-x)} - \frac{4x+7}{5(2+2x+x^2)}$$

$$= \frac{1(5)(1-x)(2+2x+x^2) + (1+x)(2+2x+x^2) - (1+x)(1-x)(4x+7)}{5(1+x)(1-x)(2+2x+x^2)}$$

$$= \frac{5(2-x^2-x^3) + (2+4x+3x^2+x^3) - (1-x^2)(4x+7)}{5(1-x^2)(2+2x+x^2)}$$

$$= \frac{5(2-x^2-x^3) + (2+4x+3x^2+x^3) - (7+4x-7x^2-4x^3)}{5(2+2x-x^2-2x^3-x^4)}$$

$$= \frac{5+5x^2}{5(2+2x-x^2-2x^3-x^4)}$$

$$= \frac{1+x^2}{2+2x-x^2-2x^3-x^4}$$

But it is clear from this example that reversing the process (working backwards from the final expression) is not easy, and we require a different method in order to find the partial fractions of a given function. To describe the method in its full generality is easy but difficult to understand, so we will apply the method to a number of commonly occurring types of function in the next section before stating the general algorithm.

2.5.1 Partial fractions

In this section we will illustrate how proper rational functions of the form $p(x)/q(x)$ may be expressed in partial fractions.

(a) Distinct linear factors

Each distinct linear factor, of the form $(x + \alpha)$, in the denominator $q(x)$ will give rise to a partial fraction of the form $\dfrac{A}{x + \alpha}$, where A is a real constant.

Example 2.29 Express in partial fractions the rational function

$$\frac{3x}{(x-1)(x+2)}$$

Solution In this case we have two distinct linear factors $(x - 1)$ and $(x + 2)$ in the denominator so the corresponding partial fractions are of the form

$$\frac{3x}{(x-1)(x+2)} = \frac{A}{x-1} + \frac{B}{x+2} = \frac{A(x+2) + B(x-1)}{(x-1)(x+2)}$$

where A and B are constants to be determined. Since both expressions are equal and their denominators are identical we must therefore make their numerators equal, yielding

$$3x = A(x + 2) + B(x - 1)$$

This identity is true for all values of x, so we can find A and B by setting first $x = 1$ and then $x = -2$. So

$$x = 1 \quad \text{gives} \quad 3 = A(3) + B(0) \quad \text{that is} \quad A = 1$$

and

$$x = -2 \quad \text{gives} \quad -6 = A(0) + B(-3) \quad \text{that is} \quad B = 2$$

Thus

$$\frac{3x}{(x - 1)(x + 2)} = \frac{1}{x - 1} + \frac{2}{x + 2}$$

When the denominator $q(x)$ of a strictly proper rational function $\dfrac{p(x)}{q(x)}$ is a product of linear factors, as in Example 2.29, there is a quick way of expressing $\dfrac{p(x)}{q(x)}$ in partial fractions.

Considering again Example 2.29, if

$$\frac{3x}{(x - 1)(x + 2)} = \frac{A}{(x - 1)} + \frac{B}{(x - 2)}$$

then to obtain A simply **cover up** the factor $(x - 1)$ in

$$\frac{3x}{(x - 1)(x + 2)}$$

and evaluate what is left at $x = 1$, giving

$$A = \frac{3(1)}{(x - 1)(1 + 2)} = 1$$

Likewise, to obtain B **cover up** the factor $(x + 2)$ in the left-hand side and evaluate what is left at $x = -2$, giving

$$B = \frac{3(-2)}{(-2 - 1)(x + 2)} = 2$$

Thus, as before,

$$\frac{3x}{(x - 1)(x + 2)} = \frac{1}{x - 1} + \frac{2}{x + 2}$$

This method of obtaining partial fractions is called the **cover up rule**.

Example 2.30   Using the cover up rule, express in partial fractions the rational function

$$\frac{2x + 1}{(x - 2)(x + 1)(x - 3)}$$

Solution   The corresponding partial fractions are of the form

$$\frac{2x + 1}{(x - 2)(x + 1)(x - 3)} = \frac{A}{(x - 2)} + \frac{B}{(x + 1)} + \frac{C}{(x - 3)}$$

Using the cover up rule

$$A = \frac{2(2) + 1}{(x - 2)(2 + 1)(2 - 3)} = -\frac{5}{3}$$

$$B = \frac{2(-1) + 1}{(-1 - 2)(x + 1)(-1 - 3)} = -\frac{1}{12}$$

$$C = \frac{2(3) + 1}{(3 - 2)(3 + 1)(x - 3)} = \frac{7}{4}$$

so that

$$\frac{2x + 1}{(x - 2)(x + 1)(x - 3)} = -\frac{\frac{5}{3}}{x - 2} - \frac{\frac{1}{12}}{x + 1} + \frac{\frac{7}{4}}{x - 3}$$

Because it is easy to make an error with this process, it is sensible to check the answers obtained. This can be done by using a 'spot' value to check that the left- and right-hand sides yield the same value. When doing this avoid using $x = 0$ or any of the special values of x that were used in finding the coefficients.

For example, taking $x = 1$ in the partial fraction expansion of Example 2.30, we have

$$\text{left-hand side} \quad = \frac{2(1) + 1}{(1 - 2)(1 + 1)(1 - 3)} = \frac{3}{4}$$

$$\text{right-hand side} \quad = -\frac{\frac{5}{3}}{1 - 2} - \frac{\frac{1}{12}}{1 + 1} + \frac{\frac{7}{4}}{1 - 3} = \frac{3}{4}$$

giving a positive check.

(b) Repeated linear factors

Each k times repeated linear factor, of the form $(x - \alpha)^k$, in the denominator $q(x)$ will give rise to a partial fraction of the form

$$\frac{A_1}{(x - \alpha)} + \frac{A_2}{(x - \alpha)^2} + \ldots + \frac{A_k}{(x - \alpha)^k}$$

where $A_1, A_2, \ldots, A_k$ are real constants.

Example 2.31

Express as partial fractions the rational function

$$\frac{3x + 1}{(x - 1)^2 (x + 2)}$$

Solution

In this case the denominator consists of the distinct linear factor $(x + 2)$ and the twice repeated linear factor $(x - 1)$. Thus, the corresponding partial fractions are of the form

$$\frac{3x + 1}{(x - 1)^2 (x + 2)} = \frac{A}{(x - 1)} + \frac{B}{(x - 1)^2} + \frac{C}{(x + 2)}$$

$$= \frac{A(x - 1)(x + 2) + B(x + 2) + C(x - 1)^2}{(x - 1)^2 (x + 2)}$$

which gives

$$3x + 1 = A(x - 1)(x + 2) + B(x + 2) + C(x - 1)^2$$

Setting $x = 1$ gives $4 = B(3)$ and $B = \frac{4}{3}$. Setting $x = -2$ gives $-5 = C(-3)^2$ and $C = -\frac{5}{9}$. To obtain A we can give x any other value, so taking $x = 0$ gives

$$1 = (-2)A + 2B + C$$

and substituting the values of B and C gives $A = \frac{5}{9}$. Hence

$$\frac{3x + 1}{(x - 1)^2 (x + 2)} = \frac{\frac{5}{9}}{x - 1} + \frac{\frac{4}{3}}{(x - 1)^2} - \frac{\frac{5}{9}}{(x + 2)}$$

(c) Irreducible quadratic factors

Each distinct irreducible quadratic factor, of the form $(ax^2 + bx + c)$, in the denominator $q(x)$ will give rise to a partial fraction of the form

$$\frac{Ax + B}{ax^2 + bx + c}$$

where A and B are real constants.

Example 2.32

Express as partial fractions the rational function

$$\frac{5x}{(x^2 + x + 1)(x - 2)}$$

Solution

In this case the denominator consists of the distinct linear factor $(x - 2)$ and the distinct irreducible quadratic factor $(x^2 + x + 1)$. Thus, the corresponding partial fractions are of the form

$$\frac{5x}{(x^2 + x + 1)(x - 2)} = \frac{Ax + B}{x^2 + x + 1} + \frac{C}{x - 2} = \frac{(Ax + B)(x - 2) + C(x^2 + x + 1)}{(x^2 + x + 1)(x - 2)}$$

giving

$$5x = (Ax + B)(x - 2) + C(x^2 + x + 1)$$

Setting $x = 2$ enables us to calculate C:

$$10 = (2A + B)(0) + C(7) \quad \text{and} \quad C = \tfrac{10}{7}$$

Here, however, we cannot select special values of x that give A and B immediately, because $x^2 + x + 1$ is an irreducible quadratic and cannot be factorized. Instead we make use of Property (i) of polynomials, described in Section 2.4.1, which stated that if two polynomials are equal in value for all values of x then the corresponding coefficients are equal. Applying this to

$$5x = (Ax + B)(x - 2) + C(x^2 + x + 1)$$

we see that the coefficient of x^2 on the right-hand side is $A + C$ while that on the left-hand side is zero. Thus

$$A + C = 0 \quad \text{and} \quad A = -C = -\tfrac{10}{7}$$

Similarly the coefficient of x^0 on the right-hand side is $-2B + C$ and that on the left-hand side is zero, and we obtain $-2B + C = 0$, which implies $B = \tfrac{1}{2}C = \tfrac{5}{7}$.
Hence

$$\frac{5x}{(x^2 + x + 1)(x - 2)} = \frac{\tfrac{5}{7} - \tfrac{10}{7}x}{x^2 + x + 1} + \frac{\tfrac{10}{7}}{x - 2}$$

Example 2.33   Express as partial fractions the rational function

$$\frac{3x^2}{(x - 1)(x + 2)}$$

Solution   In this example the numerator has the same degree as the denominator.

The first step in such examples is to divide the bottom into the top to obtain a polynomial and a strictly proper rational function. Thus

$$\frac{3x^2}{(x - 1)(x + 2)} = 3 + \frac{6 - 3x}{(x - 1)(x + 2)}$$

We then apply the partial-fraction process to the remainder, setting

$$\frac{6 - 3x}{(x - 1)(x + 2)} = \frac{A}{x - 1} + \frac{B}{x + 2} = \frac{A(x + 2) + B(x - 1)}{(x - 1)(x + 2)}$$

giving

$$6 - 3x = A(x + 2) + B(x - 1)$$

Setting first $x = 1$ and then $x = -2$ gives $A = 1$ and $B = -4$ respectively. Thus

$$\frac{3x^2}{(x - 1)(x + 2)} = 3 + \frac{1}{x - 1} - \frac{4}{x + 2}$$

Summary of method

A summary of the method for expressing a rational function in partial fractions is given in Figure 2.40.

Figure 2.40
Summary of method for expressing a rational function in partial fractions.

In general, the method for finding the partial fractions of a given function $f(x) = p(x)/q(x)$ consists in the following steps.

Step 1
If the degree of p is greater than or equal to the degree of q, divide q into p to obtain

$$f(x) = r(x) + \frac{s(x)}{q(x)}$$

where the degree of s is less than the degree of q.

Step 2
Factorize $q(x)$ fully into real linear and irreducible quadratic factors, collecting together all like factors.

Step 3
Each **linear factor** $ax + b$ in $q(x)$ will give rise to a fraction of the type

$$\frac{A}{ax + b}$$

(Here a and b are known and A is to be found.)
  Each **repeated linear factor** $(ax + b)^n$ will give rise to n fractions of the type

$$\frac{A_1}{ax + b} + \frac{A_2}{(ax + b)^2} + \frac{A_3}{(ax + b)^3} + \ldots + \frac{A_n}{(ax + b)^n}$$

Each **irreducible quadratic factor** $ax^2 + bx + c$ in $q(x)$ will give rise to a fraction of the type

$$\frac{Ax + B}{ax^2 + bx + c}$$

Each **repeated irreducible quadratic factor** $(ax^2 + bx + c)^n$ will give rise to n fractions of the type

$$\frac{A_1 x + B_1}{ax^2 + bx + c} + \frac{A_2 x + B_2}{(ax^2 + bx + c)^2} + \ldots + \frac{A_n x + B_n}{(ax^2 + bx + c)^n}$$

Put $p(x)/q(x)$ (or $s(x)/q(x)$, if that case occurs) equal to the sum of all the fractions involved.

Figure 2.40
continued

Step 4
Multiply both sides of the equation by $q(x)$ to obtain an identity involving polynomials, from which the multiplying constants of the linear combination may be found (because of Property (i) in Section 2.4.1).

Step 5
To find these coefficients, two strategies are used.

- *Strategy 1*: Choose special values of x that make finding the values of the unknown coefficients easy: for example choose x equal to the roots of $q(x) = 0$ in turn and use the 'cover up' rule.
- *Strategy 2*: Compare the coefficients of like powers of x on both sides of the identity. Starting with the highest and lowest powers usually makes it easier.

Strategy 1 may leave some coefficients undetermined. In that case we complete the process using Strategy 2.

Step 6
Lastly, check the answer either by choosing a test value for x or by putting the partial fractions over a common denominator.

2.5.2   Exercises

34 Express the following improper rational functions as the sum of a polynomial function and a strictly proper rational function

(a) $f(x) = (x^2 + x + 1)/[(x + 1)(x - 1)]$

(b) $f(x) = (x^5 - x^4 - x + 1)/(x^2 + x + 1)$

35 Express as a single fraction

(a) $\dfrac{1}{x} - \dfrac{2}{x - 2} + \dfrac{x - 1}{x^2 + 1}$

(b) $\dfrac{1}{x^3 - 3x^2 + 3x - 1} - \dfrac{1}{x^3 - x^2 - x + 1}$

(c) $\dfrac{x + 1}{x^2 + 1} + \dfrac{1}{x - 1} - \dfrac{1}{(x - 1)^2} + \dfrac{2}{x - 2}$

36 Express as partial fractions

(a) $\dfrac{1}{(x + 1)(x - 2)}$ (b) $\dfrac{2x - 1}{(x + 1)(x - 2)}$

(c) $\dfrac{x^2 - 2}{(x + 1)(x - 2)}$ (d) $\dfrac{x - 1}{(x + 1)(x - 2)^2}$

(e) $\dfrac{1}{(x + 1)(x^2 + 2x + 2)}$ (f) $\dfrac{1}{(x + 1)(x^2 - 4)}$

37 Express as partial fractions

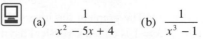

(a) $\dfrac{1}{x^2 - 5x + 4}$ (b) $\dfrac{1}{x^3 - 1}$

(c) $\dfrac{3x - 1}{x^3 - 3x - 2}$ (d) $\dfrac{x^2 - 1}{x^2 - 5x + 6}$

(e) $\dfrac{x^2 + x - 1}{(x^2 + 1)^2}$ (f) $\dfrac{18x^2 - 5x + 47}{(x^2 + 4)(x - 1)(x + 5)}$

2.5.3 Asymptotes

Sketching the graphs of rational functions gives rise to the concept of an asymptote. To illustrate, let us consider the graph of the function

$$y = f(x) = \frac{x}{1 + x} \quad (x > 0)$$

and that of its inverse

$$y = f^{-1}(x) = \frac{x}{1 - x} \quad (0 \leqslant x < 1)$$

Expressing $x/(x + 1)$ as $(x + 1 - 1)/(x + 1) = 1 - 1/(x + 1)$, we see that as x gets larger and larger $1/(x + 1)$ gets smaller and smaller, so that $x/(x + 1)$ approaches closer and closer to the value 1. This is illustrated in the graph of $y = f(x)$ shown in Figure 2.41(a). The line $y = 1$ is called a **horizontal asymptote** to the curve, and we note that the graph of $f(x)$ approaches this asymptote as $|x|$ becomes large.

Figure 2.41
Horizontal and
vertical asymptotes.

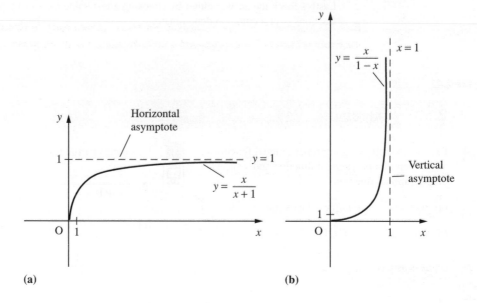

(a)

(b)

The graph of the inverse function $y = f^{-1}(x)$ is shown in Figure 2.41(b), and the line $x = 1$ is called a **vertical asymptote** to the curve.

The existence of asymptotes is a common feature of the graphs of rational functions. They feature in various engineering applications, such as in the plotting of root locus plots in control engineering. In more advanced applications of mathematics to engineering the concept of an asymptote is widely used for the purposes of making approximations. Asymptotes need not necessarily be horizontal or vertical lines; they may be sloping lines or indeed nonlinear graphs, as we shall see in Example 2.35.

Example 2.34

Sketch the graph of the function

$$y = \frac{1}{3 - x} \quad (x \neq 3)$$

and find the values of x for which

$$\frac{1}{3 - x} < 2$$

Solution We can see from the formula for y that the line $x = 3$ is a vertical asymptote of the function. As x gets closer and closer to the value $x = 3$ from the left-hand side (that is, $x < 3$), y gets larger and larger and is positive. As x gets closer and closer to $x = 3$ from the right-hand side (that is, $x > 3$), y is negative and large. As x gets larger and larger, y gets smaller and smaller for both $x > 0$ and $x < 0$, so $y = 0$ is a horizontal asymptote. Thus we obtain the sketch shown in Figure 2.42. By drawing the line $y = 2$ on the sketch, we see at once that

$$\frac{1}{3 - x} < 2$$

for $x < \frac{5}{2}$ and $x > 3$. This result was obtained algebraically in Example 1.17. Generally we use a mixture of algebraic and graphical methods to solve such problems.

Figure 2.42

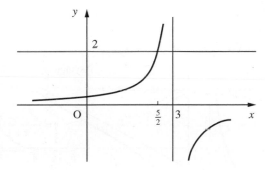

Example 2.35

Sketch the graph of the function

$$y = f(x) = \frac{x^2 - x - 6}{x + 1} \quad (x \neq -1)$$

Solution We begin the task by locating points at which the function is zero. Now $f(x) = 0$ implies that $x^2 - x - 6 = (x - 3)(x + 2) = 0$, from which we deduce that $x = 3$ and $x = -2$ are zeros of the function. Thus the graph $y = f(x)$ crosses the x axis at $x = -2$ and $x = 3$.

Next we locate the points at which the denominator of the rational function is zero, which in this case is $x = -1$. As x approaches such a point, the value of $f(x)$ becomes infinitely large in magnitude, and the value of the rational function is undefined at such a point. Thus the graph of $y = f(x)$ has a vertical asymptote at $x = -1$. (There is usually a vertical asymptote to the graph of the rational function $y = p(x)/q(x)$ at points where the denominator $q(x) = 0$.)

Next we consider the behaviour of the function as x gets larger and larger, that is as $x \to \infty$ or $x \to -\infty$. To do this, we first simplify the rational function by algebraic division, giving

$$y = f(x) = x - 2 - \frac{4}{x + 1}$$

As $x \to \pm\infty$, $4/(x + 1) \to 0$. Thus, for large values of x, both positive and negative, $4/(x + 1)$ becomes negligible compared with x, so that $f(x)$ tends to behave like $x - 2$. Thus the line $y = x - 2$ is also an asymptote to the graph of $y = f(x)$.

Having located the asymptotes, we then need to find how the graph approaches them. When x is large and positive the term $4/(x + 1)$ will be small but positive, so that $f(x)$ is slightly less than $x - 2$. Hence the graph approaches the asymptote from below. When x is large and negative the term $4/(x + 1)$ is small but negative, so the graph approaches the asymptote from above. To consider the behaviour of the function near $x = -1$, we examine the factorized form

$$y = f(x) = \frac{(x - 3)(x + 2)}{x + 1}$$

When x is slightly less than -1, $f(x)$ is positive. When x is slightly greater than -1, $f(x)$ is negative.

We are now in a position to sketch the graph of $y = f(x)$ as shown in Figure 2.43.

Figure 2.43

Graph of $y = \dfrac{x^2 - x - 6}{x + 1}$.

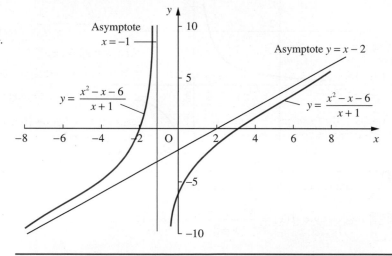

Modern computational aids have made graphing functions much easier but to obtain graphs of a reasonably good quality some preliminary analysis is always necessary. This helps to select the correct range of values for the independent variable and for the function. For example, asking a computer package to plot the function

$$y = \frac{13x^2 - 34x + 25}{x^2 - 3x + 2}$$

Figure 2.44

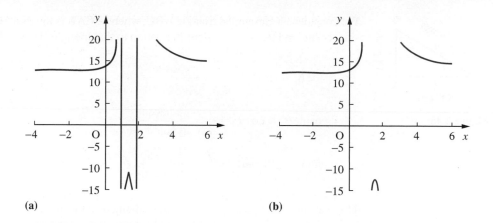

(a)   (b)

without prior analysis might result in the graph shown in Figure 2.44(a). A little analysis shows that the function is undefined at $x = 1$ and 2. Excluding these points from the range of values for x produces the more acceptable plot shown in Figure 2.44(b), although it is not clear from either plot that the graph has a horizontal asymptote $y = 13$. Clearly, much more preliminary work is needed to obtain a good quality graph of the function.

2.5.4 Exercises

38 Plot the graphs of the functions

(a) $y = \dfrac{2 + x}{1 + x}$ (b) $y = \dfrac{1}{2}\left(x + \dfrac{2}{x}\right)$

(c) $y = \dfrac{3x^4 + 12x^2 - 4}{8x^3}$ (d) $y = \dfrac{(x - 1)(x - 2)}{(x + 1)(x - 3)}$

for the domain $-3 \leqslant x \leqslant 3$. Find the points on each graph at which they intersect with the line $y = x$.

(a) $y = \dfrac{x^2 - 8x + 15}{x}$ (b) $y = \dfrac{x + 1}{x - 1}$

(c) $y = \dfrac{x^2 + 5x - 14}{x + 5}$

(*Hint*: writing (a) as

$$y = (\sqrt{x} - \sqrt{(15/x)})^2 + 2\sqrt{15} - 8$$

shows that there is a turning point at $x = \sqrt{15}$.)

39 Sketch the graphs of the functions given below locating their turning points and asymptotes.

40 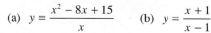 Check the sketches obtained in Questions 38 and 39 using a graphics calculator or a computer package.

2.6   Circular functions

There are two approaches to the definition of the **circular** or **trigonometric functions** and this is reflected in their double name. One approach is static in nature and the other dynamic.

2.6.1 Trigonometric ratios

The static approach began with practical problems of surveying and gave rise to the mathematical problems of triangles and their measurement that we call trigonometry.

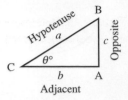

Figure 2.45

We consider a right-angled triangle ABC, where ∠CAB is the right-angle, and define the sine, cosine and tangent functions in relation to that triangle. Thus in Figure 2.45 we have

$$\text{sine } \theta° = \sin \theta° = \frac{c}{a} = \frac{\text{opposite}}{\text{hypotenuse}}$$

$$\text{cosine } \theta° = \cos \theta° = \frac{b}{a} = \frac{\text{adjacent}}{\text{hypotenuse}}$$

$$\text{tangent } \theta° = \tan \theta° = \frac{c}{b} = \frac{\text{opposite}}{\text{adjacent}}$$

The way in which these functions were defined led to their being called the 'trigonometrical ratios'. The context of the applications implied that the angles were measured in the sexagesimal system (degees, minutes, seconds): for example 35°24′41″ which today is written in the decimal form 35.41°. In modern textbooks this is shown explicitly, writing, for example, $\sin 30°$, or $\cos 35.41°$, or $\tan \theta°$, so that the independent variable θ is a pure number. For example, by considering the triangles shown in Figure 2.46(a), we can readily write down the trigonometric ratios for 30°, 45° and 60° as indicated in the table of Figure 2.46(b).

Figure 2.46

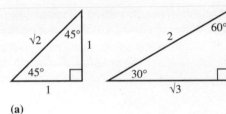

$\theta°$	$\sin \theta°$	$\cos \theta°$	$\tan \theta°$
30°	1/2	√3/2	1/√3
45°	1/√2	1/√2	1
60°	√3/2	1/2	√3

(a)   (b)

To extend trigonometry to problems involving triangles that are not necessarily right-angled, we make use of the sine and cosine rules. Using the notation of Figure 2.47 (note that it is usual to label the side opposite an angle by the corresponding lower-case letter), we have, for any triangle ABC,

Figure 2.47

The sine rule

$$\frac{a}{\sin A} = \frac{b}{\sin B} = \frac{c}{\sin C} \tag{2.14}$$

The cosine rule

$$a^2 = b^2 + c^2 - 2bc \cos A \tag{2.15}$$

or

$$b^2 = a^2 + c^2 - 2ac \cos B$$

or

$$c^2 = a^2 + b^2 - 2ab \cos C$$

Example 2.36   Consider the crank and connecting rod mechanism illustrated in Figure 2.48. Determine a functional relationship between the displacement of Q and the angle through which the crank OP has turned.

Figure 2.48
Crank and connecting
rod mechanism.

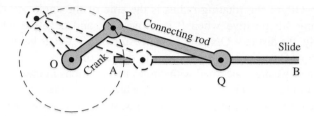

Solution   As the crank OP rotates about O, the other end of the connecting rod moves backwards and forwards along the slide AB. The displacement of Q from its initial position depends on the angle through which the crank OP has turned. A mathematical model for the mechanism replaces the crank and connecting rod, which have thickness as well as length, by straight lines, which have length only, and we consider the motion of the point Q as the line OP rotates about O, with PQ fixed in length and Q constrained to move on the line AB, as shown in Figure 2.49. We can specify the dependence of Q on the angle of rotation of OP by using some elementary trigonometry. Labelling the length of OP as r units, the length of PQ as l units, the length of OQ as y units and the angle $\angle$AOP as x degrees, and applying the cosine formula gives

Figure 2.49

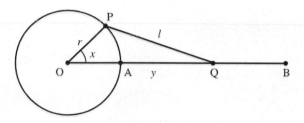

$$l^2 = r^2 + y^2 - 2yr \cos x°$$

which implies

$$(y - r \cos x°)^2 = l^2 - r^2 + r^2 \cos^2 x°$$
$$= l^2 - r^2 \sin^2 x°$$

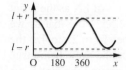

Figure 2.50

and

$$y = r \cos x° + \sqrt{(l^2 - r^2 \sin^2 x°)}$$

Thus for any angle x we can calculate the corresponding value of y. We can represent this relationship by means of a graph, as shown in Figure 2.50.

2.6.2 Circular functions

The dynamic definition of the functions arises from considering the motion of a point P around a circle as shown in Figure 2.51. Many practical mechanisms involve this mathematical model.

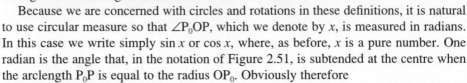

Figure 2.51

The distance OP is one unit, and the perpendicular distance NP of P from the initial position OP_0 of the rotating radius is the **sine** of the angle $\angle P_0OP$. Note that we are measuring NP positive when P is above OP_0 and negative when P is below OP_0. Similarly, the distance ON defines the **cosine** of $\angle P_0OP$ as being positive when N is to the right of O and negative when it is to the left of O.

Because we are concerned with circles and rotations in these definitions, it is natural to use circular measure so that $\angle P_0OP$, which we denote by x, is measured in radians. In this case we write simply $\sin x$ or $\cos x$, where, as before, x is a pure number. One radian is the angle that, in the notation of Figure 2.51, is subtended at the centre when the arclength P_0P is equal to the radius OP_0. Obviously therefore

$$180° = \pi \text{ radians}$$

a result we can use to convert degrees to radians and vice versa. It also follows from the definition of a radian that

(a) the length of the arc AB shown in Figure 2.52(a), of a circle of radius r, subtending an angle θ radians at the centre of the circle, is given by

$$\text{length of arc} = r\theta \tag{2.16}$$

(b) the area of the sector OAB of a circle of radius r, subtending an angle θ radians at the centre of the circle (shown shaded in Figure 2.52(b)), is given by

$$\text{area of sector} = \tfrac{1}{2}r^2\theta \tag{2.17}$$

Figure 2.52
(a) Arc of a circle.
(b) Sector of a circle.

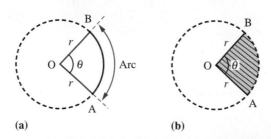

(a)   (b)

To obtain the graph of $\sin x$, we simply need to read off the values of PN as the point P moves around the circle, thus generating the graph of Figure 2.53. Note that as we continue around the circle for a second revolution (that is, as x goes from 2π to 4π) the graph produced is a replica of that produced as x goes from 0 to 2π, the same being true for subsequent intervals of 2π. By allowing P to rotate clockwise around the circle, we see that $\sin(-x) = -\sin x$, so that the graph of $\sin x$ can be extended to negative values of x, as shown in Figure 2.54.

Figure 2.53
Generating the
graph of sin x.

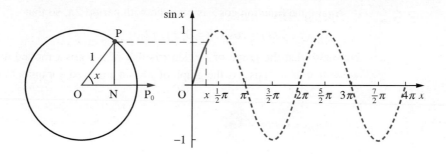

Figure 2.54
Graph of $y = \sin x$.

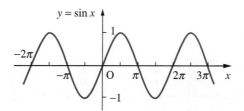

Since the graph replicates itself for every interval of 2π,

$$\sin(x + 2\pi k) = \sin x, \ k = 0, \pm 1, \pm 2, \ldots \qquad (2.18)$$

and the function $\sin x$ is said to be **periodic with period** 2π.

To obtain the graph of $y = \cos x$, we need to read off the value of ON as the point P moves around the circle. To make the plotting of the graph easier, we first rotate the circle through $90°$ anticlockwise and then proceed as for $y = \sin x$ to produce the graph of Figure 2.55. By allowing P to rotate clockwise around the circle, we see that $\cos(-x) = \cos x$, so that the graph can be extended to negative values of x, as shown in Figure 2.56.

Figure 2.55
Generating the
graph of cos x.

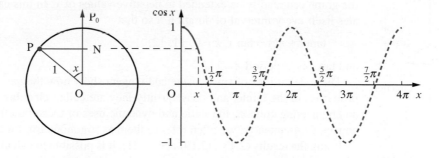

Figure 2.56
Graph of $y = \cos x$.

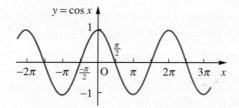

Again, the function cos x is periodic with period 2π, so that

$$\cos(x + 2\pi k) = \cos x, \; k = 0, \pm 1, \pm 2, \ldots \tag{2.19}$$

Note also that the graph of $y = \sin x$ is that of $y = \cos x$ moved $\frac{1}{2}\pi$ units to the right, while that of $y = \cos x$ is the graph of $y = \sin x$ moved $\frac{1}{2}\pi$ units to the left. Thus, from Section 2.2.3,

$$\sin x = \cos(x - \tfrac{1}{2}\pi) \tag{2.20}$$

or

$$\cos x = \sin(x + \tfrac{1}{2}\pi)$$

Figure 2.57

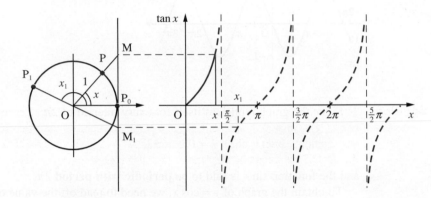

The definition of tan x is similar, and makes obvious the origin of the name 'tangent' for this function. In Figure 2.57 the rotating radius OP is extended until it cuts the tangent $P_0 M$ to the circle at the initial position P_0. The length $P_0 M$ is the **tangent** of $\angle P_0 OP$. Allowing P to move around the circle, we generate the graph shown in Figure 2.57. Again, by allowing P to move in a clockwise direction, we have $\tan(-x) = -\tan x$, and the graph can readily be extended to negative values of x. In this case the graph replicates itself every interval of duration π so that

$$\tan(x + \pi k) = \tan x, \; k = 0, \pm 1, \pm 2, \ldots \tag{2.21}$$

and tan x is of period π.

These definitions of sine, cosine and tangent show how they are associated with the properties of the circle, and consequently they are called **circular functions**. Often in an engineering context, the static and dynamic uses of these functions occur simultaneously. Consequently, we often refer to them as trigonometric functions.

Using the results (2.18), (2.19) and (2.21), it is possible to calculate the values of the trigonometric functions for angles greater than $\frac{1}{2}\pi$ using their values for angles between zero and $\frac{1}{2}\pi$. The rule is: take the acute angle that the direction makes with the initial direction, find the sine, cosine or tangent of this angle and multiply by $+1$ or -1 according to the scheme of Figure 2.58. For example

$$\cos(135°) = \cos(180° - 45°) = -\cos 45° = -\sqrt{\tfrac{1}{2}}$$

$$\sin(330°) = \sin(360° - 30°) = -\sin 30° = -\tfrac{1}{2}$$

$$\tan(240°) = \tan(180° + 60°) = \tan 60° = \sqrt{3}$$

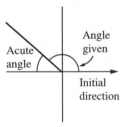

(a)

sine +	all +
cosine −	
tangent −	
tangent +	cosine +
sine −	sine −
cosine −	tangent −

(b)

Figure 2.58

If the radius OP is rotating with constant angular velocity ω (in rad s^{-1}) about O then $x = \omega t$, where t is the time (in s). The time T taken for one complete revolution is given by $\omega T = 2\pi$; that is, $T = 2\pi/\omega$. This is the **period** of the motion. In one second the radius makes $\omega/2\pi$ such revolutions. This is the **frequency**, v. Its value is given by

$$v = \text{frequency} = \frac{1}{\text{period}} = \frac{\omega}{2\pi}$$

Thus, the function $y = A \sin \omega t$, which is associated with oscillatory motion in engineering, has period $2\pi/\omega$ and **amplitude** A. The term amplitude is used to indicate the maximum distance of the graph of $y = A \sin \omega t$ from the horizontal axis.

Example 2.37   Sketch using the same set of axes the graphs of the functions

(a) $y = 2 \sin t$ (b) $y = \sin t$ (c) $y = \frac{1}{2} \sin t$

and discuss.

Solution   The graphs of the three functions are shown in Figure 2.59. The functions (a), (b) and (c) have amplitudes 2, 1 and $\frac{1}{2}$ respectively. We note that the effect of changing the amplitude is to alter the size of the 'humps' in the sine wave. Note that changing only the amplitude does not alter the points at which the graph crosses the x axis. All three functions have period 2π.

Figure 2.59

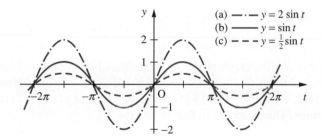

Example 2.38   Sketch using the same axes the graphs of the functions

(a) $y = \sin t$ (b) $y = \sin 2t$ (c) $y = \sin \frac{1}{2} t$

and discuss.

Solution   The graphs of the three functions (a), (b) and (c) are shown in Figure 2.60. All three have amplitude 1 and periods 2π, π and 4π respectively. We note that the effect of changing the parameter ω in $\sin \omega t$ is to 'squash' or 'stretch' the basic sine wave $\sin t$. All that happens is that the basic pattern repeats itself less or more frequently; that is, the period changes.

Figure 2.60

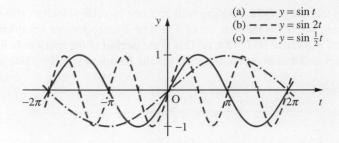

In engineering we frequently encounter the sinusoidal function

$$y = A \sin(\omega t + \alpha), \quad \omega > 0 \tag{2.22}$$

Following the discussion in Section 2.2.4, we have that the graph of this function is obtained by moving the graph of $y = A \sin \omega t$ horizontally:

$\dfrac{\alpha}{\omega}$ units to the left if α is positive

or

$\dfrac{|\alpha|}{\omega}$ units to the right if α is negative

The sine wave of (2.22) is said to 'lead' the sine wave $A \sin \omega t$ when α is positive and to 'lag' it when α is negative.

Example 2.39   Sketch the graph of $y = 3 \sin(2t + \frac{1}{3}\pi)$.

Solution   First we sketch the graph of $y = 3 \sin 2t$, which has amplitude 3 and period π, as shown in Figure 2.61(a). In this case $\alpha = \frac{1}{3}\pi$ and $\omega = 2$, so it follows that the graph of $y = 3 \sin(2t + \frac{1}{3}\pi)$ is obtained by moving the graph of $y = 3 \sin 2t$ horizontally to the left by $\frac{1}{6}\pi$ units. This is shown in Figure 2.61(b).

Figure 2.61

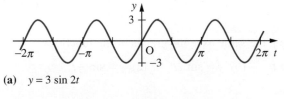

(a) $y = 3 \sin 2t$

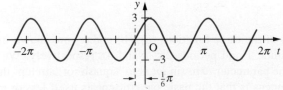

(b) $y = 3 \sin (2t + \frac{1}{3}\pi)$

2.6.3 Trigonometric identities

Other circular functions are defined in terms of the three basic functions sine, cosine and tangent. In particular, we have

$$\sec x = \frac{1}{\cos x}, \quad \text{the \textbf{secant} function}$$

$$\operatorname{cosec} x = \frac{1}{\sin x}, \quad \text{the \textbf{cosecant} function}$$

$$\cot x = \frac{1}{\tan x}, \quad \text{the \textbf{cotangent} function}$$

From the basic definitions it is possible to deduce the following trigonometric identities relating the functions.

Triangle identities

$$\cos^2 x + \sin^2 x = 1 \tag{2.23a}$$
$$1 + \tan^2 x = \sec^2 x \tag{2.23b}$$
$$1 + \cot^2 x = \operatorname{cosec}^2 x \tag{2.23c}$$

The first of these follows immediately from the use of Pythagoras' theorem in a right-angled triangle with a unit hypotenuse. Dividing (2.23a) through by $\cos^2 x$ yields identity (2.23b), and dividing through by $\sin^2 x$ yields identity (2.23c).

Compound-angle identities

$$\sin(x + y) = \sin x \cos y + \cos x \sin y \tag{2.24a}$$
$$\sin(x - y) = \sin x \cos y - \cos x \sin y \tag{2.24b}$$
$$\cos(x + y) = \cos x \cos y - \sin x \sin y \tag{2.24c}$$
$$\cos(x - y) = \cos x \cos y + \sin x \sin y \tag{2.24d}$$
$$\tan(x + y) = \frac{\tan x + \tan y}{1 - \tan x \tan y} \tag{2.24e}$$
$$\tan(x - y) = \frac{\tan x - \tan y}{1 + \tan x \tan y} \tag{2.24f}$$

Sum and product identities

$$\sin x + \sin y = 2 \sin \tfrac{1}{2}(x + y) \cos \tfrac{1}{2}(x - y) \tag{2.25a}$$
$$\sin x - \sin y = 2 \sin \tfrac{1}{2}(x - y) \cos \tfrac{1}{2}(x + y) \tag{2.25b}$$
$$\cos x + \cos y = 2 \cos \tfrac{1}{2}(x + y) \cos \tfrac{1}{2}(x - y) \tag{2.25c}$$
$$\cos x - \cos y = -2 \sin \tfrac{1}{2}(x + y) \sin \tfrac{1}{2}(x - y) \tag{2.25d}$$

From identities (2.24a), (2.24c) and (2.24e) we can obtain the double-angle formulae.

$$\sin 2x = 2 \sin x \cos x \tag{2.26a}$$

$$\cos 2x = \cos^2 x - \sin^2 x \tag{2.26b}$$

$$= 2 \cos^2 x - 1 \tag{2.26c}$$

$$= 1 - 2 \sin^2 x \tag{2.26d}$$

$$\tan 2x = \frac{2 \tan x}{1 - \tan^2 x} \tag{2.26e}$$

(Writing $x = \theta/2$ we can obtain similar identities called half-angle formulae.)

Example 2.40   Express $\cos(\pi/2 + 2x)$ in terms of $\sin x$ and $\cos x$.

Solution   Using identity (2.24c) we obtain

$$\cos(\pi/2 + 2x) = \cos \pi/2 \cos 2x - \sin \pi/2 \sin 2x$$

Since $\cos \pi/2 = 0$ and $\sin \pi/2 = 1$, we can simplify to obtain

$$\cos(\pi/2 + 2x) = -\sin 2x$$

Now using the double-angle formula (2.26a), we obtain

$$\cos(\pi/2 + 2x) = -2 \sin x \cos x$$

Example 2.41   Show that

$$\sin(A + B) + \sin(A - B) = 2 \sin A \cos B$$

and deduce that

$$\sin x + \sin y = 2 \sin \tfrac{1}{2}(x + y) \cos \tfrac{1}{2}(x - y)$$

Hence sketch the graph of $y = \sin 4x + \sin 2x$.

Solution   Using identities (2.24a) and (2.24b) we have

$$\sin(A + B) = \sin A \cos B + \cos A \sin B$$

$$\sin(A - B) = \sin A \cos B - \cos A \sin B$$

Adding these two identities gives

$$\sin(A + B) + \sin(A - B) = 2 \sin A \cos B$$

Now setting $A + B = x$ and $A - B = y$, we see that $A = \tfrac{1}{2}(x + y)$ and $B = \tfrac{1}{2}(x - y)$ so that

$$\sin x + \sin y = 2 \sin \tfrac{1}{2}(x + y) \cos \tfrac{1}{2}(x - y)$$

which is identity (2.25a). (The identities (2.25b–d) can be proved in the same manner.)

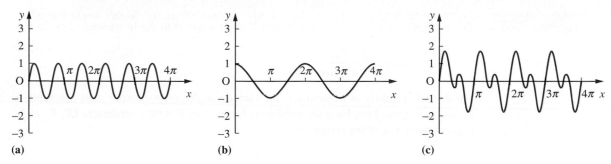

(a)   (b)   (c)

Figure 2.62

Applying the formula to

$$y = \sin 4x + \sin 2x$$

we obtain

$$y = 2 \sin 3x \cos x$$

The graphs of $y = \sin 3x$ and $y = \cos x$ are shown in Figures 2.62(a) and (b). The combination of these two graphs yields Figure 2.62(c). This type of combination of oscillations in practical situations leads to the phenomena of 'beats'.

Example 2.42   Solve the equation $2 \cos^2 x + 3 \sin x = 3$ for $0 \leqslant x \leqslant 2\pi$.

Solution   First we express the equation in terms of $\sin x$ only. This can be done by eliminating $\cos^2 x$ using the identity (2.23a), giving

$$2(1 - \sin^2 x) + 3 \sin x = 3$$

which reduces to

$$2 \sin^2 x - 3 \sin x + 1 = 0$$

This is now a quadratic equation in $\sin x$, and it is convenient to write $\lambda = \sin x$, giving

$$2\lambda^2 - 3\lambda + 1 = 0$$

Factorizing then gives $(2\lambda - 1)(\lambda - 1) = 0$

leading to the two solutions $\lambda = \frac{1}{2}$ and $\lambda = 1$

We now return to the fact that $\lambda = \sin x$ to determine the corresponding values of x.

(i) If $\lambda = \frac{1}{2}$ then $\sin x = \frac{1}{2}$. Remembering that $\sin x$ is positive for x lying in the first and second quadrants and that $\sin \frac{1}{6}\pi = \frac{1}{2}$, we have two solutions corresponding to $\lambda = \frac{1}{2}$, namely $x = \frac{1}{6}\pi$ and $x = \frac{5}{6}\pi$.

(ii) If $\lambda = 1$ then $\sin x = 1$, giving the single solution $\lambda = \frac{1}{2}\pi$.

Thus there are three solutions to the given equation, namely

$$x = \tfrac{1}{6}\pi, \quad \tfrac{1}{2}\pi \quad \text{and} \quad \tfrac{5}{6}\pi$$

Example 2.43

The path of a projectile fired with speed V at an angle α to the horizontal is given by

$$y = x \tan \alpha - \frac{1}{2} \frac{gx^2}{V^2 \cos^2 \alpha}$$

For fixed V a family of trajectories, for various angles of projection α, is obtained, as shown in Figure 2.63. Find the condition for a point P with coordinates (X, Y) to lie beyond the reach of the projectile.

Solution

Given the coordinates (X, Y), the possible angles α of launch are given by the roots of the equation

$$Y = X \tan \alpha - \frac{1}{2} \frac{gX^2}{V^2 \cos^2 \alpha}$$

Using the trigonometric identity

$$1 + \tan^2 \alpha = \frac{1}{\cos^2 \alpha}$$

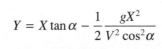

Figure 2.63
Trajectories for different launch angles.

gives

$$Y = X \tan \alpha - \frac{1}{2} \frac{gX^2}{V^2} (1 + \tan^2 \alpha)$$

Writing $T = \tan \alpha$, this may be rewritten as

$$(gX^2)T^2 - (2XV^2)T + (gX^2 + 2V^2Y) = 0$$

which is a quadratic equation in T. From (1.5), this equation will have two different real roots if

$$(2XV^2)^2 > 4(gX^2)(gX^2 + 2V^2Y)$$

but no real roots if

$$(2XV^2)^2 < 4(gX^2)(gX^2 + 2V^2Y)$$

Thus the point P(X, Y) is 'safe' if

$$V^4 < g^2X^2 + 2gV^2Y$$

The critical case where the point (X, Y) lies on the curve

$$V^4 = g^2x^2 + 2gV^2y$$

gives us the so-called 'parabola of safety', with the safety region being that above this parabola

$$y = \frac{V^2}{2g} - \frac{gx^2}{2V^2}$$

2.6.4 Amplitude and phase

Often in engineering contexts we are concerned with vibrations of parts of a structure or machine. These vibrations are a response to a periodic external force and will usually have the same frequency as that force. Usually, also, the response will lag behind the exciting force. Mathematically this is often represented by an external force of the form $F \sin \omega t$ with a response of the form $a \sin \omega t + b \cos \omega t$ where a and b are constants dependent on F, ω and the physical characteristics of the system. To find the size of the response we need to write it in the form $A \sin(\omega t + \alpha)$ where

$$A \sin(\omega t + \alpha) = a \sin \omega t + b \cos \omega t$$

This we can always do, as is illustrated in Example 2.44.

Example 2.44 Express $y = 4 \sin 3t - 3 \cos 3t$ in the form $y = A \sin(3t + \alpha)$.

Solution To determine the appropriate values of A and α, we proceed as follows.
Using the identity (2.24a), we have

$$A \sin(3t + \alpha) = A(\sin 3t \cos \alpha + \cos 3t \sin \alpha)$$

$$= (A \cos \alpha) \sin 3t + (A \sin \alpha) \cos 3t$$

Since this must equal the expression

$$4 \sin 3t - 3 \cos 3t$$

for all values of t, the respective coefficients of $\sin 3t$ and $\cos 3t$ must be the same in both expressions, so that

$$4 = A \cos \alpha \tag{2.27}$$

and

$$-3 = A \sin \alpha \tag{2.28}$$

The angle α is shown in Figure 2.64. By Pythagoras' theorem,

$$A = \sqrt{(16 + 9)} = 5$$

and clearly

$$\tan \alpha = -\tfrac{3}{4}$$

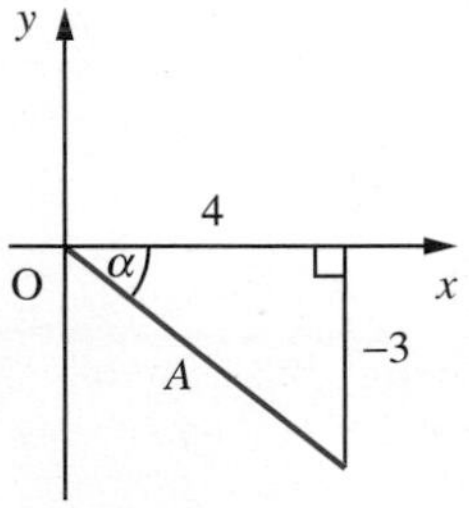

Figure 2.64
The angle α.

The value of α may now be determined using a calculator. However, care must be taken to ensure that the correct quadrant is chosen for α. Since A is taken to be positive, it follows from Figure 2.64 that α lies in the fourth quadrant. Thus, using a calculator, we have $\alpha = -0.64$ rad and

$$y = 4 \sin 3t - 3 \cos 3t = 5 \sin(3t - 0.64)$$

2.6.5 Inverse circular (trigonometric) functions

Considering the inverse of the trigonometric functions, it follows from the definition given in (2.4) that the inverse sine function $\sin^{-1}x$ (also sometimes denoted by arcsin x) is such that

$$\text{if} \quad y = \sin^{-1}x \quad \text{then} \quad x = \sin y$$

Here x should not be interpreted as an angle – rather $\sin^{-1}x$ represents the angle whose sine is x. Applying the procedures for obtaining the graph of the inverse function given in Section 2.2.3 to the graph of $y = \sin x$ (Figure 2.54) leads to the graph shown in Figure 2.65(a). As we explained in Example 2.6, when considering the inverse of $y = x^2$, the graph of Figure 2.65(a) is not representative of a function, since for each value of x in the domain $-1 \leqslant x \leqslant 1$ there are an infinite number of image values (as indicated by the points of intersection of the broken vertical line with the graph). To overcome this problem, we restrict the range of the inverse function $\sin^{-1}x$ to $-\frac{1}{2}\pi \leqslant \sin^{-1}x \leqslant \frac{1}{2}\pi$ and define the inverse sine function by

$$\text{if } y = \sin^{-1}x \text{ then } x = \sin y, \quad \text{where } -\tfrac{1}{2}\pi \leqslant y \leqslant \tfrac{1}{2}\pi \text{ and } -1 \leqslant x \leqslant 1 \qquad \textbf{(2.29)}$$

The corresponding graph is shown in Figure 2.65(b).

Similarly, in order to define the inverse cosine and inverse tangent functions $\cos^{-1}x$ and $\tan^{-1}x$ (also sometimes denoted by arccos x and arctan x), we have to restrict the ranges. This is done according to the following definitions.

$$\text{if } y = \cos^{-1}x \text{ then } x = \cos y, \quad \text{where } 0 \leqslant y \leqslant \pi \text{ and } -1 \leqslant x \leqslant 1 \qquad \textbf{(2.30)}$$

$$\text{if } y = \tan^{-1}x \text{ then } x = \tan y, \quad \text{where } -\tfrac{1}{2}\pi < y < \tfrac{1}{2}\pi \text{ and } x \text{ is any real number} \qquad \textbf{(2.31)}$$

Figure 2.65
Graph of $\sin^{-1}x$.

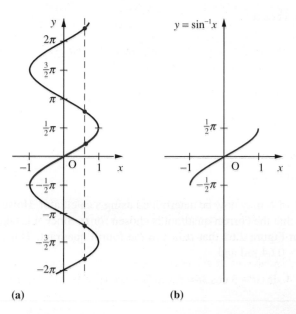

(a)   (b)

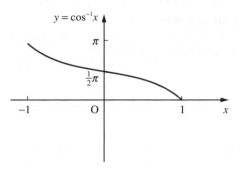

Figure 2.66 Graph of $\cos^{-1}x$.

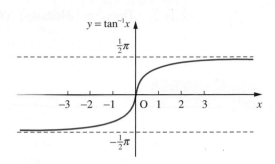

Figure 2.67 Graph of $\tan^{-1}x$.

The corresponding graphs of $y = \cos^{-1}x$ and $y = \tan^{-1}x$ are shown in Figures 2.66 and 2.67, respectively.

In some books (2.29)–(2.31) are called the *principal values* of the inverse functions. A calculator will automatically give these values.

Example 2.45   Evaluate $\sin^{-1}x$, $\cos^{-1}x$, $\tan^{-1}x$ where (a) $x = 0.35$ and (b) $x = -0.7$, expressing the answers correct to 4dp.

Solution   (a) $\sin^{-1}(0.35)$ is the angle α which lies between $-\pi/2$ and $+\pi/2$ and is such that $\sin \alpha = 0.35$. Using a calculator we have

$$\sin^{-1}(0.35) = 0.3576 \ (4\text{dp}) = 0.1138\pi$$

which clearly lies between $-\pi/2$ and $+\pi/2$.

$\cos^{-1}(0.35)$ is the angle β which lies between 0 and π and is such that $\cos \beta = 0.35$. Using a calculator we obtain

$$\cos^{-1}(0.35) = 1.2132 \ (4\text{dp}) = 0.3862\pi$$

which lies between 0 and π.

$\tan^{-1}(0.35)$ is the angle γ which lies between $-\pi/2$ and $+\pi/2$ and is such that $\tan \gamma = 0.35$. Using a calculator we have

$$\tan^{-1}(0.35) = 0.3367 \ (4\text{dp}) = 0.1072\pi$$

which lies in the correct range of values.

Notice

$$\frac{\sin^{-1}(0.35)}{\cos^{-1}(0.35)} \neq \tan^{-1}(0.35)$$

(b) $\sin^{-1}(-0.7)$ is the angle α which lies between $-\pi/2$ and $+\pi/2$ and is such that $\sin \alpha = -0.7$. Again using a calculator we obtain

$$\sin^{-1}(-0.7) = -0.7754 \ (4\text{dp})$$

which lies in the correct range of values.

$\cos^{-1}(-0.7)$ is the angle β which lies between 0 and π and is such that $\cos \beta = -0.7$.

Thus $\beta = 2.3462$, which lies in the second quadrant as expected.

$\tan^{-1}(-0.7)$ is the angle γ which lies between $-\pi/2$ and $+\pi/2$ and is such that $\tan \gamma = -0.7$. Thus $\gamma = -0.6107$, lying in the fourth quadrant, as expected.

Example 2.46 Sketch the graph of the function $y = \sin^{-1}(\sin x)$.

Solution Before beginning to sketch the graph we need to examine the algebraic properties of the function. Because of the way $\sin^{-1}$ is defined we know that for $-\pi/2 \leqslant x \leqslant \pi/2$, $\sin^{-1}(\sin x) = x$. (The function $\sin^{-1}x$ *strictly* is the inverse function of $\sin x$ with the restricted domain $-\pi/2 \leqslant x \leqslant \pi/2$.) We also know that $\sin x$ is an odd function, so that $\sin(-x) = -\sin x$. This implies that $\sin^{-1}x$ is an odd function. In fact, this is obvious from its graph (Figure 2.65(b)). Thus, $\sin^{-1}(\sin x)$ is an odd function. Lastly, since $\sin x$ is a periodic function with period 2π we conclude that $\sin^{-1}(\sin x)$ is also a periodic function of period 2π. Thus, if we can sketch the graph between 0 and π, we can obtain the graph between $-\pi$ and 0 by antisymmetry about $x = 0$ and the whole graph by periodicity elsewhere. Using Figures 2.65(a) and 2.65(b) we can obtain the graph of the function for $0 \leqslant x \leqslant \pi$ as shown in Figure 2.68 (blue). The graph between $-\pi$ and 0 is obtained by antisymmetry about the origin, as shown with the broken line in Figure 2.68, and the whole graph is obtained making use of the piece between $-\pi$ and $+\pi$ and periodicity.

Figure 2.68
Graph of
$y = \sin^{-1}(\sin x)$.

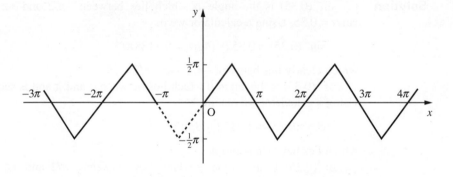

2.6.6 Polar coordinates

In some applications the position of a point P in a plane is represented by its distance r from a fixed point O and the angle θ that the line joining P to O makes with some fixed direction. The pair (r, θ) determine the point uniquely and are called the **polar coordinates** of P. If polar coordinates are chosen sharing the same origin O as rectangular cartesian coordinates and with the angle θ measured from the direction of the Ox axis then, as can be seen from Figure 2.69, the polar coordinates (r, θ) and the cartesian coordinates (x, y) of a point are related by

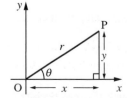

Figure 2.69

$$x = r \cos \theta, \quad y = r \sin \theta \tag{2.32}$$

and also

$$r = \sqrt{(x^2 + y^2)}, \quad \tan\theta = \frac{y}{x}$$

Note that the origin does not have a well-defined θ. Some care must be taken when evaluating θ using the above formula to ensure that it is located in the correct quadrant. The angle $\tan^{-1}(y/x)$ obtained from tables or a calculator will usually lie between $\pm\frac{1}{2}\pi$ and will give the correct value of θ if P lies in the first or fourth quadrant. If P lies in the second or third quadrant then $\theta = \tan^{-1}(y/x) + \pi$. It is sensible to use the values of $\sin\theta$ and $\cos\theta$ to check that θ lies in the correct quadrant.

Note that the angle θ is positive when measured in an anticlockwise direction and negative when measured in a clockwise direction. Many calculators have rectangular (cartesian) to polar conversion and vice versa.

Example 2.47

(a) Find the polar coordinates of the points whose cartesian coordinates are (1, 2), (−1, 3), (−1, −1), (1, −2), (1, 0), (0, 2), (0, −2).

(b) Find the cartesian coordinates of the points whose polar coordinates are (3, $\pi/4$), (2, $-\pi/6$), (2, $-\pi/2$), (5, $3\pi/4$).

Solution

(a) Using the formula (2.32) we see that:

$$(x = 1, y = 2) \equiv (r = \sqrt{5}, \theta = \tan^{-1}(2/1) = 1.107)$$

$$(x = -1, y = 3) \equiv (r = \sqrt{10}, \theta = 1.893)$$

$$(x = -1, y = -1) \equiv (r = \sqrt{2}, \theta = 5\pi/4)$$

$$(x = 1, y = -2) \equiv (r = \sqrt{5}, \theta = -1.107)$$

$$(x = 1, y = 0) \equiv (r = 1, \theta = 0)$$

$$(x = 0, y = 2) \equiv (r = 2, \theta = \pi/2)$$

$$(x = 0, y = -2) \equiv (r = 2, \theta = -\pi/2)$$

(Here answers, where appropriate, are given to 3dp.)

(b) Using the formula (2.32) we see that

$$(r = 3, \theta = \pi/4) \equiv (x = 3/\sqrt{2}, y = 3/\sqrt{2})$$

$$(r = 2, \theta = -\pi/6) \equiv (x = \sqrt{3}, y = -1)$$

$$(r = 2, \theta = -\pi/2) \equiv (x = 0, y = -2)$$

$$(r = 5, \theta = 3\pi/4) \equiv (x = -5/\sqrt{2}, y = 5\sqrt{2})$$

To plot a curve specified using polar coordinates we first look for any features, for example symmetry, which would reduce the amount of calculation, and then we draw up a table of values of r against values of θ. This is a tedious process and we usually use a graphics calculator or a computer package to perform the task. There are, however, different conventions in use about polar plotting. Some packages are designed to

Figure 2.70
(a) $r = 2a \cos \theta, 0 \leqslant \theta \leqslant \pi, r \geqslant 0$. (b) $r = 2a \cos \theta, 0 \leqslant \theta \leqslant \pi, r$ unrestricted.

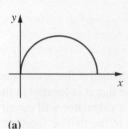

(a)

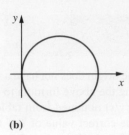

(b)

plot only points where r is positive, so that plotting $r = 2a \cos \theta$ for $0 \leqslant \theta \leqslant \pi$ would yield Figure 2.70(a) while other packages plot negative values of r treating r as a number line, so that $r = 2a \cos \theta$ for $0 \leqslant \theta \leqslant \pi$ yields Figure 2.70(b).

Example 2.48 Express the equation of the circle

$$(x - a)^2 + y^2 = a^2$$

in polar form.

Solution Expanding the squared term, the equation of the given circle becomes

$$x^2 + y^2 - 2ax = 0$$

Using the relationships (2.32), we have

$$r^2(\cos^2\theta + \sin^2\theta) - 2ar \cos \theta = 0$$

Using the trigonometric identity (2.23a),

$$r(r - 2a \cos \theta) = 0, \quad -\pi/2 < \theta \leqslant \pi/2$$

Since $r = 0$ gives the point $(0, 0)$, we can ignore this, and the equation of the circle becomes

$$r = 2a \cos \theta, \quad -\pi/2 < \theta \leqslant \pi/2$$

2.6.7 Exercises

41 Sketch for $-3\pi \leqslant x \leqslant 3\pi$ the graphs of

(a) $y = \sin 2x$ (b) $y = \sin \frac{1}{2}x$

(c) $y = \sin^2 x$ (d) $y = \sin x^2$

(e) $y = \dfrac{1}{\sin x}$ $(x \neq n\pi, n = 0, \pm 1, \pm 2, \dots)$

(f) $y = \sin\left(\dfrac{1}{x}\right)$ $(x \neq 0)$

42 Solve the following equations for $0 \leqslant x \leqslant 2\pi$:

(a) $3 \sin^2 x + 2 \sin x - 1 = 0$

(b) $4 \cos^2 x + 5 \cos x + 1 = 0$

(c) $2 \tan^2 x - \tan x - 1 = 0$

(d) $\sin 2x = \cos x$

43 By referring to an equilateral triangle, show that $\cos \frac{1}{3}\pi = \frac{1}{2}\sqrt{3}$ and $\tan \frac{1}{6}\pi = \frac{1}{3}\sqrt{3}$, and find values

for $\sin \frac{1}{3}\pi$, $\tan \frac{1}{3}\pi$, $\cos \frac{1}{6}\pi$ and $\sin \frac{1}{6}\pi$. Hence, using the double-angle formulae, find $\sin \frac{1}{12}\pi$, $\cos \frac{1}{12}\pi$ and $\tan \frac{1}{12}\pi$. Using appropriate properties from Section 2.6, calculate

(a) $\sin \frac{2}{3}\pi$ (b) $\tan \frac{7}{6}\pi$ (c) $\cos \frac{11}{6}\pi$

(d) $\sin \frac{5}{12}\pi$ (e) $\cos \frac{7}{12}\pi$ (f) $\tan \frac{11}{12}\pi$

44 Given $s = \sin \theta$, where $\frac{1}{2}\pi < \theta < \pi$, find, in terms of s,

(a) $\cos \theta$ (b) $\sin 2\theta$

(c) $\sin 3\theta$ (d) $\sin \frac{1}{2}\theta$

45 Show that

$$\frac{1 + \sin 2\theta + \cos 2\theta}{1 + \sin 2\theta - \cos 2\theta} = \cot \theta$$

46 Given $t = \tan \frac{1}{2}x$, prove that

(a) $\sin x = \dfrac{2t}{1 + t^2}$ (b) $\cos x = \dfrac{1 - t^2}{1 + t^2}$

(c) $\tan x = \dfrac{2t}{1 - t^2}$

Hence solve the equation

$$2 \sin x - \cos x = 1$$

47 In each of the following, the value of one of the six circular functions is given. Without using a calculator, find the values of the remaining five.

(a) $\sin x = \frac{1}{2}$ (b) $\cos x = -\frac{1}{2}\sqrt{3}$

(c) $\tan x = -1$ (d) $\sec x = \sqrt{2}$

(e) $\operatorname{cosec} x = -2$ (f) $\cot x = \sqrt{3}$

48   The lower edge of a mural, which is 4 m high, is 2 m above an observer's eyes, as shown in Figure 2.71.

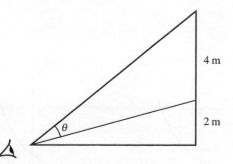

4 m

2 m

θ

Figure 2.71 Mural of Question 48.

Assuming that the best view of the mural is obtained when the angle θ subtended at the eye is a maximum, prove that the observer should stand $2\sqrt{3}$ m away from the wall.

49 Express as a product of sines and/or cosines

(a) $\sin 3\theta + \sin \theta$ (b) $\cos \theta - \cos 2\theta$

(c) $\cos 5\theta + \cos 2\theta$ (d) $\sin \theta - \sin 2\theta$

50 Express as a sum or difference of sines or cosines

(a) $\sin 3\theta \sin \theta$ (b) $\sin 3\theta \cos \theta$

(c) $\cos 3\theta \sin \theta$ (d) $\cos 3\theta \cos \theta$

51 Express in the forms $r \cos(\theta - \alpha)$ and $r \sin(\theta - \beta)$

(a) $\sqrt{3} \sin \theta - \cos \theta$ (b) $\sin \theta - \cos \theta$

(c) $\sin \theta + \cos \theta$ (d) $2 \cos \theta + 3 \sin \theta$

52 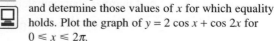 Show that $-\frac{3}{2} \leqslant 2 \cos x + \cos 2x \leqslant 3$ for all x, and determine those values of x for which equality holds. Plot the graph of $y = 2 \cos x + \cos 2x$ for $0 \leqslant x \leqslant 2\pi$.

53 Evaluate

(a) $\sin^{-1}(0.5)$ (b) $\sin^{-1}(-0.5)$

(c) $\cos^{-1}(0.5)$ (d) $\cos^{-1}(-0.5)$

(e) $\tan^{-1}(\sqrt{3})$ (f) $\tan^{-1}(-\sqrt{3})$

54 Sketch the graph of the functions

(a) $y = \sin^{-1}(\cos x)$

(b) $y = \cos^{-1}(\sin x)$

(c) $y = \cos^{-1}(\cos x)$

(d) $y = \cos^{-1}(\cos x) - \sin^{-1}(\sin x)$

55 If $\tan^{-1}x = \alpha$ and $\tan^{-1}y = \beta$, show that

$$\tan(\alpha + \beta) = \frac{x + y}{1 - xy}$$

Deduce that

$$\tan^{-1}x + \tan^{-1}y = \tan^{-1}\left(\frac{x + y}{1 - xy}\right) + k\pi$$

where $k = -1, 0, 1$ depending on the values of x and y.

2.7 # Exponential, logarithmic and hyperbolic functions

The members of this family of functions are closely interconnected. They occur in widely varied applications, from heat transfer analysis to bridge design, from transmission line modelling to the production of chemicals. Historically the exponential and logarithmic functions arose in very different contexts, the former in the calculation of compound interest and the latter in computational mathematics, but, as often happens in mathematics, the discoveries in specialized areas of applicable mathematics have found applications widely elsewhere.

2.7.1 Exponential functions

Functions of the type $f(x) = a^x$ where a is a positive constant (and x is the independent variable as usual) are called **exponential functions**.

The graphs of the exponential functions, shown in Figure 2.72, are similar. By a simple scaling of the x axis, we can obtain the same graphs for $y = 2^x$, $y = 3^x$ and $y = 4^x$, as shown in Figure 2.73. The reason for this is that we can write $3^x = 2^{kx}$ where $k \approx 1.585$ and $4^x = 2^{2x}$. Thus all exponential functions can be expressed in terms of one exponential function. The standard exponential function that is used is $y = e^x$, where e is a special number approximately equal to

$$2.718\,281\,828\,459\,045\,2\ldots$$

Figure 2.72
Graphs of exponential functions.

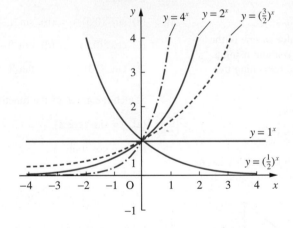

Figure 2.73
Scaled graphs of exponential functions.

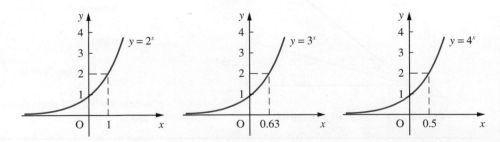

Figure 2.74
The standard
exponential
function $y = e^x$.

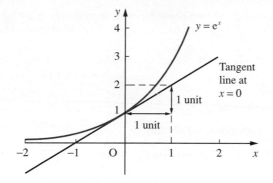

This number e is chosen because the graph of $y = e^x$ (Figure 2.74) has the property that the slope of the tangent at any point on the curve is equal to the value of the function at that point. We shall discuss this property again in Chapter 8.

We note that the following properties are satisfied by the exponential function:

$$e^{x_1}e^{x_2} = e^{x_1 + x_2}$$ (2.33a)

$$e^{x+c} = e^x e^c = Ae^x, \quad \text{where } A = e^c$$ (2.33b)

$$\frac{e^{x_1}}{e^{x_2}} = e^{x_1 - x_2}$$ (2.33c)

$$e^{kx} = (e^k)^x = a^x, \quad \text{where } a = e^k$$ (2.33d)

Often e^x is written as $\exp x$ for clarity when 'x' is a complicated expression. For example,

$$e^{(x+1)/(x+2)} = \exp\left(\frac{x+1}{x+2}\right)$$

Example 2.49   The temperature T of a body cooling in an environment, whose unknown ambient temperature is α, is given by

$$T(t) = \alpha + (T_0 - \alpha)e^{-kt}$$

where T_0 is the initial temperature of the body and k is a physical constant. To determine the value of α, the temperature of the body is recorded at two times, t_1 and t_2, where $t_2 = 2t_1$ and $T(t_1) = T_1$, $T(t_2) = T_2$. Show that

$$\alpha = \frac{T_0 T_2 - T_1^2}{T_2 - 2T_1 + T_0}$$

Solution   From the formula for $T(t)$ we have

$$T_1 - \alpha = (T_0 - \alpha)e^{-kt_1}$$

and

$$T_2 - \alpha = (T_0 - \alpha)e^{-2kt_1}$$

Squaring the first of these two equations and then dividing by the second gives

$$\frac{(T_1 - \alpha)^2}{T_2 - \alpha} = \frac{(T_0 - \alpha)^2 e^{-2kt_1}}{(T_0 - \alpha)e^{-2kt_1}}$$

This simplifies to

$$(T_1 - \alpha)^2 = (T_2 - \alpha)(T_0 - \alpha)$$

Multiplying out both sides, we obtain

$$T_1^2 - 2\alpha T_1 + \alpha^2 = T_0 T_2 - (T_0 + T_2)\alpha + \alpha^2$$

which gives

$$(T_0 - 2T_1 + T_2)\alpha = T_0 T_2 - T_1^2$$

Hence the result.

2.7.2 Logarithmic functions

From the graph of $y = e^x$, given in Figure 2.74, it is clear that it is a one-to-one function, so that its inverse function is defined. This inverse is called the **natural logarithm** function and is written as

$$y = \ln x$$

(In some textbooks it is written as $\log_e x$, while in many pure mathematics books it is written simply as $\log x$.) Using the procedures given in Section 2.2.1, its graph can be drawn as in Figure 2.75. From the definition we have

$$\text{if } y = e^x \quad \text{then} \quad x = \ln y \tag{2.34}$$

which implies that

$$\ln e^x = x, \quad e^{\ln y} = y$$

In the same way as there are many exponential functions (2^x, 3^x, 4^x, . . .), there are also many logarithmic functions. In general,

Figure 2.75
Graph of $y = \ln x$.

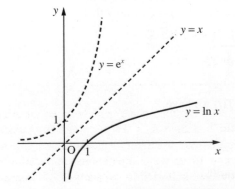

$$y = a^x \quad \text{gives} \quad x = \log_a y \tag{2.35}$$

which can be expressed verbally as 'x equals log to base a of y'. (Note that $\log_{10} x$ is often written, except in advanced mathematics books, simply as log x.) Recalling that $a^x = e^{kx}$ for some constant k, we see now that $a^x = (e^k)^x$, so that $a = e^k$ and $k = \ln a$.

From the definition of $\log_a x$ it follows that

$$\log_a(x_1 x_2) = \log_a x_1 + \log_a x_2 \tag{2.36a}$$

$$\log_a\left(\frac{x_1}{x_2}\right) = \log_a x_1 - \log_a x_2 \tag{2.36b}$$

$$\log_a x^n = n \log_a x \tag{2.36c}$$

$$x = a^{\log_a x} \tag{2.36d}$$

$$y^x = a^{x \log_a y} \tag{2.36e}$$

$$\log_a x = \frac{\log_b x}{\log_b a} \tag{2.36f}$$

Example 2.50

(a) Express $\frac{1}{3}\log_2 8 - \log_2\frac{2}{7}$ in terms of $\log_2 7$.

(b) Expand $\ln\left(\dfrac{\sqrt{(10x)}}{y^2}\right)$.

Solution

(a) $\frac{1}{3}\log_2 8 = \log_2 8^{1/3} = \log_2 2 = 1$

$\log_2\frac{2}{7} = \log_2 2 - \log_2 7 = 1 - \log_2 7$

Thus

$\frac{1}{3}\log_2 8 - \log_2\frac{2}{7} = \log_2 7$

(b) $\ln\left(\dfrac{\sqrt{(10x)}}{y^2}\right) = \ln\sqrt{(10x)} - \ln(y^2)$

$= \frac{1}{2}\ln(10x) - 2\ln y$

$= \frac{1}{2}\ln 10 + \frac{1}{2}\ln x - 2\ln y$

Despite the fact that these functions occur widely in engineering analysis, they first occurred in computational mathematics. Property (2.36a) transforms the problem of multiplying two numbers to that of adding their logarithms. The widespread use of scientific calculators has now made the computational application of logarithms largely irrelevant. They are, however, still used in the analysis of experimental data.

2.7.3 Hyperbolic functions

In applications certain combinations of exponential functions recur many times and these combinations are given special names. For example, the mathematical model for the steady state heat transfer in a straight bar leads to an expression for the temperature $T(x)$ at a point distance x from one end, given by

$$T(x) = \frac{T_0(e^{m(l-x)} - e^{-m(l-x)}) + T_1(e^{mx} - e^{-mx})}{e^{ml} - e^{-ml}}$$

where l is the total length of the bar, T_0 and T_1 are the temperatures at the ends and m is a physical constant. To simplify such expressions a family of functions, called the **hyperbolic** functions, is defined as follows:

$$\cosh x = \tfrac{1}{2}(e^x + e^{-x}), \quad \text{the \textbf{hyperbolic cosine}}$$

$$\sinh x = \tfrac{1}{2}(e^x - e^{-x}), \quad \text{the \textbf{hyperbolic sine}}$$

$$\tanh x = \frac{\sinh x}{\cosh x}, \quad \text{the \textbf{hyperbolic tangent}}$$

Thus, the expression for $T(x)$ becomes

$$T(x) = \frac{T_0 \sinh m(l - x) + T_1 \sinh mx}{\sinh ml}$$

The reason for the names of these functions is geometric. They bear the same relationship to the hyperbola as the circular functions do to the circle, as shown in Figure 2.76.

Following the pattern of the circular or trigonometric functions, other hyperbolic functions are defined as follows:

$$\text{sech } x = \frac{1}{\cosh x}, \quad \text{the \textbf{hyperbolic secant}}$$

$$\text{cosech } x = \frac{1}{\sinh x} \quad (x \neq 0), \quad \text{the \textbf{hyperbolic cosecant}}$$

$$\coth x = \frac{1}{\tanh x} \quad (x \neq 0), \quad \text{the \textbf{hyperbolic cotangent}}$$

Figure 2.76
The analogy between circular and hyperbolic functions. The circle has parametric equations $x = \cos\theta$, $y = \sin\theta$. The hyperbola has parametric equations $x = \cosh t$, $y = \sinh t$.

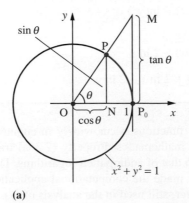

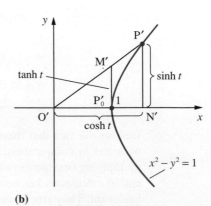

(a)   (b)

Figure 2.77
Graphs of the
hyperbolic functions.

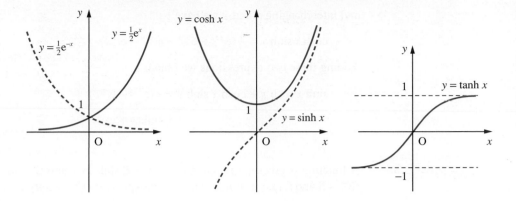

The graphs of sinh x, cosh x and tanh x are shown in Figure 2.77, where the black broken lines indicate asymptotes.

The hyperbolic functions satisfy identities analogous to those satisfied by the circular functions. From their definitions we have

$$\left.\begin{aligned} \cosh x &= \tfrac{1}{2}(e^x + e^{-x}) \\ \sinh x &= \tfrac{1}{2}(e^x - e^{-x}) \end{aligned}\right\} \tag{2.37}$$

from which we deduce

$$\cosh x + \sinh x = e^x$$
$$\cosh x - \sinh x = e^{-x}$$

and

$$(\cosh x + \sinh x)(\cosh x - \sinh x) = e^x e^{-x}$$

that is,

$$\cosh^2 x - \sinh^2 x = 1 \tag{2.38}$$

Similarly, we can show that

$$\sinh(x \pm y) = \sinh x \cosh y \pm \cosh x \sinh y \tag{2.39a}$$

$$\cosh(x \pm y) = \cosh x \cosh y \pm \sinh x \sinh y \tag{2.39b}$$

$$\tanh(x \pm y) = \frac{\tanh x \pm \tanh y}{1 \pm \tanh x \tanh y} \tag{2.39c}$$

To prove the first two of these results, it is easier to begin with the expressions on the right-hand sides and replace each hyperbolic function by its exponential form. The third result follows immediately from the previous two by dividing them. Thus

$$\sinh x \cosh y = \tfrac{1}{4}(e^x - e^{-x})(e^y + e^{-y})$$
$$= \tfrac{1}{4}(e^{x+y} + e^{x-y} - e^{-x+y} - e^{-x-y})$$

and interchanging x and y we have

$$\cosh x \sinh y = \tfrac{1}{4}(e^{x+y} + e^{y-x} - e^{-y+x} - e^{-x-y})$$

Adding these two expressions we obtain

$$\sinh x \cosh y + \cosh x \sinh y = \tfrac{1}{2}(e^{x+y} - e^{-x-y})$$

$$= \sinh(x + y)$$

Example 2.51   A function is given by $f(x) = A \cosh 2x + B \sinh 2x$ where A and B are constants and $f(0) = 5$ and $f(1) = 0$. Find A and B and express $f(x)$ as simply as possible.

Solution   Given $f(x) = A \cosh 2x + B \sinh 2x$ with the conditions $f(0) = 5, f(1) = 0$, we see that

$$A(1) + B(0) = 5$$

and

$$A \cosh 2 + B \sinh 2 = 0$$

Hence we have $A = 5$ and $B = -5 \cosh 2/\sinh 2$. Substituting into the formula for $f(x)$ we obtain

$$f(x) = 5 \cosh 2x - 5 \cosh 2 \sinh 2x/\sinh 2$$

$$= \frac{5 \sinh 2 \cosh 2x - 5 \cosh 2 \sinh 2x}{\sinh 2}$$

$$= \frac{5 \sinh(2 - 2x)}{\sinh 2}, \quad \text{using (2.39a)}$$

$$= \frac{5 \sinh 2(1 - x)}{\sinh 2}$$

Osborn's rule

In general, to obtain the formula for hyperbolic functions from the analogous identity for the circular functions, we replace each circular function by the corresponding hyperbolic function and change the sign of every product or implied product of two sines. This result is called **Osborn's rule**. Its justification will be discussed in Section 3.2.7.

Example 2.52   Verify the identity

$$\tanh 2x = \frac{2 \tanh x}{1 + \tanh^2 x}$$

using the definition of $\tanh x$. Confirm that it obeys Osborn's rule.

Solution   From the definition

$$\tanh 2x = \frac{e^{2x} - e^{-2x}}{e^{2x} + e^{-2x}}$$

and

$$1 + \tanh^2 x = 1 + \frac{(e^x - e^{-x})^2}{(e^x + e^{-x})^2} = \frac{(e^x + e^{-x})^2 + (e^x - e^{-x})^2}{(e^x + e^{-x})^2}$$

$$= \frac{2(e^{2x} + e^{-2x})}{(e^x + e^{-x})^2}$$

Thus

$$\frac{2\tanh x}{1 + \tanh^2 x} = \frac{2(e^x - e^{-x})/(e^x + e^{-x})}{2(e^{2x} + e^{-2x})/(e^x + e^{-x})^2} = \frac{(e^x - e^{-x})(e^x + e^{-x})}{e^{2x} + e^{-2x}}$$

$$= \frac{e^{2x} - e^{-2x}}{e^{2x} + e^{-2x}}$$

$$= \tanh 2x \text{ as required}$$

The formula for $\tan 2\theta$ is

$$\tan 2\theta = \frac{2\tan\theta}{1 - \tan^2\theta}$$

We see that this has an implied product of two sines ($\tan^2\theta$) so that in terms of hyperbolic functions we have, using Osborn's rule,

$$\tanh 2x = \frac{2\tanh x}{1 + \tanh^2 x}$$

which confirms the proof above.

Example 2.53   Solve the equation

$$5\cosh x + 3\sinh x = 4$$

Solution   The first step in solving problems of this type is to express the hyperbolic functions in terms of exponential functions. Thus we obtain

$$\tfrac{5}{2}(e^x + e^{-x}) + \tfrac{3}{2}(e^x - e^{-x}) = 4$$

On rearranging, this gives

$$4e^x - 4 + e^{-x} = 0$$

or

$$4e^{2x} - 4e^x + 1 = 0$$

which may be written as

$$(2e^x - 1)^2 = 0$$

from which we deduce

$$e^x = \tfrac{1}{2}$$

and hence

$$x = -\ln 2$$

2.7.4 Inverse hyperbolic functions

The inverse hyperbolic functions, illustrated in Figure 2.78, are defined in a completely natural way:

$$y = \sinh^{-1}x \quad (x \text{ in } \mathbb{R})$$
$$y = \cosh^{-1}x \quad (x \geqslant 1, y \geqslant 0)$$
$$y = \tanh^{-1}x \quad (-1 < x < 1)$$

(These are also sometimes denoted as arsinh x, arcosh x and artanh x – *not* arcsinh x, etc.) Note the restriction on the range of the inverse hyperbolic cosine to meet the condition that exactly one value of y be obtained. These functions, not surprisingly, can be expressed in terms of logarithms.

For example,

$$y = \sinh^{-1}x \quad \text{implies} \quad x = \sinh y = \tfrac{1}{2}(e^y - e^{-y})$$

Thus

$$(e^y)^2 - 2x(e^y) - 1 = 0$$

and

$$e^y = x \pm \sqrt{(x^2 + 1)}$$

Since $e^y > 0$, we can discount the negative root, and we have, on taking logarithms,

$$y = \sinh^{-1}x = \ln[x + \sqrt{(x^2 + 1)}] \tag{2.40}$$

Similarly,

$$\cosh^{-1}x = \ln[x + \sqrt{(x^2 - 1)}] \quad (x \geqslant 1) \tag{2.41}$$

Figure 2.78
Graphs of the inverse
hyperbolic functions.

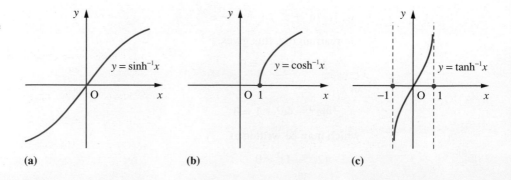

(a)   (b)   (c)

and

$$\tanh^{-1}x = \tfrac{1}{2}\ln\left(\frac{1+x}{1-x}\right) \quad (-1 < x < 1) \tag{2.42}$$

Example 2.54 Evaluate (to 4sf)

(a) $\sinh^{-1}(0.5)$ (b) $\cosh^{-1}(3)$ (c) $\tanh^{-1}(-2/5)$

using the logarithmic forms of these functions. Check your answers directly using a calculator.

Solution (a) Using formula (2.40), we have

$$\sinh^{-1}(0.5) = \ln[0.5 + \sqrt{(0.25 + 1)}]$$

$$= \ln(0.5 + 1.118\,034)$$

$$= \ln(1.618\,034)$$

$$= 0.4812$$

(b) Using formula (2.41), we have

$$\cosh^{-1}(3) = \ln(3 + \sqrt{8})$$

$$= 1.7627$$

(c) Using formula (2.42), we have

$$\tanh^{-1}(-2/5) = \frac{1}{2}\ln\left(\frac{1 - \frac{2}{5}}{1 + \frac{2}{5}}\right)$$

$$= \frac{1}{2}\ln\left(\frac{5 - 2}{5 + 2}\right)$$

$$= \frac{1}{2}\ln\frac{3}{7} = -0.4236$$

2.7.5 Exercises

56 Simplify

(a) $(e^2)^3 + e^2 \times e^3 + (e^3)^2$ (b) e^{7x}/e^{3x}

(c) $(e^3)^2$ (d) $\exp(3^2)$ (e) $\sqrt{(e^x)}$

57 Sketch the graphs of $y = e^{-2x}$ and $y = e^{-x^2}$ on the same axes. Note that $(e^{-x})^2 \neq e^{-x^2}$.

58 Find the following logarithms *without* using a calculator:

(a) $\log_2 8$ (b) $\log_2\frac{1}{4}$

(c) $\log_2\frac{1}{\sqrt{2}}$ (d) $\log_3 81$

(e) $\log_9 3$ (f) $\log_4 0.5$

59 Express in terms of $\ln x$ and $\ln y$

(a) $\ln(x^2 y)$ (b) $\ln \sqrt{(xy)}$ (c) $\ln(x^5/y^2)$

60 Express as a single logarithm

(a) $\ln 14 - \ln 21 + \ln 6$

(b) $4 \ln 2 - \frac{1}{2} \ln 25$

(c) $1.5 \ln 9 - 2 \ln 6$

(d) $2 \ln(2/3) - \ln(8/9)$

61 Simplify (a) $\exp\left\{\frac{1}{2} \ln\left[\dfrac{1-x}{1+x}\right]\right\}$ (b) $e^{2 \ln x}$

62 Sketch carefully the graphs of the functions

(a) $y = 2^x$, $y = \log_2 x$ (on the same axes)

(b) $y = e^x$, $y = \ln x$ (on the same axes)

(c) $y = 10^x$, $y = \log x$ (on the same axes)

63 Express $\ln y$ as simply as possible when

$$y = \frac{(x^2 + 1)^{3/2}}{(x^4 + 1)^{1/3}(x^4 + 4)^{1/5}}$$

64 In each of the following exercises a value of one of the six hyperbolic functions of x is given. Find the remaining five.

(a) $\cosh x = \frac{5}{4}$ (b) $\sinh x = \frac{8}{15}$

(c) $\tanh x = -\frac{7}{25}$ (d) $\operatorname{sech} x = \frac{5}{13}$

(e) $\operatorname{cosech} x = -\frac{3}{4}$ (f) $\coth x = \frac{13}{12}$

65 Use Osborn's rule to write down formulae corresponding to

(a) $\tan 3x = \dfrac{(3 - \tan^2 x)\tan x}{1 - 3 \tan^2 x}$

(b) $\cos(x + y) = \cos x \cos y - \sin x \sin y$

(c) $\cosh 2x = 1 + 2 \sinh^2 x$

(d) $\sin x - \sin y = 2 \sin \frac{1}{2}(x - y) \cos \frac{1}{2}(x + y)$

66 Prove that

(a) $\cosh^{-1} x = \ln[x + \sqrt{(x^2 - 1)}]$ $(x \geqslant 1)$

(b) $\tanh^{-1} x = \frac{1}{2} \ln\left(\dfrac{1 + x}{1 - x}\right)$ $(|x| < 1)$

67 Find to 4dp

(a) $\sinh^{-1} 0.8$

(b) $\cosh^{-1} 2$

(c) $\tanh^{-1}(-0.5)$

68 The speed V of waves in shallow water is given by

$$V^2 = 1.8L \tanh \frac{6.3d}{L}$$

where d is the depth and L the wavelength. If $d = 30$ and $L = 270$, calculate the value of V.

69 The formula

$$\lambda = \frac{\alpha t}{2} \frac{\sinh \alpha t + \sin \alpha t}{\cosh \alpha t - \cos \alpha t}$$

gives the increase in resistance of strip conductors due to eddy currents at power frequencies. Calculate λ when $\alpha = 1.075$ and $t = 1$.

70 The functions

$$f_1(x) = \frac{1}{1 + e^{-x}}, \quad f_2(x) = \frac{1}{2} \tanh \frac{1}{2} x$$

are two different forms of activating functions representing the output of a neuron in a typical neural network. Sketch the graphs of $f_1(x)$ and $f_2(x)$ and show that $f_1(x) - f_2(x) = \frac{1}{2}$.

71 The potential difference E (in V) between a telegraph line and earth is given by

$$E = A \cosh\left(x\sqrt{\frac{r}{R}}\right) + B \sinh\left(x\sqrt{\frac{r}{R}}\right)$$

where A and B are constants, x is the distance in km from the transmitting end, r is the resistance per km of the conductor and R is the insulation resistance per km. Find the values of A and B when the length of the line is 400 km, $r = 8\,\Omega$, $R = 3.2 \times 10^7\,\Omega$ and the voltages at the transmitting and receiving ends are 250 and 200 V respectively.

72 Sketch the graph of $y = e^{-x} - e^{-2x}$. Prove that the maximum of y is $\frac{1}{4}$ and find the corresponding value of x. Find the two values of x corresponding to $y = \frac{1}{40}$.

2.8   Irrational functions

The circular and exponential functions are examples of **transcendental functions**. They cannot be expressed as rational functions, that is, as the quotient of two polynomials. Other irrational functions occur in engineering, and they may be classified either as algebraic or as transcendental functions. An algebraic irrational function is a function with a formula $y = f(x)$ that is the root of a polynomial equation in y whose coefficients are rational functions of x. For example

$$y = \frac{\sqrt{(x+1)} - 1}{\sqrt{(x+1)} + 1} \quad (x \geq -1)$$

is an algebraic irrational function, since y is a root of

$$xy^2 - 2(2+x)y + x = 0$$

On the other hand, $y = |x|$, although it satisfies $y^2 = x^2$, is not a root of that equation (whose roots are $y = x$ and $y = -x$). The modulus function $|x|$ is an example of a non-algebraic irrational function.

2.8.1   Algebraic functions

In general we have an algebraic function $y = f(x)$ defined when y is the root of a polynomial equation of the form

$$a_n(x)y^n + a_{n-1}(x)y^{n-1} + \ldots + a_1(x)y + a_0(x) = 0$$

Note that here all the coefficients $a_0 \ldots a_n$ may be polynomial functions of the independent variable x. For example, consider

$$y^2 - 2xy - 8x = 0$$

This defines, for $x \geq 0$, two algebraic functions with formulae

$$y = x + \sqrt{(x^2 + 8x)} \quad \text{and} \quad y = x - \sqrt{(x^2 + 8x)}$$

One of these corresponds to $y^2 - 2xy - 8x = 0$ with $y \geq 0$ and the other to $y^2 - 2xy - 8x = 0$ with $y \leq 0$. So, when we specify a function implicitly by means of an equation we often need some extra information to define it uniquely. Often, too, we cannot obtain an explicit algebraic formula for y in terms of x and we have to evaluate the function at each point of its domain by solving the polynomial equation for y numerically.

Care has to be exercised when using algebraic functions in a larger computation in case special values of parameters produce sudden changes in value, as illustrated in Example 2.55.

Example 2.55   Sketch the graphs of the function

$$y = \sqrt{(a + bx^2 + cx^3)}/(d - x)$$

for the domain $-3 < x < 3$ where

(a) $a = 18$, $b = 1$, $c = -1$ and $d = 6$

(b) $a = 0$, $b = 1$, $c = -1$ and $d = 0$

Solution   (a) $y = \sqrt{(18 + x^2 - x^3)}/(6 - x)$

We can see that the term inside the square root is positive only when $18 + x^2 - x^3 > 0$. Since we can factorize this $(18 + x^2 - x^3) = (3 - x)(x^2 + 2x + 6)$ we deduce that y is not defined for $x > 3$. Also, for large negative values of x it behaves like $\sqrt{(-x)}$. A sketch of the graph is shown in Figure 2.79.

Figure 2.79
Graph of $y = \sqrt{(18 + x^2 - x^3)}/(6 - x)$.

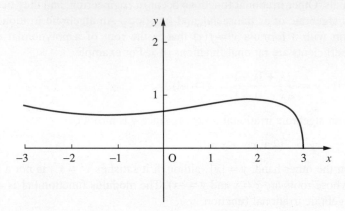

(b) $y = -\sqrt{(x^2 - x^3)}/x$

Here we can see that the function is defined for $x \leqslant 1$, $x \neq 0$. Near $x = 0$, since we can write $x = \sqrt{x^2}$ for $x > 0$ and $x = -\sqrt{x^2}$ for $x < 0$, we see that

$$y = -\sqrt{(1 - x)} \quad \text{for } x > 0$$

and

$$y = \sqrt{(1 - x)} \quad \text{for } x < 0$$

At $x = 0$ the function is not defined. The graph of the function is shown in Figure 2.80.

Figure 2.80
Graph of $y = -\sqrt{(x^2 - x^3)}/x$.

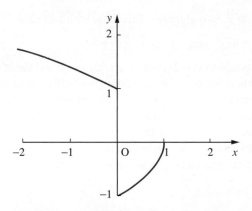

2.8.2 Implicit functions

We have seen in Section 2.8.1 that some algebraic functions are defined implicitly because we cannot obtain an algebraic formula for them. This applies to a wider class of functions where we have an equation relating the dependent and independent variables, but where finding the value of y corresponding to a given value of x requires a numerical solution of the equation. Generally we have an equation connecting x and y such as

$$f(x, y) = 0$$

Sometimes we are able to draw a curve which represents the relationship (using algebraic methods), but more commonly we have to calculate for each value of x the corresponding value of y. Most computer graphics packages have an implicit function option which will perform the task efficiently.

Example 2.56

The concentrations of two substances in a chemical process are related by the equation

$$xye^{2-y} = 2e^{x-1}, \quad 0 < x < 3, 0 < y < 3$$

Investigate this relationship graphically and discover whether it defines a function.

Solution

Separating the variables in the equation, we have

$$ye^{-y} = 2e^{-3}e^{x}/x$$

Substituting $u = e^x/x$ and $v = ye^{-y}$ reduces this equation to

$$v = 2e^{-3}u$$

so on the u–v plane the relationship is represented by a straight line. Putting the first quadrant of the four planes x–y, v–y, u–x, u–v together we obtain the diagram shown in Figure 2.81. From that diagram it is clear that the smallest value of u that occurs is at P and the largest value of v that occurs is at Q, so all the solutions of the equation lie

Figure 2.81

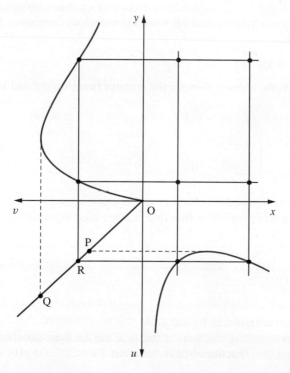

Figure 2.82

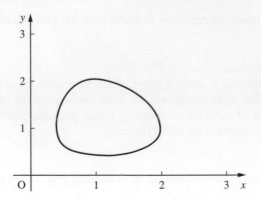

between P and Q. Any point R which lies between P and Q on the line corresponds to two values of y and two values of x. So each point R corresponds to four points of the $x–y$ plane. By considering all the points between P and Q we obtain the closed curve shown in Figure 2.82. We can see from that diagram that the equation does not define a function, since one value of x can give rise to two values of y. It is, of course, possible to specify the range of y and obtain, in this case, two functions, one for $y \geqslant 1$ and the other for $y \leqslant 1$.

2.8.3 Piecewise defined functions

Such functions often occur in the mathematical models of practical problems. For example, friction always opposes the motion of an object, so that the force F is $-R$ when the velocity v is positive and $+R$ when the velocity is negative. To represent the force, we can write

$$F = -R \, \text{sgn} \, (v)$$

where sgn is the abbreviation for the **signum function** defined by

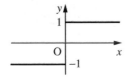

Figure 2.83
$y = \text{sgn} \, x$.

$$\text{sgn}(x) = \begin{cases} +1 & (x > 0) \\ -1 & (x < 0) \\ 0 & (x = 0) \end{cases}$$

and shown in Figure 2.83. The signum function is used in modelling relays.

The **Heaviside unit step function** is often used in modelling physical systems. It is defined by

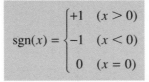

Figure 2.84
$y = H(x)$.

$$H(x) = \begin{cases} 0 & (x < 0) \\ 1 & (x \geqslant 0) \end{cases} \tag{2.43}$$

and its graph is shown in Figure 2.84.

Three other useful functions of this type are the **floor function** $\lfloor x \rfloor$, the **ceiling function** $\lceil x \rceil$ and the **fractional-part function** FRACPT (x). (In older textbooks $[x]$ is denoted by [*x*] and is sometimes called the **integer-part function**.) These are defined by

$$\lfloor x \rfloor = \text{greatest integer not greater than } x \tag{2.44}$$

$$\lceil x \rceil = \text{least integer not less than } x \tag{2.45}$$

and

$$\text{FRACPT}(x) = x - \lfloor x \rfloor \tag{2.46}$$

These definitions need to be interpreted with care. Notice, for example, that

$$\lfloor 3.43 \rfloor = 3$$

while

$$\lfloor -3.43 \rfloor = -4$$

Similarly,

$$\text{FRACPT}(3.43) = 0.43 \quad \text{and} \quad \text{FRACPT}(-3.43) = 0.57$$

The graphs of these functions are shown in Figure 2.85.

Care must be exercised when using the integer-part and fractional-part functions. Some calculators and computer implementations are different from the above definitions.

Figure 2.85
The graphs of the 'floor', 'ceiling' and 'fractional-part' functions.

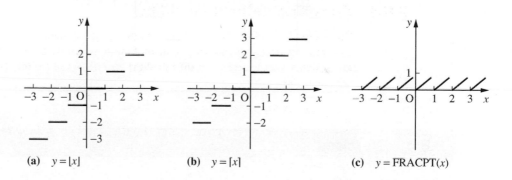

(a) $y = \lfloor x \rfloor$ (b) $y = \lceil x \rceil$ (c) $y = \text{FRACPT}(x)$

Example 2.57 Sketch the graphs of the functions with formula $y = f(x)$, where $f(x)$ is

(a) $H(x-1) - H(x-2)$ (b) $\lfloor x \rfloor - 2\lfloor \frac{1}{2}x \rfloor$

Solution (a) From the definition (2.43) of the Heaviside unit function $H(x)$ as

$$H(x) = \begin{cases} 0 & (x < 0) \\ 1 & (x \geqslant 0) \end{cases}$$

the effect of composing it with the linear function $f(x) = x - 1$ is to shift its graph one unit to the right, as shown in Figure 2.86(a). Similarly, $H(x - 2)$ has the same graph as $H(x)$, but shifted two units to the right (Figure 2.86(b)). Combining the graphs in

Figure 2.86

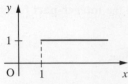

(a) $y = H(x - 1)$

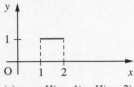

(b) $y = H(x - 2)$ (c) $y = H(x - 1) - H(x - 2)$

Figures 2.86(a) and (b), we can find the graph of their difference, $H(x - 1) - H(x - 2)$ as illustrated in Figure 2.86(c). Analytically, we can write this as

$$H(x - 1) - H(x - 2) = \begin{cases} 0 & (x < 1) \\ 1 & (1 \leqslant x < 2) \\ 0 & (x \geqslant 2) \end{cases}$$

(b) The graphs of $\lfloor x \rfloor$ and $2\lfloor \frac{1}{2}x \rfloor$ are shown in Figure 2.87. Combining these, we can find the graph of their difference.

Figure 2.87

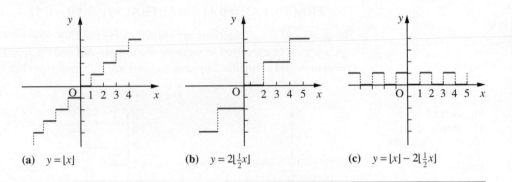

(a) $y = \lfloor x \rfloor$ (b) $y = 2\lfloor \frac{1}{2}x \rfloor$ (c) $y = \lfloor x \rfloor - 2\lfloor \frac{1}{2}x \rfloor$

2.8.4 Exercises

73 Sketch the graphs of the functions

(a) $y = \sqrt{(x^2)}$

(b) $y = \sqrt{(x^2 + x^3)}$, $x \geqslant -1$

(c) $y = x\sqrt{(1 + x)}$, $x \geqslant -1$

(d) $y = \sqrt{(1 + x)} + \sqrt{(1 - x)}$, $-1 \leqslant x \leqslant 1$

74 Sketch the curves represented by

(a) $y^2 = x(x^2 - 1)$

(b) $y^2 = (x - 1)(x - 3)/x^2$

75 Sketch the curves represented by the following equations, locating their turning points and asymptotes:

(a) $x^3 + y^3 = 6x^2$ (b) $y^2 = \dfrac{x^2}{x - 1}$

76 Sketch the graphs of

 (a) $y = |x|$

(b) $y = \frac{1}{2}(x + |x|)$

(c) $y = |x + 1|$

(d) $y = |x| + |x + 1| - 2|x + 2| + 3$

(e) $|x + y| = 1$

77 Sketch the graph of the functions $f(x)$ with formulae

(a) $f(x) = \dfrac{ax}{l}H(x)$

(b) $f(x) = \dfrac{ax}{l}[H(x) - H(x - l)]$

(c) $f(x) = \dfrac{ax}{l}H(x) - \dfrac{a}{l}(x - l)H(x - l)$

(d) $f(x) = \dfrac{ax}{l}H(x) - \dfrac{2a}{l}(x - l)H(x - l)$

78 Show that the function $g(x) = [H(x - a) - H(x - b)]\,f(x)$, $a < b$, may alternatively be expressed as

$$g(x) = \begin{cases} 0 & (x < a) \\ f(x) & (a \le x < b) \\ 0 & (x \ge b) \end{cases}$$

In other words, $g(x)$ is a function that is identical with the function $f(x)$ in the interval $[a, b]$ and zero elsewhere. Hence express as simply as possible in terms of Heaviside functions the function defined by

$$f(x) = \begin{cases} 0 & (x < 0) \\ \dfrac{ax}{l} & (0 \le x \le l) \\ \dfrac{a(2l - x)}{l} & (l \le x \le 2l) \\ 0 & (x \ge 2l) \end{cases}$$

79 Sketch the graph of the function

$$y = \begin{cases} x & (x \le 0) \\ 0 & (0 < x \le 1) \\ 1 - x & (1 < x) \end{cases}$$

Express the formula for y in terms of Heaviside functions.

80 The function INT(x) is defined as the 'nearest integer to x, with rounding up in the ambiguous case'. Sketch the graph of this function and express it in terms of $\lfloor x \rfloor$.

81 Sketch the graphs of the functions

(a) $y = \lfloor x \rfloor - \lfloor x - \tfrac{1}{2} \rfloor$

(b) $y = |\,\mathrm{FRACPT}(x) - \tfrac{1}{2}\,|$

82 It is a familiar observation that spoked wheels do not always appear to be rotating at the correct speed when seen on films. Show that if a wheel has s spokes and is rotating at n revolutions per second, and the camera operates at f frames per second, then the image of the wheel appears to rotate at N revolutions per second, where

$$N = \frac{f}{s}\left[\mathrm{FRACPT}\left(\frac{sn}{f} - \frac{1}{2} \right) - \frac{1}{2} \right]$$

Hence explain the illusion.

2.9 Numerical evaluation of functions

The introduction of calculators has greatly eased the burden of the numerical evaluation of functions. Often, however, the functions encountered in solving practical problems are not standard ones, and we have to devise methods of representing them numerically. The simplest method is to use a graph, a second method is to draw up a table of values of the function, and the third method is to give an analytical approximation to the function in terms of simpler functions. To illustrate this, consider the function e^{-x}. We can represent this by a graph as shown in Figure 2.88.

To evaluate the function for a given value of x, we read the corresponding value of y from the graph. For example, $x = 0.322$ gives $y = 0.73$ or thereabouts. Alternatively, we can tabulate the function as shown in Figure 2.89. Note that the notation $x = 0.00(0.05)0.50$ means for x from 0.00 to 0.50 in steps of 0.05.

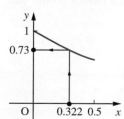

Figure 2.88
The graph of $y = e^{-x}$ for $0 \le x \le 0.5$.

Figure 2.89
Table of e^{-x} values for
$x = 0.00(0.05)0.50$.

x	0.00	0.05	0.10	0.15	0.20	0.25	0.30	0.35	0.40	0.45	0.50
e^{-x}	1.0000	0.9512	0.9048	0.8607	0.8187	0.7788	0.7408	0.7047	0.6703	0.6376	0.6065

To evaluate the function for a given value of x, we interpolate linearly within the table of values, to obtain the value of y. For example, $x = 0.322$ gives

$$y \approx 0.7408 + \frac{0.322 - 0.30}{0.35 - 0.30}(0.7047 - 0.7408)$$

$$= 0.7408 + (0.44)(-0.0361) = 0.7480 - 0.015\,884$$

$$= 0.7249$$

Another way of representing the function is to use the approximation

$$e^{-x} \approx \frac{x^2 - 6x + 12}{x^2 + 6x + 12}$$

which will be obtained in Section 7.11, Example 7.31. Setting $x = 0.322$ gives

$$y \approx \frac{(0.322 - 6)0.322 + 12}{(0.322 + 6)0.322 + 12} = \frac{10.171\,684}{14.035\,684}$$

$$= 0.724\,70 \ldots$$

The question remains as to how accurate these representations of the function are. The graphical method of representation has within it an implicit error bound. When we read the graph, we make a judgement about the number of significant digits in the answer. In the other two methods it is more difficult to assess the error – but it is also more important, since it is easy to write down more digits than can be justified. Are the answers correct to one decimal place or two, or how many? We shall discuss the accuracy of the tabular representation now and defer the algebraic approximation case until Section 7.11.

2.9.1 Tabulated functions and interpolation

To estimate the error involved in evaluating a function from a table of values as above, we need to look more closely at the process involved. Essentially the process assumes that the function behaves like a straight line between tabular points, as illustrated in Figure 2.90. Consequently it is called **linear interpolation**. The error involved depends on how closely a linear function approximates the function between tabular points, and this in turn depends on how close the tabular points are.

Figure 2.90
Linear interpolation
for e^{-x}
$(0.30 < x < 0.35)$.

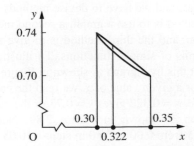

If the distance h between tabular points is sufficiently small, most functions arising from applications of mathematics behave locally like linear functions; that is to say, the error involved in approximating to the function between tabular points by a linear function is less than a rounding error. (Note that we have to use a different linear function between each consecutive pair of values of the function. We have a **piecewise-linear approximation**.) This, however, is a qualitative description of the process, and we need a quantitative description. In general, consider the function $f(x)$ with values $f_i = f(x_i)$ where $x_i = x_0 + ih$, $i = 0, 1, 2, \ldots, n$. To calculate the value $f(x)$ at a non-tabular point, where $x = x_i + \theta h$ and $0 < \theta < 1$, using linear interpolation, we have

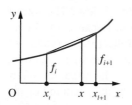

Figure 2.91

$$f(x) \approx f_i + \frac{x - x_i}{x_{i+1} - x_i}(f_{i+1} - f_i) \tag{2.47}$$

as shown in Figure 2.91.

The formula (2.47) may be written in a number of different ways, but it always gives the same numerical result. The form used will depend on the computational context. Thus we may write

$$f(x) \approx f_i + \theta(f_{i+1} - f_i), \quad \text{where } \theta = \frac{x - x_i}{x_{i+1} - x_i} \text{ and } 0 < \theta < 1 \tag{2.48}$$

or

$$f(x) \approx \frac{x - x_{i+1}}{x_i - x_{i+1}}f_i + \frac{x - x_i}{x_{i+1} - x_i}f_{i+1} \quad \text{(Lagrange's form)} \tag{2.49}$$

The difference $f_{i+1} - f_i$ between successive values in the table is often denoted by Δf_i, so that (2.47) may be rewritten as

$$f(x) \approx f_i + \theta \Delta f_i$$

Example 2.58

Use linear interpolation and the data of Figure 2.89 to estimate the value of

(a) e^{-x} where $x = 0.235$ (b) x where $e^{-x} = 0.7107$

Solution

(a) From the table of values in Figure 2.89 we see that $x = 0.235$ lies between the tabular points $x = 0.20$ and $x = 0.25$. Applying the formula (2.47) with $x_i = 0.20$, $x_{i+1} = 0.25$, $f_i = 0.8187$ and $f_{i+1} = 0.7788$ we have

$$f(0.235) \approx 0.8187 + \frac{0.235 - 0.20}{0.25 - 0.20}(0.7788 - 0.8187) = 0.7868$$

(b) From the table of values we see that $e^{-x} = 0.7107$ occurs between $x = 0.30$ and $x = 0.35$. Thus the value of x is given, using formula (2.47), by the equation

$$0.7107 \approx 0.7408 + \frac{x - 0.30}{0.35 - 0.30}(0.7047 - 0.7408)$$

Hence

$$x \approx \frac{0.7107 - 0.7408}{0.7047 - 0.7408}(0.35 - 0.30) + 0.30 = 0.3417$$

The difficulty with both the estimates obtained in Example 2.58 is that we do not know how accurate the answers are. Are they correct to 4dp or 3dp or less? The size of the error in the answer depends on the curvature of the function. Because any linear interpolation formula is, by definition, a straight line it cannot reflect the curvature of the function it is trying to model. In order to model curvature a parabola is required, that is a quadratic interpolating function. The difference between the quadratic interpolation formula and the linear formula will give us a measure of the accuracy of the linear formula. We have

function value = linear interpolation value + C_1

and

function value = quadratic interpolation value + C_2

where ideally C_2 is very much smaller than C_1. Subtracting these equations we see that

$C_1 \approx$ quadratic interpolation value – linear interpolation value

Now to determine a quadratic function we require three points. Using formula (2.10) obtained earlier, we see that the quadratic function which passes through (x_i, f_i), (x_{i+1}, f_{i+1}) and (x_{i+2}, f_{i+2}) may be expressed as

$$p(x) = \frac{(x - x_{i+1})(x - x_{i+2})f_i}{(x_i - x_{i+1})(x_i - x_{i+2})} + \frac{(x - x_i)(x - x_{i+2})f_{i+1}}{(x_{i+1} - x_i)(x_{i+1} - x_{i+2})}$$

$$+ \frac{(x - x_i)(x - x_{i+1})f_{i+2}}{(x_{i+2} - x_i)(x_{i+2} - x_{i+1})}$$

We can simplify $p(x)$, when the data points are equally spaced, by remembering that $x_{i+2} = x_i + 2h$, $x_{i+1} = x_i + h$ and $x = x_i + \theta h$, with $0 \leqslant \theta \leqslant 1$, giving

$$p(x) = \frac{(\theta - 1)(\theta - 2)}{2}f_i - \frac{\theta(\theta - 2)}{1}f_{i+1} + \frac{\theta(\theta - 1)}{2}f_{i+2}$$

This formula looks intimidatingly unlike that for linear interpolation, but, after some rearrangement, we have

$$p(x) = [f_i + \theta(f_{i+1} - f_i)] + \tfrac{1}{2}\theta(\theta - 1)(f_{i+2} - 2f_{i+1} + f_i)$$

$$= [f_i + \theta\Delta f_i] + \tfrac{1}{2}\theta(\theta - 1)(\Delta f_{i+1} - \Delta f_i)$$

where $0 < \theta < 1$. Here the term in square brackets is the linear interpolation approximation to $f(x)$, so that

$$\tfrac{1}{2}\theta(\theta - 1)(\Delta f_{i+1} - \Delta f_i)$$

is the quadratic correction for that approximation (remember: the correction is added to eliminate the error). Note that this involves the difference of two successive differences, so we may write it as $\frac{1}{2}\theta(\theta - 1)\Delta^2 f_i$, where $\Delta^2 f_i = \Delta(\Delta f_i) = \Delta f_{i+1} - \Delta f_i$.

Error in linear interpolation

We can use this to estimate the error in linear interpolation for a function. If

$$f(x) \approx f_i + \theta\Delta f_i + \tfrac{1}{2}\theta(\theta - 1)\Delta^2 f_i$$

in the interval $[x_i, x_{i+1}]$ then the error in using the linear interpolation

$$f(x) \approx f_i + \theta\Delta f_i$$

will be approximately $\frac{1}{2}\theta(\theta - 1)\Delta^2 f_i$, and an estimate of the error bound of the linear approximation is given by

$$\max_{0 \leqslant \theta \leqslant 1} \left[|\tfrac{1}{2}\theta(\theta - 1)\Delta^2 f_i| \right]$$

Now $\theta(\theta - 1) = (\theta - \tfrac{1}{2})^2 - \tfrac{1}{4}$, so that $\max\limits_{0 \leqslant \theta \leqslant 1} |\theta(\theta - 1)| = \tfrac{1}{4}$, and our estimate of the error bound is

$$\tfrac{1}{8}|\Delta^2 f_i|$$

For accurate linear interpolation we require this error bound to be less than a rounding error. That is, it must be less than $\frac{1}{2}$ unit in the least significant figure. This implies

$$\tfrac{1}{8}|\Delta^2 f_i| < \tfrac{1}{2} \text{ unit of least significant figure}$$

giving the condition

$$|\Delta^2 f_i| < 4 \text{ units of the least significant figure}$$

for linear interpolation to yield answers as accurate as those in the original table.

Thus, from the table of values of the function e^{-x} shown in Figure 2.89 we can construct the table shown in Figure 2.92. The final row shows the estimate of the maximum error incurred in linear interpolation within each interval $[x_i, x_{i+1}]$. In order to complete the table with error estimates for the intervals [0.00, 0.05] and [0.45, 0.50], we should need values of e^{-x} for $x = -0.05$ and 0.55. From the information we have in Figure 2.92 we can say that the largest error likely in using linear interpolation from this table of 11 values of e^{-x} is approximately 3 units in the fourth decimal place. Values obtained could therefore safely be quoted to 3dp.

i	0	1	2	3	4	5	6	7	8	9	10
x_i	0.00	0.05	0.10	0.15	0.20	0.25	0.30	0.35	0.40	0.45	0.50
e^{-x_i}	1.0000	0.9512	0.9048	0.8607	0.8187	0.7788	0.7408	0.7047	0.6703	0.6376	0.6065
$\tfrac{1}{8}|\Delta^2 f_i|$		0.000 29	0.000 28	0.000 26	0.000 25	0.000 24	0.000 23	0.000 21	0.000 20		

Figure 2.92 Table of values of e^{-x}, with error estimates for linear interpolation.

Critical tables

An ordinary table of values uses equally spaced values of the independent variable and tabulates the corresponding values of the dependent variable (the function values). A **critical table** gives the function values at equal intervals, usually a unit of the last decimal place, and then tabulates the limits between which the independent variable gives each value. Thus, for example, $\cos x° = 0.999$ for $1.82 \leqslant x < 3.14$ and $\cos x° = 0.998$ for $3.14 \leqslant x < 4.06$ and so on. Thus we obtain the table of values shown in Figure 2.93. If a value of the independent variable falls between two tabular values, the value of the dependent variable is that printed between these values. Thus $\cos 2.62° = 0.999$. The advantages of critical tables are that they do not require interpolation, they always give answers that are accurate to within half a unit of the last decimal place and they require less space.

Figure 2.93
A critical table
for $\cos x°$.

x	0.00		1.82		3.14		4.06		4.80		5.44
$\cos x°$		1.000		0.999		0.998		0.997		0.996	

2.9.2 Exercises

83 Tabulate the function $f(x) = \sin x$ for $x = 0.0(0.2)1.6$. From this table estimate, by linear interpolation, the value of $\sin 1.23$. Construct a table equivalent to Figure 2.92, and so estimate the error in your value of $\sin 1.23$. Use a pocket calculator to obtain a value of $\sin 1.23$ and compare this with your estimates.

84 Tabulate the function $f(x) = x^3$ for $x = 4.8(0.1)5.6$. Construct a table equivalent to Figure 2.92, and hence estimate the largest error that would be incurred in using linear interpolation in your table of values over the range [5.0, 5.4]. Construct a similar table for $x = 4.8(0.2)5.6$ (that is, for linear interpolation with twice the tabulation interval) and estimate the largest error that would be incurred by linear interpolation from this table in the range [5.0, 5.4]. What do you think the maximum error in interpolating in a similar table formed for $x = 4.8(0.05)5.6$ might be? What tabulation interval do you think would be needed to allow linear interpolation accurate to 3dp?

85 The function $f(x)$ is tabulated at unequal intervals as follows:

x	15	18	20
$f(x)$	0.2316	0.3464	0.4864

Use linear interpolation to estimate $f(17)$, $f(16.34)$ and $f^{-1}(0.3)$.

86 Assess the accuracy of the answers obtained in Question 85 using quadratic interpolation (Lagrange's formula, (2.10)).

87 Show that Lagrange's interpolation formula for cubic interpolation (see Section 2.4) is

$$f(x) = \frac{(x - x_1)(x - x_2)(x - x_3)}{(x_0 - x_1)(x_0 - x_2)(x_0 - x_3)} f_0$$

$$+ \frac{(x - x_0)(x - x_2)(x - x_3)}{(x_1 - x_0)(x_1 - x_2)(x_1 - x_3)} f_1$$

$$+ \frac{(x - x_0)(x - x_1)(x - x_3)}{(x_2 - x_0)(x_2 - x_1)(x_2 - x_3)} f_2$$

$$+ \frac{(x - x_0)(x - x_1)(x - x_2)}{(x_3 - x_0)(x_3 - x_1)(x_3 - x_2)} f_3$$

Use this formula to find a cubic polynomial that fits the function f given in the following table:

x	−1	0	1	8
$f(x)$	−1	0	1	2

Draw the graph of the cubic for $-1 < x < 8$ and compare it with the graph of $y = x^{1/3}$.

88 Construct a critical table for

$$y = \sqrt[3]{x}$$

for $y = 14.50(0.01)14.55$.

2.10   Engineering application: a design problem

Mathematics plays an important role in engineering design. We shall illustrate how some of the elementary ideas described in this chapter are used to produce optimal designs. Consider the open container shown in Figure 2.94. The base and long sides are constructed from material of thickness t cm and the short sides from material of thickness $3t$ cm. The internal dimensions of the container are l cm $\times b$ cm $\times h$ cm. The design problem is to produce a container of a given capacity that uses the least amount of material. (Mass production of such items implies that small savings on individual items produce large savings in the bulk product.) First we obtain an expression for the volume A of material used in the manufacture of the container.

Figure 2.94

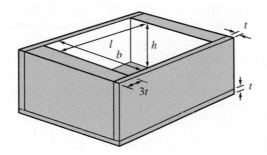

The capacity C of the box is $C(l, b, h) = lbh$. Then

$$A(l, b, h, t) = C(l + 6t, b + 2t, h + t) - C(l, b, h)$$

$$= (l + 6t)(b + 2t)(h + t) - lbh$$

$$= (lb + 6bh + 2hl)t + (2l + 6b + 12h)t^2 + 12t^3 \qquad \textbf{(2.50)}$$

For a specific design the thickness t of the material and the capacity K of the container would be specified, so, since $lbh = K$, we can define one of the variables l, b and h in terms of the other two. For example $l = K/bh$.

For various reasons, for example ease of handling, markcting display and so on, the manufacturer may impose other constraints on the design. We shall illustrate this by first considering a special case, and then look at the more general case.

Special case

Let us seek the optimal design of a container whose breadth b is four times its height h and whose capacity is $10\,000$ cm^3, using material of thickness 0.4 cm and 1.2 cm (so that $t = 0.4$). The function $f(h)$ that we wish to minimize is given by $A(l, b, h, t)$, where $t = 0.4$, $b = 4h$ and $lbh = 10\,000$ (so that $l = 2500/h^2$). Substituting these values in (2.50) gives, after some rearrangement,

$$f(h) = 9.6h^2 + 5.76h + 0.768 + 6000/h + 800/h^2$$

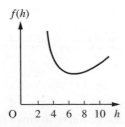

Figure 2.95

The graph of this function is shown in Figure 2.95. The graph has a minimum point *near* $h = 7$. We can obtain a better estimate for the optimal choice for h by approximating $f(h)$ locally by a quadratic function. Evaluating f at $h = 6, 7$ and 8 gives

$$f(6) = 1403.2, f(7) = 1385.0, f(8) = 1423.7$$

This shows clearly that the minimum value occurs between $h = 6$ and $h = 8$.
We approximate to $f(h)$ using a local quadratic approximation of the form

$$f(h) \simeq A(h-7)^2 + B(h-7) + C$$

Setting $h = 7$ gives $\qquad C = 1385.0$
Setting $h = 6$ gives $\quad A - B + C = 1403.2$
Setting $h = 8$ gives $\quad A + B + C = 1423.7$

Hence $C = 1385.0$, $A = 28.45$ and $B = 10.25$. The minimum of the approximating quadratic function occurs where $h - 7 = -B/(2A)$, that is, at $h = 7 - 0.18 = 6.82$. Thus the optimal choice for h is approximately 6.82 giving a value for $f(h)$ at that point of 1383.5.

The corresponding values for b and l are $b = 27.3$ and $l = 53.7$. Thus we have obtained an optimal design of the container in this special case.

General case

Here we seek the optimal design without restricting the ratio of b to h. For a container of capacity K, we have to minimize $A(l, b, h, t)$ subject to the constraint $C(l, b, h) = K$. Here

$$A(l, b, h, t) = (lb + 6bh + 2hl)t + (2l + 6b + 12h)t^2 + 12t^3$$

and

$$C(l, b, h) = lbh$$

These functions have certain algebraic symmetries that enable us to solve the problem algebraically. Consider the formula for A and set $x = 2h$ and $y = l/3$, then

$$A(l, b, h, t) = 3(by + bx + xy)t + 6(y + b + x)t^2 + 12t^3$$

$$= A^*(y, b, x, t)$$

and

$$C(l, b, h) = 3bxy/2$$

From this we can conclude that if $A^*(y, b, x, t)$ has a minimum value at (y_0, b_0, x_0) for a given value of t, then it has the same value at (x_0, b_0, y_0), (x_0, y_0, b_0), (y_0, x_0, b_0), (b_0, y_0, x_0) and (b_0, x_0, y_0). Assuming that the function has a unique minimum point, we conclude that these six points are the same, that is $b_0 = y_0 = x_0$. Thus we deduce that the minimum occurs where $l = 6h$ and $b = 2h$. Since the capacity is fixed, we have $lbh = K$, which implies that $12h^3 = K$.

Thus the optimal choice for h in the general case is $(\frac{1}{12}K)^{1/3}$.

Returning to the special case where $K = 10\,000$ and $t = 0.4$, we obtain an optimal design when

$$h = 9.41, \quad b = 18.82, \quad l = 56.46$$

using 1330.1 cm^3 of material. Note that the amount of material used is close to that used in the special case where $b = 4h$. This indicates that the design is not sensitive to small errors made during its construction.

2.11   Review exercises (1–22)

1 The functions f and g are defined by

$$f(x) = x^2 - 4 \quad (x \text{ in } [-20, 20])$$

$$g(x) = x^{1/2} \quad (x \text{ in } [0, 200])$$

Let $h(x)$ and $k(x)$ be the compositions $f \circ g(x)$ and $g \circ f(x)$ respectively. Determine $h(x)$ and $k(x)$. Is the composite function $k(x)$ defined for all x in the domain of $f(x)$? If not, then for what part of the domain of $f(x)$ is $k(x)$ defined?

2 The perimeter of an ellipse depends on the lengths of its major and minor axes, and is given by

$$\text{perimeter} = 2 \times (\text{major axis}) \times E(m)$$

where

$$m = \frac{(\text{major axis})^2 - (\text{minor axis})^2}{(\text{major axis})^2}$$

and E is the function whose graph is given in Figure 2.96.

(a) Calculate the perimeter of the ellipse whose axes are of length 10 cm and 6 cm.

(b) A fairing is to be made from sheet metal bent into the shape of an ellipse of major axis 55 cm and minor axis 13 cm, and is to be of length 2 m. Estimate the area of sheet metal required.

3 The sales volume of a product depends on its price as follows:

Price/£	1.00	1.05	1.10	1.15	1.20	1.25	1.30
Sales/000	8	7	6	5	4	3	2

The cost of production is £1 per unit. Draw up a table showing the sales revenue, the cost and the profits for each selling price, and deduce the selling price to be adopted.

4 A function f is defined by

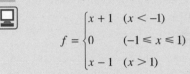

$$f = \begin{cases} x + 1 & (x < -1) \\ 0 & (-1 \leqslant x \leqslant 1) \\ x - 1 & (x > 1) \end{cases}$$

Draw the graphs of $f(x)$, $f(x - 2)$ and $f(2x)$. The function $g(x)$ is defined as $f(x + 2) - f(2x - 1)$. Draw a graph of $g(x)$.

5 The function $f(x)$ has formula $y = x^2$ for $0 \leqslant x < 1$. Sketch the graphs of $f(x)$ for $-4 < x < 4$ when

(a) $f(x)$ is periodic with period 1

(b) $f(x)$ is even and periodic with period 2

(c) $f(x)$ is odd and periodic with period 2

6 Assuming that all the numbers given are correctly rounded, calculate the positive root together with its error bound of the quadratic equation

$$1.4x^2 + 5.7x - 2.3 = 0$$

Give your answer also as a correctly rounded number.

7 Sketch the functions

(a) $x^2 - 4x + 7$ (b) $x^3 - 2x^2 + 4x - 3$

(c) $\dfrac{x + 4}{x^2 - 1}$ (d) $\dfrac{x^2 - 2x + 3}{x^2 + 2x - 3}$

8 Find the Taylor expansion of

$$x^4 + 3x^3 - x^2 + 2x - 1 \text{ about } x = 1$$

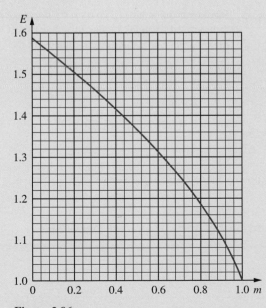

Figure 2.96

9 Find the partial fractions of

(a) $\dfrac{x+2}{(x-1)(x-4)}$ (b) $\dfrac{x^2+4}{(x+1)(x-3)}$

(c) $\dfrac{x^2-2x+3}{(x+2)^2(x-1)}$ (d) $\dfrac{x(2x-1)}{(x^2-x+1)(x+3)}$

10 Express as products of sines and/or cosines

(a) $\sin 2\theta - \sin \theta$ (b) $\cos 2\theta + \cos 3\theta$

(c) $\sin 4\theta - \sin 7\theta$

11 Express in the form $r\sin(\theta - \alpha)$

(a) $4\sin\theta - 2\cos\theta$ (b) $\sin\theta + 8\cos\theta$

(c) $\sqrt{3}\sin\theta + \cos\theta$

12 (a) From the definition of the hyperbolic sine function prove

$$\sinh 3x = 3\sinh x + 4\sinh^3 x$$

(b) Sketch the graph of $y = x^3 + x$ carefully, and show that for each value of y there is exactly one value of x. Setting $z = \tfrac{1}{2}x\sqrt{3}$, show that

$$4z^3 + 3z = \frac{3\sqrt{3}}{2}y$$

and using (a), deduce that

$$x = \frac{2}{\sqrt{3}}\sinh\left[\frac{1}{3}\sinh^{-1}\left(\frac{3\sqrt{3}}{2}y\right)\right]$$

13 The parts produced by three machines along a factory aisle (shown in Figure 2.97 as the x axis) are to be taken to a nearby bench for assembly before they undergo further processing. Each assembly takes one part from each machine. There is a fixed cost per metre for moving any of the parts. Show that if x represents the position of the assembly bench the cost $C(x)$ of moving the parts for each assembled item is given by

$$C(x) \propto d(x)$$

where $d(x) = |x+3| + |x-2| + |x-4|$

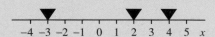

−4 −3 −2 −1   0   1   2   3   4   5 x

Figure 2.97

Draw the graph of $d(x)$ and find the optimal position of the bench.

14 Sketch the graphs of the functions

(a) $\lfloor \tfrac{1}{2}x \rfloor - \tfrac{3}{2}\lfloor \tfrac{1}{3}x \rfloor$

(b) $xH(x) - (x-1)H(x-1) + (x-2)H(x-2)$

15 Draw up a table of values of the function $f(x) = x^2 e^{-x}$ for $x = -0.1(0.1)1.1$. Determine the maximum error incurred in linearly interpolating for the function $f(x)$ in this table, and hence estimate the value of $f(0.83)$, giving your estimate to an appropriate number of decimal places.

16 By setting $t = \tan\tfrac{1}{2}x$, find the maximum value of $(\sin x)/(2 - \cos x)$.

17 (a) Show that a root x_0 of the equation

$$x^4 - px^3 + q = 0$$

is a repeated root if and only if

$$4x_0 - 3p = 0$$

(b) The stiffness of a rectangular beam varies with the cube of its height h and directly with its breadth b. Find the section of the beam that can be cut from a circular log of diameter D that has the maximum stiffness.

18 Starting at the point $(x_0, y_0) = (1, 0)$, a sequence of right-angled triangles is constructed as shown in Figure 2.98. Show that the coordinates of the vertices satisfy the recurrence relations

$$x_i = x_{i-1} - w_i y_{i-1}$$

$$y_i = w_i x_{i-1} + y_{i-1}$$

where $w_i = \tan\alpha_i^\circ$, $x_0 = 1$ and $y_0 = 0$.

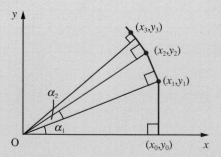

Figure 2.98

Any angle $0° < \theta° < 360°$ can be expressed in the form

$$\theta = \sum_{i=0}^{\infty} n_i\phi_i$$

where $\tan \phi_i° = 10^{-i}$ and n_i is a non-negative integer. Express $\theta = 56.5$ in this form and, using the recurrence relations above, calculate $\sin \theta°$ and $\cos \theta°$ to 5dp. (This method of calculating the trigonometric functions is used in some calculators.)

19 A mechanism consists of the linkage of three rods AB, BC and CD as shown in Figure 2.99, where

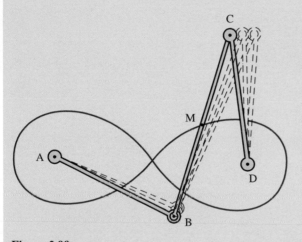

Figure 2.99

$AB = CD (= a,$ say$)$, $BC = AD = a\sqrt{2}$, and M is the midpoint. The rods are freely jointed at B and C, and are free to rotate about A and D. Using polar coordinates with their pole O at the midpoint of AD and initial line OD, show that the curve described by M as CD rotates about D is $r^2 = a^2 \cos 2\theta$. Draw a careful graph of this curve, the 'lemniscate' of Bernoulli.

Show that

(a) The cartesian coordinates of M satisfy

$$(x^2 + y^2)^2 = a^2(x^2 - y^2)$$

(b) $AM \times DM = \frac{1}{2}a^2$.

20 Show that the equation

$$r = p/\sin(\theta - \alpha)$$

represents a straight line which cuts the x axis at the angle α and whose perpendicular distance from the origin is p.

21 Use the result of Question 20 to find the polar coordinate representation of the line which passes through the points $(1, 2)$ and $(3, 3)$.

22 Show that the equation

$$r = ep/(1 + e\cos\theta)$$

where e and p are constants represents an ellipse where $0 < e < 1$, a parabola where $e = 1$ and a hyperbola where $e > 1$, the origin of the coordinate system being at a focus of the conic concerned.

3 Complex Numbers

3.1 Introduction

Complex numbers first arose in the solution of cubic equations in the sixteenth century using a method known as Cardano's solution. This gives the solution of the equation

$$x^3 + qx + r = 0$$

as

$$x = \sqrt[3]{[-\tfrac{1}{2}r + \sqrt{(\tfrac{1}{4}r^2 + \tfrac{1}{27}q^3)}]} + \sqrt[3]{[-\tfrac{1}{2}r - \sqrt{(\tfrac{1}{4}r^2 + \tfrac{1}{27}q^3)}]}$$

which may be verified by direct substitution. This solution gave difficulties when it unexpectedly involved square roots of negative numbers. For example, the equation

$$x^3 - 15x - 4 = 0$$

was known to have three roots. An obvious one is $x = 4$, but the corresponding root obtained using the formula was

$$x = \sqrt[3]{[2 + \sqrt{(-121)}]} + \sqrt[3]{[2 - \sqrt{(-121)}]}$$

Writing in 1572, Bombelli showed that

$$2 + \sqrt{(-121)} = [2 + \sqrt{(-1)}]^3$$

and

$$2 - \sqrt{(-121)} = [2 - \sqrt{(-1)}]^3$$

and so

$$x = [2 + \sqrt{(-1)}] + [2 - \sqrt{(-1)}] = 4$$

as expected. Since

$$\sqrt{(-x)} = \sqrt{(-1)}\sqrt{x}$$

where x is a positive number, the square roots of negative numbers can be represented as a number multiplied by $\sqrt{(-1)}$. Thus $\sqrt{(-121)} = 11\sqrt{(-1)}$, $\sqrt{(-4900)} = 70\sqrt{(-1)}$ and so on. Because the introduction of the special number $\sqrt{(-1)}$ simplified calculations, it quickly gained acceptance by mathematicians. Denoting $\sqrt{(-1)}$ by the letter j, we obtain the general number z where

$$z = x + \mathrm{j}y$$

Here x and y are ordinary **real numbers** and obey the Fundamental Rules of Arithmetic. (Most mathematics and physics texts use the letter i instead of j. However, we shall follow the standard engineering practice and use j.) The number z is called a **complex number**. The ordinary processes of arithmetic still apply, but become a little more complicated. As well as simplifying the process of obtaining roots as above, the introduction of $\mathrm{j} = \sqrt{(-1)}$ simplified the theory of equations, so that, for example, the quadratic equation

$$ax^2 + bx + c = 0$$

always has two roots

$$x = \frac{-b \pm \sqrt{(b^2 - 4ac)}}{2a}$$

These roots are real numbers when $b^2 \geq 4ac$ and complex numbers when $b^2 < 4ac$. Thus, any irreducible quadratic may be factorized into two complex factors. It then follows from property (ii) of the polynomial functions, given in Section 2.4.1, that any polynomial equation of degree n having real coefficients has exactly n roots which may be real or complex. This is a result known as the **Fundamental Theorem of Algebra**, which is also valid for polynomial equations having complex coefficients. Thus

$$x^7 - 7x^5 - 6x^4 + 4x^3 - 28x - 24 = 0$$

is an equation of degree seven and has the roots

$$x = -1, -2, -3, -1 - j, -1 + j, 1 - j, 1 + j$$

As has often been the case, what began as a mathematical curiosity has turned out to be of considerable practical importance, and complex numbers are invaluable in many aspects of engineering analysis. An elementary, but important, application is discussed later in this chapter.

3.2 Properties

To specify a complex number z, we use two real numbers, x and y, and write

$$z = x + jy$$

where $j = \sqrt{(-1)}$, and x is called the **real part** of z and y its **imaginary part**. This is often abbreviated to

$$z = x + jy, \quad \text{where } x = \text{Re}(z) \text{ and } y = \text{Im}(z)$$

Note that the imaginary part of z does *not* include the j. For example, if $z = 3 - j2$ then $\text{Re}(z) = 3$ and $\text{Im}(z) = -2$. If $x = 0$, the complex number is said to be **purely imaginary** and if $y = 0$ it is said to be **purely real**.

3.2.1 The Argand diagram

Geometrically, complex numbers can be represented as points on a plane similarly to the way in which real numbers are represented by points on a straight line. The number $z = x + jy$ is represented by the point P with coordinates (x, y), as shown in Figure 3.1. Such a diagram is called an **Argand diagram**, after one of its inventors. The x axis is called the **real axis** and the y axis is called the **imaginary axis**.

Figure 3.1
The Argand diagram:
$z = x + jy$.

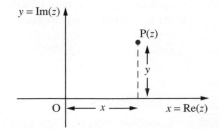

Example 3.1   Represent on an Argand diagram the complex numbers

(a) $3 + j2$ (b) $-5 + j3$ (c) $8 - j5$ (d) $-2 - j3$

Solution   (a) The number $3 + j2$ is represented by the point A(3, 2)

(b) The number $-5 + j3$ is represented by the point B(−5, 3)

(c) The number $8 - j5$ is represented by the point C(8, −5)

(d) The number $-2 - j3$ is represented by the point D(−2, −3)

as shown in Figure 3.2.

Figure 3.2

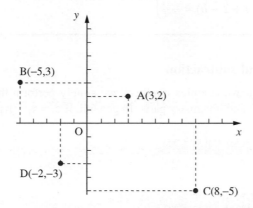

3.2.2 The arithmetic of complex numbers

(i) Equality

If two complex numbers $z_1 = x_1 + jy_1$ and $z_2 = x_2 + jy_2$ are equal then they are represented by the same point on the Argand diagram and it clearly follows that

$$x_1 = x_2 \quad \text{and} \quad y_1 = y_2$$

That is, when two complex numbers are equal we can equate their respective real and imaginary parts.

Example 3.2   If the two complex numbers

$$z_1 = (3a + 2) + j(3b - 1) \quad \text{and} \quad z_2 = (b + 1) - j(a + 2 - b)$$

are equal

(a) find the values of the real numbers a and b, and

(b) write down the real and imaginary parts of z_1 and z_2.

Solution (a) Since $z_1 = z_2$ we can equate their respective real and imaginary parts, giving

$$(3a + 2) = (b + 1) \qquad \text{or} \quad 3a - b = -1$$

and

$$(3b - 1) = -(a + 2 - b) \quad \text{or} \quad a + 2b = -1$$

Solving for a and b then gives

$$a = -\tfrac{3}{7}, \quad b = -\tfrac{2}{7}$$

(b) $\left. \begin{aligned} \mathrm{Re}(z_1) &= 3a + 2 = \tfrac{5}{7} \\ \mathrm{Re}(z_2) &= b + 1 = \tfrac{5}{7} \end{aligned} \right\}$ thus $\mathrm{Re}(z_1) = \mathrm{Re}(z_2) = \tfrac{5}{7}$

$\left. \begin{aligned} \mathrm{Im}(z_1) &= 3b - 1 = -\tfrac{13}{7} \\ \mathrm{Im}(z_2) &= -(a + 2 - b) = -\tfrac{13}{7} \end{aligned} \right\}$ thus $\mathrm{Im}(z_1) = \mathrm{Im}(z_2) = -\tfrac{13}{7}$

(ii) Addition and subtraction

To add or subtract two complex numbers, we simply perform the operations on their corresponding real and imaginary parts. In general, if $z_1 = x_1 + jy_1$ and $z_2 = x_2 + jy_2$ then

$$z_1 + z_2 = (x_1 + x_2) + j(y_1 + y_2)$$

and

$$z_1 - z_2 = (x_1 - x_2) + j(y_1 - y_2)$$

In Chapter 4, Section 4.2.5, we shall interpret complex numbers geometrically as two-dimensional vectors and illustrate how the rules for the addition of vectors can be used to represent addition of complex numbers in the Argand diagram.

Example 3.3 If $z_1 = 3 + j2$ and $z_2 = 5 - j3$ determine

(a) $z_1 + z_2$ (b) $z_1 - z_2$

Solution (a) Adding the corresponding real and imaginary parts gives

$$z_1 + z_2 = (3 + 5) + j(2 - 3) = 8 - j1$$

(b) Subtracting the corresponding real and imaginary parts gives

$$z_1 - z_2 = (3 - 5) + j(2 - (-3)) = -2 + j5$$

(iii) Multiplication

When multiplying two complex numbers the normal rules for multiplying out brackets hold. Thus, in general, if $z_1 = x_1 + jy_1$ and $z_2 = x_2 + jy_2$ then

$$z_1 z_2 = (x_1 + jy_1)(x_2 + jy_2)$$

$$= x_1 x_2 + jy_1 x_2 + jx_1 y_2 + j^2 y_1 y_2$$

Making use of the fact that $j^2 = -1$ then gives

$$z_1 z_2 = x_1 x_2 - y_1 y_2 + j(x_1 y_2 + x_2 y_1)$$

Example 3.4   If $z_1 = 3 + j2$ and $z_2 = 5 + j3$ determine $z_1 z_2$.

Solution
$$z_1 z_2 = (3 + j2)(5 + j3) = 15 + j10 + j9 + j^2 6$$
$$= 15 - 6 + j(10 + 9)$$
$$= 9 + j19$$

(iv) Division

The division of two complex numbers is less straightforward. If $z_1 = x_1 + jy_1$ and $z_2 = x_2 + jy_2$, then we use the following technique to obtain the quotient. We multiply 'top and bottom' by $x_2 - jy_2$, giving

$$\frac{z_1}{z_2} = \frac{x_1 + jy_1}{x_2 + jy_2} = \frac{(x_1 + jy_1)(x_2 - jy_2)}{(x_2 + jy_2)(x_2 - jy_2)}$$

Multiplying out 'top and bottom', we obtain

$$\frac{z_1}{z_2} = \frac{(x_1 x_2 + y_1 y_2) + j(x_2 y_1 - x_1 y_2)}{x_2^2 + y_2^2}$$

giving

$$\frac{z_1}{z_2} = \frac{(x_1 x_2 + y_1 y_2)}{x_2^2 + y_2^2} + j\frac{(x_2 y_1 - x_1 y_2)}{x_2^2 + y_2^2}$$

The number $x - jy$ is called the **complex conjugate** of $z = x + jy$ and is denoted by z^*. (Sometimes the complex conjugate is denoted with an overbar as $\bar{z}$.) Note that the complex conjugate z^* is obtained by changing the sign of the imaginary part of z. In the Argand diagram z^* is the mirror image of z in the real or x axis. The following important results are readily deduced.

$$z + z^* = 2x = 2\text{Re}(z)$$
$$z - z^* = 2jy = 2j\text{Im}(z)$$
$$zz^* = (x + jy)(x - jy) = x^2 + y^2$$

(3.1)

with the last result indicating that the product of a complex number and its complex conjugate is a real number.

Example 3.5 If $z_1 = 3 + j2$ and $z_2 = 5 + j3$ determine $\dfrac{z_1}{z_2}$.

Solution

$$\frac{z_1}{z_2} = \frac{3 + j2}{5 + j3}$$

Multiplying 'top and bottom' by the conjugate $5 - j3$ of the denominator gives

$$\frac{z_1}{z_2} = \frac{(3 + j2)(5 - j3)}{(5 + j3)(5 - j3)}$$

Multiplying out 'top and bottom' we obtain

$$\frac{3 + j2}{5 + j3} = \frac{(15 + 6) + j(10 - 9)}{(25 + 9) + j(15 - 15)} = \frac{21 + j}{34} = \tfrac{21}{34} + j\tfrac{1}{34}$$

Example 3.6 Find the real and imaginary parts of the complex number $z + 1/z$ for $z = (2 + j)/(1 - j)$.

Solution

$$z = \frac{2 + j}{1 - j} = \frac{(2 + j)(1 + j)}{(1 - j)(1 + j)} = \frac{1 + j3}{2} = \tfrac{1}{2} + j\tfrac{3}{2}$$

then

$$z^{-1} = \frac{2}{1 + j3} = \frac{2(1 - j3)}{(1 + j3)(1 - j3)} = \frac{2 - j6}{10} = \tfrac{1}{5} - j\tfrac{3}{5}$$

so that

$$z + \frac{1}{z} = (\tfrac{1}{2} + j\tfrac{3}{2}) + (\tfrac{1}{5} - j\tfrac{3}{5}) = (\tfrac{1}{2} + \tfrac{1}{5}) + j(\tfrac{3}{2} - \tfrac{3}{5}) = \tfrac{7}{10} + j\tfrac{9}{10}$$

giving

$$\mathrm{Re}\left(z + \frac{1}{z}\right) = \tfrac{7}{10} \quad \text{and} \quad \mathrm{Im}\left(z + \frac{1}{z}\right) = \tfrac{9}{10}$$

3.2.3 Modulus and argument

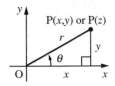

Figure 3.3
Modulus (r) and
argument (θ) of the
complex number
$z = x + jy$.

As indicated in the Argand diagram of Figure 3.3, the point P is specified uniquely if we know the length of the line OP and the angle it makes with the positive x direction. The length of OP is a measure of the size of z and is called the **modulus** of z, which is usually denoted by mod z or $|z|$. The angle between the positive real axis and OP is called the **argument** of z and is denoted by arg z. Since the polar coordinates (r, θ) and $(r, \theta + 2\pi)$ represent the same point, a convention is used to determine the argument of z uniquely, restricting its range so that $-\pi < \arg z \leqslant \pi$. (In some textbooks this is referred to as the 'principal value' of the argument.) The argument of the complex number $0 + j0$ is not defined.

Thus from Figure 3.3, $|z|$ and arg z are given by

$$|z| = r = \sqrt{(x^2 + y^2)}$$
$$\arg z = \theta \quad \text{where } \tan\theta = y/x, z \neq 0$$

(3.2)

Note that from equations (3.1)

$$zz^* = x^2 + y^2 = |z|^2$$

There are two common mistakes to avoid when calculating $|z|$ and $\arg z$ using (3.2). First note that the modulus of z is the square root of the sum of squares of x and y, *not* of x and jy. The j part of the number has been accounted for in the representation of the Argand diagram. The second common mistake is to place θ in the wrong quadrant. To avoid this, it is advisable when evaluating $\arg z$ to draw a sketch of the Argand diagram showing the location of the number.

Example 3.7 Determine the modulus and argument of

(a) $3 + j2$ (b) $1 - j$ (c) $-1 + j$ (d) $-\sqrt{6} - j\sqrt{2}$

Solution Note that the sketches of the Argand diagrams locating the positions of the complex numbers are given in Figure 3.4(a–d).

(a) $|3 + j2| = \sqrt{(3^2 + 2^2)} = \sqrt{(9 + 4)} = \sqrt{13} = 3.606$

$$\arg(3 + j2) = \tan^{-1}\left(\frac{2}{3}\right) = 0.588$$

(b) $|1 - j| = \sqrt{[1^2 + (-1)^2]} = \sqrt{2} = 1.414$

$$\arg(1 - j) = -\tan^{-1}\left(\frac{1}{1}\right) = -\tfrac{1}{4}\pi$$

(c) $|-1 + j| = \sqrt{[(-1)^2 + 1^2]} = \sqrt{2} = 1.414$

$$\arg(-1 + j) = \pi - \tan^{-1}\left(\frac{1}{1}\right) = \pi - \tfrac{1}{4}\pi = \tfrac{3}{4}\pi$$

(d) $|-\sqrt{6} - j\sqrt{2}| = \sqrt{(6 + 2)} = \sqrt{8} = 2.828$

$$\arg(-\sqrt{6} - j\sqrt{2}) = -(\pi - \tan^{-1}\tfrac{\sqrt{2}}{\sqrt{6}}) = -(\pi - \tan^{-1}\sqrt{\tfrac{1}{3}}) = -(\pi - \tfrac{1}{6}\pi) = -\tfrac{5}{6}\pi$$

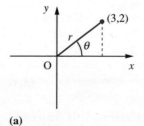

(a)

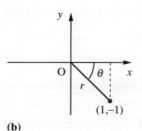

(b)

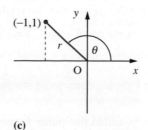

(c)

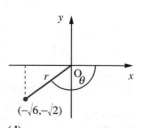

(d)

Figure 3.4

3.2.4 Exercises

1 Show in an Argand diagram the points representing the following complex numbers:

(a) $1 + j$ (b) $\sqrt{3} - j$

(c) $-3 + j4$ (d) $1 - j\sqrt{3}$

(e) $-1 + j\sqrt{3}$ (f) $-1 - j\sqrt{3}$

2 Express in the form $x + jy$ where x and y are real numbers

(a) $(5 + j3)(2 - j) - (3 + j)$ (b) $(1 - j2)^2$

(c) $\dfrac{5 - j8}{3 - j4}$ (d) $\dfrac{1 - j}{1 + j}$

(e) $\frac{1}{2}(1 + j)^2$ (f) $(3 - j2)^2$

(g) $\dfrac{1}{5 - j3} - \dfrac{1}{5 + j3}$ (h) $\dfrac{1}{2} - \dfrac{3 - j4}{5 - j8}$

3 Find the roots of the equations

(a) $x^2 + 2x + 2 = 0$ (b) $x^3 + 8 = 0$

4 Find z such that

$$zz^* + 3(z - z^*) = 13 + j12$$

5 With $z = 2 - j3$, find

(a) jz (b) z^* (c) $1/z$ (d) $(z^*)^*$

6 Find the modulus and argument of each of the complex numbers given in Question 1.

7 Find the complex numbers w, z which satisfy the simultaneous equations

$$4z + 3w = 23$$

$$z + jw = 6 + j8$$

8 For $z = x + jy$ (x and y real) satisfying

$$\frac{2z}{1 + j} - \frac{2z}{j} = \frac{5}{2 + j}$$

find x and y.

9 Given $z = 2 - j2$ is a root of

$$2z^3 - 9z^2 + 20z - 8 = 0$$

find the remaining roots of the equation.

10 Find the real and imaginary parts of z when

$$\frac{1}{z} = \frac{2}{2 + j3} + \frac{1}{3 - j2}$$

11 Find $z = z_1 + z_2 z_3 / (z_2 + z_3)$ when $z_1 = 2 + j3$, $z_2 = 3 + j4$ and $z_3 = -5 + j12$.

12 Find the values of the real numbers x and y which satisfy the equation

$$\frac{2 + x - jy}{3x + jy} = 1 + j2$$

13 Find z_3 in the form $x + jy$, where x and y are real numbers, given that

$$\frac{1}{z_3} = \frac{1}{z_1} + \frac{1}{z_1 z_2}$$

where $z_1 = 3 - j4$ and $z_2 = 5 + j2$.

3.2.5 Polar form of a complex number

Figure 3.3 shows that the relationships between (x, y) and (r, θ) are

$$x = r \cos \theta \quad \text{and} \quad y = r \sin \theta$$

Hence the complex number $z = x + jy$ can be expressed in the form

$$z = r \cos \theta + jr \sin \theta = r(\cos \theta + j \sin \theta) \tag{3.3}$$

This is called the **polar form** of the complex number. In engineering it is frequently written as $r \angle \theta$, so that

$$z = r \angle \theta = r(\cos \theta + j \sin \theta)$$

Example 3.8

Express the following complex numbers in polar form.

(a) $12 + j5$ (b) $-3 + j4$ (c) $-4 - j3$

Solution

(a) A sketch of the Argand diagram locating the position of $12 + j5$ is given in Figure 3.5(a). Thus

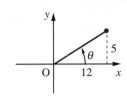

$$|12 + j5| = \sqrt{(144 + 25)} = 13$$

$$\arg(12 + j5) = \tan^{-1}\tfrac{5}{12} = 0.395$$

Thus in polar form

$$12 + j5 = 13[\cos(0.395) + j\sin(0.395)]$$

(a)

(b) A sketch of the Argand diagram locating the position of $-3 + j4$ is given in Figure 3.5(b). Thus

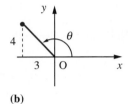

$$|-3 + j4| = \sqrt{(9 + 16)} = 5$$

$$\arg(-3 + j4) = \pi - \tan^{-1}\tfrac{4}{3} = \pi - 0.9273$$

$$= 2.214$$

(b)

Thus in polar form

$$-3 + j4 = 5[\cos(2.214) + j\sin(2.214)]$$

(c) A sketch of the Argand diagram locating the position of $-4 - j3$ is given in Figure 3.5(c). Thus

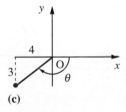

$$|-4 - j3| = \sqrt{(16 + 9)} = 5$$

$$\arg(-4 - j3) = -(\pi - \tan^{-1}\tfrac{3}{4}) = -(\pi - 0.643)$$

$$= -2.498$$

(c)

Figure 3.5

Thus in polar form

$$-4 - j3 = 5[\cos(-2.498) + j\sin(-2.498)]$$

$$= 5[\cos(2.498) - j\sin(2.498)]$$

using the results $\cos(-t) = \cos t$ and $\sin(-t) = -\sin t$.

Note: Rectangular to polar conversion can be done using a calculator and students are encouraged to check the answers in this way.

Multiplication in polar form

Let

$$z_1 = r_1(\cos\theta_1 + j\sin\theta_1) \quad \text{and} \quad z_2 = r_2(\cos\theta_2 + j\sin\theta_2)$$

then

$$z_1 z_2 = r_1 r_2(\cos\theta_1 + j\sin\theta_1)(\cos\theta_2 + j\sin\theta_2)$$

$$= r_1 r_2[(\cos\theta_1\cos\theta_2 - \sin\theta_1\sin\theta_2) + j(\sin\theta_1\cos\theta_2 + \cos\theta_1\sin\theta_2)]$$

which, on using the trigonometric identities (2.24a, c), gives

$$z_1 z_2 = r_1 r_2 [\cos(\theta_1 + \theta_2) + j\sin(\theta_1 + \theta_2)]$$

Hence

$$|z_1 z_2| = r_1 r_2 = |z_1||z_2| \tag{3.4}$$

and

$$\arg(z_1 z_2) = \theta_1 + \theta_2 = \arg z_1 + \arg z_2 \tag{3.5}$$

When using these results care must be taken to ensure that $-\pi < \arg(z_1 z_2) \le \pi$.

Example 3.9   If $z_1 = -12 + j5$ and $z_2 = -4 + j3$, determine, using (3.4) and (3.5), $|z_1 z_2|$ and $\arg(z_1 z_2)$.

Solution
$$|z_1| = \sqrt{(144 + 25)} = \sqrt{(169)} = 13$$

$$\arg(z_1) = \pi - \tan^{-1}\tfrac{5}{12} = \pi - 0.395 = 2.747$$

$$|z_2| = \sqrt{(16 + 9)} = 5$$

$$\arg(z_2) = \pi - \tan^{-1}\tfrac{3}{4} = 2.498$$

Thus from (3.4) and (3.5)

$$|z_1 z_2| = |z_1||z_2| = (13)(5) = 65$$

$$\arg(z_1 z_2) = \arg z_1 + \arg z_2 = 2.747 + 2.498$$

$$= 5.245 \ (\text{or } 300.51°)$$

However, this does not express $\arg(z_1 z_2)$ within the defined range $-\pi < \arg \le \pi$. Thus

$$\arg(z_1 z_2) = -2\pi + 5.245 = -1.038$$

Geometrical representation of multiplication by j

Since

$$z = r(\cos\theta + j\sin\theta) \quad \text{and} \quad j = 1(\cos\tfrac{1}{2}\pi + j\sin\tfrac{1}{2}\pi)$$

it follows from (3.4) that

$$jz = r[\cos(\theta + \tfrac{1}{2}\pi) + j\sin(\theta + \tfrac{1}{2}\pi)]$$

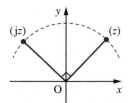

Figure 3.6
Relationship between z and jz.

Thus the effect of multiplying a complex number by j is to leave the modulus unaltered but to increase the argument by $\tfrac{1}{2}\pi$ as indicated in Figure 3.6. This property is of importance in the application of complex numbers to the theory of alternating current.

Division in polar form

Now

$$\frac{1}{\cos\theta + j\sin\theta} = \frac{1}{\cos\theta + j\sin\theta}\frac{\cos\theta - j\sin\theta}{\cos\theta - j\sin\theta}$$

$$= \frac{\cos\theta - j\sin\theta}{\cos^2\theta + \sin^2\theta}$$

$$= \cos\theta - j\sin\theta, \quad \text{since} \quad \cos^2\theta + \sin^2\theta = 1$$

Thus if

$$z_1 = r_1(\cos\theta_1 + j\sin\theta_1) \quad \text{and} \quad z_2 = r_2(\cos\theta_2 + j\sin\theta_2)$$

then

$$\frac{z_1}{z_2} = \frac{r_1(\cos\theta_1 + j\sin\theta_1)}{r_2(\cos\theta_2 + j\sin\theta_2)}$$

$$= \frac{r_1}{r_2}(\cos\theta_1 + j\sin\theta_1)(\cos\theta_2 - j\sin\theta_2) \quad \text{(from above)}$$

$$= \frac{r_1}{r_2}[(\cos\theta_1\cos\theta_2 + \sin\theta_1\sin\theta_2) + j(\sin\theta_1\cos\theta_2 - \cos\theta_1\sin\theta_2)]$$

or

$$\frac{z_1}{z_2} = \frac{r_1}{r_2}[\cos(\theta_1 - \theta_2) + j\sin(\theta_1 - \theta_2)] \tag{3.6}$$

using the trigonometric identities (2.24b, d). Hence

$$\left|\frac{z_1}{z_2}\right| = \frac{r_1}{r_2} = \frac{|z_1|}{|z_2|} \tag{3.7}$$

and

$$\arg\left(\frac{z_1}{z_2}\right) = \theta_1 - \theta_2 = \arg z_1 - \arg z_2 \tag{3.8}$$

Again some adjustment may be necessary to ensure that $-\pi < \arg(z_1/z_2) \leq \pi$.

Example 3.10 For the following pairs of complex numbers obtain z_1/z_2 and z_2/z_1.

(a) $z_1 = 4(\cos\pi/2 + j\sin\pi/2)$, $z_2 = 9(\cos\pi/3 + j\sin\pi/3)$

(b) $z_1 = \cos 3\pi/4 + j\sin 3\pi/4$, $z_2 = 2(\cos\pi/8 + j\sin\pi/8)$

Solution   (a) $|z_1| = 4$, $\arg z_1 = \pi/2$; $|z_2| = 9$, $\arg z_2 = \pi/3$

From (3.7)

$$\left|\frac{z_1}{z_2}\right| = \frac{4}{9} \quad \text{and} \quad \left|\frac{z_2}{z_1}\right| = \frac{9}{4}$$

From (3.8)

$$\arg\left(\frac{z_1}{z_2}\right) = \frac{\pi}{2} - \frac{\pi}{3} = \frac{\pi}{6} \quad \text{and} \quad \arg\left(\frac{z_2}{z_1}\right) = \frac{\pi}{3} - \frac{\pi}{2} = -\frac{\pi}{6}$$

Thus $\dfrac{z_1}{z_2} = \dfrac{4}{9}\left(\cos\dfrac{\pi}{6} + j\sin\dfrac{\pi}{6}\right)$

and $\dfrac{z_2}{z_1} = \dfrac{9}{4}\left(\cos\dfrac{\pi}{6} - j\sin\dfrac{\pi}{6}\right)$

(b) $|z_1| = 1$, $\arg z_1 = 3\pi/4$; $|z_2| = 2$, $\arg z_2 = \pi/8$

From (3.7)

$$\left|\frac{z_1}{z_2}\right| = \frac{1}{2} \quad \text{and} \quad \left|\frac{z_2}{z_1}\right| = 2$$

From (3.8)

$$\arg\left(\frac{z_1}{z_2}\right) = \frac{3\pi}{4} - \frac{\pi}{8} = \frac{5\pi}{8} \quad \text{and} \quad \arg\left(\frac{z_2}{z_1}\right) = \frac{\pi}{8} - \frac{3\pi}{4} = -\frac{5\pi}{8}$$

Thus $\dfrac{z_1}{z_2} = \dfrac{1}{2}\left(\cos\dfrac{5\pi}{8} + j\sin\dfrac{5\pi}{8}\right)$

and $\dfrac{z_2}{z_1} = 2\left(\cos\dfrac{5\pi}{8} - j\sin\dfrac{5\pi}{8}\right)$

Example 3.11   Find the modulus and argument of

$$z = \frac{(1 + j2)^2(4 - j3)^3}{(3 + j4)^4(2 - j)^3}$$

Solution $|z| = \dfrac{|1 + j2|^2|4 - j3|^3}{|3 + j4|^4|2 - j|^3}$

$$= \frac{[\sqrt{(1 + 4)}]^2[\sqrt{(16 + 9)}]^3}{[\sqrt{(9 + 16)}]^4[\sqrt{(4 + 1)}]^3} = \frac{1}{25}\sqrt{5}$$

$\arg z = 2 \arg(1 + j2) + 3 \arg(4 - j3) - 4 \arg(3 + j4) - 3 \arg(2 - j)$

$= 2(1.107) + 3(-0.643) - 4(0.927) - 3(-0.461) = -2.035$

3.2.6 Euler's formula

In Chapter 2, Section 2.7.3, we obtained the result

$$e^x = \cosh x + \sinh x$$

which links the exponential and hyperbolic functions. A similar, but more important, formula links the exponential and circular functions. It is

$$e^{j\theta} = \cos \theta + j \sin \theta \tag{3.9}$$

This formula is known as **Euler's formula**. The justification for this definition depends on the following facts.

We know from the properties of the exponential function that

$$e^{j\theta_1} e^{j\theta_2} = e^{j(\theta_1 + \theta_2)}$$

When expressed in terms of Euler's formula this becomes

$$(\cos \theta_1 + j \sin \theta_1)(\cos \theta_2 + j \sin \theta_2) = \cos(\theta_1 + \theta_2) + j \sin(\theta_1 + \theta_2)$$

which is just (3.4) with $r_1 = r_2 = 1$.

Similarly

$$\frac{e^{j\theta_1}}{e^{j\theta_2}} = e^{j(\theta_1 - \theta_2)}$$

becomes

$$\frac{\cos \theta_1 + j \sin \theta_1}{\cos \theta_2 + j \sin \theta_2} = \cos(\theta_1 - \theta_2) + j \sin(\theta_1 - \theta_2)$$

which is just (3.6) with $r_1 = r_2 = 1$.

Euler's formula enables us to write down the polar form of the complex number z very concisely:

$$z = r(\cos \theta + j \sin \theta) = re^{j\theta} = r \angle \theta \tag{3.10}$$

This is known as the **exponential form** of the complex number z.

Example 3.12   Express the following complex numbers in exponential form

(a) $2 + j3$ (b) $-2 + j$

Solution   (a) A sketch of the Argand diagram showing the position of $2 + j3$ is given in Figure 3.7(a).

$$|2 + j3| = \sqrt{(2^2 + 3^2)} = \sqrt{13}$$

$$\arg(2 + j3) = \tan^{-1}(3/2) = 0.9828$$

Thus $2 + j3 = \sqrt{13}e^{j0.9828}$

Figure 3.7

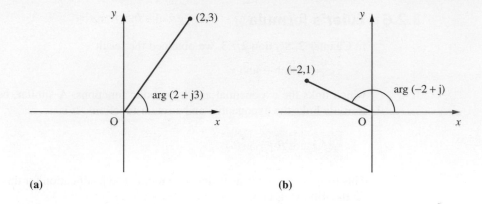

(a)   (b)

(b) A sketch of the Argand diagram showing the position of $-2 + j$ is given in Figure 3.7(b).

$$|-2 + j| = \sqrt{5}$$

$$\arg(-2 + j) = \pi - \tan^{-1}(1/2) = 2.6779$$

Thus $-2 + j = \sqrt{5}e^{j2.6779}$

Example 3.13   Express in cartesian form the complex number $e^{2+j\pi/3}$.

Solution $e^{2+j\pi/3} = e^2 e^{j\pi/3} = e^2(\cos \pi/3 + j \sin \pi/3)$

Now $e^2 = 7.3891$, $\cos \pi/3 = 0.5$ and $\sin \pi/3 = 0.8660$, so that

$$e^{2+j\pi/3} = 3.6945 + j6.3991$$

3.2.7   Relationship between circular and hyperbolic functions

Euler's formula provides the theoretical link between circular and hyperbolic functions. Since

$$e^{j\theta} = \cos \theta + j \sin \theta \quad \text{and} \quad e^{-j\theta} = \cos \theta - j \sin \theta$$

we deduce that

$$\cos \theta = \frac{e^{j\theta} + e^{-j\theta}}{2} \tag{3.11a}$$

and

$$\sin \theta = \frac{e^{j\theta} - e^{-j\theta}}{2j} \tag{3.11b}$$

In Section 2.7 we defined the hyperbolic functions by

$$\cosh x = \frac{e^x + e^{-x}}{2}$$ (3.12a)

and

$$\sinh x = \frac{e^x - e^{-x}}{2}$$ (3.12b)

Comparing (3.12a, b) with (3.11a, b), we have

$$\cosh jx = \frac{e^{jx} + e^{-jx}}{2} = \cos x$$ (3.13a)

$$\sinh jx = \frac{e^{jx} - e^{-jx}}{2} = j\sin x$$ (3.13b)

so that

$$\tanh jx = j \tan x$$ (3.13c)

Also,

$$\cos jx = \frac{e^{j^2x} + e^{-j^2x}}{2} = \frac{e^{-x} + e^x}{2} = \cosh x$$ (3.14a)

$$\sin jx = \frac{e^{j^2x} - e^{-j^2x}}{2j} = \frac{e^{-x} - e^x}{2j} = j\sinh x$$ (3.14b)

so that

$$\tan jx = j \tanh x$$ (3.14c)

These relationships provide the justification for Osborn's rule used in Section 2.7 for obtaining hyperbolic function identities from those satisfied by circular functions, since whenever a product of two sines occurs, j^2 will also occur.

Using these results we can evaluate functions such as $\sin z$, $\cos z$, $\tan z$, $\sinh z$, $\cosh z$ and $\tanh z$. For example, to evaluate

$$\cos z = \cos(x + jy)$$

we use the identity

$$\cos(A + B) = \cos A \cos B - \sin A \sin B$$

and obtain

$$\cos z = \cos x \cos jy - \sin x \sin jy$$

Using results (3.14a, b), this gives

$$\cos z = \cos x \cosh y - j \sin x \sinh y$$

Example 3.14

Find the values of

(a) $\sin[\frac{1}{4}\pi(1+j)]$ (b) $\sinh(3+j4)$

(c) $\tan(\frac{\pi}{4} - j3)$ (d) z such that $\cos z = 2$

Solution

(a) We may use the identity

$$\sin(A+B) = \sin A \cos B + \cos A \sin B$$

and obtain

$$\sin(\tfrac{1}{4}\pi + j\tfrac{1}{4}\pi) = \sin \tfrac{1}{4}\pi \cos j\tfrac{1}{4}\pi + \cos \tfrac{1}{4}\pi \sin j\tfrac{1}{4}\pi$$

Here $\sin \frac{1}{4}\pi$ and $\cos \frac{1}{4}\pi$ are evaluated as usual $(= \sqrt{\frac{1}{2}})$, while we make use of results (3.14a, b) to obtain

$$\cos j\tfrac{1}{4}\pi = \cosh \tfrac{1}{4}\pi \text{ and } \sin j\tfrac{1}{4}\pi = j \sinh \tfrac{1}{4}\pi$$

giving

$$\sin[\tfrac{1}{4}\pi(1+j)] = \sin \tfrac{1}{4}\pi \cosh \tfrac{1}{4}\pi + j \cos \tfrac{1}{4}\pi \sinh \tfrac{1}{4}\pi$$

$$= (0.7071)(1.3246) + j(0.7071)(0.8687)$$

$$= 0.9366 + j0.6142$$

(b) Using the identity

$$\sinh(A+B) = \sinh A \cosh B + \cosh A \sinh B$$

we obtain

$$\sinh(3+j4) = \sinh 3 \cosh j4 + \cosh 3 \sinh j4$$

which, on using results (3.13a, b), gives

$$\sinh(3+j4) = \sinh 3 \cos 4 + j \cosh 3 \sin 4$$

$$= (10.0179)(-0.6536) + j(10.0677)(-0.7568)$$

$$= -6.548 - j7.619$$

(c) Using the identity

$$\tan(A-B) = \frac{\tan A - \tan B}{1 + \tan A \tan B}$$

we obtain

$$\tan(\tfrac{1}{4}\pi - j3) = \frac{\tan \tfrac{1}{4}\pi - \tan j3}{1 + \tan \tfrac{1}{4}\pi \tan j3}$$

which, on using result (3.14c) and $\tan \frac{1}{4}\pi = 1$, gives

$$\tan(\tfrac{1}{4}\pi - j3) = \frac{1 - j\tanh 3}{1 + j\tanh 3} = \frac{(1 - j\tanh 3)^2}{1 + \tanh^2 3}$$

$$= \frac{1 - \tanh^2 3}{1 + \tanh^2 3} - j\frac{2\tanh 3}{1 + \tanh^2 3}$$

$$= \frac{1}{\cosh^2 3 + \sinh^2 3} - j\frac{2\sinh 3\cosh 3}{\cosh^2 3 + \sinh^2 3}$$

$$= \frac{1}{\cosh 6} + j\frac{\sinh 6}{\cosh 6}$$

$$= 0.005 - j1.000$$

(d) Writing $z = x + jy$, we have

$$2 = \cos(x + jy)$$

Expanding the right-hand side gives

$$2 = \cos x \cos jy - \sin x \sin jy$$

$$= \cos x \cosh y - \sin x \,(j \sinh y)$$

$$2 = \cos x \cosh y - j \sin x \sinh y$$

Equating real and imaginary parts of each side of this equation gives

$$2 = \cos x \cosh y$$

and

$$0 = \sin x \sinh y$$

The latter equation implies either $\sin x = 0$ or $y = 0$. If $y = 0$ then the first equation implies $2 = \cos x$, so clearly that is not a solution. The alternative, $\sin x = 0$, implies $x = 0, \pm\pi, \pm 2\pi, \pm 3\pi, \ldots$, and hence

$$2 = \cos(\pm n\pi) \cosh y, \quad n = 0, 1, 2, \ldots$$

This gives

$$2 = \cos n\pi \cosh y$$

$$= (-1)^n \cosh y$$

But $\cosh y \geq 1$, so n must be an even number. Thus the values of z such that $\cos z = 2$ are

$$z = \pm 2n\pi \pm j \cosh^{-1}2, \, n = 0, 1, 2, \ldots$$

$$= \pm 2n\pi \pm j(1.3170)$$

3.2.8 Logarithm of a complex number

Consider the equation

$$z = e^w$$

Writing $z = x + jy$ and $w = u + jv$, we have

$$x + jy = e^{u+jv} = e^u e^{jv}$$

$$= e^u(\cos v + j \sin v), \quad \text{by Euler's formula}$$

Equating real and imaginary parts,

$$x = e^u \cos v \quad \text{and} \quad y = e^u \sin v$$

Squaring both these equations and adding gives

$$x^2 + y^2 = e^{2u}(\cos^2 v + \sin^2 v) = e^{2u}$$

so that

$$u = \tfrac{1}{2} \ln(x^2 + y^2) = \ln |z|$$

Dividing the two equations,

$$\tan v = \frac{y}{x}$$

From this and $x = e^u \cos v$

$$v = \arg z + 2n\pi, \quad n = 0, \pm 1, \pm 2, \dots$$

Hence

$$v = \ln |z| + j \arg z + j2n\pi, \quad n = 0, \pm 1, \pm 2, \dots$$

We select just one of these solutions to define for us the logarithm of the complex number z, writing

$$\ln z = \ln |z| + j \arg z \tag{3.15}$$

This is sometimes called its **principal value**.

Example 3.15   Evaluate $\ln(-3 + j4)$ in the form $x + jy$.

Solution
$$|-3 + j4| = \sqrt{(9 + 16)} = 5$$

$$\arg(-3 + j4) = \pi - \tan^{-1}\tfrac{4}{3} = 2.214$$

Thus from (3.15)

$$\ln(-3 + j4) = \ln 5 + j2.214 = 1.61 + j2.214$$

3.2.9 Exercises

14 Given $z_1 = e^{j\pi/4}$ and $z_2 = e^{-j\pi/3}$, find

(a) the arguments of $z_1 z_2^2$ and z_1^3/z_2

(b) the real and imaginary parts of $z_1^2 + jz_2$

15 Given $z_1 = 2e^{j\pi/3}$ and $z_2 = 4e^{-2j\pi/3}$, find the modulus and argument of

(a) $z_1^3 z_2^2$ (b) $z_1^2 z_2^4$ (c) z_1^2/z_2^3

16 Express in polar form the complex numbers

(a) j   (b) 1

(c) −1   (d) $1 - j$

(e) $\sqrt{3} - j\sqrt{3}$ (f) $-2 + j$

(g) $-3 - j2$ (h) $7 - j5$

(i) $(2 - j)(2 + j)$ (j) $(-2 + j7)^2$

17 Using the exponential forms of $\cos\theta$ and $\sin\theta$ given in (3.11a, b), prove the following trigonometric identities:

(a) $\sin(\alpha + \beta) = \sin\alpha\cos\beta + \cos\alpha\sin\beta$

(b) $\sin^3\theta = \frac{3}{4}\sin\theta - \frac{1}{4}\sin 3\theta$

18 Express $z = (2 - j)(3 + j2)/(3 - j4)$ in the form $x + jy$ and also in polar form.

19 Writing $\tanh(u + jv) = x + jy$, with x, y, u and v real, determine x and y in terms of u and v. Hence evaluate $\tanh(2 + j\frac{1}{4}\pi)$ in the form $x + jy$.

20 Express in the form $x + jy$

(a) $\sin(\frac{5}{6}\pi + j)$ (b) $\cos(j\frac{3}{4})$

(c) $\sinh[\frac{\pi}{3}(1 + j)]$ (d) $\cosh(j\frac{\pi}{4})$

21 Solve $z = x + jy$ when

(a) $\sin z = 2$ (b) $\cos z = j\frac{3}{4}$

(c) $\sin z = 3$ (d) $\cosh z = -2$

22 Show that

(a) $\ln(5 + j12) = \ln 13 + j1.176$

(b) $\ln(-\frac{1}{2} - j\frac{1}{2}\sqrt{3}) = -j\frac{2\pi}{3}$

23 In a certain cable of length l the current I_0 at the sending end when it is raised to a potential V_0 and the other end is earthed is given by

$$I_0 = \frac{V_0}{Z_0}\tanh Pl$$

Calculate the value of I_0 when $V_0 = 100$, $Z_0 = 500 + j400$, $l = 10$ and $P = 0.1 + j0.15$.

3.3 ## Powers of complex numbers

In earlier sections we have discussed the extensions of ordinary arithmetic, including $+$, $-$, $\times$, $\div$, to complex numbers. We now extend the arithmetical operations to include the operation of powers.

3.3.1 De Moivre's theorem

From (3.10) a complex number z may be expressed in terms of its modulus r and argument θ in the exponential form

$$z = re^{j\theta}$$

Using the rules of indices and the property (2.33a) of the exponential function, we have, for any n,

$$z^n = r^n(e^{j\theta})^n = r^n e^{j(n\theta)}$$

so that

$$z^n = r^n(\cos n\theta + j \sin n\theta) \qquad (3.16)$$

This result is known as **de Moivre's theorem**.

Example 3.16   Express $1 - j$ in the form $r(\cos\theta + j\sin\theta)$ and hence evaluate $(1 - j)^{12}$.

Solution   From Example 3.7(b)

$$|1 - j| = \sqrt{2} \quad \text{and} \quad \arg(1 - j) = -\tfrac{1}{4}\pi$$

so that

$$1 - j = \sqrt{2}[\cos(-\tfrac{1}{4}\pi) + j \sin(-\tfrac{1}{4}\pi)]$$
$$= \sqrt{2}(\cos \tfrac{1}{4}\pi - j \sin \tfrac{1}{4}\pi)$$

Then

$$(1 - j)^{12} = (\sqrt{2})^{12}(\cos \tfrac{1}{4}\pi - j \sin \tfrac{1}{4}\pi)^{12}$$

which, on using de Moivre's theorem (3.16), gives

$$(1 - j)^{12} = 2^6[\cos(12 \times \tfrac{1}{4}\pi) - j \sin(12 \times \tfrac{1}{4}\pi)]$$
$$= 2^6(\cos 3\pi - j \sin 3\pi)$$
$$= 2^6(-1 - j0)$$
$$= -64$$

Most commonly, we use de Moivre's theorem to find the roots of complex numbers like $\sqrt{z}$ and $\sqrt[3]{z}$. More generally, we want to find $z^{1/n}$, the nth root, where n is a natural number. Setting $w = z^{1/n}$, we see that $z = w^n$, and by (3.16),

$$w^n = R^n(\cos n\phi + j \sin n\phi), \quad \text{where } |w| = R \text{ and } \arg w = \phi$$
$$= r(\cos\theta + j\sin\theta), \qquad \text{where } |z| = r \text{ and } \arg z = \theta$$

Comparing real and imaginary parts of this equality, we deduce that

$$r \cos\theta = R^n \cos n\phi$$

and

$$r \sin\theta = R^n \sin n\phi$$

Squaring and adding these two equations gives $r^2 = R^{2n}$; that is, $R = r^{1/n}$. Substituting this value into the equations gives

$$\cos\theta = \cos n\phi$$

and

$$\sin\theta = \sin n\phi$$

This pair of simultaneous equations has an infinite number of solutions because of the 2π-periodicity of the sine and cosine functions. Thus

$$n\phi = \theta + 2\pi k, \quad \text{where } k \text{ is an integer}$$

and

$$\phi = \frac{\theta}{n} + \frac{2\pi k}{n}, \quad \text{where } k = 0, 1, -1, 2, -2, 3, -3, \ldots$$

Substituting these values for R and ϕ into the formula for w gives

$$z^{1/n} = r^{1/n}\left[\cos\left(\frac{\theta}{n} + \frac{2\pi k}{n}\right) + j\sin\left(\frac{\theta}{n} + \frac{2\pi k}{n}\right)\right] \qquad (3.17)$$

where k is an integer. This expression yields exactly n different roots, corresponding to $k = 0, 1, 2, \ldots, n-1$. The value for $k = n$ is the same as that for $k = 0$, the value for $k = n+1$ is the same as that for $k = 1$, and so on. The n values of $z^{1/n}$ are equally spaced around a circle of radius $r^{1/n}$ whose centre is the origin of the Argand diagram. Also, the arguments increase in arithmetic progression, so that joining the roots on the circle creates a regular polygon inscribed in the latter.

Equation (3.17) may be written alternatively in the exponential form

$$z^{1/n} = r^{1/n}e^{j(\theta/n + 2\pi k/n)}, \quad k = 0, 1, 2, \ldots, n-1 \qquad (3.18)$$

Example 3.17 Given $z = -\tfrac{1}{2} + j\tfrac{1}{2}$, evaluate

(a) $z^{1/2}$ (b) $z^{1/3}$

and display the roots on an Argand diagram.

Solution We first express z in polar form.

Since $r = |z| = \sqrt{(\tfrac{1}{4} + \tfrac{1}{4})} = 2^{-1/2}$, and $\theta = \arg(z) = \pi - \tan^{-1}1 = \tfrac{3}{4}\pi$, we have

$$z = 2^{-1/2}(\cos\tfrac{3}{4}\pi + j\sin\tfrac{3}{4}\pi)$$

(a) From (3.17)

$$z^{1/2} = r^{1/2}\left[\cos\left(\frac{\theta}{2} + \frac{2\pi k}{2}\right) + j\sin\left(\frac{\theta}{2} + \frac{2\pi k}{2}\right)\right], \quad k = 0, 1$$

$$= 2^{-1/4}[\cos(\tfrac{3}{8}\pi + \pi k) + j\sin(\tfrac{3}{8}\pi + \pi k)], \qquad k = 0, 1$$

Thus we have two square roots:

$$z^{1/2} = 2^{-1/4}(\cos\tfrac{3}{8}\pi + j\sin\tfrac{3}{8}\pi) \qquad (\text{for } k = 0)$$

and

$$z^{1/2} = 2^{-1/4}(\cos\tfrac{11}{8}\pi + j\sin\tfrac{11}{8}\pi) \quad (\text{for } k = 1)$$

as shown in Figure 3.8(a). These can be evaluated numerically, giving respectively (to 4dp) $z = 0.3218 + j0.7769$ and $z = -0.3218 - j0.7769$.

Figure 3.8

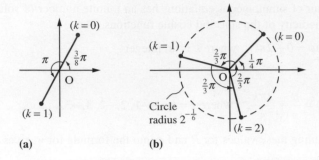

(a) (b)

(b) From (3.17)

$$z^{1/3} = r^{1/3}\left[\cos\left(\frac{\theta}{3} + \frac{2\pi k}{3}\right) + j\sin\left(\frac{\theta}{3} + \frac{2\pi k}{3}\right)\right], \quad k = 0, 1, 2$$

$$= 2^{-1/6}[\cos(\tfrac{1}{4}\pi + \tfrac{2}{3}\pi k) + j\sin(\tfrac{1}{4}\pi + \tfrac{2}{3}\pi k)], \quad k = 0, 1, 2$$

Thus we obtain three cube roots:

$$z^{1/3} = 2^{-1/6}(\cos \tfrac{1}{4}\pi + j\sin \tfrac{1}{4}\pi) \quad \text{(for } k = 0\text{)}$$

$$z^{1/3} = 2^{-1/6}(\cos \tfrac{11}{12}\pi + j\sin \tfrac{11}{12}\pi) \quad \text{(for } k = 1\text{)}$$

and

$$z^{1/3} = 2^{-1/6}(\cos \tfrac{19}{12}\pi + j\sin \tfrac{19}{12}\pi) \quad \text{(for } k = 2\text{)}$$

as shown in Figure 3.8(b). Note that the three roots are equally spaced around a circle of radius $2^{-1/6}$ with centre at the origin.

Formula (3.17) can easily be extended to deal with the general rational power z^p of z. Let $p = \dfrac{m}{n}$, where n is a natural number and m is an integer, then

$$z^p = (z^{1/n})^m$$

$$= \left\{r^{1/n}\left[\cos\left(\frac{\theta}{n} + \frac{2\pi k}{n}\right) + j\sin\left(\frac{\theta}{n} + \frac{2\pi k}{n}\right)\right]\right\}^m, \quad k = 0, 1, 2, \ldots, (n-1)$$

$$= r^{m/n}\left[\cos\left(\frac{m\theta}{n} + \frac{2\pi km}{n}\right) + j\sin\left(\frac{m\theta}{n} + \frac{2\pi km}{n}\right)\right]$$

$$= r^p[\cos(p\theta + 2\pi kp) + j\sin(p\theta + 2\pi kp)], \quad k = 0, 1, 2, \ldots, (n-1)$$

Example 3.18 Evaluate $(-\tfrac{1}{2} + j\tfrac{1}{2})^{-2/3}$ and display the roots on an Argand diagram.

Solution From Example 3.17, we can write

$$-\tfrac{1}{2} + j\tfrac{1}{2} = 2^{-1/2}(\cos \tfrac{3}{4}\pi + j\sin \tfrac{3}{4}\pi)$$

Figure 3.9

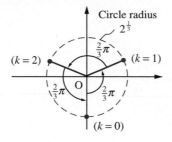

giving

$$z^{-2/3} = r^{-2/3}\left[\cos\left(-\frac{2\theta}{3} - \frac{4\pi k}{3}\right) + j\sin\left(-\frac{2\theta}{3} - \frac{4\pi k}{3}\right)\right], \quad k = 0, 1, 2$$

$$= 2^{1/3}[\cos(-\tfrac{1}{2}\pi - \tfrac{4}{3}\pi k) + j\sin(-\tfrac{1}{2}\pi - \tfrac{4}{3}\pi k)], \quad k = 0, 1, 2$$

Thus we obtain three values:

$$z^{-2/3} = 2^{1/3}[\cos(-\tfrac{1}{2}\pi) + j\sin(-\tfrac{1}{2}\pi)] \quad \text{(for } k = 0)$$

$$z^{-2/3} = 2^{1/3}(\cos\tfrac{1}{6}\pi + j\sin\tfrac{1}{6}\pi) \quad \text{(for } k = 1)$$

and

$$z^{-2/3} = 2^{1/3}(\cos\tfrac{5}{6}\pi + j\sin\tfrac{5}{6}\pi) \quad \text{(for } k = 2)$$

as shown in Figure 3.9.

Example 3.19   Solve the quadratic equation

$$z^2 + (2j - 3)z + (5 - j) = 0$$

Solution   Using formula (1.5)

$$z = \frac{-(2j - 3) \pm \sqrt{[(2j - 3)^2 - 4(5 - j)]}}{2}$$

that is,

$$z = \frac{-(2j - 3) \pm \sqrt{(-15 - j8)}}{2} \tag{3.19}$$

Now we need to determine $(-15 - j8)^{1/2}$ so first we express it in polar form. Since

$$|-15 - j8| = \sqrt{[(15)^2 + (8)^2]} = 17$$

and from Figure 3.10

$$\arg(-15 - j8) = -(\pi - \tan^{-1}\tfrac{8}{15})$$

$$= -2.6516$$

we have

$$-15 - j8 = 17[\cos(2.6516) - j\sin(2.6516)]$$

Figure 3.10
The complex
number $-15 - j8$.

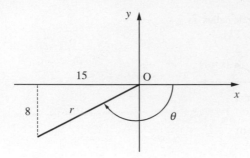

From (3.17)

$$(-15 - j8)^{1/2} = (17)^{1/2}\left[\cos\left(\frac{2.6516}{2} + \frac{2\pi k}{2}\right) - j\sin\left(\frac{2.6516}{2} + \frac{2\pi k}{2}\right)\right]$$

$$= (17)^{1/2}[\cos(1.3258 + \pi k) - j\sin(1.3258 + \pi k)], \quad k = 0, 1$$

Thus we have the two square roots

$$(-15 - j8)^{1/2} = (17)^{1/2}[\cos(1.3258) - j\sin(1.3258)] = 1 - j4 \quad \text{(for } k = 0\text{)}$$

(the reader should verify that $(1 - j4)^2 = -15 - j8$)

and

$$(-15 - j8)^{1/2} = (17)^{1/2}[\cos(4.4674) - j\sin(4.4674)] = -1 + j4 \quad \text{(for } k = 1\text{)}$$

Substituting back in (3.19) gives the roots of the quadratic as

$$z = 2 - j3 \quad \text{and} \quad 1 + j$$

3.3.2 Powers of trigonometric functions and multiple angles

Euler's formula may be used to express $\sin^n\theta$ and $\cos^n\theta$ in terms of sines and cosines of multiple angles. If $z = \cos\theta + j\sin\theta$ then

$$z^n = \cos n\theta + j\sin n\theta$$

and

$$z^{-n} = \cos n\theta - j\sin n\theta$$

so that

$$z^n + z^{-n} = 2\cos n\theta \tag{3.20a}$$

$$z^n - z^{-n} = 2j\sin n\theta \tag{3.20b}$$

Using these results, $\cos^n\theta$ and $\sin^n\theta$ can be expressed in terms of sines and cosines of multiple angles as illustrated in Example 3.20.

Example 3.20 Expand in terms of sines and cosines of multiple angles

(a) $\cos^5\theta$ (b) $\sin^6\theta$

Solution   (a) Using (3.20a) with $n = 1$,

$$(2\cos\theta)^5 = \left(z + \frac{1}{z}\right)^5 = z^5 + 5z^3 + 10z + \frac{10}{z} + \frac{5}{z^3} + \frac{1}{z^5}$$

so that

$$32\cos^5\theta = \left(z^5 + \frac{1}{z^5}\right) + 5\left(z^3 + \frac{1}{z^3}\right) + 10\left(z + \frac{1}{z}\right)$$

which, on using (3.20a) with $n = 5$, 3 and 1, gives

$$\cos^5\theta = \tfrac{1}{32}(2\cos 5\theta + 10\cos 3\theta + 20\cos\theta) = \tfrac{1}{16}(\cos 5\theta + 5\cos 3\theta + 10\cos\theta)$$

(b) Using (3.20b) with $n = 1$,

$$(2j\sin\theta)^6 = \left(z - \frac{1}{z}\right)^6 = z^6 - 6z^4 + 15z^2 - 20 + \frac{15}{z^2} - \frac{6}{z^4} + \frac{1}{z^6}$$

which, on noting that $j^6 = -1$, gives

$$-64\sin^6\theta = \left(z^6 + \frac{1}{z^6}\right) - 6\left(z^4 + \frac{1}{z^4}\right) + 15\left(z^2 + \frac{1}{z^2}\right) - 20$$

Using (3.20a) with $n = 6$, 4 and 2 then gives

$$\sin^6\theta = -\tfrac{1}{64}(2\cos 6\theta - 12\cos 4\theta + 30\cos 2\theta - 20)$$
$$= \tfrac{1}{32}(10 - 15\cos 2\theta + 6\cos 4\theta - \cos 6\theta)$$

Conversely, de Moivre's theorem may be used to expand $\cos n\theta$ and $\sin n\theta$, where n is a positive integer, as polynomials in $\cos\theta$ and $\sin\theta$. From the theorem

$$\cos n\theta + j\sin n\theta = (\cos\theta + j\sin\theta)^n$$

we obtain, writing $s = \sin\theta$ and $c = \cos\theta$ for convenience,

$$\cos n\theta + j\sin n\theta = (c + js)^n = c^n + jnc^{n-1}s + j^2\frac{n(n-1)}{2!}c^{n-2}s^2 + \ldots + j^n s^n$$

Equating real and imaginary parts yields

$$\cos n\theta = c^n - \frac{n(n-1)}{2!}c^{n-2}s^2 + \frac{n(n-1)(n-2)(n-3)}{4!}c^{n-4}s^4 + \ldots$$

and

$$\sin n\theta = nc^{n-1}s - \frac{n(n-1)(n-2)}{3!}c^{n-3}s^3 + \ldots$$

Using the trigonometric identity $\cos^2\theta = 1 - \sin^2\theta$ (so that $c^2 = 1 - s^2$), we see that

(a) $\cos n\theta$ can be expanded in terms of $(\cos\theta)^n$ for any n or in terms of $(\sin\theta)^n$ if n is even;

(b) $\sin n\theta$ can be expanded in terms of $(\sin\theta)^n$ if n is odd.

Example 3.21   Expand $\cos 4\theta$ as a polynomial in $\cos \theta$.

Solution   By de Moivre's theorem,

$$(\cos 4\theta + j \sin 4\theta) = (\cos \theta + j \sin \theta)^4 = (c + js)^4$$

$$= c^4 + j4c^3s + j^2 6c^2s^2 + j^3 4cs^3 + j^4 s^4$$

$$= c^4 + j4c^3s - 6c^2s^2 - j4cs^3 + s^4$$

Equating real parts,

$$\cos 4\theta = c^4 - 6c^2s^2 + s^4$$

which on using $s^2 = 1 - c^2$ gives

$$\cos 4\theta = c^4 - 6c^2(1 - c^2) + (1 - c^2)^2 = 8c^4 - 8c^2 + 1$$

Thus

$$\cos 4\theta = 8 \cos^4\theta - 8 \cos^2\theta + 1$$

Note that by equating imaginary parts we could have obtained a polynomial expansion for $\sin 4\theta$.

3.3.3   Exercises

24   Use de Moivre's theorem to calculate the third and fourth powers of the complex numbers

(a) $1 + j$ (b) $\sqrt{3} - j$ (c) $-3 + j4$

(d) $1 - j\sqrt{3}$ (e) $-1 + j\sqrt{3}$ (f) $-1 - j\sqrt{3}$

(The moduli and arguments of these numbers were found in Exercises 3.2.4, Question 6.)

25   Expand in terms of multiple angles

(a) $\cos^4\theta$ (b) $\sin^3\theta$

26   Use the method of Section 3.3.2 to prove the following results:

(a) $\sin 3\theta = 3 \cos^2\theta \sin \theta - \sin^3\theta$

(b) $\cos 8\theta = 128 \cos^8\theta - 256 \cos^6\theta + 160 \cos^4\theta - 32 \cos^2\theta + 1$

(c) $\tan 5\theta = \dfrac{5 \tan \theta - 10 \tan^3\theta + \tan^5\theta}{1 - 10 \tan^2\theta + 5 \tan^4\theta}$

27   Find the three values of $(8 + j8)^{1/3}$ and show them on an Argand diagram.

28   Find the following complex numbers in their polar forms:

(a) $(\sqrt{3} - j)^{1/4}$ (b) $(j8)^{1/3}$

(c) $(3 - j3)^{-2/3}$ (d) $(-1)^{1/4}$

(e) $(2 + j2)^{4/3}$ (f) $(5 - j3)^{-1/2}$

29   Obtain the four solutions of the equation

$$z^4 = 3 - j4$$

giving your answers to three decimal places.

30   Solve the quadratic equation

$$z^2 - (3 + j5)z + j8 - 5 = 0$$

31   Find the values of $z^{1/3}$, where $z = \cos 2\pi + j \sin 2\pi$. Generalize this to an expression for $1^{1/n}$. Hence solve the equations

(a) $\left[\dfrac{z - 2}{z + 2}\right]^5 = 1$ (*Hint*: first show that there are only 4 roots)

(b) $(z - 3)^6 - z^6 = 0$

3.4 Loci in the complex plane

A **locus** (plural **loci**) is the set of points that have a specified property. For example, a circle is the locus of the points in a plane that are a fixed distance, its radius, from a fixed point, its centre. The property may be specified in words or algebraically. Loci occur frequently in engineering contexts, from the design of safety guards around moving machinery to the design of aircraft wing sections. The Argand diagram representation of complex numbers as points on a plane often makes it possible to represent complicated loci very concisely in terms of a complex variable and this simplifies the engineering analysis. This occurs in a wide range of engineering problems, from water percolation through dams to the design of microelectronic devices.

3.4.1 Straight lines

There are many ways in which straight lines may be represented using complex numbers. We will illustrate these with a number of examples.

Example 3.22 Describe the locus of z given by

(a) $\mathrm{Re}(z) = 4$ (b) $\arg(z - 1 - \mathrm{j}) = \pi/4$

(c) $\left| \dfrac{z - \mathrm{j}2}{z - 1} \right| = 1$ (d) $\mathrm{Im}((1 - \mathrm{j}2)z) = 3$

Solution (a) Here $z = 4 + \mathrm{j}y$ for any real y, so that the locus is the vertical straight line with equation $x = 4$ illustrated in Figure 3.11(a).

(b) Here $z = 1 + \mathrm{j} + r(\cos \pi/4 + \mathrm{j} \sin \pi/4)$ for any positive (> 0) real number r, so that the locus is a half-line making an angle $\pi/4$ with the positive x direction with the end point $(1, 1)$ *excluded* (since arg 0 is not defined). Algebraically we can write it as $y = x$, $x > 1$, and it is illustrated in Figure 3.11(b).

(c) The equation, in this case, may be written

$$|z - \mathrm{j}2| = |z - 1|$$

Recalling the definition of modulus, we can rewrite this as

$$\sqrt{[x^2 + (y - 2)^2]} = \sqrt{[(x - 1)^2 + y^2]}$$

Squaring both sides and multiplying out, we obtain

$$x^2 + y^2 - 4y + 4 = x^2 - 2x + 1 + y^2$$

which simplifies to

$$y = \tfrac{1}{2}x + \tfrac{3}{4}$$

the equation of a straight line.

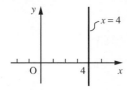

(a) Line $x = 4$

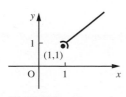

(b) Half-line $y = x$, $x > 1$

Figure 3.11

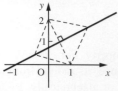

(c) Line $y = \frac{1}{2}x + \frac{3}{4}$

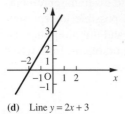

(d) Line $y = 2x + 3$

Figure 3.11
continued

Alternatively, we can interpret $|z - j2|$ as the distance on the Argand diagram from the point $0 + j2$ to the point z, and $|z - 1|$ as the distance from the point $1 + j0$ to the point z, so that

$$|z - j2| = |z - 1|$$

is the locus of points that are equidistant from the two fixed points $(0, 2)$ and $(1, 0)$, as shown in Figure 3.11(c).

(d) Writing $z = x + jy$,

$$(1 - j2)z = (1 - j2)(x + jy) = x + 2y + j(y - 2x)$$

so that $\mathrm{Im}((1 - j2)z) = 3$, implies $y - 2x = 3$.

Thus $\mathrm{Im}((1 - j2)z) = 3$ describes the straight line

$$y = 2x + 3$$

illustrated in Figure 3.11(d).

3.4.2 Circles

The simplest representation of a circle on the Argand diagram makes use of the fact that $|z - z_1|$ is the distance between the point $z = x + jy$ and the point $z_1 = a + jb$ on the diagram. Thus a circle of radius R and centre (a, b), illustrated in Figure 3.12, may be written

$$|z - z_1| = R$$

We can also write this as $z - z_1 = Re^{jt}$, where t is a parameter such that

$$-\pi < t \leqslant \pi$$

Figure 3.12
The circle $|z - z_1| = R$.

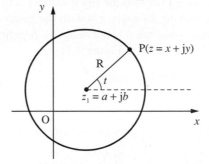

Example 3.23 Find the cartesian equation of the circle

$$|z - (2 + j3)| = 2$$

Solution Now,

$$z - (2 + j3) = (x - 2) + j(y - 3)$$

so that

$$|z - (2 + j3)| = \sqrt{[(x - 2)^2 + (y - 3)^2]}$$

and hence on the circle

$$|z - (2 + j3)| = 2$$

we have

$$\sqrt{[(x - 2)^2 + (y - 3)^2]} = 2$$

which implies

$$(x - 2)^2 + (y - 3)^2 = 4$$

indicating that the circle has centre (2, 3) and radius 2.

This may be written in the standard form

$$x^2 + y^2 - 4x - 6y + 9 = 0$$

This is not the only method of representing a circle, as is shown in the following two examples.

Example 3.24 Find the cartesian equation of the curve whose equation on the Argand diagram is

$$\left| \frac{z - j}{z - 1 - j2} \right| = \sqrt{2}$$

Solution We can interpret this equation as 'the distance between z and j is $\sqrt{2}$ times the distance between z and $(1 + j2)$', so this is different from Example 3.22(d).

Putting $z = x + jy$ into the equation gives

$$|x + j(y - 1)| = \sqrt{2}|(x - 1) + j(y - 2)|$$

Thus

$$\sqrt{[x^2 + (y - 1)^2]} = \sqrt{2}\sqrt{[(x - 1)^2 + (y - 2)^2]}$$

which implies

$$x^2 + (y - 1)^2 = 2[(x - 1)^2 + (y - 2)^2]$$

Multiplying out the brackets and collecting terms we obtain

$$x^2 + y^2 - 4x - 6y + 9 = 0 \quad \text{or} \quad (x - 2)^2 + (y - 3)^2 = 4$$

which is the equation of the circle of centre (2, 3), and radius 2.

This is a special case of a general result. If z_1 and z_2 are fixed complex numbers and k is a positive real number, then the locus of z which satisfies $\left| \dfrac{z - z_1}{z - z_2} \right| = k$ is a circle, known as the circle of Apollonius, *unless* $k = 1$. When $k = 1$, the locus is a straight line, as we saw in Example 3.22(d).

Example 3.25   Find the locus of z in the Argand diagram such that

$$\text{Re}[(z - j)/(z + 1)] = 0$$

Solution   Setting $z = x + jy$, as usual, we obtain

$$\frac{z - j}{z + 1} = \frac{x + j(y - 1)}{(x + 1) + jy} = \frac{[x + j(y - 1)][(x + 1) - jy]}{(x + 1)^2 + y^2}$$

Hence $\text{Re}[(z - j)/(z + 1)] = 0$ implies $x(x + 1) + y(y - 1) = 0$.
  Rearranging this, we have

$$x^2 + y^2 + x - y = 0$$

and

$$(x + \tfrac{1}{2})^2 + (y - \tfrac{1}{2})^2 = \tfrac{1}{2}$$

Hence the locus of z on the Argand diagram is a circle of centre $(-\tfrac{1}{2}, \tfrac{1}{2})$ and radius $\sqrt{2}/2$.

3.4.3 More general loci

In general we approach the problem of finding the locus of z on the Argand diagram using a mixture of elementary pure geometry and algebraic manipulation of expressions involving $z = x + jy$. We illustrate this in Example 3.26.

Example 3.26   Find the cartesian equation of the locus of z given by

$$|z + 1| + |z - 1| = 4$$

Solution   The defining equation here may be interpreted as the sum of the distances of the point z from the points 1 and -1 is a constant ($= 4$). By elementary considerations (Figure 3.13) we can see that the locus passes through $(2, 0)$, $(0, \sqrt{3})$, $(-2, 0)$ and $(0, -\sqrt{3})$. Results from classical geometry would identify the locus as an ellipse with foci at $(1, 0)$ and $(-1, 0)$, using the 'string property' (see Example 1.30). Using algebraic methods, however, we set $z = x + jy$ into the equation, giving

$$\sqrt{[(x + 1)^2 + y^2]} + \sqrt{[(x - 1)^2 + y^2]} = 4$$

Figure 3.13
The ellipse of
Example 3.26.

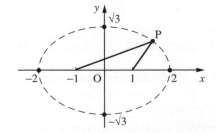

Rewriting this equation as

$$\sqrt{[(x+1)^2 + y^2]} = 4 - \sqrt{[(x-1)^2 + y^2]}$$

and squaring both sides gives

$$(x+1)^2 + y^2 = 16 - 8\sqrt{[(x-1)^2 + y^2]} + (x-1)^2 + y^2$$

This simplifies to give

$$4 - x = 2\sqrt{[(x-1)^2 + y^2]}$$

so that squaring both sides again gives

$$16 - 8x + x^2 = 4[x^2 - 2x + 1 + y^2]$$

which reduces to

$$\frac{x^2}{4} + \frac{y^2}{3} = 1$$

in the standard form of an ellipse.

3.4.4 Exercises

32 Let $z = 8 + j$ and $w = 4 + j4$. Calculate the distance on the Argand diagram from z to w and from z to $-w$.

33 Describe the locus of z when

(a) $\operatorname{Re} z = 5$ (b) $|z - 1| = 3$

(c) $\left|\dfrac{z-1}{z+1}\right| = 3$ (d) $\arg(z - 2) = \pi/4$

34 The circle $x^2 + y^2 + 4x = 0$ and the straight line $y = 3x + 2$ are taken to lie on the Argand diagram. Describe the circle and the straight line in terms of z.

35 Identify and sketch the loci on the complex plane given by

(a) $\operatorname{Re}\left(\dfrac{z+j}{z-j}\right) = 1$ (b) $\operatorname{Re}\left(\dfrac{z+j}{z-j}\right) = 2$

(c) $\left|\dfrac{z+j}{z-j}\right| = 3$ (d) $\tan\arg\left(\dfrac{z+j}{z-j}\right) = \sqrt{3}$

(e) $\operatorname{Im}(z^2) = 2$ (f) $|z + j| + |z - 1| = 2$

(g) $|z + j| - |z - 1| = \frac{1}{2}$ (h) $\arg(z + j2) = \frac{1}{4}\pi$

(i) $\arg(2z - 3) = -\frac{2}{3}\pi$ (j) $|z - j2| = 1$

36 Express as simply as possible the following loci in terms of a complex variable:

(a) $y = 3x - 2$ (b) $x^2 + y^2 + 4x = 0$

(c) $x^2 + y^2 + 2x - 4y - 4 = 0$ (d) $x^2 - y^2 = 1$

37 Find the locus of the point z in the Argand diagram which satisfies the equation

(a) $|z - 1| = 2$ (b) $|2z - 1| = 3$

(c) $|z - 2 - j3| = 4$ (d) $\arg(z) = 0$

(e) $|z - 4| = 3|z + 1|$ (f) $\arg\left(\dfrac{z-1}{z-j}\right) = \frac{1}{2}\pi$

38 Find the cartesian equation of the circle given by

$$\left|\frac{z+j}{z-1}\right| = \sqrt{2}$$

and give two other representations of the circle in terms of z.

39 Given that the argument of $(z - 1)/(z + 1)$ is $\frac{1}{4}\pi$, show that the locus of z in the Argand diagram is part of a circle of centre $(0, 1)$ and radius $\sqrt{2}$.

40 Find the cartesian equation of the locus of the point $z = x + jy$ that moves in the Argand diagram such that $|(z + 1)/(z - 2)| = 2$.

3.5 Engineering application: alternating currents in electrical networks

When an alternating current $i = I \sin \omega t$ (ω is a constant and t is the time) flows in a circuit the corresponding voltage depends on ω and on the resistance, capacitance and inductance of the circuit. (Note that the frequency of the current is $\omega/2\pi$.) For simplicity we shall separate these three elements and consider their effects individually.

For a resistor of resistance R the corresponding voltage is $v = IR \sin \omega t$. This voltage is 'in phase' with the current. It is zero at the same times as i and achieves its maxima at the same times as i, as shown in Figure 3.14. For a capacitor of capacitance C the corresponding voltage is $v = (I/\omega C) \sin(\omega t - \frac{1}{2}\pi)$, as shown in Figure 3.15. Here the voltage 'lags' behind the current by a phase of $\frac{1}{2}\pi$. For an inductor of inductance L the corresponding voltage is $v = \omega L I \sin(\omega t + \frac{1}{2}\pi)$, as shown in Figure 3.16. Here the voltage 'leads' the current by a phase of $\frac{1}{2}\pi$.

Figure 3.14
A resistor of
resistance R.

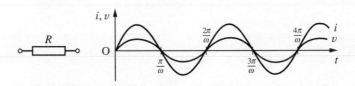

Figure 3.15
A capacitor of
capacitance C.

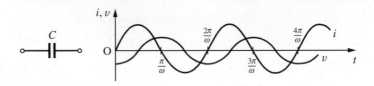

Figure 3.16
An inductor of
inductance L.

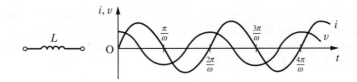

Combining these results to find v in the case of a general network is easily done using the properties of complex numbers. Remembering that $\sin \theta = \mathrm{Im}(e^{j\theta})$, we can summarize the results as

$$v = \begin{cases} \mathrm{Im}(IR e^{j\omega t}) & \text{for a resistor} \\[2mm] \mathrm{Im}\left(\dfrac{I}{\omega C} e^{j(\omega t - \pi/2)} \right) & \text{for a capacitor} \\[2mm] \mathrm{Im}(\omega L I e^{j(\omega t + \pi/2)}) & \text{for an inductor} \end{cases}$$

Now $e^{j\pi/2} = \cos \frac{1}{2}\pi + j \sin \frac{1}{2}\pi = j$ and $e^{-j\pi/2} = -j$, so we may rewrite these as

$$v = \mathrm{Im}(IZ e^{j\omega t})$$

where

$$
Z = \begin{cases}
R & \text{for a resistor} \\[2mm]
-\dfrac{j}{\omega C} & \text{for a capacitor} \\[2mm]
j\omega L & \text{for an inductor}
\end{cases}
$$

Z is called the **complex impedance** of the element, and $V = IZ$ is the **complex voltage**. For the general LCR circuit shown in Figure 3.17 the complex voltage V is the algebraic sum of the complex voltages of the individual elements; that is,

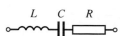

Figure 3.17
A linear LCR circuit.

$$
V = IR + j\omega L I - \frac{jI}{\omega C} = IZ
$$

where

$$
Z = R + j\omega L - \frac{j}{\omega C}
$$

The actual voltage

$$
v = \mathrm{Im}(V\mathrm{e}^{j\omega t}) = I|Z|\sin(\omega t + \phi)
$$

where

$$
|Z| = \left[R^2 + \left(L\omega - \frac{1}{C\omega} \right)^2 \right]^{1/2}
$$

is the **impedance** of the circuit and

$$
\phi = \tan^{-1}\left(\frac{L\omega - 1/C\omega}{R} \right)
$$

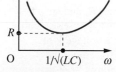

Figure 3.18
The impedance of
an LCR circuit.

is the **phase**. The impedance $|Z|$ clearly varies with ω, and the graph of this dependence is shown in Figure 3.18. The minimum value occurs when $L\omega = 1/C\omega$; that is, when $\omega = 1/\sqrt{(LC)}$. This implies that the circuit 'blocks' currents with low and high frequencies, and 'passes' currents with frequencies near $1/(2\pi\sqrt{(LC)})$.

Example 3.27   Calculate the complex impedance of the element shown in Figure 3.19 when an alternating current of frequency 100 Hz flows.

Solution   The complex impedance is the sum of the individual impedances. Thus

$$
Z = R + j\omega L
$$

15 Ω   41.3 mH

Figure 3.19
The element of
Example 3.27.

Here $R = 15\,\Omega$, $\omega = 2\pi \times 100\,\mathrm{rad\,s}^{-1}$ and $L = 41.3 \times 10^{-3}\,\mathrm{H}$, so that

$$
Z = 15 + j25.9
$$

and $|Z| = 30\,\Omega$ and $\phi = \tfrac{1}{3}\pi$.

3.5.1 Exercises

41 Calculate the complex impedance for the circuit shown in Figure 3.20 when an alternating current of frequency 50 Hz flows.

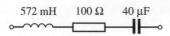

Figure 3.20

42 The complex impedance of two circuit elements in series as shown in Figure 3.21(a) is the sum of the complex impedances of the individual elements, and the reciprocal of the impedance of two elements in parallel is the sum of the reciprocals of the individual impendances, as shown in Figure 3.21(b). Use these results to calculate the complex impedance of the network shown in Figure 3.22, where $Z_1 = 1 + j\,\Omega$, $Z_2 = 5 - j5\,\Omega$ and $Z_3 = 1 + j2\,\Omega$.

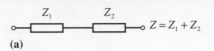

(a)

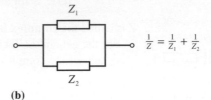

(b)

Figure 3.21

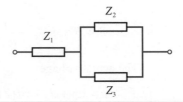

Figure 3.22

3.6 Review exercises (1–30)

1 For x and y real solve the equation

$$\frac{jy}{jx + 1} - \frac{3y + j4}{3x + y} = 0$$

2 Given $z = (2 + j)/(1 - j)$, find the real and imaginary parts of $z + z^{-1}$.

3 (a) Find the loci in the Argand diagram corresponding to the equation

$$|z - 1| = 2|z - j|$$

(b) If the point $z = x + jy$ describes the circle $|z - 1| = 1$, show that the real part of $1/(z - 2)$ is constant.

4 Writing $\ln[(x + jy + a)/(x + jy - a)] = u + jv$, show that

(a) $x^2 + y^2 - 2ax\coth u + a^2 = 0$

(b) $x = a\sinh u/(\cosh u - \cos v)$

(c) $|x + jy|^2 = a^2(\cosh u + \cos v)/(\cosh u - \cos v)$

5 A circuit consists of a resistance R_1 and an inductance L in parallel connected in series with a second resistance R_2. When a voltage V of frequency $\omega/2\pi$ is applied to the circuit the complex impedance Z is given by

$$\frac{1}{Z - R_2} = \frac{1}{R_1} + \frac{1}{j\omega L}$$

Show that if R_1 varies from zero to infinity the locus of Z on the Argand diagram is part of a circle and find its centre and radius.

6 (a) Express $\cos 6\theta$ as a polynomial in $\cos\theta$.

(b) Given $z = \cos\theta + j\sin\theta$ show, by expanding $(z + 1/z)^5(z - 1/z)^5$ or otherwise, that

$$\sin^5\theta\cos^5\theta = \frac{1}{2^9}(\sin 10\theta - 5\sin 6\theta + 10\sin 2\theta)$$

7 Show that the solutions of

$$z^4 - 3z^2 + 1 = 0$$

are given by

$$z = 2\cos 36°, \ 2\cos 72°, \ 2\cos 216°, \ 2\cos 252°$$

Hence show that

(a) $\cos 36° = \frac{1}{4}(\sqrt{5} + 1)$ (b) $\cos 72° = \frac{1}{4}(\sqrt{5} - 1)$

8 Prove that if $p(z)$ is a polynomial in z with real coefficients then $[p(z)]^* = p(z^*)$. Deduce that the roots of a polynomial equation with real coefficients occur in complex-conjugate pairs.

9 Show that

(a) $\sin^4\theta = \frac{1}{8}[\cos 4\theta - 4\cos 2\theta + 3]$

(b) $\sin^5\theta = \frac{1}{16}[\sin 5\theta - 5\sin 3\theta + 10\sin\theta]$

(c) $\cos^6\theta = \frac{1}{32}[\cos 6\theta + 6\cos 4\theta + 15\cos 2\theta + 10]$

(d) $\cos^2\theta \sin^3\theta = \frac{1}{16}[2\sin\theta + \sin 3\theta - \sin 5\theta]$

10 Prove that the statements

(a) $|z + 1| > |z - 1|$ (b) $\text{Re}(z) > 0$

are equivalent.

11 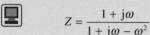 For a certain network the impedance Z is given by

$$Z = \frac{1 + j\omega}{1 + j\omega - \omega^2}$$

Sketch the variation of $|Z|$ and arg Z with the frequency ω. (Take values of $\omega \gtrless 0$.)

12 The characteristic impedance Z_0 and the propagation constant C of a transmission line are given by

$$Z_0 = \sqrt{(Z/Y)} \quad \text{and} \quad C = \sqrt{(ZY)}$$

where Z is the series impedance and Y the admittance of the line, and $\text{Re}(Z_0) > 0$ and $\text{Re}(C) > 0$. Find Z_0 and C when $Z = 0.5 + j0.3\,\Omega$ and $Y = (1 - j250) \times 10^{-8}\,\Omega$.

13 The input impedance Z of a particular network is related to the terminating impedance z by the equation

$$Z = \frac{(1 + j)z - 2 + j4}{z + 1 + j}$$

Find Z when $z = 0$, 1 and $j\,\Omega$ and sketch the variation of $|Z|$ and arg Z as z moves along the positive real axis from the origin.

14 Find the modulus and argument of

$$\frac{(3 + j4)^4 (12 - j5)^2}{(3 - j4)^2 (12 + j5)^3}$$

15 Express in the form $a + jb$, with a and b expressed to 2dp

(a) $\sin(0.2 + j0.48)$ (b) $\cosh^{-1}(j2)$

(c) $\cosh(3.8 - j5.2)$ (d) $\ln(2 + j)$

(e) $\cos(\frac{1}{4}\pi - j)$

16 Using complex numbers, show that

$$\sin^7\theta = \frac{1}{64}(35\sin\theta - 21\sin 3\theta +$$
$$7\sin 5\theta - \sin 7\theta)$$

17 Two impedances Z_1 and Z_0 are related by the equation

$$Z_1 = Z_0 \tanh(\alpha l + j\beta l)$$

where α, β and l are real. If αl is so small that we may take $\sinh \alpha l = \alpha l$, $\cosh \alpha l = 1$ and $(\alpha l)^2$ as negligible, show that

$$Z_1 = Z_0[\alpha l \sec^2 \beta l + j\tan \beta l]$$

18 In a transmission line the voltage reflection equation is given by

$$Ke^{j\theta} = \frac{Z - Z_0}{Z + Z_0}$$

where K is a real constant, $Z = R + jX$ and $Z_0 = R_0 + jX_0$. Obtain an expression for θ, the phase angle, in terms of R_0, R, X_0 and X. Hence show that if Z_0 is purely resistive (that is, real) then

$$\theta = \tan^{-1}\left[\frac{2R_0 X}{R^2 + X^2 - R_0^2}\right]$$

assuming $R_0^2 < R^2 + X^2$.

19 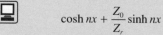 The voltage in a cable is given by the expression

$$\cosh nx + \frac{Z_0}{Z_r}\sinh nx$$

Calculate its value in the form $a + jb$, giving a and b correct to 2dp, when

$$nx = 0.40 + j0.93$$
$$Z_0 = 15 - j20 \qquad Z_r = 3 + j4$$

20 Express $Z = \cosh(0.5 + j\frac{1}{4}\pi)$ in the forms

(a) $x + jy$ (b) $re^{j\theta}$

The current in a cable is equal to the real part of the expression $e^{j0.7}/Z$. Calculate the current, giving your answer correct to 3dp.

21 Show that if the propagation constant of a cable is given by

$$X + jY = \sqrt{[(R + j\omega L)(G + j\omega C)]}$$

where R, G, ω, L and C are real, then the value of X^2 is given by

$$X^2 = \tfrac{1}{2}[RG - \omega^2 LC + \sqrt{\{(R^2 + \omega^2 L^2)}$$
$$(G^2 + \omega^2 C^2)\}]$$

22 Given $Z = (1 + j)/(3 - j4)$ obtain

(a) Z (b) $\sqrt{Z}$ (c) e^Z

(d) $\ln Z$ (e) $\sin Z$

in the form $a + jb$, a, b real, giving a and b correct to 2dp.

23 Find, in exponential form, the four values of

$$\left[\frac{7 + j24}{25}\right]^{1/4}$$

Denoting any one of these by p, show that the other three are given by $j^n p$ ($n = 1, 2, 3$).

24 Determine the six roots of the complex number $-1 + j\sqrt{3}$, in the form $re^{j\theta}$ where $-\pi < \theta \leqslant \pi$, and show that three of these are also solutions of the equation

$$\sqrt{2}Z^3 + 1 + j\sqrt{3} = 0$$

25 Find the real part of

$$\frac{(R + j\omega L)/j\omega C}{j\omega L + R + 1/j\omega C}$$

and deduce that if R^2 is negligible compared with $(\omega L)^2$ and $(LC\omega^2)^2$ is negligible compared with unity then the real part is approximately $R(1 + 2LC\omega^2)$.

26 Show that if ω is a complex cube root of unity, then $\omega^2 + \omega + 1 = 0$. Deduce that

$$(x + y + z)(x + \omega y + \omega^2 z)(x + \omega^2 y + \omega z)$$
$$= x^3 + y^3 + z^3 - 3xyz$$

Hence show that the three roots of

$$x^3 + (-3yz)x + (y^3 + z^3) = 0$$

are

$$x = -(y + z),\ -(\omega y + \omega^2 z),\ -(\omega^2 y + \omega z)$$

Use this result to obtain Cardano's solution to the cubic equation

$$x^3 + qx + r = 0$$

in the form

$$-(u + v)$$

where $u^3 = \tfrac{1}{2}r + \sqrt{[\tfrac{1}{4}r^2 + \tfrac{1}{27}q^3]}$

and $v^3 = \tfrac{1}{2}r - \sqrt{[\tfrac{1}{4}r^2 + \tfrac{1}{27}q^3]}$

Express the remaining two roots in terms of u, v and ω and find the condition that all three roots are real.

27 ABCD is a square, lettered anticlockwise, on an Argand diagram. If the points A, B represent $3 + j2$, $-1 + j4$ respectively, show that C lies on the real axis, and find the number represented by D and the length of AB.

28 If $z_1 = 3 + j2$ and $z_2 = 1 + j$, and O, P, Q, R represent the numbers 0, z_1, $z_1 z_2$, z_1/z_2 on the Argand diagram, show that RP is parallel to OQ and is half its length.

29 Show that as z describes the circle $z = be^{j\theta}$, $u + jv = z + a^2/z$ describes an ellipse ($a \neq b$). What is the image locus when $a = b$?

30 Show that as θ varies the point $z = a(h + \cos\theta) + ja(k + \sin\theta)$ describes a circle. The Joukowski transformation $u + jv = z + l^2/z$ is applied to this circle to produce an aerofoil shape in the u–v plane. Show that the coordinates of the aerofoil can be written in the form

$$\frac{u}{a} = (h + \cos\theta)$$

$$\times \left\{1 + \frac{l^2}{a^2(1 + h^2 + k^2 + 2h\cos\theta + 2k\sin\theta)}\right\}$$

$$\frac{v}{a} = (k + \sin\theta)$$

$$\times \left\{1 - \frac{l^2}{a^2(1 + h^2 + k^2 + 2h\cos\theta + 2k\sin\theta)}\right\}$$

Trace this aerofoil for the case where $h = 0.04$, $k = 0.05$ and $l^2 = 0.8a^2$, showing that it has a rounded trailing edge.

4 Vector Algebra

Chapter 4 Contents

4.1 Introduction

Much of the work of engineers and scientists involves forces. Ensuring the structural integrity of a building or a bridge involves knowing the forces acting on the system and designing the structural members to withstand them. Many have seen the dramatic pictures of the Tacoma bridge disaster (see also Section 10.10.3), when the forces acting on the bridge were not predicted accurately. To analyse such a system requires the use of Newton's laws in a situation where vector notation is essential. Similarly, in a reciprocating engine periodic forces act, and Newton's laws are used to design a crankshaft that will reduce the side forces to zero, thereby minimizing wear on the moving parts. Forces are three-dimensional quantities and provide one of the commonest examples of vectors. Associated with these forces are accelerations and velocities, which can also be represented by vectors. The use of formal mathematical notation and rules becomes progressively more important as problems become complicated and, in particular, in three-dimensional situations. Forces, velocities and accelerations all satisfy rules of addition that identify them as vectors. In this chapter we shall construct an algebraic theory for the manipulation of vectors and see how it can be applied to some simple practical problems.

The ideas behind vectors as formal quantities developed mainly during the nineteenth century, and they became a well-established tool in the twentieth century. Vectors provide a convenient and compact way of dealing with multi-dimensional situations without the problem of writing down every bit of information. They allow the principles of the subject to be developed without being obscured by complicated notation.

It is inconceivable that modern scientists and engineers could work successfully without computers. Since such machines cannot think like an engineer or scientist, they have to be told in a totally precise and formal way what to do. For instance, a robot arm needs to be given instructions on how to position itself to perform a spot weld. Three-dimensional vectors prove to be the perfect way to tell the computer how to specify the position of the workpiece of the robot arm, and a set of rules then tells the robot how to move to its working position.

Computers have put a great power at the disposal of the engineer; problems that proved to be impossible 50 years ago are now routine. With the aid of numerical algorithms, equations can often be solved very quickly. The stressing of a large structure or an aircraft wing, the lubrication of shafts and bearings, the flow of sewage in pipes and the flow past the fuselage of an aircraft are all examples of systems that were well understood in principle but could not be analysed until the necessary computer power became available. Algorithms are usually written in terms of vectors and matrices (see Chapter 5), since these form a natural setting for the numerical solution of engineering problems and are also ideal for the computer. It is vital that the manipulation of vectors be understood before embarking on more complex mathematical structures used in engineering computations.

Perhaps the most powerful influence of computers is in their graphical capabilities, which have proved invaluable in displaying the static and dynamic behaviour of systems. We accept this tool without thinking how it works. A simple example shows the complexity. How do we display a box with an open top with 'hidden' lines when we look at it from a given angle? The problem is a complicated three-dimensional one that must be analysed instantly by a computer. Vectors allow us to define lines that can be projected on to the screen, and intersections can then be computed so that the 'hidden' portion can be eliminated. Extending the analysis to a less regular shape is a formidable vector problem. Work of this type is the basis of CAD/CAM systems, which now assist engineers in all stages of the manufacturing process from design to production of a finished product. Such systems typically allow engineers to manipulate the product geometry during initial design, to produce working drawings, to generate toolpaths in the production process and generally to automate a host of previously tedious and time-consuming tasks.

The general development of the theory of vectors is closely associated with coordinate geometry, so we shall introduce a few ideas in the next section that will be used later in the chapter. The comments largely concern the two- and three-dimensional cases, but we shall mention higher-dimensional extensions where they are relevant to later work such as on the theory of matrices. While in two and three dimensions we can appeal to geometrical intuition, it is necessary to work in a much more formal way in higher dimensions, as with many other areas of mathematics.

4.2 Basic definitions and results

4.2.1 Cartesian coordinates

Setting up rectangular cartesian axes $Oxyz$ or $Ox_1x_2x_3$, we define the position of a point by **coordinates** or **components** (x, y, z) or (x_1, x_2, x_3) as indicated in Figure 4.1(a). The indicial notation is particularly important when we consider vectors in many dimensions $(x_1, x_2, \ldots, x_n)$. The axes Ox, Oy, Oz, in that order, are assumed to be right-handed in the sense of Figure 4.1(b), so that a rotation of a right-handed screw from Ox to Oy advances it along Oz, a rotation from Oy to Oz advances it along Ox and a rotation from Oz to Ox advances it along Oy. This is an accepted convention, and it will be seen to be particularly important in Section 4.2.9 when we deal with the vector product.

The length of OP in Figure 4.1(a) is obtained from Pythagoras' theorem as

$$r = (x^2 + y^2 + z^2)^{1/2}$$

The angle $\angle POx$ is the angle that OP makes with the positive x direction as in Figure 4.2. We can see that

Figure 4.1
(a) Right-handed coordinate axes.
(b) Right-hand rule.

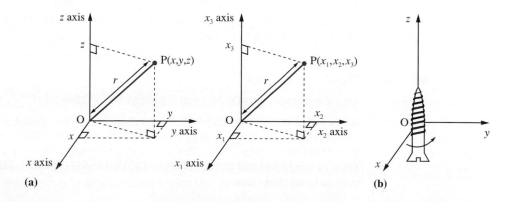

(a)   (b)

Figure 4.2
Direction cosines
of OP, $l = \cos\alpha$,
$m = \cos\beta$, $n = \cos\gamma$.

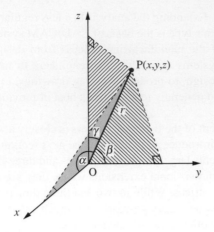

$$l = \cos\alpha = \frac{x}{r}$$

and similarly

$$m = \cos\beta = \frac{y}{r}, \quad n = \cos\gamma = \frac{z}{r}$$

The triad (l, m, n) are called the **direction cosines** of the line OP. Note that

$$l^2 + m^2 + n^2 = \frac{x^2}{r^2} + \frac{y^2}{r^2} + \frac{z^2}{r^2} = \frac{x^2 + y^2 + z^2}{r^2} = 1$$

Example 4.1 If P has coordinates $(2, -1, 3)$, find the length OP and the direction cosines of OP.

Solution $\quad OP^2 = 4 + 1 + 9, \qquad$ so that $OP = \sqrt{14}$

The direction cosines are

$$l = 2\sqrt{\tfrac{1}{14}}, \quad m = -\sqrt{\tfrac{1}{14}}, \quad n = 3\sqrt{\tfrac{1}{14}}$$

Example 4.2 A surveyor sets up his theodolite on horizontal ground, at a point O, and observes the top of a church spire, as illustrated in Figure 4.3. Relative to axes Oxyz, with Oz vertical, the surveyor measures the angles $\angle TOx = 66°$ and $\angle TOz = 57°$. The church is known to have height 35 m. Find the angle $\angle TOy$ and calculate the coordinates of T with respect to the given axes.

Solution The direction cosines

$$l = \cos 66° = 0.406\,74 \quad \text{and} \quad n = \cos 57° = 0.544\,64$$

are known and hence the third direction cosine can be computed as

$$m^2 = 1 - l^2 - n^2 = 0.537\,93$$

Thus, $m = 0.733\,44$ and hence $\angle TOy = \cos^{-1}(0.733\,44) = 42.82°$. The length $OT = r$ can now be computed from

Figure 4.3
Representation of the
axes and church spire
in Example 4.2.

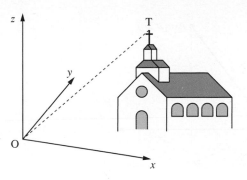

$$\cos 57° = 35/r, \quad \text{so} \quad r = 64.26 \, \text{m}$$

The remaining coordinates are obtained from

$$x/r = \cos 66° \quad \text{and} \quad y/r = \cos 42.82°$$

and hence the coordinates of T are (26.14, 47.13, 35).

4.2.2 Scalars and vectors

Quantities like distance or temperature are represented by real numbers in appropriate units, for instance 5 m or 10°C. Such quantities are called **scalars** – they obey the usual rules of real numbers and they have no direction associated with them. However, **vectors** have both a magnitude and a direction associated with them; these include force, velocity and magnetic field. To qualify as vectors, the quantities must have more than just magnitude and direction – they must also satisfy some particular rules of combination. Angular displacement in three dimensions gives an example of a quantity which has a direction and magnitude but which does not add by the addition rules of vectors, so angular displacements are *not* vectors.

We represent a vector geometrically by a line segment whose length represents the vector's magnitude in some appropriate units and whose direction represents the vector's direction, with the arrowhead indicating the sense of the vector, as shown in Figure 4.4. According to this definition, the starting point of the vector is irrelevant. In Figure 4.4, the two line segments OA and O'A' represent the same vector because their lengths are the same, their directions are the same and the sense of the arrows is the same. Thus each of these vectors is equivalent to the vector through the origin, with A given by its coordinates (a_1, a_2, a_3) as in Figure 4.5. We can therefore represent a vector in a three-dimensional space by an ordered set of three numbers or a 3-tuple. We shall see how this representation is used in Section 4.2.4.

We shall now introduce some of the basic notation and definitions for vectors. The vector of Figure 4.5 is handwritten or typewritten as a͟, a̱, $\overrightarrow{OA}$ or printed in bold-face type $\boldsymbol{a}$. Using the coordinate definition, the vector could equally be written as (a_1, a_2, a_3). There are several possible coordinate notations; the traditional one is (a_1, a_2, a_3), but in Chapter 5 on matrices we shall use an alternative standard notation.

The **modulus** or **length** or **magnitude** of a vector $\boldsymbol{a}$ is written as $|\boldsymbol{a}|$ or $|\overrightarrow{OA}|$ or a if there is no ambiguity. A vector with modulus one is called a **unit vector** and is written $\hat{\boldsymbol{a}}$, with the hat (ˆ) indicating a unit vector.

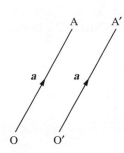

Figure 4.4
Line segments
representing a vector
a.

Figure 4.5
Representation of the
vector a by the line
segment OA.

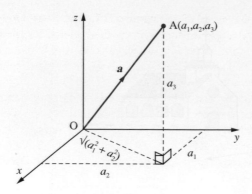

As we considered earlier, two vectors a and b are equal if and only if they have the same modulus and the same direction and sense. We write this in the usual way

$$a = b$$

If λ is a scalar and the vectors are related by $a = \lambda b$ then

if $\lambda > 0$, $\quad a$ is a vector in the same direction as b with magnitude λ times the magnitude of b;

if $\lambda < 0$, $\quad a$ is a vector in the opposite direction to b with magnitude $|\lambda|$ times the magnitude of b.

The vectors a and b are said to be **parallel** or **antiparallel** according as $\lambda > 0$ or $\lambda < 0$ respectively. (Note that we do not insert any multiplication symbol between λ and b since the common symbols $\cdot$ and $\times$ are reserved for special uses that we shall discuss later.)

The **zero** or **null vector** has zero modulus; it is written as **0** or often just as 0 when there is no ambuigity whether or not it is a vector.

4.2.3 Addition of vectors

Having introduced vectors and their basic properties, it is natural to ask if vectors can be combined. The simplest form of vector combination is addition and it is the definition of addition that finally identifies a vector. Consider the following situation. The helmsman of a small motor boat steers his vessel due east (E) at 4 knots for one hour. The path taken by the boat could be represented by the line OA in Figure 4.6. Unfortunately there is also a tidal stream running north-north-east (NNE) at $2\frac{1}{2}$ knots. Where will the boat actually be at the end of one hour?

Figure 4.6
Addition of two
vectors.

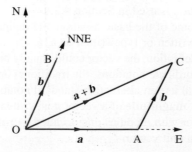

If we imagine the vessel to be steaming E for one hour through still water, and then lying still in the water and drifting with the tidal stream for one hour, we can see that it will travel from O to A in the first hour and from A to C in the second hour. If, on the other hand, the vessel steams due E through water that is simultaneously moving NNE with the tidal stream then the result will be to arrive at C after one hour. The net velocity of the boat is represented by the line OC. Putting this another way, the result of subjecting the boat to a velocity $\overrightarrow{OA}$ and a velocity $\overrightarrow{AC}$ simultaneously is the same as the result of subjecting it to a velocity $\overrightarrow{OC}$. Thus the velocity $\overrightarrow{OC}$ is the sum of the velocity $\overrightarrow{OA}$ and the velocity $\overrightarrow{AC}$.

This leads us to the **parallelogram rule** for vector addition illustrated in Figure 4.7 and stated as follows:

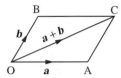

Figure 4.7
Parallelogram rule for addition of vectors.

> The sum, or resultant, of two vectors **a** and **b** is found by forming a parallelogram with **a** and **b** as two adjacent sides. The sum **a** + **b** is the vector represented by the diagonal of the parallelogram.

In Figure 4.7 the vectors $\overrightarrow{OB}$ and $\overrightarrow{AC}$ are the same, so we can rewrite the parallelogram rule as an equivalent **triangle law** (Figure 4.8), which can be stated as follows:

> If two vectors **a** and **b** are represented in magnitude and direction by the two sides of a triangle taken in order then their sum is represented in magnitude and direction by the closing third side.

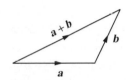

Figure 4.8
Triangle law for addition of vectors.

The triangle law for the addition of vectors can be extended to the addition of any number of vectors. If from a point O (Figure 4.9), displacements $\overrightarrow{OA}$, $\overrightarrow{AB}$, $\overrightarrow{BC}$, . . . , $\overrightarrow{LK}$ are drawn along the adjacent sides of a polygon to represent in magnitude and direction the vectors **a**, **b**, **c**, . . . , **k** respectively then the sum

$$r = a + b + c + \ldots + k$$

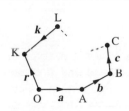

Figure 4.9
Polygon law for addition of vectors.

of these vectors is represented in magnitude and direction by the closing side OK of the polygon, the sense of the sum vector being represented by the arrow in Figure 4.9. This is referred to as the **polygon law** for the addition of vectors.

We now need to look at the usual rules of algebra for scalar quantities to check whether or not they are satisfied for vectors.

(a) Commutative law

$$a + b = b + a$$

This result is obvious from the geometrical definition, and says that order does not matter.

(b) Associative law

Geometrically, the result can be deduced using the triangle and polygon laws, as shown in Figure 4.10. We see that brackets do not matter and can be omitted.

Figure 4.10
Deduction of
associative law.

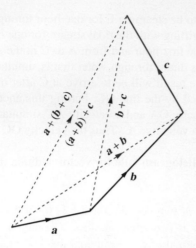

(c) ***Distributive law***

$$\lambda(a + b) = \lambda a + \lambda b$$

The result follows from similar triangles. In Figure 4.11 the side O′B′ is just λ times OB in length and in the same direction so $\overrightarrow{O'B'} = \lambda(a + b)$. The triangle law therefore gives the required result since $\overrightarrow{O'B'} = \overrightarrow{O'A'} + \overrightarrow{A'B'} = \lambda a + \lambda b$. This result just says that we can multiply brackets out by the usual laws of algebra.

Figure 4.11
Similar triangles for
the proof of the
distributive law.

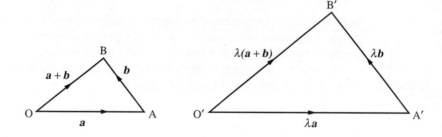

(d) ***Subtraction***
We define subtraction in the obvious way:

$$a - b = a + (-b)$$

This is illustrated geometrically in Figure 4.12, from which an important result can be deduced, namely

$$\overrightarrow{BA} = \overrightarrow{OA} - \overrightarrow{OB}$$

Figure 4.12
Subtraction of vectors.

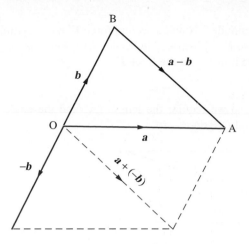

Example 4.3   From Figure 4.13, evaluate

    g in terms of a and b, f in terms of b and c

    e in terms of c and d, e in terms of f, g and h

Figure 4.13
Figure of Example 4.3.

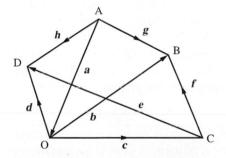

Solution   From the triangle OAB: $\overrightarrow{AB} = \overrightarrow{AO} + \overrightarrow{OB}$ and hence $g = a + b$

From the triangle OBC: $\overrightarrow{CB} = \overrightarrow{OB} - \overrightarrow{OC}$ and hence $f = b - c$

From the triangle OCD: $\overrightarrow{CD} = \overrightarrow{OD} - \overrightarrow{OC}$ and hence $e = d - c$

From the quadrilateral CBAD the polygon rule gives

$\overrightarrow{CD} + \overrightarrow{DA} + \overrightarrow{AB} + \overrightarrow{BC} = 0$ and hence $e + (-h) + g + (-f) = 0$ so $e = f - g + h$

Example 4.4   A quadrilateral OACB is defined in terms of the vectors $\overrightarrow{OA} = a$, $\overrightarrow{OB} = b$ and $\overrightarrow{OC} = b + \frac{1}{2}a$. Calculate the vector representing the other two sides $\overrightarrow{BC}$ and $\overrightarrow{CA}$.

Solution   Now $\overrightarrow{BC} = \overrightarrow{OC} - \overrightarrow{OB} = (b + \frac{1}{2}a) - b = \frac{1}{2}a$

and $\overrightarrow{CA} = \overrightarrow{OA} - \overrightarrow{OC} = a - (b + \frac{1}{2}a) = \frac{1}{2}a - b$

Example 4.5

A force F has magnitude 2 N and a second force F' has magnitude 1 N and is inclined at an angle of 60° to F as illustrated in Figure 4.14. Find the magnitude of the resultant force and the angle it makes to the force F.

Solution

Now $R = F + F'$ so we require the length OC and the angle ∠CON. The problem involves some simple trigonometry.

Figure 4.14
Figure of Example 4.5.

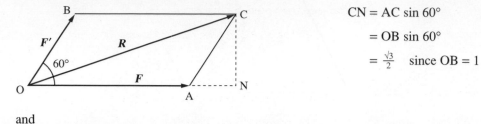

$$CN = AC \sin 60°$$
$$= OB \sin 60°$$
$$= \frac{\sqrt{3}}{2} \quad \text{since OB} = 1$$

and

$$ON = OA + AN = 2 + AC \cos 60° = 2 + OB \cos 60° = 2 + \frac{1}{2} = \frac{5}{2}$$

Thus, using Pythagoras' theorem, $OC^2 = (\frac{\sqrt{3}}{2})^2 + (\frac{5}{2})^2 = 7$ and hence the resultant has magnitude $\sqrt{7}$. The angle ∠CON is determined from $\tan \text{CON} = \dfrac{CN}{ON} = \frac{\sqrt{3}}{5}$ giving ∠CON = 19.1°.

Example 4.6

An aeroplane is flying at 400 knots in a strong NW wind of 50 knots. The plane wishes to fly due west. In which direction should the pilot fly the plane to achieve this end, and what will be his actual speed over the ground?

Solution

The resultant velocity of the plane is the vector sum of 50 knots from the NW direction and 400 knots in a direction α° north of west. In appropriate units the situation is shown in Figure 4.15(a). The vector $\overrightarrow{OA}$ represents the wind velocity and $\overrightarrow{OB}$ represents the aeroplane velocity. The resultant velocity is $\overrightarrow{OP}$, which is required to be due W.

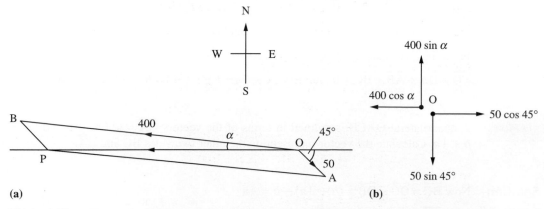

(a)

(b)

Figure 4.15 (a) The track of the aeroplane in Example 4.6. (b) Resolving the velocity into components.

Resolving the velocity into components as illustrated in Figure 4.15(b) and recognizing that the resultant velocity is in the westerly direction, we have no resultant velocity perpendicular to this direction. Thus

$$400 \sin \alpha° = 50 \sin 45°$$

so that

$$\alpha = 5.07°$$

The resultant speed due west is

$$400 \cos \alpha° - 50 \cos 45° = 363 \, knots$$

Example 4.7   If ABCD is any quadrilateral, show that $\overrightarrow{AD} + \overrightarrow{BC} = 2\overrightarrow{EF}$, where E and F are the midpoints of AB and DC respectively, and that

$$\overrightarrow{AB} + \overrightarrow{AD} + \overrightarrow{CB} + \overrightarrow{CD} = 4\overrightarrow{XY}$$

where X and Y are the midpoints of the diagonals AC and BD respectively.

Solution   Applying the polygon law for the addition of vectors to Figure 4.16,

$$\overrightarrow{EF} = \overrightarrow{EA} + \overrightarrow{AD} + \overrightarrow{DF}$$

and

$$\overrightarrow{EF} = \overrightarrow{EB} + \overrightarrow{BC} + \overrightarrow{CF}$$

Adding these two then gives

$$2\overrightarrow{EF} = \overrightarrow{EA} + \overrightarrow{AD} + \overrightarrow{DF} + \overrightarrow{EB} + \overrightarrow{BC} + \overrightarrow{CF}$$
$$= \overrightarrow{AD} + \overrightarrow{BC} + (\tfrac{1}{2}\overrightarrow{BA} + \tfrac{1}{2}\overrightarrow{CD} - \tfrac{1}{2}\overrightarrow{BA} - \tfrac{1}{2}\overrightarrow{CD})$$

since E and F are the midpoints of AB and CD respectively. Thus

$$2\overrightarrow{EF} = \overrightarrow{AD} + \overrightarrow{BC}$$

Also, by the polygon law for addition of vectors,

$$\overrightarrow{XY} = \overrightarrow{XA} + \overrightarrow{AB} + \overrightarrow{BY}$$

and

$$\overrightarrow{XY} = \overrightarrow{XC} + \overrightarrow{CB} + \overrightarrow{BY}$$

Adding and multiplying by two gives

$$4\overrightarrow{XY} = 2\overrightarrow{XA} + 2\overrightarrow{AB} + 2\overrightarrow{BY} + 2\overrightarrow{XC} + 2\overrightarrow{CB} + 2\overrightarrow{BY}$$
$$= 2\overrightarrow{AB} + 2\overrightarrow{CB} + 4\overrightarrow{BY} \quad (\text{since } \overrightarrow{XA} = -\overrightarrow{XC})$$
$$= 2\overrightarrow{AB} + 2\overrightarrow{CB} + 2\overrightarrow{BD} \quad (\text{since } \overrightarrow{BD} = 2\overrightarrow{BY})$$
$$= \overrightarrow{AB} + \overrightarrow{CB} + (\overrightarrow{AB} + \overrightarrow{BD}) + (\overrightarrow{CB} + \overrightarrow{BD})$$

so that

$$4\overrightarrow{XY} = \overrightarrow{AB} + \overrightarrow{CB} + \overrightarrow{AD} + \overrightarrow{CD}$$

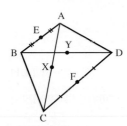

Figure 4.16
Quadrilateral of
Example 4.7.

Figure 4.17
The component form
of a vector.

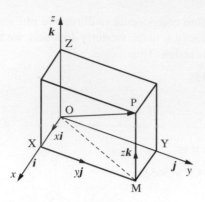

4.2.4 Cartesian components

In Section 4.2.2 we saw that vectors could be written as an ordered set of three numbers or 3-tuple. We shall now explore the properties of these ordered triples and how they relate to the geometrical definitions used in previous sections.

In Figure 4.17, we denote mutually perpendicular unit vectors in the three coordinate directions by i, j and k. (Sometimes the alternative notation $\hat{e}_1, \hat{e}_2$ and $\hat{e}_3$ is used.) The notation i, j, k is so standard that the 'hats' indicating unit vectors are usually omitted.

Applying the triangle law to the triangle OXM, we have

$$\overrightarrow{OM} = \overrightarrow{OX} + \overrightarrow{OY} = xi + yj$$

Applying the triangle law to the triangle OMP then yields

$$\overrightarrow{OP} = \overrightarrow{OM} + \overrightarrow{MP} = xi + yj + zk \tag{4.1}$$

The analysis applies to any point, so we can write any vector r in terms of its **components** x, y, z with respect to the unit vectors i, j, k as

$$r = xi + yj + zk$$

Indeed, the vector notation $r = (x, y, z)$ should be intepreted as the vector given in (4.1). In some contexts it is more convenient to use a suffix notation for the coordinates, and

$$(x_1, x_2, x_3) = x_1 i + x_2 j + x_3 k$$

is interpreted in exactly the same way. It is assumed that the three basic unit vectors are known, and all vectors in coordinate form are referred to them.

The **modulus** of a vector is just the length OP so from Figure 4.5 we have, using Pythagoras' theorem,

$$|a| = (a_1^2 + a_2^2 + a_3^2)^{1/2}$$

The basic properties of vectors follow easily from the component definition in (4.1). Two vectors $a = (a_1, a_2, a_3)$ and $b = (b_1, b_2, b_3)$ are **equal** if and only if the three components are equal, that is

$$a_1 = b_1, \quad a_2 = b_2, \quad a_3 = b_3$$

Figure 4.18
Parallelogram rule, x component.

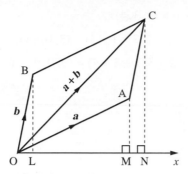

The zero vector has zero components, so

$$\mathbf{0} = (0, 0, 0)$$

The addition rule is expressed very simply in terms of vector components:

$$\mathbf{a} + \mathbf{b} = (a_1 + b_1, a_2 + b_2, a_3 + b_3)$$

The equivalence of the definitions can be deduced from Figure 4.18. We know that $\overrightarrow{OB} = \overrightarrow{AC}$, since they are equivalent displacements, and hence their x components are the same so that we have OL = MN. Thus if we take the x component of $\mathbf{a} + \mathbf{b}$

$$(\mathbf{a} + \mathbf{b})_1 = ON = OM + MN = OM + OL = a_1 + b_1$$

the y and z components can be considered in a similar manner.

If λ is a scalar and the vectors are related by $\mathbf{a} = \lambda\mathbf{b}$ then the components satisfy

$$a_1 = \lambda b_1, \quad a_2 = \lambda b_2, \quad a_3 = \lambda b_3$$

which follows from the similar triangles of Figure 4.11.

The distributive law in components is simply a restatement of the distributive law for the addition of numbers:

$$\begin{aligned}
\lambda(\mathbf{a} + \mathbf{b}) &= \lambda(a_1 + b_1, a_2 + b_2, a_3 + b_3) \\
&= (\lambda(a_1 + b_1), \lambda(a_2 + b_2), \lambda(a_3 + b_3)) \\
&= (\lambda a_1 + \lambda b_1, \lambda a_2 + \lambda b_2, \lambda a_3 + \lambda b_3) \\
&= (\lambda a_1, \lambda a_2, \lambda a_3) + (\lambda b_1, \lambda b_2, \lambda b_3) \\
&= \lambda\mathbf{a} + \lambda\mathbf{b}
\end{aligned}$$

Subtraction is again straightforward and the components are just subtracted from each other:

$$\mathbf{a} - \mathbf{b} = (a_1 - b_1, a_2 - b_2, a_3 - b_3)$$

The component form of vectors allows problems to be solved algebraically and results can be interpreted either as algebraic ideas or in a geometrical manner. Both these interpretations can be very useful in applications of vectors to engineering.

Example 4.8 Given the vectors $\mathbf{a} = (1, 1, 1)$, $\mathbf{b} = (-1, 2, 3)$ and $\mathbf{c} = (0, 3, 4)$, find

(a) $\mathbf{a} + \mathbf{b}$ (b) $2\mathbf{a} - \mathbf{b}$ (c) $\mathbf{a} + \mathbf{b} - \mathbf{c}$

(d) the unit vector in the direction of $\mathbf{c}$

Solution   (a) $a + b = (1 - 1, 1 + 2, 1 + 3) = (0, 3, 4)$

(b) $2a - b = (2 \times 1 - (-1), 2 \times 1 - 2, 2 \times 1 - 3) = (3, 0, -1)$

(c) $a + b - c = (1 - 1 + 0, 1 + 2 - 3, 1 + 3 - 4) = (0, 0, 0) = 0$

(d) $|c| = (3^2 + 4^2)^{1/2} = 5$, so

$$\hat{c} = \frac{c}{5} = (0, \tfrac{3}{5}, \tfrac{4}{5})$$

Example 4.9   Given $a = (2, -3, 1) = 2i - 3j + k$, $b = (1, 5, -2) = i + 5j - 2k$ and $c = (3, -4, 3) = 3i - 4j + 3k$

(a) find the magnitude of the vector $d = a - 2b + 3c$;

(b) write down a unit vector in the direction of d;

(c) what are the direction cosines of d?

Solution   (a) $d = a - 2b + 3c$

$$= (2i - 3j + k) - 2(i + 5j - 2k) + 3(3i - 4j + 3k)$$

$$= (2i - 3j + k) - (2i + 10j - 4k) + (9i - 12j + 9k)$$

$$= (2 - 2 + 9)i + (-3 - 10 - 12)j + (1 + 4 + 9)k$$

that is, $d = 9i - 25j + 14k$.

(b) The magnitude of d is $d = \sqrt{[9^2 + (-25)^2 + 14^2]} = \sqrt{902}$
A unit vector in the direction of d is $\hat{d}$, where

$$\hat{d} = \frac{d}{d} = \frac{9}{\sqrt{902}}i - \frac{25}{\sqrt{902}}j + \frac{14}{\sqrt{902}}k$$

(c) The direction cosines of d are $9/\sqrt{902}$, $-25/\sqrt{902}$ and $14/\sqrt{902}$.

Example 4.10   Determine whether constants α and β can be found to satisfy the vector equations

(a) $(2, 1, 0) = \alpha(-2, 0, 2) + \beta(1, 1, 1)$

(b) $(-3, 1, 2) = \alpha(-2, 0, 2) + \beta(1, 1, 1)$

and interpret the results.

Solution   (a) For the two vectors to be the same each of the components must be equal, and hence

$$2 = -2\alpha + \beta$$

$$1 = \beta$$

$$0 = 2\alpha + \beta$$

Thus the second equation gives $\beta = 1$ and both of the other two equations give the same value of α, namely $\alpha = -\frac{1}{2}$, so the equations can be satisfied.

(b) A similar argument gives

$$-3 = -2\alpha + \beta$$

$$1 = \beta$$

$$2 = 2\alpha + \beta$$

Again, the second equation gives $\beta = 1$ but the first equation leads to $\alpha = 2$ and the third to $\alpha = \frac{1}{2}$. The equations are now not consistent and no appropriate α and β can be found.

In case (a) the three vectors lie in a plane, and any vector in a plane, including the one given, can be written as the vector sum of the two vectors $(-2, 0, 2)$ and $(1, 1, 1)$ with appropriate multipliers. In case (b), however, the vector $(-3, 1, 2)$ does not lie in the plane of the two vectors $(-2, 0, 2)$ and $(1, 1, 1)$ and can, therefore, never be written as the vector sum of the two vectors $(-2, 0, 2)$ and $(1, 1, 1)$ with appropriate multipliers.

Example 4.11

A molecule XY_3 has a tetrahedral form; the position vector of the X atom is $(2\sqrt{3} + \sqrt{2}, 0, -2 + \sqrt{6})$ and those of the three Y atoms are

$$\overrightarrow{OY} = (\sqrt{3}, -2, -1), \quad \overrightarrow{OY'} = (\sqrt{3}, 2, -1), \quad \overrightarrow{OY''} = (\sqrt{2}, 0, \sqrt{6})$$

(a) Show that all of the bond lengths are equal.

(b) Show that $\overrightarrow{XY} + \overrightarrow{YY'} + \overrightarrow{Y'Y''} + \overrightarrow{Y''X} = \mathbf{0}$

Solution

(a) $\overrightarrow{XY} = \overrightarrow{OY} - \overrightarrow{OX} = (-\sqrt{3} - \sqrt{2}, -2, 1 - \sqrt{6})$ and the bond length is

$$|\overrightarrow{XY}| = [(-\sqrt{3} - \sqrt{2})^2 + (-2)^2 + (1 - \sqrt{6})^2]^{1/2} = 4$$

$\overrightarrow{YY'} = \overrightarrow{OY'} - \overrightarrow{OY} = (0, 4, 0)$ and clearly the bond length is again 4.

The other four bonds $\overrightarrow{XY'}, \overrightarrow{XY''}, \overrightarrow{Y'Y''}, \overrightarrow{Y''Y}$ are treated in exactly the same way, and each gives a bond length of 4.

(b) Now $\overrightarrow{Y'Y''} = (\sqrt{2} - \sqrt{3}, -2, \sqrt{6} + 1)$ and $\overrightarrow{Y''X} = (2\sqrt{3}, 0, -2)$ so adding the four vectors

$$\overrightarrow{XY} + \overrightarrow{YY'} + \overrightarrow{Y'Y''} + \overrightarrow{Y''X}$$

$$= (-\sqrt{3} - \sqrt{2}, -2, 1 - \sqrt{6}) + (0, 4, 0) + (\sqrt{2} - \sqrt{3}, -2, \sqrt{6} + 1) + (2\sqrt{3}, 0, -2)$$

$$= \mathbf{0}$$

and is just a verification of the polygon law.

Example 4.12

Three forces, with units of newtons,

$$\boldsymbol{F}_1 = (1, 1, 1)$$

$\boldsymbol{F}_2$ has magnitude 6 and acts in the direction $(1, 2, -2)$

$\boldsymbol{F}_3$ has magnitude 10 and acts in the direction $(3, -4, 0)$

act on a particle. Find the resultant force that acts on a particle. What additional force must be imposed on the particle to reduce the resultant force to zero?

Solution The first force is given in the usual vector form. The second two are given in an equally acceptable way but it is necessary to convert the information to the normal vector form so that the resultant can be found by vector addition. The unit vector in the given direction of F_2 is required

$$|(1, 2, -2)| = (1 + 2^2 + (-2)^2)^{1/2} = 3$$

and hence the unit vector in this direction is

$$\tfrac{1}{3}(1, 2, -2) \quad \text{and} \quad F_2 = 6(\tfrac{1}{3}, \tfrac{2}{3}, -\tfrac{2}{3}) = (2, 4, -4)$$

Similarly for F_3, the unit vector is $\tfrac{1}{5}(3, -4, 0)$ and hence $F_3 = (6, -8, 0)$. The resultant force is obtained by vector addition.

$$F = F_1 + F_2 + F_3 = (1, 1, 1) + (2, 4, -4) + (6, -8, 0) = (9, -3, -3)$$

To make the resultant zero, the force $(-9, 3, 3)$ must be added.

Example 4.13 Two geostationary satellites have known positions $(0, 0, h)$ and $(0, A, H)$ relative to a fixed set of axes on the earth's surface (which is assumed flat, with the x and y axes lying on the surface and the z axis vertical). Radar signals measure the distance of a ship from the satellites. Find the position of the ship relative to the given axes.

Solution Figure 4.19 illustrates the situation described.
The vectors

$$\overrightarrow{PR} = \overrightarrow{OR} - \overrightarrow{OP} = (a, b, 0) - (0, 0, h) = (a, b, -h)$$
$$\overrightarrow{QR} = \overrightarrow{OR} - \overrightarrow{OQ} = (a, b, 0) - (0, A, H) = (a, b - A, -H)$$

are easily calculated. If the lengths of the two vectors are known as p and q then

$$p^2 = a^2 + b^2 + h^2 \quad \text{and} \quad q^2 = a^2 + (b - A)^2 + H^2$$

Subtracting gives

$$p^2 - q^2 = A(2b - A) + h^2 - H^2$$

Figure 4.19

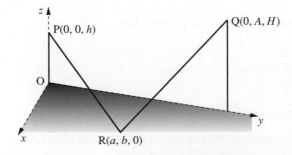

and hence

$$b = (p^2 - q^2 - h^2 + H^2 + A^2)/2A$$

Now b is known a can be calculated from

$$a = \pm\sqrt{(p^2 - b^2 - h^2)}$$

Note the ambiguity in sign; clearly the ship will need to know which side of the y axis it is lying.

In practice the axes will need to be transformed to standard latitude and longitude and the curvature of the earth will need to be taken into consideration. Can the same calculation be used for aircraft? The speed of the ship has been neglected in the calculation above but is the speed of the aircraft important?

A rather more difficult example, which may be omitted on a first reading, illustrates the use of position vectors in a practical situation.

Example 4.14   A simple robot arm is illustrated in Figure 4.20. Find the position of the elbow C in terms of the angles θ and ϕ and the vector position of the workpiece D.

Solution   We first write down the vectors

$$\overrightarrow{OA} = 2(0, 0, 1)$$
$$\overrightarrow{AB} = 0.1(\cos\theta, \sin\theta, 0)$$
$$\overrightarrow{BC} = 1.1(\sin\theta\sin\phi, -\cos\theta\sin\phi, -\cos\phi)$$

Figure 4.20
(a) Robot arm in
Example 4.14.
(b) Illustration of the
angles in Example
4.14 (exaggerated
lengths).

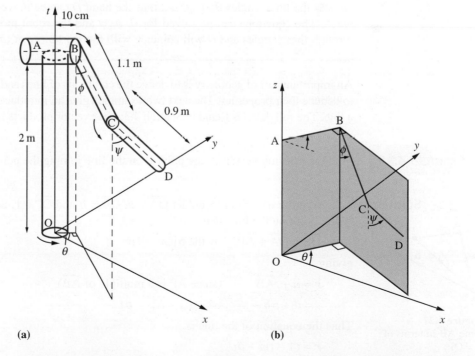

(a)   (b)

Note that all the vectors are in the form of a unit vector times the magnitude. Adding, we have

$$\overrightarrow{OC} = \overrightarrow{OA} + \overrightarrow{AB} + \overrightarrow{BC}$$
$$= (0.1 \cos \theta + 1.1 \sin \theta \sin \phi, 0.1 \sin \theta - 1.1 \cos \theta \sin \phi, 2 - 1.1 \cos \phi)$$

If we try to move the elbow to a point (a, b, c), omitting the remaining arm CD, then we find that this is not possible. The equation $c = 2 - 1.1 \cos \phi$ determines ϕ and the equation $b = 0.1 \sin \theta - 1.1 \cos \theta \sin \phi$ then determines θ. Thus we cannot also satisfy $a = 0.1 \cos \theta + 1.1 \sin \theta \sin \phi$ since we already know θ and ϕ. This result is as expected intuitively; in a three-dimensional space, physically we would expect three degrees of freedom and mathematically we would expect to have three coordinates. The robot needs the extra arm to find all points within its reach.

The additional arm is along the vector.

$$\overrightarrow{CD} = 0.9(\sin \theta \sin \psi, -\cos \theta \sin \psi, -\cos \psi)$$

and we have

$$\overrightarrow{OD} = \overrightarrow{OC} + \overrightarrow{CD}$$
$$= (0.1 \cos \theta + 1.1 \sin \theta \sin \phi + 0.9 \sin \theta \sin \psi,$$
$$0.1 \sin \theta - 1.1 \cos \theta \sin \phi - 0.9 \cos \theta \sin \psi, 2 - 1.1 \cos \phi - 0.9 \cos \psi)$$

There are now three unknowns, θ, ϕ and ψ, which can be calculated for a given (a, b, c). It is an interesting exercise in trigonometry to solve the three equations

$$a = 0.1 \cos \theta + \sin \theta (1.1 \sin \phi + 0.9 \sin \psi)$$
$$b = 0.1 \sin \theta - \cos \theta (1.1 \sin \phi + 0.9 \sin \psi)$$
$$2 - c = 1.1 \cos \phi + 0.9 \cos \psi$$

There are now three degrees of freedom, physically the three arm sections and mathematically the three angles θ, ϕ, ψ, and thus the hand D can be moved to all points within reach. The equations can be solved for θ, ϕ, ψ and the robot moves the arm sections through these angles and D will coincide with the workpiece at (a, b, c).

An important part of geometry is to define the equations of lines and planes and to be able to deduce their properties. The next two examples give an introduction to the equation of a line. The full detail is found in Section 4.3, after vector products have been introduced.

Example 4.15   Find the position vector of any point P on the line joining the points A and B.

Solution   Take an arbitrary origin O and let $\overrightarrow{OA} = a$, $\overrightarrow{OB} = b$ and $\overrightarrow{OP} = r$, as in Figure 4.21. If P is any point on the line then

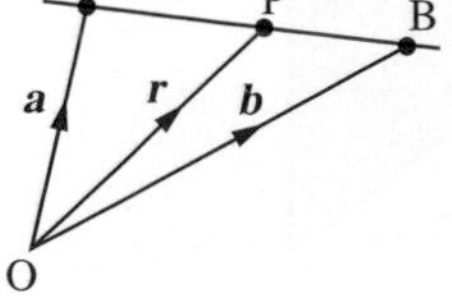

$$\overrightarrow{OP} = \overrightarrow{OA} + \overrightarrow{AP}, \quad \text{by the triangle law}$$

giving

$$r = a + t\overrightarrow{AB} \quad (\text{since } \overrightarrow{AP} \text{ is a multiple of } \overrightarrow{AB})$$
$$= a + t(b - a) \quad (\text{since } a + \overrightarrow{AB} = b)$$

Thus the equation of the line is

$$r = (1 - t)a + tb \tag{4.2}$$

Figure 4.21
Line AB in terms of
$r = \overrightarrow{OP}$.

As t varies from $-\infty$ to $+\infty$, the point P sweeps along the line, with $t = 0$ corresponding to point A and $t = 1$ to point B.

If we take the vectors in their component form $\mathbf{r} = (x, y, z)$, $\mathbf{a} = (a_1, a_2, a_3)$ and $\mathbf{b} = (b_1, b_2, b_3)$ with respect to an appropriate coordinate system then (4.2) gives

$$(x, y, z) = (1 - t)(a_1, a_2, a_3) + t(b_1, b_2, b_3)$$

Equating the individual components, we obtain

$$\frac{x - a_1}{b_1 - a_1} = \frac{y - a_2}{b_2 - a_2} = \frac{z - a_3}{b_3 - a_3} = t \qquad \textbf{(4.3)}$$

which is the usual *cartesian* equation of a line passing through the points (a_1, a_2, a_3) and (b_1, b_2, b_3).

Note that if any of the denominators is zero then equation (4.3) is interpreted so that the corresponding numerator is zero also.

Example 4.16

The line L_1 passes through the points with position vectors

$$(5, 1, 7) \quad \text{and} \quad (6, 0, 8)$$

and the line L_2 passes through the points with position vectors

$$(3, 1, 3) \quad \text{and} \quad (-1, 3, \alpha)$$

Find the value of α for which the two lines L_1 and L_2 intersect.

Solution

Using the vector form:
From Example 4.15 the equations of the two lines can be written in vector form as

$$L_1: \quad \mathbf{r} = (5, 1, 7) + t(1, -1, 1)$$
$$L_2: \quad \mathbf{r} = (3, 1, 3) + s(-4, 2, \alpha - 3)$$

These two lines intersect if t, s and α can be chosen so that the two vectors are equal, that is they have the same components. Thus

$$5 + t = 3 - 4s$$
$$1 - t = 1 + 2s$$
$$7 + t = 3 + s(\alpha - 3)$$

The first two of these equations are simultaneous equations for t and s. Solving gives $t = 2$ and $s = -1$. Putting these values into the third equation

$$9 = 3 - (\alpha - 3) \Rightarrow \alpha = -3$$

and it can be checked that the point of intersection is $(7, -1, 9)$.

Using the cartesian form:
Equation (4.3) gives the equations of the lines as

$$L_1: \quad \frac{x - 5}{6 - 5} = \frac{y - 1}{0 - 1} = \frac{z - 7}{8 - 7}$$

$$L_2: \quad \frac{x - 3}{-1 - 3} = \frac{y - 1}{3 - 1} = \frac{z - 3}{\alpha - 3}$$

The two equations for x and y are

$$x - 5 = 1 - y$$

$$\tfrac{1}{4}(3 - x) = \tfrac{1}{2}(y - 1)$$

and are solved to give $x = 7$ and $y = -1$. Putting in these values, the equations for z and α become

$$z - 7 = 2$$

$$\frac{z - 3}{\alpha - 3} = -1$$

which give $z = 9$ and $\alpha = -3$.

4.2.5 Complex numbers as vectors

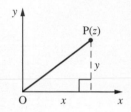

Figure 4.22
Argand diagram
representation of
$z = x + \mathrm{j}y$.

We saw in Chapter 3, Section 3.2, that a complex number $z = x + \mathrm{j}y$ can be represented geometrically by the point P in the Argand diagram as illustrated in Figure 4.22. We could equally well represent the point P by the vector $\overrightarrow{OP}$. Hence we can express the complex number z as a two-dimensional vector

$$z = \overrightarrow{OP}$$

With this interpretation of a complex number we can use the parallelogram rule to represent the addition and subtraction of complex numbers geometrically as illustrated in Figures 4.23(a, b).

Figure 4.23
(a) Addition of
complex numbers.
(b) Subtraction of
complex numbers.

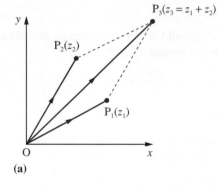

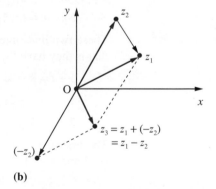

(a)

(b)

Example 4.17	A square is formed in the first and second quadrant with OP as one side of the square and $\overrightarrow{OP} = (1, 2)$. Find the coordinates of the other two vertices of the square.

Solution The situation is illustrated in Figure 4.24. Using the complex form $\overrightarrow{OP} = 1 + 2\mathrm{j}$ the side OQ is obtained by rotating OP through $\pi/2$ radians, then

$$\overrightarrow{OQ} = \mathrm{j}(1 + 2\mathrm{j}) = -2 + \mathrm{j}$$

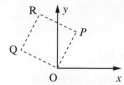

Figure 4.24
Square of Example
4.17.

The fourth point R is found by observing that $\overrightarrow{OR}$ is the vector sum of $\overrightarrow{OP}$ and $\overrightarrow{OQ}$, and hence

$$\overrightarrow{OR} = \overrightarrow{OP} + \overrightarrow{OQ} = -1 + j3$$

The four coordinates are therefore

$$(0, 0), (1, 2), (-2, 1) \text{ and } (-1, 3)$$

Example 4.18

M is the centre of a square with vertices A, B, C and D taken anticlockwise in that order. If, in the Argand diagram, M and A are represented by the complex numbers $-2 + j$ and $1 + j5$ respectively, find the complex numbers represented by the vertices B, C and D.

Solution

Applying the triangle law for addition of vectors of Figure 4.25 gives

$$\overrightarrow{MA} = \overrightarrow{MO} + \overrightarrow{OA}$$
$$= \overrightarrow{OA} - \overrightarrow{OM}$$
$$\equiv (1 + j5) - (-2 + j)$$
$$= 3 + j4$$

Since ABCD is a square,

$$MA = MB = MC = MD$$
$$\angle AMB = \angle BMC = \angle CMD = \angle DMA = \tfrac{1}{2}\pi$$

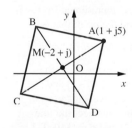

Figure 4.25
Square of Example
4.18.

Remembering that multiplying a complex number by j rotates it through $\tfrac{1}{2}\pi$ radians in an anticlockwise direction, we have

$$\overrightarrow{MB} = j\overrightarrow{MA} \equiv j(3 + j4) = -4 + j3$$

giving

$$\overrightarrow{OB} = \overrightarrow{OM} + \overrightarrow{MB} \equiv (-2 + j) + (-4 + j3) = -6 + j4$$

Likewise

$$\overrightarrow{MC} = j\overrightarrow{MB} \equiv j(-4 + j3) = -3 - j4$$

giving

$$\overrightarrow{OC} = \overrightarrow{OM} + \overrightarrow{MC} \equiv -5 - j3$$

and

$$\overrightarrow{MD} = j\overrightarrow{MC} \equiv j(-3 - j4) = 4 - j3$$

giving

$$\overrightarrow{OD} = \overrightarrow{OM} + \overrightarrow{MD} \equiv 2 - j2$$

Thus the vertices B, C and D are represented by the complex numbers $-6 + j4$, $-5 - j3$ and $2 - j2$ respectively.

4.2.6 Exercises

1 Given $a = (1, 1, 0)$, $b = (2, 2, 1)$ and $c = (0, 1, 1)$, evaluate

(a) $a + b$ (b) $a + \frac{1}{2}b + 2c$ (c) $b - 2a$

(d) $|a|$ (e) $|b|$ (f) $|a - b|$

(g) $\hat{a}$ (h) $\hat{b}$

2 If the position vectors of the points P and Q are $i + 3j - 7k$ and $5i - 2j + 4k$ respectively, find $\overrightarrow{PQ}$ and determine its length and direction cosines.

3 A particle P is acted upon by forces (measured in newtons) $F_1 = 3i - 2j + 5k$, $F_2 = -i + 7j - 3k$, $F_3 = 5i - j + 4k$ and $F_4 = -2j + 3k$. Determine the magnitude and direction of the resultant force acting on P.

4 If $a = 3i - 2j + k$, $b = -2i + 5j + 4k$, $c = -4i + j - 2k$ and $d = 2i - j + 4k$, determine α, β and γ such that

$$d = \alpha a + \beta b + \gamma c$$

5 Prove that the vectors $2i - 4j - k$, $3i + 2j - 2k$ and $5i - 2j - 3k$ can form the sides of a triangle. Find the lengths of each side of the triangle and show that it is right-angled.

6 The vector $\overrightarrow{OP}$ makes an angle of $60°$ with the positive x axis and $45°$ with the positive y axis. Find the possible angles that the vector can make with the z axis.

7 Given the points P(1, −3, 4), Q(2, 2, 1) and R(3, 7, −2), find the vectors $\overrightarrow{PQ}$ and $\overrightarrow{QR}$. Show that P, Q and R lie on a straight line and find the ratio PQ:QR.

8 A cyclist travelling east at 8 kilometres per hour finds that the wind appears to blow directly from the north. On doubling his speed it appears to blow from the north-east. Find the actual velocity of the wind.

9 Relative to a landing stage, the position vectors in kilometres of two boats A and B at noon are

$$3i + j \quad \text{and} \quad i - 2j$$

respectively. The velocities of A and B, which are constant and in kilometres per hour, are

$$10i + 24j \quad \text{and} \quad 24i + 32j$$

Find the distance between the boats t hours after noon and find the time at which this distance is a minimum.

10 If the complex numbers z_1, z_2 and z_3 are represented on the Argand diagram by the points P_1, P_2 and P_3 respectively and

$$\overrightarrow{OP_2} = 2j\overrightarrow{OP_1} \quad \text{and} \quad \overrightarrow{OP_3} = \tfrac{2}{5}j\overrightarrow{P_2P_1}$$

prove that P_3 is the foot of the perpendicular from O on to the line P_1P_2.

11 ABCD is a square, lettered anticlockwise, on an Argand diagram, with A representing $3 + j2$ and B representing $-1 + j4$. Show that C lies on the real axis and find the complex number represented by D and the length of AB.

12 A triangle has vertices A, B, C represented by $1 + j$, $2 - j$ and -1 respectively. Find the point that is equidistant from A, B and C.

13 Find the vector equation of the line through the points with position vectors $a = (2, 0, -1)$ and $b = (1, 2, 3)$. Write down the equivalent cartesian coordinate form. Does this line intersect the line through the points $c = (0, 0, 1)$ and $d = (1, 0, 1)$?

14 Given the triangle OAB, where O is the origin, and denoting the midpoints of the opposite sides as O', A' and B', show vectorially that the lines OO', AA' and BB' meet at a point. (Note that this is the result that the medians of a triangle meet at the centroid.)

15 Three weights W_1, W_2 and W_3 hang in equilibrium on the pulley system shown in Figure 4.26. The pulleys are considered to be smooth and the forces add by the rules of vector addition. Calculate θ and ϕ, the angles the ropes make with the horizontal.

16 A telegraph pole OP has three wires connected to it at P. The other ends of the wires are connected to houses at A, B and C. Axes are set up as shown in Figure 4.27. The points relative to these axes, with distances in metres, are $\overrightarrow{OP} = 8k$, $\overrightarrow{OA} = 20j + 6k$, $\overrightarrow{OB} = -i - 18j + 10k$ and $\overrightarrow{OC} = -22i + 3j + 7k$.

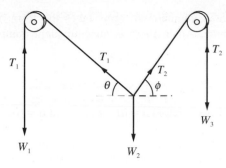

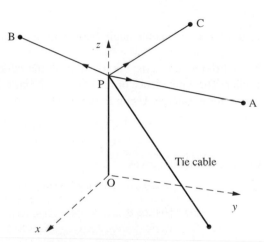

Figure 4.26 Pulley system in Question 15.

The tension in each wire is 900 N. Find the total force acting at P. A tie cable at an angle of 45° is connected to P and fixed in the ground. Where should the ground fixing be placed, and what is the tension required to ensure a zero horizontal resultant force at P?

17 A boom OB carries a load F of magnitude 500 N and is supported by cables BC and BD as shown in Figure 4.28 where the dimensions of the system are given. Determine the tensions in the cables so that equilibrium is maintained and the resultant force at the point B is along OB.

Figure 4.27 The telegraph pole of Question 16.

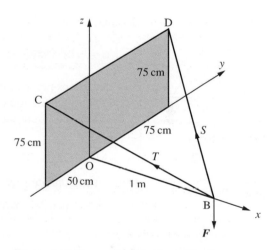

Figure 4.28 Boom supported by cables in Question 17.

4.2.7 The scalar product

A natural idea in mathematics, explored in Chapter 1, is not only to add quantities but also to multiply them together. The concept of multiplication of vectors translates into a useful tool for many engineering applications, with two different products of vectors – the 'scalar' and 'vector' products – turning out to be particularly important.

The determination of a component of a vector is a basic procedure in analysing many physical problems. For the vector a shown in Figure 4.29 the component of a in the direction of OP is just $ON = |a| \cos \theta$. The component is relevant in the physical context of work done by a force. Suppose the point of application, O, of a constant force F is moved along the vector a from O to the point A, as in Figure 4.30. The component of F in the a direction is $|F| \cos \theta$, and O is moved a distance $|a|$. The work done is defined as the product of the distance moved by the point of application and the component of the force in this direction. It is thus given by

$$\text{work done} = |F| \, |a| \cos \theta$$

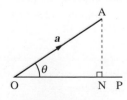

Figure 4.29
The component of a in the direction OP is $ON = |a| \cos \theta$.

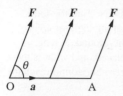

Figure 4.30
The work done by a constant force F with point of application moved from O to A is $|F||a|\cos\theta$.

The definition of the scalar product in geometrical terms takes the form of this expression for the work done by a force. Again there is an equivalent component definition, and both are now presented.

Definition

The **scalar** (or **dot** or **inner**) **product** of two vectors $a = (a_1, a_2, a_3)$ and $b = (b_1, b_2, b_3)$ is defined as follows:

In components

$$a \cdot b = a_1 b_1 + a_2 b_2 + a_3 b_3$$

Geometrically

$$a \cdot b = |a||b|\cos\theta, \quad \text{where } \theta \ (0 \leqslant \theta \leqslant \pi) \text{ is the angle between the two vectors}$$

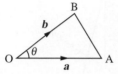

Figure 4.31
Cosine rule for a triangle; equivalence of the geometrical and component definitions of the scalar product.

Both definitions prove to be useful in different contexts, but to establish the basic rules the component definition is the simpler. The equivalence of the two definitions can easily be established from the cosine rule for a triangle. Using Figure 4.31 the cosine rule (2.15) states

$$\text{AB}^2 = \text{OA}^2 + \text{OB}^2 - 2(\text{OA})(\text{OB})\cos\theta$$

which in appropriate vector or component notation gives

$$(a_1 - b_1)^2 + (a_2 - b_2)^2 + (a_3 - b_3)^2 = (a_1^2 + a_2^2 + a_3^2) + (b_1^2 + b_2^2 + b_3^2)$$
$$- 2|a||b|\cos\theta$$

Thus

$$a_1^2 - 2a_1 b_1 + b_1^2 + a_2^2 - 2a_2 b_2 + b_2^2 + a_3^2 - 2a_3 b_3 + b_3^2$$
$$= a_1^2 \qquad + b_1^2 + a_2^2 \qquad + b_2^2 + a_3^2 \qquad + b_3^2 - 2|a||b|\cos\theta$$

and hence

$$a_1 b_1 + a_2 b_2 + a_3 b_3 = |a||b|\cos\theta \tag{4.4}$$

The first point to note is that the scalar product of two vectors gives a **number**. Secondly, it should be remembered that the scalar product is only defined as the product of two vectors and *not* between any other two quantities. For this reason, the presence of the dot ($\cdot$) is essential between the two vectors.

The basic rules are now very straightforward to establish.

(a) Commutative law

$$a \cdot b = b \cdot a$$

This rule follows immediately from the component definition, since interchanging a_i and b_i does not make any difference to the products. The rule says that 'order does not matter'.

(b) Associative law

The idea of associativity involves the product of three vectors. Since $a \cdot b$ is a scalar, it cannot be dotted with a third vector, so the idea of associativity is not applicable here.

(c) Distributive law for products with a scalar λ

$$a \cdot (\lambda b) = (\lambda a) \cdot b = \lambda (a \cdot b)$$

These results follow directly from the component definition. The implication is that scalars can be multiplied out in the normal manner.

(d) Distributive law over addition

$$a \cdot (b + c) = a \cdot b + a \cdot c$$

The proof is straightforward, since

$$a \cdot (b + c) = a_1(b_1 + c_1) + a_2(b_2 + c_2) + a_3(b_3 + c_3)$$
$$= (a_1 b_1 + a_2 b_2 + a_3 b_3) + (a_1 c_1 + a_2 c_2 + a_3 c_3)$$
$$= a \cdot b + a \cdot c$$

Thus the normal rules of algebra apply, and brackets can be multiplied out in the usual way.

(e) Powers of a

One simple point to note is that

$$a \cdot a = a_1^2 + a_2^2 + a_3^2 = |a| \, |a| \cos 0 = |a|^2$$

in agreement with Section 4.2.4. This expression is written $a^2 = a \cdot a$ and, where there is no ambiguity, $a^2 = a^2$ is also used. No other powers of vectors can be constructed, since, as in (b) above, scalar products of more than two vectors do not exist. For the standard unit vectors, i, j and k,

$$i^2 = i \cdot i = 1, \quad j^2 = j \cdot j = 1, \quad k^2 = k \cdot k = 1 \qquad (4.5)$$

(f) Perpendicular vectors

It is clear from (4.4) that if a and b are perpendicular (orthogonal) then $\cos \theta = \cos \frac{1}{2}\pi = 0$, and hence $a \cdot b = 0$, or in component notation

$$a \cdot b = a_1 b_1 + a_2 b_2 + a_3 b_3 = 0$$

However, the other way round, $a \cdot b = 0$, *does not* imply that a and b are perpendicular. There are three possibilities:

$$\text{either } a = 0 \quad \text{or} \quad b = 0 \quad \text{or} \quad \theta = \tfrac{1}{2}\pi$$

It is only when the first two possibilities have been dismissed that perpendicularity can be deduced.

The commonest mistake is to deduce from

$$\boldsymbol{a} \cdot \boldsymbol{b} = \boldsymbol{a} \cdot \boldsymbol{c}$$

that $\boldsymbol{b} = \boldsymbol{c}$. This is only one of three possible solutions – the other two being $\boldsymbol{a} = 0$ and $\boldsymbol{a}$ perpendicular to $\boldsymbol{b} - \boldsymbol{c}$. The rule to follow is that *you can't cancel vectors in the same way as scalars*.

Since the unit vectors $\boldsymbol{i}, \boldsymbol{j}$ and $\boldsymbol{k}$ are mutually perpendicular,

$$\boldsymbol{i} \cdot \boldsymbol{j} = \boldsymbol{j} \cdot \boldsymbol{k} = \boldsymbol{k} \cdot \boldsymbol{i} = 0 \qquad\qquad (4.6)$$

Using the distributive law over addition, we obtain using (4.6)

$$(a_1, a_2, a_3) \cdot (b_1, b_2, b_3) = (a_1\boldsymbol{i} + a_2\boldsymbol{j} + a_3\boldsymbol{k}) \cdot (b_1\boldsymbol{i} + b_2\boldsymbol{j} + b_3\boldsymbol{k})$$
$$= a_1b_1\boldsymbol{i} \cdot \boldsymbol{i} + a_1b_2\boldsymbol{i} \cdot \boldsymbol{j} + a_1b_3\boldsymbol{i} \cdot \boldsymbol{k} + a_2b_1\boldsymbol{j} \cdot \boldsymbol{i} + a_2b_2\boldsymbol{j} \cdot \boldsymbol{j}$$
$$+ a_2b_3\boldsymbol{j} \cdot \boldsymbol{k} + a_3b_1\boldsymbol{k} \cdot \boldsymbol{i} + a_3b_2\boldsymbol{k} \cdot \boldsymbol{j} + a_3b_3\boldsymbol{k} \cdot \boldsymbol{k}$$
$$= a_1b_1 + a_2b_2 + a_3b_3$$

which is consistent with the component definition of a scalar product.

Perpendicularity is a very important idea, which is used a great deal in both mathematics and engineering. Pressure acts on a surface in a direction perpendicular to the surface, so that the force per unit area is given by $p\hat{\boldsymbol{n}}$, where p is the pressure and $\hat{\boldsymbol{n}}$ is the unit normal. To perform many calculations, we must be able to find a vector that is perpendicular to another vector. We shall also see that many matrix methods rely on being able to construct a set of mutually orthogonal vectors. Such constructions are not only of theoretical interest, but form the basis of many practical numerical methods used in engineering. The whole of the study of Fourier series (considered in Chapter 12), which is central to much of signal processing and is heavily used by electrical engineers, is based on constructing functions that are orthogonal.

 Although computer packages are not particularly helpful with vector problems they perform the arithmetic with total accuracy. The scalar product appears in MATLAB as dot(**a**, **b**) and the length as norm(**u**); the corresponding instructions in MAPLE are innerprod(**a**, **b**) and norm(**u**, 2).

Example 4.19   Given the vectors $\boldsymbol{a} = (1, -1, 2)$, $\boldsymbol{b} = (-2, 0, 2)$ and $\boldsymbol{c} = (3, 2, 1)$, evaluate

(a) $\boldsymbol{a} \cdot \boldsymbol{c}$ (b) $\boldsymbol{b} \cdot \boldsymbol{c}$ (c) $(\boldsymbol{a} + \boldsymbol{b}) \cdot \boldsymbol{c}$

(d) $\boldsymbol{a} \cdot (2\boldsymbol{b} + 3\boldsymbol{c})$ (e) $(\boldsymbol{a} \cdot \boldsymbol{b})\boldsymbol{c}$

Solution   (a) $\boldsymbol{a} \cdot \boldsymbol{c} = (1 \times 3) + (-1 \times 2) + (2 \times 1) = 3$

(b) $\boldsymbol{b} \cdot \boldsymbol{c} = (-2 \times 3) + (0 \times 2) + (2 \times 1) = -4$

(c) $(\boldsymbol{a} + \boldsymbol{b}) \cdot \boldsymbol{c} = (-1, -1, 4) \cdot (3, 2, 1) = -3 - 2 + 4 = -1$

(note that $(\boldsymbol{a} + \boldsymbol{b}) \cdot \boldsymbol{c} = \boldsymbol{a} \cdot \boldsymbol{c} + \boldsymbol{b} \cdot \boldsymbol{c}$)

(d) $a \cdot (2b + 3c) = (1, -1, 2) \cdot [(-4, 0, 4) + (9, 6, 3)]$

$$= (1, -1, 2) \cdot (5, 6, 7) = 13$$

(note that $2(a \cdot b) + 3(a \cdot c) = 4 + 9 = 13$)

(e) $(a \cdot b)c = [(1, -1, 2) \cdot (-2, 0, 2)](3, 2, 1)$

$$= 2(3, 2, 1) = (6, 4, 2)$$

(note that $a \cdot b$ is a scalar, so $(a \cdot b)c$ is a vector parallel or antiparallel to c)

Example 4.20   Find the angle between the vectors $a = (1, 2, 3)$ and $b = (2, 0, 4)$.

Solution   By definition

$$a \cdot b = |a| \, |b| \cos \theta = a_1 b_1 + a_2 b_2 + a_3 b_3$$

We have in the left-hand side

$$(1, 2, 3) \cdot (2, 0, 4) = 2 + 0 + 12 = 14$$

Also

$$|(1, 2, 3)| = \sqrt{(1^2 + 2^2 + 3^2)} = \sqrt{14}$$

and

$$|(2, 0, 4)| = \sqrt{(2^2 + 0^2 + 4^2)} = \sqrt{20}$$

Thus, from the definition of the scalar product,

$$14 = \sqrt{(14)}\sqrt{(20)} \cos \theta$$

giving

$$\theta = \cos^{-1} \sqrt{\tfrac{7}{10}}$$

Example 4.21   Given $a = (1, 0, 1)$ and $b = (0, 1, 0)$, show that $a \cdot b = 0$, and interpret this result.

Solution $a \cdot b = (1, 0, 1) \cdot (0, 1, 0) = 0$

Since $|a| \neq 0$ and $|b| \neq 0$, the two vectors are perpendicular. We can see this result geometrically, since a lies in the x–z plane and b is parallel to the y axis.

Example 4.22   The three vectors

$$a = (1, 1, 1), \quad b = (3, 2, -3) \quad \text{and} \quad c = (-1, 4, -1)$$

are given. Show that $a \cdot b = a \cdot c$ and intepret the result.

Solution   Now $a \cdot b = 1 \times 3 + 1 \times 2 - 1 \times 3 = 2$

and $a \cdot c = 1 \times (-1) + 1 \times 4 + 1 \times (-1) = 2$

so the two scalar products are clearly equal. Certainly $b \neq c$ since they are given to be unequal and a is non-zero so the conclusion from

$$a \cdot (b - c) = 0$$

is that the vectors a and $(b - c) = (4, -2, -2)$ are perpendicular.

Example 4.23

In a triangle ABC show that the perpendiculars from the vertices to the opposite sides intersect in a point.

Solution

Let the perpendiculars AD and BE meet in O as indicated in Figure 4.32, and choose O to be the origin. Define $\overrightarrow{OA} = a$, $\overrightarrow{OB} = b$ and $\overrightarrow{OC} = c$. Then

AD perpendicular to BC implies $a \cdot (b - c) = 0$

BE perpendicular to AC implies $b \cdot (c - a) = 0$

Hence, adding,

$$a \cdot b - a \cdot c + b \cdot c - b \cdot a = 0$$

so

$$b \cdot c - a \cdot c = c \cdot (b - a) = 0$$

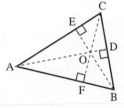

Figure 4.32
The altitudes of a triangle meet in a point (Example 4.23).

This statement implies that $b - a$ is perpendicular to c or AB is perpendicular to CF, as required. The case $b - a = 0$ is dismissed, since then the triangle would collapse. The case $c = 0$ implies that C is at O; the triangle is then right-angled and the result is trivial.

Example 4.24

It is necessary to drill to an underground pipeline in order to undertake repairs, so it is decided to aim for the nearest point from the measuring point. Relative to axes x, y in the horizontal ground and with z vertically downwards, remote measuring instruments locate two points on the pipeline at

$$(20, 20, 30) \quad \text{and} \quad (0, 15, 32)$$

with distances in metres. Find the nearest point on the pipeline from the origin O.

Solution

The situation is illustrated in Figure 4.33. The direction of the pipeline is $d = (0, 15, 32) - (20, 20, 30) = (-20, -5, 2)$. Thus any point on the pipeline will have position vector

$$r = (20, 20, 30) + t(-20, -5, 2)$$

for some t. Note that this is just the equation of the line derived in Example 4.15. At the shortest distance from O to the pipeline the vector $r = \overrightarrow{OP}$ is perpendicular to d, so $r \cdot d = 0$ gives the required condition to evaluate t. Thus

$$(-20, -5, 2) \cdot [(20, 20, 30) + t(-20, -5, 2)] = 0$$

and hence $-440 + 429t = 0$. Putting this value back into r gives

$$r = (-0.51, 14.87, 32.05)$$

Figure 4.33
Pipeline of Example
4.24.

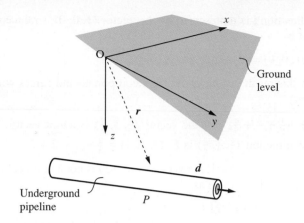

Note that the value of t is close to 1 so the optimum point is not far from the second of the points located.

Example 4.25

Find the work done by the force $F = (3, -2, 5)$ in moving a particle from a point P to a point Q having position vectors $(1, 4, -1)$ and $(-2, 3, 1)$ respectively.

Solution

Applying the triangle law to Figure 4.34, we have the displacement of the particle given by

$$r = \overrightarrow{PQ} = \overrightarrow{PO} + \overrightarrow{OQ} = \overrightarrow{OQ} - \overrightarrow{OP}$$

$$= (-2, 3, 1) - (1, 4, -1) = (-3, -1, 2)$$

Then the work done by the force F is

$$F \cdot r = (3, -2, 5) \cdot (-3, -1, 2) = -9 + 2 + 10$$

$$-3 \text{ units}$$

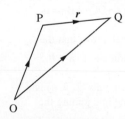

Figure 4.34
Triangle law for
Example 4.25.

The **component** of a vector in a given direction was discussed at the start of this section, and, as indicated in Figure 4.29, the component of F in the a direction is $|F| \cos \theta$. Taking $\hat{a}$ to be the unit vector in the a direction,

$$F \cdot \hat{a} = |F| \, |\hat{a}| \cos \theta = |F| \cos \theta$$

$$= \text{the component of } F \text{ in the } a \text{ direction}$$

Example 4.26

Find the component of the vector $F = (2, -1, 3)$ in

(a) the i direction

(b) the direction $(\frac{1}{3}, \frac{2}{3}, \frac{2}{3})$

(c) the direction $(4, 2, -1)$

Solution   (a) The direction i is represented by the vector $(1, 0, 0)$, so the component of F in the i direction is

$$F \cdot (1, 0, 0) = (2, -1, 3) \cdot (1, 0, 0) = 2$$

(note how this result just picks out the x component and agrees with the usual idea of a component).

(b) Since $\sqrt{(\frac{1}{9} + \frac{4}{9} + \frac{4}{9})} = 1$, the vector $(\frac{1}{3}, \frac{2}{3}, \frac{2}{3})$ is a unit vector. Thus the component of F in the direction $(\frac{1}{3}, \frac{2}{3}, \frac{2}{3})$ is $F \cdot (\frac{1}{3}, \frac{2}{3}, \frac{2}{3}) = \frac{2}{3} - \frac{2}{3} + 2 = 2$.

(c) Since $\sqrt{(16 + 4 + 1)} \neq 1$, the vector $(4, 2, -1)$ is not a unit vector. Therefore we must first compute its magnitude as

$$\sqrt{(4^2 + 2^2 + 1^2)} = \sqrt{21}$$

indicating that a unit vector in the direction of $(4, 2, -1)$ is $(4, 2, -1)/\sqrt{21}$. Thus the component of F in the direction of $(4, 2, -1)$ is

$$F \cdot (4, 2, -1)/\sqrt{21} = 3/\sqrt{21}$$

4.2.8   Exercises

18   Given that $u = (4, 0, -2)$, $v = (3, 1, -1)$, $w = (2, 1, 6)$ and $s = (1, 4, 1)$, evaluate

(a) $u \cdot v$ (b) $v \cdot s$

(c) $\hat{w}$ (d) $(v \cdot s)\hat{u}$

(e) $(u \cdot w)(v \cdot s)$ (f) $(u \cdot i)v + (w \cdot s)k$

19   Given u, v, w and s as for Question 18, find

(a) the angle between u and w

(b) the angle between v and s

(c) the value of λ for which the vectors $u + \lambda k$ and $v - \lambda i$ are perpendicular

(d) the value of μ for which the vectors $w + \mu i$ and $s - \mu i$ are perpendicular.

20 Find the work done by the force $F = (-2, -1, 3)$ in moving a particle from the point P to the point Q having position vectors $(-1, 2, 3)$ and $(1, -3, 4)$ respectively.

21 Find the resolved part in the direction of the vector $(3, 2, 1)$ of a force of 5 units acting in the direction of the vector $(2, -3, 1)$.

22 Find the value of t that makes the angle between the two vectors $a = (3, 1, 0)$ and $b = (t, 0, 1)$ equal to $45°$.

23 For any four points A, B, C and D in space, prove that

$$(\overrightarrow{DA} \cdot \overrightarrow{BC}) + (\overrightarrow{DB} \cdot \overrightarrow{CA}) + (\overrightarrow{DC} \cdot \overrightarrow{AB}) = 0$$

24 If $(c - \frac{1}{2}a) \cdot a = (c - \frac{1}{2}b) \cdot b = 0$, prove that the vector $c - \frac{1}{2}(a + b)$ is perpendicular to $a - b$.

25 Prove that the line joining the points $(2, 3, 4)$ and $(1, 2, 3)$ is perpendicular to the line joining the points $(1, 0, 2)$ and $(2, 3, -2)$.

26 Show that the diagonals of a rhombus intersect at right-angles.

27 (a) A batch of bricks weighing $10\,\text{N}$ is lifted from storage, taken to be the point $(0, 0, 0)$, against gravity to a point on the first floor of the building with coordinates $(0, 4, 5)\,\text{m}$. Gravity acts in the $(-z)$ direction and the x, y directions are at ground level. Calculate the work done in raising the bricks.

(b) A straight wall is to be built with p bricks in each layer. The weight of each brick is W newtons and it has thickness $h\,\text{m}$. Neglecting the thickness of the mortar, estimate the work done in raising the bricks from ground level to build a wall of height $nh\,\text{m}$. Show that the work done increases linearly with p but as the square of n.

28 Find the equation of a circular cylinder with the origin on the axis of the cylinder, the unit vector $\boldsymbol{a}$ along the axis and radius R.

29 A cube has corners with coordinates (0, 0, 0), (1, 0, 0), (0, 1, 0), (1, 1, 0), (0, 0, 1), (1, 0, 1), (0, 1, 1) and (1, 1, 1). Find the vectors representing the diagonals of the cube and hence find the length of the diagonals and the angle between the diagonals.

30 A lifeboat hangs from a davit as shown in Figure 4.35 with the x direction, the vertical part

of the davit and the arm of the davit being mutually perpendicular. The rope is fastened to the deck at a distance X from the davit. It is known that the maximum force in the x direction that the davit can withstand is 200 N. If the weight supported is 500 N and the pulley system is a single loop so that the tension is 250 N, then determine the maximum value that X can take.

31 A simple derrick is constructed as in Figure 4.36 with axes set up as indicated. The wires AP and BP are in tension, and the arm of the derrick, PC, is loaded with a weight W at C. The x and y components of the forces at P are always in equilibrium. Determine the range of the angle θ that will ensure that the tensions T_1 and T_2 are always positive and hence the wires will not slacken.

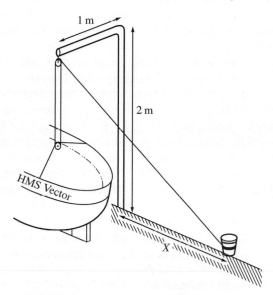

Figure 4.35 Davit in Question 30.

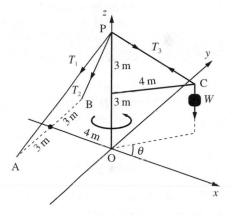

Figure 4.36 Simple derrick in Question 31.

4.2.9   The vector product

The **vector** or **cross product** was developed during the nineteenth century, its main practical use being to define the moment of a force in three dimensions. It is generally only in three dimensions that the vector product is used. The adaptation for two-dimensional vectors is of restricted scope, since for two-dimensional problems, where all vectors are confined to a plane, the direction of the vector product is always perpendicular to that plane.

Given two vectors $\boldsymbol{a}$ and $\boldsymbol{b}$, we define the vector product geometrically as

$$\boldsymbol{a} \times \boldsymbol{b} = |\boldsymbol{a}|\,|\boldsymbol{b}| \sin\theta \hat{\boldsymbol{n}} \qquad (4.7)$$

where θ is the angle between $\boldsymbol{a}$ and $\boldsymbol{b}$ ($0 \leqslant \theta \leqslant \pi$), and $\hat{\boldsymbol{n}}$ is the unit vector perpendicular to both $\boldsymbol{a}$ and $\boldsymbol{b}$ such that $\boldsymbol{a}, \boldsymbol{b}, \hat{\boldsymbol{n}}$ form a right-handed set – see Figure 4.37 and the definition at the beginning of Section 4.2.1.

Figure 4.37
Vector product $a \times b$,
right-hand rule.

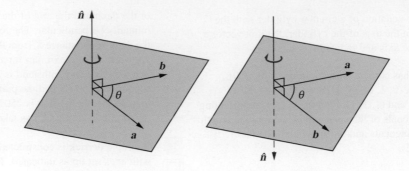

Figure 4.37
Vector product $a \times b$,
right-hand rule.

It is important to recognize that the vector product of two vectors is itself a vector. The alternative notation $a \wedge b$ is also sometimes used to denote the vector product, but this is less common since the similar wedge symbol $\wedge$ is also used for other purposes (see e.g. Chapter 6, Section 6.4.2).

One of the best-known physical applications of the vector product concerns the motion of a charged particle in a magnetic field. If a charged particle has velocity v and moves in a magnetic field H then the particle experiences a force perpendicular to both v and H, which is proportional to $v \times H$. It is this force that is used to direct the beam in a television tube. Similarly a wire moving with velocity v in a magnetic field H produces a current proportional to $v \times H$, thus converting mechanical energy into electric current, and provides the principle of the **dynamo**. For an **electric motor** the idea depends on the observation that an electric current C in a wire that lies in a magnetic field H produces a mechanical force proportional to $C \times H$. Thus electrical energy is converted to a mechanical force.

The moment or torque of a force F provides the classical application of the vector product in a mechanical context. Although moments are easy to define in two dimensions, the extension to three dimensions is not so easy. In vector notation, however, if the force passes through the point P and $\overrightarrow{OP} = r$, as illustrated in Figure 4.38, then the moment M of the force about O is simply defined as

$$M = r \times F \tag{4.8}$$

Figure 4.38
Moment of a force.

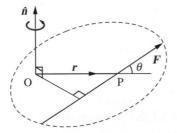

This is a vector in the direction of the normal $\hat{n}$, and moments add by the usual parallelogram law.

A further application of the vector product relates to rotating bodies. Consider a rigid body rotating with angular speed ω (in rad s^{-1}) about a fixed axis LM that passes through a fixed point O as illustrated in Figure 4.39. A point P of the rigid body having position vector r relative to O will move in a circular path whose plane is perpendicular

Figure 4.39
Angular velocity of a
rigid body.

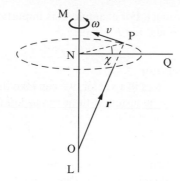

to OM and whose centre N is on OM. If NQ is a fixed direction and the angle QNP is equal to χ then

$$\text{the magnitude of angular velocity} = \frac{d\chi}{dt} = \omega$$

(Note that we have used here the idea of a derivative, which will be introduced in Chapter 8.) The velocity v of P will be in the direction of the tangent shown and will have magnitude

$$v = \text{NP}\frac{d\chi}{dt} = \text{NP}\omega$$

If we define $\boldsymbol{\omega}$ to be a vector of magnitude ω and having direction along the axis of rotation, in the sense in which the rotation would drive a right-handed screw, then

$$\boldsymbol{v} = \boldsymbol{\omega} \times \boldsymbol{r} \tag{4.9}$$

correctly defines the velocity of P in both magnitude and direction. This vector $\boldsymbol{\omega}$ is called the **angular velocity** of the rigid body.

Geometrically we have from Figure 4.40 that the area of a parallelogram ABCD is given by

$$\text{area} = h\,|\overrightarrow{\text{AB}}| = |\overrightarrow{\text{AD}}|\sin\theta\,|\overrightarrow{\text{AB}}| = |\overrightarrow{\text{AD}} \times \overrightarrow{\text{AB}}|$$

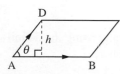

Figure 4.40
Representation of a
parallelogram.

Note also that the area of the triangle ABD is $\frac{1}{2}|\overrightarrow{\text{AD}} \times \overrightarrow{\text{AB}}|$, which corresponds to the result

$$\text{area of triangle ABD} = \tfrac{1}{2}(\text{AD})(\text{AB})\sin\theta$$

We now examine the properties of vector products in order to determine whether or not the usual laws of algebra apply.

(a) Anti-commutative law

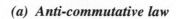

$$a \times b = -b \times a$$

This follows directly from the right-handedness of the set in the geometrical definition (4.7), since $\hat{n}$ changes direction when the order of multiplication is reversed. Thus the vector product does not commute, but rather anti-commutes, unlike the multiplication of scalars or the scalar product of two vectors. Therefore the order of multiplication matters when using the vector product. For example, it is important that the moment of a force is calculated as $M = r \times F$ and *not* $F \times r$.

(b) Non-associative multiplication

Since the vector product of two vectors is a vector, we can take the vector product with a third vector, and associativity can be tested. It turns out to *fail in general*, and

$$a \times (b \times c) \neq (a \times b) \times c$$

except in special cases, such as when $a = 0$. This can be seen to be the case from geometrical considerations using the definition (4.7). The vector $b \times c$ is perpendicular to both b and c, and is thus perpendicular to the plane containing b and c. Also, by definition, $a \times (b \times c)$ is perpendicular to $b \times c$, and is therefore in the plane of b and c. Similarly, $(a \times b) \times c$ is in the plane of a and b. Hence, in general, $a \times (b \times c)$ and $(a \times b) \times c$ are different vectors.

Since the associative law does not hold in general, we never write $a \times b \times c$, since it is ambiguous. Care must be taken to maintain the correct order, and thus brackets must be inserted when more than two vectors are involved in a vector product.

(c) Distributive law over multiplication by a scalar

The definition (4.7) shows trivially that

$$a \times (\lambda b) = \lambda(a \times b) = (\lambda a) \times b$$

and the usual algebraic rule applies.

(d) Distributive law over addition

$$a \times (b + c) = (a \times b) + (a \times c)$$

This law holds for the vector product. It can be proved geometrically using the definition (4.7). The proof, however, is rather protracted and is omitted here.

(e) Parallel vectors

It is obvious from the definition (4.7) that if a and b are parallel or antiparallel then $\theta = 0$ or π, so that $a \times b = 0$, and this includes the case $a \times a = 0$. We note, however, that if $a \times b = 0$ then we have three possible cases: either $a = 0$ or $b = 0$ or a and b are parallel. As with the scalar product, if we have $a \times b = a \times c$ then we cannot deduce that $b = c$. We first have to show that $a \neq 0$ and that a is not parallel to $b - c$.

(f) Cartesian form

From the definition (4.7), it clearly follows that the three unit vectors, i, j and k parallel to the coordinate axes satisfy

$$i \times i = j \times j = k \times k = 0$$
$$i \times j = k, \quad j \times k = i, \quad k \times i = j$$

(4.10)

Note the cyclic order of these latter equations. Using these results, we can obtain the cartesian or component form of the vector product. Taking

$$a = (a_1, a_2, a_3) = a_1 i + a_2 j + a_3 k$$

and

$$b = (b_1, b_2, b_3) = b_1 i + b_2 j + b_3 k$$

then, using rules (c), (d) and (a),

$$a \times b = (a_1 i + a_2 j + a_3 k) \times (b_1 i + b_2 j + b_3 k)$$

$$= a_1 b_1 (i \times i) + a_1 b_2 (i \times j) + a_1 b_3 (i \times k) + a_2 b_1 (j \times i) + a_2 b_2 (j \times j)$$

$$+ a_2 b_3 (j \times k) + a_3 b_1 (k \times i) + a_3 b_2 (k \times j) + a_3 b_3 (k \times k)$$

$$= a_1 b_2 k + a_1 b_3 (-j) + a_2 b_1 (-k) + a_2 b_3 i + a_3 b_1 j + a_3 b_2 (-i)$$

so that

$$a \times b = (a_2 b_3 - a_3 b_2) i + (a_3 b_1 - a_1 b_3) j + (a_1 b_2 - a_2 b_1) k \tag{4.11}$$

The cartesian form (4.11) can be more easily remembered in its determinant form (actually an accepted misuse of the determinant form)

$$a \times b = \begin{vmatrix} i & j & k \\ a_1 & a_2 & a_3 \\ b_1 & b_2 & b_3 \end{vmatrix} = i \begin{vmatrix} a_2 & a_3 \\ b_2 & b_3 \end{vmatrix} - j \begin{vmatrix} a_1 & a_3 \\ b_1 & b_3 \end{vmatrix} + k \begin{vmatrix} a_1 & a_2 \\ b_1 & b_2 \end{vmatrix}$$

$$= (a_2 b_3 - b_2 a_3) i - (a_1 b_3 - b_1 a_3) j + (a_1 b_2 - b_1 a_2) k \tag{4.12}$$

This notation is so convenient that we use it here before formally introducing determinants in the next chapter.

An alternative way to work out the cross product, which is easy to memorize, is to write the vectors (a, b, c) and (A, B, C) twice and read off the components by taking the products as indicated in Figure 4.41.

Figure 4.41
Gives the three components as $bC - cB$, $cA - aC$, $aB - bA$.

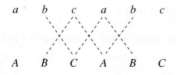

 The evaluation of the cross product is again available in the various computer packages: in MATLAB as cross(a, b) and in MAPLE as crossprod(a, b).

Example 4.27 Given the vectors $a = (2, 1, 0)$, $b = (2, -1, 1)$ and $c = (0, 1, 1)$, evaluate

(a) $a \times b$ (b) $(a \times b) \times c$ (c) $(a \cdot c)b - (b \cdot c)a$

(d) $b \times c$ (e) $a \times (b \times c)$ (f) $(a \cdot c)b - (a \cdot b)c$

Solution

(a) $\boldsymbol{a} \times \boldsymbol{b} = \begin{vmatrix} \boldsymbol{i} & \boldsymbol{j} & \boldsymbol{k} \\ 2 & 1 & 0 \\ 2 & -1 & 1 \end{vmatrix} = \boldsymbol{i} \begin{vmatrix} 1 & 0 \\ -1 & 1 \end{vmatrix} - \boldsymbol{j} \begin{vmatrix} 2 & 0 \\ 2 & 1 \end{vmatrix} + \boldsymbol{k} \begin{vmatrix} 2 & 1 \\ 2 & -1 \end{vmatrix}$

$= (1, -2, -4)$

(b) $(\boldsymbol{a} \times \boldsymbol{b}) \times \boldsymbol{c} = \begin{vmatrix} \boldsymbol{i} & \boldsymbol{j} & \boldsymbol{k} \\ 1 & -2 & -4 \\ 0 & 1 & 1 \end{vmatrix} = \boldsymbol{i} \begin{vmatrix} -2 & -4 \\ 1 & 1 \end{vmatrix} - \boldsymbol{j} \begin{vmatrix} 1 & -4 \\ 0 & 1 \end{vmatrix} + \boldsymbol{k} \begin{vmatrix} 1 & -2 \\ 0 & 1 \end{vmatrix}$

$= (2, -1, 1)$

(c) $(\boldsymbol{a} \cdot \boldsymbol{c})\boldsymbol{b} - (\boldsymbol{b} \cdot \boldsymbol{c})\boldsymbol{a} = 1\boldsymbol{b} - 0\boldsymbol{a} = (2, -1, 1)$
(Note that (b) and (c) give the same result.)

(d) $\boldsymbol{b} \times \boldsymbol{c} = (-2, -2, 2)$

(e) $\boldsymbol{a} \times (\boldsymbol{b} \times \boldsymbol{c}) = (2, -4, -2)$
(Note that (b) and (e) do not give the same result and the cross product is *not* associative.)

(f) $(\boldsymbol{a} \cdot \boldsymbol{c})\boldsymbol{b} - (\boldsymbol{a} \cdot \boldsymbol{b})\boldsymbol{c} = 1\boldsymbol{b} - 3\boldsymbol{c} = (2, -4, -2)$
(Note that (e) and (f) give the same result.)

Example 4.28   Find a unit vector perpendicular to the plane of the vectors $\boldsymbol{a} = (2, -3, 1)$ and $\boldsymbol{b} = (1, 2, -4)$.

Solution   A vector perpendicular to the plane of the two vectors is the vector product

$$\boldsymbol{a} \times \boldsymbol{b} = \begin{vmatrix} \boldsymbol{i} & \boldsymbol{j} & \boldsymbol{k} \\ 2 & -3 & 1 \\ 1 & 2 & -4 \end{vmatrix} = (10, 9, 7)$$

whose modulus is

$$|\boldsymbol{a} \times \boldsymbol{b}| = \sqrt{(100 + 81 + 49)} = \sqrt{230}$$

Hence a unit vector perpendicular to the plane of $\boldsymbol{a}$ and $\boldsymbol{b}$ is $(10/\sqrt{230}, 9/\sqrt{230}, 7/\sqrt{230})$.

Example 4.29   Find the area of the triangle having vertices at P(1, 3, 2), Q(-2, 1, 3) and R(3, -2, -1).

Solution   We have seen in Figure 4.40 that the area of the parallelogram formed with sides $\overrightarrow{PQ}$ and $\overrightarrow{PR}$ is $|\overrightarrow{PQ} \times \overrightarrow{PR}|$ so the area of the triangle PQR is $\frac{1}{2}|\overrightarrow{PQ} \times \overrightarrow{PR}|$. Now

$$\overrightarrow{PQ} = (-2 - 1, 1 - 3, 3 - 2) = (-3, -2, 1)$$

and

$$\overrightarrow{PR} = (3 - 1, -2 - 3, -1 - 2) = (2, -5, -3)$$

so that

$$\overrightarrow{PQ} \times \overrightarrow{PR} = \begin{vmatrix} i & j & k \\ -3 & -2 & 1 \\ 2 & -5 & -3 \end{vmatrix} = (11, -7, 19)$$

Hence the area of the triangle PQR is

$$\tfrac{1}{2}|\overrightarrow{PQ} \times \overrightarrow{PR}| = \tfrac{1}{2}\sqrt{(121 + 49 + 361)} = \tfrac{1}{2}\sqrt{531} \approx 11.52 \text{ square units.}$$

Example 4.30   A force of 4 units acts through the point P(2, 3, −5) in the direction of the vector (4, 5, −2). Find its moment about the point A(1, 2, −3).

What are the moments of the force about axes through A parallel to the coordinate axes?

Solution   A unit vector in the direction of the force is

$$\frac{4i + 5j - 2k}{\sqrt{(16 + 25 + 4)}} = \frac{1}{\sqrt{45}}(4, 5, -2)$$

Since the force F has a magnitude of 4 units

$$F = \frac{4}{\sqrt{45}}(4, 5, -2)$$

The position vector of P relative to A is

$$\overrightarrow{AP} = (1, 1, -2)$$

Thus from (4.8) the moment M of the force about A is

$$M = \overrightarrow{AP} \times F = \frac{4}{\sqrt{45}}\begin{vmatrix} i & j & k \\ 1 & 1 & -2 \\ 4 & 5 & -2 \end{vmatrix}$$

$$= (32/\sqrt{45}, -24/\sqrt{45}, 4/\sqrt{45})$$

The moments about axes through A parallel to the coordinate axes are $32/\sqrt{45}$, $-24/\sqrt{45}$ and $4/\sqrt{45}$.

Example 4.31   A rigid body is rotating with an angular velocity of 5 rad s^{-1} about an axis in the direction of the vector (1, 3, −2) and passing through the point A(2, 3, −1). Find the linear velocity of the point P(−2, 3, 1) of the body.

Solution   A unit vector in the direction of the axis of rotation is $\dfrac{1}{\sqrt{14}}(1, 3, -2)$. Thus the angular velocity vector of the rigid body is

$$\boldsymbol{\omega} = (5/\sqrt{14})(1, 3, -2)$$

The position vector of P relative to A is

$$\overrightarrow{AP} = (-2 - 2, 3 - 3, 1 + 1) = (-4, 0, 2)$$

Thus from (4.9) the linear velocity of P is

$$v = \boldsymbol{\omega} \times \overrightarrow{AP} = \frac{5}{\sqrt{14}} \begin{vmatrix} i & j & k \\ 1 & 3 & -2 \\ -4 & 0 & 2 \end{vmatrix}$$

$$= (30/\sqrt{14}, 30/\sqrt{14}, 60/\sqrt{14})$$

Example 4.32

A trapdoor is raised and lowered by a rope attached to one of its corners. The rope is pulled via a pulley fixed to a point A, 50 cm above the hinge as shown in Figure 4.42. If the trapdoor is uniform and of weight 20 N, what is the tension required to lift the door?

Figure 4.42
Trapdoor in Example 4.32.

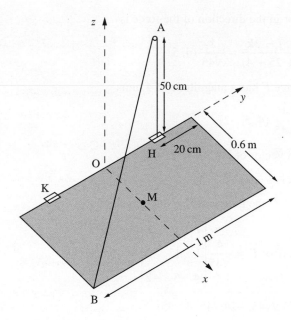

Solution

From the data given we can calculate various vectors immediately.

$$\overrightarrow{OA} = (0, 30, 50), \quad \overrightarrow{OB} = (60, -50, 0), \quad \overrightarrow{OH} = (0, 30, 0)$$

If M is the midpoint of the trapdoor then

$$\overrightarrow{OM} = (30, 0, 0)$$

The forces acting are the tension $\boldsymbol{T}$ in the rope along BA, the weight $\boldsymbol{W}$ through M in the $-z$ direction and reactions $\boldsymbol{R}$ and $\boldsymbol{S}$ at the hinges. Now

$$\overrightarrow{AB} = \overrightarrow{OB} - \overrightarrow{OA} = (60, -80, -50)$$

so that $|\overrightarrow{AB}| = 112$, and hence

$$\boldsymbol{T} = -T(60, -80, -50)/112$$

Taking moments about the hinge H, we first note that there is no moment of the reaction at H. For the remaining forces

$$M_\text{H} = \overrightarrow{HM} \times W + \overrightarrow{HB} \times T + \overrightarrow{HK} \times R$$

$$= (30, -30, 0) \times (0, 0, -20) + (60, -80, 0) \times (60, -80, -50)(-T/112) + \overrightarrow{HK} \times R$$

$$= (600, 600, 0) + T(-35.8, -26.8, 0) + \overrightarrow{HK} \times R$$

Since we require the moment about the y axis, we take the scalar product of M_H and j. The vector $\overrightarrow{HK}$ is along j, so $j \cdot (\overrightarrow{HK} \times R)$ must be zero. Thus the j component of M_H must be zero as the trapdoor just opens; that is,

$$0 = 600 - 26.8T$$

so

$$T = 22.4 \text{ N}$$

4.2.10 Exercises

32 Given $p = (1, 1, 1)$, $q = (0, -1, 2)$ and $r = (2, 2, 1)$, evaluate

(a) $p \times q$ (b) $p \times r$

(c) $r \times q$ (d) $(p \times r) \cdot q$

(e) $q \cdot (r \times p)$ (f) $(p \times r) \times q$

33 Show that the area of the triangle ABC in Figure 4.43 is $\frac{1}{2} |\overrightarrow{AB} \times \overrightarrow{AC}|$. Show that

$$\overrightarrow{AB} \times \overrightarrow{AC} = \overrightarrow{BC} \times \overrightarrow{BA} = \overrightarrow{CA} \times \overrightarrow{CB}$$

and hence deduce the sine rule

$$\frac{\sin A}{a} = \frac{\sin B}{b} = \frac{\sin C}{c}$$

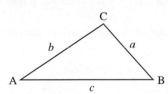

Figure 4.43
Sine rule: Section 2.6.1.

34 Prove that

$$(a - b) \times (a + b) = 2(a \times b)$$

and interpret geometrically.

35 The points A, B and C have coordinates $(1, -1, 2)$, $(9, 0, 8)$ and $(5, 0, 5)$ relative to rectangular cartesian axes. Find

(a) the vectors $\overrightarrow{AB}$ and $\overrightarrow{AC}$

(b) a unit vector perpendicular to the triangle ABC

(c) the area of the triangle ABC.

36 Use the definitions of the scalar and vector products to show that

$$|a \cdot b|^2 + |a \times b|^2 = a^2b^2$$

37 If a, b and c are three vectors such that $a + b + c = 0$, prove that

$$a \times b = b \times c = c \times a$$

and interpret geometrically.

38 A rigid body is rotating with angular velocity 6 rad s^{-1} about an axis in the direction of the vector $(3, -2, 1)$ and passing through the point A$(3, -2, 5)$. Find the linear velocity of the point P$(3, -2, 1)$ on the body.

39 A force of 4 units acts through the point P$(4, -1, 2)$ in the direction of the vector $(2, -1, 4)$. Find its moment about the point A$(3, -1, 4)$.

40 The moment of a force F acting at a point P about a point O is defined to be a vector M perpendicular to the plane containing F and the point O such that $|M| = p|F|$, where p is the perpendicular distance from O to the line of action of r. Figure 4.44 illustrates such a force F. Show that the perpendicular distance from O to the line of action of F is $|r|\sin\theta$, where r is the position vector of P. Hence deduce that $M = r \times F$. Show that the moment of F about O is the same for any point P on the line of action of F.

Forces $(1, 0, 0)$, $(1, 2, 0)$ and $(1, 2, 3)$ act through the points $(1, 1, 1)$, $(0, 1, 1)$ and $(0, 0, 1)$ respectively:

(a) find the moment of each force about the origin;

(b) find the moment of each force about the point $(1, 1, 1)$;

(c) find the total moment of the three forces about the point $(1, 1, 1)$.

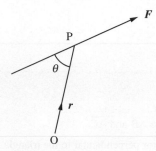

Figure 4.44
Moment of force F about O.

 41 Find a unit vector perpendicular to the plane of the two vectors $(2, -1, 1)$ and $(3, 4, -1)$. What is the sine of the angle between these two vectors?

 42 Prove that the shortest distance of a point P from the line through the points A and B is

$$\frac{|\overrightarrow{AP} \times \overrightarrow{AC}|}{|\overrightarrow{AB}|}$$

A satellite is stationary at P(2, 5, 4) and a warning signal is activated if any object comes within a distance of 3 units. Determine whether a rocket moving in a straight line passing through A(1, 5, 2) and B(3, −1, 5) activates the warning signal.

 43 The position vector r, with respect to a given origin O, of a charged particle of mass m and charge e at time t is given by

$$r = \left(\frac{Et}{B} + a\sin(\omega t)\right)i + a\cos(\omega t)j + ct k$$

where E, B, a and ω are constants. The corresponding velocity and acceleration are

$$v = \left(\frac{E}{B} + a\omega\cos(\omega t)\right)i - a\omega\sin(\omega t)j + ck$$

$$f = -a\omega^2\sin(\omega t)i - a\omega^2\cos(\omega t)j$$

For the case when $B = Bk$, show that the equation of motion

$$mf = e(Ej + v \times B)$$

is satisfied provided ω is chosen suitably.

4.2.11 Triple products

In Example 4.27 products of several vectors were computed: the product $(a \times b) \cdot c$ is called the **triple scalar product** and the product $(a \times b) \times c$ is called the **triple vector product**.

The triple scalar product is of interest because of its geometrical interpretation. Looking at Figure 4.45, we see that

$$a \times b = |a||b|\sin\theta k$$

$$= (\text{area of the parallelogram OACB})k$$

Figure 4.45
Triple scalar product
as the volume of a
parallelepiped.

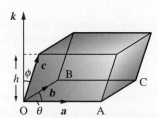

Thus, by definition,

$$(\boldsymbol{a} \times \boldsymbol{b}) \cdot \boldsymbol{c} = (\text{area of OACB})\boldsymbol{k} \cdot \boldsymbol{c}$$

$$= (\text{area of OACB})|\boldsymbol{k}||\boldsymbol{c}|\cos\phi$$

$$= (\text{area of OACB})h \quad (\text{where } h \text{ is the height of the parallelepiped})$$

$$= \text{volume of the parallelepiped}$$

Considering $(\boldsymbol{a} \times \boldsymbol{b}) \cdot \boldsymbol{c}$ to be the volume of the parallelepiped mounted on $\boldsymbol{a}, \boldsymbol{b}, \boldsymbol{c}$ has several useful consequences.

(a) If two of the vectors $\boldsymbol{a}, \boldsymbol{b}$ and $\boldsymbol{c}$ are parallel then $(\boldsymbol{a} \times \boldsymbol{b}) \cdot \boldsymbol{c} = 0$. This follows immediately since the parallelepiped collapses to a plane and has zero volume. In particular,

$$(\boldsymbol{a} \times \boldsymbol{b}) \cdot \boldsymbol{a} = 0 \quad \text{and} \quad (\boldsymbol{a} \times \boldsymbol{b}) \cdot \boldsymbol{b} = 0$$

(b) If the three vectors are coplanar then $(\boldsymbol{a} \times \boldsymbol{b}) \cdot \boldsymbol{c} = 0$. The same reasoning as in (a) gives this result.

(c) If $(\boldsymbol{a} \times \boldsymbol{b}) \cdot \boldsymbol{c} = 0$ then either $\boldsymbol{a} = 0$ or $\boldsymbol{b} = 0$ or $\boldsymbol{c} = 0$ or two of the vectors are parallel or the three vectors are coplanar.

(d) In the triple scalar product the dot $\cdot$ and the cross $\times$ can be interchanged:

$$(\boldsymbol{a} \times \boldsymbol{b}) \cdot \boldsymbol{c} = \boldsymbol{a} \cdot (\boldsymbol{b} \times \boldsymbol{c})$$

since it is easily checked that they measure the same volume mounted on $\boldsymbol{a}, \boldsymbol{b}, \boldsymbol{c}$. If we retain the same cyclic order of the three vectors then we obtain

$$\boldsymbol{a} \cdot (\boldsymbol{b} \times \boldsymbol{c}) = \boldsymbol{b} \cdot (\boldsymbol{c} \times \boldsymbol{a}) = \boldsymbol{c} \cdot (\boldsymbol{a} \times \boldsymbol{b}) \tag{4.13}$$

(e) In cartesian form the scalar triple product can be written as the determinant

$$\boldsymbol{a} \cdot (\boldsymbol{b} \times \boldsymbol{c}) = \begin{vmatrix} a_1 & a_2 & a_3 \\ b_1 & b_2 & b_3 \\ c_1 & c_2 & c_3 \end{vmatrix}$$

$$= a_1 b_2 c_3 - a_1 b_3 c_2 - a_2 b_1 c_3 + a_2 b_3 c_1 + a_3 b_1 c_2 - a_3 b_2 c_1 \tag{4.14}$$

Example 4.33 Find λ so that $\boldsymbol{a} = (2, -1, 1)$, $\boldsymbol{b} = (1, 2, -3)$ and $\boldsymbol{c} = (3, \lambda, 5)$ are coplanar.

Solution None of these vectors are zero or parallel, so by property (b) the three vectors are coplanar if $(\boldsymbol{a} \times \boldsymbol{b}) \cdot \boldsymbol{c} = 0$. Now

$$\boldsymbol{a} \times \boldsymbol{b} = (1, 7, 5)$$

so

$$(\boldsymbol{a} \times \boldsymbol{b}) \cdot \boldsymbol{c} = 3 + 7\lambda + 25$$

This will be zero, and the three vectors coplanar, when $\lambda = -4$.

For the triple vector product Example 4.27 suggests that $(a \times b) \times c$ could be written in terms of b and a, and indeed we shall show in general that

$$(a \times b) \times c = (a \cdot c)b - (b \cdot c)a \qquad (4.15)$$

We have

$$a \times b = (a_2 b_3 - a_3 b_2,\ a_3 b_1 - a_1 b_3,\ a_1 b_2 - a_2 b_1)$$

and hence

$$(a \times b) \times c = ((a_3 b_1 - a_1 b_3)c_3 - (a_1 b_2 - a_2 b_1)c_2,$$
$$(a_1 b_2 - a_2 b_1)c_1 - (a_2 b_3 - a_3 b_2)c_3,$$
$$(a_2 b_3 - a_3 b_2)c_2 - (a_3 b_1 - a_1 b_3)c_1)$$

The first component of this vector is

$$a_3 c_3 b_1 - b_3 c_3 a_1 - b_2 c_2 a_1 + a_2 c_2 b_1 = (a_1 c_1 + a_2 c_2 + a_3 c_3)b_1 - (b_1 c_1 + b_2 c_2 + b_3 c_3)a_1$$
$$= (a \cdot c)b_1 - (b \cdot c)a_1$$

Treating the second and third components similarly, we find

$$(a \times b) \times c = ((a \cdot c)b_1 - (b \cdot c)a_1,\ (a \cdot c)b_2 - (b \cdot c)a_2,\ (a \cdot c)b_3 - (b \cdot c)a_3)$$
$$= (a \cdot c)b - (b \cdot c)a$$

In a similar way we can show that

$$a \times (b \times c) = (a \cdot c)b - (a \cdot b)c \qquad (4.16)$$

We can now see why the associativity of the vector product does not hold in general. The vector in (4.15) is in the plane of b and a, while the vector in (4.16) is in the plane of b and c; hence they are not in the same planes in general, as we inferred geometrically in Section 4.2.9.

Example 4.34   If $a = (3, -2, 1)$, $b = (-1, 3, 4)$ and $c = (2, 1, -3)$, confirm that

$$a \times (b \times c) = (a \cdot c)b - (a \cdot b)c$$

Solution

$$b \times c = \begin{vmatrix} i & j & k \\ -1 & 3 & 4 \\ 2 & 1 & -3 \end{vmatrix} = (-13, 5, -7)$$

$$a \times (b \times c) = \begin{vmatrix} i & j & k \\ 3 & -2 & 1 \\ -13 & 5 & -7 \end{vmatrix} = (9, 8, -11)$$

$$(a \cdot c)b - (a \cdot b)c = [(3)(2) + (-2)(1) + (1)(-3)](-1, 3, 4)$$
$$- [(3)(-1) + (-2)(3) + (1)(4)](2, 1, -3)$$
$$= (-1, 3, 4) + 5(2, 1, -3)$$
$$= (9, 8, -11)$$

thus confirming the result

$$a \times (b \times c) = (a \cdot c)b - (a \cdot b)c$$

4.2.12 Exercises

44 Find the volume of the parallelepiped whose edges are represented by the vectors $(2, -3, 4)$, $(1, 3, -1)$, $(3, -1, 2)$.

45 Prove that the vectors $(3, 2, -1)$, $(5, -7, 3)$ and $(11, -3, 1)$ are coplanar.

46 Find the constant λ such that the three vectors $(3, 2, -1)$, $(1, -1, 3)$ and $(2, -3, \lambda)$ are coplanar.

47 Prove that the four points having position vectors $(2, 1, 0)$, $(2, -2, -2)$, $(7, -3, -1)$ and $(13, 3, 5)$ are coplanar.

48 Given $p = (1, 4, 1)$, $q = (2, 1, -1)$ and $r = (1, -3, 2)$, find

(a) a unit vector perpendicular to the plane containing p and q,

(b) a unit vector in the plane containing $p \times q$ and $p \times r$ that has zero x component.

49 Show that if a is any vector and $\hat{u}$ any unit vector then

$$a = (a \cdot \hat{u})\hat{u} + \hat{u} \times (a \times \hat{u})$$

and draw a diagram to illustrate this relation geometrically.

   The vector $(3, -2, 6)$ is resolved into two vectors along and perpendicular to the line whose direction cosines are proportional to $(1, 1, 1)$. Find these vectors.

50 Three vectors u, v, w are expressed in terms of the three vectors l, m, n in the form

$$u = u_1 l + u_2 m + u_3 n$$
$$v = v_1 l + v_2 m + v_3 n$$
$$w = w_1 l + w_2 m + w_3 n$$

Show that

$$u \cdot (v \times w) = \lambda l \cdot (m \times n)$$

and evaluate λ.

51 Forces $F_1, F_2, \ldots, F_n$ act at the points $r_1, r_2, \ldots, r_n$ respectively. The total force and the total moment about the origin O are

$$F = \Sigma F_i \quad \text{and} \quad G = \Sigma r_i \times F_i$$

Show that for any other origin O$'$ the moment is given by

$$G' = G + \overrightarrow{O'O} \times F$$

If O$'$ lies on the line

$$\overrightarrow{OO'} = r = \alpha(F \times G) + tF$$

find the constant α that ensures that G' is parallel to F. This line is called the central axis of the system of forces.

52 Extended exercise on products of four vectors.

(a) Use (4.13) to show

$$(a \times b) \cdot (c \times d) = [(a \times b) \times c] \cdot d$$

and use (4.15) to simplify the expression on the right-hand side.

(b) Use (4.15) to show that

$$(a \times b) \times (a \times c) = [a \cdot (a \times c)]b$$
$$- [b \cdot (a \times c)]a$$

and show that the right-hand side can be simplified to

$$[(a \times b) \cdot c]a$$

(c) Use (4.16) to show that

$$a \times [b \times (a \times c)] = a \times [(b \cdot c)a - (b \cdot a)c]$$

and simplify the right-hand side further. Note that the product is different from the result in (b), verifying that the position of the brackets matters in cross products.

(d) Use the result in (a) to show that

$$(l \times m) \cdot (l \times n) = l^2(m \cdot n) - (l \cdot m)(l \cdot n)$$

Take l, m and n to be unit vectors along the sides of a regular tetrahedron. Deduce that the angle between two faces of the tetrahedron is $\cos^{-1}\frac{1}{3}$.

4.3 The vector treatment of the geometry of lines and planes

4.3.1 Vector equation of a line

In Example 4.15 (equation (4.2)) the equation of the line through two points A and B with position vectors a and b, illustrated in Figure 4.46, was shown to be

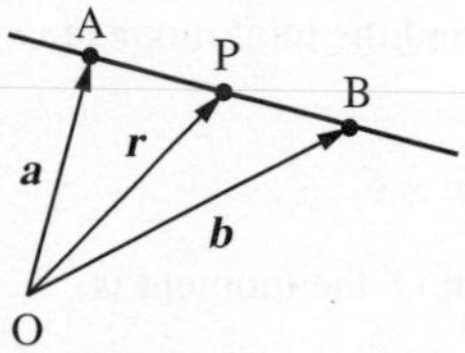

$$r = (1 - t)a + tb \qquad (4.17)$$

Since $\overrightarrow{OP} = \overrightarrow{OA} + \overrightarrow{AP} = \overrightarrow{OA} + t\overrightarrow{AB}$, we have $r = a + t(b - a)$. If we write $c = b - a$ then we have an alternative intepretation of a line through A in the direction c:

$$r = a + tc \qquad (4.18)$$

Figure 4.46
Line AB in terms of $r = \overrightarrow{OP}$.

The cartesian or component form of this equation is

$$\frac{x - a_1}{c_1} = \frac{y - a_2}{c_2} = \frac{z - a_3}{c_3}(= t) \qquad (4.19)$$

where $a = (a_1, a_2, a_3)$ and $c = (c_1, c_2, c_3)$. Alternatively, as shown by (4.3), the cartesian equation of (4.17) may be written in the form

$$\frac{x - a_1}{b_1 - a_1} = \frac{y - a_2}{b_2 - a_2} = \frac{z - a_3}{b_3 - a_3}(= t)$$

where $a = (a_1, a_2, a_3)$ and $b = (b_1, b_2, b_3)$ are two points on the line. Again, if any of the denominators is zero, then both forms of the equation of a line are interpreted as the corresponding numerator is zero.

Example 4.35 Find the equation of the lines L_1 through the points (0, 1, 0) and (1, 3, −1) and L_2 through (1, 1, 1) and (−1, −1, 1). Do the two lines intersect and, if so, at what point?

Solution From (4.17) L_1 has the equation

$$r = (0, 1 - t, 0) + (t, 3t, -t) = (t, 1 + 2t, -t)$$

and L_2 has the equation

$$\mathbf{r} = (1 - s, 1 - s, 1 - s) + (-s, -s, s) = (1 - 2s, 1 - 2s, 1)$$

Note that the cartesian equation of L_2 reduces to $x = y$; $z = 1$. The two lines intersect if it is possible to find s and t such that

$$t = 1 - 2s, \quad 1 + 2t = 1 - 2s, \quad -t = 1$$

Solving two of these equations will give the values of s and t. If these values satisfy the remaining equation then the lines intersect; however, if they do not satisfy the remaining equation then the lines do not intersect. In this particular case, the third equation gives $t = -1$ and the first equation $s = 1$. Putting these values into the second equation the left-hand side equals -1 and the right-hand side equals -1 so the equations are all satisfied and therefore the lines intersect. Substituting back into either equation, the point of intersection is $(-1, -1, 1)$.

Example 4.36

The position vectors of the points A and B are

$$(1, 4, 6) \quad \text{and} \quad (3, 5, 7)$$

Find the vector equation of the line AB and find the points where the line intersects the coordinate planes.

Solution

The line has equation

$$\mathbf{r} = (1, 4, 6) + t(2, 1, 1)$$

or in components

$$x = 1 + 2t$$

$$y = 4 + t$$

$$z = 6 + t$$

Thus the line meets the y–z plane when $x = 0$ and hence $t = -\frac{1}{2}$ and the point of intersection with the plane is $(0, \frac{7}{2}, \frac{11}{2})$.

The line meets the z–x plane when $y = 0$ and hence $t = -4$ and the point of intersection with the plane is $(-7, 0, 2)$.

The line meets the x–y plane when $z = 0$ and hence $t = -6$ and the point of intersection with the plane is $(-11, -2, 0)$.

Example 4.37

A tracking station observes an aeroplane at two successive times to be

$$(-500, 0, 1000) \quad \text{and} \quad (400, 400, 1050)$$

relative to axes x in an easterly direction, y in a northerly direction and z vertically upwards, with distances in metres. Find the equation of the path of the aeroplane. Control advises the aeroplane to change course from its present position to level flight at the current height and turn easterly through an angle of $90°$; what is the equation of the new path?

Solution The situation is illustrated in Figure 4.47. The equation of the path of the aeroplane is

$$r = (-500, 0, 1000) + t(900, 400, 50)$$

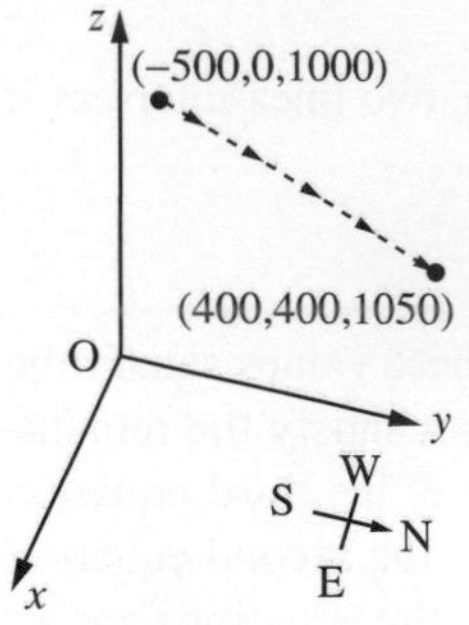

The new path starts at the point $(400, 400, 1050)$, the z component remains at the value of 1050 but the change of direction in the x–y plane implies that the aeroplane has a new direction with x and y coordinates given from

$$(900, 400, 0) \times \mathbf{k} = (400, -900, 0)$$

Thus the new path is

$$r = (400, 400, 1050) + s(400, -900, 0)$$

In cartesian coordinates the equations are

$$9x + 4y = 5200$$
$$z = 1050$$

Figure 4.47
Path of aeroplane in
Example 4.37.

Example 4.38 Find the shortest distance between the two skew lines

$$\frac{x}{3} = \frac{y - 9}{-1} = \frac{z - 2}{1} \quad \text{and} \quad \frac{x + 6}{-3} = \frac{y + 5}{2} = \frac{z - 10}{4}$$

Also determine the equation of the common perpendicular. (Note that two lines are said to be skew if they do not intersect and are not parallel.)

Solution In vector form the equations of the lines are

$$r = (0, 9, 2) + t(3, -1, 1)$$

and

$$r = (-6, -5, 10) + s(-3, 2, 4)$$

The shortest distance between the two lines will be their common perpendicular. Let P_1 and P_2 be the end points of the common perpendicular, having position vectors r_1 and r_2 respectively, where

$$r_1 = (0, 9, 2) + t_1(3, -1, 1)$$

and

$$r_2 = (-6, -5, 10) + s_2(-3, 2, 4)$$

Then the vector $\overrightarrow{P_2P_1}$ is given by

$$\overrightarrow{P_2P_1} = r_1 - r_2 = (6, 14, -8) + t_1(3, -1, 1) - s_2(-3, 2, 4) \tag{4.20}$$

Since $(3, -1, 1)$ and $(-3, 2, 4)$ are vectors in the direction of each of the lines, it follows that a vector $\mathbf{n}$ perpendicular to both lines is

$$\mathbf{n} = (-3, 2, 4) \times (3, -1, 1) = (6, 15, -3)$$

So a unit vector perpendicular to both lines is

$$\hat{\mathbf{n}} = (6, 15, -3)/\sqrt{270} = (2, 5, -1)/\sqrt{30}$$

Thus we can also express $\overrightarrow{P_2P_1}$ as

$$\overrightarrow{P_2P_1} = d\hat{n}$$

where d is the shortest distance between the two lines.

Equating the two expressions for $\overrightarrow{P_2P_1}$ gives

$$(6, 14, -8) + t_1(3, -1, 1) - s_2(-3, 2, 4) = (2, 5, -1)d/\sqrt{30}$$

Taking the scalar product throughout with the vector $(2, 5, -1)$ gives

$$(6, 14, -8) \cdot (2, 5, -1) + t_1(3, -1, 1) \cdot (2, 5, -1) - s_2(-3, 2, 4) \cdot (2, 5, -1)$$

$$= (2, 5, -1) \cdot (2, 5, -1)d/\sqrt{30}$$

which reduces to

$$90 + 0t_1 + 0s_2 = 30d/\sqrt{30}$$

giving the shortest distance between the two lines as

$$d = 3\sqrt{30}$$

To obtain the equation of the common perpendicular, we need to find the coordinates of either P_1 or P_2 – and to achieve this we need to find the value of either t_1 or s_2. We therefore take the scalar product of (4.20) with $(3, -1, 1)$ and $(-3, 2, 4)$ in turn, giving respectively

$$11t_1 + 7s_2 = 4$$

and

$$-7t_1 - 29s_2 = 22$$

which on solving simultaneously give $t_1 = 1$ and $s_2 = -1$. Hence the coordinates of the end points P_1 and P_2 of the common perpendicular are

$$\boldsymbol{r}_1 = (0, 9, 2) + 1(3, -1, 1) = (3, 8, 3)$$

and

$$\boldsymbol{r}_2 = (-6, -5, 10) - 1(-3, 2, 4) = (-3, -7, 6)$$

From (4.18) the equation of the common perpendicular is

$$\boldsymbol{r} = (3, 8, 3) + t(2, 5, -1)$$

or in cartesian form

$$\frac{x - 3}{2} = \frac{y - 8}{5} = \frac{z - 3}{-1} = t$$

 MAPLE contains a geometry package which takes a bit of time to master but which can solve many coordinate geometry problems. For the current problem the code is given: note that printing has been largely suppressed, but replace ':' by ';' at the end of statements for more information.

```
with (geom3d):
point (A, [0, 9, 2]): v := [3, -1, 1]: line (L1, [A, v]): detail (L1);
point (B, [-6, -5, 10]): w := [-3, 2, 4]: line (L2, [B, w]): detail (L2);
distance (L1, L2);                                (gives result 3√30 in the text)
```

z := Equation (L1, t): y := Equation (L2, s):
with (linalg):
m := innerprod (z − y, v): n := innerprod (z − y, w):
solve ({m, n}, {s, t});   *(gives solution t = 1 and s = −1)*
point (P, eval (z, t = 1)): point (Q, eval (y, s = −1)):
line (L3, [P, Q]): detail (L3);   *(gives the required equation of the common perpendicular)*

Example 4.39

A box with an open top and unit side length is observed from the direction (a, b, c). Determine the part of OC that is visible.

Solution

The situation is shown in Figure 4.48. The line or ray through Q parallel to the line of sight has the equation

$$\mathbf{r} = (0, 0, \alpha) + t(a, b, c)$$

where $0 \leqslant \alpha \leqslant 1$ to ensure that Q lies between O and C. The line RS has the equation

$$\mathbf{r} = (1, 0, 1) + s(0, 1, 0)$$

The ray that intersects RS must therefore satisfy

$$ta = 1, \quad tb = s, \quad \alpha = 1 - \frac{c}{a}$$

Figure 4.48
Looking for hidden lines in Example 4.39.

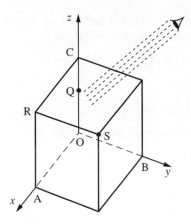

Note that if $c = 0$ then we are looking parallel to the lid and can only see the point C. If $c < 0$ then we are looking up at the box; since $\alpha > 1$, we cannot see any of side OC, so the line is hidden. If, however, $c > a$ then the solution gives α to be negative, so that all of the side OC is visible. For $0 < c < a$ the parameter α lies between 0 and 1, and only part of the line is visible. A similar analysis needs to be performed for the other sides of the lid.

What are the conditions that any or part of OA and OB can be seen?

4.3.2 Vector equation of a plane

To obtain the equation of a plane, we use the result that the line joining any two points in the plane is perpendicular to the normal to the plane as illustrated in Figure 4.49. The vector n is perpendicular to the plane, a is the position vector of a given point A in the plane and r is the position vector of any point P on the plane. The vector $\overrightarrow{AP} = r - a$ is perpendicular to n, and hence

$$(r - a) \cdot n = 0$$

so that

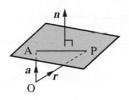

Figure 4.49
Equation of a plane;
n is perpendicular to
the plane.

$$\boxed{r \cdot n = a \cdot n \quad \text{or} \quad r \cdot n = p} \tag{4.21}$$

is the general form for the **equation of a plane** with normal n. In the particular case when n is a unit vector, p in (4.21) represents the perpendicular distance from the origin to the plane. In cartesian form we take $n = (\alpha, \beta, \gamma)$, and the equation becomes

$$\alpha x + \beta y + \gamma z = p \tag{4.22}$$

which is just a linear relation between the variables x, y and z.

Example 4.40 Find the equation of the plane through the three points

$$a = (1, 1, 1), \quad b = (0, 1, 2) \quad \text{and} \quad c = (-1, 1, -1)$$

Solution The vectors $a - b = (1, 0, -1)$ and $a - c = (2, 0, 2)$ will lie in the plane. The normal n to the plane can thus be constructed as $(a - b) \times (a - c)$, giving

$$n = (1, 0, -1) \times (2, 0, 2) = (0, -4, 0)$$

Thus from (4.21) the equation of the plane is given by

$$r \cdot n = a \cdot n$$

or

$$r \cdot (0, -4, 0) = (1, 1, 1) \cdot (0, -4, 0)$$

giving

$$r \cdot (0, -4, 0) = -4$$

or $y = 1$ in cartesian form (since $r = (x, y, z)$).

Example 4.41 A metal has a simple cubic lattice structure so that the atoms lie on the lattice points given by

$$r = a(l, m, n)$$

where a is the lattice spacing and l, m, n are integers. The metallurgist requires to identify the points that lie on two lattice planes

$$LP_1 \quad \text{through } a(0, 0, 0), \quad a(1, 1, 0) \quad \text{and} \quad a(0, 1, 2)$$

$$LP_2 \quad \text{through } a(0, 0, 2), \quad a(1, 1, 0) \quad \text{and} \quad a(0, 1, 0)$$

Solution   The direction perpendicular to LP_1 is $(1, 1, 0) \times (0, 1, 2) = (2, -2, 1)$ and hence the equation of LP_1 is

$$\boldsymbol{r} \cdot (2, -2, 1) = 0 \quad \text{or in cartesian form} \quad 2x - 2y + z = 0 \tag{4.23}$$

The direction perpendicular to LP_2 is $(1, 1, -2) \times (0, 1, -2) = (0, 2, 1)$ and hence the equation of LP_2 is

$$\boldsymbol{r} \cdot (0, 2, 1) = 2 \quad \text{or in cartesian form} \quad 2y + z = 2 \tag{4.24}$$

It is easiest to solve these equations in their cartesian form. The coordinates must be integers so take $y = m$, then z can easily be calculated from (4.24) as

$$z = 2 - 2m$$

and then x is computed from (4.23) to be $x = 2m - 1$.

Hence the required points all lie on a line and take the form

$$\boldsymbol{r} = a(2m - 1, m, 2 - 2m)$$

where m is an integer.

Example 4.42   Find the point where the plane

$$\boldsymbol{r} \cdot (1, 1, 2) = 3$$

meets the line

$$\boldsymbol{r} = (2, 1, 1) + \lambda(0, 1, 2)$$

Solution   At the point of intersection $\boldsymbol{r}$ must satisfy both equations, so

$$[(2, 1, 1) + \lambda(0, 1, 2)] \cdot (1, 1, 2) = 3$$

or

$$5 + 5\lambda = 3$$

so

$$\lambda = -\tfrac{2}{5}$$

Substituting back into the equation of the line gives the point of intersection as

$$\boldsymbol{r} = (2, \tfrac{3}{5}, \tfrac{1}{5})$$

Example 4.43   Find the equation of the line of intersection of the two planes $x + y + z = 5$ and $4x + y + 2z = 15$.

Solution   In vector form the equations of the two planes are

$$\boldsymbol{r} \cdot (1, 1, 1) = 5$$

and

$$\boldsymbol{r} \cdot (4, 1, 2) = 15$$

The required line lies in both planes, and is therefore perpendicular to the vectors $(1, 1, 1)$ and $(4, 1, 2)$, which are normal to the individual planes. Hence a vector c in the direction of the line is

$$c = (1, 1, 1) \times (4, 1, 2) = (1, 2, -3)$$

To find the equation of the line, it remains only to find the coordinates of any point on the line. To do this, we are required to find the coordinates of a point satisfying the equation of the two planes. Taking $x = 0$, the corresponding values of y and z are given by

$$y + z = 5 \quad \text{and} \quad y + 2z = 15$$

that is, $y = -5$ and $z = 10$. Hence it can be checked that the point $(0, -5, 10)$ lies in both planes and is therefore a point on the line. From (4.18) the equation of the line is

$$r = (0, -5, 10) + t(1, 2, -3)$$

or in cartesian form

$$\frac{x}{1} = \frac{y + 5}{2} = \frac{z - 10}{-3} = t$$

The MAPLE instructions to solve this example are

with (geom3d):
plane (P1, x + y + z = 5, [x, y, z]): plane (P2, 4*x + y + 2*z = 15, [x, y, z]):
intersection (L, P1, P2): detail (L);

Example 4.44   Find the perpendicular distance from the point $P(2, -3, 4)$ to the plane $x + 2y + 2z = 13$.

Solution   In vector form the equation of the plane is

$$r \cdot (1, 2, 2) = 13$$

and a vector perpendicular to the plane is

$$n = (1, 2, 2)$$

Thus from (4.18) the equation of a line perpendicular to the plane and passing through $P(2, -3, 4)$ is

$$r = (2, -3, 4) + t(1, 2, 2)$$

This will meet the plane when

$$r \cdot (1, 2, 2) = (2, -3, 4) \cdot (1, 2, 2) + t(1, 2, 2) \cdot (1, 2, 2) = 13$$

giving

$$4 + 9t = 13$$

so that

$$t = 1$$

Thus the line meets the plane at N having position vector

$$r = (2, -3, 4) + 1(1, 2, 2) = (3, -1, 6)$$

Hence the perpendicular distance is

$$PN = \sqrt{[(3 - 2)^2 + (-1 + 3)^2 + (6 - 4)^2]} = 3$$

4.3.3 Exercises

53 If A and B have position vectors (1, 2, 3) and (4, 5, 6) respectively, find

(a) the direction vector of the line through A and B,

(b) the vector equation of the line through A and B,

(c) the cartesian equation of the line.

54 Find the vector equation of the plane that passes through the points (1, 2, 3), (2, 4, 5) and (4, 5, 6). What is its cartesian equation?

55 Show that the line joining (2, 3, 4) to (1, 2, 3) is perpendicular to the line joining (1, 0, 2) to (2, 3, -2).

56 Prove that the lines $r = (1, 2, -1) + t(2, 2, 1)$ and $r = (-1, -2, 3) + s(4, 6, -3)$ intersect, and find the coordinates of their point of intersection. Also find the acute angle between the lines.

57 Find the vector equation of the plane that contains the line $r = a + \lambda b$ and passes through the point with position vector c.

58 P is a point on a straight line with position vector $r = a + tb$. Show that

$$r^2 = a^2 + 2a \cdot bt + b^2 t^2$$

By completing the square, show that r^2 is a minimum for the point P for which $t = -a \cdot b/b^2$. Show that at this point $\overrightarrow{OP}$ is perpendicular to the line $r = a + tb$. (This proves the well-known result that the shortest distance from a point to a line is the length of the perpendicular from that point to the line.)

59 The line of intersection of two planes $r \cdot n_1 = p_1$ and $r \cdot n_2 = p_2$ lies in both planes. It is therefore perpendicular to both n_1 and n_2. Give an expression for this direction, and so show that the equation of the line of intersection may be written as $r = r_0 + t(n_1 \times n_2)$, where r_0 is any vector satisfying $r_0 \cdot n_1 = p_1$ and $r_0 \cdot n_2 = p_2$. Hence find the line of intersection of the planes $r \cdot (1, 1, 1) = 5$ and $r \cdot (4, 1, 2) = 15$.

60 Find the equation of the line through the point (1, 2, 4) and in the direction of the vector (1, 1, 2). Find where this line meets the plane $x + 3y - 4z = 5$.

61 Find the acute angle between the planes $2x + y - 2z = 5$ and $3x - 6y - 2z = 7$.

62 Given that $a = (3, 1, 2)$ and $b = (1, -2, -4)$ are the position vectors of the points P and Q respectively, find

(a) the equation of the plane passing through Q and perpendicular to PQ,

(b) the distance from the point (-1, 1, 1) to the plane obtained in (a).

63 Find the equation of the line joining (1, -1, 3) to (3, 3, -1). Show that it is perpendicular to the plane $2x + 4y - 4z = 5$, and find the angle that the line makes with the plane $12x - 15y + 16z = 10$.

64 Find the shortest distance between the two lines

$$r = (4, -2, 3) + t(2, 1, -1)$$

and

$$r = (-7, -2, 1) + s(3, 2, 1)$$

65 Find the equation of the plane through the line

$$r = (1, -3, 4) + t(2, 1, 1)$$

and parallel to the line

$$r = s(1, 2, 3)$$

66 Show that the perpendicular bisectors of the sides of a triangle meet in a point.

67 Find the equation of the line through $P(-1, 0, 1)$ that cuts the line $r = (3, 2, 1) + t(1, 2, 2)$ at right-angles at Q. Also find the length PQ and the equation of the plane containing the two lines.

68 Show that the equation of the plane through the points P_1, P_2 and P_3 with position vectors r_1, r_2 and r_3 respectively takes the form

$$r \cdot [(r_1 \times r_2) + (r_2 \times r_3) + (r_3 \times r_1)] = r_1 \cdot (r_2 \times r_3)$$

4.4 Engineering application: spin-dryer suspension

Vectors are at their most powerful when dealing with complicated three-dimensional situations. Geometrical and physical intuition are often difficult to use, and it becomes necessary to work quite formally to analyse such situations. For example, the front suspension of a motor car has two struts supported by a spring-and-damper system and subject to a variety of forces and torques from both the car and the wheels. To analyse the stresses and the vibrations in the various components of the structure is non-trivial, even in a two-dimensional version; the true three-dimensional problem provides a testing exercise for even the most experienced automobile engineer. In the present text a much simpler situation is analysed to illustrate the use of vectors.

4.4.1 Point-particle model

As with the car suspension, many machines are mounted on springs to isolate vibrations. A typical, simple example is a spin-dryer, which consists of a drum connected to the casing by heavy springs. Oscillations can be very severe when spinning at high speed, and it is essential to know what forces are transmitted to the casing and hence to the mounts. Before the dynamical situation can be analysed, it is necessary to compute the restoring forces on the drum when it is displaced from its equilibrium position. This is a static problem that is best studied using vectors.

We model the spin-dryer as a heavy point particle connected to the eight corners of the casing by springs (Figure 4.50). The drum has weight W and the casing is taken to be a cube of side $2L$. The springs are all equal, having spring constant k and natural length $L\sqrt{3}$. Thus when the drum is at the midpoint of the cube the springs are neither compressed nor extended.

The particle is displaced from its central position by a small amount (a, b, c), where the natural coordinates illustrated in Figure 4.50 are used; the origin is at the centre of the cube and the axes are parallel to the sides. What is required is the total force acting on the particle arising from the weight and the springs. Clearly, this information is needed before any dynamical calculations can be performed. It will be assumed that the displacements are sufficiently small that squares $(a/L)^2$, $(b/L)^2$, $(c/L)^2$ and higher powers are neglected.

Consider a typical spring PA. The tension in the spring is assumed to obey Hooke's law that the force is along PA and has magnitude proportional to extension. In vector form this can be written as

$$T_A = k \frac{\overrightarrow{PA}}{|\overrightarrow{PA}|}(|\overrightarrow{PA}| - L\sqrt{3}) \qquad (4.25)$$

where k is the proportionality constant, $\overrightarrow{PA}/|\overrightarrow{PA}|$ is the unit vector in the direction along PA, and $|\overrightarrow{PA}| - L\sqrt{3}$ is the extension of the spring over its natural length $L\sqrt{3}$.

Figure 4.50
The particle P is
attached by equal
springs to the eight
corners of the cube.

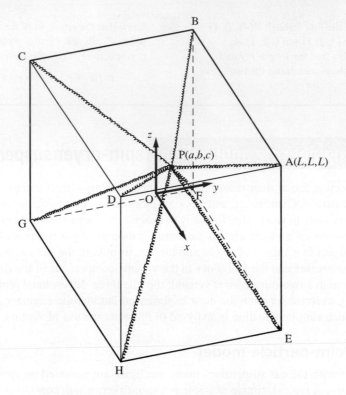

Now

$$\overrightarrow{PA} = \overrightarrow{OA} - \overrightarrow{OP} = (L - a, L - b, L - c)$$

so calculating the modulus squared gives

$$|\overrightarrow{PA}|^2 = (L - a)^2 + (L - b)^2 + (L - c)^2$$
$$= 3L^2 - 2L(a + b + c) + \text{quadratic terms}$$

Thus

$$|\overrightarrow{PA}| = \left[1 - \frac{2}{3L}(a + b + c)\right]^{1/2} L\sqrt{3}$$

and, on using the binomial expansion (see equation (7.15)) and neglecting quadratic
and higher terms, we obtain

$$|\overrightarrow{PA}| = \left[1 - \frac{1}{3L}(a + b + c)\right] L\sqrt{3}$$

Putting the information acquired back into (4.25) gives

$$T_A = kL\frac{(1 - a/L, 1 - b/L, 1 - c/L)}{[1 - (a + b + c)/3L]L\sqrt{3}}\frac{(-1)(a + b + c)L\sqrt{3}}{3L}$$

and by expanding again, using the binomial expansion to first order in a/L and so on,
we obtain

$$T_A = -\tfrac{1}{3}k(a + b + c)(1, 1, 1)$$

Similar calculations give

$$T_B = -\tfrac{1}{3}k(-a + b + c)(-1, 1, 1)$$

$$T_C = -\tfrac{1}{3}k(-a - b + c)(-1, -1, 1)$$

$$T_D = -\tfrac{1}{3}k(a - b + c)(1, -1, 1)$$

$$T_E = -\tfrac{1}{3}k(a + b - c)(1, 1, -1)$$

$$T_F = -\tfrac{1}{3}k(-a + b - c)(-1, 1, -1)$$

$$T_G = -\tfrac{1}{3}k(-a - b - c)(-1, -1, -1)$$

$$T_H = -\tfrac{1}{3}k(a - b - c)(1, -1, -1)$$

The total spring force is therefore obtained by adding these eight tensions together:

$$T = -\tfrac{8}{3}k(a, b, c)$$

The restoring force is therefore towards the centre of the cube, as expected, in the direction PO and with magnitude $\tfrac{8}{3}k$ times the length of PO.

When the weight is included, the total force is

$$F = (-\tfrac{8}{3}ka, -\tfrac{8}{3}kb, -\tfrac{8}{3}kc - W)$$

If the drum just hangs in equilibrium then $F = 0$, and hence

$$a = b = 0 \quad \text{and} \quad c = -\frac{3W}{8k}$$

Typical values are $W = 400\,\text{N}$ and $k = 10\,000\,\text{N}\,\text{m}^{-1}$, and hence

$$c = -3 \times 400/8 \times 10\,000 = -0.015\,\text{m}$$

so that the centre of the drum hangs 1.5 cm below the midpoint of the centre of the casing.

It is clear that the model used in this section is an idealized one, but it is helpful in describing how to calculate spring forces in complicated three-dimensional static situations. It also gives an idea of the size of the forces involved and the deflections. The next major step is to put these forces into the equations of motion of the drum; this, however, requires a good knowledge of calculus – and, in particular, of differential equations – so it is not appropriate at this point. You may wish to consider this problem after studying the relevant chapters later in this book. A more advanced model must include the fact that the drum is of finite size.

4.5　Engineering application: cable stayed bridge

One of the standard methods of supporting bridges is with cables. Readers will no doubt be familiar with suspension bridges such as the Golden Gate in the USA, the Humber bridge in the UK and the Tsing Ma bridge in Hong Kong with their spectacular form. Cable stayed bridges are similar in that they have towers and cables that support a roadway but they are not usually on such a grand scale as suspension bridges. They are often used when the foundations can only support a single tower at one end of the roadway. They are commonly seen on bridges over motorways and footbridges over steep narrow valleys.

In any of the situations described it is essential that information is available on the tension in the wire supports and the forces on the towers. The geometry is fully three-dimensional and quite complicated. Vectors provide a logical and efficient way of dealing with the situation.

4.5.1 A simple stayed bridge

There are many configurations that stayed bridges can take; they can have one or more towers and a variety of arrangements of stays. In Figure 4.51 a simple example of a cable stayed footbridge is illustrated. It is constructed with a central vertical pillar with four ties attached by wires to the sides of the pathway.

Figure 4.51
Model of a stayed bridge.

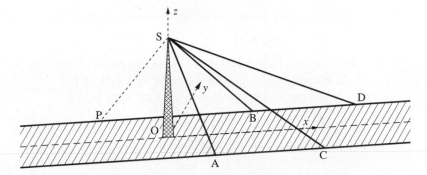

Relative to the axes, with the z axis vertical, the various points are given, in metres, as A(5, −2, 0.5), B(10, 2, 1), C(15, −2, 1.5), D(20, 2, 1) and S(0, 0, 10). Assuming the weight is evenly distributed, there is an equivalent weight of 2 tonnes at each of the four points A, B, C and D. An estimate is required of the tensions in the wires and force at the tie point S.

The vectors along the ties can easily be evaluated

$$\overrightarrow{AS} = (-5, 2, 9.5), \quad \overrightarrow{BS} = (-10, -2, 9)$$
$$\overrightarrow{CS} = (-15, 2, 8.5), \quad \overrightarrow{DS} = (-20, -2, 9)$$

The tension at S in the tie AS can be written $T_A = t_A \overrightarrow{SA}$. Assuming the whole system is in equilibrium, the vertical components at A must be equal

$$T_A \cdot k = 2 \quad \text{and hence} \quad t_A = \frac{2}{9.5}$$

and the four tensions can be computed similarly.

$$T_A = \tfrac{2}{9.5}(5, -2, -9.5) = (1.052, -0.421, -2) \quad \text{and} \quad |T_A| = 2.299 \text{ tonnes}$$
$$T_B = \tfrac{2}{9}(10, 2, -9) = (2.222, 0.444, -2) \quad \text{and} \quad |T_B| = 3.022 \text{ tonnes}$$
$$T_C = \tfrac{2}{8.5}(15, -2, -8.5) = (3.529, -0.471, -2) \quad \text{and} \quad |T_C| = 4.084 \text{ tonnes}$$
$$T_D = \tfrac{2}{9}(20, 2, -9) = (4.444, 0.444, -2) \quad \text{and} \quad |T_D| = 4.894 \text{ tonnes}$$

The total force acting at the tie point S is

$$T = T_A + T_B + T_C + T_D = (11.25, -0.004, -8)$$

Thus with straightforward addition of vectors we have been able to compute the tensions and the total force on the tower.

The question now is how to compensate for the total force on the tower and to try to ensure that it is subject to zero force or a force as small as possible. Suppose that it is decided to have just a single compensating tie wire attached to S and to one side on the pathway at P. It is assumed that on this side of the footbridge the pathway is flat and lies in the x–y plane. Where should we position the attachment of the compensating wire so that it produces zero horizontal force at S?

Let the attachment point P on the side of the footbridge be $(-a, 2, 0)$ so that the tension in the compensating cable is

$$T_P = t_P \overrightarrow{SP} = t_P(-a, 2, -10)$$

We require the y component of $(T + T_P)$ to be zero so that

$$2t_P - 0.004 = 0 \quad \text{and hence} \quad t_P = 0.002$$

which in turn gives for the x component

$$at_P = 11.248 \quad \text{and hence} \quad a = 5624 \text{ metres!}$$

Clearly the answer is ridiculous and either more than one compensating cable must be used or the y component can be neglected completely since the force in this direction is only $4\,\text{kg}$.

As a second attempt we specify the attachment wire at $P(-5, 2, 0)$. Requiring the x component of $T + T_P$ to be zero we see that

$$T + T_P = T + t_P \overrightarrow{SP} = (11.25, -0.004, -8) + t_P(-5, 2, -10)$$

gives $t_P = 2.25$. Hence the total force at S is $(0, 4.5, -30.5)$. Although the force in the x direction has been reduced to zero, an unacceptable side force on the tower in the y direction has been introduced.

In a further effort, we introduce two equal compensating wires connected to the points $P(-5, -2, 0)$ and $P'(-5, 2, 0)$. The total force at S is now

$$T + T_P + T_{P'} = T + t_P \overrightarrow{SP} + t_P \overrightarrow{SP'}$$

$$= (11.25, -0.004, -8) + t_P(-5, 2, -10) + t_P(-5, -2, -10)$$

Now choosing $t_P = 1.125$ gives a total force $(0, -0.004, -30.5)$. We now have a satisfactory resolution of the problem with the only significant force being in the downwards direction.

The different forms of stayed bridge construction will require a similar analysis to obtain an estimate of the forces involved. The example given should be viewed as illustrative.

4.6 Review exercises (1–23)

1 Given that $a = 3i - j - 4k$, $b = -2i + 4j - 3k$ and $c = i + 2j - k$, find

(a) the magnitude of the vector $a + b + c$,

(b) a unit vector parallel to $3a - 2b + 4c$,

(c) the angles between the vectors a and b and between b and c,

(d) the position vector of the centre of mass of particles of masses 1, 2 and 3 placed at points A, B and C with position vectors a, b and c respectively.

2 If the vertices X, Y and Z of a triangle have position vectors

$$x = (2, 2, 6), \quad y = (4, 6, 4) \quad \text{and} \quad z = (4, 1, 7)$$

relative to the origin O, find

(a) the midpoint of the side XY of the triangle,

(b) the area of the triangle,

(c) the volume of the tetrahedron OXYZ.

3 The vertices of a tetrahedron are the points

$$W(2, 1, 3), \quad X(3, 3, 3), \quad Y(4, 2, 4) \quad \text{and} \quad Z(3, 3, 5)$$

Determine

(a) the vectors $\overrightarrow{WX}$ and $\overrightarrow{WY}$,

(b) the area of the face WXZ,

(c) the volume of the tetrahedron WXZY,

(d) the angles between the faces WXY and WYZ.

4 Find the point P on the line L through the points

$$A(5, 1, 7) \quad \text{and} \quad B(6, 0, 8)$$

and the point Q on the line M through the points

$$C(3, 1, 3) \quad \text{and} \quad D(-1, 3, 3)$$

such that the line through P and Q is perpendicular to both lines L and M. Verify that P and Q are at a distance $\sqrt{6}$ apart, and find the point where the line through P and Q intersects the coordinate plane Oxy.

5 Given the vectors $a = (2, 1, 2)$ and $b = (-3, 0, 4)$, evaluate the unit vectors $\hat{a}$ and $\hat{b}$. Use these unit vectors to find a vector that bisects the angle between a and b.

6 According to the inverse square law, the force on a particle of mass m_1 at the point P_1 due to a particle of mass m_2 at the point P_2 is given by

$$\gamma \frac{m_1 m_2}{r^2} \hat{r} \quad \text{where } r = \overrightarrow{P_1 P_2}$$

Particles of mass $3m$, $3m$, m are fixed at the points $A(1, 0, 1)$, $B(0, 1, 2)$ and $C(2, 1, 2)$ respectively. Show that the force on the particle at A due to the presence of B and C is

$$\frac{2\gamma m^2}{\sqrt{3}}(-1, 2, 2)$$

7 Show that the vector a which satisfies the vector equation

$$a \times (i + 2j) = -2i + j + k$$

must take the form $a = (\alpha, 2\alpha - 1, 1)$. If in addition the vector a makes an angle $\cos^{-1}(\frac{1}{3})$ with the vector $(i - j + k)$ show that there are now two such vectors that satisfy both conditions.

8 The electric field at a point having position vector r, due to a charge e at R, is $e(r - R)/|r - R|^3$. Find the electric field E at the point P(2, 1, 1) given that there is a charge e at each of the points $(1, 0, 0)$, $(0, 1, 0)$ and $(0, 0, 1)$.

9 Given that $\overrightarrow{OP} = (3, 1, 2)$ and $\overrightarrow{OQ} = (1, -2, -4)$ are the position vectors of the points P and Q respectively find

(a) the equation of the plane passing through Q and perpendicular to PQ,

(b) the perpendicular distance from the point $(-1, 1, 1)$ to the plane.

10 (a) Determine the equation of the plane that passes through the points $(1, 2, -2)$, $(-1, 1, -9)$ and $(2, -2, -12)$. Find the perpendicular distance from the origin to this plane.

(b) Calculate the area of the triangle whose vertices are at the points $(1, 1, 0)$, $(1, 0, 1)$ and $(0, 1, 1)$.

11 Given $a = (-1, -3, -1)$, $b = (q, 1, 1)$ and $c = (1, 1, q)$ determine the values of q for which

(a) a is perpendicular to b

(b) $a \times (b \times c) = 0$

12 The angular momentum vector H of a particle of mass m is defined by

$$H = r \times (mv)$$

where $v = \omega \times r$.

Using the result

$$a \times (b \times c) = (a \cdot c)b - (a \cdot b)c$$

show that if r is perpendicular to ω then $H = mr^2\omega$.

Given that $m = 100$, $r = 0.1(i + j + k)$ and $\omega = 5i + 5j - 10k$ calculate

(a) $(r \cdot \omega)$ (b) H

13 A particle of mass m, charge e and moving with velocity v in a magnetic field of strength H is known to have acceleration

$$\frac{e}{mc}(v \times H)$$

where c is the speed of light. Show that the component of acceleration parallel to H is zero.

14 A force F is of magnitude 14 N and acts at the point A(3, 2, 4) in the direction of the vector $-2i + 6j + 3k$. Find the moment of the force about the point B(1, 5, −2). Find also the angle between F and $\overrightarrow{AB}$.

15 Points A, B, C have coordinates (1, 2, 1), (−1, 1, 3) and (−2, −2, −2) respectively.
   Calculate the vector product $\overrightarrow{AB} \times \overrightarrow{AC}$, the angle BAC and a unit vector perpendicular to the plane containing A, B and C. Hence obtain

(a) the equation of the plane ABC,

(b) the equation of a second plane, parallel to ABC, and containing the point D(1, 1, 1),

(c) the shortest distance between the point D and the plane containing A, B and C.

16 A plane Π passes through the three non-collinear points A, B and C having position vectors a, b and c respectively. Show that the parametric vector equation of the plane Π is

$$r = a + \lambda(b - a) + \mu(c - a)$$

The plane Π passes through the points (−3, 0, 1), (5, −8, −7) and (2, 1, −2) and the plane Θ passes through the points (3, −1, 1), (1, −2, 1) and (2, −1, 2). Find the parametric vector equation of Π and the normal vector equation of Θ, and hence show that their line of intersection is

$$r = (1, -4, -3) + t(5, 1, -3)$$

where t is a scalar variable.

17 Two skew lines L_1, L_2 have respective equations

$$\frac{x + 3}{4} = \frac{y - 3}{-1} = \frac{z - 2}{1} \quad \text{and}$$

$$\frac{x - 1}{2} = \frac{y - 5}{1} = \frac{z + 3}{2}$$

Obtain the equation of a plane through L_1 parallel to L_2 and show that the shortest distance between the lines is 6.

18 A particle P of constant mass m moves in a viscous medium under the influence of a uniform gravitational force $-mgj$, where g is a scalar constant and j is a constant unit vector. The medium offers a resistance to the motion of the

particle proportional to its momentum (with proportionality constant K). The solution of the equation of motion of P gives the velocity as

$$v = V e^{-Kt} - \frac{g}{K}(1 - e^{-Kt})j$$

where V is the velocity at time $t = 0$, and the position vector as

$$r = A - \frac{V e^{-Kt}}{K} - \frac{gt}{K}j - \frac{g e^{-Kt}}{K^2}j$$

where A is some constant vector.
   At time $t = 0$, a projectile is introduced into the atmosphere at a height H above the ground with a speed U parallel to the ground. Assuming that the atmosphere behaves as a viscous medium as described above, show that at time t the path of the projectile will be inclined at an angle α to the horizontal, where

$$\alpha = \tan^{-1}\left[\frac{g(e^{Kt} - 1)}{KU}\right]$$

Show also that the time of flight T to hitting the ground is a solution of the equation

$$T + \frac{1}{K}(e^{-KT} - 1) = \frac{HK}{g}$$

and find the horizontal distance travelled in terms of T.

19 The three vectors $a = (1, 0, 0)$, $b = (1, 1, 0)$ and $c = (1, 1, 1)$ are given. Evaluate

(a) $a \times b, b \times c, c \times a$

(b) $a \cdot (b \times c)$

For the vector $d = (2, -1, 2)$ calculate

(c) the parameters α, β, γ in the expression

$$d = \alpha a + \beta b + \gamma c$$

(d) the parameters p, q, r in the expression

$$d = p a \times b + q b \times c + r c \times a$$

and show that

$$p = \frac{c \cdot d}{a \cdot (b \times c)}, \quad q = \frac{a \cdot d}{a \cdot (b \times c)} \quad \text{and}$$

$$r = \frac{b \cdot d}{a \cdot (b \times c)}$$

20 Given the line with parametric equation

$$r = a + \lambda d$$

show that the perpendicular distance p from the origin to this line can take either of the forms

(i) $p = \dfrac{|a \times d|}{|d|}$ (ii) $p = \left| a - \dfrac{a \cdot d}{d \cdot d} d \right|$

Find the parametric equation of the straight line through the points

A(1, 0, 2)   and   B(2, 3, 0)

and determine

(a) the length of the perpendicular from the origin to the line,

(b) the point at which the line intersects the y–z plane,

(c) the coordinates of the foot of the perpendicular to the line from the point (1, 1, 1).

21 Given the three non-coplanar vectors a, b, c, and defining $v = a \cdot b \times c$, three further vectors are defined as

$$a' = b \times c/v \quad b' = c \times a/v \quad c' = a \times b/v$$

Show that

$$a = b' \times c'/v' \quad b = c' \times a'/v' \quad c = a' \times b'/v'$$

where

$$v' = a' \cdot b' \times c'$$

Deduce that

$$a \cdot a' = b \cdot b' = c \cdot c' = 1$$

$$a \cdot b' = a \cdot c' = b \cdot a' = b \cdot c' = c \cdot a'$$

$$= c \cdot b' = 0$$

If a vector is written in terms of a, b, c as

$$r = \alpha a + \beta b + \gamma c$$

evaluate α, β, γ in terms of a', b' and c'.

(*Note*: these sets of vectors are called **reciprocal sets** and are widely used in crystallography and materials science.)

22 An unbalanced machine can be approximated by two masses, 2 kg and 1.5 kg, placed at the ends A and B respectively of light rods OA and OB of lengths 0.7 m and 1.1 m. The point O lies on the axis of rotation and OAB forms a plane perpendicular to this axis; OA and OB are at right-angles. The machine rotates about the axis with an angular velocity ω, which gives a centrifugal force $mr\omega^2$ for a mass m and rod length r. Find the unbalanced force at the axis. To balance the machine a mass of 1 kg is placed at the end of a light rod OC so that C is coplanar with OAB. Determine the position of C.

23 In an automated drilling process three holes are drilled simultaneously into the centre of the faces of a block of side 0.2 m, as shown in Figure 4.52. The force exerted by each drill is 25 N, and the couples applied during the drilling are 0.2 N m in the directions indicated. Find the resultant force on the system and the moment of the forces and couples about the corner A. (Take x, y, z axes along the sides of the block, with origin at A.)

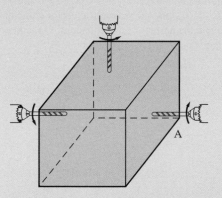

Figure 4.52 Drilling into a block, Question 23.

5 Matrix Algebra

Chapter 5 Contents

5.1 Introduction

The solution of simultaneous equations is part of elementary algebra. Many engineering problems can be formulated in terms of simultaneous equations, but in most practical situations the number of equations is extremely large and traditional methods of solution are not feasible. Even the question of whether solutions exist is not easy to answer. Setting the equations up in matrix form provides a systematic way of answering this question and also suggests practical methods of solution. Over the past 150 years or so, a large number of matrix techniques have been developed, and many have been applied to the solution of engineering and scientific problems. The advent of quantum mechanics and the matrix representation developed by Heisenberg did much to stimulate their popularity, since scientists and engineers were then able to appreciate the convenience and economy of matrix formulations.

In many problems the relationships between vector quantities can be represented by matrices. We saw in Chapter 4 that vectors in three dimensions are represented by three numbers (x_1, x_2, x_3) with respect to some coordinate system. If the coordinate system is changed, the representation of the vector changes to another triple (x_1', x_2', x_3'), related to the original through a matrix. In this three-dimensional case the matrix is a 3×3 array of numbers. Such matrices satisfy various addition and multiplication properties, which we shall develop in this chapter, and indeed it is change of axes that provides the most natural way of introducing the matrix product.

In the previous chapter we noted that forces provide an excellent example of vectors and that they have wide use in engineering. When we are dealing with a continuous medium – for instance when we try to specify the forces in a beam or an aircraft wing or the forces due to the flow of a fluid – we have to extend our ideas and define the **stress** at a point. This can be represented by a 3×3 matrix, and matrix algebra is therefore required for a better understanding of the mathematical manipulations involved.

Perhaps the major impact on engineering applications came with the advent of computers since these are ideally set up to deal with vectors and arrays (matrices), and matrix formulations of problems are therefore already in a form highly suitable for computation. Indeed, all of the widely used aspects of matrices are incorporated into most computer packages, either just for calculation or for the algebraic manipulation of matrices. Packages that are currently popular with students include MATLAB, MAPLE and MATHEMATICA but there are many others available. The use of such packages is strongly encouraged, since it takes the tedium out of the arithmetic, but they should be used to enhance understanding and not to avoid it.

Many physical problems can be modelled using differential equations, and such models form the basis of much modern science and technology. Most of these equations cannot be solved analytically because of their complexity, and it is necessary to revert to numerical solution. This almost always involves convenient vector and matrix formulations. For instance, a popular method of analysing structures is in terms of finite elements. Finite-element packages have been developed over the past 30 years or so to deal with problems having 10^5 or more variables. A major part of such packages involves setting up the data in matrix form and then solving the resulting matrix equations. They are now used to design large buildings, to stress aircraft, to determine the flow through a turbine, to study waveguides and in many other situations of great interest to engineers and scientists.

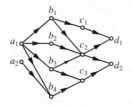

Figure 5.1
Simple electrical
network.

The analysis of circuits can be very complex. In modern VLSI systems there can be many hundreds of connections. Circuit simulation programs use matrix theory, and have proved to be a most important tool for the analysis of the electrical performance of integrated circuits. Such systems have wide use in CAD work. The detailed structure of the matrices gives information about the characteristics of the system and provides an indication of how to tackle the solution of the matrix equations. To give an example of a simple electrical network (or road system for that matter) we consider the connections in the situation illustrated in Figure 5.1. They can be represented by arrays with entries of one if a link exists and zero otherwise, thus:

$$
\begin{array}{c|cccc}
 & b_1 & b_2 & b_3 & b_4 \\
\hline
a_1 & 1 & 1 & 1 & 1 \\
a_2 & 0 & 0 & 0 & 1
\end{array}
\qquad
P = \begin{bmatrix} 1 & 1 & 1 & 1 \\ 0 & 0 & 0 & 1 \end{bmatrix}
$$

$$
\begin{array}{c|ccc}
 & c_1 & c_2 & c_3 \\
\hline
b_1 & 1 & 1 & 0 \\
b_2 & 0 & 1 & 0 \\
b_3 & 0 & 1 & 1 \\
b_4 & 0 & 0 & 1
\end{array}
\qquad
Q = \begin{bmatrix} 1 & 1 & 0 \\ 0 & 1 & 0 \\ 0 & 1 & 1 \\ 0 & 0 & 1 \end{bmatrix}
$$

$$
\begin{array}{c|cc}
 & d_1 & d_2 \\
\hline
c_1 & 1 & 0 \\
c_2 & 1 & 1 \\
c_3 & 0 & 1
\end{array}
\qquad
R = \begin{bmatrix} 1 & 0 \\ 1 & 1 \\ 0 & 1 \end{bmatrix}
$$

We now have a concise numerical way of representing the diagram in Figure 5.1. The arrays can be written in matrix form as P, Q and R, and we can perform algebraic operations on them. The product of matrices will be defined in Section 5.2.4; it is found that the entries in the product PQ give the number of different paths from a_i to c_j, and those in PQR the numbers of different paths from a_i to d_k.

The matrix product is the most interesting property in the theory, since it enables complicated sets of equations to be written in a convenient and compact way. For instance, three ores are known to contain fractions of Pb, Fe, Cu and Mn as indicated in Figure 5.2. If we mix the ores so that there are x_1 kg of ore 1, x_2 kg of ore 2 and x_3 kg of ore 3 then we can compute the amount of each element as

Figure 5.2
Table of fractions in
each kilogram of ore.

	Ore 1	Ore 2	Ore 3
Pb	0.1	0.2	0.3
Fe	0.2	0.3	0.3
Cu	0.6	0.2	0.2
Mn	0.1	0.3	0.2

$$\begin{aligned}
\text{amount of Pb} \ = A_{\text{Pb}} &= 0.1x_1 + 0.2x_2 + 0.3x_3 \\
\text{amount of Fe} \ = A_{\text{Fe}} &= 0.2x_1 + 0.3x_2 + 0.3x_3 \\
\text{amount of Cu} \ = A_{\text{Cu}} &= 0.6x_1 + 0.2x_2 + 0.2x_3 \\
\text{amount of Mn} = A_{\text{Mn}} &= 0.1x_1 + 0.3x_2 + 0.2x_3
\end{aligned}$$

(5.1)

We can rewrite the array in Figure 5.2 as a matrix

$$A = \begin{bmatrix} 0.1 & 0.2 & 0.3 \\ 0.2 & 0.3 & 0.3 \\ 0.6 & 0.2 & 0.2 \\ 0.1 & 0.3 & 0.2 \end{bmatrix}$$

and if we define the vectors

$$M = \begin{bmatrix} A_{\text{Pb}} \\ A_{\text{Fe}} \\ A_{\text{Cu}} \\ A_{\text{Mn}} \end{bmatrix} \quad \text{and} \quad X = \begin{bmatrix} x_1 \\ x_2 \\ x_3 \end{bmatrix}$$

then the equations can be written in matrix form

$$M = AX$$

with the product interpreted as in (5.1). The matrix A has 4 rows and 3 columns, so it is called a 4×3 (read '4 by 3') matrix. M and X are called column vectors; they are 4×1 and 3×1 matrices respectively.

5.2 Definitions and properties

We can look at the 'ore' problem in a different context. In the last chapter, in Section 4.3.2, we saw that the equation of a plane can be written in the form

$$\alpha x + \beta y + \gamma z = p$$

where α, β, γ and p are constants. The four planes

$$\begin{aligned}
4x + 2y + z &= 7 \\
2x + y - z &= 5 \\
x + 2y + 2z &= 3 \\
3x - 2y - z &= 0
\end{aligned}$$

(5.2)

meet in a single point. What are the coordinates of that point? Obviously they are those values of x, y and z that satisfy all four of (5.2) simultaneously.

Equations (5.2) provide an example of a mathematical problem that arises in a wide range of engineering problems: the simultaneous solution of a set of linear equations, as mentioned in the introduction. The general form of a linear equation is the sum of a set of variables, each multiplied only by a numerical factor, set equal to a constant. No variable is raised to any power or multiplied by any other variable. In this case we have four linear equations in three variables x, y and z. We shall see that there is a large body of mathematical theory concerning the solution of such equations.

As is common in mathematics, one of the first stages in solving the problem is to introduce a better notation to represent the problem. In this case we introduce the idea of an array of numbers called a **matrix**. We write

$$A = \begin{bmatrix} 4 & 2 & 1 \\ 2 & 1 & -1 \\ 1 & 2 & 2 \\ 3 & -2 & -1 \end{bmatrix}$$

and call A a 4×3 matrix; that is, a matrix with four rows and three columns. We also introduce an alternative notation for a vector, writing

$$X = \begin{bmatrix} x \\ y \\ z \end{bmatrix} \quad \text{and} \quad b = \begin{bmatrix} 7 \\ 5 \\ 3 \\ 0 \end{bmatrix}$$

We call these column vectors. Equations (5.2) can then be expressed in the form

$$AX = b$$

where the product of the matrix A and the vector X is understood to produce the left-hand sides of (5.2).

As another example let us seek the relationship between the coordinates of P in the Oxy system and the O$x'y'$ system in Figure 5.3. Trigonometry gives

$$x' = OL + LM + MN$$

$$= x \cos \theta + QM \sin \theta + MP \sin \theta$$

$$= x \cos \theta + (QM + MP) \sin \theta$$

and hence

$$x' = x \cos \theta + y \sin \theta \tag{5.3}$$

Similarly,

$$y' = -x \sin \theta + y \cos \theta \tag{5.4}$$

If we take

$$B = \begin{bmatrix} \cos\theta & \sin\theta \\ -\sin\theta & \cos\theta \end{bmatrix}, \quad X = \begin{bmatrix} x \\ y \end{bmatrix}, \quad X' = \begin{bmatrix} x' \\ y' \end{bmatrix}$$

then (5.3) and (5.4) can be written as

$$X' = BX$$

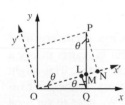

Figure 5.3
Change of axes
from Oxy to O$x'y'$.

We see that a change of axes can be written in a natural manner in matrix form, with $\boldsymbol{B}$ containing all the information about the transformation.

In each of these examples arrays $\boldsymbol{A}$ and $\boldsymbol{B}$ and vectors $\boldsymbol{X}$ and $\boldsymbol{X}'$ appear in a natural way, and the method of multiplication of the arrays and vectors is consistent. We build on this idea to define matrices generally.

5.2.1 Definitions

An array of real numbers

$$\boldsymbol{A} = \begin{bmatrix} a_{11} & a_{12} & a_{13} & \dots & a_{1n} \\ a_{21} & a_{22} & a_{23} & \dots & a_{2n} \\ \vdots & \vdots & \vdots & & \vdots \\ a_{m1} & a_{m2} & a_{m3} & \dots & a_{mn} \end{bmatrix} \tag{5.5}$$

is called an $m \times n$ **matrix**, with m rows and n columns. The entry a_{ij} denotes the **element** in the ith row and jth column. The element can be real or complex (but in this chapter we deal only with real matrices). If $m = n$ then the array is square, and $\boldsymbol{A}$ is then called a **square matrix** of order n. If the matrix has one column or one row

$$\boldsymbol{b} = \begin{bmatrix} b_1 \\ b_2 \\ \vdots \\ b_m \end{bmatrix} \quad \text{or} \quad \boldsymbol{c} = [c_1 \quad c_2 \quad \dots \quad c_n] \tag{5.6}$$

then it is called a **column vector** or a **row vector** respectively. The row vector was used in Chapter 4, Section 4.2.2, as the basic definition of a vector, but in matrix theory a vector is normally taken to be a column vector unless otherwise stated. This slight inconsistency in the notation between vector theory and matrix theory can be inconvenient, but it is so standard in the literature that we must accept it. We have to get used to vectors appearing in several different notations. It is also a common convention to use upper-case letters to represent matrices and lower-case ones for vectors. We shall adopt this convention in this chapter with one exception: the vectors

$$\begin{bmatrix} x \\ y \\ z \end{bmatrix} \quad \text{and} \quad \begin{bmatrix} x \\ y \end{bmatrix}$$

will be denoted by $\boldsymbol{X}$. (Vectors and matrices are further distinguished here by the use of a 'serif' bold face for the former (e.g. $\boldsymbol{b}$) and a 'sans serif' bold face for the latter (e.g. $\boldsymbol{A}$).) As an example of the notation used consider the matrix $\boldsymbol{A}$ and vector $\boldsymbol{b}$

$$\boldsymbol{A} = \begin{bmatrix} 0 & -1 & 2 \\ 3 & 0 & 1 \end{bmatrix} \quad \text{and} \quad \boldsymbol{b} = \begin{bmatrix} 0.15 \\ 1.11 \\ -3.01 \end{bmatrix}$$

The matrix $\boldsymbol{A}$ is a 2×3 matrix with elements $a_{11} = 0$, $a_{12} = -1$, $a_{13} = 2$, $a_{21} = 3$ and so on. The vector $\boldsymbol{b}$ is a column vector with elements $b_1 = 0.15$, $b_2 = 1.11$ and $b_3 = -3.01$.

In a square matrix of order n the diagonal containing the elements $a_{11}, a_{22}, \ldots, a_{nn}$ is called the **principal**, **main** or **leading** diagonal. The sum of the elements of the leading diagonal is called the **trace** of the square matrix $\boldsymbol{A}$, that is

$$\text{trace } \boldsymbol{A} = a_{11} + a_{22} + \ldots + a_{nn} = \sum_{i=1}^{n} a_{ii}$$

A **diagonal matrix** is a square matrix that has its only non-zero elements along the leading diagonal. (It may have zeros on the leading diagonal also.)

$$\begin{bmatrix} a_{11} & 0 & 0 & \ldots & 0 \\ 0 & a_{22} & 0 & \ldots & 0 \\ 0 & 0 & a_{33} & \ldots & 0 \\ \vdots & \vdots & \vdots & & \vdots \\ 0 & 0 & 0 & \ldots & a_{nn} \end{bmatrix}$$

An important special case of a diagonal matrix is the **unit matrix** or **identity matrix** $\boldsymbol{I}$, for which $a_{11} = a_{22} = \ldots = a_{nn} = 1$.

$$\boldsymbol{I} = \begin{bmatrix} 1 & 0 & 0 & \ldots & 0 \\ 0 & 1 & 0 & \ldots & 0 \\ 0 & 0 & 1 & \ldots & 0 \\ \vdots & \vdots & \vdots & & \vdots \\ 0 & 0 & 0 & \ldots & 1 \end{bmatrix}$$

The unit matrix can be written conveniently in terms of the Kronecker delta. This is defined as

$$\delta_{ij} = \begin{cases} 1 & \text{if} \quad i = j \\ 0 & \text{if} \quad i \neq j \end{cases}$$

The unit matrix thus has elements δ_{ij}. The notation $\boldsymbol{I}_n$ is sometimes used to denote the $n \times n$ unit matrix where its size is important.

The **zero** or **null matrix** is the matrix with every element zero, and is written as either 0 or $\boldsymbol{0}$. Sometimes a zero matrix of order $m \times n$ is written $\boldsymbol{O}_{m \times n}$.

The **transposed matrix** $\boldsymbol{A}^\mathrm{T}$ of (5.5) is the matrix with elements $b_{ij} = a_{ji}$ and is written in full as the $n \times m$ matrix

$$\boldsymbol{A}^\mathrm{T} = \begin{bmatrix} a_{11} & a_{21} & a_{31} & \ldots & a_{m1} \\ a_{12} & a_{22} & a_{32} & \ldots & a_{m2} \\ \vdots & \vdots & \vdots & & \vdots \\ a_{1n} & a_{2n} & a_{3n} & \ldots & a_{mn} \end{bmatrix}$$

This is just the matrix in (5.5) with rows and columns interchanged. We may note from (5.6) that

$$\boldsymbol{b}^T = [b_1 \quad b_2 \quad \dots \quad b_m] \quad \text{and} \quad \boldsymbol{c}^T = \begin{bmatrix} c_1 \\ c_2 \\ \vdots \\ c_n \end{bmatrix}$$

so that a column vector is transposed to a row vector and vice versa.

If a square matrix is such that $\boldsymbol{A}^T = \boldsymbol{A}$ then $a_{ij} = a_{ji}$, and the elements are therefore symmetric about the diagonal. Such a matrix is called a **symmetric matrix**; symmetric matrices play important roles in many computations. If $\boldsymbol{A}^T = -\boldsymbol{A}$, so that $a_{ij} = -a_{ji}$, the matrix is called **skew-symmetric** or **antisymmetric**. Obviously the diagonal elements of a skew-symmetric matrix satisfy $a_{ii} = -a_{ii}$ and so must all be zero.

A few examples will illustrate these definitions:

$$\boldsymbol{A} = \begin{bmatrix} 2 & 3 \\ 1 & 2 \\ 4 & 5 \end{bmatrix} \text{ is a } 3 \times 2 \text{ matrix}$$

$$\boldsymbol{A}^T = \begin{bmatrix} 2 & 1 & 4 \\ 3 & 2 & 5 \end{bmatrix} \text{ is a } 2 \times 3 \text{ matrix}$$

$$\boldsymbol{B} = \begin{bmatrix} 1 & 2 & 3 \\ 2 & 3 & 4 \\ 3 & 4 & 5 \end{bmatrix} \text{ is a symmetric } 3 \times 3 \text{ matrix}$$

trace $\boldsymbol{B} = 1 + 3 + 5 = 9$

$$\boldsymbol{C} = \begin{bmatrix} 0 & 7 & -1 \\ -7 & 0 & 4 \\ 1 & -4 & 0 \end{bmatrix} \quad \text{and} \quad \boldsymbol{C}^T = \begin{bmatrix} 0 & -7 & 1 \\ 7 & 0 & -4 \\ -1 & 4 & 0 \end{bmatrix} \text{ are skew-symmetric } 3 \times 3 \text{ matrices}$$

$$\boldsymbol{D} = \begin{bmatrix} 2 & 0 & 0 \\ 0 & 3 & 0 \\ 0 & 0 & 4 \end{bmatrix} \text{ is a } 3 \times 3 \text{ diagonal matrix}$$

trace $\boldsymbol{D} = 2 + 3 + 4 = 9$

$$\boldsymbol{I} = \begin{bmatrix} 1 & 0 & 0 & 0 \\ 0 & 1 & 0 & 0 \\ 0 & 0 & 1 & 0 \\ 0 & 0 & 0 & 1 \end{bmatrix} \text{ is the } 4 \times 4 \text{ unit matrix (sometimes written } \boldsymbol{I}_4 \text{)}$$

5.2.2 Basic operations of matrices

(a) Equality

Two matrices A and B are said to be **equal** if and only if all their elements are the same, $a_{ij} = b_{ij}$ for $1 \leqslant i \leqslant m$, $1 \leqslant j \leqslant n$, and this equality is written as

$$A = B$$

Note that this requires the two matrices to be of the same order $m \times n$.

(b) Addition

Addition of matrices is straightforward; we can only add an $m \times n$ matrix to another $m \times n$ matrix, and an element of the sum is the sum of the corresponding elements. If A has elements a_{ij} and B has elements b_{ij} then $A + B$ has elements $a_{ij} + b_{ij}$.

$$\begin{bmatrix} a_{11} & a_{12} & a_{13} & \cdots \\ a_{21} & a_{22} & a_{23} & \cdots \\ \vdots & \vdots & \vdots & \end{bmatrix} + \begin{bmatrix} b_{11} & b_{12} & b_{13} & \cdots \\ b_{21} & b_{22} & b_{23} & \cdots \\ \vdots & \vdots & \vdots & \end{bmatrix}$$

$$= \begin{bmatrix} a_{11} + b_{11} & a_{12} + b_{12} & a_{13} + b_{13} & \cdots \\ a_{21} + b_{21} & a_{22} + b_{22} & a_{23} + b_{23} & \cdots \\ \vdots & \vdots & \vdots & \end{bmatrix}$$

(c) Multiplication by a scalar

The matrix λA has elements λa_{ij}; that is, we just multiply each element by the scalar λ

$$\lambda \begin{bmatrix} a_{11} & a_{12} & a_{13} & \cdots \\ a_{21} & a_{22} & a_{23} & \cdots \\ \vdots & \vdots & \vdots & \end{bmatrix} = \begin{bmatrix} \lambda a_{11} & \lambda a_{12} & \lambda a_{13} & \cdots \\ \lambda a_{21} & \lambda a_{22} & \lambda a_{23} & \cdots \\ \vdots & \vdots & \vdots & \end{bmatrix}$$

(d) Properties of the transpose

From the definition, the transpose of a matrix is such that

$$(A + B)^{\mathrm{T}} = A^{\mathrm{T}} + B^{\mathrm{T}}$$

Similarly, we observe that

$$(A^{\mathrm{T}})^{\mathrm{T}} = A$$

so that transposing twice gives back the original matrix.

We may note as a special case of this result that for a square matrix A

$$(A^{\mathrm{T}} + A)^{\mathrm{T}} = (A^{\mathrm{T}})^{\mathrm{T}} + A^{\mathrm{T}} = A + A^{\mathrm{T}}$$

and hence $A^T + A$ must be a symmetric matrix. This proves to be a very useful result, which we shall see used in several places. Similarly, $A - A^T$ is a skew-symmetric matrix, so that any square matrix A may be expressed as the sum of a symmetric and a skew-symmetric matrix:

$$A = \tfrac{1}{2}(A + A^T) + \tfrac{1}{2}(A - A^T)$$

(e) Basic rules of addition

Because the usual rules of arithmetic are followed in the definitions of the sum of matrices and of multiplication by scalars, the

> **commutative law** $A + B = B + A$
>
> **associative law** $(A + B) + C = A + (B + C)$

and

> **distributive law** $\lambda(A + B) = \lambda A + \lambda B$

all hold for matrices.

Example 5.1   Let

$$A = \begin{bmatrix} 1 & 2 & 1 \\ 1 & 1 & 2 \\ 1 & 1 & 1 \end{bmatrix}, \quad B = \begin{bmatrix} 2 & 1 \\ 1 & 0 \\ 1 & 1 \end{bmatrix}, \quad C = \begin{bmatrix} 0 & 1 & 1 \\ 0 & 0 & 1 \\ 1 & 0 & 0 \end{bmatrix}$$

Find, where possible, (a) $A + B$, (b) $A + C$, (c) $C - A$, (d) $3A$, (e) $4B$, (f) $C + B$, (g) $3A + 2C$, (h) $A^T + A$ and (i) $A + C^T + B^T$.

Solution   (a) $A + B$ is not possible.

(b) $A + C = \begin{bmatrix} 1+0 & 2+1 & 1+1 \\ 1+0 & 1+0 & 2+1 \\ 1+1 & 1+0 & 1+0 \end{bmatrix} = \begin{bmatrix} 1 & 3 & 2 \\ 1 & 1 & 3 \\ 2 & 1 & 1 \end{bmatrix}$

(c) $C - A = \begin{bmatrix} 0-1 & 1-2 & 1-1 \\ 0-1 & 0-1 & 1-2 \\ 1-1 & 0-1 & 0-1 \end{bmatrix} = \begin{bmatrix} -1 & -1 & 0 \\ -1 & -1 & -1 \\ 0 & -1 & -1 \end{bmatrix}$

(d) $3A = \begin{bmatrix} 3 & 6 & 3 \\ 3 & 3 & 6 \\ 3 & 3 & 3 \end{bmatrix}$

(e) $4\boldsymbol{B} = \begin{bmatrix} 8 & 4 \\ 4 & 0 \\ 4 & 4 \end{bmatrix}$

(f) $\boldsymbol{C} + \boldsymbol{B}$ is not possible.

(g) $3\boldsymbol{A} + 2\boldsymbol{C} = \begin{bmatrix} 3 & 6 & 3 \\ 3 & 3 & 6 \\ 3 & 3 & 3 \end{bmatrix} + \begin{bmatrix} 0 & 2 & 2 \\ 0 & 0 & 2 \\ 2 & 0 & 0 \end{bmatrix} = \begin{bmatrix} 3 & 8 & 5 \\ 3 & 3 & 8 \\ 5 & 3 & 3 \end{bmatrix}$

(h) $\boldsymbol{A}^{\mathrm{T}} + \boldsymbol{A} = \begin{bmatrix} 1 & 1 & 1 \\ 2 & 1 & 1 \\ 1 & 2 & 1 \end{bmatrix} + \begin{bmatrix} 1 & 2 & 1 \\ 1 & 1 & 2 \\ 1 & 1 & 1 \end{bmatrix} = \begin{bmatrix} 2 & 3 & 2 \\ 3 & 2 & 3 \\ 2 & 3 & 2 \end{bmatrix}$

(Note that this matrix is symmetric.)

(i) $\boldsymbol{A} + \boldsymbol{C}^{\mathrm{T}} + \boldsymbol{B}^{\mathrm{T}}$ is not possible.

Example 5.2

A local roadside cafe serves beefburgers, eggs, chips and beans in four combination meals:

Slimmers	–	150 g chips	100 g beans	1 burger
Normal	1 egg	250 g chips	150 g beans	1 burger
Jumbo	2 eggs	350 g chips	200 g beans	2 burgers
Veggie	1 egg	200 g chips	150 g beans	–

A party orders 1 slimmer, 4 normal, 2 jumbo and 2 veggie meals. What is the total amount of materials that the kitchen staff need to cook? One of the customers sees the size of a jumbo meal and changes his order to a normal meal. How much less material will the kitchen staff need?

Solution

The meals written in matrix form are

$$
\boldsymbol{s} = \begin{bmatrix} 0 \\ 150 \\ 100 \\ 1 \end{bmatrix}, \quad
\boldsymbol{n} = \begin{bmatrix} 1 \\ 250 \\ 150 \\ 1 \end{bmatrix}, \quad
\boldsymbol{j} = \begin{bmatrix} 2 \\ 350 \\ 200 \\ 2 \end{bmatrix}, \quad
\boldsymbol{v} = \begin{bmatrix} 1 \\ 200 \\ 150 \\ 0 \end{bmatrix}
$$

and hence the kitchen requirements are

$$
\boldsymbol{s} + 4\boldsymbol{n} + 2\boldsymbol{j} + 2\boldsymbol{v} = \begin{bmatrix} 10 \\ 2250 \\ 1400 \\ 9 \end{bmatrix}
$$

The change in requirements is

$$j - n = \begin{bmatrix} 2 \\ 350 \\ 200 \\ 2 \end{bmatrix} - \begin{bmatrix} 1 \\ 250 \\ 150 \\ 1 \end{bmatrix} = \begin{bmatrix} 1 \\ 100 \\ 50 \\ 1 \end{bmatrix}$$

less materials needed.

Although this may appear to be a rather trivial example, the basic problem is identical to any production process that requires a supply of parts.

5.2.3 Exercises

1 Given the matrices

$$a = \begin{bmatrix} 1 \\ 2 \\ 0 \end{bmatrix}, \quad b = [0 \quad 1 \quad 1],$$

$$C = \begin{bmatrix} 3 & 2 & 1 \\ 1 & 2 & -1 \end{bmatrix}, \quad D = \begin{bmatrix} 5 & 6 \\ 7 & 8 \\ 9 & 10 \end{bmatrix}$$

evaluate, where possible, (a) $a + b$, (b) $b^T + a$, (c) $b + C^T$, (d) $C + D$, (e) $D^T + C$.

2 Find the values of α, β, γ that satisfy

$$\alpha \begin{bmatrix} 1 \\ 0 \\ 0 \end{bmatrix} + \beta \begin{bmatrix} 1 \\ -1 \\ 0 \end{bmatrix} + \gamma \begin{bmatrix} 0 \\ 1 \\ -1 \end{bmatrix} = \begin{bmatrix} 0 \\ 3 \\ -2 \end{bmatrix}$$

3 Given the matrix

$$A = \lambda \begin{bmatrix} 1 & 0 \\ 0 & 1 \end{bmatrix} + \mu \begin{bmatrix} 1 & 1 \\ 0 & 1 \end{bmatrix} + v \begin{bmatrix} 0 & 0 \\ 0 & 1 \end{bmatrix}$$

(a) find the value of λ, μ, v so that $A = \begin{bmatrix} 0 & -1 \\ 0 & 3 \end{bmatrix}$;

(b) show that no solution is possible if

$$A = \begin{bmatrix} 1 & -1 \\ 1 & 0 \end{bmatrix}.$$

4 Market researchers are testing customers' preferences for five products. There are four researchers who are allocated to different groups: researcher R_1 deals with men under 40, R_2 deals with men over 40, R_3 deals with women under 40 and R_4 deals with women over 40. They return their findings as a vector giving the number of customers with first preference for a particular product.

Product	R_1	R_2	R_3	R_4
a	23	32	28	39
b	34	22	33	21
c	18	21	22	17
d	9	15	10	12
e	16	10	7	11

Find the average over the whole sample. The company decides that their main target is older women so they weight the returns in the ratio 1 : 1 : 2 : 3; find the weighted average.

5.2.4 Matrix multiplication

The most important property of matrices as far as their practical applications are concerned is the multiplication of one matrix by another. We saw informally in Section 5.2 how multiplication arose and how to define multiplication of a matrix and a vector. The idea can be extended further by looking again at change of axes. Take

$$z_1 = a_{11}y_1 + a_{12}y_2, \quad y_1 = b_{11}x_1 + b_{12}x_2$$

$$z_2 = a_{21}y_1 + a_{22}y_2, \quad y_2 = b_{21}x_1 + b_{22}x_2$$

Then we can ask for the transformation from the zs to the xs. This we can do by straight substitution:

$$z_1 = (a_{11}b_{11} + a_{12}b_{21})x_1 + (a_{11}b_{12} + a_{12}b_{22})x_2$$

$$z_2 = (a_{21}b_{11} + a_{22}b_{21})x_1 + (a_{21}b_{12} + a_{22}b_{22})x_2$$

If we write the first two transformations as

$$A = \begin{bmatrix} a_{11} & a_{12} \\ a_{21} & a_{22} \end{bmatrix} \quad \text{and} \quad B = \begin{bmatrix} b_{11} & b_{12} \\ b_{21} & b_{22} \end{bmatrix}$$

then the composite transformation is written

$$AB = \begin{bmatrix} a_{11}b_{11} + a_{12}b_{21} & a_{11}b_{12} + a_{12}b_{22} \\ a_{21}b_{11} + a_{22}b_{21} & a_{21}b_{12} + a_{22}b_{22} \end{bmatrix}$$

and this is precisely how we define the matrix product.

Illustration

$z_1 = y_1 + 3y_2$
$z_2 = 2y_1 - y_2$

$y_1 = -x_1 + 2x_2$
$y_2 = 2x_1 - x_2$

Substitute to get

$z_1 = 5x_1 - x_2$
$z_2 = -4x_1 + 5x_2$

In matrix form

$$A = \begin{bmatrix} 1 & 3 \\ 2 & -1 \end{bmatrix}$$

$$B = \begin{bmatrix} -1 & 2 \\ 2 & -1 \end{bmatrix}$$

$$AB = \begin{bmatrix} 5 & -1 \\ -4 & 5 \end{bmatrix}$$

Definition

If A is an $m \times p$ matrix with elements a_{ij} and B a $p \times n$ matrix with elements b_{ij} then we define the **product** $C = AB$ as the $m \times n$ matrix with components

$$c_{ij} = \sum_{k=1}^{p} a_{ik} b_{kj} \quad \text{for } i - 1, \dots, m \quad \text{and} \quad j - 1, \dots, n$$

In pictorial form, the ith row of A is multiplied term by term with the jth column of B and the products are added to form the ijth component of C. This is commonly referred to as the 'row-by-column' method of multiplication. Clearly, in order for multiplication to be possible, A must have p columns and B must have p rows otherwise the product AB is not defined.

$$i \to \begin{bmatrix} & \vdots & \\ \cdots & c_{ij} & \cdots \\ & \vdots & \end{bmatrix} = i \to \begin{bmatrix} a_{i1} & a_{i2} & \cdots & a_{ip} \end{bmatrix} \begin{bmatrix} b_{1j} \\ b_{2j} \\ \vdots \\ b_{pj} \end{bmatrix}$$

Example 5.3   Given

$$A = \begin{bmatrix} 1 & 1 & 0 \\ 2 & 0 & 1 \end{bmatrix}, \quad B = \begin{bmatrix} 2 & 0 \\ 0 & 1 \\ 1 & 3 \end{bmatrix},$$

$$b = \begin{bmatrix} -1 \\ 2 \end{bmatrix}, \quad c = \begin{bmatrix} 1 \\ 1 \\ -1 \end{bmatrix} \quad \text{and} \quad C = \begin{bmatrix} 1 & -2 \\ -1 & 2 \\ -2 & 4 \end{bmatrix}.$$

find (a) AB, (b) BA, (c) Bb, (d) $A^T b$, (e) $c^T(A^T b)$ and (f) AC.

Solution

(a) $AB = \begin{bmatrix} 1 & 1 & 0 \\ 2 & 0 & 1 \end{bmatrix} \begin{bmatrix} 2 & 0 \\ 0 & 1 \\ 1 & 3 \end{bmatrix} = \begin{bmatrix} \text{row}\,1 \times \text{col}\,1 & \text{row}\,1 \times \text{col}\,2 \\ \text{row}\,2 \times \text{col}\,1 & \text{row}\,2 \times \text{col}\,2 \end{bmatrix}$

$= \begin{bmatrix} (1)(2) + (1)(0) + (0)(1) & (1)(0) + (1)(1) + (0)(3) \\ (2)(2) + (0)(0) + (1)(1) & (2)(0) + (0)(1) + (1)(3) \end{bmatrix} = \begin{bmatrix} 2 & 1 \\ 5 & 3 \end{bmatrix}$

(b) $BA = \begin{bmatrix} 2 & 0 \\ 0 & 1 \\ 1 & 3 \end{bmatrix} \begin{bmatrix} 1 & 1 & 0 \\ 2 & 0 & 1 \end{bmatrix} = \begin{bmatrix} 2+0 & 2+0 & 0+0 \\ 0+2 & 0+0 & 0+1 \\ 1+6 & 1+0 & 0+3 \end{bmatrix} = \begin{bmatrix} 2 & 2 & 0 \\ 2 & 0 & 1 \\ 7 & 1 & 3 \end{bmatrix}$

(Note that BA is not equal to AB.)

(c) $Bb = \begin{bmatrix} 2 & 0 \\ 0 & 1 \\ 1 & 3 \end{bmatrix} \begin{bmatrix} -1 \\ 2 \end{bmatrix} = \begin{bmatrix} -2 \\ 2 \\ 5 \end{bmatrix}$

(d) $A^T b = \begin{bmatrix} 1 & 2 \\ 1 & 0 \\ 0 & 1 \end{bmatrix} \begin{bmatrix} -1 \\ 2 \end{bmatrix} = \begin{bmatrix} 3 \\ -1 \\ 2 \end{bmatrix}$

(e) $c^T A^T b = \begin{bmatrix} 1 & 1 & -1 \end{bmatrix} \begin{bmatrix} 3 \\ -1 \\ 2 \end{bmatrix} = [0] = 0$

(Note that this matrix is the zero 1×1 matrix, which can just be written 0.)

(f) $AC = \begin{bmatrix} 1 & 1 & 0 \\ 2 & 0 & 1 \end{bmatrix} \begin{bmatrix} 1 & -2 \\ -1 & 2 \\ -2 & 4 \end{bmatrix} = \begin{bmatrix} 0 & 0 \\ 0 & 0 \end{bmatrix} = 0$

(Note that the product AC is zero even though neither A nor C is zero.)

Example 5.4   If

$$A = \begin{bmatrix} 1 & 2 & 0 \\ 1 & 1 & 0 \\ 2 & 1 & 1 \end{bmatrix} \quad \text{and} \quad X = \begin{bmatrix} x \\ y \\ z \end{bmatrix}$$

evaluate (a) $X^\mathrm{T}X$, (b) AX, (c) $X^\mathrm{T}(AX)$ and (d) $\frac{1}{2}X^\mathrm{T}[(A^\mathrm{T} + A)X]$.

Solution

(a) $X^\mathrm{T}X = \begin{bmatrix} x & y & z \end{bmatrix} \begin{bmatrix} x \\ y \\ z \end{bmatrix} = x^2 + y^2 + z^2$

(b) $AX = \begin{bmatrix} 1 & 2 & 0 \\ 1 & 1 & 0 \\ 2 & 1 & 1 \end{bmatrix} \begin{bmatrix} x \\ y \\ z \end{bmatrix} = \begin{bmatrix} x + 2y \\ x + y \\ 2x + y + z \end{bmatrix}$

(c) $X^\mathrm{T}AX = \begin{bmatrix} x & y & z \end{bmatrix} \begin{bmatrix} x + 2y \\ x + y \\ 2x + y + z \end{bmatrix} = (x^2 + 2xy) + (yx + y^2) + (2xz + yz + z^2)$

$\qquad\qquad = x^2 + y^2 + z^2 + 3xy + 2xz + yz$

(d) $\frac{1}{2}(A^\mathrm{T} + A) = \begin{bmatrix} 1 & \frac{3}{2} & 1 \\ \frac{3}{2} & 1 & \frac{1}{2} \\ 1 & \frac{1}{2} & 1 \end{bmatrix}$

and

$$\frac{1}{2}(A^\mathrm{T} + A)X = \begin{bmatrix} 1 & \frac{3}{2} & 1 \\ \frac{3}{2} & 1 & \frac{1}{2} \\ 1 & \frac{1}{2} & 1 \end{bmatrix} \begin{bmatrix} x \\ y \\ z \end{bmatrix} = \begin{bmatrix} x + \frac{3}{2}y + z \\ \frac{3}{2}x + y + \frac{1}{2}z \\ x + \frac{1}{2}y + z \end{bmatrix}$$

Therefore

$$\frac{1}{2}X^\mathrm{T}(A^\mathrm{T} + A)X = \begin{bmatrix} x & y & z \end{bmatrix} \begin{bmatrix} x + \frac{3}{2}y + z \\ \frac{3}{2}x + y + \frac{1}{2}z \\ x + \frac{1}{2}y + z \end{bmatrix}$$

$$= x^2 + y^2 + z^2 + 3xy + 2xz + yz$$

(Note that this is the same as the result of part (c).)

There are several points to note from the preceding examples. One-by-one matrices are just numbers, so the square brackets become redundant and are usually omitted. The expression X^TX just gives the square of the length of the vector X in the usual sense, namely $X^TX = x^2 + y^2 + z^2$. Similarly,

$$X^TX' = [x \quad y \quad z] \begin{bmatrix} x' \\ y' \\ z' \end{bmatrix} = xx' + yy' + zz'$$

which is the usual **scalar** or **inner product**, here written in matrix form. The expression AX gives a column vector with linear expressions as its elements. Using Example 5.4(b), we can rewrite the linear equations

$$x + 2y \quad = 3$$
$$x + y \quad = 4$$
$$2x + y + z = 5$$

as

$$\begin{bmatrix} 1 & 2 & 0 \\ 1 & 1 & 0 \\ 2 & 1 & 1 \end{bmatrix} \begin{bmatrix} x \\ y \\ z \end{bmatrix} = \begin{bmatrix} 3 \\ 4 \\ 5 \end{bmatrix}$$

which may be written in the standard matrix form for linear equations as

$$AX = b$$

It is also important to realize that if $AB = 0$ it does not follow that either A or B is zero. In Example 5.3(f) we saw that the product $AC = 0$, but neither A nor C is the zero matrix.

Properties of matrix multiplication

We now consider the basic properties of matrix multiplication. These may be proved using the definition of matrix multiplication and this is left as an exercise for the reader.

(a) Commutative law

Matrices *do not commute in general*, although they may do in special cases. In Example 5.3 we saw that $AB \neq BA$, and a further example illustrates the same result:

$$A = \begin{bmatrix} 1 & 0 \\ 0 & 0 \end{bmatrix}, \quad B = \begin{bmatrix} 1 & 2 \\ 1 & 0 \end{bmatrix}$$

$$AB = \begin{bmatrix} 1 & 2 \\ 0 & 0 \end{bmatrix}, \quad BA = \begin{bmatrix} 1 & 0 \\ 1 & 0 \end{bmatrix}$$

so again $AB \neq BA$. The products do not necessarily have the same size, as shown in Example 5.3(a and b), where AB is a 2×2 matrix while BA is 3×3. In fact, even if AB exists, it does not follow that BA does. Take, for example, the matrices

$a = [1 \quad 1 \quad 1]$ and $B = \begin{bmatrix} 1 & 2 \\ 2 & 1 \\ 1 & 1 \end{bmatrix}$. The product $aB = [4 \quad 4]$ is well defined but Ba cannot be computed since a 3×2 matrix cannot be multiplied on the right by a 1×3 matrix. Thus order matters, and we need to distinguish between AB and BA. To do this, we talk of **pre-multiplication** of B by A to form AB, and **post-multiplication** of B by A to form BA.

(b) Associative law
It can be shown that

$$A(BC) = (AB)C$$

where A is $m \times p$, B is $p \times q$ and C is $q \times n$.

Matrix multiplication is associative and we can therefore omit the brackets.

(c) Distributive law over multiplication by a scalar

$$(mA)B = A(mB) = mAB \text{ holds}$$

(d) Distributive law over addition

$$(A + B)C = AC + BC \quad \text{and} \quad A(B + C) = AB + AC$$

so we can multiply out brackets in the usual way, but making sure that the order of the products is maintained.

(e) Multiplication by unit matrices

If A is an $m \times n$ matrix and if I_m and I_n are the unit matrices of orders m and n then

$$I_m A = A I_n = A$$

Thus pre- or post-multiplication by the appropriate unit matrix leaves A unchanged.

(f) Transpose of a product

$$(AB)^{\mathrm{T}} = B^{\mathrm{T}} A^{\mathrm{T}}$$

where A is an $m \times p$ matrix and B a $p \times n$ matrix. The proof is straightforward but requires careful treatment of summation signs. Thus the transpose of the product of matrices is the product of the transposed matrices in the reverse order.

Example 5.5   Given the matrices

$$A = \begin{bmatrix} 1 & 2 & 2 \\ 0 & 1 & 1 \\ 1 & 0 & 1 \end{bmatrix}, \quad B = \begin{bmatrix} 1 & -2 & 0 \\ 1 & -1 & -1 \\ -1 & 2 & 1 \end{bmatrix}, \quad X = \begin{bmatrix} x \\ y \\ z \end{bmatrix} \quad \text{and} \quad c = \begin{bmatrix} 1 \\ 0 \\ 1 \end{bmatrix}$$

(a) find (i) AB, (ii) $(AB)^T$ and (iii) $B^T A^T$.

(b) Pre-multiply each side of the equation $BX = c$ by A.

Solution   (a) It would be a useful exercise to check these products on one of the standard soft-ware packages.

(i) $AB = \begin{bmatrix} 1 & 2 & 2 \\ 0 & 1 & 1 \\ 1 & 0 & 1 \end{bmatrix} \begin{bmatrix} 1 & -2 & 0 \\ 1 & -1 & -1 \\ -1 & 2 & 1 \end{bmatrix} = \begin{bmatrix} 1 & 0 & 0 \\ 0 & 1 & 0 \\ 0 & 0 & 1 \end{bmatrix} = I$

The MATLAB instructions

> A = [1 2 2; 0 1 1; 1 0 1]; B = [1 −2 0; 1 −1 −1; −1 2 1]; A*B

produce the correct unit matrix.

(ii) $(AB)^T = \begin{bmatrix} 1 & 0 & 0 \\ 0 & 1 & 0 \\ 0 & 0 & 1 \end{bmatrix} = I$

(iii) $B^T A^T = \begin{bmatrix} 1 & 1 & -1 \\ -2 & -1 & 2 \\ 0 & -1 & 1 \end{bmatrix} \begin{bmatrix} 1 & 0 & 1 \\ 2 & 1 & 0 \\ 2 & 1 & 1 \end{bmatrix} = \begin{bmatrix} 1 & 0 & 0 \\ 0 & 1 & 0 \\ 0 & 0 & 1 \end{bmatrix} = I$

(b) The equation $BX = c$ can be rewritten as

$$\begin{bmatrix} 1 & -2 & 0 \\ 1 & -1 & -1 \\ -1 & 2 & 1 \end{bmatrix} \begin{bmatrix} x \\ y \\ z \end{bmatrix} = \begin{bmatrix} 1 \\ 0 \\ 1 \end{bmatrix} \quad \text{or} \quad \begin{array}{r} x - 2y = 1 \\ x - y - z = 0 \\ -x + 2y + z = 1 \end{array}$$

If we now pre-multiply the equation by A we obtain

$$ABX = Ac$$

and since $AB = I$, we obtain

$$IX = \begin{bmatrix} 1 & 2 & 2 \\ 0 & 1 & 1 \\ 1 & 0 & 1 \end{bmatrix} \begin{bmatrix} 1 \\ 0 \\ 1 \end{bmatrix} = \begin{bmatrix} 3 \\ 1 \\ 2 \end{bmatrix} \quad \text{or} \quad \begin{bmatrix} x \\ y \\ z \end{bmatrix} = \begin{bmatrix} 3 \\ 1 \\ 2 \end{bmatrix}$$

and we see that we have a solution to our set of linear equations. Set up B and c in MATLAB and try $B \backslash c$.

Example 5.6   Given the three matrices

$$A = \begin{bmatrix} 1 & -1 & 1 \\ -2 & 0 & 3 \\ 0 & 1 & -2 \end{bmatrix}, \quad B = \begin{bmatrix} 0 & 0 & 1 \\ 0 & 2 & 3 \\ 1 & 2 & 3 \end{bmatrix} \quad \text{and} \quad C = \begin{bmatrix} 2 & 3 \\ -1 & 2 \\ 3 & 1 \end{bmatrix}$$

verify the associative law and the distributive law over addition.

Solution

Now $BC = \begin{bmatrix} 0 & 0 & 1 \\ 0 & 2 & 3 \\ 1 & 2 & 3 \end{bmatrix}\begin{bmatrix} 2 & 3 \\ -1 & 2 \\ -3 & 1 \end{bmatrix} = \begin{bmatrix} -3 & 1 \\ -11 & 7 \\ -9 & 10 \end{bmatrix}$ and

$$A(BC) = \begin{bmatrix} 1 & -1 & 1 \\ -2 & 0 & 3 \\ 0 & 1 & -2 \end{bmatrix}\begin{bmatrix} -3 & 1 \\ -11 & 7 \\ -9 & 10 \end{bmatrix} = \begin{bmatrix} -1 & 4 \\ -21 & 28 \\ 7 & -13 \end{bmatrix}$$

Likewise $AB = \begin{bmatrix} 1 & -1 & 1 \\ -2 & 0 & 3 \\ 0 & 1 & -2 \end{bmatrix}\begin{bmatrix} 0 & 0 & 1 \\ 0 & 2 & 3 \\ 1 & 2 & 3 \end{bmatrix} = \begin{bmatrix} 1 & 0 & 1 \\ 3 & 6 & 7 \\ -2 & -2 & -3 \end{bmatrix}$ and

$$(AB)C = \begin{bmatrix} 1 & 0 & 1 \\ 3 & 6 & 7 \\ -2 & -2 & -3 \end{bmatrix}\begin{bmatrix} 2 & 3 \\ -1 & 2 \\ -3 & 1 \end{bmatrix} = \begin{bmatrix} -1 & 4 \\ -21 & 28 \\ 7 & -13 \end{bmatrix}$$

Thus the associative law is satisfied for these three matrices. For the distributive law we need to evaluate

$$(A + B)C = \begin{bmatrix} 1 & -1 & 2 \\ -2 & 2 & 6 \\ 1 & 3 & 1 \end{bmatrix}\begin{bmatrix} 2 & 3 \\ -1 & 2 \\ -3 & 1 \end{bmatrix} = \begin{bmatrix} -3 & 3 \\ -24 & 4 \\ -4 & 10 \end{bmatrix}$$

and

$$AC + BC = \begin{bmatrix} 1 & -1 & 1 \\ -2 & 0 & 3 \\ 0 & 1 & -2 \end{bmatrix}\begin{bmatrix} 2 & 3 \\ -1 & 2 \\ -3 & 1 \end{bmatrix} + \begin{bmatrix} 0 & 0 & 1 \\ 0 & 2 & 3 \\ 1 & 2 & 3 \end{bmatrix}\begin{bmatrix} 2 & 3 \\ -1 & 2 \\ -3 & 1 \end{bmatrix}$$

$$= \begin{bmatrix} 0 & 2 \\ -13 & -3 \\ 5 & 0 \end{bmatrix} + \begin{bmatrix} -3 & 1 \\ -11 & 7 \\ -9 & 10 \end{bmatrix} = \begin{bmatrix} -3 & 3 \\ -24 & 4 \\ -4 & 10 \end{bmatrix}$$

The two matrices are equal so the distributive law is verified for the three given matrices.

Example 5.7

Show that the transformation

$$\begin{bmatrix} x' \\ y' \end{bmatrix} = \begin{bmatrix} \cos\theta & \sin\theta \\ -\sin\theta & \cos\theta \end{bmatrix} \begin{bmatrix} x \\ y \end{bmatrix}$$

with $\theta = 60°$, maps the square with corners $\begin{bmatrix} 1 \\ 1 \end{bmatrix}, \begin{bmatrix} 1 \\ 2 \end{bmatrix}, \begin{bmatrix} 2 \\ 2 \end{bmatrix}$ and $\begin{bmatrix} 2 \\ 1 \end{bmatrix}$ on to a square.

Solution

Substituting the given vectors in turn for $\begin{bmatrix} x \\ y \end{bmatrix}$ into the equation

$$\begin{bmatrix} x' \\ y' \end{bmatrix} = \begin{bmatrix} 0.5 & 0.8660 \\ -0.8660 & 0.5 \end{bmatrix} \begin{bmatrix} x \\ y \end{bmatrix}$$

we find the following vectors for $\begin{bmatrix} x' \\ y' \end{bmatrix}$

$$\begin{bmatrix} 1.366 \\ -0.366 \end{bmatrix}, \begin{bmatrix} 2.232 \\ 0.134 \end{bmatrix}, \begin{bmatrix} 2.732 \\ -0.732 \end{bmatrix} \text{ and } \begin{bmatrix} 1.866 \\ -1.232 \end{bmatrix}$$

Plotting these points on the plane as in Figure 5.4, we see that the square has been rotated through an angle of 60° about the origin. It is left as an exercise for the reader to verify the result.

This type of analysis forms the basis of manipulation of diagrams on a computer screen, and is used in many CAD/CAM situations.

Figure 5.4
Transformation of a square in Example 5.7.

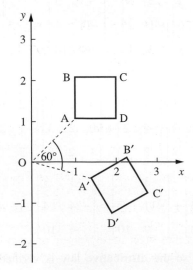

Example 5.8   A rectangular site is to be levelled, and the amount of earth that needs to be removed must be determined. A survey of the site at a regular mesh of points 10 m apart is made. The heights in metres above the level required are given in the following table.

0	0.31	0.40	0.45	0.51	0.60
0.12	0.33	0.51	0.58	0.66	0.75
0.19	0.38	0.60	0.69	0.78	0.86
0.25	0.46	0.68	0.77	0.89	0.97

It is known that the approximate volume of a cell of side x and with corner heights of a, b, c and d is

$$V = \tfrac{1}{4}x^2(a + b + c + d)$$

Write the total approximate volume in matrix form and hence estimate the volume to be removed.

Solution   Note that for the first row of cells the volume is

$$25(\quad 0 \quad +0.31 \quad +0.31 + 0.40 \quad +0.40 + 0.45 \quad +0.45 + 0.51 \quad +0.51 + 0.60$$
$$+0.12 + 0.33 \quad +0.33 + 0.51 \quad +0.51 + 0.58 \quad +0.58 + 0.66 \quad +0.66 + 0.75)$$

$$= 25[0 + 2(0.31 + 0.40 + 0.45 + 0.51) + 0.60]$$
$$+ 25[0.12 + 2(0.33 + 0.51 + 0.58 + 0.66) + 0.75]$$

The second and third rows of cells are dealt with in a similar manner, so that, when we compute the total volume, we need to multiply the corner values by 1, the other side values by 2 and the centre values by 4. In matrix form this multiplication can be performed as

$$
[1 \quad 2 \quad 2 \quad 1]
\begin{bmatrix}
0 & 0.31 & 0.40 & 0.45 & 0.51 & 0.60 \\
0.12 & 0.33 & 0.51 & 0.58 & 0.66 & 0.75 \\
0.19 & 0.38 & 0.60 & 0.69 & 0.78 & 0.86 \\
0.25 & 0.46 & 0.68 & 0.77 & 0.86 & 0.97
\end{bmatrix}
\begin{bmatrix}
1 \\ 2 \\ 2 \\ 2 \\ 2 \\ 1
\end{bmatrix}
$$

This can be checked by multiplying the matrices out. The checking can be done on one of the symbolic manipulation packages, such as MAPLE or MATHEMATICA, by putting in general symbols for the matrix and verifying that, after the matrix multiplications, the elements are multiplied by the stated factors. Performing the calculation and multiplying by the 25 gives the total volume as 816.5 m³.

A similar analysis can be applied to other situations – all that is needed is measured heights and a matrix multiplication routine on a computer to deal with the large amount of data that would be required. For other mesh shapes, or even irregular meshes, the method is similar, but the multiplying vectors will need careful calculation.

Example 5.9

A contractor makes two products P_1 and P_2. The four components required to make the products are subcontracted out and each of the components is made up from three ingredients A, B and C as follows:

Component	Units of A	Units of B	Units of C	Make-up cost and profit for subcontractor
1 requires	5	4	3	10
2 requires	2	1	1	7
3 requires	0	1	3	5
4 requires	3	4	1	2

The cost per unit of the ingredients A, B and C are a, b and c respectively. The contractor makes the product P_1 with 2 of component 1, 3 of component 2 and 4 of component 4, the make-up cost is 15; product P_2 requires 1 of component 1, 1 of component 2, 1 of component 3 and 2 of component 4, the make-up cost is 12. Find the cost to the contractor for P_1 and P_2. What is the change in costs if a increases to $(a + 1)$? It is found that the 5 units of A required for component 1 can be reduced to 4. What is the effect on the costs?

Solution

The information presented can be written naturally in matrix form. Let C_1, C_2, C_3 and C_4 be the cost the subcontractor charges the contractor for the four components, then the cost C_1 is computed as $C_1 = 5a + 4b + 3c + 10$. This expression is the first row of the matrix equation

$$\begin{bmatrix} C_1 \\ C_2 \\ C_3 \\ C_4 \end{bmatrix} = \begin{bmatrix} 5 & 4 & 3 \\ 2 & 1 & 1 \\ 0 & 1 & 3 \\ 3 & 4 & 1 \end{bmatrix} \begin{bmatrix} a \\ b \\ c \end{bmatrix} + \begin{bmatrix} 10 \\ 7 \\ 5 \\ 2 \end{bmatrix}$$

and the other three costs follow in a similar manner. Now let p_1, p_2 be the costs of producing the final products. The costs are constructed in exactly the same way as

$$\begin{bmatrix} p_1 \\ p_2 \end{bmatrix} = \begin{bmatrix} 2 & 3 & 0 & 4 \\ 1 & 1 & 1 & 2 \end{bmatrix} \begin{bmatrix} C_1 \\ C_2 \\ C_3 \\ C_4 \end{bmatrix} + \begin{bmatrix} 15 \\ 12 \end{bmatrix}$$

Substituting gives

$$\begin{bmatrix} p_1 \\ p_2 \end{bmatrix} = \begin{bmatrix} 2 & 3 & 0 & 4 \\ 1 & 1 & 1 & 2 \end{bmatrix} \begin{bmatrix} 5 & 4 & 3 \\ 2 & 1 & 1 \\ 0 & 1 & 3 \\ 3 & 4 & 1 \end{bmatrix} \begin{bmatrix} a \\ b \\ c \end{bmatrix} + \begin{bmatrix} 2 & 3 & 0 & 4 \\ 1 & 1 & 1 & 2 \end{bmatrix} \begin{bmatrix} 10 \\ 7 \\ 5 \\ 2 \end{bmatrix} + \begin{bmatrix} 15 \\ 12 \end{bmatrix}$$

or

$$\begin{bmatrix} p_1 \\ p_2 \end{bmatrix} = \begin{bmatrix} 28 & 27 & 13 \\ 13 & 14 & 9 \end{bmatrix} \begin{bmatrix} a \\ b \\ c \end{bmatrix} + \begin{bmatrix} 64 \\ 38 \end{bmatrix}$$

Thus a simple matrix formulation gives a convenient way of coding the data. If a is increased to $(a + 1)$ then multiplying out shows that p_1 increases by 28 and p_2 by 13. If the 5 in the first matrix is reduced to 4 then the costs will be

$$\begin{bmatrix} p_1 \\ p_2 \end{bmatrix} = \begin{bmatrix} 26 & 27 & 13 \\ 12 & 14 & 9 \end{bmatrix} \begin{bmatrix} a \\ b \\ c \end{bmatrix} + \begin{bmatrix} 64 \\ 38 \end{bmatrix}$$

so p_1 is reduced by $2a$ and p_2 by a.

A similar approach can be used in more complicated, realistic situations. Storing and processing the information is convenient, particularly in conjunction with a computer package or spreadsheet.

Example 5.10

The tape of a tape recorder passes the reading head at a constant speed v. On the feed reel there is a length of tape L left and it has radius R. If the rev counter is set to zero and the thickness of the tape is h find how the radius and the length of the reel vary with the number of revolutions n.

Solution

Let t_n, R_n and L_n denote the time, radius and length after n revolutions. It is given that $t_0 = 0$, $R_0 = R$ and $L_0 = L$. At stage n after one further revolution a thickness h is peeled off the radius, the length is reduced by $2\pi R_n$ and the time has advanced by $2\pi R_n/v$. A table summarizes these remarks.

Revs (n)	Time (t_n)	Radius (R_n)	Length remaining (L_n)
0	$t_0 = 0$	$R_0 = R$	$L_0 = L$
1	$t_1 = 2\pi R_0/v + t_0$	$R_1 = R_0 - h$	$L_1 = L_0 - 2\pi R_0$
2	$t_2 = 2\pi R_1/v + t_1$	$R_2 = R_1 - h$	$L_2 = L_1 - 2\pi R_1$
. . .	. . .	. . .	. . .
. . .	. . .	. . .	. . .
$n + 1$	$t_{n+1} = 2\pi R_n/v + t_n$	$R_{n+1} = R_n - h$	$L_{n+1} = L_n - 2\pi R_n$

This data can be written conveniently in matrix notation as

$$\begin{bmatrix} t_{n+1} \\ R_{n+1} \\ L_{n+1} \end{bmatrix} = \begin{bmatrix} 1 & 2\pi/v & 0 \\ 0 & 1 & 0 \\ 0 & -2\pi & 1 \end{bmatrix} \begin{bmatrix} t_n \\ R_n \\ L_n \end{bmatrix} - \begin{bmatrix} 0 \\ h \\ 0 \end{bmatrix}$$

or identifying the matrices in an obvious manner as

$$X_{n+1} = AX_n - \Delta$$

Successive substitution gives

$$X_{n+1} = A(AX_{n-1} - \Delta) - \Delta = A^2 X_{n-1} - (A + I)\Delta$$

$$= A^2(AX_{n-2} - \Delta) - (A + I)\Delta = A^3 X_{n-2} - (A^2 + A + I)\Delta$$

$$\cdots$$

$$= A^{n+1} X_0 - (A^n + A^{n-1} + \ldots + A + I)\Delta \tag{5.7}$$

The required values can now be determined by evaluating the right-hand side of this equation. Repeated products of A gives

$$A^n = \begin{bmatrix} 1 & 2\pi n/v & 0 \\ 0 & 1 & 0 \\ 0 & -2\pi n & 1 \end{bmatrix}$$

so the sum

$$A^n + A^{n-1} + \ldots + A + I = \begin{bmatrix} n+1 & 2\pi S/v & 0 \\ 0 & n+1 & 0 \\ 0 & -2\pi S & n+1 \end{bmatrix}$$

can be computed, where

$$S = 1 + 2 + 3 + \ldots + n = \tfrac{1}{2}n(n+1)$$

Writing out equation (5.7) in full gives

$$\begin{bmatrix} t_{n+1} \\ R_{n+1} \\ L_{n+1} \end{bmatrix} = \begin{bmatrix} 1 & 2\pi(n+1)/v & 0 \\ 0 & 1 & 0 \\ 0 & -2\pi(n+1) & 1 \end{bmatrix}\begin{bmatrix} t_0 \\ R_0 \\ L_0 \end{bmatrix} - \begin{bmatrix} n+1 & \pi n(n+1)/v & 0 \\ 0 & n+1 & 0 \\ 0 & -\pi n(n+1) & n+1 \end{bmatrix}\begin{bmatrix} 0 \\ h \\ 0 \end{bmatrix}$$

Thus

$$R_{n+1} = R - (n+1)h$$

$$t_{n+1} = \frac{2\pi}{v}(n+1)(R - \tfrac{1}{2}nh)$$

$$L_{n+1} = L - 2\pi(n+1)(R - \tfrac{1}{2}nh)$$

which gives the required result.

Note that the length depends on time in a linear fashion but that the relation between the radius and the time is quadratic

$$t_{n+1} = \frac{\pi}{vh}(R - R_{n+1})(R + R_{n+1} + h)$$

Example 5.11   Find the values of x that make the matrix $\mathbf{Z}^5$ a diagonal matrix, where

$$\mathbf{Z} = \begin{bmatrix} x & 0 & 0 \\ 0 & x & 1 \\ 0 & -1 & 0 \end{bmatrix}$$

Solution   Although this problem can be done by hand it is tedious and a MAPLE solution is given.

with (linalg):
Z := array ([[x, 0, 0], [0, x, 1], [0, −1, 0]]):
simplify (multiply (Z, Z, Z, Z, Z));

$$\begin{bmatrix} x^5 & 0 & 0 \\ 0 & 3x - 4x^3 + x^5 & 1 - 3x^2 + x^4 \\ 0 & 1 - 3x^2 + x^4 & 2x - x^3 \end{bmatrix}$$

evalf (solve ({% [2, 3] = 0}, {x}));
{x = 1.618}, {x = −0.618}, {x = 0.618}, {x = −1.618}

5.2.5 Exercises

5 Given the matrices

$$\mathbf{A} = \begin{bmatrix} 1 & 1 & 1 \\ 1 & 1 & 1 \end{bmatrix}, \quad \mathbf{B} = \begin{bmatrix} 1 & 1 & 0 \\ 1 & 1 & 0 \\ 0 & 0 & -1 \end{bmatrix}$$

and

$$\mathbf{C} = \begin{bmatrix} 0 & 2 \\ 1 & 1 \\ -1 & -1 \end{bmatrix}$$

evaluate $\mathbf{AB}$, $\mathbf{AC}$, $\mathbf{BC}$, $\mathbf{CA}$ and $\mathbf{BA}^{\mathrm{T}}$. Which if any of these are diagonal, unit or symmetric?

6 The matrices

$$\mathbf{A} = \begin{bmatrix} 1 & 2 & 1 \\ 3 & 0 & 2 \end{bmatrix}, \quad \mathbf{B} = \begin{bmatrix} 4 & 1 & 3 \\ 0 & 2 & 1 \end{bmatrix}$$

and

$$\mathbf{C} = \begin{bmatrix} 1 & 2 \\ 3 & 1 \\ 2 & 3 \end{bmatrix}$$

are given.

(a) Which of the following make sense: $\mathbf{AB}$, $\mathbf{AC}$, $\mathbf{BC}$, $\mathbf{AB}^{\mathrm{T}}$, $\mathbf{AC}^{\mathrm{T}}$ and $\mathbf{BC}^{\mathrm{T}}$?

(b) Evaluate those products that do exist.

(c) Evaluate $(\mathbf{A}^{\mathrm{T}}\mathbf{B})\mathbf{C}$ and $\mathbf{A}^{\mathrm{T}}(\mathbf{BC})$ and show that they are equal.

7 Given

$$\mathbf{A} = \begin{bmatrix} 1 & 2 & 3 \\ 3 & 4 & 5 \\ 5 & 6 & 7 \end{bmatrix}, \quad \mathbf{X} = \begin{bmatrix} x \\ y \\ z \end{bmatrix} \quad \text{and} \quad \mathbf{b} = \begin{bmatrix} 2 \\ 3 \\ 4 \end{bmatrix}$$

evaluate $X^{\mathrm{T}}X$ and $X^{\mathrm{T}}\mathbf{A}X$ and write out the equations given by $\mathbf{A}X = \mathbf{b}$.

8 A matrix with m rows and n columns is said to be of type $m \times n$. Give simple examples of matrices $\mathbf{A}$ and $\mathbf{B}$ to illustrate the following situations:

(a) $\mathbf{AB}$ is defined but $\mathbf{BA}$ is not,

(b) $\mathbf{AB}$ and $\mathbf{BA}$ are both defined but have different type,

(c) $\mathbf{AB}$ and $\mathbf{BA}$ are both defined and have the same type but are unequal.

9 Given

$$A = \begin{bmatrix} 1 & 3 & 2 \\ 2 & -1 & 0 \\ 1 & 4 & 1 \end{bmatrix}$$

determine a symmetric matrix C and a skew-symmetric matrix D such that

$$A = C + D$$

10 Given the matrices

$$a = [3 \quad 2 \quad -1], \quad b = \begin{bmatrix} 11 \\ 0 \\ 2 \end{bmatrix} \quad \text{and}$$

$$C = \begin{bmatrix} 4 & 1 & 1 \\ -1 & 7 & -3 \\ -1 & 3 & 5 \end{bmatrix}$$

determine the elements of G where

$$(ab)I + C^2 = C^T + G$$

and I is the unit matrix.

11 A firm allocates staff into four categories: welders, fitters, designers and administrators. It is estimated that for their three main products the time spent, in hours, on each item is given in the following matrix.

	Boiler	Water tank	Holding frame
Welder	2	0.75	1.25
Fitter	1.4	0.5	1.75
Designer	0.3	0.1	0.1
Admin	0.1	0.25	0.3

The wages, pension contributions and overheads, in £ per hour, are known to be

	Welder	Fitter	Designer	Administrator
Wages	12	8	20	10
Pension	1	0.5	2	1
O/heads	0	0	1	3

Write the problem in matrix form and use matrix products to find the total cost of producing 10 boilers, 25 water tanks and 35 frames.

12 Given

$$A = \begin{bmatrix} 1 & 1 & 1 \\ 2 & 1 & 2 \\ -2 & 1 & -1 \end{bmatrix}$$

evaluate A^2 and A^3. Verify that

$$A^3 - A^2 - 3A + I = 0$$

13 Given

$$A = \begin{bmatrix} 5 & -2 & 0 \\ -2 & 6 & 2 \\ 0 & 2 & 7 \end{bmatrix} \quad \text{and} \quad X = \begin{bmatrix} x_1 \\ x_2 \\ x_3 \end{bmatrix}$$

show that

$$X^T A X = 27 \tag{5.8}$$

implies that

$$5x_1^2 + 6x_2^2 + 7x_3^2 - 4x_1x_2 + 4x_2x_3 = 27$$

Under the transformation

$$X = BY$$

show that (5.8) becomes

$$Y^T (B^T A B) Y = 27$$

If

$$B = \begin{bmatrix} 2 & 2 & -1 \\ 2 & -1 & 2 \\ -1 & 2 & 2 \end{bmatrix} \quad \text{and} \quad Y = \begin{bmatrix} y_1 \\ y_2 \\ y_3 \end{bmatrix}$$

evaluate $B^T A B$, and hence show that

$$y_1^2 + 2y_2^2 + 3y_3^2 = 1$$

14 Figure 5.5 shows a Wheatstone bridge circuit. Kirchhoff's law states that the total current entering a junction is equal to the total current leaving it. Ohm's law in a circuit states that the imposed voltage in the circuit is the sum of current times resistance in the sections of the circuit. Write down the equations for the system and put them into the matrix form

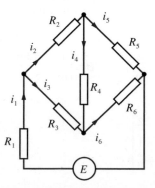

Figure 5.5 Wheatstone bridge circuit

$$A\begin{bmatrix} i_1 \\ i_2 \\ \vdots \\ i_6 \end{bmatrix} = b$$

15 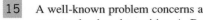 A well-known problem concerns a mythical country that has three cities, A, B and C, with a total population of 2400. At the end of each year it is decreed that all people must move to another city, half to one and half to the other. If a, b and c are the populations in the cities A, B and C respectively, show that in the next year the populations are given by

$$\begin{bmatrix} a' \\ b' \\ c' \end{bmatrix} = \begin{bmatrix} 0 & \frac{1}{2} & \frac{1}{2} \\ \frac{1}{2} & 0 & \frac{1}{2} \\ \frac{1}{2} & \frac{1}{2} & 0 \end{bmatrix} \begin{bmatrix} a \\ b \\ c \end{bmatrix}$$

Supposing that the three cities have initial populations of 600, 800 and 1000, what are the populations after 10 years and after a very long time (a package such as MATLAB is ideal for the calculations)? (Note that this example is a version of a **Markov chain** problem. Markov chains have applications in many areas of science and engineering.)

16 Find values of h, k, l and m so that $A \neq 0$, $B \neq 0$, $A^2 = A$, $B^2 = B$ and $AB = 0$, where

$$A = h\begin{bmatrix} 1 & 1 & 1 \\ 1 & 1 & 1 \\ 1 & 1 & 1 \end{bmatrix} \quad \text{and} \quad B = \begin{bmatrix} k & -l & -l \\ -l & m & m \\ -l & m & m \end{bmatrix}$$

17 The Königsberg bridge problem concerns trying to follow a path across all the bridges to the islands in a river, as shown in Figure 5.6, without going over any bridge twice. Defining three matrices in an obvious way,

$$\text{the adjacency matrix } A = \begin{bmatrix} 0 & 2 & 0 & 1 \\ 2 & 0 & 2 & 1 \\ 0 & 2 & 0 & 1 \\ 1 & 1 & 1 & 0 \end{bmatrix}$$

$$\text{the degree matrix} \quad D = \begin{bmatrix} 3 & 0 & 0 & 0 \\ 0 & 5 & 0 & 0 \\ 0 & 0 & 3 & 0 \\ 0 & 0 & 0 & 3 \end{bmatrix}$$

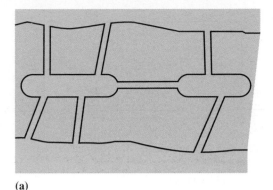

(a)

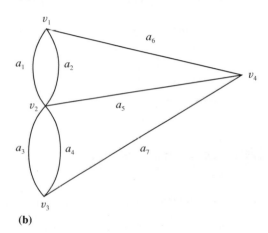

(b)

Figure 5.6 The Königsberg bridge problem.

and the vertex-arc matrix

$$B - \begin{bmatrix} 1 & 1 & 0 & 0 & 0 & 1 & 0 \\ 1 & 1 & 1 & 1 & 1 & 0 & 0 \\ 0 & 0 & 1 & 1 & 0 & 0 & 1 \\ 0 & 0 & 0 & 0 & 1 & 1 & 1 \end{bmatrix}$$

part of the solution involves showing that

$$BB^\mathrm{T} = A + D$$

Verify this result.

18 A computer screen has dimensions 20 cm × 30 cm. Axes are set up at the centre of the screen as illustrated in Figure 5.7. A box containing an arrow has dimensions 2 cm × 2 cm and is situated with its centre at the point $(-16, 10)$. It is first to be rotated through 45° in an anticlockwise direction. Find this transformation in the form

$$\begin{bmatrix} x' + 16 \\ y' - 10 \end{bmatrix} = A \begin{bmatrix} x + 16 \\ y - 10 \end{bmatrix}$$

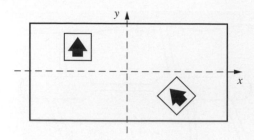

Figure 5.7 Manipulation of a computer screen in Question 18.

The rotated box is now moved to a new position with its centre at $(16, -10)$. Find the overall transformation in the form

$$\begin{bmatrix} x'' \\ y'' \end{bmatrix} = \begin{bmatrix} a \\ b \end{bmatrix} + B \begin{bmatrix} x \\ y \end{bmatrix}$$

 19 Given the matrix

$$A = \begin{bmatrix} 0 & 1 & 0 & 0 & 0 & 0 & 0 & 0 \\ 0 & 0 & 0 & 1 & 0 & 0 & 0 & 0 \\ 0 & 0 & 1 & 0 & 0 & 0 & 0 & 0 \\ 1 & 0 & 0 & 0 & 0 & 0 & 0 & 0 \\ 0 & 0 & 0 & 0 & 0 & 0 & 0 & 1 \\ 0 & 0 & 0 & 0 & 1 & 0 & 0 & 0 \\ 0 & 0 & 0 & 0 & 0 & 0 & 1 & 0 \\ 0 & 0 & 0 & 0 & 0 & 1 & 0 & 0 \end{bmatrix}$$

it is known that $A^n = I$, the unit matrix, for some n; find this value.

5.3 Determinants

The idea of a determinant is closely related to that of a square matrix and is crucial to the solution of linear equations. We shall deal here mainly with 2×2 and 3×3 determinants.

Given the square matrices

$$A = \begin{bmatrix} a_{11} & a_{12} \\ a_{21} & a_{22} \end{bmatrix} \quad \text{and} \quad B = \begin{bmatrix} a_{11} & a_{12} & a_{13} \\ a_{21} & a_{22} & a_{23} \\ a_{31} & a_{32} & a_{33} \end{bmatrix}$$

the **determinant** of A, denoted by det A or $|A|$, is given by

$$|A| = a_{11}a_{22} - a_{12}a_{21} \tag{5.9}$$

For the 3×3 matrix B

$$|B| = a_{11} \begin{vmatrix} a_{22} & a_{23} \\ a_{32} & a_{33} \end{vmatrix} - a_{12} \begin{vmatrix} a_{21} & a_{23} \\ a_{31} & a_{33} \end{vmatrix} + a_{13} \begin{vmatrix} a_{21} & a_{22} \\ a_{31} & a_{32} \end{vmatrix} \tag{5.10}$$

This is known as the expansion of the determinant along the first row.

The determinant of a 1×1 matrix, $A = [a]$, having a single entry a is simply its entry. Thus

$$|A| = a$$

It is important that this be distinguished from mod a which is also written as $|a|$.

Example 5.12   Evaluate the third-order determinant

$$\begin{vmatrix} 1 & 2 & 4 \\ -1 & 0 & 3 \\ 3 & 1 & -2 \end{vmatrix}$$

Solution   Expanding along the first row as in (5.10), we have

$$\begin{vmatrix} 1 & 2 & 4 \\ -1 & 0 & 3 \\ 3 & 1 & -2 \end{vmatrix} = 1 \begin{vmatrix} 0 & 3 \\ 1 & -2 \end{vmatrix} - 2 \begin{vmatrix} -1 & 3 \\ 3 & -2 \end{vmatrix} + 4 \begin{vmatrix} -1 & 0 \\ 3 & 1 \end{vmatrix}$$

$$= 1[(0)(-2) - (1)(3)] - 2[(-1)(-2) - (3)(3)]$$

$$+ 4[(-1)(1) - (3)(0)] \quad \text{(using (5.9))}$$

$$= 1(-3) - 2(-7) + 4(-1)$$

$$= 7$$

If we take a determinant and delete row i and column j then the determinant remaining is called the **minor** M_{ij}. In general we can take *any* row (or column) and evaluate an $n \times n$ determinant $|\boldsymbol{A}|$ as

$$|\boldsymbol{A}| = \sum_{j=1}^{n} (-1)^{i+j} a_{ij} M_{ij} \qquad \qquad (5.11)$$

The fact that the determinant is the same for *any* i requires detailed proof.
    The sign associated with a minor is given in the array

$$\begin{vmatrix} & + & - & + & \cdots \\ - & + & - & + & - & \cdots \\ + & - & + & - & + & \cdots \\ \vdots & \vdots & \vdots & \vdots & \vdots & \end{vmatrix}$$

A minor multiplied by the appropriate sign is called the **cofactor** A_{ij} of the element, so

$$A_{ij} = (-1)^{i+j} M_{ij}$$

and thus

$$|\boldsymbol{A}| = \sum_{j} a_{ij} A_{ij}$$

Example 5.13

Evaluate the minors and cofactors of the determinant

$$|A| = \begin{vmatrix} 3 & 4 & 5 \\ 6 & -4 & 2 \\ 2 & -1 & 1 \end{vmatrix}$$

associated with the first row, and hence evaluate the determinant.

Solution

$$\begin{vmatrix} 3 & 4 & 5 \\ 6 & -4 & 2 \\ 2 & -1 & 1 \end{vmatrix} \rightarrow \begin{vmatrix} -4 & 2 \\ -1 & 1 \end{vmatrix} = -4 - (-2) = -2$$

Element a_{11} has minor $M_{11} = -2$ and cofactor $A_{11} = -2$.

$$\begin{vmatrix} 3 & 4 & 5 \\ 6 & -4 & 2 \\ 2 & -1 & 1 \end{vmatrix} \rightarrow \begin{vmatrix} 6 & 2 \\ 2 & 1 \end{vmatrix} = 2$$

Element a_{12} has minor $M_{12} = 2$ and cofactor $A_{12} = -2$.

$$\begin{vmatrix} 3 & 4 & 5 \\ 6 & -4 & 2 \\ 2 & -1 & 1 \end{vmatrix} \rightarrow \begin{vmatrix} 6 & -4 \\ 2 & -1 \end{vmatrix} = 2$$

Element a_{13} has minor $M_{13} = 2$ and cofactor $A_{13} = 2$. Thus the determinant is

$$|A| = 3 \times (-2) + 4 \times (-2) + 5 \times 2 = -4$$

It may be checked that the same result is obtained by expanding along any row (or column), care being taken to incorporate the correct signs.

The properties of determinants are not always obvious, and are often quite difficult to prove in full generality. The commonly useful row operations are as follows.

(a) Two rows (or columns) equal

$$|A| = \begin{vmatrix} a_{11} & a_{12} & a_{13} \\ a_{21} & a_{22} & a_{23} \\ a_{21} & a_{22} & a_{23} \end{vmatrix} = a_{11} \begin{vmatrix} a_{22} & a_{23} \\ a_{22} & a_{23} \end{vmatrix} - a_{12} \begin{vmatrix} a_{21} & a_{23} \\ a_{21} & a_{23} \end{vmatrix} + a_{13} \begin{vmatrix} a_{21} & a_{22} \\ a_{21} & a_{22} \end{vmatrix} = 0$$

Thus if two rows (or columns) are the same, the determinant is zero.

(b) Multiple of a row by a scalar

$$|B| = \begin{vmatrix} \lambda a_{11} & \lambda a_{12} & \lambda a_{13} \\ a_{21} & a_{22} & a_{23} \\ a_{31} & a_{32} & a_{33} \end{vmatrix} = \lambda|A|$$

The proof of this follows immediately from the definition. A consequence of (a) and (b) is that if any row (or column) is a multiple of another row (or column) then the determinant is zero.

(c) Interchange of two rows (or columns)
Consider

$$|A| = \begin{vmatrix} a_{11} & a_{12} & a_{13} \\ a_{21} & a_{22} & a_{23} \\ a_{31} & a_{32} & a_{33} \end{vmatrix} \quad \text{and} \quad |B| = \begin{vmatrix} a_{21} & a_{22} & a_{23} \\ a_{11} & a_{12} & a_{13} \\ a_{31} & a_{32} & a_{33} \end{vmatrix}$$

Expanding $|A|$ by the first row,

$$|A| = a_{11}\begin{vmatrix} a_{22} & a_{23} \\ a_{32} & a_{33} \end{vmatrix} - a_{12}\begin{vmatrix} a_{21} & a_{23} \\ a_{31} & a_{33} \end{vmatrix} + a_{13}\begin{vmatrix} a_{21} & a_{22} \\ a_{31} & a_{32} \end{vmatrix}$$

and $|B|$ by the second row

$$|B| = -a_{11}\begin{vmatrix} a_{22} & a_{23} \\ a_{32} & a_{33} \end{vmatrix} + a_{12}\begin{vmatrix} a_{21} & a_{23} \\ a_{31} & a_{33} \end{vmatrix} - a_{13}\begin{vmatrix} a_{21} & a_{22} \\ a_{31} & a_{32} \end{vmatrix}$$

Thus

$$|A| = -|B|$$

so that interchanging two rows (or columns) changes the sign of the determinant.

(d) Addition rule

$$\begin{vmatrix} a_{11} + b_{11} & a_{12} + b_{12} & a_{13} + b_{13} \\ a_{21} & a_{22} & a_{23} \\ a_{31} & a_{32} & a_{33} \end{vmatrix}$$

$$= (a_{11} + b_{11})A_{11} + (a_{12} + b_{12})A_{12} + (a_{13} + b_{13})A_{13}$$

$$= (a_{11}A_{11} + a_{12}A_{12} + a_{13}A_{13}) + (b_{11}A_{11} + b_{12}A_{12} + b_{13}A_{13})$$

$$= \begin{vmatrix} a_{11} & a_{12} & a_{13} \\ a_{21} & a_{22} & a_{23} \\ a_{31} & a_{32} & a_{33} \end{vmatrix} + \begin{vmatrix} b_{11} & b_{12} & b_{13} \\ a_{21} & a_{22} & a_{23} \\ a_{31} & a_{32} & a_{33} \end{vmatrix}$$

(e) Adding multiples of rows (or columns)

Consider

$$|A| = \begin{vmatrix} a_{11} & a_{12} & a_{13} \\ a_{21} & a_{22} & a_{23} \\ a_{31} & a_{32} & a_{33} \end{vmatrix}$$

Then

$$|B| = \begin{vmatrix} a_{11} + \lambda a_{21} & a_{12} + \lambda a_{22} & a_{13} + \lambda a_{23} \\ a_{21} & a_{22} & a_{23} \\ a_{31} & a_{32} & a_{33} \end{vmatrix}$$

$$= \begin{vmatrix} a_{11} & a_{12} & a_{13} \\ a_{21} & a_{22} & a_{23} \\ a_{31} & a_{32} & a_{33} \end{vmatrix} + \lambda \begin{vmatrix} a_{21} & a_{22} & a_{23} \\ a_{21} & a_{22} & a_{23} \\ a_{31} & a_{32} & a_{33} \end{vmatrix} \quad \text{(using (d) and then (b))}$$

$$= |A| \quad \text{(since, by (a), the second determinant is zero)}$$

This means that adding multiples of rows (or columns) together makes no difference to the determinant.

(f) Transpose

$$|A^{\mathrm{T}}| = |A|$$

This just states that expanding by the first row or the first column gives the same result.

(g) Product

$$|AB| = |A||B|$$

This result is difficult to prove generally, but it can be verified rather tediously for the 2×2 or 3×3 cases. For the 2×2 case

$$|A||B| = (a_{11}a_{22} - a_{12}a_{21})(b_{11}b_{22} - b_{12}b_{21})$$

$$= a_{11}a_{22}b_{11}b_{22} - a_{11}a_{22}b_{12}b_{21} - a_{12}a_{21}b_{11}b_{22} + a_{12}a_{21}b_{12}b_{21}$$

and

$$|AB| = \begin{vmatrix} a_{11}b_{11} + a_{12}b_{21} & a_{11}b_{12} + a_{12}b_{22} \\ a_{21}b_{11} + a_{22}b_{21} & a_{21}b_{12} + a_{22}b_{22} \end{vmatrix}$$

$$= (a_{11}b_{11} + a_{12}b_{21})(a_{21}b_{12} + a_{22}b_{22}) - (a_{11}b_{12} + a_{12}b_{22})(a_{21}b_{11} + a_{22}b_{21})$$

$$= a_{11}a_{22}b_{11}b_{22} - a_{11}a_{22}b_{12}b_{21} - a_{12}a_{21}b_{11}b_{22} + a_{12}a_{21}b_{12}b_{21}$$

Example 5.14   Evaluate the 3×3 determinants

$$
\text{(a) } \begin{vmatrix} 1 & 0 & 1 \\ 0 & 1 & 2 \\ 1 & 1 & 0 \end{vmatrix}, \quad
\text{(b) } \begin{vmatrix} 1 & 0 & 1 \\ 1 & 1 & 0 \\ 0 & 1 & 2 \end{vmatrix}, \quad
\text{(c) } \begin{vmatrix} 1 & 1 & 0 \\ 0 & 1 & 1 \\ 1 & 0 & 2 \end{vmatrix}, \quad
\text{(d) } \begin{vmatrix} 1 & 0 & 1 \\ 0 & 2 & 4 \\ 3 & 3 & 0 \end{vmatrix}
$$

Solution   (a) Expand by the first row:

$$
\begin{vmatrix} 1 & 0 & 1 \\ 0 & 1 & 2 \\ 1 & 1 & 0 \end{vmatrix} = 1 \begin{vmatrix} 1 & 2 \\ 1 & 0 \end{vmatrix} - 0 \begin{vmatrix} 0 & 2 \\ 1 & 0 \end{vmatrix} + 1 \begin{vmatrix} 0 & 1 \\ 1 & 1 \end{vmatrix} = -2 - 0 - 1 = -3
$$

(b) Expand by the first column:

$$
\begin{vmatrix} 1 & 0 & 1 \\ 1 & 1 & 0 \\ 0 & 1 & 2 \end{vmatrix} = 1 \begin{vmatrix} 1 & 0 \\ 1 & 2 \end{vmatrix} - 1 \begin{vmatrix} 0 & 1 \\ 1 & 2 \end{vmatrix} + 0 \begin{vmatrix} 0 & 1 \\ 1 & 0 \end{vmatrix} = 2 + 1 + 0 = 3
$$

Note that (a) and (b) are the same determinant, but with two rows interchanged. The result confirms property (c) just stated above.

(c) Expand by the third row:

$$
\begin{vmatrix} 1 & 1 & 0 \\ 0 & 1 & 1 \\ 1 & 0 & 2 \end{vmatrix} = 1 \begin{vmatrix} 1 & 0 \\ 1 & 1 \end{vmatrix} - 0 \begin{vmatrix} 1 & 0 \\ 0 & 1 \end{vmatrix} + 2 \begin{vmatrix} 1 & 1 \\ 0 & 1 \end{vmatrix} = 1 - 0 + 2 = 3
$$

Note that the matrix associated with the determinant in (c) is just the transpose of the matrix associated with the determinant in (b).

$$
\text{(d) } \begin{vmatrix} 1 & 0 & 1 \\ 0 & 2 & 4 \\ 3 & 3 & 0 \end{vmatrix} = 2 \begin{vmatrix} 1 & 0 & 1 \\ 0 & 1 & 2 \\ 3 & 3 & 0 \end{vmatrix} = 6 \begin{vmatrix} 1 & 0 & 1 \\ 0 & 1 & 2 \\ 1 & 1 & 0 \end{vmatrix} = -18
$$

Note that we have used the multiple of a row rule on two occasions; the final determinant is the same as (a).

In MATLAB the evaluation is straightforward; the instructions

```
a = [1 0 1; 0 2 4; 3 3 0];
det(a)
```

give the result -18.

Example 5.15

Given the matrices

$$A = \begin{bmatrix} 1 & 2 & 3 \\ 2 & 3 & 4 \\ 4 & 5 & 6 \end{bmatrix} \quad \text{and} \quad B = \begin{bmatrix} 1 & 0 & 1 \\ 1 & 1 & 1 \\ 1 & 2 & 3 \end{bmatrix}$$

evaluate (a) $|A|$, (b) $|B|$ and (c) $|AB|$.

Solution

(a) $|A| = 1 \begin{vmatrix} 3 & 4 \\ 5 & 6 \end{vmatrix} - 2 \begin{vmatrix} 2 & 4 \\ 4 & 6 \end{vmatrix} + 3 \begin{vmatrix} 2 & 3 \\ 4 & 5 \end{vmatrix}$

$= 1 \times (-2) - 2 \times (-4) + 3 \times (-2) = 0$

(b) $|B| = \begin{vmatrix} 1 & 0 & 1 \\ 1 & 1 & 1 \\ 1 & 2 & 3 \end{vmatrix}$

$= \begin{vmatrix} 1 & 0 & 0 \\ 1 & 1 & 0 \\ 1 & 2 & 2 \end{vmatrix}$ (subtracting column 1 from column 3)

$= 1 \begin{vmatrix} 1 & 0 \\ 2 & 2 \end{vmatrix} = 2$ (expanding by first row)

(c) $|AB| = \begin{vmatrix} 6 & 8 & 12 \\ 9 & 11 & 17 \\ 15 & 17 & 27 \end{vmatrix}$

$= 6[(11)(27) - (17)(17)] - 8[(9)(27) - (17)(15)] + 12[(9)(17) - (11)(15)]$

$= 48 + 96 - 144 = 0$

We can use properties (a)–(e) to reduce the amount of computation involved in evaluating a determinant. We introduce as many zeros as possible into a row or column, and then expand along that row or column.

Example 5.16

Evaluate

$$D = \begin{vmatrix} 1 & 1 & 1 & 1 \\ 1 & 1+a & 1 & 1 \\ 1 & 1 & 1+b & 1 \\ 1 & 1 & 1 & 1+c \end{vmatrix}$$

Solution

$$D = \begin{vmatrix} 1 & 0 & 0 & 0 \\ 1 & a & 0 & 0 \\ 1 & 0 & b & 0 \\ 1 & 0 & 0 & c \end{vmatrix} \qquad \text{(by subtracting col. 1 from col. 2, col. 3 and col. 4)}$$

$$= 1 \begin{vmatrix} a & 0 & 0 \\ 0 & b & 0 \\ 0 & 0 & c \end{vmatrix} \qquad \text{(by expanding by the top row)}$$

$$= a \begin{vmatrix} b & 0 \\ 0 & c \end{vmatrix} = abc$$

A solution using MAPLE is given by the instructions

```
dd := array (1..4, 1..4, [[1, 1, 1, 1], [1, 1 + a, 1, 1], [1, 1, 1 + b, 1], [1, 1, 1, 1 + c]]):
det (dd);
```

A point that should be carefully noted concerns large determinants; they are extremely difficult and time-consuming to evaluate (using the basic definition (5.11) for an $n \times n$ determinant involves $n!(n - 1)$ multiplications). This is a problem even with computers – which in fact use alternative methods. If at all possible, evaluation of large determinants should be avoided. They do, however, play a central role in matrix theory.

The cofactors $A_{11}, A_{12}, \ldots$ defined earlier have the property that

$$|\mathbf{A}| = a_{11}A_{11} + a_{12}A_{12} + a_{13}A_{13}$$

Consider the expression $a_{21}A_{11} + a_{22}A_{12} + a_{23}A_{13}$. In determinant form we have

$$a_{21}A_{11} + a_{22}A_{12} + a_{23}A_{13} = \begin{vmatrix} a_{21} & a_{22} & a_{23} \\ a_{21} & a_{22} & a_{23} \\ a_{31} & a_{32} & a_{33} \end{vmatrix} = 0$$

since two rows are identical. Similarly,

$$a_{31}A_{11} + a_{32}A_{12} + a_{33}A_{13} = 0$$

Thus in general

$$\sum_k a_{ik}A_{jk} = \begin{cases} |\mathbf{A}| & \text{if } i = j \\ 0 & \text{if } i \neq j \end{cases} \tag{5.12}$$

and, expanding by columns,

$$\sum_k a_{ki}A_{kj} = \begin{cases} |\mathbf{A}| & \text{if } i = j \\ 0 & \text{if } i \neq j \end{cases} \tag{5.13}$$

A numerical example illustrates these points.

Example 5.17   Illustrate the use of cofactors in the expansion of determinants on the matrix

$$A = \begin{bmatrix} 1 & 2 & 3 \\ 6 & 5 & 4 \\ 7 & 8 & 1 \end{bmatrix}$$

Solution   The cofactors are evaluated as

$$A_{11} = \begin{vmatrix} 5 & 4 \\ 8 & 1 \end{vmatrix} = -27, \quad A_{12} = -\begin{vmatrix} 6 & 4 \\ 7 & 1 \end{vmatrix} = 22, \quad A_{13} = \begin{vmatrix} 6 & 5 \\ 7 & 8 \end{vmatrix} = 13$$

and continuing in the same way

$$A_{21} = 22, A_{22} = -20, A_{23} = 6, A_{31} = -7, A_{32} = 14 \text{ and } A_{33} = -7$$

A selection of the evaluations in equation (5.12), that is expansion by rows, is

$$a_{11}A_{11} + a_{12}A_{12} + a_{13}A_{13} = 1 \times (-27) + 2 \times 22 + 3 \times 13 = 56$$

$$a_{21}A_{11} + a_{22}A_{12} + a_{23}A_{13} = 6 \times (-27) + 5 \times 22 + 4 \times 13 = 0$$

$$a_{31}A_{21} + a_{32}A_{22} + a_{33}A_{23} = 7 \times 22 + 8 \times (-20) + 1 \times 6 = 0$$

and in equation (5.13), that is expansion by columns, is

$$a_{11}A_{12} + a_{21}A_{22} + a_{31}A_{32} = 1 \times 22 + 6 \times (-20) + 7 \times 14 = 0$$

$$a_{12}A_{12} + a_{22}A_{22} + a_{32}A_{32} = 2 \times 22 + 5 \times (-20) + 8 \times 14 = 56$$

$$a_{13}A_{11} + a_{23}A_{21} + a_{33}A_{31} = 3 \times (-27) + 4 \times 22 + 1 \times (-7) = 0$$

The other expansions in equations (5.12) and (5.13) can be verified on this example. It may be noted that the determinant of the matrix is 56.

A matrix with particularly interesting properties is the **adjoint** or **adjugate matrix**, which is defined as the transpose of the matrix of cofactors; that is,

$$\text{adj } A = \begin{bmatrix} A_{11} & A_{12} & A_{13} \\ A_{21} & A_{22} & A_{23} \\ A_{31} & A_{32} & A_{33} \end{bmatrix}^{\text{T}} \tag{5.14}$$

If we now calculate A (adj A), we have

$$[A(\text{adj } A)]_{ij} = \sum_k a_{ik}(\text{adj } A)_{kj} = \sum_k a_{ik}A_{jk}$$

$$= \begin{cases} |A| & \text{if } i = j \quad (\text{from (5.12))} \\ 0 & \text{if } i \neq j \end{cases}$$

So

$$A(\text{adj } A) = \begin{bmatrix} |A| & 0 & 0 \\ 0 & |A| & 0 \\ 0 & 0 & |A| \end{bmatrix} = |A|I \tag{5.15}$$

and we have thus discovered a matrix that when multiplied by A gives a scalar times the unit matrix. Note that adj A is available in most computer packages, either the arithmetical evaluation or symbolically. Students are encouraged to master the packages and use them to answer the exercises. The instruction adj(A); in MAPLE produces the adjoint of matrix A.

If A is a square matrix of order n then, taking determinants on both sides of (5.15),

$$|A|\,|\text{adj } A| = |A(\text{adj } A)| = ||A|I_n| = |A|^n$$

If $|A| \neq 0$, it follows that

$$|\text{adj } A| = |A|^{n-1} \tag{5.16}$$

a result known as **Cauchy's theorem**.

It is also the case that

$$\text{adj}(AB) = (\text{adj } B)(\text{adj } A) \tag{5.17}$$

so in taking the adjoint of a product the order is reversed.

An important piece of notation that has significant implications for the solution of sets of linear equations concerns whether or not a matrix has zero determinant. A square matrix A is called **non-singular** if $|A| \neq 0$ and **singular** if $|A| = 0$.

Example 5.18 Derive the adjoint of the 2×2 matrices

$$A = \begin{bmatrix} 1 & 3 \\ 2 & 8 \end{bmatrix} \quad \text{and} \quad B = \begin{bmatrix} -1 & 2 \\ -3 & -4 \end{bmatrix}$$

and verify the results in equations (5.15), (5.16) and (5.17).

Solution The cofactors are very easy to evaluate in the 2×2 case: for the matrix A

$$A_{11} = 8, A_{12} = -2, A_{21} = -3 \quad \text{and} \quad A_{22} = 1$$

and for the matrix B

$$B_{11} = -4, B_{12} = 3, B_{21} = -2 \quad \text{and} \quad B_{22} = -1$$

The adjoint or adjugate matrices can be written down immediately as

$$\text{adj } A = \begin{bmatrix} 8 & -3 \\ -2 & 1 \end{bmatrix} \quad \text{and} \quad \text{adj } B = \begin{bmatrix} -4 & -2 \\ 3 & -1 \end{bmatrix}$$

Now (5.15) gives

$$A(\text{adj } A) = \begin{bmatrix} 1 & 3 \\ 2 & 8 \end{bmatrix}\begin{bmatrix} 8 & -3 \\ -2 & 1 \end{bmatrix} = \begin{bmatrix} 2 & 0 \\ 0 & 2 \end{bmatrix} = 2I$$

$$B(\text{adj } B) = \begin{bmatrix} -1 & 2 \\ -3 & -4 \end{bmatrix}\begin{bmatrix} -4 & -2 \\ 3 & -1 \end{bmatrix} = \begin{bmatrix} 10 & 0 \\ 0 & 10 \end{bmatrix} = 10I$$

so the property is satisfied and the determinants are 2 and 10 respectively. For equation (5.16) we have $n = 2$ so

$$|\text{adj } A| = \begin{vmatrix} 8 & -3 \\ -2 & 1 \end{vmatrix} = 2 \quad \text{and} \quad |\text{adj } B| = \begin{vmatrix} -4 & -2 \\ 3 & -1 \end{vmatrix} = 10$$

as required.

Evaluating the matrices in (5.17)

$$\text{adj}(AB) = \text{adj}\begin{bmatrix} -10 & -10 \\ -26 & -28 \end{bmatrix} = \begin{bmatrix} -28 & 10 \\ 26 & -10 \end{bmatrix}$$

and

$$\text{adj } B \text{ adj } A = \begin{bmatrix} -4 & -2 \\ 3 & -1 \end{bmatrix}\begin{bmatrix} 8 & -3 \\ -2 & 1 \end{bmatrix} = \begin{bmatrix} -28 & 10 \\ 26 & -10 \end{bmatrix}$$

and the statement is clearly verified. It is left as an exercise to show that the product of the matrices the other way round, $\text{adj } A \text{ adj } B$, gives a totally different matrix.

Example 5.19   Given

$$A = \begin{bmatrix} 1 & 1 & 2 \\ 2 & 0 & 1 \\ 3 & 1 & 1 \end{bmatrix}$$

determine $\text{adj } A$ and show that $A(\text{adj } A) = (\text{adj } A)A = |A|I$.

Solution   The matrix of cofactors is

$$\begin{bmatrix} \begin{vmatrix} 0 & 1 \\ 1 & 1 \end{vmatrix} & -\begin{vmatrix} 2 & 1 \\ 3 & 1 \end{vmatrix} & \begin{vmatrix} 2 & 0 \\ 3 & 1 \end{vmatrix} \\ -\begin{vmatrix} 1 & 2 \\ 1 & 1 \end{vmatrix} & \begin{vmatrix} 1 & 2 \\ 3 & 1 \end{vmatrix} & -\begin{vmatrix} 1 & 1 \\ 3 & 1 \end{vmatrix} \\ \begin{vmatrix} 1 & 2 \\ 0 & 1 \end{vmatrix} & -\begin{vmatrix} 1 & 2 \\ 2 & 1 \end{vmatrix} & \begin{vmatrix} 1 & 1 \\ 2 & 0 \end{vmatrix} \end{bmatrix} = \begin{bmatrix} -1 & 1 & 2 \\ 1 & -5 & 2 \\ 1 & 3 & -2 \end{bmatrix}$$

so, from (5.14)

$$\text{adj } A = \begin{bmatrix} -1 & 1 & 2 \\ 1 & -5 & 2 \\ 1 & 3 & -2 \end{bmatrix}^T = \begin{bmatrix} -1 & 1 & 1 \\ 1 & -5 & 3 \\ 2 & 2 & -2 \end{bmatrix}$$

$$A(\text{adj } A) = \begin{bmatrix} 1 & 1 & 2 \\ 2 & 0 & 1 \\ 3 & 1 & 1 \end{bmatrix}\begin{bmatrix} -1 & 1 & 1 \\ 1 & -5 & 3 \\ 2 & 2 & -2 \end{bmatrix} = \begin{bmatrix} 4 & 0 & 0 \\ 0 & 4 & 0 \\ 0 & 0 & 4 \end{bmatrix}$$

$$(\text{adj } A)A = \begin{bmatrix} -1 & 1 & 1 \\ 1 & -5 & 3 \\ 2 & 2 & -2 \end{bmatrix}\begin{bmatrix} 1 & 1 & 2 \\ 2 & 0 & 1 \\ 3 & 1 & 1 \end{bmatrix} = \begin{bmatrix} 4 & 0 & 0 \\ 0 & 4 & 0 \\ 0 & 0 & 4 \end{bmatrix}$$

Since $|A| = 4$ the result then follows.

'**Spring and dashpot' systems** are used extensively in modelling mechanical engineering situations. The stability of such systems is of crucial importance and is determined by a determinantal equation. A simple example is illustrated in Figure 5.8 and involves three equal masses m connected by equal springs, with spring constants k. The middle mass is damped with damper constant c. The x_i are the displacements from the equilibrium position.

The equations of motion involve calculus and will not be written down explicitly. If the displacements are

$$x_1 = a_1 e^{st}, x_2 = a_2 e^{st} \quad \text{and} \quad x_3 = a_3 e^{st}$$

then it can be shown that the complex frequencies of vibration s are the solution of the equation

$$0 = \begin{vmatrix} ms^2 + 2k & -k & 0 \\ -k & ms^2 + cs + 2k & -k \\ 0 & -k & ms^2 + 2k \end{vmatrix}$$

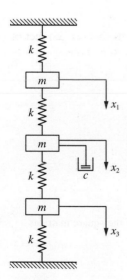

Figure 5.8
Spring and dashpot
system: displacements
measured from the
equilibrium position.

When expanded the determinant gives a sixth-order polynomial in s, which of course is difficult to solve in full generality. We know there will be six complex roots and there are techniques available to determine whether or not the roots have negative real parts and hence the system is stable.

5.3.1 Exercises

20 Find all the minors and cofactors of the determinant

$$\begin{vmatrix} 1 & 2 & 3 \\ 1 & 0 & 1 \\ 1 & 1 & 1 \end{vmatrix}$$

Hence evaluate the determinant.

21 Evaluate the determinants of the following matrices:

(a) $\begin{bmatrix} 1 & 7 \\ 4 & 9 \end{bmatrix}$ (b) $\begin{bmatrix} 1 & 4 & 3 \\ 2 & -4 & 1 \\ 3 & 2 & -6 \end{bmatrix}$

(c) $\begin{bmatrix} 2 & -1 & 3 \\ 4 & 2 & 9 \\ 1 & 3 & -4 \end{bmatrix}$

22 Determine adj A when

$$A = \begin{bmatrix} a & b \\ c & d \end{bmatrix}$$

23 Determine adj A when

$$A = \begin{bmatrix} 2 & 1 & 1 \\ 3 & 2 & 2 \\ 1 & 1 & 2 \end{bmatrix}$$

Check that $A(\text{adj}\,A) = (\text{adj}\,A)A = |A|I$.

24 Show that the matrix

$$B = \begin{bmatrix} 1 & 0 & 2 \\ 3 & 4 & 0 \\ 6 & -2 & 1 \end{bmatrix}$$

is non-singular and verify Cauchy's theorem, namely $|\text{adj}\,B| = |B|^2$.

25 If $\det(A) = 0$ deduce that $\det(A^n) = 0$ for any integer n.

26 Given

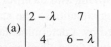

$$A = \begin{bmatrix} 2 & -1 & 0 \\ -4 & 3 & -1 \\ 1 & -1 & 1 \end{bmatrix} \text{ and}$$

$$B = \begin{bmatrix} 1 & 0 & 2 \\ 3 & 4 & 0 \\ 6 & -2 & 1 \end{bmatrix}$$

verify that $\text{adj}(AB) = (\text{adj}\,B)(\text{adj}\,A)$.

27 Find the values of λ that make the following determinants zero:

(a) $\begin{vmatrix} 2 - \lambda & 7 \\ 4 & 6 - \lambda \end{vmatrix}$

(b) $\begin{vmatrix} 1 & 3 - \lambda & 4 \\ 4 - \lambda & 2 & -1 \\ 1 & \lambda - 6 & 2 \end{vmatrix}$

(c) $\begin{vmatrix} 0 & 2 - \lambda & 0 \\ 2 - \lambda & 4 & 1 \\ 2 & -3 & \lambda - 4 \end{vmatrix}$

28 Evaluate the determinants of the square matrices

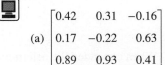

(a) $\begin{bmatrix} 0.42 & 0.31 & -0.16 \\ 0.17 & -0.22 & 0.63 \\ 0.89 & 0.93 & 0.41 \end{bmatrix}$

(b) $\begin{bmatrix} 5 & 4 & 1 & 1 \\ 4 & 5 & 1 & 1 \\ 1 & 1 & 4 & 2 \\ 1 & 1 & 2 & 4 \end{bmatrix}$

29 Show that the area of a triangle with vertices (x_1, y_1), (x_2, y_2) and (x_3, y_3) is given by the absolute value of

$$\frac{1}{2}\begin{vmatrix} 1 & x_1 & y_1 \\ 1 & x_2 & y_2 \\ 1 & x_3 & y_3 \end{vmatrix}$$

30 Show that $x + x^2 - 2x^3$ is a factor of the determinant D where

$$D = \begin{vmatrix} 0 & x & 2 & x^2 \\ -x & 0 & 1 & x^3 \\ -2 & -1 & 0 & 1 \\ -x^2 & -x^3 & -1 & 0 \end{vmatrix}$$

and hence express D as a product of linear factors.

31 Show that

$$\begin{vmatrix} x & a & b \\ x^2 & a^2 & b^2 \\ a+b & x+b & x+a \end{vmatrix}$$

$$= (b-a)(x-a)(x-b)(x+a+b)$$

Such an exercise can be solved in two lines of code of a symbolic manipulation package such as MAPLE.

32 Verify that if A is a symmetric matrix then so is adj A.

33 If A is a skew-symmetric $n \times n$ matrix, verify that adj A is symmetric or skew-symmetric according to whether n is odd or even.

5.4 The inverse matrix

In Section 5.3 we constructed adj A and saw that it had interesting properties in relation to the unit matrix. We also saw, in Example 5.5, that we had a method of solving linear equations if we could construct B such that $AB = I$. These ideas can be brought together to provide a comprehensive theory of the solution of linear equations, which we will consider in Section 5.5.

Given a square matrix A, if we can construct a matrix B such that

$$BA = AB = I$$

then **we call B the inverse of A and write it as A^{-1}**. From (5.15)

$$A(\text{adj } A) = |A|I$$

so that we have gone a long way to constructing the inverse. We have two cases:

- If A is non-singular then $|A| \neq 0$ and

$$A^{-1} = \frac{\text{adj } A}{|A|}$$

- If A is singular then $|A| = 0$ and it can be shown that the inverse A^{-1} does not exist.

If the inverse exists then it is unique. Suppose for a given A we have two inverses B and C. Then

$$AB = BA = I, \quad AC = CA = I$$

and therefore

$$AB = AC$$

Pre-multiplying by C, we have

$$C(AB) = C(AC)$$

But matrix multiplication is associative, so we can write this as

$$(CA)B = (CA)C$$

Hence

$$IB = IC \quad \text{(since } CA = I)$$

and so

$$B = C$$

The inverse is therefore unique.

It should be noted that if both A and B are square matrices then $AB = I$ if and only if $BA = I$.

Example 5.20   Find A^{-1} and B^{-1} for the matrices

(a) $A = \begin{bmatrix} 1 & 2 \\ 2 & 3 \end{bmatrix}$ and (b) $B = \begin{bmatrix} 5 & 2 & 4 \\ 3 & -1 & 2 \\ 1 & 4 & -3 \end{bmatrix}$

Solution

(a) $\text{adj}\,A = \begin{bmatrix} 3 & -2 \\ -2 & 1 \end{bmatrix}^{\mathrm{T}} = \begin{bmatrix} 3 & -2 \\ -2 & 1 \end{bmatrix}$ and $|A| = -1$

so that

$$A^{-1} = \frac{\text{adj}\,A}{|A|} = \begin{bmatrix} -3 & 2 \\ 2 & -1 \end{bmatrix}$$

(b) $\text{adj}\,B = \begin{bmatrix} -5 & 11 & 13 \\ 22 & -19 & -18 \\ 8 & 2 & -11 \end{bmatrix}^{\mathrm{T}} = \begin{bmatrix} -5 & 22 & 8 \\ 11 & -19 & 2 \\ 13 & -18 & -11 \end{bmatrix}$ and $|B| = 49$

so that

$$B^{-1} = \frac{1}{49} \begin{bmatrix} -5 & 22 & 8 \\ 11 & -19 & 2 \\ 13 & -18 & -11 \end{bmatrix}$$

In both cases it can be checked that $AA^{-1} = I$ and $BB^{-1} = I$.

Finding the inverse of a 2×2 matrix is very easy, since for

$$A = \begin{bmatrix} a & b \\ c & d \end{bmatrix}$$

$$A^{-1} = \frac{1}{ad - bc} \begin{bmatrix} d & -b \\ -c & a \end{bmatrix} \quad \text{(provided that } ad - bc \neq 0)$$

Unfortunately there is no simple extension of this result to higher-order matrices. On the other hand, in most practical situations the inverse itself is rarely required – it is the solution of the corresponding linear equations that is important. To understand the power and applicability of the various methods of solution of linear equations, the role of the inverse is essential. The consideration of the adjoint matrix provides a theoretical framework for this study, but as a practical method for finding the inverse of a matrix it is virtually useless, since, as we saw earlier, it is so time-consuming to compute determinants. However, computer packages give the inverse with a single instruction, such as inv(A) in MATLAB or inverse(A) in MAPLE.

To find the inverse of a product of two matrices, the order is reversed:

$$(AB)^{-1} = B^{-1}A^{-1} \tag{5.18}$$

(provided that A and B are invertible). To prove this, let $C = B^{-1}A^{-1}$. Then

$$C(AB) = (B^{-1}A^{-1})(AB) = B^{-1}(A^{-1}A)B = B^{-1}IB = B^{-1}B = I$$

and thus

$$C = B^{-1}A^{-1} = (AB)^{-1}$$

Since matrices do not commute in general $A^{-1}B^{-1} \neq B^{-1}A^{-1}$.

Example 5.21   Given

$$A = \begin{bmatrix} 1 & 2 \\ 2 & 1 \end{bmatrix} \quad \text{and} \quad B = \begin{bmatrix} 0 & 1 \\ 1 & 1 \end{bmatrix}$$

evaluate $(AB)^{-1}$, $A^{-1}B^{-1}$, $B^{-1}A^{-1}$ and show that $(AB)^{-1} = B^{-1}A^{-1}$.

Solution
$$A^{-1} = \begin{bmatrix} -\frac{1}{3} & \frac{2}{3} \\ \frac{2}{3} & -\frac{1}{3} \end{bmatrix}, \quad B^{-1} = \begin{bmatrix} -1 & 1 \\ 1 & 0 \end{bmatrix}$$

$$AB = \begin{bmatrix} 2 & 3 \\ 1 & 3 \end{bmatrix}, \quad (AB)^{-1} = \begin{bmatrix} 1 & -1 \\ -\frac{1}{3} & \frac{2}{3} \end{bmatrix}$$

$$A^{-1}B^{-1} = \begin{bmatrix} -\frac{1}{3} & \frac{2}{3} \\ \frac{2}{3} & -\frac{1}{3} \end{bmatrix}\begin{bmatrix} -1 & 1 \\ 1 & 0 \end{bmatrix} = \begin{bmatrix} 1 & -\frac{1}{3} \\ -1 & \frac{2}{3} \end{bmatrix}$$

$$B^{-1}A^{-1} = \begin{bmatrix} -1 & 1 \\ 1 & 0 \end{bmatrix}\begin{bmatrix} -\frac{1}{3} & \frac{2}{3} \\ \frac{2}{3} & -\frac{1}{3} \end{bmatrix} = \begin{bmatrix} 1 & -1 \\ -\frac{1}{3} & \frac{2}{3} \end{bmatrix} = (AB)^{-1}$$

Example 5.22 Given the two matrices

$$A = \begin{bmatrix} 0 & -\frac{3}{5} & 0 \\ \frac{5}{3} & 0 & -\frac{5}{3} \\ 0 & 6 & -6 \end{bmatrix} \quad \text{and} \quad T = \begin{bmatrix} 0.6 & 0.3 & 0.1 \\ 1 & 1 & 0.5 \\ 1.2 & 1.5 & 1 \end{bmatrix}$$

show that the matrix $T^{-1}AT$ is diagonal.

Solution The inverse is best computed using MATLAB or a similar package. It may be verified by direct multiplication that

$$T^{-1} = \frac{1}{6} \begin{bmatrix} 25 & -15 & 5 \\ -40 & 48 & -20 \\ 30 & -54 & 30 \end{bmatrix}$$

The further multiplications give

$$\frac{1}{6} \begin{bmatrix} 25 & -15 & 5 \\ -40 & 48 & -20 \\ 30 & -54 & 30 \end{bmatrix} \begin{bmatrix} 0 & -\frac{3}{5} & 0 \\ \frac{5}{3} & 0 & -\frac{5}{3} \\ 0 & 6 & -6 \end{bmatrix} \begin{bmatrix} 0.6 & 0.3 & 0.1 \\ 1 & 1 & 0.5 \\ 1.2 & 1.5 & 1 \end{bmatrix} = \begin{bmatrix} -1 & 0 & 0 \\ 0 & -2 & 0 \\ 0 & 0 & -3 \end{bmatrix}$$

This technique is an important one mathematically (see the companion volume *Advanced Modern Engineering Mathematics*) since it provides a method of uncoupling a system of coupled equations. Practically it is the process used to reduce a physical system to principal axes; in elasticity it provides the principal stresses in a body.

5.4.1 Exercises

34 Determine whether the following matrices are singular or non-singular:

$$\begin{bmatrix} 1 & 2 \\ 2 & 1 \end{bmatrix}, \quad \begin{bmatrix} 1 & 2 & 3 \\ 2 & 2 & 1 \\ 5 & 6 & 5 \end{bmatrix}$$

$$\begin{bmatrix} 1 & 0 & 0 & 1 \\ 0 & 1 & 0 & 1 \\ 0 & 0 & 1 & 1 \\ 0 & 0 & 0 & 1 \end{bmatrix}, \quad \begin{bmatrix} 1 & 0 & 1 \\ 0 & 1 & 0 \\ 1 & 0 & 1 \end{bmatrix}$$

35 Find the inverse of the non-singular matrices in Question 34.

36 For the matrix

$$A = \begin{bmatrix} 1 & 2 & 2 \\ 2 & 1 & 2 \\ 2 & 2 & 1 \end{bmatrix}$$

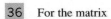
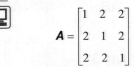

show that $A^2 - 4A - 5I = 0$ and hence that $A^{-1} = \frac{1}{5}(A - 4I)$. Calculate A^{-1} from this result. Further show that the inverse of A^2 is given by $\frac{1}{25}(21I - 4A)$ and evaluate.

37 Given

$$A = \begin{bmatrix} 1 & 0 & 2 \\ 6 & 4 & 0 \\ 6 & -2 & 1 \end{bmatrix} \quad \text{and} \quad B = \begin{bmatrix} 5 & 2 & 4 \\ 3 & -1 & 2 \\ 1 & 4 & -3 \end{bmatrix}$$

find A^{-1} and B^{-1}. Verify that $(AB)^{-1} = B^{-1}A^{-1}$.

38 Given the matrices

$$A = \begin{bmatrix} 1 & 0 & 0 & 0 \\ 0 & 0 & 1 & 0 \\ 0 & 1 & 0 & 0 \\ 0 & 0 & 0 & 1 \end{bmatrix} \quad \text{and} \quad B = \begin{bmatrix} 1 & 0 & 0 & 0 \\ 0 & 0 & 1 & 0 \\ 0 & 0 & 0 & 1 \\ 0 & 1 & 0 & 0 \end{bmatrix}$$

show that $A^2 = I$ and $B^3 = I$, and hence find A^{-1}, B^{-1} and $(AB)^{-1}$.

(*Note*: the matrices A and B in this exercise are examples of **permutation matrices**, since

$$A \begin{bmatrix} a_1 \\ a_2 \\ a_3 \\ a_4 \end{bmatrix} = \begin{bmatrix} a_1 \\ a_3 \\ a_2 \\ a_4 \end{bmatrix}$$

and the suffixes are just permuted; B has similar properties.)

5.5   Linear equations

Although matrices are of great importance in themselves, their practical importance lies in the solution of sets of linear equations. Such sets of equations occur in a wide range of scientific and engineering problems. In the first part of this section we shall consider whether or not a solution exists, and then in Sections 5.5.2 and 5.5.4 we shall look at practical methods of solution.

We now make some definitive statements about the solution of the system of simultaneous linear equations.

$$\left. \begin{aligned} a_{11}x_1 + a_{12}x_2 + \ldots + a_{1n}x_n &= b_1 \\ a_{21}x_1 + a_{22}x_2 + \ldots + a_{2n}x_n &= b_2 \\ \vdots \qquad\qquad \vdots \\ a_{n1}x_1 + a_{n2}x_2 + \ldots + a_{nn}x_n &= b_n \end{aligned} \right\}$$

(5.19)

or, in matrix notation,

$$\begin{bmatrix} a_{11} & a_{12} & \ldots & a_{1n} \\ a_{21} & a_{22} & \ldots & a_{2n} \\ \vdots & & & \\ a_{n1} & a_{n2} & \ldots & a_{nn} \end{bmatrix} \begin{bmatrix} x_1 \\ x_2 \\ \vdots \\ x_n \end{bmatrix} = \begin{bmatrix} b_1 \\ b_2 \\ \vdots \\ b_n \end{bmatrix}$$

that is,

$$AX = b$$

(5.20)

where A is the matrix of coefficients and X the vector of unknowns. If $b = 0$ the equations are called **homogeneous**, while if $b \neq 0$ they are called **nonhomogeneous** (or **inhomogeneous**). There are several cases to consider.

Case (a) **b** $\neq 0$ *and* $|\mathbf{A}| \neq 0$

We know that $\mathbf{A}^{-1}$ exists, and hence

$$\mathbf{A}^{-1}\mathbf{A}X = \mathbf{A}^{-1}b$$

so that

$$X = \mathbf{A}^{-1}b \qquad\qquad\qquad (5.21)$$

and we have a unique solution to (5.19) and (5.20).

Case (b) **b** $= 0$ *and* $|\mathbf{A}| \neq 0$

Again $\mathbf{A}^{-1}$ exists, and the homogeneous equations

$$\mathbf{A}X = 0$$

give

$$\mathbf{A}^{-1}\mathbf{A}X = \mathbf{A}^{-1}0 \quad \text{or} \quad X = 0$$

We therefore only have the **trivial solution** $X = 0$.

Case (c) **b** $\neq 0$ *and* $|\mathbf{A}| = 0$

The inverse matrix does not exist, and this is perhaps the most complicated case. We have two possibilities: either we have no solution or we have infinitely many solutions. A simple example will illustrate the situation. The equations

$$\left.\begin{array}{r}3x + 2y = 2 \\ 3x + 2y = 6\end{array}\right\}, \quad \text{or} \quad \begin{bmatrix} 3 & 2 \\ 3 & 2 \end{bmatrix}\begin{bmatrix} x \\ y \end{bmatrix} = \begin{bmatrix} 2 \\ 6 \end{bmatrix}$$

are clearly inconsistent, and no solution exists. However, in the case of

$$\left.\begin{array}{r}3x + 2y = 2 \\ 6x + 4y = 4\end{array}\right\}, \quad \text{or} \quad \begin{bmatrix} 3 & 2 \\ 6 & 4 \end{bmatrix}\begin{bmatrix} x \\ y \end{bmatrix} = \begin{bmatrix} 2 \\ 4 \end{bmatrix}$$

where one equation is a multiple of the other, we have infinitely many solutions: $x = \lambda$, $y = 1 - \frac{3}{2}\lambda$ is a solution for any value of λ.

The same behaviour is observed for problems involving more than two variables, but the situation is then much more difficult to analyse. The problem of determining whether or not a set of equations has a solution will be discussed in Section 5.6.

Case (d) **b** $= 0$ *and* $|\mathbf{A}| = 0$

As in case (c), we have infinitely many solutions. For instance, the case of two equations takes the form

$$ax + by = 0$$

$$\alpha ax + \alpha by = 0$$

so that $|\mathbf{A}| = 0$ and we find a solution $x = \lambda$, $y = -a\lambda/b$ if $b \neq 0$. If $b = 0$ then $x = 0$, $y = \lambda$ is a solution.

This case is one of the most important, since we deduce the important result that *the equation*

$$\mathbf{AX} = 0$$

has a non-trivial solution if and only if $|\mathbf{A}| = 0$.

Example 5.23   Write the five sets of equations in matrix form and decide whether they have or do not have a solution.

(a) $2x + y = 5$
    $x - 2y = -5$

(b) $2x + y = 0$
    $x - 2y = 0$

(c) $-3x + 6y = 15$
    $x - 2y = -5$

(d) $-3x + 6y = 10$
    $x - 2y = -5$

(e) $-3x + 6y = 0$
    $x - 2y = 0$

Solution   (a) In matrix form the equations are $\begin{bmatrix} 2 & 1 \\ 1 & -2 \end{bmatrix}\begin{bmatrix} x \\ y \end{bmatrix} = \begin{bmatrix} 5 \\ -5 \end{bmatrix}$. The determinant of the

matrix has the value -5 and the right-hand side is non-zero so the problem is of the type **Case (a)** and hence has a unique solution, namely $x = 1$, $y = 3$.

(b) In matrix form the equations are $\begin{bmatrix} 2 & 1 \\ 1 & -2 \end{bmatrix}\begin{bmatrix} x \\ y \end{bmatrix} = \begin{bmatrix} 0 \\ 0 \end{bmatrix}$. The determinant of the

matrix has the value -5 and the right-hand side is now zero so the problem is of the type **Case (b)** and hence only has the trivial solution, namely $x = 0$, $y = 0$.

(c) In matrix form the equations are $\begin{bmatrix} -3 & 6 \\ 1 & -2 \end{bmatrix}\begin{bmatrix} x \\ y \end{bmatrix} = \begin{bmatrix} 15 \\ -5 \end{bmatrix}$. The determinant of the

matrix is now zero and the right-hand side is non-zero so the problem is of the type **Case (c)** and hence the solution is not so easy. Essentially the first equation is just (-3) times the second equation so a solution can be computed. A bit of rearrangement soon gives $x = 2t - 5$, $y = t$ for any t, and thus there are infinitely many solutions to this set of equations.

(d) In matrix form the equations are $\begin{bmatrix} -3 & 6 \\ 1 & -2 \end{bmatrix}\begin{bmatrix} x \\ y \end{bmatrix} = \begin{bmatrix} 10 \\ -5 \end{bmatrix}$. The determinant of the

matrix is zero again and the right-hand side is non-zero so the problem is once more of the type **Case (c)** and hence the solution is not so easy. The left-hand side of the first equation is (-3) times the second equation but the right-hand side is only (-2) times the second equation so the equations are inconsistent and there is no solution to this set of equations.

(e) In matrix form the equations are $\begin{bmatrix} -3 & 6 \\ 1 & -2 \end{bmatrix}\begin{bmatrix} x \\ y \end{bmatrix} = \begin{bmatrix} 0 \\ 0 \end{bmatrix}$. The determinant of the

matrix is zero again and the right-hand side is also zero so the problem is of the type **Case (d)** and hence a non-trivial solution can be found. It can be seen that $x = 2s$ and $y = s$ gives the solution for any s.

Example 5.24

Find a solution of

$$
\begin{aligned}
x + y + z &= 6 \\
x + 2y + 3z &= 14 \\
x + 4y + 9z &= 36
\end{aligned}
$$

Solution

Expressing the equations in matrix form $AX = b$

$$
\begin{bmatrix} 1 & 1 & 1 \\ 1 & 2 & 3 \\ 1 & 4 & 9 \end{bmatrix} \begin{bmatrix} x \\ y \\ z \end{bmatrix} = \begin{bmatrix} 6 \\ 14 \\ 36 \end{bmatrix}
$$

we have

$$
|A| = \begin{vmatrix} 1 & 1 & 1 \\ 1 & 2 & 3 \\ 1 & 4 & 9 \end{vmatrix} = \begin{vmatrix} 1 & 0 & 0 \\ 1 & 1 & 2 \\ 1 & 3 & 8 \end{vmatrix} = 2 \neq 0 \quad \text{(subtracting column 1 from columns 2 and 3)}
$$

so that a solution does exist and is unique. The inverse of A can be computed as

$$
A^{-1} = \begin{bmatrix} 3 & -\frac{5}{2} & \frac{1}{2} \\ -3 & 4 & -1 \\ 1 & -\frac{3}{2} & \frac{1}{2} \end{bmatrix}
$$

and hence, from (5.21),

$$
X = \begin{bmatrix} x \\ y \\ z \end{bmatrix} = A^{-1} \begin{bmatrix} 6 \\ 14 \\ 36 \end{bmatrix} = \begin{bmatrix} 1 \\ 2 \\ 3 \end{bmatrix}
$$

so the solution is $x = 1$, $y = 2$ and $z = 3$.

Example 5.25

Find the values of k for which the equations

$$
\begin{aligned}
x + 5y + 3z &= 0 \\
5x + y - kz &= 0 \\
x + 2y + kz &= 0
\end{aligned}
$$

have a non-trivial solution.

Solution

The matrix of coefficients is

$$
A = \begin{bmatrix} 1 & 5 & 3 \\ 5 & 1 & -k \\ 1 & 2 & k \end{bmatrix}
$$

For a non-zero solution, $|A| = 0$. Hence

$$0 = |A| = \begin{vmatrix} 1 & 5 & 3 \\ 5 & 1 & -k \\ 1 & 2 & k \end{vmatrix} = 27 - 27k$$

Thus the equations have a non-trivial solution if $k = 1$; if $k \neq 1$, the only solution is $x = y = z = 0$. For $k = 1$ a simple calculation gives $x = \lambda$, $y = -2\lambda$ and $z = 3\lambda$ for any λ.

Example 5.26   Find the values of λ and the corresponding X such that

$$(A - \lambda I)X = 0$$

has a non-trivial solution, given

$$A = \begin{bmatrix} 3 & 1 \\ -2 & 0 \end{bmatrix}$$

Solution   We require

$$0 = |A - \lambda I| = \begin{vmatrix} 3 - \lambda & 1 \\ -2 & -\lambda \end{vmatrix} = -3\lambda + \lambda^2 + 2 = (\lambda - 2)(\lambda - 1)$$

Non-trivial solutions occur only if $\lambda = 1$ or 2.
   If $\lambda = 1$,

$$\begin{bmatrix} 2 & 1 \\ -2 & -1 \end{bmatrix}\begin{bmatrix} x \\ y \end{bmatrix} = 0, \quad \text{so} \quad \begin{bmatrix} x \\ y \end{bmatrix} = \alpha\begin{bmatrix} 1 \\ -2 \end{bmatrix} \quad \text{for any } \alpha$$

If $\lambda = 2$,

$$\begin{bmatrix} 1 & 1 \\ -2 & -2 \end{bmatrix}\begin{bmatrix} x \\ y \end{bmatrix} = 0, \quad \text{so} \quad \begin{bmatrix} x \\ y \end{bmatrix} = \beta\begin{bmatrix} 1 \\ -1 \end{bmatrix} \quad \text{for any } \beta$$

(Note that the problem described here is an important one. The λ and X are called **eigenvalues** and **eigenvectors**, which are introduced in Section 5.7.)

It is possible to write down the solution of a set of equations explicitly in terms of the cofactors of a matrix. However, as a method for computing the solution, this is extremely inefficient; a set of ten equations, for example, will require 4×10^8 multiplications – which takes a long time even on modern computers. The method is of great theoretical interest though. Consider the set of equations

$$\left.\begin{array}{l} a_{11}x_1 + a_{12}x_2 + a_{13}x_3 = b_1 \\ a_{21}x_1 + a_{22}x_2 + a_{23}x_3 = b_2 \\ a_{31}x_1 + a_{32}x_2 + a_{33}x_3 = b_3 \end{array}\right\} \tag{5.22}$$

Denoting the matrix of coefficients by **A** and recalling the definitions of the cofactors, we multiply the equations by A_{11}, A_{21} and A_{31} respectively, and add to give

$$(a_{11}A_{11} + a_{21}A_{21} + a_{31}A_{31})x_1 + (a_{12}A_{11} + a_{22}A_{21} + a_{32}A_{31})x_2$$
$$+ (a_{13}A_{11} + a_{23}A_{21} + a_{33}A_{31})x_3$$
$$= b_1A_{11} + b_2A_{21} + b_3A_{31}$$

Using (5.13), we obtain

$$|\boldsymbol{A}|x_1 + 0x_2 + 0x_3 = b_1A_{11} + b_2A_{21} + b_3A_{31}$$

The right-hand side can be written as a determinant, so

$$|\boldsymbol{A}|x_1 = \begin{vmatrix} b_1 & a_{12} & a_{13} \\ b_2 & a_{22} & a_{23} \\ b_3 & a_{32} & a_{33} \end{vmatrix}$$

The other x_i follow similarly, and we derive **Cramer's rule** that a solution of (5.22) is

$$x_1 = |\boldsymbol{A}|^{-1} \begin{vmatrix} b_1 & a_{12} & a_{13} \\ b_2 & a_{22} & a_{23} \\ b_3 & a_{32} & a_{33} \end{vmatrix},$$

$$x_2 = |\boldsymbol{A}|^{-1} \begin{vmatrix} a_{11} & b_1 & a_{13} \\ a_{21} & b_2 & a_{23} \\ a_{31} & b_3 & a_{33} \end{vmatrix},$$

$$x_3 = |\boldsymbol{A}|^{-1} \begin{vmatrix} a_{11} & a_{12} & b_1 \\ a_{21} & a_{22} & b_2 \\ a_{31} & a_{32} & b_3 \end{vmatrix}$$

Again it should be stressed that this rule should not be used as a computational method because of the large effort required to evaluate determinants.

Example 5.27

A function $u(x, y)$ is known to take values u_1, u_2 and u_3 at the points (x_1, y_1), (x_2, y_2) and (x_3, y_3) respectively. Find the linear interpolating function within the triangle.

Solution We assume the linear interpolating function takes the form

$$u = a + bx + cy$$

To fit the data

$$u_1 = a + bx_1 + cy_1$$
$$u_2 = a + bx_2 + cy_2 \quad \text{or in matrix form}$$
$$u_3 = a + bx_3 + cy_3$$

$$\begin{bmatrix} u_1 \\ u_2 \\ u_3 \end{bmatrix} = \begin{bmatrix} 1 & x_1 & y_1 \\ 1 & x_2 & y_2 \\ 1 & x_3 & y_3 \end{bmatrix} \begin{bmatrix} a \\ b \\ c \end{bmatrix}$$

The values of a, b and c can be obtained from Cramer's rule as

$$a = \begin{vmatrix} u_1 & x_1 & y_1 \\ u_2 & x_2 & y_2 \\ u_3 & x_3 & y_3 \end{vmatrix} / \det(\boldsymbol{A}), \quad b = \begin{vmatrix} 1 & u_1 & y_1 \\ 1 & u_2 & y_2 \\ 1 & u_3 & y_3 \end{vmatrix} / \det(\boldsymbol{A}) \quad \text{and}$$

$$c = \begin{vmatrix} 1 & x_1 & u_1 \\ 1 & x_2 & u_2 \\ 1 & x_3 & u_3 \end{vmatrix} / \det(\boldsymbol{A})$$

where $\boldsymbol{A}$ is the matrix of coefficients. The interpolation formula is now known. In finite-element analysis the evaluation of interpolation functions, such as the one described, is of great importance. Finite elements are central to many large-scale calculations in all branches of engineering.

Example 5.28

Solve the matrix equation $\boldsymbol{AX} = \boldsymbol{c}$ where

$$\boldsymbol{A} = \begin{bmatrix} 4 & 1 & 0 & 0 & 0 & 0 & 0 & 0 & 0 & 0 \\ 1 & 4 & 1 & 0 & 0 & 0 & 0 & 0 & 0 & 0 \\ 1 & 0 & 4 & 1 & 0 & 0 & 0 & 0 & 0 & 0 \\ 1 & 0 & 0 & 4 & 1 & 0 & 0 & 0 & 0 & 0 \\ 1 & 0 & 0 & 0 & 4 & 1 & 0 & 0 & 0 & 0 \\ 1 & 0 & 0 & 0 & 0 & 4 & 1 & 0 & 0 & 0 \\ 1 & 0 & 0 & 0 & 0 & 0 & 4 & 1 & 0 & 0 \\ 1 & 0 & 0 & 0 & 0 & 0 & 0 & 4 & 1 & 0 \\ 1 & 0 & 0 & 0 & 0 & 0 & 0 & 0 & 4 & 1 \\ 1 & 0 & 0 & 0 & 0 & 0 & 0 & 0 & 0 & 4 \end{bmatrix} \quad \text{and} \quad \boldsymbol{c} = \begin{bmatrix} 1 \\ 2 \\ 3 \\ 4 \\ 5 \\ 5 \\ 4 \\ 3 \\ 2 \\ 1 \end{bmatrix}$$

Solution

The solution of such a problem is beyond the scope of hand computation; Cramer's rule, evaluation of the adjoint, direct evaluation of the inverse are impracticable. Even the more practical methods in the next sections struggle with this size of problem if hand computation is tried. A computer package must be used. In MATLAB the relevant instructions are given.

```
b = zeros (10, 10);
for i = 1 : 9, b (i, i) = 4; b (i, i + 1) = 1; b (i + 1, 1) = 1; end
b (10, 10) = 4;
c = [1; 2; 3; 4; 5; 5; 4; 3; 2; 1];
b\c
```

gives the solution

0.1685   0.3258   0.5282   0.7188   0.9563   1.0063   0.8063   0.6064
0.4059   0.2079

5.5.1 Exercises

39 Find the inverse of the matrix

$$A = \begin{bmatrix} -1 & 2 & 1 \\ 0 & 1 & -2 \\ 1 & 4 & -1 \end{bmatrix}$$

and hence solve the equations

$$-x + 2y + z = 2$$
$$y - 2z = -3$$
$$x + 4y - z = 4$$

40 Show that there are two values of α for which the equations

$$\alpha x - 3y + (1 + \alpha)z = 0$$
$$2x + y - \alpha z = 0$$
$$(\alpha + 2)x - 2y + \alpha z = 0$$

have non-trivial solutions. Find the solutions corresponding to these two values of α.

41 If

$$A = \begin{bmatrix} -3 & 1 & -1 \\ 1 & -5 & 1 \\ -1 & 1 & -3 \end{bmatrix}$$

find the values of λ for which the equation $AX = \lambda X$ has non-trivial solutions.

42 Given the matrix

$$A = \begin{bmatrix} 1 & a & -1 \\ a & -2 & 2 \\ -1 & 1 & a \end{bmatrix}$$

(a) solve $|A| = 0$ for real a;

(b) if $a = 2$, find A^{-1} and hence solve

$$A \begin{bmatrix} x \\ y \\ z \end{bmatrix} = \begin{bmatrix} 1 \\ 0 \\ 2 \end{bmatrix}$$

(c) if $a = 0$, find the general solution of

$$A \begin{bmatrix} x \\ y \\ z \end{bmatrix} = \begin{bmatrix} 0 \\ 0 \\ 0 \end{bmatrix}$$

(d) if $a = 1$, show that

$$A \begin{bmatrix} x \\ y \\ z \end{bmatrix} = 2 \begin{bmatrix} x \\ y \\ z \end{bmatrix}$$

can be solved for non-zero x, y and z.

43 Use MATLAB or a similar package to find the inverse of the matrix

$$\begin{bmatrix} 6 & 2 & 1 & 0 & 0 & 0 \\ 2 & 6 & 2 & 1 & 0 & 0 \\ 1 & 2 & 6 & 2 & 1 & 0 \\ 0 & 1 & 2 & 6 & 2 & 1 \\ 0 & 0 & 1 & 2 & 6 & 2 \\ 0 & 0 & 0 & 1 & 2 & 6 \end{bmatrix}$$

and hence solve the matrix equation

$$AX = c$$

where $c^{\mathrm{T}} = [1\ 0\ 0\ 0\ 0\ 1]$.

44 In finite-element calculations the bilinear function

$$u(x, y) = a + bx + cy + dxy$$

is commonly used for interpolation over a quadrilateral and data is always stored in matrix form. If the function fits the data $u(0, 0) = u_1$, $u(p, 0) = u_2$, $u(0, q) = u_3$ and $u(p, q) = u_4$ at the four corners of a rectangle, use matrices to find the coefficients a, b, c and d.

45 In an industrial process water flows through three tanks in succession as illustrated in Figure 5.9.

The tanks have unit cross-section and have heads of water x, y and z respectively. The rate of inflow into the first tank is u, the flowrate in the tube connecting tanks 1 and 2 is $6(x - y)$, the flowrate in the tube connecting tanks 2 and 3 is $5(y - z)$ and the rate of outflow from tank 3 is $4.5z$.

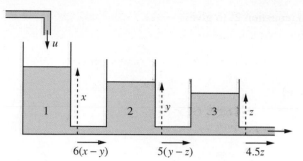

Figure 5.9 Flow though three tanks in Question 45.

Show that the equations of the system in the steady flow situation are

$$u = 6x - 6y$$

$$0 = 6x - 11y + 5z$$

$$0 = 5y - 9.5z$$

and hence find x, y and z.

46 A function is known to fit closely to the approximate function

$$f(z) = \frac{az + b}{cz + 1}$$

It is fitted to the three points $(z = 0, f = 1)$, $(z = 0.5, f = 1.128)$ and $(z = 1.3, f = 1.971)$. Show that the parameters satisfy

$$\begin{bmatrix} 1 \\ 1.128 \\ 1.971 \end{bmatrix} = \begin{bmatrix} 0 & 1 & 0 \\ 0.5 & 1 & -0.5640 \\ 1.3 & 1 & -2.562 \end{bmatrix} \begin{bmatrix} a \\ b \\ c \end{bmatrix}$$

Find a, b and c and hence the approximating function (use of a computer package is recommended). Check the value $f(1) = 1.543$. (Note that the values were chosen from tables of $\cosh z$.)

The method described here is a simple example of a powerful approximation method.

47 A cantilever beam bends under a uniform load w per unit length and is subject to an axial force P at its free end. For small deflections a numerical approximation to the shape of the beam is given by the set of equations

$$-vy_1 + y_2 \qquad\qquad = -u$$

$$y_1 - vy_2 + y_3 \qquad\quad = -4u$$

$$y_2 - vy_3 + y_4 = -9u$$

$$2y_3 - vy_4 = -16u$$

The deflections are indicated on Figure 5.10. The parameter v is defined as

$$v = 2 + \frac{PL^2}{16EI}$$

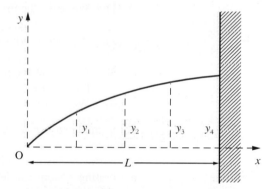

Figure 5.10 Cantilever beam in Question 47.

where EI is the flexural rigidity and L is the length of the beam. The parameter $u = wL^4/32EI$.

Use either Cramer's rule or the adjoint matrix to solve the equations when $v = 3$ and $u = 1$.

Note the immense effort required to solve this very simple problem using these methods. In later sections much more efficient methods will be described. A computer package should be used to check the results.

5.5.2 The solution of linear equations: elimination methods

The idea behind elimination techniques can be seen by considering the solution of two simultaneous equations

$$x + 2y = 4$$

$$2x + y = 5$$

Subtract $2 \times$ (equation 1) from (equation 2) to give

$$x + 2y = 4$$

$$-3y = -3$$

Divide the second equation by -3

$$x + 2y = 4$$

$$y = 1$$

From the second of these equations $y = 1$ and substituting into the first equation gives $x = 2$.

This example illustrates the basic technique for the solution of a set of linear equations by Gaussian elimination, which is very straightforward in principle. However, it needs considerable care to ensure that the calculations are carried out efficiently. Given n linear equations in the variables $x_1, x_2, \ldots, x_n$, we solve in a series of steps:

(1) We solve the first equation for x_1 in terms of $x_2, \ldots, x_n$, and eliminate x_1 from the remaining equations.

(2) We then solve the second equation of the remaining set for x_2 in terms of x_3, $\ldots, x_n$ and eliminate x_2 from the remaining equations.

(3) We repeat the process in turn on $x_3, x_4, \ldots$ until we arrive at a final equation for x_n, which we can then solve.

(4) We substitute back to get in turn $x_{n-1}, x_{n-2}, \ldots, x_1$.

For a small number of variables, say two, three or four, the method is easy to apply, and efficiency is not of the highest priority. In most science and engineering problems we are normally dealing with a large number of variables – a simple stability analysis of a vibrating system can lead to seven or eight variables, and a plate-bending problem could easily give rise to several hundred variables.

As a further example of the basic technique, we solve

$$x_1 + x_2 \qquad = 3 \tag{5.23}$$

$$2x_1 + x_2 + x_3 = 7 \tag{5.24}$$

$$x_1 + 2x_2 + 3x_3 = 14 \tag{5.25}$$

First, we eliminate x_1:

(5.23) gives $\qquad x_1 = 3 - x_2$ $\hfill$ (5.23′)

(5.24) gives $\qquad 2(3 - x_2) + x_2 + x_3 = 7,$ or $-x_2 + x_3 = 1$ $\hfill$ (5.24′)

(5.25) gives $\qquad (3 - x_2) + 2x_2 + 3x_3 = 14,$ or $x_2 + 3x_3 = 11$ $\hfill$ (5.25′)

Secondly we eliminate x_2:

(5.24′) gives $\qquad x_2 = x_3 - 1$ $\hfill$ (5.24″)

(5.25′) gives $\qquad (x_3 - 1) + 3x_3 = 11,$ or $4x_3 = 12$ $\hfill$ (5.25″)

Equation (5.25″) gives $x_3 = 3$; we put this into (5.24″) to obtain $x_2 = 2$; we then put this into (5.23′) to obtain $x_1 = 1$. Thus the values $x_1 = 1$, $x_2 = 2$ and $x_3 = 3$ give a solution to the original problem.

Equations (5.23)–(5.25) in matrix form become

$$\begin{bmatrix} 1 & 1 & 0 \\ 2 & 1 & 1 \\ 1 & 2 & 3 \end{bmatrix} \begin{bmatrix} x_1 \\ x_2 \\ x_3 \end{bmatrix} = \begin{bmatrix} 3 \\ 7 \\ 14 \end{bmatrix}$$

The elimination procedure has reduced the equations to (5.23), (5.24′) and (5.25″), which in matrix form become

$$\begin{bmatrix} 1 & 1 & 0 \\ 0 & -1 & 1 \\ 0 & 0 & 4 \end{bmatrix} \begin{bmatrix} x_1 \\ x_2 \\ x_3 \end{bmatrix} = \begin{bmatrix} 3 \\ 1 \\ 12 \end{bmatrix}$$

Essentially the elimination has brought the equations to **upper-triangular** form (that is, a form in which the matrix of coefficients has zeros in every position below the diagonal), which are then very easy to solve.

Elimination procedures rely on the manipulation of equations or, equivalently, the rows of the matrix equation. There are various **elementary row operations** used which do not alter the solution of the equations:

(a)   multiply a row by a constant,
(b)   interchange any two rows,
(c)   add or subtract one row from another.

To illustrate these, we take the matrix equation

$$\begin{bmatrix} 1 & 1 & 0 \\ 2 & 1 & 1 \\ -1 & 2 & 3 \end{bmatrix} \begin{bmatrix} x_1 \\ x_2 \\ x_3 \end{bmatrix} = \begin{bmatrix} 3 \\ 7 \\ 12 \end{bmatrix}$$

which has the solution $x_1 = 1$, $x_2 = 2$, $x_3 = 3$.

Multiplying the first row by 2 (a row operation of type (a)) yields

$$\begin{bmatrix} 2 & 2 & 0 \\ 2 & 1 & 1 \\ -1 & 2 & 3 \end{bmatrix} \begin{bmatrix} x_1 \\ x_2 \\ x_3 \end{bmatrix} = \begin{bmatrix} 6 \\ 7 \\ 12 \end{bmatrix}$$

Interchanging rows 1 and 3 (a row operation of type (b)) yields

$$\begin{bmatrix} -1 & 2 & 3 \\ 2 & 1 & 1 \\ 1 & 1 & 0 \end{bmatrix} \begin{bmatrix} x_1 \\ x_2 \\ x_3 \end{bmatrix} = \begin{bmatrix} 12 \\ 7 \\ 3 \end{bmatrix}$$

Subtracting row 1 from row 2 (a row operation of type (c)) yields

$$\begin{bmatrix} 1 & 1 & 0 \\ 1 & 0 & 1 \\ -1 & 2 & 3 \end{bmatrix} \begin{bmatrix} x_1 \\ x_2 \\ x_3 \end{bmatrix} = \begin{bmatrix} 3 \\ 4 \\ 12 \end{bmatrix}$$

In each case we see that the solution of the modified equations is still $x_1 = 1$, $x_2 = 2$, $x_3 = 3$.

Elimination procedures use repeated applications of (a), (b) and (c) in some systematic manner until the equations are processed into a required form such as the upper-triangular equations

$$
\begin{bmatrix}
a_{11} & a_{12} & a_{13} & \cdots & a_{1n} \\
0 & a_{22} & a_{23} & \cdots & a_{2n} \\
0 & 0 & a_{33} & \cdots & a_{3n} \\
\vdots & \vdots & \vdots & & \vdots \\
0 & 0 & 0 & \cdots & a_{nn}
\end{bmatrix}
\begin{bmatrix}
x_1 \\ x_2 \\ x_3 \\ \vdots \\ x_n
\end{bmatrix}
=
\begin{bmatrix}
b_1 \\ b_2 \\ b_3 \\ \vdots \\ b_n
\end{bmatrix}
\tag{5.26}
$$

The solution of the equations in upper-triangular form can be written as

$$x_n = b_n/a_{nn}$$

$$x_{n-1} = (b_{n-1} - a_{n-1,n}x_n)/a_{n-1,n-1}$$

$$x_{n-2} = (b_{n-2} - a_{n-2,n}x_n - a_{n-2,n-1}x_{n-1})/a_{n-2,n-2}$$

$$\vdots$$

$$x_1 = (b_1 - a_{1n}x_n - a_{1,n-1}x_{n-1} - \ldots - a_{12}x_2)/a_{11}$$

A pseudocode procedure implementing these equations is shown in Figure 5.11. The elementary row operations and the elimination technique are illustrated in Example 5.29.

Figure 5.11
Procedure to solve the upper-triangular system (5.26).

```
procedure uppertriangular (A, b, n → x)
{A is an n × n matrix, b and x are vectors with n elements}
  x(n)←b(n)/A(n, n)
  for j is n − 1  to 1 by −1 do
    sum←b(j)
    for i is n  to j + 1 by −1 do
      sum←sum − A(j, i) * x(i)
    endfor
    x(j)←sum/A(j, j)
  endfor
endprocedure
```

Example 5.29 Use elementary row operations and elimination to solve the set of linear equations

$$x + 2y + 3z = 10$$

$$-x + y + z = 0$$

$$y - z = 1$$

Solution In matrix form the equations are:
$$\begin{bmatrix} 1 & 2 & 3 \\ -1 & 1 & 1 \\ 0 & 1 & -1 \end{bmatrix} \begin{bmatrix} x \\ y \\ z \end{bmatrix} = \begin{bmatrix} 10 \\ 0 \\ 1 \end{bmatrix}$$

Add row 1 to row 2:
$$\begin{bmatrix} 1 & 2 & 3 \\ 0 & 3 & 4 \\ 0 & 1 & -1 \end{bmatrix} \begin{bmatrix} x \\ y \\ z \end{bmatrix} = \begin{bmatrix} 10 \\ 10 \\ 1 \end{bmatrix}$$

Divide row 2 by 3:
$$\begin{bmatrix} 1 & 2 & 3 \\ 0 & 1 & \frac{4}{3} \\ 0 & 1 & -1 \end{bmatrix} \begin{bmatrix} x \\ y \\ z \end{bmatrix} = \begin{bmatrix} 10 \\ \frac{10}{3} \\ 1 \end{bmatrix}$$

Subtract row 2 from row 3:
$$\begin{bmatrix} 1 & 2 & 3 \\ 0 & 1 & \frac{4}{3} \\ 0 & 0 & -\frac{7}{3} \end{bmatrix} \begin{bmatrix} x \\ y \\ z \end{bmatrix} = \begin{bmatrix} 10 \\ \frac{10}{3} \\ -\frac{7}{3} \end{bmatrix}$$

Divide row 3 by $(-\frac{7}{3})$:
$$\begin{bmatrix} 1 & 2 & 3 \\ 0 & 1 & \frac{4}{3} \\ 0 & 0 & 1 \end{bmatrix} \begin{bmatrix} x \\ y \\ z \end{bmatrix} = \begin{bmatrix} 10 \\ \frac{10}{3} \\ 1 \end{bmatrix}$$

The equations are now in a standard upper-triangular form for the application of the back substitution procedure described formally in Figure 5.11.

From the third row $\quad z = 1$

From the second row $\quad y = \frac{10}{3} - \frac{4}{3} z = 2$

From the first row $\quad x = 10 - 2y - 3z = 3$

It remains to undertake the operations in the example in a routine and logical manner to make the method into one of the most powerful techniques, called **elimination methods**, available for the solution of sets of linear equations. The method is available on all computer packages. Such packages are excellent at undertaking the rather tedious arithmetic and some will even illustrate the computational detail also. They are well worth mastering.

Tridiagonal or Thomas algorithm

Because of the ease of solution of upper-triangular systems, many methods use the general strategy of reducing the equations to this form. As an example of this strategy, we shall look at a **tridiagonal system**, which takes the form

$$
\begin{bmatrix}
a_1 & b_1 & 0 & 0 & 0 & 0 & \cdots & 0 \\
c_2 & a_2 & b_2 & 0 & 0 & 0 & \cdots & 0 \\
0 & c_3 & a_3 & b_3 & 0 & 0 & \cdots & 0 \\
0 & 0 & c_4 & a_4 & b_4 & 0 & \cdots & 0 \\
 & & & & & & & \vdots \\
\vdots & & & & & & & 0 \\
0 & \cdots & & 0 & c_{n-1} & a_{n-1} & b_{n-1} \\
0 & \cdots & & 0 & 0 & c_n & a_n
\end{bmatrix}
\begin{bmatrix}
x_1 \\ x_2 \\ \\ \vdots \\ \\ \\ x_{n-1} \\ x_n
\end{bmatrix}
=
\begin{bmatrix}
d_1 \\ d_2 \\ \\ \vdots \\ \\ \\ d_{n-1} \\ d_n
\end{bmatrix}
\tag{5.27}
$$

or

$$
\begin{aligned}
a_1 x_1 + b_1 x_2 & = d_1 \\
c_2 x_1 + a_2 x_2 + b_2 x_3 & = d_2 \\
c_3 x_2 + a_3 x_3 + b_3 x_4 & = d_3 \\
\ddots \qquad \ddots \qquad & \;\; \vdots \\
c_n x_{n-1} + a_n x_n & = d_n
\end{aligned}
$$

First we eliminate x_1:

$$
\begin{aligned}
x_1 + b_1' x_2 & = d_1' \\
a_2' x_2 + b_2 x_3 & = d_2' \\
c_3 x_2 + a_3 x_3 + b_3 x_4 & = d_3
\end{aligned}
$$

and so on

where

$$
b_1' = \frac{b_1}{a_1}, \quad d_1' = \frac{d_1}{a_1}, \quad a_2' = a_2 - c_2 b_1' \quad \text{and} \quad d_2' = d_2 - c_2 d_1'
$$

Next we eliminate x_2:

$$
\begin{aligned}
x_1 + b_1' x_2 & = d_1' \\
x_2 + b_2'' x_3 & = d_2'' \\
a_3'' x_3 + b_3 x_4 & = d_3'' \\
c_4 x_3 + a_4 x_4 + b_4 x_5 & = d_4
\end{aligned}
$$

and so on

where

$$
b_2'' = \frac{b_2}{a_2'}, \quad d_2'' = \frac{d_2'}{a_2'}, \quad a_3'' = a_3 - c_3 b_2'' \quad \text{and} \quad d_3'' = d_3 - c_3 d_2''
$$

We can proceed to eliminate all the variables down to the nth. We have then converted the problem to an upper-triangular form, which can be solved by the procedure in Figure 5.11. A pseudocode procedure to solve (5.27) called the **tridiagonal or Thomas**

Figure 5.12
Tridiagonal or Thomas
algorithm for the
solution of (5.27).

```
procedure tridiagonal (a, b, c, d, n→a, b, d, x)
    {a, b, c, d and x are vectors with n elements}
    for i is 1 to n − 1 do
    b(i)←b(i)/a(i)
    d(i)←d(i)/a(i)
    a(i + 1)←a(i + 1) − c(i + 1) * b(i)
    d(i + 1)←d(i + 1) − c(i + 1) * d(i)
    endfor {elimination stage}
        x(n)←d(n)/a(n)
        for i is n − 1 to 1 by −1 do
            x(i)←d(i) − b(i) * x(i + 1)
        endfor {back substitution}
endprocedure
```

algorithm is shown in Figure 5.12. The algorithm is written so that each primed value, when it is computed, replaces the previous value. Similarly the double-primed values replace the primed values. This is called **overwriting**, and reduces the storage required to implement the algorithm on a computer. It should be noted, however, that the algorithm is written for clarity and not minimum storage or maximum efficiency. The algorithm is very widely used; it is exceptionally fast and requires very little storage.

Example 5.30   Use the tridiagonal procedure to solve

$$
\begin{bmatrix} 2 & 1 & 0 & 0 \\ 1 & 2 & 1 & 0 \\ 0 & 1 & 2 & 1 \\ 0 & 0 & 1 & 2 \end{bmatrix} \begin{bmatrix} x \\ y \\ z \\ t \end{bmatrix} = \begin{bmatrix} 1 \\ 1 \\ 1 \\ -2 \end{bmatrix}
$$

Solution   The sequence of matrices is given by

$$
\begin{bmatrix} 1 & \frac{1}{2} & 0 & 0 \\ 0 & \frac{3}{2} & 1 & 0 \\ 0 & 1 & 2 & 1 \\ 0 & 0 & 1 & 2 \end{bmatrix} \begin{bmatrix} x \\ y \\ z \\ t \end{bmatrix} = \begin{bmatrix} \frac{1}{2} \\ \frac{1}{2} \\ 1 \\ -2 \end{bmatrix}, \quad
\begin{bmatrix} 1 & \frac{1}{2} & 0 & 0 \\ 0 & 1 & \frac{2}{3} & 0 \\ 0 & 0 & \frac{4}{3} & 1 \\ 0 & 0 & 1 & 2 \end{bmatrix} \begin{bmatrix} x \\ y \\ z \\ t \end{bmatrix} = \begin{bmatrix} \frac{1}{2} \\ \frac{1}{3} \\ \frac{2}{3} \\ -2 \end{bmatrix}
$$

$$
\begin{bmatrix} 1 & \frac{1}{2} & 0 & 0 \\ 0 & 1 & \frac{2}{3} & 0 \\ 0 & 0 & 1 & \frac{3}{4} \\ 0 & 0 & 0 & \frac{5}{4} \end{bmatrix} \begin{bmatrix} x \\ y \\ z \\ t \end{bmatrix} = \begin{bmatrix} \frac{1}{2} \\ \frac{1}{3} \\ \frac{1}{2} \\ -\frac{5}{2} \end{bmatrix}
$$

The elimination stage is now complete, and we substitute back to give

$$ t = -2, \; z = \tfrac{1}{2} - \tfrac{3}{4}t = 2, \; y = \tfrac{1}{3} - \tfrac{2}{3}z = -1 \quad \text{and} \quad x = \tfrac{1}{2} - \tfrac{1}{2}y = 1 $$

so that the complete solution is $x = 1$, $y = -1$, $z = 2$, $t = -2$.

Although the Thomas algorithm is efficient, the procedure in Figure 5.12 is not fool-proof, as illustrated by the simple example

$$
\begin{bmatrix} -1 & 1 & 0 \\ 1 & -1 & 1 \\ 0 & 1 & -1 \end{bmatrix} \begin{bmatrix} x \\ y \\ z \end{bmatrix} = \begin{bmatrix} -1 \\ 2 \\ 1 \end{bmatrix}
$$

After the first step we have

$$
\begin{bmatrix} 1 & -1 & 0 \\ 0 & 0 & 1 \\ 0 & 1 & -1 \end{bmatrix} \begin{bmatrix} x \\ y \\ z \end{bmatrix} = \begin{bmatrix} 1 \\ 1 \\ 1 \end{bmatrix}
$$

The next step divides by the diagonal element a_{22}. Since this element is zero, the method crashes to a halt. There is a perfectly good solution, however, since simply interchanging the last two rows,

$$
\begin{bmatrix} 1 & -1 & 0 \\ 0 & 1 & -1 \\ 0 & 0 & 1 \end{bmatrix} \begin{bmatrix} x \\ y \\ z \end{bmatrix} = \begin{bmatrix} 1 \\ 1 \\ 1 \end{bmatrix}
$$

gives an upper-triangular matrix with the obvious solution $z = 1$, $y = 2$, $x = 3$. It is clear that checks must be put in the algorithm to prevent such failures.

Gaussian elimination

Since most matrix equations are not tridiagonal, we should like to extend the idea to a general matrix

$$
\begin{bmatrix} a_{11} & a_{12} & a_{13} & \dots & a_{1n} \\ a_{21} & a_{22} & a_{23} & \dots & a_{2n} \\ \vdots & \vdots & \vdots & & \vdots \\ a_{n1} & a_{n2} & a_{n3} & \dots & a_{nn} \end{bmatrix} \begin{bmatrix} x_1 \\ x_2 \\ \vdots \\ x_n \end{bmatrix} = \begin{bmatrix} b_1 \\ b_2 \\ \vdots \\ b_n \end{bmatrix}
\tag{5.28}
$$

The result of doing this is a method known as **Gaussian elimination**. It is a little more involved than the Thomas algorithm. First we eliminate x_1:

$$
\begin{bmatrix} 1 & a'_{12} & a'_{13} & \dots & a'_{1n} \\ 0 & a'_{22} & a'_{23} & \dots & a'_{2n} \\ 0 & a'_{32} & a'_{33} & \dots & a'_{3n} \\ \vdots & \vdots & \vdots & & \vdots \\ 0 & a'_{n2} & a'_{n3} & \dots & a'_{nn} \end{bmatrix} \begin{bmatrix} x_1 \\ x_2 \\ \vdots \\ x_n \end{bmatrix} = \begin{bmatrix} b'_1 \\ b'_2 \\ \\ \vdots \\ b'_n \end{bmatrix}
$$

where

$$a'_{12} = \frac{a_{12}}{a_{11}}, \quad a'_{13} = \frac{a_{13}}{a_{11}}, \quad \ldots, \quad a'_{1n} = \frac{a_{1n}}{a_{11}}, \quad b'_1 = \frac{b_1}{a_{11}}$$

$$a'_{22} = a_{22} - a_{21}a'_{12}, \quad a'_{23} = a_{23} - a_{21}a'_{13}, \quad \ldots, \quad b'_2 = b_2 - a_{21}b'_1$$

$$a'_{32} = a_{32} - a_{31}a'_{12}, \quad a'_{33} = a_{33} - a_{31}a'_{13}, \quad \ldots, \quad b'_3 = b_3 - a_{31}b'_1$$

and so on.

Generally these can be written as

$$a'_{1j} = \frac{a_{1j}}{a_{11}}, \quad j = 1, \ldots, n \quad b'_1 = \frac{b_1}{a_{11}}$$

$$\left.\begin{array}{l} a'_{ij} = a_{ij} - a_{i1}a'_{ij} \\ b'_i = b_i - a_{i1}b'_1 \end{array}\right\}, \quad i = 2, \ldots, n \quad \text{and} \quad j = 2, \ldots, n$$

We now operate in an identical manner on the $(n-1) \times (n-1)$ submatrix, formed by ignoring row 1 and column 1, and repeat the process until the equations are of upper-triangular form. At the general step in the algorithm the equations will take the form

$$\begin{bmatrix} 1 & * & * & * & & \cdots & * \\ 0 & 1 & * & * & & \cdots & * \\ 0 & 0 & 1 & * & & \cdots & * \\ 0 & 0 & & \ddots & & & \vdots \\ \vdots & \vdots & & \ddots & & & \\ 0 & \cdots & & 0 & 1 & & \\ 0 & \cdots & & & 0 & a_{ii} & \cdots & a_{in} \\ \vdots & & & & 0 & \vdots & \ddots & \vdots \\ & & & & & \vdots & & \\ 0 & \cdots & & & 0 & a_{ni} & \cdots & a_{nn} \end{bmatrix} \begin{bmatrix} x_1 \\ x_2 \\ \\ \\ \vdots \\ \\ \\ \\ \\ x_n \end{bmatrix} = \begin{bmatrix} * \\ * \\ \\ \\ \vdots \\ \\ \\ \\ \\ * \end{bmatrix}$$

(5.29)

Again overwriting avoids the need for introducing primed symbols; the algorithm is shown in Figure 5.13.

Figure 5.13
Elimination procedure for (5.28).

```
procedure eliminate (A, b, n→A, b)
    {A is an n × n matrix and b is a vector with n elements}
    for i is 1 to n − 1 do
        {a segment will be inserted here later}
        b(i)←b(i)/A(i, i)
        for j is i to n do
            A(i, j)←A(i, j)/A(i, i)
            for k is i + 1 to n do
                A(k, j)←A(k, j) − A(k, i)*A(i, j)
            endfor
        endfor
        for k is i + 1 to n do
            b(k)←b(k) − A(k, i)*b(i)
        endfor
    endfor
endprocedure
```

The general Gaussian elimination procedure would then put the eliminate and upper-triangular procedures together in a program

```
read (file, A, b, n)
        eliminate (A, b, n→A, b)
        uppertriangular (A, b, n→x)
write (vdu, x)
```

This algorithm, sharing the merits of the Thomas algorithm, is very widely used by engineers to solve linear equations.

In Example 5.29 the basic elimination technique was illustrated but now the two procedures *eliminate* and *uppertriangular* have reduced the method to one of routine. The major problem is to perform the arithmetic accurately.

Example 5.31

Using elimination and back substitution solve the equations

$$\begin{bmatrix} 2 & 3 & 4 \\ 1 & 2 & 3 \\ 1 & 4 & 5 \end{bmatrix} \begin{bmatrix} x \\ y \\ z \end{bmatrix} = \begin{bmatrix} 1 \\ 1 \\ 2 \end{bmatrix}$$

Solution

From the method in Figure 5.13 the steps are

Divide first row by 2:
$$\begin{bmatrix} 1 & \frac{3}{2} & 2 \\ 1 & 2 & 3 \\ 1 & 4 & 5 \end{bmatrix} \begin{bmatrix} x \\ y \\ z \end{bmatrix} = \begin{bmatrix} \frac{1}{2} \\ 1 \\ 2 \end{bmatrix}$$

Subtract row 1 from row 2 and row 3:
$$\begin{bmatrix} 1 & \frac{3}{2} & 2 \\ 0 & \frac{1}{2} & 1 \\ 0 & \frac{5}{2} & 3 \end{bmatrix} \begin{bmatrix} x \\ y \\ z \end{bmatrix} = \begin{bmatrix} \frac{1}{2} \\ \frac{1}{2} \\ \frac{3}{2} \end{bmatrix}$$

Divide second row by $\frac{1}{2}$:
$$\begin{bmatrix} 1 & \frac{3}{2} & 2 \\ 0 & 1 & 2 \\ 0 & \frac{5}{2} & 3 \end{bmatrix} \begin{bmatrix} x \\ y \\ z \end{bmatrix} = \begin{bmatrix} \frac{1}{2} \\ 1 \\ \frac{3}{2} \end{bmatrix}$$

Subtract $\frac{5}{2} \times$ (row 2) from row 3:
$$\begin{bmatrix} 1 & \frac{3}{2} & 2 \\ 0 & 1 & 2 \\ 0 & 0 & -2 \end{bmatrix} \begin{bmatrix} x \\ y \\ z \end{bmatrix} = \begin{bmatrix} \frac{1}{2} \\ 1 \\ -1 \end{bmatrix}$$

Divide row 3 by (−2):
$$\begin{bmatrix} 1 & \frac{3}{2} & 2 \\ 0 & 1 & 2 \\ 0 & 0 & 1 \end{bmatrix} \begin{bmatrix} x \\ y \\ z \end{bmatrix} = \begin{bmatrix} \frac{1}{2} \\ 1 \\ \frac{1}{2} \end{bmatrix}$$

The elimination procedure is now complete and the back substitution (from Figure 5.11) is applied to the upper-triangular matrix.

From the third row $\quad\quad z = \frac{1}{2}$

From the second row $\quad\quad y = 1 - 2z = 0$

From the first row $\quad\quad\quad x = \frac{1}{2} - \frac{3}{2}y - 2z = -\frac{1}{2}$

so the solution is $x = -\frac{1}{2}$, $y = 0$, $z = \frac{1}{2}$.

Example 5.32 Solve

$$
\begin{bmatrix} 1 & 2 & 3 & 1 \\ 2 & 1 & 1 & 1 \\ 1 & 2 & 1 & 0 \\ 0 & 1 & 1 & 2 \end{bmatrix}
\begin{bmatrix} x \\ y \\ z \\ t \end{bmatrix}
=
\begin{bmatrix} 5 \\ 3 \\ 4 \\ 0 \end{bmatrix}
$$

Solution The elimination sequence is

$$
\begin{bmatrix} 1 & 2 & 3 & 1 \\ 0 & -3 & -5 & -1 \\ 0 & 0 & -2 & -1 \\ 0 & 1 & 1 & 2 \end{bmatrix}
\begin{bmatrix} x \\ y \\ z \\ t \end{bmatrix}
=
\begin{bmatrix} 5 \\ -7 \\ -1 \\ 0 \end{bmatrix},
\quad
\begin{bmatrix} 1 & 2 & 3 & 1 \\ 0 & 1 & \frac{5}{3} & \frac{1}{3} \\ 0 & 0 & -2 & -1 \\ 0 & 0 & -\frac{2}{3} & \frac{5}{3} \end{bmatrix}
\begin{bmatrix} x \\ y \\ z \\ t \end{bmatrix}
=
\begin{bmatrix} 5 \\ \frac{7}{3} \\ -1 \\ -\frac{7}{3} \end{bmatrix}
$$

$$
\begin{bmatrix} 1 & 2 & 3 & 1 \\ 0 & 1 & \frac{5}{3} & \frac{1}{3} \\ 0 & 0 & 1 & \frac{1}{2} \\ 0 & 0 & 0 & 2 \end{bmatrix}
\begin{bmatrix} x \\ y \\ z \\ t \end{bmatrix}
=
\begin{bmatrix} 5 \\ \frac{7}{3} \\ \frac{1}{2} \\ -2 \end{bmatrix}
$$

and application of the upper-triangular procedure gives $t = -1$, $z = 1$, $y = 1$ and $x = 1$.

It is clear again that if, in the algorithm shown in Figure 5.13, A(i, i) is zero at any time, the method will fail. It is, in fact, also found to be beneficial to the stability of the method to have A(i, i) as large as possible. Thus in (5.29) it is usual to perform a 'partial pivoting' so that the largest value in the column, $\max_{i\leqslant p\leqslant n} |A(p, i)|$, is chosen and the equations are swapped around to make this element the pivot. In Figure 5.13 the following segment of program would need to be inserted at the point indicated:

{find $\max_{i\leqslant p\leqslant n} |A(p, i)|$}

{interchange row i with row p_{max}}

In practical computer implementations of the algorithm the elements of the rows would not be swapped explicitly. Instead, a pointer system would be used to implement a technique known as **indirect addressing**, which allows much faster computations. The interested reader is referred to texts on computer programming techniques for a full explanation of this method.

In a hand-computation version of this elimination procedure there are methods that maintain running checks and minimize the amount of writing. In this book the emphasis is on a computer implementation, and the hand computations are provided to illustrate the principle of the method. It is a powerful learning technique to write your own programs, but the practising professional engineer will normally use procedures from a computer software library, where these are available.

 In MATLAB the instruction [L, U] = lu(A) provides in U the eliminated matrix. However the method used in MATLAB always uses partial pivoting and the method only works for a square matrix. The instruction $A\backslash b$ will give the solution to the matrix equation in one step. The MAPLE package can deal with any size matrix, so the right-hand side of the matrix equation should be appended to A and hence included in the elimination, and the instruction gausselim(A); provides the elimination but without partial pivoting. The instruction gaussjord(A); uses a much more subtle elimination process – see any advanced textbook on numerical linear algebra – and gives the solution in the most convenient form.

Example 5.33   Solve the matrix equation

$$\begin{bmatrix} 1 & 2 & 3 & 1 \\ 2 & 1 & 1 & 1 \\ 1 & 3 & 1 & 0 \\ 0 & 1 & 1 & 2 \end{bmatrix} \begin{bmatrix} x \\ y \\ z \\ t \end{bmatrix} = \begin{bmatrix} 4 \\ 3 \\ 2 \\ 1 \end{bmatrix}$$

by Gaussian elimination with partial pivoting.

Solution   The sequence is as follows. We first interchange rows 1 and 2 and eliminate:

$$\begin{bmatrix} 2 & 1 & 1 & 1 \\ 1 & 2 & 3 & 1 \\ 1 & 3 & 1 & 0 \\ 0 & 1 & 1 & 2 \end{bmatrix} \begin{bmatrix} x \\ y \\ z \\ t \end{bmatrix} = \begin{bmatrix} 3 \\ 4 \\ 2 \\ 1 \end{bmatrix} \rightarrow \begin{bmatrix} 1 & \frac{1}{2} & \frac{1}{2} & \frac{1}{2} \\ 0 & \frac{3}{2} & \frac{5}{2} & \frac{1}{2} \\ 0 & \frac{5}{2} & \frac{1}{2} & -\frac{1}{2} \\ 0 & 1 & 1 & 2 \end{bmatrix} \begin{bmatrix} x \\ y \\ z \\ t \end{bmatrix} = \begin{bmatrix} \frac{3}{2} \\ \frac{5}{2} \\ \frac{1}{2} \\ 1 \end{bmatrix}$$

We then interchange rows 2 and 3 and eliminate:

$$\begin{bmatrix} 1 & \frac{1}{2} & \frac{1}{2} & \frac{1}{2} \\ 0 & \frac{5}{2} & \frac{1}{2} & -\frac{1}{2} \\ 0 & \frac{3}{2} & \frac{5}{2} & \frac{1}{2} \\ 0 & 1 & 1 & 2 \end{bmatrix} \begin{bmatrix} x \\ y \\ z \\ t \end{bmatrix} = \begin{bmatrix} \frac{3}{2} \\ \frac{1}{2} \\ \frac{5}{2} \\ 1 \end{bmatrix} \rightarrow \begin{bmatrix} 1 & \frac{1}{2} & \frac{1}{2} & \frac{1}{2} \\ 0 & 1 & \frac{1}{5} & -\frac{1}{5} \\ 0 & 0 & \frac{11}{5} & \frac{4}{5} \\ 0 & 0 & \frac{4}{5} & \frac{11}{5} \end{bmatrix} \begin{bmatrix} x \\ y \\ z \\ t \end{bmatrix} = \begin{bmatrix} \frac{3}{2} \\ \frac{1}{5} \\ \frac{11}{5} \\ \frac{4}{5} \end{bmatrix}$$

There is no need to interchange rows at this stage, and the elimination proceeds immediately:

$$\begin{bmatrix} 1 & \frac{1}{2} & \frac{1}{2} & \frac{1}{2} \\ 0 & 1 & \frac{1}{5} & -\frac{1}{5} \\ 0 & 0 & 1 & \frac{4}{11} \\ 0 & 0 & 0 & \frac{21}{11} \end{bmatrix} \begin{bmatrix} x \\ y \\ z \\ t \end{bmatrix} = \begin{bmatrix} \frac{3}{2} \\ \frac{1}{5} \\ 1 \\ 0 \end{bmatrix}$$

Back substitution now gives $t = 0$, $z = 1$, $y = 0$ and $x = 1$.

Ill-conditioning

Elimination methods are not without their difficulties, and the following example will highlight some of them.

Example 5.34

Solve, by elimination, the equations

(a) $\begin{bmatrix} 2 & 1 \\ 1 & 0.5001 \end{bmatrix} \begin{bmatrix} x \\ y \end{bmatrix} = \begin{bmatrix} 0.3 \\ 0.6 \end{bmatrix}$ (b) $\begin{bmatrix} 2 & 1 \\ 1 & 0.4999 \end{bmatrix} \begin{bmatrix} x \\ y \end{bmatrix} = \begin{bmatrix} 0.3 \\ 0.6 \end{bmatrix}$

Solution

Keeping the calculations parallel,

(a) $\begin{bmatrix} 1 & 0.5 \\ 1 & 0.5001 \end{bmatrix} \begin{bmatrix} x \\ y \end{bmatrix} = \begin{bmatrix} 0.15 \\ 0.6 \end{bmatrix}$ (b) $\begin{bmatrix} 1 & 0.5 \\ 1 & 0.4999 \end{bmatrix} \begin{bmatrix} x \\ y \end{bmatrix} = \begin{bmatrix} 0.15 \\ 0.6 \end{bmatrix}$

$\begin{bmatrix} 1 & 0.5 \\ 0 & 0.0001 \end{bmatrix} \begin{bmatrix} x \\ y \end{bmatrix} = \begin{bmatrix} 0.15 \\ 0.45 \end{bmatrix}$ $\begin{bmatrix} 1 & 0.5 \\ 0 & -0.0001 \end{bmatrix} \begin{bmatrix} x \\ y \end{bmatrix} = \begin{bmatrix} 0.15 \\ 0.45 \end{bmatrix}$

with solution   with solution

$y = 4500,$ $x = -2249.85$ $y = -4500,$ $x = 2250.15$

In Example 5.34 simple equations that have only marginally different coefficients have wildly different solutions. This situation is totally unsatisfactory, and must be analysed carefully. To do so in full detail is not appropriate here, but the problem is clearly connected with taking differences of numbers that are almost equal: $0.5001 - 0.5 = 0.0001$.

Systems of equations that exhibit such awkward behaviour are called **ill-conditioned**. It is not straightforward to identify ill-conditioning in matrices involving many variables, but an example will illustrate the difficulties in the two-variable case. Suppose we solve

$$2x + y = 0.3$$

$$x - \alpha y = 0$$

where $\alpha = 1 \pm 0.05$ has some error in its value. We easily obtain $x = 0.3\alpha/(1 + 2\alpha)$ and $y = 0.3/(1 + 2\alpha)$, and putting in the range of α values we get $0.0983 \leqslant x \leqslant 0.1016$ and $0.0968 \leqslant y \leqslant 0.1034$. Thus an error of $\pm 5\%$ in the value of α produces an error of $\pm 2\%$ in x and an error of $\pm 3\%$ in y.

If we now try to solve

$$2x + \ y = 0.3$$

$$x + \alpha y = 0.3$$

where $\alpha = 0.4 \pm 0.05$, then we get the solution $x = 0.3(1 - \alpha)/(1 - 2\alpha)$, $y = -0.3/(1 - 2\alpha)$. Putting in the range of α values now gives $0.65 \leqslant x \leqslant 1.65$ and $-3 \leqslant y \leqslant -1$, and an error of $\pm 12\%$ in the value of α produces errors in x and y of up to 100%.

Figure 5.14 illustrates these equations geometrically. We see that a small change in the slope of the line $x - \alpha y = 0$ makes only a small difference in the solution. However, changing the slope of the line $x + \alpha y = 0.3$ makes a large difference, because the lines are nearly parallel. Identifying such behaviour for higher-dimensional problems is not at all easy. Sets of equations of this kind do occur in engineering contexts, so the difficulties outlined here should be appreciated. In each of the ill-conditioned cases we have studied, the determinant of the system is 'small':

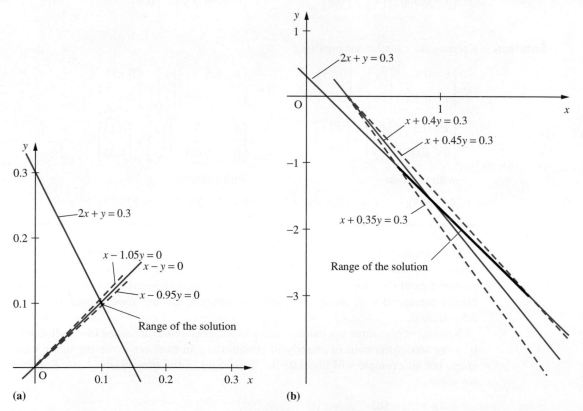

Figure 5.14 Solution of (a) $2x + y = 0.3$, $x - \alpha y = 0$ with $\alpha = 1 \pm 0.05$; and (b) $2x + y = 0.3$, $x + \alpha y = 0.3$ with $\alpha = 0.4 \pm 0.05$. The heavy black lines indicate the ranges of the solutions.

$$\begin{vmatrix} 2 & 1 \\ 1 & 0.5001 \end{vmatrix} = 0.0002, \qquad \begin{vmatrix} 2 & 1 \\ 1 & 0.4999 \end{vmatrix} = -0.0002$$

$$\begin{vmatrix} 2 & 1 \\ 1 & 0.4 \pm 0.05 \end{vmatrix} = -0.2 \pm 0.1$$

Thus the equations are 'nearly singular' – and this is one means of identifying the problem. However, the reader should refer to a more advanced book on numerical analysis to see how to identify and deal with ill-conditioning in the general case.

5.5.3 Exercises

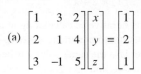

48 Use elimination with and/or without partial pivoting, to solve the equations

(a) $\begin{bmatrix} 1 & 3 & 2 \\ 2 & 1 & 4 \\ 3 & -1 & 5 \end{bmatrix} \begin{bmatrix} x \\ y \\ z \end{bmatrix} = \begin{bmatrix} 1 \\ 2 \\ 1 \end{bmatrix}$

(b) $\begin{bmatrix} 0 & 1 & 1 \\ 3 & -1 & 1 \\ 1 & 1 & -3 \end{bmatrix} \begin{bmatrix} x \\ y \\ z \end{bmatrix} = \begin{bmatrix} 6 \\ -7 \\ -13 \end{bmatrix}$

(c) $\begin{bmatrix} 1 & 2 & 4 \\ -6 & 2 & 10 \\ 2 & 8 & 7 \end{bmatrix} \begin{bmatrix} x \\ y \\ z \end{bmatrix} = \begin{bmatrix} 0 \\ 1 \\ 0 \end{bmatrix}$

49 Solve the equations

$$\begin{aligned} 4x - y &= 2 \\ -x + 4y - z &= 5 \\ -y + 4z - t &= 3 \\ -z + 4t &= 10 \end{aligned}$$

using the tridiagonal algorithm.

50 Solve the equations

$$\begin{aligned} 4x - y \quad - t &= -4 \\ -x + 4y - z \quad &= 1 \\ -y + 4z - t &= 4 \\ -x \quad - z + 4t &= 10 \end{aligned}$$

using Gaussian elimination.

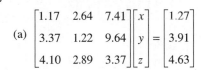

51 Solve, using Gaussian elimination with partial pivoting, the following equations:

(a) $\begin{bmatrix} 1.17 & 2.64 & 7.41 \\ 3.37 & 1.22 & 9.64 \\ 4.10 & 2.89 & 3.37 \end{bmatrix} \begin{bmatrix} x \\ y \\ z \end{bmatrix} = \begin{bmatrix} 1.27 \\ 3.91 \\ 4.63 \end{bmatrix}$

(b) $\begin{bmatrix} 3.21 & 4.18 & -2.31 \\ -4.17 & 3.63 & 4.20 \\ 1.88 & -8.14 & 0.01 \end{bmatrix} \begin{bmatrix} x \\ y \\ z \end{bmatrix} = \begin{bmatrix} 3.27 \\ -1.21 \\ 4.88 \end{bmatrix}$

(c) $\begin{bmatrix} 1 & 7 & 2 & -1 \\ 11 & 4 & -3 & 9 \\ 7 & 6 & 4 & -2 \\ 5 & 8 & -5 & 3 \end{bmatrix} \begin{bmatrix} x \\ y \\ z \\ t \end{bmatrix} = \begin{bmatrix} 12 \\ -12 \\ 7 \\ -7 \end{bmatrix}$

52 The two almost identical matrix equations are given

$$\begin{bmatrix} 0.11 & 0.19 & 0.10 \\ 0.49 & -0.31 & 0.21 \\ 1.55 & -0.70 & 0.70 \end{bmatrix} \begin{bmatrix} x \\ y \\ z \end{bmatrix} = \begin{bmatrix} 1 \\ 1 \\ 1 \end{bmatrix} \quad \text{and}$$

$$\begin{bmatrix} 0.11 & 0.19 & 0.10 \\ 0.49 & -0.31 & 0.21 \\ 1.55 & -0.70 & 0.71 \end{bmatrix} \begin{bmatrix} x \\ y \\ z \end{bmatrix} = \begin{bmatrix} 1 \\ 1 \\ 1 \end{bmatrix}$$

Use MATLAB or MAPLE to show that the solutions are wildly different. Evaluate the determinants of the two 3×3 matrices.

53 Show that a tridiagonal matrix can be written in the form

$$
\begin{bmatrix}
a_1 & b_1 & & & & \\
c_2 & a_2 & b_2 & & \text{\LARGE 0} & \\
 & c_3 & a_3 & b_3 & & \\
 & & \ddots & \ddots & \ddots & \\
\text{\LARGE 0} & & c_{n-1} & a_{n-1} & b_{n-1} \\
 & & & & c_n & a_n
\end{bmatrix}
$$

$$
=
\begin{bmatrix}
l_{11} & & & & \\
l_{21} & l_{22} & & \text{\LARGE 0} & \\
 & l_{32} & l_{33} & & \\
 & & \ddots & \ddots & \\
\text{\LARGE 0} & & & l_{n,n-1} & l_{nn}
\end{bmatrix}
$$

$$
\times
\begin{bmatrix}
1 & u_{12} & & & & \\
 & 1 & u_{23} & & \text{\LARGE 0} & \\
 & & 1 & u_{34} & & \\
 & & & \ddots & \ddots & \\
 & & & & 1 & u_{n-1,n} \\
\text{\LARGE 0} & & & & & 1
\end{bmatrix}
$$

A matrix that has zeros in every position below the diagonal is called an **upper-triangular matrix** and one with zeros everywhere above the diagonal is called a **lower-triangular matrix**. A matrix that only has non-zero elements in certain diagonal lines is called a **banded matrix**. In this case we have shown that a tridiagonal matrix can be written as the product of a lower-triangular banded matrix and an upper-triangular banded matrix.

54 The cantilever beam in Question 47 (Exercises 5.5.1) and illustrated in Figure 5.10 was solved using very inefficient methods. Solve the same equations using the Thomas algorithm.

If the displacements are not small then the equations are more complicated. They take the form

$$
\left.
\begin{aligned}
-v_1 y_1 + y_2 &= -uw_1 \\
y_1 - v_2 y_2 + y_3 &= -4uw_2 \\
y_2 - v_3 y_3 + y_4 &= -9uw_3 \\
2y_3 - v_4 y_4 &= -16u
\end{aligned}
\right\} \quad \textbf{(5.30)}
$$

where

$$
w_1 = \left[1 + 4\left(\frac{y_2}{L} \right)^2 \right]^{3/2}
$$

$$
w_2 = \left[1 + 4\left(\frac{y_3 - y_1}{L} \right)^2 \right]^{3/2}
$$

and

$$
w_3 = \left[1 + 4\left(\frac{y_4 - y_2}{L} \right)^2 \right]^{3/2}
$$

and putting $k = PL^2/16EI$

$$
v_1 = 2 + kw_1, \quad v_2 = 2 + kw_2
$$
$$
v_3 = 2 + kw_3, \quad v_4 = 2 + k
$$

Solve the equations iteratively, calculating w_i and v_i from the previous iteration and then solving the tridiagonal scheme. Use the same values as before, namely $k = 1$ and $u = 1$, and take $L = 2$. A full solution of this exercise will require the use of a computer package such as MATLAB.

55 A wire is loaded with equal weights W at nine uniformly spaced points as illustrated in Figure 5.15. The wire is sufficiently taut that the tension T may be considered to be constant. The end points are at the same level, so that $u_0 = u_{10} = 0$ and the system is symmetrical about its midpoint. The equations to determine the displacements u_i are

$$
\begin{aligned}
W &= (T/d)(2u_1 - u_2) \\
W &= (T/d)(-u_1 + 2u_2 - u_3) \\
W &= (T/d)(-u_2 + 2u_3 - u_4) \\
W &= (T/d)(-u_3 + 2u_4 - u_5) \\
W &= (T/d)(- 2u_4 + 2u_5)
\end{aligned}
$$

Taking $Wd/T = l$, calculate u_i/l for $i = 1, \ldots, 5$.

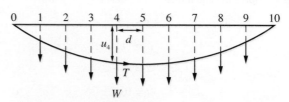

Figure 5.15 Loaded wire.

56

A ladder network is shown in Figure 5.16. The driver is an a.c. voltage of $E = E_0 e^{j\omega t}$ and the currents are taken to be $I_p = Z_p e^{j\omega t}$. The equations satisfied by the Z_p are

$$jE_0\omega = -\tfrac{1}{2}LZ_0\omega^2 + \frac{1}{C}(Z_0 - Z_1)$$

$$0 = -LZ_p\omega^2 + \frac{1}{C}(-Z_{p-1} + 2Z_p - Z_{p+1})$$

$$\text{for } p = 1, \ldots, n-1$$

$$0 = -\tfrac{1}{2}LZ_n\omega^2 + \frac{1}{C}(Z_n - Z_{n-1})$$

Take $n = 3$ and solve for Z_3. Evaluate the effective resistance in the final circuit as $|E_0/Z_3|$. Plot this

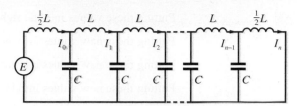

Figure 5.16 Ladder network.

resistance against ω and interpret the graph obtained. Note that the matrices contain complex numbers. (See Review exercises 7.12, Question 22.) Although it is possible to solve this exercise by hand, the use of a symbolic manipulation package, such as MAPLE, is recommended.

5.5.4 The solution of linear equations: iterative methods

An alternative and very popular way of solving linear equations is by iteration. This has the attraction of being easy to program. In practice, the availability of efficient procedures in computer libraries means that elimination methods are usually preferred for small problems. However, when the number of variables gets large, say several hundred, elimination methods struggle because the matrices can contain 10^6 or more elements. Problems of such size commonly occur in those scientific and engineering computations that require numerical solution on a mesh. Typically, in a turbine flow, we have a three-dimensional fluid flow problem that would need to be solved for three velocities and pressure on a $30 \times 30 \times 30$ mesh. The problem would require the solution of a $27\,000 \times 27\,000$ matrix equation. The saving feature of such problems is that it is very common for almost all the entries in the matrix to be zero. Matrices in which the large majority of elements are zero are called **sparse matrices**. Unless there is special structure to the equations, elimination will quickly destroy the sparseness. On the other hand, iterative methods only have to deal with the non-zero terms, so there is considerable computational saving. As usual, there is a price to pay:

(a)   it is not always easy to decide when the method has converged;

(b)   if the method takes a very large number of iterations to converge, any savings are quickly consumed.

A simple example will illustrate the way the method proceeds; in this example exact fractions will be used.

To solve the equations

$$4x + y = 2$$

$$x + 4y = -7$$

we first rearrange them as

$$x = \tfrac{1}{4}(2 - y)$$

$$y = \tfrac{1}{4}(-7 - x)$$

and start with $x = 0$, $y = 0$.

Putting these values into the right-hand side gives $\qquad x = \frac{1}{2}, \quad y = -\frac{7}{4}$

Putting these new values into the right-hand side gives $\quad x = \frac{15}{16}, \quad y = -\frac{15}{8}$

Putting these new values into the right-hand side gives $\quad x = \frac{31}{32}, \quad y = -\frac{127}{64}$

Putting these new values into the right-hand side gives $\quad x = \frac{255}{256}, y = -\frac{255}{128}$

Putting these new values into the right-hand side gives $\quad x = \frac{511}{512}, y = -\frac{2047}{1024}$

Performing the same procedure repeatedly, normally called **iteration**, gives a set of numbers that appear to be tending to the solution $x = 1$, $y = -2$.

This particular example shows the strength of the method but we are not always so fortunate, as illustrated in the next example.

Consider the tridiagonal equations in Example 5.30:

$$
\begin{aligned}
2x + y &&&= 1 \\
x + 2y + z &&&= 1 \\
y + 2z + t &&= 1 \\
z + 2t &&= -2
\end{aligned}
$$

We can rearrange these as

$$
\begin{aligned}
x &= \tfrac{1}{2}(1 - y) \\
y &= \tfrac{1}{2}(1 - x - z) \\
z &= \tfrac{1}{2}(1 - y - t) \\
t &= \tfrac{1}{2}(-2 - z)
\end{aligned}
\quad \text{or} \quad
\begin{bmatrix} x \\ y \\ z \\ t \end{bmatrix}
= \tfrac{1}{2}
\begin{bmatrix}
0 & -1 & 0 & 0 \\
-1 & 0 & -1 & 0 \\
0 & -1 & 0 & -1 \\
0 & 0 & -1 & 0
\end{bmatrix}
\begin{bmatrix} x \\ y \\ z \\ t \end{bmatrix}
+ \tfrac{1}{2}
\begin{bmatrix} 1 \\ 1 \\ 1 \\ -2 \end{bmatrix}
\qquad \textbf{(5.31)}
$$

Suppose we start with $x = y = z = t = 0$. We substitute these into the right-hand side and evaluate the new x, y, z and t; we then substitute the new values back in and repeat the process. Such iteration gives the results shown in Figure 5.17. This shows values going depressingly slowly to the solution $1, -1, 2, -2$: even after 20 iterations the values are 0.9381, -0.9767, 1.9727, -1.9856. The method just described is called the **Jacobi method**, and can be written, using superscripts as iteration counters, in the form

$$
\begin{aligned}
x^{(r+1)} &= \tfrac{1}{2}(1 - y^{(r)}) \\
y^{(r+1)} &= \tfrac{1}{2}(1 - x^{(r)} - z^{(r)}) \\
z^{(r+1)} &= \tfrac{1}{2}(1 - y^{(r)} - t^{(r)}) \\
t^{(r+1)} &= \tfrac{1}{2}(-2 - z^{(r)})
\end{aligned}
$$

Iteration	0	1	2	3	4	5	6	7	8	9	10
x	0	0.5	0.25	0.5	0.5	0.6562	0.6719	0.7734	0.7852	0.8516	0.8594
y	0	0.5	0	0	-0.3125	-0.3437	-0.5469	-0.5703	-0.7031	-0.7187	-0.8057
z	0	0.5	0.75	1.1250	1.1875	1.4375	1.4687	1.6328	1.6523	1.7598	1.7725
t	0	-1	-1.25	-1.3750	-1.5625	-1.5937	-1.7187	-1.7344	-1.8164	-1.8262	-1.8799

Figure 5.17 Iterative solution of (5.31) using Jacobi iteration.

An obvious step is to use the new values as soon as they are available. In the two-variable example the same equations

$$x = \tfrac{1}{4}(2 - y)$$
$$y = \tfrac{1}{4}(-7 - x)$$

are used and the same starting point $x = 0$, $y = 0$ is used. The iteration proceeds slightly differently:

Put the values 0, 0 in the first equation $\Rightarrow x = \tfrac{1}{2}$

Put the values $\tfrac{1}{2}$, 0 in the second equation $\Rightarrow y = -\tfrac{15}{8}$

Put the values $\tfrac{1}{2}$, $-\tfrac{15}{8}$ in the first equation $\Rightarrow x = \tfrac{31}{32}$

Put the values $\tfrac{31}{32}$, $-\tfrac{15}{8}$ in the second equation $\Rightarrow y = -\tfrac{255}{128}$

Put the values $\tfrac{31}{32}$, $-\tfrac{255}{128}$ in the first equation $\Rightarrow x = \tfrac{511}{512}$

and continue in the same way. It can be seen that already the convergence is very much faster.

In the second example we use the new values of x, y, z and t as soon as they are calculated: the method is called **Gauss–Seidel iteration**. This can be written as

$$x^{(r+1)} = \tfrac{1}{2}(1 - y^{(r)})$$
$$y^{(r+1)} = \tfrac{1}{2}(1 - x^{(r+1)} - z^{(r)})$$
$$z^{(r+1)} = \tfrac{1}{2}(1 - y^{(r+1)} - t^{(r)})$$
$$t^{(r+1)} = \tfrac{1}{2}(-2 - z^{(r+1)})$$

The calculation now yields the results shown in Figure 5.18. We see that, after the ten iterations quoted, the solution obtained by Gauss–Seidel iteration is within 4% of the actual solution whereas that obtained by Jacobi iteration still has an error of about 20%. The Gauss–Seidel method is both faster and more convenient for computer implementation. Within 20 iterations the Gauss–Seidel solution is accurate to three decimal places. A pseudocode algorithm for this method applied to the present problem is shown in Figure 5.19.

Although the two iteration methods have been described in terms of a particular example, the method is quite general. To solve

$$AX = b$$

we rewrite

$$A = D + L + U$$

Iteration	0	1	2	3	4	5	6	7	8	9	10
x	0	0.5	0.375	0.4375	0.6172	0.7480	0.8350	0.8920	0.9293	0.9537	0.9697
y	0	0.25	0.125	−0.2344	−0.4961	−0.6699	−0.7839	−0.8586	−0.9074	−0.9394	−0.9603
z	0	0.375	1.0312	1.3750	1.5918	1.7329	1.8252	1.8856	1.9251	1.9510	1.9679
t	0	−1.1875	−1.5156	−1.6875	−1.7959	−1.8665	−1.9126	−1.9426	−1.9626	−1.9755	−1.9840

Figure 5.18 Iterative solution of (5.31) using Gauss–Seidel iteration.

Figure 5.19
Algorithm to
implement the
Gauss–Seidel iteration.

```
read(vdu,eps,kmax)
k←0
x←0;y←0;z←0;t←0
repeat
   k←k + 1
   xold←x;   x←(1 − y)/2
   yold←y;        y←(1 − x − z)/2
   zold←z;        z←(1 − y − t)/2
   told←t;   t←(−2 − z)/2
until ((abs(x − xold) < eps)and(abs(y − yold) < eps)and(abs(z − zold) < eps)
      and(abs(t − told) < eps))or (k > kmax)
```

where D is diagonal, L only has non-zero elements below the diagonal and U only has non-zero elements above the diagonal, so that

$$A = \begin{bmatrix} a_{11} & & & & \\ & a_{22} & & \mathbf{0} & \\ & & a_{33} & & \\ & \mathbf{0} & & \ddots & \\ & & & & a_{nn} \end{bmatrix} + \begin{bmatrix} 0 & & & & \\ a_{21} & 0 & & \mathbf{0} & \\ a_{31} & a_{32} & 0 & & \\ \vdots & \vdots & & \ddots & \\ a_{n1} & a_{n2} & \cdots & a_{n,n-1} & 0 \end{bmatrix}$$

$$+ \begin{bmatrix} 0 & a_{12} & a_{13} & \cdots & a_{1n} \\ & 0 & a_{23} & \cdots & a_{2n} \\ & & \ddots & & \vdots \\ & \mathbf{0} & 0 & & a_{n-1,n} \\ & & & & 0 \end{bmatrix}$$

The Jacobi method is written in this notation as

$$DX^{(r+1)} = -(L + U)X^{(r)} + b$$

and the Gauss–Seidel method as

$$DX^{(r+1)} = -LX^{(r+1)} - UX^{(r)} + b$$

(Remember that $X^{(r)}$ denotes the rth iteration of X, not X raised to the power r.)

By changing the method slightly, we have been able to speed up the method, so it is natural to ask if it can be speeded up even further. A popular method for doing this is **successive over-relaxation (SOR)**. This anticipates what the x_i values might be and overshoots the values obtained by Gauss–Seidel iteration. The new value of each component of the vector $X^{(r+1)}$ is taken to be

$$wx_i^{(r+1)} + (1 - w)x_i^{(r)} \tag{5.32}$$

Iteration	0	1	2	3	4	5	6	7	8	9	10
x	0	0.7	0.273	0.5643	1.1060	1.0195	1.0226	1.0055	0.9956	0.9995	0.9995
y	0	0.21	0.0378	−0.9025	−1.0885	−1.0434	−1.0208	−0.9969	−0.9967	−0.9990	−0.9997
z	0	0.553	1.7033	1.9646	2.0930	2.0320	2.0019	1.9979	1.9971	1.9996	2.0001
t	0	−1.7871	−1.8775	−2.0242	−2.0554	−2.0002	−2.0013	−1.9981	−1.9988	−2.0002	−2.0000

Figure 5.20 Iterative solution of (5.31) with SOR factor $w = 1.4$ in (5.32).

SOR factor w	0.2	0.4	0.6	0.8	1.0	1.2	1.4	1.6	1.8
Iterations required for convergence	>50	>50	47	34	26	21	17	29	>50

Figure 5.21 Variation of rate of convergence with SOR factor w.

which is the weighted average of the previous value and the new value given by Gauss–Seidel iteration. In the two-variable example the weighted average rearranges the equations as

$$x = w[\tfrac{1}{4}(2 - y)] + (1 - w)x = x + w[\tfrac{1}{4}(2 - y - 4x)]$$
$$y = w[\tfrac{1}{4}(-7 - x)] + (1 - w)y = y + w[\tfrac{1}{4}(-7 - x - 4y)]$$

The convergence for this example is so rapid that the enhanced convergence of SOR is hardly worth the effort; an optimum value of $w = 1.05$ reduces the convergence, to six significant figures, from seven to six iterations. However, for most problems the improved convergence is significant. Note that $w = 1$ gives the Gauss–Seidel method.

If we repeat the calculation for (5.31) including (5.32) with $w = 1.4$, we obtain the results shown in Figure 5.20. It may be noted that the iterations converge even faster than the two previous methods, with a solution accurate to about 1% after ten steps. The optimum value of w is of great interest, and specialist books on numerical analysis give details of how this can be computed (for example, *An Introduction to Numerical Methods with Pascal*, L. V. Atkinson and P. J. Harley (1983), Addison-Wesley Publishers Ltd). Usually the best approach is a heuristic one – experiment with w to find a value that gives the fastest convergence. For 'one-off' problems this is hardly worth the effort so long as convergence is achieved, but in many scientific and engineering problems the same calculation may be done many hundreds of times, so the optimum value of w can reduce calculation time by half or more. For the current problem the number of iterations required to give four-decimal-place accuracy is shown in Figure 5.21.

It can be shown that outside the region $0 < w < 2$ the method will diverge but that inside it may or may not converge. The case $w < 1$ is called **under-relaxation** and $w > 1$ is called **over-relaxation**. In straightforward problems w in the range 1.2–1.8 usually gives the most rapid convergence, and this is normally the region to explore as a first guess. In the problem studied a value of $w = 1.4$ gives just about the fastest convergence, requiring only about two-thirds of the iterations required for the Gauss–Seidel method. In some physical problems, however, under-relaxation is required in order to avoid too rapid variation from iteration to iteration.

Great care must be taken with iterative methods, and convergence for some equations can be particularly difficult. Considerable experience is needed in looking at sets of equations to decide whether or not convergence can be expected, and often – even for the experienced mathematician – the answer is 'try it and see'. One simple test that will guarantee convergence is to test whether the matrix is **diagonally dominant**. This means that the magnitude of a diagonal element is larger than the sum of the magnitudes of the off-diagonal elements in that row, or $|a_{ii}| \geqslant \sum_{\substack{j=1 \\ i \neq j}}^{n} |a_{ij}|$ for each i. If the system is not diagonally dominant, the iteration method may or may not converge.

A detailed analysis of the convergence of iterative methods is not possible without a study of eigenvalues, and can be found in specialist numerical analysis books.

5.5.5 Exercises

(*Note*: All of these exercises are best solved using a computer matrix package or spreadsheet.)

57 Solve the equations in Question 49 (Exercises 5.5.3) using Jacobi iteration starting from the estimate $X = [1 \quad 1 \quad 1 \quad 1]^{\mathrm{T}}$. How accurate is the solution obtained after five iterations?

58 Solve the equations in Question 50 (Exercises 5.5.3) using Gauss–Seidel iteration, starting from the estimate $X = [1 \quad 0 \quad 0 \quad 0]^{\mathrm{T}}$. How accurate is the solution obtained after three iterations?

59 Write a computer program or set up a spreadsheet to obtain the solution, by SOR, to the equations in Question 51 (Exercises 5.5.3). Determine the optimum SOR factor for each equation.

60 Use an SOR program to solve the equations

$$x - 0.7y \qquad = -4$$

$$-0.7x + y - 0.7z = 34$$

$$-0.7y + z = -44$$

so that successive iterations differ by no more than 1 in the fourth decimal place. Find an SOR factor that produces this convergence in less than 50 iterations.

61 Show that the circuit in Figure 5.22 has equations

$$\begin{bmatrix} R_1 + R_2 + R_4 & -R_2 & -R_4 \\ -R_2 & R_3 + R_5 + R_2 & -R_5 \\ -R_4 & -R_5 & R_4 + R_5 \end{bmatrix} \begin{bmatrix} I_1 \\ I_2 \\ I_3 \end{bmatrix}$$

$$= \begin{bmatrix} 0 \\ 0 \\ E \end{bmatrix}$$

Take $R_1 = 1$, $R_2, = 2$, $R_3 = 2$, $R_4 = 2$ and $R_5 = 3$ (all in Ω) and $E = 1.5$ V. Show that the equations are diagonally dominant, and hence solve the equations by an iterative method.

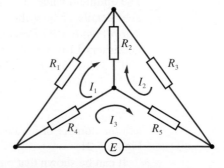

Figure 5.22 Circuit for Question 61.

62 Solve the 10×10 matrix equation in Example 5.28 using an iterative method starting from $X = [1 \quad 1 \quad 1 \quad 1 \quad 1 \quad 1 \quad 1 \quad 1 \quad 1 \quad 1]^{\mathrm{T}}$. Verify that a solution to four-figure accuracy can be obtained in less than ten iterations.

5.6 Rank

The solution of sets of linear equations has been considered in Section 5.5. Provided the determinant of a matrix is non-zero, we can obtain explicit solutions in terms of the inverse matrix. However, when we looked at cases with zero determinant the results were much less clear. The idea of the **rank** of a matrix helps to make these results more precise. Unfortunately, rank is not an easy concept, and it is usually difficult to compute. We shall take an informal approach that is not fully general but is sufficient to deal with the cases (c) and (d) of Section 5.5. The method we shall use is to take the Gaussian elimination procedure described in Figure 5.13 (Section 5.5.2) and examine the consequences for a zero-determinant situation.

If we start with the equations

$$\begin{bmatrix} 1 & 0 & 1 & 1 & 0 & 0 \\ 0 & 1 & 1 & 0 & 1 & 2 \\ 1 & 1 & 2 & 1 & 1 & 2 \\ 1 & 0 & 1 & 0 & 1 & 3 \\ 0 & 0 & 0 & 0 & 1 & 3 \\ 1 & 1 & 2 & 0 & 2 & 5 \end{bmatrix} \begin{bmatrix} x_1 \\ x_2 \\ . \\ . \\ . \\ x_6 \end{bmatrix} = \begin{bmatrix} 1 \\ 1 \\ 2 \\ 0 \\ 0 \\ 1 \end{bmatrix} \tag{5.33}$$

and proceed with the elimination, the first and second steps are quite normal:

$$\begin{bmatrix} 1 & 0 & 1 & 1 & 0 & 0 \\ 0 & 1 & 1 & 0 & 1 & 2 \\ 0 & 1 & 1 & 0 & 1 & 2 \\ 0 & 0 & 0 & -1 & 1 & 3 \\ 0 & 0 & 0 & 0 & 1 & 3 \\ 0 & 1 & 1 & -1 & 2 & 5 \end{bmatrix} \begin{bmatrix} x_1 \\ x_2 \\ . \\ . \\ . \\ x_6 \end{bmatrix} = \begin{bmatrix} 1 \\ 1 \\ 1 \\ -1 \\ 0 \\ 0 \end{bmatrix}, \quad \begin{bmatrix} 1 & 0 & 1 & 1 & 0 & 0 \\ 0 & 1 & 1 & 0 & 1 & 2 \\ 0 & 0 & 0 & 0 & 0 & 0 \\ 0 & 0 & 0 & -1 & 1 & 3 \\ 0 & 0 & 0 & 0 & 1 & 3 \\ 0 & 0 & 0 & -1 & 1 & 3 \end{bmatrix} \begin{bmatrix} x_1 \\ x_2 \\ . \\ . \\ . \\ x_6 \end{bmatrix} = \begin{bmatrix} 1 \\ 1 \\ 0 \\ -1 \\ 0 \\ -1 \end{bmatrix}$$

The next step in the elimination procedure looks for a non-zero entry in the third column on or below the diagonal element. All the entries are zero – so the procedure, as it stands, fails. To overcome the problem, we just proceed to the next column and repeat the normal sequence of operations. We interchange the third and fourth rows and perform the elimination on column 4. Finally we interchange rows 4 and 5 to give

$$\begin{bmatrix} 1 & 0 & 1 & 1 & 0 & 0 \\ 0 & 1 & 1 & 0 & 1 & 2 \\ 0 & 0 & 0 & 1 & -1 & -3 \\ 0 & 0 & 0 & 0 & 1 & 3 \\ 0 & 0 & 0 & 0 & 0 & 0 \\ 0 & 0 & 0 & 0 & 0 & 0 \end{bmatrix} \begin{bmatrix} x_1 \\ x_2 \\ . \\ . \\ . \\ x_6 \end{bmatrix} = \begin{bmatrix} 1 \\ 1 \\ 1 \\ 0 \\ 0 \\ 0 \end{bmatrix} \tag{5.34}$$

To perform the back substitution we put $x_6 = \mu$. Then

row 4   gives $x_5 = -3x_6 = -3\mu$

row 3   gives $x_4 = 1 + x_5 + 3x_6 = 1$

put $x_3 = \lambda$

row 2   gives $x_2 = 1 - x_3 - x_5 - 2x_6 = 1 - \lambda + \mu$

row 1   gives $x_1 = 1 - x_3 - x_4 = -\lambda$

Thus our solution is

$$x_1 = -\lambda, \quad x_2 = 1 - \lambda + \mu, \quad x_3 = \lambda, \quad x_4 = 1, \quad x_5 = -3\mu, \quad x_6 = \mu$$

The equations have been reduced to **echelon form**, and it is clear that the same process can be followed for any matrix.

In general we use the elementary row operations, introduced in Section 5.5.2, to manipulate the equation or matrix to **echelon form**:

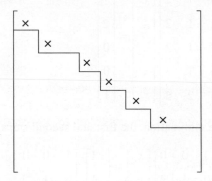

Below the line all the entries are zero, and the leading element, marked $\times$, in each row above the line is non-zero. The row operations do not change the solution to the set of equations corresponding to the matrix.

When this procedure is applied to a non-singular matrix, the method reduces to that shown in Figure 5.13, the final matrix has non-zero diagonal elements, and back substitution gives a unique solution. When the determinant is zero, as in (5.33), the elimination gives a matrix with some zeros in the diagonal and some zero rows, as in (5.34). The number of non-zero rows in the echelon form is called the **rank** of the matrix, rank $\boldsymbol{A}$; in the case of the matrix in (5.33) and that derived from it by row manipulation (5.34), we have rank $\boldsymbol{A} = 4$.

The more common definition of rank is given by the order of the largest square submatrix with non-zero determinant. A square submatrix is formed by deleting rows and columns to form a square matrix. In (5.33) the 6×6 determinant is zero and all the 5×5 submatrices have zero determinant; however, if we delete columns 3 and 6 and rows 3 and 4, we obtain

$$\begin{bmatrix} 1 & 0 & 1 & 0 \\ 0 & 1 & 0 & 1 \\ 0 & 0 & 0 & 1 \\ 1 & 1 & 0 & 2 \end{bmatrix}$$

which has determinant equal to one, hence confirming that the matrix is of rank 4. To show equivalence of the two definitions is not straightforward and is omitted here. To determine the rank of a matrix, it is very much easier to look at the echelon form.

If we find any of the rows of the echelon matrix to be zero then, for consistency, the corresponding right-hand sides of the matrix equation must also be zero. The elementary row operations reduce the equation to echelon form, so that the equations take the form

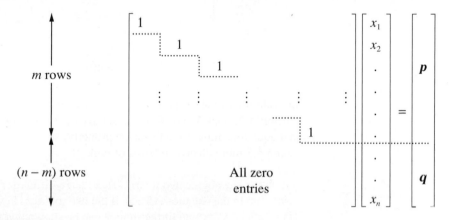

where p is a vector with m elements and q is a vector with $(n - m)$ elements. Note that each of the m non-zero rows will have a leading non-zero entry of 1 but this entry will not necessarily be on a diagonal, as illustrated for example in (5.34). Three statements follow from this reduction.

(i)   The matrix has rank $(A) = m$.
(ii)   If $q \neq 0$ then the equations are inconsistent.
(iii) If $q = 0$ then the equations are consistent and have a solution. In addition it can be shown that the solution has $(n - m)$ free parameters.

Writing the equations as

$$AX = b \tag{5.35}$$

we define the **augmented matrix** $(A : b)$ as the matrix A with the b column added to it. When reduced to echelon form the matrix and the augmented matrix take the form

$$A = \begin{bmatrix} 1 & & & & & \\ & 1 & & & & \\ & & 1 & & & \\ \vdots & \vdots & \vdots & \vdots & & \vdots \\ & & & & 1 & \\ & & \text{All zero} & & & \\ & & \text{entries} & & & \end{bmatrix}$$

$$\text{and } (A : b) = \begin{bmatrix} 1 & & & & & & \\ & 1 & & & & & \\ & & 1 & & & & \\ \vdots & \vdots & \vdots & \vdots & \vdots & & p \\ & & & & & 1 & \\ & & \text{All zero} & & & & q \\ & & \text{entries} & & & & \end{bmatrix}$$

and the solution of the equations can be written in terms of *rank*. It is easy to see from the echelon form that A and $(A:b)$ must have the same rank to ensure consistency. The original equations must have the same property, so we can state the results (c) and (d) of Section 5.5 more clearly in terms of rank.

> If A and the augmented matrix $(A:b)$ have different rank then we have no solution to the equations (5.35). If the two matrices have the same rank then a solution exists, and furthermore it can be shown that the solution will contain a number of free parameters equal to $n -$ rank A.

The calculation of rank is not easy, so, while the result is rigorous, it is not simple to apply. Reducing equations to echelon form tells us immediately the rank of the associated matrix, and gives a constructive method of solution. There is a large amount of arithmetic in the reduction, but if the solution is required then this is inevitable anyway. The numerical calculation of rank does not normally entail reduction to echelon form; rather more advanced methods such as singular value decomposition are used.

Example 5.35   Reduce the following equations to echelon form, calculate the rank of the matrices and find the solutions of the equations (if they exist):

(a) $\begin{bmatrix} 0 & 1 & 1 & 0 \\ 1 & 0 & 3 & 2 \\ 2 & 1 & 5 & 4 \\ 1 & -2 & 0 & 2 \end{bmatrix} \begin{bmatrix} x_1 \\ x_2 \\ x_3 \\ x_4 \end{bmatrix} = \begin{bmatrix} 1 \\ 3 \\ 7 \\ 2 \end{bmatrix}$

(b) $\begin{bmatrix} 1 & 0 & -1 & 1 & -1 \\ 0 & 1 & 1 & -1 & 1 \\ 1 & 1 & 0 & 0 & 0 \\ 2 & 3 & 1 & -1 & 1 \\ 2 & 2 & 0 & 0 & 0 \end{bmatrix} \begin{bmatrix} x_1 \\ x_2 \\ x_3 \\ x_4 \\ x_5 \end{bmatrix} = \begin{bmatrix} 0 \\ 1 \\ 1 \\ 3 \\ 2 \end{bmatrix}$

Solution   (a) Rows 1 and 3 are interchanged, and the elimination then proceeds as follows:

$$\begin{bmatrix} 2 & 1 & 5 & 4 \\ 1 & 0 & 3 & 2 \\ 0 & 1 & 1 & 0 \\ 1 & -2 & 0 & 2 \end{bmatrix} \begin{bmatrix} x_1 \\ x_2 \\ x_3 \\ x_4 \end{bmatrix} = \begin{bmatrix} 7 \\ 3 \\ 1 \\ 2 \end{bmatrix} \rightarrow \begin{bmatrix} 1 & \frac{1}{2} & \frac{5}{2} & 2 \\ 0 & -\frac{1}{2} & \frac{1}{2} & 0 \\ 0 & 1 & 1 & 0 \\ 0 & -\frac{5}{2} & -\frac{5}{2} & 0 \end{bmatrix} \begin{bmatrix} x_1 \\ x_2 \\ x_3 \\ x_4 \end{bmatrix} = \begin{bmatrix} \frac{7}{2} \\ -\frac{1}{2} \\ 1 \\ -\frac{3}{2} \end{bmatrix}$$

interchange row 2 and row 4

$$\rightarrow \begin{bmatrix} 1 & \frac{1}{2} & \frac{5}{2} & 2 \\ 0 & -\frac{5}{2} & -\frac{5}{2} & 0 \\ 0 & 1 & 1 & 0 \\ 0 & -\frac{1}{2} & \frac{1}{2} & 0 \end{bmatrix} \begin{bmatrix} x_1 \\ x_2 \\ x_3 \\ x_4 \end{bmatrix} = \begin{bmatrix} \frac{7}{2} \\ -\frac{3}{2} \\ 1 \\ -\frac{1}{2} \end{bmatrix}$$

eliminate elements in column 2

$$\begin{matrix}\rightarrow\\\rightarrow\\\hookrightarrow\end{matrix} \begin{bmatrix} 1 & \frac{1}{2} & \frac{5}{2} & 2 \\ 0 & 1 & 1 & 0 \\ 0 & 0 & 0 & 0 \\ 0 & 0 & 1 & 0 \end{bmatrix} \begin{bmatrix} x_1 \\ x_2 \\ x_3 \\ x_4 \end{bmatrix} = \begin{bmatrix} \frac{7}{2} \\ \frac{3}{5} \\ \frac{2}{5} \\ -\frac{1}{5} \end{bmatrix}$$

interchange row 3 and row 4

$$\rightarrow \begin{bmatrix} 1 & \frac{1}{2} & \frac{5}{2} & 2 \\ 0 & 1 & 1 & 0 \\ 0 & 0 & 1 & 0 \\ 0 & 0 & 0 & 0 \end{bmatrix} \begin{bmatrix} x_1 \\ x_2 \\ x_3 \\ x_4 \end{bmatrix} = \begin{bmatrix} \frac{7}{2} \\ \frac{3}{5} \\ -\frac{1}{5} \\ \frac{2}{5} \end{bmatrix}$$

The rank of the matrix is 3 while that of the augmented matrix ($A : b$) is 4, so the equations represented by the matrix equation (a) are not consistent. Note that the last row cannot be satisfied and hence the equations have no solution.

(b) Interchanging the first and last rows, making the pivot 1 and performing the first elimination, we obtain

$$\begin{bmatrix} 1 & 1 & 0 & 0 & 0 \\ 0 & 1 & 1 & -1 & 1 \\ 0 & 0 & 0 & 0 & 0 \\ 0 & 1 & 1 & 1 & 1 \\ 0 & -1 & -1 & 1 & -1 \end{bmatrix} \begin{bmatrix} x_1 \\ . \\ . \\ . \\ x_5 \end{bmatrix} = \begin{bmatrix} 1 \\ 1 \\ 0 \\ 1 \\ -1 \end{bmatrix} \rightarrow \begin{bmatrix} 1 & 1 & 0 & 0 & 0 \\ 0 & 1 & 1 & -1 & 1 \\ 0 & 0 & 0 & 0 & 0 \\ 0 & 0 & 0 & 0 & 0 \\ 0 & 0 & 0 & 0 & 0 \end{bmatrix} \begin{bmatrix} x_1 \\ . \\ . \\ . \\ x_5 \end{bmatrix} = \begin{bmatrix} 1 \\ 1 \\ 0 \\ 0 \\ 0 \end{bmatrix}$$

The matrix and the augmented matrix both have rank 2, so the equations are consistent and we can compute the solution

$$x_1 = 1 - \lambda, \quad x_2 = \lambda, \quad x_3 = 1 - \lambda + \mu - \nu, \quad x_4 = \mu, \quad x_5 = \nu$$

As expected, the solution contains three free parameters, since the order of the equation is 5 and the rank is 2.

In most practical problems that reduce to the solution of linear equations, it is usual that there are n independent variables to be computed from n equations. This is not always the case and the resulting matrix form is *not* square. A geometrical example of four equations and three unknowns was described in equation (5.2). The idea of a determinant is only sensible if matrices are square so the simple results about the solution of the equations cannot be used. However, the ideas of elementary row operations, reduction to echelon form and rank still hold and the existence or non-existence of solutions can be written in terms of these concepts. Some examples will illustrate the possible situations that can occur.

Underspecified sets of equations

Here there are more variables than equations.

Case (a)

Solve
$$\begin{bmatrix} 1 & 1 & 1 \\ 1 & 2 & 3 \end{bmatrix} \begin{bmatrix} x \\ y \\ z \end{bmatrix} = \begin{bmatrix} 1 \\ 2 \end{bmatrix}$$

Subtract row 1 from row 2:
$$\begin{bmatrix} 1 & 1 & 1 \\ 0 & 1 & 2 \end{bmatrix} \begin{bmatrix} x \\ y \\ z \end{bmatrix} = \begin{bmatrix} 1 \\ 1 \end{bmatrix}$$

The elimination is now complete and the back substitution starts

Put $z = t$

From row 2 $y = 1 - 2t$

From row 1 $x = 1 - y - z = t$

so the full solution is

$$x = t, \quad y = 1 - 2t, \quad z = t$$

for any t. Note that the solution has one free parameter.

Case (b)

Solve
$$\begin{bmatrix} 1 & 1 & 1 \\ 2 & 2 & 2 \end{bmatrix} \begin{bmatrix} x \\ y \\ z \end{bmatrix} = \begin{bmatrix} 1 \\ 1 \end{bmatrix}$$

Subtract $2 \times$ (row 1) from row 2:
$$\begin{bmatrix} 1 & 1 & 1 \\ 0 & 0 & 0 \end{bmatrix} \begin{bmatrix} x \\ y \\ z \end{bmatrix} = \begin{bmatrix} 1 \\ -1 \end{bmatrix}$$

and it is clear that there is no solution since the last row is inconsistent. Although this example may be seen to be almost trivial since the equations are obviously inconsistent $[x + y + z = 1$ and $2(x + y + z) = 1]$, in larger systems the situation is hardly ever obvious.

Overspecified sets of equations

Here there are more equations than variables.

Case (c)

Solve
$$\begin{bmatrix} 1 & 1 \\ 1 & 2 \\ 1 & 3 \end{bmatrix} \begin{bmatrix} x \\ y \end{bmatrix} = \begin{bmatrix} -1 \\ 0 \\ 1 \end{bmatrix}$$

Subtract row 1 from rows 2 and 3:
$$\begin{bmatrix} 1 & 1 \\ 0 & 1 \\ 0 & 2 \end{bmatrix} \begin{bmatrix} x \\ y \end{bmatrix} = \begin{bmatrix} -1 \\ 1 \\ 2 \end{bmatrix}$$

Subtract $2 \times$ (row 2) from row 3:
$$\begin{bmatrix} 1 & 1 \\ 0 & 1 \\ 0 & 0 \end{bmatrix} \begin{bmatrix} x \\ y \end{bmatrix} = \begin{bmatrix} -1 \\ 1 \\ 0 \end{bmatrix}$$

It can be observed that the equations are consistent since the last row contains all zeros. The unique solution is obtained from back substitution as $x = -2$, $y = 1$.

However, for overspecified equations the more common situation is that no solution is possible.

Case (d)

Solve
$$\begin{bmatrix} 1 & 1 \\ 1 & 2 \\ 1 & 3 \end{bmatrix} \begin{bmatrix} x \\ y \end{bmatrix} = \begin{bmatrix} 0 \\ 1 \\ -2 \end{bmatrix}$$

Subtract row 1 from rows 2 and 3:
$$\begin{bmatrix} 1 & 1 \\ 0 & 1 \\ 0 & 2 \end{bmatrix} \begin{bmatrix} x \\ y \end{bmatrix} = \begin{bmatrix} 0 \\ 1 \\ -2 \end{bmatrix}$$

Subtract $2 \times$ (row 2) from row 3:
$$\begin{bmatrix} 1 & 1 \\ 0 & 1 \\ 0 & 0 \end{bmatrix} \begin{bmatrix} x \\ y \end{bmatrix} = \begin{bmatrix} 0 \\ 1 \\ -4 \end{bmatrix}$$

The equations are now clearly inconsistent since the last row says $0 = -4$.

The existence or non-existence of solutions can be deduced from the echelon form and hence the idea of rank, and we can understand the solution of matrix equations involving non-square matrices. If A is a $p \times q$ matrix and b a $p \times 1$ *column vector, the matrix equation* $AX = b$ represents p linear equations in q variables. The rank of a matrix, being the number of non-zero rows in the echelon form of the matrix, cannot exceed p. On the other hand, the row reduction process will produce an echelon form with at most q non-zero rows. Hence the rank of a $p \times q$ matrix cannot exceed the smaller of p and q. There are two possible cases:

(i) $p < q$: Here there are more variables than equations. The rank of A must be less than the number of variables. If rank $(A : b) >$ rank A, the equations are inconsistent and there is no solution, as in case (b). If rank $(A : b) =$ rank A, as in case (a), there is a solution, which must contain $q -$ rank A free parameters.

(ii) $p > q$: Here there are more equations than variables. The rank of A cannot exceed the number of variables. If rank $(A : b) >$ rank A, as in case (d), the equations are inconsistent and there is no solution. If rank $(A : b) =$ rank A, as in case (c), some of the equations are redundant and there is a solution containing $q -$ rank A free parameters.

5.6.1 Exercises

63 Find the rank of the matrices

(a) $\begin{bmatrix} 2 & 1 & 1 & 1 \\ 4 & 2 & 2 & 3 \\ 0 & 0 & 0 & 1 \\ -2 & -1 & -1 & 0 \end{bmatrix}$, (b) $\begin{bmatrix} 1 & 1 & 1 & 1 \\ 2 & 1 & 2 & 1 \\ 0 & 1 & 0 & 1 \\ 1 & 0 & 1 & 1 \end{bmatrix}$

64 Reduce the matrices in the following equations to echelon form, determine their ranks and solve the equations, if a solution exists:

(a) $\begin{bmatrix} 1 & 2 & 3 \\ 3 & 2 & 1 \\ 1 & 1 & 1 \end{bmatrix} \begin{bmatrix} x \\ y \\ z \end{bmatrix} = \begin{bmatrix} 8 \\ 4 \\ 3 \end{bmatrix}$

(b) $\begin{bmatrix} 1 & 2 & -1 & 1 \\ 1 & 1 & 0 & 0 \\ 0 & 1 & -1 & 1 \\ 1 & 0 & 1 & -1 \end{bmatrix} \begin{bmatrix} x \\ y \\ z \\ t \end{bmatrix} = \begin{bmatrix} 0 \\ 1 \\ -1 \\ 1 \end{bmatrix}$

65 By obtaining the order of the largest square submatrix with non-zero determinant determine the rank of the matrix

$$A = \begin{bmatrix} 1 & 1 & 0 & 1 \\ 1 & 0 & 0 & 1 \\ 0 & 1 & 0 & 0 \\ 1 & 1 & 1 & 1 \end{bmatrix}$$

Reduce the matrix to echelon form and confirm your result. Check the rank of the augmented

matrix $(A : b)$, where $b^T = [-1 \quad 0 \quad -1 \quad 0]$. Does the equation $AX = b$ have a solution?

66 Solve, where possible, the following matrix equations:

(a) $\begin{bmatrix} 1 & 3 & 4 \\ -1 & 3 & 4 \end{bmatrix} \begin{bmatrix} x \\ y \\ z \end{bmatrix} = \begin{bmatrix} 1 \\ 3 \end{bmatrix}$

(b) $\begin{bmatrix} 2 & 1 \\ 4 & 6 \\ 3 & 5 \end{bmatrix} \begin{bmatrix} x \\ y \end{bmatrix} = \begin{bmatrix} 1 \\ 4 \\ -2 \end{bmatrix}$

(c) $\begin{bmatrix} 1 & 4 & 7 & -3 \\ -2 & 3 & -6 & 1 \\ 0 & 11 & 8 & -5 \end{bmatrix} \begin{bmatrix} x \\ y \\ z \\ t \end{bmatrix} = \begin{bmatrix} 1 \\ 3 \\ 5 \end{bmatrix}$

(d) $\begin{bmatrix} 2 & 1 & 4 \\ 3 & 2 & 9 \\ 4 & 1 & 3 \\ 3 & 3 & 3 \end{bmatrix} \begin{bmatrix} x \\ y \\ z \end{bmatrix} = \begin{bmatrix} 1 \\ 4 \\ -2 \\ -3 \end{bmatrix}$

67 In a fluid flow problem there are five natural parameters. These have dimensions in terms of length L, mass M and time T as follows:

$$\text{velocity} = V = LT^{-1}, \qquad \text{density} = \rho = ML^{-3}$$
$$\text{distance} = D = L, \qquad \text{gravity} = g = LT^{-2}$$

and

viscosity $= \mu = ML^{-1}T^{-1}$

To determine how many non-dimensional parameters can be constructed, seek values of p, q, r, s and t so that

$$V^p \rho^q D^r g^s \mu^t$$

is dimensionless. Write the equations for p, q, r, s and t in matrix form and show that the resulting 3×5 matrix has rank 3. Thus there are two parameters that can be chosen independently. By choosing these appropriately, show that they correspond to the Reynolds number $Re = V\rho D/\mu$ and the Froude number $Fr = Dg/V^2$.

Repeat a similar dimensional analysis when heat transfer is included.

68 Four points in a three-dimensional space have coordinates (x_i, y_i, z_i) for $i = 1, \ldots, 4$. From the rank of the matrix

$$\begin{bmatrix} x_1 & y_1 & z_1 & 1 \\ x_2 & y_2 & z_2 & 1 \\ x_3 & y_3 & z_3 & 1 \\ x_4 & y_4 & z_4 & 1 \end{bmatrix}$$

determine whether the points lie on a plane or a line or whether there are other possibilities.

69 A popular method of numerical integration – see the work in Chapter 8 – involves Gaussian integration; it is used in finite-element calculations which are well used in most of engineering. As a simple example, the numerical integral over the interval $-1 \leqslant x \leqslant 1$ is written

$$\int_{-1}^{1} f(x)dx = C_1 f(x_1) + C_2 f(x_2)$$

and the formula is made exact for the four functions $f = 1$, $f = x$, $f = x^2$ and $f = x^3$, so it must be accurate for all cubics. This leads to the four equations

$$C_1 + C_2 = 2$$
$$C_1 x_1 + C_2 x_2 = 0$$
$$C_1 x_1^2 + C_2 x_2^2 = \tfrac{2}{3}$$
$$C_1 x_1^3 + C_2 x_2^3 = 0$$

Use Gaussian elimination to reduce the equations and hence deduce that the equations are only consistent if x_1 and x_2 are chosen at the 'Gauss' points $\pm \frac{1}{\sqrt{3}}$.

5.7 The eigenvalue problem

A problem that leads to a concept of crucial importance in many branches of mathematics and its applications is that of seeking non-trivial solutions $x \neq 0$ to the matrix equation

$$\boldsymbol{A}x = \lambda x$$

This is referred to as the eigenvalue problem; values of the scalar λ for which non-trivial solutions exist are called **eigenvalues** and the corresponding solutions $x \neq 0$ are called the **eigenvectors**. We saw an example of eigenvalues in Example 5.26. Such problems arise naturally in many branches of engineering. For example, in vibrations the eigenvalues and eigenvectors describe the frequency and mode of vibration respectively, while in mechanics they represent principal stresses and the principal axes of stress in bodies subjected to external forces. Eigenvalues also play an important role in the stability analysis of dynamical systems and are central to the evaluation of energy levels in quantum mechanics.

Vibrational examples provide the most straightforward applications of eigenvalues in science and engineering. They do, however, require calculus so are a little premature in this text. The ideas are so important that it is worth considering a simple problem out of its natural order. Spring and dashpot models can be used successfully in a vast array of mechanical engineering problems; a typical example is given in the engineering application in Section 5.8. Without going into the detail of the *dynamics* of the problem, the equations (taken with $k = k_1 = k_2 = k_3$ and $l_1 = l_2 = l_3$) are quoted as

$$\frac{m}{k}\frac{\mathrm{d}^2 X_1}{\mathrm{d}t^2} = -2X_1 + X_2, \quad \frac{m}{k}\frac{\mathrm{d}^2 X_2}{\mathrm{d}t^2} = X_1 - 2X_2$$

where m is the mass of the particles and X_1 and X_2 are the displacements from their equilibrium positions.

Putting

$$X_1 = a\mathrm{e}^{pt}, \quad X_2 = b\mathrm{e}^{pt}$$

into the equations gives

$$\begin{aligned}(mp^2/k)a &= -2a + b \\ (mp^2/k)b &= a - 2b\end{aligned} \quad \text{or in matrix form} \quad \begin{bmatrix} -2 & 1 \\ 1 & -2 \end{bmatrix}\begin{bmatrix} a \\ b \end{bmatrix} = \lambda \begin{bmatrix} a \\ b \end{bmatrix}$$

where $\lambda = (mp^2/k)$. Here we have a typical eigenvalue problem. The value of λ, and hence p, determines the stability and the vibrational frequencies of the system.

5.7.1 The characteristic equation

The set of simultaneous equations

$$\mathbf{A}x = \lambda x \tag{5.36}$$

where $\mathbf{A}$ is an $n \times n$ matrix and $x = [x_1 \quad x_2 \quad \ldots \quad x_n]^\mathrm{T}$ is an $n \times 1$ column vector can be written in the form

$$(\lambda\mathbf{I} - \mathbf{A})x = 0 \tag{5.37}$$

where $\mathbf{I}$ is the identity matrix. The matrix equation (5.37) represents simply a set of homogeneous equations, and we know that a non-trivial solution exists if

$$c(\lambda) = |\lambda\mathbf{I} - \mathbf{A}| = 0 \tag{5.38}$$

Here $c(\lambda)$ is the expansion of the determinant and is a polynomial of degree n in λ, called the **characteristic polynomial** of $\mathbf{A}$. Thus

$$c(\lambda) = \lambda^n + c_{n-1}\lambda^{n-1} + c_{n-2}\lambda^{n-2} + \ldots + c_1\lambda + c_0$$

and the equation $c(\lambda) = 0$ is called the **characteristic equation** of $\mathbf{A}$. We note that this equation can be obtained just as well by evaluating $|\mathbf{A} - \lambda\mathbf{I}| = 0$; however, the form (5.38) is preferred for the definition of the characteristic equation, since the coefficient of λ^n is then always +1.

In many areas of engineering, particularly in those involving vibration or the control of processes, the determination of those values of λ for which (5.37) has a non-trivial solution (that is, a solution for which $x \neq 0$) is of vital importance. These values of λ are precisely the values that satisfy the characteristic equation, and are called the eigenvalues of $\mathbf{A}$.

Example 5.36   Find the characteristic equation and the eigenvalues of the matrix

$$A = \begin{bmatrix} -2 & 1 \\ 1 & -2 \end{bmatrix}$$

Solution   Equation (5.38) gives

$$0 = |\lambda I - A| = \begin{vmatrix} \lambda + 2 & -1 \\ -1 & \lambda + 2 \end{vmatrix} = (\lambda + 2)^2 - 1$$

so the characteristic equation is

$$\lambda^2 + 4\lambda + 3 = 0$$

The roots of this equation, namely $\lambda = -1$ and -3, give the eigenvalues.

Example 5.37   Find the characteristic equation for the matrix

$$A = \begin{bmatrix} 1 & 1 & -2 \\ -1 & 2 & 1 \\ 0 & 1 & -1 \end{bmatrix}$$

Solution   By (5.38), the characteristic equation for A is the cubic equation

$$c(\lambda) = \begin{vmatrix} \lambda - 1 & -1 & 2 \\ 1 & \lambda - 2 & -1 \\ 0 & -1 & \lambda + 1 \end{vmatrix} = 0$$

Expanding the determinant along the first column gives

$$c(\lambda) = (\lambda - 1)\begin{vmatrix} \lambda - 2 & -1 \\ -1 & \lambda + 1 \end{vmatrix} - \begin{vmatrix} -1 & 2 \\ -1 & \lambda + 1 \end{vmatrix}$$

$$= (\lambda - 1)[(\lambda - 2)(\lambda + 1) - 1] - [2 - (\lambda + 1)]$$

Thus

$$c(\lambda) = \lambda^3 - 2\lambda^2 - \lambda + 2 = 0$$

is the required characteristic equation.

For matrices of large order, determining the characteristic polynomial by direct expansion of $|\lambda I - A|$ is unsatisfactory in view of the large number of terms involved in the determinant expansion but alternative procedures are available.

5.7.2 Eigenvalues and eigenvectors

The roots of the characteristic equation (5.38) are called the eigenvalues of the matrix A (the terms latent roots, proper roots and characteristic roots are also sometimes used). By the Fundamental Theorem of Algebra, a polynomial equation of degree n has exactly n roots, so that the matrix A has exactly n eigenvalues λ_i, $i = 1, 2, \ldots, n$. These eigenvalues may be real or complex, and not necessarily distinct. Corresponding to each eigenvalue λ_i, there is a non-zero solution $x = e_i$ of (5.37); e_i is called the eigenvector of A corresponding to the eigenvalue λ_i. (Again the terms latent vector, proper vector and characteristic vector are sometimes seen, but are generally obsolete.) We note that if $x = e_i$ satisfies (5.37) then any scalar multiple $\beta_i e_i$ of e_i also satisfies (5.37), so that the eigenvector e_i may only be determined to within a scalar multiple.

Example 5.38   Verify that $\begin{bmatrix} 1 \\ 1 \end{bmatrix}$ and $\begin{bmatrix} 1 \\ -1 \end{bmatrix}$ are eigenvectors of the matrix

$$A = \begin{bmatrix} -2 & 1 \\ 1 & -2 \end{bmatrix}$$

Solution   The matrix is the same as the one given in Example 5.36 so we would expect that these eigenvectors correspond to the eigenvalues -1 and -3. To verify the fact we must check that equation (5.36) is satisfied. Now for the first column vector

$$\begin{bmatrix} -2 & 1 \\ 1 & -2 \end{bmatrix}\begin{bmatrix} 1 \\ 1 \end{bmatrix} = \begin{bmatrix} -1 \\ -1 \end{bmatrix} = -1\begin{bmatrix} 1 \\ 1 \end{bmatrix}$$

so $\begin{bmatrix} 1 \\ 1 \end{bmatrix}$ is an eigenvector corresponding to the eigenvector -1.

For the second column vector

$$\begin{bmatrix} -2 & 1 \\ 1 & -2 \end{bmatrix}\begin{bmatrix} 1 \\ -1 \end{bmatrix} = \begin{bmatrix} -3 \\ 3 \end{bmatrix} = -3\begin{bmatrix} 1 \\ -1 \end{bmatrix}$$

so $\begin{bmatrix} 1 \\ -1 \end{bmatrix}$ is an eigenvector corresponding to the eigenvector -3.

Example 5.39   Find the eigenvalues and eigenvectors of the matrix $A = \begin{bmatrix} 0 & -1 \\ 1 & 0 \end{bmatrix}$.

Solution   To find the eigenvalues use equation (5.38)

$$0 = |\lambda I - A| = \begin{vmatrix} \lambda & 1 \\ -1 & \lambda \end{vmatrix} = \lambda^2 + 1$$

This characteristic equation has two roots $\lambda = +j$ and $-j$ which are the eigenvalues, in this case complex. Note that in general eigenvalues are complex although in most of the remaining examples in this section they have been constructed to be real.

To obtain the eigenvectors use equation (5.37).

For the eigenvalue $\lambda = j$ then (5.37) gives

$$(\lambda \mathbf{I} - \mathbf{A}) \begin{bmatrix} a \\ b \end{bmatrix} = \begin{bmatrix} j & 1 \\ -1 & j \end{bmatrix} \begin{bmatrix} a \\ b \end{bmatrix} = 0$$

or in expanded form

$$ja + b = 0$$
$$-a + jb = 0$$

with solution $a = j$ and $b = 1$

and hence the eigenvector corresponding to $\lambda = j$ is $\begin{bmatrix} j \\ 1 \end{bmatrix}$.

For the eigenvalue $\lambda = -j$ then (5.37) gives

$$(\lambda \mathbf{I} - \mathbf{A}) \begin{bmatrix} c \\ d \end{bmatrix} = \begin{bmatrix} -j & 1 \\ -1 & -j \end{bmatrix} \begin{bmatrix} c \\ d \end{bmatrix} = 0$$

or in expanded form

$$-jc + d = 0$$
$$-c - jd = 0$$

with solution $c = 1$ and $d = j$

and hence the eigenvector corresponding to $\lambda = -j$ is $\begin{bmatrix} 1 \\ j \end{bmatrix}$.

Example 5.40   Determine the eigenvalues and eigenvectors for the matrix $\mathbf{A}$ of Example 5.37.

Solution

$$\mathbf{A} = \begin{bmatrix} 1 & 1 & -2 \\ -1 & 2 & 1 \\ 0 & 1 & -1 \end{bmatrix}$$

The eigenvalues λ_i of $\mathbf{A}$ satisfy the characteristic equation $c(\lambda) = 0$, and this has been obtained in Example 5.37 as the cubic

$$\lambda^3 - 2\lambda^2 - \lambda + 2 = 0$$

which can be solved to obtain the eigenvalues λ_1, λ_2 and λ_3.

Alternatively, it may be possible, using the determinant form $|\lambda \mathbf{I} - \mathbf{A}|$, or indeed (as we often do when seeking the eigenvalues) the form $|\mathbf{A} - \lambda \mathbf{I}|$, by carrying out suitable row and/or column operations to factorize the determinant.

In this case

$$|A - \lambda I| = \begin{vmatrix} 1 - \lambda & 1 & -2 \\ -1 & 2 - \lambda & 1 \\ 0 & 1 & -1 - \lambda \end{vmatrix}$$

and adding column 1 to column 3 gives

$$\begin{vmatrix} 1 - \lambda & 1 & -1 - \lambda \\ -1 & 2 - \lambda & 0 \\ 0 & 1 & -1 - \lambda \end{vmatrix} = -(1 + \lambda) \begin{vmatrix} 1 - \lambda & 1 & 1 \\ -1 & 2 - \lambda & 0 \\ 0 & 1 & 1 \end{vmatrix}$$

Subtracting row 3 from row 1 gives

$$-(1 + \lambda) \begin{vmatrix} 1 - \lambda & 0 & 0 \\ -1 & 2 - \lambda & 0 \\ 0 & 1 & 1 \end{vmatrix} = -(1 + \lambda)(1 - \lambda)(2 - \lambda)$$

Setting $|A - \lambda I| = 0$ gives the eigenvalues as $\lambda_1 = 2$, $\lambda_2 = 1$ and $\lambda_3 = -1$. The order in which they are written is arbitrary, but for consistency we shall adopt the convention of taking $\lambda_1, \lambda_2, \ldots, \lambda_n$ in decreasing order.

Having obtained the eigenvalues λ_i ($i = 1, 2, 3$), the corresponding eigenvectors e_i are obtained by solving the appropriate homogeneous equations

$$(A - \lambda_i I)e_i = 0 \tag{5.39}$$

When $i = 1$, $\lambda_i = \lambda_1 = 2$ and (5.39) is

$$\begin{bmatrix} -1 & 1 & -2 \\ -1 & 0 & 1 \\ 0 & 1 & -3 \end{bmatrix} \begin{bmatrix} e_{11} \\ e_{12} \\ e_{13} \end{bmatrix} = 0$$

that is,

$$-e_{11} + e_{12} - 2e_{13} = 0$$

$$-e_{11} + 0e_{12} + e_{13} = 0$$

$$0e_{11} + e_{12} - 3e_{13} = 0$$

leading to the solution

$$\frac{e_{11}}{-1} = \frac{-e_{12}}{3} = \frac{e_{13}}{-1} = \beta_1$$

where β_1 is an arbitrary non-zero scalar. Thus the eigenvector e_1 corresponding to the eigenvalue $\lambda_1 = 2$ is

$$e_1 = \beta_1 [1 \quad 3 \quad 1]^{\mathrm{T}}$$

As a check, we can compute

$$\boldsymbol{Ae}_1 = \beta_1 \begin{bmatrix} 1 & 1 & -2 \\ -1 & 2 & 1 \\ 0 & 1 & -1 \end{bmatrix} \begin{bmatrix} 1 \\ 3 \\ 1 \end{bmatrix} = \beta_1 \begin{bmatrix} 2 \\ 6 \\ 2 \end{bmatrix} = 2\beta_1 \begin{bmatrix} 1 \\ 3 \\ 1 \end{bmatrix} = \lambda_1 \boldsymbol{e}_1$$

and thus conclude that our calculation was correct.

When $i = 2$, $\lambda_i = \lambda_2 = 1$ and we have to solve

$$\begin{bmatrix} 0 & 1 & -2 \\ -1 & 1 & 1 \\ 0 & 1 & -2 \end{bmatrix} \begin{bmatrix} e_{21} \\ e_{22} \\ e_{23} \end{bmatrix} = 0$$

that is,

$$0e_{21} + e_{22} - 2e_{23} = 0$$

$$-e_{21} + e_{22} + e_{23} = 0$$

$$0e_{21} + e_{22} - 2e_{23} = 0$$

leading to the solution

$$\frac{e_{21}}{-3} = \frac{-e_{22}}{2} = \frac{e_{23}}{-1} = \beta_2$$

where β_2 is an arbitrary scalar. Thus the eigenvector $\boldsymbol{e}_2$ corresponding to the eigenvalue $\lambda_2 = 1$ is

$$\boldsymbol{e}_2 = \beta_2 [3 \quad 2 \quad 1]^{\mathrm{T}}$$

Again a check could be made by computing $\boldsymbol{Ae}_2$.

Finally, when $i = 3$, $\lambda_i = \lambda_3 = -1$ and we obtain from (5.39)

$$\begin{bmatrix} 2 & 1 & -2 \\ -1 & 3 & 1 \\ 0 & 1 & 0 \end{bmatrix} \begin{bmatrix} e_{31} \\ e_{32} \\ e_{33} \end{bmatrix} = 0$$

that is,

$$2e_{31} + e_{32} - 2e_{33} = 0$$

$$-e_{31} + 3e_{32} + e_{33} = 0$$

$$0e_{31} + e_{32} + 0e_{33} = 0$$

and hence

$$\frac{e_{31}}{-1} = \frac{e_{32}}{0} = \frac{e_{33}}{-1} = \beta_3$$

Here again β_3 is an arbitrary scalar, and the eigenvector e_3 corresponding to the eigenvalue λ_3 is

$$e_3 = \beta_3 [1 \quad 0 \quad 1]^T$$

The calculation can be checked as before. Thus we have found that the eigenvalues of the matrix A are 2, 1 and -1, with corresponding eigenvectors

$$\beta_1 [1 \quad 3 \quad 1]^T, \quad \beta_2 [3 \quad 2 \quad 1]^T \quad \text{and} \quad \beta_3 [1 \quad 0 \quad 1]^T$$

respectively.

Since in Example 5.40 the β_i, $i = 1, 2, 3$, are arbitrary, it follows that there are an infinite number of eigenvectors, scalar multiples of each other, corresponding to each eigenvalue. Sometimes it is convenient to scale the eigenvectors according to some convention. A convention frequently adopted is to **normalize** the eigenvectors so that they are uniquely determined up to a scale factor of ± 1. The normalized form of an eigenvector $e = [e_1 \quad e_2 \quad \ldots \quad e_n]^T$ is denoted by $\hat{e}$ and is given by

$$\hat{e} = \frac{e}{|e|}$$

where

$$|e| = \sqrt{(e_1^2 + e_2^2 + \ldots + e_n^2)}$$

For example, for the matrix A of Example 5.40, the normalized forms of the eigenvectors are

$$\hat{e}_1 = [1/\sqrt{11} \quad 3/\sqrt{11} \quad 1/\sqrt{11}]^T, \quad \hat{e}_2 = [3/\sqrt{14} \quad 2/\sqrt{14} \quad 1/\sqrt{14}]^T$$

and

$$\hat{e}_3 = [1/\sqrt{2} \quad 0 \quad 1/\sqrt{2}]^T$$

However, throughout the text, unless otherwise stated, the eigenvectors will always be presented in their 'simplest' form, so that for the matrix of Example 5.40 we take $\beta_1 = \beta_2 = \beta_3 = 1$ and write

$$e_1 = [1 \quad 3 \quad 1]^T, \quad e_2 = [3 \quad 2 \quad 1]^T \quad \text{and} \quad e_3 = [1 \quad 0 \quad 1]^T$$

In many computer packages eigenvectors are given automatically in their normalized form. For instance for the matrix in Example 5.40 the MATLAB instruction

$$[u, v] = \text{eig}(A)$$

produces the output

$$u = \begin{matrix} -0.8018 & 0.3015 & 0.7071 \\ -0.5345 & 0.9045 & 0.0000 \\ -0.2673 & 0.3015 & 0.7071 \end{matrix} \qquad v = \begin{matrix} 1.0000 & 0 & 0 \\ 0 & 2.0000 & 0 \\ 0 & 0 & -1.0000 \end{matrix}$$

with the three normalized eigenvectors given as the columns of u and the three corresponding eigenvalues given on the diagonal of v.

Example 5.41   Find the eigenvalues and eigenvectors of

$$A = \begin{bmatrix} \cos\theta & -\sin\theta \\ \sin\theta & \cos\theta \end{bmatrix}$$

Solution   Now

$$|\lambda I - A| = \begin{vmatrix} \lambda - \cos\theta & \sin\theta \\ -\sin\theta & \lambda - \cos\theta \end{vmatrix}$$

$$= \lambda^2 - 2\lambda\cos\theta + \cos^2\theta + \sin^2\theta = \lambda^2 - 2\lambda\cos\theta + 1$$

So the eigenvalues are the roots of

$$\lambda^2 - 2\lambda\cos\theta + 1 = 0$$

that is,

$$\lambda = \cos\theta \pm j\sin\theta$$

Solving for the eigenvectors as in Example 5.40, we obtain

$$e_1 = [1 \quad -j]^T \quad \text{and} \quad e_2 = [1 \quad j]^T$$

In Examples 5.39 and 5.41 we see that eigenvalues can be complex numbers, and that the eigenvectors may have complex components. This situation arises when the characteristic equation has complex (conjugate) roots.

5.7.3   Exercises

70 Obtain the characteristic polynomials of the matrices

(a) $\begin{bmatrix} 2 & -1 \\ -1 & 2 \end{bmatrix}$ (b) $\begin{bmatrix} 2 & 1 \\ 1 & 1 \end{bmatrix}$ (c) $\begin{bmatrix} 1 & 2 & 3 \\ 0 & 2 & 3 \\ 0 & 0 & 3 \end{bmatrix}$

(d) $\begin{bmatrix} 1 & 2 & 0 \\ 0 & 2 & 2 \\ 0 & 1 & 3 \end{bmatrix}$ (e) $\begin{bmatrix} 3 & 2 & 1 \\ 4 & 5 & -1 \\ 2 & 3 & 4 \end{bmatrix}$

and hence evaluate the eigenvalues of the matrices. Use a computer package to check your working.

71 Find the eigenvalues and corresponding eigenvectors of the matrices

(a) $\begin{bmatrix} 1 & 1 \\ 1 & 1 \end{bmatrix}$ (b) $\begin{bmatrix} 1 & 2 \\ 3 & 2 \end{bmatrix}$

(c) $\begin{bmatrix} 1 & 0 & -4 \\ 0 & 5 & 4 \\ -4 & 4 & 3 \end{bmatrix}$ (d) $\begin{bmatrix} 1 & 1 & 2 \\ 0 & 2 & 2 \\ -1 & 1 & 3 \end{bmatrix}$

(e) $\begin{bmatrix} 5 & 0 & 6 \\ 0 & 11 & 6 \\ 6 & 6 & -2 \end{bmatrix}$ (f) $\begin{bmatrix} 1 & -1 & 0 \\ 1 & 2 & 1 \\ -2 & 1 & -1 \end{bmatrix}$

(g) $\begin{bmatrix} 4 & 1 & 1 \\ 2 & 5 & 4 \\ -1 & -1 & 0 \end{bmatrix}$ (h) $\begin{bmatrix} 1 & -4 & -2 \\ 0 & 3 & 1 \\ 1 & 2 & 4 \end{bmatrix}$

 Use a computer package to check your working.

5.7.4   Repeated eigenvalues

In the examples considered so far the eigenvalues λ_i $(i = 1, 2, \ldots)$ of the matrix A have been distinct, and in such cases the corresponding eigenvectors can be found. The matrix A is then said to have a full set of independent eigenvectors. It is clear that the roots of the characteristic equation $c(\lambda)$ may not all be distinct; and when $c(\lambda)$ has $p \leq n$ distinct roots, $c(\lambda)$ may be factorized as

$$c(\lambda) = (\lambda - \lambda_1)^{m_1}(\lambda - \lambda_2)^{m_2} \ldots (\lambda - \lambda_p)^{m_p}$$

indicating that the root $\lambda = \lambda_i$, $i = 1, 2, \ldots, p$, is a root of order m_i, where the integer m_i is called the **algebraic multiplicity** of the eigenvalue λ_i. Clearly $m_1 + m_2 + \ldots + m_p = n$. When a matrix A has repeated eigenvalues, the question arises as to whether it is possible to obtain a full set of independent eigenvectors for A. We first consider some examples to illustrate the situation.

Example 5.42   Determine the eigenvalues and corresponding eigenvectors of the matrices

(a) $A = \begin{bmatrix} 1 & 0 \\ 0 & 1 \end{bmatrix}$ (b) $B = \begin{bmatrix} 1 & 1 \\ 0 & 1 \end{bmatrix}$

Solution   (a) The eigenvalues of A are obtained from

$$0 = |\lambda I - A| = \begin{vmatrix} \lambda - 1 & 0 \\ 0 & \lambda - 1 \end{vmatrix} = (\lambda - 1)^2$$

giving the value 1 repeated twice.
    The eigenvectors we calculate from

$$0 = (I - A)\begin{bmatrix} a \\ b \end{bmatrix} = \begin{bmatrix} 0 & 0 \\ 0 & 0 \end{bmatrix}\begin{bmatrix} a \\ b \end{bmatrix}$$

which is clearly satisfied by any values of a and b. Thus taking

$$\begin{bmatrix} a \\ b \end{bmatrix} = a\begin{bmatrix} 1 \\ 0 \end{bmatrix} + b\begin{bmatrix} 0 \\ 1 \end{bmatrix}$$

it can be seen that there are two independent eigenvectors $\begin{bmatrix} 1 \\ 0 \end{bmatrix}$ and $\begin{bmatrix} 0 \\ 1 \end{bmatrix}$. Any linear combination of the two vectors is also an eigenvector. Geometrically this corresponds to the fact that the unit matrix maps *every* vector on to itself.

(b) The eigenvalues of B are obtained from

$$0 = |\lambda I - B| = \begin{vmatrix} \lambda - 1 & -1 \\ 0 & \lambda - 1 \end{vmatrix} = (\lambda - 1)^2$$

giving the value 1 repeated twice.

The eigenvectors we calculate from

$$0 = (I - B)\begin{bmatrix} c \\ d \end{bmatrix} = \begin{bmatrix} 0 & -1 \\ 0 & 0 \end{bmatrix}\begin{bmatrix} c \\ d \end{bmatrix} = \begin{bmatrix} -d \\ 0 \end{bmatrix}$$

Thus $d = 0$ and there is *only one* eigenvector $\begin{bmatrix} 1 \\ 0 \end{bmatrix}$ and, of course, any multiple of this vector.

We note from Example 5.42 that the evaluation of eigenvectors is much more complicated when there are multiple eigenvalues. The idea of rank, introduced in Section 5.6, is required to sort out the complications but the details are left to the companion text *Advanced Modern Engineering Mathematics*.

The following two 3×3 examples illustrate similar points.

Example 5.43   Determine the eigenvalues and corresponding eigenvectors of the matrix

$$A = \begin{bmatrix} 3 & -3 & 2 \\ -1 & 5 & -2 \\ -1 & 3 & 0 \end{bmatrix}$$

Solution   We find the eigenvalues from

$$\begin{vmatrix} 3 - \lambda & -3 & 2 \\ -1 & 5 - \lambda & -2 \\ -1 & 3 & -\lambda \end{vmatrix} = 0$$

as $\lambda_1 = 4$, $\lambda_2 = \lambda_3 = 2$.

The eigenvectors are obtained from

$$(A - \lambda I)e_i = 0 \tag{5.40}$$

and when $\lambda = \lambda_1 = 4$, we obtain from (5.40)

$$e_1 = [1 \quad -1 \quad -1]^T$$

When $\lambda = \lambda_2 = \lambda_3 = 2$, (5.40) becomes

$$\begin{bmatrix} 1 & -3 & 2 \\ -1 & 3 & -2 \\ -1 & 3 & -2 \end{bmatrix}\begin{bmatrix} e_{21} \\ e_{22} \\ e_{23} \end{bmatrix} = 0$$

so that the corresponding eigenvector is obtained from the single equation

$$e_{21} - 3e_{22} + 2e_{23} = 0 \tag{5.41}$$

Clearly we are free to choose any two of the components e_{21}, e_{22} or e_{23} at will, with the remaining one determined by (5.41). Suppose we set $e_{22} = \alpha$ and $e_{23} = \beta$; then (5.41) means that $e_{21} = 3\alpha - 2\beta$, and thus

$$e_2 = [3\alpha - 2\beta \quad \alpha \quad \beta]^\text{T}$$

$$= \alpha \begin{bmatrix} 3 \\ 1 \\ 0 \end{bmatrix} + \beta \begin{bmatrix} -2 \\ 0 \\ 1 \end{bmatrix} \tag{5.42}$$

Now $\lambda = 2$ is an eigenvalue of multiplicity 2, and we seek, if possible, two independent eigenvectors defined by (5.42). Setting $\alpha = 1$ and $\beta = 0$ yields

$$e_2 = [3 \quad 1 \quad 0]^\text{T}$$

and setting $\alpha = 0$ and $\beta = 1$ gives a second vector

$$e_3 = [-2 \quad 0 \quad 1]^\text{T}$$

These two vectors are independent and of the form defined by (5.42), and it is clear that many other choices are possible. However, any other choices of the form (5.42) will be linear combinations of e_2 and e_3 as chosen above. For example, $e = [1 \quad 1 \quad 1]$ satisfies (5.42), but $e = e_2 + e_3$.

In this example, although there was a repeated eigenvalue of algebraic multiplicity 2, it was possible to construct two independent eigenvectors corresponding to this eigenvalue. Thus the matrix A has three and only three independent eigenvectors.

Example 5.44 Determine the eigenvalues and corresponding eigenvectors of the matrix

$$A = \begin{bmatrix} 1 & 2 & 2 \\ 0 & 2 & 1 \\ -1 & 2 & 2 \end{bmatrix}$$

Solution Solving $|A - \lambda I| = 0$ gives the eigenvalues as $\lambda_1 = \lambda_2 = 2$, $\lambda_3 = 1$. The eigenvector corresponding to the non-repeated or simple eigenvalue $\lambda_3 = 1$ is easily found as

$$e_3 = [1 \quad 1 \quad -1]^\text{T}$$

When $\lambda = \lambda_1 = \lambda_2 = 2$, the corresponding eigenvector is given by

$$(A - 2I)e_1 = 0$$

that is, as the solution of

$$-e_{11} + 2e_{12} + 2e_{13} = 0 \tag{i}$$

$$e_{13} = 0 \tag{ii}$$

$$-e_{11} + 2e_{12} \phantom{+ 2e_{13}} = 0 \tag{iii}$$

From (ii) we have $e_{13} = 0$, and from (i) and (ii) it follows that $e_{11} = 2e_{12}$. We deduce that there is only one independent eigenvector corresponding to the repeated eigenvalue $\lambda = 2$, namely

$$e_1 = [2 \quad 1 \quad 0]^{\mathrm{T}}$$

and in this case the matrix $\mathbf{A}$ does not possess a full set of independent eigenvectors.

We see from Examples 5.42–5.44 that if an $n \times n$ matrix $\mathbf{A}$ has repeated eigenvalues then a full set of n independent eigenvectors may or may not exist.

5.7.5 Exercises

72 Obtain the eigenvalues and corresponding eigenvectors of the matrices

(a) $\begin{bmatrix} 2 & 2 & 1 \\ 1 & 3 & 1 \\ 1 & 2 & 2 \end{bmatrix}$ (b) $\begin{bmatrix} 0 & -2 & -2 \\ -1 & 1 & 2 \\ -1 & -1 & 2 \end{bmatrix}$

(c) $\begin{bmatrix} 4 & 6 & 6 \\ 1 & 3 & 2 \\ -1 & -5 & -2 \end{bmatrix}$ (d) $\begin{bmatrix} 7 & -2 & -4 \\ 3 & 0 & -2 \\ 6 & -2 & -3 \end{bmatrix}$

73 Given that $\lambda = 1$ is a three-times repeated eigenvalue of the matrix

$$\mathbf{A} = \begin{bmatrix} -3 & -7 & -5 \\ 2 & 4 & 3 \\ 1 & 2 & 2 \end{bmatrix}$$

determine how many independent eigenvectors correspond to this value of λ. Determine a corresponding set of independent eigenvectors.

74 Given that $\lambda = 1$ is a twice-repeated eigenvalue of the matrix

$$\mathbf{A} = \begin{bmatrix} 2 & 1 & -1 \\ -1 & 0 & 1 \\ -1 & -1 & 2 \end{bmatrix}$$

determine a set of independent eigenvectors.

5.7.6 Some useful properties of eigenvalues

The following basic properties of the eigenvalues $\lambda_1, \lambda_2, \ldots, \lambda_n$ of an $n \times n$ matrix $\mathbf{A}$ are sometimes useful. The results are readily proved from either the definition of eigenvalues as the values of λ satisfying (5.36), or by comparison of corresponding characteristic polynomials (5.38). Consequently, the proofs are left to Exercise 75.

Property 1

The sum of the eigenvalues of $\mathbf{A}$ is

$$\sum_{i=1}^{n} \lambda_i = \text{trace } \mathbf{A} = \sum_{i=1}^{n} a_{ii}$$

Property 2

The product of the eigenvalues of $\boldsymbol{A}$ is

$$\prod_{i=1}^{n} \lambda_i = \det \boldsymbol{A}$$

where $\det \boldsymbol{A}$ denotes the determinant of the matrix $\boldsymbol{A}$.

Property 3

The eigenvalues of the inverse matrix $\boldsymbol{A}^{-1}$, provided it exists, are

$$\frac{1}{\lambda_1}, \quad \frac{1}{\lambda_2}, \quad \ldots, \quad \frac{1}{\lambda_n}$$

Property 4

The eigenvalues of the transposed matrix $\boldsymbol{A}^{\mathrm{T}}$ are

$$\lambda_1, \quad \lambda_2, \quad \ldots, \quad \lambda_n$$

as for the matrix $\boldsymbol{A}$.

Property 5

If k is a scalar then the eigenvalues of $k\boldsymbol{A}$ are

$$k\lambda_1, \quad k\lambda_2, \quad \ldots, \quad k\lambda_n$$

Property 6

If k is a scalar and $\boldsymbol{I}$ the $n \times n$ identity (unit) matrix then the eigenvalues of $\boldsymbol{A} \pm k\boldsymbol{I}$ are respectively

$$\lambda_1 \pm k, \quad \lambda_2 \pm k, \quad \ldots, \quad \lambda_n \pm k$$

Property 7

If k is a positive integer then the eigenvalues of $\boldsymbol{A}^k$ are

$$\lambda_1^k, \quad \lambda_2^k, \quad \ldots, \quad \lambda_n^k$$

5.7.7   Symmetric matrices

A square matrix $\boldsymbol{A}$ is said to be **symmetric** if $\boldsymbol{A}^\mathrm{T} = \boldsymbol{A}$. Such matrices form an important class and arise in a variety of practical situations. Two important results concerning the eigenvalues and eigenvectors of such matrices are:

> (i)   The eigenvalues of a real symmetric matrix are real.
> (ii) For an $n \times n$ real symmetric matrix it is always possible to find n independent eigenvectors $\boldsymbol{e}_1, \boldsymbol{e}_2, \ldots, \boldsymbol{e}_n$ that are mutually orthogonal so that $\boldsymbol{e}_i^\mathrm{T} \boldsymbol{e}_j = 0$ for $i \neq j$.

If the orthogonal eigenvectors of a symmetric matrix are normalized as

$$\hat{\boldsymbol{e}}_1, \hat{\boldsymbol{e}}_2, \ldots, \hat{\boldsymbol{e}}_n$$

then the **inner (scalar) product** is

$$\hat{\boldsymbol{e}}_i^\mathrm{T} \hat{\boldsymbol{e}}_j = \delta_{ij} \quad (i, j = 1, 2, \ldots, n)$$

where δ_{ij} is the Kronecker delta defined in Section 5.2.1.

The set of normalized eigenvectors of a symmetric matrix therefore form an orthonormal set (that is, they form a mutually orthogonal normalized set of vectors).

Example 5.45   Obtain the eigenvalues and corresponding orthogonal eigenvectors of the symmetric matrix

$$\boldsymbol{A} = \begin{bmatrix} 2 & 2 & 0 \\ 2 & 5 & 0 \\ 0 & 0 & 3 \end{bmatrix}$$

and show that the normalized eigenvectors form an orthonormal set.

Solution   The eigenvalues of $\boldsymbol{A}$ are $\lambda_1 = 6$, $\lambda_2 = 3$ and $\lambda_3 = 1$, with corresponding eigenvectors

$$\boldsymbol{e}_1 - [1 \quad 2 \quad 0]^\mathrm{T}, \quad \boldsymbol{e}_2 - [0 \quad 0 \quad 1]^\mathrm{T}, \quad \boldsymbol{e}_3 = [-2 \quad 1 \quad 0]^\mathrm{T}$$

which in normalized form are

$$\hat{\boldsymbol{e}}_1 = [1 \quad 2 \quad 0]^\mathrm{T}/\sqrt{5}, \quad \hat{\boldsymbol{e}}_2 = [0 \quad 0 \quad 1]^\mathrm{T}, \quad \hat{\boldsymbol{e}}_3 = [-2 \quad 1 \quad 0]^\mathrm{T}/\sqrt{5}$$

Evaluating the inner products, we see that, for example,

$$\hat{\boldsymbol{e}}_1^\mathrm{T} \hat{\boldsymbol{e}}_1 = \tfrac{1}{5} + \tfrac{4}{5} + 0 = 1, \quad \hat{\boldsymbol{e}}_1^\mathrm{T} \hat{\boldsymbol{e}}_3 = -\tfrac{2}{5} + \tfrac{2}{5} + 0 = 0$$

and that

$$\hat{\boldsymbol{e}}_i^\mathrm{T} \hat{\boldsymbol{e}}_j = \delta_{ij} \quad (i, j = 1, 2, 3)$$

confirming that the eigenvectors form an orthonormal set.

5.7.8 Exercises

75 Verify Properties 1–7 of Section 5.7.6.

76 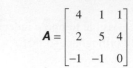 Given that the eigenvalues of the matrix

$$A = \begin{bmatrix} 4 & 1 & 1 \\ 2 & 5 & 4 \\ -1 & -1 & 0 \end{bmatrix}$$

are 5, 3 and 1:

(a) confirm Properties 1–4 of Section 5.7.6;

(b) taking $k = 2$, confirm Properties 5–7 of Section 5.7.6.

77 Determine the eigenvalues and corresponding eigenvectors of the symmetric matrix

$$A = \begin{bmatrix} -3 & -3 & -3 \\ -3 & 1 & -1 \\ -3 & -1 & 1 \end{bmatrix}$$

and verify that the eigenvectors are mutually orthogonal.

78 The 3×3 symmetric matrix A has eigenvalues 6, 3 and 2. The eigenvectors corresponding to the eigenvalues 6 and 3 are $[1 \quad 1 \quad 2]^T$ and $[1 \quad 1 \quad -1]^T$ respectively. Find an eigenvector corresponding to the eigenvalue 2.

79 Verify that the matrix

$$A = \begin{bmatrix} -\frac{3}{20} & \frac{1}{5} \\ \frac{1}{5} & \frac{3}{20} \end{bmatrix}$$

has eigenvalues $\pm\frac{1}{4}$ and corresponding eigenvectors $X = \begin{bmatrix} 2 \\ -1 \end{bmatrix}$ and $Y = \begin{bmatrix} 1 \\ 2 \end{bmatrix}$. What are the eigenvalues of A^n? Show that any vector $Z = \begin{bmatrix} a \\ b \end{bmatrix}$ can be written as $Z = \alpha X + \beta Y$ and hence deduce that $A^n Z \to 0$ as $n \to \infty$.

5.8 Engineering application: spring systems

The vibration of many mechanical systems can be modelled very satisfactorily by spring and damper systems. The shock absorbers and springs of a motor car give one of the simplest practical examples. On a more fundamental level, the vibration of the atoms or molecules of a solid can be modelled by a lattice containing atoms or molecules that interact with each other through spring forces. The model gives a detailed understanding of the structure of the solid and the strength of interactions and has practical applications in such areas as the study of impurities or 'doped' materials in semiconductor physics.

The motion of these systems demands the use of Newton's equations, which in turn require the calculus. We shall look at methods of solution in Chapters 10 and 11. In this case study we shall not consider vibrations but shall restrict our attention to the static situation. This is the first step in the solution of vibrational systems. Even here, we shall see that matrices and vectors allow a systematic approach to the more complicated situation.

5.8.1 A two-particle system

We start with the very simple situation illustrated in Figure 5.23. Two masses are connected by springs of stiffnesses k_1, k_2 and k_3 and of natural lengths l_1, l_2 and l_3 that are fixed to the walls at A and B, with distance $AB = L$. It is required to calculate the equilibrium values of x_1 and x_2. We use Hooke's law – that force is proportional to extension – to calculate the tension:

$$T_1 = k_1(x_1 - l_1)$$

$$T_2 = k_2(x_2 - x_1 - l_2)$$

$$T_3 = k_3(L - x_2 - l_3)$$

Figure 5.23
Two-particle system.

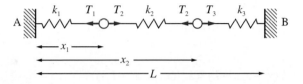

Since the forces are in equilibrium,

$$k_1(x_1 - l_1) = k_2(x_2 - x_1 - l_2)$$

$$k_2(x_2 - x_1 - l_2) = k_3(L - x_2 - l_3)$$

We have two simultaneous equations in the two unknowns, which can be written in matrix form as

$$\begin{bmatrix} k_1 + k_2 & -k_2 \\ k_2 & k_2 + k_3 \end{bmatrix} \begin{bmatrix} x_1 \\ x_2 \end{bmatrix} = \begin{bmatrix} k_1 l_1 - k_2 l_2 \\ k_2 l_2 - k_3 l_3 + k_3 L \end{bmatrix}$$

It is easy to invert 2×2 matrices, so we can compute the solution as

$$\begin{bmatrix} x_1 \\ x_2 \end{bmatrix} = \frac{1}{(k_1 + k_2)(k_2 + k_3) - k_2^2} \begin{bmatrix} k_2 + k_3 & k_2 \\ k_2 & k_1 + k_2 \end{bmatrix} \begin{bmatrix} k_1 l_1 - k_2 l_2 \\ k_2 l_2 - k_3 l_3 + k_3 L \end{bmatrix}$$

If we take the simplest situation when $k_1 = k_2 = k_3$ and $l_1 = l_2 = l_3$ then we obtain the obvious solution $x_1 = \frac{1}{3}L$, $x_2 = \frac{2}{3}L$.

5.8.2 An *n*-particle system

In the simplest situation, described in Section 5.8.1, matrix notation is convenient but not really necessary. If we try to extend the problem to many particles and many springs then such notation simplifies the statement of the problem considerably. Consider the problem illustrated in Figure 5.24. From Hooke's law

Figure 5.24
n-particle system.

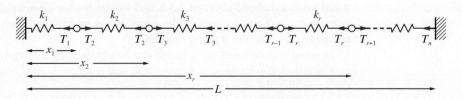

$$T_1 = k_1(x_1 - l_1)$$

$$T_2 = k_2(x_2 - x_1 - l_2)$$

$$T_3 = k_3(x_3 - x_2 - l_3)$$

$$\vdots$$

$$T_r = k_r(x_r - x_{r-1} - l_r)$$

$$\vdots$$

$$T_n = k_n(L - x_{n-1} - l_n)$$

The equilibrium equations for each 'unit' are

$$k_1(x_1 - l_1) = k_2(x_2 - x_1 - l_2)$$

$$k_2(x_2 - x_1 - l_2) = k_3(x_3 - x_2 - l_3)$$

$$\vdots$$

$$k_r(x_r - x_{r-1} - l_r) = k_{r+1}(x_{r+1} - x_r - l_{r+1})$$

$$\vdots$$

$$k_{n-1}(x_{n-1} - x_{n-2} - l_{n-1}) = k_n(L - x_{n-1} - l_n)$$

In matrix form, these become

$$
\begin{bmatrix}
k_1 + k_2 & -k_2 & & & & \\
-k_2 & k_2 + k_3 & -k_3 & & & \mathbf{0} \\
& -k_3 & k_3 + k_4 & -k_4 & & \\
& & \ddots & \ddots & & \ddots \\
\mathbf{0} & & & -k_{n-2} & k_{n-2} + k_{n-1} & -k_{n-1} \\
& & & & -k_{n-1} & k_{n-1} + k_n
\end{bmatrix}
\begin{bmatrix}
x_1 \\
x_2 \\
\vdots \\
x_{n-1}
\end{bmatrix}
$$

$$
=
\begin{bmatrix}
k_1 l_1 - k_2 l_2 \\
k_2 l_2 - k_3 l_3 \\
\vdots \\
k_{n-2} l_{n-2} - k_{n-1} l_{n-1} \\
k_{n-1} l_{n-1} - k_n l_n + k_n L
\end{bmatrix}
$$

We recognize the form of these equations immediately, since they constitute a tridiagonal system studied in Section 5.5.2, and we can use the Thomas algorithm (Figure 5.12) to solve them. Thus, by writing the equations in matrix form, we are immediately able to identify an efficient method of solution.

In some special cases the solution can be obtained by a mixture of insight and physical intuition. If we take $k_1 = k_2 = \ldots = k_n$ and $l_1 = l_2 = \ldots = l_n$ and the couplings are all the same then the equations become

$$
\begin{bmatrix}
2 & -1 & & & & & & \\
-1 & 2 & -1 & & & & \Large 0 & \\
 & -1 & 2 & -1 & & & & \\
 & & -1 & 2 & & & & \\
 & & & & \ddots & \ddots & \ddots & \\
 & \Large 0 & & & -1 & 2 & -1 & \\
 & & & & & -1 & 2
\end{bmatrix}
\begin{bmatrix}
x_1 \\ x_2 \\ \vdots \\ \\ \\ \\ x_{n-1}
\end{bmatrix}
=
\begin{bmatrix}
0 \\ 0 \\ \vdots \\ \\ \\ 0 \\ L
\end{bmatrix}
$$

We should expect all the spacings to be uniform, so we seek a solution $x_1 = \alpha$, $x_2 = 2\alpha$, $x_3 = 3\alpha$, ... The first $n - 2$ equations are satisfied identically, as expected, and the final equation in matrix formulation gives $[-(n - 2) + 2(n - 1)]\alpha = L$. Thus $\alpha = L/n$, and our intuitive solution is justified.

In a second special case where a simple solution is possible, we assume one of the couplings to be a 'rogue'. We take $k_1 = k_2 = \ldots = k_{r-1} = k_{r+1} = \ldots = k_n = k$, $k_r = k'$ and $l_1 = l_2 = \ldots = l_n = l$. If we divide all the equations in the matrix by k and write $\lambda = k'/k$ then the matrix takes the form

$$
\begin{bmatrix}
2 & -1 & & & & & & & & & \\
-1 & 2 & -1 & & & & & & \Large 0 & & \\
 & -1 & 2 & -1 & & & & & & & \\
 & & \ddots & & \ddots & & & \ddots & & & \\
 & & & & -1 & 2 & -1 & & & & \\
 & & & & & -1 & 1+\lambda & -\lambda & & & \\
 & & & & & & -\lambda & 1+\lambda & -1 & & \\
 & & & & & & & -1 & 2 & -1 & \\
 & & & & & & & & \ddots & \ddots & \ddots \\
 & \Large 0 & & & & & & & & & -1 \\
 & & & & & & & & & -1 & 2
\end{bmatrix}
\begin{bmatrix}
x_1 \\ x_2 \\ \vdots \\ \\ \\ x_{r-1} \\ x_r \\ \vdots \\ \\ \\ x_{n-1}
\end{bmatrix}
=
\begin{bmatrix}
0 \\ 0 \\ \vdots \\ \\ 0 \\ l(1-\lambda) \\ l(\lambda-1) \\ \vdots \\ \\ 0 \\ \vdots \\ L
\end{bmatrix}
$$

A reasonable assumption is that the spacings between 'good' links are all the same. Thus we try a solution of the form

$$x_1 = a, \quad x_2 = 2a, \quad \ldots, \quad x_{r-1} = (r-1)a, \quad x_r = b$$

$$x_{r+1} = b + a, \quad x_{r+2} = b + 2a, \quad \ldots, \quad x_{n-1} = b + (n - 1 - r)a$$

It can be checked that the matrix equation is satisfied except for the $(r - 1)$th, rth and $(n - 1)$th rows. These give respectively

$$-\lambda b + a(-\lambda + 1 + \lambda r) = l(1 - \lambda)$$

$$\lambda b + a(\lambda - 1 - \lambda r) = l(\lambda - 1)$$

and

$$b + a(n - r) = L$$

The first two of these are identical, so we have two equations in the two unknowns, a and b, to solve. We obtain

$$a = \frac{L - l(1 - \lambda^{-1})}{n - (1 - \lambda^{-1})}, \quad b = \frac{rL - (1 - \lambda^{-1})[L - (n - r)l]}{n - (1 - \lambda^{-1})}$$

We note that if $\lambda = 1$ then the solution reduces to the previous one, as expected.

The solution just obtained gives the deformation due to a single rogue coupling. Although this problem is of limited interest, its two- and three-dimensional extensions are of great interest in the theory of crystal lattices. It is possible to determine the deformation due to a single impurity, to compute the effect of two or more impurities and how close they have to be to interact with each other. These are problems with considerable application in materials science.

5.9 Engineering application: steady heat transfer through composite materials

5.9.1 Introduction

In many practical situations heat is transferred through several layers of different materials. Perhaps the simplest example is a double glazing unit, which comprises a layer of glass, a layer of air and another layer of glass. The thermal properties and the thicknesses of the individual layers are known but what is required is the overall thermal properties of the composite unit. How do the overall properties depend on the components? Which parameters are the most important? How sensitive is the overall heat transfer to changes in each of the components?

A second example looks at the thickness of a furnace wall. A furnace wall will comprise three layers: refractory bricks for heat resistance, insulating bricks for heat insulation and steel casing for mechanical protection. Such a furnace is enormously expensive to construct so it is important that the thickness of the wall is minimized subject to acceptable heat losses, working within the serviceable temperatures and known thickness constraints. The basic problem is again to construct a model that will give some idea how heat is transferred through such a composite material.

The basic properties of heat conduction will be discussed, and it will then be seen that matrices give a natural method of solving the theoretical equations of composite layers.

5.9.2 Heat conduction

In its full generality heat conduction forms a part of partial differential equations (see Chapter 9 of *Advanced Modern Engineering Mathematics* 2nd edition). However, for current purposes a simplified one-dimensional version is sufficient. The theory is based on the well-established **Fourier law**:

Heat transferred per unit area is proportional to the temperature gradient.

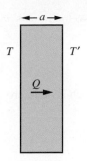

Figure 5.25
Heat transfer
through layers.

Provided a layer is not too thick and the thermal properties do not vary, then the temperature varies linearly across the solid. If Q is the amount of heat transferred per unit area from left, at temperature T, to right, at temperature T', as shown in Figure 5.25, then this law can be written mathematically

$$Q = -k\frac{T' - T}{a}$$

where k is the proportionality constant, called the thermal conductivity, a is the thickness of the layer and the minus sign is to ensure that heat is transferred from hot to cold.

For the conduction *through* an interface between two solids with good contact, as in the situation of the furnace wall, it is assumed that

(i) The temperatures at each side of the interface are equal.

(ii) The heat transferred out of the left side is equal to the heat transferred into the right side.

When the heat is transferred from solid to air, as in the situation of double glazing, it follows **Newton's law of cooling**:

> Heat transferred per unit area from solid to air is proportional to the temperature difference.

In mathematical terms, if T_s is the temperature of the solid and T_a is the temperature of the air, the heat transferred per unit area is $h(T_s - T_a)$ where h is the proportionality constant, called the heat transfer coefficient of the surface.

With the Fourier law and these interface conditions the multilayer situation can be analysed satisfactorily, provided, of course, the heat flow remains one-dimensional and steady.

5.9.3 The three-layer situation

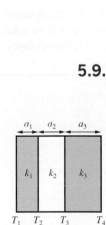

Figure 5.26
Temperature
distribution across
three layers.

Let the three layers have thicknesses a_1, a_2 and a_3 and thermal conductivities k_1, k_2 and k_3, as illustrated in Figure 5.26. At the interfaces the temperatures are taken to be T_1, T_2, T_3 and T_4. The simplest problem to study is to fix the temperatures T_1 and T_4 at the edges and determine how the temperatures T_2 and T_3 depend on the known parameters.

From the specification of the problem the temperatures at the interfaces are specified so it only remains to satisfy the heat transfer condition across the interface.

At the first interface:

$$\frac{k_1}{a_1}(T_2 - T_1) = \frac{k_2}{a_2}(T_3 - T_2)$$

and at the second interface:

$$\frac{k_2}{a_2}(T_3 - T_2) = \frac{k_3}{a_3}(T_4 - T_3)$$

It turns out to be convenient to let $u_1 = \dfrac{a_1}{k_1}$, $u_2 = \dfrac{a_2}{k_2}$ and so on. The equations then become

$$u_2(T_2 - T_1) = u_1(T_3 - T_2)$$
$$u_3(T_3 - T_2) = u_2(T_4 - T_3)$$

or in matrix form

$$\begin{bmatrix} (u_1 + u_2) & -u_1 \\ -u_3 & (u_2 + u_3) \end{bmatrix} \begin{bmatrix} T_2 \\ T_3 \end{bmatrix} = u_2 \begin{bmatrix} T_1 \\ T_4 \end{bmatrix}$$

The determinant of the matrix is easily calculated as $u_2(u_1 + u_2 + u_3)$, which is non-zero so a solution can be computed as

$$\begin{bmatrix} T_2 \\ T_3 \end{bmatrix} = \frac{1}{(u_1 + u_2 + u_3)} \begin{bmatrix} (u_2 + u_3) & u_1 \\ u_3 & (u_1 + u_2) \end{bmatrix} \begin{bmatrix} T_1 \\ T_4 \end{bmatrix}$$

Thus the temperatures T_2 and T_3 are now known, and any required properties can be deduced.

For the furnace problem described in Section 5.9.1 the following data is known:

$$T_1 = 1650 \text{ K} \quad \text{and} \quad T_4 = 300 \text{ K}$$

and

	Maximum working temperature (K)	Thermal conductivity at 100 K ($W\,m^{-1}K^{-1}$)	Thermal conductivity at 2000 K ($W\,m^{-1}K^{-1}$)
Refractory brick	1700	3.1	6.2
Insulating brick	1400	1.6	3.1
Steel	–	45.2	45.2

It may be noted that the thermal conductivity depends on the temperature but in these calculations it is assumed constant (a more sophisticated analysis is required to take these variations into account). Average values $k_1 = 5$, $k_2 = 2.5$ and $k_3 = 45.2$ are chosen. The required temperatures are evaluated as

$$T_2 = \frac{(0.4a_2 + 0.022a_3)1650 + (0.2a_1)300}{0.2a_1 + 0.4a_2 + 0.022a_3}$$

$$T_3 = \frac{(0.022a_3)1650 + (0.2a_1 + 0.4a_2)300}{0.2a_1 + 0.4a_2 + 0.022a_3}$$

A typical question that would be asked is how to minimize the thickness (or perhaps the cost) subject to appropriate constraints. For instance find

$$\min(a_1 + a_2 + a_3)$$

subject to

$$\frac{k_3}{a_3}(300 - T_3) < 50\,000 \qquad \text{(allowable heat loss at the right-hand boundary)}$$

$$T_2 < 1400 \qquad \text{(below the maximum working temperature)}$$

$$a_1 > 0.1 \qquad \text{(must have a minimum refractory thickness)}$$

The problem is a linear programming problem (see Chapter 10 of *Advanced Modern Engineering Mathematics* 2nd edition), which is beyond the scope of the present book, but it illustrates the type of question that can be answered.

A more straightforward question is to evaluate the effective conductivity of the composite. It may be noted that in the general case, the heat flow is

$$Q = -\frac{1}{u_1}(T_2 - T_1) \quad \text{which on substitution gives } Q = -\frac{T_4 - T_1}{u_1 + u_2 + u_3}$$

so the effective conductivity over the whole region is

$$k = \frac{a_1 + a_2 + a_3}{u_1 + u_2 + u_3} \quad \text{or} \quad \frac{a_1 + a_2 + a_3}{k} = \frac{a_1}{k_1} + \frac{a_2}{k_2} + \frac{a_3}{k_3}$$

5.9.4 Many-layer situation

Although matrix theory was used to solve the three-layer problem, it was unnecessary since the mathematics reduced to the solution of a pair of simultaneous equations. However, for the many-layer system it is important to approach the problem in a logical and systematic manner, and matrix theory proves to be the ideal mathematical method to use.

Consider the successive interfaces in turn and construct the heat flow equation for each of them (Figure 5.27).

$$\frac{k_1}{a_1}(T_2 - T_1) = \frac{k_2}{a_2}(T_3 - T_2)$$

$$\frac{k_2}{a_2}(T_3 - T_2) = \frac{k_3}{a_3}(T_4 - T_3)$$

$$\vdots$$

$$\frac{k_{n-1}}{a_{n-1}}(T_n - T_{n-1}) = \frac{k_n}{a_n}(T_{n+1} - T_n)$$

As in the three-layer case, it is convenient to define $u_1 = \dfrac{a_1}{k_1}, u_2 = \dfrac{a_2}{k_2}$ and so on. The equations then become

$$u_2(T_2 - T_1) = u_1(T_3 - T_2)$$

$$u_3(T_3 - T_2) = u_2(T_4 - T_3)$$

$$u_4(T_4 - T_3) = u_3(T_5 - T_4)$$

$$\vdots$$

$$u_n(T_n - T_{n-1}) = u_{n-1}(T_{n+1} - T_n)$$

Figure 5.27
n-layered problem.

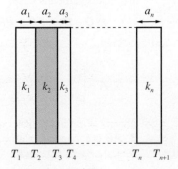

or in matrix form

$$
\begin{bmatrix}
(u_1 + u_2) & -u_1 & 0 & . & . & . & . & 0 \\
-u_3 & (u_2 + u_3) & -u_2 & 0 & . & . & . & 0 \\
0 & -u_4 & (u_3 + u_4) & -u_3 & 0 & . & . & 0 \\
. & . & . & . & . & . & . & . \\
. & . & . & . & . & . & . & . \\
. & . & . & . & . & . & . & . \\
0 & 0 & 0 & . & . & 0 & -u_n & (u_{n-1} + u_n)
\end{bmatrix}
\begin{bmatrix}
T_2 \\
. \\
. \\
. \\
. \\
. \\
T_n
\end{bmatrix}
$$

$$
=
\begin{bmatrix}
u_2 T_1 \\
0 \\
0 \\
. \\
. \\
0 \\
u_{n-1} T_{n+1}
\end{bmatrix}
$$

The matrix equation is of tridiagonal form, hence we know that there is an efficient algorithm for solution. An explicit solution, as in the three-layer case, is not so easy and requires a lot of effort. However, it is a comparatively easy exercise to prove that the effective conductivity (k) of the whole composite is obtained from the equivalent formula

$$
\frac{\Sigma a_i}{k} = \frac{a_1}{k_1} + \frac{a_2}{k_2} + \ldots + \frac{a_n}{k_n}
$$

5.10 Review exercises (1–25)

1 Given

$$
P = \begin{bmatrix} 2 & 4 & 1 \\ 0 & 2 & 1 \\ 0 & 0 & 4 \end{bmatrix}
\quad
Q = \begin{bmatrix} 6 & 2 & -1 \\ 0 & 2 & -2 \\ 0 & 0 & 4 \end{bmatrix}
$$

and

$$
R = \begin{bmatrix} 2 & 7 & -6 \\ 0 & 0 & 2 \\ 0 & 0 & 1 \end{bmatrix}
$$

(a) calculate RQ and $Q^T R^T$;

(b) calculate $Q + R$, PQ and PR, and hence verify that in this particular case

$$
P(Q + R) = PQ + PR
$$

2 Let

$$
A = \begin{bmatrix} -1 & 2 \\ 4 & 1 \end{bmatrix}
\quad \text{and} \quad
B = \begin{bmatrix} 1 & 1 \\ \lambda & \mu \end{bmatrix}
$$

where $\lambda \neq \mu$. Find all pairs of values λ, μ such that $B^{-1}AB$ is a diagonal matrix.

3 At a point in an elastic continuum the matrix representation of the infinitesimal strain tensor referred to axes $Ox_1x_2x_3$ is

$$e = \begin{bmatrix} 1 & -3 & \sqrt{2} \\ -3 & 1 & -\sqrt{2} \\ \sqrt{2} & -\sqrt{2} & 4 \end{bmatrix}$$

If i, j and k are unit vectors in the direction of the $Ox_1x_2x_3$ coordinate axes, determine the normal strain in the direction of

$$n = \tfrac{1}{2}(i - j + \sqrt{2}k)$$

and the shear strain between the directions n and

$$m = \tfrac{1}{2}(-i + j + \sqrt{2}k)$$

(Note that, using matrix notation, the normal strain is en, and the shear strain between two directions is $m^T en$.)

4 Express the determinant

$$\begin{vmatrix} \alpha & \beta & \gamma \\ \beta\gamma & \gamma\alpha & \alpha\beta \\ -\alpha + \beta + \gamma & \alpha - \beta + \gamma & \alpha + \beta - \gamma \end{vmatrix}$$

as a product of linear factors.

5 Determine the values of θ for which the system of equations

$$x + y + z = 1$$
$$x + 2y + 4z = \theta$$
$$x + 4y + 10z = \theta^2$$

possesses a solution, and for each such value find all solutions.

6 Given

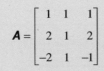

$$A = \begin{bmatrix} 1 & 1 & 1 \\ 2 & 1 & 2 \\ -2 & 1 & -1 \end{bmatrix}$$

evaluate A^2 and A^3. Verify that

$$A^3 - A^2 - 3A + I = 0$$

where I is the unit matrix of order 3. Using this result, or otherwise, find the inverse A^{-1} of A, and hence solve the equations

$$x + y + z = 3$$
$$2x + y + 2z = 7$$
$$-2x + y - z = 6$$

7 (a) If $P = \dfrac{1}{3}\begin{bmatrix} 2 & 1 & 2 \\ -2 & 2 & 1 \\ -1 & -2 & 2 \end{bmatrix}$ write down the

transpose matrix P^T. Calculate PP^T and hence show that $P^T = P^{-1}$. What does this mean about the solution of the matrix equation $Px = b$?

(b) The matrix $F = \begin{bmatrix} I_x & I_{xy} & Q_x \\ I_{xy} & I_y & Q_y \\ Q_x & Q_y & A \end{bmatrix}$ occurs in the

structural analysis of an arch. If

$$B = \begin{bmatrix} 1 & 0 & -Q_x/A \\ 0 & 1 & -Q_y/A \\ 0 & 0 & 1 \end{bmatrix}$$

find $E = BFB^T$ and show that it is a symmetric matrix.

8 (a) If the matrix $A = \begin{bmatrix} 1 & 0 & 0 \\ 1 & -1 & 0 \\ 1 & -2 & 1 \end{bmatrix}$ show that $A^2 = I$,

the unit matrix, and derive the elements of a square matrix B which satisfies

$$BA = \begin{bmatrix} 1 & 4 & 3 \\ 0 & 2 & 1 \\ -1 & 0 & 0 \end{bmatrix}$$

(b) Find suitable values for k in order that the following system of linear simultaneous equations are consistent:

$$6x + (k - 6)y = 3$$
$$2x + y = 5$$
$$(2k + 1)x + 6y = 1$$

9 Express the system of linear equations

$$3x - y + 4z = 13$$

$$5x + y - 3z = 5$$

$$x - y + z = 3$$

in the form $AX = b$, where A is a 3×3 matrix and X, b are appropriate column matrices.

(a) Find adj A, det A and A^{-1} and hence solve the system of equations.

(b) Find a matrix Y which satisfies the equation

$$AYA^{-1} = 22A^{-1} + 2A$$

(c) Find a matrix Z which satisfies the equation

$$AZ = 44I_3 - A + AA^T$$

where I_3 is the 3×3 identity matrix.

10 (a) Using the method of Gaussian elimination, find the solution of the equation

$$\begin{bmatrix} 1 & 2 & 4 & 8 \\ 2 & 7 & 13 & 25 \\ -1 & 1 & 5 & 9 \\ 2 & 1 & 11 & 24 \end{bmatrix} \begin{bmatrix} x_1 \\ x_2 \\ x_3 \\ x_4 \end{bmatrix} = \begin{bmatrix} 19 \\ 57 \\ 16 \\ 52 \end{bmatrix}$$

Hence evaluate the determinant of the matrix in the equation.

(b) Solve by the method of Gaussian elimination

$$\begin{bmatrix} 1 & 1 & -1 & 1 \\ 2 & 3 & -3 & 3 \\ -1 & 1 & 0 & 0 \\ 2 & 3 & -1 & 2 \end{bmatrix} \begin{bmatrix} x_1 \\ x_2 \\ x_3 \\ x_4 \end{bmatrix} = \begin{bmatrix} 4 \\ 11 \\ 1 \\ 13 \end{bmatrix}$$

with partial pivoting.

11 Rearrange the equations

$$x_1 - x_2 + 3x_3 = 8$$

$$4x_1 + x_2 - x_3 = 3$$

$$x_1 + 2x_2 + x_3 = 8$$

so that they are diagonally dominant to ensure convergence of the Gauss–Seidel method. Write a program on a spreadsheet or software package to obtain the solution of these equations using this method, starting from $(0, 0, 0)$. Compare your solution with that from a program when the equations are not rearranged. Use SOR, with $\omega = 1.3$, to solve the equations. Is there any improvement?

12 Find the rank of the matrix

$$\begin{bmatrix} 0 & c & b & a \\ -c & 0 & a & b \\ -b & -a & 0 & c \\ -a & -b & -c & 0 \end{bmatrix}$$

where $b \neq 0$ and $a^2 + c^2 = b^2$.

13 For a given set of discrete data points (x_i, f_i) $(i = 0, 1, 2, \ldots, n)$, show that the coefficients a_k $(k = 0, 1, \ldots, n)$ fitted to the polynomial

$$y(x) = \sum_{k=0}^{n} a_k x^k$$

are given by the solution of the equations written in the matrix form as

$$Aa = f$$

where

$$A = \begin{bmatrix} 1 & x_0 & x_0^2 & \cdots & x_0^n \\ 1 & x_1 & x_1^2 & \cdots & x_1^n \\ \vdots & \vdots & & & \vdots \\ 1 & x_n & x_n^2 & \cdots & x_n^n \end{bmatrix}$$

$$a = [a_0 \quad a_1 \quad \cdots \quad a_n]^T$$

$$f = [f_0 \quad f_1 \quad \cdots \quad f_n]^T$$

(See Question 87 in Exercises 2.9.2 of Chapter 2 for the Lagrange interpolation solution of these equations for the case $n = 3$.)

The following data is taken from the tables of the Airy function $f(x) = \text{Ai}(-x)$:

x	1	1.5	2.3	3.0	3.9
$f(x)$	0.535 56	0.464 26	0.026 70	−0.378 81	−0.147 42

Estimate from the polynomial approximation the values of $f(2.0)$ and $f(3.5)$.

14 A rotation of a set of rectangular cartesian axes $\Phi(Ox_1x_2x_3)$ to a set $\Phi'(Ox_1'x_2'x_3')$ is described by the matrix $\mathbf{L} = (l_{ij})$ $(i, j = 1, 2, 3)$, where l_{ij} is the cosine of the angle between Ox_i' and Ox_j. Show that $\mathbf{L}$ is such that

$$\mathbf{L}\mathbf{L}^{\mathrm{T}} = \mathbf{I}$$

and that the coordinates of a point in space referred to the two sets of axes are related by

$$\mathbf{X}' = \mathbf{L}\mathbf{X}$$

where $\mathbf{X}' = [x_1' \quad x_2' \quad x_3']^{\mathrm{T}}$ and $\mathbf{X} = [x_1 \quad x_2 \quad x_3]^{\mathrm{T}}$. Prove that

$$x_1'^2 + x_2'^2 + x_3'^2 = x_1^2 + x_2^2 + x_3^2$$

Describe the relationship between the axes Φ and Φ', given that

$$\mathbf{L} = \begin{bmatrix} \frac{1}{2} & 0 & \frac{1}{2}\sqrt{3} \\ 0 & 1 & 0 \\ -\frac{1}{2}\sqrt{3} & 0 & \frac{1}{2} \end{bmatrix}$$

The axes Φ' are now rotated through $45°$ about Ox_3' in the sense from Ox_1' to Ox_2' to form a new set Φ''. Show that the angle θ between the line OP and the axis Ox_1'', where P is the point with coordinates $(1, 2, -1)$ referred to the original system Φ, is

$$\theta = \cos^{-1}\left(\frac{5\sqrt{3} - 3}{12}\right)$$

15 A car is at rest on horizontal ground as shown in Figure 5.28. The weight W acts through the centre of gravity, and the springs have stiffness constants k_1 and k_2 and natural lengths a_1 and a_2. Show that the height z and the angle θ satisfy the matrix equation

$$\begin{bmatrix} -W + a_1k_1 + a_2k_2 \\ l_1k_1a_1 - l_2k_2a_2 \end{bmatrix}$$

$$= \begin{bmatrix} k_1 + k_2 & -l_1k_1 + l_2k_2 \\ l_1k_1 - l_2k_2 & -l_1^2k_1 - l_2^2k_2 \end{bmatrix}\begin{bmatrix} z \\ \theta \end{bmatrix}$$

Obtain reasonable values for the various parameters to ensure that $\theta = 0$.

16 In the circuit in Figure 5.29(a) show that the equations can be written

$$\begin{bmatrix} E_1 \\ I_1 \end{bmatrix} = \begin{bmatrix} 1 & Z_1 \\ 0 & 1 \end{bmatrix}\begin{bmatrix} E_2 \\ I_2 \end{bmatrix}$$

and that in Figure 5.29(b) they take the form

$$\begin{bmatrix} E_1 \\ I_1 \end{bmatrix} = \begin{bmatrix} 1 & 0 \\ 1/Z_2 & 1 \end{bmatrix}\begin{bmatrix} E_2 \\ I_2 \end{bmatrix}$$

Dividing the circuit in Figure 5.29(c) into blocks, with the output from one block inputting to the next block, analyse the relation between I_1, E_1 and I_2, E_2.

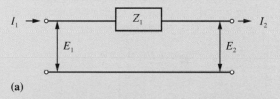

(a)

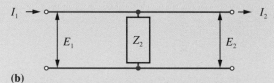

(b)

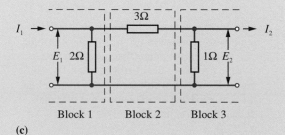

(c)

Figure 5.29

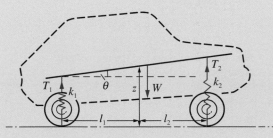

Figure 5.28 Car at rest on horizontal ground.

17
Data is fitted to a cubic

$$f = ax^3 + bx^2 + cx + d$$

with the slope of the curve given by

$$f' = 3ax^2 + 2bx + c$$

If $f_1 = f(x_1), f_2 = f(x_2), f_1' = f'(x_1)$ and $f_2' = f'(x_2)$, show that fitting the data gives the matrix equation for a, b, c and d as

$$\begin{bmatrix} f_1 \\ f_2 \\ f_1' \\ f_2' \end{bmatrix} = \begin{bmatrix} x_1^3 & x_1^2 & x_1 & 1 \\ x_2^3 & x_2^2 & x_2 & 1 \\ 3x_1^2 & 2x_1 & 1 & 0 \\ 3x_2^2 & 2x_2 & 1 & 0 \end{bmatrix} \begin{bmatrix} a \\ b \\ c \\ d \end{bmatrix}$$

Use Gaussian elimination to evaluate a, b, c and d. For the case

x	f	f'
0.4	0.327 54	0.511 73
0.8	0.404 90	−0.054 14

evaluate a, b, c and d. Plot the cubic and estimate the maximum value of f in the region $0 < x < 1$. Note that this exercise forms the basis of one of the standard methods for finding the maximum of a function $f(x)$ numerically.

18
The transformation $y = Ax$ where

$$A = \frac{1}{9} \begin{bmatrix} 8 & -1 & -4 \\ 4 & 4 & 7 \\ 1 & -8 & 4 \end{bmatrix}$$

$$y = \begin{bmatrix} y_1 \\ y_2 \\ y_3 \end{bmatrix} \quad \text{and} \quad x = \begin{bmatrix} x_1 \\ x_2 \\ x_3 \end{bmatrix}$$

takes a point with coordinates (x_1, x_2, x_3) into a point with coordinates (y_1, y_2, y_3). Show that the coordinates of the points that transform into themselves satisfy the matrix equation $Bx = 0$, where $B = A − I$, with I the identity matrix. Find the rank of B and hence deduce that for points which transform into themselves

$$[x_1 \quad x_2 \quad x_3] = \alpha[-3 \quad -1 \quad 1]$$

where α is a parameter.

Find AA^T. What is the inverse of A?
If $y_1 = 3$, $y_2 = -1$ and $y_3 = 2$, determine the values of x_1, x_2 and x_3 under this transformation.

19 (a) If
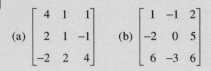

$$A = \begin{bmatrix} 1 & 0 & 1 & 0 \\ 2 & 1 & 2 & 1 \\ 1 & -2 & 2 & -2 \\ 2 & 0 & 3 & 1 \end{bmatrix}$$

verify that

$$A^{-1} = \begin{bmatrix} 6 & -2 & -1 & 0 \\ -5 & 3 & 1 & -1 \\ -5 & 2 & 1 & 0 \\ 3 & -2 & -1 & 1 \end{bmatrix}$$

(b) Use the inverse matrix given in (a) to solve the system of linear equations $Ax = b$ in which

$$b^T = [5 \quad -5 \quad -4 \quad 4]$$

20 When a body is deformed in a certain manner, the particle at point x moves to Ax, where

$$x = \begin{bmatrix} x \\ y \\ z \end{bmatrix} \quad \text{and} \quad A = \begin{bmatrix} 1 & -2 & 0 \\ -2 & 3 & 0 \\ 0 & 0 & 2 \end{bmatrix}$$

(a) Where would the point $\begin{bmatrix} 2 \\ 1 \\ 1 \end{bmatrix}$ move to?

(b) Find the point from which the particle would move to the point $\begin{bmatrix} 2 \\ 1 \\ 1 \end{bmatrix}$.

21 Find the eigenvalues and the normalized eigenvectors of the matrices

(a) $\begin{bmatrix} 4 & 1 & 1 \\ 2 & 1 & -1 \\ -2 & 2 & 4 \end{bmatrix}$ (b) $\begin{bmatrix} 1 & -1 & 2 \\ -2 & 0 & 5 \\ 6 & -3 & 6 \end{bmatrix}$

(c) $\begin{bmatrix} 5 & -2 & 0 \\ -2 & 6 & 0 \\ 0 & 2 & 7 \end{bmatrix} = \boldsymbol{C}$

In (c) write the normalized eigenvectors as the columns of the matrix $\boldsymbol{U}$ and show that $\boldsymbol{U}^{\mathrm{T}}\boldsymbol{C}\boldsymbol{U}$ is a diagonal matrix with the eigenvalues in the diagonal.

22 The vector $[1 \quad 0 \quad 1]^{\mathrm{T}}$ is an eigenvector of the symmetric matrix

$$\begin{bmatrix} 6 & -1 & 3 \\ -1 & 7 & \alpha \\ 3 & \alpha & \beta \end{bmatrix}$$

Find the values of α and β and find the corresponding eigenvalue.

23 Show that the matrix $\begin{bmatrix} -1 & 0 & 2 \\ 0 & 1 & 0 \\ 2 & 0 & -1 \end{bmatrix}$ has eigenvalues 1, 1 and −3. Find the corresponding eigenvectors. Is there a full set of three independent eigenvectors?

24 (a) Find the eigenvalues λ_1, λ_2 and the normalized eigenvectors $\boldsymbol{X}_1$, $\boldsymbol{X}_2$ of the matrix $\boldsymbol{A} = \begin{bmatrix} 2 & 1 \\ 1 & 2 \end{bmatrix}$. Check that

$$\boldsymbol{A} = \lambda_1 \boldsymbol{X}_1 \boldsymbol{X}_1^{\mathrm{T}} + \lambda_2 \boldsymbol{X}_2 \boldsymbol{X}_2^{\mathrm{T}}$$

 (b) Use MATLAB or MAPLE to repeat a similar calculation for the three eigenvalues and normalized eigenvectors of

$$\boldsymbol{B} = \begin{bmatrix} -1 & 1 & 0 \\ 1 & 0 & 1 \\ 0 & 1 & -2 \end{bmatrix}$$

(*Note*: The process described in this question calculates the spectral decomposition of a symmetric matrix.)

25 In Section 5.7.7 it was stated that a symmetric matrix $\boldsymbol{A}$ has real eigenvalues $\lambda_1, \lambda_2, \dots, \lambda_n$ (written in descending order) and corresponding orthonormal eigenvectors $\boldsymbol{e}_1, \boldsymbol{e}_2, \dots, \boldsymbol{e}_n$, that is $\boldsymbol{e}_i^{\mathrm{T}}\boldsymbol{e}_j = \delta_{ij}$. In consequence any vector can be written as

$$\boldsymbol{X} = c_1 \boldsymbol{e}_1 + c_2 \boldsymbol{e}_2 + \dots + c_n \boldsymbol{e}_n$$

Deduce that

$$\frac{\boldsymbol{X}^{\mathrm{T}}\boldsymbol{A}\boldsymbol{X}}{\boldsymbol{X}^{\mathrm{T}}\boldsymbol{X}} \leq \lambda_1 \qquad \text{(5.43)}$$

so that a lower bound of the largest eigenvalue has been found. The left-hand side of (5.43) is called the Rayleigh quotient.

It is known that the matrix $\begin{bmatrix} 0 & 1 & 0 & 0 \\ 1 & 0 & 1 & 0 \\ 0 & 1 & 0 & 1 \\ 0 & 0 & 1 & 0 \end{bmatrix}$ has a largest eigenvalue of $\frac{1}{2}(1 + \sqrt{5})$. Check that the result (5.43) holds for any vector of your choice.

6 An Introduction to Discrete Mathematics

Chapter 6 Contents

6.1 Introduction

It is ironic that the term 'discrete mathematics' is often seen as describing a new and exciting area of mathematics with applications to information technology and computing. Virtually everyone these days knows that microcomputers operate using digital electronics, and previously analogue systems such as radio and television transmissions are also turning digital. Digital systems are less prone to signal loss through dissipation, attenuation and interference through noise than traditional analogue systems. The ability of digital systems to handle the vast quantity of information required to reproduce high-resolution graphics in a very efficient and cost-effective way is a consequence of this. Another consequence of digitization is greater security due to less penetrable encryption algorithms based on the discrete mathematics of number systems. The present and the future are therefore most definitely digital, and digital systems make use of discrete mathematics. Discrete mathematics itself is remarkably old. In fact it pre-dates calculus, which might be called 'continuous mathematics'. All counting is discrete mathematics. However, it was only in the nineteenth and twentieth centuries that mathematicians like George Boole (1816–1864) gave a rigorous basis to set theory. The work of Bertrand Russell (1872–1970) and Alfred North Whitehead (1861–1947), and later Kurt Gödel (1906–1978), on logic and the foundation of mathematics, which was to have a great effect on the development of mathematics in the twentieth century, was intimately connected with questions of set theory. This material is now seen to be of great relevance to engineering. Electronic engineers have for a long time required knowledge of Boolean algebra in order to understand the principles of switching circuits. The computer is now very much part of engineering: processes are computer controlled, manufacturing by robots is now commonplace and design is computer aided. Engineers now have a duty to understand how to check the correctness of the algorithms that design, build and repair. In order to do this, branches of discrete mathematics such as propositional logic have to be part of the core curriculum for engineers and not optional extras. This chapter develops the mathematics required in a logical and systematic way, beginning with sets and applications to manufacturing, moving on to switching circuits and applications to electronics, and then to propositional calculus and applications to computing.

6.2 Set theory

The concept of a set is a relatively recent one in that it was born in the past hundred years. In the past few decades it has gained in popularity, and now forms part of school mathematics – this is natural, since the concepts involved, although they may seem unfamiliar initially, are not difficult.

Set theory is concerned with identifying one or more common characteristics among objects. We introduce basic concepts and set operations first, and then examine some applications. The largest areas of application deserve sections to themselves; however, in this section we apply set theory fundamentals to the manufacture and efficient assembly of components.

6.2.1 Definitions and notation

A **set** is a collection of objects, which are called the **elements** or **members** of the set. We shall denote sets by capital letters such as A, S and X, and elements of a set by lower-case letters such as a, s and x. The notation $\in$ is used as follows: if an element a is contained in a set S then we write

$$a \in S$$

which is read 'a belongs to S'. If b does not belong to S then the symbol $\notin$ is used:

$$b \notin S$$

read as 'b does not belong to S'.

A **finite set** is one that contains only a finite number of elements, while an **infinite set** is one consisting of an infinite number of elements. For example,

(i) the months of the year form a finite set, while
(ii) the set consisting of all integers is an infinite set.

If we wish to indicate the composition of S then there are two ways of doing this. The first method is suitable only for finite sets, and involves listing the elements of the set between open and closed braces as, for example, in

$$S = \{a, b, c, d, e, f\}$$

which denotes the set S consisting only of the six elements a, b, c, d, e and f.

The second method involves giving a rule by which all elements of the set can be determined. The notation

$$S = \{x : x \text{ has property } P\}$$

will be used to denote the set of all elements x that have the property P. For example,

(i) $S = \{N : N \in Z, N \leqslant 500\}$

is the set of integers that are less than or equal to 500, and

(ii) $S = \{x : x^2 - x - 6 = 0, x \in \mathbb{R}\}$

is the set containing only the two elements 3 and −2.

An example of an infinite set would be

$$S = \{x : 0 \leqslant x \leqslant 1, x \in \mathbb{R}\}$$

which denotes all real numbers that lie in the range 0 to 1, including 0 and 1 themselves.

Very seldom are we satisfied with the type of statement 'S is the set of all fruit' beloved of early school mathematics.

> Two sets A and B are said to be **equal** if every element of each is also an element of the other. For such sets we write $A = B$; otherwise we write $A \neq B$.

For example,

$$A = \{3, 4\} \quad \text{and} \quad B = \{x : x^2 - 7x + 12 = 0\}$$

are two equal sets.

If every element of a set A is also an element of the set B then A is said to be a **subset** of B or, alternatively, B is a **superset** of A. The statement 'A is a subset of B' is written $A \subset B$, while the statement 'B is a superset of A' is written $B \supset A$. The negations of these two statements are written as $A \not\subset B$ and $B \not\supset A$ respectively. Note that if $A \subset B$ and $B \subset A$ then $A = B$, since every element of A is an element of B and vice versa. Thus the definition of a subset does not exclude the possibility of the two sets being equal. If $A \subset B$ and $A \neq B$ then A is said to be a **proper subset** of B. In order to distinguish between a **subset** and a **proper subset**, we shall use the notation $A \subseteq B$ to denote 'A is a subset of B' and $A \subset B$ to denote 'A is a proper subset of B'. For example,

$$A = \{a, b, c\} \quad \text{is a proper subset of} \quad B = \{a, b, c, d, e, f\}$$

A set containing no elements is called the **empty** or **null** set, and is denoted by $\varnothing$. For example,

$$A = \{x : x^2 = 25, x \text{ even}\}$$

is an example of a null set, so $A = \varnothing$. It is noted that the empty set may be considered to be a subset of any set.

In most applications it is possible to define sensibly a universal set U that contains all the elements of interest. For example, when dealing with sets of integers, the universal set is the set of all integers, while in two-dimensional geometry the universal set contains all the points in the plane. In such cases we can define the complement of a set A: if all the elements of a set A are removed from the universal set U then the elements that remain in U form the **complement** of A, which is denoted by $\bar{A}$. Thus the sets A and $\bar{A}$ have no elements in common, and we may write

$$\bar{A} = \{x : x \in U, x \notin A\}$$

Relations between sets can be illustrated by schematic drawings called **Venn diagrams**, in which each set is represented as the interior of a closed region (normally drawn as a circle) of the plane. It is usual to represent the universal set by a surrounding rectangle. For example, $A \subset B$ and $\bar{A}$ are illustrated by the Venn diagrams of Figures 6.1(a) and (b) respectively.

Figure 6.1

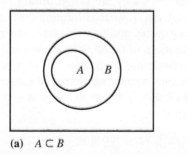

(a) $A \subset B$ (b) $\bar{A}$ (shaded)

6.2.2 Union and intersection

If A and B are two sets, related to the same universal set U, then we can combine A and B to form new sets in the following two different ways.

Figure 6.2

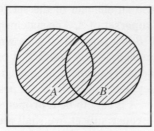
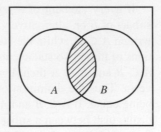

(a) $A \cup B$ (shaded)   **(b)** $A \cap B$ (shaded)

Union

The union of two sets A and B is a third set containing all the elements of A and all the elements of B. It is denoted by $A \cup B$, read as 'A union B'. Thus

$$A \cup B = \{x : x \in A \quad \text{or} \quad x \in B\}$$

where 'or' in this context is used in the inclusive sense: x is an element of A, or B, or both.

Intersection

The intersection of two sets A and B is a third set containing all the elements that belong to both A and B. It is denoted by $A \cap B$, read as 'A intersection B'. Thus

$$A \cap B = \{x : x \in A \quad \text{and} \quad x \in B\}$$

These two definitions are illustrated by the Venn diagrams of Figures 6.2(a) and (b). It is clear from the illustration that union and intersection are commutative, so that

$$A \cup B = B \cup A$$

and

$$A \cap B = B \cap A$$

If the two sets A and B have no elements in common then $A \cap B = \emptyset$: the sets A and B are said to be **disjoint**.

    Since union ($\cup$) and intersection ($\cap$) combine two sets from within the same universal set U to form a third set in U, they are called **binary** operations on U. On the other hand, operations on a single set A, such as forming the complement $\bar{A}$, are called **unary** operations on U. It is worthwhile noting at this stage the importance of the words 'or', 'and' and 'not' in the definitions of union, intersection and complementation, and we shall return to this when considering applications in later sections. It is also worth noting that the numerical solutions to the examples and exercises that follow can be checked using the package MAPLE.

Example 6.1 If $A = \{3, 4, 5, 6\}$ and $B = \{1, 5, 7, 9\}$, determine

(a) $A \cup B$ (b) $A \cap B$

Solution (a) $A \cup B = \{1, 3, 4, 5, 6, 7, 9\}$

(b) $A \cap B = \{5\}$

6.2.3 Exercises

1   Express the following sets in listed form:

$A = \{x : x < 10, x \text{ a natural number}\}$

$B = \{x : x^2 = 16, x \in \mathbb{R}\}$

$C = \{x : 4 < x < 11, x \text{ an integer}\}$

$D = \{x : 0 < x < 28, x \text{ an integer divisible by 4}\}$

2   For the sets A, B, C and D of Question 1 list the sets $A \cup B, A \cap B, A \cup C, A \cap C, B \cup D, B \cap D$ and $B \cap C$.

3   If $A = \{1, 3, 5, 7, 9\}$, $B = \{2, 4, 6, 8, 10\}$ and $C = \{1, 4, 5, 8, 9\}$, list the sets $A \cup B, A \cap C, A \cap B, B \cup C$ and $B \cap C$.

4   Illustrate the following sets using Venn diagrams:

$\bar{A} \cap \bar{B}, \bar{A} \cup \bar{B}, \overline{A \cap B}, \overline{A \cup B}, A \cap \bar{B}$

5   Given

$A = \{N : N \text{ an integer } 1 \leq N \leq 10\}$

$B = \{N : N \text{ an even integer}, N \leq 20\}$

and

$C = \{N : N = 2^n, n \text{ an integer}, 1 \leq n \leq 5\}$

determine the following:

(a) $A \cup B$ (b) $A \cap B$

(c) $A \cup C$ (d) $A \cap C$

6   For the sets defined in Question 5 check whether the following statements are true or false:

(a) $A \cap B \supseteq A \cap C$

(b) $A \cup B \supseteq C$

(c) $A \cup B \subseteq C$

7   If the universal set is the set of all integers less than or equal to 32, and A and B are as in Question 5, interpret

(a) $\bar{A}$ (b) $\overline{A \cup B}$ (c) $\bar{A} \cap \bar{B}$

(d) $\overline{A \cap B}$ (e) $\bar{A} \cup \bar{B}$

8   (a) If $A \subset B$ and $A \subset \bar{B}$, show that $A = \varnothing$.

(b) If $A \subset B$ and $C \subset D$, show that $(A \cup C) \subset (B \cup D)$ and illustrate the result using a Venn diagram.

6.2.4 Algebra of sets

In Section 6.2.2 we saw that, given two sets A and B, the operations $\cup$ and $\cap$ could be used to generate two further sets $A \cup B$ and $A \cap B$. These two new sets can then be combined with a third set C, associated with the same universal set U as the sets A and B, to form four further sets

$$C \cup (A \cup B), C \cap (A \cup B), C \cup (A \cap B), C \cap (A \cap B)$$

and the compositions of these sets are clearly indicated by the shaded regions in the Venn diagrams of Figure 6.3.

Clearly, by using various combinations of the binary operations $\cup$ and $\cap$ and the unary operation of complementation ($\bar{\ }$), many further sets can be generated. In practice, it is useful to have rules that enable us to simplify expressions involving $\cup$, $\cap$ and ($\bar{\ }$). In this section we develop such rules, which form the basis of the algebra of sets. In the next section we then proceed to show the analogy between this algebra and the algebra of switching circuits, which is widely used by practising engineers.

Given the three sets A, B and C, belonging to the same universal set U, we have already seen that the operations $\cup$ and $\cap$ are commutative, so that we have the following.

Figure 6.3

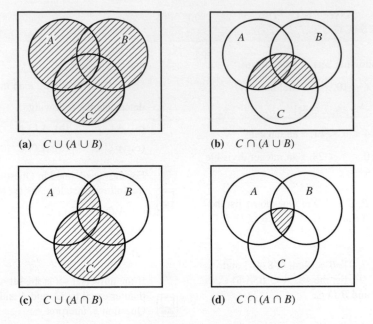

(a) $C \cup (A \cup B)$ **(b)** $C \cap (A \cup B)$

(c) $C \cup (A \cap B)$ **(d)** $C \cap (A \cap B)$

Commutative laws

$$A \cup B = B \cup A \quad \text{(union is commutative)}$$
$$A \cap B = B \cap A \quad \text{(intersection is commutative)}$$

(6.1)

It follows directly from the definitions that we have the

Idempotent laws

$$A \cup A = A \quad \text{(union is idempotent)}$$
$$A \cap A = A \quad \text{(intersection is idempotent)}$$

(6.2)

Identity laws

$$A \cup \varnothing = A \quad (\varnothing \text{ is an identity relative to union)}$$
$$A \cap U = A \quad (U \text{ is an identity relative to intersection)}$$

(6.3)

Complementary laws

$$A \cup \bar{A} = U$$
$$A \cap \bar{A} = \varnothing$$

(6.4)

In addition, it can be shown that the following associative and distributive laws hold:

Associative laws

$$A \cup (B \cup C) = (A \cup B) \cup C \quad \text{(union is associative)}$$

$$A \cap (B \cap C) = (A \cap B) \cap C \quad \text{(intersection is associative)}$$

(6.5)

Distributive laws

$$A \cup (B \cap C) = (A \cup B) \cap (A \cup C) \quad \text{(union is distributive over intersection)}$$

$$A \cap (B \cup C) = (A \cap B) \cup (A \cap C) \quad \text{(intersection is distributive over union)}$$

(6.6)

Readers should convince themselves of the validity of the results (6.5) and (6.6) by considering the Venn diagrams of Figure 6.3.

The laws expressed in (6.1)–(6.6) constitute the basic laws of the algebra of sets. This itself is a particular example of a more general logical structure called **Boolean algebra**, which is briefly defined by the statement

A class of members (equivalent to sets here) together with two binary operations (equivalent to union and intersection) and a unary operation (equivalent to complementation) is a Boolean algebra provided the operations satisfy the equivalent of the commutative laws (6.1), the identity laws (6.3), the complementary laws (6.4) and the distributive laws (6.6).

We note that it is therefore not essential to include the idempotent laws (6.2) and associative laws (6.5) in the basic rules of the algebra of sets, since these are readily deducible from the others. The reader should, at this stage, reflect on and compare the basic rules of the algebra of sets with those associated with conventional numerical algebra in which the binary operations are addition (+) and multiplication (×), and the identity elements are zero (0) and unity (1). It should be noted that in numerical algebra there is no unary operation equivalent to complementation, the idempotency laws do not hold, and that addition is not distributive over multiplication.

While the rules (6.1)–(6.6) are sufficient to enable us to simplify expressions involving $\cup$, $\cap$ and $(\bar{\ })$ the following, known as the De Morgan laws, are also useful in practice.

De Morgan laws

$$\overline{A \cup B} = \bar{A} \cap \bar{B}$$

$$\overline{A \cap B} = \bar{A} \cup \bar{B}$$

(6.7)

The first of these laws states 'the complement of the union of two sets is the intersection of the two complements', while the second states that 'the complement of the intersection of two sets is the union of the two complements'. The validity of the results is illustrated by the Venn diagrams of Figure 6.4, and they are such that they enable us to negate or invert expressions.

If we look at the pairs of laws in each of (6.1)–(6.6) and replace $\cup$ by $\cap$ and interchange $\emptyset$ and U in the first law in each pair then we get the second law in each pair.

Figure 6.4

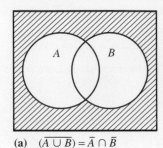
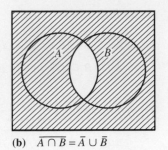

(a) $\overline{(A \cup B)} = \bar{A} \cap \bar{B}$ (b) $\overline{A \cap B} = \bar{A} \cup \bar{B}$

Conversely, if we replace ∩ and ∪ and interchange ∅ and U in the second law of each pair, we get the first law. This important observation is embedded in the **principle of duality**, which states that if any statement involving ∪, ∩ and (¯) is true for all sets then the dual statement (obtained by replacing ∪ by ∩, ∅ by U, and U by ∅) is also true for all sets. This holds for inclusion, with duality existing between ⊂ and ⊃.

Example 6.2

Using the laws (6.1)–(6.6), verify the statement

$$\overline{(A \cap B)} \cup \overline{(\bar{A} \cap \bar{B} \cap C)} \cup A = U$$

stating clearly the law used in each step.

Solution

Starting with the left-hand side, we have

$$\text{LHS} = \overline{(A \cap B)} \cup \overline{(\bar{A} \cap \bar{B} \cap C)} \cup A$$

$$= (\bar{B} \cup \bar{A}) \cup (\bar{\bar{A}} \cup \bar{\bar{B}} \cup \bar{C}) \cup A \qquad \text{(De Morgan laws)}$$

$$= (\bar{B} \cup \bar{A}) \cup (A \cup B \cup \bar{C}) \cup A \qquad (\bar{\bar{A}} = A)$$

$$= \bar{A} \cup (A \cup A) \cup (\bar{B} \cup B) \cup \bar{C} \qquad \text{(associative and commutative)}$$

$$= (\bar{A} \cup A) \cup (\bar{B} \cup B) \cup \bar{C} \qquad \text{(idempotent)}$$

$$= (U \cup U) \cup \bar{C} \qquad \text{(complementary)}$$

$$= U \cup \bar{C} \qquad \text{(idempotent)}$$

$$= U \qquad \text{(definition of union)}$$

$$= \text{RHS}$$

Example 6.3

When carrying out a survey on the popularity of three different brands X, Y and Z of washing powder, 100 users were interviewed, and the results were as follows: 30 used brand X only, 22 used brand Y only, 18 used brand Z only, 8 used brands X and Y, 9 used brands X and Z, 7 used brands Z and Y and 14 used none of the brands.

(a) How many users used brands X, Y and Z?

(b) How many users used brands X and Z but not brand Y?

Figure 6.5

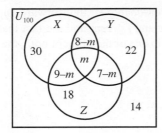

Solution   We can regard the users using brands X, Y and Z as being elements of the sets X, Y and Z respectively. If we denote the number of users using brands X, Y and Z by m then we can illustrate all the given information by the Venn diagram of Figure 6.5. We are then in a position to answer the two given questions.

(a) Since 14 users used none of the three brands, we have that $100 - 14 = 86$ users used one or more of the brands, so

number of elements of $X \cup Y \cup Z = 86$

Thus, from the Venn diagram,

$$30 + (8 - m) + m + (9 - m) + 22 + (7 - m) + 18 = 86$$

$$94 - 2m = 86$$

giving $m = 4$

indicating that 4 users use all three brands X, Y and Z.

(b) The number of users using brands X and Z and not Y is the number of elements in $(X \cap Z) \cap \overline{Y}$, which is the region indicated as having $9 - m$ elements in the Venn diagram. Thus the required answer is $9 - m = 9 - 4 = 5$ users.

Example 6.4   A company manufactures cranes. There are three basic types of crane, labelled A, B and C. Each crane is assembled from a subassembly set $\{a, b, c, d, e, f\}$ as follows:

A is assembled from $\{a, b, c, d\}$
B is assembled from $\{a, c, f\}$
C is assembled from $\{b, d, e\}$

In turn, the subassemblies are manufactured from basic components $\{p, q, r, s, t, u, v, w, x, y\}$ as follows:

a is manufactured from $\{p, q, r, s\}$
b is manufactured from $\{q, r, t, v\}$
c is manufactured from $\{p, r, s, t\}$
d is manufactured from $\{p, w, y\}$
e is manufactured from $\{u, x\}$
f is manufactured from $\{p, r, u, v, x, y\}$

(a) Give the make-up of the following subassemblies:

(i) $a \cup b$, (ii) $a \cup c \cup f$, (iii) $d \cup e$

(b) Given that A is made in Newcastle, and B and C are made in Birmingham, what components need to be available on both sites?

Solution   The solution of this problem is a reasonably straightforward application of set theory. From the definitions of a, b, c, d, e and f given, and the fact that the union of two sets contains those items that are either in one or the other or both, the following can be written down:

(i) $a \cup b \quad = \{p, q, r, s, t, v\}$

(ii) $a \cup c \cup f = \{p, q, r, s, t, u, v, x, y\}$

(iii) $d \cup e \quad = \{p, u, w, x, y\}$

This solves (a).

Now, A is made from subassemblies $\{a, b, c, d\}$, whereas B and C require $\{a, b, c, d, e, f\}$ in all of them. Inspection of those components required to make all six subassemblies reveals that subassemblies a, b, c and d do not require components u and x. Therefore only components u and x need not be made available in both sites. Using the notation of set theory, the solution to (b) is that the components that constitute

$$a \cup b \cup c \cup d$$

have to be available on both sites, or equivalently

$$\overline{a \cup b \cup c \cup d}$$

need only be available at the Birmingham site.

Comment   Of course, Example 6.4, which took much longer to state than to solve, is far too simple to represent a real situation. In a real crane manufacturing company there will be perhaps 20 basic types, and in a car production plant only a few basic types but far more than three hierarchies. However, what this example does is show how set theory can be used for sort purposes. It should also be clear that set theory, being precise, is ideally suited as a framework upon which to build a user-friendly computer program (an expert system) that can answer questions equivalent to part (b) of Example 6.4, when questioned by, for example, a managing director.

6.2.5   Exercises

9   If A, B and C are the sets $\{2, 5, 6, 7, 10\}$, $\{1, 3, 4, 7, 9\}$ and $\{2, 3, 5, 8, 9\}$ respectively, verify that

(a) $A \cap (B \cap C) = (A \cap B) \cap C$

(b) $A \cap (B \cup C) = (A \cap B) \cup (A \cap C)$

10   Using the rules of set algebra, verify the absorption rules

(a) $X \cup (X \cap Y) = X$ (b) $X \cap (X \cup Y) = X$

11   Using the laws of set algebra, simplify the following:

(a) $A \cap (\bar{A} \cup B)$ (b) $(\bar{A} \cup \bar{B}) \cap (A \cap B)$

(c) $(A \cup B) \cap (A \cup \bar{B})$ (d) $(\bar{A} \cap \bar{B}) \cup (A \cup B)$

(e) $(A \cup B \cup C) \cap (A \cup B \cup \bar{C}) \cap (A \cup \bar{B})$

(f) $(A \cup B \cup C) \cap (A \cup (B \cap C))$

(g) $(A \cap B \cap C) \cap (A \cup B \cup \bar{C}) \cup (A \cup \bar{B})$

12   Defining the difference $A - B$ between two sets A and B belonging to the same universal set U to be the set of elements of A that are not elements of B, that is $A - B = A \cap \bar{B}$, verify the following properties:

(a) $U - A = \bar{A}$ (b) $(A - B) \cup B = A \cup B$

(c) $C \cap (A - B) = (C \cap A) - (C \cap B)$

(d) $(A \cup B) \cup (B - A) = A \cup B$

Illustrate the identities using Venn diagrams.

13 If $n(X)$ denotes the number of elements of a set X, verify the following results, which are used for checking the results of opinion polls:

(a) $n(A \cup B \cup C) = n(A) + n(B) + n(C)$
$$- n(A \cap B) - n((A \cup B) \cap C)$$

(b) $n((A \cup B) \cap C) = n(A \cap C) + n(B \cap C)$
$$- n(A \cap B \cap C)$$

(c) $n(A \cup B \cup C) = n(A) + n(B) + n(C)$
$$- n(A \cap B) - n(B \cap C)$$
$$- n(C \cap A) + n(A \cap B \cap C)$$

Here the sets A, B and C belong to the same universal set U.

14 In carrying out a survey of the efficiency of lights, brakes and steering of motor vehicles, 100 vehicles were found to be defective, and the reports on them were as follows:

no. of vehicles with defective lights	= 35
no. of vehicles with defective brakes	= 40
no. of vehicles with defective steering	= 41
no. of vehicles with defective lights and brakes	= 8
no. of vehicles with defective lights and steering	= 7
no. of vehicles with defective brakes and steering	= 6

Use a Venn diagram to determine

(a) how many vehicles had defective lights, brakes and steering,

(b) how many vehicles had defective lights only.

15 On carrying out a later survey on the efficiency of the lights, brakes and steering on the 100 vehicles of Question 14, the report was as follows:

no. of vehicles with defective lights	= 42
no. of vehicles with defective brakes	= 30
no. of vehicles with defective steering	= 28
no. of vehicles with defective lights and brakes	= 8
no. of vehciles with defective lights and steering	= 10
no. of vehicles with defective brakes and steering	= 5

no. of vehciles with defective lights, brakes and steering	= 3

Use a Venn diagram to determine

(a) how many vehicles were non-defective,

(b) how many vehicles had defective lights only.

16 An analysis of 100 personal injury claims made upon a motor insurance company revealed that loss or injury in respect of an eye, an arm or a leg occurred in 30, 50 and 70 cases respectively. Claims involving the loss or injury to two of these members numbered 42. How many claims involved loss or injury to all three members? (You may assume that one or other of the three members was mentioned in each of the 100 claims.)

17 Bright Homes plc has warehouses in three different locations, L_1, L_2 and L_3, for making replacement windows. There are three different styles, called 'standard', 'executive' and 'superior':

> standard units require parts B, C and D;
> executive units require parts B, C, D and E;
> superior units require parts A, B, C and F.

The parts A, B, C, D, E and F are made from components a, b, c, d, e, f, g, h and i as follows:

> A is made from $\{a, b, c\}$
> B is made from $\{c, d, e, f\}$
> C is made from $\{c, e, f, g, h\}$
> D is made from $\{b, e, h\}$
> E is made from $\{c, h, i\}$
> F is made from $\{b, c, f, i\}$

(a) If the universal set is the set of all components $\{a, b, c, d, e, f, g, h, i\}$, write down the following:

$$\bar{C}, \quad \overline{B \cup C}, \quad \bar{B} \cap \bar{C}, \quad A \cap B \cap D,$$
$$A \cup F, \quad D \cup (E \cap F), \quad (D \cup E) \cap F$$

(b) New parts $B \cup C$, $C \cup E$ and $D \cup E \cup F$ are to be made; what are their components?

(c) Standard units are made at L_1, L_2 and L_3. Executive units are made at L_1 and L_2 only. Superior units are made at L_3 only. What basic components are needed at each location?

6.3 Switching and logic circuits

Throughout engineering, extensive use is made of switches. This is now truer than ever, since microprocessors and miniaturized electronic devices have found their way into practically every branch of engineering. A switch is either on or off: denoted by the digits 1 or 0. We shall see that the analysis of circuits containing switches provides a natural vehicle for the use of algebra of sets introduced in the last section.

6.3.1 Switching circuits

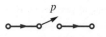

Figure 6.6
An 'on–off' switch.

Consider a simple 'on–off' switch, which we shall denote by a lower-case letter such as p and illustrate as in Figure 6.6. Such a switch is a two-state device in that it is either **closed** (or 'on') or **open** (or 'off'). We denote a closed contact by 1 and an open contact by 0, so that the variable p can only take one of the two values 1 or 0, with

$p = 1$ denoting a closed contact (or 'on' switch), so that a current is able to flow through it

and

$p = 0$ denoting an open contact (or 'off' switch), so that a current cannot flow through it

A **switching circuit** will consist of an energy source or input, for example a battery, and an output, for example a light bulb, together with a number of switches p, q, r and so on. Two switches may be combined together in two basic ways, namely by a series connection or by a parallel connection as illustrated in Figures 6.7 and 6.8 respectively.

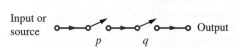

Figure 6.7 Two switches in series.

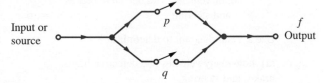

Figure 6.8 Two switches in parallel.

Associated with such a circuit is a **switching function** or **Boolean function** f of the variables contained in the circuit. This is a binary function with

$f = 1$ denoting that the entire circuit is closed

and

$f = 0$ denoting that the entire circuit is open

Clearly the states of f depend upon the states of the individual switches comprising the circuit, so we need to know how to write down an expression for f. For the series circuit of Figure 6.7 there are four possible states:

(a) p open, q open   (b) p open, q closed

(c) p closed, q open   (d) p closed, q closed

	p	q	f
Case (a)	0	0	0
Case (b)	0	1	0
Case (c)	1	0	0
Case (d)	1	1	1

Figure 6.9
Truth table for series
connection $f = p \cdot q$.

p	q	f
0	0	0
0	1	1
1	0	1
1	1	1

Figure 6.10
Truth table for parallel
connection $f = p + q$.

and it is obvious that current will flow through the circuit from input to output only if both switches p *and* q are closed. In tabular form the state of the circuit may be represented by the **truth table** of Figure 6.9.

Drawing an analogy with use of the word 'and' in the algebra of sets we write

$$f = p \cdot q$$

with $p \cdot q$ being read as 'p and q' (sometimes the dot is omitted and $p \cdot q$ is written simply as pq). Here the 'multiplication' or dot symbol is used in an analogous manner to $\cap$ in the algebra of sets.

When we connect two switches p and q in parallel, as in Figure 6.8, the state of the circuit may be represented by the truth table of Figure 6.10, and it is clear that current will flow through the circuit if either p or q is closed or if they are both closed.

Again, drawing an analogy with the use of the word 'or' in the algebra of sets, we write

$$f = p + q$$

read as 'p or q', with the $+$ symbol used in an analogous manner to $\cup$ in the algebra of sets.

So far we have assumed that the two switches p and q act independently of one another. However, two switches may be connected to one another so that

they open and close simultaneously

or

the closing (opening) of one switch will open (close) the other

This is illustrated in Figures 6.11(a) and (b) respectively. We can easily accommodate the situation of Figure 6.11(a) by denoting both switches by the same letter. To accommodate the situation of Figure 6.11(b), we define the **complement switch** $\bar{p}$ (or p') of a switch p to be a switch always in the state opposite to that of p. The action of the complement switch is summarized in the truth table of Figure 6.12.

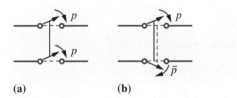

(a)   (b)

Figure 6.11 Two switches not acting independently.

p	$\bar{p}$
0	1
1	0

Figure 6.12 Truth table for complementary switch.

6.3.2 Algebra of switching circuits

We can use the operations $\cdot$, $+$ and ($\bar{\ }$) to write down the Boolean function f for complex switching circuits. The states of such circuits may then be determined by constructing truth tables.

Example 6.5

Draw up the truth table that determines the state of the switching circuit given by the Boolean function

$$f = (p \cdot \bar{q}) + (\bar{p} \cdot q)$$

Figure 6.13
Truth table for
$f = (p \cdot \bar{q}) + (\bar{p} \cdot q)$.

p	q	$\bar{p}$	$\bar{q}$	$p \cdot \bar{q}$	$\bar{p} \cdot q$	$(p \cdot \bar{q}) + (\bar{p} \cdot q)$
0	0	1	1	0	0	0
0	1	1	0	0	1	1
1	0	0	1	1	0	1
1	1	0	0	0	0	0

Solution The required truth table is shown in Figure 6.13. This circuit is interesting in that it is closed (that is, there is a current flow at the output) only if the two switches p and q are in different states. We will see later that it corresponds to the EXCLUSIVE OR function in logic circuits.

By constructing the appropriate truth table, it is readily shown that the operations $\cdot$, $+$ and $(\bar{})$ satisfy the following laws, analogous to results (6.1)–(6.6) for the algebra of sets:

Commutative laws

$$p + q = q + p, \quad p \cdot q = q \cdot p$$

Idempotent laws

$$p + p = p, \quad p \cdot p = p$$

Identity laws

$$p + 0 = p \quad \text{(0 is the identity relative to +),} \quad p + 1 = 1$$
$$p \cdot 1 = p \quad \text{(1 is the identity relative to } \cdot \text{),} \quad p \cdot 0 = 0$$

Complementary laws

$$p + \bar{p} = 1, \quad p \cdot \bar{p} = 0$$

Associative laws

$$p + (q + r) = (p + q) + r, \quad p \cdot (q \cdot r) = (p \cdot q) \cdot r$$

Distributive laws

$$p + (q \cdot r) = (p + q) \cdot (p + r), \quad p \cdot (q + r) = p \cdot q + p \cdot r$$

These rules form the basis of the algebra of switching circuits, and it is clear that it is another example of a Boolean algebra, with + and · being the two binary operations, $(\overline{})$ being the unary operation, and 0 and 1 the identity elements. It follows that the results developed for the algebra of sets carry through to the algebra of switching circuits, with equivalence between $\cup$, $\cap$, $(\overline{})$, $\varnothing$, U and +, ·, $(\overline{})$, 0, 1 respectively. Using these results, complicated switching circuits may be reduced to simpler equivalent circuits.

Example 6.6   Construct truth tables to verify the De Morgan laws for the algebra of switching circuits analogous to (6.7) for the algebra of sets.

Solution   The analogous De Morgan laws for the switching circuits are

$$\overline{p+q} = \bar{p} \cdot \bar{q} \quad \text{and} \quad \overline{p \cdot q} = \bar{p} + \bar{q}$$

the validity of which is verified by the truth tables of Figures 6.14(a) and (b).

Figure 6.14
Truth tables for De Morgan laws.

p	q	$\bar{p}$	$\bar{q}$	$p+q$	$\overline{p+q}$	$\bar{p} \cdot \bar{q}$
0	0	1	1	0	1	1
0	1	1	0	1	0	0
1	0	0	1	1	0	0
1	1	0	0	1	0	0

(a) $\overline{p+q} = \bar{p} \cdot \bar{q}$

p	q	$\bar{p}$	$\bar{q}$	$p \cdot q$	$\overline{p \cdot q}$	$\bar{p} + \bar{q}$
0	0	1	1	0	1	1
0	1	1	0	0	1	1
1	0	0	1	0	1	1
1	1	0	0	1	0	0

(b) $\overline{p \cdot q} = \bar{p} + \bar{q}$

Example 6.7   Simplify the Boolean function

$$f = p + p \cdot q \cdot r + \bar{p} \cdot \bar{q}$$

stating the law used in each step of the simplication.

Solution

$$f = p + p \cdot q \cdot r + \bar{p} \cdot \bar{q}$$

$$= p \cdot 1 + p \cdot (q \cdot r) + \bar{p} \cdot \bar{q} \qquad \text{(identity, } p \cdot 1 = p, \text{ and associative)}$$

$$= p \cdot (1 + (q \cdot r)) + \bar{p} \cdot \bar{q} \qquad \text{(distributive, } p \cdot (1 + (q \cdot r)) = p \cdot 1 + p \cdot (q \cdot r))$$

$$= p \cdot 1 + \bar{p} \cdot \bar{q} \qquad \text{(identity, } 1 + (q \cdot r) = 1)$$

$$= p + \bar{p} \cdot \bar{q} \qquad \text{(identity, } p \cdot 1 = p)$$

$$= (p + \bar{p}) \cdot (p + \bar{q}) \qquad \text{(distributive, } p + (\bar{p} \cdot \bar{q}) = (p + \bar{p}) \cdot (p + \bar{q}))$$

$$= 1 \cdot (p + \bar{q}) \qquad \text{(complementary, } p + \bar{p} = 1)$$

that is,

$$f = p + \bar{q} \qquad \text{(identity, } 1 \cdot (p + \bar{q}) = p + \bar{q})$$

Example 6.8

A machine contains three fuses p, q and r. It is desired to arrange them so that if p blows then the machine stops, but if p does not blow then the machine only stops when both q and r have blown. Derive the required fuse circuit.

Solution

In this case we can regard the fuses as being switches, with '1' representing fuse intact (current flows) and '0' representing the fuse blown (current does not flow). We are then faced with the problem of designing a circuit given a statement of its requirements. To do this, we first convert the specified requirements into logical specification in the form of a truth table. From this, the Boolean function representing the machine is written down. This may then be simplified using the algebraic rules of switching circuits to determine the simplest appropriate circuit.

Denoting the state of the machine by f (that is, $f = 1$ denotes that the machine is operating, and $f = 0$ denotes that it has stopped), the truth table of Figure 6.15 summarizes the state f in relation to the states of the individual fuses. We see from the last two columns that the machine is operating when it is in either of the three states

$$p \cdot q \cdot r \quad \text{or} \quad p \cdot q \cdot \bar{r} \quad \text{or} \quad p \cdot \bar{q} \cdot r$$

Thus it may be represented by the Boolean function

$$f = p \cdot q \cdot r + p \cdot q \cdot \bar{r} + p \cdot \bar{q} \cdot r$$

Simplifying this expression gives

$$f = (p \cdot r) \cdot (q + \bar{q}) + p \cdot q \cdot \bar{r} \qquad \text{(distributive)}$$

$$= p \cdot r + p \cdot q \cdot \bar{r} \qquad \text{(complementary)}$$

$$= p \cdot (r + q \cdot \bar{r}) \qquad \text{(distributive)}$$

$$= p \cdot ((r + q) \cdot (r + \bar{r})) \qquad \text{(distributive)}$$

$$= p \cdot (r + q) \cdot 1 \qquad \text{(complementary)}$$

$$= p \cdot (r + q) \qquad \text{(identity)}$$

Thus a suitable layout of the three fuses is as given in Figure 6.16.

In the case of this simple example we could have readily drawn the required layout from the problem specification. However, it serves to illustrate the procedure that could be adopted for a more complicated problem.

p	q	r	f	State of circuit
1	1	1	1	$p \cdot q \cdot r$
1	1	0	1	$p \cdot q \cdot \bar{r}$
1	0	1	1	$p \cdot \bar{q} \cdot r$
1	0	0	0	$p \cdot \bar{q} \cdot \bar{r}$
0	1	1	0	$\bar{p} \cdot q \cdot r$
0	1	0	0	$\bar{p} \cdot q \cdot \bar{r}$
0	0	1	0	$\bar{p} \cdot \bar{q} \cdot r$
0	0	0	0	$\bar{p} \cdot \bar{q} \cdot \bar{r}$

Figure 6.15

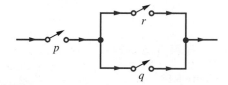

Figure 6.16

Example 6.9

In a large hall there are three electrical switches next to the three doors to operate the central lights. The three switches operate alternatively; that is, each can switch on or switch off the lights. Design a suitable switching circuit.

Solution

The light state f is either '1' (light on) or '0' (light off). Denoting the three switches by p, q and r, the state of f as it relates to the states of the three switches is given in the truth table of Figure 6.17, remembering that operating any switch turns the light off if

Figure 6.17

p	q	r	f	State of circuit
1	1	1	1	$p \cdot q \cdot r$
1	1	0	0	$p \cdot q \cdot \bar{r}$
1	0	1	0	$p \cdot \bar{q} \cdot r$
1	0	0	1	$p \cdot \bar{q} \cdot \bar{r}$
0	1	1	0	$\bar{p} \cdot q \cdot r$
0	1	0	1	$\bar{p} \cdot q \cdot \bar{r}$
0	0	1	1	$\bar{p} \cdot \bar{q} \cdot r$
0	0	0	0	$\bar{p} \cdot \bar{q} \cdot \bar{r}$

it was on and turns the light on if it was off. We arbitrarily set $p = q = r = 1$ and $f = 1$ initially. We see from the last two columns that the light is on $(f = 1)$ when the circuit is in either of the four states

$$p \cdot q \cdot r \quad \text{or} \quad p \cdot \bar{q} \cdot \bar{r} \quad \text{or} \quad \bar{p} \cdot q \cdot \bar{r} \quad \text{or} \quad \bar{p} \cdot \bar{q} \cdot r$$

Thus the required circuit is specified by the Boolean function

$$f = p \cdot q \cdot r + p \cdot \bar{q} \cdot \bar{r} + \bar{p} \cdot q \cdot \bar{r} + \bar{p} \cdot \bar{q} \cdot r$$

In this case it is not possible to simplify f any further, and in order to design the corresponding switching circuit we need to use two 1-pole, 2-way switches and one 2-pole, 2-way switch (or intermediate switch), as illustrated in Figure 6.18(a). The four possible combinations leading to 'light on' are shown in Figures 6.18(b), (c), (d) and (e) respectively.

Figure 6.18

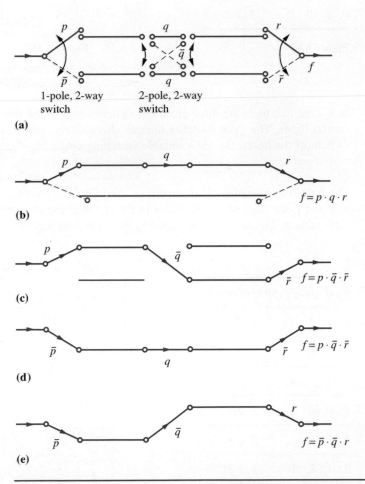

6.3.3 Exercises

18 By setting up truth tables, find the possible values of the following Boolean functions:

(a) $p \cdot (q \cdot p)$ (b) $p + (q + p)$

(c) $(p + q) \cdot (\bar{p} \cdot \bar{q})$

(d) $[(\bar{p} + \bar{q})(\bar{r} + \bar{p})] + (r + p)$

19 Figure 6.19 shows six circuits. Write down a Boolean function that represents each by using truth tables.

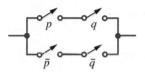

(a)

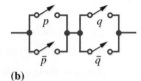

(b)

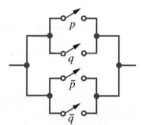

(c)

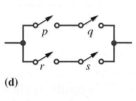

(d)

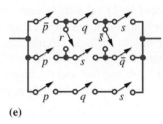

(e)

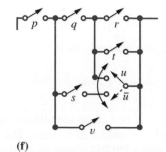

(f)

Figure 6.19

20 Use the De Morgan laws to negate the function

$$f = (p + q) \cdot (\bar{r} \cdot s) \cdot (q + \bar{t})$$

21 Give a truth table for the expression

$$f = \bar{p} \cdot q \cdot \bar{r} + \bar{p} \cdot q \cdot r + p \cdot \bar{q} \cdot \bar{r} + p \cdot q \cdot r$$

22 Simplify the following Boolean functions, stating the law used in each step of the simplification:

(a) $p \cdot (\bar{p} + p \cdot q)$ (b) $r \cdot (\overline{p + \bar{q} \cdot \bar{r}})$

(c) $(\overline{p \cdot \bar{q} + \bar{p} \cdot q})$ (d) $p + q + r + \bar{p} \cdot q$

(e) $(\overline{p \cdot q}) + (\overline{\bar{p} \cdot q \cdot r}) + p$

(f) $q + p \cdot r + p \cdot q + r$

23 Write down the Boolean functions for the switching circuits of Figure 6.20.

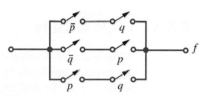

(a)

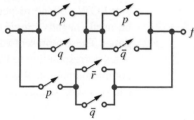

(b)

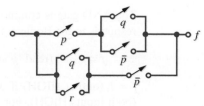

(c)

Figure 6.20

24   Draw the switching circuit corresponding to the following Boolean functions:

    (a) $f = (p + q) \cdot r + s \cdot t$ (b) $(p + q) \cdot (r + \bar{p})$

    (c) $p \cdot q + \bar{p} \cdot q$ (d) $p \cdot (q + \bar{p}) + (q + r) \cdot \bar{p}$

25   Four engineers J, F, H and D are checking a rocket. Each engineer has a switch that he or she presses in the event of discovering a fault. Show how these must be wired to a warning lamp, in the countdown control room, if the lamp is to light only under the following circumstances:

    (i)   D discovers a fault,

    (ii)   any two of J, F and H discover a fault.

26   In a public discussion a chairman asks questions of a panel of three. If to a particular question a majority of the panel answer 'yes' then a light will come on, while if to a particular question a majority of the panel answer 'no' then a buzzer will sound. The members of the panel record their answers by means of a two-position switch having position '1' for 'yes' and position '0' for 'no'. Design a suitable circuit for the discussion.

27   Design a switching circuit that can turn a lamp 'on' or 'off' at three different locations independently.

28   Design a switching circuit containing three independent contacts for a machine so that the machine is turned on when any two, but not three, of the contacts are closed.

29   The operation of a machine is monitored on a set of three lamps A, B and C, each of which at any given instant is either 'on' or 'off'. Faulty operation is indicated by each of the following conditions:

    (a)   when both A and B are off,

    (b)   when all lamps are on,

    (c)   when B is on and either A is off or C is on.

Simplify these conditions by describing as concisely as possible the state of the lamps that indicates faulty operation.

6.3.4   Logic circuits

As indicated in Section 6.3.1, a switch is a two-state device, and the algebra of switching circuits developed in Section 6.3.2 is equally applicable to systems involving other such devices. In this section we consider how the algebra may be applied to logic circuit design.

In logic circuit design the two states denoted by '1' and '0' usually denote HIGH and LOW voltage respectively (positive logic), although the opposite convention can be used (negative logic). The basic building blocks of logic circuits are called **logic gates**. These represent various standard Boolean functions. First let us consider the logic gates corresponding to the binary operation of 'and' and 'or' and the unary operation of complementation. We shall illustrate this using two inputs, although in practice more can be used.

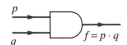

Figure 6.21
AND gate.

AND gate

The AND gate is commonly represented diagrammatically as in Figure 6.21, and corresponds to the Boolean function

$$f = p \cdot q \quad \text{(read '}p\text{ and }q\text{')}$$

$f = 1$ (output HIGH) if and only if the inputs p and q are simultaneously in state 1 (both inputs HIGH). For all other input combinations f will be zero. The corresponding truth table is as in Figure 6.9, with 1 denoting HIGH voltage and 0 denoting LOW voltage.

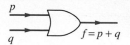

Figure 6.22
OR gate.

OR gate

The OR gate is represented diagrammatically as in Figure 6.22, and corresponds to the Boolean function

$$f = p + q \quad \text{(read '}p \text{ or } q\text{')}$$

In this case $f = 1$ (HIGH output) if either p or q or both are in state 1 (at least one input HIGH). $f = 0$ (LOW output) if and only if inputs are simultaneously 0. The corresponding truth table is as in Figure 6.10.

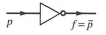

Figure 6.23
NOT gate.

NOT gate

The NOT gate is represented diagrammatically as in Figure 6.23, and corresponds to the Boolean function

$$f = \bar{p} \quad \text{(read 'not } p\text{')}$$

When the input is in state 1 (HIGH), the output is in state 0 (LOW) and vice versa. The corresponding truth table is as in Figure 6.12.

With these interpretations of $\cdot$, $+$, $(\bar{})$, 0 and 1, the rules developed in Section 6.3.2 for the algebra of switching circuits are applicable to the analysis and design of logic circuits.

Example 6.10   Build a logic circuit to represent the Boolean function

$$f = \bar{p} \cdot q + p$$

Solution   We first use a NOT gate to obtain $\bar{p}$ then an AND gate to generate $\bar{p} \cdot q$, and finally an OR gate to represent f. The resulting logic circuit is shown in Figure 6.24.

Figure 6.24
Logic circuit
$f = \bar{p} \cdot q + p$.

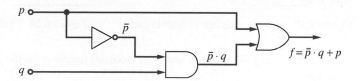

Example 6.11   Build a logic circuit to represent the Boolean function

$$f = (p + \bar{q}) \cdot (r + s \cdot q)$$

Solution   Adopting a similar procedure to the previous example leads to the logic circuit of Figure 6.25.

Figure 6.25
Logic circuit
$f = (p + \bar{q}) \cdot (r + s \cdot q)$.

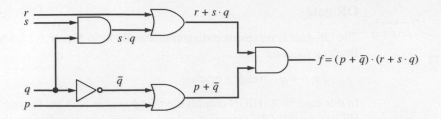

So far we have considered the three logic gates AND, OR and NOT and indicated how these can be used to build a logic circuit representative of a given Boolean function. We now introduce two further gates, which are invaluable in practice and are frequently used.

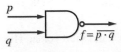

Figure 6.26
NAND gate $f = \overline{p \cdot q}$.

NAND gate

The NAND (or 'NOT AND') gate is represented diagrammatically in Figure 6.26, and corresponds to the function

$$f = \overline{p \cdot q}$$

The small circle on the output line of the gate symbol indicates negation or NOT. Thus the gate negates the AND gate, and is equivalent to the logic circuit of Figure 6.27.

The corresponding truth table is given in Figure 6.28.

p	q	$p \cdot q$	f
1	1	1	0
1	0	0	1
0	1	0	1
0	0	0	1

Figure 6.27 Equivalent circuit to NAND gate.

Figure 6.28 Truth table for NAND gate.

Note that, using De Morgan laws, the Boolean function for the NAND gate may also be written as

$$f = \overline{p \cdot q} = \bar{p} + \bar{q}$$

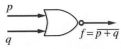

Figure 6.29
NOR gate $f = \overline{p + q}$.

NOR gate

The NOR (or 'NOT OR') gate is represented diagrammatically as in Figure 6.29, and corresponds to the Boolean function

$$f = \overline{p + q}$$

Again we have equivalence with the logic circuit of Figure 6.30, and (using the De Morgan laws) with the Boolean function

$$f = \overline{p+q} = \bar{p} \cdot \bar{q}$$

The corresponding truth table is given in Figure 6.31.

p	q	$p+q$	f
1	1	1	0
1	0	1	0
0	1	1	0
0	0	0	1

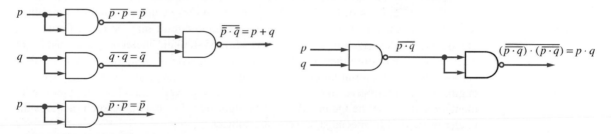

Figure 6.30 Equivalent circuit to NOR gate.

Figure 6.31 Truth table for NOR gate.

It is of interest to recognize that, using either one of the NAND or NOR gates, it is possible to build a logic circuit to represent any given Boolean function. To prove this, we have to show that, using either gate, we can implement the three basic Boolean functions $p + q$, $p \cdot q$ and $\bar{p}$. This is illustrated in Figure 6.32 for the NAND gate; the illustration for the NOR gate is left as an exercise for the reader.

Figure 6.32 Basic Boolean functions using NAND gates.

Example 6.12   Using only NOR gates, build a logic circuit to represent the Boolean function

$$f = \bar{p} \cdot q + p \cdot \bar{q}$$

Solution   The required logic circuit is illustrated in Figure 6.33.

Figure 6.33

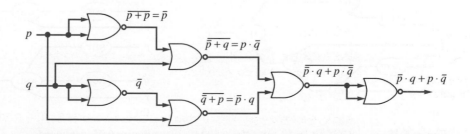

We note that the Boolean function considered in Example 6.12 is the same as that considered in Example 6.5, where its truth table was constructed, indicating that the output is in state 1 only if the two inputs are in different states. This leads us to defining a further logic gate used in practice.

EXCLUSIVE OR gate

The EXCLUSIVE OR gate is represented diagrammatically as in Figure 6.34, and corresponds to the Boolean function

$$f = \bar{p} \cdot q + p \cdot \bar{q}$$

Figure 6.34
EXCLUSIVE OR
gate.

$$p \quad q \quad f = \bar{p} \cdot q + p \cdot \bar{q}$$

As indicated above, $f = 1$ (output HIGH) only if the inputs p and q are in different states; that is, either p or q is in state 1 but *not both*. It therefore corresponds to the everyday exclusive usage of the word 'OR' where it is taken to mean 'one or the other but not both'. On the other hand, the OR gate introduced earlier is used in the sense 'one or the other or both', and could more precisely be called the INCLUSIVE OR gate.

Although present technology is such that a logic circuit consisting of thousands of logic gates may be incorporated in a single silicon chip, the design of smaller equivalent logic circuits is still an important problem. As for switching circuits, simplification of a Boolean function representation of a logic circuit may be carried out using the algebraic rules given in Section 6.3.2. More systematic methods are available for carrying out such simplification. For Boolean expressions containing not more than six variables the pictorial approach of constructing Karnaugh maps is widely used by engineers. An alternative algebraic approach, which is well suited for computer implementation, is to use the Quine–McCluskey algorithm. For details of such methods the reader is referred to specialist texts on the subject.

6.3.5 Exercises

30 Write down the Boolean function for the logic blocks of Figure 6.35. Simplify the functions as far as possible and draw the equivalent logic block.

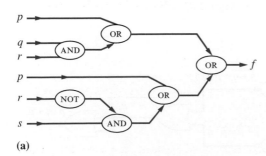

(a)

Figure 6.35

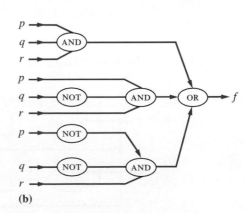

(b)

Figure 6.35 *continued*

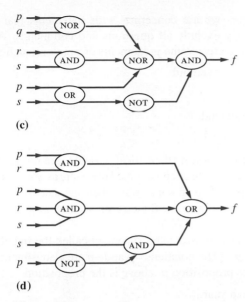

(c)

(d)

Figure 6.35 *continued*

31 Simplify the following Boolean functions and sketch the logic block corresponding to both the given and simplified functions:

(a) $(\bar{p} \cdot q + p \cdot \bar{q}) \cdot (\bar{p} + \bar{q}) \cdot (p + q)$

(b) $\bar{r} \cdot \bar{p} \cdot \bar{q} + \bar{r} \cdot \bar{p} \cdot q + r \cdot \bar{p} \cdot \bar{q}$

(c) $\bar{p} \cdot \bar{q} + r \cdot \bar{p} \cdot s + \bar{p} \cdot \bar{q} \cdot s$

(d) $(p + q) \cdot (p + r) + r \cdot (p + q \cdot r)$

(e) $(\bar{p} + \bar{q}) \cdot (\bar{p} + q) \cdot (p + q)$

6.4 Propositional logic and methods of proof

In the last section we dealt with switches that are either off or on. These lend themselves naturally to the application of set algebra. On the other hand, everyday use of English contains many statements that are neither obviously true nor false: for example 'Chilly for the time of year, isn't it?' There are, however, some statements that are immediately either true or false: for example, 'In 1988 the Summer Olympics were held in Seoul, South Korea' (true) or 'All children watch too much television' (false). Propositional logic can be used to analyse, simplify and establish the equivalence of statements. Applications of propositional logic include the efficient operation of computer-based expert systems, where the user may phrase questions differently or answer in different ways, and yet the answers are logically equivalent. Propositional logic leads naturally to the precise formulation of the proof of statements that, though important in themselves, are also the basis by which computer programs can be made more efficient. Thus we shall develop tools with a vast potential for use throughout engineering.

6.4.1 Propositions

A proposition is a statement (or sentence) for which it is immediately decidable whether it is true (T) or false (F), but not both. For example

p_1:   The year 1973 was a leap year

is a proposition readily decidable as false. Note the use of the label 'p_1: . . .', so that the overall statement is read 'p_1 is the statement. "The year 1973 was a leap year" '.

Since when considering propositions we are concerned with statements that are decidable as true or false, we obviously exclude all questions and commands. Also excluded are assertions that involve subjective value judgements or opinions such as

> r: The Director of the company is overpaid

Statements such as

> m: He was Prime Minister of England
>
> n: The number $x + 3$ is divisible by 3

that involve pronouns (he, she, and so on) or a mathematical variable, are not readily decidable as true or false, and are therefore not propositions. However, as soon as the pronoun or variable is specified (or quantified in some way) then the statements are decidable as true or false and become propositions. Statements such as m and n are examples of **predicates**.

Given any statement p, there is always an associated statement called the **negation** of p. We denote this by $\tilde{p}$, read as 'not p'. (The notations $\neg p$ and $\sim p$ are also sometimes used.) For example, the negation of the proposition p_1 above is the proposition

> $\tilde{p}_1$:   The year 1973 was not a leap year

which is decidable as true, the opposite truth value to p_1. In general the negation $\tilde{p}$ of a statement p always has precisely the opposite truth value to that of p itself. The truth values of both p and $\tilde{p}$ are given in the truth table shown in Figure 6.36.

p	$\tilde{p}$
T	F
F	T

Figure 6.36
Truth table for $\tilde{p}$.

Example 6.13

List A is a list of propositions, while list B is a list of sentences that are not propositions.

(a) Determine the truth values of the propositions in list A and state their negation statements.

(b) Explain why the sentences in list B are not propositions.

List A:

(a) Everyone can say where they were when President J. F. Kennedy was assassinated

(b) $2^n = n^2$ for some $n \in \mathbb{N}$, where $\mathbb{N}$ is the set of natural numbers

(c) The number 5 is negative

(d) $2^{89\,301} + 1$ is a prime number

(e) Air temperatures were never above 0°C in February 1935 in Bristol, UK

List B:

(a) Maths is fun

(b) Your place or mine?

(c) $y - x = x - y$

(d) Why am I reading this?

(e) Flowers are more interesting than calculus

(f) n is a prime number

(g) He won an Olympic medal

Solution   First of all, let us examine list A.

(a) This is obviously false. Besides those with poor memories or those from remote parts of the world, not everyone had been born in 1963.

(b) This is true (for $n = 2$).

(c) This is obviously false.

(d) There is no doubt that this is either true or false, but only specialists would know which (it is true).

(e) This is true, but again specialist knowledge is required before this can be verified.

All statements in list A are propositions because they are either true or false, never both. The negation predicates for list A are as follows.

(a) Not everyone can say where they were when President J. F. Kennedy was assassinated.

(b) $2^n \neq n^2$ for all $n \in \mathbb{N}$.

(c) The number 5 is not negative.

(d) $2^{89\,301} + 1$ is not a prime number.

(e) Air temperature was above 0°C at some time in February 1935 in Bristol, UK.

The sentences in list B are not propositions, for the following reasons.

(a) This is a subjective judgement. I think maths is fun (most of the time) – you probably do not!

(b) This is a question, and thus cannot be a proposition.

(c) This can easily be made into a proposition by the addition of the phrase 'for some real numbers x and y'. It is then true (whenever $x = y$).

(d) This is the same category as (b), a question.

(e) This is a subjective statement in the same category as (a).

(f) This is a predicate, since it will become a proposition once n is specified.

(g) Again, this is a predicate, since once we know who 'he' is, the statement will be certainly either true or false and hence be a proposition.

6.4.2 Compound propositions

When we combine simple statements together by such words as 'and', 'or' and so on we obtain compound statements. For example,

> m: Today is Sunday and John has gone to church

> n: Mary is 35 years old or Mary is 36 years old

constitute compound statements, with the constituent simple statements being respectively

m_1: Today is Sunday, $\qquad$ m_2: John has gone to church

n_1: Mary is 35 years old, $\qquad$ n_2: Mary is 36 years old

As for switching circuits, we can again draw an analogy between the use of the words 'or' and 'and' in English and their use in the algebra of sets to form the union $A \cup B$ and intersection $A \cap B$ of two sets A and B. Drawing on the analogy the word 'or' is used to mean 'at least one statement' and the word 'and' to mean 'both statements'. The symbolism commonly used in propositional logic is to adopt the symbol $\vee$ (analogous to $\cup$) for 'or' and the symbol $\wedge$ (analogous to $\cap$) for 'and'. Thus in symbolic form the statements m and n may be written in terms of their constituent simple statements as

$$m = m_1 \wedge m_2 \quad (m_1 \text{ and } m_2)$$

$$n = n_1 \vee n_2 \quad (n_1 \text{ or } n_2)$$

In general for two statements p and q the truth values of the compound statements

$$p \vee q \quad \text{(meaning '} p \text{ or } q \text{' and called the \textbf{disjunction} of } p, q)$$

$$p \wedge q \quad \text{(meaning '} p \text{ and } q \text{' and called the \textbf{conjunction} of } p, q)$$

p	q	$p \vee q$	$p \wedge q$
T	T	T	T
T	F	T	F
F	T	T	F
F	F	F	F

Figure 6.37
Truth table for $p \vee q$ and $p \wedge q$.

are as given in the truth table of Figure 6.37.

Here are two examples that use compound statements and also make use of $\tilde{p}$ meaning 'not p' and

$$p \rightarrow q$$

meaning p implies q. There will be more about this kind of compound statement $p \rightarrow q$ (p implies q) when we deal with proof in Section 6.4.5.

Example 6.14

Let A, B and C be the following propositions:

A: It is frosty
B: It is after 11.00 a.m.
C: Jim drives safely

(a) Translate the following statements into logical statements using the notation of this section.

  (i) It is not frosty.
  (ii) It is frosty and after 11.00 a.m.
(iii) It is not frosty, it is before 11.00 a.m. and Jim drives safely.

(b) Translate the following into English sentences:

(i) $A \wedge B$,   (ii) $\tilde{A} \rightarrow C$,   (iii) $A \wedge \tilde{B} \rightarrow \tilde{C}$,   (iv) $\tilde{A} \vee B \rightarrow C$

Solution

(a)   (i) is the negation of A, so is written $\tilde{A}$.
    (ii) is A AND B, written $A \wedge B$.
    (iii) is slightly more involved, but is a combination of NOT A, NOT B AND C, and so is written $\tilde{A} \wedge \tilde{B} \wedge C$.

(b)   (i) $A \wedge B$ is A AND B; that is, 'It is frosty and it is after 11.00 a.m.'
    (ii) $\tilde{A} \rightarrow C$ is NOT A implies C; that is, 'It is not frosty, therefore Jim drives safely'.

(iii) $A \wedge \tilde{B} \rightarrow \tilde{C}$ is A AND NOT B implies NOT C; that is, 'It is frosty and before 11.00 a.m.; therefore Jim does not drive safely.'

(iv) $\tilde{A} \vee B \rightarrow C$ is NOT A or B implies C; that is, 'It is not frosty or it is after 11.00 a.m.; therefore Jim drives safely.'

Example 6.15

(Adapted from Exercise 5.15 in K.A. Ross and C.R.B. Wright, *Discrete Mathematics*, Prentice Hall, Englewood Cliffs, NJ, 1988.) In a piece of software, we have the following three propositions:

P: The flag is set
Q: $I = 0$
R: Subroutine S is completed

Translate the following into symbols:

(a) If the flag is set then $I = 0$.

(b) Subroutine S is completed if the flag is set.

(c) The flag is set if subroutine S is not completed.

(d) Whenever $I = 0$, the flag is set.

(e) Subroutine S is completed only if $I = 0$.

(f) Subroutine S is completed only if $I = 0$ or the flag is set.

Solution

Most of the answers can be given with minimal explanation. The reader should check each and make sure each is understood before going further.

(a) $P \rightarrow Q$ (that is, P implies Q)

(b) $P \rightarrow R$ (that is, P implies R)

(c) $\tilde{R} \rightarrow P$ (that is, NOT R implies P)

Note that the logical expression is sometimes, as in (b) and (c), the 'other way round' from the English sentence. This reflects the adaptability of the English language, but can be a pitfall for the unalert student.

(d) $Q \rightarrow P$ (that is, Q implies P)

(e) $R \rightarrow Q$ (that is, R implies Q)

(f) This is really two statements owing to the presence of the (English, not logical) 'or'. 'S is completed only if $I = 0$' is written in logical symbols as (e) $R \rightarrow Q$. So including 'the flag is set' as a logical alternative gives

$$(R \rightarrow Q) \vee P$$

as the logical interpretation of (f). Alternatively, we can interpret the phrase '$I = 0$ or the flag is set' logically first as $Q \vee P$, then combine this with 'subroutine S is completed' to give

$$R \rightarrow (Q \vee P)$$

Now these two logical expressions are not the same. The sentence (f) may seem harmless; however, some extra punctuation or rephrasing is required before it is rendered unambiguous. One version could read:

(f) Subroutine S is completed only if either $I = 0$ or the flag is set (or both).

  This is $R \rightarrow (Q \vee P)$.

Another could read:

(f) Subroutine S is completed only if $I = 0$, or the flag is set (or both).

  This is $(R \rightarrow Q) \vee P$.

Part (f) highlights the fact that there is no room for sloppy thought in this branch of engineering mathematics.

6.4.3 Algebra of statements

In the same way as we used $\cup$, $\cap$ and $(\overline{})$ to generate complex expressions for sets we can use $\vee$, $\wedge$ and $\sim$ to form complex compound statements by constructing truth tables.

Example 6.16   Construct the truth table determining the truth values of the compound proposition

$$p \vee (p \wedge q)$$

Solution   The truth table is shown in Figure 6.38. Note that this verifies the analogous absorption law for set algebra of Question 10 (Exercises 6.2.5).

Figure 6.38
Truth table for
$p \vee (p \wedge q)$.

p	q	$p \wedge q$	$p \vee (p \wedge q)$
T	T	T	T
T	F	F	T
F	T	F	F
F	F	F	F

The statements are said to be **equivalent** (or more precisely **logically equivalent**) if they have the same truth values. Again, to show that two statements are equivalent, we simply need to construct the truth table for each statement and compare truth values. For example, from Example 6.16 we see that the two statements

$$p \vee (p \wedge q) \quad \text{and} \quad p$$

are equivalent. The symbolism $\equiv$ is used to denote equivalent statements, so we can write

$$p \vee (p \wedge q) \equiv p$$

By constructing the appropriate truth tables, the following laws, analogous to the results (6.1), (6.2), (6.5) and (6.6) for set algebra, are readily verified:

Commutative laws

$$p \vee q \equiv q \vee p, \quad p \wedge q \equiv q \wedge p$$

Idempotent laws

$$p \vee p \equiv p, \quad p \wedge p \equiv p$$

Associative laws

$$p \vee (q \vee r) \equiv (p \vee q) \vee r, \quad p \wedge (q \wedge r) \equiv (p \wedge q) \wedge r$$

Distributive laws

$$p \vee (q \wedge r) \equiv (p \vee q) \wedge (p \vee r), \quad p \wedge (q \vee r) \equiv (p \wedge q) \vee (p \wedge r)$$

To develop a complete parallel with the algebra of sets, we need to identify two unit elements analogous to $\varnothing$ and U, relative to $\vee$ and $\wedge$ respectively.

Relative to $\vee$, we need to identify a statement s such that

$$p \vee s \equiv p$$

for any statement p. Clearly s must have a false value under all circumstances, and an example of such a statement is

$$s \equiv q \wedge \tilde{q}$$

where q is any statement, as evidenced by the truth table of Figure 6.39(a). Such a statement that is false under all circumstances is called a **contradiction**, and its role in the algebra of statements is analogous to the role of the empty set $\varnothing$ in the algebra of sets.

Relative to $\wedge$, we need to identify a statement t such that

$$p \wedge t \equiv p$$

for any statement p. Clearly, t must have a true truth value under all circumstances, and an example of such a statement is

$$t \equiv q \vee \tilde{q}$$

for any statement q, as evidenced by the truth table of Figure 6.39(b). Such a statement that is true under all circumstances is called a **tautology** and its role in the algebra of statements is analogous to that of the universal set U in the algebra of sets.

q	$\tilde{q}$	$q \wedge \tilde{q}$
T	F	F
F	T	F

(a) Contradiction

q	$\tilde{q}$	$q \wedge \tilde{q}$
T	F	T
F	T	T

(b) Tautology

Figure 6.39

Introducing the tautology and contradiction statements t and s respectively leads to the identity and complementary laws

Identity laws

$$p \vee s \equiv p \quad (s \text{ is the identity relative to } \vee)$$
$$p \wedge t \equiv p \quad (t \text{ is the identity relative to } \wedge)$$

Complementary laws

$$p \vee \tilde{p} \equiv t, \quad p \wedge \tilde{p} \equiv s$$

analogous to (6.3) and (6.4) for set algebra.

It then follows that the algebra of statements is another example of a Boolean algebra, with $\vee$ and $\wedge$ being the two binary operations, $\tilde{}$ being the unary operation and s and t the identity elements. Consequently all the results developed for the algebra of sets carry through to the algebra of statements with equivalence between $\cup, \cap, (\bar{}), \varnothing$, U and $\vee, \wedge, \tilde{}, s, t$ respectively. These rules may then be used to reduce complex statements to simpler compound statements. These rules of the algebra of statements form the basis of propositional logic.

Example 6.17

Construct a truth table to verify the De Morgan laws for the algebra of statements analogous to (6.1) for the algebra of sets.

Solution

The analogous De Morgan laws for statements are the negations

$$\widetilde{(p \vee q)} \equiv \tilde{p} \wedge \tilde{q}, \qquad \widetilde{(p \wedge q)} \equiv \tilde{p} \vee \tilde{q}$$

whose validity is verified by the tables displayed in Figures 6.40 and 6.41.

p	q	$\tilde{p}$	$\tilde{q}$	$p \vee q$	$\widetilde{p \vee q}$	$\tilde{p} \wedge \tilde{q}$
T	T	F	F	T	F	F
T	F	F	T	T	F	F
F	T	T	F	T	F	F
F	F	T	T	F	T	T

Figure 6.40 Truth table for $\widetilde{p \vee q} \equiv \tilde{p} \wedge \tilde{q}$.

p	q	$\tilde{p}$	$\tilde{q}$	$p \wedge q$	$\widetilde{p \wedge q}$	$\tilde{p} \vee \tilde{q}$
T	T	F	F	T	F	F
T	F	F	T	F	T	T
F	T	T	F	F	T	T
F	F	T	T	F	T	T

Figure 6.41 Truth table for $\widetilde{p \wedge q} \equiv \tilde{p} \vee \tilde{q}$.

6.4.4 Exercises

32 Negate the following propositions:

(a) Fred is my brother.

(b) 12 is an even number.

(c) There will be gales next winter.

(d) Bridges collapse when design loads are exceeded.

33 Determine the truth values of the following propositions:

(a) The world is flat.

(b) $2^n + n$ is a prime number for some integer n.

(c) $a^2 = 0$ implies $a = 0$ for all $a \in \mathbb{N}$.

(d) $a + bc = (a + b)(a + c)$ for real numbers a, b and c.

34 Determine which of the following are propositions and which are not. For those that are, determine their truth values.

(a) $x + y = y + x$ for all $x, y \in \mathbb{R}$.

(b) $\mathbf{AB} = \mathbf{BA}$, where $\mathbf{A}$ and $\mathbf{B}$ are square matrices.

(c) Academics are absent-minded.

(d) I think that the world is flat.

(e) Go fetch a policeman.

(f) Every even integer greater than 4 is the sum of two prime numbers. (This is Goldbach's conjecture.)

35 Let A, B and C be the following propositions:

A: It is raining
B: The sun is shining
C: There are clouds in the sky

Translate the following into logical notation:

(a) It is raining and the sun is shining.

(b) If it is raining then there are clouds in the sky.

(c) If it is not raining then the sun is not shining and there are clouds in the sky.

(d) If there are no clouds in the sky then the sun is shining.

36 Let A, B and C be as in Question 35. Translate the following logical expressions into English sentences:

(a) $A \wedge B \rightarrow C$ (b) $(A \rightarrow C) \rightarrow B$

(c) $\tilde{A} \rightarrow (B \vee C)$ (d) $(\widetilde{A \vee B}) \wedge C$

37 Consider the ambiguous sentence

$$x^2 = y^2 \text{ implies } x = y \text{ for all } x \text{ and } y$$

(a) Make the sentence into a proposition that is true.

(b) Make the sentence into a proposition that is false.

6.4.5 Implications and proofs

A third type of compound statement of importance in propositional logic is that of **implication**, which lies at the heart of a mathematical argument. We have already met it briefly in Example 6.14, but here we give its formal definition. If p and q are two statements then we write the implication compound statement as

> If p then q

which asserts that the truth of p guarantees the truth of q. Alternatively, we say

> p implies q

and adopt the symbolism $p \rightarrow q$ (the notation $p \Rightarrow q$ is also commonly in use).

The truth table corresponding to $p \rightarrow q$ is given in Figure 6.42. From the truth table we see that $p \rightarrow q$ is false only when p is true and q is false. At first the observation that $p \rightarrow q$ is true whenever p is false may appear strange, but a simple example should

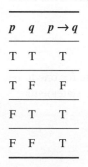

p	q	$p \rightarrow q$
T	T	T
T	F	F
F	T	T
F	F	T

Figure 6.42
Truth table for $p \rightarrow q$.

convince you. Suppose that prior to interviews for a senior management post within a company, the candidate states

If I am appointed then company profits will rise

This is clearly an implication statement $p \rightarrow q$, with the statements p and q being

p: I am appointed

q: Company profits will rise

If the candidate is not appointed (that is, p is 'false') then the statement made by the candidate is not false – independently of whether or not the company profits will rise. Hence $p \rightarrow q$ must be 'true'.

Example 6.18

Use truth tables to show that the following are tautologies:

(a) $A \rightarrow A$,   (b) $A \wedge (A \rightarrow B) \rightarrow B$

A	A → A
F	T
T	T

Figure 6.43 Truth table for $A \rightarrow A$.

A	B	A → B	A ∧ (A → B)	[A ∧ (A → B)] → B
F	F	T	F	T
F	T	T	F	T
T	F	F	F	T
T	T	T	T	T

Figure 6.44 Truth table for $[A \wedge (A \rightarrow B)] \rightarrow B$.

Solution

(a) The truth table in Figure 6.43 is easily constructed, and shows that, no matter whether A is true or false, $A \rightarrow A$ is true. It is thus a tautology.

(b) The truth table shown in Figure 6.44 can be drawn, and we see that all the entries in the last column are true and the outcome of $A \wedge (A \rightarrow B) \rightarrow B$ is always true; it is thus a tautology.

The implication statement

$q \rightarrow p$

is called the **converse** of the statement $p \rightarrow q$, and it is perfectly possible for one to be true and the other to be false. For example, if p and q are defined by the statements

p: I go for a walk in the rain

q: I get wet

then the implication statements $p \rightarrow q$ and $q \rightarrow p$ are

If I go for a walk in the rain then I get wet

and

>   If I am getting wet then I am going for a walk in the rain

respectively. The first, $p \to q$, is true but the second, $q \to p$, is false (I could be taking a shower).

An implication statement that asserts both $p \to q$ and $q \to p$ is called **double implication**, and is denoted by

> $p \leftrightarrow q$

which may be expressed verbally as

> p if and only if q

or 'p is a necessary and sufficient condition for q'. Again the notation $p \Leftrightarrow q$ is also frequently used to represent double implication.

It thus follows that $p \leftrightarrow q$ is defined to be

> $(p \to q) \wedge (q \to p)$

and its truth table is given in Figure 6.45.

Figure 6.45
Truth table for $p \leftrightarrow q$.

p	q	$p \to q$	$q \to p$	$p \leftrightarrow q$
T	T	T	T	T
T	F	F	T	F
F	T	T	F	F
F	F	T	T	T

From Figure 6.45 we see that $p \leftrightarrow q$ is true if p and q have the same truth values, and is false if p and q have different truth values. It therefore follows that

> $(p \leftrightarrow q) \leftrightarrow (p \equiv q)$

meaning that each of the statements $p \leftrightarrow q$ and $p \equiv q$ implies the other. We must be careful when interpreting implication when negation statements are involved. A commonly made mistake is to assume that if the implication

> $p \to q$

is valid then the implication

> $\tilde{p} \to \tilde{q}$

is also valid. A little thought should convince you that this is not necessarily the case. This can be confirmed by reconsidering the previous example, when the negations $\tilde{p}$ and $\tilde{q}$ would be

> $\tilde{p}$:   I do not go for a walk in the rain

> $\tilde{q}$:   I do not get wet

and $\tilde{p} \to \tilde{q}$ is

>   If I do not go for a walk in the rain then I do not get wet

Figure 6.46
The equivalence of
$p \rightarrow q$ and $\tilde{q} \rightarrow \tilde{p}$.

p	q	$\tilde{p}$	$\tilde{q}$	$p \rightarrow q$	$\tilde{q} \rightarrow \tilde{p}$
T	T	F	F	T	T
T	F	F	T	F	F
F	T	T	F	T	T
F	F	T	T	T	T

(This is obviously false, since someone could throw a bucket of water over me.) $p \rightarrow q$ and $\tilde{p} \rightarrow \tilde{q}$ so have different truth values. The construction of the two truth tables will establish this rigorously. On the other hand, the implication statements $p \rightarrow q$ and $\tilde{q} \rightarrow \tilde{p}$ are equivalent, as can be seen from the truth table in Figure 6.46. The implication $\tilde{q} \rightarrow \tilde{p}$ is called the **contrapositive form** of the implication $p \rightarrow q$.

In mathematics we need to establish beyond any doubt the truth of statements. If we denote by p a type of statement called a **hypothesis** and by q a second type of statement called a **conclusion** then the implication $p \rightarrow q$ is called a **theorem**.

In general, p can be formed from several statements; there is, however, usually only one conclusion in a theorem. A sequence of propositions that end with a conclusion, each proposition being regarded as valid, is called a **proof**. In practice, there are three ways of proving a theorem. These are direct proof, indirect proof and proof by induction. **Direct proof** is, as its name suggests, directly establishing the conclusion by a sequence of valid implementations. Here is an example of direct proof.

Example 6.19 If $a, b, c, d \in \mathbb{R}$, prove that the inverse of the 2×2 matrix

$$\mathbf{A} = \begin{bmatrix} a & b \\ c & d \end{bmatrix} \quad (ad \neq bc) \quad \text{is} \quad \frac{1}{ad - bc} \begin{bmatrix} d & -b \\ -c & a \end{bmatrix}$$

Solution This has already been done in Chapter 5, Section 5.4. In the context of propositional logic, we conveniently split the proof as follows.

H_1: If there exists a 2×2 matrix $\mathbf{B}$ such that

$$\mathbf{AB} = \mathbf{BA} = \mathbf{I}_2$$

where $\mathbf{I}_2$ is the 2×2 identity matrix, then $\mathbf{B}$ is the inverse of $\mathbf{A}$

$$H_2: \quad \begin{bmatrix} \alpha & 0 \\ 0 & \alpha \end{bmatrix} = \alpha \begin{bmatrix} 1 & 0 \\ 0 & 1 \end{bmatrix} = \alpha \mathbf{I}_2$$

$$H_3: \quad \begin{bmatrix} a & b \\ c & d \end{bmatrix}\begin{bmatrix} d & -b \\ -c & a \end{bmatrix} = \begin{bmatrix} d & -b \\ -c & a \end{bmatrix}\begin{bmatrix} a & b \\ c & d \end{bmatrix} = \begin{bmatrix} ad - bc & 0 \\ 0 & ad - bc \end{bmatrix}$$

Using H_2, we deduce that

$$\begin{bmatrix} ad - bc & 0 \\ 0 & ad - bc \end{bmatrix} = (ad - bc) \begin{bmatrix} 1 & 0 \\ 0 & 1 \end{bmatrix}$$

$$= (ad - bc)\mathbf{I}_2$$

Dividing by $ad - bc$ then gives the result.

In this proof, H_1 is a definition and hence true, H_2 and H_3 are properties of matrices established in Chapter 5. (It is possible to split H_3 into arithmetical hypotheses detailing the process of matrix multiplication.) Hence

$$H_1 \wedge H_2 \wedge H_3 \text{ implies } \mathbf{A}^{-1} = \frac{1}{ad - bc} \begin{bmatrix} d & -b \\ -c & a \end{bmatrix}$$

hence establishing that the right-hand side is the inverse of $\mathbf{A}$.

We have seen that $p \rightarrow q$ and $\tilde{q} \rightarrow \tilde{p}$ are logically equivalent. The use of this in a proof sometimes makes the arguments easier to follow, and we call this an **indirect proof**. Here is an example of this.

Example 6.20   Prove that if $a + b \geq 15$ then either $a \geq 8$ or $b \geq 8$, where a and b are integers.

Solution   Let p, q and r be the statements

$$p: \quad a + b \geq 15 \qquad q: \quad a \geq 8 \qquad r: \quad b \geq 8$$

Then the negations of these statements are

$$\tilde{p}: \quad a + b < 14 \qquad \tilde{q}: \quad a < 7 \qquad \tilde{r}: \quad b < 7$$

The statement to be proved can be put into logical notation

$$p \rightarrow (q \vee r)$$

This is equivalent to

$$(\widetilde{q \vee r}) \rightarrow \tilde{p}$$

or, using the De Morgan laws,

$$(\tilde{q} \wedge \tilde{r}) \rightarrow \tilde{p}$$

If we prove the truth of this implication statement then we have also proved that

$$p \rightarrow (q \vee r)$$

We have

$$\tilde{q} \wedge \tilde{r}: \quad a < 7 \text{ and } b < 7$$

$$\tilde{p}: \quad a + b < 14$$

Hence $\tilde{p} \wedge \tilde{r} \to \tilde{p}$ is

$a < 7$ and $b < 7$ implies $a + b < 14$ for integers a and b

which is certainly true.

We have thus proved that $p \to (q \vee r)$, as required.

Another indirect form of proof is **proof by contradiction**. Instead of proving 'p is true' we prove '$\tilde{p}$ is false'. An example of this kind of indirect proof follows.

Example 6.21   Prove that $\sqrt{2}$ is irrational.

Solution   Let p be the statement

p: $\sqrt{2}$ is irrational

then $\tilde{p}$ is the statement

$\tilde{p}$: $\sqrt{2}$ is rational

Here are the arguments establishing that $\tilde{p}$ is 'false'. If $\sqrt{2}$ is rational then there are integers m and n, with no common factor, such that

$$\sqrt{2} = \frac{m}{n}$$

Squaring this gives

$$2 = \frac{m^2}{n^2} \quad \text{or} \quad m^2 = 2n^2$$

This implies that m^2 is an even number, and therefore so is m. Hence

$m = 2k$ with k an integer

So

$$m^2 = 4k^2$$

However, since $n^2 = \frac{1}{2}m^2$, this implies

$$n^2 = 2k^2$$

and therefore n^2 is also even, which means that n is even. But if both m and n are even, they have the factor 2 (at least) in common. We thus have a contradiction, since we have assumed that m and n have no common factors. Thus $\tilde{p}$ must be false. If $\tilde{p}$ is false then p is true, and hence we have proved that $\sqrt{2}$ is irrational.

The final method of proof we shall examine is **proof by induction**. If $p_1, p_2, \ldots,$ $p_n, \ldots$ is a sequence of propositions, n is a natural number and

(a) p_1 is true (the **basis for induction**)

(b)   if p_n is true then p_{n+1} is true (the **induction hypothesis**) then p_n is true for all n **by induction**. Proof by induction is used extensively by mathematicians to establish formulae. Here is such an example.

Example 6.22

Use mathematical induction to show that

$$1 + 2 + \ldots + N = \tfrac{1}{2}N(N + 1) \tag{6.8}$$

for any natural number N.

Solution

Let us follow the routine for proof by induction.

First of all, we set $N = 1$ in the proposition (6.8):

$$1 = \tfrac{1}{2}1(1 + 1)$$

which is certainly true. Now we set $N = n$ in (6.8) and assume the statement is true:

$$1 + 2 + \ldots + n = \tfrac{1}{2}n(n + 1) \tag{6.9}$$

We now have to show that

$$1 + 2 + \ldots + n + (n + 1) = \tfrac{1}{2}(n + 1)(n + 2) \tag{6.10}$$

which is the proposition (6.8) with N replaced by $n + 1$. If we add $n + 1$ to both sides of (6.9) then the right-hand side becomes

$$\tfrac{1}{2}n(n + 1) + (n + 1)$$

which can be rewritten as

$$(\tfrac{1}{2}n + 1)(n + 1) = \tfrac{1}{2}(n + 1)(n + 2)$$

thus establishing the proof of the induction hypothesis. The truth of (6.8) then follows by induction.

6.4.6 Exercises

38 The **counterexample** is a good way of disproving assertions. (Examples can *never* be used as proof.) Find counterexamples for the following assertions:

(a) $2^n - 1$ is a prime for every $n \geqslant 2$

(b) $2^n + 3^n$ is a prime for all $n \in \mathbb{N}$

(c) $2^n + n$ is prime for every positive odd integer n

39 Give the converse and contrapositive for each of the following propositions:

(a) $A \rightarrow (B \wedge C)$

(b) If $x + y = 1$ then $x^2 + y^2 \geqslant 1$

(c) If $2 + 2 = 4$ then $3 + 3 = 9$

40 Construct the truth tables for the following:

(a) $A \wedge \tilde{A}$ (b) $\tilde{A} \vee \tilde{B}$

(c) $(A \wedge B) \rightarrow C$ (d) $\overparen{(A \wedge B) \rightarrow C)}$

41 Prove or disprove the following:

(a) $(B \rightarrow A) \leftrightarrow (A \wedge B)$

(b) $(A \wedge B) \rightarrow (A \rightarrow B)$

(c) $(A \wedge B) \rightarrow (A \vee B)$

Note that to *disprove* a tautology, only one line of a truth table is required.

42 Use contradiction to show that $\sqrt{3}$ is irrational.

43 Prove or disprove the following:

(a) The sum of two even integers is an even integer.

(b) The sum of two odd integers is an odd integer.

(c) The sum of two primes is never a prime.

(d) The sum of three consecutive integers is divisible by 3.

Indicate the methods of proof where appropriate.

44 Prove that the number of primes is infinite by contradiction.

45 Use induction to establish the following results:

(a) $\sum_{k=1}^{n} k^2 = \frac{1}{6}n(n + 1)(2n + 1)$ (n a natural number)

(b) $4 + 10 + 16 + \ldots$

$+ (6n - 2) = n(3n + 1)$ ($n \in \mathbb{N}$)

(c) $(2n + 1) + (2n + 3) + (2n + 5) + \ldots$

$+ (4n - 1) = 3n^2$ (n a natural number)

(d) $1^3 + 2^3 + \ldots$

$+ n^3 = (1 + 2 + \ldots + n)^2$ (n a natural number)

(*Hint*: use $1 + 2 + \ldots + n = \frac{1}{2}n(n + 1)$, established in the text.)

46 Prove that $11^n - 4^n$ is divisible by 7 for all natural numbers n.

47 Consider the following short procedure:

Step 1: Let $S = 1$
Step 2: Print S
Step 3: Replace S by $S + 2\sqrt{S} + 1$ and go back to step 2.

List the first four printed values of S, and prove by induction that $S = n^2$ the nth time the procedure reaches step 2.

6.5 Engineering application: expert systems

In the early 1960s many people believed that machines could be made to think and that computers that could, for instance, automatically translate text from one language to another or make accurate medical diagnoses would soon be available. The problems associated with creating machines that could undertake these tasks are well illustrated by the story (possibly apocryphal but none the less salutary) of the early language-translating machine that was asked to translate the English sentence 'The spirit is willing but the flesh is weak', into Russian. The machine's attempt was found to read, in Russian, 'The vodka is very strong but the meat has gone off'. Problems such as these and the growing appreciation of the sheer magnitude of the computing power needed to undertake these intelligent tasks (an effect often referred to as the 'combinatorial explosion') finally resulted in the realization that thinking machines were further away than some scientists had thought. Interest waned for 20 years until, in the early 1980s, advances both in our understanding of theoretical issues in computer software and in the design of computer hardware again brought the achievement of intelligent tasks by computers nearer reality.

The modern approach to producing intelligent machines (or at least machines that seem intelligent) is through 'expert systems'. The basis of an expert system is a database of facts and rules together with an 'inference engine', that is, a computer program that matches some query with the known facts and rules and determines the answer to the query. The essence of the 'intelligence' of an expert system is the way in which the inference engine is able to combine the known facts, using the given rules together with the general methods of proof that we discussed in Section 6.4, to answer queries that could not be answered by direct interrogation of the database of facts. The theoretical basis of these systems lies in propositional logic and predicate calculus. The facts and rules of the expert system's database loosely correspond to the concepts of proposition and predicate that we discussed in Section 6.4.

Expert systems that are able to answer routine queries in certain restricted areas of knowledge are now in everyday use in industry, commerce and public service. Such systems can, for instance, help tax lawyers advise clients, help geologists assess the results of seismographic tests, or advise disabled people on the benefits to which they are entitled. Nearer home, the same techniques are used in computer programs that can help with the routine drudgery of mathematics, differentiating, integrating and manipulating expressions with a speed and accuracy that humans cannot match. It is easy to envisage that expert systems that can undertake some of the work of the design engineer or design building structures and carry out the routine tasks of architecture (routeing cables and pipework within a building for instance) cannot be far away. Here we shall give more of the flavour of expert systems by an example in the domain of family relationships. Imagine that an expert system has a set of facts about the relationships in a certain family such as those shown in Figure 6.47. It is easy for a human to deduce that the family tree is that shown in Figure 6.48 (assuming, of course, that no one in the family has been married more than once and that all the children were born within wedlock). From the family tree a human could ascertain the truth of some further statements about the family. For instance, it is obvious that the statement 'Peter is the grandfather of David' is true and that the statement 'Alan is the brother of Robert' is false.

(1) Peter is the father of Robert
(2) James is the father of Alan
(3) Anne is the mother of Robert
(4) Anne is the mother of Melanie
(5) Lilian is the mother of David
(6) Robert is the father of Jennifer
(7) James is the brother of Peter
(8) Lilian is the wife of Robert
(9) Alan is the son of Martha

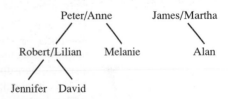

Figure 6.47 A short database of facts about family relationships.

Figure 6.48 The family tree deduced from the facts in the database.

An expert system can equally well be designed to evaluate the truth of such statements. In order to do so it needs, as well as the facts, some rules about how relationships combine. A typical set of rules is shown in Figure 6.49. If we were to ask if the statement 'Peter is a grandparent of David' is true the expert system might reason as follows:

> From fact (1) Peter is the father of Robert;
> therefore, from rule (1), Peter is a parent of Robert.

> From fact (8) Lilian is the wife of Robert;
> therefore, from rule (4), Lilian is the spouse of Robert;
> therefore, from rule (7), Robert is the spouse of Lilian.

> From fact (5) Lilian is the mother of David;
> therefore, from rule (2), Lilian is a parent of David.

Figure 6.49
A short database of
rules about family
relationships.

(1) If X is the father of Y
      then X is a parent of Y
(2) If X is the mother of Y
      then X is a parent of Y
(3) If X is a parent of Y
    and Y is a parent of Z
      then X is a grandparent of Z
(4) If X is the wife of Y
      then X is the spouse of Y
(5) If X is the husband of Y
      then X is the spouse of Y
(6) If X is the spouse of Y
    and Y is a parent of Z
      then X is a parent of Z
(7) If X is the spouse of Y
      then Y is the spouse of X

Now it has been proved that Robert is the spouse of Lilian
and that Lilian is a parent of David;
therefore, from rule (6), Robert is a parent of David.

Finally, it has been proved that Peter is a parent of Robert
and Robert is a parent of David;
therefore, from rule (3), Peter is a grandparent of David.

A little more is needed to deduce that Peter is the grandfather of David and this is left as an exercise for the reader.

Of course, the expert system needs a way of determining which rule to try to apply next in seeking to prove the truth of the query. That is the role of the part of the program called the inference engine – the inference engine attempts to prove the truth of the query by using rules in the most effective order and in such a way as to leave no possible path to a proof unexplored. In many expert systems this is achieved by using a search algorithm.

It is interesting to ask how such an expert system can prove that some assertion ('Alan is the brother of Robert' for instance) is false. Most expert systems tackle this by exhaustively trying every possible way of proving that the assertion is true. Then, if this fails, to most expert systems, it actually means merely that, given the facts and rules at the disposal of the expert systems, the assertion cannot be proved to be true. There are obviously dangers in this approach, since an incomplete database may lead an expert system to classify as false an assertion that, given more complete data, can be shown to be true. If, for instance, we were to ask the family expert system if the statement 'Alan is the cousin of Robert' is true, the expert system would allege it was not. On the other hand, if we gave the system some further, more sophisticated, rules about relationships then it would be able to deduce that the statement is actually true.

6.6   Engineering application: control

We consider a simplified model of a container for chemical reactions and design a circuit that involves four variables: upper and lower contacts for each of the temperature and pressure gauges. The control of the reaction within the container is managed using a mixing motor, a cooling-water valve, a heating device and a safety valve. We will analyse the control of the reaction given the following data and notation:

T_L = lower temperature, T_u = upper temperature

p_L = lower pressure, p_u = upper pressure

m = mixing motor, c = cooling-water valve

h = heating device, s = safety valve

$T_L = 0$, $T_u = 0$ temperature is too low

$T_L = 1$, $T_u = 0$ temperature is correct

$T_L = 1$, $T_u = 1$ temperature is too high

$p_L = 0$, $p_u = 0$ pressure is too low

$p_L = 1$, $p_u = 0$ pressure is correct

$p_L = 1$, $p_u = 1$ pressure is too high

$m = 0, 1$ mixing motor is off, on

$c = 0, 1$ cooling-water valve is off, on

$h = 0, 1$ heating is off, on

$s = 0, 1$ safety valve is closed, open

Figure 6.50 shows the container. The table in Figure 6.51 gives nine states – three initial states, three normal states and three danger states – exemplified by the pressure in the vessel. From this table we can write down that

$$s = T_L \cdot T_u \cdot p_L \cdot p_u$$

Figure 6.50

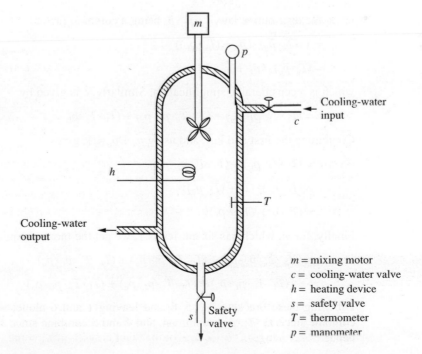

Cooling-water input

Cooling-water output

Safety valve

m = mixing motor
c = cooling-water valve
h = heating device
s = safety valve
T = thermometer
p = manometer

Figure 6.51

	T_L	T_u	p_L	p_u	h	c	m	s	Comments
Initial state (low pressure)	0	0	0	0	1	0	1	0	Gauges off; switch on motor and heater
	1	0	0	0	1	0	0	0	Correct temperature; switch off motor
	1	1	0	0	0	0	1	0	Temperature too high; heater off, motor on
Normal state (pressure acceptable)	0	0	1	0	1	0	0	0	Cold; heater on
	1	0	1	0	0	0	0	0	Normal; heater off
	1	1	1	0	0	1	1	0	Hot; motor on, cooling water in
Danger state (pressure high)	0	0	1	1	0	0	1	0	Low temperature; motor on
	1	0	1	1	0	1	1	0	Normal temperature; motor on, cooling water in
	1	1	1	1	0	1	1	1	High temperature; $c = m = s = 1$ to try to prevent an explosion!

that is, the safety valve is only open when the temperature and pressure are too high. The Boolean expressions for h, c and m are obtained by taking the union of the rows of T_u, T_L, p_u and p_L that have 1 under the columns headed h, c and m respectively. Hence

$$h = (\bar{T}_L \cdot \bar{T}_u \cdot \bar{p}_L \cdot \bar{p}_u) + (T_L \cdot \bar{T}_u \cdot \bar{p}_L \cdot \bar{p}_u) + (\bar{T}_L \cdot \bar{T}_u \cdot p_L \cdot \bar{p}_u)$$

$$= (\bar{T}_L + T_L) \cdot (\bar{T}_u \cdot \bar{p}_L \cdot \bar{p}_u) + (\bar{T}_L \cdot \bar{T}_u \cdot p_L \cdot \bar{p}_u)$$

using the distributive law, $\bar{T}_u \cdot \bar{p}_L \cdot \bar{p}_u$ being a common factor

$$= 1 \cdot (\bar{T}_u \cdot \bar{p}_u) \cdot (\bar{p}_L + (\bar{T}_L \cdot p_L))$$

$$= (\bar{T}_u \cdot \bar{p}_u) \cdot (\bar{p}_L + \bar{T}_L)$$

which is a considerable simplification. Similarly, c is given by

$$c = (T_L \cdot T_u \cdot p_L \cdot \bar{p}_u) + (T_L \cdot \bar{T}_u \cdot p_L \cdot p_u) + (T_L \cdot T_u \cdot p_L \cdot p_u)$$

Combining the first and last, and using $p_u + \bar{p}_u = 1$, gives

$$c = (T_L \cdot T_u \cdot p_L) + (T_L \cdot \bar{T}_u \cdot p_L \cdot p_u)$$

$$= (T_L \cdot p_L) \cdot (T_u + (\bar{T}_u \cdot p_u))$$

$$= (T_L \cdot p_L) \cdot (T_u + p_u)$$

Finally, for m, which has six entries as 1, we get the more complicated expression

$$m = (\bar{T}_L \cdot \bar{T}_u \cdot \bar{p}_L \cdot \bar{p}_u) + (T_L \cdot T_u \cdot \bar{p}_L \cdot \bar{p}_u) + (T_L \cdot T_u \cdot p_L \cdot \bar{p}_u)$$

$$+ (\bar{T}_L \cdot \bar{T}_u \cdot p_L \cdot p_u) + (T_L \cdot \bar{T}_u \cdot p_L \cdot p_u) + (T_L \cdot T_u \cdot p_L \cdot p_u)$$

Labelling these brackets $1, \dots, 6$, and leaving 1 and 6 alone, we note that 2 and 3 combine since $T_L \cdot T_u \cdot \bar{p}_u$ is common, and 4 and 5 combine since $\bar{T}_u \cdot p_L \cdot p_u$ is common; hence

Figure 6.52

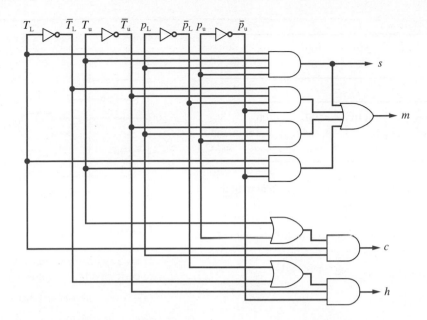

$$m = (\bar{T}_L \cdot \bar{T}_u \cdot \bar{p}_L \cdot \bar{p}_u) + (T_L \cdot T_u \cdot p_L \cdot p_u) + (T_L \cdot T_u \cdot \bar{p}_u) + (\bar{T}_u \cdot p_L \cdot p_u)$$

Hence we can draw the control of the vessel in terms of the switching circuit in Figure 6.52.

6.7 Review exercises (1–23)

1 If $U = \{1, 2, 3, 4, 5, 6, 7, 8, 9\}$, $A = \{2, 4, 6\}$, $B = \{1, 3, 5, 7\}$ and $C = \{2, 3, 4, 7, 8\}$ find the sets

(a) $\overline{A \cup B}$ (b) $C - A$ (c) $\bar{C} \cap \bar{B}$

2 Let $A = \{n \in \mathbb{N}, n \leqslant 11\}$

$B = \{n \in \mathbb{N}, n \text{ is even and } n \leqslant 20\}$

$C = \{10, 11, 12, 13, 14, 15, 17, 20\}$

Write down the sets

(a) $A \cap B$ (b) $A \cap B \cap C$

(c) $A \cup (B \cap C)$

and verify that $(A \cup B) \cap (A \cup C) = A \cup (B \cap C)$.

3 If A, B and C are defined as in Question 2, and the universal set is the set of all integers less than or equal to 20, find the following sets:

(a) $\bar{A}$ (b) $\bar{A} \cup \bar{B}$

(c) $\overline{A \cup B}$ (d) $A \cap (\bar{B} \cup \bar{C})$

Verify the De Morgan laws for A and B.

4 The sets A and B are defined by

$A = \{x : x^2 + 6 = 5x \quad \text{or} \quad x^2 + 2x = 8\}$

$B = \{2, 3, 4\}$

Which of the following statements is true?

(a) $A \neq B$

(b) $A = B$

Give reasons for your answers.

5 (a) Simplify the Boolean functions

$f = (A \cap \bar{B} \cap C) \cup (A \cap (B \cup \bar{C}))$

$g = ((\bar{A} \cup \bar{B}) \cap C) \cap ((\bar{C} \cup A) \cap B)$

(b) Draw Venn diagrams to verify that

$(A \cap \bar{B}) \cup (\bar{A} \cap B) = A \cup B$

if and only if $A \cap B = \emptyset$.

6 In an election there are three candidates and 800 voters. The voters may exercise one, two or three votes each. The following results were obtained:

Votes cast	240	400	500
Candidate	A	B	C

Voters	110	90	200	50
Candidates	B and C	A and C	A and B	A and B and C

Show that these results are inconsistent if all the voters use at least one vote.

7   Draw switching circuits to establish the truth of the following laws:

(a) $p + p \cdot q = p$

(b) $p + \bar{p} \cdot q = p + q$

(c) $p \cdot q + p \cdot r = p \cdot (q + r)$

(d) $(p + q) \cdot (p + r) = p + q \cdot r$

Use these to simplify the expression

$$s = p \cdot \bar{p} + p \cdot q + \bar{p} \cdot r + q \cdot r$$

so that s only contains two pairs of products added.

8   Write down, in set-theoretical notation, expressions corresponding to the outputs in (a) Figure 6.53 and (b) Figure 6.54.

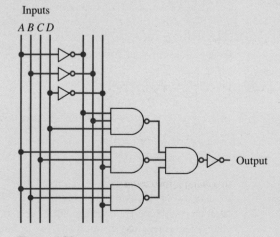

Figure 6.53

Inputs
A B C D

Output

Inputs
B C

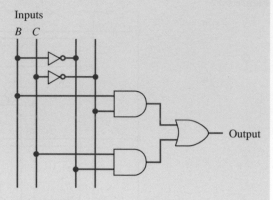

Output

Figure 6.54

9   Draw a switching circuit with inputs x, y, z and u to correspond to the following expressions:

(a) $(x \cdot y \cdot z \cdot u) + (\bar{x} \cdot \bar{y} \cdot z \cdot u) + (x \cdot \bar{u})$

(b) $(\bar{x} \cdot \bar{y}) + (\bar{z} \cdot \bar{u}) + (x \cdot y \cdot z)$

(c) $(x \cdot y \cdot z \cdot u) + (\bar{x} \cdot \bar{y} \cdot z \cdot u) + (x \cdot \bar{y} \cdot \bar{z} \cdot u)$

$\quad + (\bar{x} \cdot y \cdot z \cdot \bar{u}) + (\bar{x} \cdot y \cdot \bar{z} \cdot u)$

$\quad + (x \cdot \bar{y} \cdot z \cdot \bar{u})$

$\quad + (\bar{x} \cdot \bar{y} \cdot \bar{z} \cdot u)$

For (c) establish the output for the input states

(i) $x = y = 1, \quad z = u = 0$

(ii) $x = 1, \quad y = z = u = 0$

10   Write down truth tables for the following expressions:

(a) $p \wedge q$ (b) $p \vee q$ (c) $p \rightarrow q$

The contrapositive of the conditional statement $p \rightarrow q$ is defined as $\tilde{q} \rightarrow \tilde{p}$.

(d) Use truth tables to show that

$$\tilde{q} \rightarrow \tilde{p} \equiv p \rightarrow q$$

(e) Use truth tables to evaluate the status of the expression

$$(p \vee q) \wedge (\tilde{p} \wedge q) \rightarrow p$$

(f) By taking the contrapositive of this conditional statement and using (d) together with the De Morgan laws (see Example 6.6) show that

$$\tilde{p} \rightarrow (\tilde{p} \wedge \tilde{q}) \vee (\tilde{p} \wedge q)$$

is a tautology

11 Reduce the following Boolean expressions by taking complements:

(a) $\overline{[(\overline{p \cdot q}) \cdot p][(\overline{p \cdot q}) \cdot q]}$

(b) $\overline{(p + q + \bar{r}) \cdot (\overline{p \cdot q} + \overline{r \cdot s}) + \overline{q \cdot r \cdot s}}$

(c) $\overline{(p \cdot q \cdot r + q \cdot \bar{r} \cdot s) + (\overline{q \cdot r \cdot s} + \bar{q} \cdot \bar{r} \cdot \bar{s} + q \cdot r \cdot \bar{s})}$

12 (a) Simplify the Boolean expressions

(i) $p \cdot r + p \cdot q \cdot r + q \cdot \bar{r} \cdot s + \bar{q} \cdot r \cdot \bar{s} + p \cdot q \cdot r \cdot s$

(ii) $\overline{[(\bar{p} + q) \cdot (\bar{r} + s)] \cdot (\bar{s} + p) + r}$

(b) Show the Boolean function $p \cdot q + \bar{p} \cdot \bar{r}$ on a Venn diagram.

13 A lift (elevator) services three floors. On each floor there is a call button to call the lift. It is assumed that at the moment of call the cabin is stationary at one of the three floors. Using these six input variables, determine a control that moves the motor in the right direction for the current situation. (*Hint*: There are 24 combinations to consider.)

14 There are four people on a TV game show. Each has a 'Yes/No' button for recording opinions. The display must register 'Yes' or 'No' according to a majority vote.

(a) Derive a truth table for the above.

(b) Write down the Boolean expression for the output.

(c) Simplify this expression and suggest a suitable circuit.

(d) If there is a tie, the host has a 'casting vote'. Modify the above circuit to indicate this.

15 Consider the following logical statements:

(a) Mike never smokes dope.

(b) Rick smokes if, and only if, Mike and Vivian are present.

(c) Neil smokes under all conditions – even by himself.

(d) Vivian smokes if, and only if, Mike is not present.

The police raid: determine the state of there being no dope smoking in terms of M, R, N and V's presence (Mike, Rick, Neil and Vivian respectively).

16 Find the explicit Boolean function for the logic circuit of Figure 6.55. Show that the function simplifies to $f = q \cdot \bar{r}$ and draw two different simplified circuits which may be used to represent the circuit.

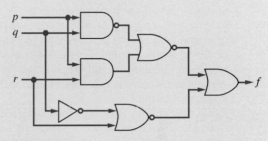

Figure 6.55

17 Which of the following statements are propositions? For those that are not, say why and suggest ways of changing them so that they become propositions. For those that are, comment on their truth value.

(a) Julius Caesar was prime minister of Great Britain.

(b) Stop hitting me.

(c) Turn right at the next roundabout.

(d) The Moon is made of green cheese.

(e) If the world is flat then $3 + 3 = 6$.

(f) If you get a degree then you will be rich.

(g) $x + y + z = 0$.

(h) The 140th decimal digit in the representation of π is 8.

(i) There are five Platonic solids.

18 (a) Draw up truth tables to represent the statements

(i) p is equivalent to q

(ii) p implies q

(b) Using the algebra of statements determine the truth of the following statements:

(i) If p implies q, and r implies q, then either r implies p or else p implies r.

(ii) If p is equivalent to q, and q is equivalent to r, then p implies r.

19 A panel light in the control room of a satellite launching site is to go on if the pressure in both the oxidizer and fuel tanks is equal to or above a required minimum value and there are 15 minutes or less to 'lift-off', or if the pressure in the oxidizer tank is equal to or above the required minimum value and the pressure in the fuel tank is below the required minimum value but there are more than 15 minutes to 'lift-off', or if the pressure in the oxidizer tank is below the required minimum value but there are more than 15 minutes to 'lift-off'. By using a truth table, write down a Boolean expression to represent the state of the panel light. Minimize the Boolean function.

20 In the control problem of Section 6.6 show that h may also be expressed as

$$h = \overline{T_u + p_u} + p_L \cdot \overline{T_L}$$

Compare the resulting control switching circuit with that of Figure 6.52.

21 Write down all subsets of the set $A = \{p, q, r, s\}$ that contain the product of *four* of p, q, r, s or their complement. Represent these on a Venn diagram. [The ideas are pursued through Karnaugh maps which are outside the scope of this text.]

22 State the converse and contrapositive of each of the following statements:

(a) If the train is late, I will not go.

(b) If you have enough money, you will retire.

(c) I cannot do it unless you are there too.

(d) If you go, so will I.

23 An island is inhabited by two tribes of vicious cannibals and, sadly, you are a prisoner of one of them. One tribe always tell the truth, the other tribe always lie. Unfortunately both tribes look identical. They will answer 'yes' or 'no' to a single question they will allow you. The God of one tribe is female, the God of the other tribe is male, and if you correctly state the sex of their God they will set you free. Use truth tables to help you formulate a question that will enable you to survive.

7 Sequences, Series and Limits

Chapter 7   Contents

7.1 Introduction

In the analysis of practical problems certain mathematical ideas and techniques appear in many different contexts. One such idea is the concept of a sequence. Sequences occur in management activities such as the determination of programmes for the maintenance of hardware or production schedules for bulk products. They also arise in investment plans and financial control. They are intrinsic to computing activities, since the most important feature of computers is their ability to perform sequences of instructions quickly and accurately. Sequences are of great importance in the numerical methods that are essential for modern design and the development of new products. As well as illustrating these basic applications, we shall show how these simple ideas lead to the idea of a limit, which is a prerequisite for a proper understanding of the calculus and numerical methods. Without that understanding, it is not possible to form mathematical models of real problems, to solve them or to interpret their solutions adequately. At the same time, we shall illustrate some of the elementary properties of the standard functions described in Chapter 2 and how they link together, and we shall look forward to further applications in more advanced engineering applications, in particular to the work on Z transforms contained in the companion text *Advanced Modern Engineering Mathematics*.

7.2 Sequences and series

7.2.1 Notation

Consider a function f whose domain is the set of whole numbers $\{0, 1, 2, 3, \ldots\}$. The set of values of the function $\{f(0), f(1), f(2), f(3), \ldots\}$ is called a **sequence**. Usually we denote the values using a subscript, so that $f(0) = f_0, f(1) = f_1, f(2) = f_2$, and so on. Often we list the elements of a sequence in order on the assumption that the first in the list is f_0, the second is f_1 and so on. For example, we may write

'Consider the sequence 1, 1, 2, 3, 5, 8, 13, 21, 34, . . .'

implying $f_0 = 1, f_5 = 8, f_8 = 34$ and so on. In this example the continuation dots . . . are used to imply that the sequence does not end. Such a sequence is called an **infinite** sequence to distinguish it from **finite** or **terminating** sequences. The finite sequence $\{f_0, f_1, \ldots, f_n\}$ is often denoted by $\{f_k\}_{k=0}^{n}$ and the infinite sequence by $\{f_k\}_{k=0}^{\infty}$. When the context makes the meaning clear, the notation is further abbreviated to $\{f_k\}$.

Example 7.1

A bank pays interest at a fixed rate of 8.5% per year, compounded annually. A customer deposits the fixed sum of £1000 into an account at the beginning of each year. How much is in the account at the beginning of each of the first four years?

Solution Let £x_n denote the amount in the account at the beginning of the $(n + 1)$th year. Then

Amount at beginning of first year $x_0 = 1000$

Amount at beginning of 2nd year $x_1 = 1000(1 + \frac{8.5}{100}) + 1000 = 2085$

Amount at beginning of 3rd year $x_2 = 2085(1 + 0.085) + 1000 = 3262.22$

Amount at beginning of 4th year $x_3 = 3262.22(1.085) + 1000 = 4539.51$

We can see that in general

$$x_n = 1.085x_{n-1} + 1000$$

This is a **recurrence relation**, which gives the value of each element of the sequence in terms of the value of the previous element.

Example 7.2 Consider the ducting of a number of cables of the same diameter d. The diameter D_n of the smallest duct with circular cross-section depends on the number n of cables to be enclosed, as shown in Figure 7.1:

$$D_0 = 0, \quad D_1 = d, \quad D_2 = 2d, \quad D_3 = (1 + 2/\sqrt{3})d, \quad D_4 = (1 + \sqrt{2})d$$

$$D_5 = \tfrac{1}{4}\sqrt{[2(5 - \sqrt{5})]}d, \quad D_6 = 3d, \quad D_7 = 3d, \ldots$$

Thus the duct diameters form a sequence of values $\{D_1, D_2, D_3, \ldots\} = \{D_n\}_{n=1}^{\infty}$.

Figure 7.1
Enclosing a
number of cables
in a circular duct.

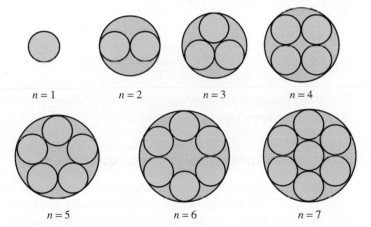

$n = 1$ $n = 2$ $n = 3$ $n = 4$

$n = 5$ $n = 6$ $n = 7$

Example 7.3 A computer simulation of the crank and connecting rod mechanism considered in Example 2.36 evaluates the position of the end Q of the connecting rod at equal intervals of the angle $x°$. Given that the displacement y of Q satisfies

$$y = r\cos x° + \sqrt{(l^2 - r^2\sin^2 x°)}$$

find the sequence of values of y where $r = 5, l = 10$ and the interval between successive values of $x°$ is $1°$.

Solution In this example the independent variable x is restricted to the sequence of values $\{0, 1, 2, \ldots, 360\}$. The corresponding sequence of values of y can be calculated from the formula

$$y_k = 5\cos k^\circ + \sqrt{(100 - 25\sin^2 k^\circ)}$$
$$= 5[\cos k^\circ + \sqrt{(4 - \sin^2 k^\circ)}]$$
$$= 5[\cos k^\circ + \sqrt{(3 + \cos^2 k^\circ)}]$$

Thus

$$\{y_k\}_{k=0}^{360} = \{15, 14.999, 14.995, 14.990, \ldots, 14.999, 15\}$$

Notice how in Example 7.3 we did not list every element of the sequence. Instead, we relied on the formula for y_k to supply the value of a particular element in the sequence. In Example 7.1 we could use the recurrence relation to determine the elements of the sequence. In Example 7.2, however, there is no formula or recurrence relation that enables us to work out the elements of the sequence. These three examples are representative of the general situation.

A **series** is an extended sum of terms. For example, a very simple series is the sum

$$1 + 2 + 3 + 4 + 5 + 6 + 7 + 8 + 9 + 10 + 11$$

When we look for a general formula for summing such series, we effectively turn it into a sequence, writing, for example, the sum to eleven terms as S_{11} and the sum to n terms as S_n, where

$$S_n = 1 + 2 + 3 + \ldots + n = \sum_{k=1}^{n} k$$

Series often occur in the mathematical analysis of practical problems and we give some important examples later in this chapter.

7.2.2 Graphical representation of sequences

Sequences, as remarked earlier, are functions whose domains are the whole numbers. We can display their properties using a conventional graph with the independent variable (now an integer n) represented as points along the positive x axis. This will show the behaviour of the sequence for low values of n but will not display the whole behaviour adequately. An alternative approach displays the terms of the sequence against the values of $1/n$. This enables us to see the whole sequence but in a rather 'telescoped' manner. When the terms of a sequence are generated by a recurrence relation a third method, known as a **cobweb diagram**, is available to us. We will illustrate these three methods in the examples below.

Example 7.4 Calculate the sequence $\left\{1 + \dfrac{(-1)^n}{n}\right\}_{n=1}^{10}$ and illustrate the answer graphically.

Solution   By means of a calculator we can obtain the terms of the sequence explicitly (to 2dp) as

$$\{0, 1.50, 0.67, 1.25, 0.80, 1.17, 0.86, 1.12, 0.89, 1.10\}$$

The graph of this function is strictly speaking the set of points

$$\{(1, 0), (2, 1.5), (3, 0.67), \ldots, (10, 1.1)\}$$

These can be displayed on a graph as isolated points but it is more helpful to the reader to join the points by straight line segments, as shown in Figure 7.2. The figure tells us that the values of the sequence oscillate about the value 1, getting closer to it as n increases.

Example 7.5   Calculate the sequence $\{n^{1/n}\}_{n=4}^{10}$ and show the points $\{(1/n, n^{1/n})\}_{n=4}^{10}$ on a graph.

Solution   Using a calculator we obtain (to 2dp)

$$\{n^{1/n}\}_{n=4}^{10} = \{1.41, 1.38, 1.35, 1.32, 1.30, 1.28, 1.26\}$$

and the set of points is

$$\{(1/n, n^{1/n})\}_{n=4}^{10} = \{(0.25, 1.41), (0.2, 1.38), \ldots, (0.1, 1.26)\}$$

In Figure 7.3 these points are displayed with a smooth curve drawn through them. The graph suggests that as n increases (i.e. $1/n$ decreases), $n^{1/n}$ approaches the value 1.

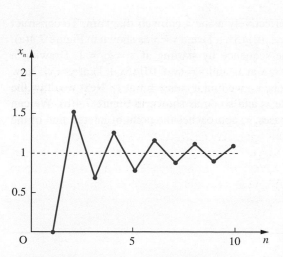

Figure 7.2 Graph of the sequence defined by $x_n = 1 + (-1)^n/n$.

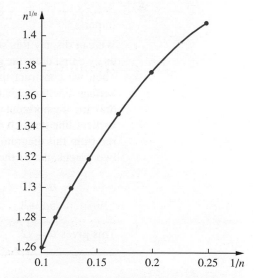

Figure 7.3 Graph of the points $\{(\frac{1}{n}, n^{1/n})\}_{n=4}^{10}$.

Figure 7.4
(a) Graphs of
$y = (x + 10)/$
$(5x + 1)$ and $y = x$.
(b) Construction of
the sequence defined
by $x_{n+1} = (x_n + 10)/$
$(5x_n + 1)$, $x_0 = 1$.

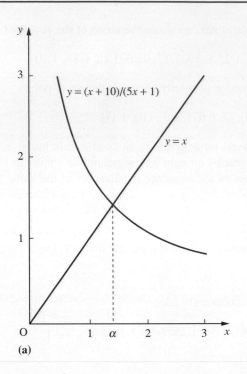

(a)

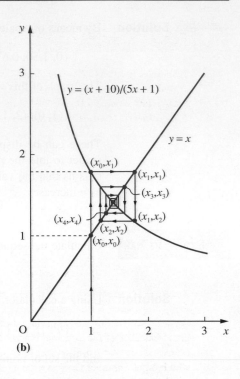

(b)

Example 7.6   Calculate the sequence $\{x_n\}_{n=0}^6$ where $x_0 = 1$ and $x_{n+1} = \dfrac{x_n + 10}{5x_n + 1}$.

Solution   Using a calculator we obtain (to 2dp) the values of the sequence

$$\{x_n\}_{n=0}^6 = \{1, 1.83, 1.16, 1.64, 1.27, 1.54, 1.33\}$$

We can display this sequence very effectively using a **cobweb diagram**. To construct this we first draw the graphs of $y = (x + 10)/(5x + 1)$ and $y = x$ as shown in Figure 7.4(a). Then we construct the points of the sequence by starting at $x = x_0 = 1$. Drawing a vertical line through $x = 1$, we cut $y = x$ at x_0 and $y = (x + 10)/(5x + 1)$ at $y = x_1$. Now drawing the horizontal line through (x_0, x_1) we find it cuts $y = x$ at x_1. Next we draw the vertical line through (x_1, x_1) to locate x_2 and so on as shown in Figure 7.4(b). We can see from this diagram that as n increases, x_n approaches the point of intersection of the two graphs, that is the value α where

$$\alpha = \frac{\alpha + 10}{5\alpha + 1} \quad (\alpha > 0)$$

This gives $\alpha = \sqrt{2}$.

As we have seen, three different methods can be used for representing sequences graphically. The choice of method will depend on the problem context.

7.2.3   Exercises

1 Triangular numbers (T_n) are defined by the number of dots that occur when arranged in equilateral triangles as shown in Figure 7.5. Show that $T_n = \frac{1}{2}n(n + 1)$ for every positive integer n.

Figure 7.5 Triangular numbers.

2 A detergent manufacturer wishes to forecast their future sales. Their market research department assesses that their 'Number One' brand has 20% of the potential market at present. They also estimate that 15% of those who bought 'Number One' in a given month will buy a different detergent in the following month and that 35% of those who bought a rival brand will buy 'Number One' in the next month. Show that their share P_n% of the market in the nth month satisfies the recurrence relation

$$P_{n+1} = 35 + 0.5P_n, \quad \text{with } P_0 = 20$$

Find the values of P_n for $n = 1, 2, 3$ and 4 and illustrate them on an appropriate diagram.

3 (a) If $x_r = r(r - 1)(2r - 5)$, calculate $\displaystyle\sum_{r=0}^{4} x_r$

  (b) If $x_r = r^{r+1} + 3(-1)^r$,   calculate $\displaystyle\sum_{r=1}^{5} x_r$

  (c) If $x_r = r^2 - 3r + 1$,   calculate $\displaystyle\sum_{r=2}^{6} x_r$

4 A precipitate at the bottom of a beaker of capacity V always retains about it a volume v of liquid. What percentage of the original solution remains about it after it has been washed n times by filling the beaker with distilled water and emptying it?

5 A certain process in statistics involves the following steps S_i ($i = 1, 2, \ldots, 6$):

S_1: Selecting a number from the set
$T = \{x_1, x_2, \ldots, x_n\}$

S_2: Subtracting 10 from it

S_3: Squaring the result

S_4: Repeating steps S_1–S_3 with the remaining numbers in T

S_5: Adding the results obtained at stage S_3 of each run through

S_6: Dividing the result of S_5 by n

Express the final outcome algebraically using Σ notation.

6   Newton's recurrence formula for determining the root of a certain equation is

$$x_{n+1} = \frac{x_n^2 - 1}{2x_n - 3}$$

Taking $x_0 = 3$ as your initial approximation, obtain the root correct to 4sf.

7 Calculate the terms of the sequence

$$\left\{ \frac{n^4}{n^4 + n^3 + 1} \right\}_{n=0}^{5}$$

and show them on graphs similar to Figures 7.2 and 7.3.

8 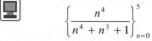 Calculate the sequence $\{x_n\}_{n=0}^{6}$ where

$$x_{n+1} = \frac{x_n + 2}{x_n + 1}, \quad x_0 = 1$$

Show the sequence using a cobweb diagram similar to Figure 7.4.

9 A steel ball-bearing drops on to a smooth hard surface from a height h. The time to the first impact is $T = \sqrt{(2h/g)}$ where g is the acceleration due to gravity. The times between successive bounces are $2eT, 2e^2T, 2e^3T, \ldots$, where e is the coefficient of restitution between the ball and the surface ($0 < e < 1$). Find the total time taken up to the fifth bounce. If $T = 1$ and $e = 0.1$, show in a diagram the times taken up to the first, second, third, fourth and fifth bounces and estimate how long the total motion lasts.

10 Consider the following puzzle: how many single, loose, smooth 30 cm bricks are necessary to form a single leaning pile with no part of the bottom brick under the top brick? Begin by considering a pile of 2 bricks. The top brick cannot project further than 15 cm without collapse. Then consider a pile of 3 bricks. Show that the top one cannot project further than 15 cm beyond the second one and that the second one cannot project further than 7.5 cm beyond the bottom brick (so that the maximum total lean is $(\frac{1}{2} + \frac{1}{4})30$ cm). Show that the maximum total lean for a pile of 4 bricks is $(\frac{1}{2} + \frac{1}{4} + \frac{1}{6})30$ cm and deduce that for a pile of n bricks it is $(\frac{1}{2} + \frac{1}{4} + \frac{1}{6} + \ldots + \frac{1}{2n+2})30$ cm. Hence solve the puzzle.

7.3 Finite sequences and series

In this section we consider some finite sequences and series that are frequently used in engineering.

7.3.1 Arithmetical sequences and series

An **arithmetical sequence** is one in which the difference between successive terms is a constant number. Thus, for example, $\{2, 5, 8, 11, 14\}$ and $\{2, 0, -2, -4, -6, -8, -10\}$ define arithmetical sequences. In general an arithmetical sequence has the form $\{a + kd\}_{k=0}^{n-1}$ where a is the first term, d is the common difference and n is the number of terms in the sequence. Thus, in the first example above, $a = 2$, $d = 3$ and $n = 5$, and in the second example, $a = 2$, $d = -2$ and $n = 7$. (The old name for such sequences was **arithmetical progressions**.) The sum of the terms of an arithmetical sequence is an **arithmetical series**. The general arithmetical series is

$$S_n = a + (a + d) + (a + 2d) + \ldots + [a + (n-1)d] = \sum_{k=0}^{n-1}(a + kd) \tag{7.1}$$

To obtain an expression for the sum of the n terms in this series, write the series in the reverse order,

$$S_n = \quad a \quad + \quad (a + d) \quad + \quad (a + 2d) \quad + \ldots + [a + (n-1)d]$$

$$S_n = [a + (n-1)d] + [a + (n-2)d] + [a + (n-3)d] + \ldots + \quad a$$

Summing the two series then gives

$$2S_n = [2a + (n-1)d] + [2a + (n-1)d] + [2a + (n-1)d] + \ldots + [2a + (n-1)d]$$

giving the sum S_n of the first n terms of an arithmetical series as

$$S_n = \tfrac{1}{2}n[2a + (n-1)d] = \tfrac{1}{2}n(\text{first term} + \text{last term}) \tag{7.2}$$

The result is illustrated geometrically for $n = 6$ in Figure 7.6, where the breadth of each rectangle is unity and the area under each shaded step is equal to a term of the series. In particular, when $a = 1$ and $d = 1$,

$$S_n = 1 + 2 + \ldots + n = \sum_{k=1}^{n} k = \tfrac{1}{2}n(n+1) \tag{7.3}$$

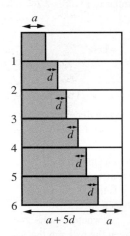

Figure 7.6

$$S_6 = \sum_{k=0}^{5}(a + kd)$$

$$= \tfrac{1}{2} \times 6 \times (2a + 5d).$$

Example 7.7   How many terms of the arithmetical series 11, 15, 19, etc. will give a sum of 341?

Solution   In this particular case the first term $a = 11$ and the common difference $d = 4$. We need to find the number of terms n such that the sum S_n is 341. Using the result in (7.2)

$$S_n = 341 = \tfrac{1}{2}n[2(11) + (n-1)(4)]$$

leading to

$$4n^2 + 18n - 682 = 0$$

or

$$(4n + 62)(n - 11) = 0$$

giving

$$n = 11 \quad \text{or} \quad n = -\tfrac{31}{2}$$

Since $n = -\tfrac{31}{2}$ is not a whole number, the number of terms required is $n = 11$.

Example 7.8   A contractor agrees to sink a well 40 metres deep at a cost of £30 for the first metre, £35 for the second metre and increasing by £5 for each subsequent metre.

(a) What is the total cost of sinking the well?

(b) What is the cost of drilling the last metre?

Solution   (a) The total cost constitutes an arithmetical series whose terms are the cost per metre. Thus, taking $a = 30$, $d = 5$ and $n = 40$ in (7.2) gives the total cost

$$S_n = £\tfrac{40}{2}[2(30) + (40 - 1)5] = £5100$$

(b) The cost of drilling the last metre is given by the 40th term of the series. Since the nth term is $a + (n - 1)d$, the cost of drilling the last metre $= 30 + (40 - 1)5 = £225$.

7.3.2   Geometric sequences and series

A **geometric sequence** is one in which the ratio of successive terms is a constant number. Thus, for example, $\{2, 4, 8, 16, 32\}$ and $\{2, -1, \tfrac{1}{2}, -\tfrac{1}{4}, \tfrac{1}{8}, -\tfrac{1}{16}, \tfrac{1}{32}\}$ define geometric sequences. In general a geometric sequence has the form $\{ar^k\}_{k=0}^{n-1}$ where a is the first term, r is the common ratio and n is the number of terms in the sequence. Thus, in the first example above, $a = 2$, $r = 2$, $n = 5$ and, in the second example, $a = 2$, $r = -\tfrac{1}{2}$ and $n = 7$. (The old name for such sequences was **geometric progressions**.) The sum of the terms of a geometric sequence is called a **geometric series**. The general geometric series has the form

$$S_n = a + ar + ar^2 + ar^3 + \ldots + ar^{n-1}$$

To obtain the sum S_n of the first n terms of the series we multiply S_n by the common ratio r, to obtain

$$rS_n = ar + ar^2 + \ldots + ar^{n-1} + ar^n$$

Subtracting this from S_n then gives

$$S_n - rS_n = a - ar^n$$

so that

$$(1 - r)S_n = a(1 - r^n)$$

Thus for $r \neq 1$, the sum of the first n terms is

$$S_n = \sum_{k=0}^{n-1} ar^k = \frac{a(1 - r^n)}{1 - r} \qquad\qquad (7.4)$$

Clearly, for the particular case of $r = 1$ the sum is $S_n = an$.

The geometric series is very important. It has many applications in practical problems as well as within mathematics.

Example 7.9

In its publicity material an insurance company guarantees that, for a fixed annual premium payable at the beginning of each year for a period of 25 years, the return will be at least equivalent to the premiums paid, together with 3% per annum compound interest. For an annual premium of £250 what is the guaranteed sum at the end of 25 years?

Solution

The first-year premium earns interest for 25 years and thus guarantees

$$£250(1 + 0.03)^{25}$$

The second-year premium earns interest for 24 years and thus guarantees

$$£250(1 + 0.03)^{24}$$

$$\vdots$$

The final-year premium earns interest for 1 year and thus guarantees

$$£250(1 + 0.03)$$

Thus, the total sum guaranteed is

$$£250[(1.03) + (1.03)^2 + \ldots + (1.03)^{25}]$$

The term inside the square brackets is a geometric series. Thus, taking $a = 1.03$, $r = 1.03$ and $n = 25$ in (7.4) gives

$$\text{Guaranteed sum} = £250\left[1.03\frac{(1.03^{25} - 1)}{(1.03 - 1)}\right] \approx £9388.$$

7.3.3 Other finite series

In addition to the arithmetical and geometric series, there are other finite series that occur in engineering applications for which an expression can be obtained for the sum of the first n terms. We shall illustrate this in Examples 7.10 and 7.11.

Example 7.10

Consider the sum-of-squares series

$$S_n = 1^2 + 2^2 + 3^2 + \ldots + n^2 = \sum_{k=1}^{n} k^2$$

Obtain an expression for the sum of this series.

Solution

There are various methods for finding the sum. A method that can be generalized makes use of the identity

$$(k + 1)^3 - k^3 = 3k^2 + 3k + 1$$

Thus

$$\sum_{k=1}^{n} [(k + 1)^3 - k^3] = \sum_{k=1}^{n} (3k^2 + 3k + 1)$$

The left-hand side equals

$$2^3 - 1^3 + 3^3 - 2^3 + 4^3 - 3^3 + \ldots + (n + 1)^3 - n^3 = (n + 1)^3 - 1$$

The right-hand side equals

$$3 \sum_{k=1}^{n} k^2 + 3 \sum_{k=1}^{n} k + \sum_{k=1}^{n} 1$$

Now

$$\sum_{k=1}^{n} k = \tfrac{1}{2}n(n + 1) \text{ from (7.3)} \quad \text{and} \quad \sum_{k=1}^{n} 1 = n$$

so that

$$(n + 1)^3 - 1 = 3 \sum_{k=1}^{n} k^2 + \frac{3n}{2}(n + 1) + n$$

whence

$$\sum_{k=1}^{n} k^2 = \tfrac{1}{6}n(n + 1)(2n + 1) \tag{7.5}$$

This method can be generalized to obtain the sum of other similar series. For example, to find the sum of cubes series $\sum_{k=1}^{n} k^3$, we would consider $(k + 1)^4 - k^4$ and so on.

| Example 7.11 | Obtain the sum of the series |

$$S_n = \frac{1}{1\cdot2} + \frac{1}{2\cdot3} + \frac{1}{3\cdot4} + \ldots + \frac{1}{n(n+1)} = \sum_{k=1}^{n} \frac{1}{k(k+1)}$$

Solution The technique for summing this series is to express the general term in its partial fractions:

$$\frac{1}{k(k+1)} = \frac{1}{k} - \frac{1}{k+1}$$

Then

$$S_n = \sum_{k=1}^{n} \frac{1}{k} - \sum_{k=1}^{n} \frac{1}{k+1}$$

$$= \left(1 + \frac{1}{2} + \frac{1}{3} + \ldots + \frac{1}{n}\right) - \left(\frac{1}{2} + \frac{1}{3} + \ldots + \frac{1}{n} + \frac{1}{n+1}\right)$$

$$= 1 - \frac{1}{n+1}$$

giving

$$S_n = \frac{n}{n+1}$$

There are many other similar series that can be summed by expressing the general term in its partial fractions. Some examples are given in Exercises 7.3.4.

7.3.4 Exercises

11 (a) Find the fifth and tenth terms of the arithmetical sequence whose first and second terms are 4 and 7. (b) The first and sixth terms of a geometric sequence are 5 and 160 respectively. Find the intermediate terms.

12 An individual starts a business and loses £150k in the first year, £120k in the second year and £90k in the third year. If the improvement continues at the same rate, find the individual's total profit or loss at the end of 20 years. After how many years would the losses be just balanced by the gains?

13 Show that

$$\frac{1}{1+\sqrt{x}}, \quad \frac{1}{1-x}, \quad \frac{1}{1-\sqrt{x}}$$

are in arithmetical progression and find the nth term of the sequence of which these are the first three terms.

14 The area of a circle of radius 1 is a transcendental number (that is, a number that cannot be obtained by the process of solving algebraic equations) denoted by the Greek letter π. To calculate its value, we may use a limiting process in which π is the limit of a sequence of known numbers. The method used by Archimedes was to inscribe in the circle a sequence of regular polygons. As the number of sides increased, so the polygon 'filled' the circle. Show, by use of the trigonometric identity $\cos 2\theta = 1 - 2\sin^2\theta$, that the area a_n of an inscribed regular polygon of n sides satisfies the equation

$$2\left(\frac{a_{2n}}{n}\right)^2 = 1 - \sqrt{\left[1 - \left(\frac{2a_n}{n}\right)^2\right]} \quad (n \geq 4)$$

Show that $a_4 = 2$ and use the recurrence relation to find a_{64}.

15 A **harmonic sequence** is a sequence with the property that every three consecutive terms (a, b and c, say) of the sequence satisfy

$$\frac{a}{c} = \frac{a-b}{b-c}$$

Prove that the reciprocals of the terms of a harmonic sequence form an arithmetical progression. Hence find the intermediate terms of a harmonic sequence of 8 terms whose first and last terms are $\frac{2}{3}$ and $\frac{2}{17}$ respectively.

16 The price of houses increases at 10% per year. Show that the price P_n in the nth year satisfies the recurrence relation

$$P_{n+1} = 1.1P_n$$

A house is currently priced at £80 000. What was its price two years ago? What will be its price in five years' time? After how many years will its price be double what it is now?

17 Evaluate each of the following sums:

(a) $1 + 2 + 3 + \ldots + 152 + 153$

(b) $1^2 + 2^2 + 3^2 + \ldots + 152^2 + 153^2$

(c) $\frac{1}{2} + \frac{1}{4} + \frac{1}{8} + \ldots + (\frac{1}{2})^{152} + (\frac{1}{2})^{153}$

(d) $2 + 6 + 18 + \ldots + 2(3)^{152} + 2(3)^{153}$

(e) $1 \cdot 2 + 2 \cdot 3 + 3 \cdot 4 + \ldots + 152 \cdot 153 + 153 \cdot 154$

(f) $\dfrac{1}{1 \cdot 2} + \dfrac{1}{2 \cdot 3} + \dfrac{1}{3 \cdot 4} + \ldots + \dfrac{1}{152 \cdot 153} + \dfrac{1}{153 \cdot 154}$

18 A certain bacterium propagates itself by subdividing, creating four additional bacteria, each identical to the parent bacterium. If the bacteria subdivide in this manner n times, then, assuming that none of the bacteria die, the number of bacteria present after each subdivision is given by the sequence $\{B_k\}_{k=0}^n$, where

$$B_k = \frac{4^{k+1} - 1}{3}$$

Three such bacteria subdivide n times and none of the bacteria die. The total number of bacteria is then 1 048 575. How many times did the bacteria divide?

19 By considering the sum

$$\sum_{k=1}^{n} [(k+1)^4 - k^4]$$

show that

$$\sum_{k=1}^{n} k^3 = [\tfrac{1}{2}n(n+1)]^2$$

20 The repayment instalment of a fixed rate, fixed period loan may be calculated by summing the *present values* of each instalment. This sum must equal the amount borrowed. The present value of an instalment £x paid after k years where $r\%$ is the rate of interest is

$$£\frac{x}{(1 + r/100)^k}$$

Thus £1000 borrowed over n years at $r\%$ satisfies the equation

$$1000 = \frac{x}{1 + r/100} + \frac{x}{(1 + r/100)^2} + \ldots$$

$$+ \frac{x}{(1 + r/100)^n}$$

Find x in terms of r and n, and compute its value when $r = 10$ and $n = 20$.

21 Consider the series

$$S_n = \tfrac{1}{2} + \tfrac{2}{4} + \tfrac{3}{8} + \ldots + \frac{n}{2^n}$$

Show that

$$\tfrac{1}{2}S_n = \tfrac{1}{4} + \tfrac{2}{8} + \tfrac{3}{16} + \ldots + \frac{n}{2^{n+1}}$$

and hence that

$$S_n - \tfrac{1}{2}S_n = \tfrac{1}{2} + \tfrac{1}{4} + \tfrac{1}{8} + \tfrac{1}{16} + \ldots + \frac{1}{2^n} + \frac{n}{2^{n+1}}$$

Hence sum the series.

22 Consider the general arithmetico-geometric series

$$S_n = a + (a+d)r + (a+2d)r^2 + \ldots$$

$$+ [a + (n-1)d]r^{n-1}$$

Show that

$$(1-r)S_n = a + dr + dr^2 + \ldots$$

$$+ dr^{n-1} - [a + (n-1)d]r^n$$

and find a simple expression for S_n.

7.4 Recurrence relations

We saw in Example 7.1 that sometimes the elements of a sequence satisfy a recurrence relation such that the value of an element x_n of a sequence $\{x_k\}$ can be expressed in terms of the values of earlier elements of the sequence. In general we may have a formula of the form

$$x_n = f(x_{n-1}, x_{n-2}, \ldots, x_1, x_0)$$

In this section we are going to consider two commonly occurring types of recurrence relation. These will provide sufficient background to make possible the solution of more difficult problems.

7.4.1 First-order linear recurrence relations with constant coefficients

These relations have the general form

$$x_{n+1} = ax_n + b_n, \quad n = 0, 1, 2, \ldots$$

where a is constant and b_n is a known sequence. The simplest case that occurs is when $b_n = 0$, when the relation reduces to

$$x_{n+1} = ax_n \tag{7.6}$$

This is called a **homogeneous relation** and every solution is a geometric sequence of the form

$$x_n = Aa^n \tag{7.7}$$

This is called the **general solution** of (7.6) since A is a constant which may be given any value. To determine the value of A we require more information about the sequence. For example, if we know the value of x_0 (say C) then $C = Aa^0$, which gives the value of A.

A slightly more difficult example is

$$x_{n+1} = ax_n + b \tag{7.8}$$

where b is a constant as well as a.

If the first term of the sequence is $x_0 = C$, as before, then

$$x_1 = aC + b$$

$$x_2 = ax_1 + b = a(aC + b) + b = Ca^2 + b(1 + a)$$

$$x_3 = ax_2 + b = a[Ca^2 + b(1 + a)] + b = Ca^3 + b(1 + a + a^2)$$

and so on.

In general, we obtain

$$x_n = Ca^n + \left(\frac{1 - a^n}{1 - a} \right) b, \quad a \neq 1$$

Rearranging, we can express this as

$$x_n = Aa^n + \frac{b}{1-a}, \quad a \neq 1 \tag{7.9}$$

where $A = C - b/(1-a)$. After the next example we will see that this solution (and that of more general problems) can be obtained more quickly by an alternative method. Notice that Aa^n is the general solution of the homogeneous relation (7.6) and that $x_n = b/(1-a)$, for all n, satisfies the full recurrence relation $x_{n+1} = ax_n + b$, so that it is a **particular solution** of the relation.

Example 7.12   Calculate the fixed annual payments £B required to amortize a debt of £D over N years, when the rate of interest is fixed at $100i\%$.

Solution   Let £d_n denote the debt after n years. Then, following the same argument as in Example 7.1, $d_0 = D$ and

$$d_{n+1} = (1+i)d_n - B$$

This is similar to the recurrence relation (7.8) but with $a = (1+i)$ an $b = -B$. Hence, using (7.9) we can write the general solution as

$$d_n = A(1+i)^n - \frac{B}{1-(1+i)} = A(1+i)^n + B/i$$

In addition, we know that $d_0 = D$ so that $D = A + B/i$ and thus the particular solution is given by

$$d_n = (D - B/i)(1+i)^n + B/i$$

We require the value of B so that the debt is zero after N years, that is $d_N = 0$. Thus

$$0 = (D - B/i)(1+i)^N + B/i$$

Solving this equation for B gives

$$B = \frac{iD(1+i)^N}{(1+i)^N - 1} = iD/[1 - (1+i)^{-N}]$$

as the required payment.

In summary, we have that the general solution to the first-order recurrence relation

$$x_{n+1} = ax_n + b$$

can be expressed as the sum of the **general solution** of the reduced relation

$$x_{n+1} = ax_n$$

and a **particular solution** of the full relation (7.8).

This is true for linear recurrence relations in general, that is, recurrence relations of the form

$$x_{n+1} = a_n x_n + a_{n-1} x_{n-1} + \ldots + a_1 x_1 + a_0$$

where the coefficients a_k are independent of the x_k but may depend on n. The property is easy to show in full generality but the same proof holds for the simplest case (7.8) above.

Suppose we can identify one particular solution p_n of (7.8) so that

$$p_{n+1} = ap_n + b$$

Now we seek a function q_n which complements p_n in such a way that

$$x_n = p_n + q_n$$

is the general solution of (7.8). Substituting x_n into this relation gives

$$p_{n+1} + q_{n+1} = ap_n + aq_n + b$$

Since $p_{n+1} = ap_n + b$, this implies that

$$q_{n+1} = aq_n$$

From (7.7), the general solution of this relation is

$$q_n = Aa^n$$

where A is a constant. Thus the general solution of (7.8) is

$$x_n = p_n + Aa^n$$

Because q_n complements p_n to form the general solution, it is usually called the **complementary solution**. As we have seen, with first-order recurrence relations, we can always find the complementary solution. Thus we are left with the task of finding the particular solution p_n. The method for finding p_n depends on the term b, as we illustrate in Example 7.13.

Indeed, the property of the general solution being the sum of a particular solution and a complementary solution applies to all linear systems, both continuous and discrete. We will meet it again in Chapter 10 when considering the general solution of linear ordinary differential equations.

Example 7.13   Find the general solutions of the recurrence relations

(a) $x_{n+1} = 3x_n + 4$ (b) $x_{n+1} = x_n + 4$

(c) $x_{n+1} = \alpha x_n + C\beta^n$ (d) $x_{n+1} = \alpha x_n + C\alpha^n$ (α, β, C given constants)

Solution   (a) First we try to find any function of n which will satisfy the relation. Since it contains the constant term 4, it is common sense to see if a constant K can be found which satisfies the relation. (Then all terms will be constants.) Setting $x_n = K$ implies $x_{n+1} = K$ and we have

$$K = 3K + 4$$

which gives $K = -2$. Thus, in this case, we can choose $p_n = -2$. Next we find the complementary solution q_n, which is the general solution of

$$x_{n+1} = 3x_n$$

From (7.7) we can see that $q_n = A3^n$ where A is a constant. Thus the general solution of (a) is

$$x_n = -2 + A3^n$$

(b) The basic steps are the same for this relation. We first find a particular solution p_n of the relation. Then we find the complementary solution q_n, so that $x_n = p_n + q_n$ is the general solution. In this case trying $x_n = K$ leads nowhere, since we obtain the inconsistent equation $K = K + 4$. Trying something a little more complicated than just a constant, we set $x_n = Kn$ and $x_{n+1} = K(n + 1)$ and we have

$$K(n + 1) = Kn + 4$$

which yields $K = 4$ and $p_n = 4n$. The general solution of $x_{n+1} = x_n$ is $q_n = A1^n$, so that the general solution of (b) is

$$x_n = 4n + A$$

(c) Since the recurrence relation has the term $C\beta^n$, it is natural to expect a solution of the form $K\beta^n$, where K is a constant, to satisfy the relation. Setting $x_n = K\beta^n$ gives

$$K\beta^{n+1} = \alpha K\beta^n + C\beta^n$$

Dividing through by β^n gives $K\beta = \alpha K + C$, from which we deduce $K = C/(\beta - \alpha)$ provided that $\beta \neq \alpha$. Thus we deduce the particular solution

$$p_n = C\beta^n/(\beta - \alpha)$$

The complementary solution q_n is the general solution of

$$x_{n+1} = \alpha x_n$$

which, using (7.7), is $q_n = A\alpha^n$. Hence the general solution of (c) is

$$x_n = C\beta^n/(\beta - \alpha) + A\alpha^n$$

(d) This is the special case of (c) where $\beta = \alpha$. If we set $p_n = K\alpha^n$, we obtain the equation $K\alpha^{n+1} = K\alpha^{n+1} + C\alpha^n$, which can only be true if $C = 0$. (We see then that p_n is the solution of $x_{n+1} = \alpha x_n$, that is, it is the complementary solution.) As in case (b), we instead seek a solution of the form $p_n = Kn\alpha^n$, so that $p_{n+1} = K(n + 1)\alpha^{n+1}$ and

$$K(n + 1)\alpha^{n+1} = \alpha Kn\alpha^n + C\alpha^n$$

This last equation gives $K = C/\alpha$. Hence the general solution of (d) is

$$x_n = Cn\alpha^{n-1} + A\alpha^n$$

where A is an arbitrary constant.

7.4.2 Second-order linear recurrence relations with constant coefficients

A second-order linear recurrence with constant coefficients has the form

$$x_{n+2} = ax_{n+1} + bx_n + c_n \tag{7.10}$$

If $c_n = 0$ for all n, then the relation is said to be **homogeneous**. As before, the solution of (7.10) can be expressed in the form

$$x_n = p_n + q_n$$

where p_n is any solution which satisfies (7.10) while q_n is the general solution of the associated homogeneous recurrence relation

$$x_{n+2} = ax_{n+1} + bx_n \tag{7.11}$$

Let α and β be the two roots of the algebraic equation

$$\lambda^2 = a\lambda + b$$

so that $\alpha^{n+2} = a\alpha^{n+1} + b\alpha^n$ and $\beta^{n+2} = a\beta^{n+1} + b\beta^n$, which imply that $y_n = \alpha^n$ and $y_n = \beta^n$ are particular solutions of (7.11). Since $(\lambda - \alpha)(\lambda - \beta) = 0$ implies $\lambda^2 = (\alpha + \beta)\lambda - \alpha\beta$ we may rewrite (7.11) as

$$x_{n+2} = (\alpha + \beta)x_{n+1} - \alpha\beta x_n$$

Rearranging the relation, we have

$$x_{n+2} - \alpha x_{n+1} = \beta(x_{n+1} - \alpha x_n)$$

Substituting $t_n = x_{n+1} - \alpha x_n$, this becomes

$$t_{n+1} = \beta t_n$$

with general solution, from (7.7), $t_n = C\beta^n$ where C is any constant.
  Thus

$$x_{n+1} - \alpha x_n = C\beta^n$$

which, using the results of Example 7.13(c) and (d), has the general solution

$$x_n = \begin{cases} C\beta^n/(\beta - \alpha) + A\alpha^n, & \alpha \neq \beta \\ Cn\alpha^{n-1} + A\alpha^n, & \alpha = \beta \end{cases}$$

Since C is any constant, we can rewrite this in the neater form

$$x_n = \begin{cases} A\alpha^n + B\beta^n, & \alpha \neq \beta \\ A\alpha^n + Bn\alpha^n, & \alpha = \beta \end{cases} \tag{7.12}$$

where A and B are arbitrary constants. Thus (7.12) gives the general solution of (7.11) where α and β are the roots of the equation

$$\lambda^2 = a\lambda + b$$

This is called the **characteristic equation** of the recurrence relation; λ is used as the unknown instead of x to avoid confusion.

Example 7.14   Find the solution of the Fibonacci recurrence relation

$$x_{n+2} = x_{n+1} + x_n$$

given $x_0 = 1$, $x_1 = 1$.

Solution   The characteristic equation of the recurrence relation is

$$\lambda^2 = \lambda + 1$$

which has roots $\lambda_1 = (1 + \sqrt{5})/2$ and $\lambda_2 = (1 - \sqrt{5})/2$.
    Hence its general solution is

$$x_n = A\left(\frac{1 + \sqrt{5}}{2}\right)^n + B\left(\frac{1 - \sqrt{5}}{2}\right)^n$$

Since $x_0 = 1$, we deduce $1 = A + B$

Since $x_1 = 1$, we deduce $1 = A\left(\frac{1 + \sqrt{5}}{2}\right) + B\left(\frac{1 - \sqrt{5}}{2}\right)$

Solving these simultaneous equations gives

$$A = (1 + \sqrt{5})/(2\sqrt{5}) \quad \text{and} \quad B = -(1 - \sqrt{5})/(2\sqrt{5})$$

and hence

$$x_n = \frac{1}{\sqrt{5}}\left[\left(\frac{1 + \sqrt{5}}{2}\right)^{n+1} - \left(\frac{1 - \sqrt{5}}{2}\right)^{n+1}\right]$$

defining the Fibonacci sequence explicitly.

We have seen that we can always find the complementary solution q_n of the recurrence relation (7.10)

$$x_{n+2} = ax_{n+1} + bx_n + c_n$$

The general solution of this relation is the sum of a particular solution p_n of the relation and its complementary solution q_n. The problem, then, is how to find one solution p_n. Here we will use methods based on experience and trial and error.

Example 7.15   Find all the solutions of

(a) $x_{n+2} = \frac{7}{2}x_{n+1} - \frac{3}{2}x_n + 12$ (b) $x_{n+2} = \frac{7}{2}x_{n+1} - \frac{3}{2}x_n + 12n$

(c) $x_{n+2} = \frac{7}{2}x_{n+1} - \frac{3}{2}x_n + 3(2^n)$

Solution   (a) First we find the general solution of the associated homogeneous relation $x_{n+2} = \frac{7}{2}x_{n+1} - \frac{3}{2}x_n$ which has characteristic equation $\lambda^2 = \frac{7}{2}\lambda - \frac{3}{2}$ with roots $\lambda = 3$ and $\lambda = \frac{1}{2}$. Thus, the complementary solution is

$$x_n = A3^n + B(\tfrac{1}{2})^n$$

Next we find a particular solution of

$$x_{n+2} = \tfrac{7}{2}x_{n+1} - \tfrac{3}{2}x_n + 12$$

We try the simplest possible function $x_n = K$ (for all n). Then, if this is a solution, we have

$$K = \tfrac{7}{2}K - \tfrac{3}{2}K + 12$$

giving $K = -12$.

Thus $p_n = -12$ and the general solution is

$$x_n = -12 + A3^n + B(\tfrac{1}{2})^n$$

(b) This has the same complementary solution as (a) so we have only to find a particular solution. We try the function $x_n = Kn + L$ where K and L are constants. Substituting into the recurrence relation gives

$$K(n + 2) + L \equiv \tfrac{7}{2}[K(n + 1) + L] - \tfrac{3}{2}[Kn + L] + 12n$$

Thus

$$Kn + 2K + L \equiv 2Kn + \tfrac{7}{2}K + 2L + 12n$$

Comparing coefficients of n gives

$$K = 2K + 12$$

so that $K = -12$.

Comparing the terms independent of n gives

$$2K + L \equiv \tfrac{7}{2}K + 2L$$

so that $L = -\tfrac{3}{2}K = 18$, and the general solution required is

$$x_n = -12n + 18 + A3^n + B(\tfrac{1}{2})^n$$

(c) This has the same complementary function as (a) so we only need to find a particular solution. To find this we try $x_n = K2^n$, giving

$$K(2^{n+2}) = \tfrac{7}{2}K(2^{n+1}) - \tfrac{3}{2}K(2^n) + 3(2^n)$$

so that

$$(4 - 7 + \tfrac{3}{2})K(2^n) = 3(2^n)$$

Hence $K = -2$ and the general solution required is

$$x_n = -2(2^n) + A3^n + B/2^n$$

When the roots of the characteristic equation are complex numbers, the general solution of the homogeneous recurrence relation has a different form, as illustrated in Example 7.16.

Example 7.16   Show that the general solution of the recurrence relation

$$x_{n+2} = 6x_{n+1} - 25x_n$$

may be expressed in the form

$$x_n = 5^n(A \cos n\theta + B \sin n\theta)$$

where θ is such that $\sin \theta = \frac{4}{5}$ and $\cos \theta = \frac{3}{5}$.

Solution   The characteristic equation

$$\lambda^2 = 6\lambda - 25$$

has the (complex) roots $\lambda = 3 + j4$ and $\lambda = 3 - j4$ so that we can write the general solution in the form

$$x_n = A(3 + j4)^n + B(3 - j4)^n$$

Now writing the complex numbers in polar form we have

$$x_n = A(re^{j\theta})^n + B(re^{-j\theta})^n$$

where $r^2 = 3^2 + 4^2$ and $\tan \theta = \frac{4}{3}$ with $0 < \theta < \pi/2$ (or $\cos \theta = \frac{3}{5}$, $\sin \theta = \frac{4}{5}$). This can be simplified to give

$$x_n = A(5^n e^{jn\theta}) + B(5^n e^{-jn\theta}) = A5^n(\cos n\theta + j \sin n\theta) + B5^n(\cos n\theta - j \sin n\theta)$$

$$= (A + B)5^n \cos n\theta + j(A - B)5^n \sin n\theta$$

Here A and B are arbitrary complex constants, so their sum and difference are also arbitrary constants and we can write

$$x_n = P5^n \cos n\theta + Q5^n \sin n\theta$$

giving the form required. (Since P and Q are constants we can replace them by A and B if we wish.)

The general result corresponding to that obtained in Example 7.16 is that if the roots of the characteristic equation can be written in the form

$$\lambda = u \pm jv$$

where u, v are real numbers, then the general solution of the homogeneous recurrence relation is

$$x_n = r^n(A \cos n\theta + B \sin n\theta)$$

where $r = \sqrt{(u^2 + v^2)}$, $\cos \theta = u/r$, $\sin \theta = v/r$ and A and B are arbitrary constants.

Recurrence relations are sometimes called **difference equations**. This name is used since we can rearrange the relations in terms of the differences of unknown sequence x_n. Thus

$$x_{n+1} = ax_n + b$$

can be rearranged as

$$\Delta x_n = (a - 1)x_n + b$$

where $\Delta x_n = x_{n+1} - x_n$.

Similarly, after some algebraic manipulation, we may write

$$x_{n+2} = ax_{n+1} + bx_n + c$$

as

$$\Delta^2 x_n = (a - 2)\Delta x_n + (a + b - 1)x_n + c$$

where

$$\Delta^2 x_n = \Delta x_{n+1} - \Delta x_n = x_{n+2} - 2x_{n+1} + x_n$$

The method for solving second-order linear recurrence relations with constant coefficients is summarized in Figure 7.7.

Figure 7.7
Summary: second-order linear recurrence relation with constant coefficients.

Homogeneous case:

$$x_{n+2} = ax_{n+1} + bx_n \tag{1}$$

(i) Solve the characteristic equation.
(ii) Write down the general solution for x_n from the table:

Roots of characteristic equation	General solution (A and B are arbitrary constants)
Real α, β and $\alpha \neq \beta$	$A\alpha^n + B\beta^n$
Real α, β and $\alpha = \beta$	$(A + Bn)\alpha^n$
Non-real α, $\beta = u \pm jv$	$(u^2 + v^2)^{n/2}(A \cos n\theta + B \sin n\theta)$ where $\cos\theta = u/(u^2 + v^2)^{1/2}$, $\sin\theta = v/(u^2 + v^2)^{1/2}$

Nonhomogeneous case:

$$x_{n+2} = ax_{n+1} + bx_n + c_n \text{ where } c_n \text{ is a known sequence.} \tag{2}$$

(i) Find the general solution of the associated homogeneous problem (1).
(ii) Find a particular solution of (2).
(iii) The general solution of (2) is the sum of (i) and (ii).

To find a particular solution to (2) substitute a likely form of particular solution into (2). If the correct form has been chosen then comparing coefficients will be enough to determine the values of the constants in the trial solution. Here are some suitable forms of particular solutions:

c_n	7	$3n + 5$	$2n^2 + 3n + 8$	$3\cos(7n) + 5\sin(7n)$	6^n	$n5^n$
p_n	C	$Cn + D$	$Cn^2 + Dn + E$	$C\cos(7n) + D\sin(7n)$	$C6^n$	$5^n(C + Dn)$

In solving problems, note that the top line of the table involves any *known* constants (these will be different from problem to problem), while the bottom line involves *unknown* constants, C, D, E, which must be determined by substituting the trial form into the nonhomogeneous relation.

An exceptional case arises when the suggested form for p_n already is present in the general solution of the associated homogeneous problem. If this happens, just multiply the suggested form by n (and if that does not work, by n repeatedly until it does).

7.4.3 Exercises

23 If a debt is amortized by equal annual payments of amount B, and if interest is charged at rate i per annum, then the debt after n years, d_n, satisfies $d_{n+1} = (1 + i)d_n - B$, where $d_0 = D$, the initial debt.

Show that $d_n = D(1 + i)^n + B\dfrac{1 - (1 + i)^n}{i}$

and deduce that to clear the debt on the Nth

payment we must take $B = \dfrac{Di}{1 - (1 + i)^{-N}}$.

If £10 000 is borrowed at an interest rate of 0.12 (= 12%) per annum, calculate (to the nearest £) the appropriate annual payment which will amortize the debt at the end of 10 years.

For this annual payment calculate the amount of the debt d_n for $n = 1, 2, \ldots, 10$ (use the recurrence rather than its solution, and record your answers to the nearest £) and calculate the first differences for this sequence. Comment briefly on the behaviour of the first differences.

24 Find the general solution of the linear recurrence relation

$$(n + 1)^2 x_{n+1} - n^2 x_n = 1, \quad \text{for } n \geqslant 1$$

(*Hint*: the coefficients are not constants. Use the substitution $z_n = n^2 x_n$ to find a constant coefficient equation for z_n. Find the general solution for z_n and hence for x_n.)

25 Evaluate the expression $2x_{n+2} - 7x_{n+1} + 3x_n$ when x_n is defined for all $n \geqslant 0$ by

(a) $x_n = 3^n$ (b) $x_n = 2^n$

(c) $x_n = 2^{-n}$ (d) $x_n = 3(-2)^n$

Which of (a) to (d) are solutions of the following recurrence relation?

$$2x_{n+2} - 7x_{n+1} + 3x_n = 0$$

26 Show, by substituting them into the recurrence relation, that $x_n = 2^n$ and $x_n = (-1)^n$ are two solutions of $x_{n+2} - x_{n+1} - 2x_n = 0$. Verify similarly that $x_n = A(2^n) + B(-1)^n$ is also a solution of the recurrence relation for all constants A and B.

27 Obtain the general solutions of

(a) $Y_{n+2} - 7Y_{n+1} + 10Y_n = 0$

(b) $u_{n+2} - u_{n+1} - 6u_n = 0$

(c) $25T_{n+2} = -T_n$

(d) $p_{n+2} - 5p_{n+1} = 5(p_{n+1} - 5p_n)$

(e) $2E_{n+2} = E_{n+1} + E_n$

28 Solve the nonhomogeneous problems (use parts of Question 27):

(a) $Y_{n+2} - 7Y_{n+1} + 10Y_n = 1, \quad Y_0 = 5/4, \quad Y_1 = 2$

(b) $2E_{n+2} - E_{n+1} - E_n = 1, \quad E_0 = 2, \quad E_1 = 0$

(c) $u_{n+2} - u_{n+1} - 6u_n = n$ (general solution only)

29 Show that the characteristic equation for the recurrence relation $x_{n+2} - 2ax_{n+1} + a^2 x_n = 0$, where a is a non-zero constant, has two equal roots $\lambda = a$.

(a) Verify (by substituting into the relation) that $x_n = (A + Bn)a^n$ is a solution for all constants A and B.

(b) Find the particular solution which satisfies $x_0 = 1$, $x_1 = 0$. (Your answer will involve a, of course.)

(c) Find the particular solution for which $x_0 = 3$, $x_{10} = 20$.

30 Let x be a constant such that $|x| < 1$. Find the solution of

$$T_{n+2} - 2xT_{n+1} + T_n = 0, \quad T_0 = 1, \quad T_1 = x$$

Find T_2, T_3 and T_4 also directly by recursion and deduce that $\cos(2\cos^{-1}x) = 2x^2 - 1$ and express $\cos(3\cos^{-1}x)$ and $\cos(4\cos^{-1}x)$ as polynomials in x.

31 A topic from information theory: imagine an information transmission system that uses an alphabet consisting of just two symbols 'dot' and 'dash', say. Messages are transmitted by first encoding them into a string of these symbols, and no other symbols (e.g. blank spaces) are allowed. Each symbol requires some length of time for its transmission. Therefore, for a fixed total time duration only a finite number of different message strings is possible. Let N_t denote the number of different message strings possible in t time units.

(a) Suppose that dot and dash each require one time unit for transmission. What is the value of N_1? Why is $N_{t+1} = 2N_t$ for all $t \geqslant 1$? Write down a simple formula for N_t for $t \geqslant 1$.

(b) Suppose instead that dot requires one unit of time for transmission while dash requires two units. What are the values of N_1 and N_2? Justify the relation $N_{t+2} = N_{t+1} + N_t$ for $t \geqslant 1$. Hence write down a formula for N_t in terms of t. (*Hint*: the general solution of Fibonacci recurrence is given in Example 7.14.)

| 7.5 | # Limit of a sequence |

In Section 7.2.1 the idea of a sequence and the associated notation were described. We shall now develop the concept of a limit of a sequence and then discuss the properties of sequences that have limits (termed convergent sequences) and methods for evaluating those limits algebraically and numerically.

7.5.1 Convergent sequences

In Example 7.6, we obtained the following sequence of approximations (working to 2dp) for $\sqrt{2}$:

$$x_0 = 1$$

$$x_1 = 1.83$$

$$x_2 = 1.16$$

$$x_3 = 1.64$$

Continuing with the process, we obtain

$$x_{22} = 1.41$$

$$x_{23} = 1.41$$

and

$$x_n = 1.41 \quad \text{for } n \geqslant 22$$

The terms x_{22} and x_{23} of the sequence are indistinguishable to two decimal places; in other words, their difference is less than a rounding error. This situation is shown clearly in Figure 7.4(b). This phenomenon occurs with many sequences, and we say that the sequence **tends to a limit** or **has a limiting value** or **converges** or **is convergent**. While it is clear in the above example what we mean by saying that the sequence converges to $\sqrt{2}$, we need a precise definition for all the cases that may occur.

In general, a sequence $\{a_k\}_{k=0}^{\infty}$ has the limiting value a as n becomes large if, given a small positive number ε (no matter how small), a_n differs from a by less than ε for all sufficiently large n. More concisely,

> $a_n \to a$ as $n \to \infty$ if, given any $\varepsilon > 0$, there is a number N such that $|a_n - a| < \varepsilon$ for all $n > N$

Here the $\to$ stands for 'tends to the value' or 'converges to the limit'. An alternative notation for $a_n \to a$ as $n \to \infty$ is

$$\lim_{n \to \infty} a_n = a$$

Diagrammatically, this means that the terms of the sequence lie between $y = a - \varepsilon$ and $y = a + \varepsilon$ for $n > N$, as shown in Figure 7.8.

Note that the limit of a sequence need not actually be an element of the sequence. For example $\{n^{-1}\}_{n=1}^{\infty}$ has limit 0, but 0 does not occur in the sequence.

Figure 7.8
Convergence
of $\{a_n\}$ to limit a.

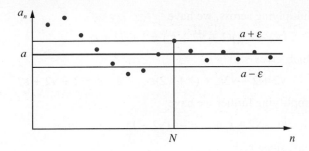

Figure 7.9
Convergence
of $\{x_n\}$ to $\sqrt{2}$.

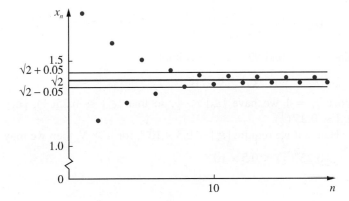

Returning to the square-root example discussed above, we have

$$x_n \to \sqrt{2} \quad \text{as} \quad n \to \infty$$

It is clear from the terms of the sequence that for an error bound of 0.05 we need $n > 8$ (see Figure 7.9). Thus $\sqrt{2} = 1.4$ (to 1dp). However, to prove convergence in the formal sense, we have to be able to say how many terms we need to take in order to obtain a specified level of precision. Suppose we need an answer correct to 10dp, or 100dp, or whatever; we must be able to give the corresponding value of N in the definition of convergence. Finding an expression for N is not often easy.

We shall illustrate the type of methods used by finding an expression for N for a classical method for calculating $\sqrt{2}$. This uses the iteration

$$x_{n+1} = \frac{2 + x_n}{1 + x_n} \quad \text{with } x_0 = 1$$

This produces the rational approximations

$$\left\{ 1, \frac{3}{2}, \frac{7}{5}, \frac{17}{12}, \frac{41}{29}, \frac{99}{70}, \dots \right\}$$

The last given approximation has an error of less than 0.0001. Suppose we require an approximation which is correct to p decimal places, then we need to find an N such that

$$|x_n - \sqrt{2}| < 0.5 \times 10^{-p}$$

for $n > N$. Writing $\varepsilon_n = x_n - \sqrt{2}$ so that $x_0 = \sqrt{2} + \varepsilon_0$, $x_1 = \sqrt{2} + \varepsilon_1, \dots, x_{n+1} = \sqrt{2} + \varepsilon_{n+1}$ (and so on) we have

$$\sqrt{2} + \varepsilon_{n+1} = \frac{2 + \sqrt{2} + \varepsilon_n}{1 + \sqrt{2} + \varepsilon_n}$$

Multiplying across, we have

$$(\sqrt{2} + \varepsilon_{n+1})(1 + \sqrt{2} + \varepsilon_n) = 2 + \sqrt{2} + \varepsilon_n$$

which gives

$$\sqrt{2} + 2 + \sqrt{2}\varepsilon_n + (1 + \sqrt{2})\varepsilon_{n+1} + \varepsilon_{n+1}\varepsilon_n = 2 + \sqrt{2} + \varepsilon_n$$

Simplifying further we have

$$(1 + \sqrt{2} + \varepsilon_n)\varepsilon_{n+1} = -\varepsilon_n(\sqrt{2} - 1)$$

Thus, since $x_n = \sqrt{2} + \varepsilon_n$

$$|\varepsilon_{n+1}| = \frac{(\sqrt{2} - 1)|\varepsilon_n|}{1 + x_n}$$

Since $x_n \geq 1$ and $\sqrt{2} < 1.5$, this implies

$$|\varepsilon_{n+1}| < \frac{0.5}{2}|\varepsilon_n| < 0.25|\varepsilon_n|$$

Since $x_0 = 1$ we have $|\varepsilon_0| < \frac{1}{2}$, so that $|\varepsilon_1| < 0.25(\frac{1}{2})$, $|\varepsilon_2| < 0.25^2(\frac{1}{2})$, ... and $|\varepsilon_n| < 0.25^n(\frac{1}{2})$.

Hence if we require $|\varepsilon_n| < 0.5 \times 10^{-p}$, for $n > N$, then we may find m such that

$$0.25^m(\tfrac{1}{2}) < 0.5 \times 10^{-p}$$

or

$$\frac{1}{4^m} < \frac{1}{10^p}$$

which implies $4^m > 10^p$.

Taking logarithms to base 10, this gives

$$m > p/\log 4$$

Then choose N to be the greatest integer not greater than m, that is, $N = \lfloor p/\log 4 \rfloor$. Thus, to guarantee 10dp, we need to evaluate at most $\lfloor 10/\log 4 \rfloor = 16$ iterations, which you may verify on your calculator.

7.5.2 Properties of convergent sequences

As we have seen in the $\sqrt{2}$ example, it is usually difficult and tedious to prove the convergence of a sequence from first principles. Normally we are able to compute the limit of a sequence from simpler sequences by means of very simple rules based on the properties of convergent sequences. These are:

(a) Every convergent sequence is bounded; that is, if $\{a_n\}_{n=0}^{\infty}$ is convergent then there is a positive number M such that $|a_n| < M$ for all n.

(b) If $\{a_n\}$ has limit a, and $\{b_n\}$ has limit b, then
   (i) $\{a_n + b_n\}$ has limit $a + b$
   (ii) $\{a_n - b_n\}$ has limit $a - b$
   (iii) $\{a_n b_n\}$ has limit ab
   (iv) $\{a_n/b_n\}$ has limit a/b, for $b_n \neq 0$, $b \neq 0$.

We illustrate the technique in Example 7.17.

Example 7.17   Find the limits of the sequence $\{x_n\}_{n=0}^{\infty}$ defined by

$$\text{(a)} \quad x_n = \frac{n}{n+1} \qquad \text{(b)} \quad x_n = \frac{2n^2 + 3n + 1}{5n^2 + 6n + 2}$$

Solution   (a) With $x_n = n/(n+1)$, we generate the sequence $\{0, \frac{1}{2}, \frac{2}{3}, \frac{3}{4}, \frac{4}{5}, \dots\}$. From these values it seems clear that $x_n \to 1$ as $n \to \infty$. This can be proved by rewriting x_n as

$$x_n = 1 - \frac{1}{n+1}$$

and we make $1/(n+1)$ as small as we please by taking n sufficiently large.
Alternatively, we write

$$x_n = \frac{1}{1 + 1/n}$$

Now $1/n \to 0$ as $n \to \infty$. Hence, by the property (b)(i), $1 + 1/n \to 1$ and so, by the property (b)(iv),

$$\frac{1}{1 + 1/n} \to 1 \quad \text{as } n \to \infty$$

as illustrated in Figure 7.10.

(b) For

$$x_n = \frac{2n^2 + 3n + 1}{5n^2 + 6n + 2}$$

the easiest approach is to divide both numerator and denominator by the highest power of n occurring and use the fact that $1/n \to 0$ as $n \to \infty$. Thus

$$x_n = \frac{2 + 3/n + 1/n^2}{5 + 6/n + 2/n^2}$$

The limits of numerator and denominator are 2 and 5 (using the property (b)(i) repeatedly), and so $x_n \to \frac{2}{5}$ as $n \to \infty$ (using (b)(iv)). This is shown clearly in Figure 7.11.

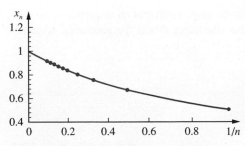

Figure 7.10 Sequence $x_n = n/(n+1)$ plotted against $1/n$.

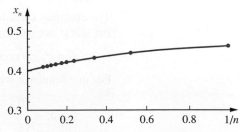

Figure 7.11 Sequence $x_n = \dfrac{2n^2 + 3n + 1}{5n^2 + 6n + 2}$ plotted against $1/n$.

Example 7.18

Show that the ratio x_n of successive terms of the Fibonacci sequence satisfies the recurrence relation

$$x_{n+1} = 1 + 1/x_n, \quad x_0 = 1$$

Calculate the first few terms of this sequence and find the value of its limit.

Solution

The Fibonacci sequence was defined in Example 7.14 as

$$f_{n+2} = f_{n+1} + f_n \quad \text{with} \quad f_0 = f_1 = 1$$

Defining $x_n = f_{n+1}/f_n$ gives $f_{n+2} = x_{n+1} \times f_{n+1}$ and $f_n = f_{n+1}/x_n$, so that the recurrence relation becomes

$$x_{n+1} f_{n+1} = f_{n+1} + f_{n+1}/x_n$$

and dividing through by f_{n+1} we have

$$x_{n+1} = 1 + 1/x_n$$

Also, $x_0 = f_1/f_0 = 1/1 = 1$.

Using the recurrence relation, we obtain the sequence

$$\{1, 2, 1.5, 1.6667, 1.6, 1.625, 1.6154, 1.6190, \ldots\}$$

The numerical results suggest a limiting value near 1.62. Indeed, the oscillatory nature of the sequence suggests $1.6154 < x_n < 1.6190$ for $n > 8$, which implies a limit value $x = 1.62$ correct to 2dp.

In this case we can check this conclusion, for if $x_n \to x$ as $n \to \infty$ then $x_{n+1} \to x$ also, and so the recurrence relation yields

$$x = 1 + \frac{1}{x}, \quad \text{with } x > 0$$

Thus $x^2 - x - 1 = 0$, which implies $x = \frac{1}{2}(1 + \sqrt{5})$ or $x = \frac{1}{2}(1 - \sqrt{5})$. Since the sequence has positive values only, it is clear that the appropriate root is $x = \frac{1}{2}(1 + \sqrt{5}) = 1.62$ (to 2dp).

This limiting value is called the **golden number**. A rectangle the ratio of whose sides is the golden number is said to be the most pleasing aesthetically, and this has often been adopted by architects as a basis of design.

7.5.3 Computation of limits

The examples considered so far tend to create the impression that all sequences converge, but this is not so. An important sequence that illustrates this is the geometric sequence

$$a_n = r^n, \quad r \text{ constant}$$

For this sequence we have

$$\lim_{n \to \infty} a_n = \begin{cases} 0 & (-1 < r < 1) \\ 1 & (r = 1) \end{cases}$$

If $r > 1$, the sequence increases without bound as $n \to \infty$, and we say it **diverges**. If $r = -1$, the sequence takes the values -1 and 1 alternately, and there is no limiting value. If $r < -1$, the sequence is unbounded and the terms alternate in sign.

Often in computational applications of sequences the limit of the sequence is not known, so that it is not possible to apply the formal definition to determine the number of terms N we need to take in order to obtain a specified level of precision. If we do not know the limit a, to which a sequence $\{a_n\}$ converges, then we cannot measure $|a_n - a|$. In the computational context, when we apply a recurrence relation to find a solution to a problem, we say that the sequence $\{a_n\}$ has converged to its limit when all subsequent terms yield the same value of the approximation required. In other words, we say that the sequence of finite terms is convergent if, for any n and $m > N$,

$$|a_n - a_m| < \varepsilon$$

where the bound ε is specified. Thus a sequence tends to a limit if all the terms of the sequence for $n > N$ are restricted to an interval that can be made arbitrarily small by choosing N sufficiently large. This is called **Cauchy's test for convergence**.

In many practical problems we need to find a numerical estimate for the limit of a sequence. A graphical method for this is to sketch the graph defined by the points $\{(1/n, a_n): n = 1, 2, 3, \ldots\}$ and then extrapolate from it, since $1/n \to 0$ as $n \to \infty$. If greater precision is required than can be obtained in this way, an effective numerical procedure is a form of repeated linear extrapolation due to Aitken. We illustrate the procedure in Example 7.19.

Example 7.19   Examine the convergence of the sequence $\{a_n\}_{n=1}^{\infty}$, $a_n = (1 + 1/n)^n$.

Solution   It can be shown that $\lim_{n\to\infty} a_n = e$, but convergence is rather slow. In fact,

$$a_1 = 2, \quad a_2 = 2.2500, \quad a_3 = 2.3704, \quad a_4 = 2.4414, \quad \ldots$$

$$a_8 = 2.5658, \quad \ldots, \quad a_{16} = 2.6379, \quad \ldots, \quad a_{32} = 2.6770, \quad \ldots$$

$$a_{64} = 2.6973, \quad \ldots, \quad \text{and} \quad e = 2.7183 \text{ to 4dp}$$

Now consider the two terms corresponding to $n = 16$ and $n = 32$ and set $x_n = 1/n$. Then

$$n = 16 \quad \text{gives} \quad x_{16} = 0.0625 \quad \text{and} \quad a_{16} = 2.6379$$

$$n = 32 \quad \text{gives} \quad x_{32} = 0.031\,25 \quad \text{and} \quad a_{32} = 2.6770$$

We wish to find the value corresponding to $x = 0$. To estimate this, we may use linear extrapolation as shown in Figure 7.12. This gives

$$b_{16,32} = \frac{x_{16}a_{32} - x_{32}a_{16}}{x_{16} - x_{32}} = 2.7161$$

Note that $b_{16,32}$ is a better estimate for e than either a_{16} or a_{32}.

Figure 7.12
Linear extrapolation
for the limit of a
sequence.

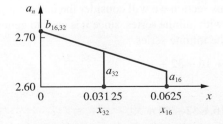

7.5.4 Exercises

32 Calculate the first six terms of each of the following sequences $\{a_n\}$ and draw a graph of a_n versus $1/n$. (Some care is needed in choosing the scale of the y axis.) What is the behaviour of a_n as $n \to \infty$?

(a) $a_n = \dfrac{n}{n^2 + 1}$ $(n \geqslant 1)$

(b) $a_n = \dfrac{3n^2 + 2n + 1}{6n^2 + 5n + 2}$ $(n \geqslant 1)$

(c) $a_n = (2n)^{1/n}$ $(n \geqslant 1)$

(d) $a_n = \left(1 + \dfrac{1}{2n}\right)^n$ $(n \geqslant 1)$

(e) $a_n = \sqrt{(1 + a_{n-1})}$, $a_1 = 1$ $(n \geqslant 2)$

(f) $a_n = n \sin \dfrac{\pi}{n}$ $(n \geqslant 1)$

33 Calculate the first six terms of each of the following sequences $\{a_n\}$ and draw a graph of a_n against n. What is the behaviour of a_n as $n \to \infty$?

(a) $a_n = \dfrac{n^2 + 1}{n + 1}$ $(n \geqslant 0)$

(b) $a_n = (\sin \tfrac{1}{2} n\pi)^n$ $(n \geqslant 1)$

(c) $a_n = 3/a_{n-1}$, $a_0 = 1$ $(n \geqslant 1)$

34 Find the least value of N such that when $n \geqslant N$,

(a) $n^2 + 2n > 100$ (b) $\dfrac{n^2}{2^n} < \dfrac{1}{1000}$

(c) $\dfrac{1}{n} - \dfrac{(-1)^n}{n^2} < 0.000\,001$

(d) $\sqrt{(n + 1)} - \sqrt{n} < \tfrac{1}{10}$

(e) $\dfrac{n^2 + 2}{n^2 - 1} - 1 < 0.01$

35 What is the long-term share of the detergent market achieved by the brand 'Number One', described in Question 2 (Exercises 7.2.3)?

36 A **linearly convergent** sequence has the property that

$$a_n - a = \lambda(a_{n-1} - a) \quad \text{for all } n$$

where λ is a constant and $a = \lim_{n \to \infty} a_n$. Show that

$$a_{n+1} - a = \lambda(a_n - a)$$

Deduce that

$$\frac{a_{n+1} - a}{a_n - a} = \frac{a_n - a}{a_{n-1} - a}$$

and show that

$$a = a_{n-1} - \frac{(a_n - a_{n-1})^2}{a_{n+1} - 2a_n + a_{n-1}}$$

This is known as **Aitken's estimate** for the limit of a sequence.

Compute the first four terms of the sequence

$$a_0 = 2, \quad a_{n+1} = \tfrac{1}{5}(3 + 4a_n^2 - a_n^3) \quad (n \geqslant 0)$$

and estimate the limit of the sequence.

7.6 Infinite series

Infinite series occur in a large variety of practical problems, from estimating the long-term effects of pollution to the stability analysis of the motions of machinery parts. They also occur in the development of computer algorithms for the numerical solution of practical problems. In this section we will consider the underlying ideas. Care has to be exercised when dealing with infinite series, since it is easy to generate fallacious results. For example, consider the infinite series

$$S = 1 - 2 + 4 - 8 + 16 - 32 + \dots$$

Then we can write

$$2S = 2 - 4 + 8 - 16 + 32 - 64 + \dots$$

and adding these two results, we obtain

$$3S = 1 \quad \text{or} \quad S = \tfrac{1}{3}$$

which is clearly wrong. Such blunders, however, are not always so glaringly obvious, so we have to develop simple methods for determining whether an infinite series sums to a finite value and for obtaining or estimating that value.

7.6.1 Convergence of infinite series

As we discussed in Section 7.2.1, series and sequences are closely connected. When the sum S_n of a series of n terms tends to a limit as $n \to \infty$, the series is **convergent**. When we can express S_n in a simple form, it is usually easy to establish whether or not the series converges. To find the sum of an infinite series, the sequence of partial sums $\{S_n\}$ is taken to the limit.

Example 7.20 Examine the following series for convergence:

(a) $1 + 3 + 5 + 7 + 9 + \ldots + (2k + 1) + \ldots$

(b) $1^2 + 2^2 + 3^2 + 4^2 + 5^2 + \ldots + k^2 + \ldots$

(c) $1 + \dfrac{1}{2} + \dfrac{1}{4} + \dfrac{1}{8} + \dfrac{1}{16} + \ldots + \dfrac{1}{2^k} + \ldots$

(d) $\dfrac{1}{1 \cdot 2} + \dfrac{1}{2 \cdot 3} + \dfrac{1}{3 \cdot 4} + \dfrac{1}{4 \cdot 5} + \ldots + \dfrac{1}{(k + 1)(k + 2)} + \ldots$

Solution (a) This is an arithmetic series so we can write its finite sum as a simple formula

$$S_n = \sum_{k=0}^{n-1} (2k + 1) = 1 + 3 + 5 + \ldots + (2n - 1) = n^2 \quad (n \text{ terms})$$

It is clear from this that $S_n \to \infty$ as $n \to \infty$ and the series does not converge to a limit. It is a **divergent** series.

(b) As we saw in Example 7.10

$$S_n = 1^2 + 2^2 + 3^2 + \ldots + n^2 = \tfrac{1}{6} n(n + 1)(2n + 1) \quad (n \text{ terms})$$

As n becomes large, so does S_n, and $S_n \to \infty$ as $n \to \infty$. Hence the series is divergent.

(c) $S_n = 1 + \tfrac{1}{2} + \tfrac{1}{4} + \ldots + \dfrac{1}{2^{n-1}} \quad (n \text{ terms})$

This is a geometric series with common ratio $\tfrac{1}{2}$. Using the formula (7.4) with $a = 1$ and $r = \tfrac{1}{2}$ gives

$$S_n = \frac{1 - \dfrac{1}{2^n}}{1 - \dfrac{1}{2}} = 2\left(1 - \dfrac{1}{2^n}\right)$$

As $n \to \infty$, $\dfrac{1}{2^n} \to 0$, so that $S_n \to 2$. Hence the series converges to the sum 2.

(d) We showed in Example 7.11 that

$$S_n = \frac{1}{1 \cdot 2} + \frac{1}{2 \cdot 3} + \ldots + \frac{1}{n(n+1)} = 1 - \frac{1}{n+1}$$

As $n \to \infty$, $1/(n+1) \to 0$, so that $S_n \to 1$. Hence the series converges to the sum 1.

Among the elementary series, the geometric series is the most important.

$$S_n = a + ar + ar^2 + \ldots + ar^{n-1} \quad (n \text{ terms})$$

$$= \frac{a(1 - r^n)}{1 - r}$$

$$= \frac{a}{1 - r} - \frac{ar^n}{1 - r}$$

Since $r^n \to 0$ as $n \to \infty$ when $|r| < 1$, we conclude that $S_n \to a/(1 - r)$ where $|r| < 1$ and the series is convergent. Where $|r| \geq 1$, the series is divergent. These results are used in many applications and the sum of the infinite series is

$$S = a + ar + ar^2 + ar^3 + \ldots = \frac{a}{1 - r}, \quad |r| < 1 \tag{7.13}$$

7.6.2 Tests for convergence of positive series

The convergence or divergence of the series discussed in Example 7.20 was established by considering the behaviour of the partial sum S_n as $n \to \infty$. In many cases, however, it is not possible to express S_n in a closed form. When this occurs, the convergence or divergence of the series is established by means of a test. Two tests are commonly used.

(a) Comparison test

Suppose we have a series, $\sum_{k=0}^{\infty} c_k$, of positive terms ($c_k \geq 0$, all k) which is known to be convergent. If we have another series, $\sum_{k=0}^{\infty} u_k$, of positive terms such that $u_k \leq c_k$ for all k then $\sum_{k=0}^{\infty} u_k$ is convergent also.

Also, if $\sum_{k=0}^{\infty} c_k$ diverges and $u_k \geq c_k \geq 0$ for all k, then $\sum_{k=0}^{\infty} u_k$ also diverges.

Example 7.21 Examine for convergence the series

(a) $1 + \dfrac{1}{1!} + \dfrac{1}{2!} + \dfrac{1}{3!} + \dfrac{1}{4!} + \ldots + \dfrac{1}{n!} + \ldots$ (the **factorial series**)

(b) $1 + \dfrac{1}{2} + \dfrac{1}{3} + \dfrac{1}{4} + \ldots + \dfrac{1}{n} + \ldots$ (the **harmonic series**)

Solution   (a) We can establish the convergence of the series (a) by considering its partial sum

$$A_n = 1 + \frac{1}{1!} + \frac{1}{2!} + \frac{1}{3!} + \frac{1}{4!} + \ldots + \frac{1}{n!}$$

Each term of this series is less than or equal to the corresponding term of the series

$$C_n = 1 + 1 + \frac{1}{2} + \frac{1}{2^2} + \frac{1}{2^3} + \ldots + \frac{1}{2^{n-1}}$$

This geometric series may be summed to give

$$C_n = 3 - \frac{1}{2^{n-1}}$$

Thus

$$A_n < 3 - \frac{1}{2^{n-1}}$$

which implies that, since all the terms of the series are positive numbers, A_n tends to a limit less than 3 as $n \to \infty$. Thus the series is convergent.

(b) The divergence of the series (b) is similarly established.

$$1 + \tfrac{1}{2} + \tfrac{1}{3} + \tfrac{1}{4} + \tfrac{1}{5} + \tfrac{1}{6} + \tfrac{1}{7} + \tfrac{1}{8} + \tfrac{1}{9} + \ldots$$

Collecting together successive groups of two, four, eight, . . . terms, we have

$$1 + \tfrac{1}{2} + (\tfrac{1}{3} + \tfrac{1}{4}) + (\tfrac{1}{5} + \tfrac{1}{6} + \tfrac{1}{7} + \tfrac{1}{8}) + (\tfrac{1}{9} + \ldots + \tfrac{1}{16}) + (\tfrac{1}{17} + \ldots$$

which may be compared with the series

(c) $1 + \tfrac{1}{2} + (\tfrac{1}{4} + \tfrac{1}{4}) + (\tfrac{1}{8} + \tfrac{1}{8} + \tfrac{1}{8} + \tfrac{1}{8}) + (\tfrac{1}{16} + \ldots + \tfrac{1}{16}) + (\tfrac{1}{32} + \ldots$

Each term of the rearranged (b) is greater than or at least equal to the corresponding term of the series (c), and so the 'sum' of the series (b) is greater than the 'sum' of the series (c), which is

$$1 + \tfrac{1}{2} + \tfrac{1}{2} + \tfrac{1}{2} + \tfrac{1}{2} + \ldots$$

on summing the terms in brackets and which is clearly divergent.

Note that the harmonic series is divergent despite the fact that its nth term tends to zero as $n \to \infty$.

(b) d'Alembert's ratio test

Suppose we have a series of positive terms, $\sum_{k=0}^{\infty} u_k$, and also $\lim\limits_{n \to \infty} \dfrac{u_{n+1}}{u_n} = l$ exists.

Then the series is convergent if $l < 1$ and divergent if $l > 1$. If $l = 1$, we are not able to decide, using this test, whether the series converges or diverges.

The proof of this result is straightforward. Assume that $\lim\limits_{n \to \infty} \dfrac{u_{n+1}}{u_n} = l < 1$ and choose r to be any number between l and 1. Then since the values of u_{n+1}/u_n, when n is sufficiently large, differ from l by as little as we please, we have

$$\frac{u_{n+1}}{u_n} < r$$

for $n \geqslant N$. Thus

$$u_{N+1} < ru_N, \quad u_{N+2} < r^2 u_N, \ldots$$

Thus, from and after the term u_N of the series, the terms do not exceed those of the convergent geometric series

$$u_N(1 + r + r^2 + r^3 + \ldots)$$

Hence $\sum_{k=0}^{\infty} u_k$ converges.

It is left as an exercise for the reader to show that the series diverges when $l > 1$.

Example 7.22   Use d'Alembert's test to determine whether the following series are convergent.

(a) $\displaystyle\sum_{k=0}^{\infty} \frac{2^k}{k!}$ (b) $\displaystyle\sum_{k=0}^{\infty} \frac{2^k}{(k+1)^2}$

Solution   (a) Let $u_k = \dfrac{2^k}{k!}$, then

$$\frac{u_{n+1}}{u_n} = \frac{2^{n+1}}{(n+1)!} \Bigg/ \frac{2^n}{n!} = \frac{2}{n+1}$$

which tends to zero as $n \to \infty$. Thus $l = 0$ and the series is convergent.

(b) Here

$$l = \lim_{n \to \infty}\left[\frac{2^{n+1}}{(n+2)^2} \Bigg/ \frac{2^n}{(n+1)^2}\right] = 2$$

so that the series diverges.

A necessary condition for convergence of all series is that the terms of the series must tend to zero as $n \to \infty$. Thus a simple test for divergence is:

If $u_n \to u \neq 0$ as $n \to \infty$, then $\sum_{k=0}^{\infty} u_k$ is divergent

Notice, however, that $u_n \to 0$ as $n \to \infty$ does not guarantee that $\sum_{k=0}^{\infty} u_k$ is convergent. To prove that, we need more information. (Recall, for example, the harmonic series $1 + \frac{1}{2} + \frac{1}{3} + \frac{1}{4} + \ldots$, of Example 7.21, which is divergent.)

| **Example 7.23** | Show that the series $\frac{1}{2} + \frac{2}{3} + \frac{3}{4} + \dots$ is divergent. |

Solution Here $u_k = \dfrac{k}{k+1}$, so that d'Alembert's ratio test does not give a conclusion (since $l = 1$). However, we note that $u_n = 1 - \dfrac{1}{n+1}$, so that $u_n \to 1$ as $n \to \infty$, from which we conclude that $\sum_{k=1}^{\infty} u_k$ diverges.

7.6.3 The absolute convergence of general series

In practical problems, we are concerned with series which may have both positive and negative terms. **Absolutely convergent** series are a special case of such series. Consider the general series

$$S = \sum_{k=0}^{\infty} u_k$$

which may have both positive and negative terms u_k. If the associated series

$$T = \sum_{k=0}^{\infty} |u_k|$$

is convergent then S is convergent and is said to be **absolutely convergent**. If it is impossible to obtain a value for the limit of the partial sum T_n, we must use some other test to determine the convergence (or divergence) of T. A simple test for absolute convergence of a series $\sum_{k=1}^{\infty} u_k$ is a natural extension of d'Alembert's ratio test.

If $\displaystyle\lim_{n \to \infty} \left| \frac{u_{n+1}}{u_n} \right| < 1$ then $\displaystyle\sum_{k=0}^{\infty} u_k$ is absolutely convergent

If $\displaystyle\lim_{n \to \infty} \left| \frac{u_{n+1}}{u_n} \right| > 1$ then $\displaystyle\sum_{k=0}^{\infty} u_k$ is divergent

If $\displaystyle\lim_{n \to \infty} \left| \frac{u_{n+1}}{u_n} \right| = 1$ then no conclusion is possible

Absolutely convergent series have the following useful properties:

(a) the insertion of brackets into the series does not alter its sum;

(b) the rearrangement of the series does not alter its sum;

(c) the product of two absolutely convergent series $A = \sum a_n$ and $B = \sum b_n$ is an absolutely convergent series C, where

$$C = a_1 b_1 + (a_2 b_1 + a_1 b_2) + (a_3 b_1 + a_2 b_2 + a_1 b_3) + (a_4 b_1 + a_3 b_2 + a_2 b_3 + a_1 b_4) + \dots$$

There are convergent series that are not absolutely convergent; that is $\sum_{k=1}^{\infty} u_k$ converges but $\sum_{k=0}^{\infty} |u_k|$ diverges. The most common series of this type are alternating series. Here the u_k alternate in sign. If, in addition, the terms decrease in size and tend to zero,

$$|u_n| < |u_{n-1}| \quad \text{for all } n, \quad \text{with } u_n \to 0 \quad \text{as} \quad n \to \infty$$

then the series converges. Thus

$$\sum_{k=1}^{\infty} (-1)^{k+1} \frac{1}{k} = 1 - \tfrac{1}{2} + \tfrac{1}{3} - \tfrac{1}{4} + \tfrac{1}{5} - \tfrac{1}{6} + \dots$$

which we write as

$$\sum_{k=1}^{\infty} (-1)^{k+1} \frac{1}{k} = 1 + (-\tfrac{1}{2}) + (\tfrac{1}{3}) + (-\tfrac{1}{4}) + (\tfrac{1}{5}) + (-\tfrac{1}{6}) + \dots$$

converges. Its sum is $\ln 2$, as we shall show in Section 7.7. The associated series of positive terms, $\sum_{k=1}^{\infty}(1/k)$, diverges of course (see Example 7.21b).

7.6.4 Exercises

37 Decide which of the following geometric series are convergent.

(a) $2 + \tfrac{2}{3} + \tfrac{2}{9} + \tfrac{2}{27} + \dots + \dfrac{2}{3^k} + \dots$

(b) $4 - 2 + 1 - \tfrac{1}{2} + \dots + \dfrac{(-1)^k 4}{2^k} + \dots$

(c) $10 + 11 + \tfrac{121}{10} + \tfrac{1331}{100} + \dots + 10(\tfrac{11}{10})^k + \dots$

(d) $1 - \tfrac{5}{4} + \tfrac{25}{16} - \tfrac{125}{64} + \dots + (\tfrac{-5}{4})^k + \dots$

38 Show that if

$$T_n = a + 2ar + 3ar^2 + 4ar^3 + \dots + nar^{n-1}$$

then $(1 - r)T_n = a + ar + ar^2 + \dots + ar^{n-1} - nar^n$
Deduce that

$$T_n = \frac{a(1 - r^n)}{(1 - r)^2} - \frac{nar^n}{1 - r}$$

Show that if $|r| < 1$, then $T_n \to a/(1 - r)^2$ as $n \to \infty$. Hence sum the infinite series

$$1 + \tfrac{2}{3} + \tfrac{1}{3} + \tfrac{4}{27} + \tfrac{5}{81} + \dots + \dfrac{k}{3^{k-1}} + \dots$$

39 For each of the following series find the sum of the first N terms, and, by letting $N \to \infty$, show that the infinite series converges and state its sum.

(a) $\tfrac{2}{1 \cdot 3} + \tfrac{2}{3 \cdot 5} + \tfrac{2}{5 \cdot 7} + \dots$

(b) $\tfrac{1}{1} + \tfrac{2}{2} + \tfrac{3}{2^2} + \tfrac{4}{2^3} + \tfrac{5}{2^4} + \dots$

(c) $\tfrac{1}{1 \cdot 2 \cdot 3} + \tfrac{1}{2 \cdot 3 \cdot 4} + \tfrac{1}{3 \cdot 4 \cdot 5} + \dots$

40 Which of the following series are convergent?

(a) $\displaystyle\sum_{k=1}^{\infty} (-1)^k$ (b) $-\tfrac{2}{3} + \tfrac{3}{4} - \tfrac{4}{5} + \dots$

(c) $\displaystyle\sum_{k=0}^{\infty} \frac{1}{3^k + 1}$

41 By comparison with the series $\sum_{k=2}^{\infty}[1/k(k-1)]$ and $\sum_{k=2}^{\infty}[1/k(k+1)]$, show that $S = \sum_{k=2}^{\infty}(1/k^2)$ is convergent and $\tfrac{1}{2} < S < 1$. (In fact, $\sum_{k=1}^{\infty}(1/k^2) = S + 1 = \tfrac{1}{6}\pi^2$.)

42 Show that $0.\dot{5}\dot{7}$ (that is, $0.575\,757 \dots$) may be expressed as $57 \times 10^{-2} + 57 \times 10^{-4} + \dots$, and so $0.\dot{5}\dot{7} = 57\sum_{r=1}^{\infty}100^{-r}$. Hence express $0.\dot{5}\dot{7}$ as a rational number. Use a similar method to express as rational numbers:

(a) $0.\dot{4}1\dot{3}$ (b) $0.101\,010 \dots$

(c) $0.999\,999 \dots$ (d) $17.231\,723\,172\,3 \dots$

43 Consider the series $\sum_{r=1}^{\infty} k^{-p}$. By means of the inequalities $(p > 0)$

$$\frac{1}{2^p} + \frac{1}{3^p} < \frac{2}{2^p}$$

$$\frac{1}{4^p} + \frac{1}{5^p} + \frac{1}{6^p} + \frac{1}{7^p} < \frac{4}{4^p}$$

$$\frac{1}{8^p} + \frac{1}{9^p} + \frac{1}{10^p} + \frac{1}{11^p} + \frac{1}{12^p} + \frac{1}{13^p}$$

$$+ \frac{1}{14^p} + \frac{1}{15^p} < \frac{8}{8^p}$$

and so on, deduce that the series is convergent for $p > 1$. Show that it is divergent for $p \leqslant 1$.

44 Two attempts to evaluate the sum $\sum_{k=1}^{\infty} k^{-4}$ are made on a computer working to 8 digits. The first evaluates the sum

$$1 + \frac{1}{2^4} + \frac{1}{3^4} + \frac{1}{4^4} + \ldots + \frac{1}{72^4}$$

from the left; the second evaluates it from the right. The first method yields the result 1.082 320 2, the second 1.082 322 1. Which is the better approximation and why?

45 Show that

$$\sum_{k=1}^{2n} \frac{(-1)^{k+1}}{k^4} = \sum_{k=1}^{2n} \frac{1}{k^4} - \frac{1}{8}\sum_{k=1}^{n} \frac{1}{k^4}$$

and deduce that

$$\sum_{k=1}^{\infty} \frac{(-1)^{k+1}}{k^4} = \frac{7}{8}\sum_{k=1}^{\infty} \frac{1}{k^4}$$

Deduce that the modulus of error in the estimate for the sum $\sum_{k=1}^{\infty} k^{-4}$ obtained by computing $\frac{8}{7}\sum_{k=1}^{\infty}(-1)^k k^{-4}$ is less than $\frac{8}{7}(N+1)^{-4}$.

7.7 Power series

Power series frequently occur in the solution of practical problems, as we shall see in Chapter 9, Sections 9.5.2, 9.9 and elsewhere. Often they are used to determine the sensitivity of systems to small changes in design parameters, to examine whether such systems are stable when small variations occur (as they always will in real life). The basic mathematics involved in power series is a natural extension of the series considered earlier.

A series of the type

$$a_0 + a_1 x + a_2 x^2 + a_3 x^3 + \ldots + a_n x^n + \ldots$$

where the $a_0, a_1, a_2, \ldots$ are independent of x is called a **power series**.

7.7.1 Convergence of power series

Power series will, in general, converge for certain values of x and diverge elsewhere. Applying d'Alembert's ratio test to the above series, we see that it is absolutely convergent when

$$\lim_{n \to \infty} \left| \frac{a_{n+1}x^{n+1}}{a_n x^n} \right| < 1$$

Thus the series converges if

$$|x| \lim_{n \to \infty} \left| \frac{a_{n+1}}{a_n} \right| < 1$$

that is, if

$$|x| < \lim_{n \to \infty} \left| \frac{a_n}{a_{n+1}} \right|$$

Denoting $\lim_{n \to \infty} |a_n/a_{n+1}|$ by r, we see that the series is absolutely convergent for $-r < x < r$ and divergent for $x < -r$ and $x > r$. The limit r is called the **radius of convergence** of the series. The behaviour at $x = \pm r$ has to be determined by other methods.

The various cases that occur are shown in Example 7.24.

Example 7.24   Find the radius of convergences of the series

(a) $\displaystyle\sum_{n=1}^{\infty} \frac{x^n}{n}$ (b) $\displaystyle\sum_{n=1}^{\infty} n^n x^n$

Solution   (a) Here $a_n = 1/n$, so that $|a_n/a_{n+1}| = (n+1)/n$ and $r = 1$. Thus the domain of absolute convergence of the series is $-1 < x < 1$. The series diverges for $|x| > 1$ and for $x = 1$. At $x = -1$ the series is

$$-1 + \tfrac{1}{2} - \tfrac{1}{3} + \tfrac{1}{4} + \ldots$$

which is convergent to $\ln \tfrac{1}{2}$ (see Section 7.6.3 and formula (7.14) below). Thus the series

$$\sum_{n=1}^{\infty} \frac{x^n}{n} = x + \tfrac{1}{2}x^2 + \tfrac{1}{3}x^3 + \tfrac{1}{4}x^4 + \ldots$$

is convergent for $-1 \leqslant x < 1$.

(b) Here $a_n = n^n$ and

$$\left| \frac{a_n}{a_{n+1}} \right| = \frac{n^n}{(n+1)^{n+1}} = \left(\frac{n}{n+1} \right)^n \frac{1}{n+1}$$

Now

$$\left(\frac{n}{n+1} \right)^n = \frac{1}{(1 + 1/n)^n} \to e^{-1} \quad \text{as } n \to \infty \text{ (see Example 7.19)}$$

and

$$\frac{1}{n+1} \to 0 \quad \text{as } n \to \infty$$

so that $a_n/a_{n+1} \to 0$ as $n \to \infty$. Thus the series converges only at $x = 0$, and diverges elsewhere.

7.7.2 Special power series

Power series may be added, multiplied and divided within their common domains of convergence (provided the denominator is non-zero within this common domain) to give power series that are convergent, and these properties are often exploited to express a given power series in terms of standard series and to obtain power series expansions of complicated functions.

Four elementary power series that are of widespread use are

(a) The geometric series

$$\frac{1}{1+x} = 1 - x + x^2 - x^3 + \ldots + (-1)^n x^n + \ldots \quad (-1 < x < 1) \tag{7.14}$$

(b) The binomial series

$$(1+x)^r = 1 + \binom{r}{1}x + \binom{r}{2}x^2 + \binom{r}{3}x^3 + \ldots + \binom{r}{n}x^n + \ldots \quad (-1 < x < 1) \tag{7.15}$$

where

$$\binom{r}{n} = \frac{r(r-1)\ldots(r-n+1)}{1\cdot 2\cdot 3\cdot\ldots\cdot n}$$

is the binomial coefficient.

In series (7.15) r is any real number. When r is a positive integer, N say, the series terminates at the term x^N and we have the binomial expansion discussed in Chapter 1. When r is not a positive integer, the series does not terminate.

We can see that setting $r = -1$ gives

$$(1+x)^{-1} = \frac{1}{1+x} = 1 + \frac{(-1)}{1}x + \frac{(-1)(-2)}{1\cdot 2}x^2 + \frac{(-1)(-2)(-3)}{1\cdot 2\cdot 3}x^3 + \ldots$$

which simplifies to the geometric series

$$\frac{1}{1+x} = 1 - x + x^2 - x^3 + \ldots$$

Similarly,

$$(1+x)^{-2} = 1 + \frac{(-2)}{1}x + \frac{(-2)(-3)}{1\cdot 2}x^2 + \frac{(-2)(-3)(-4)}{1\cdot 2\cdot 3}x^3 + \ldots$$

which simplifies to the arithmetical–geometric series

$$\frac{1}{(1+x)^2} = 1 - 2x + 3x^2 - 4x^3 + \ldots$$

(Compare Exercises 7.6.4, Question 38.) So the geometric series may be thought of as a special case of the binomial series.

(c) The exponential series

$$e^x = 1 + \frac{x}{1!} + \frac{x^2}{2!} + \frac{x^3}{3!} + \ldots + \frac{x^n}{n!} + \ldots \quad \text{(all } x\text{)} \tag{7.16}$$

We saw in Example 7.19 that the number e is defined by

$$e = \lim_{n \to \infty} \left(1 + \frac{1}{n}\right)^n$$

Similarly, the function e^x is defined by

$$e^x = \lim_{n \to \infty} \left(1 + \frac{x}{n}\right)^n$$

(or, equivalently, by $\lim_{n \to \infty} (1 + \frac{1}{n})^{nx}$).

Using the binomial expansion, we have

$$e^x = \lim_{n \to \infty} \left\{1 + \frac{n}{1}\left(\frac{x}{n}\right) + \frac{n(n-1)}{1 \cdot 2}\left(\frac{x}{n}\right)^2 + \frac{n(n-1)(n-2)}{1 \cdot 2 \cdot 3}\left(\frac{x}{n}\right)^3 + \ldots\right\}$$

$$= \lim_{n \to \infty} \left\{1 + \frac{1}{1}x + \frac{(1 - \frac{1}{n})}{1 \cdot 2}x^2 + \frac{(1 - \frac{1}{n})(1 - \frac{2}{n})}{1 \cdot 2 \cdot 3}x^3 + \ldots\right\}$$

$$= 1 + \frac{x}{1!} + \frac{x^2}{2!} + \frac{x^3}{3!} + \ldots$$

So we see the connection between the binomial and exponential series.

(d) The logarithmic series

$$\ln(1 + x) = x - \frac{x^2}{2} + \frac{x^3}{3} - \frac{x^4}{4} + \ldots + (-1)^n \frac{x^{n+1}}{n+1} + \ldots \quad (-1 < x \leqslant 1) \tag{7.17}$$

The logarithmic function is the inverse function of the exponential function, so that

$$y = \ln(1 + x)$$

implies $1 + x = e^y = \lim_{n \to \infty} \left(1 + \frac{y}{n}\right)^n$.

Unscrambling the limit to solve for y gives

$$y = \lim_{n \to \infty} \{n[(1 + x)^{1/n} - 1]\}$$

Using the binomial expansion again gives

$$y = \lim_{n \to \infty} \left\{n\left[\frac{\frac{1}{n}}{1}x + \frac{\frac{1}{n}(\frac{1}{n} - 1)}{1 \cdot 2}x^2 + \frac{\frac{1}{n}(\frac{1}{n} - 1)(\frac{1}{n} - 2)}{1 \cdot 2 \cdot 3}x^3 + \ldots\right]\right\}$$

$$= x - \frac{x^2}{2} + \frac{x^3}{3} - \frac{x^4}{4} + \ldots$$

Thus we see the connection between the binomial and logarithmic series. Note that taking $x = 1$ in series (7.17) we have the result

$$\sum_{k=1}^{\infty} \frac{1}{k}(-1)^{k+1} = \ln 2$$

used in Section 7.6.3.

A summary of the standard series introduced together with some other useful series deduced from them is given in Figure 7.13. Note that using the series expansions for e^x, $\sin x$ and $\cos x$ given in the figure we can demonstrate the validity of Euler's formula

$$e^{jx} = \cos x + j \sin x$$

introduced in equation (3.9). The radius of convergence of all these series may be determined using d'Alembert's test.

Figure 7.13
Table of some useful series.

$$\frac{1}{1+x} = 1 - x + x^2 - x^3 + \ldots + (-1)^n x^n + \ldots \quad (-1 < x < 1)$$

$$\frac{1}{1-x} = 1 + x + x^2 + x^3 + \ldots + x^n + \ldots \quad (-1 < x < 1)$$

$$(1+x)^r = 1 + \binom{r}{1}x + \binom{r}{2}x^2 + \binom{r}{3}x^3 + \ldots + \binom{r}{n}x^n + \ldots \quad (-1 < x < 1)$$

$$\ln(1+x) = x - \frac{x^2}{2} + \frac{x^3}{3} - \frac{x^4}{4} + \ldots + (-1)^n \frac{x^{n+1}}{n+1} + \ldots \quad (-1 < x \leq 1)$$

$$-\ln(1-x) = x + \frac{x^2}{2} + \frac{x^3}{3} + \frac{x^4}{4} + \ldots + \frac{x^{n+1}}{n+1} + \ldots \quad (-1 \leq x < 1)$$

$$\ln\frac{1+x}{1-x} = 2\left(x + \frac{x^3}{3} + \frac{x^5}{5} + \ldots + \frac{x^{2n+1}}{2n+1} \ldots\right) \quad (-1 < x < 1)$$

$$e^x = 1 + \frac{x}{1!} + \frac{x^2}{2!} + \frac{x^3}{3!} + \ldots + \frac{x^n}{n!} + \ldots \quad (\text{all } x)$$

$$e^{-x} = 1 - \frac{x}{1!} + \frac{x^2}{2!} - \frac{x^3}{3!} + \frac{x^4}{4!} + \ldots + (-1)^n \frac{x^n}{n!} + \ldots \quad (\text{all } x)$$

$$\cosh x = 1 + \frac{x^2}{2!} + \frac{x^4}{4!} + \frac{x^6}{6!} + \ldots + \frac{x^{2n}}{(2n)!} + \ldots \quad (\text{all } x)$$

$$\sinh x = x + \frac{x^3}{3!} + \frac{x^5}{5!} + \frac{x^7}{7!} + \ldots + \frac{x^{2n+1}}{(2n+1)!} + \ldots \quad (\text{all } x)$$

$$\cos x = 1 - \frac{x^2}{2!} + \frac{x^4}{4!} - \frac{x^6}{6!} + \ldots + (-1)^n \frac{x^{2n}}{(2n)!} + \ldots \quad (\text{all } x)$$

$$\sin x = x - \frac{x^3}{3!} + \frac{x^5}{5!} - \frac{x^7}{7!} + \ldots + (-1)^n \frac{x^{2n+1}}{(2n+1)!} + \ldots \quad (\text{all } x)$$

(*Note*: in the last two series x is an angle measured in radians.)

Example 7.25   Obtain the power series expansions of

(a) $\dfrac{1}{\sqrt{(1-x^2)}}$ (b) $\dfrac{1}{(1-x)(1+3x)}$ (c) $\dfrac{\ln(1+x)}{1+x}$

Solution   (a) Using the binomial series (7.15) with $n = -\tfrac{1}{2}$ gives

$$\frac{1}{\sqrt{(1+x)}} = (1+x)^{-1/2}$$

$$= 1 + \frac{(-\tfrac{1}{2})}{1!}x + \frac{(-\tfrac{1}{2})(-\tfrac{3}{2})}{2!}x^2 + \frac{(-\tfrac{1}{2})(-\tfrac{3}{2})(-\tfrac{5}{2})}{3!}x^3 + \ldots \quad (-1 < x < 1)$$

Now replacing x with $-x^2$ gives the required result

$$\frac{1}{\sqrt{(1-x^2)}} = 1 + \frac{(\tfrac{1}{2})}{1!}x^2 + \frac{(\tfrac{1}{2})(\tfrac{3}{2})}{2!}x^4 + \frac{(\tfrac{1}{2})(\tfrac{3}{2})(\tfrac{5}{2})}{3!}x^6 + \ldots \quad (-1 < x < 1)$$

$$= 1 + \frac{1}{2}x^2 + \frac{1\cdot 3}{2\cdot 4}x^4 + \frac{1\cdot 3\cdot 5}{2\cdot 4\cdot 6}x^6 + \ldots \quad (-1 < x < 1)$$

(b) Expressed in partial fractions

$$\frac{1}{(1-x)(1+3x)} = \frac{\tfrac{1}{4}}{1-x} + \frac{\tfrac{3}{4}}{1+3x}$$

From the table of Figure 7.13

$$\frac{1}{1-x} = 1 + x + x^2 + x^3 + \ldots + x^n + \ldots \quad (-1 < x < 1)$$

and replacing x by $3x$ in (7.14) gives

$$\frac{1}{1+3x} = 1 - (3x) + (3x)^2 - (3x)^3 + \ldots + (-1)^n(3x)^n + \ldots \quad (-\tfrac{1}{3} < x < \tfrac{1}{3})$$

Thus

$$\frac{1}{(1-x)(1+3x)}$$

$$= \tfrac{1}{4}[1 + x + x^2 + x^3 + \ldots] + \tfrac{3}{4}[1 - 3x + 9x^2 - 27x^3 + \ldots] \quad (-\tfrac{1}{3} < x < \tfrac{1}{3})$$

$$= 1 - 2x + 7x^2 - 20x^3 + \ldots + \tfrac{1}{4}(1 + (-1)^n 3^{n+1})x^n + \ldots \quad (-\tfrac{1}{3} < x < \tfrac{1}{3})$$

(c) Using the series for $\ln(1+x)$ and $(1+x)^{-1}$ from (7.17) and (7.14),

$$\frac{\ln(1+x)}{1+x} = (x - \tfrac{1}{2}x^2 + \tfrac{1}{3}x^3 - \tfrac{1}{4}x^4 + \ldots)(1 - x + x^2 - x^3 + \ldots)$$

$$= x - (1 + \tfrac{1}{2})x^2 + (1 + \tfrac{1}{2} + \tfrac{1}{3})x^3 - (1 + \tfrac{1}{2} + \tfrac{1}{3} + \tfrac{1}{4})x^4 + \ldots$$

$$(-1 < x < 1)$$

The inverse process of expressing the sum of a power series in terms of the elementary functions is often difficult or impossible, but when it can be achieved it usually results in dramatic simplification of a practical problem.

Example 7.26   Sum the series

(a) $1 + 2x + 3x^2 + 4x^3 + 5x^4 + \ldots$ (b) $1 + \dfrac{x}{2!} + \dfrac{x^2}{4!} + \dfrac{x^3}{6!} + \dfrac{x^4}{8!} + \ldots$

Solution   (a) Set

$$S = 1 + 2x + 3x^2 + 4x^3 + 5x^4 + \ldots$$

Then

$$xS = x + 2x^2 + 3x^3 + 4x^4 + \ldots$$

and subtracting this from S gives

$$(1 - x)S = 1 + x + x^2 + x^3 + \ldots$$

which is a geometric series. Thus

$$(1 - x)S = \frac{1}{1 - x} \quad (-1 < x < 1)$$

giving

$$S = \frac{1}{(1 - x)^2} \quad (-1 < x < 1)$$

(b) Summing this series relies on recognizing its similarity to the series for the hyperbolic cosine:

$$\cosh x = 1 + \frac{x^2}{2!} + \frac{x^4}{4!} + \frac{x^6}{6!} + \ldots \quad (-\infty < x < \infty)$$

Replacing x by $\sqrt{x}$ gives

$$\cosh \sqrt{x} = 1 + \frac{x}{2!} + \frac{x^2}{4!} + \frac{x^3}{6!} + \ldots \quad (-\infty < x < \infty)$$

and thus the series is summed.

7.7.3 Exercises

46 For what values of x are the following series convergent?

(a) $\displaystyle\sum_{n=1}^{\infty} nx^n$

(b) $\displaystyle\sum_{n=0}^{\infty} (-1)^n \frac{x^{2n}}{(2n+1)!}$

(c) $\displaystyle\sum_{n=1}^{\infty} \frac{x^n}{n(n+1)}$

(d) $\displaystyle\sum_{n=1}^{\infty} n^2 x^n$

47 From known series deduce the following:

(a) $\dfrac{1}{1+x^2} = 1 - x^2 + x^4 - x^6 + \ldots$

(b) $\frac{1}{2}\ln\dfrac{1+x}{1-x} = x + \frac{1}{3}x^3 + \frac{1}{5}x^5 + \frac{1}{7}x^7 + \ldots$

(c) $\dfrac{1}{(1+x)^2} = 1 - 2x + 3x^2 - 4x^3 + 5x^4 - \ldots$

(d) $\sqrt{(1-x)} = 1 - \frac{1}{2}x - \frac{1}{8}x^2 + \frac{1}{16}x^3 - \frac{5}{128}x^4 - \ldots$

(e) $\dfrac{1}{(1-2x)(2+x)} = \frac{1}{2} + \frac{3}{4}x + \frac{13}{8}x^2 + \frac{51}{16}x^3 + \ldots$

(f) $\dfrac{1}{(1-x)(1+x^2)} = 1 + x + x^4 + x^5 + \ldots$

In each case give the general term and the radius of convergence.

48 Calculate the binomial coefficients

(a) $\dbinom{5}{2}$ (b) $\dbinom{-2}{3}$

(c) $\dbinom{1/2}{3}$ (d) $\dbinom{-1/2}{4}$

49 From known series deduce the following (the general term is not required):

(a) $\tan x = x + \frac{1}{3}x^3 + \frac{2}{15}x^5 + \ldots$

(b) $\cos^2 x = 1 - \dfrac{2x^2}{2!} + \dfrac{2^3 x^4}{4!} - \dfrac{2^5 x^6}{6!} + \ldots$

(c) $e^x \cos x = 1 + x - \dfrac{2x^3}{3!} - \dfrac{2^2 x^4}{4!} - \dfrac{2^2 x^5}{5!} + \ldots$

(d) $\ln(1+\sin x) = x - \frac{1}{2}x^2 + \frac{1}{6}x^3 + \frac{1}{12}x^4 + \ldots$

50 Show that

$$\frac{1}{1-x} = 1 + x + x^2 + \ldots + x^{n-1} + \frac{x^n}{1-x} \quad (x \neq 1)$$

Hence derive a polynomial approximation to $(1-x)^{-1}$ with an error that, in modulus, is less than 0.5×10^{-4} for $0 \leqslant x \leqslant 0.25$.

Using nested multiplication, calculate from your approximation the reciprocal of 0.84 to 4dp, and compare your answer with the value given by your calculator. How many multiplications are needed in this case?

51 Find the sums of the following power series:

(a) $\displaystyle\sum_{k=0}^{\infty} (-1)^k 2^k x^{2k}$

(b) $1 + \frac{1}{2}x + \dfrac{1 \cdot 3}{2 \cdot 4}x^2 + \dfrac{1 \cdot 3 \cdot 5}{2 \cdot 4 \cdot 6}x^3 + \dfrac{1 \cdot 3 \cdot 5 \cdot 7}{2 \cdot 4 \cdot 6 \cdot 8}x^4 + \ldots$

(c) $\displaystyle\sum_{k=1}^{\infty} \frac{x^k}{k(k+1)}$

(d) $\frac{1}{2}x^2 + \frac{2}{3}x^3 + \frac{3}{4}x^4 + \frac{4}{5}x^5 + \ldots$

52 A regular polygon of n sides is inscribed in a circle of unit diameter. Show that its perimeter p_n is given by

$$p_n = n \sin \frac{\pi}{n}$$

Using the series expansion for sine, prove that

$$\pi = p_n + \frac{\pi^3}{3!}\frac{1}{n^2} - \frac{\pi^5}{5!}\frac{1}{n^4} + \ldots$$

and deduce that

$$\pi = \frac{1}{3}(4p_{2n} - p_n) + \frac{1}{4}\frac{\pi^5}{5!}\frac{1}{n^4} + \ldots$$

Given $p_{12} = 3.1058$ and $p_{24} = 3.1326$, use this result to obtain a better estimate of π.

7.8   Functions of a real variable

So far in this chapter we have concentrated on sequences and series. The terms of a sequence may be seen as defining a function whose domain is a subset of integers, such as N. We now turn to the fundamental properties that are essential to mathematical modelling and problem-solving, but we shall also be developing some basic mathematics that is necessary for later chapters.

7.8.1   Limit of a function of a real variable

The notion of limit can be extended in a natural way to include functions of a real variable:

> A function $f(x)$ is said to approach a limit l as x approaches the value a if, given any small positive quantity ε, it is possible to find a positive number δ such that $|f(x) - l| < \varepsilon$ for all x satisfying $0 < |x - a| < \delta$.

Less formally, this means that we can make the value of $f(x)$ as close as we please to l by taking x sufficiently close to a. Note that, using the formal definition, there is no need to evaluate $f(a)$; indeed, $f(a)$ may or may not equal l. The limiting value of f as $x \to a$ depends only on nearby values!

Example 7.27   Using a calculator, examine the values of $f(x)$ near $x = 0$ where

$$f(x) = \frac{x}{1 - \sqrt{(1 + x)}}, \quad x \neq 0$$

What is the value of $\lim_{x \to 0} f(x)$?

Solution   Note that $f(x)$ is not defined where $x = 0$. At nearby values of x we can calculate $f(x)$, and some values are shown in Figure 7.14.

Figure 7.14
Values of $f(x)$ to 6dp.

x	−0.1	−0.01	−0.001	0.001	0.01	0.1
$f(x)$	−1.948 683	−1.994 987	−1.999 500	−2.000 500	−2.004 988	−2.048 809

It seems that as x gets close to the value of 0, $f(x)$ gets close to the value of −2. Indeed, it can be proved that for $0 < |x| < 2\varepsilon - \varepsilon^2$, $|f(x) + 2| < \varepsilon$, so that

$$\lim_{x \to 0} f(x) = -2.$$

The elementary rules for limits (listed in Section 7.5.2) carry over from those of sequences, and these enable us to evaluate many limits by reduction to standard cases. Some common standard limits are

(i) $\displaystyle\lim_{x \to a} \frac{x^r - a^r}{x - a} = ra^{r-1}$, where r is a real number

(ii) $\displaystyle\lim_{x \to 0} \frac{\sin x}{x} = 1$, where x is in radians

(iii) $\displaystyle\lim_{h \to 0} (1 + xh)^{1/h} = e^x$

These results can be deduced from the results of Section 7.7.2. For instance, consider $x^r - a^r$. Since $x \to a$, set $x = a + h$. Then as $x \to a$, $h \to 0$. We have

$$x^r - a^r = a^r\left(1 + \frac{h}{a}\right)^r - a^r \quad (a \neq 0)$$

Expanding $(1 + h/a)^r$ by the binomial series (7.15), we have

$$x^r - a^r = \frac{r}{1!}ha^{r-1} + \frac{r(r-1)}{2!}h^2a^{r-2} + \frac{r(r-1)(r-2)}{3!}h^3a^{r-3} + \dots$$

But $x - a = h$, so

$$\frac{x^r - a^r}{x - a} = ra^{r-1} + \frac{r(r-1)}{2!}ha^{r-2} + \dots$$

and letting $h \to 0$ yields the result (i)

$$\lim_{x \to a} \frac{x^r - a^r}{x - a} = ra^{r-1}$$

(When $a = 0$, the result is obtained trivially.)

The result (ii) is obtained even more simply. The series expansion

$$\sin x = x - \frac{x^3}{3!} + \frac{x^5}{5!} - \dots$$

gives

$$\lim_{x \to 0} \frac{\sin x}{x} = 1$$

A geometric interpretation of (ii) is given in Figure 7.15. OAB is a sector of a circle of unit radius with angle x (measured in radians). Then

the area of $\triangle OBD <$ area of sector OBA $<$ area of $\triangle OCA$

Algebraically, we have

$$\tfrac{1}{2}\sin x \cos x < \tfrac{1}{2}x < \tfrac{1}{2}\tan x$$

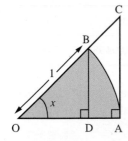

Figure 7.15
Geometric
interpretation of
$\displaystyle\lim_{x \to 0} \frac{\sin x}{x} = 1$.

Considering $x > 0$, we may write this as

$$1 < \frac{\sin x}{x} < \frac{1}{\cos x}$$

As $x \to 0$, $\cos x \to 1$, so that $\dfrac{\sin x}{x} \to 1$ also.

The result (iii) is obtained from a binomial series:

$$(1 + xh)^{1/h} = 1 + \frac{1}{1!}\frac{1}{h}xh + \frac{1}{2!}\frac{1}{h}\left(\frac{1}{h} - 1\right)(xh)^2 + \frac{1}{3!}\frac{1}{h}\left(\frac{1}{h} - 1\right)\left(\frac{1}{h} - 2\right)(xh)^3 + \ldots$$

$$= 1 + \frac{x}{1!} + \frac{x^2(1 - h)}{2!} + \frac{x^3(1 - h)(1 - 2h)}{3!} + \ldots$$

and, as $h \to 0$,

$$(1 + xh)^{1/h} \to 1 + \frac{x}{1!} + \frac{x^2}{2!} + \frac{x^3}{3!} + \ldots = e^x$$

Example 7.28   Evaluate the following limits:

(a) $\displaystyle\lim_{x \to 0} \frac{\sqrt{(1 + x^2)} - 1}{x^2}$ (b) $\displaystyle\lim_{x \to 0} \frac{1 - \cos x}{x^2}$

Solution   (a) Method 1: Expand $\sqrt{(1 + x^2)}$ by the binomial series (7.15), giving

$$\sqrt{(1 + x^2)} = (1 + x^2)^{1/2} = 1 + \tfrac{1}{2}x^2 - \tfrac{1}{8}x^4 + \ldots$$

so that

$$\frac{\sqrt{(1 + x^2)} - 1}{x^2} = \frac{\tfrac{1}{2}x^2 - \tfrac{1}{8}x^4 + \ldots}{x^2} = \tfrac{1}{2} - \tfrac{1}{8}x^2 + \ldots$$

Thus

$$\lim_{x \to 0} \frac{\sqrt{(1 + x^2)} - 1}{x^2} = \tfrac{1}{2}$$

Method 2: Multiply numerator and denominator by $\sqrt{(1 + x^2)} + 1$, giving

$$\frac{[\sqrt{(1 + x^2)} - 1][\sqrt{(1 + x^2)} + 1]}{x^2[\sqrt{(1 + x^2)} + 1]} = \frac{(1 + x^2) - 1}{x^2[\sqrt{(1 + x^2)} + 1]} = \frac{1}{\sqrt{(1 + x^2)} + 1}$$

Now let $x \to 0$, to obtain

$$\lim_{x \to 0} \frac{\sqrt{(1 + x^2)} - 1}{x^2} = \tfrac{1}{2}$$

(b) Method 1: Replace $\cos x$ by its power series expansion,

$$\cos x = 1 - \frac{x^2}{2!} + \frac{x^4}{4!} - \frac{x^6}{6!} - \ldots$$

giving

$$\frac{1 - \cos x}{x^2} = \frac{x^2/2! - x^4/4! + x^6/6! - \ldots}{x^2} = \frac{1}{2!} - \frac{x^2}{4!} + \frac{x^4}{6!} - \ldots$$

Thus

$$\lim_{x \to 0} \frac{1 - \cos x}{x^2} = \tfrac{1}{2}$$

Method 2: Using the half-angle formula for $\cos x$, we have

$$1 - \cos x = 2 \sin^2 \tfrac{1}{2} x$$

so

$$\frac{1 - \cos x}{x^2} = \frac{2 \sin^2 \tfrac{1}{2} x}{x^2} = 2\left(\frac{\sin \tfrac{1}{2} x}{x}\right)^2 = \frac{1}{2}\left(\frac{\sin \varphi}{\varphi}\right)^2 \quad \text{where } \varphi = \tfrac{1}{2} x$$

On letting $x \to 0$, we have $\varphi \to 0$ and $(\sin \varphi)/\varphi \to 1$, so that

$$\frac{1 - \cos x}{x^2} \to \tfrac{1}{2}$$

7.8.2 One-sided limits

In some applications we have to use one-sided limits, for example

$$\lim_{x \to 0+} \sqrt{x} = 0 \quad \text{(as } x \text{ tends to zero 'from above')}$$

In this example, $\lim_{x \to 0-} \sqrt{x}$ (as x tends to zero 'from below') does not exist, since no negative numbers are in the domain of $\sqrt{x}$. When we write

$$\lim_{x \to a} f(x) = l$$

we mean that

$$\lim_{x \to a-} f(x) = \lim_{x \to a+} f(x) = l$$

Example 7.29   Sketch the graph of the function $f(x)$ where

$$f(x) = \frac{\sqrt{(x^2 - x^3)}}{x}, \quad x \neq 0 \text{ and } x < 1$$

and show that $\lim_{x \to 0} f(x)$ does not exist.

Figure 7.16
Graph of
$$y = \frac{\sqrt{(x^2 - x^3)}}{x}.$$

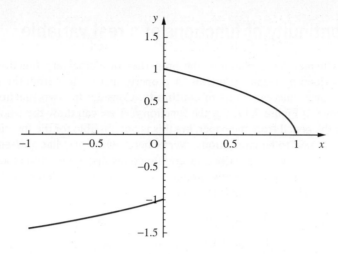

Solution   Notice that the function is not defined for $x = 0$. A sketch of the function is given for $-1 \leq x \leq 1$, $x \neq 0$ in Figure 7.16. From that diagram we see that $f(x) \to -1$ as $x \to 0$ from below and $f(x) \to +1$ as $x \to 0$ from above. Since the existence of a limit requires the same value whether we approach from above or below we deduce that $\lim_{x \to 0} f(x)$ does not exist.

7.8.3   Exercises

53   Evaluate the following limits:

(a) $\displaystyle\lim_{x \to 0} \frac{\sqrt{(1 + x)} - \sqrt{(1 - x)}}{3x}$

(b) $\displaystyle\lim_{x \to 0} \frac{\cos x \sin x - x}{x^3}$

(c) $\displaystyle\lim_{x \to 0} \frac{\sin^{-1} 2x}{x}$

(d) $\displaystyle\lim_{x \to \pi/2} (\sec x - \tan x)$

54   Show that

$$\lim_{x \to \infty} f(x) = \lim_{y \to 0+} f\left(\frac{1}{y}\right)$$

Hence find

(a) $\displaystyle\lim_{x \to \infty} \frac{3x^2 - x - 2}{x^2 - 1}$

(b) $\displaystyle\lim_{x \to \infty} x(\sqrt{(1 + x^2)} - x)$

55   Evaluate the following limits:

(a) $\displaystyle\lim_{x \to 0-} \tanh \frac{1}{x}$ (b) $\displaystyle\lim_{x \to 0+} \tanh \frac{1}{x}$

(c) $\displaystyle\lim_{x \to n-} \lfloor x \rfloor$ $(n \in \mathbb{Z})$

(d) $\displaystyle\lim_{x \to n+} \lfloor x \rfloor$ $(n \in \mathbb{Z})$

56   Draw (carefully) graphs of

(a) xe^{-x} (b) $x^2 e^{-x}$ (c) $x^3 e^{-x}$

for $0 \leq x \leq 5$. Use the series expansion of e^x to prove that $x^n e^{-x} \to 0$ as $x \to \infty$ for all $n \in \mathbb{Z}$.

57   Use a calculator to evaluate the function $f(x) = x^x$

for $x = 1, 0.1, 0.01, \ldots, 0.000\,000\,001$. What do these calculations suggest about $\lim_{x \to 0+} f(x)$?

Since $x^x = e^{x \ln x}$, the value of this limit is related to $\lim_{x \to 0+} (x \ln x)$. By setting $x = e^{-y}$ and using the results of Question 56, prove that $x^x \to 1$ as $x \to 0+$.

7.9 Continuity of functions of a real variable

In Chapter 2 we examined the properties of elementary functions. Often these were described by means of graphs. A property that is clear from the graphical representation of a function is that of **continuity**. Consider the two functions whose graphs are shown in Figure 7.17. For the function $f(x)$ we can draw the whole curve without lifting the pencil from the paper, but this is not possible for the function $g(x)$. The function $f(x)$ is said to be **continuous everywhere**, while $g(x)$ has a **discontinuity** at $x = 0$. In Section 2.8.3, we described several functions that are used to model practical problems and that have points of discontinuity similar to the function $g(x)$. The most important of these is Heaviside's unit function

$$H(x) = \begin{cases} 0 & (x < 0) \\ 1 & (x \geq 0) \end{cases}$$

which has a discontinuity at $x = 0$.

Figure 7.17
Graphs of
the functions
(a) $f(x) = x(x^2 - 1)$
and (b) $g(x) =$
$\tan^{-1}(1/x)$, $x \neq 0$.

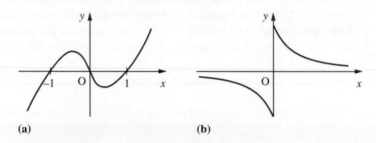

(a)   (b)

The formal mathematical definition of continuity for a function $f(x)$ defined in the neighbourhood of a point $x = x_0$ and at the point itself is that

$$f(x) \to f(x_0) \quad \text{as } x \to x_0$$

A function with this property is said to be **continuous at** $x = x_0$.

Continuous functions have some very special properties, which we shall now list.

7.9.1 Properties of continuous functions

If $f(x)$ is continuous in the interval $[a, b]$ then it has the following properties.

(a) $f(x)$ is a bounded function: there are numbers m and M such that

$$m < f(x) < M \quad \text{for all } x \in [a, b]$$

Any numbers satisfying this relation are called a **lower bound** and an **upper bound** respectively.

(b) $f(x)$ has a largest and a least value on $[a, b]$. The least value of $f(x)$ on $[a, b]$ is called the **minimum** of $f(x)$ on $[a, b]$, the largest value is the **maximum** of $f(x)$ on $[a, b]$ and the difference between the two is called the **oscillation** of $f(x)$ on $[a, b]$. This is illustrated in Figure 7.18.

(c) $f(x)$ takes every value between its least and its largest value somewhere between $x = a$ and $x = b$. This property is known as the **intermediate value theorem**.

Figure 7.18
The oscillation
of a function.

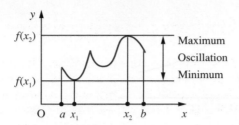

(d) If $a \leqslant x_1 \leqslant x_2 \leqslant x_3 \leqslant \ldots \leqslant x_n < b$, there is an $X \in [a, b]$ such that

$$f(X) = \frac{f(x_1) + f(x_2) + \ldots + f(x_n)}{n}$$

This property is known as the **average value theorem**.

(e) Given $\varepsilon > 0$, the interval $[a, b]$ can be divided into a number of intervals in each of which the oscillation of the function is less than ε.

(f) Given $\varepsilon > 0$, there is a subdivision of $[a, b]$, $a = x_0 < x_1 < x_2 < \ldots < x_n = b$, such that in each subinterval (x_i, x_{i+1})

$$\left| f(x) - \left[f_i + (x - x_i) \frac{f_{i+1} - f_i}{x_{i+1} - x_i} \right] \right| < \varepsilon, \, f_i = f(x_i)$$

That is, by making a subtabulation that is sufficiently fine, we can represent $f(x)$ locally by linear interpolation to within any prescribed error bound.

(g) Given $\varepsilon > 0$, $f(x)$ can be approximated on the interval $[a, b]$ by a polynomial of suitable degree such that

$$|f(x) - p_n(x)| < \varepsilon \quad \text{for } x \in [a, b]$$

This is known as **Weierstrass' theorem**. Note, however, that the theorem does not tell us how to obtain $p_n(x)$.

The properties of limits listed in Section 7.5.2 enable us to determine the continuity of functions formed by combining continuous functions. Thus if $f(x)$ and $g(x)$ are continuous functions then so are the functions

(a) $af(x)$, where a is a constant
(b) $f(x) + g(x)$
(c) $f(x)g(x)$
(d) $f(x)/g(x)$, except where $g(x) = 0$

Also the composite function $f(g(x))$ is continuous at x_0 if $g(x)$ is continuous at x_0 and $f(x)$ is continuous at $x = g(x_0)$.

Some of the properties of continuous functions are illustrated in Example 7.30 and in Exercises 7.9.4.

Example 7.30

Show that $f(x) = 2x/(1 + x^2)$ for $x \in \mathbb{R}$ is continuous on its whole domain. Find its maximum and minimum values and show that it attains every value between these extrema.

Figure 7.19
Graph of $2x/(1 + x^2)$.

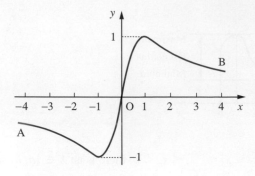

Solution The graph of the function is shown in Figure 7.19, from which we can see that the part shown is a continuous curve. That is to say, we can put a pencil at point A at the left-hand end of the graph and trace along the whole length of the curve to reach the point B at the right-hand end without lifting the pencil from the page. We can prove this more formally as follows. Select any point x_0 of the domain of the function. Then we have to show that $|f(x) - f(x_0)|$ can be made as small as we please by taking x sufficiently close to x_0. Now

$$\frac{2x}{1 + x^2} - \frac{2x_0}{1 + x_0^2} = \frac{2x(1 + x_0^2) - 2x_0(1 + x^2)}{(1 + x^2)(1 + x_0^2)} = \frac{2(1 - xx_0)(x - x_0)}{(1 + x^2)(1 + x_0^2)}$$

$$\to 0 \quad \text{as } x \to x_0$$

This implies that $f(x)$ is continuous at x_0, and since x_0 is any point of the domain, it follows that $f(x)$ is continuous for all x.

   The proof that the function takes every value between its least and largest values actually determines those values in this example. Let $y = f(x)$. Then, given y in the range of $f(x)$, we can find the corresponding values of x. Thus

$$y = \frac{2x}{1 + x^2} \quad \text{gives} \quad yx^2 - 2x + y = 0$$

and so

$$x = \frac{1 \pm \sqrt{(1 - y^2)}}{y}, \, y \neq 0$$

This gives two values of x for each $y \in (-1, 1)$, $y \neq 0$. Clearly $y = 0$ is also attained for $x = 0$. The maximum and minimum values for y are 1 and -1 respectively, and the corresponding values of x are 1 and -1. Thus $f(x)$ is a continuous function on its domain, and it attains its maximum and minimum values and every value in between.

7.9.2 Continuous and discontinuous functions

The technique used to show that $f(x)$ is a continuous function in Example 7.30 can be used to show that polynomials, rational functions (except where the denominator is zero) and many transcendental functions are continuous on their domains. We frequently make use of the properties of continuous functions unconsciously in problem-solving! For example, in solving equations we trap the root between two points x_1 and x_2 where $f(x_1) < 0$ and $f(x_2) > 0$ and conclude that the root we seek lies between x_1 and x_2. The

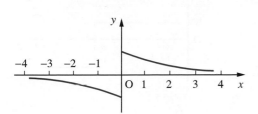

Figure 7.20 Graph of $\tan^{-1}(1/x)$, $x \neq 0$.

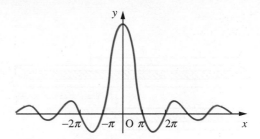

Figure 7.21 Graph of sinc $x = (\sin x)/x$.

need for continuity here is shown by the graph of $y = \tan^{-1}(1/x)$ (Figure 7.20). There is no value of x corresponding to $y = 0$, despite the facts that $\tan^{-1}(1/0.01)$ is positive and $\tan^{-1}[1/(-0.01)]$ is negative.

Similarly, when locating the maximum or minimum value of a function $y = f(x)$, in many practical situations we would be content with a solution that yields a value close to the true optimum value, and property (e) above tells us we can make that value as close as we please. Sometimes we use the continuity idea to fill in 'gaps' in function definitions. A simple example of this is $f(x) = (\sin x)/x$ for $x \neq 0$. This function is defined everywhere except at $x = 0$. We can extend it to include $x = 0$ by insisting that it be continuous at $x = 0$. Since $(\sin x)/x \rightarrow 1$ as $x \rightarrow 0$, defining $f(x)$ as

$$f(x) = \begin{cases} \dfrac{\sin x}{x} & (x \neq 0) \\ 1 & (x = 0) \end{cases} \qquad\qquad \textbf{(7.18)}$$

yields a function with no 'gaps' in its domain. The function $f(x)$ in (7.18) is known as the **sinc function**; that is,

$$\text{sinc } x = \begin{cases} \dfrac{\sin x}{x} & (x \neq 0) \\ 1 & (x = 0) \end{cases}$$

and its graph is drawn in Figure 7.21. This function has important applications in engineering, particularly in digital signal analysis. See the chapter on Fourier transforms in the companion text *Advanced Modern Engineering Mathematics*.

Of course it is not always possible to fill in 'gaps' in function definitions. The function

$$g(x) = \frac{x}{\sin x} \qquad (x \neq n\pi, \, n = 0, \pm 1, \pm 2, \ldots)$$

can have its domain extended to include the points $x = n\pi$, but it will always have a discontinuity at those points (except perhaps $x = 0$). Thus

$$f(x) = \begin{cases} \dfrac{x}{\sin x} & (x \neq n\pi, \, n = 0, \pm 1, \pm 2, \ldots) \\ 1 & (x = 0) \\ 0 & (x = n\pi, \, n = \pm 1, \pm 2, \ldots) \end{cases}$$

yields a function that is defined everywhere but is discontinuous at an infinite set of points.

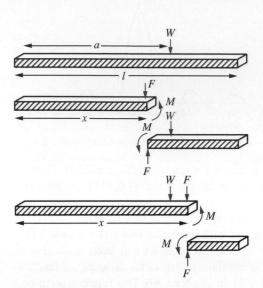

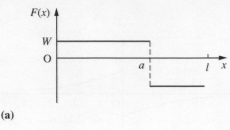

(a)

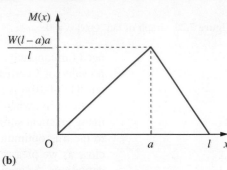

(b)

Figure 7.22 A beam, hinged at both ends, carrying a point load.

Figure 7.23 (a) The shear force and (b) the bending moment for a freely hinged beam.

In the analysis of practical problems we frequently use functions that have different formulae on different parts of their domain. For example, consider a beam of length l that is freely hinged at both ends and carries a concentrated load W at $x = a$, as shown in Figure 7.22. Then the shear force F is given by

$$F(x) = \begin{cases} W - Wa/l & (0 < x < a) \\ -Wa/l & (a \leqslant x < l) \end{cases}$$

and is sketched in Figure 7.23(a).

The bending moment M is

$$M(x) = \begin{cases} W(l - a)x/l & (0 < x \leqslant a) \\ W(l - x)a/l & (a \leqslant x < l) \end{cases}$$

and is sketched in Figure 7.23(b).

Notice here that F has a finite discontinuity at $x = a$ while M is continuous there.

7.9.3 Numerical location of zeros

Many practical engineering problems may involve the determination of the points at which a function takes a specific value (often zero) or the points at which it takes its maximum or minimum values. There are many different numerical procedures for solving such problems and we shall illustrate the technique by considering its application to the analysis of structural vibration.

This is a very common problem in engineering. To avoid resonance effects, it is necessary to calculate the natural frequencies of vibration of a structure. For a beam built in at one end and simply supported at the other as shown in Figure 7.24 the natural frequencies are given by

Figure 7.24
A beam built in at one end and simply supported at the other.

$$\frac{\theta^2}{2\pi l^2}\sqrt{\frac{EI}{\rho}}$$

where l is the length of the beam, E is Young's modulus, I is the moment of inertia of the beam about its neutral axis, ρ is its density and θ satisfies the equation

$$\tan\theta = \tanh\theta$$

We can find approximate values for θ that satisfy the above equation by means of a graph, as shown in Figure 7.25. From the diagram it is clear that the roots occur just before the points $\theta = 0$, $\frac{5}{4}\pi$, $\frac{9}{4}\pi$, $\frac{13}{4}\pi$, ... Using a calculator, we can compare the values of $\tan\theta$ and $\tanh\theta$, to produce the table of Figure 7.26, which gives us the estimate for the root near $\theta = \frac{5}{4}\pi$ as 3.925 ± 0.005.

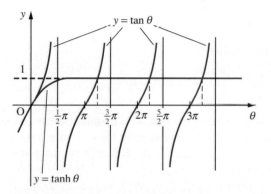

θ	$\tan\theta$	$\tanh\theta$
3.90	0.9474	0.9992
3.91	0.9666	0.9992
3.92	0.9861	0.9992
3.93	1.0060	0.9992

Figure 7.25 The roots of the equation $\tan\theta = \tanh\theta$.

Figure 7.26 Table of values.

If we require a more precise answer than this provides, we can resort to a finer subtabulation. In some problems this can be very tedious and time-consuming. A better strategy is to use an **interval-halving** or **bisection method**. We know that the root lies between $\theta_1 = 3.92$ and $\theta_2 = 3.93$. We work out the value of the functions at the midpoint of this interval, $\theta_3 = 3.925$, and determine whether the root lies between θ_1 and θ_3 or between θ_3 and θ_2. The process is then repeated on the subinterval that contains the root, and so on until sufficient precision is obtained.

The process is set out in tabular form in Figure 7.27. Note the renaming of the end points of the root-bracketing interval at each step, so that the interval under scrutiny is always denoted by $[\theta_1, \theta_2]$. After five applications we have $\theta = 3.926\,72 \pm (0.005/2^5)$.

Figure 7.27
Solution of
$\tan\theta - \tanh\theta = 0$
by the bisection
method.

θ_1	$f(\theta_1)$	θ_2	$f(\theta_2)$	θ_m	$f(\theta_m)$
3.92	−0.013 098	3.93	0.006 808	3.925	−0.003 195
3.925	−0.003 195	3.93	0.006 808	3.927 5	0.001 794
3.925	−0.003 195	3.927 5	0.001 794	3.926 25	−0.000 703
3.926 25	−0.000 703	3.927 5	0.001 794	3.926 875	0.000 545
3.926 25	−0.000 703	3.926 875	0.000 545	3.926 562 5	−0.000 079

A refinement of the bisection method is the **method of false position** (also known as *regula falsi*). To solve the equation $f(x) = 0$, given x_1 and x_2 such that $f(x_1) > 0$ and $f(x_2) < 0$ and $f(x)$ is continuous in (x_1, x_2), the bisection method takes the point

Figure 7.28
Solution of
$\tan \theta - \tanh \theta = 0$
by *regula falsa*.

θ_1	$f(\theta_1)$	θ_2	$f(\theta_2)$	$\dfrac{\theta_1 f(\theta_2) - \theta_2 f(\theta_1)}{f(\theta_2) - f(\theta_1)}$	$f\left(\dfrac{\theta_1 f(\theta_2) - \theta_2 f(\theta_1)}{f(\theta_2) - f(\theta_1)}\right)$
3.92	−0.013 098	3.93	0.006 808	3.926 580	−0.000 045
3.926 580	−0.000 045	3.93	0.006 808	3.926 602	−0.000 000

$\frac{1}{2}(x_1 + x_2)$ as the next estimate of the root. The method of false position uses linear interpolation to derive the next estimate of the root. The straight line joining the points $(x_1, f(x_1))$ and $(x_2, f(x_2))$ is given by

$$\frac{y - f(x_1)}{f(x_2) - f(x_1)} = \frac{x - x_1}{x_2 - x_1}$$

This line cuts the x axis where

$$x = \frac{x_1 f(x_2) - x_2 f(x_1)}{f(x_2) - f(x_1)}$$

so this is the new estimate of the root. This method usually converges more rapidly than the bisection method. The computation of the root of $\tan \theta - \tanh \theta = 0$ in the interval (3.92, 3.93) is shown in Figure 7.28. Notice how, as a result of the first step, the estimate of the root is $\theta = 3.926\,580$ and $f(3.926\,580) = 0.000\,045$. The root is now bracketed in the interval (3.926 580, 3.93), and the method is repeated. In two steps we have an estimate of the root giving a value of $f(\theta) < 10^{-6}$. This obviously converges much faster than the bisection method.

Both the bisection method and the method of false position are **bracketing methods** – the root is known to lie in an interval of steadily decreasing size. As such, they are guaranteed to converge to a solution. An alternative method of solution for an equation $f(x) = 0$ is to devise a scheme producing a convergent sequence whose limit is the root of the equation. Such **fixed point iteration methods** are based on a relation of the form $x_{n+1} = g(x_n)$. If $\lim_{n \to \infty} x_n = \alpha$, say, then evidently $\alpha = g(\alpha)$. The simplest way to devise an iterative scheme for the solution of an equation $f(x) = 0$ is to find some rearrangement of the equation in the form $x = g(x)$. Then, if the scheme $x_{n+1} = g(x_n)$ converges, the limit will be a root of $f(x) = 0$.

We can arrange the equation $\tan \theta = \tanh \theta$ in the form

$$\theta = \tan^{-1}(\tanh \theta) + k\pi \quad (k = 0, \pm1, \pm2, \dots)$$

If we take $k = 1$ and $\theta_0 = \frac{5}{4}\pi$ we obtain, using the iteration scheme,

$$\theta_n = \tan^{-1}(\tanh \theta_{n-1}) + \pi$$

the sequence

$$\theta_0 = 3.926\,991, \quad \theta_1 = 3.926\,603, \quad \theta_2 = 3.926\,602, \quad \theta_3 = 3.926\,602$$

and the root is $\theta = 3.926\,602$ to 6dp. (Taking other values of k will, of course, give schemes that converge to other roots of $\tan \theta = \tanh \theta$.)

The disadvantage of such iterative schemes is that not all of them converge. We shall return to this topic in Section 9.4.2.

7.9.4   Exercises

58 Draw sketches and discuss the continuity of

(a) $\dfrac{|x|}{x}$ (b) $\dfrac{x-1}{2-x}$

(c) $\tanh\dfrac{1}{x}$ (d) $\lfloor 1 - x^2 \rfloor$

59 Find upper and lower bounds obtained by

(a) $2x^2 - 4x + 7$ $(0 \leqslant x \leqslant 2)$

(b) $-x^2 + 4x - 1$ $(0 \leqslant x \leqslant 3)$

in the appropriate domains. Draw sketches to illustrate your answers.

60 Use the intermediate value theorem to show that the equation

$$x^3 + 10x^2 + 8x - 50 = 0$$

has roots between 1 and 2, between −4 and −3 and between −9 and −8. Find the root between 1 and 2 to 2dp using the bisection method.

61 Show that the equation $3^x = 3x$ has a root in the interval (0.7, 0.9). Use the intermediate value theorem and the method of *regula falsa* to find this root to 3dp.

62 Show that the equation

$$x^3 - 3x + 1 = 0$$

has three roots α, β and γ, where $\alpha < -1$, $0 < \beta < 1$ and $\gamma > 1$. For which of these is the iterative scheme

$$x_{n+1} = \tfrac{1}{3}(x_n^3 + 1)$$

convergent? Calculate the roots to 3dp.

63 The cubic equation $x^3 + 2x - 2 = 0$ can be written as

(a) $x = 1 - \tfrac{1}{2}x^3$ (b) $x = \dfrac{2}{2 + x^2}$

(c) $x = (2 - 2x)^{1/3}$

Determine which of the corresponding iteration processes converges most rapidly to find the real root of the equation. Hence calculate the root to 3dp.

64 Show that the iteration

$$x_{n+1} = \frac{1}{3}\left(2x_n + \frac{a}{x_n^2}\right)$$

converges to the limit $a^{1/3}$. Use the formula with $a = 157$ and $x_0 = 5$ to compare x_1 and x_2.

Show that the error ε_n in the nth iterate is given by $\varepsilon_{n+1} \approx \varepsilon_n^2/x_{n-1}$, where $x_n = a^{1/3} + \varepsilon_n$. Hence estimate the error in x_1 obtained above.

65 The periods of natural vibrations of a cantilever are given by

$$\frac{2\pi l^2}{\theta^2}\sqrt{\frac{\rho}{EI}}$$

where l, E, I and ρ are physical constants dependent on the shape and material of the cantilever and θ is a root of the equation

$$\cosh\theta\cos\theta = -1$$

Examine this equation graphically. Estimate its lowest root α_0 and obtain an approximation for the kth root α_k. Compare the two iterations:

$$\theta_{n+1} - \cosh^{-1}(-\sec\theta_n)$$

and

$$\theta_{n+1} = \cos^{-1}(-\operatorname{sech}\theta_n)$$

Which should be used to find an improved approximation to x_0?

7.10   Engineering application: insulator chain

The voltage V_k at the kth pin of the insulator chain shown in Figure 7.29 satisfies the recurrence relation

$$V_{k+2} - \left(2 + \frac{C_2}{C_1}\right)V_{k+1} + V_k = 0$$

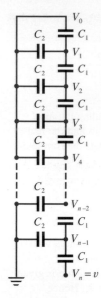

Figure 7.29

with $V_0 = 0$ and $V_n = v$, the amplitude of the voltage applied at the head of the chain. The characteristic equation for this recurrence relation is

$$\lambda^2 - \left(2 + \frac{C_2}{C_1}\right)\lambda + 1 = 0$$

which has real roots

$$\lambda_{1,2} = 1 + \frac{C_2}{2C_1} \pm \sqrt{\left[\frac{C_2}{C_1}\left(1 + \frac{C_2}{4C_1}\right)\right]}$$

Thus, the general solution is

$$V_k = A\lambda_1^k + B\lambda_2^k$$

Applying the condition $V_0 = 0$ gives

$$A + B = 0$$

Applying the condition $V_n = v$ gives

$$A\lambda_1^n + B\lambda_2^n = v$$

Hence $B = -A$ and $A = v/(\lambda_1^n - \lambda_2^n)$ and

$$V_k = \frac{v(\lambda_1^k - \lambda_2^k)}{\lambda_1^n - \lambda_2^n}$$

In a typical insulator chain $C_2/C_1 = 0.1$ and $n = 10$. It is left to the reader to calculate V_k/v for $k = 1, 2, \ldots, 9$.

7.11 Engineering application: approximating functions and Padé approximants

In Section 2.9.1 we introduced linear and quadratic interpolation as a means of obtaining estimates of the values of functions in between known values. Often in engineering applications it is of considerable importance to obtain good approximations to functions. In this section we shall show how what we have learned about power series representation can be used to produce a type of approximate representation of a function widely used by engineers, for example, when approximating exponentials by rational functions in modelling time delays in control systems. The approach is attributed to Padé and is based on the matching of series expansion.

Example 7.31 Obtain an approximation to the function e^{-x} in the form

$$e^{-x} \approx \frac{a + bx + cx^2}{A + Bx + Cx^2}$$

and find an estimate for the error.

Solution Assuming an exact match at $x = 0$, we deduce at once that $a = A$. Also, we know that $1/e^x = e^{-x}$, and assuming a similar relation for the approximation

$$\frac{A - Bx + Cx^2}{a - bx + cx^2} \equiv \frac{a + bx + cx^2}{A + Bx + Cx^2}$$

This holds if we choose $A = a$ (as above), $B = -b$ and $C = c$, giving

$$e^{-x} \approx \frac{A - Bx + Cx^2}{A + Bx + Cx^2}$$

We can see from this that it would be possible to express both sides of the equation as power series in x (at least in a restricted domain). We can rewrite the approximation to make it exact:

$$(A + Bx + Cx^2)e^{-x} = (A - Bx + Cx^2) + px^3 + qx^4 + rx^5 + \dots$$

where $p, q, \dots$ are to be found.

Replacing e^{-x} by its power series representation, we have

$$(A + Bx + Cx^2)(1 - x + \tfrac{1}{2}x^2 - \tfrac{1}{6}x^3 + \tfrac{1}{24}x^4 - \tfrac{1}{120}x^5 + \dots)$$
$$= A - Bx + Cx^2 + px^3 + qx^4 + rx^5 + \dots$$

Multiplying out the left-hand side and collecting terms, we obtain

$$A + (B - A)x + (\tfrac{1}{2}A - B + C)x^2 + (-\tfrac{1}{6}A + \tfrac{1}{2}B - C)x^3$$
$$+ (\tfrac{1}{24}A - \tfrac{1}{6}B + \tfrac{1}{2}C)x^4 + (-\tfrac{1}{120}A + \tfrac{1}{24}B - \tfrac{1}{6}C)x^5 + \dots$$
$$= A - Bx + Cx^2 + px^3 + qx^4 + rx^5 + \dots$$

Comparing the coefficients of like powers of x on either side of this equation gives

$$A = A$$
$$B - A = -B$$
$$\tfrac{1}{2}A - B + C = C$$
$$-\tfrac{1}{6}A + \tfrac{1}{2}B - C = p$$
$$\tfrac{1}{24}A - \tfrac{1}{6}B + \tfrac{1}{2}C = q$$
$$-\tfrac{1}{120}A + \tfrac{1}{24}B - \tfrac{1}{6}C = r$$

and so on.

We see from this that there is not a unique solution for A, B and C, but that we may choose them (or some of them) arbitrarily. Taking $A = 1$ gives $B = \tfrac{1}{2}$ and $\tfrac{1}{12} - C = p$. Setting $p = 0$ will make the error term smaller near $x = 0$, so we adopt that choice, giving $C = \tfrac{1}{12}$. This gives $q = 0$ and $r = -\tfrac{1}{720}$. Thus

$$(1 + \tfrac{1}{2}x + \tfrac{1}{12}x^2)e^{-x} = (1 - \tfrac{1}{2}x + \tfrac{1}{12}x^2) - \tfrac{1}{720}x^5 + \dots$$

so that

$$e^{-x} = \frac{1 - \tfrac{1}{2}x + \tfrac{1}{12}x^2}{1 + \tfrac{1}{2}x + \tfrac{1}{12}x^2} - \frac{\tfrac{1}{720}x^5 + \dots}{1 + \tfrac{1}{2}x + \tfrac{1}{12}x^2}$$

$$= \frac{12 - 6x + x^2}{12 + 6x + x^2} - \tfrac{1}{720}x^5 + \dots$$

The principal term of the error, $-\frac{1}{720}x^5$, enables us to decide the domain of usefulness of the approximation. For example, if we require an approximation correct to 4dp, we need $\frac{1}{720}x^5$ to be less then $\frac{1}{2} \times 10^{-4}$. Thus the approximation

$$e^{-x} \approx \frac{12 - 6x + x^2}{12 + 6x - x^2}$$

yields answers correct to 4dp for $|x| < 0.51$.

This particular approximation is used by control engineers in order to enable them to apply linear systems techniques to the analysis and design of systems characterizing a time delay in their dynamics. Since the degree of both the numerator and denominator is 2, this is referred to as the (2, 2) Padé approximant.

As an extended exercise, the reader should obtain the following (1, 1) and (3, 3) Padé approximants:

$$e^{-x} \approx \frac{2 - x}{2 + x} \quad \text{and} \quad e^{-x} \approx \frac{120 - 60x + 12x^2 - x^3}{120 + 60x + 12x^2 + x^3}$$

7.12 Review exercises (1–23)

1 There are two methods of assessing the value of a wasting asset. The first assumes that it decreases each year by a fixed amount; the second assumes that it depreciates by a fixed percentage.

A piece of equipment costs £1000 and has a 'lifespan' of six years after which its scrap value is £100. Estimate the value of the equipment by both methods for the intervening years.

2 A machine that costs £1000 has a working life of three years, after which it is valueless and has to be replaced. It saves the owner £500 per year while it is in use. Show that the true total saving £S to the owner over the three years is

$$S = 500\left[\frac{1}{1 + r/100} + \frac{1}{(1 + r/100)^2} + \frac{1}{(1 + r/100)^3} \right] - 1000$$

where $r\%$ is the current rate of interest. Estimate S for $r = 5, 10, 15$ and 20. When does the machine truly save the owner money?

3 An economic model for the supply $S(P)$ and demand $D(P)$ of a product at a market price of P is given by

$$D(P) = 2 - P$$

$$S(P) = \tfrac{1}{2} + \tfrac{1}{2}P$$

and

$$D(P_{t+1}) = S(P_t)$$

(so that supply lags behind demand by one time unit). Show that

$$P_{t+1} - 1 = -\tfrac{1}{2}(P_t - 1)$$

and deduce that

$$P_t = 1 + (-\tfrac{1}{2})^t(P_0 - 1)$$

Find the particular solution of the recurrence relation corresponding to $P_0 = 0.8$ and sketch it in a cobweb diagram. What is the steady-state price of the product?

4 Show that

$$\sum_{k=1}^{n} \frac{1}{T_k} = \frac{T_{n-1} + T_n}{T_n}$$

where T_k is the kth triangular number. (See Question 1 in Exercises 7.2.3.)

5 Find the general solutions of the following linear recurrence relations.

(a) $f_{n+2} - 5f_{n+1} + 6f_n = 0$ (b) $f_{n+2} - 4f_{n+1} + 4f_n = 0$

(c) $f_{n+2} - 5f_{n+1} + 6f_n = 4^n$ (d) $f_{n+2} - 5f_{n+1} + 6f_n = 3^n$

6 Suppose that consumer spending in period t, C_t, is related to personal income two periods earlier, I_{t-2}, by

$$C_t = 0.875I_{t-2} - 0.2C_{t-1} \quad (t \geq 2)$$

Deduce that if personal income increases by a factor 1.05 each period, that is

$$I_{t+1} = 1.05I_t$$

then $I_t = 1.05^t I_0$ and hence

$$C_t = (C_1 - 0.7I_0)(-0.2)^{t-1} + 0.7I_0(1.05)^{t-1}$$

Describe the behaviour of C_t in the long run.

7 An economist believes that the price P_t of a seasonal commodity in period t, satisfies the recurrence relation

$$P_{t+2} = 2(P_{t+1} - P_t) + C \quad (t \geq 0)$$

where C is a positive constant.
   Show that

$$P_t = A(1 + j)^t + B(1 - j)^t + C$$

where A and B are complex conjugate constants. Noting that $1 \pm j = \sqrt{2}(\cos\frac{\pi}{4} \pm j\sin\frac{\pi}{4})$, explain why the economist is mistaken.

8 The cobweb model applied to agricultural commodities assumes that current supply depends on prices in the previous season. If P_t denotes market price in any period and Q_{St}, Q_{Dt} supply and demand in that period, then

$$Q_{Dt} = 180 - 0.75P_t$$

$$Q_{St} = -30 + 0.3P_{t-1} \quad \text{where } P_0 = 220$$

   Find the market price and comment on its form.

9 Solve for National Income, Y_t, the set of recurrence relations

$$Y_t = 1 + C_t + I_t$$

$$C_t = \tfrac{1}{2}Y_{t-1}$$

$$I_t = 2(C_t - C_{t-1})$$

Comment on your solution.

10 A sequence is defined by

$$a_k = 1 + \frac{1}{2} + \frac{1}{3} + \ldots + \frac{1}{k} - \ln k \quad (k = 1, 2, \ldots)$$

Given $a_{10} = 0.626\,383$, $a_{16} = 0.608\,140$ and $a_{20} = 0.602\,009$ estimate $\gamma = \lim_{n\to\infty} a_n$, using repeated linear extrapolation. (γ is known as **Euler's constant**.)

11 Discuss the convergence of

(a) $\dfrac{2}{1^2} + \dfrac{3}{2^2} + \dfrac{4}{3^2} + \dfrac{5}{4^2} + \ldots$

(b) $\displaystyle\sum_{k=1}^{\infty} \frac{k^p}{k!} \quad$ (all p)

(c) $\frac{1}{11} - \frac{2}{13} + \frac{3}{15} - \frac{4}{17} + \ldots$

(d) $1 - \frac{1}{3} + \frac{1}{5} - \frac{1}{7} + \ldots$

12 Express the following recurring decimal numbers in the form p/q where p and q are integers:

(a) $1.231\,231\,23\ldots$ (b) $0.429\,429\,429\ldots$

(c) $0.101\,101\,101\ldots$ (d) $0.517\,251\,72\ldots$

13 Determine which of the following series are convergent:

(a) $\displaystyle\sum_{n=0}^{\infty} \frac{1}{n^2+1}$ (b) $\displaystyle\sum_{n=1}^{\infty} \frac{n+2}{n^2}$

(c) $\displaystyle\sum_{n=1}^{\infty} \frac{n-1}{2n^5-1}$ (d) $\displaystyle\sum_{n=1}^{\infty} \frac{n-1}{n^2+n-3}$

14 A rational function $f(x)$ has the following power series representation for $-1 < x < 1$:

$$f(x) = 1^2x + 2^2x^2 + 3^2x^3 + 4^2x^4 + \ldots$$

Find a closed-form expression for $f(x)$.

15 Find the values of a and b such that

$$\tan x = \frac{ax}{1 + bx^2} + cx^5 + O(x^7)$$

giving the value of c. (The series for $\tan x$ is given in Question 49(a) in Exercises 7.7.3.)
   For what values of x will the approximation

$$\tan x = \frac{ax}{1 + bx^2}$$

be valid to 4dp? Use the approximation to calculate $\tan 0.29$ and $\tan 0.295$, and compare your answers with the values given by your calculator. Comment on your results.

[Here $O(x^7)$ mean terms involving powers of n greater than or equal to 7.]

16 The function $f(x) = \sinh^{-1}x$ has the power series expansion

$$\sinh^{-1}x = x - \frac{1}{2}\frac{x^3}{3} + \frac{1 \cdot 3}{2 \cdot 4}\frac{x^5}{5} - \frac{1 \cdot 3 \cdot 5}{2 \cdot 4 \cdot 6}\frac{x^7}{7} + \dots$$

Obtain polynomial approximations for $\sinh^{-1}x$ for $-0.5 < x < 0.5$ such that the truncation error is less than (a) 0.005 and (b) 0.000 05.

17 A chord of a circle is half a mile long and supports an arc whose length is 1 foot longer (1 mile = 5280 feet). Show that the angle θ subtended by the arc at the centre of the circle satisfies

$$\sin \tfrac{1}{2}\theta = \frac{1320}{2641}\theta$$

Use the series expansion for sine to obtain an approximate solution of this equation, and estimate the maximum height of the arc above its chord.

18 A machine is purchased for £3600. The annual running cost of the machine is initially £1800, but rises annually by 10%. After x years its secondhand value is £3600e$^{-0.35x}$. Show that the average annual cost £C (including depreciation) after x years is given by

$$C = \frac{3600(1 - e^{-0.35x})}{x} + 90(19 + x)$$

Show graphically that the machine should be replaced after about 4 years, and use an iterative method to refine this estimate.

19 The rate of discharge from a circular pipe is proportional to $[(\theta - \sin\theta)^3/\theta]^{1/2}$, where θ is the angle subtended at the centre by the wetted perimeter. Find the value of θ that maximizes the discharge rate.

20 Consider the sequence ϕ_n defined by

$$\phi_n = \frac{1}{2}\left[\left(1 + \frac{1}{n}\right)^n + \left(1 - \frac{1}{n}\right)^{-n}\right]$$

Show that $\phi_n \to e$ as $n \to \infty$. Using the power series expansions of $\ln(1 + x)$ and e^x, show that

$$\left(1 + \frac{1}{n}\right)^n = \exp\left[n \ln\left(1 + \frac{1}{n}\right)\right]$$

$$= e\left(1 - \frac{1}{2n} + \frac{11}{24n^2} - \frac{7}{16n^3} + \dots\right)$$

and deduce that

$$\phi_n = e\left(1 + \frac{11}{24n^2} + \dots\right)$$

Evaluate ϕ_{64} and ϕ_{128} (without using the y^x key of your calculator), and use extrapolation to estimate the value of e.

21 A beam of weight W per unit length is simply supported at the same level at $(N + 1)$ equidistant points, the extreme supports being at the ends of the beam. The bending moment M_k at the kth support satisfies the recurrence relation

$$M_{k+2} + 4M_{k+1} + M_k = \tfrac{1}{2}Wa^2$$

where a is the distance between the supports and $M_0 = 0$ and $M_N = 0$. (This is a consequence of Clapeyron's theorem of three moments.) Show that if the sequences $\{A_k\}_{k=0}^{N}$ and $\{B_k\}_{k=0}^{N}$ are calculated by the recurrences

$$A_0 = A_1 = 0$$

$$A_{k+2} + 4A_{k+1} + A_k = 1 \quad (k = 0, 1, \dots, N - 2)$$

and

$$B_0 = 0, \quad B_1 = 1$$

$$B_{k+2} + 4B_{k+1} + B_k = 0 \quad (k = 0, 1, \dots, N - 2)$$

then the solution of the bending-moment problem is given by

$$M_k = \tfrac{1}{2}Wa^2A_k + M_1B_k \quad (k = 0, \dots, N)$$

with $\tfrac{1}{2}Wa^2A_N + M_1B_N = 0$ determining the value of M_1.

Perform the calculation for the case where $N = 8$, $a = 1$ and $W = 25$.

22 A complex voltage E is applied to the ladder network of Figure 7.30. Show that the (complex) mesh currents I_k satisfy the equations

$$\tfrac{1}{2}L\omega jI_0 - \frac{j}{C\omega}(I_0 - I_1) = E$$

$$L\omega jI_k - \frac{j}{C\omega}(I_k - I_{k+1}) + \frac{j}{C\omega}(I_{k-1} - I_k) = 0$$

$$(k = 1, \dots, N - 1) \quad \textbf{(7.19)}$$

$$\tfrac{1}{2}L\omega jI_n + \frac{j}{C\omega}(I_{n-1} - I_n) = 0$$

(See Section 3.5 for the application of complex numbers to alternating circuits.)

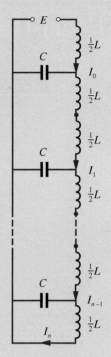

Figure 7.30

Show that $I_k = A(e^\theta)^k = Ae^{k\theta}$ satisfies (7.19) provided that $\cosh\theta = 1 - \frac{1}{2}LC\omega^2$. Note that this equation yields two values for θ, so that in general I_k may be written as

$$I_k = Ae^{k\theta} + Be^{-k\theta}$$

where A and B are independent of k. Using the special equations for I_0 and I_n, obtain the values of A and B and prove that

$$I_k = jEC\omega\,\frac{\cosh(n-k)\theta}{\sinh\theta\,\sinh n\theta}$$

23 A lightweight beam of length l is clamped horizontally at both ends. It carries a concentrated load W at a distance a from one end ($x = 0$). The shear force F and bending moment M at the point x on the beam are given by

$$F = \begin{cases} \dfrac{W(l-a)^2(l+2a)}{l^3} & (0 < x < a) \\[2ex] \dfrac{-Wa^2(3l-2a)}{l^3} & (a < x < l) \end{cases}$$

and

$$M = \begin{cases} \dfrac{W(l-a)^2[al - x(l+2a)]}{l^3} & (0 < x < a) \\[2ex] \dfrac{Wa^2[al - 2l^2 + x(3l-2a)]}{l^3} & (a < x < l) \end{cases}$$

Draw the graphs of these functions. Use Heaviside functions to obtain single formulae for M and F.

8 Differentiation and Integration

8.1 Introduction

Many of the practical situations that engineers have to analyse involve quantities that are varying. Whether it is the temperature of a coolant, the voltage on a transmission line or the torque on a turbine blade, the mathematical tools for performing such analyses are the same. One of the most successful of these is **calculus**, which involves two fundamental operations: differentiation and integration. Historically, integration was discovered first, and indeed some of the ideas and results date back over 2000 years to when the Greeks developed the **method of exhaustion** to evaluate the area of a region bounded on one side by a curve – a method used by Archimedes (287–212 BC) to obtain the exact formula for the area of a circle. Differentiation was discovered very much later, during the seventeenth century, in relation to the problem of determining the tangent at an arbitrary point on a curve. Its characteristic features were probably first used by Fermat in 1638 to find the maximum and minimum points of some special functions. He noticed that tangents must be horizontal at some points, and developed a method for finding them by slightly changing the variable in a single algebraic equation and then letting the change 'disappear'. The connection between the two processes of determining the area under a curve and obtaining a tangent at a point on a curve was first realized in 1663 by Barrow, who was Newton's professor at Cambridge University. However, it was Newton (1642–1727) and Leibniz (1646–1716), working independently, who fully recognized the implications of this relationship. This led them to develop the calculus as a way of dealing with change and motion. Exploitation of their work resulted in an era of tremendous mathematical activity, much of which was motivated by the desire to solve applied problems, particularly by Newton, whose accomplishments were immense and included the formulation of the laws of gravitation. The calculus was put on a firmer mathematical basis in the nineteenth century by Cauchy and Riemann. It remains today one of the most powerful mathematical tools used by engineers, and in this chapter and the next we shall review its basic ideas and techniques and show their application both in the formulation of mathematical models of practical problems and in their solution.

In recent years we have seen significant developments in symbolic algebra packages, such as MATHEMATICA, MACSYMA, MAPLE, REDUCE, AXIOM and DERIVE, which are capable of performing algebraic manipulation, including the calculation of derivatives and integrals. To the inexperienced, this development may appear to eliminate the need for engineers to be able to carry out even basic operations in calculus by hand. This, however, is far from the truth. If engineers are to apply the powerful techniques associated with the calculus to the design and analysis of industrial problems then it is essential that they have a sound grounding of differentiation and integration. First, this allows effective formulation, comprehension and analysis of mathematical models. Secondly, it provides the basis for understanding symbolic algebra packages, particularly when specific forms of results are desired. In order to acquire this understanding it is necessary to have a certain degree of fluency in the manipulation of associated basic techniques. It is the objective of this chapter and the next to provide the minimum requirements for this. At the same time, students should be given the opportunity to develop their skills in the use of a symbolic algebra package and, whenever appropriate, be encouraged to check their answers to the exercises using such a

 package.

8.2 Differentiation

Here we shall introduce the concept of differentiation and illustrate its role in some problem-solving and modelling situations.

8.2.1 Rates of change

Consider an object moving along a straight line with constant velocity u (in $m\,s^{-1}$). The distance s (in metres) travelled by the object in time t (in seconds) is given by the formula $s = ut$. The distance–time graph of this motion is the straight line shown in Figure 8.1. Note that the velocity u is the rate of change of distance with respect to time, and that on the distance–time graph it is the gradient (slope) of the straight line representing the relationship between the distance travelled and the time elapsed. This, of course, is a special case where the velocity is constant and the distance travelled is a linear function of time. Even when the velocity varies with time, however, it is still given by the gradient of the distance–time graph, although it then varies from point to point along the curve.

Consider the distance–time graph shown in Figure 8.2(a). Suppose we wish to find the velocity at the time $t = t_1$. To do that we can enlarge that piece of the graph near $t = t_1$, as shown in Figure 8.2(b) and (c). We recall that continuous functions have the property that locally they may be approximated by linear functions. We see that as we increase the magnification the graph takes on the appearance of a straight line through (t_1, s_1). The gradient of that straight line (in the limit) gives us the gradient of the graph at (t_1, s_1). Consider that section of the graph contained in the rectangle whose sides

Figure 8.1
Distance–time graph
for constant velocity u.

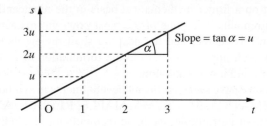

Figure 8.2
(a) Distance–time
graph, (b) enlargement
of outer rectangle
surrounding (t_1, s_1)
and (c) enlargement
of inner rectangle
surrounding (t_1, s_1).

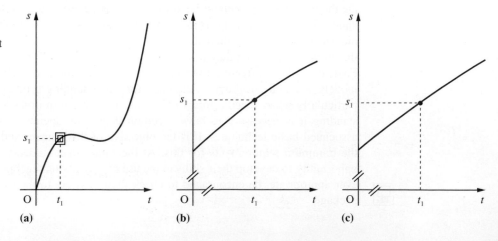

Figure 8.3
Section of the
distance–time graph.

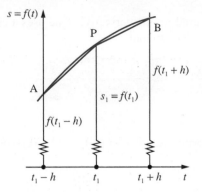

parallel to the s axis are $t = t_1 + h$ and $t = t_1 - h$ where h is a positive (small) number, as shown in Figure 8.3. If we denote the function relating distance and time by $f(t)$, then $s_1 = f(t_1)$ and we can approximate the gradient of the function $f(t)$ by the gradients of either of the chords AP or BP. Thus

$$\text{gradient} \simeq \frac{f(t_1 + h) - f(t_1)}{h} \simeq \frac{f(t_1 - h) - f(t_1)}{(-h)}.$$

As h becomes smaller and smaller (corresponding to greater and greater magnifications) these approximations become better and better so that in the limit ($h \to 0$) they cease being approximations and become exact. Thus we may write

$$(\text{gradient of } f(t) \text{ at } t = t_1) = \lim_{h \to 0} \frac{f(t_1 + h) - f(t_1)}{h}$$

$$= \lim_{h \to 0} \frac{f(t_1 - h) - f(t_1)}{(-h)}$$

Here we specified $h > 0$, which means that the former limit is the limit from above and the latter is the limit from below of the expression

$$\frac{f(t_1 + \Delta t) - f(t)}{\Delta t}$$

where $\Delta t \to 0$. So provided that the right-hand and left-hand limits have the same value, the gradient of the function $f(t)$ is defined at $t = t_1$ by

$$\lim_{\Delta t \to 0} \frac{f(t_1 + \Delta t) - f(t)}{\Delta t}$$

(Here we have used the composite symbol Δt to indicate a small change in the value of t. It may be positive or negative.)

8.2.2   Definition of a derivative

Formally we define the derivative of the function $f(x)$ at the point x to be

$$\lim_{\Delta x \to 0} \frac{f(x + \Delta x) - f(x)}{\Delta x} = \lim_{\Delta x \to 0} \frac{\Delta f}{\Delta x}$$

Two kinds of notation are used for the derivative. One uses a composite symbol, $\dfrac{df}{dx}$, and the other uses a prime, $f'(x)$, so that

$$\frac{df}{dx} = f'(x) = \lim_{\Delta x \to 0} \frac{\Delta f}{\Delta x} = \lim_{\Delta x \to 0} \frac{f(x + \Delta x) - f(x)}{\Delta x} \tag{8.1}$$

Example 8.1

Using the definition of a derivative given in (8.1), find $f'(x)$ when $f(x)$ is

(a) x^2 (b) $\dfrac{1}{x}$ (c) $mx + c$ (m, c constants)

Solution   (a) $f(x) = x^2$

$$f(x + \Delta x) = (x + \Delta x)^2 = x^2 + 2x\,\Delta x + (\Delta x)^2$$

so that

$$\frac{\Delta f}{\Delta x} = \frac{f(x + \Delta x) - f(x)}{\Delta x} = \frac{2x\,\Delta x + (\Delta x)^2}{\Delta x} = 2x + \Delta x$$

Thus, from (8.1), the derivative of $f(x)$ is

$$\frac{df}{dx} = f'(x) = \lim_{\Delta x \to 0} \frac{\Delta f}{\Delta x} = \lim_{\Delta x \to 0} (2x + \Delta x) = 2x$$

so that

$$\frac{d}{dx}(x^2) = 2x$$

(b) $f(x) = \dfrac{1}{x}$

$$f(x + \Delta x) = \frac{1}{x + \Delta x}$$

so that

$$\frac{\Delta f}{\Delta x} = \frac{f(x + \Delta x) - f(x)}{\Delta x} = \left[\frac{\dfrac{1}{x + \Delta x} - \dfrac{1}{x}}{\Delta x} \right] = \left[\frac{x - x - \Delta x}{\Delta x (x + \Delta x) x} \right]$$

$$= \left[\frac{-1}{x^2 + x\,\Delta x} \right]$$

Thus, from (8.1), the derivative of $f(x)$ is

$$\frac{df}{dx} = \lim_{\Delta x \to 0} \frac{\Delta f}{\Delta x} = \lim_{\Delta x \to 0} \left[\frac{-1}{x^2 + x\,\Delta x} \right] = -\frac{1}{x^2}$$

so that

$$\frac{\mathrm{d}}{\mathrm{d}x}(x^{-1}) = -1x^{-2}$$

(c) $f(x) = mx + c$

$$f(x + \Delta x) = m(x + \Delta x) + c$$

so that

$$\frac{\Delta f}{\Delta x} = \frac{f(x + \Delta x) - f(x)}{\Delta x} = \frac{m\Delta x}{\Delta x} = m$$

Thus, from (8.1), the derivative of $f(x)$ is

$$\frac{\mathrm{d}f}{\mathrm{d}x} = \lim_{\Delta x \to 0} \frac{\Delta f}{\Delta x} = m$$

so the gradient of the function $f(x) = mx + c$ is the same as that of the straight line $y = mx + c$, as we would expect.

8.2.3   Interpretation as the slope of a tangent

The definition is illustrated graphically in Figure 8.4, where Δx denotes a small incremental change in the independent variable x and Δf is the corresponding incremental change in $f(x)$. The slope of the line segment PQ is

$$\frac{\Delta f}{\Delta x} = \frac{f(x + \Delta x) - f(x)}{\Delta x}$$

In the limit as $\Delta x \to 0$ the point Q $\to$ P, and the line segment becomes the tangent to the curve at P, whose slope is given by the derivative

$$\frac{\mathrm{d}f}{\mathrm{d}x} = \lim_{\Delta x \to 0} \frac{\Delta f}{\Delta x}$$

Figure 8.4
Illustration of
derivative.

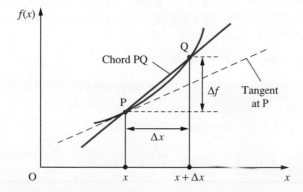

Summary

If $y = f(x)$ then the derivative of $f(x)$ is defined by

$$\frac{dy}{dx} = \frac{df}{dx} = f'(x) = \lim_{\Delta x \to 0} \frac{\Delta f}{\Delta x} = \lim_{\Delta x \to 0} \frac{f(x + \Delta x) - f(x)}{\Delta x}$$

The derivative may be interpreted as

(a) the rate of change of the function $f(x)$ with respect to x, or

(b) the slope of the tangent at the point (x, y) on the graph of $y = f(x)$.

Example 8.2   Consider the function $f(x) = 25x - 5x^2$. Find

(a) the derivative of $f(x)$ from first principles,

(b) the rate of change of $f(x)$ at $x = 1$,

(c) the equation of the tangent to the graph of $f(x)$ at the point $(1, 20)$.

Solution   (a) $f(x) = 25x - 5x^2$

$$f(x + \Delta x) = 25(x + \Delta x) - 5(x + \Delta x)^2$$

$$= 25x + 25\Delta x - 5x^2 - 10x\Delta x - 5(\Delta x)^2$$

so that

$$\frac{\Delta f}{\Delta x} = \frac{f(x + \Delta x) - f(x)}{\Delta x} = \frac{25\Delta x - 10x\Delta x - 5(\Delta x)^2}{\Delta x} = 25 - 10x - 5\Delta x$$

Thus the derivative of $f(x)$ is

$$\frac{df}{dx} = f'(x) = \lim_{\Delta x \to 0} \frac{\Delta f}{\Delta x} = \lim_{\Delta x \to 0} (25 - 10x - 5\Delta x) = 25 - 10x$$

(b) The rate of change of $f(x)$ at $x = 1$ is $f'(1) = 15$.

(c) The slope of the tangent to the graph of $f(x)$ at $(1, 20)$ is $f'(1) = 15$. Remembering from equation (1.11) that the equation of a line passing through a point (x_1, y_1) and having slope m is

$$y - y_1 = m(x - x_1)$$

we have the equation of the tangent to the graph of $f(x)$ at $(1, 20)$ is

$$y - 20 = 15(x - 1)$$

or

$$y = 15x + 5$$

8.2.4 Differentiable functions

The formal definition of the derivative of $f(x)$ implies that the right-hand and left-hand limits are equal. In some cases this does not happen. For example, the function $f(x) = \sqrt{(1 + \sin x)}$ is such that

$$\lim_{\Delta x \to 0-} \frac{f(3\pi/2 + \Delta x) - f(3\pi/2)}{\Delta x} = \frac{-1}{\sqrt{2}}$$

$$\lim_{\Delta x \to 0+} \frac{f(3\pi/2 + \Delta x) - f(3\pi/2)}{\Delta x} = \frac{1}{\sqrt{2}}$$

Clearly the derivative of the function is not defined at $x = 3\pi/2$ (the two limits above are sometimes referred to as 'left-hand' and 'right-hand' derivatives, respectively).

The graph of $y = \sqrt{(1 + \sin x)}$ is shown in Figure 8.5, and it is clear that at $x = 3\pi/2$ a unique tangent cannot be drawn to the graph of the function. This is not surprising since from the interpretation of the derivative as the slope of the tangent, it follows that for a function $f(x)$ to be **differentiable** at $x = a$, the graph of $f(x)$ must have a unique, non-vertical well-defined tangent at $x = a$. Otherwise the limit

$$\lim_{\Delta x \to 0} \frac{f(a + \Delta x) - f(a)}{\Delta x}$$

does not exist. We say that a function $f(x)$ is differentiable if it is differentiable at all points in its domain. For practical purposes it is sufficient to interpret a differentiable function as one having a smooth continuous graph with no sharp corners. Engineers frequently refer to such functions as being 'well behaved'. Clearly the function having the graph shown in Figure 8.6(a) is differentiable at all points except $x = x_1$ and $x = x_2$, since a unique tangent cannot be drawn at these points. Similarly, the function having the graph shown in Figure 8.6(b) is differentiable at all points except at $x = 0$.

Figure 8.5
The graph of
$y = \sqrt{(1 + \sin x)}$.

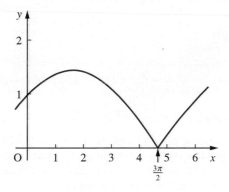

Figure 8.6

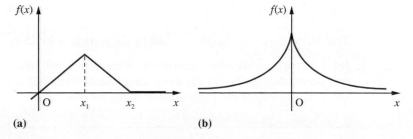

(a)   (b)

8.2.5 Speed, velocity and acceleration

Considering the motion of the object in Section 8.2.1 enables us to distinguish between the terms **speed** and **velocity**. In everyday usage we talk of speed rather than velocity, and always regard it as being positive or zero. As we saw in Chapter 4, velocity is a vector quantity and has a direction associated with it, while speed is a scalar quantity, being the magnitude or modulus of the velocity. When s and v are measured horizontally, the object will have a positive velocity when travelling to the right and a negative velocity when travelling to the left. Throughout its motion, the speed of the particle will be positive or zero. Likewise, **acceleration** a, being the rate of change of velocity with respect to time, is a vector quantity and is determined by

$$a(t) = \frac{dv}{dt}$$

Example 8.3 A particle is thrown vertically upwards into the air. Its height s (in m) above the ground after time t (in seconds) is given by

$$s = 25t - 5t^2$$

(a) What height does the particle reach?

(b) What is its velocity when it returns to hit the ground?

(c) What is its acceleration?

Solution Since velocity v is rate of change of distance s with time t we have

$$v = \frac{ds}{dt}$$

In this particular example

$$s(t) = 25t - 5t^2 \tag{8.2}$$

so, from Example 8.2,

$$v(t) = \frac{ds}{dt} = 25 - 10t \tag{8.3}$$

(a) When the particle reaches its maximum height, it will be momentarily at rest, so that its velocity will be momentarily zero. From (8.3) this will occur when $t = 25/10 = 2.5$. Then, from (8.2), the height reached at this instant is

$$s(2.5) = 25 \times \tfrac{5}{2} - 5 \times (\tfrac{5}{2})^2 = \tfrac{125}{4}$$

That is, the maximum height reached by the particle is 31.25 m.

(b) First we need to find the time at which the particle will return to hit the ground. This will occur when the height s is again zero, which from (8.2) is when $t = 5$. Then, from (8.3), the velocity of the particle when it hits the ground is

$$v(5) = -25$$

That is, when it returns to hit the ground, the particle will be travelling at $25\,\mathrm{m\,s}^{-1}$, with the negative sign indicating that it is travelling downwards, since s and v are measured upwards.

(c) The acceleration, a, is the rate of change of velocity with respect to time. Thus, from (8.3)

$$a(t) = \frac{\mathrm{d}v}{\mathrm{d}t} = -10$$

that is, $a \approx -g = -9.806\,65\ (\mathrm{m\,s}^{-2})$, the acceleration due to gravity.

8.2.6 Mathematical modelling using derivatives

We have seen that the gradient of a tangent to the graph $y = f(x)$ can be expressed as a derivative, but derivatives have much wider application than just this. Any quantity that can be expressed as a limit of the form (8.1) can be represented by a derivative, and such quantities arise in many practical situations. Because gradients of tangents to graphs can be expressed as derivatives, it follows that we can always interpret a derivative geometrically as the slope of a tangent to a graph.

Example 8.4

In a suspension bridge a roadway, of length $2l$, is suspended by vertical hangers from cables carried by towers at the ends of the span, as illustrated in Figure 8.7(a). The lowest points of the cables are a distance h below the top of the supporting towers. Find an equation which represents the line shape of the cables.

Solution

To solve this problem we have to make some simplifying assumptions. We assume that the roadway is massive compared to the cables, so that the weight W of the roadway is the dominant factor in determining the shape of the cables. Secondly, we assume

Figure 8.7
(a) Schematic diagram for a suspension bridge. (b) Forces acting on the cable between A and B.

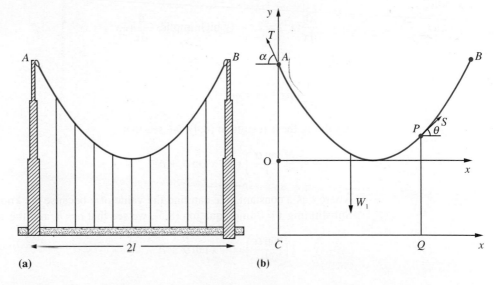

that the weight of the roadway is uniformly distributed along its length, and if the hangers are equally spaced they can be adjusted in length so that they carry equal vertical loads.

We solve this problem using elementary statics because at each point P on the cable the forces are in equilibrium. Figure 8.7(b) shows the forces acting on the part of the cable between A and P. These are the weight W_1 of the roadway between C and Q ($W_1 = Wx/2l$, where $x = CQ$), the tension T in the cable acting at the angle α at A and the tension S in the cable acting at the angle θ at P.

Resolving forces horizontally, we have $T\cos\alpha = S\cos\theta$.

Resolving forces vertically, we have $T\sin\alpha + S\sin\theta = \dfrac{Wx}{2l}$.

Eliminating S between these equations gives $T\sin\alpha + T\cos\alpha\tan\theta = \dfrac{Wx}{2l}$.

Also, we know that the total weight W of the roadway is supported by the tensions at A and B, so that $2T\sin\alpha = W$. Hence, substituting, we obtain

$$\frac{W}{2\sin\alpha}\sin\alpha + \frac{W}{2\sin\alpha}\cos\alpha\tan\theta = \frac{Wx}{2l}$$

giving

$$\tan\theta = x\frac{\tan\alpha}{l} - \tan\alpha$$

Now $\tan\theta$ is the slope of the curve at P (the tensions act along the direction of the tangent at each point of the curve), so that

$$\frac{dy}{dx} = x\frac{\tan\alpha}{l} - \tan\alpha \tag{8.4}$$

using the coordinate system shown in Figure 8.7(b). This equation involving the derivative of y is called a **differential equation**. To solve this equation we have to find the function whose derivative is the right-hand side of equation (8.4). In this case we can make use of the results of Example 8.1, since we know that

$$\frac{d}{dx}(x^2) = 2x \quad \text{(which implies } \frac{d}{dx}(\tfrac{1}{2}x^2) = x)$$

and

$$\frac{d}{dx}(mx + c) = m$$

Applying these results to (8.4), we see that

$$y = \tfrac{1}{2}\left(\frac{\tan\alpha}{l}\right)x^2 - x\tan\alpha + c \tag{8.5}$$

where c is a constant. We can find the value of c because we know that $y = h$ at $x = 0$. Substituting $x = 0$ into equation (8.5), we see that $c = h$, and the solution becomes

$$y = \tfrac{1}{2}\left(\frac{\tan\alpha}{l}\right)x^2 - x\tan\alpha + h \tag{8.6}$$

But we also know that $y = 0$ where $x = l$. This enables us to find the value of $\tan \alpha$. Substituting $x = l$ into equation (8.6), we have

$$0 = \tfrac{1}{2}l \tan \alpha - l \tan \alpha + h$$

which implies $\tan \alpha = 2h/l$. Thus the shape of the supporting cable is given by

$$y = \frac{hx^2}{l^2} - \frac{2hx}{l} + h$$
$$= h(x - l)^2/l^2$$

indicating that the points of attachment of the hangers to the cable lie on a parabolic curve.

Example 8.5

Suppose that a tank initially contains 80 litres of pure water. At a given instant (taken to be $t = 0$) a salt solution containing 0.25 kg of salt per litre flows into the tank at the rate of 8 litres min^{-1}. The liquid in the tank is kept homogeneous by constant stirring. Also, at time $t = 0$ liquid is allowed to flow out from the tank at the rate of 12 litres min^{-1}. Show that the amount of salt $x(t)$ (in kg) in the tank at time t (min) $\geqslant 0$ is determined by the mathematical model

$$\frac{\mathrm{d}x(t)}{\mathrm{d}t} + \frac{3x(t)}{20 - t} = 2 \quad (t < 20)$$

Solution

The situation is illustrated in Figure 8.8. Since $x(t)$ denotes the amount of salt in the tank at time $t \geqslant 0$, the rate of increase of the amount of salt in the tank is $\mathrm{d}x/\mathrm{d}t$, and is given by

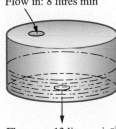

Flow in: 8 litres min^{-1}

Flow out: 12 litres min^{-1}

Figure 8.8
Water tank of
Example 8.5.

$$\frac{\mathrm{d}x}{\mathrm{d}t} = \text{rate of inflow of salt} - \text{rate of outflow of salt} \tag{8.7}$$

The rate of inflow of salt is $(0.25 \text{ kg litre}^{-1})(8 \text{ litres min}^{-1}) = 2 \text{ kg min}^{-1}$.

The rate of outflow of salt is $c \times (\text{rate of outflow of liquid}) = c \times 12 \text{ litres min}^{-1}$

$$= 12c \quad (\text{in kg min}^{-1})$$

where $c(t)$ is the concentration of salt in the tank (in kg litre^{-1}). The concentration at time t is given by

$$c(t) = \frac{\text{amount of salt in the tank at time } t}{\text{volume of liquid in the tank at time } t}$$

After time t (in min) $8t$ litres have entered the tank and $12t$ litres have left. Also, at $t = 0$ there were 80 litres in the tank. Therefore the volume V of liquid in the tank at time t is given by

$$V(t) = 80 - (12t - 8t) = (80 - 4t)$$

(Note that $V(t) \geqslant 0$ only if $t \leqslant 20$ min; after this time the liquid will flow out as quickly as it flows in and none will accumulate in the tank.) Thus the concentration $c(t)$ is given by

$$c(t) = \frac{x(t)}{V(t)} = \frac{x(t)}{80 - 4t}$$

so that

$$\text{rate of outflow of salt} = 12 \times \frac{x(t)}{80 - 4t} = \frac{3x(t)}{20 - t}$$

Substituting back into (8.7) gives the rate of increase as

$$\frac{dx}{dt} = 2 - \frac{3x}{20 - t}$$

or

$$\frac{dx(t)}{dt} + \frac{3x(t)}{20 - t} = 2$$

This is another example of a differential equation, and in Chapter 10 (Review exercises 10.13, Question 9) we shall show how it can be solved to give the quantity $x(t)$ of salt in the tank at time t.

Example 8.6

A radio telescope has the shape of a paraboloid of revolution. Show that all the radio waves arriving in a direction parallel to its axis of symmetry are reflected to pass through the same point on that axis of symmetry.

Solution

The diagram in Figure 8.9 shows a section of the paraboloid through its axis of symmetry. We choose the coordinate system such that the equation of the parabola shown is $y = x^2$. Let AP represent the path of a radio signal travelling parallel to the y axis. At P it is reflected to pass through the point B on the y axis. The laws of reflection state that $\angle APN = \angle BPN$, where PN is the normal to the curve at P. Now given the coordinates (a, a^2) of the point P we have to find the coordinates $(0, b)$ of the point B. From the diagram we can see that if $\angle PTQ = \theta$, then $\angle PQT = \pi/2 - \theta$, which implies that $\angle ONP = \theta$. Since AP is parallel to NB, we see that $\angle APN = \theta$ and hence $\angle BPN = \theta$. This implies that $\angle PBN = \pi - 2\theta$. With all of these angles known we can calculate the coordinates of B. From the diagram

$$\tan \angle NBP = \frac{a}{a^2 - b}$$

Figure 8.9
Section of paraboloid through axis of symmetry.

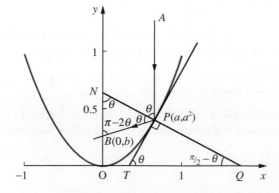

Since $\angle NBP = \pi - 2\theta$, this implies $\tan 2\theta = \dfrac{a}{b - a^2}$. Also

$$\tan \theta = \left(\frac{dy}{dx}\right)_{x=a} = 2a$$

and since $\tan 2\theta = \dfrac{2 \tan \theta}{1 - \tan^2 \theta}$ we obtain

$$\frac{4a}{1 - 4a^2} = \frac{a}{b - a^2}$$

This gives $b = \frac{1}{4}$. Notice that the value of b is independent of a. Thus all the reflected rays pass through $(0, \frac{1}{4})$. As was indicated in Section 1.4.4, this property is important in many engineering design projects.

Example 8.7   Show that the shear force F acting in a beam is related to the bending moment M by

$$F = \frac{dM}{dx}$$

Solution   A beam is a horizontal structural member which carries loads. These induce forces and stresses inside the beam in transmitting the loads to the supports. For design safety two internal quantities are used, the shear force F and the bending moment M. At each point along the beam the forces are in equilibrium. We find the shear force F at a point distance x from the left-hand end of the beam by considering the vertical equilibrium of forces for the left-hand portion of the beam, and we find the bending moment M by looking at the balance of moments of force for that left-hand portion (see Figure 8.10(a)). The force F is the sum of the forces acting vertically on AP, and M is the sum of moments.

Consider the small element of the beam of length Δx between P and Q shown in Figure 8.10(b). Then examining the balance of moments about Q we see that

$$M(x + \Delta x) = M(x) + \Delta x F(x)$$

so that

$$M(x + \Delta x) - M(x) = \Delta x F(x)$$

Figure 8.10
(a) Horizontal beam.
(b) Element of the beam.

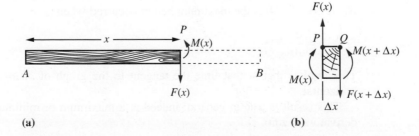

(a)   (b)

giving

$$F(x) = \frac{M(x + \Delta x) - M(x)}{\Delta x}$$

Now letting $\Delta x \to 0$, we obtain

$$F(x) = \frac{dM}{dx}$$

as required.

We saw in Section 7.9.2 that for a freely hinged beam with a point load W at $x = a$

$$M(x) = \begin{cases} W(l - a)x/l & 0 < x \leq a \\ W(l - x)a/l & a \leq x < l \end{cases}$$

$$F(x) = \begin{cases} W - Wa/l & 0 < x < a \\ -Wa/l & a \leq x < l \end{cases}$$

It is left to the reader to verify that these satisfy the equation relating $M(x)$ and $F(x)$.

8.2.7 Maximum and minimum values

In Section 2.3.3 we saw how to find the maximum or minimum value of a quadratic function by the algebraic method of completing the square. The technique was illustrated in Example 2.19. Unfortunately there are no algebraic methods for finding such optimal values in general. The example discussed in Section 2.10 gives an indication of the difficulty of approaching such problems in that way. The ideas of calculus provide us with another approach to such problems.

In Example 8.3 we discussed an object which was projected vertically under gravity. Its height $s(t)$, velocity $v(t)$ and acceleration $a(t)$ were given by

$$s(t) = 25t - 5t^2$$

$$v(t) = \frac{ds}{dt} = 25 - 10t$$

$$a(t) = \frac{dv}{dt} = -10$$

We saw then that it reached its maximum height 31.25 m when $t = 2.5$. For $0 \leq t < 2.5$, $s(t)$ was an increasing function of t and for $t > 2.5$ it was a decreasing function. In terms of its velocity, we can express this as saying that its velocity was positive for $0 \leq t < 2.5$ and negative for $t > 2.5$. At $t = 2.5$ the object was momentarily at rest, that is, $v(2.5) = 0$. Thus the maximum height occurred when

$$v = \frac{ds}{dt} = 0$$

This implies that at that time the tangent to the graph of *distance* against *time* was horizontal.

This result is true in general: indeed at a maximum or minimum of a function, its derivative is zero.

So a *necessary* condition for $y = f(x)$ to have a maximum or minimum at the point $x = a$ is that $f'(a) = 0$. This is confirmed by Figures 8.11(a) and (b), since at a maximum or minimum value of the function its graph has a horizontal tangent.

Unfortunately this is not a *sufficient* condition. There are some functions that have horizontal tangents at non-maximal/minimal values, as shown in Figure 8.11(c). In many elementary applications, however, the problem context will indicate whether such a point is a maximum or a minimum or neither. This is discussed fully in Sections 9.2.1 and 9.5.9.

Figure 8.11
(a) Local maximum.
(b) Local minimum.
(c) Neither maximum nor minimum.

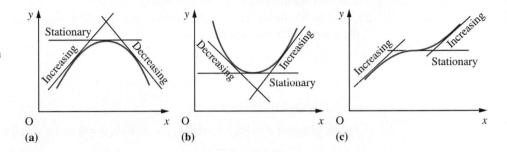

(a) (b) (c)

Example 8.8

An open box, illustrated in Figure 8.12(a), is made from an A4 sheet of card using the folds shown in Figure 8.12(b). Find the dimensions of the tray which maximize its capacity.

Figure 8.12
(a) The open box.
(b) The net of an open box used commercially.

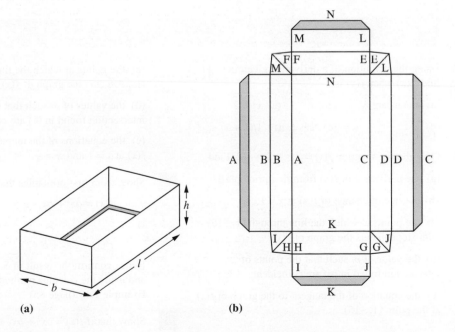

(a) (b)

Solution Boxes like these are used commercially for food sales. Packaging is expensive so that manufacturers often try to design a container that has the biggest capacity for a standard size of cardboard. An A4 sheet has size 210×297 mm.

Allowing 10 mm flaps as stiffeners, shaded in the diagram, and denoting the length, breadth and height by l, b and h (in mm) respectively, we have

$$l + 4h + 20 = 297$$

$$b + 4h + 20 = 210$$

and the capacity C is $l \times b \times h \, \text{mm}^3$. Thus

$$C(h) = (277 - 4h)(190 - 4h)h = 52\,630x - 1868x^2 + 16x^3$$

The maximum capacity C^* occurs where $C'(h) = 0$.

It can be shown from first principles, see Question 5 of Exercises 8.2.8 below, that the general cubic function

$$f(x) = ax^3 + bx^2 + cx + d$$

has derivative

$$f'(x) = 3ax^2 + 2bx + c$$

In this example $a = 16$, $b = -1868$, $c = 52\,630$, $d = 0$ and $x = h$. Thus

$$C'(h) = 52\,630 - 3736h + 48h^2$$

so that the value h^* of h which yields the maximum capacity is $h^* = 18.47$. We can verify that it is a maximum by showing that

$$C'(18.4) > 0 \quad \text{and} \quad C'(18.5) < 0$$

8.2.8 Exercises

1 Using the definition of a derivative given in (8.1), find $f'(x)$ when $f(x)$ is

(a) a constant K (b) x (c) $x^2 - 2$

(d) x^3 (e) $\sqrt{x}$ (f) $1/(1 + x)$

2 Consider the function $f(x) = 2x^2 - 5x - 12$. Find

(a) the derivative of $f(x)$ from first principles,

(b) the rate of change of $f(x)$ at $x = 1$,

(c) the points at which the line through $(1, -15)$ with slope m cuts the graph of $f(x)$,

(d) the value of m such that the points of intersection found in (c) are coincident,

(e) the equation of the tangent to the graph of $f(x)$ at the point $(1, -15)$.

3 Consider the function $f(x) = 2x^3 - 3x^2 + x + 3$. Find

(a) the derivative of $f(x)$ from first principles,

(b) the rate of change of $f(x)$ at $x = 1$,

(c) the points at which the line through $(1, 3)$ with slope m cuts the graph of $f(x)$,

(d) the values of m such that two of the points of intersection found in (c) are coincident,

(e) the equations of the tangents to the graph of $f(x)$ at $x = 1$ and $x = \frac{1}{4}$.

4 Show from first principles that the derivative of

$$f(x) = ax^2 + bx + c$$

is

$$f'(x) = 2ax + b$$

Hence confirm the result of Section 2.3.3 (page 84) and using the calculus method verify the results of Example 2.19 (page 85).

5 Show that if $f(x) = ax^3 + bx^2 + cx + d$, then

$$f(x + \Delta x) = ax^3 + bx^2 + cx + d + (3ax^2 + 2bx \\ + c)\Delta x + (3ax + b)(\Delta x)^2 + a(\Delta x)^3$$

Deduce the formula for $f'(x)$ given in Example 8.8.

6 Gas escapes from a spherical balloon at $2\,\mathrm{m^3\,min^{-1}}$. How fast is the surface area shrinking when the radius equals $12\,\mathrm{m}$? (The surface area of a sphere of radius r is $4\pi r^2$.)

7 A tank is initially filled with 1000 litres of brine, containing $0.15\,\mathrm{kg}$ of salt per litre. Fresh brine containing $0.25\,\mathrm{kg}$ of salt per litre runs into the tank at the rate of $4\,\mathrm{litres\,s^{-1}}$, and the mixture (kept uniform by vigorous stirring) runs out at the same rate. Show that if Q (in kg) is the amount of salt in the tank at time t (in s) then

$$\frac{\mathrm{d}Q}{\mathrm{d}t} = 1 - \frac{Q}{250}$$

8 The displacement–time graph for a vehicle is given by

$$s(t) = \begin{cases} t, & 0 \leqslant t \leqslant 1 \\ t^2 - t + 1, & 1 \leqslant t \leqslant 2 \\ 3t - 3, & 2 \leqslant t \leqslant 3 \\ 9 - t, & 3 \leqslant t \leqslant 9 \end{cases}$$

Obtain the formula for the velocity–time graph.

9 The bending moment $M(x)$ for a beam of length l is given by $M(x) = W(2x - l)^3/8l^2$, $0 \leqslant x \leqslant l$. Find the formula for the shear force F. (See Example 8.7.)

10 A small weight is dragged across a horizontal plane by a string PQ of length a, the end P being attached to the weight while the end Q is made to move steadily along a fixed line perpendicular to the original position of PQ. Choosing the coordinate axes so that Oy is that fixed line and Ox passes through the initial position of P as shown in Figure 8.13, show that the curve $y = y(x)$ described by P is such that

$$\frac{\mathrm{d}y}{\mathrm{d}x} = -\frac{\sqrt{(a^2 - x^2)}}{x}$$

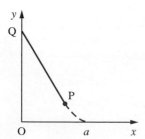

Figure 8.13

11 Consider the function $f(x) = \sqrt{(1 + \sin x)}$. Show that $f(3\pi/2 \pm h) = \sqrt{2}\sin\frac{1}{2}h$ ($h > 0$) and deduce that $f'(x)$ does not exist at $x = 3\pi/2$.

12 The limiting tension in a rope wound round a capstan (that is, the tension when the rope is about to slip) depends on the angle of wrap θ as shown in Figure 8.14. Show that an increase $\Delta\theta$ in the angle of wrap produces a corresponding increase ΔT in the value of the limiting tension such that

$$\Delta T \approx \mu T \Delta\theta$$

where μ is the coefficient of friction. Deduce $\mathrm{d}T/\mathrm{d}\theta$.

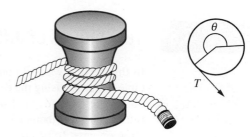

Figure 8.14

13 A chemical dissolves in water at a rate jointly proportional to the amount undissolved and to the difference between the concentration in the solution and that in the saturated solution. Initially none of the chemical is dissolved in the water. Show that the amount $x(t)$ of undissolved chemical satisfies the differential equation

$$\frac{\mathrm{d}x}{\mathrm{d}t} = kx(M - x_0 + x)$$

where k is a constant, M is the amount of the chemical in the saturated solution and $x_0 = x(0)$.

14 The rate at which a solute diffuses through a membrane is proportional to the area and to the concentration difference across the membrane. A solution of concentration C flows down a tube with constant velocity v. The solute diffuses through the wall of the tube into an ambient solution of the same solute of a lower fixed concentration C_0. If the tube has constant circular cross-section of radius r, show that at distance x along the tube the concentration $C(x)$ satisfies the differential equation

$$\frac{\mathrm{d}C}{\mathrm{d}x} = -\frac{2k}{rv}(C - C_0)$$

where k is a constant.

15   A lecture theatre having volume $1000\,m^3$ is designed to seat 200 people. The air is conditioned continuously by an inflow of fresh air at the constant rate V (in $m^3\,min^{-1}$). An average person generates $980\,cm^3$ of CO_2 per minute, while fresh air contains 0.04% of CO_2 by volume. Show that the percentage concentration x of CO_2 by volume in the lecture theatre at time t (in min) after the audience enters satisfies the differential equation

$$1000\frac{dx}{dt} = 19.6 + 0.04V - Vx(t)$$

16   Consider the chemical reaction

$$A + B \rightarrow X$$

Let x be the amount of product X, and a and b the initial amounts of A and B (with x, a and b in mol). The rate of reaction is proportional to the product of the uncombined amounts of A and B remaining. Express this relationship in terms of $\frac{dx}{dt}$, x, a and b.

8.3   Techniques of differentiation

In this section we will obtain the derivatives of some basic functions from 'first principles', that is, using the definition of a derivative given in (8.1), and will show how we obtain the derivatives of other functions using the basic results and some elementary rules. The rules themselves may be derived from the basic definition of the differentiation process. In practice, we make use of a very few basic facts, which, together with the rules, enable us to differentiate a wide variety of functions.

8.3.1   Rules of differentiation

To enable us to exploit the basic derivatives as we obtain them, we will first obtain the rules which make that exploitation possible. These rules you should know 'by heart'.

Rule 1 (constant multiplication rule)

If $y = f(x)$ and k is a constant then

$$\frac{d}{dx}(ky) = k\frac{dy}{dx} = kf'(x)$$

Rule 2 (sum rule)

If $u = f(x)$ and $v = g(x)$ then

$$\frac{d}{dx}(u + v) = \frac{du}{dx} + \frac{dv}{dx} = f'(x) + g'(x)$$

Rule 3 (product rule)

If $u = f(x)$ and $v = g(x)$ then

$$\frac{d}{dx}(uv) = u\frac{dv}{dx} + v\frac{du}{dx} = f(x)g'(x) + g(x)f'(x)$$

Rule 4 (quotient rule)

If $u = f(x)$ and $v = g(x)$ then

$$\frac{d}{dx}\left(\frac{u}{v}\right) = \frac{v(du/dx) - u(dv/dx)}{v^2} = \frac{g(x)f'(x) - f(x)g'(x)}{[g(x)]^2}$$

Rule 5 (composite-function or chain rule)

If $z = g(x)$ and $y = f(z)$ then

$$\frac{dy}{dx} = \frac{dy}{dz}\frac{dz}{dx} = f'(z)g'(x)$$

When z is a linear function of x (that is, $y = f(ax + b)$, with a and b constants),

$$\frac{dy}{dx} = af'(ax + b)$$

Rule 6 (inverse-function rule)

If $y = f^{-1}(x)$ then $x = f(y)$ and

$$\frac{dy}{dx} = \frac{1}{dx/dy} = \frac{1}{f'(y)}$$

Verification of rules

Rule 1 follows directly from the definition given in (8.1), for if

$$g(x) = kf(x), \quad k \text{ constant}$$

$$g(x + \Delta x) = kf(x + \Delta x)$$

and

$$\Delta g = g(x + \Delta x) - g(x) = k[f(x + \Delta x) - f(x)]$$

so that

$$\frac{dg}{dx} = \lim_{\Delta x \to 0}\frac{\Delta g}{\Delta x} = \lim_{\Delta x \to 0} k\left[\frac{f(x + \Delta x) - f(x)}{\Delta x}\right] = kf'(x)$$

using the properties of limits given in Section 7.5.2.

Likewise, for *Rule 2*, if

$$h(x) = f(x) + g(x)$$

$$h(x + \Delta x) = f(x + \Delta x) + g(x + \Delta x)$$

and

$$\Delta h = h(x + \Delta x) - h(x) = [f(x + \Delta x) - f(x)] + [g(x + \Delta x) - g(x)]$$

8.3.2 Derivative of x^r

Using the definition of a derivative given in (8.1) and following the procedure of Example 8.1, we can proceed to obtain the derivative of the power function $f(x) = x^r$ when r is a real number.

Since $f(x) = x^r$ we have

$$f(x + \Delta x) = (x + \Delta x)^r$$

Using the binomial series (7.15) from Section 7.7.2 we have

$$(x + \Delta x)^r = x^r \left(1 + \frac{\Delta x}{x} \right)^r, \quad x \neq 0$$

$$= x^r \left[1 + r\frac{\Delta x}{x} + \frac{1}{2}r(r-1)\left(\frac{\Delta x}{x}\right)^2 + \dots \right]$$

$$= x^r + rx^{r-1}\Delta x + \frac{1}{2}r(r-1)x^{r-2}(\Delta x)^2 + \dots$$

so that

$$\frac{\Delta f}{\Delta x} = \frac{f(x + \Delta x) - f(x)}{\Delta x} = rx^{r-1} + \frac{1}{2}r(r-1)x^{r-2}(\Delta x) + \dots$$

Now letting $\Delta x \to 0$ we have that

$$\frac{df}{dx} = \frac{d(x^r)}{dx} = \lim_{\Delta x \to 0} \frac{\Delta f}{\Delta x} = rx^{r-1}$$

leading to the general result

$$\frac{d}{dx}(x^r) = rx^{r-1}, r \in \mathbb{R} \tag{8.10}$$

Note that the solutions of Example 8.1 satisfy this general result. Note also that (8.10) implies that if k is a constant then $\dfrac{dk}{dx} = 0$, which is as expected since the derivative measures the rate of change of the function.

Example 8.9   Using result (8.10) find $f'(x)$ when $f(x)$ is

(a) $\sqrt{x}$ (b) $\dfrac{1}{x^5}$ (c) $\dfrac{1}{\sqrt[3]{x}}$

Solution   (a) Taking $r = \frac{1}{2}$ in (8.10) gives

$$\frac{d}{dx}(\sqrt{x}) = \frac{d}{dx}(x^{1/2}) = \tfrac{1}{2}x^{-1/2} = \frac{1}{2\sqrt{x}}$$

(b) Taking $r = -5$ in (8.10) gives

$$\frac{d}{dx}\left(\frac{1}{x^5}\right) = \frac{d}{dx}(x^{-5}) = -5x^{-6} = -\frac{5}{x^6}$$

(c) Taking $r = -\frac{1}{3}$ in (8.10) gives

$$\frac{d}{dx}\left(\frac{1}{\sqrt[3]{x}}\right) = \frac{d}{dx}(x^{-1/3}) = -\frac{1}{3}x^{-4/3} = -\frac{1}{3}\left(\frac{1}{\sqrt[3]{x^4}}\right)$$

Using the result

$$\frac{d}{dx}(x^r) = rx^{r-1}$$

given in equation (8.10), together with the rules developed in Section 8.3.1, we proceed in the following sections to find the derivatives of a range of algebraic functions.

8.3.3 Polynomial functions

It is a simple matter to find the derivative of the polynomial function

$$f(x) = a_0 + a_1 x + a_2 x^2 + a_3 x^3 + \ldots + a_{n-1} x^{n-1} + a_n x^n = \sum_{r=0}^{n} a_r x^r \qquad \textbf{(8.11)}$$

where n is a non-negative integer and the coefficients a_r, $r = 0, 1, \ldots, n$, are real numbers.

Using the constant multiplication rule together with the sum rule we may differentiate term by term to give

$$f'(x) = a_1 + 2a_2 x + 3a_3 x^2 + \ldots + (n-1)a_{n-1} x^{n-2} + na_n x^{n-1} = \sum_{r=1}^{n} ra_r x^{r-1}$$

Example 8.10   If $y = 2x^4 - 2x^3 - x^2 + 3x - 2$, find $\dfrac{dy}{dx}$.

Solution   Differentiating term by term, using the sum rule, gives

$$\frac{dy}{dx} = \frac{d}{dx}(2x^4) - \frac{d}{dx}(2x^3) - \frac{d}{dx}(x^2) + \frac{d}{dx}(3x) - \frac{d}{dx}(2)$$

which on using the constant multiplication rule gives

$$\frac{dy}{dx} = 2\frac{d}{dx}(x^4) - 2\frac{d}{dx}(x^3) - \frac{d}{dx}(x^2) + 3\frac{d}{dx}(x) - \frac{d}{dx}(2)$$

$$= 2(4x^3) - 2(3x^2) - (2x) + 3(1) - 0$$

so that

$$\frac{dy}{dx} = 8x^3 - 6x^2 - 2x + 3$$

Example 8.11

The distance s metres moved by a body in t seconds is given by

$$s = 2t^3 - 1.5t^2 - 6t + 12$$

Determine the velocity and acceleration after 2 seconds.

Solution

$$s = 2t^3 - 1.5t^2 - 6t + 12$$

The velocity v (m s^{-1}) is given by $v = \dfrac{ds}{dt}$ so that

$$v = \frac{ds}{dt} = 2(3t^2) - 1.5(2t) - 6(1) = 6t^2 - 3t - 6$$

When $t = 2$ seconds

$$v = 6(4) - 3(2) - 6 = 12$$

so that the velocity after 2 seconds is $12 \, \text{m s}^{-1}$.

The acceleration f (m s^{-2}) is given by $f = \dfrac{dv}{dt}$ so that

$$f = \frac{dv}{dt} = \frac{d}{dt}(6t^2 - 3t - 6) = 12t - 3$$

When $t = 2$ seconds

$$f = 12(2) - 3 = 21$$

so that the acceleration after 2 seconds is $21 \, \text{m s}^{-2}$.

Sometimes polynomial functions are not expressed in the standard form of (8.11), $f(x) = (2x + 5)^3$ and $f(x) = (3x - 1)^2(x + 2)^3$ being such examples. Such cases will be considered in Section 8.3.5 where applications of the chain rule will be discussed.

The derivatives of polynomial functions can be evaluated numerically by a simple extension of the method of synthetic division (or nested multiplication) which is used for evaluating the function itself. We saw in Section 2.4.3 that

$$f(x) = a_n x^n + a_{n-1} x^{n-1} + \ldots + a_1 x + a_0$$

could be written as $f(x) = g(x)(x - c) + f(c)$ where

$$g(x) = b_{n-1} x^{n-1} + b_{n-2} x^{n-2} + \ldots + b_1 x + b_0$$

and where the coefficients $b_{n-1}, \ldots, b_0$ were generated in the process of nested multiplication. Differentiating $f(x)$ with respect to x using the product rule gives

$$f'(x) = g'(x)(x - c) + g(x)(1)$$

so that

$$f'(c) = g'(c)(0) + g(c)(1) = g(c)$$

Thus we can evaluate $f'(c)$ by applying the nested multiplication method again but this time to $g(x)$.

Example 8.12 Evaluate $f(2)$ and $f'(2)$ for the polynomial function

$$f(x) = 2x^4 - 2x^3 - x^2 + 3x - 2$$

Solution

$$
\begin{array}{r|rrrrr}
 & 2 & -2 & -1 & 3 & -2 \\
\times 2 & 0 & 4 & 4 & 6 & 18 \\
\hline
 & 2 & 2 & 3 & 9 & 16 = f(2) \\
\times 2 & 0 & 4 & 12 & 30 & \\
\hline
 & 2 & 6 & 15 & 39 = f'(2) &
\end{array}
$$

This method of evaluating the function and its derivative is very efficient and is often used in computer packages which require finding the roots of polynomial equations.

8.3.4 Rational functions

As we saw in Section 2.5, rational functions have the general form

$$f(x) = \frac{p(x)}{q(x)}$$

where $p(x)$ and $q(x)$ are polynomials. To obtain the derivatives of such functions we make use of the constant multiplication, sum and quotient rules as illustrated in Example 8.13.

Example 8.13 Find the derivative of the following functions of x:

(a) $\dfrac{3x + 2}{2x^2 + 1}$ (b) $\dfrac{2x + 3}{x^2 + x + 1}$

(c) $x^3 + 2x^2 - \dfrac{1}{x} + \dfrac{1}{x^2} + 3, \quad x \neq 0$

Solution (a) Taking $u = 3x + 2$ and $v = 2x^2 + 1$ gives

$$\frac{\mathrm{d}u}{\mathrm{d}x} = 3 \quad \text{and} \quad \frac{\mathrm{d}v}{\mathrm{d}x} = 4x$$

so, from the quotient rule,

$$
\frac{\mathrm{d}}{\mathrm{d}x}\left[\frac{3x+2}{2x^2+1}\right] = \frac{v\left(\dfrac{\mathrm{d}u}{\mathrm{d}x}\right) - u\left(\dfrac{\mathrm{d}v}{\mathrm{d}x}\right)}{v^2}
$$

$$
= \frac{(2x^2+1)3 - (3x+2)4x}{(2x^2+1)^2}
$$

$$
= \frac{-(6x^2 + 8x - 3)}{(2x^2+1)^2}
$$

(b) Taking $u = 2x + 3$ and $v = x^2 + x + 1$

$$\frac{du}{dv} = 2 \quad \text{and} \quad \frac{dv}{dx} = 2x + 1$$

so, from the quotient rule,

$$\frac{d}{dx}\left[\frac{2x + 3}{x^2 + x + 1}\right] = \frac{(x^2 + x + 1)(2) - (2x + 3)(2x + 1)}{(x^2 + x + 1)^2}$$

$$= -\frac{(2x^2 + 6x + 1)}{(x^2 + x + 1)^2}$$

(c) In this case we can express the function as

$$y = x^3 + 2x^2 - x^{-1} + x^{-2} + 3, \quad x \neq 0$$

and differentiate term by term to give

$$\frac{dy}{dx} = 3x^2 + 2(2x) - (-x^{-2}) + (-2x^{-3}) + 0 = 3x^2 + 4x + x^{-2} - 2x^{-3}$$

$$= 3x^2 + 4x + \frac{1}{x^2} - \frac{2}{x^3}, \quad x \neq 0$$

8.3.5 Other algebraic functions

In dealing with more elaborate algebraic functions we make extensive use of the chain rule. Adapting Figure 2.12 (Section 2.2.4), the chain rule (Rule 5 of Section 8.3.1)

$$\frac{dy}{dx} = \frac{dy}{dz} \cdot \frac{dz}{dx}$$

may be represented as in Figure 8.17.

Figure 8.17
The chain rule of
differentiation.

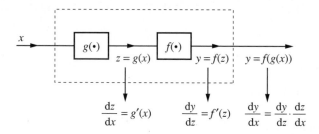

Example 8.14 Find $\dfrac{dy}{dx}$ when y is

(a) $(5x^2 + 11)^9$ (b) $\sqrt{(3x^2 + 1)}$

Solution (a) In this case we could expand out $(5x^2 + 11)^9$ and treat it as a polynomial of degree 18. However, it is advantageous to view it as a composite function, as represented in Figure 8.18(a). Thus, taking

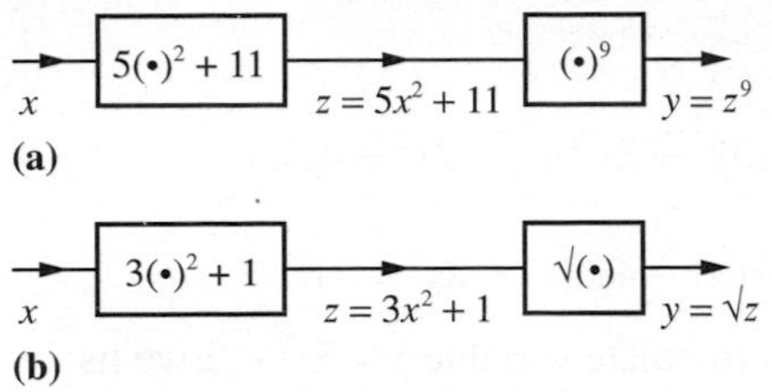

Figure 8.18
(a) Representation
of $y = (5x^2 + 11)^9$.
(b) Representation
of $y = \sqrt{(3x^2 + 1)}$.

(a)

(b)

$$y = z^9 \quad \text{and} \quad z = 5x^2 + 11$$

$$\frac{dy}{dz} = 9z^8 \quad \text{and} \quad \frac{dz}{dx} = 10x$$

so, by the chain rule,

$$\frac{dy}{dx} = \frac{dy}{dz}\frac{dz}{dx} = 9(5x^2 + 11)^8(10x) = 90x(5x^2 + 11)^8$$

(b) The composite function $y = \sqrt{(3x^2 + 1)}$ may be represented as in Figure 8.18(b). Thus, taking

$$y = \sqrt{z} = z^{1/2} \quad \text{and} \quad z = 3x^2 + 1$$

$$\frac{dy}{dz} = \tfrac{1}{2}z^{-1/2} = \frac{1}{2\sqrt{z}} \quad \text{and} \quad \frac{dz}{dx} = 6x$$

so, by the chain rule,

$$\frac{dy}{dx} = \frac{dy}{dz}\frac{dz}{dx} = \frac{1}{2\sqrt{(3x^2 + 1)}}6x = \frac{3x}{\sqrt{(3x^2 + 1)}}$$

It is usual to refer to z as the **intermediate** (or auxiliary) variable, and once the process has been understood the schematic representation, by a block diagram, stage is dispensed with.

Example 8.15   Find $\dfrac{dy}{dx}$ when y is

(a) $(3x^3 - 2x^2 + 1)^5$ (b) $\dfrac{1}{(5x^2 - 2)^7}$

(c) $(x^2 + 1)^3\sqrt{(x - 1)}$ (d) $\dfrac{\sqrt{(2x + 1)}}{(x^2 + 1)^3}$

Solution   (a) Introducing the intermediate variable $z = 3x^3 - 2x^2 + 1$ we have

$$y = z^5 \quad \text{and} \quad z = 3x^3 - 2x^2 + 1$$

$$\frac{dy}{dz} = 5z^4 \quad \text{and} \quad \frac{dz}{dx} = 9x^2 - 4x$$

so, by the chain rule,

$$\frac{dy}{dx} = \frac{dy}{dz}\frac{dz}{dx} = 5(3x^3 - 2x^2 + 1)^4(9x^2 - 4x)$$

$$= 5x(9x - 4)(3x^3 - 2x^2 + 1)^4$$

(b) Introducing the intermediate variable $z = 5x^2 - 2$ we have

$$y = \frac{1}{z^7} = z^{-7} \qquad \text{and} \quad z = 5x^2 - 2$$

$$\frac{dy}{dz} = -7z^{-8} = -\frac{7}{z^8} \quad \text{and} \quad \frac{dz}{dx} = 10x$$

so, by the chain rule,

$$\frac{dy}{dx} = \frac{dy}{dz}\frac{dz}{dx} = -\frac{7}{(5x^2 - 2)^8}10x = -\frac{70x}{(5x^2 - 2)^8}$$

(c) In this case we are dealing with the product $y = uv$ where $u = (x^2 + 1)^3$ and $v = \sqrt{(x - 1)}$. Then by the product rule

$$\frac{dy}{dx} = u\frac{dv}{dx} + v\frac{du}{dx}$$

To find $\dfrac{du}{dx}$ we introduce the intermediate variable $z = x^2 + 1$ giving $u = z^3$ and $z = x^2 + 1$, so, by the chain rule,

$$\frac{du}{dx} = \frac{du}{dz}\frac{dz}{dx} = 3(x^2 + 1)^2 2x = 6x(x^2 + 1)^2$$

Likewise, to find $\dfrac{dv}{dx}$ we introduce the intermediate variable $w = x - 1$ giving $v = \sqrt{w} = w^{1/2}$ and $w = x - 1$, so by the chain rule

$$\frac{dv}{dx} = \frac{1}{2\sqrt{(x - 1)}}$$

It then follows from the product rule that

$$\begin{array}{cccc} u & \dfrac{dv}{dx} & v & \dfrac{du}{dx} \\ \downarrow & \downarrow & \downarrow & \downarrow \end{array}$$

$$\frac{dy}{dx} = (x^2 + 1)^3 \frac{1}{2\sqrt{(x - 1)}} + \sqrt{(x - 1)}\,6x(x^2 + 1)^2$$

$$= \frac{(x^2 + 1)^2}{2\sqrt{(x - 1)}}[(x^2 + 1) + 12x(x - 1)]$$

$$= \frac{(x^2 + 1)^2(13x^2 - 12x + 1)}{2\sqrt{(x - 1)}}$$

(d) In this case we are dealing with the quotient

$$y = \frac{u}{v}$$

where $u = \sqrt{(2x + 1)}$ and $v = (x^2 + 1)^3$. Then by the quotient rule

$$\frac{dy}{dx} = \frac{v\left(\dfrac{du}{dx}\right) - u\left(\dfrac{dv}{dx}\right)}{v^2}$$

To find $\dfrac{du}{dx}$ we introduce the intermediate variable $z = 2x + 1$, giving $u = z^{1/2}$ and $z = 2x + 1$, so by the chain rule

$$\frac{du}{dx} = \frac{du}{dz}\frac{dz}{dx} = \frac{1}{\sqrt{(2x + 1)}}$$

To find $\dfrac{dv}{dx}$ we introduce the intermediate variable $w = x^2 + 1$, giving $v = w^3$ and $w = x^2 + 1$, so by the chain rule

$$\frac{dv}{dx} = \frac{dv}{dw}\frac{dw}{dx} = 3(x^2 + 1)^2 2x = 6x(x^2 + 1)^2$$

Then, by the quotient rule,

$$\begin{array}{ccccc} v & \dfrac{du}{dx} & u & \dfrac{dv}{dx} & v^2 \\ \downarrow & \downarrow & \downarrow & \downarrow & \end{array}$$

$$\frac{dy}{dx} = \frac{(x^2 + 1)^3 \dfrac{1}{\sqrt{(2x + 1)}} - \sqrt{(2x + 1)}\,6x(x^2 + 1)^2}{(x^2 + 1)^6} = \frac{1 - 6x - 11x^2}{(x^2 + 1)^4 \sqrt{(2x + 1)}}$$

8.3.6 Exercises

17 Differentiate the function f where $f(x)$ is

(a) $5x^2 - 2x + 1$ (b) $4x^3 + x - 8$

(c) $x^{24} + 3$ (d) $(5x + 3)^9$

(e) $(4x - 2)^7$ (f) $(1 - 3x)^6$

(g) $(3x^2 - x + 1)^3$

(h) $(4x^3 - 2x + 1)^6$

(i) $(1 + x - x^4)^5$

18 Differentiate the function f where $f(x)$ is

(a) $(2x + 4)^7(3x - 2)^5$ (b) $(5x + 1)^3(3 - 2x)^4$

(c) $(\frac{1}{2}x + 2)^2(x + 3)^4$

(d) $(x^2 + x - 2)(3x^2 - 5x + 1)$

(e) $(x^4 - 3x + 1)(6x^2 + 5)$

(f) $(x^2 + x + 1)^2(x^3 + 2x^2 + 1)^4$

(g) $(x^5 + 2x + 1)^3(2x^2 + 3x - 1)^4$

(h) $(2x + 1)^3(7 - x)^5$ (i) $(x^2 + 4x + 1)(3x + 1)^5$

19 An open water conduit is to be cut in the shape of an isosceles trapezium and lined with material which is available in a standard width of 1 metre as shown in Figure 8.19.

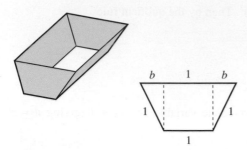

Figure 8.19 Water conduit of Question 19.

To achieve maximum potential capacity, the designer has to maximize the area of cross-section $A(b)$. Show that

$$A(b) = [(1 + b)^3(1 - b)]^{1/2}$$

and that this is maximized when $b = 0.5$.

20 A carton is made from a sheet of A4 card (210 mm × 297 mm) using the net shown in Figure 8.20. Find the dimensions that yield the largest capacity.

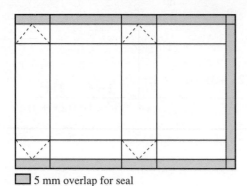

▭ 5 mm overlap for seal

Figure 8.20 Net used in Question 20.

21 Differentiate:

(a) $\sqrt{(1 + 2x)}$ (b) $(x + 2)\sqrt{x}$ (c) $x\sqrt{(x + 2)}$

22 Differentiate:

(a) $x\sqrt{(4 + x^2)}$ (b) $x\sqrt{(9 - x^2)}$

(c) $(x + 1)\sqrt{(x^2 + 2x + 3)}$ (d) $x^{2/3} - x^{1/4}$

(e) $\sqrt[3]{(x^2 + 1)}$ (f) $x(2x - 1)^{1/3}$

23 Differentiate:

(a) $1/(x + 3)^2$ (b) $x/(x + 1)$

(c) $(x - 3)/(x - 2)$ (d) $1/(x^2 - 4x + 1)$

(e) $\left(\sqrt{x} + \dfrac{1}{\sqrt{x}}\right)^2$ (f) $x/(x^2 + 5x + 6)$

(g) $x/\sqrt{(x^2 - 1)}$ (h) $(2x + 1)^2/(3x^2 + 1)^3$

24 A fruit juice manufacturer wishes to design a carton that has a square face as shown in Figure 8.21(a). The carton is to contain 1 litre of juice and is made from a rectangular sheet of waxed cardboard by folding it into a rectangular tube and sealing down the edge and then folding and sealing the top and bottom. To make the carton airtight and robust for handling, an overlap of at least 0.5 cm is needed. The net for the carton is shown in Figure 8.21(b).

Show that the amount $A(h)$ cm^2 of card used is given by

$$A(h) = \left[h + \frac{1000}{h^2} + 1\right]\left[2h + \frac{2000}{h^2} + 0.5\right]$$

Verify that

$$A(h) = 2\left[h + \frac{1000}{h^2} + \tfrac{5}{8}\right]^2 - \tfrac{9}{32}$$

and hence find the value h^* of h which minimizes $A(h)$.

(a)

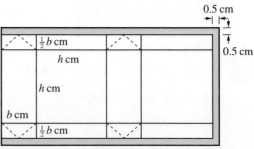

(b)

Figure 8.21 Carton of Question 24.

8.3.7 Differentiation of circular functions

Taking $f(x) = \sin x$ and using the sum identity (2.25b) gives from the formal definition (8.1)

$$f'(x) = \lim_{\Delta x \to 0} \frac{\sin(x + \Delta x) - \sin x}{\Delta x} = \lim_{\Delta x \to 0} \frac{\cos(x + \frac{1}{2}\Delta x)\sin(\frac{1}{2}\Delta x)}{\frac{1}{2}\Delta x}$$

Since, from Section 7.8.1,

$$\lim_{\Delta x \to 0} \frac{\sin(\frac{1}{2}\Delta x)}{\frac{1}{2}\Delta x} = 1 \quad \text{and} \quad \lim_{\Delta x \to 0} \cos(x + \frac{1}{2}\Delta x) = \cos x$$

we have

$$f'(x) = \frac{d}{dx}(\sin x) = \cos x \tag{8.12}$$

Likewise, using the sum identity (2.25d), we have

$$\frac{\cos(x + \Delta x) - \cos x}{\Delta x} = \frac{-\sin(x + \frac{1}{2}\Delta x)\sin(\frac{1}{2}\Delta x)}{\frac{1}{2}\Delta x}$$

from which we deduce using (8.1) that

$$\frac{d}{dx}(\cos x) = -\sin x \tag{8.13}$$

Since $\tan x = \sin x/\cos x$, we take $u = \sin x$ and $v = \cos x$, giving

$$\frac{du}{dx} = \cos x \quad \text{and} \quad \frac{dv}{dx} = -\sin x$$

Then, from the quotient rule,

$$\frac{d}{dx}(\tan x) = \frac{(\cos x)(\cos x) - (\sin x)(-\sin x)}{\cos^2 x}$$

$$= \frac{\cos^2 x + \sin^2 x}{\cos^2 x} = \frac{1}{\cos^2 x}$$

That is,

$$\frac{d}{dx}(\tan x) = \sec^2 x \tag{8.14}$$

Since $\sec x = 1/\cos x$, we take $u = 1$ and $v = \cos x$ in the quotient rule to give

$$\frac{d}{dx}(\sec x) = \frac{(\cos x)(0) - (1)(-\sin x)}{\cos^2 x} = \frac{1}{\cos x}\frac{\sin x}{\cos x}$$

That is,

$$\frac{d}{dx}(\sec x) = \sec x \tan x \qquad (8.15)$$

Since $\operatorname{cosec} x = 1/\sin x$, following the same procedure as above we obtain

$$\frac{d}{dx}(\operatorname{cosec} x) = -\operatorname{cosec} x \cot x \qquad (8.16)$$

Since $\cot x = \cos x/\sin x$, taking $u = \cos x$ and $v = \sin x$ and using the quotient rule gives

$$\frac{d}{dx}(\cot x) = -\operatorname{cosec}^2 x \qquad (8.17)$$

Taking $y = \sin^{-1}x$, we have $x = \sin y$, so that

$$\frac{dx}{dy} = \cos y$$

Then, from the inverse-function rule,

$$\frac{dy}{dx} = \frac{1}{\cos y}$$

Using the identity $\cos^2 y = 1 - \sin^2 y$, this simplifies to

$$\frac{d}{dx}(\sin^{-1}x) = \frac{1}{\sqrt{(1 - x^2)}}, \quad |x| < 1 \qquad (8.18)$$

(Note that we have taken the positive square root, since from Figure 2.65(b) the derivative must be positive.)

Taking $y = \cos^{-1}x$, we have $x = \cos y$, so that

$$\frac{dy}{dx} = \frac{1}{dx/dy} = -\frac{1}{\sin y}$$

which, using the identity $\sin^2 y = 1 - \cos^2 y$, reduces to

$$\frac{d}{dx}(\cos^{-1}x) = -\frac{1}{\sqrt{(1 - x^2)}}, \quad |x| < 1 \qquad (8.19)$$

(Note from Figure 2.66 that the derivative is negative.)

Taking $y = \tan^{-1}x$, we have $x = \tan y$, so that

$$\frac{dy}{dx} = \frac{1}{dx/dy} = \frac{1}{\sec^2 y}$$

Using the identity $1 + \tan^2 y = \sec^2 y$, this reduces to

$$\frac{d}{dx}(\tan^{-1}_\bullet x) = \frac{1}{1 + x^2} \qquad (8.20)$$

Summary

$$\frac{d}{dx}(\sin x) = \cos x, \quad \frac{d}{dx}(\cos x) = -\sin x$$

$$\frac{d}{dx}(\tan x) = \sec^2 x, \quad \frac{d}{dx}(\sec x) = \sec x \tan x$$

$$\frac{d}{dx}(\operatorname{cosec} x) = -\operatorname{cosec} x \cot x, \quad \frac{d}{dx}(\cot x) = -\operatorname{cosec}^2 x$$

$$\frac{d}{dx}(\sin^{-1} x) = \frac{1}{\sqrt{(1-x^2)}}, \quad |x| < 1$$

$$\frac{d}{dx}(\cos^{-1} x) = \frac{-1}{\sqrt{(1-x^2)}}, \quad |x| < 1$$

$$\frac{d}{dx}(\tan^{-1} x) = \frac{1}{1+x^2}$$

Example 8.16 Find $\dfrac{dy}{dx}$ when y is given by:

(a) $\sin(2x + 3)$ (b) $x^2 \cos x$ (c) $\dfrac{\sin 2x}{x^2 + 2}$

(d) $\sec 6x$ (e) $x \tan 2x$ (f) $\sin^{-1} 6x$

(g) $x^2 \cos^{-1} x$ (h) $\tan^{-1} \dfrac{2x}{1+x^2}$

Solution (a) Introducing the intermediate variable $z = 2x + 3$, we have $y = \sin z$ and $z = 2x + 3$, so by the chain rule

$$\frac{dy}{dx} = \frac{dy}{dz}\frac{dz}{dx} = \cos(2x + 3)2 = 2\cos(2x + 3)$$

Note that this result could have been written using the particular linear case of the composite-function rule given in Section 8.3.1.

(b) Taking $u = x^2$ and $v = \cos x$ gives

$$\frac{du}{dx} = 2x \quad \text{and} \quad \frac{dv}{dx} = -\sin x$$

so by the product rule

$$\frac{dy}{dx} = x^2 \sin x + 2x \cos x$$

(c) Taking $u = \sin 2x$ and $v = x^2 + 2$ gives

$$\frac{du}{dx} = 2\cos 2x \quad \text{and} \quad \frac{dv}{dx} = 2x$$

so by the quotient rule

$$\frac{dy}{dx} = \frac{2(x^2 + 2)\cos 2x - 2x\sin 2x}{(x^2 + 2)^2}$$

(d) Introducing the intermediate variable $z = 6x$ we have

$y = \sec z$ and $z = 6x$ so by the chain rule

$$\frac{dy}{dx} = \frac{dy}{dz}\frac{dz}{dx} = 6\sec 6x \tan 6x$$

(e) Taking $u = x$ and $v = \tan 2x$ gives

$$\frac{du}{dx} = 1 \quad \text{and} \quad \frac{dv}{dx} = 2\sec^2 2x$$

where the chain rule, with intermediate variable $z = 2x$, has been used to find $\dfrac{dv}{dx}$. Then by the product rule

$$\frac{dy}{dx} = 2x\sec^2 2x + \tan 2x$$

(f) Introducing the intermediate variable $z = 6x$ we have $y = \sin^{-1}z$ and $z = 6x$

$$\frac{dy}{dz} = \frac{1}{\sqrt{(1 - z^2)}} \quad \text{and} \quad \frac{dz}{dx} = 6, \quad |z| < 1$$

so by the chain rule

$$\frac{dy}{dx} = \frac{dy}{dz}\frac{dz}{dx} = \frac{6}{\sqrt{(1 - 36x)}}, \quad |x| < \tfrac{1}{6}$$

(g) Taking $u = x^2$ and $v = \cos^{-1}x$ gives

$$\frac{du}{dx} = 2x \quad \text{and} \quad \frac{dv}{dx} = -\frac{1}{\sqrt{(1 - x^2)}}$$

so by the quotient rule

$$\frac{dy}{dx} = -\frac{x^2}{\sqrt{(1 - x^2)}} + 2x\cos^{-1}x$$

(h) Introducing the intermediate variable $z = \dfrac{2x}{1 + x^2}$ we have

$$y = \tan^{-1}z \quad \text{and} \quad z = \frac{2x}{1 + x^2}$$

$$\frac{dy}{dz} = \frac{1}{1 + z^2} \quad \text{and} \quad \frac{dz}{dx} = \frac{(1 + x^2)2 - 2x(2x)}{(1 + x^2)^2} = \frac{2(1 - x^2)}{(1 + x^2)^2}$$

so from the chain rule

$$\frac{dy}{dx} = \frac{dy}{dz}\frac{dz}{dx} = \frac{1}{1 + \left(\dfrac{2x}{1 + x^2}\right)^2} \cdot \frac{2(1 - x^2)}{(1 + x^2)^2} = \frac{2(1 - x^2)}{(1 + x^2)^2 + (2x)^2}$$

$$= \frac{2(1 - x^2)}{x^4 + 6x^2 + 1}$$

8.3.8   Extended form of the chain rule

Sometimes there are more than two component functions involved in a composite function. For example, consider the composite function

$$y = f(w), \quad w = g(z), \quad z = h(x)$$

which may be represented schematically by the block diagram of Figure 8.22. To obtain the derivative $\dfrac{dy}{dx}$ we first consider y as a composite function of h and the 'dotted box', giving, on applying the chain rule,

$$\frac{dy}{dx} = \frac{dy}{dz}\frac{dz}{dx}$$

Figure 8.22
Composite function containing three component functions.

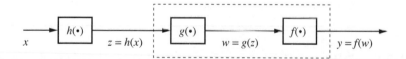

Reapplying the chain rule, this time with z as the domain variable, gives

$$\frac{dy}{dz} = \frac{dy}{dw}\frac{dw}{dz}$$

which on back substitution gives

$$\frac{dy}{dx} = \frac{dy}{dw}\frac{dw}{dz}\frac{dz}{dx}$$

as the extended form of the chain rule.

Example 8.17   Find $\dfrac{dy}{dx}$ when y is given by

(a) $\sin^2(x^2 + 1)$ (b) $\cos^{-1}\sqrt{(1 - x^2)}$

Solution (a) Introducing the intermediate variables $z = x^2 + 1$ and $w = \sin z$, then

$$y = w^2, \qquad w = \sin z, \qquad z = x^2 + 1$$

$$\frac{dy}{dw} = 2w, \quad \frac{dw}{dz} = \cos z, \quad \frac{dz}{dx} = 2x$$

so by the extended chain rule

$$\frac{dy}{dx} = \frac{dy}{dw}\frac{dw}{dz}\frac{dz}{dx} = (2w)(\cos z)(2x)$$

Since $z = x^2 + 1$ and $w = \sin z = \sin(x^2 + 1)$,

$$\frac{dy}{dx} = 4x\sin(x^2 + 1)\cos(x^2 + 1)$$

(b) Introducing the intermediate variables $z = 1 - x^2$ and $w = \sqrt{z}$, then

$$y = \cos^{-1}w, \qquad w = z^{1/2}, \qquad z = 1 - x^2$$

$$\frac{dy}{dw} = -\frac{1}{\sqrt{(1 - w^2)}}, \quad \frac{dw}{dz} = \tfrac{1}{2}z^{-1/2}, \quad \frac{dz}{dx} = -2x$$

Since $z = 1 - x^2$ and $w = \sqrt{(1 - x^2)}$ we have, by the extended chain rule,

$$\frac{dy}{dx} = \frac{dy}{dw}\frac{dw}{dz}\frac{dz}{dx} = \left(-\frac{1}{\sqrt{[1 - (1 - x^2)]}}\right)\left(\frac{1}{2\sqrt{(1 - x^2)}}\right)(-2x) = \frac{1}{\sqrt{(1 - x^2)}}$$

Here we have assumed $0 < x < 1$. If $-1 < x < 0$ the derivative is $-1/\sqrt{(1 - x^2)}$. The function has no derivative at $x = 0$.

8.3.9 Differentiation of exponential and related functions

The formal definition (8.1) gives the derivative of e^x as

$$\frac{d}{dx}(e^x) = \lim_{\Delta x \to 0}\frac{e^{x+\Delta x} - e^x}{\Delta x} = \lim_{\Delta x \to 0}\frac{e^x(e^{\Delta x} - 1)}{\Delta x}$$

$$= e^x\lim_{\Delta x \to 0}\frac{1 + \Delta x + (\Delta x)^2/2! + (\Delta x)^3/3! + \ldots - 1}{\Delta x}$$

(using (7.16))

$$= e^x\lim_{\Delta x \to 0}[1 + \tfrac{1}{2}\Delta x + \tfrac{1}{6}(\Delta x)^2 + \ldots]$$

so that

$$\frac{d}{dx}(e^x) = e^x \tag{8.21}$$

Thus the exponential function (to base e) has the special property that it is its own derivative.

Taking $y = \ln x$, we have $x = e^y$ so that

$$\frac{dx}{dy} = e^y$$

Then, from the inverse-function rule,

$$\frac{dy}{dx} = \frac{1}{e^y} = \frac{1}{x}$$

That is

$$\frac{d}{dx}(\ln x) = \frac{1}{x}, \quad x > 0 \qquad\qquad (8.22)$$

Example 8.18 Find $\dfrac{dy}{dx}$ when y is given by:

(a) $x^2 e^x$ (b) $3e^{-2x}$ (c) $\dfrac{\ln x}{x^2}$

(d) $\ln(x^2 + 1)$ (e) $e^{-x}(\sin x + \cos x)$

Solution (a) Taking $u = x^2$ and $v = e^x$

$$\frac{du}{dx} = 2x \quad\text{and}\quad \frac{dv}{dx} = e^x$$

Then by the product rule

$$\frac{dy}{dx} = x^2 e^x + 2x e^x = x(x+2)e^x$$

(b) Introducing the intermediate variable $z = -2x$ then

$$y = 3e^z \quad\text{and}\quad z = -2x$$

$$\frac{dy}{dz} = 3e^z \quad\text{and}\quad \frac{dz}{dx} = -2$$

so by the chain rule

$$\frac{dy}{dx} = \frac{dy}{dz}\frac{dz}{dx} = (3e^{-2x})(-2) = -6e^{-2x}$$

(c) Taking $u = \ln x$ and $v = x^2$ gives

$$\frac{du}{dx} = \frac{1}{x} \quad\text{and}\quad \frac{dv}{dx} = 2x \quad (x \neq 0)$$

so by the quotient rule

$$\frac{d}{dx}\left(\frac{\ln x}{x^2}\right) = \frac{(1/x)x^2 - (\ln x)(2x)}{x^4} = \frac{1 - 2\ln x}{x^3}$$

(d) Introducing the intermediate variable $z = x^2 + 1$ then

$$y = \ln z \quad \text{and} \quad z = x^2 + 1$$

$$\frac{dy}{dz} = \frac{1}{z} \quad \text{and} \quad \frac{dz}{dx} = 2x$$

so by the chain rule

$$\frac{dy}{dx} = \frac{dy}{dz}\frac{dz}{dx} = \frac{1}{x^2 + 1}(2x) = \frac{2x}{x^2 + 1}$$

(e) Taking $u = e^{-x}$ and $v = \sin x + \cos x$

$$\frac{du}{dx} = -e^{-x} \quad \text{and} \quad \frac{dv}{dx} = \cos x - \sin x$$

Then by the product rule

$$\frac{dy}{dx} = e^{-x}(\cos x - \sin x) + (\sin x + \cos x)(-e^{-x})$$

$$= -2e^{-x}\sin x$$

The hyperbolic functions, introduced in Section 2.7.3, are closely related to the exponential function and their derivatives are readily deduced. From their definitions

$$\frac{d}{dx}(\sinh x) = \frac{d}{dx}\left[\frac{e^x - e^{-x}}{2}\right] = \tfrac{1}{2}(e^x + e^{-x}) = \cosh x \tag{8.23a}$$

$$\frac{d}{dx}(\cosh x) = \frac{d}{dx}\left[\frac{e^x + e^{-x}}{2}\right] = \tfrac{1}{2}(e^x - e^{-x}) = \sinh x \tag{8.23b}$$

$$\frac{d}{dx}(\tanh x) = \frac{d}{dx}\left[\frac{\sinh x}{\cosh x}\right] = \frac{(\cosh x)(\cosh x) - (\sinh x)(\sinh x)}{\cosh^2 x}$$

$$= \frac{1}{\cosh^2 x} = \operatorname{sech}^2 x \tag{8.23c}$$

$$\frac{d}{dx}(\operatorname{sech} x) = \frac{d}{dx}\left[\frac{1}{\cosh x}\right] = \frac{-\sinh x}{\cosh^2 x} = -\operatorname{sech} x \tanh x \tag{8.23d}$$

$$\frac{d}{dx}(\operatorname{cosech} x) = \frac{d}{dx}\left[\frac{1}{\sinh x}\right] = -\operatorname{cosech} x \coth x \tag{8.23e}$$

$$\frac{d}{dx}(\coth x) = \frac{d}{dx}\left[\frac{\cosh x}{\sinh x}\right] = -\operatorname{cosech}^2 x \tag{8.23f}$$

Following the same procedure as for the inverse circular functions in Section 8.3.7 the following derivatives of the inverse hyperbolic functions are readily obtained.

$$\frac{d}{dx}(\sinh^{-1}x) = \frac{1}{\sqrt{(1 + x^2)}}$$ (8.24a)

$$\frac{d}{dx}(\cosh^{-1}x) = \frac{1}{\sqrt{(x^2 - 1)}}, \quad x > 1$$ (8.24b)

$$\frac{d}{dx}(\tanh^{-1}x) = \frac{1}{1 - x^2}, \quad |x| < 1$$ (8.24c)

Example 8.19 Find $\dfrac{dy}{dx}$ when y is given by

(a) $\tanh 2x$ (b) $\cosh^2 x$ (c) $e^{-3x} \sinh 3x$ (d) $\sinh^{-1}\left[\dfrac{3x}{4}\right]$

Solution (a) Introducing the intermediate variable $z = 2x$ gives

$$y = \tanh z \qquad \text{and} \quad z = 2x$$

$$\frac{dy}{dz} = \text{sech}^2 z \quad \text{and} \quad \frac{dz}{dx} = 2$$

so by the chain rule

$$\frac{dy}{dx} = 2\,\text{sech}^2(2x)$$

(b) Introducing the intermediate variable $z = \cosh x$ gives

$$y = z^2 \qquad \text{and} \quad z = \cosh x$$

$$\frac{dy}{dz} = 2z \quad \text{and} \quad \frac{dz}{dx} = \sinh x$$

so by the chain rule

$$\frac{dy}{dx} = 2\cosh x \sinh x = \sinh 2x$$

(c) Taking $u = e^{-3x}$ and $v = \sinh 3x$

gives using the chain rule

$$\frac{du}{dx} = -3e^{-3x} \quad \text{and} \quad \frac{dv}{dx} = 3\cosh 3x$$

so by the product rule

$$\frac{dy}{dx} = (e^{-3x})(3\cosh 3x) + (\sinh 3x)(-3e^{-3x}) = 3e^{-3x}(\cosh 3x - \sinh 3x)$$

(d) Introducing the intermediate variable $z = \frac{3}{4}x$ gives

$$y = \sinh^{-1}z \qquad \text{and} \qquad z = \tfrac{3}{4}x$$

$$\frac{dy}{dz} = \frac{1}{\sqrt{(1 + z^2)}} \qquad \text{and} \qquad \frac{dz}{dx} = \tfrac{3}{4}$$

so by the chain rule

$$\frac{dy}{dx} = \frac{3}{4} \cdot \frac{1}{\sqrt{(1 + \frac{9}{16}x^2)}} = \frac{3}{\sqrt{(16 + 9x^2)}}$$

8.3.10 Exercises

25 Differentiate with respect to x:

(a) $\sin(3x - 2)$ (b) $\cos^4 x$

(c) $\cos^2 3x$ (d) $\sin 2x \cos 3x$

(e) $x \sin x$ (f) $\sqrt{(2 + \cos 2x)}$

(g) $a\cos(x + \theta)$ (h) $\tan 4x$

26 Differentiate with respect to x:

(a) $\sin^{-1}(x/2)$ (b) $\cos^{-1}(5x)$

(c) $\sqrt{(1 + x^2)}\tan^{-1}x$ (d) $\sin^{-1}((x - 1)/2)$

(e) $\tan^{-1}3x$ (f) $\sqrt{(1 - x^2)}\sin^{-1}x$

27 Differentiate with respect to x:

(a) e^{2x} (b) $e^{-x/2}$

(c) $\exp(x^2 + x)$ (d) $x^2 e^{5x}$

(e) $(3x + 2)e^{-x}$ (f) $e^x/(1 + e^x)$

(g) $\sqrt{(1 + e^x)}$ (h) e^{ax+b}

28 Differentiate with respect to x:

(a) $\ln(2x + 3)$ (b) $\ln(x^2 + 2x + 3)$

(c) $\ln[(x - 2)/(x - 3)]$ (d) $\frac{1}{x}\ln x$

(e) $\ln[(2x + 1)/(1 - 3x)]$ (f) $\ln[(x + 1)x]$

29 Differentiate with respect to x:

(a) $\sinh 3x$ (b) $\tanh 4x$ (c) $x^3 \cosh 2x$

(d) $\ln[\cosh \tfrac{1}{2}x]$ (e) $\cos x \cosh x$ (f) $1/\cosh x$

30 Differentiate with respect to x:

(a) $\sinh^{-1}2x$ (b) $\cosh^{-1}(2x^2 - 1)$

(c) $\tanh^{-1}(1/x)$ (d) $\sqrt{(1 + x^2)}\sinh^{-1}x$

(e) $\sqrt{(4 - x^2)} - 2\cosh^{-1}(2/x)$

(f) $\tanh^{-1}x/(1 + x^2)$

31 Draw a careful sketch of $y = e^{-ax}\sin \omega x$ where a and ω are positive constants. What is the ratio of the heights of successive maxima of the function?

32 The line AB joins the points $A(a, 0)$, $B(0, b)$ on the x and y axes respectively and passes through the point $(8, 27)$. Find the positions of A and B which minimize the length of AB.

33 Sketch the curve $y = e^{-x^2}$. Find the rectangle inscribed under the curve having one edge on the x axis, which has maximum area.

34 Show that $y = 9e^{-9t}/(10 - e^{-9t})$ satisfies the differential equation

$$\frac{dy}{dt} = -y(9 + y)$$

8.3.11 Parametric and implicit differentiation

The chain rule is used with the inverse-function rule to evaluate derivatives when a function is specified **parametrically**.

In general, if a function is defined by $y = f(x)$, where $x = g(t)$ and $y = h(t)$ and t is a parameter, then

$$\frac{dy}{dx} = \frac{dy}{dt} \Big/ \frac{dx}{dt} \quad \text{or} \quad \frac{dy}{dx} = \frac{dy}{dt}\frac{dt}{dx}$$

(8.25)

Example 8.20   The function $y = f(x)$ is defined by $x = t^3$, $y = t^2$ ($t \in \mathbb{R}$). Find dy/dx.

Solution   The graph of $f(x)$ is shown in Figure 8.23. There are many ways in which dy/dx may be evaluated. The simplest uses the result (8.25). In this case

$$\frac{dy}{dt} = 2t \quad \text{and} \quad \frac{dx}{dt} = 3t^2$$

so that

$$\frac{dy}{dx} = \frac{dy}{dt} \Big/ \frac{dx}{dt} = \frac{2}{3t} \quad (t \neq 0)$$

This gives the result in terms of t. In terms of x, it may be written as

$$\frac{dy}{dx} = \tfrac{2}{3}x^{-1/3} \quad (x \neq 0)$$

In terms of x and y, we have

$$\frac{dy}{dx} = \frac{2y}{3x} \quad (x \neq 0)$$

Note from Figure 8.23 that the graph does not have a well-defined tangent at $x = 0$, so the derivative does not exist at this point; that is, the function is not differentiable at $x = 0$.

We can also obtain these results directly. Eliminating t between the defining equations for x and y, we have

$$y = x^{2/3}$$

Differentiating with respect to x gives

$$\frac{dy}{dx} = \tfrac{2}{3}x^{-1/3}$$

Figure 8.23
The graph of
$\{(x, y): y = t^2, x = t^3,$
$-\infty < t < \infty\}$.

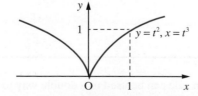

The chain rule may also be used to differentiate functions expressed in an implicit form. For example, the function of Example 8.20 may be expressed implicitly, by eliminating t, as

$$y^3 = x^2$$

To obtain the derivative dy/dx, we use the method known as **implicit differentiation**. In this method we treat y as an unknown function of x and differentiate both sides term by term with respect to x. This gives

$$\frac{d}{dx}(y^3) = \frac{d}{dx}(x^2)$$

Now y^3 is a composite function of x, with y being the intermediate variable, so the chain rule gives

$$\frac{d}{dx}(y^3) = \frac{d}{dy}(y^3)\frac{dy}{dx} = 3y^2\frac{dy}{dx}$$

Then, substituting back, we have

$$3y^2\frac{dy}{dx} = 2x$$

giving

$$\frac{dy}{dx} = \frac{2x}{3y^2} = \frac{2y}{3x} \quad \text{(on substituting for } y^3\text{)}$$

Example 8.21 Find $\dfrac{dy}{dx}$ when $x^2 + y^2 + xy = 1$.

Solution Differentiating both sides, term by term, gives

$$\frac{d}{dx}(x^2) + \frac{d}{dx}(y^2) + \frac{d}{dx}(xy) = \frac{d}{dx}(1)$$

Recognizing that y is a function of x and taking care over the product term xy, the chain rule gives

$$2x + \frac{d}{dy}(y^2)\frac{dy}{dx} + x\frac{dy}{dx} + y = 0$$

$$2x + 2y\frac{dy}{dx} + x\frac{dy}{dx} + y = 0$$

leading to

$$\frac{dy}{dx} = -\frac{(2x + y)}{(x + 2y)}$$

The implicit differentiation rule can be used in a double way to obtain derivatives of functions of the form $f(x)^{g(x)}$ as illustrated in Example 8.22.

Example 8.22   Find the derivative of the function

$$f(x) = (\sin x)^x \quad (x \in (0, \pi))$$

Solution   The simplest way of dealing with this is first to take logarithms. Thus $y = (\sin x)^x$ gives

$$\ln y = x \ln \sin x$$

Then differentiating implicitly with respect to x, remembering that y is a function of x, gives

$$\frac{1}{y}\frac{dy}{dx} = \ln \sin x + x\frac{\cos x}{\sin x} = \ln \sin x + x \cot x$$

and so

$$\frac{dy}{dx} = (\ln \sin x + x \cot x)(\sin x)^x$$

Sometimes the technique used in Example 8.22 is described as **logarithmic differentiation**. Implicit differentiation is useful in calculating the slopes of tangents and normals to curves specified implicitly, such as in Example 8.21. Having obtained the slope of the tangent at the point (x, y) as $\dfrac{dy}{dx}$, the slope of the normal to the curve at the corresponding point may be inferred from Figure 8.24.

Figure 8.24
Relationship between slopes of the tangent and normal to a plane curve.

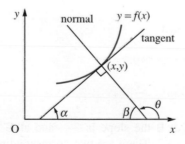

Slope of tangent $= \tan \alpha$
Slope of normal $= \tan \theta = -\tan \beta$
$= -\tan (90 - \alpha)$
$= -\dfrac{1}{\text{slope of tangent}}$

Example 8.23   Find the equations of the tangent and normal to the curve having equation $x^2 + y^2 - 3xy + 4 = 0$ at the point $(2, 4)$.

Solution   Differentiating the equation implicitly with respect to x

$$2x + 2y\frac{dy}{dx} - 3x\frac{dy}{dx} - 3y = 0$$

gives

$$\frac{dy}{dx} = \frac{3y - 2x}{2y - 3x}$$

This represents the slope of the tangent at the point (x, y) on the curve. Thus the slope of the tangent at the point $(2, 4)$ is

$$\left[\frac{dy}{dx}\right]_{(2,4)} = \frac{12 - 4}{8 - 6} = 4$$

Remembering from equation (1.11) that the equation of a line passing through a point (x, y) and having slope m is $y - y_1 = m(x - x_1)$ we have that the equation of the tangent to the graph at $(2, 4)$ is

$$(y - 4) = 4(x - 2) \quad \text{or} \quad y = 4x - 4$$

The slope of the normal at $(2, 4)$ is $-\frac{1}{4}$ so it has equation

$$y - 4 = -\tfrac{1}{4}(x - 2) \quad \text{or} \quad 4y = 18 - x$$

Example 8.24   Find the slope of the tangents to the circle

$$x^2 + y^2 - 2x + 4y - 20 = 0$$

at the points A(1, 3), B(4, 2) and C(−2, −6).

Solution   The circle defined by the equation is shown in Figure 8.25, together with the three points A, B and C. Clearly this equation does not define a function in general, but near specific points we can restrict it so that it behaves locally like a function. To compute the slopes of the tangents, we differentiate the equation defining the curve with respect to x implicitly, and then we insert the x and y coordinates of the points. Thus in this example we have

$$2x + 2y\frac{dy}{dx} - 2 + 4\frac{dy}{dx} = 0$$

giving

$$\frac{dy}{dx} = \frac{1 - x}{2 + y} \quad (y \neq -2)$$

Then at A the slope is zero, at B the slope is $-\frac{3}{4}$, and at C the slope is $-\frac{3}{4}$. Note that dy/dx is not defined at $y = -2$. There are two corresponding points: $(-4, -2)$ and $(6, -2)$. At these points the curve has a vertical tangent.

Figure 8.25
Graph of the
circle $x^2 + y^2 - 2x$
$+ 4y - 20 = 0$.

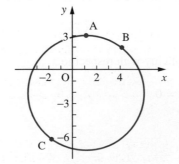

8.3.12 Exercises

35 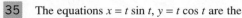 The equations $x = t \sin t$, $y = t \cos t$ are the parametric equations for a spiral. Find $\dfrac{dy}{dx}$ in terms of t.

36 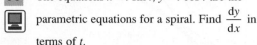 A curve is defined parametrically by the equations

$$x = 2 \cos \theta + \cos 2\theta$$

$$y = 2 \sin \theta - \sin 2\theta$$

Draw a sketch of the curve for $0 \leqslant \theta \leqslant 2\pi$. Find the equation of the tangent to the curve at the point where $\theta = \pi/4$.

37 Find the equation of the tangent, at the point $(0, 4)$, to the curve defined by

$$y^3 x + y + 7x^4 = 4$$

38 Find the value of $\dfrac{dy}{dx}$ at the point $(1, -1)$ on the curve given by the equation

$$x^3 - y^3 - xy - x = 0$$

39 Differentiate with respect to x:

(a) 10^x (b) 2^{-x} (c) $\dfrac{(x-1)^{7/2}(x+1)^{1/2}}{x^2+2}$

40 Use logarithmic differentiation to prove that

$$\frac{d}{dx}(y_1 y_2 \ldots y_n)$$

$$= \sum_{k=1}^{n} (y_1 y_2 \ldots y_{k-1} y_{k+1} \ldots y_n) y_k'$$

Hence differentiate $x^3 e^{-2x} \sin \pi x$.

8.3.13 Higher derivatives

The derivative df/dx of function $f(x)$ is itself a function and may be differentiable. The derivative of a derivative is called the **second derivative**, and is written as

$$\frac{d^2 f}{dx^2} \quad \text{or} \quad f''(x) \quad \text{or} \quad f^{(2)}(x)$$

This may in turn be differentiated, yielding **third derivatives** and so on. In general, the **nth derivative** is written as

$$\frac{d^n f}{dx^n} \quad \text{or} \quad f^{(n)}(x)$$

The second derivative $d^2 f/dx^2$ represents the rate of change of df/dx as x increases; geometrically, this gives us information as to how the slope of the tangent to the graph of $y = f(x)$ is changing with increasing x.

- If $d^2 f/dx^2 > 0$ then df/dx is increasing as x increases, and the tangent rotates in an anticlockwise direction as we move along the horizontal axis, as illustrated in Figure 8.26(a).
- If $d^2 f/dx^2 < 0$ then df/dx is decreasing as x increases, and the tangent rotates in a clockwise direction as we move along the horizontal axis, as illustrated in Figure 8.26(b).

Also note that when $d^2 f/dx^2 > 0$, the graph of $y = f(x)$ is 'concave up', and when $d^2 f/dx^2 < 0$ the graph is 'concave down'. Thus the sign of $d^2 f/dx^2$ relates to the concavity of the graph; we shall use this information in Chapter 9 to define a point of inflection.

Figure 8.26
Rates of change of
$\dfrac{dy}{dx}$ as x increases.

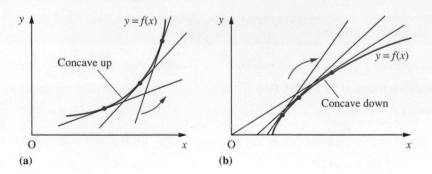

(a)

(b)

Example 8.25

Find the second derivative of the functions given by

(a) $y = e^{-x} \sin 2x$ (b) $y = \dfrac{\ln x}{x}$

Solution

(a) This simply requires two differentiations. Applying the product rule, we have

$$\frac{dy}{dx} = (e^{-x})(2\cos 2x) + (\sin 2x)(-e^{-x}) = e^{-x}(2\cos 2x - \sin 2x)$$

Applying the rule again we have

$$\frac{d^2 y}{dx^2} = (e^{-x})(-4\sin 2x - 2\cos 2x) + (2\cos 2x - \sin 2x)(-e^{-x})$$

$$= -e^{-x}(3\sin 2x + 4\cos 2x)$$

(b) Again this simply requires two differentiations. Applying the quotient rule, we have

$$\frac{dy}{dx} = \frac{(1/x)x - \ln x}{x^2} = \frac{1 - \ln x}{x^2}$$

Applying the rule again, we obtain

$$\frac{d^2 y}{dx^2} = \frac{-(1/x)x^2 - (1 - \ln x)(2x)}{x^4} = \frac{2\ln x - 3}{x^3} \quad (x \neq 0)$$

When determining the second derivative using parametric or implicit differentiation care must be taken to ensure correct use of the chain rule. The approach is illustrated in Example 8.26.

Example 8.26

Find $\dfrac{d^2 y}{dx^2}$ when y is given by

(a) $y = t^2, x = t^3$ (b) $x^2 + y^2 - 2x + 4y - 20 = 0$

Solution

(a) Here $y = t^2$ and $x = t^3$ gives, as in Example 8.20,

$$\frac{dy}{dx} = \tfrac{2}{3}\frac{1}{t} \quad (t \neq 0)$$

Differentiating again, using the chain rule, gives

$$\frac{d^2y}{dx^2} = \frac{d}{dx}\left(\frac{dy}{dx}\right) = \frac{d}{dt}\left(\frac{dy}{dx}\right)\frac{dt}{dx} \quad \text{(this is an important step)}$$

$$= \frac{d}{dt}\left(\frac{dy}{dx}\right)\bigg/\frac{dx}{dt}$$

$$= \frac{\frac{2}{3}(-1/t^2)}{3t^2} = -\frac{2}{9}\frac{1}{t^4}$$

(b) Here x and y are related by the equation

$$x^2 + y^2 - 2x + 4y - 20 = 0$$

so that as in Example 8.24

$$2x + 2y\frac{dy}{dx} - 2 + 4\frac{dy}{dx} = 0$$

and

$$(y + 2)\frac{dy}{dx} + x - 1 = 0$$

Differentiating a second time gives

$$\left(\frac{dy}{dx}\right)\frac{dy}{dx} + (y + 2)\frac{d^2y}{dx^2} + 1 = 0$$

using the product rule and remembering that

$$\frac{d}{dx}\left(\frac{dy}{dx}\right) = \frac{d^2y}{dx^2}$$

After rearrangement, we have

$$\frac{d^2y}{dx^2} = -\frac{1 + (dy/dx)^2}{y + 2} \quad (y \neq -2)$$

and substituting

$$\frac{dy}{dx} = \frac{1 - x}{2 + y}$$

into the right-hand side gives eventually

$$\frac{d^2y}{dx^2} = -\frac{x^2 + y^2 - 2x + 4y + 5}{(2 + y)^3}$$

This may be further simplified, using the original equation, to give

$$\frac{d^2y}{dx^2} = -\frac{25}{(2 + y)^3} \quad (y \neq -2)$$

Further results for higher derivatives are developed in Exercises 8.3.15. These are on the whole straightforward extensions of previous work. One result that sometimes causes blunders is the extension of the inverse-function rule to higher derivatives.

We know that

$$\frac{dx}{dy} = 1 \Big/ \frac{dy}{dx}$$

To find the second derivative of x with respect to y needs a little care:

$$\frac{d^2x}{dy^2} = \frac{d}{dy}\left(\frac{dx}{dy}\right) = \frac{d}{dy}\left[\left(\frac{dy}{dx}\right)^{-1}\right]$$

$$= \frac{d}{dx}\left[\left(\frac{dy}{dx}\right)^{-1}\right]\frac{dx}{dy} \quad \text{(using the chain rule)}$$

$$= \left[-\frac{d}{dx}\left(\frac{dy}{dx}\right)\Big/\left(\frac{dy}{dx}\right)^2\right]\left(1\Big/\frac{dy}{dx}\right) = -\frac{d^2y}{dx^2}\Big/\left(\frac{dy}{dx}\right)^3$$

8.3.14 Curvature of plane curves

The **curvature** κ of a plane curve, having equation $y = f(x)$, at any point is the rate at which the curve is bending or curving away from the tangent at that point. In other words, the curvature measures the rate at which the tangent to the curve changes as it moves along the curve.

Take two points P and Q on the curve $y = f(x)$ and a distance Δs apart measured along the curve. Then, with the notation of Figure 8.27(a), the average curvature of the curve PQ is $\Delta\theta/\Delta s$. We then define the curvature κ of the curve at the point P to be the absolute value of the average curvature as Q approaches P. That is,

$$\kappa = \left|\lim_{\Delta s \to 0}\frac{\Delta\theta}{\Delta s}\right| = \left|\frac{d\theta}{ds}\right| \tag{8.26}$$

Figure 8.27
Curvature and
radius of curvature.

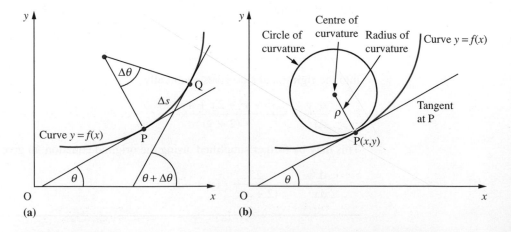

(a) (b)

If we now construct a circle, as shown in Figure 8.27(b), so that it

- has the same tangent at P as $y = f(x)$,
- lies on the same side of the tangent as $y = f(x)$ and
- has the same curvature κ as $y = f(x)$ at P

then this is called the **circle of curvature** at P. Its radius ρ is called the **radius of curvature** at P and is given by

$$\rho = \text{radius of curvature} = \frac{1}{\kappa}$$

The centre of the circle is called the **centre of curvature** at P.

In order to obtain the curvature of a curve given by an equation of the form $y = f(x)$, we must obtain a more usable formula than (8.26). Since

$$\tan \theta = \text{slope of the tangent at P} = \frac{dy}{dx}$$

differentiating with respect to s, using the chain rule, gives

$$\sec^2 \theta \frac{d\theta}{ds} = \frac{d^2 y}{dx^2} \frac{dx}{ds}$$

so that

$$\frac{d\theta}{ds} = \frac{d^2 y/dx^2}{[1 + (dy/dx)^2]} \frac{dx}{ds} \tag{8.27}$$

using the trigonometric identity $1 + \tan^2 \theta = \sec^2 \theta$.

We shall see in Section 8.7.6 that

$$\frac{ds}{dx} = \sqrt{\left[1 + \left(\frac{dy}{dx}\right)^2\right]}$$

so, from the inverse-function rule,

$$\frac{dx}{ds} = \frac{1}{ds/dx} = \frac{1}{\sqrt{[1 + (dy/dx)^2]}}$$

which on substituting into (8.27) gives the formula

$$\kappa = \left|\frac{d\theta}{ds}\right| = \frac{|d^2 y/dx^2|}{[1 + (dy/dx)^2]^{3/2}} \tag{8.28}$$

If we denote the coordinates of the centre of curvature by (X, Y) then it follows from Figure 8.27(b) that

$$X = x - \rho \sin \theta, \quad Y = y + \rho \cos \theta$$

Since $\tan \theta = dy/dx$, it follows that

$$\sin \theta = \frac{dy/dx}{\sqrt{[1 + (dy/dx)^2]}}, \quad \cos \theta = \frac{1}{\sqrt{[1 + (dy/dx)^2]}}$$

Using these results together with $\rho = 1/\kappa$, with κ from (8.28), gives the coordinates of the centre of curvature as

$$X = x - \frac{dy}{dx}\left[1 + \left(\frac{dy}{dx}\right)^2\right]\bigg/\frac{d^2y}{dx^2}, \quad Y = y + \left[1 + \left(\frac{dy}{dx}\right)^2\right]\bigg/\frac{d^2y}{dx^2} \tag{8.29}$$

Although these results have been deduced for the curve of Figure 8.27(b), which at the point P(x, y) has positive slope $(dy/dx > 0)$ and is concave upwards $(d^2y/dx^2 > 0)$, it can be shown that these are valid in all cases.

It is left as an exercise for the reader to show that if the curve $y = f(x)$ is given in parametric form

$$x = g(t), \quad y = h(t)$$

then the curvature κ is given by

$$\kappa = \left|\frac{dg}{dt}\frac{d^2h}{dt^2} - \frac{d^2g}{dt^2}\frac{dh}{dt}\right|\bigg/\left[\left(\frac{dg}{dt}\right)^2 + \left(\frac{dh}{dt}\right)^2\right]^{3/2} \tag{8.30}$$

8.3.15 Exercises

41 Find $\dfrac{d^2y}{dx^2}$ when y is given by:

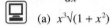

(a) $x^3\sqrt{(1 + x^2)}$

(b) $\ln(x^2 + x + 1)$

(c) $y^3x + y + 7x^4 = 4$

(d) $x^3 - y^3 - xy - x = 0$

42 Find $\dfrac{d^2y}{dx^2}$ when x and y are given by:

(a) $x = t \sin t$ and $y = t \cos t$

(b) $x = 2 \cos t + \cos 2t$ and $y = 2 \sin t - \sin 2t$

43 If $y = 3e^{2x} \cos(2x - 3)$, verify that

$$\frac{d^2y}{dx^2} - 4\frac{dy}{dx} + 8y = 0$$

44 If $y = (\sin^{-1}x)^2$, prove that

$$(1 - x^2)\left(\frac{dy}{dx}\right)^2 = 4y$$

and deduce that

$$(1 - x^2)\frac{d^2y}{dx^2} - x\frac{dy}{dx} - 2 = 0$$

45 (a) If $y = x^2 + 1/x^2$, find dy/dx and d^2y/dx^2. Hence show that

$$x^2\frac{d^2y}{dx^2} + 4x\frac{dy}{dx} + 2y = 12x^2$$

(b) If $x = \tan t$ and $y = \cot t$, show that

$$\frac{d^2y}{dx^2} + 2y\frac{dy}{dx} = 0$$

46 If $x = a(\theta - \sin \theta)$ and $y = a(1 - \cos \theta)$, find dy/dx and d^2y/dx^2.

47 The equation of a curve is

$$xy^3 - 2x^2y^2 + x^4 - 1 = 0$$

Show that the tangent to the curve at the point (1, 2) has a slope of unity. Hence write down the equation of the tangent to the curve at this point. What are the coordinates of the points at which this tangent crosses the coordinate axes?

48 A **cycloid** is a curve traced out by a point p on the rim of a wheel as it rolls along the ground. Using the coordinate system shown in Figure 8.28, show that the curve has the parametric representation

$$x = a(\theta - \sin \theta), \quad y = a(1 - \cos \theta)$$

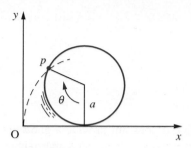

Figure 8.28

where θ is the angle through which the wheel has turned.

Draw a sketch of the curve.

Find the gradient of the curve at a general point (x, y).

If the wheel rotates at a constant speed, with $\theta = \omega t$, where ω is constant and t is the time, show that the speed V of the point on the rim is given by

$$V(t) = 2a\omega |\sin \tfrac{1}{2}\omega t|$$

49 Find dy/dx in terms of t for the curve with parametric representation

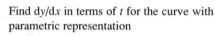

$$x = \frac{1-t}{1+2t} \qquad y = \frac{1-2t}{1+t}$$

Show that

$$\frac{d^2 y}{dx^2} = -\frac{2}{3}\left(\frac{1+2t}{1+t}\right)^3$$

and find a similar expression for d^2x/dy^2.

50 Find the slope of the tangent to the lemniscate

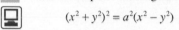

$$(x^2 + y^2)^2 = a^2(x^2 - y^2)$$

at the point (x, y). (See Review exercises 2.11, Question 19.)

51 Confirm that the point $(1, 1)$ lies on the curve with equation $x^3 - y^2 + xy - x^2 = 0$ and find the values of dy/dx and d^2y/dx^2 at that point.

52 Use logarithmic differentiation to differentiate

(a) $(\ln x)^x$ (b) $x^{\ln x}$

(c) $(1 - x^2)^{1/2}(2x^2 + 3)^{-4/3}$

53 Find $f^{(4)}(x)$ and $f^{(n)}(x)$ for the following functions $f(x)$:

(a) e^{3x} (b) $\ln(x + 2)$ (c) $\dfrac{1}{1 - x^2}$

54 Find the fourth derivative of $f(x) = \sin(ax + b)$ and verify that $f^{(n)}(x) = a^n \sin(ax + b + \tfrac{1}{2}n\pi)$.

55 Prove that

$$\frac{d^n}{dx^n}(e^{ax} \sin bx) = (a^2 + b^2)^{n/2} e^{ax} \sin(bx + n\theta)$$

where $\cos \theta = a/\sqrt{(a^2 + b^2)}$, $\sin \theta = b/\sqrt{(a^2 + b^2)}$.

56 If $y = u(x)v(x)$, prove that

(a) $y^{(2)}(x) = u^{(2)}(x)v(x) + 2u^{(1)}(x)v^{(1)}(x) + u(x)v^{(2)}(x)$

(b) $y^{(3)}(x) = u^{(3)}(x)v(x) + 3u^{(2)}(x)v^{(1)}(x)$
$$+ 3u^{(1)}(x)v^{(2)}(x) + u(x)v^{(3)}(x)$$

Hence prove **Leibniz' theorem** for the nth derivative of a product:

$$y^{(n)}(x) = u^{(n)}(x)v(x) + \binom{n}{1}u^{(n-1)}(x)v^{(1)}(x)$$

$$+ \binom{n}{2}u^{(n-2)}(x)v^{(2)}(x) + \ldots + u(x)v^{(n)}(x)$$

57 Use Leibniz' theorem (Question 56) to find the following:

(a) $\dfrac{d^5}{dx^5}(x^2 \sin x)$ (put $u = \sin x$, $v = x^2$)

(b) $\dfrac{d^4}{dx^4}(xe^{-x})$ (c) $\dfrac{d^3}{dx^3}[x^2(3x + 1)^{12}]$

58 Using logarithmic differentiation, find the derivatives of

(a) $x^3 e^{-2x} \ln x$ (b) $\dfrac{1}{x}e^x \sin 2x$

59 Find the radius of curvature at the point $(2, 8)$ on the curve $y = x^3$.

60 Show that the radius of curvature at the origin to the curve

$$x^3 + y^3 + 2x^2 - 4y + 3x = 0$$

is $\frac{125}{64}$.

61 Find the radius of curvature and the coordinates of the centre of curvature of the curve

$$y = (11 - 4x)/(3 - x)$$

at the point $(2, 3)$.

62 Find the radius of curvature at the point where $\theta = \tfrac{1}{3}\pi$ on the curve defined parametrically by

$$x = 2 \cos \theta, \quad y = \sin \theta$$

Numerical differentiation

Although the formula

$$f'(x) = \lim_{\Delta x \to 0} \frac{f(x + \Delta x) - f(x)}{\Delta x} = \lim_{\Delta x \to 0} \frac{\Delta f}{\Delta x}$$

provides the definition of the derivative of $f(x)$, it does not provide a good basis for evaluating $f'(x)$ numerically. This is because it provides a one-sided approximation of the gradient at x as shown in Figure 8.29. When we set $\Delta x = h\ (> 0)$, we obtain the slope of the chord PR. When we set $\Delta x = -h\ (< 0)$, we obtain the slope of the chord QP. Clearly the chord QR offers a better approximation to the tangent at P. A second reason why the formal definition of a derivative yields a poor approximation is that the evaluation of derivatives involves the division of a small quantity Δf by a second small quantity Δx. This process magnifies the rounding errors involved in calculating Δf from the values of $f(x)$, a process that worsens as $\Delta x \to 0$. This phenomenon is called **ill-conditioning**. Generally speaking, numerical differentiation is a process in which accuracy is lost and the 'noise' caused by experimental error is magnified.

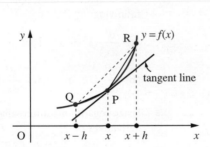

Figure 8.29 Approximations to the tangent at P.

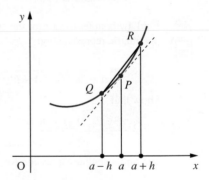

Figure 8.30 Chord approximation.

8.4.1 The chord approximation

This method uses the slope of a chord QR symmetrically disposed about x to approximate the slope of the tangent at x, as shown in Figure 8.30. Thus

$$f'(x) \approx \frac{f(x + h) - f(x - h)}{2h} = \phi(h)$$

Thus when the function is specified graphically, a value of h is chosen, and at a series of points along the curve the quotient $\phi(h)$ is calculated. When the function is given as a table of values, we do not have control of the value of h, but the same approximation is used using the tabular interval as h. Consequently, to estimate the value of the derivative $f'(a)$ at $x = a$ we use the approximation

$$\frac{f(a + h) - f(a - h)}{2h} = \phi(h)$$

as the basis for an extrapolation. For almost all functions commonly occurring in engineering applications

$$f'(a) = \phi(h) + \text{terms involving powers of } h \text{ greater than or equal to } h^2$$

For example, considering $f(x) = x^3$

$$\phi(h) = \frac{(a+h)^3 - (a-h)^3}{2h} = \frac{6a^2h + 2h^3}{2h} = 3a^2 + h^2$$

Similarly, for $f(x) = x^4$

$$\phi(h) = 4a^3 + 4ah^2$$

In general, we may write

$$f'(a) = \phi(h) + Ah^2 + \text{terms involving higher powers of } h$$

where A is independent of h. Interval-halving gives

$$f'(a) = \phi(\tfrac{1}{2}h) + \tfrac{1}{4}A'h^2 + \text{terms involving higher powers of } h$$

where $A' \approx A$. Hence we obtain a better estimate for $f'(a)$ by extrapolation, eliminating the terms involving h^2:

$$f'(a) \approx \tfrac{1}{3}[4\phi(\tfrac{1}{2}h) - \phi(h)] \tag{8.31}$$

We illustrate this technique in Example 8.27.

Example 8.27   Estimate $f'(0.5)$, where $f(x)$ is given by the table

x	0.1	0.2	0.3	0.4	0.5	0.6	0.7	0.8	0.9
$f(x)$	0.0998	0.1987	0.2955	0.3894	0.4794	0.5646	0.6442	0.7174	0.7833

Solution   Using the data provided, taking $h = 0.4$ and 0.2, we obtain

$$\phi(0.4) = \frac{0.7833 - 0.0998}{0.8} = 0.8544$$

and

$$\phi(0.2) = \frac{0.6442 - 0.2955}{0.4} = 0.8718$$

Hence, by extrapolation, we have, using (8.31),

$$f'(0.5) \approx (4 \times 0.8718 - 0.8544)/3 = 0.8776$$

The tabulated function is actually $\sin x$, so that in this illustrative example we can compare the estimate with the true value $\cos 0.5$, and we find that the answer is correct to 4dp.

In general, any numerical procedure is subject to two types of error. One is due to the accumulation of rounding errors within a calculation, while the other is due to the nature of the approximation formula (the truncation error). In this example the truncation error is of order h^2 for $\phi(h)$, but we do not have an estimate for the truncation error for the extrapolated estimate for $f'(a)$. This will be discussed in the next chapter following the introduction of the Taylor series (see Exercises 9.5.6, Question 36). The effect of the rounding errors on the answer can be assessed, however, and, using the methods of Chapter 1, we see that the maximum effect of the rounding errors on the answer in Example 8.27 is $\pm 2.5 \times 10^{-4}$.

8.4.2 Exercises

 63 Use the chord approximation to obtain two estimates for $f'(1.2)$ using $h = 0.2$ and $h = 0.1$ where $f(x)$ is given in the table below.

x	1.0	1.1	1.2	1.3	1.4
$f(x)$	1.000	1.008	1.061	1.192	1.414

Use extrapolation to obtain an improved approximation.

 64 Use your calculator (in radian mode) to calculate the quotient $\{f(x+h) - f(x-h)\}/(2h)$ for $f(x) = \sin x$, where $x = 0(0.1)1.0$ and $h = 0.001$. Compare your answers with $\cos x$.

 65 Consider the function $f(x) = xe^x$, tabulated below:

x	0.96	0.97	0.98	0.99	1.00
$f(x)$	2.5072	2.5588	2.6112	2.6643	2.7183

x	1.01	1.02	1.03	1.04
$f(x)$	2.7731	2.8287	2.8851	2.9424

(a) Find, *exactly*, $f'(1)$ and $f''(1)$.

(b) Use the tabulated values and the formula

$$f'(a) \simeq (f(a+h) - f(a-h))/2h$$

to estimate $f'(1)$, for various h. Compute the errors involved and comment on the results.

(c) Repeat (b) for $f''(1)$ using

$$f''(a) \simeq (f(a+h) - 2f(a) + f(a-h))/h^2$$

 66 Use the following table of $f(x) = (e^x - e^{-x})/2$ to estimate $f'(1.0)$ by means of an extrapolation method.

x	0.2	0.6	0.8	1.2	1.4	1.8
$f(x)$	0.2013	0.6367	0.8881	1.5095	1.9043	2.9422

Compare your answer with $(e + e^{-1})/2 = 1.5431$ correct to 4dp.

67 Investigate the effect of using a smaller value for h in Example 8.27. Show that $\phi(0.1)$ gives a poorer estimate for $f'(0.5)$ and the error bound for the consequent extrapolation $[4\phi(0.1) - \phi(0.2)]/3$ is 7×10^{-4}.

8.5 Integration

In this section we shall introduce the concept of integration and illustrate its role in problem-solving and modelling situations.

8.5.1 Basic ideas and definitions

Consider an object moving along a line with constant velocity u (in m s^{-1}). The distance s (in m) travelled by the object between times t_1 and t_2 (in s) is given by

$$s = u(t_2 - t_1)$$

Figure 8.31
Velocity–time graph
for an object moving
with constant velocity
u. The shaded area
shows the distance
travelled by the object
between times t_1 and t_2.

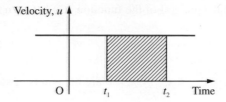

Figure 8.32
A velocity–time
graph and two
piecewise-constant
approximations to it.

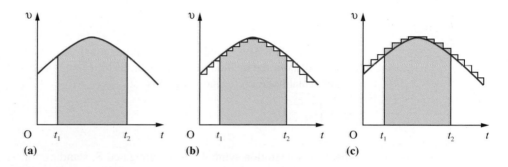

This is the area 'under' the graph of the velocity function between $t = t_1$ and $t = t_2$, as shown in Figure 8.31. This, of course, deals with the special case where the velocity is a constant function. However, even when the velocity varies with time, the area under the velocity graph still gives the distance travelled. Consider the velocity graph shown in Figure 8.32(a). We can approximate the velocity–time graph by a series of small horizontal lines that lie either entirely below the curve (as in Figure 8.32(b)) or entirely above it (Figure 8.32(c)). An object moving such that its velocity–time graph is (b) would always be slower at a particular time than an object with velocity–time graph (a), so that the distance it covers is less than that of the object with graph (a). Similarly, an object with velocity–time graph (c) will cover a greater distance than an object with graph (a). Thus

distance with graph (b) < distance with graph (a) < distance with graph (c)

In cases (b) and (c), because the velocities are piecewise-constant, the distances covered are represented by the areas under the graphs between $t = t_1$ and $t = t_2$. So we have

area under graph (b) < area under graph (a) < area under graph (c)

If the horizontal steps of graphs (b) and (c) are made very small, the difference between the areas for the approximating graphs (b) and (c) becomes very small. In other words, the distance for graph (a) is just the area under the graph between $t = t_1$ and $t = t_2$.

This is one of many practical problems that involve this process of area evaluation at some stage in their solution. This process is called **integration**: the summing together of all the parts that make up a given area. The area under the graph is called the **integral** of the function. For some functions it is possible to obtain formulae for their integrals; for others we have to be content with numerical approximations.

Formally, we define the integral of the function $f(x)$ between $x = a$ and $x = b$ to be

$$\lim_{\substack{n \to \infty \\ \Delta x \to 0}} \sum_{r=1}^{n} f(x_r^*)\Delta x_{r-1}$$

where $a = x_0 < x_1 < x_2 < \ldots < x_{n-1} < x_n = b$ are the points of subdivision of the interval $[a, b]$,

$$\Delta x_{r-1} = x_r - x_{r-1}, \, \Delta x = \max(\Delta x_0, \Delta x_1, \ldots, \Delta x_{n-1}) \text{ and } x_{r-1} \leq x_r^* \leq x_r$$

Here we have used the special notation

$$\lim_{\substack{n \to \infty \\ \Delta x \to 0}}$$

to emphasize that $n \to \infty$ and $\Delta x \to 0$ simultaneously. The value of the integral is independent of both the method of subdivision of $[a, b]$ and the choices of x_r^*.

The usual notation for the integral is

$$\int_a^b f(x)\mathrm{d}x$$

where the integration symbol $\int$ is an elongated S, standing for 'summation'. The $\mathrm{d}x$ is called the **differential** of x, and a and b are called the **limits of integration**. The function $f(x)$ being integrated is the **integrand**.

The process is illustrated in Figure 8.33, where the area under the graph of $f(x)$ for $x \in [a, b]$ has been subdivided into n vertical strips (by which we strictly mean that the area has been approximated by the n vertical strips). The area of a typical strip is given by

$$f(x_r^*)(x_r - x_{r-1}) = f(x_r^*)\Delta x_{r-1}$$

where $x_{r-1} \leq x_r^* \leq x_r$ and $\Delta x_{r-1} = x_r - x_{r-1}$. Thus the area under the graph can be approximated by

$$\sum_{r=1}^{n} f(x_r^*)\Delta x_{r-1}$$

This approximation becomes closer to the exact area as the number of strips is increased and their widths decreased. In the limiting case as $n \to \infty$ and $\Delta x \to 0$ this leads to the exact area being given by

$$A = \int_a^b f(x)\,\mathrm{d}x$$

Figure 8.33

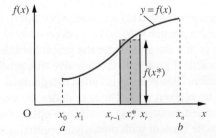

so that

$$\int_a^b f(x)\mathrm{d}x = \lim_{\substack{n\to\infty \\ \Delta x \to 0}} \sum_{r=1}^n f(x_r^*)\Delta x_{r-1} \tag{8.32}$$

In line with the definition of an integral, we note that if the graph of $f(x)$ is below the x axis then the summation involves products of negative ordinates with positive widths, so that areas below the x axis must be interpreted as being negative.

Example 8.28   By considering the area under the graph of $y = x + 3$, evaluate the integral $\int_{-5}^5 (x + 3)\mathrm{d}x$.

Solution   The area under the graph is shown hatched in Figure 8.34, with the area A_1 being negative and the area A_2 positive. In each case the areas are triangular, so that

$$A_1 = -\tfrac{1}{2} \times 2 \times 2 = -2$$

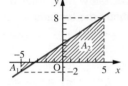

and

$$A_2 = \tfrac{1}{2} \times 8 \times 8 = 32$$

Thus

$$\int_{-5}^5 (x + 3)\mathrm{d}x = A_1 + A_2 = -2 + 32 = 30$$

Figure 8.34

We have seen that the area under the graph $y = f(x)$ can be expressed as an integral, but integrals have a much wider application. Any quantity that can be expressed in the form of the limit of a sum as in (8.32) can be represented by an integral, and this occurs in many practical situations. Because areas can be expressed as integrals, it follows that we can always interpret an integral geometrically as an area under a graph.

Example 8.29   A beam of length l is freely hinged at both ends and carries a distributed load $w(x)$ where

$$w(x) = \begin{cases} 4Wx/l^2 & 0 \leqslant x \leqslant l/2 \\ 4W(l - x)/l^2 & l/2 \leqslant x \leqslant l \end{cases}$$

Show that the total load is W and find the shear force at a point on the beam.

Solution   To find the total load on the beam we divide the interval $(0, l)$ into n subintervals of length Δx, so that $x_k = k\Delta x$ and $\Delta x = l/n$. Then the load on the subinterval (x_k, x_{k+1}) is $w(x_k^*)\Delta x$ where $w(x_k^*)$ is the average value of $w(x)$ in that subinterval. The total load on the beam is the sum of all such elementary loads, and we have

$$\text{total load} = \sum_{k=0}^{n-1} w(x_k^*)\Delta x$$

Figure 8.35
Non-uniform
load on a beam.

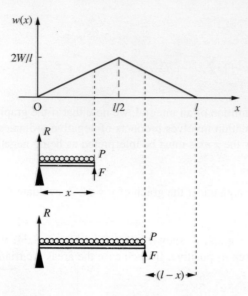

This formula, while it is exact, is not very useful since we do not know the values of the x_k^*'s. By proceeding to the limit, however, $x_k^* \to x_k$ and we obtain the formula

$$\text{total load} = \int_0^l w(x)\,dx$$

Now the integral $\int_0^l w(x)\,dx$ is the area under the curve $y = w(x)$ between $x = 0$ and $x = l$, and by considering the graph of $w(x)$ shown in Figure 8.35 we see that this is W. From the symmetry of the loading and the end conditions we see that the reactions at the supports at both ends are equal (to R, say). Then

$$2R = W$$

for equilibrium. To find the shear force F we have to consider the vertical equilibrium of the portion of the beam to the left of P. Thus

$$R + F = \text{load between } A \text{ and } P$$

Consideration of the areas under the graph of $y = w(x)$ for x shows that

$$R + F = \begin{cases} \frac{1}{2}x(4Wx/l^2) & 0 \leqslant x \leqslant l/2 \\ W - \frac{1}{2}(l - x)[4W(l - x)/l^2] & l/2 \leqslant x \leqslant l \end{cases}$$

This simplifies as

$$R + F = \begin{cases} 2Wx^2/l^2 & 0 \leqslant x \leqslant l/2 \\ W - 2W(l - x)^2/l^2 & l/2 \leqslant x \leqslant l \end{cases}$$

Thus

$$F = \begin{cases} 2Wx^2/l^2 - W/2 & 0 \leqslant x \leqslant l/2 \\ W/2 - 2W(l - x)^2/l^2 & l/2 \leqslant x \leqslant l \end{cases}$$

8.5.2 Exercises

68 Two hot-rodders, Alan and Brian, compete in a drag race. Each accelerates at a constant rate from a standing start. Alan covers the last quarter of the course in 3 s, while Brian covers the last third in 4 s. Who wins and by what time margin?

69 Show that the area under the graph of the constant function $f(x) = 1$ between $x = a$ and $x = b$ $(a < b)$ is given by $b - a$.

70 Show that the area under the graph of the linear function $f(x) = x$ between $x = a$ and $x = b$ $(a < b)$ is given by $\frac{1}{2}(b^2 - a^2)$.

71 Draw the graph of the function $f(x) = 2x - 1$ for $-3 < x < 3$. By considering the area under the graph, evaluate the integral $\int_{-3}^{3}(2x - 1)\mathrm{d}x$.

72 Using n strips of equal width, show that the area under the graph $y = x^2$ between $x = 0$ and $x = c$ satisfies the inequality

$$h^3 \sum_{r=1}^{n-1} r^2 < \text{area} < h^3 \sum_{r=1}^{n} r^2$$

and deduce

(a) $\displaystyle\int_{0}^{c} x^2\,\mathrm{d}x = \frac{1}{3}c^3$ (b) $\displaystyle\int_{a}^{b} x^2\,\mathrm{d}x = \frac{1}{3}(b^3 - a^3)$

(c) $\displaystyle\int_{a}^{b} x^{1/2}\,\mathrm{d}x = \frac{2}{3}(b^{3/2} - a^{3/2})$

(Recall that $\sum_{r=1}^{n} r^2 = \frac{1}{6}n(n + 1)(2n + 1)$.)

73 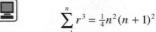 Using the method of Question 72 and the fact that

$$\sum_{r=1}^{n} r^3 = \frac{1}{4}n^2(n + 1)^2$$

show that

$$\int_{a}^{b} x^3\,\mathrm{d}x = \frac{1}{4}(b^4 - a^4)$$

74 A cylinder of length l and diameter D is constructed such that the density of the material comprising it varies as the distance from the base. Show that the mass of the cylinder is given by

$$\int_{0}^{l} \frac{1}{4}KD^2\pi x\,\mathrm{d}x$$

where K is a proportionality constant.

75 A beam of length l is freely hinged at both ends and carries a distributed load $w(x)$ where

$$w = \begin{cases} 4W/l & 0 \leq x \leq l/4 \\ 0 & l/4 < x \leq l \end{cases}$$

Find the shear force at a point on the beam.

76 A hemi-spherical vessel has internal radius 0.5 m. It is initially empty. Water flows in at a constant rate of 1 litre per second. Find an expression for the depth of the water after t seconds.

8.5.3 Definite and indefinite integrals

We have seen that the area under the graph $y = f(x)$ between $x = a$ and $x = b$ is given by the integral

$$\int_{a}^{b} f(x)\mathrm{d}x$$

Clearly, this area depends on the values of a and b as well as on the function $f(x)$. Thus the integral of a function $f(x)$ may be regarded as a function of a and b. If we replace the number b by the variable x, we obtain a function, F say, that is the area under the graph between a and x, as shown in Figure 8.36. This type of integral is called an

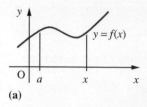
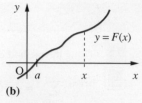

Figure 8.36 (a) Graph of $y = f(x)$. (b) Graph of $\int_a^x f(t)dt$.

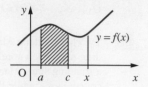

Figure 8.37

indefinite integral to distinguish it from integrals with fixed a and b, which are called **definite integrals**. We have defined F by the relation

$$F(x) = \int_a^x f(t)dt$$

Notice here that the dummy variable t, used as the integrator, is chosen to be different from the variable x on which the function F depends.

If a different lower limit is chosen, a different function is obtained, say G:

$$G(x) = \int_c^x f(t)dt$$

By interpreting an integral as the area under a curve, we see from Figure 8.37 that this new function differs from F only by a constant. This follows since

$$F(x) - G(x) = \int_a^x f(t)dt - \int_c^x f(t)dt = \int_a^c f(t)dt$$

which is a definite integral having a constant value representing the area under the graph between a and c, shown shaded in Figure 8.37.

For example, using the definition of an integral, we can show that

$$\int_a^b (t^2 - 1)dt = \tfrac{1}{3}(b^3 - a^3) - (b - a)$$

so that

$$\int_a^x (t^2 - 1)dt = \tfrac{1}{3}(x^3 - a^3) - (x - a) = \tfrac{1}{3}x^3 - x + (a - \tfrac{1}{3}a^3)$$

Giving a the values 1 and 2 leads to the two functions

$$F(x) = \int_1^x (t^2 - 1)dt = \tfrac{1}{3}x^3 - x + \tfrac{2}{3}$$

and

$$G(x) = \int_2^x (t^2 - 1)dt = \tfrac{1}{3}x^3 - x - \tfrac{2}{3}$$

In fact, all indefinite integrals of $f(x) = x^2 - 1$ are of the general form

Figure 8.38
(a) Graph of $f(x)$.
(b) Graph of
$F(x) = \int_0^x f(t)\mathrm{d}(t)$.

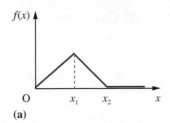

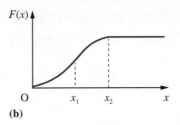

(a)

(b)

$\frac{1}{3}x^3 - x + \text{constant}$

When the lower limit is not specified, we denote the indefinite integral by

$$\int f(x)\mathrm{d}x \quad \text{or} \quad \int^x f(t)\mathrm{d}t$$

and include the constant as an arbitrary **constant of integration**. Thus

$$\int (x^2 - 1)\mathrm{d}x = \tfrac{1}{3}x^3 - x + c$$

where c is the arbitrary constant of integration.

It is important to recognize that an indefinite integral is itself a function, while a definite integral is a number.

We noted in Section 8.2.4 that a function could only be differentiated at points where its graph had a unique tangent, and that, for example, the function represented by the graph of Figure 8.6(a), reproduced as Figure 8.38(a), is not differentiable at domain values $x = x_1$ and $x = x_2$. However, such functions are integrable, with the corresponding indefinite integrals being functions having 'smooth' graphs. For example, the graph of the indefinite integral $F(x)$ of the function $f(x)$ shown in Figure 8.38(a) has the form shown in Figure 8.38(b). For this reason, engineers often refer to integration as being a 'smoothing' process and an integrator is frequently incorporated within a system design in order to ensure 'smoother' operation.

We can express definite integrals in terms of indefinite integrals. Setting $g(x) = \int f(x)\mathrm{d}x$, we see that

$$\int_a^b f(x)\mathrm{d}x = g(b) - g(a)$$

This is often denoted by

$$\int_a^b f(x)\mathrm{d}x = [g(x)]_a^b$$

a notation introduced by Fourier. Thus, for example,

$$\int_1^5 (x^2 - 1)\mathrm{d}x = [\tfrac{1}{3}x^3 - x + c]_1^5 = [\tfrac{125}{3} - 5 + c] - [\tfrac{1}{3} - 1 + c] = 37\tfrac{1}{3}$$

When evaluating definite integrals, the constant of integration can be omitted, since it cancels out in the arithmetic.

8.5.4 The Fundamental Theorem of Calculus

From Questions 69, 70 and 72 (Exercises 8.5.2) we have

$$\int_a^b 1\,dx = b - a, \qquad \text{giving} \qquad \int 1\,dx = x + \text{constant}$$

$$\int_a^b x\,dx = \tfrac{1}{2}(b^2 - a^2), \quad \text{giving} \quad \int x\,dx = \tfrac{1}{2}x^2 + \text{constant}$$

$$\int_a^b x^2\,dx = \tfrac{1}{3}(b^3 - a^3), \quad \text{giving} \quad \int x^2\,dx = \tfrac{1}{3}x^3 + \text{constant}$$

The comparable results for differentiation are

$$\frac{d}{dx}(k) = 0, \quad k \text{ constant}$$

$$\frac{d}{dx}(x) = 1$$

$$\frac{d}{dx}(x^2) = 2x$$

Using the sum and constant multiplication rules for differentiation from Section 8.3.1,

$$\frac{d}{dx}[f(x) + k] = \frac{d}{dx}f(x) + \frac{d}{dx}(k) = \frac{d}{dx}f(x), \quad k \text{ constant}$$

$$\frac{d}{dx}[kf(x)] = k\frac{d}{dx}f(x)$$

the above results may be combined to give

$$\frac{d}{dx}\left(\int 1\,dx\right) = \frac{d}{dx}(x + \text{constant}) = 1$$

$$\frac{d}{dx}\left(\int x\,dx\right) = x$$

$$\frac{d}{dx}\left(\int x^2\,dx\right) = x^2$$

These results suggest a more general result:

> The process of differentiation is the inverse of that of integration.

This conjecture is also supported by elementary applications of the processes. We obtained the distance travelled by an object by integrating its velocity function. We obtained the velocity of an object by differentiating its distance function. The general result is called the **Fundamental Theorem of Integral and Differential Calculus**, and may be stated in the form of the following theorem.

Theorem 8.1 The indefinite integral $F(x)$ of a continuous function $f(x)$ always possesses a derivative $F'(x)$, and, moreover, $F'(x) = f(x)$.

Proof The formula for $F(x)$ may be written as

$$F(x) = \int_a^x f(t)\,dt, \quad \text{where } a \text{ is a constant}$$

The quotient

$$\frac{F(x+h) - F(x)}{h}$$

may be written in terms of $f(x)$ as

$$\frac{F(x+h) - F(x)}{h} = \frac{\displaystyle\int_a^{x+h} f(t)\,dt - \int_a^x f(t)\,dt}{h} = \frac{1}{h}\int_x^{x+h} f(t)\,dt$$

Consider the case when h is positive. The function $f(x)$ is continuous, and so it is bounded on $[x, x+h]$. Suppose it attains its upper bound at x_1, as shown in Figure 8.39, and its lower bound at x_2. Then by considering the area under the graph, we see that

Figure 8.39

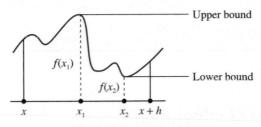

$$hf(x_2) \leq \int_x^{x+h} f(t)\,dt \leq hf(x_1)$$

which implies that

$$f(x_2) \leq \frac{1}{h}\int_x^{x+h} f(t)\,dt \leq f(x_1)$$

or equivalently

$$f(x_2) \leq \frac{F(x+h) - F(x)}{h} \leq f(x_1)$$

As $h \to 0$, $x_2 \to x$ and $x_1 \to x$, and we obtain the result

$$F'(x) = f(x)$$

(The proof when h is negative is similar.)

end of theorem

This theorem is of fundamental importance, and is used repeatedly in practical problem-solving using calculus.

8.5.5 Exercise

77 Using the Fundamental Theorem of Integral and Differential Calculus, evaluate the following integrals:

(a) $\int x^6 \, dx$, noting that $\dfrac{d}{dx} x^7 = 7x^6$

(b) $\int e^{3x} \, dx$, noting that $\dfrac{d}{dx} e^{3x} = 3e^{3x}$

(c) $\int \sin 5x \, dx$, noting that $\dfrac{d}{dx} \cos 5x = -5 \sin 5x$

(d) $\int (2x + 1)^3 \, dx$, noting that $\dfrac{d}{dx}(2x + 1)^4 = 8(2x + 1)^3$

(e) $\int \sec^2 3x \, dx$, noting that $\dfrac{d}{dx}(\tan 3x) = 3 \sec^2 3x$

(f) $\int \dfrac{2}{x} \, dx$, noting that $\dfrac{d}{dx} \ln x = \dfrac{1}{x}$

(g) $\int \dfrac{3}{x^2} \, dx$, noting that $\dfrac{d}{dx}\left(\dfrac{1}{x}\right) = -\dfrac{1}{x^2}$

(h) $\int \cos 2x \, dx$, noting that $\dfrac{d}{dx} \sin 2x = 2 \cos 2x$

(i) $\int \sec 4x \tan 4x \, dx$, noting that $\dfrac{d}{dx} \sec 4x = 4 \sec 4x \tan 4x$

(j) $\int \sqrt{(4x - 1)} \, dx$, noting that $\dfrac{d}{dx}(4x - 1)^{3/2} = 6(4x - 1)^{1/2}$

8.6 Techniques of integration

In this section we consider some of the methods available for determining the integrals of functions. Again we shall concentrate on developing techniques, leaving problem-solving applications for later in both this chapter and the rest of the book.

8.6.1 Integration as antiderivative

Applying the Fundamental Theorem of Calculus to some of the standard derivatives deduced in Section 8.3, we deduce the integrals given in Figure 8.40. Note that we have used the notation

$$\ln|x| = \begin{cases} \ln x, & x > 0 \\ \ln(-x), & x < 0 \end{cases}$$

To help extend the number of functions that can be integrated analytically, using the results of Figure 8.40, the following rules may be used.

Figure 8.40
Some standard
integrals.

$f(x)$	$\int f(x)\mathrm{d}x$ *Here c is a constant of integration*
$x^n \quad (n \neq -1)$	$\dfrac{x^{n+1}}{n+1} + c$
$\dfrac{1}{x}$	$\left.\begin{array}{l}\ln x \quad + c \ (x>0)\\ \ln(-x) + c \ (x<0)\end{array}\right\} = \ln\lvert x\rvert + c$
$\sin x$	$-\cos x + c$
$\cos x$	$\sin x + c$
e^x	$\mathrm{e}^x + c$
$\sec^2 x$	$\tan x + c$
$\dfrac{1}{\sqrt{(1-x^2)}}, \lvert x\rvert < 1$	$\sin^{-1} x + c$
$\dfrac{1}{1+x^2}$	$\tan^{-1} x + c$

Rule 1 (scalar-multiplication rule)
If k is a constant then

$$\int kf(x)\mathrm{d}x = k\int f(x)\mathrm{d}x$$

Rule 2 (sum rule)

$$\int [f(x) \pm g(x)]\mathrm{d}x = \int f(x)\mathrm{d}x \pm \int g(x)\mathrm{d}x$$

Rule 3 (linear composite rule)
If a and b are constants and $F'(x) = f(x)$ then

$$\int f(ax+b)\mathrm{d}x = \frac{1}{a}F(ax+b) + \text{constant}, \quad a \neq 0$$

Rule 4 (inverse-function rule)
If $y = f^{-1}(x)$, so that $x = f(y)$, then

$$\int f^{-1}(x)\mathrm{d}x = xy - \int f(y)\mathrm{d}y$$

Rules 1–3 follow directly from the definition of an integral, while rule 4 may be demonstrated graphically as illustrated in Figure 8.41.

Figure 8.41
Illustration of
$\int f^{-1}(x)\,dx =$
$xy - \int f(y)\,dy.$

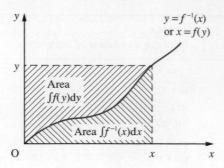

Example 8.30

Using the inverse-function rule, obtain the integrals of

(a) $\sin^{-1}x$ (b) $\ln x$

Solution   (a) If $y = \sin^{-1}x$ then $x = \sin y$ and

$$\int \sin^{-1}x\,dx = xy - \int \sin y\,dy$$

$$= xy + \cos y + \text{constant}$$

which, on using the identity $\sin^2 y + \cos^2 y = 1$, gives

$$\int \sin^{-1}x\,dx = x\sin^{-1}x + \sqrt{(1 - x^2)} + \text{constant}$$

since $\cos y \geqslant 0$ on the range of $\sin^{-1}x$.

(b) If $y = \ln x$ then $x = e^y$, and

$$\int \ln x\,dx = xy - \int e^y\,dy = xy - e^y + \text{constant}$$

$$= x \ln x - x + \text{constant}$$

since $e^{\ln x} = x$.

Example 8.31

Find the indefinite integrals of

(a) $6x^4 + 4x - \dfrac{3}{x}$ (b) $(2 - x)\sqrt{x}$ (c) $\sqrt{(5x + 2)}$ (d) $\dfrac{x + 1}{x}$

Solution   (a) Using the scalar-multiplication and sum rules,

$$\int \left(6x^4 + 4x - \frac{3}{x}\right)dx = 6\int x^4\,dx + 4\int x\,dx - 3\int \frac{1}{x}\,dx$$

$$= \tfrac{6}{5}x^5 + 2x^2 - 3\ln|x| + \text{constant}$$

using the standard integrals of Figure 8.40.

(b) Looking at the function, we see that because it involves a square root, its domain is restricted to values of $x \geqslant 0$. Multiplying through the brackets and using the scalar-multiplication and sum rules, we have

$$\int (2 - x)\sqrt{x}\,dx = 2 \int x^{1/2}\,dx - \int x^{3/2}\,dx$$

$$= \tfrac{4}{3}x^{3/2} - \tfrac{2}{5}x^{5/2} + \text{constant} \quad (x \geqslant 0)$$

(c) Examining the function, we see in this case that its domain is restricted to values of x greater than or equal to $-\tfrac{2}{5}$. We note that the formula is the square root of a linear function, and so we use the linear composite rule to obtain its integral. Thus, since

$$\int \sqrt{x}\,dx = \tfrac{2}{3}x^{3/2} + \text{constant}$$

we obtain

$$\int \sqrt{(5x + 2)}\,dx = \tfrac{1}{5}[\tfrac{2}{3}(5x + 2)^{3/2}] + \text{constant}$$

$$= \tfrac{2}{15}(5x + 2)^{3/2} + \text{constant} \quad (x \geqslant -\tfrac{2}{5})$$

(d) In this case we see that the function is defined except at $x = 0$. Expressing $(x + 1)/x$ as $1 + 1/x$ and using the sum rule, we obtain

$$\int \frac{x + 1}{x}\,dx = \int \left(1 + \frac{1}{x}\right)dx = \int 1\,dx + \int \frac{1}{x}\,dx$$

$$= \begin{cases} x + \ln x + \text{constant} & (x > 0) \\ x + \ln(-x) + \text{constant} & (x < 0) \end{cases}$$

$$= x + \ln|x| + \text{constant}$$

We have seen in Example 8.31 how a carefully chosen rearrangement of the integrand makes it possible to evaluate non-standard integrals. This technique is widely used in finding integrals of products of sines and cosines and of quadratic functions and their square roots. The first case uses the trigonometric sum identities (Section 2.6.3) and the second uses 'completion of the square'.

Example 8.32 Find the indefinite integrals of

(a) $\sin(5x + 1)\cos(x + 2)$ (b) $\dfrac{1}{x^2 + 10x + 50}$

Solution (a) First we express the product as the sum of two sine terms:

$$\sin(5x + 1)\cos(x + 2) = \tfrac{1}{2}[\sin(6x + 3) + \sin(4x - 1)]$$

Then we evaluate the integral, using the scalar-multiplication, sum and linear composite rules:

$$\int \sin(5x+1)\cos(x+2)\,dx = \int \frac{1}{2}[\sin(6x+3)+\sin(4x-1)]\,dx$$

$$= \frac{1}{2}\int \sin(6x+3)\,dx + \frac{1}{2}\int \sin(4x-1)\,dx$$

$$= -\frac{1}{12}\cos(6x+3) - \frac{1}{8}\cos(4x-1) + \text{constant}$$

recalling that

$$\int \sin x\,dx = -\cos x + \text{constant}$$

(b) Rewriting the quadratic term by completing the square (Section 1.3.1) gives

$$\int \frac{1}{x^2+10x+50}\,dx = \int \frac{1}{(x+5)^2+5^2}\,dx = \int \frac{1}{25[(\frac{1}{5}x+1)^2+1]}\,dx$$

Recalling that

$$\int \frac{1}{x^2+1}\,dx = \tan^{-1}x + \text{constant}$$

we have, using the linear composite and scalar-multiplication rules,

$$\int \frac{1}{x^2+10x+50}\,dx = \frac{1}{25}\frac{\tan^{-1}(\frac{1}{5}x+1)}{1/5} + \text{constant} = \frac{1}{5}\tan^{-1}\frac{x+5}{5} + \text{constant}$$

Finally, we consider the use of partial fractions in evaluating integrals of rational functions. Partial fractions are so frequently used to evaluate such integrals that one talks of the **partial fraction method of integration**.

Example 8.33 Using partial fractions, evaluate the integrals

(a) $\displaystyle\int \frac{6}{x^2-2x-8}\,dx$ (b) $\displaystyle\int \frac{9}{(x-1)(x+2)^2}\,dx$

Solution (a) Factorizing the denominator as $x^2-2x-8 = (x+2)(x-4)$, we can express the integrand in terms of its partial fractions:

$$\frac{6}{x^2-2x-8} = \frac{6}{(x+2)(x-4)} = \frac{-1}{x+2} + \frac{1}{x-4}$$

Thus

$$\int \frac{6}{x^2 - 2x - 8}\,\mathrm{d}x = \int \frac{-1}{x+2}\,\mathrm{d}x + \int \frac{1}{x-4}\,\mathrm{d}x$$

$$= -\ln|x+2| + \ln|x-4| + \text{constant}$$

$$= \ln\left|\frac{x-4}{x+2}\right| + \text{constant}$$

(b) In partial fractions we have

$$\frac{9}{(x-1)(x+2)^2} = \frac{1}{x-1} + \frac{-1}{x+2} + \frac{-3}{(x+2)^2}$$

Then

$$\int \frac{9}{(x-1)(x+2)^2}\,\mathrm{d}x = \int \frac{1}{x-1}\,\mathrm{d}x - \int \frac{1}{x+2}\,\mathrm{d}x - \int \frac{3}{(x+2)^2}\,\mathrm{d}x$$

$$= \ln|x-1| - \ln|x+2| + 3(x+2)^{-1} + \text{constant}$$

$$= \ln\left|\frac{x-1}{x+2}\right| + \frac{3}{x+2} + \text{constant}$$

Note in the preceding examples how the problems are systematically reduced to simpler ones by rearranging the integrand and carefully applying the basic rules together with standard integrals. The ability to do this comes only with practice of the type offered by the exercises at the end of this section.

In addition to the rules given earlier, two further results follow immediately from the basic definition of an integral. These are

$$\int_a^b f(x)\,\mathrm{d}x = -\int_b^a f(x)\,\mathrm{d}x$$

and

$$\int_a^b f(x)\,\mathrm{d}x = \int_a^c f(x)\,\mathrm{d}x + \int_c^b f(x)\,\mathrm{d}x$$

(Thus we may break the interval $[a, b]$ into convenient subintervals if the function is defined piecewise as illustrated in Example 8.34.)

Example 8.34   Evaluate

(a) $\displaystyle\int_{-1}^2 |x|\,\mathrm{d}x$ (b) $\displaystyle\int_0^{10} H(x-5)\,\mathrm{d}x$

where H is the Heaviside step function given by (2.43).

Figure 8.42

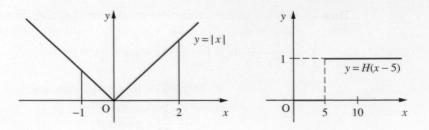

Solution   The areas involved are illustrated in Figure 8.42.

(a) Since

$$|x| = \begin{cases} -x & (x \leqslant 0) \\ x & (x \geqslant 0) \end{cases}$$

we split the integral at $x = 0$ and write

$$\int_{-1}^{2} |x|\,dx = \int_{-1}^{0} -x\,dx + \int_{0}^{2} x\,dx = [-\tfrac{1}{2}x^2]_{-1}^{0} + [\tfrac{1}{2}x^2]_{0}^{2} = \tfrac{5}{2}$$

(b) Since $H(x-5)$ has a discontinuity at $x = 5$, we write

$$\int_{0}^{10} H(x-5)\,dx = \int_{0}^{5} H(x-5)\,dx + \int_{5}^{10} H(x-5)\,dx$$

$$= \int_{0}^{5} 0\,dx + \int_{5}^{10} 1\,dx = 5$$

These results can be readily confirmed by inspection of the relevant areas.

We see from this last example that it is sometimes possible to integrate functions even if they have discontinuities. This is possible provided that there are only a finite number of finite discontinuities within the domain of integration and that elsewhere the function is continuous and bounded. To illustrate this, consider the function $f(x)$ illustrated in Figure 8.43 where

$$y = f(x) = \begin{cases} f_1(x) & (a \leqslant x < b_1) \\ f_2(x) & (b_1 < x < b_2) \\ f_3(x) & (b_2 < x \leqslant b) \end{cases}$$

Figure 8.43
Piecewise-continuous
function.

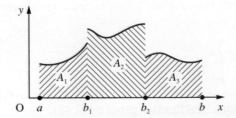

Such a function is called a **piecewise-continuous function**. Interpreting the integral as the area under the curve, we have

$$\int_a^b f(x)\mathrm{d}x = A_1 + A_2 + A_3$$

but in this case we interpret the individual areas as

$$\int_a^b f(x)\mathrm{d}x = \int_a^{b_1^-} f_1(x)\mathrm{d}x + \int_{b_1^+}^{b_2^-} f_2(x)\mathrm{d}x + \int_{b_2^+}^b f_3(x)\mathrm{d}x$$

where, as before, b_1^- signifies approaching b_1 from the left and b_1^+ signifies approaching b_1 from the right (see Section 7.8.1). It is in this sense that we evaluated $\int_0^{10} H(x-5)\mathrm{d}x$ in Example 8.34, and – strictly speaking – we should have written

$$\int_0^{10} H(x-5)\mathrm{d}x = \int_0^{5^-} H(x-5)\mathrm{d}x + \int_{5^+}^{10} H(x-5)\mathrm{d}x$$

and, since

$$H(x-5) = \begin{cases} 0 & (x < 5) \\ 1 & (x \geqslant 5) \end{cases}$$

$$\int_0^{10} H(x-5)\mathrm{d}x = \int_0^{5^-} 0\,\mathrm{d}x + \int_{5^+}^{10} 1\,\mathrm{d}x = 5$$

Example 8.35

As shown in Example 8.7, the bending moment M and shear force F acting in a beam satisfy the differential equation

$$F = \frac{\mathrm{d}M}{\mathrm{d}x}$$

In Example 8.29, we showed that for a continuously non-uniformly loaded beam which is freely hinged at both ends the shear force F is given by

$$F(x) = \begin{cases} 2Wx^2/l^2 - W/2 & 0 \leqslant x \leqslant l/2 \\ W/2 - 2W(l-x)^2/l^2 & l/2 \leqslant x \leqslant l \end{cases}$$

Given that $M = 0$ at $x = 0$, find an expression for $M(x)$ at a general point.

Solution

Since $\dfrac{\mathrm{d}M}{\mathrm{d}x} = F(x)$ with $M(0) = 0$ we deduce by the Fundamental Theorem

$$M(x) = \int_0^x F(t)\mathrm{d}t$$

In evaluating this integral we have to remember that $F(x)$ is defined separately on $(0, l/2)$ and $(l/2, l)$.

For $x < l/2$, we have

$$M(x) = \int_0^x (2Wt^2/l^2 - W/2)\,dt$$

For $x > l/2$, we have

$$M(x) = \int_0^{l/2} (2Wt^2/l^2 - W/2)\,dt + \int_{l/2}^x (W/2 - 2W(l - t)^2/l^2)\,dt$$

Thus

$$M(x) = \begin{cases} \dfrac{Wx}{6l^2}(4x^2 - 3l^2) & 0 \leqslant x \leqslant l/2 \\[3mm] \dfrac{W(l - x)}{6l^2}(4(l - x)^2 - 3l^2) & l/2 \leqslant x \leqslant l \end{cases}$$

8.6.2 Exercises

78 Find the indefinite integrals of

(a) $3x^{2/3}$

(b) $\sqrt{(2x)}$

(c) $2x^3 - 2x^2 + \dfrac{1}{x} - 2$

(d) $2e^x + 3\cos 2x$

(e) $x^2 + 3e^x - \dfrac{1}{x^2}$

(f) $(2x + 1)^3$

(g) $(1 - 2x)^{1/3}$

(h) $(2x^2 + 1)^3$

(i) $\cos(2x + 1)$

(j) 2^x (*Hint*: $2 = e^{\ln 2}$)

(k) $\sin 3x \cos 5x$

(l) $\cos 7x \cos 5x$

79 Using partial fractions, integrate

(a) $\dfrac{x}{x^2 - 3x - 4}$

(b) $\dfrac{x}{(x - 2)^2}$

(c) $\dfrac{1}{x(x + 1)}$

(d) $\dfrac{x}{x^2 + 2x + 1}$

(e) $\dfrac{1}{x^2 - 1}$

(f) $\dfrac{1}{x^2(x - 1)}$

(g) $\dfrac{1}{x(x - 1)(x - 2)}$

(h) $\dfrac{1}{1 + x - 2x^2}$

(i) $\dfrac{2x^3}{x^3 - 1}$

(j) $\dfrac{3x^3 - 3x^2 + 4x - 2}{x(x - 1)(x^2 + 1)}$

(k) $\dfrac{9}{(x - 1)(x + 2)^2}$

(l) $\dfrac{x^2 - 2x + 3}{(x - 1)(x^2 - x - 1)}$

80 Evaluate

(a) $\displaystyle\int_2^3 \dfrac{x\,dx}{\sqrt{(x + 1)}}$

(b) $\displaystyle\int_0^1 x(x - 1)^{11}\,dx$

(c) $\displaystyle\int_0^\pi \sin 5x \sin 6x\,dx$

(d) $\displaystyle\int_0^\pi \sin^2 5x\,dx$

(e) $\displaystyle\int_1^2 \left(x^{3/2} - \dfrac{1}{x^2}\right)dx$

(f) $\displaystyle\int_0^{\pi/2} \sin x\,dx$

(g) $\displaystyle\int_0^2 \dfrac{dx}{\sqrt{(3 + 2x - x^2)}}$

(h) $\displaystyle\int_0^1 \dfrac{2\,dx}{(x + 1)^2(x^2 + 1)}$

(*Hint*: Replace the x in (a) by $(x + 1) - 1$ and in (b) by $(x - 1) + 1$.)

81 Express $12/(x - 3)(x + 1)$ in partial fractions and hence show that

$$\int_4^6 \dfrac{12}{(x - 3)(x + 1)}\,dx = 3\ln \tfrac{15}{7}$$

82 Find the indefinite integrals of

(a) x^{-2}

(b) $(x + 1)^{-1/3}$

(c) $\dfrac{4x^3 - 7x^2 + 1}{x^2}$

(d) $\sin x + \cos x$

(e) $\sin^2 x$

(f) $\cos^2 x$

(g) $\cosh^2 x$

(h) $\sinh(5x + 1)$

(i) $\dfrac{1}{9 - 16x^2}$

(j) $\dfrac{1}{\sqrt{(2x - x^2)}}$

(k) $\dfrac{1}{\sqrt{(1 - 9x^2)}}$

(l) $\dfrac{1}{\sqrt{(4 - x^2)}}$

(m) $\dfrac{1}{\sqrt{(1 - x - x^2)}}$

(n) $\dfrac{1}{\sqrt{[x(1 - x)]}}$

(o) $\dfrac{1}{\sqrt{(5 + 4x - x^3)}}$

(p) $\dfrac{1}{x^2 + 6x + 13}$

83 Evaluate

(a) $\displaystyle\int_0^3 |x - 2|\,dx$

(b) $\displaystyle\int_0^5 (x - 2)H(x - 2)\,dx$

(c) $\displaystyle\int_0^3 \lfloor x \rfloor\,dx$

(d) $\displaystyle\int_0^3 \mathrm{FRACPT}(x)\,dx$

(e) $\displaystyle\int_0^3 x\lfloor x \rfloor\,dx$

8.6.3 Further analytical methods of integration

The product rule for differentiation

$$\frac{\mathrm{d}}{\mathrm{d}x}(uv) = \frac{\mathrm{d}u}{\mathrm{d}x}v + u\frac{\mathrm{d}v}{\mathrm{d}x}$$

may also be used for integration after a little rearrangement. From the above we have

$$u\frac{\mathrm{d}v}{\mathrm{d}x} = \frac{\mathrm{d}}{\mathrm{d}x}(uv) - v\frac{\mathrm{d}u}{\mathrm{d}x}$$

and on integrating we have

$$\int u\frac{\mathrm{d}v}{\mathrm{d}x}\,\mathrm{d}x = uv - \int v\frac{\mathrm{d}u}{\mathrm{d}x}\,\mathrm{d}x$$

We may use this result to determine an integral when the integrand is the product of the two functions. The method is called **integration by parts**. The procedure is to choose one term of the product to be u and the other to be $\mathrm{d}v/\mathrm{d}x$. We then calculate $\mathrm{d}u/\mathrm{d}x$ and v, and the hope is that the resulting integral on the right-hand side is easier than the one we started with. We shall illustrate the method with a few examples.

Example 8.36   Find the indefinite integrals of

(a) $x^2 \cos x$ (b) $x \ln x$ (c) $e^x \sin 2x$

Solution   (a) Since differentiation reduces the squared term to a linear one, leading to some simplification, we choose

$$u = x^2 \quad \text{and} \quad \frac{\mathrm{d}v}{\mathrm{d}x} = \cos x$$

so that

$$\frac{\mathrm{d}u}{\mathrm{d}x} = 2x \quad \text{and} \quad v = \sin x$$

Note: There is no need to introduce a constant of integration when determining v. Substituting in the formula for integration by parts gives

$$
\begin{array}{cccc}
u & dv/dx & u & v & v & du/dx \\
\downarrow & \downarrow & \downarrow & \downarrow & \downarrow & \downarrow
\end{array}
$$

$$
\int x^2 \cos x \, dx = x^2(\sin x) - \int (\sin x)(2x)dx = x^2 \sin x - 2\int x \sin x \, dx
$$

We now apply the same technique to the last integral, taking

$$
u = x \quad \text{and} \quad \frac{dv}{dx} = \sin x
$$

to give

$$
\int x \sin x \, dx = (x)(-\cos x) - \int (-\cos x)(1)dx = -x \cos x + \sin x + \text{constant}
$$

Substituting back gives

$$
\int x^2 \cos x \, dx = x^2 \sin x - 2(-x \cos x + \sin x) + \text{constant}
$$

$$
= x^2 \sin x - 2 \sin x + 2x \cos x + \text{constant}
$$

(b) With this integral, we set

$$
u = \ln x \quad \text{and} \quad \frac{dv}{dx} = x
$$

giving

$$
\frac{du}{dx} = \frac{1}{x} \quad \text{and} \quad v = \tfrac{1}{2}x^2
$$

Integration by parts then gives

$$
\int x \ln x \, dx = (\tfrac{1}{2}x^2) \ln x - \int (\tfrac{1}{2}x^2)\left(\frac{1}{x}\right)dx = \tfrac{1}{2}x^2 \ln x - \int \tfrac{1}{2}x \, dx
$$

$$
= \tfrac{1}{2}x^2 \ln x - \tfrac{1}{4}x^2 + \text{constant}
$$

(c) In this case it is not obvious that any choice of u and v will result in a simpler integral. Setting

$$
u = \sin 2x \quad \text{and} \quad \frac{dv}{dx} = e^x
$$

(only because integrating $\sin 2x$ will mean dividing by 2 and getting clumsy fractions!) gives

$$
\frac{du}{dx} = 2 \cos 2x \quad \text{and} \quad v = e^x
$$

Integration by parts then gives

$$\int e^{x}\sin 2x\,dx = e^{x}\sin 2x - \int e^{x}(2\cos 2x)\,dx$$

$$= e^{x}\sin 2x - 2\int e^{x}\cos 2x\,dx$$

which has produced no simplification at all. We repeat the process, however, on the last integral, taking care to integrate the part we integrated the first time and to differentiate the part we differentiated the first time. Thus we take

$$u = \cos 2x \quad \text{and} \quad \frac{dv}{dx} = e^{x}$$

giving

$$\int e^{x}\cos 2x\,dx = e^{x}\cos 2x - \int e^{x}(-2\sin 2x)\,dx$$

$$= e^{x}\cos 2x + 2\int e^{x}\sin 2x\,dx$$

Substituting in the previous expression, we obtain

$$\int e^{x}\sin 2x\,dx = e^{x}\sin 2x - 2\left(e^{x}\cos 2x + 2\int e^{x}\sin 2x\,dx \right)$$

Hence

$$5\int e^{x}\sin 2x\,dx = e^{x}(\sin 2x - 2\cos 2x)$$

so

$$\int e^{x}\sin 2x\,dx = \tfrac{1}{5}e^{x}(\sin 2x - 2\cos 2x) + \text{constant}$$

The composite-function rule for differentiation

$$\frac{d}{dx}[f(g(x))] = f'(g(x))g'(x)$$

can be used to evaluate some integrals. Reversing the differentiation process, we may write

$$\int f'(g(x))g'(x)\,dx = f(g(x)) + \text{constant}$$

The key step here is identifying the function $g(x)$. This will not be unique: different choices of $g(x)$ may differ by a constant. To make the process of manipulation easier to follow, it is usual to set $t = g(x)$, so that the integral becomes

$$\int f'(g(x))g'(x)dx = \int f'(t)\frac{dt}{dx}dx = \int f'(t)dt = f(t) + \text{constant}$$

$$= f(g(x)) + \text{constant} \quad \text{(on back substitution)}$$

This technique for evaluating integrals is called the **substitution method**; we shall illustrate its use with a number of examples.

Example 8.37   Find the indefinite integrals

(a) $\displaystyle\int 2x\sqrt{(x^2 + 3)}\,dx$ (b) $\displaystyle\int \frac{x + 1}{x^2 + 2x + 2}\,dx$

Solution   (a) Comparison with the general form above suggests that we take

$$g(x) = x^2 + 3, \quad \text{with } g'(x) = 2x$$

Setting $t = x^2 + 3$, the integral becomes

$$\int 2x\sqrt{(x^2 + 3)}\,dx = \int \frac{dt}{dx}\sqrt{t}\,dx = \int t^{1/2}dt$$

$$= \tfrac{2}{3}t^{3/2} + \text{constant} = \tfrac{2}{3}(x^2 + 3)^{3/2} + \text{constant}$$

(b) Comparison with the general form suggests that we choose

$$g(x) = x^2 + 2x + 2, \quad \text{with } g'(x) = 2x + 2$$

This necessitates a slight modification of the integral giving

$$\int \frac{x + 1}{x^2 + 2x + 2}\,dx = \tfrac{1}{2}\int \frac{2x + 2}{x^2 + 2x + 2}\,dx = \tfrac{1}{2}\int \frac{1}{t}\,dt$$

where $t = x^2 + 2x + 2$ and $dt = (2x + 2)dx$. Thus

$$\int \frac{x + 1}{x^2 + 2x + 2}\,dx = \tfrac{1}{2}\ln t + \text{constant} = \tfrac{1}{2}\ln(x^2 + 2x + 2) + \text{constant}$$

This example is a special case of a commonly occurring form when the integrand can be written as

$$\frac{\text{derivative of denominator}}{\text{denominator}}$$

so that the integral is the logarithm of the denominator.

The choice of substitution $t = g(x)$ that results in a convenient simplification of the integral is not always dictated by the form of the composite-function rule given above. Sometimes the integrand contains terms that suggest an initial substitution, at least, in the hope of simplification.

Example 8.38

Find the indefinite integral

$$\int \frac{1}{2 + \sqrt{(1-x)}} \, dx$$

Solution

The source of the difficulty with this integral is the square-root term in the denominator. We try to simplify the integral by the substitution $t = \sqrt{(1-x)}$. Thus $x = 1 - t^2$ and $dx/dt = -2t$, giving

$$\int \frac{1}{2 + \sqrt{(1-x)}} \, dx = \int \frac{1}{2 + t} \frac{dx}{dt} \, dt = \int \frac{1}{2+t}(-2t) \, dt$$

$$= \int \frac{-2t}{2+t} \, dt = 2 \int \left(\frac{2}{2+t} - 1 \right) dt$$

$$= 4 \ln(2 + t) - 2t + \text{constant}$$

$$= 4 \ln[2 + \sqrt{(1-x)}] - 2\sqrt{(1-x)} + \text{constant}$$

The choice of such substitutions is not always immediately obvious. We shall consider a further example and then give a list of substitutions commonly used to simplify integrals.

Example 8.39

Find the indefinite integral $\int \sqrt{(1 - x^2)} \, dx$, $0 \leqslant x \leqslant 1$.

Solution

Based on our experience with Example 8.38, we are tempted to try to remove the square-root term using the substitution

$$u = \sqrt{(1 - x^2)}$$

Then $u^2 = 1 - x^2$ and $2u = -2x \, dx/du$, so that

$$\frac{dx}{du} = -\frac{u}{x} = -\frac{u}{\sqrt{(1 - u^2)}}$$

giving

$$\int \sqrt{(1 - x^2)} \, dx = -\int \frac{u^2 \, du}{\sqrt{(1 - u^2)}}$$

which leaves us with an integral more complicated than the one with which we started.

Thus in this case the simple substitution does not work, and we need to look for a more sophisticated substitution, bearing in mind that what we wish to do is to remove the awkward square-root term $\sqrt{(1-x^2)}$. Noting that $\cos^2\theta = 1 - \sin^2\theta$ we try the substitution $x = \sin\theta$ so that $\dfrac{dx}{d\theta} = \cos\theta$ giving

$$\int \sqrt{(1-x^2)}dx = \int \sqrt{(1-\sin^2\theta)}\frac{dx}{d\theta}d\theta = \int \cos\theta\cos\theta\,d\theta$$

$$= \int \cos^2\theta\,d\theta$$

which looks simpler than the original integral but is not immediately integrable.

Using the double-angle trigonometric identity

$$\cos 2\theta = 2\cos^2\theta - 1$$

we obtain

$$\int \sqrt{(1-x^2)}dx = \int \tfrac{1}{2}(1 + \cos 2\theta)d\theta$$

$$= \tfrac{1}{2}\theta + \tfrac{1}{4}\sin 2\theta + \text{constant}$$

This gives the answer in terms of θ rather than the original variable x. Since $\theta = \sin^{-1}x$, back substitution gives

$$\int \sqrt{(1-x^2)}dx = \tfrac{1}{2}\sin^{-1}x + \tfrac{1}{4}\sin(2\sin^{-1}x) + \text{constant}$$

or, since $\sin 2\theta = 2\sin\theta\cos\theta = 2\sin\theta\sqrt{(1-\sin^2\theta)}$, we may write this in the alternative form

$$\int \sqrt{(1-x^2)}dx = \tfrac{1}{2}\sin^{-1}x + \tfrac{1}{2}x\sqrt{(1-x^2)} + \text{constant}$$

Figure 8.44 shows a number of substitutions that are often used in the evaluation of $\int f(x)dx$. This list is not exhaustive. There are many special cases, some of which are given in Exercises 8.6.4.

When using substitution methods with definite integrals, it is usually best to change the limits of the integral when the integrating variable is changed. This saves returning to the original variable, which can sometimes be very tedious. In general, setting $x = g(t)$ gives

$$\int_a^b f(x)dx = \int_{g^{-1}(a)}^{g^{-1}(b)} f(g(t))g'(t)dt$$

$$= \int_{t_a}^{t_b} h(t)dt, \quad \text{where } a = g(t_a),\ b = g(t_b) \text{ and } h(t) = f(g(t))g'(t)$$

Figure 8.44
Substitutions for
evaluation of $\int f(x)\mathrm{d}x$.

If $f(x)$ contains	try	
$\sqrt{(a^2 - x^2)}$	$x = a\sin\theta,$	$\dfrac{\mathrm{d}x}{\mathrm{d}\theta} = a\cos\theta$
	or $x = a\tanh u,$	$\dfrac{\mathrm{d}x}{\mathrm{d}u} = a\,\mathrm{sech}^2 u$
$\sqrt{(a^2 + x^2)}$	$x = a\sinh u,$	$\dfrac{\mathrm{d}x}{\mathrm{d}u} = a\cosh u$
	or $x = a\tan\theta,$	$\dfrac{\mathrm{d}x}{\mathrm{d}\theta} = a\sec^2\theta$
$\sqrt{(x^2 - a^2)}$	$x = a\cosh u$	$\dfrac{\mathrm{d}x}{\mathrm{d}u} = a\sinh u$
	or $x = a\sec\theta$	$\dfrac{\mathrm{d}x}{\mathrm{d}\theta} = a\sec\theta\tan\theta$
Circular functions	$s = \sin x,$	$\dfrac{\mathrm{d}s}{\mathrm{d}x} = \cos x$
	or $c = \cos x,$	$\dfrac{\mathrm{d}c}{\mathrm{d}x} = -\sin x$
	or $t = \tan\frac{1}{2}x,$	
	$\left(\sin x = \dfrac{2t}{1+t^2},\quad \cos x = \dfrac{1-t^2}{1+t^2},\quad \dfrac{\mathrm{d}x}{\mathrm{d}t} = \dfrac{2}{1+t^2}\right)$	
Hyperbolic functions	$u = \mathrm{e}^x,$	$\dfrac{\mathrm{d}u}{\mathrm{d}x} = \mathrm{e}^x$
	or $s = \sinh x,$	$\dfrac{\mathrm{d}s}{\mathrm{d}x} = \cosh x$
	or $c = \cosh x,$	$\dfrac{\mathrm{d}c}{\mathrm{d}x} = \sinh x$
	or $t = \tanh\frac{1}{2}x,$	$\dfrac{\mathrm{d}t}{\mathrm{d}x} = \frac{1}{2}\,\mathrm{sech}^2\frac{1}{2}x$

Example 8.40   Using the substitution $u = \sqrt{(x + 2)}$, evaluate the definite integral

$$\int_{-2}^{2} \frac{\sqrt{(x + 2)}}{x + 6}\,\mathrm{d}x.$$

Solution   Setting $u = \sqrt{(x + 2)}$, or $u^2 = x + 2$, gives $2u\,\mathrm{d}u = \mathrm{d}x$. Regarding limits, when $x = -2$, $u = 0$ and when $x = 2$, $u = \sqrt{4} = 2$.

Making the substitution gives

$$\int_{-2}^{2} \frac{\sqrt{(x + 2)}}{x + 6}\,\mathrm{d}x = \int_{0}^{2} \frac{u}{u^2 + 4}\,2u\,\mathrm{d}u = \int_{0}^{2} \frac{2u^2}{u^2 + 4}\,\mathrm{d}u = \int_{0}^{2} 2 - \frac{8}{u^2 + 4}\,\mathrm{d}u$$

$$= \left[2u - 4\tan^{-1}\frac{u}{2}\right]_{0}^{2} = 4 - \pi$$

8.6.4 Exercises

84 Use integration by parts to find the indefinite integrals of

(a) $x \sin x$ (b) xe^{3x} (c) $x^3 \ln x$

(d) $e^{-2x} \sin 3x$ (e) $x \tan^{-1}x$ (f) $x \cos 2x$

85 Show that

$$\int f(x)\,\mathrm{d}x = xf(x) - \int xf'(x)\,\mathrm{d}x$$

Use this result to integrate

(a) $\sin^{-1}x$ (b) $\ln x$

(c) $\cosh^{-1}x$ (d) $\tan^{-1}x$

86 Using the substitution method, integrate the following functions:

(a) $x\sqrt{(1+x^2)}$ (b) $\cos x \sin^3 x$

(c) $\dfrac{x}{(1+x^2)^2}$ (d) $\dfrac{x}{\sqrt{(x^2-1)}}$

(e) $\dfrac{2x+3}{x^2+3x+2}$ (f) $\dfrac{1}{x\ln x}$

87 Find the values of the constants a and b such that

$$\frac{3x+2}{x^2+2x+5} = \frac{a(2x+2)}{x^2+2x+5} + \frac{b}{x^2+2x+5}$$

and hence find its integral. (Note that $(\mathrm{d}/\mathrm{d}x)(x^2+2x+5) = 2x+2$.)

88 Use the technique of Question 87 to integrate

(a) $\dfrac{x+1}{x^2+4x+5}$ (b) $\dfrac{2x+3}{\sqrt{(5+4x-x^2)}}$

(c) $\dfrac{\sin x}{\sin x + \cos x}$

89 Use the given substitutions to integrate the following functions:

(a) $x^3\sqrt{(1+x^2)}$,   with $t = \sqrt{(1+x^2)}$

(b) $\sin^3 x \cos^5 x$,   with $t = \sin x$

(c) $\dfrac{x}{(1+x^2)^2}$,   with $x = \tan t$

(d) $\dfrac{3}{x\sqrt{(x^2+9)}}$,   with $t = \dfrac{1}{x}$

(e) $\dfrac{x}{\sqrt{(4-x^2)}}$,   with $x = 2\sin t$

(f) $\dfrac{1}{3+\sqrt{x}}$,   with $t = \sqrt{x}$

90 Use an appropriate substitution to integrate the following functions:

(a) $\dfrac{1}{1+\sqrt{(1+x)}}$ (b) $\sin^2 x \cos^3 x$ (c) $\sin\sqrt{x}$

91 Show that $t = \tan\frac{1}{2}x$ implies

$$\sin x = \frac{2t}{1+t^2}, \qquad \cos x = \frac{1-t^2}{1+t^2}$$

and

$$\mathrm{d}x = \frac{2}{1+t^2}\,\mathrm{d}t$$

Hence integrate

(a) $\operatorname{cosec} x$ (b) $\sec x$

(c) $\dfrac{1}{3+4\sin x}$ (d) $\dfrac{1}{5\sin x + 12\cos x}$

92 Using the substitution $u = \sqrt{x}-1$, evaluate the definite integral

$$\int_4^9 \frac{\mathrm{d}x}{(\sqrt{x}-1)\sqrt{x}}$$

93 Using integration by parts, evaluate the definite integrals

(a) $\displaystyle\int_0^{\pi/2} x^2 \sin x\,\mathrm{d}x$ (b) $\displaystyle\int_1^3 x^2 \ln x\,\mathrm{d}x$

(c) $\displaystyle\int_0^1 xe^{3x}\,\mathrm{d}x$

94 Evaluate the following definite integrals using the given substitutions:

(a) $\displaystyle\int_1^4 \frac{e^{\sqrt{x}}}{\sqrt{x}}\,\mathrm{d}x$,   with $u = \sqrt{x}$

(b) $\displaystyle\int_{-2}^2 \frac{x+6}{\sqrt{(x+2)}}\,\mathrm{d}x$,   with $u = \sqrt{(x+2)}$

(c) $\displaystyle\int_0^{\sqrt{3}} \frac{\tan^{-1}x}{1+x^2}\,dx$, with $u = \tan^{-1}x$

(d) $\displaystyle\int_{1/6}^{1/2} \frac{dx}{(5+6x)^3}$, with $u = 5 + 6x$

95 In Question 10 (Exercises 8.2.8) the equation of the path of P was found to be such that

$$\frac{dy}{dx} = \frac{\sqrt{(a^2 - x^2)}}{x}, \quad \text{with } y = 0 \text{ at } x = a$$

Use the substitution $x = a\,\text{sech}\,u$ to integrate this differential equation and show that

$$y = \ln\left[\frac{a + \sqrt{(a^2 - x^2)}}{x}\right] - \sqrt{(a^2 - x^2)}$$

This curve is called a **tractrix**.

8.7 Exploitation of integration

In this section we consider some other situations in which integration is widely used.

8.7.1 Volume of a solid of revolution

Imagine rotating the plane area A under the graph of the function $f(x)$, $x \in [a, b]$, of Figure 8.45 through a complete revolution about the x axis. The result would be to generate a solid having the x axis as axis of symmetry as shown in Figure 8.46(a): this is called a **solid of revolution**. If we wish to determine the volume of this solid, we proceed as in Section 8.5.1 and subdivide the rotating area into n vertical strips. When a typical strip within the subinterval $[x_{r-1}, x_r]$ is rotated through a revolution about the x axis, it will generate a thin disc of radius $f(x_r^*)$ (with $x_{r-1} < x_r^* < x_r$) and thickness Δx_{r-1} as shown in Figure 8.46(b). The volume of the disc is given by

$$\Delta V_r = \pi [f(x_r^*)]^2 \Delta x_{r-1}$$

Figure 8.45
Plane area rotated.

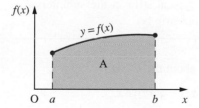

Figure 8.46
Solid of revolution.

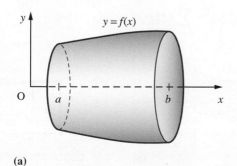

(a)

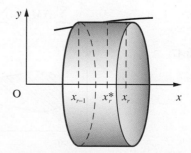

(b)

Thus the volume of the solid can be approximated by

$$V \approx \sum_{r=1}^{n} \Delta V_r = \pi \sum_{r=1}^{n} [f(x_r^*)]^2 \Delta x_{r-1}$$

Again this approximation is closer to the exact volume as the number of strips is increased. Thus in the limiting case as $n \to \infty$ and $\Delta x \to 0$, $\Delta x = \max_{r} \Delta x_r$, it leads to the volume being given by

$$V = \lim_{\substack{n \to \infty \\ \Delta x \to 0}} \pi \sum_{r=1}^{n} [f(x_r^*)]^2 \Delta x_{r-1} = \pi \int_a^b [f(x)]^2 \mathrm{d}x \qquad (8.33)$$

8.7.2   Centroid of a plane area

Consider the plane region of Figure 8.47(a) bounded between the graphs of the two continuous functions $f(x)$ and $g(x)$ on the interval $x \in [a, b]$, with $g(x) \leqslant f(x)$ on the interval. The area A of this region is clearly given by

$$A = \text{area under the graph of } f(x) - \text{area under the graph of } g(x)$$

$$= \int_a^b f(x)\mathrm{d}x - \int_a^b g(x)\mathrm{d}x$$

That is

$$A = \int_a^b [f(x) - g(x)]\mathrm{d}x \qquad (8.34)$$

We now wish to find the coordinates $(\bar{x}, \bar{y})$ of the centroid of this area. To do this, we take moments of area about the x and y axes in turn. As before, we subdivide the region into n strips, with a typical strip in the subinterval $[x_{r-1}, x_r]$ being shown in Figure 8.47(b). The area of the strip is

$$\Delta A_r = [f(x_r^*) - g(x_r^*)]\Delta x_{r-1}$$

and the moment of this area about the y axis is

$$\Delta M_{y_r} = x_r^* \Delta A_r = x_r^* [f(x_r^*) - g(x_r^*)]\Delta x_{r-1}$$

Thus the sum of the moments of the n strips about the y axis is

$$\sum_{r=1}^{n} \Delta M_{y_r} = \sum_{r=1}^{n} x_r^* \Delta A_r = \sum_{r=1}^{n} x_r^* [f(x_r^*) - g(x_r^*)]\Delta x_{r-1}$$

Figure 8.47

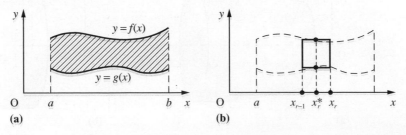

(a)   (b)

Proceeding to the limit $n \to \infty$, $\Delta x \to 0$, $\Delta x = \max_r \Delta x_r$, we have the moment of the plane area about the y axis being given by

$$M_y = \lim_{\substack{n \to \infty \\ \Delta x \to 0}} \sum_{r=1}^{n} x_r^*[f(x_r^*) - g(x_r^*)]\Delta x_{r-1} = \int_a^b x[f(x) - g(x)]\,\mathrm{d}x$$

Since the x coordinate of the centroid of the plane area is $\bar{x}$, it follows that the moment of the area about the y axis is also given by

$$M_y = A\bar{x}$$

Equating, we have

$$\bar{x} = \frac{1}{A}\int_a^b x[f(x) - g(x)]\,\mathrm{d}x \tag{8.35}$$

where the area A is given by (8.34).

Likewise, taking moments about the x axis,

$$M_x = A\bar{y} = \lim_{\substack{n \to \infty \\ \Delta x \to 0}} \left[\sum_{r=1}^{n} \tfrac{1}{2}f(x_r^*)f(x_r^*)\Delta x_{r-1} - \sum_{r=1}^{n} \tfrac{1}{2}g(x_r^*)g(x_r^*)\Delta x_{r-1} \right]$$

$$= \lim_{\substack{n \to \infty \\ \Delta x \to 0}} \tfrac{1}{2}\sum_{r=1}^{n}\{[f(x_r^*)]^2 - [g(x_r^*)]^2\}\Delta x_{r-1} = \tfrac{1}{2}\int_a^b \{[f(x)]^2 - [g(x)]^2\}\,\mathrm{d}x$$

giving

$$\bar{y} = \frac{1}{2A}\int_a^b \{[f(x)]^2 - [g(x)]^2\}\,\mathrm{d}x \tag{8.36}$$

where A again is given by (8.34).

In the particular case when $g(x)$ is the x axis, we find that the centroid of the plane area bounded by $f(x)$ ($x \in [a, b]$) and the x axis has coordinates

$$\bar{x} = \frac{1}{A}\int_a^b xf(x)\,\mathrm{d}x, \quad \bar{y} = \frac{1}{2A}\int_a^b [f(x)]^2\,\mathrm{d}x \tag{8.37}$$

8.7.3 Centre of gravity of a solid of revolution

Proceeding as in Section 8.7.2, we can obtain the coordinates $(\bar{X}, \bar{Y})$ of the centre of gravity of the solid of revolution generated by $f(x)$ ($x \in [a, b]$) and shown in Figure 8.46. By symmetry, it lies on the x axis, so that

$$\bar{Y} = 0$$

Taking moments about the y axis gives

$$V\bar{X} = \lim_{\substack{n \to \infty \\ \Delta x \to 0}} \pi \sum_{r=1}^{n} x_r^*[f(x_r^*)]^2 \Delta x_{r-1} = \pi \int_a^b x[f(x)]^2\,\mathrm{d}x \tag{8.38}$$

giving

$$\overline{X} = \frac{\pi}{V} \int_a^b x[f(x)]^2 \, \mathrm{d}x \tag{8.39}$$

where the volume V is given by (8.33).

8.7.4 Mean values

In many engineering applications we need to know the mean value of a continuously varying quantity. When dealing with a sequence of values we can compute the mean value simply by adding the values together and then dividing by the number of values taken. When dealing with a continuously varying quantity, we cannot do that directly. Using integration, however, we are able to calculate the mean value.

Consider the function $f(x)$ on the interval $[a, b]$ and divide the interval into n equal strips of width h so that $nh = b - a$. Now evaluate the function at the midpoint of each strip. Formally, let $x_k = a + kh$ be the points of subdivision, so that the points of evaluation are $f(x_k^*)$ where $x_k^* = x_k + h/2$. Then the mean value (m.v.) of $f(x)$ on $[a, b]$ is approximately

$$\mathrm{m.v.}(f(x)) \approx \frac{1}{n} \sum_{k=0}^{n-1} f(x_k^*) = \frac{1}{b-a} \sum_{k=0}^{n-1} f(x_k^*)h$$

Now allowing $n \to \infty$ (with $h \to 0$), the summation becomes an integral and the approximation becomes exactly true. Thus

$$\mathrm{m.v.}(f(x)) = \frac{1}{b-a} \int_a^b f(x)\,\mathrm{d}x \tag{8.40}$$

The graphical representation of this makes the situation quite clear. In Figure 8.48, the sum of the shaded areas above the line $y = $ (mean value) is equal to the sum of the shaded areas below it, so that the area of the rectangle ABCD is the same as the area between the curve and the x axis.

Figure 8.48
Mean value
of a function
$y = f(x), x \in [a, b]$.

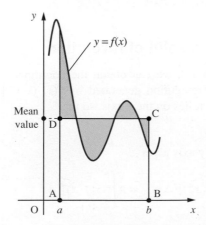

8.7.5 Root mean square values

In some contexts the computation of the mean value of a function is not useful, for example the mean of an alternating current is zero but that does not imply it is not dangerous! To deal with such situations we use the root mean square (r.m.s.) of the function $f(x)$. Literally this is the square root of the mean value of $[f(x)]^2$. Thus we can write

$$[\text{r.m.s.}(f(x))]^2 = \frac{1}{b-a} \int_a^b [f(x)]^2 \, dx$$

(8.41)

Although the obvious applications of root mean square values are in electrical engineering, they also occur in the application of statistics to engineering contexts (as standard deviations of continuously distributed random variables). They also occur in the design of gyroscopes and in mechanics, where the 'radius of gyration' is in effect the root mean of moments about an axis.

8.7.6 Arclength and surface area

In many practical problems we are required to work out the length of a curve or the surface area generated by rotating a curve. The formula for the length s of a curve with formula $y = f(x)$ between two points corresponding to $x = a$ and $x = b$ is obtained using the basic idea of integration. Let Δs_k be the element of arclength between $x = x_k$ and $x = x_{k+1}$. Then for a curve that is concave upwards as in Figure 8.49 we deduce that

$$\Delta x_k \sec \theta_k \leq \Delta s_k \leq \Delta x_k \sec \theta_{k+1}$$

where θ_k and θ_{k+1} are the angles of slope made by the tangents to the curve at P_k and P_{k+1}. Thus the length s of the curve between $x = a$ and $x = b$ satisfies the inequality

$$\sum_{k=0}^{n-1} \Delta x_k \sec \theta_k \leq s = \sum_{k=0}^{n-1} \Delta s_k \leq \sum_{k=0}^{n-1} \Delta x_k \sec \theta_{k+1}$$

Letting $n \to \infty$ and max $\Delta x_k \to 0$ yields the inequality

$$\int_a^b \sec \theta \, dx \leq s \leq \int_a^b \sec \theta \, dx$$

Figure 8.49
(a) Curve $y = f(x)$.
(b) Element of arclength.

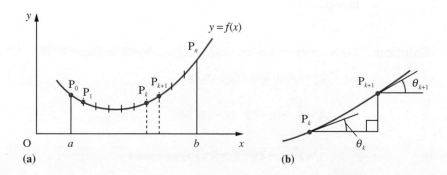

(a)   (b)

from which we deduce that

$$s = \int_a^b \sec\theta \, dx$$

A similar analysis for curves that are concave downwards yields the same result.

We can express $\sec\theta$ in terms of dy/dx by means of the identity

$$\sec^2\theta = 1 + \tan^2\theta$$

Here $\tan\theta = dy/dx$, so, using the convention that s increases with x, we obtain

$$\sec\theta = \sqrt{\left[1 + \left(\frac{dy}{dx}\right)^2\right]}$$

so that the length of the curve is

$$s = \int_a^b \sqrt{\left[1 + \left(\frac{dy}{dx}\right)^2\right]} dx \tag{8.42}$$

The surface area S generated by s when it is rotated through 2π radians about the x axis is calculated in a similar way. The element of arc Δs_k generates an element of surface area ΔS_k, where

$$\Delta S_k = 2\pi \bar{y}_k \Delta s_k$$

where $\bar{y}_k$ is the average value of y between $y_k = f(x_k)$ and $y_{k+1} = f(x_{k+1})$. Thus the total surface area is given by

$$S = \int_a^b 2\pi y \sqrt{\left[1 + \left(\frac{dy}{dx}\right)^2\right]} dx \tag{8.43}$$

Example 8.41

The area enclosed between the curve $y = \sqrt{(x-2)}$ and the ordinates $x = 2$ and $x = 5$ is rotated through 2π radians about the x axis. Calculate

(a) the rotating area and the coordinates of its centroid;

(b) the volume of the solid of revolution generated and the coordinates of its centre of gravity.

Solution

The rotating area is the shaded region shown in Figure 8.50.

(a) The rotating area is given by

$$A = \int_2^5 y \, dx = \int_2^5 (x-2)^{1/2} \, dx$$

$$= [\tfrac{2}{3}(x-2)^{3/2}]_2^5 = 2\sqrt{3} \text{ square units}$$

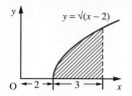

Figure 8.50

If we denote the coordinates of the centroid of the area by $(\bar{x}, \bar{y})$ then, from (8.37),

$$\bar{x} = \frac{1}{A}\int_2^5 xy\,dx = \frac{1}{A}\int_2^5 x(x-2)^{1/2}\,dx = \frac{1}{A}\int_2^5 [(x-2)^{3/2} + 2(x-2)^{1/2}]\,dx$$

$$= \frac{1}{A}\left[\frac{2}{5}(x-2)^{5/2} + \frac{4}{3}(x-2)^{3/2}\right]_2^5 = \frac{1}{A}\left[\frac{2}{5}(3)^{5/2} + \frac{4}{3}(3)^{3/2}\right] = \frac{1}{A}\frac{38}{5}\sqrt{3}$$

Inserting the value $A = 2\sqrt{3}$ obtained earlier gives $\bar{x} = \frac{19}{5}$. .

Likewise, from (8.37),

$$\bar{y} = \frac{1}{A}\int_2^5 \tfrac{1}{2}y^2\,dx = \frac{1}{A}\int_2^5 \tfrac{1}{2}(x-2)\,dx$$

$$= \frac{1}{A}[\tfrac{1}{4}(x-2)^2]_2^5 = \frac{9}{4A}$$

Inserting $A = 2\sqrt{3}$ then gives $\bar{y} = \frac{3}{8}\sqrt{3}$ so that the coordinates of the centroid are $(\frac{19}{5}, \frac{3}{8}\sqrt{3})$.

(b) From (8.33) the volume V of the solid of revolution formed is

$$V = \pi\int_2^5 y^2\,dx = \pi\int_2^5 (x-2)\,dx$$

$$= \pi[\tfrac{1}{2}x^2 - 2x]_2^5 = \tfrac{9}{2}\pi \text{ cubic units}$$

If we denote the coordinates of the centre of gravity of the solid of revolution by $(\bar{X}, \bar{Y})$ then, from (8.38) and (8.39),

$$\bar{Y} = 0$$

and

$$\bar{X} = \frac{\pi}{V}\int_2^5 xy^2\,dx = \frac{\pi}{V}\int_2^5 x(x-2)\,dx$$

$$= \frac{\pi}{V}[\tfrac{1}{3}x^3 - x^2]_2^5 = \frac{\pi}{V}[(\tfrac{125}{3} - 25) - (\tfrac{8}{3} - 4)]$$

$$= \frac{18\pi}{V}$$

Inserting the value $V = \frac{9}{2}\pi$ obtained earlier gives $\bar{X} = 4$ so that the coordinates of the centre of gravity are $(4, 0)$.

Example 8.42 An electric current i is given by the expression

$$i = I\sin\theta$$

where I is a constant. Find the root mean square value of the current over the interval $0 \leqslant \theta \leqslant 2\pi$.

Solution   Using (8.41) the r.m.s. value of the given current is given by

$$(\text{r.m.s.}\,i)^2 = \frac{1}{2\pi - 0}\int_0^{2\pi} I^2 \sin^2\theta\,d\theta$$

$$= \frac{I^2}{2\pi}\int_0^{2\pi} \tfrac{1}{2}(1 - \cos 2\theta)d\theta = \frac{I^2}{4\pi}[\theta - \tfrac{1}{2}\sin 2\theta]_0^{2\pi} = \frac{I^2}{4\pi}2\pi = \tfrac{1}{2}I^2$$

so that

$$\text{r.m.s. current} = \sqrt{(\tfrac{1}{2}I^2)} = I/\sqrt{2}$$

Example 8.43   A parabolic reflector is formed by rotating the part of the curve $y = \sqrt{x}$ between $x = 0$ and $x = 1$ about the x axis. What is the surface area of the reflector?

Solution   The parabolic reflector is shown in Figure 8.51. Since $y = x^{1/2}$,

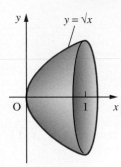

$$\frac{dy}{dx} = \frac{1}{2}x^{-1/2} = \frac{1}{2\sqrt{x}}$$

so that, using (8.43), the surface area S of the reflector is given by

$$S = 2\pi \int_0^1 y \sqrt{\left[1 + \left(\frac{dy}{dx}\right)^2\right]}\,dx = 2\pi \int_0^1 \sqrt{x}\sqrt{\left(1 + \frac{1}{4x}\right)}\,dx$$

$$= 2\pi \int_0^1 \sqrt{x}\,\frac{\sqrt{(4x + 1)}}{2\sqrt{x}}\,dx = \pi \int_0^1 \sqrt{(4x + 1)}\,dx$$

Figure 8.51
Parabolic reflector.

$$= \pi\,[\tfrac{1}{4}\tfrac{2}{3}(4x + 1)^{3/2}]_0^1 = \tfrac{1}{6}\pi\,(5^{3/2} - 1) \text{ square units}$$

Example 8.44   The curve described by the cable of the suspension bridge shown in Figure 8.52 is given by

$$y = \frac{hx^2}{l^2} - \frac{2h}{l}x + h$$

where x is the distance measured from one end of the bridge. What is the length of the cable?

Figure 8.52
Suspension bridge.

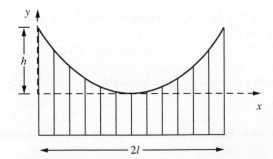

Solution   Here the equation of the curve is

$$y = h\left(\frac{x}{l} - 1\right)^2 \quad \text{so that} \quad \frac{dy}{dx} = \frac{2h}{l}\left(\frac{x}{l} - 1\right)$$

Using (8.42), the length s of the cable is

$$s = \int_0^{2l} \sqrt{\left[1 + \frac{4h^2}{l^2}\left(\frac{x}{l} - 1\right)^2\right]}\,dx$$

This integral can be simplified by putting

$$t = \frac{2h}{l}\left(\frac{x}{l} - 1\right)$$

Thus

$$s = \frac{l^2}{2h}\int_{-2h/l}^{2h/l} \sqrt{(1 + t^2)}\,dt = \frac{l^2}{h}\int_0^{2h/l} \sqrt{(1 + t^2)}\,dt \quad \text{(from symmetry)}$$

This can be further simplified by putting $t = \sinh u$, giving

$$s = \frac{l^2}{h}\int_0^{\sinh^{-1}(2h/l)} \cosh^2 u\,du = \frac{l^2}{2h}\int_0^{\sinh^{-1}(2h/l)} (\cosh 2u + 1)\,du$$

$$= \frac{l^2}{2h}\left[\tfrac{1}{2}\sinh 2u + u\right]_0^{\sinh^{-1}(2h/l)} = \frac{l^2}{2h}\left[\sinh u \cosh u + u\right]_0^{\sinh^{-1}(2h/l)}$$

$$= \frac{l^2}{2h}\left[\frac{2h}{l}\sqrt{\left(1 + \frac{4h^2}{l^2}\right)} + \sinh^{-1}\left(\frac{2h}{l}\right)\right]$$

That is,

$$s = \sqrt{(l^2 + 4h^2)} + \frac{l^2}{2h}\sinh^{-1}\left(\frac{2h}{l}\right)$$

Example 8.45   Find the equation of the curve described by a heavy cable hanging, without load, under gravity, from two equally high points.

Solution   Consider the cable illustrated in Figure 8.53. Let T be the tension acting at a point P that is a horizontal distance x from the axis of symmetry, as shown, and let the tangent to the curve at P make an angle θ to the horizontal. If s is the length of the curve between A and P, and T_0 is the tension at A, then resolving the forces acting on the length of cable between A and P horizontally and vertically gives

$$T_0 = T\cos\theta \quad \text{and} \quad s\rho g = T\sin\theta$$

where ρ is the line density of the cable and g is the acceleration due to gravity. Dividing these equations, we obtain

Figure 8.53
Heavy hanging cable.

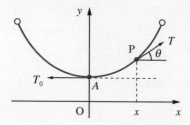

$$\tan \theta = \frac{s}{c}$$

where $c = T_0/\rho g$. In terms of x and y, this equation, using (8.42), implies that the co-ordinates of P satisfy

$$y'(x) = \frac{1}{c} \int_0^x \sqrt{[1 + (y'(t))^2]} \, dt$$

To solve this equation to obtain the equation of the curve, we first differentiate it with respect to x, giving

$$y''(x) = \frac{1}{c} \sqrt{[1 + (y'(x))^2]}$$

with $dy/dx = 0$ at $x = 0$. This may be rewritten as

$$\frac{d^2y/dx^2}{\sqrt{[1 + (dy/dx)^2]}} = \frac{1}{c}$$

and integrating with respect to x, using the substitution $dy/dx = \sinh u$, and remembering that $(d/dx)(dy/dx) = d^2y/dx^2$, gives

$$\sinh^{-1}\left(\frac{dy}{dx}\right) = \frac{x}{c} + A$$

Since $dy/dx = 0$ at $x = 0$, we deduce that $A = 0$ and

$$\frac{dy}{dx} = \sinh\frac{x}{c}$$

This is easy to integrate, giving

$$y = c\cosh\frac{x}{c} + B$$

The value of B is fixed by the value of y at $x = 0$. This may be chosen quite arbitrarily without changing the shape of the curve. Choosing $y(0) = c$ gives a neat answer (with $B = 0$):

$$y = c\cosh\frac{x}{c}$$

Note that this curve, called the **catenary**, is different from the shape of the cable of a suspension bridge, which is a parabola. The catenary has many applications, including the design of roofs and arches.

8.7.7   Exercises

96 Find the volume generated when the plane figure bounded by the curve $xy = x^3 + 3$, the x axis and the ordinates at $x = 1$ and $x = 2$ is rotated about the x axis through one complete revolution.

97 Express the length of the arc of the curve $y = \sin x$ from $x = 0$ to $x = \pi$ as an integral. Also find the volume of the solid generated by revolving the region bounded by the x axis and this arc about the x axis through 2π radians.

98 (a) Sketch the curve whose equation is

$$y = (x - 2)(x - 1)$$

Show that the volume generated when the finite area between the curve and the x axis is rotated through 2π radians about the x axis is $\pi/30$.

(b) Show that the curved surface generated by the revolution about the x axis of the portion of the curve $y^2 = 4ax$ included between the origin and the ordinate $x = 3a$ is $\frac{56}{3}\pi a^2$.

99 A curve is represented parametrically by

$$x(t) = 3t - t^3, \quad y(t) = 3t^2 \quad (0 \leq t \leq 1)$$

Find the volume and surface area of the solid of revolution generated when the curve is rotated about the x axis through 2π radians.

100 The electrical resistance R (in Ω) of a rheostat at a temperature θ (in °C) is given by $R = 38(1 + 0.004\theta)$.

Find the average resistance of the rheostat as the temperature varies uniformly from 10°C to 40°C.

101 The area enclosed between the x axis, the curve $y = x(2 - x)$ and the ordinates $x = 1$ and $x = 2$ is rotated through 2π radians about the x axis. Calculate

(a) the rotating area and the coordinates of its centroid;

(b) the volume of the solid of revolution formed and the coordinates of its centre of gravity.

102 Show that the area enclosed between the x axis, the curve $4y = x^2 - 2\ln x$ and the coordinates $x = 1$ and $x = 3$ is $\frac{1}{6}(19 - 9\ln 3)$.

103 The speed V of a rocket at a time t after launch is given by

$$V = at^2 + b$$

where a and b are constants. The average speed over the first second was $10\,\text{m s}^{-1}$, and that over the next second was $50\,\text{m s}^{-1}$. Determine the values of a and b. What was the average speed over the third second?

104 Find the centroid of the area bounded by $y^2 = 4x$ and $y = 2x$ and also the centroid of the volume obtained by revolving this area about the x axis.

8.8   Numerical evaluation of integrals

In many practical problems the functions that have to be integrated are often specified by a graph or by a table of values. Even when the function is given analytically, it often cannot be integrated to give an answer in terms of simple functions. Also, in many engineering and scientific problems it is often known in advance that the value of an integral is only required to a certain precision, and the use of an approximate method can avoid considerable unwanted labour. In all these cases we have to evaluate the integrals numerically. There are many ways of doing this, varying from the simplest square-counting for working out the area under a graph to sophisticated computer procedures. In this section we shall develop a simple numerical method known as the trapezium rule, which is the basis of many computer algorithms, and a hand computation method known as Simpson's rule.

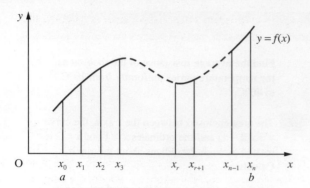

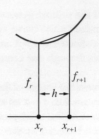

Figure 8.54 Slicing up an area into vertical strips of equal width.

Figure 8.55 Trapezium approximation to area of strip.

8.8.1 The trapezium rule

The simplest methods return to the initial ideas about integration introduced in Section 8.5.1. As indicated in Figure 8.54, they involve slicing up the area to be found into a number of strips of equal width, approximating the area of each strip in some way; the sum of these approximations then gives the final numerical result.

The points of subdivision of the domain of integration $[a, b]$ are labelled $x_0, x_1, \ldots, x_n$, where $x_0 = a$, $x_n = b$, $x_r = x_0 + rh$ $(r = 0, 1, 2, \ldots, n)$, and the width of each strip is $h = (b - a)/n$. The value of the integrand $f(x)$ at these points is, as usual, denoted by $f_r = f(x_r)$. A basic method for numerical integration approximates the area of each strip by the area of the trapezium formed when the upper end is replaced by the chord of the graph, as shown in Figure 8.55.

By the sum rule of integration

$$\int_a^b f(x)\mathrm{d}x = \int_{x_0}^{x_1} f(x)\mathrm{d}x + \int_{x_1}^{x_2} f(x)\mathrm{d}x + \ldots + \int_{x_{n-1}}^{x_n} f(x)\mathrm{d}x$$

$$= \sum_{r=0}^{n-1} \int_{x_r}^{x_{r+1}} f(x)\mathrm{d}x$$

From Figure 8.55 we can see that the approximate area of the rth strip is

$$\tfrac{1}{2}(f_r + f_{r+1})h$$

so that

$$\int_a^b f(x)\mathrm{d}x \approx \sum_{r=0}^{n-1} \tfrac{1}{2}(f_r + f_{r+1})h = \tfrac{1}{2}h \sum_{r=0}^{n-1} (f_r + f_{r+1})$$

$$= \tfrac{1}{2}h[(f_0 + f_1) + (f_1 + f_2) + \ldots + (f_{n-1} + f_n)]$$

That is,

$$\int_a^b f(x)\mathrm{d}x \approx h(\tfrac{1}{2}f_0 + f_1 + f_2 + \ldots + f_{n-1} + \tfrac{1}{2}f_n) \tag{8.44}$$

This approximation method is called the **trapezium rule**. As we shall see below, the best method for using it is given in formula (8.45).

Example 8.46   Evaluate the integral $\int_1^2 (1/x)\mathrm{d}x$ to 5dp, using the trapezium rule.

Solution   This integral is one of the standard integrals given in Figure 8.40, and so can be evaluated analytically. Its value is $\ln 2 = 0.693\,147$ to 6dp. This enables us, in this illustrative example, to check our methods. Usually, of course, the value of the integral is not known beforehand, and assessing the accuracy of the estimate obtained using the trapezium rule is an important aspect of the evaluation.

The first decision to be made in the numerical procedure is that of how many strips should be used; that is, what value n should have. A large number of strips may yield a good approximation to each strip, but will involve a lot of calculation, with the possibility of consequent rounding error accumulation. A small number of strips will obviously involve a large error in the approximation to the area of each strip. We shall investigate the situation.

First of all, we shall introduce the notation $T(h)$ to denote the approximation to the value of the integral given by the trapezium rule using strips of width h. Obviously, ignoring the possible effects of rounding errors, we expect

$$\lim_{h \to 0} T(h) = \int_1^2 \frac{1}{x}\mathrm{d}x$$

Taking $n = 1$ gives $h = (2 - 1)/n = 1$, $x_0 = 1$ and $x_1 = 2$. This gives the estimate

$$\int_1^2 \frac{1}{x}\mathrm{d}x = \tfrac{1}{2}(1)(f_0 + f_1) = T(1)$$

Here $f_0 = 1$ and $f_1 = 0.5$, so that $T(1) = 0.75$. This estimate for the value of the integral has an error of $0.75 - 0.693 = +0.057$.

Taking $n = 2$ gives $h = 0.5$, $x_0 = 1$, $x_2 = 2$ and $x_1 = 1.5$. Note that x_0 and x_2 are the two points used before, but now relabelled. This gives the estimate

$$T(0.5) = (0.5)[f_1 + \tfrac{1}{2}(f_0 + f_2)]$$

where $f_0 = 1$, $f_1 = 0.666\,667$ and $f_2 = 0.5$, so that $T(0.5) = 0.708\,333$. This estimate has an error of $+0.015$, so by doubling the number of strips, we have reduced the error by a factor of nearly four.

Taking $n = 4$ gives $h = 0.25$, $x_0 = 1$, $x_4 = 2$, $x_1 = 1.25$, $x_2 = 1.5$ and $x_3 = 1.75$. Note that three of these points were used in the previous calculation. This value of n gives the estimate

$$T(0.25) = (0.25)[f_1 + f_2 + f_3 + \tfrac{1}{2}(f_0 + f_4)]$$

where $f_0 = 1$, $f_1 = 0.8$, $f_2 = 0.666\,667$, $f_3 = 0.571\,429$ and $f_4 = 0.5$, so that $T(0.25) = 0.697\,024$. This estimate has an error of $+0.004$, so by doubling the number of strips, we have again reduced the error by a factor of four.

Continuing this process, with $n = 8$, we obtain the estimate $T(0.125) = 0.694\,122$, with an error of $+0.001$.

Figure 8.56
Points at which
integrand is evaluated.

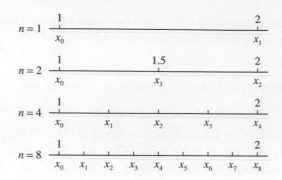

Based on these four calculations, we can estimate the values of n and h that will give an answer correct to 5dp; that is, with an absolute error less than 0.000005. If we continue the process of doubling the number of strips, reducing the error by a factor of four each time, we shall obtain an answer with the required accuracy when $n = 128$. With this large number of strips, we clearly need to organize the calculation to do it as economically as possible. Looking back at the previous calculations, we see that at each new value of n we almost double the number of points at which the integrand has to be evaluated, but as can be seen from Figure 8.56, at half of these points it has been evaluated in previous calculations.

Taking into account the effect of interval-halving on h, we can reduce the amount of calculation to evaluate $T(h)$ by making use of the result obtained for $T(2h)$:

$$T(h) = h[f_1 + f_2 + f_3 + \ldots + f_{n-1} + \tfrac{1}{2}(f_0 + f_n)], \quad h = \frac{b-a}{n}$$

$$T(2h) = (2h)[f_2 + f_4 + \ldots + f_{n-2} + \tfrac{1}{2}(f_0 + f_n)]$$

Here $f_0, f_2, f_4, \ldots, f_n$ were all calculated previously, in the evaluation of $T(2h)$. Rearranging, we have

$$T(h) = h(f_1 + f_3 + f_5 + \ldots + f_{n-1}) + \tfrac{1}{2}(2h)[f_2 + f_4 + \ldots + f_{n-2} + \tfrac{1}{2}(f_0 + f_n)]$$

Thus

$$T(h) = h(f_1 + f_3 + f_5 + \ldots + f_{n-1}) + \tfrac{1}{2}T(2h) \tag{8.45}$$

(remembering that if h is the strip width for n intervals then $2h$ is the strip width for $\tfrac{1}{2}n$ intervals).

This formula enables us to perform the calculations economically, but we can exploit it in a more subtle way.

We have seen that halving the strip width reduces the error by a factor of approximately four. This means that the error is proportional to h^2. In fact, this behaviour is typical of the application of the trapezium rule to the evaluation of many kinds of integrals and we can use it to obtain a more accurate estimate of the value of the integral. Since the error is proportional to h^2, we can write

$$T(h) - \int_1^2 \frac{1}{x}\,dx = Ah^2$$

where $h = 1/n$ and A is some number that, in general, will depend upon n but will remain bounded as n becomes large. A similar formula holds for $T(2h)$:

$$T(2h) - \int_1^2 \frac{1}{x} \, dx = 4A'h^2$$

where h has the same value as before and $A' \approx A$. These two formulae enable us to estimate the error in the approximation for the integral. Subtracting them gives

$$3Ah^2 \approx T(2h) - T(h)$$

so that the approximation $T(h)$ to the integral has an error estimate of $\frac{1}{3}[T(2h) - T(h)]$. Thus in the calculation above the estimated error for $T(0.125)$ is

$$\tfrac{1}{3}(0.697\,024 - 0.694\,122) = +0.000\,967$$

as we found before. This means that we can estimate the error in the usual situation of not knowing (unlike in this example) the true value of the integral. It also enables us to obtain a better approximation. Subtracting the estimated error from $T(h)$ gives the improved approximation (Richardson's extrapolation)

$$\int_1^2 \frac{1}{x} \, dx \approx T(h) - \tfrac{1}{3}[T(2h) - T(h)]$$

Alternatively we may write

$$\int_1^2 \frac{1}{x} \, dx \approx \tfrac{1}{3}[4T(h) - T(2h)]$$

Using the values for $T(0.25)$ and $T(0.125)$ obtained above, we have

$$\int_1^2 \frac{1}{x} \, dx \approx 0.694\,122 - 0.000\,967 = 0.693\,15$$

which is correct to 5dp. In general, of course, we could not know how good an approximation this extrapolated value is, and the usual practice is to continue interval-halving until two successive extrapolated values agree to the accuracy required. Not all integrals will converge as quickly as in this example. For example $\int_0^1 \sqrt{x} \, dx$ requires a large number of evaluations to achieve reasonable accuracy. The reason for the slow convergence of the approximation to $\int_0^1 \sqrt{x} \, dx$ compared with that of $\int_1^2 (1/x) \, dx$ is readily seen from Figure 8.57.

Figure 8.57
(a) Graph of $y = \sqrt{x}$.
(b) Graph of $y = 1/x$.

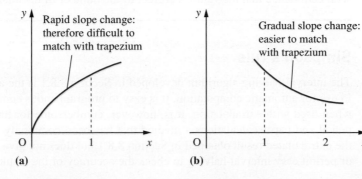

Example 8.47

Evaluate the integral $\int_0^1 \sqrt{(1 + x^2)}\,dx$ to 5dp, using the trapezium rule and extrapolation.

Solution

As before, we begin with just one strip, so that $h = 1$ and $T(1) = \frac{1}{2}(f_0 + f_1)$, where $f_0 = f(x_0) = f(0) = 1.000\,000$ and $f_1 = f(x_1) = f(1) = 1.414\,214$. Thus $T(1) = 1.207\,107$. Next we set $h = \frac{1}{2}$, and we calculate one new value of the integrand at $x = \frac{1}{2}$, giving a new $f_1 = \frac{1}{2}\sqrt{5} = 1.118\,034$ and

$$T(0.5) = hf_1 + \tfrac{1}{2}T(1)$$

$$= 0.5 \times 1.118\,034 + 0.603\,554$$

$$= 1.162\,570$$

An estimate for the error in $T(0.5)$ is

$$\tfrac{1}{3}[T(1) - T(0.5)] = 0.014\,846$$

and a better approximation for the value of the integral is given by

$$1.162\,570 - 0.014\,846 = 1.147\,724$$

Next we interval-halve again, giving $h = 0.25$, and calculate new values of the integrand (at $x = 0.25$ and $x = 0.75$):

$$f_1 = f(0.25) = 1.030\,776 \quad \text{and} \quad f_3 = f(0.75) = 1.25$$

Thus

$$T(0.25) = h(f_1 + f_3) + \tfrac{1}{2}T(0.5) = 1.151\,479$$

with an error estimate of $\frac{1}{3}[T(0.5) - T(0.25)] = 0.003\,697$ and an extrapolated value

$$1.151\,479 - 0.003\,697 = 1.147\,782$$

At this stage we can see that the value of the integral is 1.148 to 3dp. We continue interval-halving, giving: for $h = 0.125$, $T(0.125) = 1.148\,714$, with an error estimate of 0.000\,922 and an extrapolated value 1.147\,793; and for $h = 0.0625$, $T(0.0625) = 1.148\,714$, with an error estimate of 0.000\,230 and an extrapolated value 1.147\,793. Thus the extrapolated values agree to 6dp, so that we can write

$$\int_0^1 \sqrt{(1 + x^2)}\,dx = 1.147\,79$$

with confidence that the value is correct to the number of decimal places given.

8.8.2 Simpson's rule

The interval-halving algorithm developed in Section 8.8.1 is the appropriate algorithm to use for automatic computation. It is easy to program and is computationally efficient when used with extrapolation. It is, however, cumbersome for hand computation. For pencil and paper calculations a method that has been commonly used is equivalent to the extrapolated result obtained in Section 8.8.1 but does not give any estimate of error or permit easy interval-halving to check the accuracy of the result.

The trapezium rule approximation to $\int_a^b f(x)dx$ using one strip is

$$T_1 = \tfrac{1}{2}(b-a)[f(a)+f(b)]$$

and that using two strips is

$$T_2 = \frac{b-a}{4}[f(a) + 2f\left(\frac{a+b}{2}\right) + f(b)]$$

The extrapolation based on these two estimates is

$$S = [4T_2 - T_1]/3$$

$$= \frac{(b-a)}{6}[f(a) + 4f\left(\frac{a+b}{2}\right) + f(b)]$$

The formula provides the basic approximation for the area under the curve between $x = a$ and $x = b$. It can be shown to be the area under the parabola which passes through the three points $(a, f(a))$, $((a+b)/2, f((a+b)/2))$ and $(b, f(b))$.

Now consider the interval $[a, b]$ divided into n equal strips of width h where n is an even number. Then we may write

$$\int_a^b f(x)dx = \int_{x_0}^{x_2} f(x)\,dx + \int_{x_2}^{x_4} f(x)dx + \int_{x_4}^{x_6} f(x)dx + \ldots + \int_{x_{n-2}}^{x_n} f(x)dx$$

where $x_k = a + kh$.

Applying the basic formula to each of the integrals on the right-hand side yields the approximation

$$\int_a^b f(x)dx \approx \frac{h}{3}[f_0 + 4f_1 + f_2] + \frac{h}{3}[f_2 + 4f_3 + f_4] + \frac{h}{3}[f_4 + 4f_5 + f_6]$$

$$+ \ldots + \frac{h}{3}[f_{n-2} + 4f_{n-1} + f_n]$$

$$\int_a^b f(x)dx \approx \frac{h}{3}[f_0 + 4f_1 + 2f_2 + 4f_3 + 2f_4 + \ldots + 2f_{n-2} + 4f_{n-1} + f_n] \qquad (8.46)$$

or in words

> The integral is approximately one-third the step size times the sum of four times the odd ordinates plus twice the even ordinates plus first and last ordinates.

This is referred to as **Simpson's rule** and a pencil and paper calculation would be set out as shown in Example 8.48.

Example 8.48

Figure 8.58 shows a longitudinal section PQ of rough ground through which a straight horizontal road is to be cut. The width of the road is to be 10 m, and the sides of the cutting and embankment slope at 2 horizontal to 1 vertical. Estimate the net volume of earth removed in making the road.

Figure 8.58
Cross-section with
distances above or
below datum at 200 m
intervals (not to scale).

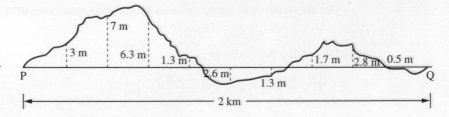

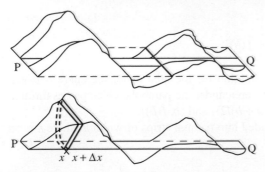

Figure 8.59 Volume of soil to be removed in road
construction.

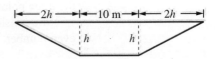

Figure 8.60 Cross-section of cutting
with sides sloping at 1 in 2.

Solution   In this case we are not dealing with a solid of revolution, and so cannot use (8.33) to
find the volume. Instead, we slice the volume up, estimate the volume of each slice and
then add all the individual volumes together as illustrated in Section 8.5.1. The volume
above the datum PQ is counted as positive and that below the datum as negative, so that
infill on site is accounted for automatically.

   Consider the 'slice' between the points at distances x and $x + \Delta x$ from P as shown
in Figure 8.59. The volume of this slice is $\bar{A}\Delta x$ where $\bar{A}$ is the average cross-sectional
area between x and $x + \Delta x$. The cross-sectional area A depends on the height h of the
soil above the datum line PQ. This relationship is given by

$$A = (2h + 10)h$$

as shown in Figure 8.60.

   The height h depends on the distance x along the road, so that we can construct a
table of values for A as a function of x, as shown in Figure 8.61.

Figure 8.61
Cross-sectional area
versus distance.

x	0	200	400	600	800	1000	1200	1400	1600	1800	2000
h	0.0	3.0	7.0	6.3	1.3	−2.6	−1.3	1.7	2.8	−0.5	0.0
A	0.0	48.0	168.0	142.4	16.4	−39.5	−16.4	22.8	43.7	−5.5	0.0

   The total volume V of soil removed from the site is the sum of the volumes of the
individual slices:

$$V = \sum A(\bar{x})\Delta x$$

where $A(\bar{x})$ is given by

$$A(\bar{x}) = \bar{A} \quad (x \le \bar{x} \le x + \Delta x)$$

Letting the number of slices tend to infinity while making their thicknesses all tend to zero gives V in the form of an integral:

$$V = \int_0^{2000} A(x)\,dx$$

This provides us with a mathematical model for the amount of soil to be removed: the next step is to evaluate the integral. In this example the integrand is known only from a table of values, so we have no alternative but to evaluate it numerically.

Using Simpson's rule with 10 strips of width 200 m, the calculation is shown in Figure 8.62 and we obtain the estimate $7.3 \times 10^4\,\text{m}^3$. If a better estimate is required, more data will have to be collected.

Figure 8.62
Simpson's rule
'paper and pencil'
calculation.

Odds	Evens	First and Last
48.0	168.0	0.0
142.4	16.4	0.0
−39.5	−16.4	0.0
22.8	43.7	423.4
−5.5	211.7 × 2	672.8
168.2 × 4		1096.2 × $\frac{200}{3}$
		73 080.0

8.8.3 Exercises

105 Use the trapezium rule to evaluate $\int_0^{0.8} e^{-x^2}\,dx$. Take the step size h equal to 0.8, 0.4, 0.2, 0.1 in turn and use extrapolation to improve the accuracy of your answer.

106 Use the trapezium rule, with interval-halving and extrapolation, to evaluate

$$\int_0^1 \log(\cosh x)\,dx \quad \text{to 4dp}$$

107 An ellipse has parametric equations $x = \cos t$, $y = \frac{1}{2}\sqrt{3}\sin t$. Show that the length of its circumference is given by

$$2\int_0^{\pi/2} \sqrt{(3 + \sin^2 t)}\,dt$$

This integral cannot be evaluated in terms of elementary functions. Use the trapezium rule with interval-halving to evaluate it to 6dp.

108 The capacity of a battery is measured by $\int i\,dt$, where i is the current. Estimate, using Simpson's rule, the capacity of a battery whose current was

measured over an 8 h period with the results shown below:

Time/h	0	1	2	3	4	5	6	7	8
Current/A	25.2	29.0	31.8	36.5	33.7	31.2	29.6	27.3	28.6

109 The speed $V(t)\,\text{m s}^{-1}$ of a vehicle at time t s is given by the table below. Use Simpson's rule to estimate the distance travelled over the eight seconds.

t	0	1	2	3	4	5	6	7	8
$V(t)$	0	0.63	2.52	5.41	9.02	13.11	16.72	18.75	20.15

110 Use Simpson's rule with $h = 0.1$ to estimate

$$\int_0^1 \sqrt{(1 + x^3)}\,dx$$

(Notice that by this method you have no way of knowing how accurate your estimate is.)

8.9   Engineering application: design of prismatic channels

The mean velocity V of flow in straight prismatic channels is proportional to $(A/p)^r$, where A is the cross-sectional area of the flow, p is the wetted perimeter and r is approximately a constant ($\frac{7}{12}$). Given the channel section for minimum flows (that is, A_0 and p_0), the objective is to design a channel such that V has the same value for all larger discharges.

Assume a symmetric channel cross-section as shown in Figure 8.63, where A_0 and p_0 are the minimum flow values of A and p. Let the shape of the channel be given by $x = f(y)$. (Note that in this application y, the height of the surface above the datum line, is the independent variable.) Then we want to find the function $f(y)$ such that the mean flow velocity is independent of y. This implies that

Figure 8.63
Channel cross-section.

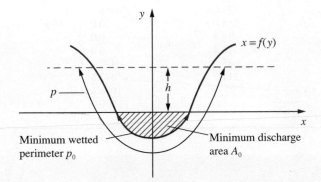

$$\frac{A}{p} = \frac{A_0}{p_0}$$

The area A is given by the integral of $f(y)$. Thus

$$A = A_0 + 2\int_0^h x\,\mathrm{d}y$$

where $x = f(y)$ and $h > 0$.

Using (8.42), the wetted perimeter p is given by

$$p = p_0 + 2\int_0^h \sqrt{\left[1 + \left(\frac{\mathrm{d}x}{\mathrm{d}y}\right)^2\right]}\,\mathrm{d}y \quad (h > 0)$$

Since $A/A_0 = p/p_0$, we deduce that

$$1 + \frac{2}{A_0}\int_0^h x\,\mathrm{d}y = 1 + \frac{2}{p_0}\int_0^h \sqrt{\left[1 + \left(\frac{\mathrm{d}x}{\mathrm{d}y}\right)^2\right]}\,\mathrm{d}y$$

Rearranging the integrals under a common integral sign gives

$$\int_0^h \left\{\frac{x}{A_0} - \frac{1}{p_0}\sqrt{\left[1 + \left(\frac{\mathrm{d}x}{\mathrm{d}y}\right)^2\right]}\right\}\,\mathrm{d}y = 0 \quad (h > 0)$$

Since this is true for all $h > 0$, it implies that the integrand must be identically zero. Thus $x = f(y)$ satisfies the differential equation

$$\frac{x}{A_0} = \frac{1}{p_0} \sqrt{\left[1 + \left(\frac{dx}{dy}\right)^2\right]}$$

which, assuming $\dfrac{dx}{dy} \geqslant 0$, implies

$$\frac{dx}{dy} = \sqrt{\left[\left(\frac{p_0 x}{A_0}\right)^2 - 1\right]} \tag{8.47}$$

Integrating with respect to y then gives

$$\int \frac{dx}{\sqrt{[(p_0 x/A_0)^2 - 1]}} = \int 1\, dy$$

Using the substitution $\cosh u = (p_0 x/A_0)$ on the left-hand side gives

$$\frac{A_0}{p_0} \cosh^{-1}\left(\frac{p_0 x}{A_0}\right) = y + c$$

If the channel has width $2b$ where $y = 0$, we can obtain the value of the constant of integration c as

$$c = \frac{A_0}{p_0} \cosh^{-1}\left(\frac{p_0 b}{A_0}\right)$$

and deduce the formula for a suitable channel shape as

$$y = \frac{A_0}{p_0}\left[\cosh^{-1}\left(\frac{p_0 x}{A_0}\right) - \cosh^{-1}\left(\frac{p_0 b}{A_0}\right)\right]$$

This solution, however, is not unique and we note that the differential equation (8.47) is also satisfied by

$$x = \frac{A_0}{p_0}$$

As an exercise, use this information to show that the general solution may take the form of either of the cross-sections shown in Figures 8.64(a) and (b). Notice that the line shape in Figure 8.64(b) does not have $\dfrac{dx}{dy} \geqslant 0$.

Figure 8.64

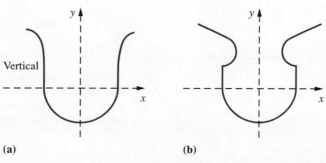

Vertical

(a)   (b)

8.10 Review exercises (1–29)

1 Differentiate the following expressions, giving your answers as simply as possible.

(a) e^{x^2+x}

(b) $\dfrac{x^3}{(3-x)^2}$

(c) $\sin(5x-1)$

(d) $(\tan x)^x$

(e) $\cos^{-1}\sqrt{(1-x^2)}$

(f) $\dfrac{1}{\sqrt{(x+1)}}$

(g) $\sin^{-1}\dfrac{1}{\sqrt{(1+x^2)}}$

(h) $\dfrac{1}{(x-1)(x+2)}$

(i) $\sin(3x+1)$

(j) $x^3 \ln x$

(k) $\dfrac{x^3}{(3-x^2)}$

(l) $\tan^{-1}(e^{-2x})$

(m) $\sqrt{(1+\cosh x)}$

(n) $(x^2+1)\sin 2x$

(o) $\dfrac{x-1}{(x+2)^2}$

(p) $e^{\sqrt{x}}$

(q) $\ln \tan x$

(r) $\dfrac{(2x-1)^{3/2}}{(x+1)^5}$

(s) $x \sin x$

(t) e^{x^2}

(u) 2^x

(v) $\dfrac{\cos x}{1+\sin x}$

(w) $\sin^{-1}\left(\dfrac{1}{x}\right)$

(x) $x^3 \cos 2x$

(y) $\sqrt{(x^3+x+3)}$

(z) $\dfrac{e^{-x}}{1+x}$

2 Evaluate

(a) $\displaystyle\int x^{1/2} \ln x \, dx$

(b) $\displaystyle\int \dfrac{(2x+3)\,dx}{x^2+2x+2}$

(c) $\displaystyle\int_0^3 \dfrac{1}{x}\lfloor x \rfloor dx$

(d) $\displaystyle\int_{1/2}^1 \dfrac{x \sin^{-1}x \, dx}{\sqrt{(1-x^2)}}$

(set $x = \sin t$)

(e) $\displaystyle\int \dfrac{x \, dx}{(x-1)(x-2)}$

(f) $\displaystyle\int \tan^4 x \, dx$

(g) $\displaystyle\int_0^1 \sqrt{(4-3x^2)}dx$

(h) $\displaystyle\int \dfrac{dx}{\sqrt{(4-9x^2)}}$

(i) $\displaystyle\int \dfrac{x^2 \, dx}{\sqrt{(x^3-1)}}$

(j) $\displaystyle\int \dfrac{(x^2+1)dx}{x+1}$

(k) $\displaystyle\int \dfrac{dx}{x^2+6x+13}$

(l) $\displaystyle\int \sqrt{x} \sin \sqrt{x} dx$

(m) $\displaystyle\int_0^2 \mathrm{FRACPT}(x)dx$

(n) $\displaystyle\int_0^1 \sinh^2 x \, dx$

(o) $\displaystyle\int (1-3x)^9 dx$

(p) $\displaystyle\int \sin 3x \sin 2x \, dx$

(q) $\displaystyle\int \ln 2x \, dx$

(r) $\displaystyle\int xe^{-x^2/2}\,dx$

(s) $\displaystyle\int \dfrac{dx}{\sqrt{(4x^2-9)}}$

(t) $\displaystyle\int_{-1}^4 \dfrac{(3x-1)dx}{\sqrt{(4+3x-x^2)}}$

(u) $\displaystyle\int_0^1 \dfrac{x \, dx}{(x+1)(x^2+1)}$

(v) $\displaystyle\int (4-3x)^4 dx$

(w) $\displaystyle\int \cos 2x \cos 3x \, dx$

(x) $\displaystyle\int \sin^{-1}x \, dx$

(y) $\displaystyle\int x^2 e^{-x} dx$

(z) $\displaystyle\int \dfrac{dx}{1+x+x^2}$

3 Find the equation of the tangent and normal at the point $(1, 4)$ to the curve whose equation is
$$y = 2x^4 - 3x^3 + 5x^2 + 3x - 3$$

4 Find the equation of the tangent to the curve $x^2 - 3xy + 2y^2 = 3$ at the point $(1, 2)$ and the equation of the normal to the curve $y = x^3 - x^2$ at the point $(1, 0)$. Find the distance of the point of intersection of these lines from the point $(-1, 2)$.

5 Using partial fractions, show that

(a) $\displaystyle\int_2^4 \dfrac{2x+3}{x(x-1)(x+2)}dx = \tfrac{3}{2}\ln 3 - \tfrac{4}{3}\ln 2$

(b) $\displaystyle\int_1^2 \dfrac{6x^2 dx}{(x+1)^2(2x-1)} = 3\ln 3 - \tfrac{8}{3}\ln 2 - \tfrac{1}{3}$

6 Working to 5dp, evaluate $\int_0^1(1+x^2)^{-1}dx$ using the trapezium rule with five ordinates. Evaluate the integral by direct integration and comment on the accuracy of the numerical method.

7 The parametric equations of a curve are

$$x = at^2, \quad y = 2at$$

If ρ is the radius of curvature and (h, k) is its centre of curvature, prove that

(a) $\dfrac{d^2y}{dx^2} = -\dfrac{1}{2at^3}$ (b) $\rho = 2a(1 + t^2)^{3/2}$

(c) $h = a(2 + 3t^2), \quad k = -2at^3$

8 (a) Using the substitution $u = x + 1$, evaluate

$$\int_3^8 x\sqrt{(x+1)}\,dx$$

(b) Using the substitution $u = \sqrt{x} + 6$, evaluate

$$\int_0^1 \dfrac{dx}{2(x + 6\sqrt{x})}$$

(c) The region R is bounded by the x axis, the line $x = \tfrac{9}{2}$ and the curve with parametric equations

$$x = a\cos t, \quad y = b\sin t \quad (0 \le t \le \tfrac{1}{3}\pi)$$

where a and b are positive constants. Let A, $\bar{x}$ and I_y denote respectively the area of R, the x coordinate of the centroid of R and the second moment of area of R about the y axis. Prove that

$$I_y = \tfrac{1}{4}a^2A + \tfrac{3}{8}a\bar{x}A$$

9 A curve has parametric equations

$$x = 2t + \sin 2t, \quad y = \cos 2t$$

Show that

$$\dfrac{dy}{dx} = -\tan t$$

Find d^2y/dx^2 and d^2x/dy^2 in terms of t, and demonstrate that

$$\dfrac{d^2y}{dx^2} \ne 1 \bigg/ \dfrac{d^2x}{dy^2}$$

10 Verify that the point $(-1, 1)$ lies on the curve

$$y(y - 3x) = y^3 - 3x^3$$

and find the values of dy/dx and d^2y/dx^2 there. What is the radius of curvature at that point?

11 Sketch the curve whose equation is

$$y^2 = x(x - 1)^2$$

and find the area enclosed by the loop.

12 Sketch the curve whose parametric representation is

$$x = a\sin^3 t, \quad y = b\cos^3 t \quad (0 \le t \le 2\pi)$$

Find the area enclosed.

13 Sketch the curve whose polar equation is

$$r = 1 + \cos\theta$$

Show that the tangent to the curve at the point $r = \tfrac{3}{2}$, $\theta = \tfrac{1}{3}\pi$ is parallel to the line $\theta = 0$. Find the total area enclosed by the curve.

14 A curve is specified in polar coordinates (r, θ) in the form $r = f(\theta)$. Show that the sectorial area bounded by the line $\theta = \alpha$, $\theta = \beta$ and the curve $r = f(\theta)$ $(\alpha \le \theta \le \beta)$ is given by

$$\dfrac{1}{2}\int_\alpha^\beta [f(\theta)]^2\,d\theta$$

Also show that the angle ϕ between the tangent to the curve at any point P and the polar line OP is given by

$$\cot\phi = \dfrac{1}{r}\dfrac{dr}{d\theta}$$

15 Find the length of the arc of the parabola $y = x^2$ that lies between $(-1, 1)$ and $(1, 1)$.

16 The parametric equations

$$x = t^2 - 1, \quad y = t^3 - t$$

describe a closed curve as t increases from -1 to 1. Sketch the curve and find the area enclosed.

17 (a) Find the area of the region bounded by the x axis and one arch of the cycloid

$$\begin{aligned} x &= a(\theta - \sin\theta), \\ y &= a(1 - \cos\theta) \; (0 \le \theta \le 2\pi) \end{aligned}$$

where a is a positive constant.

(b) Show that the radius of curvature of the cycloid defined in (a) at the point O is given by

$$\rho = 2\sqrt{2}a(1 - \cos\theta)^{1/2}$$

What is the maximum value of ρ?

(c) Discuss the nature of the radius of curvature when $\theta = 0$ or $\theta = 2\pi$.

(d) Determine the length of one arch of the cycloid.

18   Consider the integral

$$I_n = \int_0^{\pi/4} \tan^n x \, dx$$

where n is an integer. Using the trigonometric identity $1 + \tan^2 x = \sec^2 x$, show that

$$I_n + I_{n-2} = \int_0^{\pi/4} \tan^{n-2} x \sec^2 x \, dx$$

and hence obtain the recurrence relation

$$I_n = \frac{1}{n-1} - I_{n-2}$$

Use this to find

(a) $\displaystyle \int_0^{\pi/4} \tan^6 x \, dx$ (b) $\displaystyle \int_0^{\pi/4} \tan^7 x \, dx$

(Recurrence relations of this type are often called **reduction formulae**, since they provide a systematic way of reducing the value of the parameter n so that a difficult integral may be reduced to an easier one.)

19   Use integration by parts (writing the integrand as $\sin\theta \sin^{n-1}\theta$) to show that

$$I_n = \int_0^{\pi/2} \sin^n \theta \, d\theta$$

satisfies the reduction formula

$$nI_n = (n-1)I_{n-2}$$

Hence prove that

$$I_{2k+1} = \frac{2k}{2k+1} \frac{2k-2}{2k-1} \cdots \frac{2}{3}$$

and

$$I_{2k} = \frac{2k-1}{2k} \frac{2k-3}{2k-2} \cdots \frac{1}{2} \frac{\pi}{2}$$

These results are known as **Wallis' formulae**.
   Use them to show that

(a) $\displaystyle \int_0^{\pi/2} \sin^5 x \, dx = \frac{8}{15}$ (b) $\displaystyle \int_0^{\pi/2} \cos^6 x \, dx = \frac{5}{32}\pi$

20   Consider the integral

$$I_{m,n} = \int_0^{\pi/2} \cos^m x \sin^n x \, dx$$

Show that $I_{m,n}$ satisfies the reduction formula

$$I_{m,n} = \frac{n-1}{m+n} I_{m,n-2}$$

21   Reduction formulae of the type discussed in Questions 18–20 are iteration formulae – and, like other iteration formulae, when they are used, attention must be paid to their numerical properties. This is illustrated by considering the integral

$$I_n = \int_0^1 x^n e^{x-1} \, dx$$

Prove that

$$I_n = 1 - nI_{n-1} \quad (n > 0)$$

with $I_0 = 1 - e^{-1}$.
   Evaluate I_0 on your calculator and use the reduction formula to calculate I_n, $n = 1, 2, \ldots, 10$.
   Since

$$0 < x^{n+1} e^{x-1} < x^n e^{x-1} < x^n \quad (0 < x < 1)$$

we know that

$$0 < I_{n+1} < I_n < \frac{1}{n+1} \quad (n = 0, 1, 2, 3, \ldots)$$

Compare this with your results, and explain the discrepancy.
   Since the iteration diverges when used for n increasing (that is, on setting $n = 1, 2, 3, \ldots$ in turn), it will converge when used for n decreasing (say $n = 50, 49, 48, \ldots$). Since $0 < I_{19} < \frac{1}{20}$, try using the iteration with $n = 19, 18, 17, \ldots$ to obtain I_{10}. Continue the iteration backwards to find, eventually, I_0.

22   A solid of revolution is generated by rotating the area between the y axis, the line $y = 1$ and the parabola $y = x^2$ about the y axis. Find its volume and its surface area.

23   The numerical procedures developed in this chapter for evaluating integrals have all used strips of equal width. An alternative procedure is to specify the number of tabular points to be used but

not their position. It is possible to find tabular points within the domain of integration for the most accurate evaluation of the integral for the given number of points. Consider the two-point formula

$$\int_{-h}^{h} f(x)\,dx \approx h[af(\alpha h) + bf(\beta h)]$$

where a, b, α and β are constants to be found. Symmetry about $x = 0$ implies $\beta = -\alpha$. If the formula evaluates all quadratic functions exactly, prove that

$$2h = h(a + b)$$

$$0 = h(a\alpha h - b\alpha h)$$

$$\tfrac{2}{3}h^2 = h(a\alpha^2 h^2 + b\alpha^2 h^2)$$

Deduce that $a = b = 1$ and $\alpha = 1/\sqrt{3}$.

24 The symbols T_n and S_n are defined as the estimates of the integral

$$I = \int_0^1 (1 + 2x)^{-1}\,dx$$

using n intervals with the trapezium and Simpson's rules respectively. Calculate T_1, T_2, T_4, S_2 and S_4, working to 3dp only. Verify that your numerical results satisfy

$$S_{2n} = \tfrac{1}{3}(4T_{2n} - T_n)$$

for $n = 1$ and 2. Prove this result.

25 (a) A curve is represented parametrically by

$$x(t) = 3t - t^3, \quad y(t) = 3t^2 \quad (0 \leqslant t \leqslant 1)$$

Find the volume and the surface area of the solid of revolution generated when the curve is rotated about the x axis through 2π radians.

(b) Find the position of the centroid of the plane figure bounded by the curve $y = 5 \sin 2x$, $y = 0$ and $x = \tfrac{1}{6}\pi$.

26 When a homogeneous bar of constant cross-sectional area A (see Figure 8.65) is under uniformly distributed tensile stress, the elongation in the direction of the stress for a material obeying Hooke's law is given by

$$\text{stress} = E \times \text{strain}$$

where E is Young's modulus, the stress is the applied force per unit area and the strain is the

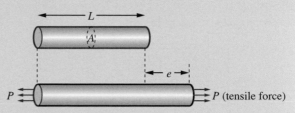

Figure 8.65

ratio of the elongation to the unstretched length of the bar. That is,

$$E\,\frac{e}{L} = \frac{P}{A}$$

Consider a bar of circular cross-section whose diameter varies along its length as shown in Figure 8.66, so that

$$A = A_0 + kx^2, \quad k = \frac{A_1 - A_0}{L^2}$$

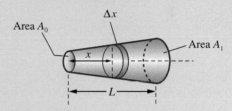

Figure 8.66

By considering the elongation of an element of thickness Δx of the bar, show that the total elongation of the bar under the tensile force P is

$$l = \int_0^L \frac{P\,dx}{E(A_0 + kx^2)}$$

Show that

$$l = \frac{4PL}{\pi d_0 d_2 E} \cos^{-1}\left(\frac{d_0}{d_1}\right)$$

where d_0 and d_1 are the end diameters of the bar, $d_0 < d_1$ and $d_2^2 = d_1^2 - d_0^2$.

27 Figure 8.67 shows an old cylindrical borehole that has been filled in part with silt and in part with water. Before the hole can be redrilled, the water has to be pumped to the surface. We wish to estimate the work required for this purpose.

(a) As a first approximation, assume that the silting has been uniform – as indicated in Figure 8.67 –

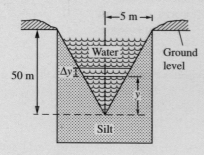

Figure 8.67

and that the water thus forms a right-circular cone of base radius 5 m and height 50 m. Hence, by considering the small element of water shown, show that an estimate of the work (in J) required to raise the water to ground level is

$$W_1 = 10^3 \times \pi g \int_0^{50} \left(\frac{y}{10}\right)^2 (50 - y)\mathrm{d}y$$

Evaluate W_1 in the form $k_1\pi g$, giving k_1 correct to 3sf.

(b) Surveying suggests that, while the water–silt boundary may still be regarded as having cylindrical symmetry about the axis of the original borehole, a more accurate profile can be obtained from the data below.

Depth below ground level (50 − y)/m	Radius of water/m
0	5
5	4.7
10	4.3
15	4.1
20	3.9
25	3.3
30	2.8
35	2.0
40	1.2
45	0.3
50	0

Use this data, with Simpson's rule, to obtain a second approximation W_2 to the work required. Give your answer in the form $k_2\pi g$, with k_2 given to 3sf.

28 Draw the graph of the function $f(x)$ defined by

$$f(x) = \int_0^x \{\lfloor x \rfloor - \tfrac{1}{2} - \lfloor x - \tfrac{1}{2} \rfloor\}\mathrm{d}x$$

for the interval $-5 \leqslant x \leqslant 5$.

29 An even function $f(x)$ of period 2π is given on the interval $[0, \pi]$ by the formula

$$y = x/\pi$$

(a) Using the even-ness property of the function, draw the graph of the function for $-\pi \leqslant x \leqslant \pi$.

(b) Using the periodicity property of the function, draw the graph of the function for $-4\pi \leqslant x \leqslant 4\pi$.

(c) Draw also the graph of the function $g(x) = \tfrac{1}{2} - \tfrac{1}{2}\cos x$, for $-4\pi \leqslant x \leqslant 4\pi$.

The function $h(x) = \tfrac{1}{2} + a\cos x$ is used as an approximation to $f(x)$ by choosing the value for the constant a which makes the total squared error, $[h(x) - f(x)]^2$, over $[0, \pi]$ a minimum, that is the value of a which minimizes

$$E(a) = \int_0^\pi [h(x) - f(x)]^2\,\mathrm{d}x$$

Show that

$$E(a) = \frac{\pi}{2}\left[a^2 + \frac{8a}{\pi^2} + \tfrac{1}{6}\right]$$

and that $E(a)$ is a minimum when $a = -4/\pi^2$. Draw a graph of the difference, $h(x) - f(x)$, between the approximation and the original function, for $0 \leqslant x \leqslant \pi$. What is its period?

9 Further Calculus

Chapter 9 Contents

9.1 Introduction

In Chapter 8 we discussed the fundamental ideas and concepts of integral and differential calculus and applied them to various practical problems. We also developed techniques for solving problems using calculus. In this chapter we shall use calculus to solve problems in optimization, extend the techniques developed in Chapter 8 to deal with a wide range of problems and develop the theory to enable us to understand the numerical techniques widely used in practical problem-solving.

9.2 Applications to optimization problems

In many industrial situations the role of management is to make decisions that will lead to the most effective use of the resources available. These decisions seldom affect the whole operation in one sweeping decision, but are usually a chain of small decisions: organizing stock control, designing a product, pricing it, servicing equipment and so on. Effective management seeks to optimize the constituent parts of the whole operation. A wide variety of mathematical techniques is used to solve such optimization problems, as discussed in a preliminary way in Section 8.2.7. Here, and later in Section 9.5.9, we consider methods based on the methods and concepts of calculus.

9.2.1 Optimal values

As we saw in Section 8.2.7 the basic idea is that the **optimal value** of a differentiable function $f(x)$ (that is, its **maximum** or **minimum value**) generally occurs where its derivative is zero; that is, where

$$f'(x) = 0$$

As can be seen from Figure 9.1, this is a necessary condition, since at a maximum or minimum value of the function its graph has a horizontal tangent. Figure 9.1 does, however, show that these extremal values are generally only local maximum or minimum values, corresponding to turning points on the graph, so some care must be exercised in using the horizontal tangent as a test for an optimal value. In seeking the extremal values of a function it is also necessary to check the end points (if any) of the domain of the function.

Figure 9.1
Maximum and
minimum values.

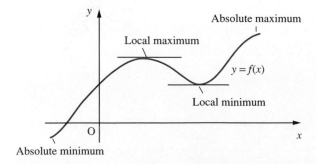

Figure 9.2
Graph with horizontal tangents.

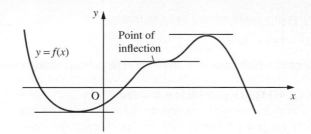

Figure 9.2 gives another illustration of why care must be exercised: at some **points of inflection** – that is, points where the graph crosses its own tangent – the tangent may be horizontal.

A third reason for caution is that a function may have an optimal value at a point where its derivative does not exist. A simple example of this is given by $f(x) = x^{2/3}$, whose graph is shown in Figure 9.3.

Having determined the **critical or stationary points** where $f'(x) = 0$, we need to be able to determine their character or nature; that is, whether they correspond to a local maximum, a local minimum or a point of inflection of the function $f(x)$. We can do this by examining values of $f'(x)$ close to and on either side of the critical point. From Figure 9.4 we see that

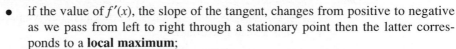

Figure 9.3
Graph of $f(x) = x^{2/3}$, with minimum at $x = 0$.

- if the value of $f'(x)$, the slope of the tangent, changes from positive to negative as we pass from left to right through a stationary point then the latter corresponds to a **local maximum**;
- if the value of $f'(x)$ changes from negative to positive as we pass from left to right through a stationary point then the latter corresponds to a **local minimum**;
- if $f'(x)$ does not change sign as we pass through a stationary point then the latter corresponds to a **point of inflection**.

Figure 9.4
Change in slope on passing through a turning point.

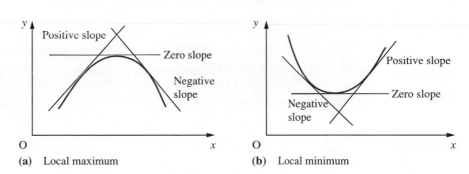

(a)   Local maximum   (b)   Local minimum

Example 9.1   Determine the stationary points of the function

$$f(x) = 4x^3 - 21x^2 + 18x + 6$$

and examine their nature.

Solution   The derivative is

$$f'(x) = 12x^2 - 42x + 18 = 6(2x - 1)(x - 3)$$

Stationary points occur when $f'(x) = 0$; that is,

$$6(2x - 1)(x - 3) = 0$$

the solutions of which are $x = \frac{1}{2}$ and $x = 3$. The corresponding values of the function are

$$f(\tfrac{1}{2}) = 4(\tfrac{1}{8}) - 21(\tfrac{1}{4}) + 18(\tfrac{1}{2}) + 6 = \tfrac{41}{4}$$

and

$$f(3) = 4(27) - 21(9) + 18(3) + 6 = -21$$

so that the stationary points of $f(x)$ are

$$(\tfrac{1}{2}, \tfrac{41}{4}) \quad \text{and} \quad (3, -21)$$

In order to investigate their nature, we use the procedure outlined above.

(a) Considering the point $(\tfrac{1}{2}, \tfrac{41}{4})$: if x is a little less than $\tfrac{1}{2}$ then $2x - 1 < 0$ and $x - 3 < 0$, so that

$$f'(x) = 6(2x - 1)(x - 3) = (\text{negative})(\text{negative}) = (\text{positive})$$

while if x is a little greater than $\tfrac{1}{2}$ then $2x - 1 > 0$ and $x - 3 < 0$, so that

$$f'(x) = (\text{positive})(\text{negative}) = (\text{negative})$$

Thus $f'(x)$ changes from (positive) to (negative) as we pass through the point so that $(\tfrac{1}{2}, \tfrac{41}{4})$ is a local maximum.

(b) Considering the point $(3, -21)$: if x is a little less than 3 then $2x - 1 > 0$ and $x - 3 < 0$, so that

$$f'(x) = (\text{positive})(\text{negative}) = (\text{negative})$$

while if x is a little greater than 3 then $2x - 1 > 0$ and $x - 3 > 0$, so that

$$f'(x) = (\text{positive})(\text{positive}) = (\text{positive})$$

Thus $f'(x)$ changes from (negative) to (positive) as we pass through the point so that $(3, -21)$ is a local minimum.

This information may now be used to sketch a graph of $f(x)$ as illustrated in Figure 9.5.

Figure 9.5
Graph of $f(x) = 4x^3 - 21x^2 + 18x + 6$.

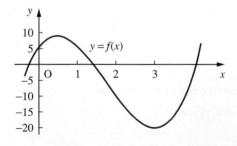

An alternative approach to determining the nature of a stationary point is to calculate the value of the second derivative $f''(x)$ at the point. Recall from Section 8.3.13 that $f''(x)$ determines the rate of change of $f'(x)$. Suppose that $f(x)$ has a stationary point at

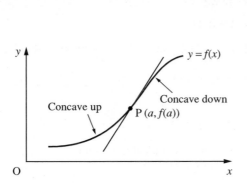

Figure 9.6 A point of inflection at $(a, f(a))$.

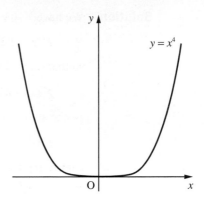

Figure 9.7 Graph of $f(x) = x^4$, illustrating the local minimum at $x = 0$.

$x = a$, so that $f'(a) = 0$. Then, provided $f''(a)$ is defined, either $f''(a) < 0$, $f''(a) = 0$ or $f''(a) > 0$.

If $f''(a) < 0$ then $f'(x)$ is decreasing at $x = a$; and since $f'(a) = 0$, it follows that $f'(x) > 0$ for values of x just less than a and $f'(x) < 0$ for values of x just greater than a. We therefore conclude that $x = a$ corresponds to a local maximum. Note that this concurs with our observation in Section 8.3.13 that the sign of $f''(x)$ determines the concavity of the graph of $f(x)$. Since the graph is concave down at a local maximum, $f''(a) \leqslant 0$. The equality case is discussed further in Section 9.5.9.

Similarly, we can argue that if $f''(a) > 0$ then the stationary point $x = a$ corresponds to a local minimum. Again this concurs with our observation that the graph is concave up at a local minimum.

Summarizing, we have

- the function $f(x)$ has a local maximum at $x = a$ provided $f'(a) = 0$ and $f''(a) < 0$;
- the function $f(x)$ has a local minimum at $x = a$ provided $f'(a) = 0$ and $f''(a) > 0$.

If $f''(a) = 0$, we cannot assume that $x = a$ corresponds to a point of inflection, and we must revert to considering the sign of $f'(x)$ on either side of the stationary point. As mentioned earlier, at a point of inflection the graph crosses its own tangent, or, in other words, the concavity of the graph changes. Since the concavity is determined by the sign of $f''(x)$, it follows that $f''(x) = 0$ at a point of inflection and that $f''(x)$ changes sign as we pass through the point. Note, as illustrated by the graph of Figure 9.6, that it is not necessary for $f'(x) = 0$ at a point of inflection. If, as illustrated in Figure 9.2, $f'(x) = 0$ at a point of inflection then it is a **stationary point of inflection**. It does not follow, however, that if $f'(a) = 0$ and $f''(a) = 0$ then $x = a$ is a point of inflection. An example of when this is not the case is $y = x^4$, which, as illustrated in Figure 9.7, has a local minimum at $x = 0$ even though both $\mathrm{d}y/\mathrm{d}x$ and $\mathrm{d}^2y/\mathrm{d}x^2$ are zero at $x = 0$. It is for this reason that we must take care and revert to considering the sign of $f'(x)$ on either side. We shall return to reconsider these conditions in Section 9.5.9 following consideration of Taylor series.

Example 9.2

Using the second derivative, confirm the nature of the stationary points of the function

$$f(x) = 4x^3 - 21x^2 + 18x + 6$$

determined in Example 9.1.

Solution We have

$$f'(x) = 12x^2 - 42x + 18$$

so that

$$f''(x) = 24x - 42$$

At the stationary point $(\frac{1}{2}, \frac{41}{4})$

$$f''(\tfrac{1}{2}) = 12 - 42 = -30 < 0$$

confirming that it corresponds to a local maximum.
   At the stationary point $(3, -21)$

$$f''(3) = 72 - 42 = 30 > 0$$

confirming that it corresponds to a local minimum.
   Note also that $f''(x) = 0$ at $x = \frac{7}{4}$ and that $f''(x) < 0$ for $x < \frac{7}{4}$ and $f''(x) > 0$ for $x > \frac{7}{4}$. Thus $(\frac{7}{4}, -\frac{43}{8})$ is a point of inflection (but not a stationary point of inflection), which is clearly identifiable in the graph of Figure 9.5.

In many applications we know for practical reasons that a particular problem has a minimum (or maximum) solution. If the equation $f'(x) = 0$ is satisfied by only one sensible value of x then that value must determine the unique minimum (or maximum) we are seeking. We will illustrate using three simple examples.

Example 9.3   A manufacturer has to supply N items per month at a uniform daily rate. Each time a production run is started it costs £c_1, the 'set-up' cost. In addition, each item costs £c_2 to manufacture. To avoid unnecessarily high production costs, the manufacturer decides to produce a large quantity q in one run and store it until the contract calls for delivery. The cost of storing each item is £c_3 per month. What is the optimal size of a production run?

Solution   As the contract calls for a monthly supply of N items, we need to look for a production run size that will minimize the total monthly cost to the manufacturer.
   The costs the manufacturer incurs are the production costs and the storage costs. The production cost for a production run of q items is

$$£(c_1 + c_2 q)$$

This production run will satisfy the contract for q/N months, so the monthly production cost will be

$$£\frac{c_1 + c_2 q}{q/N} = £\left(\frac{c_1}{q} + c_2\right)N$$

To this must be added the monthly storage cost, which will be £$\frac{1}{2}q c_3$, since the stock is depleted at a uniform rate and the average stock size is $\frac{1}{2}q$. Thus the total monthly cost £C is given by

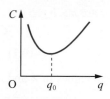

Figure 9.8
Monthly cost versus
run size.

$$C = \left(\frac{c_1}{q} + c_2\right)N + \tfrac{1}{2}qc_3$$

which has a graph similar to that shown in Figure 9.8.

To find the value q_0 of q that minimizes C, we differentiate the expression for C with respect to q and set the derivative equal to zero:

$$\frac{dC}{dq} = \frac{-c_1N}{q^2} + \tfrac{1}{2}c_3$$

and

$$\frac{dC}{dq} = 0 \quad \text{implies} \quad \frac{-c_1N}{q_0^2} + \tfrac{1}{2}c_3 = 0$$

and hence

$$q_0 = \sqrt{\left(\frac{2c_1N}{c_3}\right)}$$

This quantity is called the **economic lot size**.

Optimization plays an important role in design, and in Example 9.4 we illustrate this by applying it to the relatively easy problem of designing a milk carton.

Example 9.4   A milk retailer wishes to design a milk carton that has a square cross-section, as illustrated in Figure 9.9(a), and is to contain two pints of milk (2 pints ≡ 1.136 litres). The carton is to be made from a rectangular sheet of waxed cardboard, by folding into a square tube and sealing down the edge, and then folding and sealing the top and bottom. To make the resulting carton airtight and robust for handling, an overlap of at least 5 mm is needed. The procedure is illustrated in Figure 9.9(b). As the milk retailer will be using a large number of such cartons, there is a requirement to use the design that is least expensive to produce. In particular the retailer desires the design that minimizes the amount of waxed cardboard used.

Figure 9.9
The construction of a
milk carton.

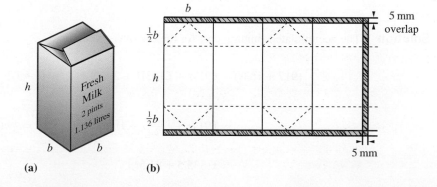

(a)   (b)

Solution If, as illustrated in Figure 9.9(a), the final dimensions of the container are $h \times b \times b$ (all in mm) then the area of waxed cardboard required is

$$A = (4b + 5)(h + b + 10) \tag{9.1}$$

Since the capacity of the carton is fixed at two pints (1.136 litres), the values of h and b must be such that

$$\text{volume} = hb^2 = 1\,136\,000\,\text{mm}^3 \tag{9.2}$$

Substituting (9.2) back into (9.1) gives

$$A = (4b + 5)\left(\frac{1\,136\,000}{b^2} + b + 10\right)$$

To find the value of b that minimizes A, we differentiate A with respect to b to obtain $A'(b)$ and then set $A'(b) = 0$. Differentiating gives

$$A'(b) = 8b + 45 - \frac{4\,544\,000}{b^2} - \frac{11\,360\,000}{b^3}$$

so the required value of b is given by the root of the equation

$$8b^4 + 45b^3 - 4\,544\,000b - 11\,360\,000 = 0$$

A straightforward tabulation of this polynomial, or use of a suitable software package, yields a root at $b = 81.8$. From (9.2) the corresponding value of h is $h = 169.8$. Thus the optimal design of the milk carton will have dimensions $81.8\,\text{mm} \times 81.8\,\text{mm} \times 169.8\,\text{mm}$.

Optimization problems also occur in programmes for replacing equipment and machinery in industry. We will illustrate this by a more commonplace decision: the best policy for replacing a car.

Example 9.5 For a particular model of car, bought in 1987 for £4750, the second-hand value after t years is given fairly accurately by the formula

$$\text{price} = £e^{8.41 - 0.189t}$$

The running costs of the car increase as the car gets older, so after t years the annual running cost is £$(917 + 163t)$. When should it be replaced?

Solution The accumulated running cost for the car over t years is

$$£\sum_{r=0}^{t-1}(917 + 163r) = £917t + £163[1 + 2 + 3 + \ldots + (t - 1)]$$

$$= £917t + £\tfrac{163}{2}(t - 1)t \quad \left(\text{using } \sum_{r=1}^{n} r = \tfrac{1}{2}n(n + 1)\right)$$

$$= £(835.5 + 81.5t)t$$

The total cost of the car clearly includes depreciation as well as running costs, so the average annual cost £C of the car is given by

$$C = \frac{4750 - e^{8.41-0.189t} + (835.5 + 81.5t)t}{t}$$

To find the optimal time for replacing the car, we find the value of t that minimizes C. Differentiating C with respect to t gives

$$C'(t) = -\frac{1}{t^2}(4750 - e^{8.41-0.189t}) + \frac{1}{t}(0.189e^{8.41-0.189t}) + 81.5$$

Setting $C'(t) = 0$ gives

$$e^{8.41-0.189t} = \frac{4750 - 81.5t^2}{1 + 0.189t}$$

Solving this numerically gives $t = 3.1$. Since car tax and insurance are usually paid on an annual basis, it is probably best to sell the car after 3 years!

9.2.2 Exercises

1 Find the stationary values of the following functions and determine their nature. In each case also find the point of inflection and sketch a graph of the function.

(a) $f(x) = 2x^3 - 5x^2 + 4x - 1$

(b) $f(x) = x^3 + 6x^2 - 15x + 51$

2 Find the stationary values of the following functions, distinguishing carefully between them. In each case sketch a graph of the function.

(a) $f(x) = \dfrac{3x}{(x-1)(x-4)}$

(b) $f(x) = 2e^{-x}(x-1)^3$

(c) $f(x) = x^2e^{-x}$

3 Consider the can shown in Figure 9.10, which has capacity 500 ml. The cost of manufacture is proportional to the amount of metal used, which in turn is proportional to the surface area of the can. Ignoring the overlaps necessary for the manufacture of the can, find the diameter and height of the can which minimizes its cost.

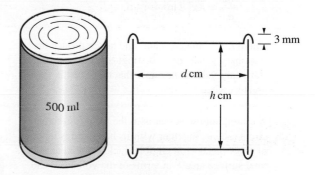

Figure 9.10 Can of Questions 3 and 4.

4 Consider again the can shown in Figure 9.10. Allowing for an overlap of 6 mm top and bottom surfaces to give a rim of 3 mm on the can, show that the area A mm² of metal used is given by

$$A(d) = \pi(d^2 + 3.6d + 1.44)/2 + 2000/d$$

where d cm is the diameter of the can.

Show that the value of d^* which minimizes the area of the can satisfies the equation

$$\pi d^2(d + 1.8) = 2000$$

Calculate d^* and the corresponding value of the height of the can.

5 A wire of length l metres is bent so as to form the boundary of a sector of a circle of radius r metres and angle θ radians. Show that

$$\theta = \frac{l - 2r}{r}$$

and prove that the area of the sector is greatest when the radius is $l/4$.

6 A cone of semi-vertical angle θ is inscribed in a sphere of radius a. Show that the volume of the cone is

$$V = \tfrac{8}{3}\pi a^3 \sin^2\theta \cos^4\theta$$

Hence prove that the cone of maximum volume that can be inscribed in a sphere of given radius is $\tfrac{8}{27}$th of the volume of the sphere.

7 In an underwater telephone cable the ratio of the radius of the core to the thickness of the protective sheath is denoted by x. The speed v at which a signal is transmitted is proportional to $x^2 \ln(1/x)$. Show that

$$\frac{dv}{dx} = Kx\left[2\ln\left(\frac{1}{x}\right) - 1\right]$$

where K is some constant, and hence deduce the stationary values of v. Distinguish between these stationary values and show that the speed is greatest when $x = 1/\sqrt{e}$.

8 A closed hollow vessel is in the form of a right-circular cone, together with its base, and is made of sheet metal of negligible thickness. Express the total surface area S in terms of the volume V and the semi-vertical angle θ of the cone. Show that for a given volume the total area of the surface is a minimum if $\sin\theta = \tfrac{1}{3}$. Find the value of S if $V = \tfrac{8}{3}\pi a^3$.

9 A numerical method which is more efficient than repeated subtabulation for obtaining the optimal solution is the following **bracketing method**. The initial tabulation locates an interval in which the solution occurs. The optimal solution is then estimated by optimizing a suitable quadratic approximation.

Consider again the milk carton problem, Example 9.4. Calculate $A(70)$, $A(80)$ and $A(90)$ and deduce that a minimum occurs in $[70, 90]$. Next find numbers p, q and r such that

$$C(b) = p(b - 80)^2 + q(b - 80) + r$$

satisfies $C(70) = A(70)$, $C(80) = A(80)$ and $C(90) = A(90)$. The minimum of C occurs at $80 - q/(2p)$. Show that this yields the estimate $b = 82.2$. Evaluate $A(82.2)$ and deduce that the solution lies in the interval $[80, 90]$. Next repeat the process using the values $A(80)$, $A(82.2)$ and $A(90)$ and show that the solution lies in the interval $[80, 83.1]$. Apply the method once more to obtain an improved estimate of the solution.

10 A pipeline is to be laid from a point A on one bank of a river of width 1 unit to a point B 2 units downstream on the opposite bank, as shown in Figure 9.11. Because it costs more to lay the pipe under water than on dry land, it is proposed to take it in a straight line across the river to a point C and then along the river bank to B. If it costs $\alpha\%$ more to lay a given length of pipe under the river than along the bank, write down a formula for the cost of the pipeline, specifying the domain of the function carefully. What recommendation would you make about the position of C when (a) $\alpha = 25$, (b) $\alpha = 10$?

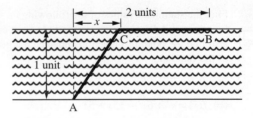

Figure 9.11

11 A manufacturer found that the sales figure for a certain item depended on the selling price. The market research department found that the maximum number of items that could be sold was 20 000 and that the number actually sold decreased by 100 for every 1p increase in price. The total cost of production of the items consisted of a set-up cost of £200 plus 50p per item manufactured. What price should be adopted to maximize profits, and how many items are produced?

12 Cross-current extraction methods are used in many chemical processes. Solute is extracted from a stream of solvent by repeated washings with water. The solvent stream is passed consecutively through a sequence of extractors, in each of which a cross-current of wash water, flowing at a determined rate, carries out some of the solute. The aim is to choose the individual wash flowrates in such a way as to extract as much solute as possible by the end, the total flow of wash water being fixed.

Consider the three-state extractor process shown in Figure 9.12, where c, x, y and z are the solute concentrations in the main stream, and αx, αy and αz are the solute concentrations in the effluent wash-water streams, with α a constant. The solute balance equations for the extractors are

$$Q(c - x) = u\alpha x$$

$$Q(x - y) = v\alpha y$$

$$Q(y - z) = w\alpha z$$

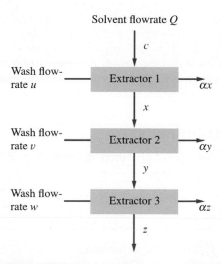

Solvent flowrate Q

Figure 9.12

The total wash-water flowrate is W, so that

$$u + v + w = W$$

We wish to find u, v and w such that the outflow concentration z is minimized.

This is an example of **dynamic programming**. The key to its solution is the **Principle of Optimality**, which states that an optimal programme has the property that, whatever the initial state and decisions, the remaining decisions must constitute an optimal policy with respect to the state resulting from the initial decision. This means we solve the problem first for a one-extractor process, then a two-extractor process, then a three-extractor process, and so on.

For a one-stage process x is minimized when $u = W$, giving $x^* = Qc/(Q + \alpha W)$.

For a two-stage process $y = Qx/(Q + \alpha v)$, where $x = \frac{1}{2}W$ with $v = \frac{1}{2}W$, giving $y^* = Q^2 c/(Q + \frac{1}{2}\alpha W)^2$.

For a three-stage process, $z = Q^2 x/[Q + \frac{1}{2}\alpha(W - u)]^2$, where $x = Qc/(Q + \alpha u)$. Show that z is minimized when $u = \frac{1}{3}W$, with $v = w = \frac{1}{3}W$, giving $z^* = Q^3 c/(Q + \frac{1}{3}\alpha W)^3$.

Generalize your answers to the case when n extractors are used.

13 The management of resources often requires a chain of decisions similar to that described in Question 12. Consider the harvesting policy for a large forest. The profit produced from the sale of felled timber is proportional to the square root of the volume sold, while the volume of standing timber increases in proportion to itself year on year. Use the technique outlined in Question 12 to produce a 10-year harvesting programme for a forest.

9.3 **Improper integrals**

When we considered the definite integral $\int_a^b f(x)\,dx$ in Chapter 8 and showed its equivalence with an area under a curve, it was assumed that the integrand $f(x)$ was continuous, or at least piecewise-continuous, over the closed domain of integration $[a, b]$. To illustrate a possible consequence of this not being the case, consider the apparent definite integral $\int_{-1}^{1}(1/x^2)\,dx$. If we proceed in a mechanistic way and follow the usual procedure, we should write

$$\int_{-1}^{1} \frac{1}{x^2}\,dx = \left[\frac{-1}{x}\right]_{-1}^{1} = -2$$

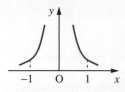

Figure 9.13
Graph of $f(x) = 1/x^2$.

However, if we plot the graph of $f(x) = 1/x^2$, as in Figure 9.13, it is clear that this is not correct, since it implies that the area under a curve that lies entirely above the x axis is negative. So where have we gone wrong? The answer lies in the fact that $f(x) = 1/x^2$ has an **infinite discontinuity** or **singularity** (that is, it is unbounded) at $x = 0$. As a consequence, the region under the curve over the domain of integration $[-1, 1]$ is unbounded, and our integration process was invalid.

In this section we consider the conditions under which the integral $\int_a^b f(x)\,dx$ exists when either

(a) the integrand $f(x)$ becomes unbounded (that is, $f(x)$ has an infinite discontinuity) at some point within the domain of integration, or

(b) the domain of integration is infinite (that is, either a or b or both are infinite).

Such integrals are called **improper integrals**, and are encountered in many contexts in engineering. For example, the period of a simple pendulum released from rest with angle α is given by

$$\int_0^\alpha \frac{4}{\sqrt{(\cos x - \cos \alpha)}}\,dx$$

The integrand is infinite at $x = \alpha$; yet we know that the answer is meaningful from elementary physics. Further examples are met when Laplace transforms are introduced in Chapter 11.

9.3.1 Integrand with an infinite discontinuity

Suppose that the lower limit $x = a$ is the only point of infinite discontinuity of $f(x)$ in $[a, b]$. Then we define

$$\int_a^b f(x)\,dx = \lim_{X \to a^+} \int_X^b f(x)\,dx \tag{9.3}$$

provided that the limit exists. Otherwise $\int_a^b f(x)\,dx$ has no meaning.

Similarly, if the upper limit $x = b$ is the only point of infinite discontinuity in $[a, b]$, we define

$$\int_a^b f(x)\,dx = \lim_{X \to b^-} \int_a^X f(x)\,dx \tag{9.4}$$

provided that the limit exists. Otherwise $\int_a^b f(x)\,dx$ has no meaning.

Example 9.6   Evaluate the following, if they are defined:

(a) $\displaystyle\int_0^1 x^{-2/3}\,dx$ (b) $\displaystyle\int_0^1 \frac{dx}{\sqrt{(1 - x^2)}}$ (c) $\displaystyle\int_0^1 \ln x\,dx$ (d) $\displaystyle\int_0^1 \frac{dx}{x^2}$

Solution   (a) Here the integral has an infinite discontinuity at the lower limit $x = 0$, and we consider

$$\lim_{X \to 0^+} \int_X^1 x^{-2/3} \, dx = \lim_{X \to 0^+} [3x^{1/3}]_X^1 = \lim_{X \to 0^+} (3 - 3X^{1/3}) = 3$$

Since the limit exists, it follows from (9.3) that

$$\int_0^1 x^{-2/3} \, dx = 3$$

(b) Here the discontinuity in the integrand occurs at the upper limit $x = 1$, and so we consider

$$\lim_{X \to 1^-} \int_0^X \frac{dx}{\sqrt{(1 - x^2)}} = \lim_{X \to 1^-} [\sin^{-1} x]_0^X = \lim_{X \to 1^-} (\sin^{-1} X) = \tfrac{1}{2}\pi$$

Since the limit exists, it follows from (9.4) that

$$\int_0^1 \frac{dx}{\sqrt{(1 - x^2)}} = \tfrac{1}{2}\pi$$

(c) Again the integrand has an infinite discontinuity at the lower limit $x = 0$, and so we consider

$$\lim_{X \to 0^+} \int_X^1 \ln x \, dx = \lim_{X \to 0^+} [x \ln x - x]_X^1 \quad \text{(integrating by parts)}$$

$$= \lim_{X \to 0^+} (X - X \ln X - 1)$$

$$= -1 \quad \text{(since } X \ln X \to 0 \text{ as } X \to 0^+, \text{ question 57, Section 7.8.3)}$$

Since the limit exists, it follows from (9.3) that

$$\int_0^1 \ln x \, dx = -1$$

(d) In this case the integrand has an infinite discontinuity at the lower limit $x = 0$, and we consider the limit

$$\lim_{X \to 0^+} \int_X^1 \frac{dx}{x^2} = \lim_{X \to 0^+} \left[\frac{-1}{x} \right]_X^1 = \lim_{X \to 0^+} \left(\frac{1}{X} - 1 \right)$$

This becomes infinite as $X \to 0$ and so the integral has no meaning.

If the integrand $f(x)$ has an infinite discontinuity at $x = c$, where $a < c < b$, then we define

$$\int_a^b f(x) \, dx = \lim_{X \to 0^+} \int_a^{c-X} f(x) \, dx + \lim_{X \to 0^+} \int_{c+X}^b f(x) \, dx \tag{9.5}$$

provided that both limits on the right-hand side exist. Otherwise $\int_a^b f(x) \, dx$ is not defined.

Example 9.7	Confirm that $\int_{-1}^{1}(1/x^2)\,dx$ is not defined.

Solution This is the apparent integral considered in the introductory discussion, where we saw that following the usual integration techniques in a mechanistic sense led to a ridiculous answer. In this case the integrand has an infinite discontinuity at $x = 0$, so, following (9.5), we consider the two limits

$$\lim_{X \to 0^+} \int_{-1}^{-X} \frac{dx}{x^2} \quad \text{and} \quad \lim_{X \to 0^+} \int_{X}^{1} \frac{dx}{x^2}$$

From the solution to Example 9.6(d) it is clear that both these tend to infinity, so that neither limit exists and the integral $\int_{-1}^{1}(1/x^2)\,dx$ is not defined.

9.3.2 Infinite integrals

The second case, where the domain of integration is infinite, is dealt with in a similar manner. We define

$$\int_{a}^{\infty} f(x)\,dx = \lim_{X \to \infty} \int_{a}^{X} f(x)\,dx \tag{9.6}$$

if that limit exists. Otherwise $\int_{a}^{\infty} f(x)\,dx$ has no meaning.

Example 9.8	Evaluate the following:

(a) $\displaystyle\int_{1}^{\infty} x^{-3/2}\,dx$ (b) $\displaystyle\int_{0}^{\infty} \frac{dx}{1 + x^2}$ (c) $\displaystyle\int_{0}^{\infty} e^{-x} \sin x\,dx$

Solution (a) $\displaystyle\int_{1}^{\infty} x^{-3/2}\,dx = \lim_{X \to \infty} \int_{1}^{X} x^{-3/2}\,dx = \lim_{X \to \infty} [-2x^{-1/2}]_{1}^{X} = \lim_{X \to \infty} (2 - 2X^{-1/2}) = 2$

(b) $\displaystyle\int_{0}^{\infty} \frac{dx}{1 + x^2} = \lim_{X \to \infty} \int_{0}^{X} \frac{dx}{1 + x^2} = \lim_{X \to \infty} [\tan^{-1} x]_{0}^{X} = \lim_{X \to \infty} (\tan^{-1} X) = \tfrac{1}{2}\pi$

(c) $\displaystyle\int_{0}^{\infty} e^{-x} \sin x\,dx = \lim_{X \to \infty} \int_{0}^{X} e^{-x} \sin x\,dx$

$$= \lim_{X \to \infty} [-\tfrac{1}{2}e^{-x}(\cos x + \sin x)]_{0}^{X} \quad \text{(integration by parts)}$$

$$= \lim_{X \to \infty} [\tfrac{1}{2} - \tfrac{1}{2}e^{-X}(\cos X + \sin X)]$$

$$= \tfrac{1}{2}$$

The indefinite integral is obtained using integration by parts, as in Example 8.36(c). It can be verified by direct differentiation.

14 Evaluate the following improper integrals.

(a) $\displaystyle\int_0^1 (-x\ln x)\,dx$

(b) $\displaystyle\int_0^\infty xe^{-x^3}\,dx$

(c) $\displaystyle\int_0^\infty x^2 e^{-2x}\,dx$

(d) $\displaystyle\int_{-\infty}^\infty e^{3x}\exp(-e^x)\,dx$

(e) $\displaystyle\int_0^1 x^2(1-x^3)^{-1/2}\,dx$

(f) $\displaystyle\int_0^1 (x-1)/\!\sqrt{x}\,dx$

(g) $\displaystyle\int_0^{\frac{\pi}{2}} \frac{\sin x}{\sqrt{\cos x}}\,dx$

(h) $\displaystyle\int_0^{\frac{\pi}{2}} \cos x \sin^{-1/3}x\,dx$

(i) $\displaystyle\int_0^\infty \frac{x}{1+x^4}\,dx$

9.4 Some theorems with applications to numerical methods

There are a number of theorems involving integration and differentiation that are useful in understanding why certain numerical methods are better than others and in devising new methods. They are also useful in the more mundane tasks of assessing the effect of data error when evaluating functions and probing the accuracy of analytical approximations to functions. We shall now briefly consider such theorems and indicate their potential uses. Deriving the results is not easy and the reader may prefer to omit the proofs. The results, however, have many practical implications and should be studied carefully.

9.4.1 Rolle's theorem and the first mean value theorems

The simplest result is the following

Theorem 9.1 **Rolle's theorem**

If the function $f(x)$ is continuous on the domain $[a, b]$ and differentiable on (a, b) with $f(a) = f(b)$ then there is at least one point $x = c$ in (a, b) such that $f'(c) = 0$.

end of theorem

The validity of this theorem can be easily illustrated geometrically as shown in Figure 9.14, since what the theorem tells us is that it is possible to find at least one point on the curve $y = f(x)$ between the values $x = a$ and $x = b$ where the tangent is parallel to the x axis; that is, there must exist at least one maximum or minimum between $x = a$ and $x = b$.

In Chapter 7, Section 7.9.1, we discussed the properties of continuous functions. All continuous functions are integrable, and this fact enables us to calculate the mean value of a continuous function over a given domain, say $[a, b]$. The mean value is given by

Figure 9.14
Four examples of
Rolle's theorem.

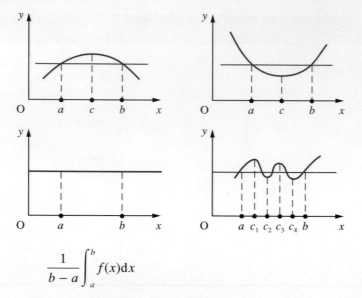

$$\frac{1}{b-a}\int_a^b f(x)\mathrm{d}x$$

Clearly the mean value of $f(x)$ lies between its maximum and minimum values on the
domain $[a, b]$ and, from the intermediate value theorem (Property (c), Section 7.9.1),
we deduce that there is a point $x = c$ in the interval $[a, b]$ such that

$$f(c) = \text{mean value of } f(x) = \frac{1}{b-a}\int_a^b f(x)\mathrm{d}x$$

This result is referred to as the first mean value theorem of integral calculus and may
be stated as follows.

Theorem 9.2

The first mean value theorem of integral calculus

If the function $f(x)$ is continuous over the domain $[a, b]$ then there exists at least one
point $x = c$, with $a < c < b$, such that

$$f(c) = \frac{1}{b-a}\int_a^b f(x)\mathrm{d}x$$

end of theorem

This theorem is illustrated geometrically in Figure 9.15(a).

Figure 9.15
The first mean value
theorems: (a) $f(c_i)$ is
the mean value of
$f(x)$ ($a \le x \le b$);
(b) the chord PQ is
parallel to the tangents
at $x = c_i$.

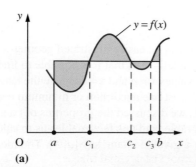

(a)

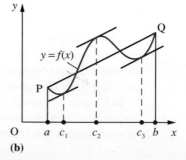

(b)

If $f(x)$ is a differentiable function then

$$\int_a^b f'(x)\mathrm{d}x = f(b) - f(a)$$

Applying Theorem 9.2 to $f'(x)$ gives

$$\int_a^b f'(x)\mathrm{d}x = (b-a)f'(c), \quad \text{with } a < c < b$$

and hence, by equating the two values of $\int_a^b f'(x)\mathrm{d}x$,

$$\frac{f(b) - f(a)}{b - a} = f'(c)$$

This result is referred to as the first mean value theorem of differential calculus, and may be stated as follows.

Theorem 9.3

First mean value theorem of differential calculus

If the function $f(x)$ is continuous on the domain $[a, b]$ and differentiable on (a, b) then there exists at least one point $x = c$, with $a < c < b$, such that

$$\frac{f(b) - f(a)}{b - a} = f'(c)$$

end of theorem

It is this theorem that is normally referred to as the first mean value theorem. Geometrically, it implies that at some point on the interval $[a, b]$ the slope of the tangent to the graph of $f(x)$ is parallel to the chord between the end points $x = a$ and $x = b$ of the graph, as shown in Figure 9.15(b).

An immediate application of Theorem 9.3 is in the estimation of the effect of rounding errors in the independent variable x on the calculated value of the dependent variable $y = f(x)$. If ε_x is the error bound for x then the error bound for y is ε_y, where

$$\varepsilon_y = \max_{x - \varepsilon_x < x^* < x + \varepsilon_x} |f(x^*) - f(x)|$$

Applying the first mean value theorem with $a = x$, $b = x^*$ gives

$$|f(x^*) - f(x)| = |x^* - x| \, |f'(c)|$$

with c lying between x and x^*. Since $f'(c) \approx f'(x)$, we have

$$\varepsilon_y \approx \max_{x - \varepsilon_x < x^* < x + \varepsilon_x} |f'(x)(x^* - x)| = |f'(x)|\varepsilon_x \tag{9.7}$$

We illustrate this by Example 9.9.

Example 9.9

Show that

$$\Delta(\sin x) \approx \cos x \, \Delta x$$

and hence estimate an error bound for $\sin a$, where $a = 1.935$ (3dp). Compare the error interval obtained with $[\sin 1.9355, \sin 1.9345]$. Express $\sin a$ as a correctly rounded number with the maximum number of decimal places.

Solution   The difference $\Delta(\sin x)$ is given by

$$\Delta(\sin x) = \sin(x + \Delta x) - \sin x$$

Since $(d/dx) \sin x = \cos x$, application of Theorem 9.3 gives

$$\frac{\sin(x + \Delta x) - \sin x}{(x + \Delta x) - x} = \cos X, \quad \text{with } x < X < x + \Delta x$$

which reduces to

$$\Delta(\sin x) = \cos X \, \Delta x, \quad \text{with } x < X < x + \Delta x$$

If Δx is small then $x \approx X$ and $\cos X \approx \cos x$, so that

$$\Delta(\sin x) \approx \cos x \, \Delta x$$

as required.

Setting $x = a$ gives $\Delta(\sin a) \approx \cos a \, \Delta a$, and hence, using (9.7), an error bound estimate for $\sin a$ is

$$\varepsilon_{\sin a} = |\cos a| \, \varepsilon_a$$

In this example $a = 1.935$ and $\varepsilon_a = 0.0005$, so that

$$\varepsilon_{\sin a} = |\cos 1.935| (0.0005) = |-0.3562| (0.0005) = 0.000\,18$$

Thus

$$\sin a = \sin 1.935 \pm 0.000\,18 = 0.934\,41 \pm 0.000\,18$$

which spans the interval $[0.934\,23, 0.934\,59]$.

Now $\sin 1.9355 = 0.934\,23$ and $\sin 1.9345 = 0.934\,59$, so that in this example the estimate of the error interval and the error interval are the same to 5dp.

Thus

$$\sin a = 0.9344 \pm 0.0002$$

or

$$\sin a = 0.93$$

9.4.2 Convergence of iterative schemes

In Chapter 7, Section 7.9.3, the solution of equations by iteration was discussed. We now consider the convergence of such iterative schemes. As before, suppose that an iteration for the root $x = \alpha$ of the equation $f(x) = 0$ is given by

$$x_{n+1} = g(x_n) \quad (n = 0, 1, 2, \ldots)$$

where $\alpha = g(\alpha)$. The usual practice is to stop the iteration when the difference $|x_{n+1} - x_n|$ between two successive iterates is sufficiently small; that is, when it is less than half-a-unit of the least significant figure required in the answer.

There are two separate issues here: one concerns the convergence of the iteration formula to the root, and the other concerns the 'stopping' mechanism. In practical computation, the rule of stopping an iteration is important because it vitally affects the accuracy of the estimate of the root of the equation.

Convergence process

To examine the convergence of the iteration to the root α, we estimate $|x_{n+1} - \alpha|$ as $n \to \infty$. Now

$$x_{n+1} = g(x_n) \quad \text{and} \quad \alpha = g(\alpha)$$

so that

$$x_{n+1} - \alpha = g(x_n) - g(\alpha)$$

Using the mean value theorem 9.3, this may be written as

$$x_{n+1} - \alpha = (x_n - \alpha)g'(X_n)$$

where X_n lies in the interval (x_n, α), assuming $x_n < \alpha$. Writing $\varepsilon_n = x_n - \alpha$, we obtain

$$|\varepsilon_{n+1}| \leqslant r|\varepsilon_n|$$

where $r = |g'(x)|_{\max}$ in the neighbourhood of $x = \alpha$. By comparison with the geometric sequence, we deduce that $\varepsilon_n \to 0$, as $n \to \infty$, if $0 < r < 1$ and that, provided we start near $x = \alpha$, the iteration converges if $|g'(x)| < 1$ near $x = \alpha$. Note that the more horizontal the graph of $g(x)$ near the root, the smaller r is and hence the more rapid is the convergence. We will discuss this further in Section 9.5.7.

Stopping process

The 'stopping' rule can be investigated similarly. The rule says that the iteration is stopped when $|x_{n+1} - x_n| < \varepsilon$, where ε is the maximum acceptable error. We therefore seek a relationship between $|x_{n+1} - \alpha|$ and $|x_{n+1} - x_n|$.

Now we can rewrite $x_{n+1} - \alpha$ as

$$x_{n+1} - \alpha = (x_{n+1} - x_{n+2}) + (x_{n+2} - x_{n+3}) + (x_{n+3} - x_{n+4}) + \ldots + (x_{n+k} - \alpha)$$

Since $x_n \to \alpha$ as $n \to \infty$, it follows that $x_{n+k} \to \alpha$ as $k \to \infty$, since all the previous terms on the right-hand side tend to zero. We may therefore write

$$x_{n+1} - \alpha = \sum_{k=1}^{\infty} (x_{n+k} - x_{n+k+1}) \tag{9.8}$$

Using the first mean value theorem 9.3, we have

$$x_{n+1} - x_{n+2} = g(x_n) - g(x_{n+1})$$

$$= (x_n - x_{n+1})g'(X_n)$$

where X_n lies in the interval (x_n, x_{n+1}), assuming $x_n < x_{n+1}$. By repeated application of this result, we have

$$x_{n+2} - x_{n+3} = (x_{n+1} - x_{n+2})g'(X_{n+1}) = (x_n - x_{n+1})g'(X_n)g'(X_{n+1})$$

$$\vdots$$

leading to

$$x_{n+k} - x_{n+k+1} = (x_n - x_{n+1})g'(X_n)g'(X_{n+1}) \ldots g'(X_{n+k-1}) \tag{9.9}$$

If, as before, $|g'(x)| < r < 1$ in the neighbourhood of $x = \alpha$ then we obtain from (9.8)

$$|x_{n+1} - \alpha| \leq \sum_{k=1}^{\infty} |x_{n+k} - x_{n+k+1}|$$

$$\leq \sum_{k=1}^{\infty} |x_n - x_{n+1}| r^k \qquad \text{(using (9.9) with } |g'(x)| < r\text{)}$$

$$= |x_n - x_{n+1}| \sum_{k=1}^{\infty} r^k$$

$$= \frac{r}{1-r} |x_n - x_{n+1}|$$

using the expression for the sum of a geometric progression given in (7.12). Hence

$$|x_{n+1} - \alpha| < \frac{r\varepsilon}{1-r}$$

Thus $|x_{n+1} - \alpha| < \varepsilon$ provided that $r < \frac{1}{2}$, and the 'stopping' rule is valid provided that $|g'(x)| < \frac{1}{2}$ near the root $x = \alpha$. In many practical problems it is necessary to estimate r by

$$|x_{n+1} - x_n|/|x_n - x_{n-1}| = |g(x_n) - g(x_{n-1})|/|x_n - x_{n-1}|$$

Clearly, the smaller the value of r, the more rapid is the convergence. Note, however, that this discussion has ignored the effects of rounding errors on the computation so that the result above has been shown only for exact arithmetic.

Example 9.10   Show that the iteration

$$\theta_{n+1} = \tan^{-1}(\tanh \theta_n), \quad \text{with } \theta_0 = \tfrac{5}{4}\pi \approx 3.9$$

considered in Section 7.9.3 is convergent to the root near $\theta = 3.9$ of the equation $\tan \theta = \tanh \theta$ (see Figure 9.16).

Solution   Here the iteration function has formula

$$g(\theta) = \tan^{-1}(\tanh \theta)$$

Figure 9.16
Roots of the equation
$\tan \theta = \tanh \theta$.

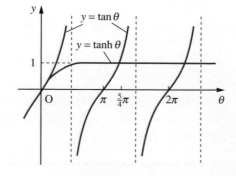

with derivative

$$g'(\theta) = \frac{1}{1 + \tanh^2\theta}\,\text{sech}^2\theta$$

$$= \frac{1}{\cosh^2\theta + \sinh^2\theta} = \frac{1}{\cosh 2\theta}$$

Near $\theta = 3.9$, $\cosh 2\theta \approx 1220$, so $|g'(\theta)|$ is small (in fact $r < 0.004$) and the method converges.

Example 9.11

A spherical wooden ball floats in water as illustrated in Figure 9.17. Its diameter is 10 cm and its density is $0.8\,\text{g cm}^{-3}$. Find the depth h cm to which it sinks.

Figure 9.17
Floating ball of
Example 9.11.

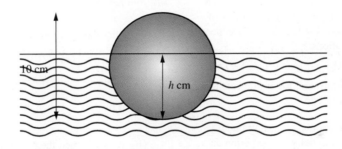

Solution

Archimedes shouted '$\varepsilon\upsilon\rho\eta\kappa\alpha$!' when he realized that the weight of a floating body must balance the weight of water it displaces. In this case we have the weight of the ball is

$$\tfrac{4}{3}\pi(5)^3 \times 0.8\,\text{g}$$

The volume of a zone depth h of a sphere of radius r is

$$\tfrac{1}{3}\pi h^2(3r - h)$$

so the weight of water displaced is

$$\tfrac{1}{3}\pi h^2(15 - h)\,\text{g}$$

Hence by Archimedes' principle we have

$$\tfrac{4}{3}\pi \times 125 \times \tfrac{4}{5} = \tfrac{1}{3}\pi h^2(15 - h)$$

that is

$$400 = h^2(15 - h)$$

Graphing $y = (x - 15)x^2 + 400$ shows that there is a root near $x = 7$. To find the root more accurately we can construct an iteration. For example

$$h_{n+1} = [(h_n^3 + 400)/15]^{1/2}$$

Starting with $h_0 = 7.00$, we obtain the iterates given in the table below.

n	0	1	2	3	4	5	6	7	8	9	10
h_n	7.00	7.04	7.06	7.08	7.10	7.11	7.11	7.12	7.12	7.12	7.12

With this set of iterates we would be tempted to conclude that the root is 7.11 or 7.12. In fact the correct answer is 7.13. This example shows the importance of the size of the derivative of the iteration function. In this case it is 0.7 near the root and there is danger of premature termination of the process. Clearly it is not of vital importance here but the example illustrates the danger of using an iteration without due care.

9.4.3 Exercises

15 By means of sketches of the graphs $y = 1/x$ and $y = \tan x$, show that the equation $x \tan x = 1$ has a root between $x = 0$ and $x = \frac{1}{2}\pi$ and an infinity of roots near $x = k\pi$, where $k = 1, 2, 3, \ldots$. Deduce which of the two iterations

(a) $x_{n+1} = \cot x_n$ (b) $x_{n+1} = \tan^{-1}(1/x_n) + k\pi$

is convergent to the roots, and use it to locate the smallest positive root to 6dp.

16 If $\alpha = f(\alpha)$ but the iteration $x_{n+1} = f(x_n)$ fails to converge to the root α, under what condition on $f(x)$ will the iteration $x_{n+1} = f^{-1}(x_n)$ converge?

17 Show the cubic equation $x^3 - 2x - 1 = 0$ has a root near $x = 2$. Prove that the iteration

$$x_{n+1} = \tfrac{1}{2}(x_n^3 - 1)$$

fails to converge to that root. Devise a simple iteration formula for the root of the equation, and use it to find the root to 6dp.

18 The equation $f(x) = 0$ has a root at $x = \alpha$. Show that rewriting the equation as $x = x + \lambda f(x)$, where λ is a constant, yields a convergent iteration for α if $\lambda = -1/f'(x_0)$ and x_0 is sufficiently close to α.

Use this method to devise an iteration for the root near $x = 2$ of the equation $x^3 - 2x - 1 = 0$.

19 Consider the iteration defined by

$$x_{n+1} = \tfrac{1}{3}(x_n^3 + 2)$$

Show that

(a) if $0 < x_0 < 1$ then the iteration tends to a limit as $n \to \infty$;

(b) if $x_0 > 1$ then the iteration is divergent. Explain this behaviour.

20 Consider the iteration

$$x_{n+1} = \frac{2 + 30x_n - x_n^2}{30}, \quad x_0 = 1.5$$

Working to 2dp, obtain the first three iterates. Then continue to obtain the following six iterates. From the numerical evidence what do you estimate as the limit of the sequence?

Assuming that the sequence has a limit near 1.5, obtain its value algebraically and then explain the phenomena observed above.

9.5 Taylor's theorem and related results

A question that frequently arises in both engineering and mathematical problem-solving is the behaviour of a solution when one (or more) of the parameters in the problem statement is changed. This occurs in sensitivity analysis when we examine solutions for their dependence on errors in the original data. It is also relevant to analysing the equilibrium of structures. One of the mathematical tools for such analyses is Taylor's theorem. In this section we shall develop the theorem and then use it to solve problems in design and numerical methods.

9.5.1   Taylor polynomials and Taylor's theorem

In Section 2.9.1 we discussed the use of interpolating functions to approximate functions specified by a table of values. The simplest case was linear interpolation. With this, we require a different formula between successive tabular points. Another approach to the problem of function approximation is to construct a polynomial that, together with its derivatives, takes the same values as those of the function and its derivatives at a particular point in the domain. That is, we seek a polynomial $p(x)$ such that

$$p(a) = f(a), \quad p'(a) = f'(a), \quad p''(a) = f''(a), \dots$$

The idea is illustrated by Example 9.12.

Example 9.12   Find a polynomial approximation to the function $f(x)$ such that

$$f(0) = 3, \quad f'(0) = 4, \quad f''(0) = -10 \quad \text{and} \quad f'''(0) = 12$$

Solution   In this example we have information about the value of the function and its first three derivatives at $x = 0$. This means that we can form an approximating polynomial of degree 3

$$p(x) = a + bx + cx^2 + dx^3$$

and determine the values of a, b, c and d from the information given.

Setting $p(0) = f(0)$ gives $a = 3$.

Differentiating gives

$$p'(x) = b + 2cx + 3dx^2$$

and on setting $p'(0) = f'(0) = 4$, we have $b = 4$.

Differentiating again gives

$$p''(x) = 2c + 6dx$$

and on setting $p''(0) = f''(0) = -10$, we have $c = -5$.

Differentiating again gives

$$p'''(x) = 6d$$

and on setting $p'''(0) = f'''(0) = 12$, we have $d = 2$.

Thus the approximating polynomial is

$$p(x) = 3 + 4x - 5x^2 + 2x^3$$

The technique used in Example 9.12 can be applied at points other than $x = 0$, as shown in Example 9.13.

Example 9.13   Find a polynomial approximation to $f(x)$ such that

$$f(1) = 4, \quad f'(1) = 0, \quad f''(1) = 2 \quad \text{and} \quad f'''(1) = 12$$

Solution   Because the information concerns the value of the function and its derivatives at the point $x = 1$, we look for a polynomial in powers of $x - 1$. So in this case we are seeking an approximation in the form

$$p(x) = a + b(x - 1) + c(x - 1)^2 + d(x - 1)^3$$

Setting $x = 1$ in $p(x)$ and its derivatives gives, in turn,

$$p(1) = a = 4 \qquad p'(1) = b = 0$$

$$p''(1) = 2c = 2 \qquad p'''(1) = 6d = 12$$

Thus the required approximation is

$$p(x) = 4 + 0(x - 1) + 1(x - 1)^2 + 2(x - 1)^3$$
$$= 4 + (x - 1)^2 + 2(x - 1)^3$$

Such polynomial approximations to functions are called **Taylor polynomials**. In general, we can write the nth-degree Taylor polynomial approximation to the function $f(x)$, given the value of the function and its derivatives at $x = a$, in the form

$$f(x) \approx p_n(x)$$

where

$$p_n(x) = f(a) + \frac{x - a}{1!}f'(a) + \frac{(x - a)^2}{2!}f''(a) + \frac{(x - a)^3}{3!}f'''(a) + \dots$$

$$+ \frac{(x - a)^n}{n!}f^{(n)}(a) \qquad\qquad \textbf{(9.10)}$$

Clearly, $p_n(a) = f(a)$, and also the first n derivatives of $p_n(x)$ match the first n derivatives of $f(x)$ at $x = a$.

The approximation of $f(x)$ given in (9.10) can be made exact by writing

$$f(x) = p_n(x) + R_n(x) \qquad\qquad \textbf{(9.11)}$$

where $R_n(x)$ is the **remainder**. The remainder term can be expressed in many different forms, with the simplest, known as **Lagrange's form**, being

$$R_n(x) = \frac{(x - a)^{n+1}}{(n + 1)!}f^{(n+1)}(a + \theta h)$$

where $h = x - a$ and $0 < \theta < 1$.

The result (9.11) constitutes **Taylor's theorem**, which may be stated as follows.

Theorem 9.4   **Taylor's theorem**

If $f(x)$, $f'(x)$, $\dots$, $f^{(n)}(x)$ exist and are continuous on the closed domain $[a, x]$ and $f^{(n+1)}(x)$ exists on the open domain (a, x) then there exists a number θ, with $0 < \theta < 1$, such that

$$f(x) = f(a) + \frac{x-a}{1!}f'(a) + \frac{(x-a)^2}{2!}f''(a) + \ldots$$

$$+ \frac{(x-a)^n}{n!}f^{(n)}(a) + \frac{(x-a)^{n+1}}{(n+1)!}f^{(n+1)}(a+\theta h) \tag{9.12}$$

where $h = x - a$.

Taylor's theorem is in fact a natural extension of the first mean value theorem (Theorem 9.3), and it is sometimes referred to as the ***n*th mean value theorem**. It may be proved by repeated use of Rolle's theorem (Theorem 9.1), but, since the proof does not add to our understanding of how to apply the result to the solution of engineering problems, it is not developed here.

9.5.2 Taylor and Maclaurin series

An alternative form of the Taylor polynomial (9.12) is obtained when we replace x in the expansion by $a + x$. Then we obtain a polynomial in x, rather than $x - a$, namely

$$f(x + a) = f(a) + \frac{x}{1!}f'(a) + \frac{x^2}{2!}f''(a) + \frac{x^3}{3!}f'''(a) + \ldots$$

$$+ \frac{x^n}{n!}f^{(n)}(a) + R_n(x) \tag{9.13}$$

where

$$R_n(x) = \frac{x^{n+1}}{(n+1)!}f^{(n+1)}(a+\theta x), \quad \text{with } 0 < \theta < 1$$

Equation (9.13) is called the **Taylor polynomial expansion of** $f(x)$ **about** $x = a$.

The remainder $R_n(x)$ represents the error involved in approximating $f(x)$ by the polynomial

$$f(a) + \frac{x}{1!}f'(a) + \frac{x^2}{2!}f''(a) + \ldots + \frac{x^n}{n!}f^{(n)}(a)$$

If $R_n(x) \to 0$ as $n \to \infty$ then we may represent $f(x)$ by the power series

$$f(x + a) = f(a) + \frac{x}{1!}f'(a) + \frac{x^2}{2!}f''(a) + \ldots = \sum_{n=0}^{\infty} \frac{x^n}{n!}f^{(n)}(a) \tag{9.14}$$

The power series (9.14) is called the **Taylor series expansion of** $f(x)$ **about** $x = a$. We saw in Section 7.3.2 that a power series may have a restricted domain of convergence. Similarly, $R_n(x)$ may tend to zero as $n \to \infty$ only for a restricted interval of values of x or not at all. In that case the power series given by (9.14) will only represent the function $f(x)$ in that interval of convergence.

Setting $a = 0$ in (8.16) leads to the special case

$$f(x) = f(0) + \frac{x}{1!}f'(0) + \frac{x^2}{2!}f''(0) + \ldots = \sum_{n=0}^{\infty} \frac{x^n}{n!} f^{(n)}(0) \qquad (9.15)$$

which is known as the **Maclaurin series expansion of** $f(x)$.

Example 9.14   Find the Maclaurin series expansion of $e^x \sin x$.

Solution   Since $f(x) = e^x \sin x$,

$$f'(x) = e^x(\sin x + \cos x)$$

This may be rewritten (see Section 2.6.4) as

$$f'(x) = \sqrt{2}e^x \sin(x + \tfrac{1}{4}\pi)$$

so the process of differentiation is equivalent to multiplying by $\sqrt{2}$ and adding $\tfrac{1}{4}\pi$ to the argument of the sine function. Thus we can write the second derivative directly as

$$f''(x) = (\sqrt{2})^2 e^x \sin(x + 2 \times \tfrac{1}{4}\pi) = 2e^x \cos x$$

and so on for higher derivatives, giving in general

$$f^{(k)}(x) = (\sqrt{2})^k e^x \sin(x + \tfrac{1}{4}k\pi)$$

Putting $x = 0$ gives $f(0) = 0, f^{(1)}(0) = 1, f^{(2)}(0) = 2, f^{(3)}(0) = 2, f^{(4)}(0) = 0, f^{(5)}(0) = -4$, $f^{(6)}(0) = -8, \ldots$, which, on substituting into (9.15), gives

$$e^x \sin x = 0 + x(1) + \frac{1}{2!}x^2(2) + \frac{1}{3!}x^3(2) + \frac{1}{4!}x^4(0) + \frac{1}{5!}x^5(-4) + \ldots$$

$$= x + x^2 + \tfrac{1}{3}x^3 - \tfrac{1}{30}x^5 + \ldots$$

It remains to show that $R_n(x) \to 0$ as $n \to \infty$. Since

$$R_n(x) = \frac{x^{n+1}}{(n+1)!} f^{(n+1)}(\theta x), \quad \text{with } 0 < \theta < 1$$

we have in this particular example

$$R_n(x) = \frac{x^{n+1}}{(n+1)!} (\sqrt{2})^{n+1} e^{\theta x} \sin[\theta x + \tfrac{1}{4}(n+1)\pi]$$

$$= \frac{(x\sqrt{2})^{n+1}}{(n+1)!} e^{\theta x} \sin[\theta x + \tfrac{1}{4}(n+1)\pi], \quad \text{with } 0 < \theta < 1$$

Now $(x\sqrt{2})^{n+1}/(n+1)! \to 0$ as $n \to \infty$, and $|\sin[\theta x + \tfrac{1}{4}(n+1)\pi]| \leqslant 1$, and so

$$R_n(x) \to 0 \quad \text{as } n \to \infty \quad \text{for all } x$$

Thus the Maclaurin expansion of $e^x \sin x$ is

$$e^x \sin x = x + x^2 + \tfrac{1}{3}x^3 - \tfrac{1}{30}x^5 + \ldots$$

(a) $(1 + x)^r = 1 + rx + \dfrac{r(r-1)x^2}{2!} + \dfrac{r(r-1)(r-2)x^3}{3!} + \ldots + \dfrac{r(r-1)\ldots(r-n+1)}{n!}x^n + \ldots$ $(-1 < x < 1, r \in \mathbb{R})$

(b) $e^x = 1 + \dfrac{x}{1!} + \dfrac{x^2}{2!} + \dfrac{x^3}{3!} + \ldots + \dfrac{x^n}{n!} + \ldots$ (all x)

(c) $\sin x = x - \dfrac{x^3}{3!} + \dfrac{x^5}{5!} - \ldots + \dfrac{(-1)^n x^{2n+1}}{(2n+1)!} + \ldots$ (all x)

(d) $\cos x = 1 - \dfrac{x^2}{2!} + \dfrac{x^4}{4!} - \ldots + \dfrac{(-1)^n x^{2n}}{(2n)!} + \ldots$ (all x)

(e) $\ln(1+x) = x - \dfrac{x^2}{2} + \dfrac{x^3}{3} - \dfrac{x^4}{4} + \ldots + \dfrac{(-1)^n x^{n+1}}{n+1} + \ldots$ $(-1 < x \leqslant 1)$

(f) $\tan x = x + \dfrac{x^3}{3} + \dfrac{2x^5}{15} + \dfrac{17x^7}{315} + \ldots$ $(-\tfrac{1}{2}\pi < x < \tfrac{1}{2}\pi)$

(g) $\sinh x = x + \dfrac{x^3}{3!} + \dfrac{x^5}{5!} + \ldots + \dfrac{x^{2n+1}}{(2n+1)!} + \ldots$ (all x)

(h) $\cosh x = 1 + \dfrac{x^2}{2!} + \dfrac{x^4}{4!} + \ldots + \dfrac{x^{2n}}{(2n)!} + \ldots$ (all x)

Figure 9.18 Some standard Maclaurin series expansions.

In practice it is rarely the case that we obtain the Maclaurin series expansion of a function by direct calculation of the derivatives as in Example 9.14. More commonly, we obtain such series by the manipulation of known standard Maclaurin series as we did in Section 7.7.2. Most of the standard series were given in Figure 7.13. For convenience, we reproduce some of them in Figure 9.18.

Example 9.15 Using the Maclaurin series expansions of e^x and $\sin x$, confirm the Maclaurin series expansion of $e^x \sin x$ obtained in Example 9.14.

Solution From entries (b) and (c) of Figure 9.18

$$e^x = 1 + \frac{x}{1!} + \frac{x^2}{2!} + \frac{x^3}{3!} + \ldots \quad \text{(all } x)$$

$$\sin x = x - \frac{x^3}{3!} + \frac{x^5}{5!} - \ldots \quad \text{(all } x)$$

As indicated in Section 7.7.2, we can multiply two power series within their common domain of convergence, giving in this case

$$e^x \sin x = \left(1 + \frac{x}{1!} + \frac{x^2}{2!} + \frac{x^3}{3!} + \frac{x^4}{4!} + \ldots\right)\left(x - \frac{x^3}{3!} + \frac{x^5}{5!} - \ldots\right)$$

$$= x + x^2 + x^3(\tfrac{1}{2} - \tfrac{1}{6}) + x^4(\tfrac{1}{6} - \tfrac{1}{6}) + x^5(\tfrac{1}{120} + \tfrac{1}{24} - \tfrac{1}{12}) + \ldots$$

$$= x + x^2 + \tfrac{1}{3}x^3 - \tfrac{1}{30}x^5 + \ldots \quad \text{(all } x)$$

which is the series obtained in Example 9.14.

Example 9.16   Obtain the binomial expansion of $(1 - x^2)^{-1/2}$ and deduce a power series expansion for $\sin^{-1}x$.

Solution   From entry (a) of Figure 9.17.

$$(1 + x)^r = 1 + rx + \frac{r(r - 1)x^2}{2!} + \frac{r(r - 1)(r - 2)x^3}{3!} + \ldots \quad (|x| < 1)$$

To obtain the expansion of $(1 - x^2)^{-1/2}$, we need to set $r = -\frac{1}{2}$ and replace x by $-x^2$. We shall do this in two steps. First setting $r = -\frac{1}{2}$ gives

$$(1 + x)^{-1/2} = 1 + \frac{-\frac{1}{2}}{1}x + \frac{(-\frac{1}{2})(-\frac{3}{2})}{1 \cdot 2}x^2 + \frac{(-\frac{1}{2})(-\frac{3}{2})(-\frac{5}{2})}{1 \cdot 2 \cdot 3}x^3 + \frac{(-\frac{1}{2})(-\frac{3}{2})(-\frac{5}{2})(-\frac{7}{2})}{1 \cdot 2 \cdot 3 \cdot 4}x^4 + \ldots$$

$$= 1 - \tfrac{1}{2}x + \frac{1 \cdot 3}{2 \cdot 4}x^2 - \frac{1 \cdot 3 \cdot 5}{2 \cdot 4 \cdot 6}x^3 + \frac{1 \cdot 3 \cdot 5 \cdot 7}{2 \cdot 4 \cdot 6 \cdot 8}x^4 + \ldots \quad (|x| < 1)$$

Then, replacing x by $-x^2$, we have

$$(1 - x^2)^{-1/2} = 1 - \tfrac{1}{2}(-x^2) + \frac{1 \cdot 3}{2 \cdot 4}(-x^2)^2 - \frac{1 \cdot 3 \cdot 5}{2 \cdot 4 \cdot 6}(-x^2)^3 + \frac{1 \cdot 3 \cdot 5 \cdot 7}{2 \cdot 4 \cdot 6 \cdot 8}(-x^2)^4 + \ldots$$

giving the required binomial expansion

$$(1 - x^2)^{-1/2} = 1 + \tfrac{1}{2}x^2 + \frac{1 \cdot 3}{2 \cdot 4}x^4 + \frac{1 \cdot 3 \cdot 5}{2 \cdot 4 \cdot 6}x^6 + \frac{1 \cdot 3 \cdot 5 \cdot 7}{2 \cdot 4 \cdot 6 \cdot 8}x^8 + \ldots$$

$$= 1 + \tfrac{1}{2}x^2 + \tfrac{3}{8}x^4 + \tfrac{5}{16}x^6 + \tfrac{35}{128}x^8 + \ldots \quad (|x| < 1) \qquad \textbf{(9.16)}$$

Now

$$\int_0^x \frac{dt}{\sqrt{(1 - t^2)}} = \sin^{-1}x$$

and so, integrating the series (9.16) term by term, we obtain

$$\sin^{-1}x = x + \tfrac{1}{6}x^3 + \tfrac{3}{40}x^5 + \tfrac{5}{112}x^7 + \ldots \quad (|x| < 1)$$

Notice that in Example 9.16 we have integrated a power series to obtain the expansion of another function. In general, we may integrate and differentiate power series within their domains of absolute convergence.

Example 9.17   A continuous belt passes over two wheels of diameter D and d with their centres separated by a distance l. Show that, if the belt is sufficiently tight for any sag to be negligible, the length L of belt required is

$$L = 2l \cos \alpha + \tfrac{1}{2}(D + d)\pi + (D - d)\alpha$$

where

$$\sin \alpha = \frac{1}{2l}(D - d)$$

Figure 9.19

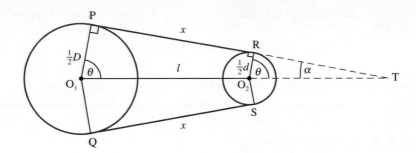

If α is such that α^3 and higher powers may be ignored, show that L is given approximately by

$$L \approx 2l + \frac{(D-d)^2}{4l} + 1.57(D+d)$$

Solution The situation is illustrated in Figure 9.19. From the geometry, the belt will leave the wheels tangentially, and so, because the radius of a circle is perpendicular to its tangent, we have right-angles at P and R as shown. Extending PR and O_1O_2 to meet at T as illustrated, it follows that the triangles PTO_1 and RTO_2 are similar, so that

$$\frac{RO_2}{PO_1} = \frac{TO_2}{TO_1} \quad \text{giving} \quad \frac{d}{D} = \frac{TO_2}{TO_2 + l}$$

so that

$$TO_2 = \frac{dl}{D-d}$$

Thus the angle α subtended at T is given by

$$\sin \alpha = \frac{\frac{1}{2}d}{TO_2} = \frac{D-d}{2l}$$

Also, $RT = TO_2 \cos \alpha$ and $x + RT = (l + TO_2)\cos \alpha$, so that

$$x = l \cos \alpha$$

Using the result that the length of an arc of a circle of radius r subtended by an angle θ is $r\theta$, we find the length of the belt L to be

$$L = 2x + \tfrac{1}{2}D(2\pi - 2\theta) + \tfrac{1}{2}d(2\theta)$$
$$= 2l \cos \alpha + D[\pi - (\tfrac{1}{2}\pi - \alpha)] + d(\tfrac{1}{2}\pi - \alpha)$$

That is,

$$L = 2l \cos \alpha + \tfrac{1}{2}(D+d)\pi + (D-d)\alpha$$

Using the Maclaurin series expansion of $\cos \alpha$, neglecting terms in α^3 and higher powers of α, gives

$$L \approx 2l(1 - \tfrac{1}{2}\alpha^2) + \tfrac{1}{2}(D+d)\pi + (D-d)\alpha$$

Also the Maclaurin expansion of $\sin \alpha$ gives

$$\sin \alpha \approx \alpha \quad \text{so that} \quad \alpha \approx \frac{D-d}{2l}$$

Substituting back and taking $\pi \approx 3.14$ gives the required approximation:

$$L \approx 2l + \frac{(D-d)^2}{4l} + 1.57(D+d)$$

See Question 31 (Exercises 9.5.4) for error considerations.

9.5.3 L'Hôpital's rule

Sometimes we need to find limits of the form

$$\lim_{x \to a} \frac{f(x)}{g(x)}$$

where $f(a) = g(a) = 0$. Even though such a limit may be defined, it cannot be found by substituting $x = a$, since this produces the indeterminate form $0/0$. Using Taylor's theorem 9.4, we can formulate a rule for obtaining such limits if they exist.

Using Taylor's series, we may write

$$\frac{f(x)}{g(x)} = \frac{f(a) + (x-a)f'(a) + \frac{1}{2}(x-a)^2 f''(a) + \dots}{g(a) + (x-a)g'(a) + \frac{1}{2}(x-a)^2 g''(a) + \dots}$$

$$= \frac{f'(a) + \frac{1}{2}(x-a)f''(a) + \dots}{g'(a) + \frac{1}{2}(x-a)g''(a) + \dots} \quad \text{since } f(a) = g(a) = 0, \, x \neq a$$

Hence

$$\lim_{x \to a} \frac{f(x)}{g(x)} = \frac{f'(a)}{g'(a)}$$

provided $g'(a) \neq 0$. This is known as **L'Hôpital's rule**.

It may be that $f'(a)/g'(a)$ is also indeterminate. Consequently, when applying L'Hôpital's rule to obtain the limit

$$\lim_{x \to a} \frac{f(x)}{g(x)}$$

we must repeat the process of differentiating $f(x)$ and $g(x)$ each time we have the indeterminate form $0/0$ at $x = c$. If, however, at any stage in the process one or other of the derivatives is non-zero at $x = a$ then we must stop the process, since the rule will no longer apply. In such cases the limit is either zero or infinite or does not exist; for example, $\lim_{x \to 0} \dfrac{1}{x}$ does not exist.

Example 9.18   Using L'Hôpital's rule, obtain the limits

(a) $\displaystyle\lim_{x \to 0} \frac{\sin x - x}{x^3}$ (b) $\displaystyle\lim_{x \to 0} \frac{1 - \cos x}{x + x^2}$

Solution   (a) Since $(\sin x - x)/x^3$ takes the indeterminate form $0/0$ at $x = 0$, we apply L'Hôpital's rule to give

$$\lim_{x \to 0} \frac{\sin x - x}{x^3} = \lim_{x \to 0} \frac{\cos x - 1}{3x^2} \qquad \text{(again } 0/0 \text{ at } x = 0)$$

$$= \lim_{x \to 0} \frac{-\sin x}{6x} \qquad \text{(again } 0/0 \text{ at } x = 0)$$

$$= \lim_{x \to 0} \frac{-\cos x}{6} = -\tfrac{1}{6}$$

so that

$$\lim_{x \to 0} \frac{\sin x - x}{x^3} = -\tfrac{1}{6}$$

(b) Since $(1 - \cos x)/(x + x^2)$ takes the form $0/0$ at $x = 0$, we apply L'Hôpital's rule to give

$$\lim_{x \to 0} \frac{1 - \cos x}{x + x^2} = \lim_{x \to 0} \frac{\sin x}{1 + 2x} = 0$$

Note that in this case the limit is zero since $(\sin x)/(1 + 2x)$ takes the form $0/1$ at $x = 0$. If we mistakenly proceeded to apply the rule once again, we should obtain

$$\lim_{x \to 0} \frac{1 - \cos x}{x + x^2} = \lim_{x \to 0} \frac{\sin x}{1 + 2x} = \lim_{x \to 0} \frac{\cos x}{2} = \frac{1}{2}$$

an incorrect answer, since the rule was not applicable. The reader may have noticed that both of these limits can be readily evaluated using Maclaurin series.

9.5.4   Exercises

21   Show that if $f(x) = e^{\cos x}$ then

$$f'(x) = -f(x) \sin x$$

and find $f(0)$ and $f'(0)$. Differentiating the expression for $f'(x)$, obtain $f''(x)$ in terms of $f(x)$ and $f'(x)$, and find $f''(0)$. Repeating the process, obtain $f^{(n)}(0)$ for $n = 3, 4, 5$ and 6, and hence obtain the Maclaurin polynomial of degree six for $f(x)$. Confirm your answer by obtaining the series using the Maclaurin expansions of e^x and $\cos x$.

22   A function $y = y(x)$ satisfies the equation

$$\frac{dy}{dx} = y - x + 1$$

with $y = 1$ when $x = 0$. By repeated differentiation, show that $y^{(n)}(0) = 1$ ($n \geqslant 2$), and find the Maclaurin series for y.

23   An alternative approach to Question 22 uses the method of successive approximation, rewriting the equation as

$$y_{n+1}(x) = 1 + \int_0^x [y_n(t) - t + 1] \, dt,$$

with $y_0(x) = y(0) = 1$

Putting $y_0(x) = 1$ into the integral, show that

$$y_1(x) = 1 + 2x - \tfrac{1}{2}x^2$$

$$y_2(x) = 1 + 2x + \tfrac{1}{2}x^2 - \tfrac{1}{6}x^3$$

and find y_3 and y_4.

24   Show that the binomial expansion of $(1 + x)^{-1}$ is

$$(1 + x)^{-1} = 1 - x + x^2 - x^3 + \ldots \quad (-1 < x < 1)$$

Hence find the Maclaurin series expansion of $\tan^{-1} x$.

25 Use the series for $\sin x$ and $\cos x$ to obtain the Maclaurin series for $\tan x$ as far as the term in x^7. Deduce the series for $\ln \cos x$.

26 Show that

$$\coth x = \frac{1}{x}(1 + \tfrac{1}{3}x^2 - \tfrac{1}{45}x^4 + \tfrac{2}{945}x^6 - \dots)$$

27 The field strength H of a magnet at a point on the axis at a distance x from its centre is given by

$$H = \frac{M}{2l}\left[\frac{1}{(x-l)^2} - \frac{1}{(x+l)^2}\right]$$

where $2l$ is the length of the magnet and M is its moment. Show that if l is very small compared with x then

$$H \approx \frac{2M}{x^3}$$

28 Using the Maclaurin series expansions of e^x and $\cos x$, show that

$$\lim_{x \to 0}\left(\frac{e^x + e^{-x} - 2}{2\cos 2x - 2}\right) = -\tfrac{1}{4}$$

29 Show that

$$\ln\left(\frac{\sin x}{x}\right) \approx -\tfrac{1}{6}x^2 - \tfrac{1}{180}x^4$$

if powers of x greater than x^5 are neglected.

30 By expanding e^{-x^2} as a Maclaurin series, show that

$$\int_0^{1/2} e^{-x^2}\,dx \approx 0.461$$

31 Considering the problem of Example 9.17, for what values of l does the approximation

$$L \approx 2l + \frac{(D-d)^2}{4l} + 1.57(D+d)$$

have a percentage error of less than 5% when $D = 10$ and $d = 8$?
What is the percentage error bound for this approximation?

32 Using L'Hôpital's rule, find the following limits:

(a) $\displaystyle\lim_{x \to 2}\frac{x^3 - 3x - 2}{x^3 - 8}$ (b) $\displaystyle\lim_{x \to 0}\frac{1 - (1-x)^{1/4}}{x}$

(c) $\displaystyle\lim_{x \to \pi}\frac{\sin 3x}{\sin 2x}$ (d) $\displaystyle\lim_{x \to 1}\left(\frac{3}{x^3 - 1} - \frac{1}{x-1}\right)$

(e) $\displaystyle\lim_{x \to 0}\frac{x\cos x - \sin x}{x^3}$ (f) $\displaystyle\lim_{x \to \pi/2}\frac{1 - \sin x}{\ln \sin x}$

33 Consider again the design of the milk carton discussed in Example 9.4. Show that if the overlap used in its construction is x mm instead of 5 mm, the objective function that must be minimized is

$$f(b) = (4b + x)\left(\frac{1136\,000}{b^2} + b + 2x\right)$$

Show that when $x = 0$, the optimal value for b is $b_0^* = 10(568)^{1/3}$. The optimal value b^* depends on x. Obtain the Maclaurin series expansion for b^* as far as the term in x^2 and discuss the effect of the overlap size on the design of the carton. (*Hint*: let $b^* \approx b_0 + b_1 x + b_2 x^2$.)

9.5.5 Interpolation revisited

In Chapter 2, Section 2.9.1, we developed the idea of linear interpolation and showed that the approximation

$$f(x) \simeq f_i + \frac{x - x_i}{x_{i+1} - x_i}(f_{i+1} - f_i)$$

gave a value for $f(x)$ which was as accurate as the original data when $|\Delta^2 f_i|$ is less than 4 units of the least significant figure. In many applications, it is easier to express this condition in terms of the second derivative rather than the second difference.
Now

$$\Delta^2 f_i = f(x_i + h) - 2f(x_i) + f(x_i - h)$$

Replacing $f(x_i + h)$ and $f(x_i - h)$ by their Taylor expansions about $x = x_i$, we have (after some cancelling of terms)

$$\Delta^2 f_i = h^2 f''(x_i) + \frac{h^4}{12} f''''(x_i) + \dots$$

The leading term provides a good estimate for $\Delta^2 f_i$ so that the condition for accurate linear interpolation becomes

$$h^2 |f''(x)| < 4 \text{ units of the least significant figure}$$

This enables us to choose an appropriate tabular interval, as is shown in Example 9.19.

Example 9.19   The function $f(x) = e^{-x}$ is to be tabulated to 4dp on the interval $[0, 0.5]$. Find the maximum tabular interval such that the resulting table is suitable for linear interpolation to 4dp; that is, to yield an interpolated value which is as accurate as the tabulated value.

Solution   Here we require that

$$h^2 |f''(x)| < 4 \times 0.0001$$

Since $f(x) = e^{-x}$ we deduce that $f''(x) = e^{-x}$. On the interval $[0, 0.5]$, the maximum value of e^{-x} occurs at $x = 0$, where $e^0 = 1$. Thus we need the largest value of h such that

$$h^2 < 4 \times 0.0001$$

Hence $h < 0.02$, so that the largest tabular interval is 0.02.

9.5.6   Exercises

34   A table for e^x is required for use with linear interpolation to 6dp. It is tabulated for values of x from $x = 0$ to $x = X$ at intervals of 0.001. What is the largest possible value of X?

35   A table for $\tan x$ is required for use with linear interpolation to 6dp. It is tabulated for values of x from $x = 0$ to $x = 1$ at intervals of h rad. What is the largest possible value of h?

36   In Section 8.4 we discussed the process of numerical differentiation using the approximation

$$\phi(h) = \frac{f(a + h) - f(a - h)}{2h}$$

Using the Taylor series for $f(a + h)$ and $f(a - h)$ about $x = a$, show that

$$f'(a) = \phi(h) - \frac{h^2}{3!} f^{(3)}(a) - \frac{h^4}{5!} f^{(5)}(a) - \dots$$

and deduce that

$$f'(a) = \tfrac{1}{3}[4\phi(\tfrac{1}{2}h) - \phi(h)] + \tfrac{1}{4}\frac{h^4}{5!} f^{(5)}(a) + \dots$$

Writing $\psi(h) = \tfrac{1}{3}[4\phi(\tfrac{1}{2}h) - \phi(h)]$, show that $\tfrac{1}{15}[16\psi(\tfrac{1}{2}h) - \psi(h)]$ yields an approximation to $f'(a)$ with truncation error $O(h^6)$. Apply this extrapolation procedure to find $f'(1)$ when $f(x) = \cosh x$, taking $h = 0.4, 0.2$ and 0.1, working to as many decimal places as your calculator will permit.

9.5.7 The convergence of iterations revisited

In Section 9.4.1 we analysed the convergence of an iteration $x_{n+1} = g(x_n)$ for the root α of an equation $f(x) = 0$. We can use the Taylor expansion to analyse the **rate of convergence** of such schemes. Setting $x_n = \alpha + \varepsilon_n$, so that ε_n is the error after n iterations, we have

$$\alpha + \varepsilon_{n+1} = g(\alpha + \varepsilon_n)$$

Expanding $g(\alpha + \varepsilon_n)$ about $x = \alpha$, using the Taylor series (9.14), gives

$$g(\alpha + \varepsilon_n) = \alpha + \varepsilon_{n+1} = g(\alpha) + \frac{\varepsilon_n}{1!}g'(\alpha) + \frac{\varepsilon_n^2}{2!}g''(\alpha) + \frac{\varepsilon_n^3}{3!}g'''(\alpha) + \ldots \qquad \textbf{(9.17)}$$

Since α is a root of the equation $f(x) = 0$, we have $\alpha = g(\alpha)$ and (9.17) simplifies to

$$\varepsilon_{n+1} = \frac{\varepsilon_n}{1!}g'(\alpha) + \frac{\varepsilon_n^2}{2!}g''(\alpha) + \frac{\varepsilon_n^3}{3!}g'''(\alpha) + \ldots \qquad \textbf{(9.18)}$$

If $g'(\alpha) \neq 0$ then ε_{n+1} is proportional to ε_n, and we have a first-order process. If $g'(\alpha) = 0$ and $g''(\alpha) \neq 0$ then ε_{n+1} is proportional to ε_n^2, and we have a second-order process, and so on.

Example 9.20

The equation $x \tan x = 4$ has an infinite number of roots. To find the root near $x = 1$, we may use the iteration

$$x_{n+1} = \tan^{-1}\left(\frac{4}{x_n}\right)$$

Show that this is a first-order process. Starting with $x_0 = 1$, find x_3 and assess its accuracy.

Solution

Here $g(x) = \tan^{-1}(4/x)$, so that

$$g'(x) = \frac{-4}{x^2 + 16}$$

Which is non-zero for all x, i.e. $g'(\alpha) \neq 0$. Thus the iteration is a first-order process. Starting with $x_0 = 1$, we obtain, working to 4dp, the following table.

n	x_n	$4/x_n$	$\tan^{-1}(4/x_n)$
0	1.0000	4.0000	1.3258
1	1.3258	3.0170	1.2507
2	1.2507	3.1982	1.2678
3	1.2678		

From (9.18) we can assess the accuracy of x_n using

$$\varepsilon_{n+1} = \varepsilon_n g'(\alpha) + \ldots$$

and approximating ε_n by $x_n - x_{n+1}$ and α by x_3. Thus in this case we have

$$\varepsilon_3 \approx g'(x_3)(x_2 - x_3) = \frac{-4}{16 + (1.2678)^2}(-0.0171) = 0.0039$$

so that the root is 1.26 to 3sf.

9.5.8 Newton–Raphson procedure

One of the most popular techniques used by engineers for solving non-linear equations is the **Newton–Raphson procedure**. The basic idea is that if x_0 is an approximation to the root $x = \alpha$ of the equation $f(x) = 0$ then a closer approximation will be given by the point $x = x_1$ where the tangent to the graph at $x = x_0$ cuts the x axis, as shown in Figure 9.20.

Figure 9.20
The Newton–Raphson root-finding method.

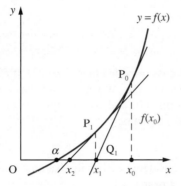

From the definition of the derivative

$$f'(x_0) = \text{slope of } P_0 Q_1 = \frac{f(x_0)}{x_0 - x_1}$$

which can be rearranged to give

$$x_1 = x_0 - \frac{f(x_0)}{f'(x_0)}$$

Taking x_1 as the new approximation to the root $x = \alpha$ and repeating the procedure, as illustrated in Figure 9.20, we obtain the closer aproximation

$$x_2 = x_1 - \frac{f(x_1)}{f'(x_1)}$$

and so on. In general, we may write

$$x_{n+1} = x_n - \frac{f(x_n)}{f'(x_n)} \quad (n = 0, 1, 2, \dots) \tag{9.19}$$

Equation (9.19) is known as the Newton–Raphson iteration procedure for obtaining an approximation to the root of $f(x) = 0$. Note that if $f'(x_n) = 0$ then (9.19) cannot be used to obtain x_{n+1}. This is because the tangent to the graph of $y = f(x)$ at $x = x_n$ will be parallel to the horizontal x axis.

Comparing with the general iteration $x_{n+1} = g(x_n)$, we see that in the case of the Newton–Raphson procedure (9.19) the iteration function is

$$g(x) = x - \frac{f(x)}{f'(x)}$$

which, using the quotient rule, has derivative

$$g'(x) = 1 - \frac{[f'(x)]^2 - f(x)f''(x)}{[f'(x)]^2} = \frac{f(x)f''(x)}{[f'(x)]^2}$$

Since α is a root of $f(x) = 0$, we have $f(\alpha) = 0$, giving

$$g'(\alpha) = 0$$

so the procedure is not a first-order process. Differentiating again and substituting $x = \alpha$, we obtain

$$g''(\alpha) = \frac{f''(\alpha)}{f'(\alpha)}$$

and we have a second-order process provided that $f'(\alpha) \neq 0$. If $f''(\alpha) = 0$, $f'(\alpha) \neq 0$ then we have a third- or higher-order process. When $f(x) = 0$ has a repeated root at $x = \alpha$, $g'(\alpha)$ has the indeterminate form 0/0, and the analysis fails. Repeated roots cause numerical as well as theoretical problems.

Example 9.21   The equation $x \tan x = 4$ was considered earlier in Example 9.20. Apply the Newton–Raphson method to find the root near $x = 1$.

Solution   First, we rewrite the equation in the more convenient (for differentiation) form

$$x \sin x - 4 \cos x = 0$$

Then taking $f(x) = x \sin x - 4 \cos x$ we have $f'(x) = x \cos x + 5 \sin x$. Using the iteration

$$x_{n+1} = x_n - f(x_n)/f'(x_n), \quad x_0 = 1$$

gives the values (to 9dp)

1.000 000 000
1.277 976 731
1.264 600 951
1.264 591 571
1.264 591 571

so that after four iterations we obtain an answer correct to 9dp.

Example 9.22   Find the root of

$$8.0000x^4 + 0.4500x^3 - 4.5440x - 0.1136 = 0$$

near $x = 0.8$ to 4sf.

Figure 9.21
Iteration for the root
of the equation
$8.0000x^4 + 0.4500x^3 - 4.5440x - 0.1136 = 0$.

n	x_n	$f(x_n)$	$f'(x_n)$	$-f_n/f'_n$
0	0.8000	−0.241 600	12.7040	0.019 018
1	0.8190	0.011 436	13.9408	−0.000 820
2	0.8182	0.000 340	13.8876	−0.000 022
3	0.8182			

Solution   In this particular example

$$f(x) = 8.0000x^4 + 0.4500x^3 - 4.5440x - 0.1136$$

giving

$$f'(x) = 32.0000x^3 + 1.3500x^2 - 4.5440$$

When iterating for the root using the Newton–Raphson procedure (9.19), it is usual to present the calculations in tabular form as shown in Figure 9.21 for this particular example. To 4sf the root is given by $x = 0.8182$. When using the Newton–Raphson method, it is recommended that the iteration formula is *not* tidied up into a single expression but is left in the 'approximation minus error' format. Tidying up may lead to ill-conditioning of the numerical procedure.

9.5.9 Optimization revisited

In Section 9.2 we indicated that we would return to reconsider the conditions for determining the nature of stationary points following the introduction of the Taylor series.

If a minimum value of a differentiable function $f(x)$ occurs at $x = a$ then the difference $f(a + h) - f(a)$ will be positive for all small h. However, from the Taylor series (9.14)

$$f(a + h) - f(a) = hf'(a) + \frac{1}{2!}h^2 f''(a) + \frac{1}{3!}h^3 f'''(a) + \ldots$$

and the sign of the expression on the right-hand side depends on the sign of h. It will change sign as h changes sign unless $f'(a) = 0$, in which case the sign depends on the sign of $f''(a)$. Thus a necessary condition for the minimum to occur at $x = a$ is that $f'(a) = 0$, and a necessary and sufficient condition for a minimum of $f(x)$ at $x = a$ is $f'(a) = 0$ and $f''(a) > 0$. Similarly, the maxima of differentiable functions occur when $f'(a) = 0$ and $f''(a) < 0$. If $f'(a) = 0$ and $f''(a) = 0$, we may have a maximum or minimum value or a point of inflection. If $f'(a) = f''(a) = 0$, a necessary condition for a minimum or maximum at $x = a$ is $f'''(a) = 0$, and so on. However, it is important to remember that a function may have an optimal value at a point where its derivative does not exist, as illustrated in Figure 9.22. A numerical scheme for locating the optimal point of a function using the Newton–Raphson procedure can be established. The resulting iteration

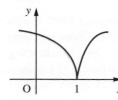

Figure 9.22
$y = (x - 1)^{2/3}$ has a
minimum at $x = 1$ but
it is not differentiable
here.

$$x_{n+1} = x_n - \frac{f'(x_n)}{f''(x_n)}$$

is, however, rarely used in practice. Generally, bracketing methods are used similar to that described in Question 9 (Exercises 9.2.2).

9.5.10 Exercises

37 Given below are three methods for calculating $\sqrt{2}$ by iteration. Find the order of each process and discuss their numerical properties.

(a) $x_{n+1} = 1 + 1/(1 + x_n)$ (b) $x_{n+1} = \frac{1}{2}(x_n + 2/x_n)$

(c) $x_{n+1} = (3x_n^4 + 12x_n^2 - 4)/(8x_n^3)$

38 Use the Newton–Raphson iteration procedure to find the real root of $x^3 - 6x^2 + 9x + 1 = 0$ to 4dp.

39 Use the Newton–Raphson method to find the two positive roots of $x^4 - 4x^3 - 12x^2 + 32x + 28 = 0$.

40 The iteration $x_{n+1} = x_n(3 - 3ax_n + a^2x_n^2)$ may be used to calculate the reciprocal of a, that is, to solve $ax = 1$. Show that this is a third-order process with $\varepsilon_{n+1} = a^2\varepsilon_n^3$. Apply the iteration with $a = 1.735$, starting with $x_0 = 0.5$, and prove that x_2 is correct to 8dp.

9.5.11 Numerical integration

A remarkable mathematical result that follows from the Taylor series is known as the **Euler–Maclaurin formula**:

$$\int_a^b f(x)\mathrm{d}x = \frac{b-a}{2}[f(b) + f(a)] - \frac{(b-a)^2}{12}[f'(b) - f'(a)]$$

$$+ \frac{(b-a)^4}{720}[f^{(3)}(b) - f^{(3)}(a)] - \frac{(b-a)^6}{30\,240}[f^{(5)}(b) - f^{(5)}(a)]\ldots$$

Subdividing the interval $[a, b]$ into n equal strips of width h, we have

$$\int_a^b f(x)\mathrm{d}x = \sum_{r=0}^{n-1} \int_{x_r}^{x_{r+1}} f(x)\mathrm{d}x, \quad x_r = a + rh$$

Applying the formula to each term in the summation, we obtain the trapezium rule together with a power series expansion of the truncation error in terms of h:

$$\int_a^b f(x)\mathrm{d}x = \frac{1}{2}h(f_0 + 2f_1 + 2f_2 + \ldots + 2f_{n-1} + f_n) - \frac{1}{12}h^2(f_n' - f_0')$$

$$+ \frac{1}{720}h^4(f_n^{(3)} - f_0^{(3)}) - \frac{1}{30\,240}h^6(f_n^{(5)} - f_0^{(5)}) + \ldots$$

$$= T(h) + \alpha_1 h^2 + \alpha_2 h^4 + \alpha_3 h^6 + \ldots \tag{9.20}$$

where $T(h)$ is the trapezium approximation to the integral using n strips of width h with $nh = b - a$, and the α's are independent of h. From this we see that the principal term of the **global truncation** error for the approximation is $\frac{1}{12}h^2[f'(b) - f'(a)]$ which, using the first mean value theorem 9.3, may be written $\frac{1}{12}h^2(b - a)f''(c)$ where $a < c < b$.

This analysis makes no allowance for the effect of rounding errors in the values of f_i $(i = 0, 1, \ldots, n)$. A simple estimate of these is

$$h(\tfrac{1}{2} + \underbrace{1 + 1 + \ldots + 1}_{n-1 \text{ terms}} + \tfrac{1}{2}) \times (\tfrac{1}{2} \text{ unit of the least significant figure})$$

$$= nh(\tfrac{1}{2} \text{ unit of the least significant figure})$$

$$= (b - a)(\tfrac{1}{2} \text{ unit of the least significant figure})$$

This result assumes a fixed number of decimal places in the values of the integrand, and is suitable for calculator work. For computers, when h is small and n large, there is the problem of loss of significant digits when adding a large number of almost-equal numbers.

Example 9.23 In Example 8.46 the integral $\int_1^2 (1/x)\mathrm{d}x$ was estimated using the trapezium rule with $h = \frac{1}{4}$ and tabulating the integrand to 6dp. Estimate an error bound for the answer obtained.

Solution Here $f(x) = 1/x$, $a = 1$ and $b = 2$. The global error is given by

$$\tfrac{1}{12}(b - a)h^2 f''(X), \quad \text{with } a \leqslant X \leqslant b$$

so that in this example it is

$$\tfrac{1}{12}(1)(0.25)^2 \frac{2}{X^3}, \quad \text{with } 1 \leqslant X \leqslant 2$$

The largest possible value this can take is when $X = 1$, so we obtain an estimate for the truncation error of 0.010. The rounding-error effect, 0.000 000 5, is negligible compared with this. The error bound we have now calculated safely overestimates the actual error 0.004 obtained in the calculation.

Returning to the full Euler–Maclaurin expansion (9.20), using $2n$ strips of width $\frac{1}{2}h$, we obtain

$$\int_a^b f(x)\mathrm{d}x = T(\tfrac{1}{2}h) + \tfrac{1}{4}\alpha_1 h^2 + \tfrac{1}{16}\alpha_2 h^4 + \tfrac{1}{64}\alpha_3 h^6 + \ldots \tag{9.21}$$

Eliminating the α_1 terms from (9.20) and (9.21) (by subtracting the former from $4 \times$ the latter, and dividing the result by 3) gives

$$\int_a^b f(x)\mathrm{d}x = \tfrac{1}{3}[4T(\tfrac{1}{2}h) - T(h)] - \tfrac{1}{4}\alpha_2 h^4 - \tfrac{5}{16}\alpha_3 h^6 - \ldots$$

Thus the estimate $\frac{1}{3}[4T(\frac{1}{2}h) - T(h)]$ is more accurate than either $T(\frac{1}{2}h)$ or $T(h)$ taken separately. This implies that the truncation error for Simpson's rule is proportional to h^4, which explains why it is a good method for hand computation (as opposed to automatic computation).

9.5.12 Exercises

41   Simpson's rule for the numerical evaluation of an integral is

$$\int_a^b f(x)\mathrm{d}x \approx \frac{b - a}{n}(f_0 + 4f_1 + 2f_2 + \ldots$$

$$+ 2f_{n-2} + 4f_{n-1} + f_n)$$

where n is an even number. The global truncation error is

$$\frac{(b - a)^5}{180n^4} f^{(4)}(c), \quad \text{with } a < c < b$$

If $f(x) = \ln\cosh x$ and $a = 0$, $b = 0.5$, show that $|f^{(4)}(x)| < 2$ for $0 \leqslant x \leqslant 0.5$ and deduce that the global truncation error will be less than $1/(2880n^4)$.

If $f(x)$ is tabulated to 4dp, show that the accumulated rounding error using the formula is less than $1/40\,000$, and find n such that, using the formula, the integral $\int_0^{0.5} \ln \cosh x \, dx$ would be evaluated correctly to 4dp.

 42 (a) Use the trapezium rule with $h = 0.25$ to evaluate $\int_0^1 \sqrt{x} \, dx$. Compare your answer with the exact value, $\frac{2}{3}$.

(b) Put $x = t^2$ in the integral and again evaluate it using the trapezium rule with four strips. Compare your answer with the exact value and with the answer found in (a).

(c) Examine the global truncation errors in both cases and draw some general conclusions.

 43 The trapezium-rule estimate for $\int_0^1 e^{x^2} dx$ with $h = 0.25$ is $1.490\,68$ to 5dp. Estimate the size of the global truncation error in this approximation and show that

$$1.40 \le \int_0^1 e^{x^2} dx < 1.48$$

What value of h will give an answer correct to 4dp?

 44 Show that the composite trapezium rule with step length h yields the approximation

$$\int_0^1 e^x dx \approx \tfrac{1}{2}h(e-1)\coth\left(\frac{h}{2}\right)$$

Using the series expansion for $\coth x$

$$\coth x = \frac{1}{x}(1 + \tfrac{1}{3}x^2 - \tfrac{1}{45}x^4 + \tfrac{2}{945}x^6 - \dots)$$

obtain the approximation

$$\int_0^1 e^x dx \approx (e-1)(1 + \tfrac{1}{12}h^2 - \tfrac{1}{720}h^4$$

$$+ \tfrac{1}{30\,240}h^6 - \dots)$$

Compare this answer with the Euler–Maclaurin theorem.

9.6 Calculus of vectors

In mechanics the vectors describing a dynamic system are time-dependent. Such vectors may be integrated and differentiated in a natural extension of the same processes for scalar quantities. In this section we briefly introduce the relevant definitions.

9.6.1 Differentiation and integration of vectors

The formal definition gives the derivative of a vector $\boldsymbol{v}(t)$ as

$$\frac{d\boldsymbol{v}}{dt} = \lim_{\Delta t \to 0} \frac{\boldsymbol{v}(t + \Delta t) - \boldsymbol{v}(t)}{\Delta t}$$

so if $\boldsymbol{v} = (v_1(t), v_2(t), v_3(t))$ then

$$\frac{d\boldsymbol{v}}{dt} = \left(\frac{dv_1}{dt}, \frac{dv_2}{dt}, \frac{dv_3}{dt}\right)$$

For example, the position vector $\boldsymbol{r}(t) = (x(t), y(t), z(t))$ of a particle may be differentiated with respect to time t to give its velocity $\boldsymbol{v}(t)$ as

$$\boldsymbol{v}(t) = \frac{d\boldsymbol{r}}{dt} = \left(\frac{dx}{dt}, \frac{dy}{dt}, \frac{dz}{dt}\right)$$

Differentiating again gives the acceleration of the particle as

$$\boldsymbol{f}(t) = \frac{d\boldsymbol{v}}{dt} = \frac{d^2\boldsymbol{r}}{dt^2} = \left(\frac{d^2x}{dt^2}, \frac{d^2y}{dt^2}, \frac{d^2z}{dt^2}\right)$$

When differentiating a vector with respect to time, it is conventional to use a 'dot' notation and write

$$\frac{\mathrm{d}\boldsymbol{r}}{\mathrm{d}t} = \dot{\boldsymbol{r}} \quad \text{and} \quad \frac{\mathrm{d}^2\boldsymbol{r}}{\mathrm{d}t^2} = \ddot{\boldsymbol{r}}$$

The usual rules of differentiation may be deduced from this definition.

(a) $\dfrac{\mathrm{d}}{\mathrm{d}t}[\boldsymbol{u}(t) + \boldsymbol{v}(t)] = \dfrac{\mathrm{d}\boldsymbol{u}}{\mathrm{d}t} + \dfrac{\mathrm{d}\boldsymbol{v}}{\mathrm{d}t}$

(b) $\dfrac{\mathrm{d}}{\mathrm{d}t}[\lambda(t)\boldsymbol{v}(t)] = \dfrac{\mathrm{d}\lambda}{\mathrm{d}t}\boldsymbol{v}(t) + \lambda(t)\dfrac{\mathrm{d}\boldsymbol{v}}{\mathrm{d}t},$ where $\lambda(t)$ is a scalar function

(c) $\dfrac{\mathrm{d}}{\mathrm{d}t}[\boldsymbol{u}(t) \cdot \boldsymbol{v}(t)] = \dfrac{\mathrm{d}\boldsymbol{u}}{\mathrm{d}t} \cdot \boldsymbol{v}(t) + \boldsymbol{u}(t) \cdot \dfrac{\mathrm{d}\boldsymbol{v}}{\mathrm{d}t}$

(d) $\dfrac{\mathrm{d}}{\mathrm{d}t}[\boldsymbol{u}(t) \times \boldsymbol{v}(t)] = \dfrac{\mathrm{d}\boldsymbol{u}}{\mathrm{d}t} \times \boldsymbol{v}(t) + \boldsymbol{u}(t) \times \dfrac{\mathrm{d}\boldsymbol{v}}{\mathrm{d}t},$ note importance of order

Example 9.24   Sketch the curve

$$\boldsymbol{r} = \sin t\,\boldsymbol{i} + \cos t\,\boldsymbol{j}$$

Calculate

(a) $\dfrac{\mathrm{d}\boldsymbol{r}}{\mathrm{d}t}$ (b) $\dfrac{\mathrm{d}^2\boldsymbol{r}}{\mathrm{d}t^2}$ (c) $\left|\dfrac{\mathrm{d}\boldsymbol{r}}{\mathrm{d}t}\right|$ (d) $\dfrac{\mathrm{d}}{\mathrm{d}t}(|\boldsymbol{r}|)$

Solution   A sketch of the curve is shown in Figure 9.23. It is a circle with centre at the origin and of unit radius.

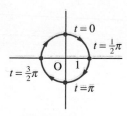

Figure 9.23

(a) $\dfrac{\mathrm{d}\boldsymbol{r}}{\mathrm{d}t} = \dfrac{\mathrm{d}}{\mathrm{d}t}(\sin t)\boldsymbol{i} + \dfrac{\mathrm{d}}{\mathrm{d}t}(\cos t)\boldsymbol{j} = \cos t\,\boldsymbol{i} - \sin t\,\boldsymbol{j}$

(b) $\dfrac{\mathrm{d}^2\boldsymbol{r}}{\mathrm{d}t^2} = \dfrac{\mathrm{d}}{\mathrm{d}t}(\cos t)\boldsymbol{i} - \dfrac{\mathrm{d}}{\mathrm{d}t}(\sin t)\boldsymbol{j} = -\sin t\,\boldsymbol{i} - \cos t\,\boldsymbol{j}$

(c) $\left|\dfrac{\mathrm{d}\boldsymbol{r}}{\mathrm{d}t}\right| = (\cos^2 t + \sin^2 t)^{1/2} = 1$

(d) $|\boldsymbol{r}| = (\sin^2 t + \cos^2 t)^{1/2} = 1$

so that

$$\frac{\mathrm{d}}{\mathrm{d}t}(|\boldsymbol{r}|) = \frac{\mathrm{d}}{\mathrm{d}t}(1) = 0$$

Note that

$$\frac{\mathrm{d}}{\mathrm{d}t}(|\boldsymbol{r}|) \neq \left|\frac{\mathrm{d}\boldsymbol{r}}{\mathrm{d}t}\right|$$

In the same way, the integration of a vector $\mathbf{v}(t)$ with respect to the variable t is usually performed in terms of its components:

$$\int \mathbf{v}(t)dt = \int (v_1(t), v_2(t), v_3(t)) \, dt$$

$$= \left(\int v_1(t) \, dt, \int v_2(t) \, dt, \int v_3(t) \, dt \right)$$

Of course, the arbitrary constant of integration is now a vector constant $\mathbf{c} = (c_1, c_2, c_3)$.

Example 9.25   Given

$$\frac{d^2\mathbf{r}}{dt^2} = -g\mathbf{k} \quad \text{with} \quad \mathbf{r}(0) = 0 \quad \text{and} \quad \dot{\mathbf{r}}(0) = \mathbf{V}$$

find $\mathbf{r}(t)$. Obtain the locus of the point P, such that $\overrightarrow{OP} = \mathbf{r}$, in terms of x and z when $\mathbf{V} = (u, 0, v)$.

Solution   This is the equation of motion of a projectile under gravity. Integrating the equation once gives

$$\frac{d\mathbf{r}}{dt} = -gt\mathbf{k} + \mathbf{c}$$

Since $\dot{\mathbf{r}}(0) = \mathbf{V}$, we have

$$\mathbf{c} = \mathbf{V} \quad \text{and} \quad \frac{d\mathbf{r}}{dt} = -gt\mathbf{k} + \mathbf{V}$$

Integrating a second time gives

$$\mathbf{r}(t) = \mathbf{V}t - \tfrac{1}{2}gt^2\mathbf{k} + \mathbf{a}$$

Since $\mathbf{r}(0) = 0$, we have $\mathbf{a} = 0$, giving

$$\mathbf{r} = \mathbf{V}t - \tfrac{1}{2}gt^2\mathbf{k}$$

Now $\mathbf{r} = (x, y, z)$, so that when $\mathbf{V} = (u, 0, v)$, we have

$$(x, y, z) = (u, 0, v)t + (0, 0, -\tfrac{1}{2}gt^2)$$
$$= (ut, 0, vt) + (0, 0, -\tfrac{1}{2}gt^2)$$
$$= (ut, 0, vt - \tfrac{1}{2}gt^2)$$

Thus

$$x = ut, \quad y = 0 \quad \text{and} \quad z = vt - \tfrac{1}{2}gt^2$$

Substituting $t = x/u$ into the equation for z gives, after some rearrangement,

$$z = \tfrac{1}{2}\frac{v^2}{g} - \frac{g}{2u^2}\left(x - \frac{uv}{g} \right)^2$$

This is a parabola with vertex at $(uv/g, 0, v^2/2g)$.

9.6.2 Exercises

45 If $r = (t, t^2, t^3)$, find $\dot{r}(t)$ and $\ddot{r}(t)$.

46 Given the vector

$$r = (1 + t)i + t^2j + \tfrac{2}{3}t^3k$$

evaluate dr/dt and write it in the form

$$\frac{dr}{dt} = f(t)\hat{T}(t)$$

where $\hat{T}$ is the unit tangent direction. Calculate $d\hat{T}/dt$ in its simplest form and show that it is perpendicular to $\hat{T}$.

47 In polar coordinates (r, θ), the unit vectors $\hat{r}$ and $\hat{\theta}$ are defined as in Figure 9.24. Show that

$$\hat{r} = \cos\theta\, i + \sin\theta\, j$$

$$\hat{\theta} = -\sin\theta\, i + \cos\theta\, j$$

Hence from the definition $r = r\hat{r}$ show that

$$\frac{dr}{dt} = \frac{dr}{dt}\hat{r} + r\omega\hat{\theta} \quad \text{where} \quad \omega = \frac{d\theta}{dt}$$

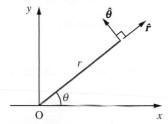

Figure 9.24

Deduce that

$$\frac{d\hat{r}}{dt} = \omega\hat{\theta} \quad \text{and} \quad \frac{d\hat{\theta}}{dt} = -\omega\hat{r}$$

and

$$\frac{d^2r}{dt^2} = \left(\frac{d^2r}{dt^2} - r\omega^2\right)\hat{r} + \left(2\omega\frac{dr}{dt} + r\frac{d\omega}{dt}\right)\hat{\theta}$$

48 Show that if the vector $a(t) = f(t)i + g(t)j$ has constant magnitude, then a and $\dfrac{da}{dt}$ are perpendicular.

49 A curve is given parametrically by $r(t) = f(t)i + g(t)j$. Show that, if s is the length of an arc measured from a fixed point P_0 on the curve so that s increases as t increases, then

$$\left|\frac{dr}{dt}\right| = \frac{ds}{dt}$$

Deduce that $\dfrac{dr}{ds}$ is a unit tangent vector to the curve at $r(t)$ and that (using the result of Question 48), $\dfrac{dr}{ds}$ and $\dfrac{d^2r}{ds^2}$ are perpendicular. Show that

$$\left|\frac{d^2r}{ds^2}\right| = |\kappa|$$

where κ is the curvature of the curve at that point.

9.7 Partial differentiation

The ideas of calculus apply to functions of several variables as well as to functions of one variable. Of course, because more variables are involved, the notation and technical detail are more complicated but the essential ideas are the same. In the remainder of this chapter we will explore the extension of the process and ideas of differentiation to functions of several independent variables. As we shall see below, the rate of change of the function with respect to its variables can be expressed in terms of the rates of change of the function with respect to each of the independent variables separately.

9.7.1 Partial derivatives

Given a function of one variable, $f(x)$, we recall from Section 8.2.2 that the derivative was defined by

$$\frac{df}{dx} = \lim_{\Delta x \to 0} \frac{\Delta f}{\Delta x} = \lim_{\Delta x \to 0} \left[\frac{f(x + \Delta x) - f(x)}{\Delta x} \right]$$

and that this was a measure of the rate of change of the value of the function $f(x)$ with respect to its variable (or argument) x. For a function of several variables it is also useful to know how the function changes when one, some or all of the variables change. To achieve this we define the **partial derivatives** of a function.

First, we consider a function $f(x, y)$ of the two variables x and y. The partial derivative, $\frac{\partial f}{\partial x}$, of $f(x, y)$ with respect to x is its derivative with respect to x treating the value of y as being constant. Thus

$$\frac{\partial f}{\partial x} = \left[\frac{df}{dx} \right]_{y=\text{const}} = \lim_{\Delta x \to 0} \left[\frac{f(x + \Delta x, y) - f(x, y)}{\Delta x} \right]$$

Likewise, the partial derivative, $\frac{\partial f}{\partial y}$, of $f(x, y)$ with respect to y is its derivative with respect to y treating the value of x as being constant, so that

$$\frac{\partial f}{\partial y} = \left[\frac{df}{dy} \right]_{x=\text{const}} = \lim_{\Delta y \to 0} \left[\frac{f(x, y + \Delta y) - f(x, y)}{\Delta y} \right]$$

The process of obtaining the partial derivatives is called **partial differentiation**. Note the use of 'curly dees', which is to distinguish between partial differentiation and ordinary differentiation. In writing, care must be taken to distinguish between $\frac{df}{dx}$, $\frac{\Delta f}{\Delta x}$ and $\frac{\partial f}{\partial x}$, all of which have different meanings.

A concise notation is sometimes used for partial derivatives; as an alternative to the 'curly dee', we write

$$f_x = \frac{\partial f}{\partial x} \quad \text{and} \quad f_y = \frac{\partial f}{\partial y}$$

It should be noted, however, that subscripts often have other connotations, so care should be taken in using them in this way.

If we write $z = f(x, y)$ then the partial derivatives may also be written as

$$\frac{\partial z}{\partial x}, \frac{\partial z}{\partial y} \quad \text{or} \quad z_x, z_y$$

Summary

The **partial derivatives** of the function $z = f(x, y)$ with respect to the variables x and y respectively are given by

$$\frac{\partial f}{\partial x} = f_x = \frac{\partial z}{\partial x} = z_x = \lim_{\Delta x \to 0} \left[\frac{f(x + \Delta x, y) - f(x, y)}{\Delta x} \right] \tag{9.22}$$

$$\frac{\partial f}{\partial y} = f_y = \frac{\partial z}{\partial y} = z_y = \lim_{\Delta y \to 0} \left[\frac{f(x, y + \Delta y) - f(x, y)}{\Delta y} \right] \tag{9.23}$$

Finding partial derivatives is no more difficult than finding derivatives of functions of one variable, with the constant multiplication, sum, product and quotient rules having counterparts for partial derivatives. Note, however, that, despite the notation, partial derivatives do not behave like fractions. For example, $\dfrac{\partial x}{\partial z} \neq 1 \bigg/ \left(\dfrac{\partial z}{\partial x}\right)$ (see Exercises 9.7.7, Question 75).

Example 9.26 Find $\dfrac{\partial f}{\partial x}$ and $\dfrac{\partial f}{\partial y}$ where $f(x, y)$ is given by

(a) $3x^2 + 2xy + y^3$ (b) $(y^2 + x)e^{-xy}$

Solution (a) $f(x, y) = 3x^2 + 2xy + y^3$

To find $\dfrac{\partial f}{\partial x}$, we differentiate $f(x, y)$ with respect to x regarding y as a constant. Thus we obtain

$$\frac{\partial f}{\partial x} = \frac{\partial}{\partial x}(3x^2) + \frac{\partial}{\partial x}(2xy) + \frac{\partial}{\partial x}(y^3)$$

$$= 3\frac{d}{dx}(x^2) + 2y\frac{d}{dx}(x) + 0 \quad \text{(Note: term in brackets involves } x \text{ only)}$$

$$= 6x + 2y$$

Similarly,

$$\frac{\partial f}{\partial y} = \frac{\partial}{\partial y}(3x^2) + \frac{\partial}{\partial y}(2xy) + \frac{\partial}{\partial y}(y^3)$$

$$= 0 + 2x\frac{d}{dy}(y) + \frac{d}{dy}(y^3)$$

$$= 2x + 3y^2$$

(b) $f(x, y) = (y^2 + x)e^{-xy}$

Using the product rule, differentiating with respect to x, regarding y as a constant, gives

$$\frac{\partial f}{\partial x} = (e^{-xy})\frac{\partial}{\partial x}(y^2 + x) + (y^2 + x)\frac{\partial}{\partial x}(e^{-xy})$$

$$= (e^{-xy})(1) + (y^2 + x)(-ye^{-xy})$$

$$= (1 - y^3 - xy)e^{-xy}$$

Similarly,

$$\frac{\partial f}{\partial y} = (e^{-xy})\frac{\partial}{\partial y}(y^2 + x) + (y^2 + x)\frac{\partial}{\partial y}(e^{-xy})$$

$$= (e^{-xy})(2y) + (y^2 + x)(-xe^{-xy})$$

$$= (2y - xy^2 - x^2)e^{-xy}$$

Example 9.27

Find $\partial f/\partial x$ and $\partial f/\partial y$ when $f(x, y)$ is

(a) $xy^2 + 3xy - x + 2$ (b) $\sin(x^2 - 3y)$

Solution

(a) Taking $f(x, y) = xy^2 + 3xy - x + 2$ and differentiating with respect to x, keeping y fixed, gives

$$\frac{\partial f}{\partial x} = f_x = y^2 + 3y - 1$$

Differentiating with respect to y, keeping x fixed, gives

$$\frac{\partial f}{\partial y} = f_y = 2xy + 3x$$

(b) Taking $f(x, y) = \sin(x^2 - 3y)$ and applying the composite-function rule, we obtain

$$\frac{\partial f}{\partial x} = \cos(x^2 - 3y)\frac{\partial}{\partial x}(x^2 - 3y) = \cos(x^2 - 3y)\,2x$$

$$= 2x\cos(x^2 - 3y)$$

and

$$\frac{\partial f}{\partial y} = \cos(x^2 - 3y)\frac{\partial}{\partial y}(x^2 - 3y) = -3\cos(x^2 - 3y)$$

Although we have introduced partial derivatives in the context of functions of two variables, the concept may be readily extended to obtain the partial derivatives of a function of as many variables as we please. Thus for a function $f(x_1, x_2, \ldots, x_n)$ of n variables the partial derivative with respect to x_i is given by

$$f_{x_i} = \frac{\partial f}{\partial x_i} = \lim_{\Delta x_i \to 0} \frac{f(x_1, x_2, \ldots, x_{i-1}, x_i + \Delta x_i, x_{i+1}, \ldots, x_n) - f(x_1, x_2, \ldots, x_n)}{\Delta x_i}$$

and is obtained by differentiating the function with respect to x_i with all the other $n - 1$ variables kept constant.

Example 9.28

Find the partial derivatives of

$$f(x, y, z) = xyz^2 + 3xy - z$$

with respect to x, y and z.

Solution

Differentiating $f(x, y, z)$ with respect to x, keeping y and z fixed, gives

$$f_x = \frac{\partial f}{\partial x} = yz^2 + 3y$$

Differentiating $f(x, y, z)$ with respect to y, keeping x and z fixed, gives

$$f_y = \frac{\partial f}{\partial y} = xz^2 + 3x$$

Differentiating $f(x, y, z)$ with respect to z, keeping x and y fixed, gives

$$f_z = \frac{\partial f}{\partial z} = xy(2z) + 0 - 1 = 2xyz - 1$$

9.7.2 Directional derivatives

Consider a function of two variables $z = f(x, y)$. This may be represented as a surface in three dimensions as shown in Figure 9.25.

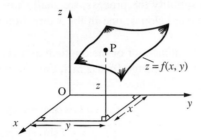

Figure 9.25 Surface $z = f(x, y)$.

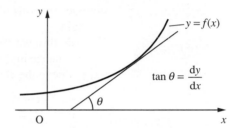

Figure 9.26 Tangent to the graph of $y = f(x)$.

We recall from Chapter 8 that the derivative of a function $f(x)$ of one variable measures the slope of the tangent to the graph of the function, as illustrated in Figure 9.26. In the case of a function of two variables, because $z = f(x, y)$ defines a surface in three dimensions, there is no unique meaning of 'slope' unless we specify the direction in which it is to be measured. In general, the slope will be different for different directions. Now consider two points P and Q on the surface $z = f(x, y)$, as shown in Figure 9.27 and let P′ and Q′ be their projections on the x–y plane. To simplify, set P′Q′ = l; then the coordinates of P′ and Q′ are given by

$$(x, y, 0) \quad \text{and} \quad (x + l\cos\alpha, y + l\sin\alpha, 0)$$

Figure 9.27
Directional derivative.

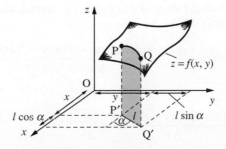

respectively, where α is the angle that P'Q' makes with the positive x direction. The slope of the line PQ is then

$$\frac{f(x + l\cos\alpha, y + l\sin\alpha) - f(x, y)}{l}$$

and the slope of the surface at P in the direction of $\overrightarrow{PQ}$ is the limit of this quotient as $l \to 0$. Denoting this slope by $m_\alpha(x, y)$, we have

$$m_\alpha(x, y) = \lim_{l \to 0} \frac{f(x + l\cos\alpha, y + l\sin\alpha) - f(x, y)}{l}$$

Here the subscript α indicates the direction with respect to which the slope is measured, and the (x, y) shows the point at which it is evaluated. Essentially, we have reduced the problem of a function of two variables to a function of one variable by fixing the direction along which we allow x and y to vary. It would be very clumsy to have to perform the calculation this way every time we wish to work out the rate of change or slope of the function. To simplify the process, we shall show how to represent the slope m_α in terms of two standard slopes: one in the x direction and the other in the y direction.

To do this, we rearrange the numerator of the quotient as a sum of terms, one showing the change in $f(x, y)$ due to the change $l\cos\alpha$ in x, the other showing the change in $f(x, y)$ due to the change $l\sin\alpha$ in y. Thus

$$f(x + l\cos\alpha, y + l\sin\alpha) - f(x, y) = [f(x + l\cos\alpha, y + l\sin\alpha) - f(x, y + l\sin\alpha)]$$

$$+ [f(x, y + l\sin\alpha) - f(x,y)]$$

and
$$m_\alpha(x, y) = \lim_{l \to 0} \frac{f(x + l\cos\alpha, y + l\sin\alpha) - f(x, y + l\sin\alpha)}{l\cos\alpha}\cos\alpha$$

$$+ \lim_{l \to 0} \frac{f(x, y + l\sin\alpha) - f(x, y)}{l\sin\alpha}\sin\alpha$$

$$= p(x, y)\cos\alpha + q(x, y)\sin\alpha$$

where $p(x, y)$ and $q(x, y)$ are the values of the respective limits

$$p(x, y) = \lim_{l \to 0} \frac{f(x + l\cos\alpha, y + l\sin\alpha) - f(x, y + l\sin\alpha)}{l\cos\alpha}$$

$$q(x, y) = \lim_{l \to 0} \frac{f(x, y + l\sin\alpha) - f(x, y)}{l\sin\alpha}$$

Examining the numerator of $p(x, y)$, we see that the 'y value' in both terms is the same, $y + l\sin\alpha$, and also that $l\sin\alpha \to 0$ as $l \to 0$. In contrast, the 'x value' in the terms differs by $l\cos\alpha$. Denoting this by Δx we may write

$$p(x, y) = \lim_{\Delta x \to 0} = \frac{f(x + \Delta x, y + \Delta x\tan\alpha) - f(x, y + \Delta x\tan\alpha)}{\Delta x}$$

which simpifies to

$$p(x, y) = \lim_{\Delta x \to 0} \frac{f(x + \Delta x, y) - f(x, y)}{\Delta x} = \frac{\partial f}{\partial x} \tag{9.24}$$

In the same way,

$$q(x, y) = \lim_{\Delta y \to 0} \frac{f(x, y + \Delta y) - f(x, y)}{\Delta y} = \frac{\partial f}{\partial y} \qquad (9.25)$$

and we may then write the slope in the direction at an angle α to the x axis as

$$m_\alpha(x, y) = \frac{\partial f}{\partial x} \cos \alpha + \frac{\partial f}{\partial y} \sin \alpha \qquad (9.26)$$

Example 9.29 Find the partial derivatives of $f(x, y) = x^2 y^3 + 3y + x$ with respect to x and y, and the slope of the function in the direction at an angle α to the x axis.

Solution To find the partial derivative of $f(x, y)$ with respect to x, we differentiate $f(x, y)$ with respect to x, keeping y constant. Thus

$$\frac{\partial f}{\partial x} = 2xy^3 + 1$$

Similarly, we obtain the partial derivative with respect to y by differentiating $f(x, y)$ with respect to y, keeping x constant. Thus

$$\frac{\partial f}{\partial y} = 3x^2 y^2 + 3$$

The general expression for the slope of the surface $z = f(x, y)$ in the direction at an angle α to the x axis is

$$m_\alpha(x, y) = \frac{\partial f}{\partial x} \cos \alpha + \frac{\partial f}{\partial y} \sin \alpha$$

So for this function we have

$$m_\alpha(x, y) = (2xy^3 + 1)\cos \alpha + (3x^2 y^2 + 3)\sin \alpha$$

Since in evaluating $\partial f/\partial x$ we consider only the variation of $f(x, y)$ in the x direction, $\partial f/\partial x$ gives the slope of the surface $z = f(x, y)$ at the point (x, y) in the x direction ($\alpha = 0$ in (9.26)). Similarly, $\partial f/\partial y$ gives the slope in the y direction ($\alpha = \frac{1}{2}\pi$ in (9.26)). This is illustrated in Figure 9.28.

Thus if we know $\partial f/\partial x$ and $\partial f/\partial y$, we can calculate the slope $m_\alpha(x, y)$ of the function in any given direction using (9.26). This is called the **directional derivative** of $f(x, y)$, and may be regarded as the projection of the vector $(\partial f/\partial x, \partial f/\partial y)$ onto the direction represented by the unit vector $(\cos \alpha, \sin \alpha)$, so that $(\cos \alpha, \sin \alpha)$ is a unit vector in the direction of the required derivative. Thus we may express $m_\alpha(x, y)$ as the scalar product

$$m_\alpha(x, y) = \left(\frac{\partial f}{\partial x}, \frac{\partial f}{\partial y} \right) \cdot (\cos \alpha, \sin \alpha)$$

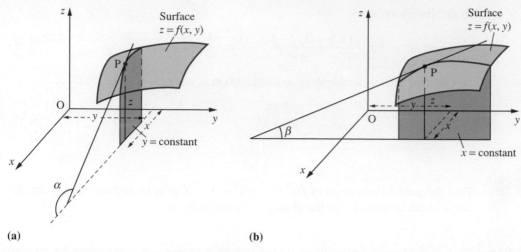

Figure 9.28 Geometrical illustration of partial derivatives (a) $\dfrac{\partial f}{\partial x} = \tan \alpha$ and (b) $\dfrac{\partial f}{\partial y} = \tan \beta$.

9.7.3 Exercises

50 Obtain from first principles the partial derivatives $\partial f/\partial x$ and $\partial f/\partial y$ of the function $f(x, y)$ at the point $(1, 2)$, where

$$f(x, y) = 2x^2 - xy + y^2$$

51 Obtain from first principles the partial derivatives $\partial f/\partial x$ and $\partial f/\partial y$ of the function $f(x, y)$ at the general point (x, y) where

$$f(x, y) = x \cos y$$

52 Find $\partial f/\partial x$ and $\partial f/\partial y$ when $f(x, y)$ is

(a) $x^3 y + 2x^2 + 9y^2 + xy + 10$

(b) $(x + y^2)^3$ (c) $(3x^2 + y^2 + 2xy)^{1/2}$

53 Find $\partial f/\partial x$ and $\partial f/\partial y$ when $f(x, y)$ is

(a) $e^{xy} \cos x$ (b) $\dfrac{x}{x^2 + y^2}$ (c) $\dfrac{x + y}{x^2 + 2y^2 + 6}$

54 Find the gradient of $f(x, y) = x^2 + 2y^2 - 3x + 2y$ at the point (x, y) in the direction making an angle α with the positive x direction. What is the value of the gradient at $(2, -1)$ when $\alpha = \frac{1}{6}\pi$? What values of α give the largest gradient at $(2, -1)$?

The level curve of $f(x, y)$ through $(2, -1)$ is given by $f(x, y) = f(2, -1)$. This defines the relationship between x and y on the curve. Show that the tangent to the level curve at $(2, -1)$ is

perpendicular to the direction of maximum gradient at that point and parallel to the direction of zero gradient.

55 Find $\partial z/\partial x$ and $\partial z/\partial y$ when $z(x, y)$ satisfies

(a) $x^2 + y^2 + z^2 = 10$ (b) $xyz = x - y + z$

56 Show that $z = x^2 y^2/(x^2 + y^2)$ satisfies the differential equation

$$x\frac{\partial z}{\partial x} + y\frac{\partial z}{\partial y} = 2z$$

57 Find f_x, f_y and f_z when $f(x, y, z)$ is

(a) $x^2 y + 3yxz - 2z^3 x^2 y$ (b) $e^{2z} \cos xy$

58 Show that

$$f(x, y, z) = (x^2 + y^2 + z^2)^{-1/2}$$

satisfies

$$xf_x + yf_y + zf_z = -f(x, y, z)$$

59 Show that

$$f(x, y, z) = x + \frac{x - y}{y - z}$$

satisfies

$$f_x + f_y + f_z = 1$$

9.7.4 The chain rule

As can be seen from Examples 9.26–9.28, the rules and results of ordinary differentiation carry over to partial differentiation. In particular, the composite-function rule still holds, but in a modified form. Consider the two-variable case where $z = f(x, y)$ and x and y are themselves functions of two independent variables, s and t. Then z itself is also a function of s and t, say $F(s, t)$, and we can find its derivatives using a composite-function rule that gives the rates of change of z with respect to s and t in terms of the rates of change of z with respect to x and y and the rates of change of x and y with respect to s and t. Thus

$$\frac{\partial z}{\partial s} = \frac{\partial z}{\partial x}\frac{\partial x}{\partial s} + \frac{\partial z}{\partial y}\frac{\partial y}{\partial s} \quad \text{and} \quad \frac{\partial z}{\partial t} = \frac{\partial z}{\partial x}\frac{\partial x}{\partial t} + \frac{\partial z}{\partial y}\frac{\partial y}{\partial t} \tag{9.27}$$

or, in vector–matrix form,

$$\begin{bmatrix} \dfrac{\partial z}{\partial s} & \dfrac{\partial z}{\partial t} \end{bmatrix} = \begin{bmatrix} \dfrac{\partial z}{\partial x} & \dfrac{\partial z}{\partial y} \end{bmatrix} \begin{bmatrix} \dfrac{\partial x}{\partial s} & \dfrac{\partial x}{\partial t} \\ \dfrac{\partial y}{\partial s} & \dfrac{\partial y}{\partial t} \end{bmatrix}$$

This result is often called the **chain rule**. The proof is straightforward. Consider $\partial z/\partial s$, given by

$$\frac{\partial z}{\partial s} = \lim_{\Delta s \to 0} \frac{F(s + \Delta s, t) - F(s, t)}{\Delta s}$$

The point $(s + \Delta s, t)$ in the s–t plane will correspond to the point $(x + \Delta x, y + \Delta y)$ in the x–y plane, while (s, t) corresponds to (x, y). Thus

$$\frac{\partial z}{\partial s} = \lim_{\Delta s \to 0} \frac{f(x + \Delta x, y + \Delta y) - f(x, y)}{\Delta s}$$

$$= \lim_{\Delta s \to 0} \frac{f(x + \Delta x, y + \Delta y) - f(x, y + \Delta y)}{\Delta x}\frac{\Delta x}{\Delta s}$$

$$+ \lim_{\Delta s \to 0} \frac{f(x, y + \Delta y) - f(x, y)}{\Delta y}\frac{\Delta y}{\Delta s}$$

$$= \frac{\partial f}{\partial x}\frac{\partial x}{\partial s} + \frac{\partial f}{\partial y}\frac{\partial y}{\partial s}$$

We can similarly prove the result for $\partial z/\partial t$.

It may happen, of course, that x and y are functions of one variable only or of three variables or more. In all these cases the chain rule still applies when the functions involved are differentiable.

Example 9.30

Find $\partial T/\partial r$ and $\partial T/\partial \theta$ when

$$T(x, y) = x^3 - xy + y^3$$

and

$$x = r\cos\theta \quad \text{and} \quad y = r\sin\theta$$

Solution   By the chain rule (9.27),

$$\frac{\partial T}{\partial r} = \frac{\partial T}{\partial x}\frac{\partial x}{\partial r} + \frac{\partial T}{\partial y}\frac{\partial y}{\partial r}$$

In this example

$$\frac{\partial T}{\partial x} = 3x^2 - y \quad \text{and} \quad \frac{\partial T}{\partial y} = -x + 3y^2$$

and

$$\frac{\partial x}{\partial r} = \cos\theta \quad \text{and} \quad \frac{\partial y}{\partial r} = \sin\theta$$

so that

$$\frac{\partial T}{\partial r} = (3x^2 - y)\cos\theta + (-x + 3y^2)\sin\theta$$

Substituting for x and y in terms of r and θ gives

$$\frac{\partial T}{\partial r} = 3r^2(\cos^3\theta + \sin^3\theta) - 2r\cos\theta\sin\theta$$

Similarly,

$$\frac{\partial T}{\partial\theta} = (3x^2 - y)(-r\sin\theta) + (-x + 3y^2)r\cos\theta$$

$$= 3r^3(\sin\theta - \cos\theta)\cos\theta\sin\theta + r^2(\sin^2\theta - \cos^2\theta)$$

Example 9.31   Find dH/dt when

$$H(t) = \sin(3x - y)$$

and

$$x = 2t^2 - 3 \quad \text{and} \quad y = \tfrac{1}{2}t^2 - 5t + 1$$

Solution   We note that x and y are functions of t only, so that the chain rule (9.27) becomes

$$\frac{dH}{dt} = \frac{\partial H}{\partial x}\frac{dx}{dt} + \frac{\partial H}{\partial y}\frac{dy}{dt}$$

Note the mixture of partial and ordinary derivatives. H is a function of the one variable t, but its dependence is expressed through the two variables x and y.

Substituting for the derivatives involved, we have

$$\frac{dH}{dt} = 3[\cos(3x - y)]4t - [\cos(3x - y)](t - 5)$$

$$= (11t + 5)\cos(3x - y)$$

$$= (11t + 5)\cos(\tfrac{11}{2}t^2 + 5t - 10)$$

Example 9.32	The base radius r cm of a right-circular cone increases at $2\,\mathrm{cm\,s^{-1}}$ and its height h cm at $3\,\mathrm{cm\,s^{-1}}$. Find the rate of increase in its volume when $r = 5$ and $h = 15$.

Solution The volume V of a cone having base radius r and height h is

$$V = \tfrac{1}{3}\pi r^2 h$$

We wish to determine dV/dt given dr/dt and dh/dt. Applying the chain rule (9.27) gives

$$\frac{dV}{dt} = \frac{\partial V}{\partial r}\frac{dr}{dt} + \frac{\partial V}{\partial h}\frac{dh}{dt}$$

Now

$$\frac{\partial V}{\partial r} = \tfrac{2}{3}\pi rh, \quad \frac{\partial V}{\partial h} = \tfrac{1}{3}\pi r^2, \quad \frac{dr}{dt} = 2, \quad \frac{dh}{dt} = 3$$

so that

$$\frac{dV}{dt} = \tfrac{4}{3}\pi rh + \pi r^2$$

When $r = 5$ cm and $h = 15$ cm, the rate of increase in volume is

$$\frac{dV}{dt} = (\tfrac{4}{3}\pi \times 5 \times 15 + \pi \times 5^2)\,\mathrm{cm^3\,s^{-1}} = 125\pi\,\mathrm{cm^3\,s^{-1}}$$

9.7.5 Exercises

60 Find dz/dt when

(a) $z^2 = x^2 + y^2$, $x = t^2 + 1$ and $y = t - 1$

(b) $z = x^2 t^2$ and $x^2 + 3xt + 2t^2 = 1$

61 Find $\partial f/\partial s$ and $\partial f/\partial t$ when $f(x, y) = e^x \cos y$, $x = s^2 - t^2$ and $y = 2st$.

62 Show that if $u = xy$, $v = xy$ and $z = f(u, v)$ then

(a) $x\dfrac{\partial z}{\partial x} - y\dfrac{\partial z}{\partial y} = (x - y)\dfrac{\partial z}{\partial u}$

(b) $\dfrac{\partial z}{\partial x} - \dfrac{\partial z}{\partial y} = (y - x)\dfrac{\partial z}{\partial v}$

63 Show that if $z = x^n f(u)$, where $u = y/x$, then

$$x\frac{\partial z}{\partial x} + y\frac{\partial z}{\partial y} = nz$$

Verify this result for $z = x^4 + 2y^4 + 3xy^3$.

64 Show that, if f is a function of the independent variables x and y, and the latter are changed to independent variables u and v where $u = e^{y/x}$ and $v = x^2 + y^2$, then

(a) $x\dfrac{\partial f}{\partial x} + y\dfrac{\partial f}{\partial y} = 2v\dfrac{\partial f}{\partial v}$

(b) $x^3\dfrac{\partial f}{\partial y} - x^2 y\dfrac{\partial f}{\partial x} = uv\dfrac{\partial f}{\partial u}$

65   In a right-angled triangle a cm and b cm are the sides containing the right-angle. a is increasing at $2\,\mathrm{cm\,s^{-1}}$ and b is increasing at $3\,\mathrm{cm\,s^{-1}}$. Calculate the rate of change of (a) the area and (b) the hypotenuse when $a = 5$ and $b = 3$.

66   Show that the total surface area S of a closed cone of base radius r cm and perpendicular height h cm is given by

$$S = \pi r^2 + \pi r \sqrt{(r^2 + h^2)}$$

If r and h are each increasing at the rate of $0.25\,\mathrm{cm\,s^{-1}}$, find the rate at which S is increasing at the instant when $r = 3$ and $h = 4$.

67   A particle moves such that its position at time t is given by $\boldsymbol{r} = (t, t^2, t^3)$. Find the rate of change of the distance of the particle from the origin.

68   Find $\partial f / \partial s$ and $\partial f / \partial t$ where

$$f(x, y) = x^2 + 2y^2$$

and $x = \mathrm{e}^{-s} + \mathrm{e}^{-t}$ and $y = \mathrm{e}^{-s} - \mathrm{e}^{-t}$.

9.7.6 Successive differentiation

Consider the function $f(x, y)$ with partial derivatives $\partial f / \partial x$ and $\partial f / \partial y$. In general, these partial derivatives will themselves be functions of x and y, and thus may themselves be differentiated to yield second derivatives. We write

$$\frac{\partial}{\partial x}\left(\frac{\partial f}{\partial x}\right) = \frac{\partial^2 f}{\partial x^2} = f_{xx}$$

$$\frac{\partial}{\partial y}\left(\frac{\partial f}{\partial x}\right) = \frac{\partial^2 f}{\partial y \partial x} = \frac{\partial}{\partial y}(f_x) = f_{xy}$$

$$\frac{\partial}{\partial x}\left(\frac{\partial f}{\partial y}\right) = \frac{\partial^2 f}{\partial x \partial y} = \frac{\partial}{\partial x}(f_y) = f_{yx}$$

and

$$\frac{\partial}{\partial y}\left(\frac{\partial f}{\partial y}\right) = \frac{\partial^2 f}{\partial y^2} = f_{yy}$$

There are some functions for which the mixed second derivatives are not equal, that is

$$\frac{\partial^2 f}{\partial x \partial y} \neq \frac{\partial^2 f}{\partial y \partial x}$$

and the order of differentiation is therefore important, but for most of the functions that occur in engineering problems, when the second derivatives are usually continuous functions, these mixed derivatives are the same in value. In a similar manner we can define higher-order partial derivatives

$$\frac{\partial^{m+n} f}{\partial x^m \partial y^n}$$

Example 9.33   Find the second partial derivatives of $f(x, y) = x^2 y^3 + 3y + x$.

Solution   We found in Example 9.29 that

$$\frac{\partial f}{\partial x} = 2xy^3 + 1 \quad \text{and} \quad \frac{\partial f}{\partial y} = 3x^2 y^2 + 3$$

Differentiating again, we obtain

$$\frac{\partial}{\partial x}\left(\frac{\partial f}{\partial x}\right) = \frac{\partial^2 f}{\partial x^2} = 2y^3, \quad \frac{\partial}{\partial y}\left(\frac{\partial f}{\partial x}\right) = \frac{\partial^2 f}{\partial y \partial x} = 6xy^2$$

$$\frac{\partial}{\partial y}\left(\frac{\partial f}{\partial y}\right) = \frac{\partial^2 f}{\partial y^2} = 6x^2 y, \quad \frac{\partial}{\partial x}\left(\frac{\partial f}{\partial y}\right) = \frac{\partial^2 f}{\partial x \partial y} = 6xy^2$$

Note that in this example

$$\frac{\partial^2 f}{\partial x \partial y} = \frac{\partial^2 f}{\partial y \partial x}$$

Example 9.34 Find the second partial derivatives of

$$f(x, y, z) = xyz^2 + 3xy - z$$

Solution In Example 9.28 we obtained the first partial derivatives as

$$f_x = \frac{\partial f}{\partial x} = yz^2 + 3y, \quad f_y = \frac{\partial f}{\partial y} = xz^2 + 3x, \quad f_z = \frac{\partial f}{\partial z} = 2xyz - 1$$

Differentiating again, we obtain

$$\frac{\partial}{\partial x}\left(\frac{\partial f}{\partial x}\right) = \frac{\partial^2 f}{\partial x^2} = f_{xx} = 0, \quad \frac{\partial}{\partial y}\left(\frac{\partial f}{\partial x}\right) = \frac{\partial^2 f}{\partial y \partial x} = f_{xy} = z^2 + 3$$

$$\frac{\partial}{\partial z}\left(\frac{\partial f}{\partial x}\right) = \frac{\partial^2 f}{\partial z \partial x} = f_{xz} = 2yz, \quad \frac{\partial}{\partial x}\left(\frac{\partial f}{\partial y}\right) = \frac{\partial^2 f}{\partial x \partial y} = f_{yx} = z^2 + 3$$

$$\frac{\partial}{\partial y}\left(\frac{\partial f}{\partial y}\right) = \frac{\partial^2 f}{\partial y^2} = f_{yy} = 0, \quad \frac{\partial}{\partial z}\left(\frac{\partial f}{\partial y}\right) = \frac{\partial^2 f}{\partial z \partial y} = f_{yz} = 2xz$$

$$\frac{\partial}{\partial x}\left(\frac{\partial f}{\partial z}\right) = \frac{\partial^2 f}{\partial x \partial z} = f_{zx} = 2yz, \quad \frac{\partial}{\partial y}\left(\frac{\partial f}{\partial z}\right) = \frac{\partial^2 f}{\partial y \partial z} = f_{zy} = 2xz$$

$$\frac{\partial}{\partial z}\left(\frac{\partial f}{\partial z}\right) = \frac{\partial^2 f}{\partial z^2} = f_{zz} = 2xy$$

Note that, as expected,

$$f_{xy} = f_{yx}, \quad f_{xz} = f_{zx} \quad \text{and} \quad f_{yz} = f_{zy}$$

Example 9.35 $f(x, y)$ is a function of two variables x and y that we wish to change to variables s and t, where

$$s = x^2 - y^2, \quad t = xy$$

Determine f_{xx} and f_{yy} in terms of s, t, f_s, f_t, f_{ss}, f_{tt} and f_{st}. Show that

$$f_{xx} + f_{yy} = \sqrt{(s^2 + 4t^2)}(4f_{ss} + f_{tt})$$

Solution Using the chain rule,

$$f_x = \frac{\partial f}{\partial x} = \frac{\partial f}{\partial s}\frac{\partial s}{\partial x} + \frac{\partial f}{\partial t}\frac{\partial t}{\partial x} = 2x\frac{\partial f}{\partial s} + y\frac{\partial f}{\partial t}$$

$$f_y = \frac{\partial f}{\partial y} = \frac{\partial f}{\partial s}\frac{\partial s}{\partial y} + \frac{\partial f}{\partial t}\frac{\partial t}{\partial y} = -2y\frac{\partial f}{\partial s} + x\frac{\partial f}{\partial t}$$

Differentiating f_x with respect to x gives

$$f_{xx} = \frac{\partial}{\partial x}\left(2x\frac{\partial f}{\partial s} + y\frac{\partial f}{\partial t}\right)$$

$$= 2\frac{\partial f}{\partial s} + 2x\frac{\partial}{\partial x}\left(\frac{\partial f}{\partial s}\right) + y\frac{\partial}{\partial x}\left(\frac{\partial f}{\partial t}\right) \quad \text{(using the product rule)}$$

Repeated use of the chain rule as indicated above leads to

$$f_{xx} = 2\frac{\partial f}{\partial s} + 2x\left[\frac{\partial}{\partial s}\left(\frac{\partial f}{\partial s}\right)\frac{\partial s}{\partial x} + \frac{\partial}{\partial t}\left(\frac{\partial f}{\partial s}\right)\frac{\partial t}{\partial x}\right] + y\left[\frac{\partial}{\partial s}\left(\frac{\partial f}{\partial t}\right)\frac{\partial s}{\partial x} + \frac{\partial}{\partial t}\left(\frac{\partial f}{\partial t}\right)\frac{\partial t}{\partial x}\right]$$

$$= 2f_s + 2x(2xf_{ss} + yf_{st}) + y(2xf_{ts} + yf_{tt})$$

which, on assuming $f_{st} = f_{ts}$, gives

$$f_{xx} = 2f_s + 4x^2f_{ss} + y^2f_{tt} + 4xyf_{st} \tag{9.28}$$

Following a similar procedure, we can determine f_{yy}. Differentiating f_y with respect to y gives

$$f_{yy} = \frac{\partial}{\partial y}(-2yf_s + xf_t)$$

$$= -2f_s - 2y\frac{\partial}{\partial y}(f_s) + x\frac{\partial}{\partial y}(f_t)$$

$$= -2f_s - 2y\left[\frac{\partial}{\partial s}(f_s)\frac{\partial s}{\partial y} + \frac{\partial}{\partial t}(f_s)\frac{\partial t}{\partial y}\right] + x\left[\frac{\partial}{\partial s}(f_t)\frac{\partial s}{\partial y} + \frac{\partial}{\partial t}(f_t)\frac{\partial t}{\partial y}\right]$$

$$= -2f_s - 2y(-2yf_{ss} + xf_{st}) + x(-2yf_{ts} + xf_{tt})$$

giving

$$f_{yy} = -2f_s + 4y^2f_{ss} + x^2f_{tt} - 4xyf_{st} \tag{9.29}$$

Adding (9.28) and (9.29), we obtain

$$f_{xx} + f_{yy} = 4(x^2 + y^2)f_{ss} + (x^2 + y^2)f_{tt}$$

$$= (x^2 + y^2)(4f_{ss} + f_{tt})$$

$$= \surd[(x^2 - y^2)^2 + 4x^2y^2](4f_{ss} + f_{tt})$$

which leads to the required result

$$f_{xx} + f_{yy} = \surd(s^2 + 4t^2)(4f_{ss} + f_{tt})$$

9.7.7 Exercises

69 Verify that

$$f(x, y) = \frac{x}{x^2 + y^2}$$

satisfies the equation

$$\frac{\partial^2 f}{\partial x^2} + \frac{\partial^2 f}{\partial y^2} = 0$$

70 Find the value of the constant a if $V(x, y) = x^3 + axy^2$ satisfies

$$\frac{\partial^2 V}{\partial x^2} + \frac{\partial^2 V}{\partial y^2} = 0$$

71 Verify that

$$\frac{\partial^2 f}{\partial x \partial y} = \frac{\partial^2 f}{\partial y \partial x}$$

in the cases

(a) $f(x, y) = x^2 \cos y$ (b) $f(x, y) = \sinh x \cos y$

72 Show that

$$V(x, y, z) = \frac{1}{z} \exp\left(-\frac{x^2 + y^2}{4z}\right)$$

satisfies the differential equation

$$\frac{\partial^2 V}{\partial x^2} + \frac{\partial^2 V}{\partial y^2} = \frac{\partial V}{\partial z}$$

73 Prove that $z = xf(x + y) + yF(x + y)$, where f and F are arbitrary functions, satisfies the equation

$$z_{xx} + z_{yy} = 2z_{xy}$$

74 Show that, if $z = xe^{Kxy}$, where K is a constant, then

$$xz_x - yz_y = z \quad \text{and} \quad xz_{xx} - yz_{xy} = 0$$

75 If $u = ax + by$ and $v = bx - ay$, where a and b are constants, obtain $\partial u/\partial x$ and $\partial v/\partial y$. By expressing x and y in terms of u and v, obtain $\partial x/\partial u$ and $\partial y/\partial v$ and deduce that

$$\frac{\partial u}{\partial x} \frac{\partial x}{\partial u} = \frac{a^2}{a^2 + b^2}$$

$$\frac{\partial v}{\partial y} \frac{\partial y}{\partial v} = \frac{a^2 + b^2}{a^2}$$

Show also that

$$\frac{\partial^2 f}{\partial x \partial y} = ab\left(\frac{\partial^2 f}{\partial u^2} - \frac{\partial^2 f}{\partial v^2}\right) + (b^2 - a^2)\frac{\partial^2 f}{\partial u \partial v}$$

76 Find the values of the constants a and b such that $u = x + ay$, $v = x + by$ transforms

$$9\frac{\partial^2 f}{\partial x^2} - 9\frac{\partial^2 f}{\partial x \partial y} + 2\frac{\partial^2 f}{\partial y^2} = 0$$

into

$$\frac{\partial^2 f}{\partial u \partial v} = 0$$

77 Regarding u and v as functions of x and y and defined by the equations

$$x = e^u \cos v, \qquad y = e^u \sin v$$

show that

(a) $\dfrac{\partial u}{\partial x} \dfrac{\partial x}{\partial u} = \cos^2 v = \dfrac{\partial v}{\partial y} \dfrac{\partial y}{\partial v}$

(b) $\dfrac{\partial^2 z}{\partial x^2} + \dfrac{\partial^2 z}{\partial y^2} = e^{-2u}\left(\dfrac{\partial^2 z}{\partial u^2} - \dfrac{\partial^2 z}{\partial v^2}\right)$

where z is a twice-differentiable function of u and v.

9.7.8 The total differential and small errors

Consider a function $u = f(x, y)$ of two variables x and y. Let Δx and Δy be increments in the values of x and y. Then the corresponding increment in u is given by

$$\Delta u = f(x + \Delta x, y + \Delta y) - f(x, y)$$

We rewrite this as two terms: one showing the change in u due to the change in x, and the other showing the change in u due to the change in y. Thus

$$\Delta u = [f(x + \Delta x, y + \Delta y) - f(x, y + \Delta y)] + [f(x, y + \Delta y) - f(x, y)]$$

Dividing the first bracketed term by Δx and the second by Δy gives

$$\Delta u = \frac{f(x + \Delta x, y + \Delta y) - f(x, y + \Delta x)}{\Delta x}\Delta x + \frac{f(x, y + \Delta y) - f(x, y)}{\Delta y}\Delta y$$

From the definition of the partial derivative, we may approximate this expression by

$$\Delta u \approx \frac{\partial f}{\partial x}\Delta x + \frac{\partial f}{\partial y}\Delta y$$

We define the **differential** du by the equation

$$du = \frac{\partial f}{\partial x}\Delta x + \frac{\partial f}{\partial y}\Delta y \tag{9.30}$$

By setting $f(x, y) = f_1(x, y) = x$ and $f(x, y) = f_2(x, y) = y$ in turn in (9.30), we see that

$$dx = \frac{\partial f_1}{\partial x}\Delta x + \frac{\partial f_1}{\partial y}\Delta y = \Delta x \quad \text{and} \quad dy = \Delta y$$

so that for the independent variables increments and differentials are equal. For the dependent variable we have

$$du = \frac{\partial f}{\partial x}dx + \frac{\partial f}{\partial y}dy \tag{9.31}$$

We see that the differential du is an approximation to the change Δu in $u = f(x, y)$ resulting from small changes Δx and Δy in the independent variables x and y; that is,

$$\Delta u \approx du = \frac{\partial f}{\partial x}dx + \frac{\partial f}{\partial y}dy = \frac{\partial f}{\partial x}\Delta x + \frac{\partial f}{\partial y}\Delta y \tag{9.32}$$

a result illustrated in Figure 9.29.

This extends to functions of as many variables as we please, provided that the partial derivatives exist. For example, for a function of three variables (x, y, z) defined by $u = f(x, y, z)$ we have

$$\Delta u \approx du = \frac{\partial f}{\partial x}dx + \frac{\partial f}{\partial y}dy + \frac{\partial f}{\partial z}dz$$

$$= \frac{\partial f}{\partial x}\Delta x + \frac{\partial f}{\partial y}\Delta y + \frac{\partial f}{\partial z}\Delta z$$

Figure 9.29
Total differential.

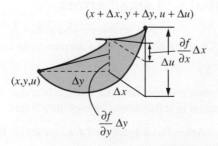

The differential of a function of several variables is often called a **total differential**, emphasizing that it shows the variation of the function with respect to small changes in *all* the independent variables.

Example 9.36   Find the total differential of $u(x, y) = x^2 y^3$.

Solution   Taking partial derivatives we have

$$\frac{\partial u}{\partial x} = 2xy^3 \quad \text{and} \quad \frac{\partial u}{\partial y} = 3x^2 y^2$$

Hence, using (9.31)

$$du = 2xy^3 \, dx + 3x^2 y^2 dy$$

All physical measurements are subject to error, and a calculated quantity usually depends on several measurements. It is very important to know the degree of accuracy that can be relied upon in a quantity that has been calculated. The total differential can be used to estimate error bounds for quantities calculated from experimental results or from data that is subject to errors. This is illustrated in Example 9.37.

Example 9.37   The volume $V \, \text{cm}^3$ of a circular cylinder of radius $r \, \text{cm}$ and height $h \, \text{cm}$ is given by $V = \pi r^2 h$. If $r = 3 \pm 0.01$ and $h = 5 \pm 0.005$ find the greatest possible error in the calculation of V and compare it with the estimate obtained using the total differential.

Solution   The total differential is

$$dV = \frac{\partial V}{\partial r} \, dr + \frac{\partial V}{\partial h} \, dh = 2\pi rh \, dr + \pi r^2 \, dh$$

Then from (9.32)

$$\Delta V \approx dV = 2\pi rh \, dr + \pi r^2 \, dh = \pi r(2h \, \Delta r + r \, \Delta h)$$

When $r = 3$ and $h = 5$, we are given that $\Delta r = \pm 0.01$ and $\Delta h = \pm 0.005$, so that

$$\Delta V \approx \pm 3\pi (10 \times 0.01 + 3 \times 0.005)$$

giving

$$\Delta V \approx \pm 0.345\pi$$

(It should be noted that $\Delta r, \Delta h, \Delta V$ represent maximum errors.) The calculated volume V is subject to a maximum positive error of

$$\{(3.01)^2(5.005) - 45\}\pi = 0.3458\pi$$

and a maximum negative error of

$$\{(2.99)^2(4.995) - 45\}\pi = -0.3442\pi$$

Thus the approximation gives a good guide to the accuracy of the result.

Example 9.38

Two variables, x and y, are related by $y = ae^{-bx}$, where a and b are constants. The values of a and b are determined from experimental data and have relative error bounds p and q respectively. What is the relative error bound for a value of y calculated using the formula with these values of a and b?

Solution

Note that in this example it is assumed that the value of x is known exactly. We are given $y = ae^{-bx}$, where a and b are approximations with errors Δa and Δb, which are unknown but are such that

$$\left|\frac{\Delta a}{a}\right| \le p \quad \text{and} \quad \left|\frac{\Delta b}{b}\right| \le q$$

The formula for the total differential gives

$$dy = \frac{\partial y}{\partial a}da + \frac{\partial y}{\partial b}db$$

For the independent variables a and b the increments and the differentials are the same quantity, so that $da = \Delta a$ and $db = \Delta b$. Also, from the given formula for y, we have

$$\frac{\partial y}{\partial a} = e^{-bx} \quad \text{and} \quad \frac{\partial y}{\partial b} = -xae^{-bx}$$

Thus, from (9.30),

$$dy = e^{-bx}\Delta a - xae^{-bx}\Delta b$$

and division by y gives

$$\frac{dy}{y} = \frac{\Delta a}{a} - bx\frac{\Delta b}{b}$$

Hence

$$\left|\frac{dy}{y}\right| \le \left|\frac{\Delta a}{a}\right| + |bx|\left|\frac{\Delta b}{b}\right| \le p + |bx|q$$

Since $\Delta y \approx dy$, we obtain an estimate for the relative error bound for y as $p + |bx|q$.

9.7.9 Exercises

78 The function z is defined by

$$z(x, y) = x^2y - 3y$$

Find Δz and dz when $x = 4$, $y = 3$, $\Delta x = -0.01$ and $\Delta y = 0.02$.

79 An open box has internal dimensions $2\,\text{m} \times 1.25\,\text{m} \times 0.75\,\text{m}$. It is made of sheet metal 4 mm thick.

Find the actual volume of metal used and compare it with the approximate volume found using the differential of the capacity of the box.

80 The angle of elevation of the top of a tower is found to be $30° \pm 0.5°$ from a point $300 \pm 0.1\,\text{m}$ on a horizontal line through the base of the tower. Estimate the height of the tower.

81 The equations

$$x + 2y + 3z + 4u = -3$$

$$x^2 + y^2 + z^2 + u^2 = 10$$

$$x^3 + y^3 + z^3 + u^3 = 0$$

define u as a function of y if x and z are eliminated. Find du/dy when $x = 1$, $y = -1$, $z = 2$, $u = -2$.

82 The acceleration f of a piston is given by

$$f = r\omega^2\left(\cos\theta + \frac{r}{L}\cos 2\theta\right)$$

When $\theta = \frac{1}{6}\pi$ radians and when $r/L = \frac{1}{2}$, calculate the approximate percentage error in the calculated value of f if the values of both r and ω are 1% too small.

83 The area of a triangle ABC is calculated using the formula

$$S = \tfrac{1}{2}bc\sin A$$

and it is known that b, c and A are measured correctly to within 1%. If the angle A is measured as 45°, prove that the percentage error in the calculated value of S is not more than about 2.8%.

84 The angular deflection θ of a beam of electrons in a cathode-ray tube due to a magnetic field is given by

$$\theta = K\frac{HL}{V^{1/2}}$$

where H is the intensity of the magnetic field, L is the length of the electron path, V is the accelerating voltage and K is a constant. If errors of up to ±0.2% are present in each of the measured H, L and V, what is the greatest possible percentage error in the calculated value of θ (assume that K is known accuratcly)?

85 In a coal processing plant the flow V of slurry along a pipe is given by

$$V = \frac{\pi p r^4}{8\eta l}$$

If r and l both increase by 5%, and p and η decrease by 10% and 30% respectively, find the approximate percentage change in V.

9.7.10 Exact differentials

Differentials sometimes arise naturally when modelling practical problems. An example in fluid dynamics is given in Section 9.10. When this occurs, it is often possible to analyse the problem further by testing to see if the expression in which the differentials occur is a total differential. Consider the equation

$$P(x, y)\,dx + Q(x, y)\,dy = 0$$

connecting x, y and their differentials. The left-hand side of this equation is said to be an **exact differential** if there is a function $f(x, y)$ such that

$$df = P(x, y)\,dx + Q(x, y)\,dy$$

Now we know that

$$df = \frac{\partial f}{\partial x}\,dx + \frac{\partial f}{\partial y}\,dy$$

so if $f(x, y)$ exists then

$$P(x, y) = \frac{\partial f}{\partial x} \quad \text{and} \quad Q(x, y) = \frac{\partial f}{\partial y}$$

For functions with continuous second derivatives we have

$$\frac{\partial^2 f}{\partial x \partial y} = \frac{\partial^2 f}{\partial y \partial x}$$

Thus if $f(x, y)$ exists then

$$\frac{\partial P}{\partial y} = \frac{\partial Q}{\partial x} \qquad\qquad (9.33)$$

This gives us a test for the existence of $f(x, y)$, but does not tell us how to find it! The technique for finding $f(x, y)$ is shown in Example 9.39.

Example 9.39   Show that

$$(6x + 9y + 11)\mathrm{d}x + (9x - 4y + 3)\mathrm{d}y$$

is an exact differential and find the relationship between y and x given

$$\frac{\mathrm{d}y}{\mathrm{d}x} = -\frac{6x + 9y + 11}{9x - 4y + 3}$$

and the condition $y = 1$ when $x = 0$.

Solution   In this example

$$P(x, y) = 6x + 9y + 11 \quad \text{and} \quad Q(x, y) = 9x - 4y + 3$$

First we test whether the expression is an exact differential. In this example

$$\frac{\partial P}{\partial y} = 9 \quad \text{and} \quad \frac{\partial Q}{\partial x} = 9$$

so from (9.33) we have an exact differential. Thus we know that there is a function $f(x, y)$ such that

$$\frac{\partial f}{\partial x} = 6x + 9y + 11, \quad \frac{\partial f}{\partial y} = 9x - 4y + 3 \qquad\qquad \textbf{(9.34), (9.35)}$$

Integrating (9.34) with respect to x, keeping y constant (that is, reversing the partial differentiation process), we have

$$f(x, y) = 3x^2 + 9xy + 11x + g(y) \qquad\qquad \textbf{(9.36)}$$

Note that the 'constant' of integration is a function of y. You can check that this expression for $f(x, y)$ is correct by differentiating it partially with respect to x. But we also know from (9.35) the partial derivative of $f(x, y)$ with respect to y, and this enables us to find $g'(y)$. Differentiating (9.36) partially with respect to y and equating it to (9.35), we have

$$\frac{\partial f}{\partial y} = 9x + \frac{\mathrm{d}g}{\mathrm{d}y} = 9x - 4y + 3$$

(Note that since g is a function of y only we use $\mathrm{d}g/\mathrm{d}y$ rather than $\partial g/\partial y$.) Thus

$$\frac{\mathrm{d}g}{\mathrm{d}y} = -4y + 3$$

so, on integrating,

$$g(y) = -2y^2 + 3y + C$$

Substituting back into (9.36) gives

$$f(x, y) = 3x^2 + 9xy + 11x - 2y^2 + 3y + C$$

Now we are given that

$$\frac{dy}{dx} = -\frac{6x + 9y + 11}{9x - 4y + 3}$$

which implies that

$$(6x + 9y + 11)dx + (9x - 4y + 3)dy = 0$$

which in turn implies that

$$3x^2 + 9xy + 11x - 2y^2 + 3y + A = 0$$

The arbitrary constant A is fixed by applying the given condition $y = 1$ when $x = 0$, giving $A = -1$. Thus x and y satisfy the equation

$$3x^2 + 9xy + 11x - 2y^2 + 3y = 1$$

9.7.11 Exercises

86 Determine which of the following are exact differentials of a function, and find, where appropriate, the corresponding function.

(a) $(y^2 + 2xy + 1)dx + (2xy + x^2)dy$

(b) $(2xy^2 + 3y\cos 3x)dx + (2x^2y + \sin 3x)dy$

(c) $(6xy - y^2)dx + (2xe^y - x^2)dy$

(d) $(z^3 - 3y)dx + (12y^2 - 3x)dy + 3xz^2dz$

87 Find the value of the constant λ such that

$$(y\cos x + \lambda \cos y)dx + (x\sin y + \sin x + y)dy$$

is the exact differential of a function $f(x, y)$. Find the corresponding function $f(x, y)$ that also satisfies the condition $f(0, 1) = 0$.

88 Show that the differential

$$g(x, y) = (10x^2 + 6xy + 6y^2)dx$$
$$+ (9x^2 + 4xy + 15y^2)dy$$

is not exact, but that a constant m can be chosen so that

$$(2x + 3y)^m g(x, y)$$

is equal to dz, the exact differential of a function $z = f(x, y)$. Find $f(x, y)$.

9.8 Taylor's theorem for functions of two variables

In this section we extend Taylor's theorem for one variable (Theorem 9.4) to a function of two variables and apply it to unconstrained and constrained optimization problems.

9.8.1 Taylor's theorem

First we consider a function of two variables. Suppose $f(x, y)$ is a function all of whose nth-order partial derivatives exist and are continuous on some circular domain D with centre (a, b). Then, if $(a + h, b + k)$ lies in D, we have

$$f(a + h, b + k) = f(a, b) + \frac{1}{1!}\left(h\frac{\partial}{\partial x} + k\frac{\partial}{\partial y}\right)f(a, b) + \frac{1}{2!}\left(h\frac{\partial}{\partial x} + k\frac{\partial}{\partial y}\right)^2 f(a, b)$$

$$+ \ldots + \frac{1}{(n - 1)!}\left(h\frac{\partial}{\partial x} + k\frac{\partial}{\partial y}\right)^{n-1} f(a, b)$$

$$+ \frac{1}{n!}\left(h\frac{\partial}{\partial x} + k\frac{\partial}{\partial y}\right)^n f(a + \theta h, b + \theta k) \tag{9.37}$$

where $0 < \theta < 1$. Here we have introduced the notation

$$\left(h\frac{\partial}{\partial x} + k\frac{\partial}{\partial y}\right)^r f(a, b)$$

to represent the value of the expression

$$h^r\frac{\partial^r f}{\partial x^r} + \binom{r}{1}h^{r-1}k\frac{\partial^r f}{\partial x^{r-1}\partial y} + \binom{r}{2}h^{r-2}k^2\frac{\partial^r f}{\partial x^{r-2}\partial y^2} + \ldots$$

$$+ \binom{r}{r-1}hk^{r-1}\frac{\partial^r f}{\partial x\partial y^{r-1}} + k^r\frac{\partial^r f}{\partial y^r}$$

at the point (a, b).

This result is obtained by repeated use of the chain rule. Setting $x = a + ht$ and $y = b + kt$, where $0 \leqslant t \leqslant 1$, we obtain

$$g(t) = f(a + ht, b + kt)$$

which is a function of one variable, so that, from Theorem 9.4, it has a Taylor expansion

$$g(t) = g(0) + \frac{t}{1!}g'(0) + \frac{t^2}{2!}g''(0) + \ldots + \frac{t^{n-1}}{(n - 1)!}g^{(n-1)}(0) + \frac{t^n}{n!}g^{(n)}(\theta t)$$

where $0 \leqslant \theta \leqslant 1$. The derivatives of g are found using the chain rule:

$$g' = \frac{dg}{dt} = \frac{dx}{dt}\frac{\partial f}{\partial x} + \frac{dy}{dt}\frac{\partial f}{\partial y} = h\frac{\partial f}{\partial x} + k\frac{\partial f}{\partial y} = \left(h\frac{\partial}{\partial x} + k\frac{\partial}{\partial y}\right)f$$

$$g'' = \frac{d^2g}{dt^2} = \frac{d}{dt}\left(h\frac{\partial}{\partial x} + k\frac{\partial}{\partial y}\right)f = \left(h\frac{\partial}{\partial x} + k\frac{\partial}{\partial y}\right)\left(h\frac{\partial}{\partial x} + k\frac{\partial}{\partial y}\right)f$$

$$= \left(h\frac{\partial}{\partial x} + k\frac{\partial}{\partial y}\right)^2 f$$

and, in general,

$$g^{(r)} = \frac{d^r g}{dt^r} = \left(h\frac{\partial}{\partial x} + k\frac{\partial}{\partial y} \right)^r f \quad (r = 0, 1, 2, \ldots, n)$$

Putting $t = 1$ into the Taylor expansion of g gives the required result.

The same method can be used to extend the result to as many variables as we please. For the function $f(\mathbf{x})$, where $\mathbf{x} = (x_1, x_2, \ldots, x_n)$, we have

$$f(\mathbf{a} + \mathbf{h}) = f(\mathbf{a}) + \sum_{i=1}^{n} h_i \frac{\partial f}{\partial x_i}(\mathbf{a}) + \frac{1}{2!} \left(\sum_{i=1}^{n} h_i \frac{\partial}{\partial x_i} \right)^2 f(\mathbf{a}) + \ldots$$

$$+ \frac{1}{(m-1)!} \left(\sum_{i=1}^{n} h_i \frac{\partial}{\partial x_i} \right)^{m-1} f(\mathbf{a}) + \frac{1}{m!} \left(\sum_{i=1}^{n} h_i \frac{\partial}{\partial x_i} \right)^{m} f(\mathbf{a} + \theta\mathbf{h}) \quad \text{(9.38)}$$

where $0 \leqslant \theta \leqslant 1$, provided that all the partial derivatives exist and are continuous.

By setting $h = x - a$ and $k = y - b$ in (9.37), we have the following alternative form of the Taylor expansion:

$$f(x, y) = f(a, b) + \frac{1}{1!} \left[(x - a)\frac{\partial}{\partial x} + (y - b)\frac{\partial}{\partial y} \right] f(a, b)$$

$$+ \frac{1}{2!} \left[(x - a)\frac{\partial}{\partial x} + (y - b)\frac{\partial}{\partial y} \right]^2 f(a, b)$$

$$+ \ldots$$

$$+ \frac{1}{n!} \left[(x - a)\frac{\partial}{\partial x} + (y - b)\frac{\partial}{\partial y} \right]^n f(a + \theta(x - a), b + \theta(y - b))$$

$$\text{(9.39)}$$

which is referred to as the **Taylor expansion** of $f(x, y)$ about the point (a, b).

Example 9.40 Obtain the Taylor series of the function $f(x, y) = \sin xy$ about the point $(1, \frac{1}{3}\pi)$, neglecting terms of degree three and higher.

Solution From (9.39) the required series is

$$f(x, y) = f(1, \tfrac{1}{3}\pi) + \frac{1}{1!} \left[(x - 1)\frac{\partial}{\partial x} + (y - \tfrac{1}{3}\pi)\frac{\partial}{\partial y} \right] f(1, \tfrac{1}{3}\pi)$$

$$+ \frac{1}{2!} \left[(x - 1)\frac{\partial}{\partial x} + (y - \tfrac{1}{3}\pi)\frac{\partial}{\partial y} \right]^2 f(1, \tfrac{1}{3}\pi) \ldots$$

Since $f(x, y) = \sin xy$, $f(1, \tfrac{1}{3}\pi) = \dfrac{\sqrt{3}}{2}$. Also,

$$\frac{\partial f}{\partial x} = y\cos xy \qquad\qquad \text{giving} \qquad \left(\frac{\partial f}{\partial x}\right)_{(1,\pi/3)} = \tfrac{1}{6}\pi$$

$$\frac{\partial f}{\partial y} = x\cos xy \qquad\qquad \text{giving} \qquad \left(\frac{\partial f}{\partial y}\right)_{(1,\pi/3)} = \tfrac{1}{2}$$

$$\frac{\partial^2 f}{\partial x^2} = -y^2\sin xy \qquad\qquad \text{giving} \qquad \left(\frac{\partial^2 f}{\partial x^2}\right)_{(1,\pi/3)} = -\tfrac{1}{18}\pi^2\sqrt{3}$$

$$\frac{\partial^2 f}{\partial x\,\partial y} = \cos xy - xy\sin xy \quad \text{giving} \qquad \left(\frac{\partial^2 f}{\partial x\,\partial y}\right)_{(1,\pi/3)} = \tfrac{1}{2} - \tfrac{1}{6}\pi\sqrt{3}$$

$$\frac{\partial^2 f}{\partial y^2} = -x^2\sin xy \qquad\qquad \text{giving} \qquad \left(\frac{\partial^2 f}{\partial y^2}\right)_{(1,\pi/3)} = -\tfrac{1}{2}\sqrt{3}$$

Hence, neglecting terms of degree three and higher,

$$\sin xy \approx \frac{\sqrt{3}}{2} + \tfrac{1}{6}\pi(x-1) + \tfrac{1}{2}(y - \tfrac{1}{3}\pi) - \tfrac{1}{36}\pi^2\sqrt{3}(x-1)^2$$

$$+ (\tfrac{1}{2} - \tfrac{1}{6}\pi\sqrt{3})(x-1)(y - \tfrac{1}{3}\pi) - \tfrac{1}{4}\sqrt{3}(y - \tfrac{1}{3}\pi)^2$$

9.8.2 Optimization of unconstrained functions

In Section 9.2 we considered the problem of determining the maximum and minimum values of a function $f(x)$ of one variable. We now turn our attention to obtaining the maximum and minimum values of a function $f(x, y)$ of two variables. Geometrically $z = f(x, y)$ represents a surface in three-dimensional space, with z being the height of the surface above the x–y plane. Suppose that $f(x, y)$ has a local maximum value at the point (a, b) as illustrated in Figure 9.30(a). Then for all possible (small) values of h and k

$$f(a, b) > f(a + h, b + k)$$

so that the difference (increment)

$$\Delta f = f(a + h, b + k) - f(a, b)$$

is negative. Then, provided that the partial derivatives exist and are continuous, using Taylor's theorem we can express Δf in terms of the partial derivatives of $f(x, y)$ evaluated at (a, b):

$$\Delta f = \left(h\frac{\partial f}{\partial x} + k\frac{\partial f}{\partial y}\right)_{(a,b)} + \frac{1}{2!}\left(h^2\frac{\partial^2 f}{\partial x^2} + 2hk\frac{\partial^2 f}{\partial x\,\partial y} + k^2\frac{\partial^2 f}{\partial y^2}\right)_{(a,b)} + \dots$$

where h and k may be negative or positive numbers. Since h and k are small, the sign of Δf depends on the sign of

$$\left(h\frac{\partial f}{\partial x} + k\frac{\partial f}{\partial y}\right)_{(a,b)}$$

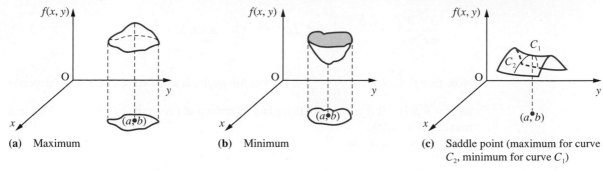

(a) Maximum (b) Minimum (c) Saddle point (maximum for curve C_2, minimum for curve C_1)

Figure 9.30

That is, the sign of Δf depends on the values of h and k. But for a maximum value of $f(x, y)$ at (a, b) the sign of Δf must be negative whatever the values of h and k. This implies that for a maximum to occur at (a, b), $\partial f/\partial x$ and $\partial f/\partial y$ must be zero there.

If $f(x, y)$ has a local minimum at (a, b), as illustrated in Figure 9.30(b), then

$$f(a, b) < f(a + h, b + k)$$

and, using the above argument, we find that for a local minimum to occur at (a, b), $\partial f/\partial x$ and $\partial f/\partial y$ must again be zero.

Thus a first necessary condition for a maximum or a minimum is

$$\frac{\partial f}{\partial x} = \frac{\partial f}{\partial y} = 0 \quad \text{at } (a, b)$$

In terms of differentials, this means that

$$\mathrm{d}f = 0 \quad \text{at } (a, b)$$

Points at which this occurs are called **stationary points** of the function and the values of the function at those points are called its **stationary values**. When this condition is satisfied, we have

$$\Delta f = \frac{1}{2!} \left(h^2 \frac{\partial^2 f}{\partial x^2} + 2hk \frac{\partial^2 f}{\partial x \partial y} + k^2 \frac{\partial^2 f}{\partial y^2} \right)_{(a,b)} + \ldots$$

Putting

$$R = \frac{\partial^2 f}{\partial x^2}, \quad S = \frac{\partial^2 f}{\partial x \partial y} \quad \text{and} \quad T = \frac{\partial^2 f}{\partial y^2}$$

we deduce that the sign of Δf depends on the sign of the second differential

$$\mathrm{d}^2 f = Rh^2 + 2Shk + Tk^2$$

Rearranging, we have, provided that $R \neq 0$,

$$\mathrm{d}^2 f = \frac{1}{R} (R^2 h^2 + 2RShk + RTk^2) = \frac{1}{R} [(Rh + Sk)^2 + (RT - S^2)k^2]$$

If $RT - S^2 > 0$, the sign of $\mathrm{d}^2 f$ is independent of the values of h and k; while if $RT - S^2 < 0$, its sign depends on those values. Thus a second necessary condition for a maximum or minimum value to occur at (a, b) is that

$$\frac{\partial^2 f}{\partial x^2}\frac{\partial^2 f}{\partial y^2} - \left(\frac{\partial^2 f}{\partial x \partial y}\right)^2 = f_{xx}f_{yy} - f_{xy}^2 \geqslant 0 \quad \text{at } (a, b)$$

Note that $f_{xx}f_{yy} - f_{xy}^2 = \begin{vmatrix} f_{xx} & f_{xy} \\ f_{yx} & f_{yy} \end{vmatrix}$. If strict inequality is satisfied, the sign of Δf depends on $R = \partial^2 f/\partial x^2$. If $\partial^2 f/\partial x^2 > 0$, there is a minimum at (a, b). If $\partial^2 f/\partial x^2 < 0$, there is a maximum at (a, b).

By expressing $\mathrm{d}^2 f$ as

$$\mathrm{d}^2 f = \frac{1}{T}[(TK + Sh)^2 + (RT - S^2)h^2], \quad T \neq 0$$

we could equally well have deduced that there is a minimum at (a, b) if $\partial^2 f/\partial y^2 > 0$ and a maximum at (a, b) if $\partial^2 f/\partial y^2 < 0$, assuming the above strict inequality.

If $\quad \dfrac{\partial^2 f}{\partial x^2}\dfrac{\partial^2 f}{\partial y^2} - \left(\dfrac{\partial^2 f}{\partial x \partial y}\right)^2 < 0 \quad \text{at } (a, b)$

then the sign of Δf depends on the values of h and k, and along some paths through (a, b) the function has a maximum value while along other paths it has a minimum value. Such a point is called a **saddle point**, as illustrated in Figure 9.30(c).

Summary

(1) A necessary condition for the function $f(x, y)$ to have a stationary value at (a, b) is that

$$\frac{\partial f}{\partial x} = 0 \quad \text{and} \quad \frac{\partial f}{\partial y} = 0 \quad \text{at } (a, b)$$

(2) If $\quad \dfrac{\partial^2 f}{\partial x^2}\dfrac{\partial^2 f}{\partial y^2} - \left(\dfrac{\partial^2 f}{\partial x \partial y}\right)^2 > 0 \quad \text{and} \quad \dfrac{\partial^2 f}{\partial x^2} \text{ or } \dfrac{\partial^2 f}{\partial y^2} < 0 \quad \text{at } (a, b)$

then the stationary point is a local maximum.

(3) If $\quad \dfrac{\partial^2 f}{\partial x^2}\dfrac{\partial^2 f}{\partial y^2} - \left(\dfrac{\partial^2 f}{\partial x \partial y}\right)^2 > 0 \quad \text{and} \quad \dfrac{\partial^2 f}{\partial x^2} \text{ or } \dfrac{\partial^2 f}{\partial y^2} > 0 \quad \text{at } (a, b)$

then the stationary point is a local minimum.

(4) If $\quad \dfrac{\partial^2 f}{\partial x^2}\dfrac{\partial^2 f}{\partial y^2} - \left(\dfrac{\partial^2 f}{\partial x \partial y}\right)^2 < 0 \quad \text{at } (a, b)$

then the stationary point is a saddle point.

(5) If $\quad \dfrac{\partial^2 f}{\partial x^2}\dfrac{\partial^2 f}{\partial y^2} - \left(\dfrac{\partial^2 f}{\partial x \partial y}\right)^2 = 0 \quad \text{at } (a, b)$

we cannot draw a conclusion, and the point may be a maximum, minimum or saddle point. Further investigation is required, and it may be necessary to consider the third-order terms in the Taylor series.

Example 9.41   Find the stationary points of the function

$$f(x, y) = 2x^3 + 6xy^2 - 3y^3 - 150x$$

and determine their nature.

Solution $\dfrac{\partial f}{\partial x} = 6x^2 + 6y^2 - 150$ and $\dfrac{\partial f}{\partial y} = 12xy - 9y^2$

For a stationary point both of these partial derivatives are zero, which gives

$$x^2 + y^2 = 25$$

and

$$y(4x - 3y) = 0$$

From the second equation we see that either $y = 0$ or $4x = 3y$. Putting $y = 0$ in the first equation gives $x = \pm 5$, so that the points $(5, 0)$ and $(-5, 0)$ are solutions of the equations. Putting $x = \frac{3}{4}y$ into the first equation gives $y = \pm 4$, so that the points $(3, 4)$ and $(-3, -4)$ are also solutions of the equation. Thus the function has stationary points at $(5, 0)$, $(-5, 0)$, $(3, 4)$ and $(-3, -4)$.

Next we have to classify these points as maxima or minima or saddle points. Working out the second derivatives, we have

$$\frac{\partial^2 f}{\partial x^2} = 12x, \quad \frac{\partial^2 f}{\partial y^2} = 12x - 18y \quad \text{and} \quad \frac{\partial^2 f}{\partial x \partial y} = 12y$$

and we can complete the following table.

Point	$\dfrac{\partial^2 f}{\partial x^2}$	$\dfrac{\partial^2 f}{\partial y^2}$	$\dfrac{\partial^2 f}{\partial x \partial y}$	$\dfrac{\partial^2 f}{\partial x^2}\dfrac{\partial^2 f}{\partial y^2} - \left(\dfrac{\partial^2 f}{\partial x \partial y}\right)^2$	*Nature*	*Value*
$(5, 0)$	60	60	0	positive	minimum	−500
$(-5, 0)$	−60	−60	0	positive	maximum	500
$(3, 4)$	36	−36	48	negative	saddle point	−300
$(-3, -4)$	−36	36	−48	negative	saddle point	300

The situation is shown quite clearly on the contour plot (level curves) of the function shown in Figure 9.31. Looking at the figure, we see that the contours distinguish clearly between saddle points and other stationary points, as indicated in Figure 9.32.

Figure 9.31
Contour plot of
$f(x, y) = 2x^3 + 6xy^2$
$- 3y^3 - 150x$.

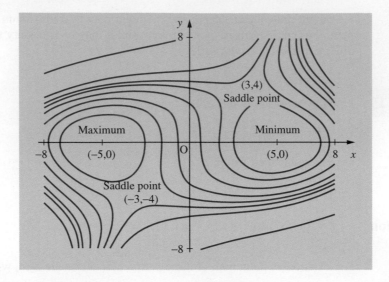

Figure 9.32
Nature of stationary
points: (a) saddle and
(b) maximum or
minimum.

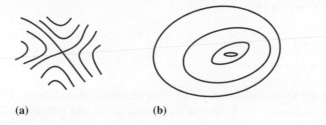

(a)   (b)

The process indicated above can be extended to functions of as many variables as we please. At a stationary point the first differential df, is zero, so that all the first partial derivatives are zero there. If, at that stationary point, the second differential d^2f is negative for all small changes in the independent variables then we have a maximum. If it is positive, we have a minimum. If it is zero, further analysis is required. However, the general conditions for this to occur are extremely complicated both to write down and to apply.

9.8.3 Exercises

89 Find the stationary values (and their classification) of

(a) $x^3 - 15x^2 - 20y^2 + 5$

(b) $2 - x^2 - xy - y^2$

(c) $2x^2 + y^2 + 3xy - 3y - 5x + 2$

(d) $x^3 + y^2 - 3(x + y) + 1$

(e) $xy^2 - 2xy - 2x^2 - 3x$

(f) $x^3y^2(1 - x - y)$

(g) $x^2 + y^2 + \dfrac{2}{x} + \dfrac{2}{y}$

90 Prove that $(x + y)/(x^2 + 2y^2 + 6)$ has a maximum at $(2, 1)$ and a minimum at $(-2, -1)$.

91 Show that

$$f(x, y) = x^3 + y^3 - 2(x^2 + y^2) + 3xy$$

has stationary values at $(0, 0)$ and $(\tfrac{1}{3}, \tfrac{1}{3})$ and investigate their nature.

92 A manufacturer produces an article in batches of N items. Each production run has a set-up cost of £100 and each item costs an additional £0.05 to produce. The weekly storage costs are a basic rental of £50 plus an additional £0.10 per item stored. Assuming that there is a steady sale of n items per week, so that the average number of items stored is $\frac{1}{2}N$, and that, when the store is exhausted, it is immediately replenished by a new production run, show that the weekly cost £K is given by

$$K = 50 + 0.05N + 0.05n + \frac{100n}{N}$$

The weekly demand n is a function of the selling price £p, and

$$n = 5000 - 10\,000p$$

Show that the weekly profit £P is

$$P = (5000 - 10\,000p)\left(p - 0.05 - \frac{100}{N}\right)$$
$$- 0.05N - 50$$

If the manufacturer is able to decide both the batch size N and the price £p, show that a maximum weekly profit is realized where $p = 0.3$, and find the corresponding values of N, n and P.

93 The gravitational attraction at the point (x, y) in the x–y plane due to point masses in the plane is

$$G(x, y) = \frac{1}{x} + \frac{4}{y} + \frac{9}{4 - x - y}$$

Show that $G(x, y)$ has a stationary value of 9.

94 Find constants a and b such that

$$\int_0^\pi [\sin x - (ax^2 + bx)]^2 \, dx$$

is a minimum.

95 A rectangular tank open at the top has a volume of $4\,\text{m}^3$. If the base measurements (in m) are x by y, show that the surface area (in m^2) is given by

$$A = xy + \frac{8}{y} + \frac{8}{x}$$

and find the dimensions of the tank for A to be a minimum.

96 A flat circular metal plate has a shape defined by the region $x^2 + y^2 \leq 1$. The plate is heated so that the temperature T at any point (x, y) on it is given by

$$T = x^2 + 2y^2 - x$$

Find the temperatures at the hottest and coldest points on the plate and the points where they occur. (*Hint*: consider the level curves of T.)

97 A metal channel is formed by turning up the sides of width x of a rectangular sheet of metal through an angle θ. If the sheet is $200\,\text{mm}$ wide, determine the values of x and θ for which the cross-section of the channel will be a maximum.

9.8.4 Optimization of constrained functions

As we have seen in Exercises 9.8.3, Questions 92 and 95–97, there are frequent situations in engineering applications when we wish to obtain the stationary values of functions of more than one variable and for which the variables themselves are subject to one or more constraint conditions. The general theory for these applications is discussed in the companion text *Advanced Modern Engineering Mathematics*. Here we will show the technique for solving such problems.

Example 9.42   Obtain the extremum value of the function

$$f(x, y) = 2x^2 + 3y^2$$

subject to the constraint $2x + y = 1$.

Solution   In this particular example it is easy to eliminate one of the two variables x and y. Eliminating y, we can write $f(x, y)$ as

$$f(x, y) = f(x) = 2x^2 + 3(1 - 2x)^2 = 14x^2 - 12x + 3$$

We can now apply the techniques used for functions of one variable to obtain the extremum value. Differentiating gives

$$f'(x) = 28x - 12 \quad \text{and} \quad f''(x) = 28$$

An extremal value occurs when $f'(x) = 0$; that is, $x = \frac{3}{7}$, and, since $f''(\frac{3}{7}) > 0$, this corresponds to a minimum value. Thus the extremum is a minimum $f_{min} = \frac{3}{7}$ at $x = \frac{3}{7}$, $y = \frac{1}{7}$.

In Example 9.42 we were fortunate in being able to use the constraint equation to eliminate one of the variables. In practice, however, it is often difficult, or even imposs-ible, to do this, and we have to retain all the original variables. Let us consider the general problem of obtaining the stationary points of $f(x, y, z)$ subject to the constraint $g(x, y, z) = 0$. We shall refer to such points as **conditional stationary points**.

At stationary points of $f(x, y, z)$ we have

$$df = \frac{\partial f}{\partial x} dx + \frac{\partial f}{\partial y} dy + \frac{\partial f}{\partial z} dz = 0 \tag{9.40}$$

This implies that the vector $(\partial f/\partial x, \partial f/\partial y, \partial f/\partial z)$ is perpendicular to the vector (dx, dy, dz). Since $g(x, y, z) = 0$

$$dg = \frac{\partial g}{\partial x} dx + \frac{\partial g}{\partial y} dy + \frac{\partial g}{\partial z} dz = 0 \tag{9.41}$$

Thus, the vector $(\partial g/\partial x, \partial g/\partial y, \partial g/\partial z)$ is also perpendicular to the vector (dx, dy, dz). This implies that the vector $(\partial f/\partial x, \partial f/\partial y, \partial f/\partial z)$ is parallel to the vector $(\partial g/\partial x, \partial g/\partial y, \partial g/\partial z)$ and that we can find a number λ such that

$$\left(\frac{\partial f}{\partial x}, \frac{\partial f}{\partial y}, \frac{\partial f}{\partial z} \right) - \lambda \left(\frac{\partial g}{\partial x}, \frac{\partial g}{\partial y}, \frac{\partial g}{\partial z} \right) = (0, 0, 0) \tag{9.42}$$

Geometrically this means that the level surface of the objective function $f(x, y, z)$ touches the constraint surface $g(x, y, z) = 0$ at the stationary point.

> This can be neatly summarized by writing $\phi(x, y, z) = f(x, y, z) - \lambda g(x, y, z)$. Then $f(x, y, z)$ will have a stationary point subject to the constraint $g(x, y, z) = 0$ when
>
> $$\frac{\partial \phi}{\partial x} = \frac{\partial \phi}{\partial y} = \frac{\partial \phi}{\partial z} = 0 \quad \text{and} \quad g(x, y, z) = 0 \tag{9.43}$$

This gives four equations to determine $(x, y, z; \lambda)$ for the stationary point. The scalar multiplier λ is called a **Lagrange multiplier** and the function $\phi(x, y, z)$ is called the **auxiliary function**.

Example 9.43　　Rework Example 9.42 using the method of Lagrange multipliers.

Solution　　Here we need to obtain the extremum of the function

$$f(x, y) = 2x^2 + 3y^2$$

Figure 9.33
Level curves of
$f(x, y) = 2x^2 + 3y^2$.

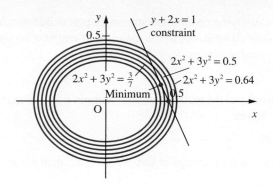

subject to the constraint

$$g(x, y) = 2x + y - 1 = 0$$

The auxiliary function is

$$\phi(x, y, z) = f(x, y) - \lambda g(x, y)$$
$$= 2x^2 + 3y^2 - \lambda(2x + y - 1)$$

and we find that the conditional extrema of $f(x, y)$ are given by

$$\frac{\partial \phi}{\partial x} = \frac{\partial \phi}{\partial y} = 0, \quad g(x, y) = 0$$

that is

$$\frac{\partial \phi}{\partial x} = 4x - 2\lambda = 0 \qquad \text{(9.44)}$$

$$\frac{\partial \phi}{\partial y} = 6y - \lambda = 0 \qquad \text{(9.45)}$$

$$g(x, y) = 2x + y - 1 = 0 \qquad \text{(9.46)}$$

Solving (9.44)–(9.46) gives

$$\lambda = \tfrac{6}{7}, \quad x = \tfrac{3}{7} \quad \text{and} \quad y = \tfrac{1}{7}$$

so that the conditional extremal value of $f(x, y)$ is $\tfrac{3}{7}$ and occurs at $x = \tfrac{3}{7}, y = \tfrac{1}{7}$.

It is clear from the level curves of $f(x, y)$, shown in Figure 9.33, that the function has a minimum at $(\tfrac{3}{7}, \tfrac{1}{7})$. In general, however, to determine the nature of the conditional stationary point, we have to resort to Taylor's theorem and consider the sign of the difference $f(x + h, y + k) - f(x, y)$. Taking a point near $(\tfrac{3}{7}, \tfrac{1}{7})$, say $(\tfrac{3}{7} + h, \tfrac{1}{7} + k)$, that still satisfies the constraint $2x + y - 1 = 0$, we have $2h + k = 0$, so that $k = -2h$. Hence a near point satisfying the constraint is $(\tfrac{3}{7} + h, \tfrac{1}{7} - 2h)$, and thus

$$f(\tfrac{3}{7} + h, \tfrac{1}{7} - 2h) - f(\tfrac{3}{7}, \tfrac{1}{7}) = 2(\tfrac{3}{7} + h)^2 + 3(\tfrac{1}{7} - 2h)^2 - [2(\tfrac{3}{7})^2 + 3(\tfrac{1}{7})^2]$$

$$= 14h^2 > 0$$

Since this is positive, it follows that the point is a minimum, confirming the result of Example 9.42.

In general, classifying conditional stationary points into maxima, minima or saddle points can be very difficult, but in the majority of engineering applications this can be done using physical reasoning.

Example 9.44   Find the dimensions of the rectangular box, without a top, of maximum capacity whose surface area is $108\,\mathrm{m}^2$.

Solution   If the dimensions of the rectangular box are $x \times y \times z$ then we are required to maximize

$$f(x, y, z) = xyz$$

subject to the constraint

$$xy + 2xz + 2yz = 108 \tag{9.47}$$

The auxiliary function is

$$\phi(x, y, z) = xyz + \lambda(xy + 2xz + 2yz - 108)$$

and the equations we have to solve are

$$\frac{\partial \phi}{\partial x} = yz + \lambda(y + 2z) = 0 \tag{9.48}$$

$$\frac{\partial \phi}{\partial y} = xz + \lambda(x + 2z) = 0 \tag{9.49}$$

$$\frac{\partial \phi}{\partial z} = xy + \lambda(2x + 2y) = 0 \tag{9.50}$$

together with (9.47).
    Taking $x \times$ (9.48) $+ y \times$ (9.49) $+ z \times$ (9.50) gives

$$3xyz + \lambda(2xy + 4xz + 4yz) = 0$$

or

$$\lambda(xy + 2xz + 2yz) + \tfrac{3}{2}xyz = 0 \tag{9.51}$$

Then, from (9.51) and (9.47),

$$108\lambda + \tfrac{3}{2}xyz = 0$$

or

$$\lambda = -\tfrac{1}{72}xyz$$

Substituting into (9.48)–(9.50) in succession and dividing throughout by common factors gives

$$1 - \tfrac{1}{72}x(y + 2z) = 0 \tag{9.52}$$
$$1 - \tfrac{1}{72}y(x + 2z) = 0 \tag{9.53}$$
$$1 - \tfrac{1}{72}z(2x + 2y) = 0 \tag{9.54}$$

Subtracting (9.53) from (9.52) gives

$$\tfrac{1}{36}yz - \tfrac{1}{36}xz = 0 \quad \text{or} \quad y = x \quad \text{(since clearly } z \neq 0\text{)}$$

Figure 9.34
Level surface of
objective function
touches constraint
surface at (6,6,3).

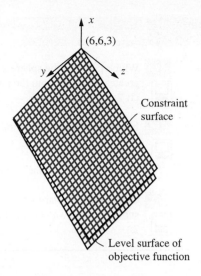

Putting this into (9.54), we have

$$1 - \tfrac{1}{18}yz = 0, \quad \text{or} \quad yz = 18$$

Substituting this and $x = y$ into (9.52) gives

$$1 - \tfrac{1}{72}y^2 - \tfrac{1}{2} = 0$$

that is

$$y^2 = 36, \quad \text{or} \quad y = 6 \text{ (since } y > 0)$$

It then follows that $x = 6$, $z = 3$. Thus the required dimensions are $6\,\text{m} \times 6\,\text{m} \times 3\,\text{m}$, and it follows from physical considerations that this corresponds to the maximum volume, since the minimum volume is zero (Figure 9.34).

The Lagrange multiplier method outlined above may be extended to a function of any number of variables. It also extends naturally to situations where there is more than one constraint equation by introducing the equivalent number of Lagrange multipliers. In general, if $f(x_1, x_2, \ldots, x_n)$ is a function of n variables subject to m ($< n$) constraints

$$g_i(x_1, \ldots, x_n) = 0 \quad (i = 1, 2, \ldots, m)$$

then, to determine the constrained stationary values of $f(x_1, x_2, \ldots, x_n)$, the procedure is to set up the auxiliary function

$$\phi(x_1, x_2, \ldots, x_n) = f(x_1, x_2, \ldots, x_n) + \lambda_1 g_1 + \ldots + \lambda_m g_m$$

and solve the resulting $m + n$ equations

$$\frac{\partial \phi}{\partial x_j} = \frac{\partial f}{\partial x_j} + \sum_{i=1}^{m} \lambda_i \frac{\partial g_i}{\partial x_j} = 0 \quad (j = 1, 2, \ldots, n)$$

$$\frac{\partial \phi}{\partial \lambda_i} = g_i = 0 \quad (i = 1, 2, \ldots, m)$$

9.8.5 Exercises

98 Find the extremum of $x^2 - 2y^2 + 2xy + 4x$ subject to the constraint $2x = y$ and verify that it is a maximum value.

99 Find the extremum of $3x^2 + 2y^2 + 6z^2$ subject to the constraint $x + y + z = 1$ and verify that it is a minimum value.

100 The equation $5x^2 + 6xy + 5y^2 - 8 = 0$ represents an ellipse whose centre is at the origin. By considering the extrema of $x^2 + y^2$, obtain the lengths of the semi-axes.

101 Which point on the sphere $x^2 + y^2 + z^2 = 1$ is at the greatest distance from the point having coordinates $(1, 2, 2)$?

102 Find the maximum and minimum values of

$$f(x, y) = 4x + y + y^2$$

where (x, y) lies on the circle $x^2 + y^2 + 2x + y = 1$.

103 Obtain the stationary value of $2x + y + 2z + x^2 - 3z^2$ subject to the two constraints $x + y + z = 1$ and $2x - y + z = 2$.

9.9 Engineering application: deflection of a built-in column

In this section we consider an example in which the techniques developed in Section 9.5 may be used to solve an engineering problem.

The deflection $y(x)$ of a column buckling under its own weight satisfies the differential equation

$$EI\frac{d^3y}{dx^3} + wx\frac{dy}{dx} = 0 \qquad\qquad (9.55)$$

where E is Young's modulus, I the second moment of area of the cross-section and w is the weight per unit run of the column. The deflection of the built-in column shown in Figure 9.35 also satisfies the conditions

$$\frac{d^2y}{dx^2} = 0 \quad \text{at } x = 0$$

and

$$y = 0 \quad \text{and} \quad \frac{dy}{dx} = 0 \quad \text{at } x = l$$

where l is the length of the column. We need to find the greatest height attainable for the column without collapse.

To make the algebraic manipulations easier, we first simplify the differential equation. Putting $x = ct$ gives

$$\frac{dy}{dx} = \frac{dy}{dt}\frac{dt}{dx} = \frac{1}{c}\frac{dy}{dt}$$

$$\frac{d^2y}{dx^2} = \frac{d}{dt}\left(\frac{1}{c}\frac{dy}{dt}\right)\frac{dt}{dx} = \frac{1}{c^2}\frac{d^2y}{dt^2}$$

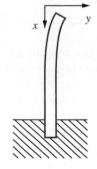

Figure 9.35
Deflection of a column.

and

$$\frac{d^3y}{dx^3} = \frac{d}{dt}\left(\frac{1}{c^2}\frac{d^2y}{dt^2}\right)\frac{dt}{dx} = \frac{1}{c^3}\frac{d^3y}{dt^3}$$

which on substituting into (9.55) transforms it to

$$\frac{EI}{c^3}\frac{d^3y}{dt^3} + wt\frac{dy}{dt} = 0$$

so that choosing $c^3 = EI/w$ and setting $f(t) = dy/dt$ simplifies the equation further to

$$\frac{d^2f}{dt^2} + tf = 0 \tag{9.56}$$

with the conditions

$$\frac{df}{dt} = 0 \quad \text{at } t = 0 \quad \text{and} \quad f(t) = 0 \quad \text{at } t = l(EI/w)^{-1/3} = T$$

Assuming that $f(t)$ has a Maclaurin series expansion, we may write it as

$$f(t) = a_0 + a_1 t + a_2 t^2 + a_3 t^3 + \ldots + a_n t^n + \ldots$$

Differentiating this, we have

$$f'(t) = a_1 + 2a_2 t + 3a_3 t^2 + \ldots + na_n t^{n-1} + (n+1)a_{n+1}t^n + \ldots$$

and

$$f''(t) = 2a_2 + 6a_3 t + 12a_4 t^2 + \ldots + n(n-1)a_n t^{n-2} + \ldots$$

Since $f'(0) = 0$, we deduce at once that $a_1 = 0$. Since $f(t)$ satisfies the differential equation (9.56), we deduce on substitution that

$$2a_2 + 6a_3 t + 12a_4 t^2 + \ldots + n(n-1)a_n t^{n-2} + \ldots$$
$$= -a_0 t - a_1 t^2 - a_2 t^3 - \ldots - a_n t^{n+1} - \ldots$$

This expression is true for all values of t, with $0 < t < T$, so we deduce from Property (i) of polynomials given in Section 2.4.1 that the coefficients of each power of t on each side of the equation are equal. That is,

$$2a_2 = 0 \qquad \text{(coefficient of } t^0\text{)}$$

$$6a_3 = -a_0 \qquad \text{(coefficient of } t^1\text{)}$$

$$12a_4 = -a_1 \qquad \text{(coefficient of } t^2\text{)}$$

and so on. In general, the coefficient of t^r (obtained by setting $n - 2 = r$ on the left-hand side and $n + 1 = r$ on the right-hand side) yields

$$(r + 2)(r + 1)a_{r+2} = -a_{r-1}$$

This recurrence relation enables us to calculate a_{r+3} in terms of a_r as

$$a_{r+3} = \frac{-a_r}{(r+3)(r+2)} \quad (r = 0, 1, 2, \ldots) \tag{9.57}$$

Thus

$$a_3 = \frac{-a_0}{3 \cdot 2} \quad (r = 0), \quad a_4 = \frac{-a_1}{4 \cdot 3} \quad (r = 1)$$

$$a_5 = \frac{-a_2}{5 \cdot 4} \quad (r = 2), \quad a_6 = \frac{-a_3}{6 \cdot 5} \quad (r = 3)$$

$$a_7 = \frac{-a_4}{7 \cdot 6} \quad (r = 4), \quad a_8 = \frac{-a_5}{8 \cdot 7} \quad (r = 5)$$

and so on.

Since we deduced earlier, using the condition $f'(0) = 0$, that $a_1 = 0$, some terms can be eliminated immediately, and we have

$$a_4 = 0, \quad a_7 = 0, \quad a_{10} = 0, \quad a_{13} = 0, \quad \ldots$$

Since $a_2 = 0$ (from the coefficient of t^0), we have

$$a_5 = 0, \quad a_8 = 0, \quad \ldots$$

We are therefore left with

$$f(t) = a_0 + a_3 t^3 + a_6 t^6 + a_9 t^9 + \ldots$$

Substituting for $a_3, a_6, a_9, \ldots$ in terms of a_0, using (9.57) gives

$$f(t) = a_0 \left(1 - \frac{1}{3!} t^3 + \frac{1 \cdot 4}{6!} t^6 - \frac{1 \cdot 4 \cdot 7}{9!} t^9 + \frac{1 \cdot 4 \cdot 7 \cdot 10}{12!} t^{12} - \ldots \right)$$

So far we have only applied the condition at $t = 0$. Now we apply the condition at $t = T$, namely $f(T) = 0$. This gives

$$a_0 \left(1 - \frac{1}{3!} T^3 + \frac{4}{6!} T^6 - \frac{4 \cdot 7}{9!} T^9 \ldots \right) = 0$$

so that either

$$a_0 = 0 \quad \text{or} \quad 1 - \frac{1}{3!} T^3 + \frac{4}{6!} T^6 - \frac{4 \cdot 7}{9!} T^9 + \ldots = 0$$

This means that there is no deflection ($a_0 = 0$) unless

$$1 - \frac{1}{3!} T^3 + \frac{4}{6!} T^6 - \frac{4 \cdot 7}{9!} T^9 + \ldots = 0$$

The smallest value of T that satisfies this equation gives the critical height of the column. At that height the value of a_0 becomes arbitrary (and non-zero), and the column buckles. A first approximation to the critical value of T can be found by solving the quadratic equation (in T^3)

$$1 - \frac{1}{3!} T^3 + \frac{4}{6!} T^6 = 1 - \tfrac{1}{6} T^3 + \tfrac{1}{180} (T^3)^2 = 0$$

giving $T^3 = 8.292$. This may be refined using the Newton–Raphson procedure (9.19), eventually giving the critical length L in terms of E, I and w:

$$L = 1.99 (EI/w)^{1/3}$$

The detailed calculation is left as an exercise for the reader.

9.10 Engineering application: streamlines in fluid dynamics

As we mentioned in Section 9.7.10, differentials often occur in mathematical modelling of practical problems. An example occurs in fluid dynamics. Consider the case of steady-state incompressible fluid flow in two dimensions. Using rectangular cartesian coordinates (x, y) to describe a point in the fluid, let u and v be the velocities of the fluid in the x and y directions respectively. Then by considering the flow in and flow out of a small rectangle, as shown in Figure 9.36, per unit time, we obtain a differential relationship between $u(x, y)$ and $v(x, y)$ that models the fact that no fluid is lost or gained in the rectangle; that is, the fluid is conserved.

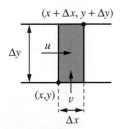

The velocity of the fluid $\boldsymbol{q}$ is a vector point function. The values of its components u and v depend on the spatial coordinates x and y. The flow into the small rectangle in unit time is

$$u(x, \bar{y})\Delta y + v(\bar{x}, y)\Delta x$$

Figure 9.36
Flow through rectangular element.

where $\bar{x}$ lies between x and $x + \Delta x$, and $\bar{y}$ lies between y and $y + \Delta y$. Similarly, the flow out of the rectangle is

$$u(x + \Delta x, \tilde{y})\Delta y + v(\tilde{x}, y + \Delta y)\Delta x$$

where $\tilde{x}$ lies between x and $x + \Delta x$ and $\tilde{y}$ lies between y and $y + \Delta y$. Because no fluid is created or destroyed within the rectangle, we may equate these two expressions, giving

$$u(x, \bar{y})\Delta y + v(\bar{x}, y)\Delta x = u(x + \Delta x, \tilde{y})\Delta y + v(\tilde{x}, y + \Delta y)\Delta x$$

Rearranging, we have

$$\frac{u(x + \Delta x, \tilde{y}) - u(x, \bar{y})}{\Delta x} + \frac{v(\tilde{x}, y + \Delta y) - v(\bar{x}, y)}{\Delta y} = 0$$

Letting $\Delta x \to 0$ and $\Delta y \to 0$ gives the **continuity equation**

$$\frac{\partial u}{\partial x} + \frac{\partial v}{\partial y} = 0$$

The fluid actually flows along paths called **streamlines** so that there is no flow across a streamline. Thus from Figure 9.37 we deduce that

$$v\,\Delta x = u\,\Delta y$$

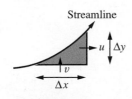

and hence

$$v\,\mathrm{d}x - u\,\mathrm{d}y = 0$$

Figure 9.37
Streamline.

The condition for this expression to be an exact differential is

$$\frac{\partial}{\partial y}(v) = \frac{\partial}{\partial x}(-u)$$

or

$$\frac{\partial u}{\partial x} + \frac{\partial v}{\partial y} = 0$$

This is satisfied for incompressible flow since it is just the continuity equation, so that we deduce that there is a function $\psi(x, y)$, called the **stream function**, such that

$$v = \frac{\partial \psi}{\partial x} \quad \text{and} \quad u = -\frac{\partial \psi}{\partial y}$$

It follows that if we are given u and v, as functions of x and y, that satisfy the continuity equation then we can find the equations of the streamlines given by $\psi(x, y) = \text{constant}$.

Example 9.45

Find the stream function $\psi(x, y)$ for the incompressible flow that is such that the velocity q at the point (x, y) is

$$(-y/(x^2 + y^2), x/(x^2 + y^2))$$

Solution

From the definition of the stream function, we have

$$u(x, y) = -\frac{\partial \psi}{\partial y} \quad \text{and} \quad v(x, y) = \frac{\partial \psi}{\partial x}$$

provided that

$$\frac{\partial u}{\partial x} + \frac{\partial v}{\partial y} = 0$$

Here we have

$$u = \frac{-y}{x^2 + y^2} \quad \text{and} \quad v = \frac{x}{x^2 + y^2}$$

so that

$$\frac{\partial u}{\partial x} = \frac{2xy}{(x^2 + y^2)^2} \quad \text{and} \quad \frac{\partial v}{\partial y} = -\frac{2yx}{(x^2 + y^2)^2}$$

confirming that

$$\frac{\partial u}{\partial x} + \frac{\partial v}{\partial y} = 0$$

Integrating

$$\frac{\partial \psi}{\partial y} = -u(x, y) = \frac{y}{x^2 + y^2}$$

with respect to y, keeping x constant, gives

$$\psi(x, y) = \tfrac{1}{2} \ln(x^2 + y^2) + g(x)$$

Differentiating partially with respect to x gives

$$\frac{\partial \psi}{\partial x} = \frac{x}{x^2 + y^2} + \frac{dg}{dx}$$

Figure 9.38
A vortex.

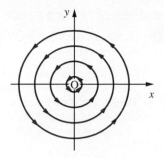

Since it is known that

$$\frac{\partial \psi}{\partial x} = v(x, y) = \frac{x}{x^2 + y^2}$$

we have

$$\frac{\mathrm{d}g}{\mathrm{d}x} = 0$$

which on integrating gives

$$g(x) = C$$

where C is a constant. Substituting back into the expression obtained for $\psi(x, y)$, we have

$$\psi(x, y) = \tfrac{1}{2} \ln(x^2 + y^2) + C$$

A streamline of the flow is given by the equation $\psi(x, y) = k$, where k is a constant. After a little manipulation this gives

$$x^2 + y^2 = a^2 \quad \text{and} \quad \ln a = k - C$$

and the corresponding streamlines are shown in Figure 9.38. This is an example of a **vortex**.

9.11 Review exercises (1–39)

1 Find the turning points on the curve

$$y = 2x^3 - 5x^2 + 4x - 1$$

and determine their nature. Find the point of inflection and sketch the graph of the curve.

2 Use the Newton–Raphson method to find the root of

$$e^x - x^2 + 3x - 2 = 0$$

in the interval $0 \leqslant x \leqslant 1$. Start with $x = 0.5$ and give the root correct to 4dp.

3 The deflection at the midpoint of a uniform beam of length l, flexural rigidity EI and weight per unit length w, subject to an axial force P, is

$$d = \frac{w}{m^2 P}(\sec \tfrac{1}{2}ml - 1) - \frac{wl^2}{8P}$$

where $m^2 = P/EI$. On making the substitution $\theta = \tfrac{1}{2}ml$, show that

$$d = \frac{wl^4}{32EI} \frac{2 \sec \theta - 2 - \theta^2}{\theta^4}$$

As the force P is relaxed, the deflection should reduce to that of a beam sagging under its own weight. By first representing $\sec\theta$ by its Maclaurin series expansion, show that

$$\lim_{\theta \to 0} d = \frac{5wl^4}{384EI}$$

4 Using the Maclaurin series expansion of e^x, determine the Maclaurin series expansion of $x/(e^x - 1)$ as far as the term in x^4, and hence obtain the approximation

$$\int_0^1 \frac{x}{e^x - 1}\, dx \approx \frac{311}{400}$$

5 Use L'Hôpital's rule to find

$$\lim_{x \to 1} \frac{\ln x}{x^2 - 1}$$

6 Determine

$$\lim_{x \to 2} \frac{2 \sin kx - x \sin 2k}{2(4 - x^2)}$$

where k is a constant.

7 Show that the equation

$$x^3 - 2x - 5 = 0$$

has a root in the neighbourhood of $x = 2$ and find it to three significant figures using the Newton–Raphson method.

8 The turning moment T on the crankshaft of an engine is given by

$$T = 6 + 2.5 \sin 2\theta - 3.8 \cos 2\theta$$

Find the maximum and minimum values of T for $0 \le \theta \le 2\pi$.

9 The deflection of a beam of length L is given by

$$y = wx^2 \frac{(L - x)^2}{EI} \quad (0 \le x \le L)$$

where w, E and I are constants. Determine

(a) the maximum deflection,

(b) the points along the beam at which points of inflection lie.

10 (a) Obtain the Maclaurin series expansions of $\sinh x$ and $\cosh x$.

(b) A telegraph wire is stretched between two poles at the same height and a distance $2l$ apart. The sag at the midpoint is h. If the axes are taken as shown in Figure 9.39, it can be shown that the equation of the curve followed by the wire is

$$y = c \cosh \frac{x}{c}$$

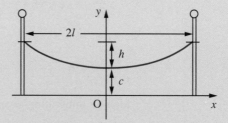

Figure 9.39 Telegraph wire of Question 10.

where c is an undetermined constant (see Example 8.45).

(i) Show that the length $2s$ of the wire is given by

$$2s = 2c \sinh \frac{l}{c}$$

(ii) If the wire is taut, so that h/c is small, it can be shown that l/c is also small. Ignoring powers of l/c higher than the second, show that

$$\frac{h^2}{l^2} \approx \frac{1}{4}\left(\frac{l}{c}\right)^2$$

Hence show that the length of the wire is approximately

$$2l\left[1 + \frac{2}{3}\left(\frac{h}{l}\right)^2\right]$$

11 A running track is set out in the form of a rectangle, of length L and width W, with two semicircular areas, of radius $\frac{1}{2}W$, adjoined at each end of the rectangle. If the perimeter of the whole track is fixed at 400 m, determine the values of L and W that maximize the area of the rectangle.

12 Find the maximum and minimum values of y where

$$y = \frac{x^2}{(x-2)(x-6)}$$

justifying your answers. Sketch the curve, indicating the stationary points and any asymptotes.

13 Prove that

$$\int_0^\infty \operatorname{sech} x \, dx = \pi$$

and deduce $\int_0^1 \operatorname{sech}^{-1} x \, dx$.

14 Evaluate

(a) $\displaystyle\int_1^\infty \frac{1}{x^3} \, dx$ (b) $\displaystyle\int_0^\infty \frac{1}{x^2 + 2x + 2} \, dx$

(c) $\displaystyle\int_0^\infty x e^{-4x} \, dx$ (d) $\displaystyle\int_1^\infty \frac{\ln x}{x^3} \, dx$

(e) $\displaystyle\int_0^\infty e^{-2x} \cos x \, dx$ (f) $\displaystyle\int_0^\infty e^{-2x} \cosh x \, dx$

15 Evaluate

(a) $\displaystyle\int_0^8 x^{-1/3} \, dx$ (b) $\displaystyle\int_{3/2}^6 \frac{1}{\sqrt{(2x-3)}} \, dx$

(c) $\displaystyle\int_0^1 \ln x \, dx$

stating in each case the value of x for which the integrand becomes unbounded.

16 The function $F(r)$ is defined by

$$F(r) = \int_0^{\pi/2} \sin^r x \, dx \quad r > -1$$

By considering $d(\cos x \sin^{r-1} x)/dx$, or otherwise, show that

$$(r+1)\int \sin^r x \, dx = \cos x \sin^{r+1} x + (r+2)\int \sin^{r+2} x \, dx$$

and deduce that $(r+1)F(r) = (r+2)F(r+2)$.
   Show that $F(-\frac{1}{2}) = \frac{21}{5}F(\frac{7}{2})$. Tabulate $f(x) = \sin^{7/2} x$ for $x = 0(\frac{1}{8}\pi)\frac{1}{2}\pi$ to 3dp and use the values to obtain three approximations to $F(3.5)$

using the trapezium rule with strips of width $\frac{1}{2}\pi$, $\frac{1}{4}\pi$ and $\frac{1}{8}\pi$ respectively. Hence obtain an approximation to $F(-0.5)$.

17 Light sources are placed at two fixed points Q and R which are 1 metre apart. The source at R is twice as intense as that at Q. The total illumination at a point P on the line QR distant x metres from Q is $cf(x)$ where c is a positive constant and

$$f(x) = \frac{1}{x^2} + \frac{2}{(1-x)^2} \quad 0 < x < 1$$

Evaluate $f(0.3), f(0.4)$ and $f(0.5)$ and find the quadratic function

$$g(x) = A(x-0.4)^2 + B(x-0.4) + C$$

which passes through $(0.3, f(0.3))$, $(0.4, f(0.4))$ and $(0.5, f(0.5))$. Use this function to estimate the value of x at which the minimum of $f(x)$ occurs. Compare your result with that obtained by calculus methods.

18 Use the Taylor series to show that the principal term of the truncation error of the approximation

$$f''(a) \approx [f(a+h) - 2f(a) + f(a-h)]/h^2$$

is $\frac{1}{12}h^2 f^{(4)}(a)$.
   Consider the function $f(x) = x e^x$. Estimate $f''(1)$ using the approximation above with $h = 0.01$, and $h = 0.02$. Compare your answer with the true value.

19 A particle moves in three-dimensional space such that its position at time t (seconds) is given by the vector $(4\cos t, 4\sin t, 3)$ where distance is measured in metres. Find the magnitude of its velocity and acceleration.

20 The acceleration $\mathbf{a}$ (m s^{-2}) of a particle at time t (s) is given by $\mathbf{a} = (1+t)\mathbf{i} + t^2\mathbf{j} + 2\mathbf{k}$. At $t = 0$ its displacement $\mathbf{r}$ is zero and its velocity $\mathbf{v}$ (m s^{-1}) is $\mathbf{i} - \mathbf{j}$. Find its displacement at time t.

21 The temperature gradient u at a point in a solid is

$$u(x, t) = t^{-1/2} e^{-x^2/4kt}$$

where k is a constant. Verify that

$$\frac{\partial^2 u}{\partial x^2} = \frac{1}{k} \frac{\partial u}{\partial t}$$

22 Show that the surfaces defined by

$$z^2 = \tfrac{1}{2}(x^2 + y^2) - 1$$

and

$$z = 1/xy$$

intersect, and that they do so orthogonally.

23 The height h of the top of a pylon is calculated by measuring its angle of elevation α at a point a distance s horizontally from the base of the pylon. Find the error in h due to small errors in s and α. If s and α are taken as 20 m and 30° respectively when the correct values are 19.8 m and 30.2°, find the error and the proportional error in the calculated height.

24 The resistance of a length of wire is given by

$$R = \frac{k\rho L}{D^2}$$

where k is a constant. L is increasing at a rate of 0.4% min^{-1}, ρ is increasing at a rate of 0.01% min^{-1} and D is decreasing at a rate of 0.1% min^{-1}. At what percentage rate is the resistance R increasing?

25 The deflection H of a metal structure can be calculated using the formula

$$H = \sqrt{\left(\frac{I\rho^4 D^2 L^{3/2}}{20g} \right)}$$

where I, ρ, D and L are the moment of inertia, density, diameter and length respectively, and g is the acceleration due to gravity. If the value of H is to remain unaltered when I increases by 0.1%, ρ by 0.2% and D decreases by 0.3%, what percentage change in L is required?

26 In the calculation of the power in an a.c. circuit using the formula $W = EI \cos \phi$, errors of +1% in I, −0.7% in E and +2% in ϕ occur. Find the percentage error in the calculated value of W when $\phi = \tfrac{1}{3}\pi$ rad.

27 (a) Prove that $u = x^3 - 3xy^2$ satisfies

$$\frac{\partial^2 u}{\partial x^2} + \frac{\partial^2 u}{\partial y^2} = 0$$

(b) Given

$$u = x^2 \tan^{-1}\left(\frac{y}{x}\right) - y^2 \tan^{-1}\left(\frac{x}{y}\right)$$

evaluate

$$x\frac{\partial u}{\partial x} + y\frac{\partial u}{\partial y}$$

in terms of u.

28 Verify that $z = \ln \sqrt{(x^2 - y^2)}$ satisfies the equation

$$\left(\frac{\partial z}{\partial x}\right)^2 + \frac{\partial^2 z}{\partial y \partial x} + \left(\frac{\partial z}{\partial y}\right)^2 = \frac{1}{(x - y)^2}$$

29 (a) Find the value of the positive constant c for which the function

$$y = \frac{k}{2\pi} \sin\left(\frac{\pi x}{k}\right) \sin\left(\frac{2\pi t}{k}\right)$$

satisfies the equation

$$c^2 \frac{\partial^2 y}{\partial x^2} = \frac{\partial^2 y}{\partial t^2}$$

(b) V is a function of the independent variables x and y. Given that $x = r \cos \theta$ and $y = r \sin \theta$, find $\partial V / \partial \theta$ and $\partial V / \partial r$ in terms of $\partial V / \partial x$ and $\partial V / \partial y$, and hence show that

$$\frac{\partial V}{\partial y} = \frac{1}{r}\left(r \sin \theta \frac{\partial V}{\partial r} + \cos \theta \frac{\partial V}{\partial \theta} \right)$$

and

$$\frac{\partial V}{\partial x} = \frac{1}{r}\left(r \cos \theta \frac{\partial V}{\partial r} - \sin \theta \frac{\partial V}{\partial \theta} \right)$$

30 A curve C in three dimensions is given parametrically by $(x(t), y(t), z(t))$, where t is a real parameter, with $a \leqslant t \leqslant b$. Show that the equation of the tangent line at a point P on this curve where $t = t_0$ is given by

$$\frac{x - x_0}{x_0'} = \frac{y - y_0}{y_0'} = \frac{z - z_0}{z_0'}$$

where $x_0 = x(t_0)$, $x_0' = x'(t_0)$, and so on.

Hence find the equation of the tangent line to the circular helix

$$x = a \cos t, \quad y = a \sin t, \quad z = at$$

at $t = \frac{1}{4}\pi$ and show that the length of the helix between $t = 0$ and $t = \frac{1}{2}\pi$ is $\pi a/\sqrt{2}$.

31 Show that $u = f(x + y) + g(x - y)$ satisfies the differential equation

$$\frac{\partial^2 u}{\partial x^2} - \frac{\partial^2 u}{\partial y^2} = 0$$

32 Show that if

$$\phi(x, t) = \frac{f(z)}{\sqrt{t}} \quad \text{and} \quad z = \frac{x}{2\sqrt{t}}$$

then

$$\frac{\partial \phi}{\partial t} = -\frac{zf'(z) + f(z)}{2t\sqrt{t}}$$

and find a similar expression for $\partial^2\phi/\partial x^2$.
Deduce that if

$$\frac{\partial^2 \phi}{\partial x^2} = \frac{1}{k}\frac{\partial \phi}{\partial t}$$

then

$$kf''(z) + 2zf'(z) + 2f(z) = 0$$

33 Water waves move in the direction of the x axis with speed c. Their height h at time t is given by

$$h(t) = a\sin(x - ct)$$

where a is a constant. A small cork floats on the water and is blown by the wind in the direction of the x axis with constant velocity U. Show that the vertical acceleration of the cork at time t is given by

$$\frac{\mathrm{d}^2 h}{\mathrm{d}t^2} = -(U - c)^2 h$$

34 The components of velocity of an inviscid incompressible fluid in the x and y directions are u and v respectively, where

$$u = \frac{x^2 - y^2}{(x^2 + y^2)^2} \quad \text{and} \quad v = \frac{2xy}{(x^2 + y^2)^2}$$

Find the stream function $\psi(x, y)$ such that

$$\mathrm{d}\psi = v\,\mathrm{d}x - u\,\mathrm{d}y$$

and verify that it satisfies Laplace's equation

$$\frac{\partial^2 \psi}{\partial x^2} + \frac{\partial^2 \psi}{\partial y^2} = 0$$

35 Show that the function

$$f(x, y) = x^2 y^2 - 5x^2 - 8xy - 5y^2$$

has one maximum and four saddle points. Sketch the part of the surface $z = f(x, y)$ that lies in the first quadrant.

36 Determine the position and nature of the stationary points on the surface

$$z = \mathrm{e}^{-(x+y)}(3x^2 + y^2)$$

37 A trough of capacity $1\,\mathrm{m}^3$ is to be made from sheet metal in the shape shown in Figure 9.40. Calculate the dimensions that use the least amount of metal. *Hint*: Set $y = xY$ and $z = xZ$ and show that the area of sheet metal needed is

$$\frac{2(1 + Y\cos\theta)Y\sin\theta + (2Y + 1)Z}{[(1 + Y\cos\theta)YZ\sin\theta]^{2/3}}$$

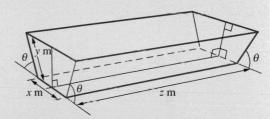

Figure 9.40 Trough of Question 37.

38 Find the critical points of the function

$$z = 12xy - 3xy^2 - x^3$$

and identify the character of each point.

39 Find the local maxima and minima of the function

$$f(x, y) = y^2 - 8x + 17$$

subject to the constraint

$$x^2 + y^2 = 9$$

10 Introduction to Ordinary Differential Equations

Chapter 10   Contents

10.1 Introduction

The essential role played by mathematical models in both engineering analysis and engineering design has been noted earlier in this book. It often happens that, in creating a mathematical model of a physical system, we need to express such relationships as 'the acceleration of A is directly proportional to B' or 'changes in D produce proportionate changes in E with constant of proportionality F'. Such statements naturally give rise to equations involving derivatives and integrals of the variables in the model as well as the variables themselves. Equations which introduce derivatives are called **differential equations**, those which introduce integrals are called **integral equations** and those which introduce both are called **integro-differential equations**. Generally speaking, integral and integro-differential equations are rather more difficult to solve than purely differential ones. This chapter starts with a discussion of the general characteristics of differential equations and then deals with ways of solving first-order differential equations. It is concluded by an examination of the solution of differential equations of second and higher orders.

Firstly, though, we will give some example of engineering problems which give rise to differential equations. As we will see, two of these (Sections 10.2.2 and 10.2.3) give rise to first-order equations and the other two (Sections 10.2.1 and 10.2.4) give rise to second-order equations. There are many other systems of engineering importance which give rise to second- or even higher-order differential equations. Section 10.10 develops an important engineering application of second-order constant-coefficient differential equations – the analysis of vibrations or oscillations. This analysis has very wide applications in engineering practice. A complete appreciation of this material requires that readers are already familiar with the material in Sections 10.8 and 10.9 but those who wish to have some understanding of the engineering importance of second-order constant-coefficient differential equations before beginning their study of the mathematical material may read quickly through that section before beginning their detailed study of the chapter.

10.2 Engineering examples

10.2.1 The take-off run of an aircraft

Aeronautical engineers need to be able to predict the length of runway that an aircraft will require to take off safely. To do this, a mathematical model of the forces acting on the aircraft during the take-off run is constructed, and the relationships holding between the forces are identified. Figure 10.1 shows an aircraft and the forces acting on it. If the mass of the aircraft is m, gravity causes a downward force mg. There is a ground reaction force through the wheels, denoted by G, and an aerodynamic lift force L. The engines provide a thrust T, which is opposed by an aerodynamic drag D and a rolling resistance from contact with the ground R. Since the aircraft is rolling along the runway, it is not accelerating vertically, so the vertical forces are in balance and the vertical equation of motion yields

$$L + G = mg \tag{10.1}$$

Figure 10.1
Forces on an
aircraft during
the take-off run.

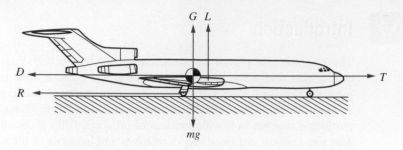

On the other hand, the aircraft is accelerating along the runway, so the horizontal equation of motion is

$$T - D - R = m\frac{d^2s}{dt^2} \qquad (10.2)$$

where s is the distance the aircraft has travelled along the runway.

We know from experimental evidence that both aerodynamic lift and aerodynamic drag forces on a body vary roughly as the square of the velocity of the airflow relative to the body. We shall therefore choose to model the lift and drag forces as proportional to velocity squared. The rolling resistance is also known to be roughly proportional to the reaction force between ground and aircraft. Thus we make the modelling assumptions

$$L = \alpha v^2, \quad D = \beta v^2 \quad \text{and} \quad R = \mu G$$

Substituting for L, D and R in (10.1) and (10.2) and eliminating G results in the equation

$$m\frac{d^2s}{dt^2} - (\mu\alpha - \beta)v^2 + \mu mg = T$$

or, replacing v by ds/dt,

$$m\frac{d^2s}{dt^2} - (\mu\alpha - \beta)\left(\frac{ds}{dt}\right)^2 = T - \mu mg \qquad (10.3)$$

Thus our model of the aircraft travelling along the runway provides an equation relating the first and second time derivatives of the distance travelled by the aircraft, the thrust provided by the engines and various constants – the model is expressed as a differential equation for the distance s travelled along the runway. The model is not yet really complete, since we have not specified how the thrust varies. The thrust could, of course, vary with time (the pilot could open or close the throttles during the take-off run), and may also vary with the forward speed of the aircraft. On the other hand, we could just assume that thrust is constant. Also, the constants m, μ, α and β need to be determined. This information might be provided by measurements on the aircraft or on scale models of it, by other calculations or by engineers' estimates.

Once the model is complete it could be used, for instance, to predict the length of runway needed by the aircraft to attain flying speed. Flying speed is, of course, the speed at which the lift (αv^2) is equal to the weight (mg) of the aircraft. For a real aircraft our model would probably need to be made more elaborate, including, for instance, the angle of attack of the wing, which would change during the take-off run as the balance between aerodynamic and ground forces changed and as the pilot (or autopilot) changed the control surface settings.

10.2.2 Domestic hot-water supply

The second example involves modelling the heating of water in a hot-water storage tank. Figure 10.2 shows schematically an 'indirect' domestic hot-water tank. In this design of a hot-water system the central heating boiler, or other primary source of heat, supplies hot water to a calorifier (which takes the form of a coiled pipe) inside the hot-water storage tank. The main mass of water in the tank is then heated by the hot water passing through the calorifier coil. We wish to calculate how quickly the hot water in the tank will heat up.

Figure 10.2
An 'indirect'
hot-water tank.

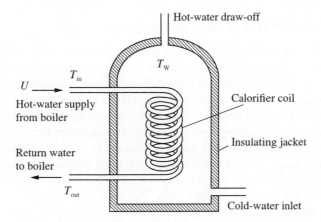

We shall assume that, to a good approximation, during heating convection ensures that the main mass of water in the tank is well mixed and at a uniform temperature T_w. The heating water flows into the calorifier at a speed U at a temperature T_{in}. The outflow from the calorifier is at temperature T_{out}. The cross-sectional area of the calorifier tube is A. The mass flowrate of heating water through the calorifier is therefore ρAU, where ρ is the density of water, and the rate of heat loss from the heating water is $\rho AU(T_{in} - T_{out})c$, where c is the specific heat of water. The heat capacity of the main mass of water in the tank is ρVc, where V is the volume of the tank, and so the rate of gain of heat in the main mass of water is given by

$$\rho Vc \frac{dT_w}{dt}$$

The tank is well insulated, so, to a first approximation, we shall assume that the heat loss from the external shell of the tank is negligible. The rate of heat gain of the main mass of water is therefore equal to the rate of heat loss from the heating water; that is,

$$AU(T_{in} - T_{out}) = V\frac{dT_w}{dt} \tag{10.4}$$

where it is assumed that no hot water is being drawn off.

We should also expect that the difference in temperature of the heating water flowing in and that flowing out of the calorifier will be greater the cooler the mass of water in the tank. If we assume direct proportionality of these two quantities, we may express this modelling assumption as

$$T_{in} - T_{out} = \alpha(T_{in} - T_w) \tag{10.5}$$

where α is a constant of proportionality. Eliminating T_{out} between (10.4) and (10.5) leads to the equation

$$V\frac{dT_w}{dt} + AU\alpha T_w = AU\alpha T_{in} \tag{10.6}$$

Thus we have a differential equation relating the temperature of the water in the tank and its derivative with respect to time to the temperature of the heating water supplied by the boiler. The equation also involves various constants determined by the characteristics of the system.

10.2.3 Hydro-electric power generation

Our third example is drawn from the sphere of hydraulic engineering. Figure 10.3 shows a cross-section through a hydro-power generation plant. Water, retained behind a dam, is drawn off through a conduit and drives a generator. In order to control the power generated, there is also a control valve in series with the generator. The conduit from the dam to the generator is typically quite long and of considerable cross-section, so that it contains many tonnes of water. Hence, when the control valve is opened or closed, the power generated does not increase or decrease instantaneously. Because of the large mass of water in the conduit that must be accelerated or decelerated, the system may take several minutes or even tens of minutes to attain its new equilibrium flowrate and power generation level. We wish to predict the behaviour of the system when the control valve setting is changed.

Figure 10.3
A hydro-electric
generation plant.

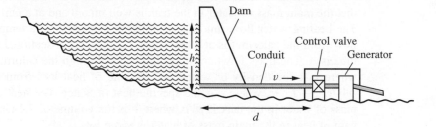

The pressure at the entry to the conduit will be atmospheric plus $\rho g h$, where ρ is the density of the water in the dam and h is the depth of the entry below the water surface. It is known that for flow in pipes, to a good approximation, the volume flowrate is proportional to the pressure differential between the ends of the pipe. We shall express this as

$$Q = \alpha \Delta p_1$$

where Q is the volume flowrate through the conduit, α is a constant and Δp_1 is the pressure difference between the two ends of the conduit. It is also known that the pressure loss across a turbine such as the generator in this case is proportional to the discharge (volume flow through the turbine), so we can write

$$\Delta p_2 = \beta Q$$

where Δp_2 is the pressure loss across the generator and β is a characteristic of the generator. The discharge of the turbine must, of course, be equal to the flowrate through the conduit feeding the turbine. In a similar way, the pressure differential across a control valve is also proportional to its discharge, so we have

$$\Delta p_3 = \gamma Q$$

where Δp_3 is the pressure loss across the valve and γ is a constant whose value will vary with the setting of the control valve. The total pressure differential between the entry to the conduit and the exit from the control valve is ρgh. Hence the pressure differential between the ends of the conduit is $\rho gh - \Delta p_2 - \Delta p_3$. If this exceeds Δp_1, the pressure differential needed to maintain the flow through the conduit at its current level, then the mass of water in the conduit will accelerate and the volume flowrate through the system will increase; if it is less than Δp_1 then the mass of water will decelerate and the volume flow will decrease. The net force on the mass of water in the conduit is the excess pressure differential multiplied by the cross-sectional area of the conduit, A, say. The mass of water is $\rho A d$, where d is the length of the conduit, and Q, the volume flowrate, is vA, where v is the velocity of the water in the conduit. Thus we can write

$$(\rho gh - \Delta p_1 - \Delta p_2 - \Delta p_3)A = \rho dA \frac{dv}{dt}$$

Assuming that the cross-sectional area of the conduit is constant and substituting for Δp_1, Δp_2 and Δp_3, we can rewrite this as

$$\left(\rho gh - \frac{Q}{\alpha} - \beta Q - \gamma Q\right)A = \rho d \frac{dQ}{dt}$$

that is,

$$\frac{\rho d}{A}\frac{dQ}{dt} + \left(\frac{1}{\alpha} + \beta + \gamma\right)Q = \rho gh \tag{10.7}$$

We find that this simple model of the hydro-power generation system results in an equation involving the volume flowrate through the system and its time derivative and, of course, various constants expressing physical characteristics of the system. One of these constants, γ, is determined by the setting of the valve controlling the whole system.

10.2.4 Simple electrical circuits

The fourth example comes from electrical engineering. A resistor, an inductor and a capacitor are connected in a series circuit with a switch and battery as shown in Figure 10.4. The switch is a spring-biased one that, when released, moves immediately on to contact B. While the switch is held against contact A, a current flows in the circuit. When it is released, the circuit must eventually become quiescent, with no current flowing. What is the manner of the decay to the quiescent state?

We know from experiment that the relation $V = iR$ holds between the potential difference across the resistor and the current flowing through a pure resistor of resistance R. In the same way, we know that for a pure capacitor of capacitance C we have $V = q/C$, where V is the potential difference across the capacitor and q is the charge on it, and that for a pure inductor of inductance L we have $V = L\,di/dt$. If we assume that the circuit components are a pure resistor, inductor and capacitor respectively, and that the switch and the wires joining the components have negligible resistance, capacitance

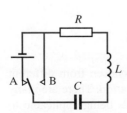

Figure 10.4
An inductor, capacitor, resistor (LCR) electrical circuit.

and inductance, then, when the switch is in contact with B, the total potential difference around the circuit must be zero and we have

$$L\frac{di}{dt} + Ri + \frac{q}{C} = 0$$

This differential equation appears to relate two different quantities: the current i flowing in the circuit and the charge q on the capacitor. Of course, these two quantities are not independent. If the current is flowing then the charge on the capacitor must be increasing or decreasing (depending on the direction in which the current is flowing). The principle of conservation of charge tells us that the current is equal to the rate of change of charge; that is, we must have

$$i = \frac{dq}{dt} \tag{10.8}$$

We can use this in one of two ways: either to eliminate q, in which case we obtain the integro-differential equation

$$L\frac{di}{dt} + Ri + \frac{1}{C}\int i\,dt = 0$$

or to eliminate i, in which case we obtain the differential equation

$$L\frac{d^2q}{dt^2} + R\frac{dq}{dt} + \frac{1}{C}q = 0$$

Alternatively, differentiating either of these equations with respect to time, we obtain

$$L\frac{d^2i}{dt^2} + R\frac{di}{dt} + \frac{1}{C}i = 0 \tag{10.9}$$

The equations are, of course, equivalent, but the final form is probably the most usual and most tractable of the three.

Thus we have found that a simple analysis of an *LCR* electrical circuit results in a differential equation for one of the variables: either the charge on the capacitor in the circuit or the current in the circuit. Once the equation has been solved to yield one of these, the other can be obtained from (10.8).

10.3 The classification of differential equations

In Section 10.2 we created mathematical models of problems chosen from different areas of engineering science. Each gave rise to a differential equation. There are many techniques for solving differential equations – different methods being applicable to different kinds of equation – so, before we go on to study these methods, it is necessary to understand the various categories and classifications of differential equations. We shall then be in a position to recognize the overall characteristics of an equation and identify which techniques will be useful in its solution.

10.3.1 Ordinary and partial differential equations

We have already noted that the characteristic defining a differential equation is that it involves not only algebraic combinations of the variables occurring in the equation but also derivatives of those variables with respect to other variables. We have met, in Chapter 8, the idea of differentiation of a function with respect to a single variable and in Chapter 9, the idea of partial differentiation. Differential equations may involve either ordinary or partial derivatives. Those equations involving ordinary derivatives are called **ordinary differential equations** while those involving partial derivatives are called **partial differential equations**. The equations (10.3), (10.6), (10.7) and (10.9) derived in the examples in Section 10.2 are all ordinary differential equations, and in this chapter we shall be concentrating our attention on such equations. This is not because partial differential equations are not important to engineers – on the contrary, they have many applications in engineering science – but because there are significant differences in the methods and techniques used for their solution. Partial differential equations are considered in the companion text *Advanced Modern Engineering Mathematics*.

Example 10.1

$$\frac{\partial f}{\partial x} + \frac{\partial f}{\partial y} = 4x^2 + 2y$$

is a partial differential equation whereas

$$\frac{d^2 f}{dx^2} - 4x\frac{df}{dx} = \cos 2x$$

is an ordinary differential equation.

10.3.2 Independent and dependent variables

The next type of classification we must understand is that of the variables occurring in a differential equation. The variables with respect to which differentiation occurs are called **independent variables** while those that are differentiated are **dependent variables**. This terminology reflects the fact that what a differential equation actually expresses is the way in which the dependent variable (or variables) depends on the independent variable (or variables). Ordinary differential equations have only one independent variable, whereas partial differential equations have two or more. A single ordinary differential equation will usually have one independent variable and one dependent variable. In much the same way as algebraic equations may occur in sets that must be solved simultaneously, we can also have sets of coupled ordinary differential equations. In this case there will be a single independent variable but more than one dependent variable.

Example 10.2

In the partial differential equation

$$\frac{\partial f}{\partial x} + \frac{\partial f}{\partial y} - 4x^2 + 2y$$

the independent variables are x and y and the dependent variable is f. In the ordinary differential equation

$$\frac{d^2 f}{dx^2} - 4x\frac{df}{dx} = \cos 2x$$

the independent variable is x and the dependent variable is f. In the pair of coupled ordinary differential equations

$$4\frac{dx}{dt} + 3\frac{dy}{dt} - x + 2y = \cos t$$

$$6\frac{dx}{dt} - 2\frac{dy}{dt} - 2x + y = 2\sin t$$

the independent variable is t and the dependent variables are x and y.

10.3.3 The order of a differential equation

Another classification of differential equations is in terms of their order. The **order of a differential equation** is the degree of the highest derivative that occurs in the equation. In the case of partial differential equations the degree of a mixed derivative is the total of the degrees of differentiation with respect to each of the independent variables. The order of an equation is not affected by any power to which the derivatives may be raised.

Example 10.3

$$\frac{\partial f}{\partial x} + \frac{\partial f}{\partial y} = 4x^2 + 2y$$

is a first-order partial differential equation.

$$\frac{\partial^3 f}{\partial x \partial y^2} + \frac{\partial^2 f}{\partial x^2} - x\frac{\partial f}{\partial y} = 4x^2 + 2y$$

is a third-order partial differential equation.

$$\frac{d^2 f}{dx^2} - 4x\frac{df}{dx} = \cos 2x$$

is a second-order ordinary differential equation. The coupled ordinary differential equations

$$4\frac{dx}{dt} + 3\frac{dy}{dt} - x + 2y = \cos t$$

$$6\frac{dx}{dt} - 2\frac{dy}{dt} - 2x + y = 2\sin t$$

are both first-order equations as is the equation

$$\left(\frac{dx}{dt}\right)^2 + 4\frac{dx}{dt} = 0$$

despite the term in $(dx/dt)^2$.

10.3.4 Linear and nonlinear differential equations

Differential equations are also classified as linear or nonlinear. We may informally define **linear equations** as those in which the dependent variable or variables and their derivatives do not occur as products, raised to powers or in nonlinear functions. We shall meet a more formal definition of a linear differential equation in Section 10.8. **Nonlinear equations** are those that are not linear. Linear equations are an important category, since they have useful simplifying properties. Many of the nonlinear equations that occur in engineering science cannot be solved easily as they stand, but can be solved, for practical engineering purposes, by the process of replacing them with linear equations that are a close approximation – at least in some region of interest – and then studying the solution of the linear approximation. We shall see more of this later.

Example 10.4

$$\frac{\partial f}{\partial x} + \frac{\partial f}{\partial y} = 4x^2 + 2y$$

and

$$\frac{\partial^3 f}{\partial x \partial y^2} = 4x^2 + 2y$$

are both linear partial differential equations.

$$\frac{\mathrm{d}^2 f}{\mathrm{d}x^2} - 4x\frac{\mathrm{d}f}{\mathrm{d}x} = \cos 2x$$

and the coupled differential equations

$$4\frac{\mathrm{d}x}{\mathrm{d}t} + 3\frac{\mathrm{d}y}{\mathrm{d}t} - x + 2y = \cos t$$

$$6\frac{\mathrm{d}x}{\mathrm{d}t} - 2\frac{\mathrm{d}y}{\mathrm{d}t} - 2x + y = 2\sin t$$

are linear ordinary differential equations.

$$\left(\frac{\mathrm{d}x}{\mathrm{d}t}\right)^2 + 4\frac{\mathrm{d}x}{\mathrm{d}t} = 0$$

$$\frac{\mathrm{d}^2 x}{\mathrm{d}t^2} + x\frac{\mathrm{d}x}{\mathrm{d}t} = 4\sin t$$

$$4\frac{\mathrm{d}x}{\mathrm{d}t} + \sin x = 0$$

are all nonlinear differential equations, the first because the derivative $\mathrm{d}x/\mathrm{d}t$ is squared, the second because of the product between the dependent variable x and its derivative, and the third because of the nonlinear function, $\sin x$, of the dependent variable.

10.3.5 Homogeneous and nonhomogeneous equations

There is a further classification that can be applied to linear equations: the distinction between homogeneous and nonhomogeneous equations. In all the examples we have presented so far the differential equations have been arranged so that all terms containing the dependent variable occur on the left-hand side of the equality sign, and those terms that involve only the independent variable and constant terms occur on the right-hand side. This is a standard way of arranging terms, and aids in the identification of equations. Specifically, when linear equations are arranged in this way, those in which the right-hand side is zero are called **homogeneous equations** and those in which it is non-zero are **nonhomogeneous equations**. Expressed another way, each term in a homogeneous equation involves the dependent variable or one of its derivatives. In a nonhomogeneous equation there is at least one term that does not contain the independent variable or any of its derivatives.

Example 10.5

$$\frac{\partial f}{\partial x} + \frac{\partial f}{\partial y} = 4x^2 + 2y$$

is a nonhomogeneous partial differential equation, whereas

$$\frac{\partial^2 f}{\partial x \partial y} = 0$$

is a homogeneous one. The equations

$$\frac{dx}{dt} + 4x = 0$$

and

$$4\frac{dx}{dt} + (\sin t)x = 0$$

are both homogeneous ordinary differential equations, while

$$\frac{d^2 x}{dt^2} + t\frac{dx}{dt} = 4\sin t$$

and

$$\frac{d^2 f}{dx^2} - 4x\frac{df}{dx} = \cos 2x$$

are both nonhomogeneous ordinary differential equations.

Example 10.6   Classify the equations (10.3), (10.6), (10.7) and (10.9) derived in the engineering examples of Section 10.2.

Solution   (a) Equation (10.3) is a second-order nonlinear ordinary differential equation whose dependent variable is s and whose independent variable is t.

(b) Equation (10.6) is a first-order linear nonhomogeneous ordinary differential equation whose dependent variable is T_w and whose independent variable is t.

(c) Equation (10.7) is a first-order linear nonhomogeneous ordinary differential equation whose dependent variable is Q and whose independent variable is t.

(d) Equation (10.9) is a second-order linear homogeneous ordinary differential equation whose dependent variable is i and whose independent variable is t.

10.3.6 Exercises

1 State the order of each of the following differential equations and name the dependent and independent variables. Classify each equation as partial or ordinary and as linear homogeneous, linear nonhomogeneous or nonlinear differential equations.

(a) $\dfrac{dx}{dt} + 2x = 0$

(b) $\dfrac{\partial y}{\partial t} + \dfrac{\partial x}{\partial t} = 0$

(c) $\dfrac{d^2x}{dt^2} + 2\dfrac{dx}{dt} + 3x = 0$

(d) $\dfrac{\partial x}{\partial t} + c\dfrac{\partial^2 y}{\partial t^2} = 0$

(e) $\left(\dfrac{dx}{dt}\right)^2 + x = 0$

(f) $\dfrac{\partial y}{\partial t}\dfrac{\partial x}{\partial t} - x = 0$

(g) $\dfrac{dx}{dt} + 2x = t^2$

(h) $\dfrac{d^2x}{dt^2} + \dfrac{dx}{dt} - 4x = \cos t + e^t$

2 Classify the following differential equations as partial or ordinary and as linear homogeneous, linear nonhomogeneous or nonlinear differential equations, state their order and name the dependent and independent variables.

(a) $\dfrac{\partial^2 f}{\partial x^2} + y\dfrac{\partial f}{\partial x}\dfrac{\partial f}{\partial y} + \sin y = 0$

(b) $\dfrac{d^2p}{dz^2}\dfrac{dp}{dz} + (\sin z)p = \ln z$

(c) $\dfrac{\partial^2 h}{\partial x \partial y} + x^2h = 0$

(d) $\dfrac{d^2s}{dt^2} + (\sin t)\dfrac{ds}{dt} + (t + \cos t)s = e^t$

(e) $\dfrac{\partial x}{\partial u} + v\dfrac{\partial x}{\partial v} = uv$

(f) $\left(\dfrac{d^3p}{dy^3}\right)^{1/2} + 4\dfrac{d^2p}{dy^2} - 6\dfrac{dp}{dy} + 8p = 0$

(g) $\dfrac{dr}{dz} + z^2 = 0$

(h) $\dfrac{\partial f}{\partial x}\dfrac{\partial f}{\partial y}\dfrac{\partial f}{\partial z} + xyzf = 0$

(i) $\dfrac{dx}{dt} = f(t)x$

(j) $\dfrac{dx}{dt} = f(t)x + g(t)$

(k) $a(x)\dfrac{\partial^2 f}{\partial x^2} + b(x, y)\dfrac{\partial^2 f}{\partial x \partial y} + c(y)\dfrac{\partial^2 f}{\partial y^2} = p(x, y)$

(l) $\dfrac{d^3p}{dq^3} + \dfrac{d^2p}{dq^2}p + 4q^2 = 0$

(m) $\dfrac{d^2x}{dy^2} = \dfrac{y}{x^2 - 1}$

(n) $\dfrac{\partial^2 y}{\partial p \partial t} + (\sin t)\dfrac{\partial y}{\partial p} + (\cos p)\dfrac{\partial y}{\partial t} = 0$

(o) $(\sin z)\dfrac{dy}{dz} + \dfrac{\cos z}{z}y = 0$

10.4 Solving differential equations

So far we have said that differential equations are equations which express relationships between a dependent variable and the derivatives of that variable with respect to one or more independent variables. We are now going to study some methods of solving differential equations. First, though, we should give some thought to exactly what form we expect that solution to take.

When we solve an algebraic equation we expect the solution to be a number (e.g. the solution of the equation $4x + 9 = 7$ is $x = -\frac{1}{2}$) or, perhaps, a set of numbers (e.g. the solution of a cubic polynomial equation like $x^3 - 5x^2 + 8x - 12 = 0$ is that x is one of a set of three real or complex numbers). Again, equations involving vectors and matrices have solutions that are constant vectors or one of a set of constant vectors. Differential equations, on the other hand, are equations involving not a simple scalar or vector variable but a function and its derivatives. The solution of a differential equation is, therefore, not a single value (or one from a set of values) but a function (or a family of functions). With this in mind let us proceed.

10.4.1 Solution by inspection

The solution to some differential equations is obvious.

Example 10.7   Faced with the differential equation

$$\frac{dx}{dt} = -4x \tag{10.10}$$

we might recall that if $x(t) = e^{-4t}$ then

$$\frac{dx}{dt} = -4e^{-4t} = -4x$$

In other words, the function $x(t) = e^{-4t}$ is a solution of the differential equation.

Example 10.8   The differential equation

$$\frac{d^2x}{dt^2} + \lambda^2 x = 0 \tag{10.11}$$

may be solved by recollecting that

$$\frac{d^2}{dt^2}(\sin \alpha t) = -\alpha^2 \sin \alpha t$$

Clearly, the function $x(t) = \sin \lambda t$ will satisfy the differential equation.

Many differential equations can be solved by inspection in a similar manner to Examples 10.7 and 10.8. Solution by inspection requires the recognition of the equation and its connection to a familiar result in differentiation. It is therefore dependent upon experience and inspiration, and for this reason is only practical for solving the simplest differential equations.

10.4.2 General and particular solutions

Examples 10.7 and 10.8 also illustrate a pitfall of solving equations in this way. The function $x(t) = e^{-4t}$ is certainly a solution of the equation in Example 10.7, but so is the function $x(t) = Ae^{-4t}$, where A is an arbitrary constant. The function $x(t) = \sin \lambda t$ is certainly a solution of the equation in Example 10.8, but so is the function $x(t) = A \sin \lambda t + B \cos \lambda t$, where A and B are arbitrary constants. Differential equations in general have this property – the most general function that will satisfy the differential equation contains one or more arbitrary constants. Such a function is known as the **general solution** of the differential equation. Giving particular numerical values to the constants in the general solution results in a **particular solution** of the equation. The general solution normally contains a number of arbitrary constants equal to the order of the differential equation.

When solving differential equations, we should, as a rule, seek the most general solution that is compatible with the constraints imposed by the problem. If we do not do this, we run the risk of neglecting some feature of the problem which, when translated into the terms of the engineering problem that gave rise to the differential equation, may have serious implications for the performance, efficiency or even safety of the engineering equipment or system being analysed.

10.4.3 Boundary and initial conditions

The arbitrary constants in the general solution of a differential equation can often be determined by the application of other conditions.

Example 10.9   Find the function $x(t)$ that satisfies the differential equation

$$\frac{\mathrm{d}x}{\mathrm{d}t} = -4x$$

and that has the value 2.5 when $t = 0$.

Solution   We know by inspection that the general solution is Ae^{-4t}. Applying the condition on $x(t)$ at $t = 0$, we find that we must have $Ae^{-0} = 2.5$, that is $A = 2.5$, so the solution to the differential equation that also satisfies the condition $x(0) = 2.5$ is $x(t) = 2.5e^{-4t}$.

Additional conditions on the solution of a differential equation such as that in Example 10.9 are called **boundary conditions**. In the special case in which all the boundary conditions are given at the same value of the independent variable the boundary conditions are called **initial conditions**. In many circumstances it is convenient to consider a differential equation as incomplete until the boundary conditions have been specified. A differential equation together with its boundary conditions is referred to as a **boundary-value problem**, unless the boundary conditions satisfy the requirements for being initial conditions, in which case the differential equation together with its boundary conditions is referred to as an **initial-value problem**.

Example 10.10   Find the function $x(t)$ that satisfies the initial-value problem

$$\frac{d^2x}{dt^2} + \lambda^2 x = 0 \quad x(0) = 4, \quad \frac{dx}{dt}(0) = 3, \quad \lambda \neq 0$$

Solution   We know by inspection that the general solution of this differential equation is

$$x(t) = A \sin \lambda t + B \cos \lambda t$$

and so

$$\frac{dx}{dt} = \lambda A \cos \lambda t - \lambda B \sin \lambda t$$

Applying the initial conditions gives rise to the equations

$$0A + 1B = 4$$

$$\lambda A + 0B = 3$$

and hence to the solution

$$x(t) = \frac{3}{\lambda} \sin \lambda t + 4 \cos \lambda t$$

Example 10.11   Find the function $x(t)$ that satisfies the boundary-value problem

$$\frac{d^2x}{dt^2} + \lambda^2 x = 0 \quad x(0) = 4, \quad \frac{dx}{dt}\left(\frac{\pi}{\lambda}\right) = 3, \quad \lambda \neq 0$$

Solution   We know by inspection that the general solution of the differential equation is

$$x(t) = A \sin \lambda t + B \cos \lambda t$$

and so

$$\frac{dx}{dt} = \lambda A \cos \lambda t - \lambda B \sin \lambda t$$

Applying the boundary conditions gives rise to the equations

$$0A + 1B = 4$$

$$-\lambda A + 0B = 3$$

and hence to the solution

$$x(t) = -\frac{3}{\lambda} \sin \lambda t + 4 \cos \lambda t$$

Obviously, since a first-order differential equation has only one arbitrary constant in its solution, only one boundary condition is needed to determine the constant, and so the boundary condition of a first-order equation can always be treated as an initial condition. For higher-order equations (and for sets of coupled first-order equations) the distinction between initial-value and boundary-value problems is an important one, not least because, generally speaking, initial-value problems are easier to solve than boundary-value problems.

10.4.4 Analytical and numerical solution

We have seen that some differential equations are so simple that they can be solved by inspection, given a reasonable knowledge of differentiation. There are many differential equations that are not amenable to solution in this way. For some of these we may be able, by the use of more complex mathematical techniques, to find a solution that expresses a functional relationship between the dependent and independent variables. We say that such equations have an **analytical solution**. In the case of other equations we may not be able to find a solution in such a form – either because no suitable mathematical technique for finding the solution exists or because there is no analytical solution. In these cases the only way of solving the equation is by the use of numerical techniques, leading to a **numerical solution**.

An analytical solution is almost always preferable to a numerical one. This is chiefly because an analytical solution is a mathematical function, and so the numerical value of the dependent variable can be computed for any value of the independent variable. In contrast with this, a numerical solution takes the form of a table giving the values of the dependent variable at a discrete set of values of the independent variable. The value of the dependent variable corresponding to any value of the independent variable not included in that discrete set can only be computed by interpolation from the table (or by repeating the whole numerical solution process, making sure the desired value of the independent variable is included in the solution set).

If the differential equation being solved contains parameters (such as the constant λ in Example 10.8) then an analytical solution of the equation will contain that parameter. The behaviour of the solution of the equation as the parameter value changes can be readily understood. For a numerical solution the parameter must be given a specific numerical value before the solution is computed. The numerical solution will then be valid only for that value of the parameter. If the behaviour of the solution as the parameter value is changed is of interest then the equation must be solved repeatedly using different parameter values.

When we obtain an analytical solution of a differential equation without its associated boundary conditions, the arbitrary constants in the solution are effectively parameters of the solution. A numerical solution to a differential equation cannot be obtained unless the boundary conditions are specified. This is one reason why it is sometimes convenient to refer to the whole problem (differential equation and boundary conditions) as a unit rather than consider the differential equation separately from its boundary conditions.

Another reason for preferring an analytical solution to a numerical one when such a solution is available is that the work required to obtain a numerical solution is generally much greater than that required to obtain an analytical one. On the other hand, most of this greater quantity of work can be delegated to a computer (and this may sometimes be considered to be an argument for numerical solutions being preferable to analytical ones).

Finally, it should be pointed out that this somewhat simplified overview of the contrast between analytical and numerical solutions of differential equations is becoming increasingly blurred by the availability of computerized symbolic manipulation systems (often known as computer algebra systems). We shall, in the remainder of this chapter, be studying methods for both the numerical and analytical solution of ordinary differential equations.

10.4.5 Exercises

3 Give the general solution of the following differential equations. In each case state how many arbitrary constants you expect to find in the general solution. Are your expectations confirmed in practice?

(a) $\dfrac{dx}{dt} = 4t^2$

(b) $\dfrac{d^2x}{dt^2} = t^3 - 2t$

(c) $\dfrac{d^2x}{dt^2} = e^{4t}$

(d) $\dfrac{dx}{dt} = -6x$

(e) $\dfrac{d^3x}{dt^3} = \dfrac{2}{t^3} + \sin 5t$

(f) $\dfrac{d^2x}{dt^2} = 8x$

4 For each of the following differential equation problems state how many arbitrary constants you would expect to find in the most general solution satisfying the problem. Find the solution and check whether your expectation is confirmed.

(a) $\dfrac{d^2x}{dt^2} = 4t, \quad x(0) = 2$

(b) $\dfrac{d^2x}{dt^2} = \sin 2t, \quad x(\tfrac{1}{4}\pi) = 2, \quad x(\tfrac{3}{4}\pi) = 2$

(c) $\dfrac{dx}{dt} = 4$

(d) $\dfrac{dx}{dt} + 2t = 0, \quad x(1) = 1$

(e) $\dfrac{d^2x}{dt^2} = 2e^{-2t}, \quad x(0) = a$

(f) $\dfrac{dx}{dt} - 2\sin 2t = 0$

(g) $\dfrac{dx}{dt} = 2x, \quad x(0) = 1$

(h) $\dfrac{d^2x}{dt^2} - x = 0, \quad x(0) = 0, \quad x(1) = 1$

5 State which of the following problems are **under-determined** (that is, have insufficient boundary conditions to determine all the arbitrary constants in the general solution) and which are **fully determined**. In the case of fully determined problems state which are boundary-value problems and which are initial-value problems. (Do not attempt to solve the differential equations.)

(a) $4x\dfrac{d^2x}{dt^2} + \left(2t^2 - \dfrac{1}{x}\right)\dfrac{dx}{dt} - 4x^2t = 0, \quad x(0) = 4$

(b) $\left(\dfrac{d^3x}{dt^3}\right)^2 + t\dfrac{d^2x}{dt^2} - x\left(\dfrac{dx}{dt}\right)^2 = 0$

$x(0) = 0, \quad \dfrac{dx}{dt}(0) = 1, \quad x(2) = 0$

(c) $\left(\dfrac{dx}{dt}\right)^2 - x^2 = \sin t, \quad x(0) = a$

(d) $\dfrac{d^4x}{dt^4} + 4\dfrac{d^3x}{dt^3} - 2\dfrac{d^2x}{dt^2} + \dfrac{dx}{dt} - 4x = e^t$

$x(0) = 1, \quad x(2) = 0$

(e) $\dfrac{d^2x}{dt^2} - 2t\dfrac{dx}{dt} = t^2 - 4, \quad x(0) = 1, \quad x(2) = 0$

(f) $\dfrac{d^2x}{dt^2} + 2x\left(\dfrac{dx}{dt}\right)^2 - \dfrac{x}{t} = 0$

$x(1) = 0, \quad \dfrac{dx}{dt}(1) = 4$

(g) $\left(\dfrac{d^2x}{dt^2}\right)^2 + 2t = 0, \quad \dfrac{dx}{dt}(2) = 1$

(h) $\dfrac{d^3x}{dt^3}\dfrac{dx}{dt} + x\dfrac{d^2x}{dt^2} = 2t^2$

$x(0) = 0, \quad \dfrac{dx}{dt}(0) = 0$

(i) $\left(\dfrac{d^3 x}{dt^3}\right)^{1/2} + t\dfrac{d^2 x}{dt^2} + x\dfrac{dx}{dt} - \dfrac{x}{t} = 0$

$x(1) = 1, \quad \dfrac{dx}{dt}(1) = 0, \quad \dfrac{d^2 x}{dt^2}(3) = 0$

(j) $\dfrac{dx}{dt} = (x - t)^2, \quad x(4) = 2$

(k) $\dfrac{d^2 x}{dt^2} - 4\dfrac{dx}{dt} + 4x = \cos t, \quad x(1) = 0, \quad x(3) = 0$

(l) $\dfrac{1}{t}\dfrac{d^3 x}{dt^3} - t^2\left(\dfrac{dx}{dt}\right)^2 + x\left(\dfrac{dx}{dt}\right)^{1/2} - (t^2 + 4)x = 0$

$x(0) = 0, \quad \dfrac{dx}{dt}(0) = U, \quad \dfrac{d^2 x}{dt^2}(0) = 0$

6 A uniform horizontal beam OA, of length a and weight w per unit length, is clamped horizontally at O and freely supported at A. The transverse displacement y of the beam is governed by the differential equation

$$EI\dfrac{d^2 y}{dx^2} = \tfrac{1}{2}w(a - x)^2 - R(a - x)$$

where x is the distance along the beam measured from O, R is the reaction at A, and E and I are physical constants. At O the boundary conditions are $y(0) = 0$ and $\dfrac{dy}{dx}(0) = 0$. Solve the differential equation. What is the boundary condition at A? Use this boundary condition to determine the reaction R. Hence find the maximum transverse displacement of the beam.

10.5　First-order ordinary differential equations

For the next three sections of this chapter we are going to concentrate our attention on the solution of first-order differential equations. This is not as restrictive as it might at first sight seem, since higher-order differential equations can, using a technique that we shall meet later in this chapter, be expressed as sets of coupled first-order differential equations. Some of the methods used for the solution of first-order equations, particularly the numerical techniques, are also applicable to such sets of coupled first-order equations, and thus may be used to solve higher-order differential equations.

10.5.1　A geometrical perspective

Most first-order differential equations can be expressed in the form

$$\dfrac{dx}{dt} = f(t, x) \tag{10.12}$$

It is true that there are some nonlinear equations which cannot be reduced to this form, but they are relatively uncommon in engineering applications, and the treatment of such oddities is beyond the scope of this text. Expressing the equation in this form means that, for any point in the t–x plane for which $f(t, x)$ is defined, we can compute the value of dx/dt at that point. If we then do this for a grid of points in the t–x plane, we can draw a picture such as Figure 10.5. At each point a short line segment with gradient dx/dt is drawn. Such a diagram is called the **direction field** of the differential equation. Obviously, there is a gradient direction at every point of the t–x plane, but it is equally obviously only practical to draw in a finite number of them as we have done in Figure 10.5. The equation whose direction field is drawn in Figure 10.5 is in fact

$$\dfrac{dx}{dt} = x(1 - x)t$$

Figure 10.5
The direction field
for the equation
$dx/dt = x(1 - x)t$.

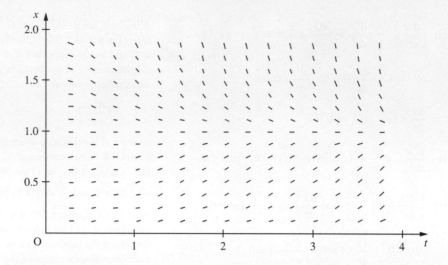

Figure 10.6
Solutions of
$dx/dt = x(1 - x)t$
superimposed on
its direction field.

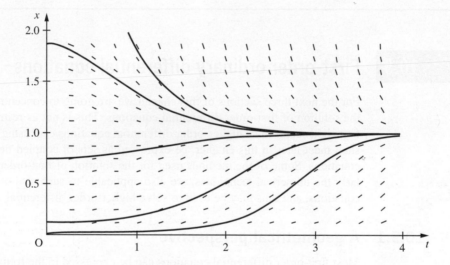

but the same process could be carried out for any equation expressible in the form
(10.12).

A solution of the differential equation is a function relating x and t (that is, a curve
in the t–x plane) which satisfies the differential equation. Since the solution function
satisfies the differential equation, the solution curve has the property that its gradient is
the same as the direction of the direction field of the equation at every point on the
curve; in other words, the direction field consists of line segments that are tangential to
the solution curves. With this insight, it is then fairly easy to infer what the solution
curves of the equation whose direction field is shown in Figure 10.5 must look like.
Some typical solution curves are shown in Figure 10.6.

By continuing this process, we could cover the whole t–x plane with an infinite
number of different solution curves. Each solution curve is a particular solution of
the differential equation. Since we are considering first-order equations, we expect
the general solution to contain one unknown constant. Giving a specific value to that
constant derives, from the general solution, one or other of the particular solution

curves. In other words, the general solution, with its unknown constant, represents a **family of solution curves**. The curves drawn in Figure 10.6 are particular members of that family.

Example 10.12   Sketch the direction field of the differential equation

$$\frac{\mathrm{d}x}{\mathrm{d}t} = -\tfrac{1}{2}x$$

Verify that $x(t) = Ce^{-t/2}$ is the general solution of the differential equation. Find the particular solution that satisfies $x(0) = 2$ and sketch it on the direction field. Do the same with the solution for which $x(3) = -1$.

Solution   The direction field is shown in Figure 10.7. Substituting the function $x(t) = Ce^{-t/2}$ into the equation immediately verifies that it is a solution. The initial condition $x(0) = 2$ implies $C = 2$. The condition $x(3) = -1$ implies $C = -e^{3/2}$. Both of these curves are shown on Figure 10.7, and are readily seen to be in the direction of the direction field at every point.

Figure 10.7
The direction field and some solution curves of $\mathrm{d}x/\mathrm{d}t = -x/2$.

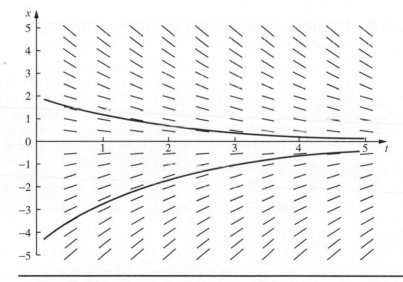

Sketching the direction field of an equation is not normally used as a way of solving a differential equation (although, as we shall see later, one of the simplest techniques for the numerical solution of ordinary differential equations may be interpreted as following lines through a direction field). It is, however, a very valuable aid to understanding the nature of the equation and its solutions. The sketching of direction fields is made very much simpler by the use of computers and particularly computer graphics. In cases of difficulty or uncertainty about the solution of a differential equation, sketching the direction field often greatly illuminates the problem.

10.5.2 Exercises

7 Sketch the direction field of the differential equation

$$\frac{dx}{dt} = -2t$$

Find the solution of the equation. Sketch the particular solutions for which $x(0) = 2$, and for which $x(2) = -3$, and check that these are consistent with your direction field.

8 Sketch the direction field of the differential equation

$$\frac{dx}{dt} = t - x$$

Verify that $x = t - 1 + Ce^{-t}$ is the solution of the equation. Sketch the solution curve for which $x(0) = 2$, and that for which $x(4) = 0$, and check that these are consistent with your direction field.

9 Draw the direction field of the equation

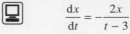

$$\frac{dx}{dt} = -\frac{2x}{t - 3}$$

Sketch some of the solution curves suggested by the direction field. Verify that the general solution of the equation is $x = C/(t - 3)^2$ and check that the members of this family resemble the solution curves you have sketched on the direction field.

10 Draw the direction field of the equation

$$\frac{dx}{dt} = \frac{1 - t}{t}x$$

Sketch some of the solution curves suggested by the direction field. Verify that the general solution of the equation is $x = Cte^{-t}$ and check that the members of this family resemble the solution curves you have sketched on the direction field.

10.5.3 Elementary analytical solution methods: separable equations

So far we have only solved differential equations such as (10.10) and (10.11) whose solution is immediately obvious. We are now going to introduce some techniques that allow us to solve somewhat more difficult equations. These techniques are basically ways of manipulating differential equations into forms in which their solutions become obvious. The first method applies to equations that take what is known as a separable form. If the function $f(t, x)$ in the first-order differential equation

$$\frac{dx}{dt} = f(t, x)$$

is such that the equation can be manipulated (by algebraic operations) into the form

$$g(x)\frac{dx}{dt} = h(t) \tag{10.13}$$

then the equation is called a **separable equation**. We may find an expression for the solution of such equations by the following argument.

Let $G(x) = \displaystyle\int g(x)\,dx$ and $H(t) = \displaystyle\int h(t)\,dt.$

Then $G'(x) = g(x)$ and $H'(t) = h(t).$

Define the function $k(t, x)$ by $k(t, x) = G(x) - H(t)$. Now, since x is a function of t,

$$\frac{dk}{dt} = \frac{d}{dt}G(x) - \frac{d}{dt}H(t) = G'(x)\frac{dx}{dt} - H'(t) = g(x)\frac{dx}{dt} - h(t) \tag{10.14}$$

Hence, from the differential equation (10.13), we have

$$g(x)\frac{dx}{dt} = h(t) \quad \Rightarrow \quad \frac{dk}{dt} = 0 \quad \Rightarrow k(t, x) = \text{constant} = C, \text{ say}$$

That is,

$$G(x) - H(t) = C \quad \Rightarrow \int g(x)dx - \int h(t)dt = C \quad \Rightarrow \int g(x)dx = \int h(t)dt$$

(Note that the constant C does not appear explicitly in the final expression since each of the indefinite integrals will involve an arbitrary constant of integration.) Thus we have shown that the differential equation (10.13) leads to

$$\int g(x)dx = \int h(t)dt \tag{10.15}$$

If the functions $g(x)$ and $h(t)$ can be integrated then the equation is solved.

Example 10.13   Solve the equation

$$\frac{dx}{dt} = 4xt, \quad x > 0$$

Solution   This equation can be written as

$$\frac{1}{x}\frac{dx}{dt} = 4t$$

and so is a separable equation. The solution is given by

$$\int \frac{dx}{x} = \int 4t\,dt$$

That is,

$$\ln x = 2t^2 + C$$

or

$$x = e^{2t^2 + C} = e^{2t^2}e^C$$

$$= C'e^{2t^2}, \quad \text{where } C' = e^C$$

Note: The cases $x < 0$ and $x = 0$ can be solved by allowing C' to be negative and zero respectively.

Note that a constant of integration has been introduced. We might expect such constants as a result of the integration of both left- and right-hand sides. However, if two constants had been introduced, they could then have been combined into one constant either on the left- or the right-hand side of the equation, so only one constant is actually necessary.

The solution of a separable differential equation is sometimes described in the following way. Multiply both sides of (10.13) by dt and cancel the dt's on the left-hand side. This produces the equation

$$g(x)dx = h(t)dt$$

Write an integral sign in front of both sides of this equation, and (10.15) is obtained. While this procedure appears to produce the correct result and may be useful as an *aide mémoire*, it is not mathematically rigorous. The reader is warned that the use of similar lines of argument in other contexts may produce fallacious results.

Some differential equations, while not being in separable form, can be transformed, by means of a substitution, into separable equations. The best known example of this is a differential equation of the form

$$\frac{dx}{dt} = f\left(\frac{x}{t}\right) \tag{10.16}$$

If the substitution $y = x/t$ is made then, since $x = yt$ and therefore, by the rule for differentiation of a product,

$$\frac{dx}{dt} = t\frac{dy}{dt} + y$$

we obtain

$$t\frac{dy}{dt} + y = f(y)$$

That is,

$$\frac{1}{f(y) - y}\frac{dy}{dt} = \frac{1}{t}$$

which is an equation of separable form.

Example 10.14   Solve the equation

$$t^2\frac{dx}{dt} = x^2 + xt, \quad t > 0, x \neq 0$$

Solution   Dividing both sides of the equation by t^2 results in

$$\frac{dx}{dt} = \frac{x^2}{t^2} + \frac{x}{t}$$

which is of the form (10.16). Making the substitution $y = x/t$ results in

$$\frac{1}{y^2}\frac{dy}{dt} = \frac{1}{t}$$

which is of separable form. The solution of this equation is given by

$$\int \frac{dy}{y^2} = \int \frac{dt}{t}$$

that is,

$$-\frac{1}{y} = \ln t + C \quad \text{or} \quad y = \frac{-1}{\ln t + C} = \frac{x}{t}$$

so

$$x = \frac{-t}{\ln t + C}$$

Note: The requirement $t > 0$ and $x \neq 0$ means that it is valid to divide throughout by t, and later by y, in the solution process. Solutions can be obtained without these restrictions and this is left as an exercise for the reader.

10.5.4 Exercises

11 Find the general solutions of the following differential equations:

(a) $\dfrac{dx}{dt} = kx$ 　(b) $\dfrac{dx}{dt} = 6xt^2$

(c) $\dfrac{dx}{dt} = \dfrac{bx}{t}$ 　(d) $\dfrac{dx}{dt} = \dfrac{a}{xt}$

12 Find the solutions of the following initial-value problems:

(a) $\dfrac{dx}{dt} = \dfrac{\sin t}{x^2}$, 　$x(0) = 4$

(b) $t^2 \dfrac{dx}{dt} = \dfrac{1}{x}$, 　$x(4) = 9$

13 Find the general solutions of the following differential equations:

(a) $xt\dfrac{dx}{dt} = x^2 + t^2$

(b) $x^2 \dfrac{dx}{dt} = \dfrac{t^3 + x^3}{t}$

(c) $t\dfrac{dx}{dt} = \dfrac{x^2 + xt}{t}$

14 Find the solution of the following initial-value problem:

$$x^3 t \frac{dx}{dt} = t^4 + x^4, \quad x(1) = 4$$

15 Find the general solutions of the following differential equations:

(a) $\sqrt{t}\dfrac{dx}{dt} = \sqrt{x}$ 　(b) $\dfrac{dx}{dt} = (1 + \sin t)\cot x$

(c) $\dfrac{dx}{dt} = xte^{t^2}$ 　(d) $x^2 \dfrac{dx}{dt} = e^t$

(e) $\dfrac{dx}{dt} = ax(x-1)$ 　(f) $x\dfrac{dx}{dt} = \sin t$

16 Find the solutions of the following initial-value problems:

(a) $\dfrac{dx}{dt} = \dfrac{t^2 + 1}{x + 2}$, 　$x(0) = -2$

(b) $t(t-1)\dfrac{dx}{dt} = x(x+1)$, 　$x(2) = 2$

(c) $\dfrac{dx}{dt} = (x^2 - 1)\cos t$, 　$x(0) = 2$

(d) $\dfrac{dx}{dt} = e^{x+t}$, 　$x(0) = a$

(e) $\dfrac{dx}{dt} = \dfrac{4\ln t}{x^2}$, 　$x(1) = 0$

17 Find the general solutions of the following differential equations:

(a) $2xt\dfrac{dx}{dt} = -x^2 - t^2$ 　(b) $t\dfrac{dx}{dt} = x + t\sin^2\left(\dfrac{x}{t}\right)$

(c) $t\dfrac{dx}{dt} = \dfrac{3t^2 - x^2}{t - 2x}$ 　(d) $t\dfrac{dx}{dt} = x + t\tan\left(\dfrac{x}{t}\right)$

(e) $\dfrac{dx}{dt} = \dfrac{x + t}{x - t}$ 　(f) $t\dfrac{dx}{dt} = x + te^{x/t}$

18 Find the solutions of the following initial-value problems:

(a) $\dfrac{\mathrm{d}x}{\mathrm{d}t} = \dfrac{x^3 - xt^2}{t^3}, \quad x(1) = 2$

(b) $xt\dfrac{\mathrm{d}x}{\mathrm{d}t} = 2(x^2 + t^2), \quad x(2) = -1$

(c) $t\dfrac{\mathrm{d}x}{\mathrm{d}t} = te^{-x/t} + x, \quad x(2) = 4$

(d) $xt\dfrac{\mathrm{d}x}{\mathrm{d}t} = t^2 e^{-x^2/t^2} + x^2, \quad x(1) = 2$

(e) $t^2\dfrac{\mathrm{d}x}{\mathrm{d}t} = x^2 + 2xt, \quad x(1) = 4$

19 Show that, by making the substitution $y = at + bx + c$, equations of the form

$$\dfrac{\mathrm{d}x}{\mathrm{d}t} = f(at + bx + c)$$

can be reduced to separable form. Hence find the general solutions of the following differential equations:

(a) $\dfrac{\mathrm{d}x}{\mathrm{d}t} = \dfrac{t - x + 2}{t - x + 3}$ (b) $2\dfrac{\mathrm{d}x}{\mathrm{d}t} = -\dfrac{(t + 2x)}{t + 2x + 1}$

(c) $\dfrac{\mathrm{d}x}{\mathrm{d}t} = \dfrac{1 - 2x - t}{4x + 2t}$ (d) $\dfrac{\mathrm{d}x}{\mathrm{d}t} = \dfrac{x - t + 2}{x - t + 1}$

(e) $\dfrac{\mathrm{d}x}{\mathrm{d}t} = 2t + x + 2$ (f) $2\dfrac{\mathrm{d}x}{\mathrm{d}t} = 2x - t + 5$

(g) $\dfrac{\mathrm{d}x}{\mathrm{d}t} = 4t^2 + 4xt + x^2 - 2$

20 A chemical reaction is governed by the differential equation

$$\dfrac{\mathrm{d}x}{\mathrm{d}t} = K(5 - x)^2$$

where $x(t)$ is the concentration of the chemical at time t. The initial concentration is zero and the concentration at time $5\,\mathrm{s}$ is found to be 2. Determine the reaction rate constant K and find the concentration at time $10\,\mathrm{s}$ and $50\,\mathrm{s}$. What is the ultimate value of the concentration?

21 A skydiver's vertical velocity is governed by the differential equation

$$m\dfrac{\mathrm{d}v}{\mathrm{d}t} = mg - Kv^2$$

where K is the skydiver's coefficient of drag. If the skydiver leaves her aeroplane at time $t = 0$ with zero vertical velocity find at what time she reaches half her final velocity.

22 A chemical A is formed by an irreversible reaction from chemicals B and C. Assuming that the amounts of B and C are adequate to sustain the reaction, the amount of A formed at time t is governed by the differential equation

$$\dfrac{\mathrm{d}A}{\mathrm{d}t} = K(1 - \alpha A)^7$$

If no A is present at time $t = 0$ find an expression for the amount of A present at time t.

10.5.5 Elementary analytical solution methods: exact equations

Some first-order differential equations are of a form (or can be manipulated into a form) that is called **exact**. Since such equations can be solved readily, it would be useful to be able to recognize them or, better still, to have a test for them. In this section we shall see how exact equations are solved, and develop a test that allows us to recognize them.

The solution of exact equations depends on the following observation: if $h(t, x)$ is a function of the variables x and t, and the variable x is itself a function of t, then, by the chain rule of differentiation,

$$\dfrac{\mathrm{d}h}{\mathrm{d}t} = \dfrac{\partial h}{\partial x}\dfrac{\mathrm{d}x}{\mathrm{d}t} + \dfrac{\partial h}{\partial t}$$

Now if a first-order differential equation is of the form

$$p(t, x)\dfrac{\mathrm{d}x}{\mathrm{d}t} + q(t, x) = 0 \tag{10.17}$$

and a function $h(t, x)$ can be found such that

$$\frac{\partial h}{\partial x} = p(t, x) \quad \text{and} \quad \frac{\partial h}{\partial t} = q(t, x) \tag{10.18}$$

then (10.17) is equivalent to the equation

$$\frac{dh}{dt} = 0$$

and the solution must be

$$h(t, x) = C$$

Example 10.15 Solve the differential equation

$$2xt\frac{dx}{dt} + x^2 - 2t = 0$$

Solution If $h(t, x) = x^2 t - t^2$ then

$$\frac{\partial h}{\partial x} = 2xt \quad \text{and} \quad \frac{\partial h}{\partial t} = x^2 - 2t$$

so the differential equation takes the form

$$\frac{d}{dt}(x^2 t - t^2) = 0$$

and the solution is

$$x^2 t - t^2 = C$$

Assuming $t > 0$ and $C > 0$ the solution can be written as

$$x = \pm\sqrt{\left(t + \frac{C}{t}\right)}$$

Thus we can solve equations of the form (10.17) provided that we can guess a function $h(t, x)$ that satisfies the conditions (10.18). If such a function is not immediately obvious, there are two possibilities: first there is no such function, and, secondly, there is such a function but we don't see what it is. We shall now develop a test that enables us to answer the question of whether an appropriate function $h(t, x)$ exists and a procedure that enables us to find such a function if it does exist. If

$$\frac{\partial h}{\partial x} = p(t, x) \quad \text{and} \quad \frac{\partial h}{\partial t} = q(t, x)$$

then

$$\frac{\partial p}{\partial t} - \frac{\partial^2 h}{\partial x \partial t} = \frac{\partial q}{\partial x}$$

so, for a function $h(t, x)$ satisfying (10.18) to exist, the functions $p(t, x)$ and $q(t, x)$ must satisfy

$$\frac{\partial p}{\partial t} = \frac{\partial q}{\partial x} \qquad \qquad \textbf{(10.19)}$$

If $p(t, x)$ and $q(t, x)$ do not satisfy this condition then there is no point in seeking a function $h(t, x)$ satisfying (10.18).

If $p(t, x)$ and $q(t, x)$ do satisfy (10.19), how do we find the function $h(t, x)$ that satisfies (10.18) and thus solve the equation (10.17)? It may be that, as in Example 10.15, the function is obvious. If not, it can be obtained by solving the two equations (10.18) independently and then comparing the answers, as in Example 10.16.

Example 10.16   Solve the differential equation

$$(\ln \sin t - 3x^2)\frac{dx}{dt} + x \cot t + 4t = 0$$

Solution   First, since

$$\frac{\partial}{\partial t}(\ln \sin t - 3x^2) = \cot t = \frac{\partial}{\partial x}(x \cot t + 4t)$$

an appropriate function $h(t, x)$ may exist. Now

$$\frac{\partial h}{\partial x} = \ln \sin t - 3x^2 \quad \text{gives} \quad h = x \ln \sin t - x^3 + C_1(t)$$

and

$$\frac{\partial h}{\partial t} = x \cot t + 4t \quad \text{gives} \quad h = x \ln \sin t + 2t^2 + C_2(x)$$

where $C_1(t)$ and $C_2(x)$ are arbitrary functions of t and x respectively. These functions play the same role in the integration of partial differential relations as the arbitrary constants do in the integration of ordinary differential relations. Comparing the two results, we see that

$$h(t, x) = x \ln \sin t - x^3 + 2t^2$$

satisfies (10.18) and so the solution of the differential equation is

$$x \ln \sin t - x^3 + 2t^2 = C$$

Equations that are not already of exact form can sometimes be put into exact form by multiplying each term of the equation by some function, known as an **integrating factor**. There is no general method for determining integrating factors for arbitrary differential equations, but they are known for some special classes of differential equation. We shall meet one such class of differential equation and its integrating factor in Section 10.5.7.

Example 10.17   The differential equation

$$\frac{\mathrm{d}x}{\mathrm{d}t} + \frac{x}{2t} = \frac{1}{x}$$

is not an exact equation (the condition (10.19) does not hold), but it can be manipulated, by multiplying each term by $2xt$ and rearranging the terms, into the form of the equation in Example 10.15.

10.5.6 Exercises

23 For each of the following differential equations determine whether they are exact equations and, if so, find the general solutions:

(a) $x\dfrac{\mathrm{d}x}{\mathrm{d}t} + t = 0$

(b) $x\dfrac{\mathrm{d}x}{\mathrm{d}t} - t = 0$

(c) $(x + t)\dfrac{\mathrm{d}x}{\mathrm{d}t} + x - t = 0$

(d) $(x - t^2)\dfrac{\mathrm{d}x}{\mathrm{d}t} - 2xt = 0$

(e) $(x - t)\dfrac{\mathrm{d}x}{\mathrm{d}t} - x + t - 1 = 0$

(f) $(2x + t)\dfrac{\mathrm{d}x}{\mathrm{d}t} + x + 2t = 0$

24 Find the solution of the following initial-value problems:

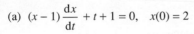

(a) $(x - 1)\dfrac{\mathrm{d}x}{\mathrm{d}t} + t + 1 = 0, \quad x(0) = 2$

(b) $(2x + t)\dfrac{\mathrm{d}x}{\mathrm{d}t} + x - t = 0, \quad x(0) = -1$

(c) $(2 - xt^2)\dfrac{\mathrm{d}x}{\mathrm{d}t} - x^2 t = 0, \quad x(1) = 2$

(d) $\cos t\dfrac{\mathrm{d}x}{\mathrm{d}t} - x\sin t + 1 = 0, \quad x(0) = 2$

25 For each of the following differential equations determine whether they are exact, and, if so, find the general solution:

(a) $(x + t)\dfrac{\mathrm{d}x}{\mathrm{d}t} - x + t = 0$

(b) $\sqrt{t}\dfrac{\mathrm{d}x}{\mathrm{d}t} - xt = 0$

(c) $[\sin(x + t) + x\cos(x + t)]\dfrac{\mathrm{d}x}{\mathrm{d}t} + x\cos(x + t) = 0$

(d) $\sin(xt)\dfrac{\mathrm{d}x}{\mathrm{d}t} + \cos xt = 0$

(e) $(1 + te^{xt})\dfrac{\mathrm{d}x}{\mathrm{d}t} + xe^{xt} = 0$

(f) $2(x + \sqrt{t})\dfrac{\mathrm{d}x}{\mathrm{d}t} + \dfrac{x}{\sqrt{t}} + 1 = 0$

(g) $te^{-xt}\dfrac{\mathrm{d}x}{\mathrm{d}t} - xe^{xt} = 0$

(h) $\dfrac{t}{x + t}\dfrac{\mathrm{d}x}{\mathrm{d}t} + \dfrac{t}{x + t} + \ln(x + t) = 0$

26 Find the solutions of the following initial-value problems:

(a) $\cos(x + t)\left(\dfrac{\mathrm{d}x}{\mathrm{d}t} + 1\right) + 1 = 0, \quad x(0) = \tfrac{1}{2}\pi$

(b) $3(x + 2t)^{1/2}\dfrac{\mathrm{d}x}{\mathrm{d}t} + 6(x + 2t)^{1/2} + 1 = 0,$

$x(-1) = 6$

(c) $x(x^2 - t^2)\dfrac{\mathrm{d}x}{\mathrm{d}t} - t(x^2 - t^2) + 1 = 0, \quad x(0) = -1$

(d) $\dfrac{1}{x + t}\dfrac{\mathrm{d}x}{\mathrm{d}t} + \dfrac{1}{x + t} - \dfrac{1}{t^2} = 0, \quad x(2) = 2$

27 What conditions on the constants a, b, e and f must be satisfied for the differential equation

$$(ax + bt)\frac{\mathrm{d}x}{\mathrm{d}t} + ex + ft = 0$$

to be exact, and what is the solution of the equation when they are satisfied?

28 What conditions on the functions $g(t)$ and $h(t)$ must be satisfied for the differential equation

$$g(t)\frac{\mathrm{d}x}{\mathrm{d}t} + h(t)x = 0$$

to be exact, and what is the solution of the equation when they are satisfied?

29 For what value of k is the function $(x + t)^k$ an integrating factor for the differential equation

$$[(x + t)\ln(x + t) + x]\frac{\mathrm{d}x}{\mathrm{d}t} + x = 0?$$

30 For what value of k is the function t^k an integrating factor for the differential equation

$$(t^2\cos xt)\frac{\mathrm{d}x}{\mathrm{d}t} + 3\sin xt + xt\cos xt = 0?$$

10.5.7 Further analytical solution methods: linear equations

In Section 10.3.4 we defined linear differential equations. The most general first-order linear differential equation must have the form

$$\frac{\mathrm{d}x}{\mathrm{d}t} + p(t)x = r(t) \tag{10.20}$$

where $p(t)$ and $r(t)$ are arbitrary functions of the independent variable t. We shall first see how to solve the slightly simpler equation

$$\frac{\mathrm{d}x}{\mathrm{d}t} + p(t)x = 0 \tag{10.21}$$

If we multiply this equation throughout by a function $g(t)$, the resulting equation

$$g(t)\frac{\mathrm{d}x}{\mathrm{d}t} + g(t)p(t)x = 0$$

will be exact if

$$\frac{\partial g}{\partial t} = \frac{\partial}{\partial x}(gpx)$$

Since g and p are functions of t only, this reduces to

$$\frac{\mathrm{d}g}{\mathrm{d}t} = gp$$

which is a separable equation with solution

$$\int\frac{\mathrm{d}g}{g} = \int p(t)\mathrm{d}t$$

That is,

$$\ln g = \int p(t)\mathrm{d}t$$

or

$$g(t) = e^{k(t)}, \quad \text{where } k(t) = \int p(t)\mathrm{d}t$$

Hence, multiplying (10.21) by $g(t)$, we obtain

$$e^{k(t)}\frac{\mathrm{d}x}{\mathrm{d}t} + p(t)e^{k(t)}x = 0$$

or

$$\frac{\mathrm{d}}{\mathrm{d}t}(e^{k(t)}x) = 0, \quad \text{since} \quad \frac{\mathrm{d}}{\mathrm{d}t}(e^{k(t)}) = p(t)e^{k(t)}$$

Hence

$$e^{k(t)}x = C$$

that is

$$x = Ce^{-k(t)}$$

This technique can, in fact, be used on the full equation (10.20). In that case, multiplying by the integrating factor $g(t)$, we obtain

$$e^{k(t)}\frac{\mathrm{d}x}{\mathrm{d}t} + p(t)e^{k(t)}x = e^{k(t)}r(t)$$

or

$$\frac{\mathrm{d}}{\mathrm{d}t}(e^{k(t)}x) = e^{k(t)}r(t)$$

That is,

$$e^{k(t)}x = \int e^{k(t)}r(t)\mathrm{d}t + C$$

or

$$x = e^{-k(t)}\left[\int e^{k(t)}r(t)\mathrm{d}t + C\right] \tag{10.22}$$

Thus (10.22) is an analytical solution of (10.20). The form of the solution can be simplified considerably if $\int p(t)\mathrm{d}t$ has a simple analytical form, as in Examples 10.18 and 10.19.

Example 10.18 Solve the first-order linear differential equation

$$\frac{dx}{dt} + tx = t$$

Solution The integrating factor for this equation is $e^{\int t\,dt}$; that is, $e^{t^2/2}$. Multiplying both sides of the equation by this factor we obtain

$$e^{t^2/2}\frac{dx}{dt} + te^{t^2/2}x = te^{t^2/2}$$

or

$$\frac{d}{dt}(e^{t^2/2}x) = te^{t^2/2}$$

That is,

$$e^{t^2/2}x = \int te^{t^2/2}\,dt = e^{t^2/2} + C$$

or

$$x(t) = 1 + Ce^{-t^2/2}$$

Note: In evaluating $\int t\,dt$ for the integrating factor we have taken the constant of integration to be zero. Any other value of the constant of integration would also produce a valid (but more complicated!) integrating factor.

Example 10.19 Solve the first-order linear initial-value problem

$$t\frac{dx}{dt} + x = t^2, \quad x(2) = \tfrac{1}{3}$$

Solution First, by dividing throughout by t, we write the equation in the standard form (10.20):

$$\frac{dx}{dt} + \frac{1}{t}x = t$$

The integrating factor for this equation is $e^{\int (1/t)\,dt}$; that is, $e^{\ln t}$, or simply t. Multiplying both sides of the equation by this factor, we obtain

$$t\frac{dx}{dt} + x = t^2$$

or

$$\frac{d}{dt}(tx) = t^2$$

That is

$$tx = \tfrac{1}{3}t^3 + C$$

The initial value $x(2) = \frac{1}{3}$, so that

$$\frac{2}{3} = \frac{8}{3} + C, \quad \text{giving } C = -2$$

Therefore

$$x(t) = \frac{1}{3}t^2 - \frac{2}{t}$$

10.5.8 Exercises

31 Find the solution of the following differential equations:

(a) $\dfrac{dx}{dt} + 3x = 2$ (b) $\dfrac{dx}{dt} - 4x = t$

(c) $\dfrac{dx}{dt} + 2x = e^{-4t}$ (d) $\dfrac{dx}{dt} + tx = -2t$

32 Find the solution of the following initial-value problems:

(a) $\dfrac{dx}{dt} - 2x = 3$, $\quad x(0) = 2$

(b) $\dfrac{dx}{dt} + 3x = t$, $\quad x(0) = 1$

(c) $\dfrac{dx}{dt} - \dfrac{x}{t} = t^2 - 3$, $\quad x(1) = -1$

33 Find the solutions of the following differential equations:

(a) $\dfrac{dx}{dt} - x = t + 2t^2$ (b) $\dfrac{dx}{dt} - 4tx = t^3$

(c) $\dfrac{dx}{dt} + \dfrac{2x}{t} = \cos t$ (d) $t\dfrac{dx}{dt} + 4x = e^t$

(e) $\dfrac{dx}{dt} - (2\cot 2t)x = \cos t$

(f) $\dfrac{dx}{dt} + 6t^2x = t^2 + 2t^5$ (g) $\dfrac{dx}{dt} - \dfrac{x}{t^2} = \dfrac{4}{t^2}$

34 Find the solutions of the following initial-value problems:

(a) $\dfrac{dx}{dt} - 2t(2x - 1) = 0$, $\quad x(0) = 0$

(b) $\dfrac{dx}{dt} = -x\ln t$, $\quad x(1) = 2$

(c) $\dfrac{dx}{dt} + 5x - t = e^{-2t}$, $\quad x(-1) = 0$

(d) $t^2\dfrac{dx}{dt} - 1 + x = 0$, $\quad x(2) = 2$

(e) $\dfrac{dx}{dt} - \dfrac{1 - 2x}{t} = 4t + e^t$, $\quad x(1) = 0$

(f) $\dfrac{dx}{dt} + (x - U)\sin t = 0$, $\quad x(\pi) = 2U$

35 Solve (10.6), which arose from the model of the heating of the water in a domestic hot-water storage tank developed in Section 10.2.2. If the water in the tank is initially at 10°C and T_{in} is 80°C, what is the ratio of the times taken for the water in the tank to reach 60°C, 70°C and 75°C?

36 Solve (10.7), which arose from the model of a hydro-electric power station developed in Section 10.2.3. The setting of the control valve is represented in the model by the value of the parameter γ. Derive an expression for the discharge $Q(t)$ following a sudden increase in the valve opening such that the parameter γ changes from γ_0 to $\frac{1}{2}\gamma_0$.

10.6 Numerical solution of first-order ordinary differential equations

Having met, in the last few sections, some techniques that may yield analytical solutions for first-order ordinary differential equations, we are now going to see how first-order ordinary differential equations can be solved numerically. In this chapter we will only study the simplest such method, Euler's method. Many more sophisticated (but also more complex) methods exist which yield solutions more efficiently, but space precludes their inclusion in this introductory treatment.

10.6.1 A simple solution method: Euler's method

In Section 10.5.1 we met the concept of the direction field of a differential equation

$$\frac{\mathrm{d}x}{\mathrm{d}t} = f(t, x)$$

We noted that solutions of the differential equation are curves in the t–x plane to which the direction field lines are tangential at every point. This immediately suggests that a curve representing a solution can be obtained by sketching on the direction field a curve that is always tangential to the lines of the direction field. In Figure 10.8 a way of systematically constructing an approximation to such a curve is shown.

Starting at some point (t_0, x_0), a straight line with gradient equal to the value of the direction field at that point, $f(t_0, x_0)$, is drawn. This line is followed to a point with abscissa $t_0 + h$. The ordinate at this point is $x_0 + hf(t_0, x_0)$, which we shall call X_1. The value of the direction field at this new point is calculated, and another straight line from this point with the new gradient is drawn. This line is followed as far as the point with abscissa $t_0 + 2h$. The process can be repeated any number of times, and a curve in the t–x plane consisting of a number of short straight line segments is constructed. The curve is completely defined by the points at which the line segments join, and these can obviously be described by the equations

Figure 10.8
The construction of a numerical solution of the equation $\mathrm{d}x/\mathrm{d}t = f(t, x)$.

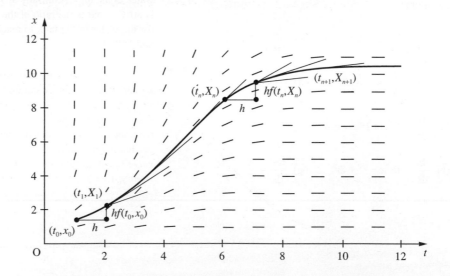

$$t_1 = t_0 + h, \qquad X_1 = x_0 + hf(t_0, x_0)$$

$$t_2 = t_1 + h, \qquad X_2 = X_1 + hf(t_1, X_1)$$

$$t_3 = t_2 + h, \qquad X_3 = X_2 + hf(t_2, X_2)$$

$$\vdots \qquad\qquad \vdots$$

$$t_{n+1} = t_n + h, \qquad X_{n+1} = X_n + hf(t_n, X_n)$$

These define, mathematically, the simplest method for integrating first-order differential equations. It is called **Euler's method**. Solutions are constructed step by step, starting from some given starting point (t_0, x_0). For a given t_0 each different x_0 will give rise to a different solution curve. These curves are all solutions of the differential equation, but each corresponds to a different initial condition.

The solution curves constructed using this method are obviously not exact solutions but only approximations to solutions, because they are only tangential to the direction field at certain points. Between these points, the curves are only approximately tangential to the direction field. Intuitively, we expect that, as the distance for which we follow each straight line segment is reduced, the curve we are constructing will become a better and better approximation to the exact solution. The increment h in the independent variable t along each straight-line segment is called the **step size** used in the solution. In Figure 10.9 three approximate solutions of the initial-value problem

$$\frac{\mathrm{d}x}{\mathrm{d}t} = x^2 t \mathrm{e}^{-t}, \quad x(0) = 0.91 \tag{10.23}$$

for step sizes $h = 0.05$, 0.025 and 0.0125 are shown. These steps are sufficiently small that the curves, despite being composed of a series of short straight lines, give the illusion of being smooth curves. The equation (10.23) actually has an analytical solution, which can be obtained by separation:

$$x = \frac{1}{(1 + t)\mathrm{e}^{-t} + C}$$

The analytical solution to the initial-value problem is also shown in Figure 10.9 for comparison. It can be seen that, as we expect intuitively, the smaller the step size the more closely the numerical solution approximates the analytical solution.

Figure 10.9
The Euler-method
solutions of
$\mathrm{d}x/\mathrm{d}t = x^2 t \mathrm{e}^{-t}$ for
$h = 0.05$, 0.025
and 0.0125.

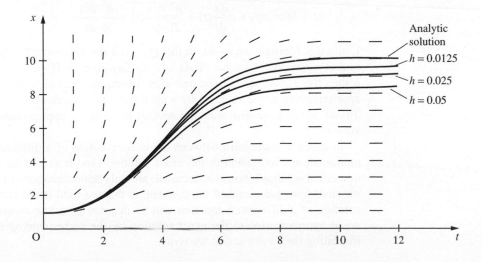

Example 10.20

The function $x(t)$ satisfies the differential equation

$$\frac{dx}{dt} = \frac{x+t}{xt}$$

and the initial condition $x(1) = 2$. Use Euler's method to obtain an approximation to the value of $x(2)$ using a step size of $h = 0.1$.

Solution

The solution is obtained step by step as set out in Figure 10.10. The approximation $X(2) = 3.1162$ results.

Figure 10.10
Computational results for Example 10.20.

t	X	$X+t$	Xt	$h\dfrac{X+t}{Xt}$
1.0000	2.0000	3.0000	2.0000	0.1500
1.1000	2.1500	3.2500	2.3650	0.1374
1.2000	2.2874	3.4874	2.7449	0.1271
1.3000	2.4145	3.7145	3.1388	0.1183
1.4000	2.5328	3.9328	3.5459	0.1109
1.5000	2.6437	4.1437	3.9656	0.1045
1.6000	2.7482	4.3482	4.3971	0.0989
1.7000	2.8471	4.5471	4.8400	0.0939
1.8000	2.9410	4.7410	5.2939	0.0896
1.9000	3.0306	4.9306	5.7581	0.0856
2.0000	3.1162			

10.6.2 Analysing Euler's method

We have introduced Euler's method via an intuitive argument from a geometrical understanding of the problem. Euler's method can be seen in another light – as an application of Taylor series. The Taylor series given in Section 9.5.2 applied to a function $x(t)$ gives

$$x(t+h) = x(t) + h\frac{dx}{dt}(t) + \frac{h^2}{2!}\frac{d^2x}{dt^2}(t) + \frac{h^3}{3!}\frac{d^3x}{dt^3}(t) + \dots \tag{10.24}$$

Using this formula, we could, in theory, given the value of $x(t)$ and all the derivatives of x at t, compute the value of $x(t+h)$ for any given h. If we choose a small value for h then the Taylor series truncated after a finite number of terms will provide a good approximation to the value of $x(t+h)$. Euler's method can be interpreted as using the Taylor series truncated after the second term as an approximation to the value of $x(t+h)$.

In order to distinguish between the exact solution of a differential equation and a numerical approximation to the exact solution (and it should be appreciated that all numerical solutions, however accurate, are only approximations to the exact solution), we shall now make explicit the convention that we used in the last section. The exact solution of a differential equation will be denoted by a lower-case letter and a numerical approximation to the exact solution by the corresponding capital letter. Thus, truncating the Taylor series, we write

$$X(t + h) = x(t) + h\frac{dx}{dt}(t) = x(t) + hf(t, x) \tag{10.25}$$

Applying this truncated Taylor series, starting at the point (t_0, x_0) and denoting $t_0 + nh$ by t_n, we obtain

$$X(t_1) = X(t_0 + h) = x(t_0) + hf(t_0, x_0)$$

$$X(t_2) = X(t_1 + h) = X(t_1) + hf(t_1, X_1)$$

$$X(t_3) = X(t_2 + h) = X(t_2) + hf(t_2, X_2)$$

and so on

which is just the Euler-method formula obtained in Section 10.6.1. As an additional abbreviated notation, we shall adopt the convention that $x(t_0 + nh)$ is denoted by x_n, $X(t_0 + nh)$ by X_n, $f(t_n, x_n)$ by f_n, and $f(t_n, X_n)$ by F_n. Hence we may express Euler's method, in general terms, as the recursive rule

$$X_0 = x_0$$

$$X_{n+1} = X_n + hF_n \quad (n \geq 0)$$

The advantage of viewing Euler's method as an application of Taylor series in this way is that it gives us a clue to obtaining more accurate methods for the numerical solution of differential equations. It also enables us to analyse in more detail how accurate Euler's method may be expected to be. We can abbreviate (10.24) to

$$x(t + h) = x(t) + hf(t, x) + O(h^2)$$

where $O(h^2)$ covers all the terms involving powers of h greater than or equal to h^2. Combining this with (10.25), we see that

$$X(t + h) = x(t + h) + O(h^2) \tag{10.26}$$

(Note that in obtaining this result we have used the fact that signs are irrelevant in determining the order of terms; that is, $-O(h^p) = O(h^p)$.) Equation (10.26) expresses the fact that at each step of the Euler process the value of $X(t + h)$ obtained has an error of order h^2, or, to put it another way, the formula used is accurate as far as terms of order h. For this reason Euler's method is known as a **first-order method**. The exact size of the error is, as we intuitively expected, dependent on the size of h, and decreases as h decreases. Since the error is of order h^2, we expect that halving h, for instance, will reduce the error at each step by a factor of four.

This does not, unfortunately, mean that the error in the solution of the initial-value problem is reduced by a factor of four. To understand why this is so, we argue as follows. Starting from the point (t_0, x_0) and using Euler's method with a step size h to obtain a value of $X(t_0 + 4)$, say, requires $4/h$ steps. At each step an error of order h^2 is incurred. The total error in the value of $X(t_0 + 4)$ will be the sum of the errors incurred at each step, and so will be $4/h$ times the value of a typical step error. Hence the total error is of the order of $(4/h)O(h^2)$; that is, the total error is $O(h)$. From this argument we should expect that if we compare solutions of a differential equation obtained using Euler's method with different step sizes, halving the step size will halve the error in the solution. Examination of Figure 10.9 confirms that this expectation is roughly correct in the case of the solutions presented there.

Example 10.21   Let X_a denote the approximation to the solution of the initial-value problem

$$\frac{\mathrm{d}x}{\mathrm{d}t} = \frac{x^2}{t+1}, \quad x(0) = 1$$

obtained using Euler's method with a step size $h = 0.1$, and X_b that obtained using a step size of $h = 0.05$. Compute the values of $X_a(t)$ and $X_b(t)$ for $t = 0.1, 0.2, \ldots, 1.0$. Compare these values with the values of $x(t)$, the exact solution of the problem. Compute the ratio of the errors in X_a and X_b.

Solution   The exact solution, which may be obtained by separation, is

$$x = \frac{1}{1 - \ln(t+1)}$$

The numerical solutions X_a and X_b and their errors are shown in Figure 10.11. Of course, in this figure the values of X_a are recorded at every step whereas those of X_b are only recorded at alternate steps.

Again, the final column of Figure 10.11 shows that our expectations about the effects of halving the step size when using Euler's method to solve a differential equation are confirmed. The ratio of the errors is not, of course, exactly one-half, because there are some higher-order terms in the errors, which we have ignored.

Figure 10.11
Computational results for Example 10.21.

t	X_a	X_b	$x(t)$	$\lvert x - X_a \rvert$	$\lvert x - X_b \rvert$	$\dfrac{\lvert x - X_b \rvert}{\lvert x - X_a \rvert}$
0.000 00	1.000 00	1.000 00	1.000 00			
0.100 00	1.100 00	1.102 50	1.105 35	0.005 35	0.002 85	0.53
0.200 00	1.210 00	1.216 03	1.222 97	0.012 97	0.006 95	0.54
0.300 00	1.332 01	1.342 94	1.355 68	0.023 67	0.012 75	0.54
0.400 00	1.468 49	1.486 17	1.507 10	0.038 61	0.020 92	0.54
0.500 00	1.622 52	1.649 52	1.681 99	0.059 47	0.032 47	0.55
0.600 00	1.798 03	1.837 91	1.886 81	0.088 78	0.048 90	0.55
0.700 00	2.000 08	2.057 92	2.130 51	0.130 42	0.072 59	0.56
0.800 00	2.235 40	2.318 57	2.425 93	0.190 53	0.107 36	0.56
0.900 00	2.513 01	2.632 51	2.792 16	0.279 15	0.159 65	0.57
1.000 00	2.845 39	3.018 05	3.258 89	0.413 50	0.240 84	0.58

10.6.3   Using numerical methods to solve engineering problems

In Example 10.21 the errors in the values of X_a and X_b are quite large (up to about 14% in the worst case). While carrying out computations with large errors such as these is quite useful for illustrating the mathematical properties of computational methods, in engineering computations we usually need to keep errors very much smaller. Exactly how small they must be is largely a matter of engineering judgement. The engineer must decide how accurately a result is needed for a given engineering purpose. It is then up to that engineer to use the mathematical techniques and knowledge available to carry out the computations to the desired accuracy. The engineering decision about the required accuracy will usually be based on the use that is to be made of the result. If, for instance, a preliminary design study is being carried out then a relatively approximate answer will often suffice, whereas for final design work much more accurate answers will normally be required. It must be appreciated that demanding greater accuracy than is actually needed for the engineering purpose in hand will usually carry a penalty in time, effort or cost.

Let us imagine that, for the problem posed in Example 10.21, we had decided we needed the value of $x(1)$ accurate to 1%. In the cases in which we should normally resort to numerical solution we should not have the analytical solution available, so we must ignore that solution. We shall suppose then that we had obtained the values of $X_a(1)$ and $X_b(1)$ and wanted to predict the step size we should need to use to obtain a better approximation to $x(1)$ accurate to 1%. Knowing that the error in $X_b(1)$ should be approximately one-half the error in $X_a(1)$ suggests that the error in $X_b(1)$ will be roughly the same as the difference between the errors in $X_a(1)$ and $X_b(1)$, which is the same as the difference between $X_a(1)$ and $X_b(1)$; that is, 0.172 66. One percent of $X_b(1)$ is roughly 0.03, that is, roughly one-sixth of the error in $X_b(1)$. Hence we expect that a step size roughly one-sixth of that used to obtain X_b will suffice; that is, a step size $h = 0.008\,33$. In practice, of course, we shall round to a more convenient non-recurring decimal quantity such as $h = 0.008$. This procedure is closely related to the Aitken extrapolation procedure introduced in Section 7.5.3 for estimating limits of convergent sequences and series.

Example 10.22 Compute an approximation $X(1)$ to the value of $x(1)$ satisfying the initial-value problem

$$\frac{dx}{dt} = \frac{x^2}{t+1}, \quad x(0) = 1$$

by using Euler's method with a step size $h = 0.008$.

Solution It is worth commenting here that the calculations performed in Example 10.21 could reasonably be carried out on any hand-held calculator, but this new calculation requires 125 steps. To do this is on the boundaries of what might reasonably be done on a hand-held calculator, and is more suited to a computer. Repeating the calculation with a step size $h = 0.008$ produces the result $X(1) = 3.213\,91$.

We had estimated from the evidence available (that is, values of $X(1)$ obtained using step sizes $h = 0.1$ and 0.05) that the step size $h = 0.008$ should provide a value of $X(1)$ accurate to approximately 1%. Comparison of the value we have just computed with the exact solution shows that it is actually in error by approximately 1.4%. This does not quite meet the target of 1% that we set ourselves. This example therefore serves, first, to illustrate how, given two approximations to $x(1)$ derived using Euler's method with different step sizes, we can estimate the step size needed to compute an approximation within a desired accuracy, and, secondly, to emphasize that the estimate of the appropriate step size is only an *estimate*, and will not *guarantee* an approximate solution to the problem meeting the desired accuracy criterion. If we had been more conservative and rounded the estimated step size down to, say, 0.005, we should have obtained $X(1) = 3.230\,43$, which is in error by only 0.9% and would have met the required accuracy criterion.

Figure 10.12
A poorly structured
algorithm for Example
10.21.

```
x1←1
x2←1
write(printer,0,1,1,1)
for i is 1 to 10 do
    x1←x1 + 0.1*x1*x1/((i − 1)*0.1 + 1)
    x2←x2 + 0.05*x2*x2/((i − 1)*0.1 + 1)
    x2←x2 + 0.05*x2*x2/((i − 1)*0.1 + 1.05)
    x←1/(1 − ln(i*0.1 + 1))
    write(printer,0.1*i,x1,x2,x,x − x1,x − x2,(x − x2)/(x − x1))
endfor
```

Figure 10.13
A better structured
algorithm for Example
10.21.

```
initial_time←0
final_time←1
initial_x←1
step←0.1
t←initial_time
x1←initial_x
x2←initial_x
h1←step
h2←step/2
write(printer,initial_time,x1,x2,initial_x)
repeat
    euler(t,x1,h1,1→x1)
    euler(t,x2,h2,2→x2)
    t←t + h
    x←exact_solution(t,initial_time,initial_x)
    write(printer,t,x1,x2,x,abs(x − x1),abs(x − x2),abs((x − x2)/(x− x1)))
until t ≥ final_time

procedure euler(t_old,x_old,step,number→x_new)
    temp_x←x_old
    for i is 0 to number − 1 do
        temp_x←temp_x + step*derivative(t_old + step*i,temp_x)
    endfor
    x_new←temp_x
endprocedure

procedure derivative(t,x → derivative)
    derivative←x*x/(t+1)
endprocedure

procedure exact_solution(t,t0,x0→exact_solution)
    c←ln(t0 + 1) + 1/x0
    exact_solution←1/(c − ln(t + 1))
endprocedure
```

 Since we have mentioned in Example 10.22 the use of computers to undertake the repetitious calculations involved in the numerical solution of differential equations, it is also worth commenting briefly on the writing of computer programs to implement those numerical solution methods. While it is perfectly possible to write informal, unstructured programs to implement algorithms such as Euler's method, a little attention to planning and structuring a program well will usually be amply rewarded – particularly in terms of the reduced probability of introducing 'bugs'. Another reason for careful structuring is that, in this way, parts of programs can often be written in fairly general terms and can be re-used later for other problems. The two pseudocode algorithms in Figures 10.12 and 10.13 will both produce the table of results in Example 10.21. The pseudocode program of Figure 10.12 is very specific to the problem posed, whereas that of Figure 10.13 is more general, better structured, and more expressive of the structure of mathematical problems. It is generally better to aim at the style of Figure 10.13.

10.6.4 Exercises

37 Find the value of $X(0.3)$ for the initial-value problem

$$\frac{dx}{dt} = x - 2t, \quad x(0) = 1$$

using Euler's method with steps of $h = 0.1$.

38 Find the value of $X(0.25)$ for the initial-value problem

$$\frac{dx}{dt} = xt, \quad x(0) = 2$$

using Euler's method with steps of $h = 0.05$.

39 Find the value of $X(1)$ for the initial-value problem

$$\frac{dx}{dt} - \frac{x}{2\sqrt{(t + x)}}, \quad x(0.5) = 1$$

using Euler's method with step size $h = 0.1$.

40 Find the value of $X(0.5)$ for the initial-value problem

$$\frac{dx}{dt} = \frac{4 - t}{t + x}, \quad x(0) = 1$$

using Euler's method with step size $h = 0.05$.

41 Denote the Euler-method solution of the initial-value problem

$$\frac{dx}{dt} = \frac{xt}{t^2 + 2}, \quad x(1) = 2$$

using step size $h = 0.1$ by $X_a(t)$, and that using $h = 0.05$ by $X_b(t)$. Find the values of $X_a(2)$ and $X_b(2)$. Estimate the error in the value of $X_b(2)$, and

suggest a value of step size that would provide a value of $X(2)$ accurate to 0.1%. Find the value of $X(2)$ using this step size. Find the exact solution of the initial-value problem, and determine the actual magnitude of the errors in $X_a(2)$, $X_b(2)$ and your final value of $X(2)$.

42 Denote the Euler-method solution of the initial-value problem

$$\frac{dx}{dt} = \frac{1}{xt}, \quad x(1) = 1$$

using step size $h = 0.1$ by $X_a(t)$, and that using $h = 0.05$ by $X_b(t)$. Find the values of $X_a(2)$ and $X_b(2)$. Estimate the error in the value of $X_b(2)$, and suggest a value of step size that would provide a value of $X(2)$ accurate to 0.2%. Find the value of $X(2)$ using this step size. Find the exact solution of the initial-value problem, and determine the actual magnitude of the errors in $X_a(2)$, $X_b(2)$ and your final value of $X(2)$.

43 Denote the Euler-method solution of the initial-value problem

$$\frac{dx}{dt} = \frac{1}{\ln x}, \quad x(1) = 1.2$$

using step size $h = 0.05$ by $X_a(t)$, and that using $h = 0.025$ by $X_b(t)$. Find the values of $X_a(1.5)$ and $X_b(1.5)$. Estimate the error in the value of $X_b(1.5)$, and suggest a value of step size that would provide a value of $X(1.5)$ accurate to 0.25%. Find the value of $X(1.5)$ using this step size. Find the exact solution of the initial-value problem, and determine the actual magnitude of the errors in $X_a(1.5)$, $X_b(1.5)$ and your final value of $X(1.5)$.

10.7 Engineering application: analysis of damper performance

In this section we shall carry out a modest engineering design exercise that will illustrate the modelling of an engineering problem using first-order differential equations and the solution of that problem using the techniques we have met so far in this chapter.

A small engineering company produces, among other artefacts, hydraulic dampers for specialized applications. One of the test rigs used by the company to check the quality and consistency of the operational characteristics of its output is illustrated in Figure 10.14. A carriage carrying a mass, which can be altered to suit the damper under test, is projected along a track of very low friction at a carefully controlled speed. At the end of the track the carriage impacts into a buffer which is connected to the damper under test. Immediately prior to impact the carriage passes through a pair of photocells whose output is used to measure the carriage speed accurately. The mass of the buffer is very small compared with the mass of the carriage and test weight. The time/displacement history of the damper as it is compressed by the impact of the carriage is recorded digitally. The apparatus can produce time/displacement graphs and time/compression speed graphs for dampers on test.

In order to interpret the test results, the company needs to know how a damper should, in theory, behave under such a test. The simplest classical model of a damper assumes that the resistance of the damper is proportional to the velocity of compression. Since the mass of the buffer and damper components is small compared with the mass of the test apparatus carriage, it is reasonable to assume that, on impact, the moving components of the buffer and damper accelerate instantaneously to the velocity of the carriage, with negligible loss of speed on the part of the carriage. Since the track is of very low friction, it will be assumed that the only force decelerating the carriage is that provided by the damper (this also means assuming the carriage is not moving sufficiently fast for air resistance to have a significant effect). With these assumptions, the equation of motion of the carriage is

$$m\frac{\mathrm{d}v}{\mathrm{d}t} = -kv, \quad v(0) = U \tag{10.27}$$

where m is the mass of the carriage, $v(t)$ is its speed and k is the damper constant. Time is measured from the moment of impact and U is the impact speed of the carriage. The damper constant describes the force produced by the damper per unit speed of compression (and, for double-acting dampers, extension). The design engineer can adjust this constant by altering the internal design and dimensions of the damper. Equation (10.27) can be solved on sight, or by separation. The solution is

$$v = C\mathrm{e}^{-\lambda t}, \quad \text{with } \lambda = k/m$$

Figure 10.14
The damper test apparatus.

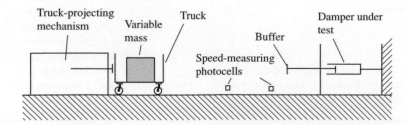

which, upon substituting in the initial conditions, becomes

$$v = Ue^{-\lambda t} \tag{10.28}$$

Writing $v = dx/dt$, where x is the compression of the damper and is taken as zero initially, this equation can be expressed as

$$\frac{dx}{dt} = Ue^{-\lambda t}, \quad x(0) = 0$$

This can be integrated directly, giving the solution, after substitution of the initial condition,

$$x = \frac{U}{\lambda}(1 - e^{-\lambda t}) \tag{10.29}$$

The velocity and displacement curves predicted by this model, (10.28) and (10.29), show that as $t \to \infty$, $v \to 0$ and $x \to U/\lambda$. Neither v nor x actually ever achieve these limits! This does not seem very realistic, since it is observed in tests that, after a finite and fairly short time (short at least when compared with infinity), the carriage comes to rest and the compression reaches a definite final value. The behaviour predicted by the simple model and the behaviour observed in tests do not quite agree. One possible explanation of this mismatch is the presence in the damper of friction between the components. Such friction would produce an additional resistance in the damper that does not vary with the speed of compression. The force resisting compression might therefore be better modelled as $kv + b$, where b is some constant force, rather than just kv. The compression of such a damper would be described by the equation

$$m\frac{dv}{dt} = -kv - b, \quad v(0) = U \tag{10.30}$$

Equation (10.30) is a linear first-order equation whose solution is

$$v = Ce^{-\lambda t} - \frac{b}{\lambda m}$$

or, substituting in the initial conditions,

$$v = Ue^{-\lambda t} - \frac{b}{\lambda m}(1 - e^{-\lambda t}) \tag{10.31}$$

This can be integrated again to provide displacement as a function of time:

$$x = \frac{1}{\lambda}\left(U + \frac{b}{\lambda m}\right)(1 - e^{-\lambda t}) - \frac{bt}{\lambda m} \tag{10.32}$$

Equation (10.31) predicts that the compression velocity of the damper will be zero when

$$t = \frac{1}{\lambda}\ln\left(\frac{b + \lambda Um}{b}\right) \tag{10.33}$$

at which time the compression of the damper will be

$$x = \frac{U}{\lambda} - \frac{b}{\lambda^2 m}\ln\left(\frac{b + \lambda Um}{b}\right) \tag{10.34}$$

This model therefore seems more realistic.

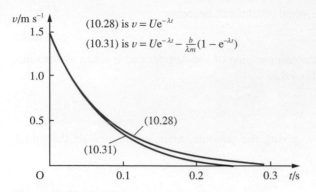

Figure 10.15 The predicted velocity–time curves for the damper test, both with and without the constant friction term.

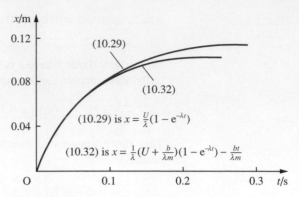

Figure 10.16 The predicted displacement–time curves for the damper test, both with and without the constant friction term.

Figures 10.15 and 10.16 show the velocity and displacement curves represented by (10.28) and (10.29) and (10.31) and (10.32) for a test in which the carriage carries a mass of 2 kg and travels at $1.5\,\text{m s}^{-1}$ at impact, the damper has a damping constant $25\,\text{N s m}^{-1}$ and the constant frictional force in the damper amounts to 1.5 N.

The company perceives that one of the disadvantages of the classical hydraulic damper is, as may be inferred from Figure 10.15, that the largest force, and hence the largest deceleration of the damped object, is produced early in the history of the impact, when the velocity is largest. This means that the object, whatever it is, must be able to withstand this high deceleration. If a damper could be designed that produced a more even force over the deceleration process, the maximum deceleration experienced by an object being stopped from a given speed in a given distance would be reduced. The company's designers think they may have a solution to this problem – they have devised a new pattern of damper with a patent internal mechanism such that the damping constant increases as the damper operates. The effect of this mechanism is that, during any given operating cycle, the damping constant may be expressed as $k(1 + at)$, where t is the time elapsed in the operating cycle. The internal mechanism is such that in a short time after an operating cycle the effective damper constant returns to its initial state and the damper is ready for another operating cycle.

A model of this new design of damper is provided by the equation

$$m\frac{\mathrm{d}v}{\mathrm{d}t} = -k(1 + at)v - b, \quad v(0) = U \tag{10.35}$$

This is a linear first-order differential equation. Applying the appropriate solution method gives the solution as

$$v = -\frac{b}{m}\mathrm{e}^{-\lambda g(t)} \int \mathrm{e}^{\lambda g(t)}\,\mathrm{d}t, \quad \text{where } g(t) = t + \tfrac{1}{2}at^2$$

Unfortunately there is no analytical solution of the integral. Two options are open to the company's mathematical engineers. Either the integral could be solved numerically by any appropriate numerical integration method (see Section 8.8) or, alternatively, (10.35) could be solved numerically. In Section 10.9.3 we shall see how (10.35) could be formulated as a second-order equation and solved to provide both (v, t) and (x, t)

Figure 10.17
Computational results
for the damper design
problem.

t	V_a	V_b	$V_a - V_b$	*(10.31)*
0.000	1.500 00	1.500 00		1.500 00
0.020	1.154 76	1.154 77	0.000 01	1.154 78
0.040	0.885 92	0.885 94	0.000 01	0.885 95
0.060	0.676 58	0.676 60	0.000 02	0.676 62
0.080	0.513 57	0.513 59	0.000 02	0.513 61
0.100	0.386 63	0.386 65	0.000 02	0.386 67
0.120	0.287 79	0.287 81	0.000 02	0.287 82
0.140	0.210 82	0.210 84	0.000 01	0.210 85
0.160	0.150 89	0.150 90	0.000 01	0.150 91
0.180	0.104 21	0.104 23	0.000 01	0.104 24
0.200	0.067 87	0.067 88	0.000 01	0.067 89
0.220	0.039 57	0.039 58	0.000 01	0.039 59
0.240	0.017 54	0.017 54	0.000 01	0.017 55
0.260	0.000 38	0.000 38	0.000 01	0.000 39
0.280	−0.012 98	−0.012 98	0.000 01	−0.012 97

curves simultaneously. This latter is what the company's engineers would probably choose to do. For the moment though, we must be content with solving (10.35) numerically to provide a (v, t) curve.

The company's engineers would wish to devise a numerical method for integrating the equation that will allow them to predict the performance of the damper for different combinations of the operational parameters U, m, k, a and b. Hence the task is to write a program that can be validated against some test cases and then be used with considerable confidence in other circumstances. If the value of a is taken to be 0 then the program to solve (10.35) should produce the same results as the analytical solution (10.31) of (10.30). This provides an appropriate test for the adequacy of the method and step size chosen. A program written to integrate equation (10.35) by Euler's method produced the results in the table in Figure 10.17. Several test runs of the program were undertaken using different step sizes, and results using $h = 0.000\,01$ (V_a) and $h = 0.000\,005$ (V_b) together with analytical solution (10.31) are shown in the figure.

It can be seen that the results using a step size of $h = 0.000\,005$ are in agreement with the analytical solution to at least 4 decimal places, and the agreement between the two numerical solutions V_a and V_b is good to 4dp. This agreement suggests that the accuracy of the numerical solution is adequate.

It therefore seems that a step size of $h = 0.000\,005$ will produce results that are accurate to at least 4dp and probably more. Using this step size, the (v, t) traces shown in Figure 10.18 were produced. First, for comparison, the predicted result of a test on a standard damper described by (10.30) is shown. Secondly, the predicted result of a test with a new model of damper with a parameter $a = 4$ is shown. It can be seen that the modified damper stops the carriage in a shorter time than the original model. The velocity–time trace is also slightly straighter, indicating that the design objective of making the deceleration more nearly uniform has been, at least in part, achieved. The third trace shown is for a new model damper with the basic damper constant k reduced to 17.5 and the parameter a kept at 4. This damper is able to halt the carriage in the same time as the original unmodified damper, but, in so doing, the maximum deceleration is somewhat smaller. This is the advantage of the new design that the company hope to exploit in the market.

Figure 10.18
Comparison of
velocity–time curves
for the damper test.

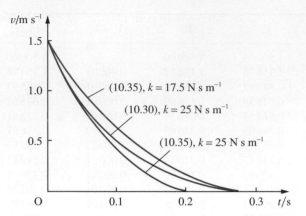

In this section we have seen how differential equations and numerical solution methods can be used to provide an analytical tool that the company can now use as a routine design tool for predicting the performance of a new model damper with any given combination of parameters. Such a tool is an invaluable aid to the designer, whose task will usually be to specify appropriate parameters to meet an operational requirement specified by a client, for instance something like 'to be capable of halting a mass M travelling at velocity U within a time T while subjecting it to a deceleration of no more than D'.

It should also be commented here that we have completed the numerical work in this example using Euler's method. In practice it would be far better to use a more sophisticated method, which would yield a solution of equivalent accuracy while using a much larger time step and therefore much less computing effort. Although the difference for a single computation would be very small (and, therefore, considerably outweighed by the additional programming effort of implementing a more complex method), if we were undertaking a large number of comparative runs or creating a design tool which would be used by many engineers over a long period of time then such issues would be important.

10.8 Linear differential equations

Having dealt, in the last three sections, with first-order differential equations we will now turn our attention to differential equations of higher orders. To begin with we will restrict our attention to linear differential equations.

In Section 10.3.4 we defined the concept of linearity and mentioned that the solutions of linear equations have important simplifying properties. In this section we are going to study these simplifying properties in more detail. Before we do so, however, it is helpful to define some new notation.

10.8.1 Differential operators

We are familiar from Section 2.2 with the idea that a function is a mapping from a set known as the domain of the function to another set, the codomain of the function. The functions we have met so far have been ones whose domain and codomain have been familiar sets such as the set of all real numbers (or perhaps some subset of that set), the set of integers or the set of complex numbers. There is, though, no reason why a function should not be defined to have a domain and codomain consisting of functions. Such functions are called **operators**. This name captures the idea that operators are functions that transform one function into another function. If f is a function and ϕ is an operator then $\phi[f]$ is another function.

Example 10.23

Let the set A be the set of functions on the real numbers; that is, functions whose domain and codomain are both the real numbers. The operator ϕ has domain A and is defined by

$$\phi[f(t)] = f(t)^2$$

In other words, the effect of an operator ϕ on a function f is defined by specifying the function $\phi[f(t)]$. Thus for the ϕ defined here

$$\phi[3t^2 - 2t + 4] = 9t^4 - 12t^3 + 28t^2 - 16t + 16$$

and

$$\phi[\sin t - t] = \sin^2 t - 2t \sin t + t^2$$

Example 10.24

The operator ϕ is defined by

$$\phi[f(t)] = tf(t)^2 - 4f(t) + t^2$$

Then $tg(t)^4 - 4g(t)^2 + t^2$ may be expressed as $\phi[g(t)^2]$ and $te^{2t} - 4e^t + t^2$ may be expressed as $\phi[e^t]$.

Where no ambiguity is likely to result, it is permissible and conventionally acceptable to write $\phi[f(t)]$ as $\phi f(t)$; that is, to omit the square brackets.

We may view the operation of differentiation as transforming a differentiable function to another function, its derivative. When we are going to take this view, we often write the differentiation symbol separately from the function on which it will operate; for instance, we write

$$\frac{\mathrm{d}x}{\mathrm{d}t} \quad \text{as} \quad \frac{\mathrm{d}}{\mathrm{d}t}[x] \quad \text{or} \quad \frac{\mathrm{d}^2 x}{\mathrm{d}t^2} \quad \text{as} \quad \frac{\mathrm{d}^2}{\mathrm{d}t^2}[x]$$

This notation is already familiar in those contexts in which we habitually write such expressions as

$$\frac{\mathrm{d}}{\mathrm{d}t}[f(t)g(t)] = \frac{\mathrm{d}f}{\mathrm{d}t}g + f\frac{\mathrm{d}g}{\mathrm{d}t}$$

In such contexts we refer to the symbol d/dt as a differential operator.

Example 10.25 Let the operator ϕ be defined by

$$\phi[f(t)] = \frac{\mathrm{d}}{\mathrm{d}t}f(t)$$

Then we have

$$\phi[t^2] = 2t, \quad \phi[\sin t] = \cos t, \quad \phi[4t^3 - \tan t] = 12t^2 - \sec^2 t, \quad \text{and so on}$$

Using this notation, a differential equation may be expressed as an operator equation.

Example 10.26 Let the operator L be defined by

$$L[f(t)] = \frac{\mathrm{d}^2 f}{\mathrm{d}t^2} - (\sin t)\frac{\mathrm{d}f}{\mathrm{d}t} + e^t f$$

The differential equation

$$\frac{\mathrm{d}^2 f}{\mathrm{d}t^2} - (\sin t)\frac{\mathrm{d}f}{\mathrm{d}t} + e^t f = t^4$$

may, using the operator notation, be written as

$$L[f(t)] = t^4$$

In Section 10.3.5 we introduced the concept of homogeneous and nonhomogeneous linear differential equations and mentioned the convention whereby differential equations are usually written with the terms involving the dependent variable on the left-hand side and those not involving it on the right-hand side. When written in this way, a homogeneous equation can be characterized as an equation of the form

$$L[x(t)] = 0$$

and a nonhomogeneous one as an equation of the form

$$L[x(t)] = f(t)$$

where L is the differential operator of the equation.

10.8.2 Linear differential equations

Returning now to linear and nonlinear equations, we see that linear ones can be more precisely and compactly defined as those for which the operator satisfies

$$L[ax_1 + bx_2] = aL[x_1] + bL[x_2] \tag{10.36}$$

for all functions x_1 and x_2 and all constants a and b.

Example 10.27

The equation

$$\frac{d^2x}{dt^2} + 4t\frac{dx}{dt} - (\sin t)x = \cos t$$

is a linear differential equation. Identify the operator of the equation and show that (10.36) holds for this operator.

Solution

The operator is

$$L \equiv \frac{d^2}{dt^2} + 4t\frac{d}{dt} - \sin t$$

Hence we have

$$L[ax_1 + bx_2] = \frac{d^2}{dt^2}[ax_1 + bx_2] + 4t\frac{d}{dt}[ax_1 + bx_2] - (\sin t)(ax_1 + bx_2)$$

$$= a\frac{d^2x_1}{dt^2} + b\frac{d^2x_2}{dt^2} + 4t\left(a\frac{dx_1}{dt} + b\frac{dx_2}{dt}\right) - (a\sin t)x_1 - (b\sin t)x_2$$

$$= a\left[\frac{d^2x_1}{dt^2} + 4t\frac{dx_1}{dt} - (\sin t)x_1\right] + b\left[\frac{d^2x_2}{dt^2} + 4t\frac{dx_2}{dt} - (\sin t)x_2\right]$$

$$= aL[x_1] + bL[x_2]$$

Equation (10.36) is the strict mathematical definition of linearity for any type of operator, and the definition we gave earlier in Section 10.3.4 is considerably less satisfactory mathematically. The formal definition of a linear differential equation is therefore any differential equation whose differential operator is linear in the sense of (10.36).

We said before that linear differential equations are an important subcategory of differential equations because they have particularly useful simplifying properties. The most important simplifying property can be summed up in the following principle:

> **Linearity principle:** if x_1 and x_2 are both solutions of the homogeneous linear differential equation $L[x] = 0$ then so is $ax_1 + bx_2$, where a and b are arbitrary constants.

This result follows directly from the definition of a linear operator. Since x_1 and x_2 are solutions of the differential equation, we have

$$L[x_1] = 0 \quad \text{and} \quad L[x_2] = 0$$

Since the equation is linear, we have

$$L[ax_1 + bx_2] = aL[x_1] + bL[x_2] = 0$$

Therefore $ax_1 + bx_2$ is a solution of the equation $L[x] = 0$.

Example 10.28

The well-known result that the general solution of the equation

$$\frac{d^2 x}{dt^2} + \lambda^2 x = 0$$

is

$$x = A \sin \lambda t + B \cos \lambda t$$

can be interpreted in the light of the linearity principle. Let $x_1 = \sin \lambda t$ and $x_2 = \cos \lambda t$. Then x_1 and x_2 are solutions of the differential equation. The equation is linear, so we know that $Ax_1 + Bx_2$ is also a solution.

Example 10.29

Find the general solution of the equation

$$\frac{d^4 x}{dt^4} - \lambda^4 x = 0$$

Solution

By inspection, we can see that $\sin \lambda t$, $\cos \lambda t$, $\sinh \lambda t$ and $\cosh \lambda t$ are all solutions of the equation. Therefore, since the equation is linear, the general solution is

$$x = A \sin \lambda t + B \cos \lambda t + C \sinh \lambda t + D \cosh \lambda t$$

In Example 10.29 we have implicitly used our expectation, introduced in Section 10.4.2, that the general solution of a pth-order differential equation contains p arbitrary constants. Since the equation is a fourth-order one, once we have found four solutions, we assemble them with four arbitrary constants and we have the general solution. Is this always the case? Not quite – we need an additional constraint on the solutions, as is shown by Example 10.30.

Example 10.30

Find the general solution of the differential equation

$$\frac{d^3 x}{dt^3} = 0$$

Solution

The functions $t + 4$, $2t$ and 7 are all solutions of the differential equation. Because the equation is linear, the function

$$x = A(t + 4) + 2Bt + 7C$$

is also a solution. Is it the general solution? By gathering like powers of the variable t, it can be rewritten as

$$x = (A + 2B)t + (4A + 7C)$$

and replacing the constant $A + 2B$ by a constant D and $4A + 7C$ by a constant E we have

$$x = Dt + E$$

Our proposed solution only really has two arbitrary constants in it, not the three we should expect in the general solution of the equation. Of course, if we notice that t^2 is also a solution of the equation, we can then apply the linearity principle to demonstrate that

$$x = Ft^2 + A(t+4) + 2Bt + 7C = Ft^2 + Dt + E$$

is also a solution and, since it now has the expected number of arbitrary constants and cannot be rewritten in any way with fewer constants, it is the general solution.

In order to resolve this problem, we need the idea of linear independence.

> The functions $f_1(t), f_2(t), \ldots, f_p(t)$ are said to be **linearly dependent** if a set of numbers $k_1, k_2, \ldots, k_p$, which are not all zero, can be found such that
>
> $$k_1 f_1(t) + k_2 f_2(t) + \ldots + k_p f_p(t) = 0$$
>
> that is,
>
> $$\sum_{j=1}^{p} k_j f_j(t) = 0$$
>
> The functions are linearly independent if no such set of numbers exists.

The essential difference between a set of linearly dependent functions and a set of linearly independent ones is that for a linearly dependent set there are functions in the set that can be written as linear combinations of some or all of the remaining functions. For a linearly independent set this is not possible. We can see that the three solutions that we first used in Example 10.30 are linearly dependent solutions. In effect this means that one of them is just a disguised form of the other two, and so we don't really have three solutions at all, only two. The additional constraint that we mentioned immediately after Example 10.29 is just that the solutions must be linearly independent. This gives us the following principle:

> **General solution of a linear homogeneous equation:** if $x_1, x_2, \ldots, x_p$ are all solutions of the pth-order homogeneous linear differential equation
>
> $$L[x] = 0$$
>
> and $x_1, x_2, \ldots, x_p$ are also linearly independent then the general solution of the differential equation is
>
> $$x = A_1 x_1 + A_2 x_2 + \ldots + A_p x_p$$

A formal proof of this result is not straightforward, and is not given here. We may, however, argue for its plausibility in the following way. Since the equation is linear, repeated application of the linearity principle shows that $A_1 x_1 + A_2 x_2 + \ldots + A_p x_p$ is a solution of $L[x] = 0$. The expression $A_1 x_1 + A_2 x_2 + \ldots + A_p x_p$ has p arbitrary constants and, since $x_1, x_2, \ldots, x_p$ are linearly independent, there is no way of rewriting the expression to reduce the number of arbitrary constants. Hence $A_1 x_1 + A_2 x_2 + \ldots + A_p x_p$ has the characteristics of the general solution of the differential equation.

The relatively simple structure of the general solution of a homogeneous linear differential equation has now been exposed. The general solution of a nonhomogeneous equation is only slightly more complex. It is given by the following result:

General solution of a linear nonhomogeneous equation: let

$$L[x] = f(t)$$

be a nonhomogeneous linear differential equation. If x^* is *any* solution of this equation and x_c is a solution of the equivalent homogeneous equation

$$L[x] = 0$$

then $x^* + x_c$ is also a solution of the nonhomogeneous equation.

This result is relatively straightforward to prove. By definition of x^* and x_c, we have

$$L[x^*] = f(t) \quad \text{and} \quad L[x_c] = 0$$

Since L is a linear operator, we have

$$L[x^* + x_c] = L[x^*] + L[x_c] = f(t) + 0 = f(t)$$

Hence $x^* + x_c$ is a solution of $L[x] = f(t)$.

It follows from this that finding the general solution of a nonhomogeneous linear differential equation can be reduced to the problem of finding any solution of the nonhomogeneous equation and adding to it the general solution of the equivalent homogeneous equation. The resulting expression is a solution of the nonhomogeneous equation containing the appropriate number of arbitrary constants, and so is the general solution. The first part of the solution (the 'any solution' of the nonhomogeneous equation, x^*) is known as a **particular integral** and the second part of the solution (the general solution of the equivalent homogeneous equation, x_c) is called the **complementary function**. The reader should note the similarity in structure with the general solution of linear recurrence relations developed in Section 7.4.

Example 10.31   Find the general solution of the differential equation

$$\frac{d^2x}{dt^2} + \lambda^2 x = 4t^3, \quad \lambda > 0$$

Solution   A particular integral of the equation is

$$x = \frac{4}{\lambda^2}t^3 - \frac{24}{\lambda^4}t$$

which you can check by direct substitution.

The complementary function is the general solution of the equation

$$\frac{d^2x}{dt^2} + \lambda^2 x = 0$$

that is,

$$x = A\sin\lambda t + B\cos\lambda t$$

Hence the general solution of

$$\frac{d^2x}{dt^2} + \lambda^2 x = 4t^3$$

is

$$x = \frac{4}{\lambda^2} t^3 - \frac{24}{\lambda^4} t + A \sin \lambda t + B \cos \lambda t$$

| **Example 10.32** | Find the general solution of the differential equation |

$$\frac{d^2 x}{dt^2} - p^2 x = \sin 2t, \quad p > 0$$

Solution A particular integral of the equation is

$$x = -\frac{\sin 2t}{4 + p^2}$$

which again can be checked by direct substitution.
  The complementary function is the general solution of the equation

$$\frac{d^2 x}{dt^2} - p^2 x = 0$$

that is,

$$x = A e^{pt} + B e^{-pt}$$

Hence the general solution of

$$\frac{d^2 x}{dt^2} - p^2 x = \sin 2t$$

is

$$x = -\frac{\sin 2t}{4 + p^2} + A e^{pt} + B e^{-pt}$$

10.8.3 Exercises

44 For each of the following differential equations write down the differential operator L that would enable the equation to be expressed to $L[x(t)] = 0$:

(a) $\dfrac{dx}{dt} + t^2 x = 0$ (b) $\dfrac{dx}{dt} = 6xt^2$

(c) $\dfrac{dx}{dt} - kx = 0$

45 Which of the following two sets are linearly dependent and which are linearly independent?

(a) $\{1, t, t^2, t^3, t^4, t^5, t^6\}$

(b) $\{1 + t, t^2, t^2 - t, 1 - t^2\}$

46 For each of the following sets of linearly dependent functions find $k_1, k_2, \ldots$ such that $k_1 f_1 + k_2 f_2 + \ldots = 0$.

(a) $\{t + 1, t, 2\}$ (b) $\{t^2 - 1, t^2 + 1, t - 1, t + 1\}$

47 For each of the following differential equations write down the differential operator L that would enable the equation to be expressed as $L[x(t)] = 0$:

(a) $\dfrac{dx}{dt} = f(t)x$

(b) $\dfrac{d^3 x}{dt^3} + (\sin t)\dfrac{d^2 x}{dt^2} + 4t^2 x = 0$

(c) $\dfrac{d^2x}{dt^2} + (\sin t)\dfrac{dx}{dt} = (t + \cos t)x$

(d) $(\sin t)\dfrac{dx}{dt} = \dfrac{\cos t}{t}x$

(e) $\dfrac{dx}{dt} = \dfrac{bx}{t}$ (f) $\dfrac{dx}{dt} = xte^{t^2}$

(g) $\dfrac{d}{dt}\left(t^2\dfrac{dx}{dt}\right) = t\dfrac{d}{dt}(xt)$

(h) $\dfrac{d}{dt}\left[\dfrac{1}{t}\dfrac{d}{dt}(t^2x)\right] = xt$

48 Which of the following sets of functions are linearly dependent and which are linearly independent?

(a) $\{\sin t + 2\cos t, \sin t - 2\cos t, 2\sin t + \cos t, 2\sin t - \cos t\}$

(b) $\{\sin t, \cos t, \sin 2t, \cos 2t, \sin 3t, \cos 3t\}$

(c) $\{1 + 2t, 2t - 3t^2, 3t^2 + 4t^3, 4t^3 - 5t^4\}$

(d) $\{1 + 2t, 2t - 3t^2, 3t^2 + 4t^3, 4t^3\}$

(e) $\{1, 1 + 2t, 2t - 3t^2, 3t^2 + 4t^3, 4t^3\}$

(f) $\{\ln a, \ln b, \ln ab\}$

(g) $\{e^s, e^t, e^{s+t}\}$

(h) $\{e^t, e^{2t} - e^t, e^{3t} - e^{2t}, e^{2t}\}$

(i) $\{f(t), f(t) - g(t), f(t) + g(t)\}$

(j) $\{1 - 2t^2, t - 3t^3, 2t^2 - 4t^4, 3t^3 - 5t^5\}$

(k) $\{1, 1 + t, 1 + t + t^2, 1 + t + t^2 + t^3\}$

49 For each of the following sets of linearly dependent functions find $k_1, k_2, \ldots$ such that $k_1f_1 + k_2f_2 + \ldots = 0$:

(a) $\{\sin t, \cos t + \sin t, \cos 2t - \sin t, \cos t - \cos 2t\}$

(b) $\{t + t^3, t - t^2, t^2 + 2t^3, t^2 - t^3\}$

(c) $\{\ln t, \ln 2t, \ln 4t^2\}$

(d) $\{f(t) + g(t), f(t)(1 + f(t)), g(t) - f(t), f(t)^2 - g(t)\}$

(e) $\{1 + t + 2t^2, t - 2t^2 + 3t^3, 1 + t - 2t^2, t - 2t^2 - 3t^3, t^3\}$

50 Determine which members of the given sets are solutions of the following differential equations. Hence, in each case, write down the general solution of the differential equation.

(a) $\dfrac{d^4x}{dt^4} = 0$ $\{1, t, t^2, t^3, t^4, t^5, t^6\}$

(b) $\dfrac{d^2x}{dt^2} - p^2x = 0$ $\{e^{pt}, e^{-pt}, \cos pt, \sin pt\}$

(c) $\dfrac{d^4x}{dt^4} - p^4x = 0$

$\{e^{pt}, e^{-pt}, \cos pt, \sin pt, \cosh pt, \sinh pt\}$

(d) $\dfrac{d^2x}{dt^2} + 2\dfrac{dx}{dt} = 0$

$\{\cos 2t, \sin 2t, e^{-2t}, e^{2t}, t^2, t, 1\}$

(e) $\dfrac{d^3x}{dt^3} + 4\dfrac{dx}{dt} = 0$

$\{\cos 2t, \sin 2t, e^{-2t}, e^{2t}, t^2, t, 1\}$

(f) $\dfrac{d^2x}{dt^2} + 2\dfrac{dx}{dt} + x = 0$

$\{e^t, e^{-t}, e^{2t}, e^{-2t}, te^t, te^{-t}, te^{2t}, te^{-2t}\}$

(g) $\dfrac{d^3x}{dt^3} - \dfrac{d^2x}{dt^2} - \dfrac{dx}{dt} + x = 0$

$\{e^t, e^{-t}, e^{2t}, e^{-2t}, te^t, te^{-t}, te^{2t}, te^{-2t}\}$

51 The operators L and M are defined by

$$L = \dfrac{d^2}{dt^2} - 4t\dfrac{d}{dt} + 6t^2$$

and

$$M = \dfrac{1}{t}\dfrac{d}{dt} - e^t$$

Find L[M[x(t)]]. Hence write down the operator LM. Find M[L[x(t)]]. Is LM = ML?

52 The operators L and M are defined by

$$L = f_1(t)\dfrac{d}{dt} + g_1(t)$$

and

$$M = f_2(t)\dfrac{d}{dt} + g_2(t)$$

Find expressions for the operators LM and ML. Under what conditions on f_1, g_1, f_2 and g_2 is LM = ML? What conditions do you think linear differential operators must satisfy in order to be commutative?

10.9 Linear constant-coefficient differential equations

10.9.1 Linear homogeneous constant-coefficient equations

One class of linear equation that arises relatively frequently in engineering practice is the linear constant-coefficient equation. These are linear equations in which the coefficients of the dependent variable and its derivatives do not depend on the independent variable but are constants. In view of the frequency with which such equations arise, and the fundamental importance of the problems that give rise to such equations, it is perhaps fortunate that they are relatively easy to solve.

We shall demonstrate the method of solution of such equations by considering, first of all, the second-order linear homogeneous constant-coefficient equation. The most general form this can take is

$$a\frac{d^2x}{dt^2} + b\frac{dx}{dt} + cx = 0, \quad a \neq 0 \tag{10.37}$$

Now the solution of the first-order linear homogeneous constant-coefficient equation

$$a\frac{dx}{dt} + bx = 0, \quad a \neq 0$$

is

$$x = Ae^{mt}, \quad \text{where } am + b = 0$$

Let us, by analogy, try the function $x(t) = e^{mt}$ as a solution of the second-order equation (10.37). Then direct substitution gives

$$am^2e^{mt} + bme^{mt} + ce^{mt} = 0$$

That is,

$$(am^2 + bm + c)e^{mt} = 0$$

Thus e^{mt} is a solution of the equation provided that

$$am^2 + bm + c = 0 \tag{10.38}$$

Suppose the roots of this quadratic equation are m_1 and m_2. Then e^{m_1t} and e^{m_2t} are solutions of the differential equation. Since it is a linear homogeneous equation, the general solution must be

$$x(t) = Ae^{m_1t} + Be^{m_2t} \tag{10.39}$$

provided that $m_1 \neq m_2$.

The form of the solution to (10.37) is deceptively simple. We know that the roots of a quadratic equation will take one of three forms:

(a) two different real numbers;
(b) a pair of complex-conjugate numbers;
(c) a repeated root (which must be real).

In the first case the solution is expressed as in (10.39). In the second case the roots may be written as

$$m_1 = \phi + j\psi \quad \text{and} \quad m_2 = \phi - j\psi$$

where ϕ and ψ are real, so that the solution is

$$
\begin{aligned}
x(t) &= Ae^{(\phi+j\psi)t} + Be^{(\phi-j\psi)t} \\
&= e^{\phi t}(Ae^{j\psi t} + Be^{-j\psi t}) \\
&= e^{\phi t}[A(\cos \psi t + j \sin \psi t) + B(\cos \psi t - j \sin \psi t)] \\
&= e^{\phi t}[(A + B)\cos \psi t + j(A - B)\sin \psi t]
\end{aligned}
$$

using Euler's formula (3.9). Writing $A + B = C$ and $j(A - B) = D$, we have

$$x(t) = e^{\phi t}(C \cos \psi t + D \sin \psi t)$$

In the third case the two roots m_1 and m_2 are equal, say, to k; therefore the solution (10.39) reduces to

$$x(t) = Ae^{kt} + Be^{kt} = Ce^{kt}$$

In this case the two solutions are not linearly independent, so we do not yet have the complete solution of (10.37). The complete solution, in this case, can be obtained by using the trial solution $x(t) = t^p e^{mt}$. In order for (10.38) to have a repeated root $m = k$, the constants in (10.37) must be such that (10.37) is of the form

$$a\frac{d^2x}{dt^2} - 2ak\frac{dx}{dt} + ak^2x = 0$$

Substituting the trial solution into this equation gives

$$a[p(p - 1)t^{p-2}e^{mt} + 2mpt^{p-1}e^{mt} + m^2t^p e^{mt}] - 2ak(pt^{p-1}e^{mt} + mt^p e^{mt}) + ak^2 t^p e^{mt} = 0$$

That is,

$$p(p - 1) + 2mpt + m^2t^2 - 2k(pt + mt^2) + k^2t^2 = 0$$

or

$$p(p - 1) + 2(m - k)pt + (m - k)^2 t^2 = 0$$

This equation is satisfied, for all values of t, if $m = k$ and $p = 1$ or $p = 0$. Hence te^{kt} and e^{kt} are two solutions of the differential equation. These are linearly independent functions, so the general solution in the case of two equal roots is

$$x(t) = Ate^{kt} + Be^{kt} = (At + B)e^{kt}$$

Evidently the solutions of the equation (10.38) that arise from substituting the trial solution into the differential equation (10.37) determine the form of the solution to the latter. Equation (10.38) is an important adjunct to the original equation, and is known as the **characteristic equation** of the differential equation (10.37). It is sometimes referred to as the **auxiliary equation**.

Summary

Summarizing, we have that if the two roots m_1 and m_2 of the characteristic equation are

- real and distinct then the corresponding solution is

$$x(t) = Ae^{m_1 t} + Be^{m_2 t}$$

- both equal to k then the corresponding solution is

$$x(t) = (At + B)e^{kt}$$

- complex conjugates $\phi \pm j\psi$ then the corresponding solution is

$$x(t) = e^{\phi t}(C \cos \psi t + D \sin \psi t)$$

Example 10.33   Find the general solution of the equation

$$\frac{d^2 x}{dt^2} - 9\frac{dx}{dt} + 6x = 0$$

Solution   The characteristic equation is

$$m^2 - 9m + 6 = 0$$

The roots of this equation are $m = 4.5 \pm \frac{1}{2}\sqrt{57}$, or, to 2dp, $m_1 = 12.05$ and $m_2 = -3.05$. Thus the solution is

$$x(t) = Ae^{12.05t} + Be^{-3.05t}$$

Example 10.34   Find the general solution of the equation

$$2\frac{d^2 x}{dt^2} - 3\frac{dx}{dt} + 5x = 0$$

Solution   The characteristic equation is

$$2m^2 - 3m + 5 = 0$$

The roots of this equation are $m = \frac{1}{4}(3 \pm j\sqrt{31})$, or, to 2dp, $m_1 = 0.75 + j1.39$ and $m_2 = 0.75 - j1.39$. Thus the solution is

$$x(t) = e^{0.75t}(A \cos 1.39t + B \sin 1.39t)$$

Example 10.35   Find the solution of the initial-value problem

$$\frac{d^2 x}{dt^2} + 6\frac{dx}{dt} + 9x = 0, \quad x(0) = 1, \quad \frac{dx}{dt}(0) = 2$$

Solution   The characteristic equation is

$$m^2 + 6m + 9 = (m + 3)^2 = 0$$

This equation has a repeated root $m = -3$. Thus the solution is

$$x(t) = (At + B)e^{-3t}$$

Now substituting in the initial conditions gives

$$B = 1, \quad -3B + A = 2$$

Hence $x(t) = (5t + 1)e^{-3t}$.

Notice, in Example 10.35, that the two initial conditions allow us to determine the values of the two arbitrary constants in the general solution of the second-order differential equation.

We have thus far demonstrated a technique that will solve any second-order linear homogeneous constant-coefficient equation. The technique extends quite satisfactorily to higher-order homogeneous constant-coefficient equations. When the same trial solution e^{mt} is substituted into a pth-order equation, it gives rise to a characteristic equation that is a polynomial equation of degree p in m, for instance

$$a_p m^p + a_{p-1} m^{p-1} + a_{p-2} m^{p-2} + \ldots + a_1 m + a_0 = 0$$

We know from the theory of polynomial equations (see Section 3.1) that such an equation has p roots. These may be real or complex, with the complex ones occurring in conjugate pairs. The roots may also be simple or repeated. These various possibilities are dealt with just as for a second-order equation. The only additional complexity over and above the solution of the second-order equation lies in the possibility of roots being repeated more than twice. In the case of a root $m = k$ of multiplicity n, the technique employed above can be used to show that the corresponding elementary solutions are $e^{kt}, te^{kt}, t^2 e^{kt}, \ldots, t^{n-1} e^{kt}$.

Example 10.36   Find the general solution of the equation

$$\frac{d^3 x}{dt^3} - 2\frac{d^2 x}{dt^2} - 5\frac{dx}{dt} + 6x = 0$$

Solution   The characteristic equation is

$$m^3 - 2m^2 - 5m + 6 = (m - 1)(m + 2)(m - 3) = 0$$

This equation has roots $m = 1, -2, 3$. Thus the solution is

$$x(t) = Ae^t + Be^{-2t} + Ce^{3t}$$

Example 10.37   Find the general solution of the equation

$$2\frac{d^4 x}{dt^4} + 3\frac{d^3 x}{dt^3} - 22\frac{d^2 x}{dt^2} - 73\frac{dx}{dt} - 60x = 0$$

Solution   The characteristic equation is

$$2m^4 + 3m^3 - 22m^2 - 73m - 60 = 0$$

that is,

$$(m - 4)(2m + 3)(m^2 + 4m + 5) = 0$$

The roots are therefore $m = 4, -3/2, -2 \pm j$. Thus the solution is

$$x(t) = Ae^{4t} + Be^{-3t/2} + e^{-2t}(C \cos t + D \sin t)$$

Example 10.38   Find the general solution of the equation

$$\frac{d^4x}{dt^4} + \frac{d^3x}{dt^3} - 3\frac{d^2x}{dt^2} - 5\frac{dx}{dt} - 2x = 0$$

Solution   The characteristic equation is

$$m^4 + m^3 - 3m^2 - 5m - 2 = 0$$

that is,

$$(m - 2)(m^3 + 3m^2 + 3m + 1) = (m - 2)(m + 1)^3 = 0$$

The roots are therefore $m = 2$ and $m = -1$ repeated three times. Thus the solution is

$$x(t) = Ae^{2t} + (Bt^2 + Ct + D)e^{-t}$$

10.9.2   Exercises

53   Find the general solution of the following differential equations:

(a) $2\dfrac{d^2x}{dt^2} - 5\dfrac{dx}{dt} + 3x = 0$

(b) $\dfrac{d^2x}{dt^2} + 2\dfrac{dx}{dt} + 5x = 0$

(c) $\dfrac{d^2x}{dt^2} + 3\dfrac{dx}{dt} - 4x = 0$

(d) $\dfrac{d^2x}{dt^2} - 4\dfrac{dx}{dt} + 13x = 0$

54   Solve the following initial-value problems:

(a) $5\dfrac{d^2x}{dt^2} - 3\dfrac{dx}{dt} - 2x = 0, \ x(0) = -1, \ \dfrac{dx}{dt}(0) = 1$

(b) $\dfrac{d^2x}{dt^2} - 6\dfrac{dx}{dt} + 10x = 0, \ x(0) = 2, \ \dfrac{dx}{dt}(0) = 0$

(c) $\dfrac{d^2x}{dt^2} - 4\dfrac{dx}{dt} + 3x = 0, \ x(0) = 0, \ \dfrac{dx}{dt}(0) = 1$

55   Find the general solutions of the following differential equations:

(a) $4\dfrac{d^2x}{dt^2} - 2\dfrac{dx}{dt} + 7x = 0$

(b) $\dfrac{d^2x}{dt^2} + 6\dfrac{dx}{dt} - 4x = 0$

(c) $3\dfrac{d^2x}{dt^2} + 3\dfrac{dx}{dt} + 3x = 0$

(d) $\dfrac{d^2x}{dt^2} - 8\dfrac{dx}{dt} + 16x = 0$

(e) $9\dfrac{d^3x}{dt^3} - 9\dfrac{d^2x}{dt^2} - 4\dfrac{dx}{dt} + 4x = 0$

(f) $\dfrac{d^3x}{dt^3} - \dfrac{d^2x}{dt^2} + 7\dfrac{dx}{dt} + 9x = 0$

(g) $\dfrac{d^3x}{dt^3} - 2\dfrac{d^2x}{dt^2} + 3\dfrac{dx}{dt} = 0$

56 Show that the characteristic equation of the differential equation

$$\frac{d^4x}{dt^4} - 4\frac{d^3x}{dt^3} + 11\frac{d^2x}{dt^2} - 14\frac{dx}{dt} + 10x = 0$$

is

$$(m^2 - 2m + 2)(m^2 - 2m + 5) = 0$$

and hence find the general solution of the equation.

57 Solve the following initial-value problems:

(a) $2\frac{d^2x}{dt^2} - 2\frac{dx}{dt} + 3x = 0$, $x(0) = 1$, $\frac{dx}{dt}(0) = 0$

(b) $\frac{d^2x}{dt^2} - 4\frac{dx}{dt} + 4x = 0$, $x(1) = 0$, $\frac{dx}{dt}(1) = 2$

(c) $\frac{d^2x}{dt^2} + 5\frac{dx}{dt} + 8x = 0$, $x(0) = 1$, $\frac{dx}{dt}(0) = -2$

(d) $9\frac{d^2x}{dt^2} + 6\frac{dx}{dt} + x = 0$, $x(-3) = 2$, $\frac{dx}{dt}(-3) = \frac{1}{2}$

(e) $\frac{d^3x}{dt^3} - 6\frac{d^2x}{dt^2} + 11\frac{dx}{dt} - 6x = 0$,

$x(0) = 1$, $\frac{dx}{dt}(0) = 0$, $\frac{d^2x}{dt^2}(0) = 1$

(f) $\frac{d^3x}{dt^3} + 6\frac{d^2x}{dt^2} + 12\frac{dx}{dt} + 8x = 0$,

$x(1) = 1$, $\frac{dx}{dt}(1) = 1$, $\frac{d^2x}{dt^2}(1) = 0$

58 Show that the characteristic equation of the differential equation

$$\frac{d^4x}{dt^4} - 2\frac{d^3x}{dt^3} + 3\frac{d^2x}{dt^2} - 2\frac{dx}{dt} + x = 0$$

is

$$(m^2 - m + 1)^2 = 0$$

and hence find the general solution of the equation.

59 Show that the characteristic equation of the differential equation

$$\frac{d^4x}{dt^4} - \frac{d^3x}{dt^3} - 9\frac{d^2x}{dt^2} - 11\frac{dx}{dt} - 4x = 0$$

is

$$(m^3 + 3m^2 + 3m + 1)(m - 4) = 0$$

and hence find the general solution of the equation.

10.9.3 Linear nonhomogeneous constant-coefficient equations

Having dealt with linear homogeneous constant-coefficient differential equations, much of the groundwork for linear nonhomogeneous constant-coefficient differential equations is already covered. The general form of such an equation of pth order is

$$L_p[x] = f(t)$$

where L_p is a pth-order linear differential operator. We have seen above that the general solution of this equation takes the form of the sum of a particular integral and the complemetary function. The complementary function is the general solution of the equation

$$L_p[x] = 0$$

Hence the complementary function may be found by the methods of the last section, and, in order to complete the treatment of the nonhomogeneous equation, we need only to discuss the finding of the particular integral.

There is no general mathematical theory that will guarantee to produce a particular integral by routine manipulation – rather finding a particular integral relies on recall of empirical rules or on intellectual inspiration. We shall proceed by first giving some examples.

Example 10.39   Find the general solution of the equation

$$\frac{d^2x}{dt^2} + 5\frac{dx}{dt} - 9x = t^2$$

Solution   First we shall seek a particular integral. Try the polynomial

$$x(t) = At^2 + Bt + C$$

Then direct substitution gives

$$2A + 5(2At + B) - 9(At^2 + Bt + C) = t^2$$

that is,

$$-9At^2 + (10A - 9B)t + 2A + 5B - 9C = t^2$$

Equating coefficients of the various powers of t on the left- and right-hand sides of this equation leads to a set of three linear equations for the unknowns A, B and C:

$$-9A \qquad\qquad = 1$$
$$10A - 9B \qquad = 0$$
$$2A + 5B - 9C = 0$$

This set of equations has solution

$$A = -\tfrac{1}{9}, \quad B = -\tfrac{10}{81}, \quad C = -\tfrac{68}{729}$$

so the particular integral is

$$x(t) = -\tfrac{1}{9}t^2 - \tfrac{10}{81}t - \tfrac{68}{729}$$

The method of Section 10.9.1 provides the complementary function, which is

$$Ae^{-(5-\sqrt{61})t/2} + Be^{-(5+\sqrt{61})t/2}$$

so the general solution of the equation is

$$x(t) = -\tfrac{1}{9}t^2 - \tfrac{10}{81}t - \tfrac{68}{729} + Ae^{-(5-\sqrt{61})t/2} + Be^{-(5+\sqrt{61})t/2}$$

Example 10.40   Find the general solution of the equation

$$\frac{d^2x}{dt^2} + 5\frac{dx}{dt} - 9x = \cos 2t$$

Solution   As in Example 10.39, we first seek a particular integral. In this case the right-hand-side function is $\cos 2t$. If we considered as a trial function $x(t) = A \cos 2t$, we should find that the left-hand side produced $\cos 2t$ and $\sin 2t$ terms. This suggests that the trial function should be

$$x(t) = A \cos 2t + B \sin 2t$$

Then direct substitution gives

$$-4A\cos 2t - 4B\sin 2t + 5(-2A\sin 2t + 2B\cos 2t) - 9(A\cos 2t + B\sin 2t)$$
$$= \cos 2t$$

that is,

$$(-13A + 10B)\cos 2t - (10A + 13B)\sin 2t = \cos 2t$$

Equating coefficients of $\cos 2t$ and $\sin 2t$ on the left- and right-hand sides of this equation leads to two linear equations for the unknowns A and B:

$$-13A + 10B = 1$$
$$10A + 13B = 0$$

so

$$A = -\tfrac{13}{269} \quad \text{and} \quad B = \tfrac{10}{269}$$

and the particular integral is

$$x(t) = \tfrac{1}{269}(10\sin 2t - 13\cos 2t)$$

The complementary function is the same as for Example 10.39,

$$Ae^{-(5-\sqrt{61})t/2} + Be^{-(5+\sqrt{61})t/2}$$

so the general solution of the equation is

$$x(t) = \tfrac{1}{269}(10\sin 2t - 13\cos 2t) + Ae^{-(5-\sqrt{61})t/2} + Be^{-(5+\sqrt{61})t/2}$$

Example 10.41   Find the general solution of the equation

$$\frac{d^2 x}{dt^2} + 5\frac{dx}{dt} - 9x = e^{4t}$$

Solution   Again we first seek a particular integral. In this case the right-hand-side function is e^{4t}. Since all derivatives of e^{4t} are multiples of e^{4t}, the trial function Ae^{4t} seems suitable. Then direct substitution gives

$$16Ae^{4t} + 20Ae^{4t} - 9Ae^{4t} = e^{4t}$$

Equating coefficients of e^{4t} on the left- and right-hand sides of this equation yields

$$27A = 1$$

so the particular integral is

$$x(t) = \tfrac{1}{27}e^{4t}$$

The complementary function is the same as for Example 10.39,

$$Ae^{-(5-\sqrt{61})t/2} + Be^{-(5+\sqrt{61})t/2}$$

so the general solution of the equation is

$$x(t) = \tfrac{1}{27}e^{4t} + Ae^{-(5-\sqrt{61})t/2} + Be^{-(5+\sqrt{61})t/2}$$

Figure 10.19
Trial functions for
particular integrals.

Right-hand-side function	Trial function	Unknown coefficients
Polynomial in t of degree p, for example $$6t^3 + 4t^2 - 2t + 5$$	Polynomial in t of degree p, for example $$At^3 + Bt^2 + Ct + D$$	Coefficients of the polynomial, for example $$A, B, C \text{ and } D$$
Exponential function of t, for example $$e^{-3t}$$	Exponential function of t with the same exponent, for example $$A\,e^{-3t}$$	Coefficient of the exponential function, for example $$A$$
Sine or cosine of a multiple of t, for example $$\sin 5t$$	Linear combination of sine and cosine of the same multiple of t, for example $$A \sin 5t + B \cos 5t$$	Coefficients of sine and cosine terms, for example $$A \text{ and } B$$

Examples 10.39–10.41 show how to deal with the most common right-hand-side functions. In each case a trial solution function is chosen to match the right-hand-side function. The trial solution function contains unknown coefficients, which are determined by substituting the trial solution into the differential equation and matching the left- and right-hand sides of the equation. Because of the unknown coefficients in the trial solution, this method of determining a particular integral is known as the **method of undetermined coefficients**. Figure 10.19 summarizes the standard trial functions which are used.

Although the examples that we have shown above all involve the solution of second-order equations, the trial solutions used to find particular integrals that are given in Figure 10.19 apply to linear nonhomogeneous constant-coefficient differential equations of any order.

If the right-hand side is a linear sum of more than one of these functions then the appropriate trial function is the sum of the trial functions for the terms making up the right-hand side. This can be seen from the properties of linear equations expressed in the following principle:

If L *is a linear differential operator and* x_1 *is a solution of the equation*

$$L[x(t)] = f_1(t)$$

and x_2 *is a solution of the equation*

$$L[x(t)] = f_2(t)$$

then $x_1 + x_2$ *is a solution of the equation*

$$L[x(t)] = f_1(t) + f_2(t)$$

This result can readily be proved as follows. Since L is a linear operator

$$L[x_1 + x_2] = L[x_1] + L[x_2]$$
$$= f_1(t) + f_2(t)$$

Hence $x_1 + x_2$ is a solution of $L[x] = f_1(t) + f_2(t)$.

Example 10.42

Find the general solution of the equation

$$\frac{d^2x}{dt^2} + 5\frac{dx}{dt} - 9x = e^{-2t} + 2 - t$$

Solution

First we find the particular integral. Since the right-hand side is the sum of an exponential and a polynomial of degree one, the trial function for this equation is

$$Ae^{-2t} + B + Ct$$

So, by direct substitution

$$4Ae^{-2t} + 5(-2Ae^{-2t} + C) - 9(Ae^{-2t} + B + Ct) = e^{-2t} + 2 - t$$

Equating coefficients of e^{-2t}, 1 and t on the left- and right-hand sides of this equation yields

$$-15A \qquad = 1$$
$$-9B + 5C = 2$$
$$-9C = -1$$

so the particular integral is

$$x(t) = -\tfrac{1}{15}e^{-2t} - \tfrac{13}{81} + \tfrac{1}{9}t$$

The complementary function is the same as for Example 10.39,

$$Ae^{-(5-\sqrt{61})t/2} + Be^{-(5+\sqrt{61})t/2}$$

so the general solution of the equation is

$$x(t) = -\tfrac{1}{15}e^{-2t} - \tfrac{13}{81} + \tfrac{1}{9}t + Ae^{-(5-\sqrt{61})t/2} + Be^{-(5+\sqrt{61})t/2}$$

The solution of linear nonhomogeneous constant-coefficient differential equations of order higher than two follows directly from the method for second-order equations. Finding a particular integral is the same whatever the degree of the equation. The principle that the solution is constructed from a particular integral added to the complementary function requires that the differential operator be linear, but is valid for an operator of any degree. Hence completing the solution of the higher-order nonhomogeneous equation only requires that the derived homogeneous equation can be solved – and we learnt how to do that in Section 10.9.1.

There is one complication that we have not yet mentioned. This is illustrated in Example 10.43.

Example 10.43

Find the general solution of the equation

$$\frac{d^2x}{dt^2} + \frac{dx}{dt} - 2x = e^{-2t}$$

Solution

Substituting in the appropriate trial solution Ae^{-2t} produces the result

$$4Ae^{-2t} - 2Ae^{-2t} - 2Ae^{-2t} = e^{-2t}$$

This equation has no solution for A.

The problem in Example 10.43 lies in the fact that the right-hand side of the equation consists of a function that is also a solution of the equivalent homogeneous equation. In such cases we must multiply the appropriate trial function for the particular integral by t. If the right-hand-side function corresponds to a function that is a repeated root of the characteristic equation then the trial function must be multiplied by t^n, where n is the multiplicity of the root of the characteristic equation.

Example 10.44

Find the general solution of the equation

$$\frac{d^4x}{dt^4} - 2\frac{d^3x}{dt^3} + 5\frac{d^2x}{dt^2} - 8\frac{dx}{dt} + 4x = e^t$$

Solution

Substituting in the appropriate trial solution Ae^t produces the result

$$Ae^t - 2Ae^t + 5Ae^t - 8Ae^t + 4Ae^t = e^t$$

for which, as in Example 10.43, there is no solution for A. The characteristic equation for the homogeneous equation is

$$m^4 - 2m^3 + 5m^2 - 8m + 4 = 0$$

that is,

$$(m - 1)^2(m^2 + 4) = 0$$

so the general solution of the homogeneous equation is

$$x(t) = (At + B)e^t + C\cos 2t + D\sin 2t$$

The right-hand side of the equation, e^t, is the function corresponding to the double root $m = 1$, so the standard trial function for this right-hand side, Ee^t, must be multiplied by t^2. Substituting this trial function, we obtain

$$E(t^2e^t + 8te^t + 12e^t) - 2E(t^2e^t + 6te^t + 6e^t)$$

$$+ 5E(t^2e^t + 4te^t + 2e^t) - 8E(t^2e^t + 2tc^t) + 4Et^2e^t = e^t$$

that is,

$$10Ee^t = e^t$$

Hence the solution of the differential equation is

$$x(t) = \tfrac{1}{10}t^2e^t + (At + B)e^t + C\cos 2t + D\sin 2t$$

Examples 10.39–10.44 have all found general solutions to problems with no boundary conditions given. Obviously values could be determined for the constants to fit the general solution to given boundary or initial conditions. Each boundary condition allows the value of one constant to be fixed. Hence, in general, the number of boundary conditions needed to completely determine the solution is equal to the order of the differential equation.

10.9.4 Exercises

 60 Find the general solution of the following differential equations:

(a) $\dfrac{d^2x}{dt^2} - 2\dfrac{dx}{dt} - 3x = t$

(b) $\dfrac{d^2x}{dt^2} - 2\dfrac{dx}{dt} - 5x = t^2 - 2t$

(c) $\dfrac{d^2x}{dt^2} - \dfrac{dx}{dt} - x = 5e^t$

 61 Find the general solutions of the following differential equations:

(a) $\dfrac{d^2x}{dt^2} - 3\dfrac{dx}{dt} + 4x = \cos 4t - 2\sin 4t$

(b) $9\dfrac{d^2x}{dt^2} - 12\dfrac{dx}{dt} + 4x = e^{-3t}$

(c) $2\dfrac{d^2x}{dt^2} + 4\dfrac{dx}{dt} - 7x = 7\cos 2t$

(d) $\dfrac{d^2x}{dt^2} + \dfrac{dx}{dt} + 4x = 5t - 7$

(e) $16\dfrac{d^2x}{dt^2} + 8\dfrac{dx}{dt} + x = t + 6$

(f) $\dfrac{d^2x}{dt^2} - 8\dfrac{dx}{dt} + 16x = -3\sin 3t$

(g) $\dfrac{d^2x}{dt^2} - 4\dfrac{dx}{dt} + 7x = e^{-5t}$

(h) $3\dfrac{d^2x}{dt^2} + 3\dfrac{dx}{dt} - x = t^2 + e^{-2t}$

(i) $\dfrac{d^2x}{dt^2} + 2\dfrac{dx}{dt} - 3x = 5e^{-3t} + \sin 2t$

(j) $\dfrac{d^2x}{dt^2} + 16x = 1 + 2\sin 4t$

(k) $\dfrac{d^2x}{dt^2} - 4\dfrac{dx}{dt} = 7 - 3e^{4t}$

 62 Show that the characteristic equation of the differential equation

$$\frac{d^4x}{dt^4} - 3\frac{d^3x}{dt^3} - 5\frac{d^2x}{dt^2} + 9\frac{dx}{dt} - 2x = 0$$

is

$$(m^2 + m - 2)(m^2 - 4m + 1) = 0$$

and hence find the general solutions of the equations

(a) $\dfrac{d^4x}{dt^4} - 3\dfrac{d^3x}{dt^3} - 5\dfrac{d^2x}{dt^2} + 9\dfrac{dx}{dt} - 2x = \cos 2t$

(b) $\dfrac{d^4x}{dt^4} - 3\dfrac{d^3x}{dt^3} - 5\dfrac{d^2x}{dt^2} + 9\dfrac{dx}{dt} - 2x = e^{2t} + e^{-2t}$

(c) $\dfrac{d^4x}{dt^4} - 3\dfrac{d^3x}{dt^3} - 5\dfrac{d^2x}{dt^2} + 9\dfrac{dx}{dt} - 2x = t^2 - 1 + e^{-t}$

 63 Show that the characteristic equation of the differential equation

$$\frac{d^3x}{dt^3} - 9\frac{d^2x}{dt^2} + 27\frac{dx}{dt} - 27x = 0$$

is

$$(m - 3)^3 = 0$$

and hence find the general solutions of the equations

(a) $\dfrac{d^3x}{dt^3} - 9\dfrac{d^2x}{dt^2} + 27\dfrac{dx}{dt} - 27x = \cos t - \sin t + t$

(b) $\dfrac{d^3x}{dt^3} - 9\dfrac{d^2x}{dt^2} + 27\dfrac{dx}{dt} - 27x = e^t$

(c) $\dfrac{d^3x}{dt^3} - 9\dfrac{d^2x}{dt^2} + 27\dfrac{dx}{dt} - 27x = e^{3t} + t$

10.10 Engineering application: second-order linear constant-coefficient differential equations

In this section we are going to show how simple mathematical models of a variety of engineering systems give rise to second-order linear constant-coefficient differential equations. We shall also investigate the major features of the solutions of such models.

10.10.1 Free oscillations of elastic systems

If a wooden plank or a metal beam is attached firmly to a rigid foundation at one end with its other end projecting and unsupported as shown in Figure 10.20 then the imposition of a force on the free end, or equivalently the placing of a heavy object on it, will cause the plank or beam to bend under the load. The greater the force or load, the greater will be the deflection. If the load is moderate then the plank or beam will spring back to its original position when the load is removed. If the load is great enough, the plank will eventually break. The metal beam, on the other hand, may either deform permanently (so that it does not return to its original position when the load is removed) or fracture, depending on the type of metal. Experiments on planks or beams such as described here have revealed that for beams made of a wide variety of materials there is commonly a range of loads for which the deflection of the beam is roughly proportional to the load applied (Figures 10.20a, b). When the load becomes large enough, however, there is usually a region in which the deflection increases either less rapidly or more rapidly than the load (Figure 10.20c), and finally a load beyond which the beam either breaks or is permanently deformed (Figure 10.20d).

Figure 10.20
The deflection of a cantilever by a load.

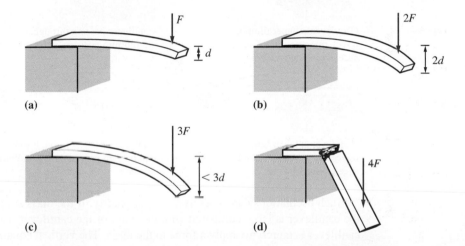

(a)

(b)

(c)

(d)

A beam that is fixed rigidly at one end and designed to support a load of some sort on the other end is called a cantilever. There are many common everyday and engineering applications of cantilevers. One with which most readers will be familiar is a diving springboard. Engineering applications include such things as warehouse hoists, the wings of aircraft and some types of bridges. For most of these applications the cantilever is designed to operate with small deflections; that is, the size and material of construction of the cantilever will be chosen by the designer so that, under the greatest anticipated load, the deflection of the cantilever will be small. Within this regime, the deflection of the tip of the cantilever will be proportional to the load applied. In the notation of Figure 10.20, we can write

$$d = \frac{1}{k}F \tag{10.40}$$

where d is the deflection of the cantilever, F is the load applied and k is a constant. Equation (10.40) essentially expresses a mathematical model of the cantilever, albeit a very simple one. The model is valid for applied loads such that the deflection of the cantilever remains within the linear range (where the deflection is proportional to load), and would not be valid for larger loads leading to nonlinear deflections, permanent distortions and breakages.

Equation (10.40) can also be used to investigate the dynamic behaviour of cantilevers. So far, we have assumed that the cantilever is in equilibrium under the applied load. Such situations, in which the cantilever is not moving, are called **static**. The term **dynamic** is conventionally used to describe situations and analyses in which the deflection of the cantilever is not constant in time. When the deflection of a cantilever is either greater than or less than the static deflection under the same load, the cantilever exerts a net force accelerating the mass back towards its equilibrium position. As a result, the deflection of the cantilever oscillates about the static equilibrium position. The situation is illustrated in Figure 10.21.

Figure 10.21
The dynamic behaviour of a loaded cantilever.

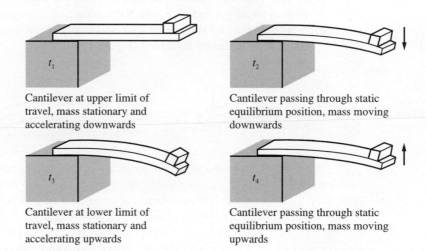

t_1

Cantilever at upper limit of travel, mass stationary and accelerating downwards

t_2

Cantilever passing through static equilibrium position, mass moving downwards

t_3

Cantilever at lower limit of travel, mass stationary and accelerating upwards

t_4

Cantilever passing through static equilibrium position, mass moving upwards

Such oscillations can be analysed fairly readily. If the mass supported on the end of the cantilever is large compared with the mass of the cantilever itself, the effect of the cantilever is merely to apply a force to the mass. The vertical equation of motion of the mass is then

$$m\frac{\mathrm{d}^2x}{\mathrm{d}t^2} = mg - F$$

where x is the instantaneous deflection of the tip of the cantilever below the horizontal, m is the mass and F is the upward force exerted on the mass by the cantilever due to its bending. But the restoring force, provided the deflection of the cantilever remains small enough at all times during the motion, is given by (10.40). Thus the motion is governed by the equation

$$m\frac{\mathrm{d}^2x}{\mathrm{d}t^2} = mg - kx$$

This equation, rearranged in the form

$$\frac{\mathrm{d}^2x}{\mathrm{d}t^2} + \frac{kx}{m} = g \tag{10.41}$$

is recognizable as a second-order linear nonhomogeneous constant-coefficient equation. In the static case, when the load is not moving, the solution of this equation is $x = mg/k$. This is, of course, also a particular integral for (10.41). The complementary function for (10.41) is

$$x = A \cos \omega t + B \sin \omega t, \quad \text{where } \omega^2 = k/m$$

The complete solution of (10.41) is therefore

$$x = \frac{mg}{k} + A \cos \omega t + B \sin \omega t \qquad \text{(10.42)}$$

The constants A and B could of course be determined if suitable initial conditions were provided. What is at least as important – if not more so for the engineer – is to understand the physical meaning of the solution (10.42). This is more easily done if (10.42) is slightly rearranged. Taking $C = (A^2 + B^2)^{1/2}$ and $\tan \delta = B/A$, so that

$$A = C \cos \delta \quad \text{and} \quad B = C \sin \delta$$

(10.42) becomes

$$x = \frac{mg}{k} + C \cos(\omega t - \delta) \qquad \text{(10.43)}$$

In physical terms this equation implies that the deflection of the cantilever takes the form of periodic oscillations of angular frequency ω and constant amplitude C about the position of static equilibrium of the cantilever (the position at which $kx = mg$).

The interested reader can check the accuracy of this description by constructing a cantilever from a flexible wooden or plastic ruler (the flexible plastic type is the most effective). The ruler should be held firmly by one end so that it projects over the edge of a desk or table, and the free end loaded with a sufficient mass of plasticine or other suitable material. The static equilibrium position is easily found. If the end of the ruler is displaced from this position and released, the plasticine-loaded end will be found to vibrate up and down around the equilibrium position. If the mass of plasticine is increased, the frequency of the vibration will be found to decrease as predicted by the relation $\omega^2 = k/m$.

A cantilever is not the only engineering system that gives rise to linear constant-coefficient equations. The pendulum shown in Figure 10.22 can be analysed thus. It is of length l and carries a mass m at its free end. If the mass of the pendulum arm is very small compared with m then, resolving forces at right-angles to the pendulum arm, the equation of motion of the mass is

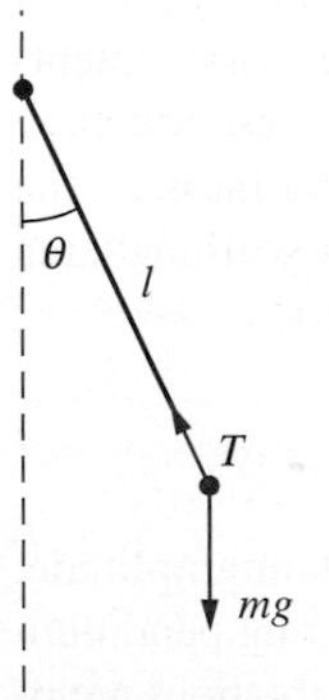

$$ml \frac{d^2\theta}{dt^2} = -mg \sin \theta$$

This is a second-order nonlinear differential equation, but if the displacement from the equilibrium position (in which the pendulum hangs stationary and vertically below the pivot) is small then $\sin \theta \approx \theta$ and the equation becomes

$$\frac{d^2\theta}{dt^2} + \frac{g}{l}\theta = 0 \qquad \text{(10.44)}$$

The solution of this equation is

$$\theta = A \cos \omega t + B \sin \omega t, \quad \text{with } \omega^2 = g/l \qquad \text{(10.45)}$$

Figure 10.22
A pendulum.

Figure 10.23
A floating buoy.

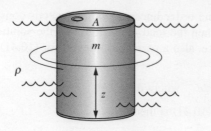

In other words, the pendulum's displacement from its equilibrium position oscillates sinusoidally with a frequency that decreases as the pendulum increases in length but is independent of the mass of the pendulum bob.

The buoy (or floating oil drum or similar) shown in Figure 10.23 also gives rise to a second-order linear constant-coefficient equation. Suppose the immersed depth of the buoy is z. Its mass (which is concentrated near the bottom of the buoy in order that it should float upright and not tip over) is m. We know, by Archimedes' principle, that the water in which the buoy floats exerts an upthrust on the buoy equal to the weight of the water displaced by the latter. If the cross-sectional area of the buoy is A and the density of the water is ρ, the upthrust will be $\rho A z g$. Hence the equation of motion is

$$m\frac{d^2z}{dt^2} = mg - \rho A z g$$

that is,

$$\frac{d^2z}{dt^2} + \frac{\rho A g}{m}z = g \tag{10.46}$$

Equation (10.46) has particular integral $z = m/\rho A$ and complementary function

$$z = A\cos\omega t + B\sin\omega t, \quad \text{with } \omega^2 = \rho A g/m$$

so the complete solution is

$$z = \frac{m}{\rho A} + A\cos\omega t + B\sin\omega t, \quad \text{with } \omega^2 = \frac{\rho A g}{m} \tag{10.47}$$

As in the case of the cantilever, the particular integral of the equation corresponds to the static equilibrium solution (when the buoy is floating just sufficiently immersed that the upthrust exerted by the water equals the weight of the buoy), and the complementary function describes oscillations of the buoy about this position. In this case the buoy oscillates with constant amplitude and a frequency that decreases as the mass of the buoy increases and increases as the density of the water and/or the cross-sectional area of the buoy increases.

10.10.2 Free oscillations of damped elastic systems

Equations (10.42), (10.45) and (10.47) all describe oscillations of constant amplitude. In reality, in all the situations described, a vibrating cantilever, an oscillating pendulum and a bobbing buoy, experience leads us to expect that the oscillations or vibrations are of decreasing amplitude, so that the motion eventually decays away and the system finally comes to rest in its static equilibrium position. This suggests that the mathematical

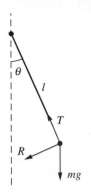

Figure 10.24
A pendulum with
air resistance.

models of the situation that we constructed in Section 10.10.1, and which are represented by (10.41), (10.44) and (10.46), are inadequate in some way.

What has been ignored in each case is the effect of dissipation of energy. Suppose, in the case of the pendulum, the motion of the pendulum were opposed by air resistance. The work which the pendulum does against the air resistance represents a continuous loss of energy, as a result of which the amplitude of oscillation of the pendulum decreases until it finally comes to rest. The situation is illustrated in Figure 10.24. The forces acting on the pendulum mass are gravity, air resistance (which opposes motion) and the tension in the pendulum arm. Resolving these forces perpendicular to the pendulum arm results in the equation of motion.

$$ml\frac{d^2\theta}{dt^2} = -R - mg\sin\theta$$

If the air resistance is assumed to be proportional to the speed of the pendulum mass then, since the speed of the mass is $l(d\theta/dt)$, we have

$$R = kl\frac{d\theta}{dt}$$

Hence

$$ml\frac{d^2\theta}{dt^2} = -kl\frac{d\theta}{dt} - mg\sin\theta$$

or, assuming θ is small so that $\sin\theta \approx \theta$ and rearranging the terms,

$$\frac{d^2\theta}{dt^2} + \frac{k}{m}\frac{d\theta}{dt} + \frac{g}{l}\theta = 0 \qquad\qquad\textbf{(10.48)}$$

This is a second-order linear constant-coefficient differential equation. It should be noted that the assumption that air resistance is proportional to speed is not the only possible assumption. For very slow-moving objects air resistance may well be more nearly constant, while for very fast-moving objects air resistance is usually taken to be proportional to the square of speed, which is a much better description of reality for fast-moving objects. For objects moving at modest speeds, however, the assumption that air resistance is proportional to speed is commonly adopted.

In the case of the cantilever and the buoy, also, we might assume that there is a resistance to motion that is proportional to the speed of motion. Again these are not the only possible assumptions, but they are ones that, under appropriate circumstances, are reasonable. The guiding principle when modelling physical systems such as these is to identify the physical source of the resistance and try to describe its behaviour. This is not a problem of mathematics but rather one of mathematical modelling, in which engineers must use their knowledge of physics and engineering as well as of mathematics in order to arrive at an appropriate mathematical description of reality.

Constructing models of a whole host of other engineering situations also leads to equations similar to (10.48). Basically, any situation in which the motion of some mass is caused by the sum of a force opposing displacement that is proportional to the displacement from some fixed position and a force that resists motion and is proportional to the speed of motion gives rise to an equation of the form

$$m\frac{d^2x}{dt^2} = -\mu\frac{dx}{dt} - \lambda x$$

that is,

$$\frac{d^2x}{dt^2} + p\frac{dx}{dt} + qx = 0 \qquad (10.49)$$

where

$$p = \frac{\mu}{m} \quad \text{and} \quad q = \frac{\lambda}{m}$$

We must have $p > 0$ and $q > 0$, because the two forces oppose displacement and motion respectively. We know from Section 10.10.1 that the solution of (10.49) is

$$x(t) = Ae^{m_1 t} + Be^{m_2 t}$$

where m_1 and m_2 are the roots of the characteristic equation

$$m^2 + pm + q = 0$$

For reasons that will become apparent, it is convenient to put (10.49) into the standard form

$$\frac{d^2x}{dt^2} + 2\zeta\omega\frac{dx}{dt} + \omega^2 x = 0 \qquad (10.50)$$

where, because $p, q > 0$, so are ζ and ω. The characteristic equation is then $m^2 + 2\zeta\omega m + \omega^2 = 0$, whose roots are

$$m = \begin{cases} -\zeta\omega \pm (\zeta^2 - 1)^{1/2}\omega & (\zeta > 1) \\ -\omega \text{ (twice)} & (\zeta = 1) \\ -\zeta\omega \pm j(1 - \zeta^2)^{1/2}\omega & (0 < \zeta < 1) \end{cases}$$

and the solution of (10.50) is therefore

$$x = A\exp\{-[\zeta - (\zeta^2 - 1)^{1/2}]\omega t\} + B\exp\{-[\zeta + (\zeta^2 - 1)^{1/2}]\omega t\} \; (\zeta > 1) \qquad (10.51a)$$

$$x = e^{-\omega t}(At + B) \qquad\qquad\qquad\qquad\qquad\qquad (\zeta = 1) \qquad (10.51b)$$

$$x = e^{-\zeta\omega t}\{A\cos[(1 - \zeta^2)^{1/2}\omega t] + B\sin[(1 - \zeta^2)^{1/2}\omega t]\} \qquad (0 < \zeta < 1) \quad (10.51c)$$

The first point to note about these solutions is that, since $\zeta > 0$ and $\omega > 0$, we have $x \to 0$ as $t \to \infty$ in all cases. Figure 10.25 shows the typical form of the solution (10.51) for various values of ζ. Variation of ω will only change the scale along the horizontal axis. For $0 < \zeta < 1$ the solution takes an oscillatory form with decaying amplitude. For $\zeta > 1$ the solution has the form of an exponential decay. The larger ζ, the slower is the final decay, since the exponential coefficient $\zeta - (\zeta^2 - 1)^{1/2} \to 0$ as $\zeta \to \infty$. In Figure 10.25 the envelopes of the oscillatory solutions are shown as broken lines. If the envelope of the oscillatory decay is compared with the solutions for $\zeta > 1$ it is quickly apparent that the most rapid decay is when $\zeta = 1$. It is now apparent why we chose to take (10.50) as the standard form for the description of second-order damped systems. The parameter ω is the **natural frequency** of the system, that is, the frequency with which it would oscillate in the absence of damping, and the parameter ζ is the **damping**

Figure 10.25
The motion of damped
second-order systems.

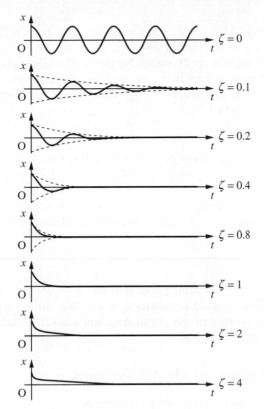

parameter of the system. When $\zeta = 1$, the decay of the motion of the system to its equilibrium state is as fast as is possible. For this reason, $\zeta = 1$ is referred to as **critical damping**. When $\zeta < 1$, the motion described by the equation decays to its equilibrium state in an oscillatory manner, passing through the equilibrium position on a number of occasions before coming to rest. For this reason, the motion is described as **under-damped**. When $\zeta > 1$, the motion described by the equation decays to the equilibrium position in a direct manner, but less rapidly than for a critically damped system. In this case the motion is described as **over-damped**.

For the under-damped case an engineering rule of thumb that is commonly used is that when $\zeta = 0.3$ the system shows three *discernible* overshoots before settling down. That is not to say that there are only three overshoots – on the contrary, there are an infinite number – but by the fourth and subsequent overshoots the amplitude of the oscillations has decayed to less than 2% of the initial amplitude. When $\zeta = 0.5$, there are two discernible overshoots (the third and subsequent ones have amplitude less than $\frac{1}{2}$% of initial); and when $\zeta = 0.7$, there is only one significant overshoot.

Another rule of thumb relates to the envelope containing the response. The response is contained within an envelope defined by the function $e^{-\zeta\omega t}$. Now $e^{-3} = 0.0498$ and $e^{-4.5} = 0.0111$; so when $\zeta\omega t = 3$, the amplitude of the response will have fallen to approximately 5% of its original amplitude; and when $\zeta\omega t = 4.5$, it will have fallen to roughly 1% of its original amplitude. For this reason, $t = 1/\zeta\omega$ is called the **decay time** of the system, and engineers use the rule of thumb that response falls to 5% in three decay times and 1% in four and a half decay times.

Example 10.45   A pendulum of mass 4 kg, length 2 m and an air resistance coefficient of $5 \, \mathrm{N \, s \, m^{-1}}$ is released from an initial position in which it makes an angle of $20°$ with the vertical. Assuming that this angle is small enough for the small-angle approximation to be made in the equation of motion, how many oscillations will be obviously observable before the pendulum comes to rest, and how long will it take for the amplitude of the motion to have fallen to less than $1°$?

Solution   The motion of the pendulum is described by (10.48). Comparing this with (10.50), we see that $\omega = (g/l)^{1/2} = 2.215 \, \mathrm{rad \, s^{-1}}$ and $2\zeta\omega = k/m = 1.25$; that is, $\zeta = 0.282$. Hence, since $\zeta \approx 0.3$, we expect to see three obvious discernible overshoots (one and a half complete cycles of oscillation). The decay time for the pendulum is $1/\zeta\omega = 2m/k = 1.6 \, \mathrm{s}$, so we expect the amplitude of oscillation of the pendulum to fall to 5% of its initial amplitude in 4.8 s.

It is evident from the preceding paragraphs and from Example 10.45 that the natural frequency ω and the damping parameter ζ of a system are a very convenient way of summarizing the properties of any physical system whose oscillations are described by a damped second-order equation.

10.10.3   Forced oscillations of elastic systems

In Sections 10.10.1 and 10.10.2 we have examined the behaviour of elastic systems undergoing oscillations in which the system is free to choose its own frequency of oscillation. In many situations elastic systems are driven by some external force at a frequency imposed by the latter.

A familiar example of such a situation is the vibration of lamp posts in strong winds. The lightweight tubular metal lamp posts that have frequently been installed by highway authorities since the 1960s are a form of cantilever. The vertical post is rigidly mounted in the ground, and carries at its top a lamp apparatus. The lamp is effectively a concentrated mass, though it would probably not be sufficiently massive to allow the assumption, which we made for the cantilevers treated in Sections 10.10.1 and 10.10.2, that the mass of the post is small compared with the mass of the lamp. Nonetheless, if the top of the lamp post were to be pulled to one side and released, the post would certainly vibrate. The frequency of that vibration would be a function of the stiffness (restoring force per unit lateral tip displacement) of the lamp post and the mass of both the post and the lamp apparatus carried at the top. The stiffer the lamp post, the higher would be the frequency of vibration. The more massive the post and the lamp apparatus, the lower the frequency.

When the wind blows past the lamp post, aerodynamic effects (known as vortex shedding) result in an oscillating side-force on the lamp post. The frequency of this side-force is a function of the wind speed and the diameter of the lamp post. There is no reason why the frequency of oscillation of the wind-induced side-force should coincide with the frequency of the oscillations that result if the top of the lamp post is displaced sideways and released to vibrate freely. Under the influence of the oscillating side-force, such lamp posts commonly vibrate from side to side in time with the oscillating side-force. As the wind speed changes, so the frequency of the side-force and therefore of the lamp post's vibrations changes. Other types of lamp post, notably the reinforced concrete type and the older cast-iron lamp posts, do not seem to exhibit this behaviour. This can be explained in terms of their greater stiffness, as we shall see later.

Oscillations of elastic systems in which the system is free to adopt its own natural frequency of vibration are called **free vibrations**, while those caused by oscillating external forces (and in which the system must vibrate at the frequency of the external forcing) are called **forced vibrations**.

Other large structures can also be forced to oscillate by the wind blowing past them, just like lamp posts. Large modern factory chimneys made of steel or aluminium sections bolted together and stayed by wires exhibit this type of vibration, as do the suspension cables and hangers of suspension bridges and the overhead power transmission lines of electricity grid systems. The legs of offshore oil rigs can be forced to vibrate by ocean currents and waves. The wings of an aircraft (which, being mounted rigidly in the fuselage of the aircraft, are also a form of cantilever) may vibrate under aerodynamic loads, particularly from atmospheric turbulence. Large pieces of static industrial machinery are usually bolted down to the ground. If such fastening is subjected to a large load, it will usually give a little, so the attachment of the machinery to the floor must be considered as elastic. If the machinery, when in operation, produces an internal side-load (such as an out-of-balance rotor would produce) then the machinery is seen to rock from side to side on its mountings at the frequency of the internally generated side-loading. This effect can often be observed in the rocking vibrations of a car engine when it is idling in a stationary car. It is well known that bodies of men or women marching are ordered to break step when passing over bridges. If they did not, the regular footfalls of the whole group would create a periodic force on the bridge. The dangers of such regular forces will become apparent in our analysis. All these situations are similar in nature to the forced vibrations of the lamp post under the influence of the wind. In most of them the oscillations induced by the side-force are potentially disastrous, and must be understood by the engineer so that engineering artefacts may be designed to avoid the destructive effects of forced vibrations.

A simple model of the vibrations of a lamp post can be constructed as shown in Figure 10.26. The lamp apparatus, of mass m, is displaced from its equilibrium position by a distance x. The structure of the cantilever results in a restoring force S and air resistance in a restoring force R. The wind load (which, remember, is not a force in the direction of the wind but rather an oscillatory side-force) is W. If the displacement x is small and the displacement velocity is not too great then we may reasonably assume

$$S = kx \quad \text{and} \quad R = \lambda \frac{\mathrm{d}x}{\mathrm{d}t}$$

Making the somewhat unrealistic assumption that the mass of the lamp post itself is small compared with the mass of the lamp apparatus, the equation of motion of the lamp is seen to be

$$m\frac{\mathrm{d}^2 x}{\mathrm{d}t^2} = -\lambda \frac{\mathrm{d}x}{\mathrm{d}t} - kx + W$$

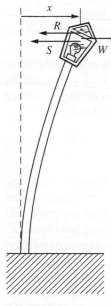

Figure 10.26
The forces acting on a vibrating lamp post.

The wind-induced force W is oscillatory, so we shall assume that it is of the form

$$W = W_0 \cos \Omega t$$

Hence the equation of motion becomes

$$m\frac{d^2x}{dt^2} + \lambda\frac{dx}{dt} + kx = W_0\cos\Omega t \tag{10.52}$$

which is a second-order linear nonhomogeneous constant-coefficient differential equation. In order to facilitate the interpretation of the result, we shall replace (10.52) with the equivalent equation

$$\frac{d^2x}{dt^2} + 2\zeta\omega\frac{dx}{dt} + \omega^2 x = F\cos\Omega t \tag{10.53}$$

The particular integral for (10.53) is obtained by assuming the form $A\cos\Omega t + B\sin\Omega t$, and is found to be

$$\frac{(\omega^2 - \Omega^2)F\cos\Omega t + 2\zeta\omega\Omega F\sin\Omega t}{(\omega^2 - \Omega^2)^2 + 4\zeta^2\omega^2\Omega^2} \tag{10.54a}$$

or equivalently

$$\frac{F}{[(\omega^2 - \Omega^2)^2 + 4\zeta^2\omega^2\Omega^2]^{1/2}}\cos(\Omega t - \delta) \tag{10.54b}$$

where

$$\delta = \tan^{-1}\left(\frac{2\zeta\omega\Omega}{\omega^2 - \Omega^2}\right)$$

The complementary function is of course the solution of the homogeneous equivalent of (10.52), which is just (10.50). The complementary function is therefore given by (10.51). The motion of a damped second-order system in response to forcing by a force $F\cos\Omega t$ is therefore the sum of (10.51) and (10.54a) or (10.54b). In Section 10.10.2 we saw that (10.51) is, for positive ζ, always a decaying function of time. The complementary function for (10.53) therefore represents a motion that decays to nothing with time, and is therefore called a **transient solution**. The particular integral, on the other hand, does not decay, but continues at a steady amplitude for as long as the forcing remains. The long-term response of a damped second-order system to forcing by a force $F\cos\Omega t$ is therefore to oscillate at the forcing frequency Ω with amplitude

$$A(\Omega) = \frac{1}{[(\omega^2 - \Omega^2)^2 + 4\zeta^2\omega^2\Omega^2]^{1/2}} \tag{10.55}$$

times the amplitude of the forcing term. This is called the **steady-state response** of the system. Evidently, the amplitude of the steady-state response changes as the frequency Ω of the forcing changes. In Figure 10.27 the form of the response amplitude $A(\Omega)$ as a function of Ω is shown for a range of values of ζ. Obviously, the characteristics of the response of a damped second-order system to forcing depend crucially on the damping. For lightly damped systems (ζ near to 0) the response has a definite maximum near to ω, the natural frequency of the system. For more heavily damped systems the peak response is smaller, and for large enough ζ the peak disappears altogether.

The significance of this is that systems subjected to an oscillatory external force at a frequency near to the natural frequency of the system will, unless they are sufficiently heavily damped, respond with large-amplitude motion. This phenomenon is known as

Figure 10.27
The response of
damped second-order
systems to sinusoidal
forcing.

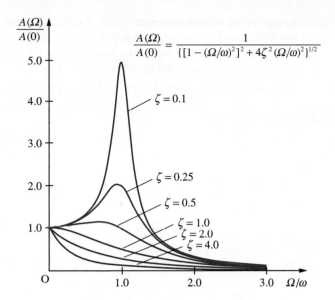

$$\frac{A(\Omega)}{A(0)} = \frac{1}{\{[1 - (\Omega/\omega)^2]^2 + 4\zeta^2 (\Omega/\omega)^2\}^{1/2}}$$

resonance. Resonance can cause catastrophic failure of the structure of a system. The history of engineering endeavour contains many examples of structures that have failed because they have been subjected to some external exciting force with a frequency near to one of the natural frequencies of vibration of the structure. Perhaps the most famous example of such a failure is the collapse in 1941 of the suspension bridge at Tacoma Narrows in the USA. This failure, due to wind-induced oscillations, was recorded on film and provides a salutary lesson for all engineers. Similar forces have destroyed factory chimneys, power transmission lines and aircraft.

It should now be obvious why the amplitude of oscillation of the tubular metal lamp post varies with wind speed. The natural frequency of the lamp post is determined by its structure, and is therefore fixed. The frequency of the vortex shedding, and so of the oscillatory side-force, is directly proportional to the wind speed. Hence, as the wind speed increases, so does the frequency of external forcing of the lamp post. As the forcing frequency approaches the natural frequency of the lamp post, the amplitude of the lamp post's vibrations increases. When the wind speed increases sufficiently, the forcing frequency exceeds the natural frequency, and the amplitude of the oscillations decreases again. The same explanation applies to the Tacoma Narrows bridge. The bridge, once constructed, stood for some months without serious difficulty. The failure was the result of the first storm in which wind speeds rose sufficiently to excite the bridge structure at one of its natural frequencies. (Since the structure of a suspension bridge is much more complex than that of a simple cantilever, such a bridge has many natural frequencies, corresponding to different modes of vibration.)

10.10.4 Oscillations in electrical circuits

In Section 10.2.4 we analysed a simple electrical circuit composed of a resistor, a capacitor and an inductor. In that case we considered what happened when a switch was thrown in a circuit containing a d.c. voltage source. If an alternating voltage signal is applied to a similar circuit the equation governing the resulting oscillations also turns out to be a second-order linear differential equation.

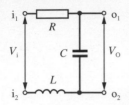

Figure 10.28
An *LCR* electrical
circuit.

Consider the circuit shown in Figure 10.28. Suppose a voltage V_i is applied across the input terminals i_1 and i_2. The voltage drop across the inductor is $L(di/dt)$, that across the capacitor is $\int (i/C)dt$ and that across the resistor is Ri. Kirchhoff's laws (or the principle of conservation of charge) tell us that the current in each component must be the same. The voltage across the output terminals o_1 and o_2 is $\int (i/C)dt = V_o$. Hence we have

$$L\frac{di}{dt} + Ri + \frac{1}{C}\int i\,dt = V_i$$

with

$$V_o = \frac{1}{C}\int i\,dt$$

That is,

$$LC\frac{d^2V_o}{dt^2} + RC\frac{dV_o}{dt} + V_o = V_i$$

or

$$\frac{d^2V_o}{dt^2} + \frac{R}{L}\frac{dV_o}{dt} + \frac{V_o}{LC} = \frac{V_i}{LC} \tag{10.56}$$

This is a second-order linear nonhomogeneous constant-coefficient differential equation. If the signal V_i is of the form $V\cos\Omega t$ then we essentially have forced oscillations of a second-order system again. If we write

$$\omega^2 = \frac{1}{LC}, \quad 2\zeta\omega = \frac{R}{L} \quad \text{and} \quad F = \frac{V}{LC}$$

then (10.56) takes the standard form of (10.53), and we can infer that the voltage V_o will be sinusoidal with amplitude

$$\frac{A(\Omega)}{LC}V$$

where

$$A(\Omega) = \frac{1}{[(\omega^2 - \Omega^2)^2 + 4\zeta^2\omega^2\Omega^2]^{1/2}}$$

Thus when a sinusoidal voltage waveform is applied to the input terminals of the circuit, the voltage appearing at the output terminals is also a sinusoidal waveform, but one whose amplitude, relative to the input waveform amplitude, depends on the frequency of the input. A circuit that has this property is of course called a **filter**.

The form of $A(\Omega)$ will depend on ω and ζ, which in turn are determined by the values of L, R and C. The latter could be chosen so that ζ is small. In that case the circuit provides a large output when the input frequency Ω is near some frequency ω (which is determined by the choice of L and C) and a smaller output otherwise. This is a **tuned circuit** or a **bandpass filter**. If L, R and C are chosen so that ζ is larger (say near unity) then the circuit provides a larger output for small Ω and a smaller output for larger Ω. Such a circuit is a **low-pass filter**.

In this section, we have seen how problems in two very different areas of engineering – one mechanical and the other electrical – both give rise to very similar equations.

Our knowledge of the form of the solutions of the equation is applicable to either area. This is a good example of the unifying properties of mathematics in engineering science. There are many other applications of the theory of the solution of second-order linear constant-coefficient differential equations in engineering.

It is also worth commenting here that filters of the type that we have described in this section are called **passive filters** since they use only inductors, resistors and capacitors – components that are referred to as **passive components**. Modern practice in electrical engineering involves the use of **active components** such as operational amplifiers in filter design, such filters being known as **active filters**. The analysis of the operation of active filters is more complex than that of passive filters. While, for many applications, active filters have displaced passive filters in modern practice, there are also many applications in which passive filters remain the norm.

10.10.5 Exercises

64 Find the damping parameters and natural frequencies of the systems governed by the following second-order linear constant-coefficient differential equations:

(a) $\dfrac{d^2x}{dt^2} + 6\dfrac{dx}{dt} + 9x = 0$

(b) $\dfrac{d^2x}{dt^2} + 4\dfrac{dx}{dt} + 7x = 0$

65 Determine the values of the appropriate parameters needed to give the systems governed by the following second-order linear constant-coefficient differential equations the damping parameters and natural frequencies stated:

(a) $\dfrac{d^2x}{dt^2} + 2a\dfrac{dx}{dt} + bx = 0$, $\zeta = 0.5$, $\omega = 2$

(b) $\dfrac{d^2x}{dt^2} + p\dfrac{dx}{dt} + qx = 0$, $\zeta = 1.4$, $\omega = 0.5$

(c) $\dfrac{d^2x}{dt^2} + \beta\dfrac{dx}{dt} + \gamma x = 0$, $\zeta = 1$, $\omega = 1.1$

66 Find the damping parameters and natural frequencies of the systems governed by the following second-order linear constant-coefficient differential equations:

(a) $\dfrac{d^2x}{dt^2} + 2a\dfrac{dx}{dt} + 16p^2x = 0$

(b) $2\dfrac{d^2x}{dt^2} + 14\dfrac{dx}{dt} + \dfrac{1}{\alpha}x = 0$

(c) $2.41\dfrac{d^2x}{dt^2} + 1.02\dfrac{dx}{dt} + 7.63x = 0$

(d) $\dfrac{1}{\eta}\dfrac{d^2x}{dt^2} + 40\dfrac{dx}{dt} + 25\eta x = 0$

(e) $1.88\dfrac{d^2x}{dt^2} + 4.71\dfrac{dx}{dt} + 0.48x = 0$

67 Determine the values of the appropriate parameters needed to give the systems governed by the following second-order linear constant-coefficient differential equations the damping parameters and natural frequencies stated:

(a) $\dfrac{d^2x}{dt^2} + \alpha\dfrac{dx}{dt} + \beta x = 0$, $\zeta = 0.5$, $\omega = \pi$

(b) $\dfrac{d^2x}{dt^2} + a\dfrac{dx}{dt} + bx = 0$, $\zeta = 0.1$, $\omega = 2\pi$

(c) $4\dfrac{d^2x}{dt^2} + q\dfrac{dx}{dt} + rx = 0$, $\zeta = 1$, $\omega = 1$

(d) $a\dfrac{d^2x}{dt^2} + b\dfrac{dx}{dt} + 14x = 0$, $\zeta = 2$, $\omega = 2\pi$

68 The function $A(\Omega)$ is as given by (10.55) and shown in Figure 10.27. Show that $A(\Omega)$ has a simple maximum point when $\zeta < \sqrt{\tfrac{1}{2}}$. Let the value of Ω for which this maximum occurs be $\Omega_{\max}$. Find $\Omega_{\max}$ as a function of ζ and ω, and also find $A(\Omega_{\max})$.

For $\zeta > \sqrt{\tfrac{1}{2}}$, $A(\Omega)$ has no maximum, but does have a single point of inflection. Show, by consideration of Figure 10.27, that $|dA/d\Omega|$ is a maximum at the point of inflection. Let Ω_c be the value of Ω for which the point of inflection occurs. Show that Ω_c satisfies the equation

$$3\Omega^6 + 5\beta\omega^2\Omega^4 + (4\beta^2 - 3)\omega^4\Omega^2 - \beta\omega^6 = 0$$

where $\beta = 2\zeta^2 - 1$. Hence show that for $\zeta = \sqrt{\frac{1}{2}}$ the greatest value of $|dA/d\Omega|$ occurs when $\Omega = \omega$ and is $1/(\sqrt{2}\omega^3)$. Also find the greatest values of $|dA/d\Omega|$ when $\zeta = \sqrt{(\frac{1}{2} + \frac{1}{6}\sqrt{3})}$ and when $\zeta = 1$.

Show that $|d^2A(0)/d\Omega^2|$ is minimized when $\zeta = \sqrt{\frac{1}{2}}$. The two values of ζ that minimize the maxima of $|dA/d\Omega|$ and $|d^2A(0)/d\Omega^2|$ respectively are important, particularly in control theory, since, in different senses, they maximize the flatness of the response function $A(\Omega)$.

69 An underwater sensor is mounted below the keel of the fast patrol boat shown in Figure 10.29. The supporting bracket is of cylindrical cross-section (diameter 0.04 m), and so is subject to an oscillating side-force due to vortex shedding. The bracket is of negligible mass compared with the sensor itself, which has a mass of 4 kg. The bracket has a tip displacement stiffness of $25\,000\,\mathrm{N\,m^{-1}}$. The frequency of the oscillating side-force is SU/d, where U is the speed of the vessel through the water, d is the diameter of the supporting bracket and S is the Strouhal number for vortex shedding from a circular cylinder. S has the value 0.20 approximately. At what speed will the frequency of the side-force coincide with the natural frequency of the sensor and mounting?

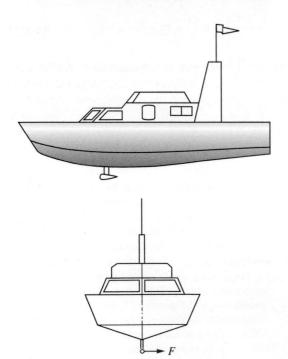

Figure 10.29 An underwater sensor mounting.

70 The piece of machinery shown in Figure 10.30 is mounted on a solid foundation in such a way that the mounting may be characterized as a rigid pivot and two stiff springs as shown. A damper is connected between the machine and an adjacent strong point. The mass of the machine is 500 kg, the length $a = 1$ m, the length $b = 1.2$ m and the spring stiffness is $8000\,\mathrm{N\,m^{-1}}$. The moment of inertia of the machine about the pivot point is $2ma^2$. The machine generates internally a side-force F that may be approximated as $F_0 \cos 2\pi ft$. As the machine runs up to speed, the frequency f increases from 0 to 6 Hz. What is the minimum damper coefficient that will prevent the machine from vibrating with any amplitude greater than twice its zero-frequency amplitude $A(0)$ during a run-up?

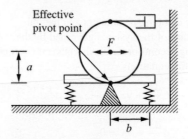

Figure 10.30 A compliantly mounted piece of machinery.

71 Figure 10.31 shows a radio tuner circuit. Show that the natural frequency and damping parameters of the circuit are $1/\sqrt{(LC)}$ and

$$\frac{1}{2}\left(\frac{L}{C}\right)^{1/2}\left(\frac{1}{R_1} + \frac{1}{R_2}\right)$$

respectively. If $R_1 = 300\,\Omega$ and $R_2 = 50\,\Omega$ what value should L have, and over what range should C be adjustable in order that the circuit have a damping factor of $\zeta = 0.1$ and can be tuned to the medium waveband (505–1605 kHz)?

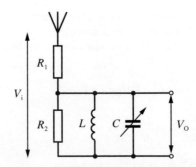

Figure 10.31 A radio tuner circuit.

| 10.11 | # Numerical solution of second- and higher-order differential equations |

Obviously, the classes of second- and higher-order differential equations that can be solved analytically, while representing an important subset of the totality of such equations, are relatively restricted. Just as for first-order equations, those for which no analytical solution exists can still be solved by numerical means. The numerical solution of second- and higher-order equations does not, in fact, need any significant new mathematical theory or technique.

10.11.1   Numerical solution of coupled first-order equations

In Section 10.6 we met Euler's method for the numerical solution of equations of the form

$$\frac{dx}{dt} = f(t, x)$$

that is, first-order differential equations involving a single dependent variable and a single independent variable. In Section 10.3 we noted that it was possible to have sets of coupled first-order equations, each involving the same independent variable but with more than one dependent variable. An example of this type of equation set is

$$\frac{dx}{dt} = x - y^2 + xt \qquad (10.57a)$$

$$\frac{dy}{dt} = 2x^2 + xy - t \qquad (10.57b)$$

This is a pair of differential equations in the dependent variables x and y with the independent variable t. The derivative of each of the dependent variables depends not only on itself and on the independent variable t, but also on the other dependent variable. Neither of the equations can be solved in isolation or independently of the other – both must be solved simultaneously, or side by side. A pair of coupled differential equations such as (10.57) may be characterized as

$$\frac{dx}{dt} = f_1(t, x, y) \qquad (10.58a)$$

$$\frac{dy}{dt} = f_2(t, x, y) \qquad (10.58b)$$

For a set of p such equations it is convenient to denote the dependent variables not by $x, y, z, \ldots$ but by $x_1, x_2, x_3, \ldots, x_p$ and to denote the set of equations by

$$\frac{dx_i}{dt} = f_i(t, x_1, x_2, \ldots, x_p) \quad (i = 1, 2, \ldots, p)$$

or equivalently, using vector notation,

$$\frac{d}{dt}[\mathbf{x}] = \mathbf{f}(t, \mathbf{x})$$

where $x(t)$ is a vector function of t given by

$$x(t) = [x_1(t) \quad x_2(t) \quad \ldots \quad x_p(t)]^T$$

$f(t, x)$ is a vector-valued function of the scalar variable t and the vector variable x.

Euler's method for the solution of a single differential equation takes the form

$$X_{n+1} = X_n + hf(t_n, X_n)$$

If we were to try to apply this method to (10.58a), we should obtain

$$X_{n+1} = X_n + hf_1(t_n, X_n, Y_n)$$

In other words, the value of X_{n+1} depends not only on t_n and X_n but also on Y_n. In the same way, we would obtain

$$Y_{n+1} = Y_n + hf_2(t_n, X_n, Y_n)$$

for Y_{n+1}. In practice, this means that to solve two coupled differential equations, we must advance the solution of both equations simultaneously in the manner shown in Example 10.46.

Example 10.46   Find the value of $X(1.4)$ satisfying the following initial-value problem:

$$\frac{dx}{dt} = x - y^2 + xt, \quad x(1) = 0.5$$

$$\frac{dy}{dt} = 2x^2 + xy - t, \quad y(1) = 1.2$$

using Euler's method with time step $h = 0.1$.

Solution   The right-hand sides of the two equations will be denoted by $f_1(t, x, y)$ and $f_2(t, x, y)$ respectively, so

$$f_1(t, x, y) = x - y^2 + xt \quad \text{and} \quad f_2(t, x, y) = 2x^2 + xy - t$$

The initial condition is imposed at $t = 1$, so t_n will denote $1 + nh$, X_n will denote $X(1 + nh)$, and Y_n will denote $Y(1 + nh)$. Then we have

$$X_1 = x_0 + hf_1(t_0, x_0, y_0) \qquad\qquad Y_1 = y_0 + hf_2(t_0, x_0, y_0)$$
$$= 0.5 + 0.1f_1(1, 0.5, 1.2) \qquad\qquad = 1.2 + 0.1f_2(1, 0.5, 1.2)$$
$$= 0.4560 \qquad\qquad\qquad\qquad = 1.2100$$

for the first step. The next step is therefore

$$X_2 = X_1 + hf_1(t_1, X_1, Y_1) \qquad\qquad Y_2 = Y_1 + hf_2(t_1, X_1, Y_1)$$
$$= 0.4560 \qquad\qquad\qquad\qquad = 1.2100$$
$$+ 0.1f_1(1.1, 0.4560, 1.2100) \qquad\qquad + 0.1f_2(1.1, 0.4560, 1.2100)$$
$$= 0.4054 \qquad\qquad\qquad\qquad = 1.1968$$

and the third step is

$$X_3 = 0.4054 \qquad\qquad Y_3 = 1.1968$$

$$+ 0.1f_1(1.2, 0.4054, 1.1968) \qquad + 0.1f_2(1.2, 0.4054, 1.1968)$$

$$= 0.3513 \qquad\qquad = 1.1581$$

Finally, we obtain

$$X_4 = 0.3513 + 0.1f_1(1.3, 0.3513, 1.1581)$$

$$= 0.2980$$

Hence we have $X(1.4) = 0.2980$.

It should be obvious from Example 10.46 that the main drawback of extending Euler's method to sets of differential equations is the additional labour and tedium of the computations. Intrinsically, the computations are no more difficult, merely much more laborious – a prime example of a problem ripe for computerization.

10.11.2 State-space representation of higher-order systems

The solution of differential equation initial-value problems of order greater than one can be reduced to the solution of a set of first-order differential equations. This is achieved by a simple transformation, illustrated by Example 10.47.

Example 10.47   The initial-value problem

$$\frac{d^2x}{dt^2} + x^2t\frac{dx}{dt} - xt^2 = \tfrac{1}{2}t^2, \quad x(0) = 1.2, \quad \frac{dx}{dt}(0) = 0.8$$

can be transformed into two coupled first-order differential equations by introducing an additional variable

$$y = \frac{dx}{dt}$$

With this definition, we have

$$\frac{d^2x}{dt^2} = \frac{dy}{dt}$$

and so the differential equation becomes

$$\frac{dy}{dt} + x^2ty - xt^2 = \tfrac{1}{2}t^2$$

Thus the original differential equation can be replaced by a pair of coupled first-order differential equations, together with initial conditions:

$$\frac{dx}{dt} = y, \quad x(0) = 1.2$$

$$\frac{dy}{dt} = -x^2 ty + xt^2 + \tfrac{1}{2}t^2, \quad y(0) = 0.8$$

This process can be extended to transform a *p*th-order initial-value problem into a set of *p* first-order equations, each with an initial condition. Once the original equation has been transformed in this way, its solution by numerical methods is just the same as if it had been a set of coupled equations in the first place.

Example 10.48

Find the value of $X(0.2)$ satisfying the initial-value problem

$$\frac{d^3x}{dt^3} + xt\frac{d^2x}{dt^2} + t\frac{dx}{dt} - t^2x = 0, \quad x(0) = 1, \quad \frac{dx}{dt}(0) = 0.5, \quad \frac{d^2x}{dt^2}(0) = -0.2$$

using Euler's method with step size $h = 0.05$.

Solution

Since this is a third-order equation, we need to introduce two new variables:

$$y = \frac{dx}{dt} \quad \text{and} \quad z = \frac{dy}{dt} = \frac{d^2x}{dt^2}$$

Then the equation is transformed into a set of three first-order differential equations

$$\frac{dx}{dt} = y \qquad\qquad x(0) = 1$$

$$\frac{dy}{dt} = z \qquad\qquad y(0) = 0.5$$

$$\frac{dz}{dt} = -xtz - ty + t^2x \quad z(0) = -0.2$$

Applied to the set of differential equations

$$\frac{dx}{dt} = f_1(t, x, y, z)$$

$$\frac{dy}{dt} = f_2(t, x, y, z)$$

$$\frac{dz}{dt} = f_3(t, x, y, z)$$

the Euler scheme is of the form

$$X_{n+1} = X_n + hf_1(t_n, X_n, Y_n, Z_n)$$
$$Y_{n+1} = Y_n + hf_2(t_n, X_n, Y_n, Z_n)$$
$$Z_{n+1} = Z_n + hf_3(t_n, X_n, Y_n, Z_n)$$

In this case, therefore, we have

$$X_0 = x_0 = 1$$

$$Y_0 = y_0 = 0.5$$

$$Z_0 = z_0 = -0.2$$

$$f_1(t_0, X_0, Y_0, Z_0) = Y_0 = 0.5000$$

$$f_2(t_0, X_0, Y_0, Z_0) = Z_0 = -0.2000$$

$$f_3(t_0, X_0, Y_0, Z_0) = -X_0 t_0 Z_0 - t_0 Y_0 + t_0^2 X_0$$

$$= -1.0000 \times 0 \times (-0.2000) - 0 \times 0.5000 + 0^2 \times 1.0000$$

$$= 0.0000$$

$$X_1 = 1.0000 + 0.05 \times 0.5000 = 1.0250$$

$$Y_1 = 0.5000 + 0.05 \times (-0.2000) = 0.4900$$

$$Z_1 = -0.2000 + 0.05 \times 0.0000 = -0.2000$$

$$f_1(t_1, X_1, Y_1, Z_1) = Y_1 = 0.4900$$

$$f_2(t_1, X_1, Y_1, Z_1) = Z_1 = -0.2000$$

$$f_3(t_1, X_1, Y_1, Z_1) = -X_1 t_1 Z_1 - t_1 Y_1 + t_1^2 X_1$$

$$= -1.0250 \times 0.05 \times (-0.2000) - 0.05 \times 0.4900$$

$$+ 0.05^2 \times 1.0250 = -0.0117$$

$$X_2 = 1.0250 + 0.05 \times 0.4900 = 1.0495$$

$$Y_2 = 0.4900 + 0.05 \times (-0.2000) = 0.4800$$

$$Z_2 = -0.2000 + 0.05 \times (-0.0117) = -0.2005$$

Proceeding similarly we have

$$X_3 = 1.0495 + 0.05 \times 0.4800 = 1.0735$$

$$Y_3 = 0.4800 + 0.05 \times (-0.2005) = 0.4700$$

$$Z_3 = -0.2005 + 0.05 \times (-0.0165) = -0.2013$$

$$X_4 = 1.0735 + 0.05 \times 0.4700 = 1.0970$$

$$Y_4 = 0.4700 + 0.05 \times (-0.2013) = 0.4599$$

$$Z_4 = -0.2013 + 0.05 \times (-0.0139) = -0.2018$$

 Hence $X(0.2) = X_4 = 1.0970$. It should be obvious by now that computations like these are sufficiently tedious to justify the effort of writing a computer program to carry out the actual arithmetic. The essential point for the reader to grasp is not the mechanics but the principle whereby methods for the solution of first-order differential equations (and this includes the more sophisticated methods as well as Euler's method) can be extended to the solution of sets of equations and hence to higher-order equations.

10.11.3 Exercises

72 Transform the following initial-value problems into sets of first-order differential equations with appropriate initial conditions:

(a) $\dfrac{d^2x}{dt^2} + 6(x^2 - t)\dfrac{dx}{dt} - 4xt = 0,$

$x(0) = 1, \quad \dfrac{dx}{dt}(0) = 2$

(b) $\dfrac{d^2x}{dt^2} - \sin\left(\dfrac{dx}{dt}\right) + 4x = 0,$

$x(0) = 0, \quad \dfrac{dx}{dt}(0) = 0$

73 Find the value of $X(0.3)$ for the initial-value problem

$\dfrac{d^2x}{dt^2} + x^2\dfrac{dx}{dt} + x = \sin t,$

$x(0) = 0, \quad \dfrac{dx}{dt}(0) = 1$

using Euler's method with step size $h = 0.1$.

74 Transform the following initial-value problems into sets of first-order differential equations with appropriate initial conditions:

(a) $\dfrac{d^2x}{dt^2} + 4(x^2 - t^2)^{1/2} = 0,$

$x(1) = 2, \quad \dfrac{dx}{dt}(1) = 0.5$

(b) $\dfrac{d^3x}{dt^3} + t\dfrac{d^2x}{dt^2} + 6e^t\dfrac{dx}{dt} - x^2t = e^{2t},$

$x(0) = 1, \quad \dfrac{dx}{dt}(0) = 2, \quad \dfrac{d^2x}{dt^2}(0) = 0$

(c) $\dfrac{d^3x}{dt^3} + t\dfrac{d^2x}{dt^2} + x^2 = \sin t,$

$x(1) = 1, \quad \dfrac{dx}{dt}(1) = 0, \quad \dfrac{d^2x}{dt^2}(1) = -2$

(d) $\left(\dfrac{d^3x}{dt^3}\right)^{1/2} + t\dfrac{d^2x}{dt^2} + x^2t^2 = 0,$

$x(2) = 0, \quad \dfrac{dx}{dt}(2) = 0, \quad \dfrac{d^2x}{dt^2}(2) = 2$

(e) $\dfrac{d^4x}{dt^4} + x\dfrac{d^2x}{dt^2} + x^2 = \ln t, \quad x(0) = 0,$

$\dfrac{dx}{dt}(0) = 0, \quad \dfrac{d^2x}{dt^2}(0) = 4, \quad \dfrac{d^3x}{dt^3}(0) = -3$

(f) $\dfrac{d^4x}{dt^4} + \left(\dfrac{dx}{dt} - 1\right)\dfrac{d^3x}{dt^3} + \dfrac{dx}{dt} - (xt)^{1/2}$

$= t^2 + 4t - 5,$

$x(0) = a, \quad \dfrac{dx}{dt}(0) = 0, \quad \dfrac{d^2x}{dt^2}(0) = b, \quad \dfrac{d^3x}{dt^3}(0) = 0$

75 Use Euler's method to compute an approximation $X(0.65)$ to the solution $x(0.65)$ of the initial-value problem

$\dfrac{d^3x}{dt^3} + \dfrac{d^2x}{dt^2}(x - t) + \left(\dfrac{dx}{dt}\right)^2 - x^2 = 0,$

$x(0.5) = -1, \quad \dfrac{dx}{dt}(0.5) = 1, \quad \dfrac{d^2x}{dt^2}(0.5) = 2$

using a step size of $h = 0.05$.

76 Write a computer program to solve the initial-value problem

$\dfrac{d^2x}{dt^2} + x^2\dfrac{dx}{dt} + x = \sin t,$

$x(0) = 0, \quad \dfrac{dx}{dt}(0) = 1$

using Euler's method. Use your program to find the value of $X(0.4)$ using steps of $h = 0.01$ and $h = 0.005$. Hence estimate the accuracy of your value of $X(0.4)$ and estimate the step size that would be necessary to obtain a value of $X(0.4)$ accurate to 4dp.

77 A water treatment plant deals with a constant influx Q of polluted water with pollutant concentration s_0. The treatment tank contains bacteria which consume the pollutant and protozoa which feed on the bacteria, thus keeping the bacteria from increasing too rapidly and overwhelming the system. If the concentration of the bacteria and the protozoa are denoted by b and p the system is governed by the differential equations

$$\frac{ds}{dt} = r(s_0 - s) - \alpha m \frac{bs}{1 + s}$$

$$\frac{db}{dt} = -rb + m \frac{bs}{1 + s} - \beta n \frac{bp}{1 + p}$$

$$\frac{dp}{dt} = -rp + n \frac{bp}{1 + p}$$

Write a program to solve these equations numerically.

Measurements have determined that the (biological) parameters α, m, β and n have the values 0.5, 1.0, 0.8 and 0.1 respectively. The parameter r is a measure of the inflow rate of polluted water and s_0 is the level of pollutant. Using the initial conditions $s(0) = 0$, $b(0) = 0.2$ and $p(0) = 0.05$ determine the final steady level of pollutant if $r = 0.05$ and $s_0 = 0.4$. What effect does doubling the inflow rate (r) have?

10.12 Qualitative analysis of second-order differential equations

Sometimes it is easier or more convenient to discover the qualitative properties of the solutions of a differential equation than to solve it completely. In some cases this qualitative knowledge is just as useful as a complete solution. In other cases the qualitative knowledge is more illuminating than a quantitative solution, particularly if the only quantitative solutions that can be derived are numerical ones. One technique that is very useful in this context is the **phase-plane plot**.

10.12.1 Phase-plane plots

The second-order nonlinear differential equation

$$\frac{d^2 x}{dt^2} + \mu(x^2 - 1)\frac{dx}{dt} + \lambda x = 0$$

is known as the Van der Pol oscillator. It has properties that are typical of many non linear oscillators. The equation has no simple analytical solution, so, if we wish to investigate its properties, we must resort to a numerical computation. The equation can readily be recast in state-space form as described in Section 10.11.2 and solved by Euler's method described in Section 10.6.

Figure 10.32 shows displacement and velocity plots for a Van der Pol oscillator with $\lambda = 40$ and $\mu = 3$. The initial conditions used were $x(0) = 0.05$ and $(dx/dt)(0) = 0$. It can be seen that initially the amplitude of the displacement oscillations grows quite rapidly, but after about three cycles this rapid growth stops and the displacement curve appears to settle into a periodically repeating pattern. Similar comments could be made about the velocity curve. Is the Van der Pol oscillator tending towards some fixed cyclical pattern?

This question can be answered much more easily if the displacement and velocity curves are plotted in a different way. Instead of plotting each individually against time, we plot velocity against displacement as in Figure 10.33. Such a plot is called a phase-plane plot. Figure 10.33(a) shows the same data as plotted in Figure 10.32. Time increases in the direction shown by the arrows, the plot starting at the point (0.05, 0) and spiralling outwards. From this plot it is easy to see that the fourth and fifth cycles of the oscillations are nearly indistinguishable. Continuing the computations for a larger

Figure 10.32
Displacement and
velocity traces for a
Van der Pol oscillator.

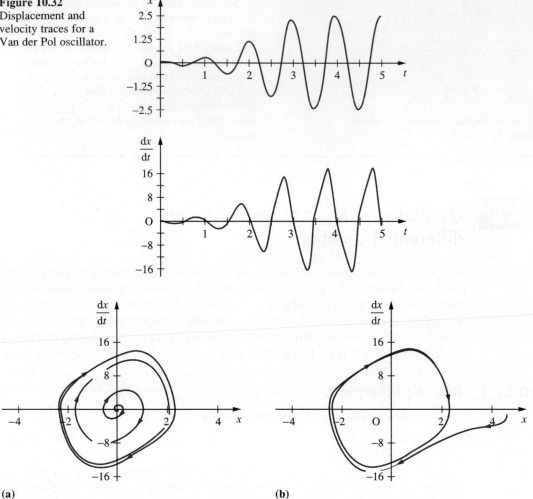

(a)   (b)

Figure 10.33 Phase-plane plots for Van der Pol oscillators – two different initial conditions.

number of cycles would confirm that, after an initial period, the oscillations settle down into a cyclical pattern. The pattern is called a **limit cycle**. The Van der Pol oscillator has the property that the limit cycle is independent of the initial conditions chosen (but depends on the parameters μ and λ). Figure 10.33(b) shows a phase-plane plot of the oscillations of the Van der Pol oscillator, starting from the initial condition $(4.5, 0)$. The interested reader may wish to explore the Van der Pol oscillator further – perhaps by writing a computer program to solve the equation and plotting solution paths in the phase plane for a number of other initial conditions. Exploration of this type will confirm that the limit cycle is independent of initial conditions, and exploration of other values of μ and λ will show how the limit cycle varies as these parameters change.

Other equations will of course produce different solution paths in the phase plane. The second-order linear constant-coefficient equation

$$\frac{d^2x}{dt^2} + \mu\frac{dx}{dt} + \lambda x = 0$$

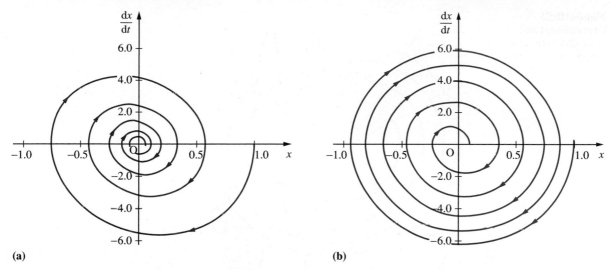

(a)   (b)

Figure 10.34 Phase-plane plots for some second-order oscillators.

yields a phase-plane plot like that shown in Figure 10.34(a). In that particular case the parameters have the values $\mu = 1.5$ and $\lambda = 40$. Other values of μ and λ that result in decaying oscillatory solutions of the equation yield similar spiral phase-plane plots tending towards the origin as $t \to \infty$. Such a plot is typical of any system whose behaviour is oscillatory and decaying. For instance, Figure 10.34(b) shows the phase-plane plot of the nonlinear second-order equation

$$\frac{\mathrm{d}^2 x}{\mathrm{d}t^2} + \mu \operatorname{sgn}\left(\frac{\mathrm{d}x}{\mathrm{d}t}\right) + \lambda x = 0 \tag{10.59}$$

with $\mu = 3$ and $\lambda = 40$ (recall that the function $\operatorname{sgn}(x)$ takes the value 1 if $x \geq 0$ and -1 if $x < 0$). The general characters of Figures 10.34(a, b) are similar. The difference between the two equations is manifest in the difference between the pattern of changing spacing of successive turns of the spirals.

The utility of phase-plane plotting is not restricted to enhancing the understanding of numerical solutions of differential equations. Second-order differential equations which can be expressed in the form

$$\frac{\mathrm{d}^2 x}{\mathrm{d}t^2} = f\left(x, \frac{\mathrm{d}x}{\mathrm{d}t}\right)$$

arise in mathematical models of many engineering systems. An equation of this form can be expressed as

$$\frac{\mathrm{d}v}{\mathrm{d}x} = \frac{f(x, v)}{v}, \quad \text{where } v = \frac{\mathrm{d}x}{\mathrm{d}t}$$

The derivative $\mathrm{d}v/\mathrm{d}x$ is of course just the gradient of the solution path in the phase plane. Hence we can sketch the path in the phase plane of the solutions of a second-order differential equation of this type without actually obtaining the solution. This provides a useful qualitative insight into the form of solution that might be expected.

Figure 10.35
The phase-plane
direction field
for (10.59).

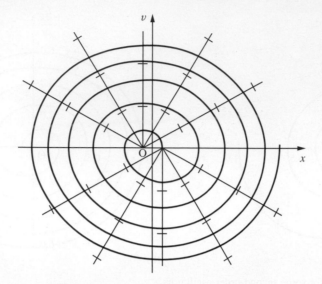

As an example, consider (10.59). This may be expressed as

$$\frac{dv}{dx} = -\frac{\mu \operatorname{sgn}(v) + \lambda x}{v}$$

Thus the gradient of the solution path in the phase plane is equal to k for all points on the curve

$$v = -\frac{\lambda}{k}x - \frac{\mu \operatorname{sgn}(v)}{k}, \quad k \neq 0$$

These curves are of course a family of straight lines. Hence we can construct a diagram similar to the direction-field diagrams described in Section 10.5.1. The phase-plane direction-field diagram is shown, with the solution path from Figure 10.34(b) superimposed upon it, in Figure 10.35.

This technique can also be used for equations for which the lines of constant gradient in the phase plane are not straight. Example 10.49 illustrates this.

Example 10.49   Draw a phase-plane direction field for the equation

$$\frac{d^2x}{dt^2} + 1.5\left(\frac{dx}{dt}\right)^3 + 40x = 0 \tag{10.60}$$

Hence sketch the solution path of the equation that starts from the initial conditions $x = 1$, $dx/dt = 0$.

Solution   Equation (10.60) can be expressed as

$$v\frac{dv}{dx} = -1.5v^3 - 40x$$

so the curve on which the solution-path gradient is equal to k is given by

Figure 10.36
The phase-plane direction field for (10.60).

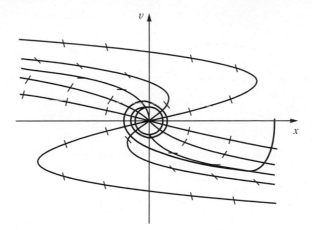

$$x = -\tfrac{1}{40}(kv + 1.5v^3)$$

Thus, as shown in Figure 10.36, the curves of constant solution-path gradient are in this case cubic functions of v. The solution path of the equation starting from the point $(1,0)$ is sketched.

10.12.2 Exercises

78 Draw phase-plane direction fields for the following equations and sketch the form you would expect the solution paths to take, starting from the points $(x, v) = (1, 0), (0, 1), (-1, 0)$ and $(0, -1)$ in each case:

(a) $\dfrac{d^2x}{dt^2} + \dfrac{dx}{dt} + x^3 = 0$

(b) $\dfrac{d^2x}{dt^2} + \dfrac{dx}{dt} + \text{sgn}(x) = 0$

(c) $\dfrac{d^2x}{dt^2} + \dfrac{dx}{dt} + x^2\,\text{sgn}(x) = 0$

(d) $\dfrac{d^2x}{dt^2} + \text{sgn}\left(\dfrac{dx}{dt}\right) + 2\,\text{sgn}(x) = 0$

79 For each of the problems in Question 78 solve the differential equation numerically and check that the solutions you obtain are similar to your sketch solutions.

10.13 Review exercises (1–35)

1 Classify each of the following as partial or ordinary and as linear homogeneous, linear nonhomogeneous or nonlinear differential equations, state the order of the equations and name the dependent and independent variables:

(a) $\dfrac{d^2x}{dt^2} + x\dfrac{dx}{dt} + x^2 = 0$

(b) $\dfrac{\partial^3 h}{\partial t\,\partial v^2} + \dfrac{\partial^3 h}{\partial t^2} + a\dfrac{\partial h}{\partial v} = 0$

(c) $\dfrac{dz}{dx} + 4z^2 = \sin x$

(d) $\dfrac{d^3p}{ds^3} + 4s\dfrac{d^2p}{ds^2} + s^2 = \cos as$

2 Classify the following differential equation problems as under-determined, fully determined or over-determined, and solve them where possible:

(a) $\dfrac{d^2x}{dt^2} = t, \quad x(0) = 1$

(b) $\dfrac{d^3x}{dt^3} - t = 0, \quad x(0) = 0, \quad x(1) = 0, \quad x(2) = 0$

(c) $\dfrac{dx}{dt} = \sin t, \quad x(0) = 0, \quad \dfrac{dx}{dt}(0) = 1$

(d) $\dfrac{d^2x}{dt^2} = e^{4t}, \quad x(0) = 0, \quad \dfrac{dx}{dt}(1) = 0$

3 Sketch the direction field of the differential equation

$$\frac{dx}{dt} = ax(1 - x^2)$$

and sketch the form of solution suggested by the direction field. Solve the equation and confirm that the solution supports the inferences you made from the direction field.

4 Solve the following differential equation problems.

(a) $\dfrac{dx}{dt} + \dfrac{\cos t}{\sin x} = 0, \quad x(0) = -\pi$

(b) $t\dfrac{dx}{dt} - e^{-x} = 0, \quad x(1) = 2$

(c) $\dfrac{dx}{dt} = xt^2, \quad x(2) = 1$

(d) $t\dfrac{dx}{dt} = \dfrac{t}{\sin(x/t)} + x, \quad x(1) = 1$

(e) $\dfrac{dx}{dt} = \dfrac{8t - x}{2x + t}, \quad x(0) = 2$

(f) $t\dfrac{dx}{dt} + x\ln t = x(\ln x + 1), \quad x(1) = 2$

(g) $t\dfrac{dx}{dt} = x - t, \quad x(1) = 3$

(h) $\dfrac{dx}{dt} = \dfrac{x - 7t}{x - t}, \quad x(0) = 2$

5 For each of the following problems, determine which are exact differentials, and hence solve the differential equations where possible:

(a) $2xt^2\dfrac{dx}{dt} = a - 2x^2t, \quad x(1) = 2$

(b) $(2xt + 2t + t^2)\dfrac{dx}{dt} + x^2 + 2tx = 0, \quad x(2) = 2$

(c) $(t\cos xt)\dfrac{dx}{dt} + x\cos xt + 1 = 0, \quad x(\pi) = 0$

(d) $(t\cos xt)\dfrac{dx}{dt} - x\cos xt = 0, \quad x(\pi) = 0$

(e) $te^{xt}\dfrac{dx}{dt} + 1 + xe^{xt} = 0, \quad x(2) = 4$

6 Solve the following differential equation problems:

(a) $\dfrac{dx}{dt} - 2x = t, \quad x(0) = 2$

(b) $\dfrac{dx}{dt} + 2tx = (t - \tfrac{1}{2})e^{-t}, \quad x(0) = 1$

(c) $\dfrac{dx}{dt} + 3x = e^{2t}, \quad x(0) = 2$

(d) $\dfrac{dx}{dt} + x\sin t = \sin t, \quad x(\pi) = e$

7 Solve the differential equation

$$\frac{dx}{dt} = \left(\frac{xt}{x^2 + t^2}\right)^{1/2}, \quad x(0) = 1$$

to find the value of $X(0.4)$ using Euler's method with step size 0.1 and 0.05. By comparing these two estimates of $x(0.4)$, estimate the accuracy of the better of the two values that you have obtained and also the step size you would need to use in order to calculate an estimate of $x(0.4)$ accurate to 2dp.

8 Solve the differential equation

$$\frac{dx}{dt} = \sin t^2, \quad x(0) = 2$$

to find the value of $X(0.25)$ using Euler's method with steps of size 0.05 and 0.025. By comparing these two estimates of $x(0.25)$, estimate the accuracy of the better of the two values that you have obtained and also the step size you would need to use in order to calculate an estimate of $x(0.25)$ accurate to 3dp.

9 Solve the differential equation

$$\frac{dx}{dt} + \frac{3x}{20 - t} = 2$$

obtained in Example 8.5 to determine the amount $x(t)$ of salt in the tank at time t minutes. Initially the tank contains pure water.

10 An open vessel is in the shape of a right circular cone of semi-vertical angle 45° with axis vertical and apex downwards. At time $t = 0$ the vessel is empty. Water is pumped in at a constant rate $p\,\mathrm{m^3\,s^{-1}}$ and escapes through a small hole at the vertex at a rate $ky\,\mathrm{m^3\,s^{-1}}$ where k is a positive constant and y is the depth of water in the cone.

Given that the volume of a circular cone is $\pi r^2 h/3$, where r is the radius of the base and h its vertical height, show that

$$\pi y^2 \frac{\mathrm{d}y}{\mathrm{d}t} = p - ky$$

Deduce that the water level reaches the value $y = p/(2k)$ at time

$$t = \frac{\pi p^2}{k^3}\left(\ln 2 - \frac{5}{8}\right)$$

11 Stefan's law states that the rate of change of temperature of a body due to radiation of heat is

$$\frac{\mathrm{d}T}{\mathrm{d}t} = -k(T^4 - T_0^4)$$

where T is the temperature of the body, T_0 is the temperature of the surrounding medium (both measured in K) and k is a constant. Show that the solution of this differential equation is

$$2\tan\left(\frac{T}{T_0}\right) + \ln\left(\frac{T + T_0}{T - T_0}\right) = 4T_0^3(kt + C)$$

Show that, when the temperature difference between the body and its surroundings is small, Stefan's law can be approximated by Newton's law of cooling

$$\frac{\mathrm{d}T}{\mathrm{d}t} = -\alpha(T - T_0)$$

and find α in terms of k and T_0.

12 A motor under load generates heat internally at a constant rate H and radiates heat, in accordance with Newton's law of cooling, at a rate $k\theta$, where k is a constant and θ is the temperature difference of the motor over its surroundings. With suitable non-dimensionalization of time the temperature of the motor is given by the differential equation

$$\frac{\mathrm{d}\theta}{\mathrm{d}t} = H - k\theta$$

Given that $\theta = 0$ and $\mathrm{d}\theta/\mathrm{d}t = 10$ when $t = 0$ and $\theta = 60$ when $t = 10$ show that

(a) the ultimate rise in temperature is $\theta = 10/k$

(b) k is a solution of the equation $\mathrm{e}^{-10k} = 1 - 6k$

(c) $t = 10 + \dfrac{1}{k}\ln\left(\dfrac{10 - 60k}{10 - k\theta}\right)$

13 A linear cam is to be made whose rate of rise (as it moves in the negative x direction) at the point (x, y) on the profile is equal to one half of the gradient of the line joining (x, y) to a fixed point on the cam (x_0, y_0). Show that the cam profile is a solution of the differential equation

$$\frac{\mathrm{d}y}{\mathrm{d}x} = \frac{y - y_0}{2(x - x_0)}$$

and hence find its equation. Sketch the cam profile.

14 Radioactive elements decay at a constant rate per unit mass of the element. Show that such decays obey equations of the form

$$\frac{\mathrm{d}m}{\mathrm{d}t} = -km$$

where k is the decay rate of the element and m is the mass of the element present. The half life of an element is the time taken for one half of any given mass of the element to decay. Find the relationship between the decay constant k and the half life of an element.

15 In Section 10.2.4 we showed that the equation governing the current flowing in a series LRC electrical circuit is (equation 10.9)

$$L\frac{\mathrm{d}^2 i}{\mathrm{d}t^2} + R\frac{\mathrm{d}i}{\mathrm{d}t} + \frac{1}{C}i = 0$$

Show, by a similar method, that the equation governing the current flowing in a series LR circuit containing a voltage source E is

$$L\frac{\mathrm{d}i}{\mathrm{d}t} + Ri = E$$

At time $t = 0$ a switch is closed applying a d.c. potential of V to an initially quiescent series LR circuit consisting of an inductor L and a resistor R. Show that the current flowing in the circuit is

$$i(t) = \frac{V}{R}(1 - \mathrm{e}^{-Rt/L})$$

and hence find the time needed for the current to reach 95% of its final value.

762 INTRODUCTION TO ORDINARY DIFFERENTIAL EQUATIONS

16 The tread of a car tyre wears more rapidly as it becomes thinner. The tread-wear rate, measured in mm per 10 000 miles, may be modelled as

$$a + b(d - t)^2$$

where d is the initial tread depth, t is the current tread depth and a and b are constants. A tyre company takes measurements on a new design of tyre whose initial tread depth is 8 mm. When the tyre is new its wear rate is found to be 1.03 mm per 10 000 miles run and when the tread depth is reduced to 4 mm the wear rate is 3.43 mm per 10 000 miles. Assuming that a tyre is discarded when the tread depth has been reduced to 2 mm what is its estimated life?

17 Express each of the following differential equations in the form

$$L[x(t)] = f(t)$$

(a) $\dfrac{d^2x}{dt^2} + (\sin t)\dfrac{dx}{dt} - 9x + \cos t = 0$

(b) $\dfrac{d^3x}{dt^3} + t\dfrac{d^2x}{dt^2} + t^2\dfrac{dx}{dt} - 4t\dfrac{dx}{dt} + e^t + x = 0$

(c) $\dfrac{dx}{dt} = e^t + e^{-t}x$

(d) $\dfrac{d^2x}{dt^2} = \cos \Omega t - 4x$

(e) $t^2\dfrac{d^3x}{dt^3} + \ln(t^2 + 4) = \dfrac{1}{t^2 + 2t + 4}\dfrac{dx}{dt}$

18 For each of the following pairs of operators calculate the operator LM − ML; hence state which of the pairs are commutative (that is, satisfy LMx(t) = MLx(t)):

(a) $L = \dfrac{d}{dt} + \sin t, \quad M = \dfrac{d}{dt} - \cos t$

(b) $L = \dfrac{d}{dt} + 4, \quad M = \dfrac{d}{dt} + 9$

(c) $L = \dfrac{d}{dt} + \sin t + 2, \quad M = \dfrac{d}{dt} + \sin t - 2$

(d) $L = \dfrac{d^2}{dt^2} + 2t^2 - 9, \quad M = \dfrac{d^2}{dt^2} + 2t^2 + t$

19 What conditions must the functions $f(t)$ and $g(t)$ satisfy in order for the following operator pairs to be commutative?

(a) $L = \dfrac{d}{dt} + f(t), \quad M = \dfrac{d}{dt} + g(t)$

(b) $L = \dfrac{d^2}{dt^2} + f(t), \quad M = \dfrac{d^2}{dt^2} + g(t)$

20 Find the general solution of the following differential equations:

(a) $\dfrac{d^2x}{dt^2} - 3\dfrac{dx}{dt} + 2x = \sin t$

(b) $\dfrac{d^3x}{dt^3} - 7\dfrac{dx}{dt} + 6x = t$

(c) $\dfrac{d^3x}{dt^3} - 7\dfrac{dx}{dt} + 6x = e^{2t}$

(d) $\dfrac{dx}{dt} - 4x = e^{4t}$

(e) $\dfrac{d^2x}{dt^2} + 3\dfrac{dx}{dt} + \frac{13}{4}x = t^2$

(f) $\dfrac{d^2x}{dt^2} + 3\dfrac{dx}{dt} + \frac{13}{4}x = \sin t$

(g) $\dfrac{d^3x}{dt^3} - 5\dfrac{d^2x}{dt^2} + 2\dfrac{dx}{dt} + 8x = t^2 - t$

(h) $\dfrac{d^2x}{dt^2} - 2\dfrac{dx}{dt} + 5x = e^{-t}$

(i) $\dfrac{d^3x}{dt^3} - 5\dfrac{d^2x}{dt^2} + 2\dfrac{dx}{dt} + 8x = e^{2t} + e^t$

(j) $\dfrac{d^2x}{dt^2} - 2\dfrac{dx}{dt} + 5x = t + e^t \cos 2t$

21 Solve the following initial-value problems:

(a) $\dfrac{d^2x}{dt^2} + 2\dfrac{dx}{dt} + 5x = 1, \quad x(0) = 0, \quad \dfrac{dx}{dt}(0) = 0$

(b) $3\dfrac{d^2x}{dt^2} - 2\dfrac{dx}{dt} - x = 2t - 1,$

$x(0) = 7, \quad \dfrac{dx}{dt}(0) = 2$

(c) $\dfrac{d^2x}{dt^2} + 2\dfrac{dx}{dt} + x = 4\cos 2t,$

$x(0) = 0, \quad \dfrac{dx}{dt}(0) = 2$

(d) $\dfrac{d^2x}{dt^2} - \dfrac{dx}{dt} = -2e^{2t},\quad x(0) = 0,\quad \dfrac{dx}{dt}(0) = 1$

(e) $\dfrac{d^2x}{dt^2} - 3\dfrac{dx}{dt} + 2x = 2e^{-4t},$

$x(0) = 0,\quad \dfrac{dx}{dt}(0) = 1$

(f) $\dfrac{d^3x}{dt^3} + 5\dfrac{d^2x}{dt^2} + 17\dfrac{dx}{dt} + 13x = 1,$

$x(0) = 1,\quad \dfrac{dx}{dt}(0) = 1,\quad \dfrac{d^2x}{dt^2}(0) = 0$

22 Find the damping parameters and natural frequencies of the systems governed by the following second-order linear constant-coefficient differential equations:

(a) $\dfrac{d^2x}{dt^2} + 7\dfrac{dx}{dt} + 2x = 0$

(b) $\dfrac{d^2x}{dt^2} + p\dfrac{dx}{dt} + p^{1/2}x = 0$

(c) $\dfrac{d^2x}{dt^2} + 2aq\dfrac{dx}{dt} + \tfrac{1}{2}qx = 0$

(d) $\dfrac{d^2x}{dt^2} + 14\dfrac{dx}{dt} + 2\alpha x = 0$

23 Determine the values of the appropriate parameters needed to give the systems governed by the following second-order linear constant-coefficient differential equations the damping parameters and natural frequencies stated:

(a) $\dfrac{d^2x}{dt^2} + \dfrac{a}{2}\dfrac{dx}{dt} + bx = 0,\quad \zeta = 0.25,\quad \omega = 2$

(b) $\dfrac{d^2x}{dt^2} + a\dfrac{dx}{dt} + bx = 0,\quad \zeta = 2,\quad \omega = \pi$

(c) $a\dfrac{d^2x}{dt^2} + 4\dfrac{dx}{dt} + cx = 0,\quad \zeta = 0.5,\quad \omega = 2$

(d) $p\dfrac{d^2x}{dt^2} + q^2\dfrac{dx}{dt} + 6x = 0,\quad \zeta = 1.2,\quad \omega = 0.2$

24 Show that by making the substitution

$$v = \dfrac{dx}{dt}$$

the equation

$$\dfrac{d^2x}{dt^2} + \dfrac{dx}{dt} = 1$$

may be expressed as

$$\dfrac{dv}{dt} + v = 1$$

Show that the solution of this equation is $v = 1 + Ce^{-t}$ and hence find $x(t)$.

This technique is a standard method for solving second-order differential equations in which the dependent variable itself does not appear explicitly. Apply the same method to obtain the solutions of the differential equations:

(a) $\dfrac{d^2x}{dt^2} = 4\dfrac{dx}{dt} + e^{-2t}$

(b) $\dfrac{d^2x}{dt^2} - \left(\dfrac{dx}{dt}\right)^2 = 1$

(c) $t\dfrac{d^2x}{dt^2} = 2\dfrac{dx}{dt}$

25 Using the method introduced in Question 24, find the solutions of the following initial-value problems:

(a) $\dfrac{d^2x}{dt^2} + k\dfrac{dx}{dt} = t^2,\quad x(0) = 0,\quad \dfrac{dx}{dt}(0) = 1$

(b) $\dfrac{d^2x}{dt^2} = \left(\dfrac{dx}{dt}\right)^2 e^{-kt},\quad x(0) = 0,\quad \dfrac{dx}{dt}(0) = U$

(c) $(t^2 + 4)\dfrac{d^2x}{dt^2} = 2t\dfrac{dx}{dt},\quad x(1) = 0,\quad \dfrac{dx}{dt}(1) = 2$

(d) $\dfrac{d^2x}{dt^2} + 4\dfrac{dx}{dt} = \sin t,\quad x(\pi) = 0,\quad \dfrac{dx}{dt}(\pi) = 1$

26 Show that by making the substitution

$$v = \dfrac{dx}{dt}$$

and noting that

$$\dfrac{d^2x}{dt^2} = \dfrac{dv}{dt} = \dfrac{dv}{dx}\dfrac{dx}{dt} = v\dfrac{dv}{dx}$$

the equation

$$\dfrac{d^2x}{dt^2} = x\dfrac{dx}{dt}$$

may be expressed as

$$v\dfrac{dv}{dx} = xv$$

Show that the solution of this equation is
$v = \frac{1}{2}x^2 + C$ and hence find $x(t)$.

This technique is a standard method for solving second-order differential equations in which the independent variable does not appear explicitly. Apply the same method to obtain the solutions of the differential equations:

(a) $\dfrac{d^2x}{dt^2} = p\dfrac{dx}{dt}$

(b) $\dfrac{d^2x}{dt^2} = \left(\dfrac{dx}{dt}\right)^2$

(c) $\dfrac{d^2x}{dt^2} = \left(\dfrac{dx}{dt}\right)^2\left(2x - \dfrac{1}{x}\right)$

27 Using the method introduced in Question 26, find the solutions of the following initial-value problems:

(a) $x\dfrac{d^2x}{dt^2} = p\left(\dfrac{dx}{dt}\right)^2$, $x(0) = 4$, $\dfrac{dx}{dt}(0) = 1$

(b) $\dfrac{d^2x}{dt^2} = \dfrac{dx}{dt}e^x$, $x(1) = 1$, $\dfrac{dx}{dt}(1) = 0$

(c) $\dfrac{d^2x}{dt^2} = x^2\dfrac{dx}{dt}$, $x(0) = 2$, $\dfrac{dx}{dt}(0) = \dfrac{8}{3}$

(d) $\dfrac{d^2x}{dt^2} + \dfrac{1}{2}\left(\dfrac{dx}{dt}\right)^2 = x$, $x(0) = 1$, $\dfrac{dx}{dt}(0) = 0$

28 Equation (10.3), arising from the model of the take-off run of an aircraft developed in Section 10.2.1, can be solved by the techniques introduced in Exercises 24 and 26. Assuming that the thrust is constant find the speed of the aircraft both as a function of time and of distance run along the ground. The take-off speed of the aircraft is denoted by V_2. Find expressions for the length of runway required and the time taken by the aircraft to become airborne in terms of take-off speed.

29 Find the values of $X(t)$ for t up to 2, where $X(t)$ is the solution of the differential equation problem

$$\dfrac{d^3x}{dt^3} + \left(\dfrac{d^2x}{dt^2}\right)^2 + 4\left(\dfrac{dx}{dt}\right)^2 - xt = \sin t,$$

$$x(1) = 0.2, \quad \dfrac{dx}{dt}(1) = 1, \quad \dfrac{d^2x}{dt^2}(1) = 0$$

using Euler's method with step size $h = 0.025$. Repeat the computation with $h = 0.0125$. Hence

estimate the accuracy of the value of $X(2)$ given by your solution.

30 The end of a chain, coiled near the edge of a horizontal surface, falls over the edge. If the friction between the chain and the horizontal surface is negligible and the chain is inextensible then, when a length x of chain has fallen, the equation of motion is

$$\dfrac{d}{dt}(mxv) = mgx$$

where m is the mass per unit length of the chain, g is gravitational acceleration and v is the velocity of the falling length of the chain. If the mass per unit length of the chain is constant show that this equation can be expressed as

$$xv\dfrac{dv}{dx} + v^2 = gx$$

and, by putting $y = v^2$, show that $v = \sqrt{(2gx/3)}$.

31 A simple mass spring system, subject to light damping, is vibrating under the action of a periodic force $F\cos pt$. The equation of motion is

$$\dfrac{d^2x}{dt^2} + 2\dfrac{dx}{dt} + 4x = F\cos pt$$

where F and p are constants.

Solve the differential equation for the displacement $x(t)$. Show that one part of the solution tends to zero as $t \to \infty$ and show that the amplitude of the steady-state solution is

$$F[(4 - p^2)^2 + 4p^2]^{-1/2}.$$

Hence show that resonance occurs when $p = \sqrt{2}$.

32 An alternating emf of $E\sin \omega t$ volt is supplied to a circuit containing an inductor of L henry, a resistor of R ohm and a capacitor of C farad in series. The differential equation satisfied by the current i amp and the charge q coulomb on the capacitor is

$$L\dfrac{di}{dt} + Ri + \dfrac{q}{C} = E\sin \omega t$$

Using $i = dq/dt$ obtain a second-order differential equation satisfied by i. Find the resistance if it is just large enough to prevent natural oscillations. For this value of R and $\omega = (LC)^{-1/2}$ prove that

$$i = \dfrac{E}{2K}(\sin \omega t - \omega t\,e^{-\omega t})$$

where $K^2 = L/C$, when the current and charge on the capacitor are both zero at time $t = 0$.

The following three questions are intended to be open-ended – there is no single 'correct' answer. They should be approached in an enquiring frame of mind, with the objective of disovering, by use of mathematical knowledge and technique, something more about how the physical world functions. The questions are designed to use primarily mathematical knowledge introduced in this chapter.

33 A truck of mass m moves along a horizontal test track subject only to a force resisting motion that is proportional to its speed. At time $t = 0$ the truck passes a reference point moving with speed U. Find the velocity of the truck both as a function of time and as a function of displacement from the reference point. Find the displacement of the truck from the reference point as a function of time. Repeat these calculations for similar trucks subject to resistance forces proportional to

(a) square root of speed,

(b) square of speed,

(c) cube of speed.

How long does the truck take to come to rest in each case? Draw plots of velocity against displacement in each case. Explain, in qualitative terms, the behaviour of the truck under each type of resistance.

How would you model mathematically a truck that is subject to a small constant resistance plus a resistance proportional to its speed? How far would such a truck travel before coming to rest, and how long would it take to do so? Can you repeat these calculations for trucks subject to a small constant resistance plus a resistance proportional to speed squared or speed cubed?

What general conclusions can you draw about the type of terms that it is sensible to use in mathematical models of engineering systems to describe resistance to motion?

34 Figure 10.37 shows a system that serves as a simplified model of the phenomenon of 'tool chatter'. The mass A rests on a moving belt and is connected to a rigid support by a spring. The coefficient of sliding friction between the belt and the mass is less than the coefficient of static friction. When the spring is uncompressed, the

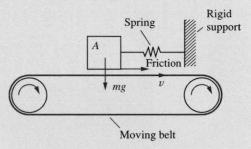

Figure 10.37 Diagram of a model of the 'tool chatter' phenomenon.

mass moves to the right with the belt. As it does so, the spring is compressed until the force exerted by the spring exceeds the maximum static frictional force available. The mass then starts to slide. The spring force slows the mass, brings it to rest, and then accelerates it back along the belt so that it moves leftwards. As it does so, the compression in the spring is reduced, the force of sliding friction slows the mass to rest, and then accelerates it so that its velocity is directed to the right. When its velocity matches that of the belt, sliding ceases and static friction takes over again.

Thus the mass undergoes a cyclic process of being pushed forwards by static friction until the spring is sufficiently compressed and then being flung backwards by the stored energy in the spring until the energy is dissipated. Analyse the model, determining such quantities as how the amplitude and frequency of motion of the mass depend on the coefficients and static friction and the other physical parameters.

35 The second-order linear nonhomogeneous constant-coefficient differential equation

$$\frac{d^2x}{dt^2} + 2\zeta\omega\frac{dx}{dt} + \omega^2x = F\cos\Omega t$$

(often referred to as a **forced harmonic oscillator**) has a response $A(\Omega)F\cos(\Omega t - \delta)$, where $A(\Omega)$ is often called the **frequency response** (strictly it is the *amplitude response* or *gain spectrum*) and is given by (10.55) and shown in Figure 10.27. How does the frequency response of the second-order nonlinear nonhomogeneous constant-coefficient differential equation

$$\frac{d^2x}{dt^2} + 2\zeta\omega\left|\frac{dx}{dt}\right|\frac{dx}{dt} + \omega^2x = F\cos\Omega t$$

differ from that of the linear one?

11 Introduction to Laplace Transforms

Chapter 11 Contents

11.1 Introduction

Laplace transform methods have a key role to play in the modern approach to the analysis and design of engineering systems. The stimulus for developing these methods was the pioneering work of the English electrical engineer Oliver Heaviside (1850–1925) in developing a method for the systematic solution of ordinary differential equations with constant coefficients. Heaviside was concerned with solving practical problems, and his method was based mainly on intuition, lacking mathematical rigour: consequently it was frowned upon by theoreticians at the time. However, Heaviside himself was not concerned with rigorous proofs, and was satisfied that his method gave the correct results. Using his ideas, he was able to solve important practical problems that could not be dealt with using classical methods. This led to many new results in fields such as the propagation of currents and voltages along transmission lines.

Because it worked in practice, Heaviside's method was widely accepted by engineers. As its power for problem-solving became more and more apparent, the method attracted the attention of mathematicians, who set out to justify it. This provided the stimulus for rapid developments in many branches of mathematics, including improper integrals, asymptotic series and transform theory. Research on the problem continued for many years before it was eventually recognized that an integral transform developed by the French mathematician Pierre Simon de Laplace (1749–1827) almost a century before provided a theoretical foundation for Heaviside's work. It was also recognized that the use of this integral transform provided a more systematic alternative for investigating differential equations than the method proposed by Heaviside. It is this alternative approach that is the basis of the **Laplace transform method**.

We have already come across instances where a mathematical transformation has been used to simplify the solution of a problem. For example, the logarithm is used to simplify multiplication and division problems. To multiply or divide two numbers, we transform them into their logarithms, add or subtract these, and then perform the inverse transformation (that is, the antilogarithm) to obtain the product or quotient of the original numbers. The purpose of using a transformation is to create a new domain in which it is easier to handle the problem being investigated. Once results have been obtained in the new domain, they can be inverse-transformed to give the desired results in the original domain.

The Laplace transform is an example of a class called **integral transforms**, and it takes a function $f(t)$ of one variable t (which we shall refer to as **time**) into a function $F(s)$ of another variable s (the **complex frequency**). Another integral transform widely used by engineers is the **Fourier transform**, which is dealt with in the companion text *Advanced Modern Engineering Mathematics*. The attraction of the Laplace transform is that it transforms *differential* equations in the t (time) domain into *algebraic* equations in the s (frequency) domain. Solving differential equations in the t domain therefore reduces to solving algebraic equations in the s domain. Having done the latter for the desired unknowns, their values as functions of time may be found by taking inverse transforms. Another advantage of using the Laplace transform for solving differential equations is that initial conditions play an essential role in the transformation process, so they are automatically incorporated into the solution. This constrasts with the classical approach considered in Chapter 10, where the initial conditions are only introduced when the unknown constants of integration are determined. The Laplace transform is

Figure 11.1
Schematic
representation of a
system.

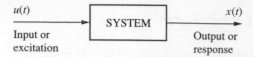

$u(t)$

Input or
excitation

SYSTEM

$x(t)$

Output or
response

therefore an ideal tool for solving initial-value problems such as those occurring in the investigation of electrical circuits and mechanical vibrations.

The Laplace transform finds particular application in the field of signals and linear systems analysis. A distinguishing feature of a system is that when it is subjected to an excitation (input), it produces a response (output). When the input $u(t)$ and output $x(t)$ are functions of a single variable t, representing time, it is normal to refer to them as **signals**. Schematically, a system may be represented as in Figure 11.1. The problem facing the engineer is that of determining the system output $x(t)$ when it is subjected to an input $u(t)$ applied at some instant of time, which we can take to be $t = 0$. The relationship between output and input is determined by the laws governing the behaviour of the system. If the system is linear and time-invariant then the output is related to the input by a linear differential equation with constant coefficients, and we have a standard initial-value problem, which is amenable to solution using the Laplace transform.

While many of the problems considered in this chapter can be solved by the classical approach of Chapter 10, the Laplace transform leads to a more unified approach and provides the engineer with greater insight into system behaviour. In practice, the input signal $u(t)$ may be a discontinuous or periodic function, or even a pulse, and in such cases the use of the Laplace transform has distinct advantages over the classical approach. Also, more often than not, an engineer is interested not only in system analysis but also in system synthesis or design. Consequently, an engineer's objective in studying a system's response to specific inputs is frequently to learn more about the system with a view to improving or controlling it so that it satisfies certain specifications. It is in this area that the use of the Laplace transform is attractive, since by considering the system response to particular inputs, such as a sinusoid, it provides the engineer with powerful graphical methods for system design that are relatively easy to apply and widely used in practice.

In modelling the system by a differential equation, it has been assumed that both the input and output signals can vary at any instant of time; that is, they are functions of a continuous time variable (note that this does not mean that the signals themselves have to be continuous functions of time). Such systems are called **continuous-time systems**, and it is for investigating these that the Laplace transform is best suited. With the introduction of computer control into system design, signals associated with a system may only change at discrete instants of time. In such cases the system is said to be a **discrete-time system**, and is modelled by a difference equation rather than a differential equation. Such systems are dealt with using the z transform considered in the companion text, *Advanced Modern Engineering Mathematics*.

In this chapter we restrict our consideration to simply introducing the Laplace transform and to illustrating its use in solving differential equations. Its more extensive role in engineering applications is dealt with in the companion text.

There is some overlap in the material covered in this chapter and in Chapter 10, particularly in relation to the modelling aspects of applications to electrical circuits and mechanical vibrations. This overlap has been included so that the two approaches to solving differential equations can be studied independently of each other.

11.2 The Laplace transform

11.2.1 Definition and notation

We define the Laplace transform of a function $f(t)$ by the expression

$$\mathscr{L}\{f(t)\} = \int_0^\infty e^{-st}f(t)\,dt \tag{11.1}$$

where s is a complex variable and e^{-st} is called the **kernel** of the transformation.

It is usual to represent the Laplace transform of a function by the corresponding capital letter, so that we write

$$\mathscr{L}\{f(t)\} = F(s) = \int_0^\infty e^{-st}f(t)\,dt \tag{11.2}$$

An alternative notation in common use is to denote $\mathscr{L}\{f(t)\}$ by $\bar{f}(s)$ or simply $\bar{f}$.

Before proceeding, there are a few observations relating to the definition (11.2) worthy of comment.

(a) The symbol $\mathscr{L}$ denotes the **Laplace transform operator**; when it operates on a function $f(t)$, it transforms it into a function $F(s)$ of the complex variable s. We say the operator transforms the function $f(t)$ in the t domain (usually called the **time domain**) into the function $F(s)$ in the s domain (usually called the **complex frequency domain**, or simply the **frequency domain**). This relationship is depicted graphically in Figure 11.2, and it is usual to refer to $f(t)$ and $F(s)$ as a **Laplace transform pair**, written as $\{f(t), F(s)\}$.

(b) Because the upper limit in the integral is infinite, the domain of integration is infinite. Thus the integral is an example of an **improper integral**, as introduced in Chapter 9, Section 9.3; that is,

$$\int_0^\infty e^{-st}f(t)\,dt = \lim_{T\to\infty}\int_0^T e^{-st}f(t)\,dt$$

This immediately raises the question of whether or not the integral converges, an issue we shall consider in Section 11.2.3.

Figure 11.2
The Laplace transform operator.

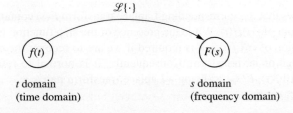

$\mathscr{L}\{\cdot\}$

$f(t)$ $F(s)$

t domain s domain
(time domain)   (frequency domain)

(c) Because the lower limit in the integral is zero, it follows that when taking the Laplace transform, the behaviour of $f(t)$ for negative values of t is ignored or suppressed. This means that $F(s)$ contains information on the behaviour of $f(t)$ only for $t \geq 0$, so that the Laplace transform is not a suitable tool for investigating problems in which values of $f(t)$ for $t < 0$ are relevant. In most engineering applications this does not cause any problems, since we are then concerned with physical systems for which the functions we are dealing with vary with time t. An attribute of physical realizable systems is that they are **non-anticipatory** in the sense that there is no output (or response) until an input (or excitation) is applied. Because of this causal relationship between the input and output, we define a function $f(t)$ to be **causal** if $f(t) = 0$ $(t < 0)$. In general, however, unless the domain is clearly specified, a function $f(t)$ is normally intepreted as being defined for all real values, both positive and negative, of t. Making use of the Heaviside unit step function $H(t)$ (see also Chapter 2, Section 2.8.3), where

$$H(t) = \begin{cases} 0 & (t < 0) \\ 1 & (t \geq 0) \end{cases}$$

we have

$$f(t)H(t) = \begin{cases} 0 & (t < 0) \\ f(t) & (t \geq 0) \end{cases}$$

Thus the effect of multiplying $f(t)$ by $H(t)$ is to convert it into a causal function. Graphically, the relationship between $f(t)$ and $f(t)H(t)$ is as shown in Figure 11.3.

Figure 11.3
Graph of $f(t)$ and its causal equivalent function.

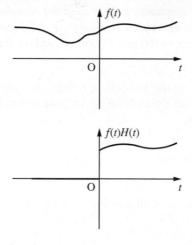

It follows that the corresponding Laplace transform $F(s)$ contains full information on the behaviour of $f(t)H(t)$. While the presence of the step function is not necessary for the determination of $\mathcal{L}\{f(t)\}$, it is required if we are to recover the original function from a knowledge of its transform. Consequently, it is normal to refer to $\{f(t)H(t), F(s)\}$ rather than $\{f(t), F(s)\}$ as being a Laplace transform pair.

(d) If the behaviour of $f(t)$ for $t < 0$ is of interest then we need to use the alternative **two-sided** or **bilateral Laplace transform** of the function $f(t)$, defined by

$$\mathscr{L}_B\{f(t)\} = \int_{-\infty}^{\infty} e^{-st} f(t) \, dt \tag{11.3}$$

The Laplace transform defined by (11.2), with lower limit zero, is sometimes referred to as the **one-sided** or **unilateral Laplace transform** of the function $f(t)$. In this chapter we shall concern ourselves only with the latter transform, and refer to it simply as the Laplace transform of the function $f(t)$. Note that when $f(t)$ is a causal function,

$$\mathscr{L}_B\{f(t)\} = \mathscr{L}\{f(t)\}$$

11.2.2 Transforms of simple functions

In this section we obtain the Laplace transformations of some simple functions.

Example 11.1 Determine the Laplace transform of the function

$$f(t) = c$$

where c is a constant.

Solution Using the definition (11.2),

$$\mathscr{L}(c) = \int_0^{\infty} e^{-st} c \, dt = \lim_{T \to \infty} \int_0^T e^{-st} c \, dt$$

$$= \lim_{T \to \infty} \left[-\frac{c}{s} e^{-st} \right]_0^T = \frac{c}{s} \left(1 - \lim_{T \to \infty} e^{-sT} \right)$$

Taking $s = \sigma + j\omega$, where σ and ω are real,

$$\lim_{T \to \infty} e^{-sT} = \lim_{T \to \infty} (e^{-(\sigma + j\omega)T}) = \lim_{T \to \infty} e^{-\sigma T} (\cos \omega T + j \sin \omega T)$$

A finite limit exists provided that $\sigma = \text{Re}(s) > 0$, when the limit is zero. Thus, provided that $\text{Re}(s) > 0$, the Laplace transform is

$$\mathscr{L}(c) = \frac{c}{s}, \quad \text{Re}(s) > 0$$

so that

$$\left. \begin{array}{l} f(t) = c \\[2mm] F(s) = \dfrac{c}{s} \end{array} \right\} \quad \text{Re}(s) > 0 \tag{11.4}$$

constitute an example of a Laplace transform pair.

Example 11.2   Determine the Laplace transform of the ramp function

$$f(t) = t$$

Solution   From the definition (11.2),

$$\mathcal{L}\{t\} = \int_0^\infty e^{-st}t\,dt = \lim_{T\to\infty}\int_0^T e^{-st}t\,dt$$

$$= \lim_{T\to\infty}\left[-\frac{t}{s}e^{-st} - \frac{e^{-st}}{s^2}\right]_0^T$$

$$= \frac{1}{s^2} - \lim_{T\to\infty}\frac{Te^{-sT}}{s} - \lim_{T\to\infty}\frac{e^{-sT}}{s^2}$$

Following the same procedure as in Example 11.1, limits exist provided that $\text{Re}(s) > 0$, when

$$\lim_{T\to\infty}\frac{Te^{-sT}}{s} = \lim_{T\to\infty}\frac{e^{-sT}}{s^2} = 0$$

Thus, provided that $\text{Re}(s) > 0$,

$$\mathcal{L}\{t\} = \frac{1}{s^2}$$

giving us the Laplace transform pair

$$\left.\begin{array}{l} f(t) = t \\[2mm] F(s) = \dfrac{1}{s^2} \end{array}\right\} \quad \text{Re}(s) > 0 \tag{11.5}$$

Example 11.3   Determine the Laplace transform of the one-sided exponential function

$$f(t) = e^{kt}$$

Solution   The definition (11.2) gives

$$\mathcal{L}\{e^{kt}\} = \int_0^\infty e^{-st}e^{kt}\,dt = \lim_{T\to\infty}\int_0^T e^{-(s-k)t}\,dt$$

$$= \lim_{T\to\infty}\frac{-1}{s-k}[e^{-(s-k)t}]_0^T$$

$$= \frac{1}{s-k}\left(1 - \lim_{T\to\infty}e^{-(s-k)T}\right)$$

Writing $s = \sigma + j\omega$, where σ and ω are real, we have

$$\lim_{T\to\infty}e^{-(s-k)T} = \lim_{T\to\infty}e^{-(\sigma-k)T}e^{j\omega T}$$

If k is real, then, provided that $\sigma = \mathrm{Re}(s) > k$, the limit exists, and is zero. If k is complex, say $k = a + jb$, then the limit will also exist, and be zero, provided that $\sigma > a$ (that is, $\mathrm{Re}(s) > \mathrm{Re}(k)$). Under these conditions, we then have

$$\mathcal{L}\{e^{kt}\} = \frac{1}{s - k}$$

giving us the Laplace transform pair

$$\left. \begin{array}{l} f(t) = e^{kt} \\[2mm] F(s) = \dfrac{1}{s - k} \end{array} \right\} \quad \mathrm{Re}(s) > \mathrm{Re}(k) \qquad (11.6)$$

Example 11.4   Determine the Laplace transforms of the sine and cosine functions

$$f(t) = \sin at, \quad g(t) = \cos at$$

where a is a real constant.

Solution   Since

$$e^{jat} = \cos at + j \sin at$$

we may write

$$f(t) = \sin at = \mathrm{Im}(e^{jat})$$

$$g(t) = \cos at = \mathrm{Re}(e^{jat})$$

Using this formulation, the required transforms may be obtained from the result

$$\mathcal{L}\{e^{kt}\} = \frac{1}{s - k}, \quad \mathrm{Re}(s) > \mathrm{Re}(k)$$

of Example 11.3.

Taking $k = ja$ in this result gives

$$\mathcal{L}\{e^{jat}\} = \frac{1}{s - ja}, \quad \mathrm{Re}(s) > 0$$

or

$$\mathcal{L}\{e^{jat}\} = \frac{s + ja}{s^2 + a^2}, \quad \mathrm{Re}(s) > 0$$

Thus, equating real and imaginary parts and assuming s is real,

$$\mathcal{L}\{\sin at\} = \mathrm{Im}\,\mathcal{L}\{e^{jat}\} = \frac{a}{s^2 + a^2}$$

$$\mathcal{L}\{\cos at\} = \mathrm{Re}\,\mathcal{L}\{e^{jat}\} = \frac{s}{s^2 + a^2}$$

These results also hold when s is complex, giving us the Laplace transform pairs

$$\mathcal{L}\{\sin at\} = \frac{a}{s^2 + a^2}, \quad \mathrm{Re}(s) > 0 \tag{11.7}$$

$$\mathcal{L}\{\cos at\} = \frac{s}{s^2 + a^2}, \quad \mathrm{Re}(s) > 0 \tag{11.8}$$

11.2.3 Existence of the Laplace transform

Clearly, from the definition (11.2), the Laplace transform of a function $f(t)$ exists if and only if the improper integral in the definition converges for at least some values of s. The examples of Section 11.2.2 suggest that this relates to the boundedness of the function, with the factor e^{-st} in the transform integral acting like a convergence factor in that the allowed values of $\mathrm{Re}(s)$ are those for which the integral converges. In order to be able to state sufficient conditions on $f(t)$ for the existence of $\mathcal{L}\{f(t)\}$, we first introduce the definition of a function of exponential order.

Definition 11.1

A function $f(t)$ is said to be of **exponential order** as $t \to \infty$ if there exists a real number σ and positive constants M and T such that

$$|f(t)| < Me^{\sigma t}$$

for all $t > T$.

What this definition tells us is that a function $f(t)$ is of exponential order if it does not grow faster than some exponential function of the form $Me^{\sigma t}$. Fortunately, most functions of practical significance satisfy this requirement, and are therefore of exponential order. There are, however, functions that are not of exponential order, an example being e^{t^2}, since this grows more rapidly than $Me^{\sigma t}$ as $t \to \infty$, whatever the values of M and σ.

Example 11.5   The function $f(t) = e^{3t}$ is of exponential order, with $\sigma \geqslant 3$.

Example 11.6   Show that the function $f(t) = t^3$ $(t \geqslant 0)$ is of exponential order.

Solution   Since

$$e^{\alpha t} = 1 + \alpha t + \tfrac{1}{2}\alpha^2 t^2 + \tfrac{1}{6}\alpha^3 t^3 + \dots$$

it follows that for any $\alpha > 0$

$$t^3 < \frac{6}{\alpha^3}e^{\alpha t}$$

so that t^3 is of exponential order, with $\sigma > 0$.

It follows from Examples 11.5 and 11.6 that the choice of σ in Definition 11.1 is not unique for a particular function. For this reason, we define the greatest lower bound σ_c of the set of possible values of σ to be the **abscissa of convergence** of $f(t)$. Thus, in the case of the function $f(t) = e^{3t}$, $\sigma_c = 3$, while in the case of the function $f(t) = t^3$, $\sigma_c = 0$.

Returning to the definition of the Laplace transform given by (11.2), it follows that if $f(t)$ is a continuous function and is also of exponential order with abscissa of convergence σ_c, so that

$$|f(t)| < Me^{\sigma t}, \quad \sigma > \sigma_c$$

then, taking $T = 0$ in Definition 11.1,

$$|F(s)| = \left| \int_0^\infty e^{-st} f(t) \mathrm{d}t \right| \leqslant \int_0^\infty |e^{-st}||f(t)|\,\mathrm{d}t$$

Writing $s = \sigma + j\omega$, where σ and ω are real, since $|e^{-j\omega t}| = 1$, we have

$$|e^{-st}| = |e^{-\sigma t}||e^{-j\omega t}| = |e^{-\sigma t}| = e^{-\sigma t}$$

so that

$$|F(s)| \leqslant \int_0^\infty e^{-\sigma t}|f(t)|\,\mathrm{d}t \leqslant M \int_0^\infty e^{-\sigma t}e^{\sigma_d t}\,\mathrm{d}t, \quad \sigma_d > \sigma_c$$

$$= M \int_0^\infty e^{-(\sigma\ \sigma_d)t}\,\mathrm{d}t$$

This last integral is finite whenever $\sigma = \mathrm{Re}(s) > \sigma_d$. Since σ_d can be chosen arbitrarily such that $\sigma_d > \sigma_c$ we conclude that $F(s)$ exists for $\sigma > \sigma_c$. Thus a continuous function $f(t)$ of exponential order, with abscissa of convergence σ_c, has a Laplace transform

$$\mathscr{L}\{f(t)\} = F(s), \quad \mathrm{Re}(s) > \sigma_c$$

where the region of convergence is as shown in Figure 11.4.

In fact, the requirement that $f(t)$ be continuous is not essential, and may be relaxed to $f(t)$ being piecewise-continuous, as defined in Chapter 8, Section 8.6.1; that is, $f(t)$ must have only a finite number of finite discontinuities, being elsewhere continuous and bounded.

We conclude this section by stating a theorem that ensures the existence of a Laplace transform.

Figure 11.4
Region of convergence for $\mathscr{L}\{f(t)\}$; σ_c is the abscissa of convergence for $f(t)$.

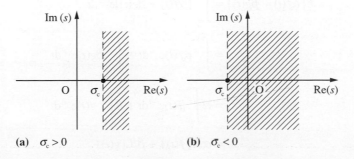

(a) $\sigma_c > 0$ (b) $\sigma_c < 0$

Theorem 11.1	**Existence of Laplace transform**

If the causal function $f(t)$ is piecewise-continuous on $[0, \infty]$ and is of exponential order, with abscissa of convergence σ_c, then its Laplace transform exists, with region of convergence $\mathrm{Re}(s) > \sigma_c$ in the s domain; that is,

$$\mathcal{L}\{f(t)\} = F(s) = \int_0^\infty \mathrm{e}^{-st} f(t)\,\mathrm{d}t, \quad \mathrm{Re}(s) > \sigma_c$$

end of theorem

The conditions of this theorem are *sufficient* for ensuring the existence of the Laplace transform of a function. They do not, however, constitute *necessary* conditions for the existence of such a transform, and it does not follow that if the conditions are violated then a transform does not exist. In fact, the conditions are more restrictive than necessary, since there exist functions with infinite discontinuities that possess Laplace transforms.

11.2.4 Properties of the Laplace transform

In this section we consider some of the properties of the Laplace transform that will enable us to find further transform pairs $\{f(t), F(s)\}$ without having to compute them directly using the definition. Further properties will be developed in later sections when the need arises.

Property 11.1: The linearity property

A fundamental property of the Laplace transform is its linearity, which may be stated as follows:

> If $f(t)$ and $g(t)$ are functions having Laplace transforms and if α and β are any constants then
>
> $$\mathcal{L}\{\alpha f(t) + \beta g(t)\} = \alpha \mathcal{L}\{f(t)\} + \beta \mathcal{L}\{g(t)\}$$

As a consequence of this property, we say that the Laplace transform operator $\mathcal{L}$ is a **linear operator**. A proof of the property follows readily from the definition (11.2), since

$$\mathcal{L}\{\alpha f(t) + \beta g(t)\} = \int_0^\infty [\alpha f(t) + \beta g(t)]\mathrm{e}^{-st}\,\mathrm{d}t$$

$$= \int_0^\infty \alpha f(t)\mathrm{e}^{-st}\,\mathrm{d}t + \int_0^\infty \beta g(t)\mathrm{e}^{-st}\,\mathrm{d}t$$

$$= \alpha \int_0^\infty f(t)\mathrm{e}^{-st}\,\mathrm{d}t + \beta \int_0^\infty g(t)\mathrm{e}^{-st}\,\mathrm{d}t$$

$$= \alpha \mathcal{L}\{f(t)\} + \beta \mathcal{L}\{g(t)\}$$

Regarding the region of convergence, if $f(t)$ and $g(t)$ have abscissae of convergence σ_f and σ_g respectively, and $\sigma_1 > \sigma_f$, $\sigma_2 > \sigma_g$, then

$$|f(t)| < M_1 e^{\sigma_1 t}, \quad |g(t)| < M_2 e^{\sigma_2 t}$$

It follows that

$$|\alpha f(t) + \beta g(t)| \leqslant |\alpha| \, |f(t)| + |\beta| \, |g(t)|$$
$$\leqslant |\alpha| M_1 e^{\sigma_1 t} + |\beta| M_2 e^{\sigma_2 t}$$
$$\leqslant (|\alpha| M_1 + |\beta| M_2) e^{\sigma t}$$

where $\sigma = \max(\sigma_1, \sigma_2)$, so that the abscissa of convergence of the linear sum $\alpha f(t) + \beta g(t)$ is less than or equal to the maximum of those for $f(t)$ and $g(t)$.

This linearity property may clearly be extended to a linear combination of any finite number of functions.

Example 11.7 Determine $\mathcal{L}\{3t + 2e^{3t}\}$.

Solution Using the results given in (11.5) and (11.6),

$$\mathcal{L}\{t\} = \frac{1}{s^2}, \quad \text{Re}(s) > 0$$

$$\mathcal{L}\{e^{3t}\} = \frac{1}{s - 3}, \quad \text{Re}(s) > 3$$

so, by the linearity property,

$$\mathcal{L}\{3t + 2e^{3t}\} = 3\mathcal{L}\{t\} + 2\mathcal{L}\{e^{3t}\}$$

$$= \frac{3}{s^2} + \frac{2}{s - 3}, \quad \text{Re}(s) > \max\{0, 3\}$$

$$= \frac{3}{s^2} + \frac{2}{s - 3}, \quad \text{Re}(s) > 3$$

Example 11.8 Determine $\mathcal{L}\{5 - 3t + 4\sin 2t - 6e^{4t}\}$.

Solution Using the results given in (11.4)–(11.7),

$$\mathcal{L}\{5\} = \frac{5}{s}, \quad \text{Re}(s) > 0 \qquad \mathcal{L}\{t\} = \frac{1}{s^2}, \quad \text{Re}(s) > 0$$

$$\mathcal{L}\{\sin 2t\} = \frac{2}{s^2 + 4}, \quad \text{Re}(s) > 0 \qquad \mathcal{L}\{e^{4t}\} = \frac{1}{s - 4}, \quad \text{Re}(s) > 4$$

so, by the linearity property,

$$\mathcal{L}\{5 - 3t + 4\sin 2t - 6e^{4t}\} = \mathcal{L}\{5\} - 3\mathcal{L}\{t\} + 4\mathcal{L}\{\sin 2t\} - 6\mathcal{L}\{e^{4t}\}$$

$$= \frac{5}{s} - \frac{3}{s^2} + \frac{8}{s^2 + 4} - \frac{6}{s - 4}, \quad \text{Re}(s) > \max\{0, 4\}$$

$$= \frac{5}{s} - \frac{3}{s^2} + \frac{8}{s^2 + 4} - \frac{6}{s - 4}, \quad \text{Re}(s) > 4$$

The first shift property is another property that enables us to add more combinations to our repertoire of Laplace transform pairs. As with the linearity property, it will prove to be of considerable importance in our later discussions, particularly when considering the inversion of Laplace transforms.

Property 11.2: The first shift property

The property is contained in the following theorem, commonly referred to as the **first shift theorem** or sometimes as the **exponential modulation theorem**.

Theorem 11.2

The first shift theorem

If $f(t)$ is a function having Laplace transform $F(s)$, with $\text{Re}(s) > \sigma_c$, then the function $e^{at}f(t)$ also has a Laplace transform, given by

$$\mathcal{L}\{e^{at}f(t)\} = F(s - a), \quad \text{Re}(s) > \sigma_c + \text{Re}(a)$$

Proof

A proof of the theorem follows directly from the definition of the Laplace transform, since

$$\mathcal{L}\{e^{at}f(t)\} = \int_0^\infty e^{at}f(t)e^{-st}\,dt = \int_0^\infty f(t)e^{-(s-a)t}\,dt$$

Then, since

$$\mathcal{L}\{f(t)\} = F(s) = \int_0^\infty f(t)e^{-st}\,dt, \quad \text{Re}(s) > \sigma_c$$

we see that the last integral above is in structure exactly the Laplace transform of $f(t)$ itself, except that $s - a$ takes the place of s, so that

$$\mathcal{L}\{e^{at}f(t)\} = F(s - a), \quad \text{Re}(s - a) > \sigma_c$$

or

$$\mathcal{L}\{e^{at}f(t)\} = F(s - a), \quad \text{Re}(s) > \sigma_c + \text{Re}(a)$$

end of theorem

An alternative way of expressing the result of Theorem 11.2, which may be found more convenient in application, is as

$$\mathcal{L}\{e^{at}f(t)\} = [\mathcal{L}\{f(t)\}]_{s \to s-a} = [F(s)]_{s \to s-a}$$

In other words, the theorem says that the Laplace transform of e^{at} times a function $f(t)$ is equal to the Laplace transform of $f(t)$ itself, with s replaced by $s - a$.

Example 11.9 Determine $\mathcal{L}\{te^{-2t}\}$.

Solution From the result given in (11.5),

$$\mathcal{L}\{t\} = F(s) = \frac{1}{s^2}, \quad \mathrm{Re}(s) > 0$$

so, by the first shift theorem,

$$\mathcal{L}\{te^{-2t}\} = F(s + 2) = [F(s)]_{s \to s+2}, \quad \mathrm{Re}(s) > 0 - 2$$

that is,

$$\mathcal{L}\{te^{-2t}\} = \frac{1}{(s + 2)^2}, \quad \mathrm{Re}(s) > -2$$

Example 11.10 Determine $\mathcal{L}\{e^{-3t}\sin 2t\}$.

Solution From the result (11.7),

$$\mathcal{L}\{\sin 2t\} = F(s) = \frac{2}{s^2 + 4}, \quad \mathrm{Re}(s) > 0$$

so, by the first shift theorem,

$$\mathcal{L}\{e^{-3t}\sin 2t\} = F(s + 3) = [F(s)]_{s \to s+3}, \quad \mathrm{Re}(s) > 0 \quad 3$$

that is,

$$\mathcal{L}\{e^{-3t}\sin 2t\} = \frac{2}{(s + 3)^2 + 4} = \frac{2}{s^2 + 6s + 13}, \quad \mathrm{Re}(s) > -3$$

The function $e^{-3t}\sin 2t$ in Example 11.10 is a member of a general class of functions called **damped sinusoids**. These play an important role in the study of engineering systems, particularly in the analysis of vibrations. For this reason, we add the following two general members of the class to our standard library of Laplace transform pairs:

$$\mathcal{L}\{e^{-kt}\sin at\} = \frac{a}{(s + k)^2 + a^2}, \quad \mathrm{Re}(s) > -k \qquad \textbf{(11.9)}$$

$$\mathcal{L}\{e^{-kt}\cos at\} = \frac{s + k}{(s + k)^2 + a^2}, \quad \mathrm{Re}(s) > -k \qquad \textbf{(11.10)}$$

where in both cases k and a are real constants.

Property 11.3: Derivative-of-transform property

This property relates operations in the time domain to those in the transformed s domain, but initially we shall simply look upon it as a method of increasing our repertoire of Laplace transform pairs. The property is also sometimes referred to as the **multiplication-by-t** property. A statement of the property is contained in the following theorem.

Theorem 11.3 **Derivative of transform**

If $f(t)$ is a function having Laplace transform

$$F(s) = \mathcal{L}\{f(t)\}, \quad \text{Re}(s) > \sigma_c$$

then the functions $t^n f(t)$ ($n = 1, 2, \ldots$) also have Laplace transforms, given by

$$\mathcal{L}\{t^n f(t)\} = (-1)^n \frac{\mathrm{d}^n F(s)}{\mathrm{d}s^n}, \quad \text{Re}(s) > \sigma_c$$

Proof By definition,

$$\mathcal{L}\{f(t)\} = F(s) = \int_0^\infty e^{-st} f(t) \mathrm{d}t$$

so that

$$\frac{\mathrm{d}^n F(s)}{\mathrm{d}s^n} = \frac{\mathrm{d}^n}{\mathrm{d}s^n} \int_0^\infty e^{-st} f(t) \mathrm{d}t$$

Owing to the convergence properties of the improper integral involved, we can interchange the operations of differentiation and integration and differentiate with respect to s under the integral sign. Thus

$$\frac{\mathrm{d}^n F(s)}{\mathrm{d}s^n} = \int_0^\infty \frac{\partial^n}{\partial s^n} [e^{-st} f(t)] \mathrm{d}t$$

which, on carrying out the repeated differentiation, gives

$$\frac{\mathrm{d}^n F(s)}{\mathrm{d}s^n} = (-1)^n \int_0^\infty e^{-st} t^n f(t) \mathrm{d}t$$

$$= (-1)^n \mathcal{L}\{t^n f(t)\}, \quad \text{Re}(s) > \sigma_c$$

the region of convergence remaining unchanged.

end of theorem

In other words, Theorem 11.3 says that differentiating the transform of a function with respect to s is equivalent to multiplying the function itself by $-t$. As with the previous properties, we can now use this result to add to our list of Laplace transform pairs.

Example 11.11   Determine $\mathcal{L}\{t \sin 3t\}$.

Solution   Using the result (11.7),

$$\mathcal{L}\{\sin 3t\} = F(s) = \frac{3}{s^2 + 9}, \quad \operatorname{Re}(s) > 0$$

so, by the derivative theorem,

$$\mathcal{L}\{t \sin 3t\} = -\frac{dF(s)}{ds} = \frac{6s}{(s^2 + 9)^2}, \quad \operatorname{Re}(s) > 0$$

Example 11.12   Determine $\mathcal{L}\{t^2 e^t\}$.

Solution   From the result (11.6),

$$\mathcal{L}\{e^t\} = F(s) = \frac{1}{s - 1}, \quad \operatorname{Re}(s) > 1$$

so, by the derivative theorem,

$$\mathcal{L}\{t^2 e^t\} = (-1)^2 \frac{d^2 F(s)}{ds^2} = (-1)^2 \frac{d^2}{ds^2}\left(\frac{1}{s - 1}\right)$$

$$= (-1)\frac{d}{ds}\left(\frac{1}{(s - 1)^2}\right)$$

$$= \frac{2}{(s - 1)^3}, \quad \operatorname{Re}(s) > 1$$

Note that the result is easier to deduce using the first shift theorem.

Example 11.13   Determine $\mathcal{L}\{t^n\}$, where n is a positive integer.

Solution   Using the result (11.4),

$$\mathcal{L}\{1\} = \frac{1}{s}, \quad \operatorname{Re}(s) > 0$$

so, by the derivative theorem,

$$\mathcal{L}\{t^n\} = (-1)^n \frac{d^n}{ds^n}\left(\frac{1}{s}\right) = \frac{n!}{s^{n+1}}, \quad \operatorname{Re}(s) > 0$$

11.2.5 Table of Laplace transforms

It is appropriate at this stage to draw together the results proved to date for easy access. This is done in the form of two short tables. Figure 11.5(a) lists some Laplace transform pairs and Figure 11.5(b) lists the properties already considered.

Figure 11.5
(a) Table of Laplace transform pairs.
(b) Some properties of the Laplace transform.

(a)

$f(t)$	$\mathcal{L}\{f(t)\} = F(s)$	Region of convergence
$c,$ c a constant	$\dfrac{c}{s}$	$\text{Re}(s) > 0$
t	$\dfrac{1}{s^2}$	$\text{Re}(s) > 0$
$t^n,$ n a positive integer	$\dfrac{n!}{s^{n+1}}$	$\text{Re}(s) > 0$
$e^{kt},$ k a constant	$\dfrac{1}{s - k}$	$\text{Re}(s) > \text{Re}(k)$
$\sin at,$ a a real constant	$\dfrac{a}{s^2 + a^2}$	$\text{Re}(s) > 0$
$\cos at,$ a a real constant	$\dfrac{s}{s^2 + a^2}$	$\text{Re}(s) > 0$
$e^{-kt}\sin at,$ k and a real constants	$\dfrac{a}{(s + k)^2 + a^2}$	$\text{Re}(s) > -k$
$e^{-kt}\cos at,$ k and a real constants	$\dfrac{s + k}{(s + k)^2 + a^2}$	$\text{Re}(s) > -k$

(b) $\mathcal{L}\{f(t)\} = F(s),$ $\text{Re}(s) > \sigma_1$ and $\mathcal{L}\{g(t)\} = G(s),$ $\text{Re}(s) > \sigma_2$

Linearity: $\mathcal{L}\{\alpha f(t) + \beta g(t)\} = \alpha F(s) + \beta G(s),$ $\text{Re}(s) > \max(\sigma_1, \sigma_2)$

First shift theorem: $\mathcal{L}\{e^{at}f(t)\} = F(s - a),$ $\text{Re}(s) > \sigma_1 + \text{Re}(a)$

Derivative of transform:

$$\mathcal{L}\{t^n f(t)\} = (-1)^n \frac{d^n F(s)}{ds^n} \quad (n = 1, 2, \ldots), \quad \text{Re}(s) > \sigma_1$$

11.2.6   Exercises

1   Use the definition of the Laplace transform to obtain the transforms of $f(t)$ when $f(t)$ is given by

(a) $\cosh 2t$ (b) t^2 (c) $3 + t$ (d) te^{-t}

stating the region of convergence in each case.

2   What are the abscissae of convergence for the following functions?

(a) e^{5t} (b) e^{-3t}

(c) $\sin 2t$ (d) $\sinh 3t$

(e) $\cosh 2t$ (f) t^4

(g) $e^{-5t} + t^2$ (h) $3\cos 2t - t^3$

(i) $3e^{2t} - 2e^{-2t} + \sin 2t$ (j) $\sinh 3t + \sin 3t$

3   Using the results shown in Figure 11.5, obtain the Laplace transforms of the following functions, stating the region of convergence:

(a) $5 - 3t$ (b) $7t^3 - 2\sin 3t$

(c) $3 - 2t + 4\cos 2t$ (d) $\cosh 3t$

(e) $\sinh 2t$ (f) $5e^{-2t} + 3 - 2\cos 2t$

(g) $4te^{-2t}$ (h) $2e^{-3t}\sin 2t$

(i) $t^2 e^{-4t}$ (j) $6t^3 - 3t^2 + 4t - 2$

(k) $2\cos 3t + 5\sin 3t$ (l) $t\cos 2t$

(m) $t^2\sin 3t$ (n) $t^2 - 3\cos 4t$

(o) $t^2 e^{-2t} + e^{-t}\cos 2t + 3$

11.2.7 The inverse transform

The symbol $\mathcal{L}^{-1}\{F(s)\}$ denotes a causal function $f(t)$ whose Laplace transform is $F(s)$; that is,

$$\text{if}\quad \mathcal{L}\{f(t)\} = F(s) \quad \text{then}\quad f(t) = \mathcal{L}^{-1}\{F(s)\}$$

This correspondence between the functions $F(s)$ and $f(t)$ is called the **inverse Laplace transformation**, $f(t)$ being the **inverse transform** of $F(s)$, and $\mathcal{L}^{-1}$ being referred to as the **inverse Laplace transform operator**. These relationships are depicted in Figure 11.6.

Figure 11.6
The Laplace transform and its inverse.

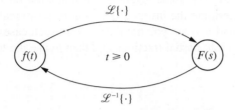

As was pointed out in observation (c) of Section 11.2.1, the Laplace transform $F(s)$ only determines the behaviour of $f(t)$ for $t \geqslant 0$. Thus $\mathcal{L}^{-1}\{F(s)\} = f(t)$ only for $t \geqslant 0$. When writing $\mathcal{L}^{-1}\{F(s)\} = f(t)$, it is assumed that $t \geqslant 0$ so strictly speaking, we should write

$$\mathcal{L}^{-1}\{F(s)\} = f(t)H(t) \tag{11.11}$$

Example 11.14

Since

$$\mathcal{L}\{e^{at}\} = \frac{1}{s - a}$$

it follows that

$$\mathcal{L}^{-1}\left\{\frac{1}{s - a}\right\} = e^{at}$$

Example 11.15

Since

$$\mathcal{L}\{\sin \omega t\} = \frac{\omega}{s^2 + \omega^2}$$

it follows that

$$\mathcal{L}^{-1}\left\{\frac{\omega}{s^2 + \omega^2}\right\} = \sin \omega t$$

The linearity property for the Laplace transform (Property 11.1) states that if α and β are any constants then

$$\mathscr{L}\{\alpha f(t) + \beta g(t)\} = \alpha \mathscr{L}\{f(t)\} + \beta \mathscr{L}\{g(t)\} = \alpha F(s) + \beta G(s)$$

It then follows from the above definition that

$$\mathscr{L}^{-1}\{\alpha F(s) + \beta G(s)\} = \alpha f(t) + \beta g(t) = \alpha \mathscr{L}^{-1}\{F(s)\} + \beta \mathscr{L}^{-1}\{G(s)\}$$

so that the inverse Laplace transform operator $\mathscr{L}^{-1}$ is also a **linear operator**.

11.2.8 Evaluation of inverse transforms

The most obvious way of finding the inverse transform of the function $F(s)$ is to make use of a table of transforms such as that given in Figure 11.5. Sometimes it is possible to write down the inverse transform directly from the table, but more often than not it is first necessary to carry out some algebraic manipulation on $F(s)$. In particular, we frequently need to determine the inverse transform of a rational function of the form $p(s)/q(s)$, where $p(s)$ and $q(s)$ are polynomials in s. In such cases the procedure is first to resolve the function into partial fractions and then to use the table of transforms.

Example 11.16   Find

$$\mathscr{L}^{-1}\left\{\frac{1}{(s+3)(s-2)}\right\}$$

Solution   First $1/(s+3)(s-2)$ is resolved into partial fractions, giving

$$\frac{1}{(s+3)(s-2)} = \frac{-\frac{1}{5}}{s+3} + \frac{\frac{1}{5}}{s-2}$$

Then, using the result $\mathscr{L}^{-1}\{1/(s+a)\} = e^{-at}$ together with the linearity property, we have

$$\mathscr{L}^{-1}\left\{\frac{1}{(s+3)(s-2)}\right\} = -\tfrac{1}{5}\mathscr{L}^{-1}\left\{\frac{1}{s+3}\right\} + \tfrac{1}{5}\mathscr{L}^{-1}\left\{\frac{1}{s-2}\right\} = -\tfrac{1}{5}e^{-3t} + \tfrac{1}{5}e^{2t}$$

Example 11.17   Find

$$\mathscr{L}^{-1}\left\{\frac{s+1}{s^2(s^2+9)}\right\}$$

Solution   Resolving $(s+1)/s^2(s^2+9)$ into partial fractions gives

$$\frac{s+1}{s^2(s^2+9)} = \frac{\frac{1}{9}}{s} + \frac{\frac{1}{9}}{s^2} - \frac{1}{9}\frac{s+1}{s^2+9}$$

$$= \frac{\frac{1}{9}}{s} + \frac{\frac{1}{9}}{s^2} - \frac{1}{9}\frac{s}{s^2+3^2} - \frac{1}{27}\frac{3}{s^2+3^2}$$

Using the results in Figure 11.5, together with the linearity property, we have

$$\mathscr{L}^{-1}\left\{\frac{s+1}{s^2(s^2+9)}\right\} = \tfrac{1}{9} + \tfrac{1}{9}t - \tfrac{1}{9}\cos 3t - \tfrac{1}{27}\sin 3t$$

11.2.9 Inversion using the first shift theorem

In Theorem 11.2 we saw that if $F(s)$ is the Laplace transform of $f(t)$ then, for a scalar a, $F(s - a)$ is the Laplace transform of $e^{at}f(t)$. This theorem normally causes little difficulty when used to obtain the Laplace transforms of functions, but it does frequently lead to problems when used to obtain inverse transforms. Expressed in the inverse form, the theorem becomes

$$\mathcal{L}^{-1}\{F(s - a)\} = e^{at}f(t)$$

The notation

$$\mathcal{L}^{-1}\{[F(s)]_{s \to s-a}\} = e^{at}[f(t)]$$

where $F(s) = \mathcal{L}\{f(t)\}$ and $[F(s)]_{s \to s-a}$ denotes that s in $F(s)$ is replaced by $s - a$, may make the relation clearer.

Example 11.18   Find

$$\mathcal{L}^{-1}\left\{\frac{1}{(s + 2)^2}\right\}$$

Solution

$$\frac{1}{(s + 2)^2} = \left[\frac{1}{s^2}\right]_{s \to s+2}$$

and, since $1/s^2 = \mathcal{L}\{t\}$, the shift theorem gives

$$\mathcal{L}^{-1}\left\{\frac{1}{(s + 2)^2}\right\} = te^{-2t}$$

Example 11.19   Find

$$\mathcal{L}^{-1}\left\{\frac{2}{s^2 + 6s + 13}\right\}$$

Solution

$$\frac{2}{s^2 + 6s + 13} = \frac{2}{(s + 3)^2 + 4} = \left[\frac{2}{s^2 + 2^2}\right]_{s \to s+3}$$

and, since $2/(s^2 + 2^2) = \mathcal{L}\{\sin 2t\}$, the shift theorem gives

$$\mathcal{L}^{-1}\left\{\frac{2}{s^2 + 6s + 13}\right\} = e^{-3t}\sin 2t$$

Example 11.20   Find

$$\mathcal{L}^{-1}\left\{\frac{s + 7}{s^2 + 2s + 5}\right\}$$

Solution

$$\frac{s+7}{s^2+2s+5} = \frac{s+7}{(s+1)^2+4} = \frac{(s+1)}{(s+1)^2+4} + 3\frac{2}{(s+1)^2+4}$$

$$= \left[\frac{s}{s^2+2^2}\right]_{s\to s+1} + 3\left[\frac{2}{s^2+2^2}\right]_{s\to s+1}$$

Since $s/(s^2+2^2) = \mathcal{L}\{\cos 2t\}$ and $2/(s^2+2^2) = \mathcal{L}\{\sin 2t\}$, the shift theorem gives

$$\mathcal{L}^{-1}\left\{\frac{s+7}{s^2+2s+5}\right\} = e^{-t}\cos 2t + 3e^{-t}\sin 2t$$

Example 11.21   Find

$$\mathcal{L}^{-1}\left\{\frac{1}{(s+1)^2(s^2+4)}\right\}$$

Solution   Resolving $1/(s+1)^2(s^2+4)$ into partial fractions gives

$$\frac{1}{(s+1)^2(s^2+4)} = \frac{\frac{2}{25}}{s+1} + \frac{\frac{1}{5}}{(s+1)^2} - \frac{1}{25}\frac{2s+3}{s^2+4}$$

$$= \frac{\frac{2}{25}}{s+1} + \frac{1}{5}\left[\frac{1}{s^2}\right]_{s\to s+1} - \frac{2}{25}\frac{s}{s^2+2^2} - \frac{3}{50}\frac{2}{s^2+2^2}$$

Since $1/s^2 = \mathcal{L}\{t\}$, the shift theorem, together with the results in Figure 11.5, gives

$$\mathcal{L}^{-1}\left\{\frac{1}{(s+1)^2(s^2+4)}\right\} = \tfrac{2}{25}e^{-t} + \tfrac{1}{5}e^{-t}t - \tfrac{2}{25}\cos 2t - \tfrac{3}{50}\sin 2t$$

11.2.10   Exercise

4   Find $\mathcal{L}^{-1}\{F(s)\}$ when $F(s)$ is given by

(a) $\dfrac{1}{(s+3)(s+7)}$

(b) $\dfrac{s+5}{(s+1)(s-3)}$

(c) $\dfrac{s-1}{s^2(s+3)}$

(d) $\dfrac{2s+6}{s^2+4}$

(e) $\dfrac{1}{s^2(s^2+16)}$

(f) $\dfrac{s+8}{s^2+4s+5}$

(g) $\dfrac{s+1}{s^2(s^2+4s+8)}$

(h) $\dfrac{4s}{(s-1)(s+1)^2}$

(i) $\dfrac{s+7}{s^2+2s+5}$

(j) $\dfrac{3s^2-7s+5}{(s-1)(s-2)(s-3)}$

(k) $\dfrac{5s-7}{(s+3)(s^2+2)}$

(l) $\dfrac{s}{(s-1)(s^2+2s+2)}$

(m) $\dfrac{s-1}{s^2+2s+5}$

(n) $\dfrac{s-1}{(s-2)(s-3)(s-4)}$

(o) $\dfrac{3s}{(s-1)(s^2-4)}$

(p) $\dfrac{36}{s(s^2+1)(s^2+9)}$

(q) $\dfrac{2s^2+4s+9}{(s+2)(s^2+3s+3)}$

(r) $\dfrac{1}{(s+1)(s+2)(s^2+2s+10)}$

11.3 ## Solution of differential equations

We first consider the Laplace transforms of derivatives and integrals, and then apply these to the solution of differential equations.

11.3.1 Transforms of derivatives

If we are to use Laplace transform methods to solve differential equations, we need to find convenient expressions for the Laplace transforms of derivatives such as df/dt, d^2f/dt^2 or, in general, d^nf/dt^n. By definition,

$$\mathcal{L}\left\{\frac{df}{dt}\right\} = \int_0^\infty e^{-st}\frac{df}{dt}\,dt$$

Integrating by parts, we have

$$\mathcal{L}\left\{\frac{df}{dt}\right\} = [e^{-st}f(t)]_0^\infty + s\int_0^\infty e^{-st}f(t)\,dt$$

$$= -f(0) + sF(s)$$

that is,

$$\boxed{\mathcal{L}\left\{\frac{df}{dt}\right\} = sF(s) - f(0)} \tag{11.12}$$

In taking the Laplace transform of a derivative we have assumed that $f(t)$ is continuous at $t = 0$, so that $f(0^-) = f(0) = f(0^+)$. In the companion text *Advanced Modern Engineering Mathematics* there are occasions when $f(0^-) \neq f(0^+)$ and we have to revert to a more generalized calculus to resolve the problem.

The advantage of using the Laplace transform when dealing with differential equations can readily be seen, since it enables us to replace the operation of differentiation in the time domain by a simple algebraic operation in the s domain.

Note that to deduce the result (11.12), we have assumed that $f(t)$ is continuous, with a piecewise-continuous derivative df/dt, for $t \geqslant 0$ and that it is also of exponential order as $t \to \infty$.

Likewise, if both $f(t)$ and df/dt are continuous on $t \geqslant 0$ and are of exponential order as $t \to \infty$, and d^2f/dt^2 is piecewise-continuous for $t \geqslant 0$, then

$$\mathcal{L}\left\{\frac{d^2f}{dt^2}\right\} = \int_0^\infty e^{-st}\frac{d^2f}{dt^2}\,dt = \left[e^{-st}\frac{df}{dt}\right]_0^\infty + s\int_0^\infty e^{-st}\frac{df}{dt}\,dt$$

$$= -\left[\frac{df}{dt}\right]_{t=0} + s\mathcal{L}\left\{\frac{df}{dt}\right\}$$

which, on using (11.11), gives

$$\mathcal{L}\left\{\frac{d^2f}{dt^2}\right\} = -\left[\frac{df}{dt}\right]_{t=0} + s[sF(s) - f(0)]$$

leading to the result

$$\mathcal{L}\left\{\frac{\mathrm{d}^2 f}{\mathrm{d}t^2}\right\} = s^2 F(s) - sf(0) - \left[\frac{\mathrm{d}f}{\mathrm{d}t}\right]_{t=0} = s^2 F(s) - sf(0) - f^{(1)}(0) \qquad \textbf{(11.13)}$$

Clearly, provided that $f(t)$ and its derivatives satisfy the required conditions, this procedure may be extended to obtain the Laplace transform of $f^{(n)}(t) = \mathrm{d}^n f/\mathrm{d}t^n$ in the form

$$\mathcal{L}\{f^{(n)}(t)\} = s^n F(s) - s^{n-1} f(0) - s^{n-2} f^{(1)}(0) - \ldots - f^{(n-1)}(0)$$

$$= s^n F(s) - \sum_{i=1}^{n} s^{n-i} f^{(i-1)}(0) \qquad \textbf{(11.14)}$$

a result that may be readily proved by induction.

Again it is noted that in determining the Laplace transform of $f^{(n)}(t)$ we have assumed that $f^{(n-1)}(t)$ is continuous.

11.3.2 Transforms of integrals

In some applications the behaviour of a system may be represented by an **integro-differential equation**, which is an equation containing both derivatives and integrals of the unknown variable. For example, the current i in a series electrical circuit consisting of a resistance R, an inductance L and capacitance C, and subject to an applied voltage E, is given by

$$L\frac{\mathrm{d}i}{\mathrm{d}t} + iR + \frac{1}{C}\int_0^t i(\tau)\mathrm{d}\tau = E$$

To solve such equations directly, it is convenient to be able to obtain the Laplace transform of integrals such as $\int_0^t f(\tau)\mathrm{d}\tau$.

Writing

$$g(t) = \int_0^t f(\tau)\mathrm{d}\tau$$

we have

$$\frac{\mathrm{d}g}{\mathrm{d}t} = f(t), \quad g(0) = 0$$

Taking Laplace transforms,

$$\mathcal{L}\left\{\frac{\mathrm{d}g}{\mathrm{d}t}\right\} = \mathcal{L}\{f(t)\}$$

which, on using (11.12), gives

$$sG(s) = F(s)$$

or

$$\mathcal{L}\{g(t)\} = G(s) = \frac{1}{s}F(s) = \frac{1}{s}\mathcal{L}\{f(t)\}$$

leading to the result

$$\mathcal{L}\left\{\int_0^t f(\tau)\mathrm{d}\tau\right\} = \frac{1}{s}\mathcal{L}\{f(t)\} = \frac{1}{s}F(s) \qquad\qquad \textbf{(11.15)}$$

Example 11.22   Obtain

$$\mathcal{L}\left\{\int_0^t (\tau^3 + \sin 2\tau)\mathrm{d}\tau\right\}$$

Solution   In this case $f(t) = t^3 + \sin 2t$, giving

$$F(s) = \mathcal{L}\{f(t)\} = \mathcal{L}\{t^3\} + \mathcal{L}\{\sin 2t\}$$

$$= \frac{6}{s^4} + \frac{2}{s^2 + 4}$$

so, by (11.15),

$$\mathcal{L}\left\{\int_0^t (\tau^3 + \sin 2\tau)\mathrm{d}\tau\right\} = \frac{1}{s}F(s) = \frac{6}{s^5} + \frac{2}{s(s^2 + 4)}$$

11.3.3   Ordinary differential equations

Having obtained expressions for the Laplace transforms of derivatives, we are now in a position to use Laplace transform methods to solve ordinary linear differential equations with constant coefficients, which were introduced in Chapter 10. To illustrate this, consider the general second-order linear differential equation

$$a\frac{\mathrm{d}^2 x}{\mathrm{d}t^2} + b\frac{\mathrm{d}x}{\mathrm{d}t} + cx = u(t) \quad (t \geqslant 0) \qquad\qquad \textbf{(11.16)}$$

subject to the initial conditions $x(0) = x_0$, $\dot{x}(0) = v_0$ where as usual a dot denotes differentiation with respect to time, t. Such a differential equation may model the dynamics of some system for which the variable $x(t)$ determines the **response** of the system to the **forcing** or **excitation** term $u(t)$. The terms **system input** and **system output** are also frequently used for $u(t)$ and $x(t)$ respectively. Since the differential equation is linear and has constant coefficients, a system characterized by such a model is said to be a **linear time-invariant system**.

Taking Laplace transforms of each term in (11.16) gives

$$a\mathscr{L}\left\{\frac{d^2x}{dt^2}\right\} + b\mathscr{L}\left\{\frac{dx}{dt}\right\} + c\mathscr{L}\{x\} = \mathscr{L}\{u(t)\}$$

which on using (11.12) and (11.13) leads to

$$a[s^2X(s) - sx(0) - \dot{x}(0)] + b[sX(s) - x(0)] + cX(s) = U(s)$$

Rearranging, and incorporating the given initial conditions, gives

$$(as^2 + bs + c)X(s) = U(s) + (as + b)x_0 + av_0$$

so that

$$X(s) = \frac{U(s) + (as + b)x_0 + av_0}{as^2 + bs + c} \tag{11.17}$$

Equation (11.17) determines the Laplace transform $X(s)$ of the response, from which, by taking the inverse transform, the desired time response $x(t)$ may be obtained.

Before considering specific examples, there are a few observations worth noting at this stage.

(a) As we have already noted in Section 11.3.1, a distinct advantage of using the Laplace transform is that it enables us to replace the operation of differentiation by an algebraic operation. Consequently, by taking the Laplace transform of each term in a differential equation, it is converted into an algebraic equation in the variable s. This may then be rearranged using algebraic rules to obtain an expression for the Laplace transform of the response; the desired time response is then obtained by taking the inverse transform.

(b) The Laplace transform method yields the complete solution to the linear differential equation, with the initial conditions automatically included. This contrasts with the classical approach adopted in Chapter 10, in which the general solution consists of two components, the **complementary function** and the **particular integral**, with the initial conditions determining the undetermined constants associated with the complementary function. When the solution is expressed in the general form (11.17), upon inversion the term involving $U(s)$ leads to a particular integral while that involving x_0 and v_0 gives a complementary function. A useful side issue is that an explicit solution for the transient is obtained that reflects the initial conditions.

(c) The Laplace transform method is ideally suited for solving initial-value problems; that is, linear differential equations in which all the initial conditions $x(0)$, $\dot{x}(0)$, and so on, at time $t = 0$ are specified. The method is less attractive for boundary-value problems, when the conditions on $x(t)$ and its derivatives are not all specified at $t = 0$, but some are specified at other values of the independent variable. It is still possible, however, to use the Laplace transform method by assigning arbitrary constants to one or more of the initial conditions and then determining their values using the given boundary conditions.

(d) It should be noted that the denominator of the right-hand side of (11.17) is the left-hand side of (11.16) with the operator d/dt replaced by s. The denominator equated to zero also corresponds to the auxiliary equation or characteristic equation used in the classical approach. Given a specific initial-value problem, the process of obtaining a solution using Laplace transform methods is fairly straightforward, and is illustrated by Example 11.23.

Example 11.23   Solve the differential equation

$$\frac{d^2x}{dt^2} + 5\frac{dx}{dt} + 6x = 2e^{-t} \quad (t \geqslant 0)$$

subject to the initial conditions $x = 1$ and $dx/dt = 0$ at $t = 0$.

Solution   Taking Laplace transforms

$$\mathscr{L}\left\{\frac{d^2x}{dt^2}\right\} + 5\mathscr{L}\left\{\frac{dx}{dt}\right\} + 6\mathscr{L}\{x\} = 2\mathscr{L}\{e^{-t}\}$$

leads to the transformed equation

$$[s^2X(s) - sx(0) - \dot{x}(0)] + 5[sX(s) - x(0)] + 6X(s) = \frac{2}{s+1}$$

which on rearrangement gives

$$(s^2 + 5s + 6)X(s) = \frac{2}{s+1} + (s+5)x(0) + \dot{x}(0)$$

Incorporating the given initial conditions $x(0) = 1$ and $\dot{x}(0) = 0$ leads to

$$(s^2 + 5s + 6)X(s) = \frac{2}{s+1} + s + 5$$

That is,

$$X(s) = \frac{2}{(s+1)(s+2)(s+3)} + \frac{s+5}{(s+3)(s+2)}$$

Resolving the rational terms into partial fractions gives

$$X(s) = \frac{1}{s+1} - \frac{2}{s+2} + \frac{1}{s+3} + \frac{3}{s+2} - \frac{2}{s+3}$$

$$= \frac{1}{s+1} + \frac{1}{s+2} - \frac{1}{s+3}$$

Taking inverse transforms gives the desired solution

$$x(t) = e^{-t} + e^{-2t} - e^{-3t} \quad (t \geqslant 0)$$

In principle the procedure adopted in Example 11.23 for solving a second-order linear differential equation with constant coefficients is readily carried over to higher-order differential equations. A general nth-order linear differential equation may be written as

$$a_n\frac{d^nx}{dt^n} + a_{n-1}\frac{d^{n-1}x}{dt^{n-1}} + \dots + a_0x = u(t) \quad (t \geqslant 0) \tag{11.18}$$

where $a_n, a_{n-1}, \ldots, a_0$ are constants, with $a_n \neq 0$. This may be written in the more concise form

$$q(\mathrm{D})x(t) = u(t) \qquad\qquad\qquad (11.19)$$

where D denotes the operator $\mathrm{d}/\mathrm{d}t$ and $q(\mathrm{D})$ is the polynomial

$$q(\mathrm{D}) = \sum_{r=0}^{n} a_r \mathrm{D}^r$$

The objective is then to determine the response $x(t)$ for a given forcing function $u(t)$ subject to the given set of initial conditions

$$\mathrm{D}^r x(0) = \left[\frac{\mathrm{d}^r x}{\mathrm{d}t^r}\right]_{t=0} = c_r \quad (r = 0, 1, \ldots, n-1)$$

Taking Laplace transforms in (11.19) and proceeding as before leads to

$$X(s) = \frac{p(s)}{q(s)}$$

where

$$p(s) = U(s) + \sum_{r=0}^{n-1} c_r \sum_{i=r+1}^{n} a_i s^{i-r-1}$$

Then, in principle, by taking the inverse transform, the desired response $x(t)$ may be obtained as

$$x(t) = \mathcal{L}^{-1}\left\{\frac{p(s)}{q(s)}\right\}$$

For high-order differential equations the process of performing this inversion may prove to be rather tedious, and matrix methods may be used as indicated in Chapter 6 of the companion text, *Advanced Modern Engineering Mathematics*.

To conclude this section, further worked examples are developed in order to help consolidate understanding of this method for solving linear differential equations.

Example 11.24 Solve the differential equation

$$\frac{\mathrm{d}^2 x}{\mathrm{d}t^2} + 6\frac{\mathrm{d}x}{\mathrm{d}t} + 9x = \sin t \quad (t \geq 0)$$

subject to the initial conditions $x = 0$ and $\mathrm{d}x/\mathrm{d}t = 0$ at $t = 0$.

Solution Taking the Laplace transforms

$$\mathcal{L}\left\{\frac{\mathrm{d}^2 x}{\mathrm{d}t^2}\right\} + 6\mathcal{L}\left\{\frac{\mathrm{d}x}{\mathrm{d}t}\right\} + 9\mathcal{L}\{x\} = \mathcal{L}\{\sin t\}$$

leads to the equation

$$[s^2 X(s) - sx(0) - \dot{x}(0)] + 6[sX(s) - x(0)] + 9X(s) = \frac{1}{s^2 + 1}$$

which on rearrangement gives

$$(s^2 + 6s + 9)X(s) = \frac{1}{s^2 + 1} + (s + 6)x(0) + \dot{x}(0)$$

Incorporating the given initial conditions $x(0) = \dot{x}(0) = 0$ leads to

$$X(s) = \frac{1}{(s^2 + 1)(s + 3)^2}$$

Resolving into partial fractions gives

$$X(s) = \frac{3}{50}\frac{1}{s + 3} + \frac{1}{10}\frac{1}{(s + 3)^2} + \frac{2}{25}\frac{1}{s^2 + 1} - \frac{3}{50}\frac{s}{s^2 + 1}$$

that is,

$$X(s) = \frac{3}{50}\frac{1}{s + 3} + \frac{1}{10}\left[\frac{1}{s^2}\right]_{s \to s+3} + \frac{2}{25}\frac{1}{s^2 + 1} - \frac{3}{50}\frac{s}{s^2 + 1}$$

Taking inverse transforms, using the shift theorem, leads to the desired solution

$$x(t) = \frac{3}{50}e^{-3t} + \frac{1}{10}te^{-3t} + \frac{2}{25}\sin t - \frac{3}{50}\cos t \quad (t \geqslant 0)$$

Example 11.25   Solve the differential equation

$$\frac{d^3x}{dt^3} + 5\frac{d^2x}{dt^2} + 17\frac{dx}{dt} + 13x = 1 \quad (t \geqslant 0)$$

subject to the initial conditions $x = dx/dt = 1$ and $d^2x/dt^2 = 0$ at $t = 0$.

Solution   Taking Laplace transforms

$$\mathscr{L}\left\{\frac{d^3x}{dt^3}\right\} + 5\mathscr{L}\left\{\frac{d^2x}{dt^2}\right\} + 17\mathscr{L}\left\{\frac{dx}{dt}\right\} + 13\mathscr{L}\{x\} = \mathscr{L}\{1\}$$

leads to the equation

$$s^3X(s) - s^2x(0) - s\dot{x}(0) - \ddot{x}(0) + 5[s^2X(s) - sx(0) - \dot{x}(0)]$$

$$+ 17[sX(s) - x(0)] + 13X(s) = \frac{1}{s}$$

which on rearrangement gives

$$(s^3 + 5s^2 + 17s + 13)X(s) = \frac{1}{s} + (s^2 + 5s + 17)x(0) + (s + 5)\dot{x}(0) + \ddot{x}(0)$$

Incorporating the given initial conditions $x(0) = \dot{x}(0) = 1$ and $\ddot{x}(0) = 0$ leads to

$$X(s) = \frac{s^3 + 6s^2 + 22s + 1}{s(s^3 + 5s^2 + 17s + 13)}$$

Clearly $s + 1$ is a factor of $s^3 + 5s^2 + 17s + 13$, and by algebraic division we have

$$X(s) = \frac{s^3 + 6s^2 + 22s + 1}{s(s + 1)(s^2 + 4s + 13)}$$

Resolving into partial fractions,

$$X(s) = \frac{\frac{1}{13}}{s} + \frac{\frac{8}{5}}{s+1} - \frac{1}{65}\frac{44s+7}{s^2+4s+13}$$

$$= \frac{\frac{1}{13}}{s} + \frac{\frac{8}{5}}{s+1} - \frac{1}{65}\frac{44(s+2)-27(3)}{(s+2)^2+3^2}$$

Taking inverse transforms, using the shift theorem, leads to the solution

$$x(t) = \tfrac{1}{13} + \tfrac{8}{5}e^{-t} - \tfrac{1}{65}e^{-2t}(44\cos 3t - 27\sin 3t) \quad (t \geqslant 0)$$

11.3.4 Exercise

5 Using Laplace transform methods, solve for $t \geqslant 0$ the following differential equations, subject to the specified initial conditions. (Readers are encouraged to check their solutions using an appropriate software package.)

(a) $\dfrac{dx}{dt} + 3x = e^{-2t}$ subject to $x = 2$ at $t = 0$

(b) $3\dfrac{dx}{dt} - 4x = \sin 2t$ subject to $x = \tfrac{1}{3}$ at $t = 0$

(c) $\dfrac{d^2x}{dt^2} + 2\dfrac{dx}{dt} + 5x = 1$

   subject to $x = 0$ and $\dfrac{dx}{dt} = 0$ at $t = 0$

(d) $\dfrac{d^2y}{dt^2} + 2\dfrac{dy}{dt} + y = 4\cos 2t$

   subject to $y = 0$ and $\dfrac{dy}{dt} = 2$ at $t = 0$

(e) $\dfrac{d^2x}{dt^2} - 3\dfrac{dx}{dt} + 2x = 2e^{-4t}$

   subject to $x = 0$ and $\dfrac{dx}{dt} = 1$ at $t = 0$

(f) $\dfrac{d^2x}{dt^2} + 4\dfrac{dx}{dt} + 5x = 3e^{-2t}$

   subject to $x = 4$ and $\dfrac{dx}{dt} = -7$ at $t = 0$

(g) $\dfrac{d^2x}{dt^2} + \dfrac{dx}{dt} - 2x = 5e^{-t}\sin t$

   subject to $x = 1$ and $\dfrac{dx}{dt} = 0$ at $t = 0$

(h) $\dfrac{d^2y}{dt^2} + 2\dfrac{dy}{dt} + 3y = 3t$

   subject to $y = 0$ and $\dfrac{dy}{dt} = 1$ at $t = 0$

(i) $\dfrac{d^2x}{dt^2} + 4\dfrac{dx}{dt} + 4x = t^2 + e^{-2t}$

   subject to $x = \tfrac{1}{2}$ and $\dfrac{dx}{dt} = 0$ at $t = 0$

(j) $9\dfrac{d^2x}{dt^2} + 12\dfrac{dx}{dt} + 5x = 1$

   subject to $x = 0$ and $\dfrac{dx}{dt} = 0$ at $t = 0$

(k) $\dfrac{d^2x}{dt^2} + 8\dfrac{dx}{dt} + 16x = 16\sin 4t$

   subject to $x = -\tfrac{1}{2}$ and $\dfrac{dx}{dt} = 1$ at $t = 0$

(l) $9\dfrac{d^2y}{dt^2} + 12\dfrac{dy}{dt} + 4y = e^{-t}$

   subject to $y = 1$ and $\dfrac{dy}{dt} = 1$ at $t = 0$

(m) $\dfrac{d^3x}{dt^3} - 2\dfrac{d^2x}{dt^2} - \dfrac{dx}{dt} + 2x = 2 + t$

   subject to $x = 0$, $\dfrac{dx}{dt} = 1$ and $\dfrac{d^2x}{dt^2} = 0$ at $t = 0$

(n) $\dfrac{d^3x}{dt^3} + \dfrac{d^2x}{dt^2} + \dfrac{dx}{dt} + x = \cos 3t$

   subject to $x = 0$, $\dfrac{dx}{dt} = 1$ and $\dfrac{d^2x}{dt^2} = 1$ at $t = 0$

11.3.5 Simultaneous differential equations

In engineering we frequently encounter systems whose characteristics are modelled by a set of simultaneous linear differential equations with constant coefficients. The method of solution is essentially the same as that adopted in Section 11.3.3 for solving a single differential equation in one unknown. Taking Laplace transforms throughout, the system of simultaneous differential equations is transformed into a system of simultaneous algebraic equations, which are then solved for the transformed variables; inverse transforms then give the desired solutions.

Example 11.26 Solve for $t \geqslant 0$ the simultaneous first-order differential equations

$$\frac{dx}{dt} + \frac{dy}{dt} + 5x + 3y = e^{-t} \tag{11.20}$$

$$2\frac{dx}{dt} + \frac{dy}{dt} + x + y = 3 \tag{11.21}$$

subject to the initial conditions $x = 2$ and $y = 1$ at $t = 0$.

Solution Taking Laplace transforms in (11.20) and (11.21) gives

$$sX(s) - x(0) + sY(s) - y(0) + 5X(s) + 3Y(s) = \frac{1}{s+1}$$

$$2[sX(s) - x(0)] + sY(s) - y(0) + X(s) + Y(s) = \frac{3}{s}$$

Rearranging and incorporating the given initial conditions $x(0) = 2$ and $y(0) = 1$ leads to

$$(s+5)X(s) + (s+3)Y(s) = 3 + \frac{1}{s+1} = \frac{3s+4}{s+1} \tag{11.22}$$

$$(2s+1)X(s) + (s+1)Y(s) = 5 + \frac{3}{s} = \frac{5s+3}{s} \tag{11.23}$$

Hence, by taking Laplace transforms, the pair of simultaneous differential equations (11.20) and (11.21) in $x(t)$ and $y(t)$ has been transformed into a pair of simultaneous algebraic equations (11.22) and (11.23) in the transformed variables $X(s)$ and $Y(s)$. These algebraic equations may now be solved simultaneously for $X(s)$ and $Y(s)$ using standard algebraic techniques.

Solving first for $X(s)$ gives

$$X(s) = \frac{2s^2 + 14s + 9}{s(s+2)(s-1)}$$

Resolving into partial fractions,

$$X(s) = -\frac{\frac{9}{2}}{s} - \frac{\frac{11}{6}}{s+2} + \frac{\frac{25}{3}}{s-1}$$

which on inversion gives

$$x(t) = -\tfrac{9}{2} - \tfrac{11}{6}e^{-2t} + \tfrac{25}{3}e^{t} \quad (t \geqslant 0) \tag{11.24}$$

Likewise, solving for $Y(s)$ gives

$$Y(s) = \frac{s^3 - 22s^2 - 39s - 15}{s(s + 1)(s + 2)(s - 1)}$$

Resolving into partial fractions,

$$Y(s) = \frac{\frac{15}{2}}{s} + \frac{\frac{1}{2}}{s + 1} + \frac{\frac{11}{2}}{s + 2} - \frac{\frac{25}{2}}{s - 1}$$

which on inversion gives

$$y(t) = \tfrac{15}{2} + \tfrac{1}{2}\mathrm{e}^{-t} + \tfrac{11}{2}\mathrm{e}^{-2t} - \tfrac{25}{2}\mathrm{e}^{t} \quad (t \geqslant 0)$$

Thus the solution to the given pair of simultaneous differential equations is

$$\left.\begin{array}{l} x(t) = -\tfrac{9}{2} - \tfrac{11}{6}\mathrm{e}^{-2t} + \tfrac{25}{3}\mathrm{e}^{t} \\[2mm] y(t) = \tfrac{15}{2} + \tfrac{1}{2}\mathrm{e}^{-t} + \tfrac{11}{2}\mathrm{e}^{-2t} - \tfrac{25}{2}\mathrm{e}^{t} \end{array}\right\} \quad (t \geqslant 0)$$

Note: When solving a pair of first-order simultaneous differential equations such as (11.20) and (11.21), an alternative approach to obtaining the value of $y(t)$ having obtained $x(t)$ is to use (11.20) and (11.21) directly.

Eliminating $\mathrm{d}y/\mathrm{d}t$ from (11.20) and (11.21) gives

$$2y = \frac{\mathrm{d}x}{\mathrm{d}t} - 4x - 3 + \mathrm{e}^{-t}$$

Substituting the solution obtained in (11.24) for $x(t)$ gives

$$2y = (\tfrac{11}{3}\mathrm{e}^{-2t} + \tfrac{25}{3}\mathrm{e}^{t}) - 4(-\tfrac{9}{2} - \tfrac{11}{6}\mathrm{e}^{-2t} + \tfrac{25}{3}\mathrm{e}^{t}) - 3 + \mathrm{e}^{-t}$$

leading as before to the solution

$$y = \tfrac{15}{2} + \tfrac{1}{2}\mathrm{e}^{-t} + \tfrac{11}{2}\mathrm{e}^{-2t} - \tfrac{25}{2}\mathrm{e}^{t}$$

A further alternative is to express (11.22) and (11.23) in matrix form and solve for $X(s)$ and $Y(s)$ using Gaussian elimination.

In principle, the same procedure as used in Example 11.26 can be employed to solve a pair of higher-order simultaneous differential equations or a larger system of differential equations involving more unknowns. However, the algebra involved can become quite complicated, and matrix methods are usually preferred.

11.3.6 Exercise

6 Using Laplace transform methods, solve for $t \geqslant 0$ the following simultaneous differential equations subject to the given initial conditions. (Readers are encouraged to check their solutions using an appropriate software package.)

(a) $2\dfrac{\mathrm{d}x}{\mathrm{d}t} - 2\dfrac{\mathrm{d}y}{\mathrm{d}t} - 9y = \mathrm{e}^{-2t}$

$2\dfrac{\mathrm{d}x}{\mathrm{d}t} + 4\dfrac{\mathrm{d}y}{\mathrm{d}t} + 4x - 37y = 0$

subject to $x = 0$ and $y = \tfrac{1}{4}$ at $t = 0$

(b) $\dfrac{\mathrm{d}x}{\mathrm{d}t} + 2\dfrac{\mathrm{d}y}{\mathrm{d}t} + x - y = 5\sin t$

$2\dfrac{\mathrm{d}x}{\mathrm{d}t} + 3\dfrac{\mathrm{d}y}{\mathrm{d}t} + x - y = \mathrm{e}^{t}$

subject to $x = 0$ and $y = 0$ at $t = 0$

(c) $\dfrac{\mathrm{d}x}{\mathrm{d}t} + \dfrac{\mathrm{d}y}{\mathrm{d}t} + 2x + y = \mathrm{e}^{-3t}$

$\dfrac{\mathrm{d}y}{\mathrm{d}t} + 5x + 3y = 5\mathrm{e}^{-2t}$

subject to $x = -1$ and $y = 4$ at $t = 0$

(d) $3\dfrac{\mathrm{d}x}{\mathrm{d}t} + 3\dfrac{\mathrm{d}y}{\mathrm{d}t} - 2x = \mathrm{e}^{t}$

$\dfrac{\mathrm{d}x}{\mathrm{d}t} + 2\dfrac{\mathrm{d}y}{\mathrm{d}t} - y = 1$

subject to $x = 1$ and $y = 1$ at $t = 0$

(e) $3\dfrac{\mathrm{d}x}{\mathrm{d}t} + \dfrac{\mathrm{d}y}{\mathrm{d}t} - 2x = 3\sin t + 5\cos t$

$2\dfrac{\mathrm{d}x}{\mathrm{d}t} + \dfrac{\mathrm{d}y}{\mathrm{d}t} + y = \sin t + \cos t$

subject to $x = 0$ and $y = -1$ at $t = 0$

(f) $\dfrac{\mathrm{d}x}{\mathrm{d}t} + \dfrac{\mathrm{d}y}{\mathrm{d}t} + y = t$

$\dfrac{\mathrm{d}x}{\mathrm{d}t} + 4\dfrac{\mathrm{d}y}{\mathrm{d}t} + x = 1$

subject to $x = 1$ and $y = 0$ at $t = 0$

(g) $2\dfrac{\mathrm{d}x}{\mathrm{d}t} + 3\dfrac{\mathrm{d}y}{\mathrm{d}t} + 7x = 14t + 7$

$5\dfrac{\mathrm{d}x}{\mathrm{d}t} - 3\dfrac{\mathrm{d}y}{\mathrm{d}t} + 4x + 6y = 14t - 14$

subject to $x = y = 0$ at $t = 0$

(h) $\dfrac{\mathrm{d}^2 x}{\mathrm{d}t^2} = y - 2x$

$\dfrac{\mathrm{d}^2 y}{\mathrm{d}t^2} = x - 2y$

subject to $x = 4$, $y = 2$, $\mathrm{d}x/\mathrm{d}t = 0$ and $\mathrm{d}y/\mathrm{d}t = 0$ at $t = 0$

(i) $5\dfrac{\mathrm{d}^2 x}{\mathrm{d}t^2} + 12\dfrac{\mathrm{d}^2 y}{\mathrm{d}t^2} + 6x = 0$

$5\dfrac{\mathrm{d}^2 x}{\mathrm{d}t^2} + 16\dfrac{\mathrm{d}^2 y}{\mathrm{d}t^2} + 6y = 0$

subject to $x = \frac{7}{4}$, $y = 1$, $\mathrm{d}x/\mathrm{d}t = 0$ and $\mathrm{d}y/\mathrm{d}t = 0$ at $t = 0$

(j) $2\dfrac{\mathrm{d}^2 x}{\mathrm{d}t^2} - \dfrac{\mathrm{d}^2 y}{\mathrm{d}t^2} - \dfrac{\mathrm{d}x}{\mathrm{d}t} - \dfrac{\mathrm{d}y}{\mathrm{d}t} = 3y - 9x$

$2\dfrac{\mathrm{d}^2 x}{\mathrm{d}t^2} - \dfrac{\mathrm{d}^2 y}{\mathrm{d}t^2} + \dfrac{\mathrm{d}x}{\mathrm{d}t} + \dfrac{\mathrm{d}y}{\mathrm{d}t} = 5y - 7x$

subject to $x = \mathrm{d}x/\mathrm{d}t = 1$ and $y = \mathrm{d}y/\mathrm{d}t = 0$ at $t = 0$

11.4 Engineering applications: electrical circuits and mechanical vibrations

To illustrate the use of Laplace transforms, we consider here their application to the analysis of electrical circuits and vibrating mechanical systems. Since initial conditions are automatically taken into account in the transformation process, the Laplace transform is particularly attractive for examining the transient behaviour of such systems. Although electrical circuits and mechanical vibrations were considered in Chapter 10, we shall review here the modelling aspects in each case. This is to enable the two chapters to be studied independently of each other.

11.4.1 Electrical circuits

Passive electrical circuits are constructed of three basic elements: **resistors** (having resistance R, measured in ohms Ω), **capacitors** (having capacitance C, measured in farads F) and **inductors** (having inductance L, measured in henries H), with the associated

Figure 11.7
Constituent elements
of an electrical circuit.

$i(t) \longrightarrow$

(a) Resistor

(b) Capacitor

(c) Inductor

variables being **current** $i(t)$ (measured in amperes A) and **voltage** $v(t)$ (measured in volts V). The current flow in the circuit is related to the charge $q(t)$ (measured in coulombs C) by the relationship

$$i = \frac{dq}{dt}$$

Conventionally, the basic elements are represented symbolically as in Figure 11.7.

The relationship between the flow of current $i(t)$ and the voltage drops $v(t)$ across these elements at time t are

voltage drop across resistor $= Ri$ (Ohm's law)

voltage drop across capacitor $= \dfrac{1}{C} \displaystyle\int i \, dt = \dfrac{q}{C}$

The interaction between the individual elements making up an electrical circuit is determined by **Kirchhoff's laws**:

Law 1
The algebraic sum of all the currents entering any junction (or node) of a circuit is zero.

Law 2
The algebraic sum of the voltage drops around any closed loop (or path) in a circuit is zero.

Use of these laws leads to circuit equations, which may then be analysed using Laplace transform techniques.

Example 11.27

The *LCR* circuit of Figure 11.8 consists of a resistor R, a capacitor C and an inductor L connected in series together with a voltage source $e(t)$. Prior to closing the switch at time $t = 0$, both the charge on the capacitor and the resulting current in the circuit are zero. Determine the charge $q(t)$ on the capacitor and the resulting current $i(t)$ in the circuit at time t given that $R = 160\,\Omega$, $L = 1\,\mathrm{H}$, $C = 10^{-4}\,\mathrm{F}$ and $e(t) = 20\,\mathrm{V}$.

Figure 11.8
LCR circuit of
Example 11.27.

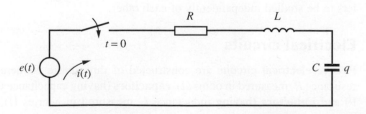

Solution   Applying Kirchhoff's second law to the circuit of Figure 11.8 gives

$$Ri + L\frac{di}{dt} + \frac{1}{C}\int i\,dt = e(t) \tag{11.25}$$

or, using $i = dq/dt$,

$$L\frac{d^2q}{dt^2} + R\frac{dq}{dt} + \frac{1}{C}q = e(t)$$

Substituting the given values for L, R, C and $e(t)$ gives

$$\frac{d^2q}{dt^2} + 160\frac{dq}{dt} + 10^4q = 20$$

Taking Laplace transforms throughout leads to the equation

$$(s^2 + 160s + 10^4)Q(s) = [sq(0) + \dot{q}(0)] + 160q(0) + \frac{20}{s}$$

where $Q(s)$ is the transform of $q(t)$. We are given that $q(0) = 0$ and $\dot{q}(0) = i(0) = 0$, so that this reduces to

$$(s^2 + 160s + 10^4)Q(s) = \frac{20}{s}$$

that is,

$$Q(s) = \frac{20}{s(s^2 + 160s + 10^4)}$$

Resolving into partial fractions gives

$$Q(s) = \frac{\frac{1}{500}}{s} - \frac{1}{500}\frac{s + 160}{s^2 + 160s + 10^4}$$

$$= \frac{1}{500}\left[\frac{1}{s} - \frac{(s + 80) + \frac{4}{3}(60)}{(s + 80)^2 + (60)^2}\right]$$

$$= \frac{1}{500}\left[\frac{1}{s} - \left[\frac{s + \frac{4}{3} \times 60}{s^2 + 60^2}\right]_{s \to s+80}\right]$$

Taking inverse transforms, making use of the shift theorem (Theorem 11.2), gives

$$q(t) = \tfrac{1}{500}(1 - e^{-80t}\cos 60t - \tfrac{4}{3}e^{-80t}\sin 60t)$$

The resulting current $i(t)$ in the circuit is then given by

$$i(t) = \frac{dq}{dt} = \tfrac{1}{3}e^{-80t}\sin 60t$$

Note that we could have determined the current by taking Laplace transforms in (11.25). Substituting the given values for L, R, C and $e(t)$ and using (11.15) leads to the transformed equation

$$160I(s) + sI(s) + \frac{10^4}{s}I(s) = \frac{20}{s}$$

that is,

$$I(s) = \frac{20}{(s^2 + 80)^2 + 60^2} \quad (= sQ(s) \quad \text{since} \quad q(0) = 0)$$

which, on taking inverse transforms, gives as before

$$i(t) = \tfrac{1}{3}e^{-80t}\sin 60t$$

Example 11.28

In the parallel network of Figure 11.9 there is no current flowing in either loop prior to closing the switch at time $t = 0$. Deduce the currents $i_1(t)$ and $i_2(t)$ flowing in the loops at time t.

Figure 11.9
Parallel circuit of
Example 11.28.

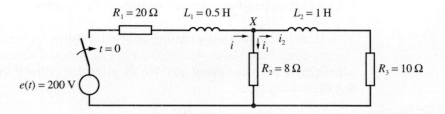

Solution

Applying Kirchhoff's first law to node X gives

$$i = i_1 + i_2$$

Applying Kirchhoff's second law to each of the two loops in turn gives

$$R_1(i_1 + i_2) + L_1\frac{\mathrm{d}}{\mathrm{d}t}(i_1 + i_2) + R_2i_1 = 200$$

$$L_2\frac{\mathrm{d}i_2}{\mathrm{d}t} + R_3i_2 - R_2i_1 = 0$$

Substituting the given values for the resistances and inductances gives

$$\left.\begin{array}{l}\dfrac{\mathrm{d}i_1}{\mathrm{d}t} + \dfrac{\mathrm{d}i_2}{\mathrm{d}t} + 56i_1 + 40i_2 = 400 \\[4mm] \dfrac{\mathrm{d}i_2}{\mathrm{d}t} - 8i_1 + 10i_2 = 0\end{array}\right\} \tag{11.26}$$

Taking Laplace transforms and incorporating the initial conditions $i_1(0) = i_2(0) = 0$ leads to the transformed equations

$$(s + 56)I_1(s) + (s + 40)I_2(s) = \frac{400}{s} \tag{11.27}$$

$$-8I_1(s) + (s + 10)I_2(s) = 0 \tag{11.28}$$

Hence

$$I_2(s) = \frac{3200}{s(s^2 + 74s + 880)} = \frac{3200}{s(s + 59.1)(s + 14.9)}$$

Resolving into partial fractions gives

$$I_2(s) = \frac{3.64}{s} + \frac{1.22}{s + 59.1} - \frac{4.86}{s + 14.9}$$

which, on taking inverse transforms, leads to

$$i_2(t) = 3.64 + 1.22e^{-59.1t} - 4.86e^{-14.9t}$$

From (11.26),

$$i_1(t) = \tfrac{1}{8}\left(10i_2 + \frac{di_2}{dt}\right)$$

that is,

$$i_1(t) = 4.55 - 7.49e^{-59.1t} + 2.98e^{-14.9t}$$

Note that as $t \to \infty$, the currents $i_1(t)$ and $i_2(t)$ approach the constant values 4.55 and 3.64A respectively. (Note that $i(0) = i_1(0) + i_2(0) \neq 0$ due to rounding errors in the calculation.)

Example 11.29   A voltage $e(t)$ is applied to the primary circuit at time $t = 0$, and mutual induction M drives the current $i_2(t)$ in the secondary circuit of Figure 11.10. If, prior to closing the switch, the currents in both circuits are zero, determine the induced current $i_2(t)$ in the secondary circuit at time t when $R_1 = 4\,\Omega$, $R_2 = 10\,\Omega$, $L_1 = 2\,\text{H}$, $L_2 = 8\,\text{H}$, $M = 2\,\text{H}$ and $e(t) = 28 \sin 2t\,\text{V}$.

Figure 11.10
Circuit of Example
11.29.

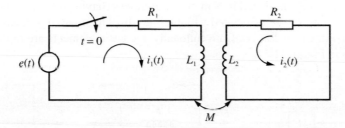

Solution   Applying Kirchhoff's second law to the primary and secondary circuits respectively gives

$$R_1 i_1 + L_1 \frac{di_1}{dt} + M\frac{di_2}{dt} = e(t)$$

$$R_2 i_2 + L_2 \frac{di_2}{dt} + M\frac{di_1}{dt} = 0$$

Substituting the given values for the resistances, inductances and applied voltage leads to

$$2\frac{di_1}{dt} + 4i_1 + 2\frac{di_2}{dt} = 28 \sin 2t$$

$$2\frac{di_1}{dt} + 8\frac{di_2}{dt} + 10i_2 = 0$$

Taking Laplace transforms and noting that $i_1(0) = i_2(0) = 0$ leads to the equations

$$(s + 2)I_1(s) + sI_2(s) = \frac{28}{s^2 + 4} \tag{11.29}$$

$$sI_1(s) + (4s + 5)I_2(s) = 0 \tag{11.30}$$

Solving for $I_2(s)$ yields

$$I_2(s) = -\frac{28s}{(3s + 10)(s + 1)(s^2 + 4)}$$

Resolving into partial fractions gives

$$I_2(s) = -\frac{\frac{45}{17}}{3s + 10} + \frac{\frac{4}{5}}{s + 1} + \frac{7}{85}\frac{s - 26}{s^2 + 4}$$

Taking inverse Laplace transforms gives the current in the secondary circuit as

$$i_2(t) = \tfrac{4}{5}e^{-t} - \tfrac{15}{17}e^{-10t/3} + \tfrac{7}{85}\cos 2t - \tfrac{91}{85}\sin 2t$$

As $t \to \infty$, the current will approach the sinusoidal response

$$i_2(t) = \tfrac{7}{85}\cos 2t - \tfrac{91}{85}\sin 2t$$

11.4.2 Mechanical vibrations

Mechanical translational systems may be used to model many situations, and involve three basic elements: **masses** (having mass M, measured in kg), **springs** (having spring stiffness K, measured in $\mathrm{N\,m^{-1}}$) and **dampers** (having damping coefficient B, measured in $\mathrm{N\,s\,m^{-1}}$). The associated variables are **displacement** $x(t)$ (measured in m) and **force** $F(t)$ (measured in N). Conventionally, the basic elements are represented symbolically as in Figure 11.11.

Figure 11.11
Constituent elements of a translational mechanical system.

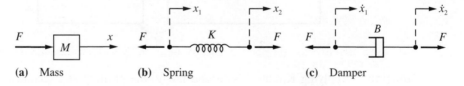

(a) Mass (b) Spring (c) Damper

Assuming we are dealing with ideal springs and dampers (that is, assuming that they behave linearly), the relationships between the forces and displacements at time t are

mass: $F = M\dfrac{\mathrm{d}^2 x}{\mathrm{d}t^2} = M\ddot{x}$ (Newton's law)

spring: $F = K(x_2 - x_1)$ (Hooke's law)

damper: $F = B\left(\dfrac{\mathrm{d}x_2}{\mathrm{d}t} - \dfrac{\mathrm{d}x_1}{\mathrm{d}t}\right) = B(\dot{x}_2 - \dot{x}_1)$

Using these relationships leads to the system equations, which may then be analysed using Laplace transform techniques.

Example 11.30

The mass of the mass–spring–damper system of Figure 11.12(a) is subjected to an externally applied periodic force $F(t) = 4 \sin \omega t$ at time $t = 0$. Determine the resulting displacement $x(t)$ of the mass at time t, given that $x(0) = \dot{x}(0) = 0$, for the two cases

(a) $\omega = 2$ (b) $\omega = 5$

In the case $\omega = 5$, what would happen to the response if the damper were missing?

Figure 11.12
Mass–spring–damper
system of Example
11.30.

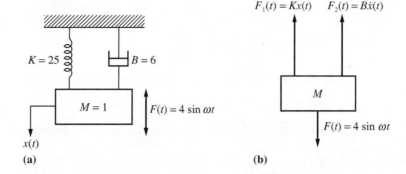

Solution As indicated in Figure 11.12(b), the forces acting on the mass M are the applied force $F(t)$ and the restoring forces F_1 and F_2 due to the spring and damper respectively. Thus, by Newton's law,

$$M\ddot{x}(t) = F(t) - F_1(t) - F_2(t)$$

Since $M = 1$, $F(t) = 4 \sin \omega t$, $F_1(t) = Kx(t) = 25x(t)$ and $F_2(t) = B\dot{x}(t) = 6\dot{x}(t)$, this gives

$$\ddot{x}(t) + 6\dot{x}(t) + 25x(t) = 4 \sin \omega t \tag{11.31}$$

as the differential equation representing the motion of the system.

Taking Laplace transforms throughout in (11.31) gives

$$(s^2 + 6s + 25)X(s) = [sx(0) + \dot{x}(0)] + 6x(0) + \frac{4\omega}{s^2 + \omega^2}$$

where $X(s)$ is the transform of $x(t)$. Incorporating the given initial conditions $x(0) = \dot{x}(0) = 0$ leads to

$$X(s) = \frac{4\omega}{(s^2 + \omega^2)(s^2 + 6s + 25)} \tag{11.32}$$

In case (a), with $\omega = 2$, (11.32) gives

$$X(s) = \frac{8}{(s^2 + 4)(s^2 + 6s + 25)}$$

which, on resolving into partial fractions, leads to

$$X(s) = \tfrac{4}{195} \frac{-4s + 14}{s^2 + 4} + \tfrac{2}{195} \frac{8s + 20}{s^2 + 6s + 25}$$

$$= \tfrac{4}{195} \frac{-4s + 14}{s^2 + 4} + \tfrac{2}{195} \frac{8(s + 3) - 4}{(s + 3)^2 + 16}$$

Taking inverse Laplace transforms gives the required response

$$x(t) = \tfrac{4}{195}(7 \sin 2t - 4 \cos 2t) + \tfrac{2}{195}e^{-3t}(8 \cos 4t - \sin 4t) \tag{11.33}$$

In case (b), with $\omega = 5$, (11.32) gives

$$X(s) = \frac{20}{(s^2 + 25)(s^2 + 6s + 25)} \tag{11.34}$$

that is,

$$X(s) = \frac{-\tfrac{2}{15}s}{s^2 + 25} + \tfrac{1}{15}\frac{2(s + 3) + 6}{(s + 3)^2 + 16}$$

which, on taking inverse Laplace transforms, gives the required response

$$x(t) = -\tfrac{2}{15}\cos 5t + \tfrac{1}{15}e^{-3t}(2 \cos 4t + \tfrac{3}{2}\sin 4t) \tag{11.35}$$

If the damping term were missing then (11.34) would become

$$X(s) = \frac{20}{(s^2 + 25)^2} \tag{11.36}$$

By Theorem 11.3,

$$\mathscr{L}\{t \cos 5t\} = -\frac{d}{ds}\mathscr{L}\{\cos 5t\} = -\frac{d}{ds}\left(\frac{s}{s^2 + 25}\right)$$

that is,

$$\mathscr{L}\{t \cos 5t\} = -\frac{1}{s^2 + 25} + \frac{2s^2}{(s^2 + 25)^2} = \frac{1}{s^2 + 25} - \frac{50}{(s^2 + 25)^2}$$

$$= \tfrac{1}{5}\mathscr{L}\{\sin 5t\} - \frac{50}{(s^2 + 25)^2}$$

Thus, by the linearity property (11.10),

$$\mathscr{L}\{\tfrac{1}{5}\sin 5t - t \cos 5t\} = \frac{50}{(s^2 + 25)^2}$$

so that taking inverse Laplace transforms in (11.36) gives the response as

$$x(t) = \tfrac{2}{25}(\sin 5t - 5t \cos 5t)$$

Because of the term $t \cos 5t$, the response $x(t)$ is unbounded as $t \to \infty$. This arises because in this case the applied force $F(t) = 4 \sin 5t$ is in **resonance** with the system (that is, the vibrating mass), whose natural oscillating frequency is $5/2\pi$ Hz, equal to that of the applied force. Even in the presence of damping, the amplitude of the system response is maximized when the applied force is approaching resonance with the system. (This is left as an exercise for the reader.) In the absence of damping we have the limiting case of **pure resonance**, leading to an unbounded response. As noted in Chapter 10, Section 10.10.3, resonance is of practical importance, since, for example, it can lead to large and strong structures collapsing under what appears to be a relatively small force.

Example 11.31

Consider the mechanical system of Figure 11.13(a), which consists of two masses $M_1 = 1$ and $M_2 = 2$, each attached to a fixed base by a spring, having constants $K_1 = 1$ and $K_3 = 2$ respectively, and attached to each other by a third spring having constant $K_2 = 2$. The system is released from rest at time $t = 0$ in a position in which M_1 is displaced 1 unit to the left of its equilibrium position and M_2 is displaced 2 units to the right of its equilibrium position. Neglecting all frictional effects, determine the positions of the masses at time t.

Figure 11.13
Two-mass system of Example 11.31.

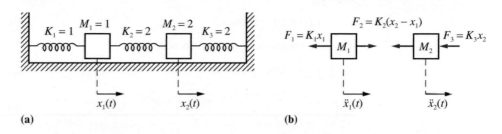

(a)   (b)

Solution

Let $x_1(t)$ and $x_2(t)$ denote the displacements of the masses M_1 and M_2 respectively from their equilibrium positions. Since frictional effects are neglected, the only forces acting on the masses are the restoring forces due to the springs, as shown in Figure 11.13(b). Applying Newton's law to the motions of M_1 and M_2 respectively gives

$$M_1\ddot{x}_1 = F_2 - F_1 = K_2(x_2 - x_1) - K_1 x_1$$

$$M_2\ddot{x}_2 = -F_3 - F_2 = -K_3 x_2 - K_2(x_2 - x_1)$$

which, on substituting the given values for M_1, M_2, K_1, K_2 and K_3, gives

$$\ddot{x}_1 + 3x_1 - 2x_2 = 0 \tag{11.37}$$

$$2\ddot{x}_2 + 4x_2 - 2x_1 = 0 \tag{11.38}$$

Taking Laplace transforms leads to the equations

$$(s^2 + 3)X_1(s) - 2X_2(s) = sx_1(0) + \dot{x}_1(0)$$

$$-X_1(s) + (s^2 + 2)X_2(s) = sx_2(0) + \dot{x}_2(0)$$

Since $x_1(t)$ and $x_2(t)$ denote displacements to the right of the equilibrium positions, we have $x_1(0) = -1$ and $x_2(0) = 2$. Also, the system is released from rest, so that $\dot{x}_1(0) = \dot{x}_2(0) = 0$. Incorporating these initial conditions, the transformed equations become

$$(s^2 + 3)X_1(s) - 2X_2(s) = -s \tag{11.39}$$

$$-X_1(s) + (s^2 + 2)X_2(s) = 2s \tag{11.40}$$

Hence

$$X_2(s) = \frac{2s^3 + 5s}{(s^2 + 4)(s^2 + 1)}$$

Resolving into partial fractions gives

$$X_2(s) = \frac{s}{s^2 + 1} + \frac{s}{s^2 + 4}$$

which, on taking inverse Laplace transforms, leads to the response

$$x_2(t) = \cos t + \cos 2t$$

Substituting for $x_2(t)$ in (11.38) gives

$$x_1(t) = 2x_2(t) + \ddot{x}_2(t)$$

$$= 2\cos t + 2\cos 2t - \cos t - 4\cos 2t$$

that is,

$$x_1(t) = \cos t - 2\cos 2t$$

Thus the positions of the masses at time t are

$$x_1(t) = \cos t - 2\cos 2t$$

$$x_2(t) = \cos t + \cos 2t$$

11.4.3 Exercises

7 Use the Laplace transform technique to find the transforms $I_1(s)$ and $I_2(s)$ of the respective currents flowing in the circuit of Figure 11.14, where $i_1(t)$ is that through the capacitor and $i_2(t)$ that through the resistance. Hence, determine $i_2(t)$. (Initially, $i_1(0) = i_2(0) = q_1(0) = 0$.) Sketch $i_2(t)$ for large values of t.

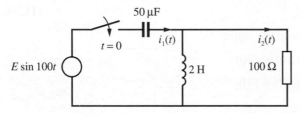

Figure 11.14 Circuit of Question 7.

8 At time $t = 0$, with no currents flowing, a voltage $v(t) = 10 \sin t$ is applied to the primary circuit of a transformer that has a mutual inductance of 1 H, as shown in Figure 11.15. Denoting the current flowing at time t in the secondary circuit by $i_2(t)$, show that

$$\mathcal{L}\{i_2(t)\} = \frac{10s}{(s^2 + 7s + 6)(s^2 + 1)}$$

and deduce that

$$i_2(t) = -e^{-t} + \tfrac{12}{37}e^{-6t} + \tfrac{25}{37}\cos t + \tfrac{35}{37}\sin t$$

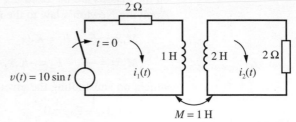

Figure 11.15 Circuit of Question 8.

9 In the circuit of Figure 11.16 there is no energy stored (that is, there is no charge on the capacitors and no current flowing in the inductances) prior to the closure of the switch at time $t = 0$. Determine $i_1(t)$ for $t > 0$ for a constant applied voltage $E_0 = 10$ V.

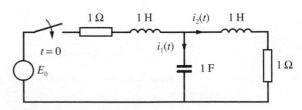

Figure 11.16 Circuit of Question 9.

10 Determine the displacements of the masses M_1 and M_2 in Figure 11.13 at time $t > 0$ when

$$M_1 = M_2 = 1$$

$$K_1 = 1, K_2 = 3 \quad \text{and} \quad K_3 = 9$$

What are the natural frequencies of the system?

11 When testing the landing-gear unit of a space vehicle, drop tests are carried out. Figure 11.17 is a schematic model of the unit at the instant when it first touches the ground. At this instant the spring is fully extended and the velocity of the mass is $\sqrt{(2gh)}$, where h is the height from which the unit has been dropped. Obtain the equation representing the displacement of the mass at time $t > 0$ when $M = 50\,\text{kg}$, $B = 180\,\text{N s m}^{-1}$ and $K = 474.5\,\text{N m}^{-1}$, and investigate the effects of different dropping heights h. (g is the acceleration due to gravity, and may be taken as $9.8\,\text{m s}^{-2}$.)

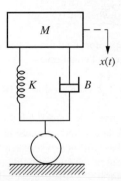

Figure 11.17 Landing-gear of Question 11.

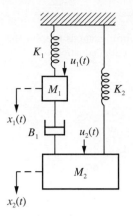

Figure 11.18 Mechanical system of Question 12.

12 Consider the mass–spring–damper system of Figure 11.18, which may be subject to two input forces $u_1(t)$ and $u_2(t)$. Show that the displacements $x_1(t)$ and $x_2(t)$ of the two masses are given by

$$x_1(t) = \mathcal{L}^{-1}\left\{ \frac{M_2 s^2 + B_1 s + K_2}{\Delta} U_1(s) + \frac{B_1 s}{\Delta} U_2(s) \right\}$$

$$x_2(t) = \mathcal{L}^{-1}\left\{ \frac{B_1 s}{\Delta} U_1(s) + \frac{M_1 s^2 + B_1 s + K_1}{\Delta} U_2(s) \right\}$$

where

$$\Delta = (M_1 s^2 + B_1 s + K_1)(M_2 s^2 + B_1 s + K_2) - B_1^2 s^2$$

11.5 Review exercises (1–18)

1 Solve, using Laplace transforms, the following differential equations:

(a) $\dfrac{d^2 x}{dt^2} + 4\dfrac{dx}{dt} + 5x = 8\cos t$

subject to $x = \dfrac{dx}{dt} = 0$ at $t = 0$

(b) $5\dfrac{d^2 x}{dt^2} - 3\dfrac{dx}{dt} - 2x = 6$

subject to $x = 1$ and $\dfrac{dx}{dt} = 1$ at $t = 0$

2 (a) Find the inverse Laplace transform of

$$\frac{1}{(s + 1)(s + 2)(s^2 + 2s + 2)}$$

(b) A voltage source $V\mathrm{e}^{-t}\sin t$ is applied across a series LCR circuit with $L = 1$, $R = 3$ and $C = \frac{1}{2}$. Show that the current $i(t)$ in the circuit satisfies the differential equation

$$\frac{d^2 i}{dt^2} + 3\frac{di}{dt} + 2i = V\mathrm{e}^{-t}\sin t$$

Find the current $i(t)$ in the circuit at time $t \geq 0$ if $i(t)$ satisfies the initial conditions $i(0) = 1$ and $(\mathrm{d}i/\mathrm{d}t)(0) = 2$.

3 Use Laplace transform methods to solve the simultaneous differential equations

$$\frac{d^2x}{dt^2} - x + 5\frac{dy}{dt} = t$$

$$\frac{d^2y}{dt^2} - 4y - 2\frac{dx}{dt} = -2$$

subject to $x = y = \dfrac{dx}{dt} = \dfrac{dy}{dt} = 0$ at $t = 0$.

4 Solve the differential equation

$$\frac{d^2x}{dt^2} + 2\frac{dx}{dt} + 2x = \cos t$$

subject to the initial conditions $x = x_0$ and $dx/dt = x_1$ at $t = 0$. Identify the steady-state and transient solutions. Find the amplitude and phase shift of the steady-state solution.

5 Resistors of 5 and $20\,\Omega$ are connected to the primary and secondary coils of a transformer with inductances as shown in Figure 11.19. At time $t = 0$, with no current flowing, a voltage $E = 100\,V$ is applied to the primary circuit. Show that subsequently the current in the secondary circuit is

$$\frac{20}{\sqrt{41}}(e^{-(11+\sqrt{41})t/2} - e^{-(11-\sqrt{41})t/2})$$

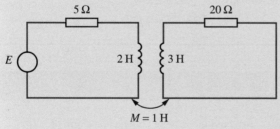

Figure 11.19 Circuit of Question 5.

6 (a) Find the Laplace transforms of

   (i) $\cos(\omega t + \phi)$ (ii) $e^{-\omega t}\sin(\omega t + \phi)$

(b) Using Laplace transform methods, solve the differential equation

$$\frac{d^2x}{dt^2} + 4\frac{dx}{dt} + 8x = \cos 2t$$

given that $x = 2$ and $dx/dt = 1$ when $t = 0$.

7 (a) Find the inverse Laplace transform of

$$\frac{s - 4}{s^2 + 4s + 13}$$

(b) Solve using Laplace transforms the differential equation

$$\frac{dy}{dt} + 2y = 2(2 + \cos t + 2\sin t)$$

given that $y = -3$ when $t = 0$.

8 Using Laplace transforms, solve the simultaneous differential equations

$$\frac{dx}{dt} + 5x + 3y = 5\sin t - 2\cos t$$

$$\frac{dy}{dt} + 3y + 5x = 6\sin t - 3\cos t$$

where $x = 1$ and $y = 0$ when $t = 0$.

9 The charge q on a capacitor in an inductive circuit is given by the differential equation

$$\frac{d^2q}{dt^2} + 300\frac{dq}{dt} + 2 \times 10^4 q = 200\sin 100t$$

and it is also known that both q and dq/dt are zero when $t = 0$. Use the Laplace transform method to find q. What is the phase difference between the steady-state component of the current dq/dt and the applied emf $200\sin 100t$ to the nearest half-degree?

10 Use Laplace transforms to find the value of x given that

$$4\frac{dx}{dt} + 6x + y = 2\sin 2t$$

$$\frac{d^2x}{dt^2} + x - \frac{dy}{dt} = 3e^{-2t}$$

and that $x = 2$ and $dx/dt = -2$ when $t = 0$.

11 (a) Use Laplace transforms to solve the differential equation

$$\frac{d^2\theta}{dt^2} + 8\frac{d\theta}{dt} + 16\theta = \sin 2t$$

given that $\theta = 0$ and $d\theta/dt = 0$ when $t = 0$.

(b) Using Laplace transforms, solve the simultaneous differential equations

$$\frac{di_1}{dt} + 2i_1 + 6i_2 = 0$$

$$i_1 + \frac{di_2}{dt} - 3i_2 = 0$$

given that $i_1 = 1$, $i_2 = 0$ when $t = 0$.

12   The terminals of a generator producing a voltage V are connected through a wire of resistance R and a coil of inductance L (and negligible resistance). A capacitor of capacitance C is connected in parallel with the resistance R as shown in Figure 11.20. Show that the current i flowing through the resistance R is given by

$$LCR\frac{d^2i}{dt^2} + L\frac{di}{dt} + Ri = V$$

Suppose that

 (i) $V = 0$ for $t < 0$ and $V = E$ (constant) for $t \geqslant 0$

(ii) $L = 2R^2C$

(iii) $CR = 1/2n$

and show that the equation reduces to

$$\frac{d^2i}{dt^2} + 2n\frac{di}{dt} + 2n^2i = 2n^2\frac{E}{R}$$

Hence, assuming that $i = 0$ and $di/dt = 0$ when $t - 0$, use Laplace transforms to obtain an expression for i in terms of t.

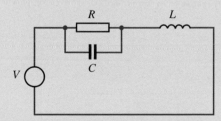

Figure 11.20 Circuit of Question 12.

13 Show that the currents in the coupled circuits of Figure 11.21 are determined by the simultaneous differential equations

$$L\frac{di_1}{dt} + R(i_1 - i_2) + Ri_1 = E$$

$$L\frac{di_2}{dt} + Ri_2 - R(i_1 - i_2) = 0$$

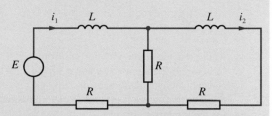

Figure 11.21 Circuit of Question 13.

Find i_1 in terms of t, L, E and R, given that $i_1 = 0$ and $di_1/dt = E/L$ at $t = 0$, and show that $i_1 \simeq \frac{2}{3}E/R$ for large t. What does i_2 tend to for large t?

14   A system consists of two unit masses lying in a straight line on a smooth surface and connected together to two fixed points by three springs. When a sinusoidal force is applied to the system, the displacements $x_1(t)$ and $x_2(t)$ of the respective masses from their equilibrium positions satisfy the equations

$$\frac{d^2x_1}{dt^2} = x_2 - 2x_1 + \sin 2t$$

$$\frac{d^2x_2}{dt^2} = -2x_2 + x_1$$

Given that the system is initially at rest in the equilibrium position ($x_1 = x_2 - 0$), use the Laplace transform method to solve the equations for $x_1(t)$ and $x_2(t)$.

15   (a) Obtain the inverse Laplace transforms of

 (i) $\dfrac{s + 4}{s^2 + 2s + 10}$ (ii) $\dfrac{s - 3}{(s - 1)^2(s - 2)}$

(b) Use Laplace transforms to solve the differential equation

$$\frac{d^2y}{dt^2} + 2\frac{dy}{dt} + y = 3te^{-t}$$

given that $y = 4$ and $dy/dt = 2$, when $t = 0$.

16 (a) Determine the inverse Laplace transform of

$$\frac{5}{s^2 - 14s + 53}$$

(b) The equation of motion of the moving coil of a galvanometer when a current i is passed through it is of the form

$$\frac{d^2\theta}{dt^2} + 2K\frac{d\theta}{dt} + n^2\theta = \frac{n^2 i}{K}$$

where θ is the angle of deflection from the 'no-current' position and n and K are positive constants. Given that i is a constant and $\theta = 0 = d\theta/dt$ when $t = 0$, obtain an expression for the Laplace transform of θ.

In constructing the galvanometer, it is desirable to have it critically damped (that is, $n = K$). Use the Laplace transform method to solve the differential equation in this case, and sketch the graph of θ against t for positive values of t.

17 Two cylindrical water tanks are connected as shown in Figure 11.22. Initially there are 250 litres in the top tank and 50 litres in the bottom tank. At time $t = 0$ the valve between the two tanks and the valve at the bottom of the lower tank are opened. The flowrate through each of these valves is proportional to the volume of water in the tank immediately above the valve, the constant of proportionality being 0.1 for both valves. Denoting the volume in the top tank by v_1 and the volume in the bottom tank by v_2 show that the following differential equations are satisfied.

$$\frac{dv_1}{dt} = -0.1v_1$$

$$\frac{dv_2}{dt} + 0.1v_2 = 0.1v_1$$

(a) Use Laplace transforms to determine v_1 and v_2.

(b) Find the time taken for the volume of water in the top tank to reach 10% of its starting value.

18 In order to transport sensitive equipment a crate is installed inside a truck on damped springs, as shown in Figure 11.23. The suspension system of the truck, including the tyres, may be modelled as a damped spring. The various spring and damper constants are indicated in the figure. The masses of the crate and truck are M_1 and M_2 respectively and their displacements from equilibrium are respectively $x_1(t)$ and $x_2(t)$. The vertical displacement of the truck as it traverses a bumpy road may be modelled by applying a force $u(t)$ to the truck.

Show that the motion of the crate and truck may be modelled by the differential equations

$$M_1\ddot{x}_1 = K_1(x_2 - x_1) + B_1(\dot{x}_2 - \dot{x}_1)$$

$$M_2\ddot{x}_2 = u - (K_1 + K_2)x_2 + K_1x_1$$
$$- (B_1 + B_2)\dot{x}_2 + B_1\dot{x}_1$$

For the particular case where $M_1 = 1$, $M_2 = 3$, $K_1 = 2$, $K_2 = 1$, $B_1 = 3$, $B_2 = 2$ and $u(t) = \sin t$, and the initial conditions at time $t = 0$ are $x_1 = x_2 = \dot{x}_1 = 0$, $\dot{x}_2 = 2$, show that the Laplace transform of $x_1(t)$ is

$$X_1(s) = \frac{18s^3 + 12s^2 + 21s + 14}{(3s^4 + 14s^3 + 15s^2 + 7s + 2)(s^2 + 1)}$$

(*Note*: Using an appropriate software package, such as MATLAB/SIMULINK, the model developed may be used as the basis for simulation studies of various scenarios.)

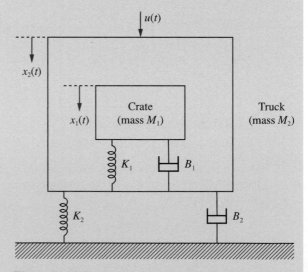

Figure 11.23 Transport crate of Question 18.

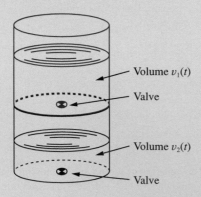

Figure 11.22 Cylindrical tanks of Question 17.

12 Introduction to Fourier Series

Chapter 12 Contents

12.1 Introduction

The representation of a function in the form of a series is fairly common practice in mathematics. Probably the most familiar expansions are power series of the form

$$f(x) = \sum_{n=0}^{\infty} a_n x^n$$

in which the resolved components or **base set** comprise the power functions

$$1, x, x^2, x^3, \ldots, x^n, \ldots$$

For example, we recall that the exponential function may be represented by the infinite series

$$e^x = 1 + x + \frac{x^2}{2!} + \frac{x^3}{3!} + \ldots + \frac{x^n}{n!} + \ldots = \sum_{n=0}^{\infty} \frac{x^n}{n!}$$

There are frequently advantages in expanding a function in such a series, since the first few terms of a good approximation are easy to deal with. For example, term-by-term integration or differentiation may be applied or suitable function approximations can be made.

Power functions comprise only one example of a base set for the expansions of functions: a number of other base sets may be used. In particular, a **Fourier series** is an expansion of a periodic function $f(t)$ of period $T = 2\pi/\omega$ in which that base set is the set of sine functions, giving an expanded representation of the form

$$f(t) = A_0 + \sum_{n=1}^{\infty} A_n \sin(n\omega t + \phi_n)$$

Although the idea of expanding a function in the form of such a series had been used by Bernoulli, D'Alembert and Euler (*c.* 1750) to solve problems associated with the vibration of strings, it was Joseph Fourier (1768–1830) who developed the approach to a stage where it was generally useful. Fourier, a French mathematician, was interested in heat-flow problems: given an initial temperature at all points of a region, he was concerned with determining the change in the temperature distribution over time. When Fourier postulated in 1807 that an arbitrary function $f(x)$ could be represented by a trigonometric series of the form

$$\sum_{n=0}^{\infty} (A_n \cos nkx + B_n \sin nkx)$$

the result was considered so startling that it met considerable opposition from the leading mathematicians of the time, notably Laplace, Poisson and, more significantly, Lagrange, who is regarded as one of the greatest mathematicians of all time. They questioned his work because of its lack of rigour, and it was probably this opposition that delayed the publication of Fourier's work, his classic text *Théorie Analytique de la Chaleur* (The Analytical Theory of Heat) not appearing until 1822. This text has since become the source for the modern methods of solving practical problems associated with partial differential equations subject to prescribed boundary conditions. In addition to heat flow, this class of problems includes structural vibrations, wave propagation and diffusion, which are discussed in the companion text *Advanced Modern Engineering Mathematics*. The task of giving Fourier's work a more rigorous mathematical underpinning was undertaken later by Dirichlet (*c*. 1830) and subsequently Riemann, his successor at the University of Göttingen.

In addition to its use in solving boundary-value problems associated with partial differential equations, Fourier series analysis is central to many other applications in engineering, such as the analysis and design of oscillating and nonlinear systems. This chapter is intended to provide only an introduction to Fourier series, with a more detailed treatment, including consideration of frequency spectra, oscillating and nonlinear systems, and generalized Fourier series, being given in *Advanced Modern Engineering Mathematics*.

12.2 Fourier series expansion

In this section we develop the Fourier series expansion of periodic functions and discuss how closely they approximate the functions. We also indicate how symmetrical properties of the function may be taken advantage of in order to reduce the amount of mathematical manipulation involved in determining the Fourier series. First, for continuity, we review the properties of periodic functions considered in Section 2.2.5.

12.2.1 Periodic functions

A function $f(t)$ is said to be **periodic** if its image values are repeated at regular intervals in its domain. Thus the graph of a periodic function can be divided into 'vertical strips' that are replicas of each other, as illustrated in Figure 12.1. The interval between two successive replicas is called the **period** of the function. We therefore say that a function $f(t)$ is periodic with period T if, for all its domain values t,

$$f(t + mT) = f(t)$$

for any integer m.

To provide a measure of the number of repetitions per unit of t, we define the **frequency** of a periodic function to be the reciprocal of its period, so that

$$\text{frequency} = \frac{1}{\text{period}} = \frac{1}{T}$$

Figure 12.1
A periodic function
with period T.

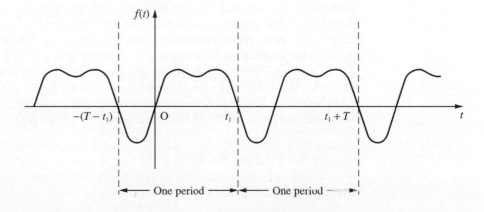

The term **circular frequency** is also used in engineering, and is defined by

$$\text{circular frequency} = 2\pi \times \text{frequency} = \frac{2\pi}{T}$$

and is measured in radians per second. It is common to drop the term 'circular' and refer to this simply as the frequency when the context is clear.

12.2.2 Fourier's theorem

This theorem states that a periodic function that satisfies certain conditions can be expressed as the sum of a number of sine functions of different amplitudes, phases and periods. That is, if $f(t)$ is a periodic function with period T then

$$f(t) = A_0 + A_1 \sin(\omega t + \phi_1) + A_2 \sin(2\omega t + \phi_2) + \ldots$$
$$+ A_n \sin(n\omega t + \phi_n) + \ldots \tag{12.1}$$

where the A's and ϕ's are constants and $\omega = 2\pi/T$ is the frequency of $f(t)$. The term $A_1 \sin(\omega t + \phi_1)$ is called the **first harmonic** or the **fundamental mode**, and it has the same frequency ω as the parent function $f(t)$. The term $A_n \sin(n\omega t + \phi_n)$ is called the **nth harmonic,** and it has frequency $n\omega$, which is n times that of the fundamental. A_n denotes the **amplitude** of the nth harmonic and ϕ_n is its **phase angle**, measuring the lag or lead of the nth harmonic with reference to a pure sine wave of the same frequency.
Since

$$A_n \sin(n\omega t + \phi_n) \equiv (A_n \cos \phi_n) \sin n\omega t + (A_n \sin \phi_n) \cos n\omega t$$

$$\equiv b_n \sin n\omega t + a_n \cos n\omega t$$

where

$$b_n = A_n \cos \phi_n, \quad a_n = A_n \sin \phi_n \tag{12.2}$$

the expansion (12.1) may be written as

$$f(t) = \tfrac{1}{2}a_0 + \sum_{n=1}^{\infty} a_n \cos n\omega t + \sum_{n=1}^{\infty} b_n \sin n\omega t \tag{12.3}$$

where $a_0 = 2A_0$ (we shall see later that taking the first term as $\frac{1}{2}a_0$ rather than a_0 is a convenience that enables us to make a_0 fit a general result). The expansion (12.3) is called the **Fourier series expansion** of the function $f(t)$, and the a's and b's are called the **Fourier coefficients**. In electrical engineering it is common practice to refer to a_n and b_n respectively as the **in-phase** and **phase quadrature components** of the nth harmonic, this terminology arising from the use of the phasor notation $e^{jn\omega t} = \cos n\omega t + j \sin n\omega t$. Clearly, (12.1) is an alternative representation of the Fourier series with the amplitude and phase of the nth harmonic being determined from (12.2) as

$$A_n = \sqrt{(a_n^2 + b_n^2)}, \quad \phi_n = \tan^{-1}\left(\frac{a_n}{b_n}\right)$$

with care being taken over choice of quadrant.

12.2.3 The Fourier coefficients

Before proceeding to evaluate the Fourier coefficients, we first state the following integrals, in which $T = 2\pi/\omega$:

$$\int_d^{d+T} \cos n\omega t \, dt = \begin{cases} 0 & (n \neq 0) \\ T & (n = 0) \end{cases} \tag{12.4}$$

$$\int_d^{d+T} \sin n\omega t \, dt = 0 \quad \text{(all } n) \tag{12.5}$$

$$\int_d^{d+T} \sin m\omega t \sin n\omega t \, dt = \begin{cases} 0 & (m \neq n) \\ \frac{1}{2}T & (m = n \neq 0) \end{cases} \tag{12.6}$$

$$\int_d^{d+T} \cos m\omega t \cos n\omega t \, dt = \begin{cases} 0 & (m \neq n) \\ \frac{1}{2}T & (m = n \neq 0) \end{cases} \tag{12.7}$$

$$\int_d^{d+T} \cos m\omega t \sin n\omega t \, dt = 0 \quad \text{(all } m \text{ and } n) \tag{12.8}$$

The results (12.4)–(12.8) constitute the **orthogonality relations** for sine and cosine functions, and show that the set of functions

$$\{1, \cos \omega t, \cos 2\omega t, \ldots, \cos n\omega t, \sin \omega t, \sin 2\omega t, \ldots, \sin n\omega t\}$$

is an orthogonal set of functions on the interval $d \leq t \leq d + T$. The choice of d is arbitrary in these results, it only being necessary to integrate over a period of duration T.

Integrating the series (12.3) with respect to t over the period $t = d$ to $t = d + T$, and using (12.4) and (12.5), we find that each term on the right-hand side is zero except for the term involving a_0; that is, we have

$$\int_d^{d+T} f(t) dt = \frac{1}{2}a_0 \int_d^{d+T} dt + \sum_{n=1}^{\infty} \left(a_n \int_d^{d+T} \cos n\omega t \, dt + b_n \int_d^{d+T} \sin n\omega t \, dt \right)$$

$$= \frac{1}{2}a_0(T) + \sum_{n=1}^{\infty} [a_n(0) + b_n(0)]$$

$$= \frac{1}{2}Ta_0$$

Thus

$$\frac{1}{2}a_0 = \frac{1}{T} \int_d^{d+T} f(t) dt$$

and we can see that the constant term $\frac{1}{2}a_0$ in the Fourier series expansion represents the mean value of the function $f(t)$ over one period. For an electrical signal it represents the bias level or d.c. (direct current) component. Hence

$$a_0 = \frac{2}{T} \int_d^{d+T} f(t) dt \tag{12.9}$$

To obtain this result, we have assumed that term-by-term integration of the series (12.3) is permissible. This is indeed so because of the convergence properties of the series – its validity is discussed in detail in more advanced texts.

To obtain the Fourier coefficient a_n ($n \neq 0$), we multiply (12.3) throughout by $\cos m\omega t$ and integrate with respect to t over the period $t = d$ to $t = d + T$, giving

$$\int_d^{d+T} f(t) \cos m\omega t \, dt = \tfrac{1}{2} a_0 \int_d^{d+T} \cos m\omega t \, dt + \sum_{n=1}^{\infty} a_n \int_d^{d+T} \cos n\omega t \cos m\omega t \, dt$$

$$+ \sum_{n=1}^{\infty} b_n \int_d^{d+T} \cos m\omega t \sin n\omega t \, dt$$

Assuming term-by-term integration to be possible, and using (12.4), (12.7) and (12.8), we find that, when $m \neq 0$, the only non-zero integral on the right-hand side is the one that occurs in the first summation when $n = m$. That is, we have

$$\int_d^{d+T} f(t) \cos m\omega t \, dt = a_m \int_d^{d+T} \cos m\omega t \cos m\omega t \, dt = \tfrac{1}{2} a_m T$$

giving

$$a_m = \frac{2}{T} \int_d^{d+T} f(t) \cos m\omega t \, dt$$

which, on replacing m by n, gives

$$a_n = \frac{2}{T} \int_d^{d+T} f(t) \cos n\omega t \, dt \qquad \textbf{(12.10)}$$

The value of a_0 given in (12.9) may be obtained by taking $n = 0$ in (12.10), so that we may write

$$a_n = \frac{2}{T} \int_d^{d+T} f(t) \cos m\omega t \, dt \quad (n = 0, 1, 2, \dots) \qquad \textbf{(12.11)}$$

This explains why the constant term in the Fourier series expansion was taken as $\tfrac{1}{2} a_0$ and not a_0, since this ensures compatibility of the results (12.9) and (12.10). Although a_0 and a_n satisfy the same formula, it is usually safer to work them out separately.

Finally, to obtain the Fourier coefficients b_n, we multiply (12.3) throughout by $\sin m\omega t$ and integrate with respect to t over the period $t = d$ to $t = d + T$, giving

$$\int_d^{d+T} f(t) \sin m\omega t \, dt = \tfrac{1}{2} a_0 \int_d^{d+T} \sin m\omega t \, dt$$

$$+ \sum_{n=1}^{\infty} \left(a_n \int_d^{d+T} \sin m\omega t \cos n\omega t \, dt + b_n \int_t^{d+T} \sin m\omega t \sin n\omega t \, dt \right)$$

Assuming term-by-term integration to be possible, and using (12.5), (12.6) and (12.8), we find that the only non-zero integral on the right-hand side is the one that occurs in the second summation when $m = n$. That is, we have

$$\int_{d}^{d+T} f(t) \sin m\omega t \, \mathrm{d}t = b_m \int_{d}^{d+T} \sin m\omega t \sin m\omega t \, \mathrm{d}t = \tfrac{1}{2} b_m T$$

giving, on replacing m by n,

$$b_n = \frac{2}{T} \int_{d}^{d+T} f(t) \sin n\omega t \, \mathrm{d}t \quad (n = 1, 2, 3, \dots) \tag{12.12}$$

The equations (12.11) and (12.12) giving the Fourier coefficients are known as **Euler's formulae**.

Summary

In summary, we have shown that if a periodic function $f(t)$ of period $T = 2\pi/\omega$ can be expanded as a Fourier series then that series is given by

$$f(t) = \tfrac{1}{2} a_0 + \sum_{n=1}^{\infty} a_n \cos n\omega t + \sum_{n=1}^{\infty} b_n \sin n\omega t \tag{12.3}$$

where the coefficients are given by Euler's formulae

$$a_n = \frac{2}{T} \int_{d}^{d+T} f(t) \cos n\omega t \, \mathrm{d}t \quad (n = 0, 1, 2, \dots) \tag{12.11}$$

$$b_n = \frac{2}{T} \int_{d}^{d+T} f(t) \sin n\omega t \, \mathrm{d}t \quad (n = 1, 2, 3, \dots) \tag{12.12}$$

The limits of integration in Euler's formulae may be specified over any period, so that the choice of d is arbitrary, and may be made in such a way as to help in the calculation of a_n and b_n. In practice, it is common to specify $f(t)$ over either the period $-\tfrac{1}{2}T < t < \tfrac{1}{2}T$ or the period $0 < t < T$, leading respectively to the limits of integration being $-\tfrac{1}{2}T$ and $\tfrac{1}{2}T$ (that is, $d = -\tfrac{1}{2}T$) or 0 and T (that is, $d = 0$).

It is also worth noting that an alternative approach may simplify the calculation of a_n and b_n. Using the formula

$$\mathrm{e}^{jn\omega t} = \cos n\omega t + j \sin n\omega t$$

we have

$$a_n + jb_n = \frac{2}{T} \int_{d}^{d+T} f(t) \mathrm{e}^{jn\omega t} \, \mathrm{d}t \tag{12.13}$$

Evaluating this integral and equating real and imaginary parts on each side gives the values of a_n and b_n. This approach is particularly useful when only the amplitude $|a_n + jb_n|$ of the nth harmonic is required.

12.2.4 Functions of period 2π

If the period T of the periodic function $f(t)$ is taken to be 2π then $\omega = 1$, and the series (12.3) becomes

$$f(t) = \tfrac{1}{2}a_0 + \sum_{n=1}^{\infty} a_n \cos nt + \sum_{n=1}^{\infty} b_n \sin nt \qquad (12.14)$$

with the coefficients given by

$$a_n = \frac{1}{\pi} \int_{d}^{d+2\pi} f(t) \cos nt \, dt \quad (n = 0, 1, 2, \dots) \qquad (12.15)$$

$$b_n = \frac{1}{\pi} \int_{d}^{d+2\pi} f(t) \sin nt \, dt \quad (n = 1, 2, \dots) \qquad (12.16)$$

While a unit frequency may rarely be encountered in practice, consideration of this particular case reduces the amount of mathematical manipulation involved in determining the coefficients a_n and b_n. Also, there is no loss of generality in considering this case, since if we have a function $f(t)$ of period T, we may write $t_1 = 2\pi t/T$, so that

$$f(t) \equiv f\left(\frac{Tt_1}{2\pi}\right) \equiv F(t_1)$$

where $F(t_1)$ is a function of period 2π. That is, by a simple change of variable, a periodic function $f(t)$ of period T may be transformed into a periodic function $F(t_1)$ of period 2π. Thus, in order to develop an initial understanding and to discuss some of the properties of Fourier series, we shall first consider functions of period 2π, returning to functions of period other than 2π in Section 12.2.10.

Example 12.1 Obtain the Fourier series expansion of the periodic function $f(t)$ of period 2π defined by

$$f(t) = t \quad (0 < t < 2\pi), \qquad f(t) = f(t + 2\pi)$$

Solution A sketch of the function $f(t)$ over the interval $-4\pi < t < 4\pi$ is shown in Figure 12.2. Since the function is periodic we only need to sketch it over one period, the pattern being repeated for other periods. Using (12.15) to evaluate the Fourier coefficients a_0 and a_n gives

$$a_0 = \frac{1}{\pi} \int_{0}^{2\pi} f(t) dt = \frac{1}{\pi} \int_{0}^{2\pi} t \, dt = \frac{1}{\pi} \left[\frac{t^2}{2} \right]_{0}^{2\pi} = 2\pi$$

Figure 12.2
Sawtooth wave
of Example 12.1.

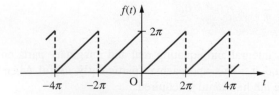

and

$$a_n = \frac{1}{\pi} \int_0^{2\pi} f(t) \cos nt \, dt \quad (n = 1, 2, \dots)$$

$$= \frac{1}{\pi} \int_0^{2\pi} t \cos nt \, dt$$

which, on integration by parts, gives

$$a_n = \frac{1}{\pi} \left[t \frac{\sin nt}{n} + \frac{\cos nt}{n^2} \right]_0^{2\pi} = \frac{1}{\pi} \left(\frac{2\pi}{n} \sin 2n\pi + \frac{1}{n^2} \cos 2n\pi - \frac{\cos 0}{n^2} \right) = 0$$

since $\sin 2n\pi = 0$ and $\cos 2n\pi = \cos 0 = 1$. Note the need to work out a_0 separately from a_n in this case. The formula (12.16) for b_n gives

$$b_n = \frac{1}{\pi} \int_0^{2\pi} f(t) \sin nt \, dt \quad (n = 1, 2, \dots)$$

$$= \frac{1}{\pi} \int_0^{2\pi} t \sin nt \, dt$$

which, on integration by parts, gives

$$b_n = \frac{1}{\pi} \left[-\frac{t}{n} \cos nt + \frac{\sin nt}{n^2} \right]_0^{2\pi}$$

$$= \frac{1}{\pi} \left(-\frac{2\pi}{n} \cos 2n\pi \right) \quad (\text{since } \sin 2n\pi = \sin 0 = 0)$$

$$= -\frac{2}{n} \quad (\text{since } \cos 2n\pi = 1)$$

Hence from (12.14) the Fourier series expansion of $f(t)$ is

$$f(t) = \pi - \sum_{n=1}^{\infty} \frac{2}{n} \sin nt$$

or, in expanded form,

$$f(t) = \pi - 2 \left(\sin t + \frac{\sin 2t}{2} + \frac{\sin 3t}{3} + \dots + \frac{\sin nt}{n} + \dots \right)$$

Example 12.2

A periodic function $f(t)$ with period 2π is defined by

$$f(t) = t^2 + t \quad (-\pi < t < \pi), \qquad f(t) = f(t + 2\pi)$$

Sketch a graph of the function $f(t)$ for values of t from $t = -3\pi$ to $t = 3\pi$ and obtain a Fourier series expansion of the function.

Figure 12.3
Graph of the function
$f(t)$ of Example 12.2.

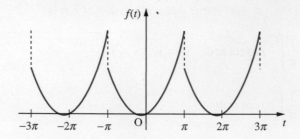

Solution A graph of the function $f(t)$ for $-3\pi < t < 3\pi$ is shown in Figure 12.3. From (12.15) we have

$$a_0 = \frac{1}{\pi} \int_{-\pi}^{\pi} f(t)\,dt = \frac{1}{\pi} \int_{-\pi}^{\pi} (t^2 + t)\,dt = \tfrac{2}{3}\pi^2$$

and

$$a_n = \frac{1}{\pi} \int_{-\pi}^{\pi} f(t) \cos nt\,dt \quad (n = 1, 2, 3, \dots)$$

$$= \frac{1}{\pi} \int_{-\pi}^{\pi} (t^2 + t) \cos nt\,dt$$

which, on integration by parts, gives

$$a_n = \frac{1}{\pi}\left[\frac{t^2}{n} \sin nt + \frac{2t}{n^2} \cos nt - \frac{2}{n^3} \sin nt + \frac{t}{n} \sin nt + \frac{1}{n^2} \cos nt \right]_{-\pi}^{\pi}$$

$$= \frac{1}{\pi} \frac{4\pi}{n^2} \cos n\pi \quad \left(\text{since } \sin n\pi = 0 \text{ and } \left[\frac{1}{n^2} \cos nt\right]_{-\pi}^{\pi} = 0 \right)$$

$$= \frac{4}{n^2}(-1)^n \qquad (\text{since } \cos n\pi = (-1)^n)$$

From (12.16)

$$b_n = \frac{1}{\pi} \int_{-\pi}^{\pi} f(t) \sin nt\,dt \quad (n = 1, 2, 3, \dots)$$

$$= \frac{1}{\pi} \int_{-\pi}^{\pi} (t^2 + t) \sin nt\,dt$$

which, on integration by parts, gives

$$b_n = \frac{1}{\pi}\left[-\frac{t^2}{n} \cos nt + \frac{2t}{n^2} \sin nt + \frac{2}{n^3} \cos nt - \frac{t}{n} \cos nt + \frac{1}{n^2} \sin nt \right]_{-\pi}^{\pi}$$

$$= -\frac{2}{n} \cos n\pi = -\frac{2}{n}(-1)^n \quad (\text{since } \cos n\pi = (-1)^n)$$

Hence from (12.14) the Fourier series expansion of $f(t)$ is

$$f(t) = \frac{1}{3}\pi^2 + \sum_{n=1}^{\infty} \frac{4}{n^2}(-1)^n \cos nt - \sum_{n=1}^{\infty} \frac{2}{n}(-1)^n \sin nt$$

or, in expanded form,

$$f(t) = \frac{1}{3}\pi^2 + 4\left(-\cos t + \frac{\cos 2t}{2^2} - \frac{\cos 3t}{3^2} + \ldots\right) + 2\left(\sin t - \frac{\sin 2t}{2} + \frac{\sin 3t}{3} \ldots\right)$$

To illustrate the alternative approach, using (12.13) gives

$$a_n + jb_n = \frac{1}{\pi}\int_{-\pi}^{\pi} f(t)e^{jnt}\,dt = \frac{1}{\pi}\int_{-\pi}^{\pi} (t^2 + t)e^{jnt}\,dt$$

$$= \frac{1}{\pi}\left(\left[\frac{t^2 + t}{jn}e^{jnt}\right]_{-\pi}^{\pi} - \int_{-\pi}^{\pi} \frac{2t + 1}{jn}e^{jnt}\,dt\right)$$

$$= \frac{1}{\pi}\left[\frac{t^2 + t}{jn}e^{jnt} - \frac{2t + 1}{(jn)^2}e^{jnt} + \frac{2e^{jnt}}{(jn)^3}\right]_{-\pi}^{\pi}$$

Since

$$e^{jn\pi} = \cos n\pi + j\sin n\pi = (-1)^n$$
$$e^{-jn\pi} = \cos n\pi - j\sin n\pi = (-1)^n$$

and

$$1/j = -j$$

$$a_n + jb_n = \frac{(-1)^n}{\pi}\left(-j\frac{\pi^2 + \pi}{n} + \frac{2\pi + 1}{n^2} + \frac{j2}{n^3} + j\frac{\pi^2 - \pi}{n} - \frac{1 - 2\pi}{n^2} - \frac{j2}{n^3}\right)$$

$$= (-1)^n\left(\frac{4}{n^2} - j\frac{2}{n}\right)$$

Equating real and imaginary parts gives, as before,

$$a_n = \frac{4}{n^2}(-1)^n, \quad b_n = -\frac{2}{n}(-1)^n$$

A periodic function $f(t)$ may be specified in a piecewise fashion over a period, or, indeed, it may only be piecewise-continuous over a period, as illustrated in Figure 12.4. In order to calculate the Fourier coefficients in such cases, it is necessary to break up the range of integration in Euler's formulae to correspond to the various components of the function. For example, for the function shown in Figure 12.4, $f(t)$ is defined in the interval $-\pi < t < \pi$ by

$$f(t) = \begin{cases} f_1(t) & (-\pi < t < -p) \\ f_2(t) & (-p < t < q) \\ f_3(t) & (q < t < \pi) \end{cases}$$

Figure 12.4
Piecewise-continuous
function over a period.

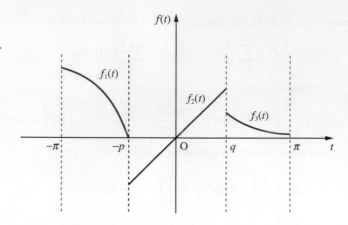

and is periodic with period 2π. Euler's formulae (12.15) and (12.16) for the Fourier coefficients become

$$a_n = \frac{1}{\pi}\left[\int_{-\pi}^{-p} f_1(t)\cos nt\,dt + \int_{-p}^{q} f_2(t)\cos nt\,dt + \int_{q}^{\pi} f_3(t)\cos nt\,dt\right]$$

$$b_n = \frac{1}{\pi}\left[\int_{-\pi}^{-p} f_1(t)\sin nt\,dt + \int_{-p}^{q} f_2(t)\sin nt\,dt + \int_{q}^{\pi} f_3(t)\sin nt\,dt\right]$$

Example 12.3   A periodic function $f(t)$ of period 2π is defined within the period $0 \leq t \leq 2\pi$ by

$$f(t) = \begin{cases} t & (0 \leq t \leq \tfrac{1}{2}\pi) \\ \tfrac{1}{2}\pi & (\tfrac{1}{2}\pi \leq t \leq \pi) \\ \pi - \tfrac{1}{2}t & (\pi \leq t \leq 2\pi) \end{cases}$$

Sketch a graph of $f(t)$ for $-2\pi \leq t \leq 3\pi$ and find a Fourier series expansion of it.

Solution   A graph of the function $f(t)$ for $-2\pi \leq t \leq 3\pi$ is shown in Figure 12.5. From (12.15),

$$a_0 = \frac{1}{\pi}\int_0^{2\pi} f(t)dt = \frac{1}{\pi}\left[\int_0^{\pi/2} t\,dt + \int_{\pi/2}^{\pi} \tfrac{1}{2}\pi\,dt + \int_{\pi}^{2\pi}\left(\pi - \tfrac{1}{2}t\right)dt\right] = \frac{5}{8}\pi$$

Figure 12.5
Graph of the function
$f(t)$ of Example 12.3.

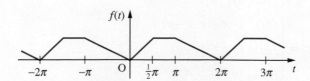

and

$$a_n = \frac{1}{\pi} \int_0^{2\pi} f(t) \cos nt \, dt \quad (n = 1, 2, 3, \ldots)$$

$$= \frac{1}{\pi} \left[\int_0^{\pi/2} t \cos nt \, dt + \int_{\pi/2}^{\pi} \frac{1}{2}\pi \cos nt \, dt + \int_{\pi}^{2\pi} \left(\pi - \frac{1}{2}t \right) \cos nt \, dt \right]$$

$$= \frac{1}{\pi} \left(\left[\frac{t}{n}\sin nt + \frac{\cos nt}{n^2} \right]_0^{\pi/2} + \left[\frac{\pi}{2n}\sin nt \right]_{\pi/2}^{\pi} + \left[\frac{2\pi - t}{2}\frac{\sin nt}{n} - \frac{\cos nt}{2n^2} \right]_{\pi}^{2\pi} \right)$$

$$= \frac{1}{\pi} \left(\frac{\pi}{2n}\sin\frac{1}{2}n\pi + \frac{1}{n^2}\cos\frac{1}{2}n\pi - \frac{1}{n^2} - \frac{\pi}{2n}\sin\frac{1}{2}n\pi - \frac{1}{2n^2} + \frac{1}{2n^2}\cos n\pi \right)$$

$$= \frac{1}{2\pi n^2} \left(2\cos\tfrac{1}{2}n\pi - 3 + \cos n\pi \right)$$

that is,

$$a_n = \begin{cases} \dfrac{1}{\pi n^2}[(-1)^{n/2} - 1] & \text{(even } n) \\[2mm] -\dfrac{2}{\pi n^2} & \text{(odd } n) \end{cases}$$

From (12.16),

$$b_n = \frac{1}{\pi} \int_0^{2\pi} f(t) \sin nt \, dt \quad (n = 1, 2, 3, \ldots)$$

$$= \frac{1}{\pi} \left[\int_0^{\pi/2} t \sin nt \, dt + \int_{\pi/2}^{\pi} \frac{1}{2}\pi \sin nt \, dt + \int_{\pi}^{2\pi} \left(\pi - \frac{1}{2}t \right) \sin nt \, dt \right]$$

$$= \frac{1}{\pi} \left(\left[-\frac{t}{n}\cos nt + \frac{1}{n^2}\sin nt \right]_0^{\pi/2} + \left[-\frac{\pi}{2n}\cos nt \right]_{\pi/2}^{\pi} \right.$$

$$\left. + \left[\frac{t - 2\pi}{2n}\cos nt - \frac{1}{2n^2}\sin nt \right]_{\pi}^{2\pi} \right)$$

$$= \frac{1}{\pi} \left(-\frac{\pi}{2n}\cos\tfrac{1}{2}n\pi + \frac{1}{n^2}\sin\frac{1}{2}n\pi - \frac{\pi}{2n}\cos n\pi + \frac{\pi}{2n}\cos\frac{1}{2}n\pi + \frac{\pi}{2n}\cos n\pi \right)$$

$$= \frac{1}{\pi n^2}\sin\frac{1}{2}n\pi$$

$$= \begin{cases} 0 & \text{(even } n) \\[2mm] \dfrac{(-1)^{(n-1)/2}}{\pi n^2} & \text{(odd } n) \end{cases}$$

Hence from (12.14) the Fourier series expansion of $f(t)$ is

$$f(t) = \frac{5}{16}\pi - \frac{2}{\pi}\left(\cos t + \frac{\cos 3t}{3^2} + \frac{\cos 5t}{5^2} + \dots\right)$$

$$- \frac{2}{\pi}\left(\frac{\cos 2t}{2^2} + \frac{\cos 6t}{6^2} + \frac{\cos 10t}{10^2} + \dots\right)$$

$$+ \frac{1}{\pi}\left(\sin t - \frac{\sin 3t}{3^2} + \frac{\sin 5t}{5^2} - \frac{\sin 7t}{7^2} + \dots\right)$$

12.2.5 Even and odd functions

Noting that a particular function possesses certain symmetrical properties enables us both to tell which terms are absent from a Fourier series expansion of the function and to simplify the expressions determining the remaining coefficients. In this section we consider even and odd function symmetries, while in Section 12.2.6 we shall consider symmetry due to even and odd harmonics.

First we review the properties of even and odd functions, considered in Section 2.2.5, that are useful for determining the Fourier coefficients. If $f(t)$ is an even function then $f(t) = f(-t)$ for all t, and the graph of the function is symmetrical about the vertical axis as illustrated in Figure 12.6(a). From the definition of integration, it follows that if $f(t)$ is an even function then

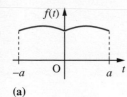

(a)

$$\int_{-a}^{a} f(t)\,dt = 2\int_{0}^{a} f(t)\,dt$$

If $f(t)$ is an odd function then $f(t) = -f(-t)$ for all t, and the graph of the function is symmetrical about the origin; that is, there is opposite-quadrant symmetry, as illustrated in Figure 12.6(b). It follows that if $f(t)$ is an odd function then

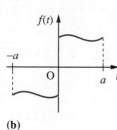

(b)

$$\int_{-a}^{a} f(t)\,dt = 0$$

Figure 12.6
Graphs of (a) an even function and (b) an odd function.

The following properties of even and odd functions are also useful for our purposes:

(a)   the *sum* of two (or more) *odd* functions is an *odd* function;
(b)   the *product* of two *even* functions is an *even* function;
(c)   the *product* of two *odd* functions is an *even* function;
(d)   the *product* of an *odd* and an *even* function is an *odd* function;
(e)   the *derivative* of an *even* function is an *odd* function;
(f)   the *derivative* of an *odd* function is an *even* function.

(Noting that t^{even} is even and t^{odd} is odd helps one to remember (a)–(f).)

Using these properties, and taking $d = -\frac{1}{2}T$ in (12.11) and (12.12), we have the following:

(i) If $f(t)$ is an *even* periodic function of period T then

$$a_n = \frac{2}{T}\int_{-T/2}^{T/2} f(t)\cos n\omega t\,dt = \frac{4}{T}\int_{0}^{T/2} f(t)\cos n\omega t\,dt$$

using property (b), and

$$b_n = \frac{2}{T} \int_{-T/2}^{T/2} f(t) \sin n\omega t \, dt = 0$$

using property (d).

Thus the Fourier series expansion of an even periodic function $f(t)$ with period T consists of cosine terms only and, from (12.3), is given by

$$f(t) = \tfrac{1}{2}a_0 + \sum_{n=1}^{\infty} a_n \cos n\omega t \qquad \textbf{(12.17)}$$

with

$$a_n = \frac{4}{T} \int_{0}^{T/2} f(t) \cos n\omega t \quad (n = 0, 1, 2, \ldots) \qquad \textbf{(12.18)}$$

(ii) If $f(t)$ is an *odd* periodic function of period T then

$$a_n = \frac{2}{T} \int_{-T/2}^{T/2} f(t) \cos n\omega t \, dt = 0$$

using property (d), and

$$b_n = \frac{2}{T} \int_{-T/2}^{T/2} f(t) \sin n\omega t \, dt = \frac{4}{T} \int_{0}^{T/2} f(t) \sin n\omega t \, dt$$

using property (c).

Thus the Fourier series expansion of an odd periodic function $f(t)$ with period T consists of sine terms only and, from (12.3), is given by

$$f(t) = \sum_{n=1}^{\infty} b_n \sin n\omega t \qquad \textbf{(12.19)}$$

with

$$b_n = \frac{4}{T} \int_{0}^{T/2} f(t) \sin n\omega t \, dt \quad (n = 1, 2, 3, \ldots) \qquad \textbf{(12.20)}$$

Example 12.4   A periodic function $f(t)$ with period 2π is defined within the period $-\pi < t < \pi$ by

$$f(t) = \begin{cases} -1 & (-\pi < t < 0) \\ 1 & (0 < t < \pi) \end{cases}$$

Find its Fourier series expansion.

Figure 12.7
Square wave of
Example 12.4.

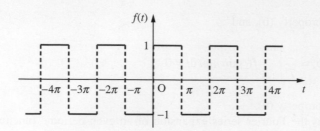

Solution   A sketch of the function $f(t)$ over the interval $-4\pi < t < 4\pi$ is shown in Figure 12.7. Clearly $f(t)$ is an odd function of t, so that its Fourier series expansion consists of sine terms only. Taking $T = 2\pi$, that is $\omega = 1$, in (12.19) and (12.20), the Fourier series expansion is given by

$$f(t) = \sum_{n=1}^{\infty} b_n \sin nt$$

with

$$b_n = \frac{2}{\pi} \int_0^{\pi} f(t) \sin nt \, dt \quad (n = 1, 2, 3, \dots)$$

$$= \frac{2}{\pi} \int_0^{\pi} 1 \sin nt \, dt = \frac{2}{\pi} \left[-\frac{1}{n} \cos nt \right]_0^{\pi}$$

$$= \frac{2}{n\pi}(1 - \cos n\pi) = \frac{2}{n\pi}[1 - (-1)^n]$$

$$= \begin{cases} 4/n\pi & (\text{odd } n) \\ 0 & (\text{even } n) \end{cases}$$

Thus the Fourier series expansion of $f(t)$ is

$$f(t) = \frac{4}{\pi}\left(\sin t + \frac{1}{3}\sin 3t + \frac{1}{5}\sin 5t + \dots \right) = \frac{4}{\pi} \sum_{n=1}^{\infty} \frac{\sin(2n-1)t}{2n-1} \tag{12.21}$$

Example 12.5   A periodic function $f(t)$ with period 2π is defined as

$$f(t) = t^2 \quad (-\pi < t < \pi), \qquad f(t) = f(t + 2\pi)$$

Obtain a Fourier series expansion for it.

Solution   A sketch of the function $f(t)$ over the interval $-3\pi < t < 3\pi$ is shown is Figure 12.8. Clearly, $f(t)$ is an even function of t, so that its Fourier series expansion consists of cosine terms only. Taking $T = 2\pi$, that is $\omega = 1$, in (12.17) and (12.18) the Fourier series expansion is given by

$$f(t) = \tfrac{1}{2}a_0 + \sum_{n=1}^{\infty} a_n \cos nt$$

Figure 12.8
The function $f(t)$
of Example 12.5.

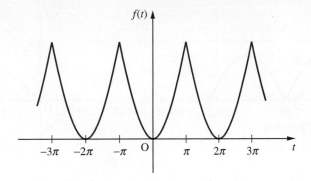

with

$$a_0 = \frac{2}{\pi} \int_0^\pi f(t)\,dt = \frac{2}{\pi} \int_0^\pi t^2\,dt = \frac{2}{3}\pi^2$$

and

$$a_n = \frac{2}{\pi} \int_0^\pi f(t) \cos nt\,dt \quad (n = 1, 2, 3, \dots)$$

$$= \frac{2}{\pi} \int_0^\pi t^2 \cos nt\,dt$$

$$= \frac{2}{\pi} \left[\frac{t^2}{n} \sin nt + \frac{2t}{n^2} \cos nt - \frac{2}{n^3} \sin nt \right]_0^\pi$$

$$= \frac{2}{\pi} \left(\frac{2\pi}{n^2} \cos n\pi \right) = \frac{4}{n^2}(-1)^n$$

since $\sin n\pi = 0$ and $\cos n\pi = (-1)^n$. Thus the Fourier series expansion of $f(t) = t^2$ is

$$f(t) = \frac{1}{3}\pi^2 + 4 \sum_{n=1}^{\infty} \frac{(-1)^n}{n^2} \cos nt \tag{12.22}$$

or, writing out the first few terms,

$$f(t) = \tfrac{1}{3}\pi^2 - 4\cos t + \cos 2t - \tfrac{4}{9}\cos 3t + \dots$$

12.2.6 Even and odd harmonics

In this section we consider types of symmetry that can be identified in order to eliminate terms from the Fourier series expansion having even values of n (including $n = 0$) or odd values of n.

(a) If a periodic function $f(t)$ is such that

$$f(t + \tfrac{1}{2}T) = f(t)$$

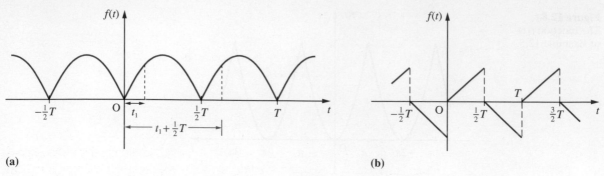

Figure 12.9 Functions having Fourier series with (a) only even harmonics and (b) only odd harmonics.

then it has period $T/2$ and frequency $\omega = 2(2\pi/T)$ so only even harmonics are present in its Fourier series expansion. For even n we have

$$a_n = \frac{4}{T} \int_0^{T/2} f(t) \cos n\omega t \, dt \tag{12.23}$$

$$b_n = \frac{4}{T} \int_0^{T/2} f(t) \sin n\omega t \, dt \tag{12.24}$$

An example of such a function is given in Figure 12.9(a).

(b) If a periodic function $f(t)$ with period T is such that

$$f(t + \tfrac{1}{2}T) = -f(t)$$

then only odd harmonics are present in its Fourier series expansion. For odd n

$$a_n = \frac{4}{T} \int_0^{T/2} f(t) \cos n\omega t \, dt \tag{12.25}$$

$$b_n = \frac{4}{T} \int_0^{T/2} f(t) \sin n\omega t \, dt \tag{12.26}$$

An example of such a function is shown in Figure 12.9(b).

The square wave of Example 12.4 is such that $f(t + \pi) = -f(t)$, so that, from (b), its Fourier series expansion consists of only odd harmonics. Since it is also an odd function, it follows that its Fourier series expansion consists only of odd-harmonic sine terms, which is confirmed by the result (12.21).

Example 12.6   Obtain the Fourier series expansion of the rectified sine wave

$$f(t) = |\sin t|$$

Solution   A sketch of the wave over the interval $-\pi < t < 2\pi$ is shown in Figure 12.10. Clearly, $f(t + \pi) = f(t)$ so that only even harmonics are present in the Fourier series expansion.

Figure 12.10
Rectified wave
$f(t) = |\sin t|$.

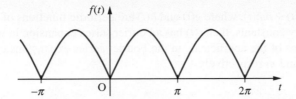

Since the function is also an even function of t, it follows that the Fourier series expansion will consist only of even-harmonic cosine terms. Taking $T = 2\pi$, that is $\omega = 1$, in (12.23), the coefficients of the even harmonics are given by

$$a_n = \frac{2}{\pi}\int_0^\pi f(t)\cos nt \quad (\text{even } n) \quad = \frac{2}{\pi}\int_0^\pi \sin t \cos nt \, dt$$

$$= \frac{1}{\pi}\int_0^\pi [\sin(n+1)t - \sin(n-1)t]dt$$

$$= \frac{1}{\pi}\left[-\frac{\cos(n+1)t}{n+1} + \frac{\cos(n-1)t}{n-1}\right]_0^\pi$$

Since both $n+1$ and $n-1$ are odd when n is even,

$$\cos(n+1)\pi = \cos(n-1)\pi = -1$$

so that

$$a_n = \frac{1}{\pi}\left[\left(\frac{1}{n+1} - \frac{1}{n-1}\right) - \left(-\frac{1}{n+1} + \frac{1}{n-1}\right)\right] = -\frac{4}{\pi}\frac{1}{n^2-1}$$

Thus the Fourier series expansion of $f(t)$ is

$$f(t) = \frac{1}{2}a_0 + \sum_{\substack{n=2 \\ (n \text{ even})}}^\infty a_n \cos nt = \frac{2}{\pi} - \frac{4}{\pi}\sum_{\substack{n=2 \\ (n \text{ even})}}^\infty \frac{1}{n^2-1}\cos nt$$

$$= \frac{2}{\pi} - \frac{4}{\pi}\sum_{n=1}^\infty \frac{1}{4n^2-1}\cos 2nt$$

or, writing out the first few terms,

$$f(t) = \frac{2}{\pi} - \frac{4}{\pi}\left(\frac{1}{3}\cos 2t + \frac{1}{15}\cos 4t + \frac{1}{35}\cos 6t + \dots\right)$$

12.2.7 Linearity property

The linearity property as applied to Fourier series may be stated in the form of the following theorem.

Theorem 12.1

If $f(t) = lg(t) + mh(t)$, where $g(t)$ and $h(t)$ are periodic functions of period T and l and m are arbitrary constants, then $f(t)$ has a Fourier series expansion in which the coefficients are the sums of the coefficients in the Fourier series expansions of $g(t)$ and $h(t)$ multiplied by l and m respectively.

Proof

Clearly $f(t)$ is periodic with period T. If the Fourier series expansions of $g(t)$ and $h(t)$ are

$$g(t) = \tfrac{1}{2}a_0 + \sum_{n=1}^{\infty} a_n \cos n\omega t + \sum_{n=1}^{\infty} b_n \sin n\omega t$$

$$h(t) = \tfrac{1}{2}\alpha_0 + \sum_{n=1}^{\infty} \alpha_n \cos n\omega t + \sum_{n=1}^{\infty} \beta_n \sin n\omega t$$

then, using (12.11) and (12.12), the Fourier coefficients in the expansion of $f(t)$ are

$$A_n = \frac{2}{T}\int_{d}^{d+T} f(t)\cos n\omega t\, dt = \frac{2}{T}\int_{d}^{d+T} [lg(t) + mh(t)]\cos n\omega t\, dt$$

$$= \frac{2l}{T}\int_{d}^{d+T} g(t)\cos n\omega t\, dt + \frac{2m}{T}\int_{d}^{d+T} h(t)\cos n\omega t\, dt$$

$$= la_n + m\alpha_n$$

and

$$B_n = \frac{2}{T}\int_{d}^{d+T} f(t)\sin n\omega t\, dt = \frac{2l}{T}\int_{d}^{d+T} g(t)\sin n\omega t\, dt + \frac{2m}{T}\int_{d}^{d+T} h(t)\sin n\omega t\, dt$$

$$= lb_n + m\beta_n$$

confirming that the Fourier series expansion of $f(t)$ is

$$f(t) = \tfrac{1}{2}(la_0 + m\alpha_0) + \sum_{n=1}^{\infty} (la_n + m\alpha_n)\cos n\omega t + \sum_{n=1}^{\infty} (lb_n + m\beta_n)\sin n\omega t$$

end of theorem

Example 12.7

Suppose that $g(t)$ and $h(t)$ are periodic functions of period 2π and are defined within the period $-\pi < t < \pi$ by

$$g(t) = t^2, \quad h(t) = t$$

Determine the Fourier series expansions of both $g(t)$ and $h(t)$ and use the linearity property to confirm the expansion obtained in Example 12.2 for the periodic function $f(t)$ defined within the period $-\pi < t < \pi$ by $f(t) = t^2 + t$.

Solution   The Fourier series of $g(t)$ is given by (12.22) as

$$g(t) = \frac{1}{3}\pi^2 + 4\sum_{n=1}^{\infty} \frac{(-1)^n}{n^2} \cos nt$$

Recognizing that $h(t) = t$ is an odd function of t, we find, taking $T = 2\pi$ and $\omega = 1$ in (12.19) and (12.20), that its Fourier series expansion is

$$h(t) = \sum_{n=1}^{\infty} b_n \sin nt$$

where

$$b_n = \frac{2}{\pi} \int_0^{\pi} h(t) \sin nt \, dt \quad (n = 1, 2, 3, \dots)$$

$$= \frac{2}{\pi} \int_0^{\pi} t \sin nt \, dt = \frac{2}{\pi} \left[-\frac{t}{n} \cos nt + \frac{\sin nt}{n^2} \right]_0^{\pi}$$

$$= -\frac{2}{n}(-1)^n$$

recognizing again that $\cos n\pi = (-1)^n$ and $\sin n\pi = 0$. Thus the Fourier series expansion of $h(t) = t$ is

$$h(t) = -2\sum_{n=1}^{\infty} \frac{(-1)^n}{n} \sin nt \tag{12.27}$$

Using the linearity property, we find, by combining (12.12) and (12.27), that the Fourier series expansion of $f(t) = g(t) + h(t) = t^2 + t$ is

$$f(t) = \frac{1}{3}\pi^2 + 4\sum_{n=1}^{\infty} \frac{(-1)^n}{n^2} \cos nt - 2\sum_{n=1}^{\infty} \frac{(-1)^n}{n} \sin nt$$

which conforms to the series obtained in Example 12.2.

12.2.8   Convergence of the Fourier series

So far we have concentrated our attention on determining the Fourier series expansion corresponding to a given periodic function $f(t)$. In reality, this is an exercise in integration, since we merely have to compute the coefficients a_n and b_n using Euler's formulae (12.11) and (12.12) and then substitute these values into (12.3). We have not yet considered the question of whether or not the Fourier series thus obtained is a valid representation of the periodic function $f(t)$. It should not be assumed that the existence of the coefficients a_n and b_n in itself implies that the associated series converges to the function $f(t)$.

A full discussion of the convergence of a Fourier series is beyond the scope of this book and we shall confine ourselves to simply stating a set of conditions which ensures that $f(t)$ has a convergent Fourier series expansion. These conditions, known as **Dirichlet's conditions**, may be stated in the form of Theorem 12.2.

Theorem 12.2 Dirichlet's conditions

If $f(t)$ is a bounded periodic function that in any period has

(a) a finite number of isolated maxima and minima, and

(b) a finite number of points of finite discontinuity

then the Fourier series expansion of $f(t)$ converges to $f(t)$ at all points where $f(t)$ is continuous and to the average of the right- and left-hand limits of $f(t)$ at points where $f(t)$ is discontinuous (that is, to the mean of the discontinuity).

end of theorem

Example 12.8

Give reasons why the functions

(a) $\dfrac{1}{3-t}$ (b) $\sin\left(\dfrac{1}{t-2}\right)$

do not satisfy Dirichlet's conditions in the interval $0 < t < 2\pi$.

Solution (a) The function $f(t) = 1/(3-t)$ has an infinite discontinuity at $t = 3$, which is within the interval, and therefore does not satisfy the condition that $f(t)$ must only have *finite* discontinuities within a period (i.e. it is bounded).

(b) The function $f(t) = \sin[1/(t-2)]$ has an infinite number of maxima and minima in the neighbourbood of $t = 2$, which is within the interval, and therefore does not satisfy the requirement that $f(t)$ must have only a finite number of isolated maxima and minima within one period.

The conditions of Theorem 12.2 are sufficient to ensure that a representative Fourier series expansion of $f(t)$ exists. However, they are not necessary conditions for convergence, and it does not follow that a representative Fourier series does not exist if they are not satisfied. Indeed, necessary conditions on $f(t)$ for the existence of a convergent Fourier series are not yet known. In practice, this does not cause any problems, since for almost all conceivable practical applications the functions that are encountered satisfy the conditions of Theorem 12.2 and therefore have representative Fourier series.

Another issue of importance in practical applications is the rate of convergence of a Fourier series, since this is an indication of how many terms must be taken in the expansion in order to obtain a realistic approximation to the function $f(t)$ it represents. Obviously, this is determined by the coefficients a_n and b_n of the Fourier series and the manner in which these decrease as n increases.

In an example, such as Example 12.1, in which the function $f(t)$ is only piecewise-continuous, exhibiting jump discontinuities, the Fourier coefficients decrease as $1/n$, and it may be necessary to include a large number of terms to obtain an adequate approximation to $f(t)$. In an example, such as Example 12.3, in which the function is a continuous function but has discontinuous first derivatives (owing to the sharp corners), the Fourier coefficients decrease as $1/n^2$, and so one would expect the series to converge more rapidly. Indeed, this argument applies in general, and we may summarize as follows:

(a)   if $f(t)$ is only piecewise-continuous then the coefficients in its Fourier series representation decrease as $1/n$;

(b)   if $f(t)$ is continuous everywhere but has discontinuous first derivatives then the coefficients in its Fourier series representation decrease as $1/n^2$;

(c)   if $f(t)$ and all its derivatives up to that of the rth order are continuous but the $(r + 1)$th derivative is discontinuous then the coefficients in its Fourier series representation decrease as $1/n^{r+2}$.

These observations are not surprising, since they simply tell us that the smoother the function $f(t)$, the more rapidly will its Fourier series representation converge.

To illustrate some of these issues related to convergence we return to Example 12.4, in which the Fourier series (12.21) was obtained as a representation of the square wave of Figure 12.7.

Since (12.21) is an infinite series, it is clearly not possible to plot a graph of the result. However, by considering finite partial sums, it is possible to plot graphs of approximations to the series. Denoting the sum of the first N terms in the infinite series by $f_N(t)$, that is

$$f_N(t) = \frac{4}{\pi} \sum_{n=1}^{N} \frac{\sin(2n - 1)t}{2n - 1} \tag{12.28}$$

the graphs of $f_N(t)$ for $N = 1, 2, 3$ and 20 are as shown in Figure 12.11. It can be seen that at points where $f(t)$ is continuous the approximation of $f(t)$ by $f_N(t)$ improves as N increases, confirming that the series converges to $f(t)$ at all such points. It can also be seen that at points of discontinuity of $f(t)$, which occur at $t = \pm n\pi$ $(n = 0, 1, 2, \ldots)$, the series converges to the mean value of the discontinuity, which in this particular example is $\frac{1}{2}(-1 + 1) = 0$. As a consequence, the equality sign in (12.21) needs to be interpreted carefully. Although such use may be acceptable, in the sense that the

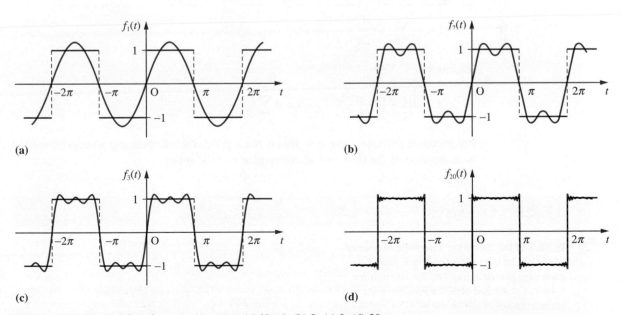

Figure 12.11 Plots of $f_N(t)$ for a square wave; (a) $N = 1$; (b) 2; (c) 3; (d) 20.

series converges to $f(t)$ for values of t where $f(t)$ is continuous, this is not so at points of discontinuity. To overcome this problem, the symbol ~ (read as 'behaves as' or 'represented by') rather than = is frequently used in the Fourier series representation of a function $f(t)$, so that (12.21) is often written as

$$f(t) \sim \sum_{n=1}^{\infty} \frac{\sin(2n-1)t}{2n-1}$$

In the companion text *Advanced Modern Engineering Mathematics* it is shown that the Fourier series converges to $f(t)$ in the sense that the integral of the square of the difference between $f(t)$ and $f_N(t)$ is minimized and tends to zero as $N \to \infty$.

We note that convergence of the Fourier series is slowest near a point of discontinuity, such as the one that occurs at $t = 0$. Although the series does converge to the mean value of the discontinuity (namely zero) at $t = 0$, there is, as indicated in Figure 12.11(d), an undershoot at $t = 0-$ (that is, just to the left of $t = 0$) and an overshoot at $t = 0+$ (that is, just to the right of $t = 0$). This non-smooth convergence of the Fourier series leading to the occurrence of an undershoot and an overshoot at points of discontinuity of $f(t)$ is a characteristic of all Fourier series representing discontinuous functions, not only that of the square wave of Example 12.4, and is known as **Gibbs' phenomenon** after the American physicist J. W. Gibbs (1839–1903). The magnitude of the undershoot/overshoot does not diminish as $N \to \infty$ in (12.28), but simply gets 'sharper' and 'sharper', tending to a spike. In general, the magnitude of the undershoot and overshoot together amount to about 18% of the magnitude of the discontinuity (that is, the difference in the values of the function $f(t)$ to the left and right of the discontinuity). It is important that the existence of this phenomenon be recognized, since in certain practical applications these spikes at discontinuities have to be suppressed by using appropriate smoothing factors.

Theoretically, we can use the series (12.21) to obtain an approximation to π. This is achieved by taking $t = \frac{1}{2}\pi$, when $f(t) = 1$; (12.21) then gives

$$1 = \frac{4}{\pi} \sum_{n=1}^{\infty} \frac{\sin\frac{1}{2}(2n-1)\pi}{2n-1}$$

leading to

$$\pi = 4(1 - \tfrac{1}{3} + \tfrac{1}{5} - \tfrac{1}{7} + \dots) = 4 \sum_{n=1}^{\infty} \frac{(-1)^{n+1}}{2n-1}$$

For practical purposes, however, this is not a good way of obtaining an approximation to π, because of the slow rate of convergence of the series.

12.2.9 Exercises

1 In each of the following a periodic function $f(t)$ of period 2π is specified over one period. In each case sketch a graph of the function for $-4\pi \le t \le 4\pi$ and obtain a Fourier series representation of the function.

(a) $f(t) = \begin{cases} -\pi & (-\pi < t < 0) \\ t & (0 < t < \pi) \end{cases}$

(b) $f(t) = \begin{cases} t + \pi & (-\pi < t < 0) \\ 0 & (0 < t < \pi) \end{cases}$

(c) $f(t) = 1 - \dfrac{t}{\pi}$ $(0 \leqslant t \leqslant 2\pi)$

(d) $f(t) = \begin{cases} 0 & (-\pi \leqslant t \leqslant -\frac{1}{2}\pi) \\ 2\cos t & (-\frac{1}{2}\pi \leqslant t \leqslant \frac{1}{2}\pi) \\ 0 & (\frac{1}{2}\pi \leqslant t \leqslant \pi) \end{cases}$

(e) $f(t) = \cos\frac{1}{2}t$ $(-\pi < t < \pi)$

(f) $f(t) = |t|$ $(-\pi < t < \pi)$

(g) $f(t) = \begin{cases} 0 & (-\pi \leqslant t \leqslant 0) \\ 2t - \pi & (0 < t \leqslant \pi) \end{cases}$

(h) $f(t) = \begin{cases} -t + e^t & (-\pi \leqslant t < 0) \\ t + e^t & (0 \leqslant t < \pi) \end{cases}$

2 Obtain the Fourier series expansion of the periodic function $f(t)$ of period 2π defined over the period $0 \leqslant t \leqslant 2\pi$ by

$$f(t) = (\pi - t)^2 \quad (0 \leqslant t \leqslant 2\pi)$$

Use the Fourier series to show that

$$\frac{1}{12}\pi^2 = \sum_{n=1}^{\infty} \frac{(-1)^{n+1}}{n^2}$$

3 The charge $q(t)$ on the plates of a capacitor at time t is as shown in Figure 12.12. Express $q(t)$ as a Fourier series expansion.

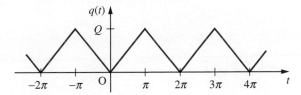

Figure 12.12 Plot of the charge $q(t)$ in Question 3.

4 The clipped response of a half-wave rectifier is the periodic function $f(t)$ of period 2π defined over the period $0 \leqslant t \leqslant 2\pi$ by

$$f(t) = \begin{cases} 5\sin t & (0 \leqslant t \leqslant \pi) \\ 0 & (\pi \leqslant t \leqslant 2\pi) \end{cases}$$

Express $f(t)$ as a Fourier series expansion.

5 Show that the Fourier series representing the periodic function $f(t)$, where

$$f(t) = \begin{cases} \pi^2 & (-\pi < t < 0) \\ (t - \pi)^2 & (0 < t < \pi) \end{cases}$$

$$f(t + 2\pi) = f(t)$$

is

$$f(t) = \frac{2}{3}\pi^2 + \sum_{n=1}^{\infty} \left[\frac{2}{n^2}\cos nt + \frac{(-1)^n}{n}\pi\sin nt \right]$$

$$- \frac{4}{\pi}\sum_{n=1}^{\infty} \frac{\sin(2n-1)t}{(2n-1)^3}$$

Use this result to show that

(a) $\displaystyle\sum_{n=1}^{\infty} \frac{1}{n^2} = \frac{1}{6}\pi^2$ (b) $\displaystyle\sum_{n=1}^{\infty} \frac{(-1)^{n+1}}{n^2} = \frac{1}{12}\pi^2$

6 A periodic function $f(t)$ of period 2π is defined within the domain $0 \leqslant t \leqslant \pi$ by

$$f(t) = \begin{cases} t & (0 \leqslant t \leqslant \frac{1}{2}\pi) \\ \pi - t & (\frac{1}{2}\pi \leqslant t \leqslant \pi) \end{cases}$$

Sketch a graph of $f(t)$ for $-2\pi < t < 4\pi$ for the two cases where

(a) $f(t)$ is an even function

(b) $f(t)$ is an odd function

Find the Fourier series expansion that represents the even function for all values of t, and use it to show that

$$\frac{1}{8}\pi^2 = \sum_{n=1}^{\infty} \frac{1}{(2n-1)^2}$$

7 A periodic function $f(t)$ of period 2π is defined within the period $0 \leqslant t \leqslant 2\pi$ by

$$f(t) = \begin{cases} 2 - t/\pi & (0 \leqslant t \leqslant \pi) \\ t/\pi & (\pi \leqslant t \leqslant 2\pi) \end{cases}$$

Draw a graph of the function for $-4\pi \leqslant t \leqslant 4\pi$ and obtain its Fourier series expansion.

By replacing t by $t - \frac{1}{2}\pi$ in your answer, show that the periodic function $f(t - \frac{1}{2}\pi) - \frac{3}{2}$ is represented by a sine series of odd harmonics.

12.2.10 Functions of period T

Although all the results have been related to periodic functions having period T, all the examples we have considered so far have involved periodic functions of period 2π. This was done primarily for ease of manipulation in determining the Fourier coefficients while becoming acquainted with Fourier series. As mentioned in Section 12.2.4, functions having unit frequency (that is, of period 2π) are rarely encountered in practice, and in this section we consider examples of periodic functions having periods other than 2π.

Example 12.9

A periodic function $f(t)$ of period 4 (that is, $f(t + 4) = f(t)$) is defined in the range $-2 < t < 2$ by

$$f(t) = \begin{cases} 0 & (-2 < t < 0) \\ 1 & (0 < t < 2) \end{cases}$$

Sketch a graph of $f(t)$ for $-6 \leqslant t \leqslant 6$ and obtain a Fourier series expansion for the function.

Solution

A graph of $f(t)$ for $-6 \leqslant t \leqslant 6$ is shown is Figure 12.13. Taking $T = 4$ in (12.11) and (12.12), we have

$$a_0 = \tfrac{1}{2} \int_{-2}^{2} f(t)\,\mathrm{d}t = \frac{1}{2}\left(\int_{-2}^{0} 0\,\mathrm{d}t + \int_{0}^{2} 1\,\mathrm{d}t \right) = 1$$

$$a_n = \tfrac{1}{2} \int_{-2}^{2} f(t)\cos \tfrac{1}{2}n\pi t \,\mathrm{d}t \quad (n = 1, 2, 3, \dots)$$

$$= \tfrac{1}{2}\left(\int_{-2}^{0} 0\,\mathrm{d}t + \int_{0}^{2} \cos \tfrac{1}{2}n\pi t \,\mathrm{d}t \right) = 0$$

and

$$b_n = \tfrac{1}{2} \int_{-2}^{2} f(t)\sin \tfrac{1}{2}n\pi t \,\mathrm{d}t \quad (n = 1, 2, 3, \dots)$$

$$= \frac{1}{2}\left(\int_{-2}^{0} 0\,\mathrm{d}t + \int_{0}^{2} \sin \frac{1}{2}n\pi t \,\mathrm{d}t \right) = \frac{1}{n\pi}(1 - \cos n\pi) = \frac{1}{n\pi}[1 - (-1)^n]$$

$$= \begin{cases} 0 & (\text{even } n) \\ 2/n\pi & (\text{odd } n) \end{cases}$$

Figure 12.13
The function $f(t)$
of Example 12.9.

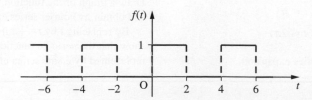

Thus, from (12.10), the Fourier series expansion of $f(t)$ is

$$f(t) = \frac{1}{2} + \frac{2}{\pi}\left(\sin\frac{1}{2}\pi t + \frac{1}{3}\sin\frac{3}{2}\pi t + \frac{1}{5}\sin\frac{5}{2}\pi t + \ldots\right)$$

$$= \frac{1}{2} + \frac{2}{\pi}\sum_{n=1}^{\infty}\frac{1}{2n-1}\sin\frac{1}{2}(2n-1)\pi t$$

Example 12.10 A periodic function $f(t)$ of period 2 is defined by

$$f(t) = \begin{cases} 3t & (0 < t < 1) \\ 3 & (1 < t < 2) \end{cases}$$

$$f(t+2) = f(t)$$

Sketch a graph of $f(t)$ for $-4 \leq t \leq 4$ and determine a Fourier series expansion for the function.

Solution A graph of $f(t)$ for $-4 \leq t \leq 4$ is shown in Figure 12.14. Taking $T = 2$ in (12.11) and (12.12), we have

$$a_0 = \frac{2}{2}\int_0^2 f(t)\,dt = \int_0^1 3t\,dt + \int_1^2 3\,dt = \tfrac{9}{2}$$

$$a_n = \frac{2}{2}\int_0^2 f(t)\cos\frac{n\pi t}{1}\,dt \quad (n = 1, 2, 3, \ldots)$$

$$= \int_0^1 3t\cos n\pi t\,dt + \int_1^2 3\cos n\pi t\,dt$$

$$= \left[\frac{3t\sin n\pi t}{n\pi} + \frac{3\cos n\pi t}{(n\pi)^2}\right]_0^1 + \left[\frac{3\sin n\pi t}{n\pi}\right]_1^2$$

$$= \frac{3}{(n\pi)^2}(\cos n\pi - 1)$$

$$= \begin{cases} 0 & (\text{even } n) \\ -6/(n\pi)^2 & (\text{odd } n) \end{cases}$$

Figure 12.14
The function $f(t)$
of Example 12.10.

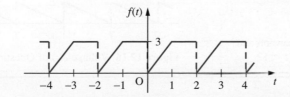

and

$$b_n = \frac{2}{2}\int_0^2 f(t)\sin\frac{n\pi t}{1}\,dt = \int_0^1 3t\sin n\pi t\,dt + \int_1^2 3\sin n\pi t\,dt \quad (n = 1, 2, 3, \dots)$$

$$= \left[-\frac{3\cos n\pi t}{n\pi} + \frac{3\sin n\pi t}{(n\pi)^2}\right]_0^1 + \left[-\frac{3\cos n\pi t}{n\pi}\right]_1^2 = -\frac{3}{n\pi}\cos 2n\pi = -\frac{3}{n\pi}$$

Thus, from (12.10), the Fourier series expansion of $f(t)$ is

$$f(t) = \frac{9}{4} - \frac{6}{\pi^2}\left(\cos\pi t + \frac{1}{9}\cos 3\pi t + \frac{1}{25}\cos 5\pi t + \dots\right)$$

$$- \frac{3}{\pi}\left(\sin\pi t + \frac{1}{2}\sin 2\pi t + \frac{1}{3}\sin 3\pi t + \dots\right)$$

$$= \frac{9}{4} - \frac{6}{\pi^2}\sum_{n=1}^{\infty}\frac{\cos(2n-1)\pi t}{(2n-1)^2} - \frac{3}{\pi}\sum_{n=1}^{\infty}\frac{\sin n\pi t}{n}$$

12.2.11 Exercises

8 Find a Fourier series expansion of the periodic function

$$f(t) = t \quad (-l < t < l)$$

$$f(t + 2l) = f(t)$$

9 A periodic function $f(t)$ of period $2l$ is defined over one period by

$$f(t) = \begin{cases} -\dfrac{K}{l}(l+t) & (-l < t < 0) \\[2mm] \dfrac{K}{l}(l-t) & (0 < t < l) \end{cases}$$

Determine its Fourier series expansion and illustrate graphically for $-3l < t < 3l$.

10 A periodic function of period 10 is defined within the period $-5 < t < 5$ by

$$f(t) = \begin{cases} 0 & (-5 < t < 0) \\ 3 & (0 < t < 5) \end{cases}$$

Determine its Fourier series expansion and illustrate graphically for $-12 < t < 12$.

11 Passing a sinusoidal voltage $A\sin\omega t$ through a half-wave rectifier produces the clipped sine wave shown in Figure 12.15. Determine a Fourier series expansion of the rectified wave.

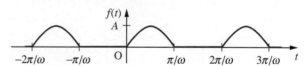

Figure 12.15 Rectified sine wave of Question 11.

12 Obtain a Fourier series expansion of the periodic function

$$f(t) = t^2 \quad (-T < t < T)$$

$$f(t + 2T) = f(t)$$

and illustrate graphically for $-3T < t < 3T$.

13 Determine a Fourier series representation of the periodic voltage $e(t)$ shown in Figure 12.16.

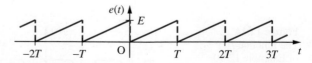

Figure 12.16 Voltage $e(t)$ of Question 13.

Functions defined over a finite interval

One of the requirements of Fourier's theorem is that the function to be expanded be periodic. Therefore a function $f(t)$ that is not periodic cannot have a Fourier series representation that converges to it *for all values* of t. However, we can obtain a Fourier series expansion that represents a *non-periodic* function $f(t)$ that is defined only over a finite time interval $0 \leq t \leq \tau$. This is a facility that is frequently used to solve problems in practice, particularly boundary-value problems involving partial differential equations, such as the consideration of heat flow along a bar or the vibrations of a string. Various forms of Fourier series representations of $f(t)$, valid only in the interval $0 \leq t \leq \tau$, are possible, including series consisting of cosine terms only or series consisting of sine terms only. To obtain these, various periodic extensions of $f(t)$ are formulated.

12.3.1 Full-range series

Suppose the given function $f(t)$ is defined only over the finite time interval $0 \leq t \leq \tau$. Then, to obtain a full-range Fourier series representation of $f(t)$ (that is, a series consisting of both cosine and sine terms), we define the **periodic extension** $\phi(t)$ of $f(t)$ by

$$\phi(t) = f(t) \quad (0 < t < \tau)$$

$$\phi(t + \tau) = \phi(t)$$

The graphs of a possible $f(t)$ and its periodic extension $\phi(t)$ are shown in Figures 12.17(a) and (b) respectively.

Provided that $f(t)$ satisfies Dirichlet's conditions in the interval $0 \leq t \leq \tau$, the new function $\phi(t)$, of period τ, will have a convergent Fourier series expansion. Since, within the particular period $0 < t < \tau$, $\phi(t)$ is identical with $f(t)$, it follows that this Fourier series expansion of $\phi(t)$ will be representative of $f(t)$ within this interval.

Figure 12.17
Graphs of a function defined only over
(a) a finite interval $0 \leq t \leq \tau$ and (b) its periodic extension.

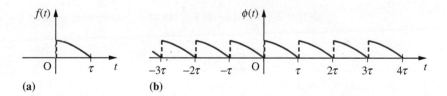

(a) (b)

Example 12.11

Find a full-range Fourier series expansion of $f(t) = t$ valid in the finite interval $0 < t < 4$. Draw graphs of both $f(t)$ and the periodic function represented by the Fourier series obtained.

Solution

Define the periodic function $\phi(t)$ by

$$\phi(t) = f(t) = t \quad (0 < t < 4)$$

$$\phi(t + 4) = \phi(t)$$

Figure 12.18
The functions $f(t)$ and
$\phi(t)$ of Example 12.11.

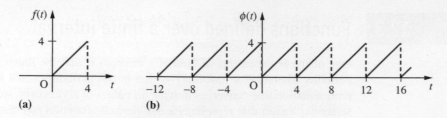

(a) (b)

Then the graphs of $f(t)$ and its periodic extension $\phi(t)$ are as shown in Figures 12.18(a) and (b) respectively. Since $\phi(t)$ is a periodic function with period 4, it has a convergent Fourier series expansion. Taking $T = 4$ in (12.11) and (12.12), the Fourier coefficients are determined as

$$a_0 = \tfrac{1}{2} \int_0^4 f(t)\,dt = \tfrac{1}{2} \int_0^4 t\,dt = 4$$

$$a_n = \tfrac{1}{2} \int_0^4 f(t) \cos \tfrac{1}{2} n\pi t\,dt \quad (n = 1, 2, 3, \dots)$$

$$= \frac{1}{2} \int_0^4 t \cos \frac{1}{2} n\pi t\,dt = \frac{1}{2} \left[\frac{2t}{n\pi} \sin \frac{1}{2} n\pi t + \frac{4}{(n\pi)^2} \cos \frac{1}{2} n\pi t \right]_0^4 = 0$$

and

$$b_n = \tfrac{1}{2} \int_0^4 f(t) \sin \tfrac{1}{2} n\pi t\,dt \quad (n = 1, 2, 3, \dots)$$

$$= \frac{1}{2} \int_0^4 t \sin \frac{1}{2} n\pi t\,dt = \frac{1}{2} \left[-\frac{2t}{n\pi} \cos \frac{1}{2} n\pi t + \frac{4}{(n\pi)^2} \sin \frac{1}{2} n\pi t \right]_0^4 = -\frac{4}{n\pi}$$

Thus, by (12.10), the Fourier series expansion of $\phi(t)$ is

$$\phi(t) = 2 - \frac{4}{\pi} \left(\sin \frac{1}{2}\pi t + \frac{1}{2} \sin \pi t + \frac{1}{3} \sin \frac{3}{2}\pi t + \frac{1}{4} \sin 2t + \frac{1}{5} \sin \frac{5}{2}\pi t + \dots \right)$$

$$= 2 - \frac{4}{\pi} \sum_{n=1}^{\infty} \frac{1}{n} \sin \frac{1}{2} n\pi t$$

Since $\phi(t) = f(t)$ for $0 < t < 4$, it follows that this Fourier series is representative of $f(t)$ within this interval, so that

$$f(t) = t = 2 - \frac{4}{\pi} \sum_{n=1}^{\infty} \frac{1}{n} \sin \frac{1}{2} n\pi t \quad (0 < t < 4) \tag{12.29}$$

It is important to appreciate that this series converges to t only within the interval $0 < t < 4$. For values of t outside this interval it converges to the periodic extended function $\phi(t)$. Again convergence is to be interpreted in the sense of Theorem 12.2, so that at the end points $t = 0$ and $t = 4$ the series does not converge to t but to the mean of the discontinuity in $\phi(t)$, namely the value 2.

12.3.2 Half-range cosine and sine series

Rather than develop the periodic extension $\phi(t)$ of $f(t)$ as in Section 12.3.1, it is possible to formulate periodic extensions that are either even or odd functions, so that the resulting Fourier series of the extended periodic functions consist either of cosine terms only or sine terms only.

For a function $f(t)$ defined only over the finite interval $0 \leqslant t \leqslant \tau$ its **even periodic extension** $F(t)$ is the even periodic function defined by

$$F(t) = \begin{cases} f(t) & (0 < t < \tau) \\ f(-t) & (-\tau < t < 0) \end{cases}$$

$$F(t + 2\tau) = f(t)$$

As an illustration, the even periodic extension $F(t)$ of the function $f(t)$ shown in Figure 12.17(a) (redrawn in Figure 12.19a) is shown in Figure 12.19(b).

Figure 12.19
(a) A function $f(t)$.
(b) Its even periodic extension $F(t)$.

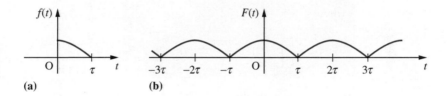

(a)   (b)

Provided that $f(t)$ satisfies Dirichlet's conditions in the interval $0 < t < \tau$, since it is an even function of period 2τ, it follows from Section 12.2.5 that the even periodic extension $F(t)$ will have a convergent Fourier series representation consisting of cosine terms only and given by

$$F(t) - \tfrac{1}{2}u_0 + \sum_{n=1}^{\infty} u_n \cos \frac{n\pi t}{\tau} \qquad (12.30)$$

where

$$a_n = \frac{2}{\tau} \int_0^{\tau} f(t) \cos \frac{n\pi t}{\tau} \, dt \quad (n = 0, 1, 2, \dots) \qquad (12.31)$$

Since, within the particular interval $0 < t < \tau$, $F(t)$ is identical with $f(t)$, it follows that the series (12.30) also converges to $f(t)$ within this interval.

For a function $f(t)$ defined only over the finite interval $0 \leqslant t \leqslant \tau$, its **odd periodic extension** $G(t)$ is the odd periodic function defined by

$$G(t) = \begin{cases} f(t) & (0 < t < \tau) \\ -f(-t) & (-\tau < t < 0) \end{cases}$$

$$G(t + 2\tau) = G(t)$$

Again, as an illustration, the odd periodic extension $G(t)$ of the function $f(t)$ shown in Figure 12.17(a) (redrawn in Figure 12.20a) is shown in Figure 12.20(b).

Figure 12.20
(a) A function $f(t)$.
(b) Its odd periodic
extension $G(t)$.

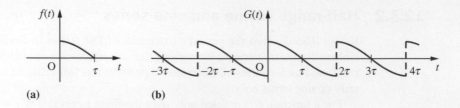

(a)   (b)

Provided that $f(t)$ satisfies Dirichlet's conditions in the interval $0 < t < \tau$, since it is an odd function of period 2τ, it follows from Section 12.2.5 that the odd periodic extension $G(t)$ will have a convergent Fourier series representation consisting of sine terms only and given by

$$G(t) = \sum_{n=1}^{\infty} b_n \sin \frac{n\pi t}{\tau} \qquad (12.32)$$

where

$$b_n = \frac{2}{\tau} \int_0^{\tau} f(t) \sin \frac{n\pi t}{\tau}\, dt \quad (n = 1, 2, 3, \dots) \qquad (12.33)$$

Again, since, within the particular interval $0 < t < \tau$, $G(t)$ is identical with $f(t)$, it follows that the series (12.32) also converges to $f(t)$ within this interval.

We note that both the even and odd periodic extensions $F(t)$ and $G(t)$ are of period 2τ, which is twice the length of the interval over which $f(t)$ is defined. However, the resulting Fourier series (12.30) and (12.32) are based only on the function $f(t)$, and for this reason are called the **half-range Fourier series expansions** of $f(t)$. In particular, the even half-range expansion $F(t)$, (12.30), is called the **half-range cosine series expansion** of $f(t)$, while the odd half-range expansion $G(t)$, (12.32), is called the **half-range sine series expansion** of $f(t)$.

Example 12.12

For the function $f(t) = t$ defined only in the interval $0 < t < 4$, and considered in Example 12.11, obtain

(a) a half-range cosine series expansion,

(b) a half-range sine series expansion.

Draw graphs of $f(t)$ and of the periodic functions represented by the two series obtained for $-20 < t < 20$.

Solution

(a) Half-range cosine series. Define the periodic function $F(t)$ by

$$F(t) = \begin{cases} f(t) = t & (0 < t < 4) \\ f(-t) = -t & (-4 < t < 0) \end{cases}$$

$$F(t + 8) = F(t)$$

Then, since $F(t)$ is an even periodic function with period 8, it has a convergent Fourier series expansion given by (12.30). Taking $\tau = 4$ in (12.31), we have

$$a_0 = \frac{2}{4} \int_0^4 f(t)\,dt = \frac{1}{2} \int_0^4 t\,dt = 4$$

$$a_n = \frac{2}{4} \int_0^4 f(t) \cos\frac{1}{4}n\pi t\,dt \quad (n = 1, 2, 3, \dots)$$

$$= \frac{1}{2} \int_0^4 t \cos\frac{1}{4}n\pi t\,dt = \frac{1}{2} \left[\frac{4t}{n\pi} \sin\frac{1}{4}n\pi t + \frac{16}{(n\pi)^2} \cos\frac{1}{4}n\pi t \right]_0^4$$

$$= \frac{8}{(n\pi)^2} (\cos n\pi - 1) = \begin{cases} 0 & (\text{even } n) \\ -16/(n\pi)^2 & (\text{odd } n) \end{cases}$$

Then, by (12.30), the Fourier series expansion of $F(t)$ is

$$F(t) = 2 - \frac{16}{\pi^2} \left(\cos\frac{1}{4}\pi t + \frac{1}{3^2} \cos\frac{3}{4}\pi t + \frac{1}{5^2} \cos\frac{5}{4}\pi t + \dots \right)$$

or

$$F(t) = 2 - \frac{16}{\pi^2} \sum_{n=1}^{\infty} \frac{1}{(2n-1)^2} \cos\frac{1}{4}(2n-1)\pi t$$

Since $F(t) = f(t)$ for $0 < t < 4$, it follows that this Fourier series is representative of $f(t)$ within this interval. Thus the half-range cosine series expansion of $f(t)$ is

$$f(t) = t = 2 - \frac{16}{\pi^2} \sum_{n=1}^{\infty} \frac{1}{(2n-1)^2} \cos\frac{1}{4}(2n-1)\pi t \quad (0 < t < 4) \tag{12.34}$$

(b) Half-range sine series. Define the periodic function $G(t)$ by

$$G(t) = \begin{cases} f(t) = t & (0 < t < 4) \\ -f(-t) = t & (-4 < t < 0) \end{cases}$$

$$G(t + 8) = G(t)$$

Then, since $G(t)$ is an odd periodic function with period 8, it has a convergent Fourier series expansion given by (12.32). Taking $\tau = 4$ in (12.33), we have

$$b_n = \frac{2}{4} \int_0^4 f(t) \sin\frac{1}{4}n\pi t\,dt \quad (n = 1, 2, 3, \dots)$$

$$= \frac{1}{2} \int_0^4 t \sin\frac{1}{4}n\pi t\,dt = \frac{1}{2} \left[-\frac{4t}{n\pi} \cos\frac{1}{4}n\pi t + \frac{16}{(n\pi)^2} \sin\frac{1}{4}n\pi t \right]_0^4$$

$$= -\frac{8}{n\pi} \cos n\pi = -\frac{8}{n\pi}(-1)^n$$

Thus, by (12.32), the Fourier series expansion of $G(t)$ is

$$G(t) = \frac{8}{\pi}\left(\sin\frac{1}{4}\pi t - \frac{1}{2}\sin\frac{1}{2}\pi t + \frac{1}{3}\sin\frac{3}{4}\pi t - \ldots \right)$$

or

$$G(t) = \frac{8}{\pi}\sum_{n=1}^{\infty}\frac{(-1)^{n+1}}{n}\sin\frac{1}{4}n\pi t$$

Since $G(t) = f(t)$ for $0 < t < 4$, it follows that this Fourier series is representative of $f(t)$ within this interval. Thus the half-range sine series expansion of $f(t)$ is

$$f(t) = t = \frac{8}{\pi}\sum_{n=1}^{\infty}\frac{(-1)^{n+1}}{n}\sin\frac{1}{4}n\pi t \quad (0 < t < 4) \tag{12.35}$$

Graphs of the given function $f(t)$ and of the even and odd periodic expansions $F(t)$ and $G(t)$ are given in Figures 12.21(a), (b) and (c) respectively.

It is important to realize that the three different Fourier series representations (12.29), (12.34) and (12.35) are representative of the function $f(t) = t$ only within the defined interval $0 < t < 4$. Outside this interval the three Fourier series converge to the three different functions $\phi(t)$, $F(t)$ and $G(t)$, illustrated in Figures 12.18(b), 12.21(b) and 12.21(c) respectively.

Figure 12.21
The functions $f(t)$, $F(t)$ and $G(t)$ of Example 12.12.

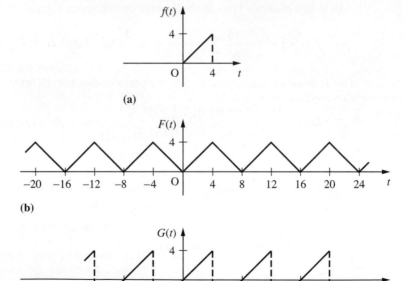

(a)

(b)

(c)

12.3.3 Exercises

14 Show that the half-range Fourier sine series expansion of the function $f(t) = 1$, valid for $0 < t < \pi$, is

$$f(t) = \frac{4}{\pi} \sum_{n=1}^{\infty} \frac{\sin(2n - 1)t}{2n - 1} \quad (0 < t < \pi)$$

Sketch the graphs of both $f(t)$ and the periodic function represented by the series expansion for $-3\pi < t < 3\pi$.

15 Determine the half-range cosine series expansion of the function $f(t) = 2t - 1$, valid for $0 < t < 1$. Sketch the graphs of both $f(t)$ and the periodic function represented by the series expansion for $-2 < t < 2$.

16 The function $f(t) = 1 - t^2$ is to be represented by a Fourier series expansion over the finite interval $0 < t < 1$. Obtain a suitable

(a) full-range series expansion,

(b) half-range sine series expansion,

(c) half-range cosine series expansion.

Draw graphs of $f(t)$ and of the periodic functions represented by each of the three series for $-4 < t < 4$.

17 A function $f(t)$ is defined by

$$f(t) = \pi t - t^2 \quad (0 \leqslant t \leqslant \pi)$$

and is to be represented by either a half-range Fourier sine series or a half-range Fourier cosine series. Find both of these series and sketch the graphs of the functions represented by them for $-2\pi < t < 2\pi$.

18 A tightly stretched flexible uniform string has its ends fixed at the points $x = 0$ and $x = l$. The midpoint of the string is displaced a distance a, as shown in Figure 12.22. If $f(x)$ denotes the displaced profile of the string, express $f(x)$ as a Fourier series expansion consisting only of sine terms.

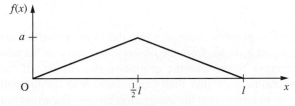

Figure 12.22 Displaced string of Question 18.

19 Repeat Question 18 for the case where the displaced profile of the string is as shown in Figure 12.23.

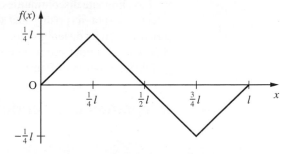

Figure 12.23 Displaced string of Question 19.

20 A function $f(t)$ is defined on $0 \leqslant t \leqslant \pi$ by

$$f(t) = \begin{cases} \sin t & (0 \leqslant t < \tfrac{1}{2}\pi) \\ 0 & (\tfrac{1}{2}\pi \leqslant t \leqslant \pi) \end{cases}$$

Find a half-range Fourier series expansion of $f(t)$ on this interval. Sketch a graph of the function represented by the series for $-2\pi \leqslant t \leqslant 2\pi$.

21 A function $f(t)$ is defined on the interval $-l \leqslant x \leqslant l$ by

$$f(x) = \frac{A}{l}(|x| - l)$$

Obtain a Fourier series expansion of $f(x)$ and sketch a graph of the function represented by the series for $-3l \leqslant x \leqslant 3l$.

22 The temperature distribution $T(x)$ at a distance x, measured from one end, along a bar of length L is given by

$$T(x) = Kx(L - x) \quad (0 \leqslant x \leqslant L),$$
$$K = \text{constant}$$

Express $T(x)$ as a Fourier series expansion consisting of sine terms only.

23 Find the Fourier series expansion of the function $f(t)$ valid for $-1 < t < 1$, where

$$f(t) = \begin{cases} 1 & (-1 < t < 0) \\ \cos \pi t & (0 < t < 1) \end{cases}$$

To what value does this series converge when $t = 1$?

Differentiation and integration of Fourier series

It is inevitable that the desire to obtain the derivative or the integral of a Fourier series will arise in some applications. Since the smoothing effects of the integration process tend to eliminate discontinuities, whereas the process of differentiation has the opposite effect, it is not surprising that the integration of a Fourier series is more likely to be possible than its differentiation. We shall not pursue the theory in depth here; rather we shall state, without proof, two theorems concerned with the term-by-term integration and differentiation of Fourier series, and make some observations on their use.

12.4.1 Integration of a Fourier series

Theorem 12.3

A Fourier series expansion of a periodic function $f(t)$ that satisfies Dirichlet's conditions may be integrated term by term, and the integrated series converges to the integral of the function $f(t)$.

end of theorem

According to this theorem, if $f(t)$ satisfies Dirichlet's conditions in the interval $-\pi \leqslant t \leqslant \pi$ and has a Fourier series expansion

$$f(t) = \tfrac{1}{2}a_0 + \sum_{n=1}^{\infty} (a_n \cos nt + b_n \sin nt)$$

then for $-\pi \leqslant t_1 < t \leqslant \pi$

$$\int_{t_1}^{t} f(t)\mathrm{d}t = \int_{t_1}^{t} \tfrac{1}{2}a_0 \,\mathrm{d}t + \sum_{n=1}^{\infty} \int_{t_1}^{t} (a_n \cos nt + b_n \sin nt)\mathrm{d}t$$

$$= \tfrac{1}{2}a_0(t - t_1) + \sum_{n=1}^{\infty}\left[\frac{b_n}{n}(\cos nt_1 - \cos nt) + \frac{a_n}{n}(\sin nt - \sin nt_1)\right]$$

Because of the presence of the term $\tfrac{1}{2}a_0 t$ on the right-hand side, this is clearly not a Fourier series expansion of the integral on the left-hand side. However, the result can be rearranged to be a Fourier series expansion of the function

$$g(t) = \int_{t_1}^{t} f(t)\mathrm{d}t - \tfrac{1}{2}a_0 t$$

Example 12.13 serves to illustrate this process. Note also that the Fourier coefficients in the new Fourier series are $-b_n/n$ and a_n/n, so, from the observations made in Section 12.2.8, the integrated series converges faster than the original series for $f(t)$. If the given function $f(t)$ is piecewise-continuous, rather than continuous, over the interval $-\pi \leqslant t \leqslant \pi$ then care must be taken to ensure that the integration process is carried out properly over the various subintervals. Again, Example 12.14 serves to illustrate this point.

Example 12.13   From Example 12.5, the Fourier series expansion of the function

$$f(t) = t^2 \quad (-\pi \leqslant t \leqslant \pi), \qquad f(t + 2\pi) = f(\pi)$$

is

$$t^2 = \frac{1}{3}\pi^2 + 4\sum_{n=1}^{\infty} \frac{(-1)^n \cos nt}{n^2} \quad (-\pi \leqslant t \leqslant \pi)$$

Integrating this result between the limits $-\pi$ and t gives

$$\int_{-\pi}^{t} t^2 \, dt = \int_{-\pi}^{t} \frac{1}{3}\pi^2 \, dt + 4\sum_{n=1}^{\infty} \int_{-\pi}^{t} \frac{(-1)^n \cos nt}{n^2} \, dt$$

that is,

$$\frac{1}{3}t^3 = \frac{1}{3}\pi^2 t + 4\sum_{n=1}^{\infty} \frac{(-1)^n \sin nt}{n^3} \quad (-\pi \leqslant t \leqslant \pi)$$

Because of the term $\frac{1}{3}\pi^2 t$ on the right-hand side, this is clearly not a Fourier series expansion. However, rearranging, we have

$$t^3 - \pi^2 t = 12\sum_{n=1}^{\infty} \frac{(-1)^n \sin nt}{n^2}$$

and now the right-hand side may be taken to be the Fourier series expansion of the function

$$g(t) = t^3 - \pi^2 t \quad (-\pi \leqslant t \leqslant \pi)$$

$$g(t + 2\pi) = g(t)$$

Example 12.14   Integrate term by term the Fourier series expansion obtained in Example 12.4 for the square wave

$$f(t) = \begin{cases} -1 & (-\pi < t < 0) \\ 1 & (0 < t < \pi) \end{cases}$$

$$f(t + 2\pi) = f(t)$$

illustrated in Figure 12.7.

Solution   From (12.21), the Fourier series expansion for $f(t)$ is

$$f(t) = \frac{4}{\pi} \frac{\sin(2n - 1)t}{2n - 1}$$

We now need to integrate between the limits $-\pi$ and t and, owing to the discontinuity in $f(t)$ at $t = 0$, we must consider separately values of t in the intervals $-\pi < t < 0$ and $0 < t < \pi$.

Case (i), interval $-\pi < t < 0$. Integrating (12.21) term by term, we have

$$\int_{-\pi}^{t} (-1)dt = \frac{4}{\pi} \sum_{n=1}^{\infty} \int_{-\pi}^{t} \frac{\sin(2n-1)t}{(2n-1)} dt$$

that is,

$$-(t + \pi) = -\frac{4}{\pi} \sum_{n=1}^{\infty} \left[\frac{\cos(2n-1)t}{(2n-1)^2} \right]_{-\pi}^{t}$$

$$= -\frac{4}{\pi} \left[\sum_{n=1}^{\infty} \frac{\cos(2n-1)t}{(2n-1)^2} + \sum_{n=1}^{\infty} \frac{1}{(2n-1)^2} \right]$$

It can be shown that

$$\sum_{n=1}^{\infty} \frac{2}{(2n-1)^2} = \frac{1}{8}\pi^2$$

(see Exercises 12.2.9, Question 6), so that the above simplifies to

$$-t = \frac{1}{2}\pi - \frac{4}{\pi} \sum_{n=1}^{\infty} \frac{\cos(2n-1)t}{(2n-1)^2} \quad (-\pi < t < 0) \tag{12.36}$$

Case (ii), interval $0 < t < \pi$. Integrating (12.21) term by term, we have

$$\int_{-\pi}^{0} (-1)dt + \int_{0}^{t} 1 dt = \frac{4}{\pi} \sum_{n=1}^{\infty} \int_{-\pi}^{t} \frac{\sin(2n-1)t}{(2n-1)} dt$$

giving

$$t = \frac{1}{2}\pi - \frac{4}{\pi} \sum_{n=1}^{\infty} \frac{\cos(2n-1)t}{(2n-1)^2} \quad (0 < t < \pi) \tag{12.37}$$

Taking (12.36) and (12.37) together, we find that the function

$$g(t) = |t| = \begin{cases} -t & (-\pi < t < 0) \\ t & (0 < t < \pi) \end{cases}$$

$$g(t + 2\pi) = g(t)$$

has a Fourier series expansion

$$g(t) = |t| = \frac{1}{2}\pi - \frac{4}{\pi} \sum_{n=1}^{\infty} \frac{\cos(2n-1)t}{(2n-1)^2}$$

12.4.2 Differentiation of a Fourier series

Theorem 12.4

If $f(t)$ is a periodic function that satisfies Dirichlet's conditions then its derivative $f'(t)$, wherever it exists, may be found by term-by-term differentiation of the Fourier series of $f(t)$ if and only if the function $f(t)$ is continuous everywhere and the function $f'(t)$ has a Fourier series expansion (that is, $f'(t)$ satisfies Dirichlet's conditions).

end of theorem

It follows from Theorem 12.4 that if the Fourier series expansion of $f(t)$ is differentiable term by term then $f(t)$ must be periodic at the end points of a period (owing to the condition that $f(t)$ must be continuous everywhere). Thus, for example, if we are dealing with a function $f(t)$ of period 2π and defined in the range $-\pi < t < \pi$ then we must have $f(-\pi) = f(\pi)$. To illustrate this point, consider the Fourier series expansion of the function

$$f(t) = t \quad (-\pi < t < \pi)$$

$$f(t + 2\pi) = f(t)$$

which, from Example 12.7, is given by

$$f(t) = 2(\sin t - \tfrac{1}{2}\sin 2t + \tfrac{1}{3}\sin 3t - \tfrac{1}{4}\sin 4t + \ldots)$$

Differentiating term by term, we have

$$f'(t) = 2(\cos t - \cos 2t + \cos 3t - \cos 4t + \ldots)$$

If this differentiation process is valid then $f'(t)$ must be equal to unity for $-\pi < t < \pi$. Clearly this is not the case, since the series on the right-hand side does not converge for any value of t. This follows since the nth term of the series is $2(-1)^{n+1}\cos nt$ and does not tend to zero as $n \to \infty$.

If $f(t)$ is continuous everywhere and has a Fourier series expansion

$$f(t) = \tfrac{1}{2}a_0 + \sum_{n=1}^{\infty}(a_n \cos nt + b_n \sin nt)$$

then, from Theorem 12.4, provided that $f'(t)$ satisfies the required conditions, its Fourier series expansion is

$$f'(t) = \sum_{n=1}^{\infty}(nb_n \cos nt - na_n \sin nt)$$

In this case the Fourier coefficients of the derived expansion are nb_n and na_n, so, in contrast to the integrated series, the derived series will converge more slowly than the original series expansion for $f(t)$.

Example 12.15

Consider the process of differentiating term by term the Fourier series expansion of the function

$$f(t) = t^2 \quad (-\pi \leqslant t \leqslant \pi), \qquad f(t + 2\pi) = f(t)$$

Solution   From Example 12.5, the Fourier series expansion of $f(t)$ is

$$t^2 = \frac{1}{3}\pi^2 + 4\sum_{n=1}^{\infty} \frac{(-1)^n \cos nt}{n^2} \quad (-\pi \leqslant t \leqslant \pi)$$

Since $f(t)$ is continuous within and at the end points of the interval $-\pi \leqslant t \leqslant \pi$, we may apply Theorem 12.4 to obtain

$$t = 2\sum_{n=1}^{\infty} \frac{(-1)^{n+1} \sin nt}{n} \quad (-\pi \leqslant t \leqslant \pi)$$

which conforms with the Fourier series expansion obtained for the function

$$f(t) = t \quad (-\pi < t < \pi), \qquad f(t + 2\pi) = f(t)$$

in Example 12.7.

12.4.3   Exercises

24   Show that the periodic function

$$f(t) = t \quad (-T < t < T)$$
$$f(t + 2T) = f(t)$$

has a Fourier series expansion

$$f(t) = \frac{2T}{\pi}\left(\sin\frac{\pi t}{T} - \frac{1}{2}\sin\frac{2\pi t}{T} + \frac{1}{3}\sin\frac{3\pi t}{T}\right.$$
$$\left. - \frac{1}{4}\sin\frac{4\pi t}{T} + \ldots\right)$$

By term-by-term integration of this series, show that the periodic function

$$g(t) = t^2 \quad (-T < t < T)$$
$$g(t + 2T) = g(t)$$

has a Fourier series expansion

$$g(t) = \frac{1}{3}T^2 - \frac{4T^2}{\pi^2}\left(\cos\frac{\pi t}{T} - \frac{1}{2^2}\cos\frac{2\pi t}{T}\right.$$
$$\left. + \frac{1}{3^2}\cos\frac{3\pi t}{T} - \frac{1}{4^2}\cos\frac{4\pi t}{T} + \ldots\right)$$

(*Hint*: A constant of integration must be introduced; it may be evaluated as the mean value over a period.)

25   The periodic function

$$h(t) = \pi^2 - t^2 \quad (-\pi < t < \pi)$$
$$h(t + 2\pi) = h(t)$$

has a Fourier series expansion

$$h(t) = \frac{2}{3}\pi^2 + 4\left(\cos t - \frac{1}{2^2}\cos 2t\right.$$
$$\left. + \frac{1}{3^2}\cos 3t \ldots\right)$$

By term-by-term differentiation of this series, confirm the series obtained for $f(t)$ in Question 24 for the case when $T = \pi$.

26   (a) Suppose that the derivative $f'(t)$ of a periodic function $f(t)$ of period 2π has a Fourier series expansion

$$f'(t) = \tfrac{1}{2}A_0 + \sum_{n=1}^{\infty} A_n \cos nt + \sum_{n=1}^{\infty} B_n \sin nt$$

Show that

$$A_0 = \frac{1}{n}[f(\pi_-) - f(-\pi_+)]$$
$$A_n = (-1)^n A_0 + nb_n$$
$$B_n = -na_n$$

where a_0, a_n and b_n are the Fourier coefficients of the function $f(t)$.

(b) In Example 12.7 we saw that the periodic function

$$f(t) = t^2 + t \quad (-\pi < t < \pi)$$
$$f(t + 2\pi) = f(t)$$

has a Fourier series expansion

$$f(t) = \tfrac{1}{3}\pi^2 + \sum_{n=1}^{\infty} \frac{4}{n^2}(-1)^n \cos nt$$

$$- \sum_{n=1}^{\infty} \frac{2}{n}(-1)^n \sin nt$$

Differentiate this series term by term, and explain why it is not a Fourier expansion of the periodic function

$$g(t) = 2t + 1 \quad (-\pi < t < \pi)$$

$$g(t + 2\pi) = g(t)$$

(c) Use the results of (a) to obtain the Fourier series expansion of $g(t)$ and confirm your solution by direct evaluation of the coefficients using Euler's formulae.

12.5   Engineering application: analysis of a slider–crank mechanism

Figure 12.24(a) represents a slider–crank mechanism. The crank OP rotates about O, and P is connected to A, which is constrained so that it slides along OQ, A special case when OP = 1 m and PA = 3 m is shown in Figure 12.24(b). The distance OA is x when the angle AOP is θ and is given by

$$x(\theta) = \cos\theta + \sqrt{(9 - \sin^2\theta)}$$

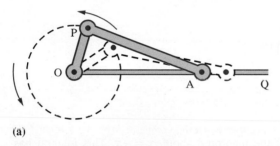

(a)   (b)

Figure 12.24 Slider–crank mechanism.

It is clear both from the basic geometry of the mechanism and from the formula for x that $x(\theta)$ is an even periodic function. This implies it can be represented by a Fourier series of the form

$$x(\theta) = a_0 + a_1 \cos\theta + a_2 \cos 2\theta + a_3 \cos 3\theta + \dots$$

Such mechanisms usually form parts of larger pieces of equipment, so it is important to know the sizes of the coefficients a_k during the design process to avoid dangerous motions due to resonance.

The process of obtaining the values of the coefficients a_k is called harmonic analysis. Truncating the Fourier series, we can obtain an approximation to $x(\theta)$ in the form

$$x(\theta) \approx a_0 + a_1 \cos\theta + a_2 \cos 2\theta + a_3 \cos 3\theta + a_4 \cos 4\theta$$

We wish to determine the values of $a_0, \dots, a_4$ so that we obtain the best approximation possible. We achieve this by choosing $a_0, \dots, a_4$ in such a way that the total squared error over a complete period is a minimum. Because $x(\theta)$ is an even function, this simplifies to finding the values of $a_0, \dots, a_4$ that minimize the integral

$$I(a_0, a_1, a_2, a_3, a_4) = \int_0^\pi [a_0 + a_1 \cos\theta + a_2 \cos 2\theta + a_3 \cos 3\theta + a_4 \cos 4\theta - x(\theta)]^2 \, d\theta$$

Thus we want $a_0, \ldots, a_4$ such that

$$\frac{\partial I}{\partial a_k} = 0 \quad (k = 0, 1, \ldots, 4)$$

Taking the case $k = 0$, this yields

$$\int_0^\pi 2[a_0 + a_1 \cos\theta + a_2 \cos 2\theta + a_3 \cos 3\theta + a_4 \cos 4\theta - x(\theta)] d\theta = 0$$

which reduces to

$$\int_0^\pi [a_0 - x(\theta)] d\theta = 0$$

on using the integration properties of $\cos k\theta$ on $(0, \pi)$ for $k = 1, \ldots, 4$. Thus

$$\int_0^\pi a_0 \, d\theta = \int_0^\pi x(\theta) d\theta$$

giving

$$a_0 = \frac{1}{\pi} \int_0^\pi [\cos\theta + \sqrt{(9 - \sin^2\theta)}] d\theta$$

$$= \frac{2}{\pi} \int_0^{\pi/2} \sqrt{(9 - \sin^2\theta)} d\theta$$

on using the symmetry properties of the integrand about $x = \frac{1}{2}\pi$. This integral has to be evaluated numerically, and, using the trapezium rule, we obtain the value $a_0 = 2.9148$.

Similarly, $\partial I / \partial a_1 = 0$ gives

$$\int_0^\pi 2[a_0 + a_1 \cos\theta + a_2 \cos 2\theta + a_3 \cos 3\theta + a_4 \cos 4\theta - x(\theta)] \cos\theta \, d\theta = 0$$

which reduces to

$$\int_0^\pi [a_1 \cos^2\theta - x(\theta) \cos\theta] d\theta = 0$$

Thus

$$\int_0^\pi a_1 \cos^2\theta \, d\theta = \int_0^\pi x(\theta) \cos\theta \, d\theta$$

which gives

$$\tfrac{1}{2}\pi a_1 = \int_0^\pi [\cos^2\theta + \cos\theta\sqrt{(9 - \sin^2\theta)}]d\theta = \tfrac{1}{2}\pi$$

on using the symmetry properties of the integrand. Thus $a_1 = 1$.

Continuing in the same fashion, we obtain

$$\int_0^\pi a_2 \cos^2 2\theta\, d\theta = \int_0^\pi x(\theta)\cos 2\theta\, d\theta$$

$$\int_0^\pi a_3 \cos^2 3\theta\, d\theta = \int_0^\pi x(\theta)\cos 3\theta\, d\theta$$

and

$$\int_0^\pi a_4 \cos^2 4\theta\, d\theta = \int_0^\pi x(\theta)\cos 4\theta\, d\theta$$

from which we deduce

$$a_2 = \frac{2}{\pi}\int_0^\pi \cos 2\theta\sqrt{(9 - \sin^2\theta)}d\theta$$

$$a_3 = \frac{2}{\pi}\int_0^\pi \cos 3\theta\sqrt{(9 - \sin^2\theta)}d\theta = 0$$

$$a_4 = \frac{2}{\pi}\int_0^\pi \cos 4\theta\sqrt{(9 - \sin^2\theta)}d\theta$$

Calculating the integrals for a_2 and a_4 numerically, we obtain the 'least-squares approximation' for $x(\theta)$ in the form

$$x(\theta) \approx 2.9148 + \cos\theta + 0.0858\cos 2\theta - 0.0006\cos 4\theta$$

This example can be generalized to show that the truncated Fourier series provides the 'best' approximation to a periodic function.

12.6   Review exercises (1–21)

1   A periodic function $f(t)$ is defined by

$$f(t) = \begin{cases} t^2 & (0 \leqslant t < \pi) \\ 0 & (\pi < t \leqslant 2\pi) \end{cases}$$

$$f(t + 2\pi) = f(t)$$

Obtain a Fourier series expansion of $f(t)$ and deduce that

$$\tfrac{1}{6}\pi^2 = \sum_{r=1}^\infty \frac{1}{r^2}$$

2   Determine the full-range Fourier series expansion of the even function $f(t)$ of period 2π defined by

$$f(t) = \begin{cases} \tfrac{2}{3}t & (0 \leqslant t \leqslant \tfrac{1}{3}\pi) \\ \tfrac{1}{3}(\pi - t) & (\tfrac{1}{3}\pi \leqslant t \leqslant \pi) \end{cases}$$

To what value does the series converge at $t = \tfrac{1}{3}\pi$?

3   A function $f(t)$ is defined for $0 \leqslant t \leqslant \tfrac{1}{2}T$ by

$$f(t) = \begin{cases} t & (0 \leqslant t \leqslant \tfrac{1}{4}T) \\ \tfrac{1}{2}T - t & (\tfrac{1}{4}T \leqslant t \leqslant \tfrac{1}{2}T) \end{cases}$$

Sketch odd and even functions that have a period T and are equal to $f(t)$ for $0 \leqslant t \leqslant \frac{1}{2}T$.

(a) Find the half-range Fourier sine series of $f(t)$.

(b) To what value will the series converge for $t = -\frac{1}{4}T$?

(c) What is the sum of the following series?

$$S = \sum_{r=1}^{\infty} \frac{1}{(2r-1)^2}$$

4 The magnetomotive force, y, in the air gap of an alternator can be represented approximately by a graph of the form shown in Figure 12.25. Find a Fourier series for y, explaining beforehand, with reasons, any special characteristics you would expect to find.

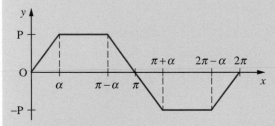

Figure 12.25 Data for $f(t)$ in Question 4.

5 Prove that if $g(x)$ is an odd function and $f(x)$ an even function of x, the product $g(x)[c + f(x)]$ is an odd function if c is a constant.

A periodic function with period 2π is defined by

$$F(\theta) = \frac{1}{12}\theta(\pi^2 - \theta^2)$$

in the interval $-\pi \leqslant \theta \leqslant \pi$. Show that the Fourier series representation of the function is

$$F(\theta) = \sum_{n=1}^{\infty} \frac{(-1)^{n+1}}{n^3} \sin n\theta$$

6 A repeating waveform of period 2π is described by

$$f(t) = \begin{cases} \pi + t & (-\pi \leqslant t \leqslant -\frac{1}{2}\pi) \\ -t & (-\frac{1}{2}\pi \leqslant t \leqslant \frac{1}{2}\pi) \\ t - \pi & (\frac{1}{2}\pi \leqslant t \leqslant \pi) \end{cases}$$

Sketch the waveform over the range $t = -2\pi$ to $t = 2\pi$ and find the Fourier series representation of $f(t)$, making use of any properties of the waveform that you can identify before any integration is performed.

7 A function $f(x)$ is periodic of period 2π and is defined by

$$f(x) = \begin{cases} -2x & (-\pi < x \leqslant 0) \\ 2x & (0 < x \leqslant \pi) \end{cases}$$

Sketch a graph of $f(x)$ from -2π to 3π and prove that

$$f(x) = \pi - \frac{8}{\pi} \sum_{n=0}^{\infty} \frac{1}{(2n+1)^2} \cos(2n+1)x$$

Hence show that

$$\frac{1}{8}\pi^2 = 1 + \sum_{n=1}^{\infty} \frac{1}{(2n+1)^2}$$

8 A function $f(x)$ of period 2π is defined in the interval $-\pi \leqslant x \leqslant \pi$ by

$$f(x) = \begin{cases} \frac{1}{2}\pi + x & (-\pi \leqslant x \leqslant 0) \\ \frac{1}{2}\pi - x & (0 \leqslant x \leqslant \pi) \end{cases}$$

Sketch a graph of $f(x)$ over the interval $-3\pi \leqslant x \leqslant 3\pi$. Express $f(x)$ as a Fourier series and from this deduce a numerical series for π.

9 A periodic function of period 2π is defined for $0 \leqslant x \leqslant 2\pi$ by

$$f(x) = \begin{cases} x & (0 \leqslant x \leqslant \frac{1}{2}\pi) \\ \frac{1}{2}\pi & (\frac{1}{2}\pi < x \leqslant \pi) \\ -\frac{1}{2}\pi & (\pi < x \leqslant \frac{3}{2}\pi) \\ x - 2\pi & (\frac{3}{2}\pi \leqslant x \leqslant 2\pi) \end{cases}$$

Sketch $f(x)$ for $-2\pi \leqslant x \leqslant 4\pi$ and show that its Fourier series representation is

$$f(x) = \left(1 + \frac{2}{\pi}\right)\sin x - \frac{1}{2}\sin 2x$$
$$+ \frac{1}{3}\left(1 - \frac{2}{3\pi}\right)\sin 3x - \frac{1}{4}\sin 4x + \dots$$

Express this series in a general form.

10 A waveform is defined by $V(t) = 10e^{-3t}$ for $0 \leqslant t < 0.4$ and $V(t) = V(t - 0.4)$ for all t. Sketch the graphs of V, dV/dt and $\int_0^t V \, dt$.

Express V as a Fourier series and show that the amplitude of the nth harmonic is about $2.22/n$.

11 A function $f(x)$ is defined in the interval $-1 \leqslant x \leqslant 1$ by

$$f(x) = \begin{cases} 1/2\varepsilon & (-\varepsilon < x < \varepsilon) \\ 0 & (-1 \leqslant x < -\varepsilon; \; \varepsilon < x \leqslant 1) \end{cases}$$

Sketch a graph of $f(x)$ and show that a Fourier series expansion of $f(x)$ valid in the interval $-1 \leqslant x \leqslant 1$ is given by

$$f(x) = \tfrac{1}{2} + \sum_{n=1}^{\infty} \frac{\sin n\pi\varepsilon}{n\pi\varepsilon} \cos n\pi x$$

12 Show that the half-range Fourier sine series for the function

$$f(t) = \left(1 - \frac{t}{\pi}\right)^2 \quad (0 \leqslant t \leqslant \pi)$$

is

$$f(t) = \sum_{n=1}^{\infty} \frac{2}{n\pi}\left\{1 - \frac{2}{n^2\pi^2}[1-(-1)^n]\right\} \sin nt$$

13 Find a half-range Fourier sine and Fourier cosine series for $f(x)$ valid in the interval $0 < x < \pi$ when $f(x)$ is defined by

$$f(x) = \begin{cases} x & (0 \leqslant x \leqslant \tfrac{1}{2}\pi) \\ \pi - x & (\tfrac{1}{2}\pi \leqslant x \leqslant \pi) \end{cases}$$

Sketch the graph of the Fourier series obtained for $-2\pi < x \leqslant 2\pi$.

14 A function $f(x)$ is periodic of period 2π and is defined by $f(x) = e^x \; (-\pi < x < \pi)$. Sketch the graph of $f(x)$ from $x = -2\pi$ to $x = 2\pi$ and prove that

$$f(x) = \frac{2\sinh\pi}{\pi}\left[\frac{1}{2} + \sum_{n=1}^{\infty} \frac{(-1)^n}{1+n^2}(\cos nx - n\sin nx)\right]$$

15 A function $f(t)$ is defined on $0 < t < \pi$ by

$$f(t) = \pi - t$$

Find

(a) a half-range Fourier sine series, and

(b) a half-range Fourier cosine series

for $f(t)$ valid for $0 < t < \pi$.
   Sketch the graphs of the functions represented by each series for $-2\pi < t < 2\pi$.

16 A periodic function $f(t)$ of period 2 is defined in the interval $-1 < t < 1$ by

$$f(t) = 1 - t^2$$

Sketch a graph of $f(t)$ for $-3 < t < 3$ and obtain a Fourier series expansion for it.

17 (a) Without actually finding the series state what terms you would expect to find in the Fourier series for the following periodic functions of period 2π.

   (i) $f(t) = \sin^2 t, \quad -\pi \leqslant t \leqslant \pi$

   (ii) $f(t) = 3e^{-t}, \quad -\pi \leqslant t \leqslant \pi$

   (iii) $f(t) = \begin{cases} 0, & -\pi < t < 0 \\ 1, & 0 < t < \pi \end{cases}$

(b) Find, up to and including the term in $\cos 4t$, the Fourier half-range cosine series for the function defined by

$$f(t) = \begin{cases} t^2, & 0 < t < \pi/2 \\ 0, & \pi/2 < t < \pi \end{cases}$$

18 (a) A periodic function $f(t)$, of period 2π, is defined in $-\pi \leqslant t \leqslant \pi$ by

$$f(t) = \begin{cases} -t & (-\pi \leqslant t \leqslant 0) \\ t & (0 \leqslant t \leqslant \pi) \end{cases}$$

Obtain a Fourier series expansion for $f(t)$.

(b) By formally differentiating the series obtained in (a), obtain the Fourier series expansion of the periodic square wave

$$g(t) = \begin{cases} -1 & (-\pi < t < 0) \\ 0 & (t = 0) \\ 1 & (0 < t < \pi) \end{cases}$$

$$g(t + 2\pi) = g(t)$$

Check the validity of your result by determining directly the Fourier series expansion of $g(t)$.

19 The periodic waveform $f(t)$ shown in Figure 12.26 may be written as

$$f(t) = 1 + g(t)$$

where $g(t)$ represents an odd function.

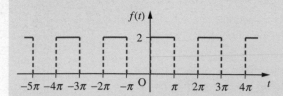

Figure 12.26 Waveform $f(t)$ of Question 19.

(a) Sketch the graph of $g(t)$.

(b) Obtain the Fourier series expansion for $g(t)$, and hence write down the Fourier series expansion for $f(t)$.

20 Show that the Fourier series

$$\frac{1}{2}\pi - \frac{4}{\pi}\sum_{n=1}^{\infty}\frac{\cos(2n-1)t}{(2n-1)^2}$$

represents the function $f(t)$, of period 2π, given by

$$f(t) = \begin{cases} t & (0 \leqslant t \leqslant \pi) \\ -t & (-\pi \leqslant t \leqslant 0) \end{cases}$$

Deduce that, apart from a transient component (that is, a complementary function that dies away as $t \to \infty$), the differential equation

$$\frac{\mathrm{d}x}{\mathrm{d}t} + x = f(t)$$

has the solution

$$x = \frac{1}{2}\pi - \frac{4}{\pi}\sum_{n=1}^{\infty}\frac{\cos(2n-1)t + (2n-1)\sin(2n-1)t}{(2n-1)^2[1+(2n-1)^2]}$$

21 Show that if $f(t)$ is a periodic function of period 2π and

$$f(t) = \begin{cases} t/\pi & (0 < t < \pi) \\ (2\pi - t)/\pi & (\pi < t < 2\pi) \end{cases}$$

then

$$f(t) = \frac{1}{2} - \frac{4}{\pi^2}\sum_{n=0}^{\infty}\frac{\cos(2n+1)t}{(2n+1)^2}$$

Show also that, when ω is not an integer,

$$y = \frac{1}{2\omega^2}(1 - \cos\omega t)$$

$$- \frac{4}{\pi^2}\sum_{n=1}^{\infty}\frac{\cos(2n+1)t - \cos\omega t}{(2n+1)^2[\omega^2 - (2n+1)^2]}$$

satisfies the differential equation

$$\frac{\mathrm{d}^2y}{\mathrm{d}t^2} + \omega^2 y = f(t)$$

subject to the initial conditions $y = \mathrm{d}y/\mathrm{d}t = 0$ at $t = 0$.

13 Data Handling and Probability Theory

Chapter 13 Contents

13.1 Introduction

Many events in our lives are subject to chance – by which we mean that they are not entirely predictable. To some extent, we can choose where we live and what sort of work we do, but even so we cannot be sure what sort of neighbours or workmates we shall have: noisy, generous, friendly and so on. In a similar way, experiments in all branches of science and engineering involve unpredictable outcomes that may be expressed either as a quality such as 'turned green' or 'exploded', or numerically in terms of mass, resistance or any standard unit. In contrast with everyday life, an 'experiment' is repeated many times, so that the limited predictability of the various outcomes can emerge as a pattern within the disorder. The subject of statistics is about extracting that pattern and drawing useful conclusions from it, and the theoretical foundation for this is contained in the theory of probability.

Engineers, in particular, are immersed in data throughout their working lives. The term 'data' is used rather loosely to refer to numerical information of all kinds, including for example the specification of a machine or part. For our present purposes, however, we shall use **data** to refer to the set of measured outcomes of an experiment. Engineering is a discipline founded upon experiment, and engineers need to know how to process their experimental data and how to assess the results of others' experiments.

The aim of statistics is to extract useful information from the data. This information can take many forms. If the aim of an experiment is to assist with the making of a decision then the people conducting the experiment will have in mind a question to which they would like the answer, and ideally the question (and its possible answers) will be expressed as simply and clearly as possible in ordinary language. On the other hand, the aim of an experiment may be to calibrate an instrument or to measure some unknown quantity, in which case the conclusions of the experiment will be numerical.

Sometimes all that is needed is to plot the data in a suitable way that makes the message clear. The information is then conveyed in graphical form to the reader. Unfortunately, it very often turns out that the data is rather ambiguous, the conclusions are not obvious and the data must be analysed in a more mathematical way. In this case the conclusion (which relates directly to the purpose of the experiment) cannot be stated with 100% confidence. This issue is taken up in the companion text *Advanced Modern Engineering Mathematics*, where the mathematical methods of statistics are introduced. In the present chapter we shall see how the data may be plotted to good effect, and then go on to cover the essential probability theory without which the proper statistical practices (in engineering and elsewhere) would be impossible.

13.2 The raw material of statistics

13.2.1 Experiments and sampling

A statistician requires data to work on, and data is usually obtained by experiment – but not any old experiment will do. The most common type of statistical experiment involves taking a **sample** from a **population** and drawing some conclusions about the whole population from the results for the sample. In general, in statistical work the population that is the object of study is assumed to be very large and rather uniform with respect to certain characteristics of interest. The sample that is drawn for investigation is much smaller. The size of the sample governs the confidence with which statements about the characteristics of the population can be made.

Ideally, the entire population would be studied, but this may be impractical for reasons of expense, ethics or destructiveness of tests:

(a) *Expense*: the population may be too large or the cost per individual may be high.

(b) *Ethics*: in medical experiments involving animals or people the aim is to use the smallest sample size that is compatible with obtaining a dependable result.

(c) *Destructiveness*: destructive testing of components, for example breaking stress or lifetime, obviously precludes using the whole population.

The quality of the sample is also important. Imagine an opinion poll in which all the people interviewed were professional engineers. The results would be of interest to someone investigating the voting intentions of this particular group, but such a poll might be a poor indicator of the result of the next general election. Now imagine an opinion poll conducted in a large hall, with a microphone passed from person to person. The intimidating nature of this situation would prevent many respondents from giving a truthful answer, particularly if the poll involved politically, socially or morally sensitive issues. These two examples demonstrate the fundamental requirements of any sampling experiment, including an opinion poll: the sample must be **representative** and successive observations must be **independent**.

13.2.2 Histograms of data

After gathering the data together, the first step is often to display it graphically using a histogram or pie chart. Computer packages are often very useful for this. For example, Figure 13.1 contains some data describing the performance of two prototype car

Figure 13.1
Car engine test data.

	Engine A				Engine B			
Time	*Temp.*	*Time*	*Temp.*	*Time*	*Temp.*	*Time*	*Temp.*
27.7	24	24.1	7	24.9	13	24.3	17
24.3	25	23.1	14	21.4	19	24.5	16
23.7	18	23.4	16	24.1	18	26.1	18
22.1	15	23.1	9	27.5	19	27.7	14
21.8	19	24.1	14	27.5	21	24.3	19
24.7	16	28.6	23	25.7	17	26.1	5
23.4	17	20.2	14	24.9	17	24.0	17
21.6	14	25.7	18	23.3	19	24.9	18
24.5	18	24.6	18	22.5	21	26.7	23
26.1	20	24.0	12	28.5	12	27.3	28
24.8	15	24.9	18	25.9	17	23.9	18
23.7	15	21.9	20	26.9	13	23.1	10
25.0	22	25.1	16	27.7	17	25.5	25
26.9	18	25.7	16	25.4	23	24.9	22
23.7	19	23.5	11	25.3	30	25.9	16

engines: a series of running times (in minutes at constant speed on 1 litre of standard fuel) and ambient temperatures at the times of the tests, for each engine. Two questions that are easy to state, and which might be answerable from this data, are

(a)   Is the fuel consumption of one engine different from that of the other?
(b)   Does fuel consumption depend upon ambient temperature?

These questions are actually related, as can be seen in *Advanced Modern Engineering Mathematics*, where this example is discussed at some length. For the moment, we shall see what can be learned just by plotting the data.

The first thing to observe is that the measured running times are rather erratic, even taking temperature into account. The six tests of engine A at 18°C produced results ranging from 23.7 min to 26.9 min. This situation is typical, and is not necessarily the result of sloppy experimental practice or inaccurate equipment (though such failings should not be condoned where they are easily avoided). There are practical limitations on the design and conduct of experiments that preclude making measurements to ultimate precision, and mean that certain causal factors that might influence the results are not measured at all. In this series of engine tests the actual quantity of fuel would have varied a little around 1 litre, the condition of the engine oil would have been different from one test to another, and so on.

Figures 13.2(a) and (b) are **histograms** of the running times for engines A and B respectively. The data has been grouped into classes, and the height of each bar indicates the number in the class. Values falling on a boundary are counted in the upper class. The width of each bar is the same, and is chosen to reveal the overall shape of the data. A histogram with too many small classes is very erratic, whereas one with too few large classes has no structure. It is typical for a histogram to span the data with about eight to ten classes.

Figure 13.2
Histograms of running times: (a) engine A; (b) engine B; (c) superimposed.

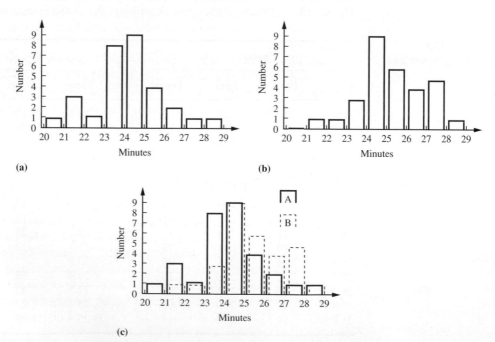

Figure 13.3
Histograms of
temperatures: (a)
engine A; (b) engine
B; (c) superimposed.

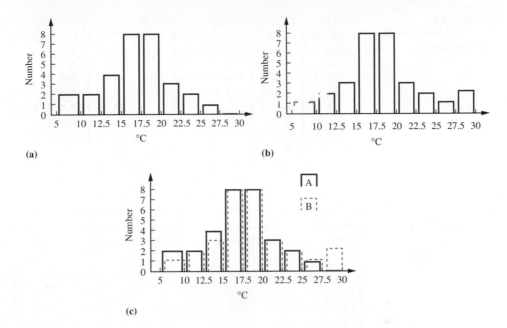

(a)

(b)

(c)

Figure 13.2(c) shows the two histograms superimposed. It is fairly clear that there is a difference here, and that the running times for engine B tend to be longer than those for engine A. However, just from the histograms, it is difficult to be precise about the amount of the difference, or to assess the confidence with which one could state that a difference exists.

Figure 13.3 contains corresponding histograms for the temperatures. This time the results are much more similar. It is easy to imagine that if a relatively small subset of the sample had given different results from those obtained then there would have been no difference at all between the histograms. This difference could therefore be attributed to chance.

Somewhere in between these two situations is one for which a difference just exists but is not obvious. It is in dealing with this type of situation (which is quite common) that the powerful mathematical discipline of statistics is important. No analysis of the data will definitely settle the question of whether or not the populations differ, but the extent of the evidence for a difference can be assessed, and this may be invaluable if a decision has to be made.

13.2.3 Alternative types of plot

Histograms are the most common types of data plot, but there are many others. For example, when data is grouped into classes, there is inevitably some loss of information. This can be avoided by using a **stem-and-leaf plot**, which is similar to a histogram except that the individual values are retained. The idea is for the leading digits in the sample values to form a **stem** (one per class), with the remaining digits entering the bar as a **leaf**. The length of the bar is simply the number of sample values with that stem. Figure 13.4 contains stem-and-leaf plots for the running time data for engines A and B (Figure 13.1). The * in the stem shows where the leaf digit goes. The similarity of these plots to the histograms in Figure 13.2 is clear.

Figure 13.4
Stem-and-leaf plots of running times.

A:

20. *	2		1
21. *	8 6 9		3
22. *	1		1
23. *	7 4 7 7 1 4 1 5		8
24. *	3 7 5 8 1 1 6 0 9		9
25. *	0 7 1 7		4
26. *	1 9		2
27. *	7		1
28. *	6		1

B:

20. *			0
21. *	4		1
22. *	5		1
23. *	3 9 1		3
24. *	9 1 9 3 5 3 0 9 9		9
25. *	7 9 4 3 5 9		6
26. *	9 1 1 7		4
27. *	5 5 7 7 3		5
28. *	5		1

Figure 13.5
Cumulative percentages for temperature data.

Class range	Number of observations	Cumulative number at upper boundary	Cumulative percentage
0–9.9	2	2	6.7
10.0–12.4	2	4	13.3
12.5–14.9	4	8	26.7
15.0–17.4	8	16	53.3
17.5–19.9	8	24	80.0
20.0–22.4	3	27	90.0
22.5–24.9	2	29	96.7
25.0–27.4	1	30	100.0

The main disadvantages of stem-and-leaf plots are that they are less suitable for large samples, they are more difficult to superimpose to detect differences, and there is less flexibility in choosing the classes. The stem can be split or several stems conjoined, but the main constraint is that ten is divisible only by two and five. (Try drawing a stem-and-leaf plot for the temperature data in Figure 13.1.)

Another useful device is the **cumulative percentage plot**, which shows for any value what proportion of observations were less than that value. This can be drawn up from the original data, but is more easily inferred from a histogram of classes by successively adding the class sizes and dividing by the total number of observations. Figure 13.5 contains such a table for the temperature data for engine A in Figure 13.1. The cumulative percentage is plotted in Figure 13.6(a). The S shape is typical for plots like this. Sometimes a special kind of graph paper is used for which the probability scale is nonlinear, as shown in Figure 13.6(b). This is known as **normal probability paper**, and is useful for testing whether the data has a particularly important kind of profile called the normal distribution, which will be introduced in Section 13.5.3. If the data is normal, the plot should fit a straight line.

The presentation of data (often using sophisticated graphics) is very important in communicating results, and is often referred to as **descriptive statistics**. In *Advanced Modern Engineering Mathematics* we introduce **inferential statistics**. This means using mathematical methods to analyse the data with a view to answering certain important questions. Inferential statistics is more powerful than descriptive statistics because of the capacity to extract conclusions and quote the confidence with which they are asserted. In the rest of this chapter the necessary theory of probability will be covered so that the statistical methods can be built upon it.

Figure 13.6
Cumulative percentage
plots: (a) linear scale;
(b) normal scale.

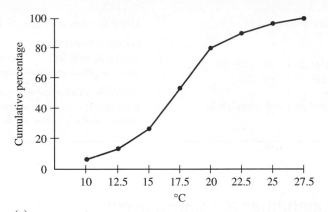

(a)

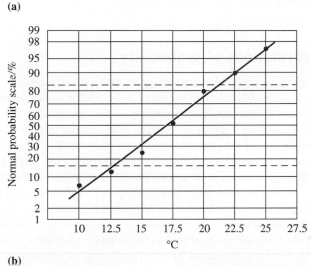

(b)

13.2.4 Exercises

1 A sample of 52 spoken sentences have the
following lengths in words:

7, 3, 8, 6, 10, 6, 2, 9, 5, 8, 2, 7, 1, 8, 5, 4,
12, 9, 3, 6, 2, 8, 2, 10, 7, 4, 11, 9, 8, 2, 6, 1,
3, 11, 7, 8, 1, 4, 2, 9, 7, 3, 8, 5, 1, 9, 2, 11,
6, 7, 3, 8

Draw a histogram of the lengths from 1 to 12
words. What do you notice about this histogram?

2 The following data consists of percentage marks
achieved by students sitting an examination:

47, 51, 75, 58, 70, 73, 63, 60, 60, 54, 60,
67, 50, 60, 74, 69, 51, 67, 49, 66, 61, 46,

66, 57, 55, 60, 62, 36, 52, 67, 62, 51, 62, 62,
59, 52, 75, 44, 75, 56, 52, 64, 63, 59, 54, 57,
68, 53, 43, 64, 39, 58, 68, 66, 72, 46, 58, 52,
50, 45

Draw histograms with (a) class boundaries at
intervals of five, and (b) your own choice of
class boundaries.

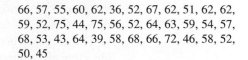

3 Construct stem-and-leaf plots for the data in
Question 2: (a) using * as a placeholder for the
second digit, and (b) using * as a placeholder for
0, 1, 2, 3 and 4 in the second digit and + as a
placeholder for 5, 6, 7, 8 and 9.

4 Figures for a well's daily production of oil in barrels are as follows:

$$214, 203, 226, 198, 243, 225, 207, 203, 208,$$
$$200, 217, 202, 208, 212, 205, 220$$

Construct a stem-and-leaf plot with stem labels $19*, 20*, \ldots, 24*$.

 5 Using the data in Figure 13.1,

 (a) draw two histograms of temperatures for engine A, first with class boundaries at even numbers, then with boundaries at multiples of five;

(b) draw a cumulative percentage plot for the running time data for engine A and compare it with a similar plot for engine B.

13.3 Probabilities of random events

13.3.1 Interpretations of probability

The theory of probability underlies the methods of inference used in statistical situations, and the concept of probability can be related to the histogram of data. The height of each bar determines the proportion of the sample that fell into the corresponding class. One way to think of probability is to assume that as a larger and larger sample is taken (ignoring the practical objections raised in Section 13.2.1), the histogram will stabilize and the class proportions will converge to the 'true' probability figures. This concept of probability is of an objective quantity that applies to each observation and measures (in a relative way) how likely it is to fall into the corresponding class. Like the speed of sound and the density of gold, it is known only imperfectly because of our limited capacity to do experiments.

An alternative concept of probability that is important in decision-making and expert systems involves **degree of belief**. This is highly subjective, because it will depend upon the individual (or group) concerned and will vary with past experience. This seems unscientific at first sight, and there is much resistance to this notion, but there are many situations where experiments are unrepeatable in principle and no 'large-sample proportion' approach is applicable. The outcome of an election is uncertain, and it is not unreasonable to say that some outcomes are 'more probable' than others, but the actual election can take place only once. It seems that one is forced into a subjective view of the uncertainties, but the probability figures that emerge must obey certain rules in order to be consistent. Advocates of subjective probability have shown that these rules are the same as those obeyed by the sample proportions.

The formal theory of probability admits a number of 'interpretations', of which these objective and subjective interpretations are by far the most important. For engineering students it is most appropriate to keep the first interpretation – that of probability as an idealized proportion – in mind when studying the theory.

13.3.2 Sample space and events

The first step is to introduce some terminology that allows us to be clear in describing what is observed in an experiment. The language used is that of set theory, introduced in Chapter 6, because this provides a natural way of describing how observed events combine and separate.

Let the set of all possible outcomes of an experiment be called the **sample space** and denote it by S. An **event** is any subset of S. One (initially unspecified) outcome is considered to be the **actual outcome** and an event is said to **occur** if it contains the actual outcome.

Example 13.1

If the experiment is to roll an ordinary six-faced die and observe the numerical value of the outcome then the sample space will be the set $\{1, 2, 3, 4, 5, 6\}$. The event 'the outcome is an even number' will be represented by the set $\{2, 4, 6\}$. If the actual outcome is a 5 then the event 'the outcome is an even number' has not occurred, because $5 \notin \{2, 4, 6\}$.

Example 13.2

If the experiment is to toss a fair coin twice then the sample space may be represented as the set $\{HH, HT, TH, TT\}$. The event 'both tosses yield the same result' is represented by the set $\{HH, TT\}$. If the actual outcome is TT then the event 'both tosses yield the same result' has occurred, because $TT \in \{HH, TT\}$.

The sample space S therefore contains everything that can occur, and may be discrete or continuous. It is **discrete** if the possible outcomes can be written as a list: for instance, for somebody's birthday $S = \{1\text{ January}, 2\text{ January}, \ldots, 31\text{ December}\}$. The list does not need to be finite. **Continuous** sample spaces arise when experiments involve measurements of some continuous variable such as a person's height or the voltage in a circuit. Then brackets (rather than curly brackets) are used to denote an open interval such as $S = (0, 10)$, and square brackets to denote a closed interval (see Section 1.2). Of course we can only measure to limited precision in practice, so the possible values could be listed, but this is a rather arbitrary technical limitation.

Events are what we observe, in general. It is not always necessary – and for a continuous observation it is impossible – to know the actual outcome. If you want to send someone a birthday card then it is sufficient to know that her birthday occurs 'about the end of June'. Events range from S itself (the certain event) to the empty set $\emptyset$ (the impossible event). Most interesting events are in-between: neither certain nor impossible, but with a reasonable chance of containing the actual outcome.

The usual operations of set theory (Section 6.2) apply to events. If A and B are events then so are the following:

(a) *union*: $A \cup B$ corresponding to 'A or B occurs';
(b) *intersection*: $A \cap B$ corresponding to 'A and B occur';
(c) *complement*: $S - A$ corresponding to 'not-A occurs'.

The complement of A is also written $\bar{A}$.

Using this language, it is possible to describe situations in which at least one of two events occurs (union), or where both events occur (intersection), or where an event fails to occur (complement). Since so much of our everyday experience is structured in this way, this should be a fairly natural starting-point for the theory.

13.3.3 Axioms of probability

The next step is to associate a real number $P(A)$ with each event $A \subseteq S$, called the **probability** of that event. (Strictly speaking, assigning probabilities to arbitrary subsets of a continuous sample space is not possible, but there are ways around this and we shall ignore this technical limitation here.) These numbers must satisfy the rules prompted by the interpretations discussed in Section 13.3.1. The following three rules are referred to as the **axioms of probability**, and lay the foundation for the whole theory:

(1) The certain event S has probability one: $P(S) = 1$.
(2) All probabilities are non-negative: $P(A) \geqslant 0$.
(3) Addition rule: if A and B are disjoint events (so that $A \cap B = \varnothing$) then

$$P(A \cup B) = P(A) + P(B)$$

If probability is regarded as an idealized proportion then clearly its maximum value must be one, it must be non-negative, and the addition rule describes how proportions behave in exclusive situations: for instance, if 5% of units of a brand of power supply produce a voltage that is too low and 8% produce a voltage that is too high then the proportion that produces a voltage that is either too low or too high must be 13%. The three axioms are therefore exactly what we should intuitively expect. What is remarkable is that they are also sufficient. Further rules of probability follow from the axioms:

(4) Complement rule: $P(S - A) = 1 - P(A)$.
(5) $P(\varnothing) = 0$.
(6) If $A \subseteq B$ then $P(A) \leqslant P(B)$.
(7) General addition rule:

$$P(A \cup B) = P(A) + P(B) - P(A \cap B)$$

The complement rule (4) follows immediately from axioms (1) and (3) using the fact that a set does not intersect with its complement:

$$P(A \cup \bar{A}) = P(S) = 1 = P(A) + P(\bar{A})$$

The general addition rule (7) can be illustrated using a Venn diagram as in Figure 13.7. Imagine the probability as a unit mass, spread out (unevenly) over S. Because of the overlap between A and B, adding their probabilities makes the probability of the intersection contribute twice to the total, so this has to be subtracted to compensate.

Figure 13.7
Venn diagram illustrating general addition rule.

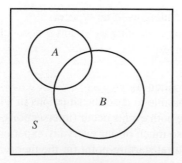

Example 13.3   A fair six-sided die is tossed. Find the probability of the event 'even number or number less than four'.

Solution   The sample space is $S = \{1, 2, 3, 4, 5, 6\}$, with all values equally likely. Since $P(1) + P(2) + P(3) + P(4) + P(5) + P(6)$ must sum to one, we must have

$$P(1) = P(2) = \ldots = P(6) = \tfrac{1}{6}$$

Figure 13.8
Sample space for
Example 13.3.

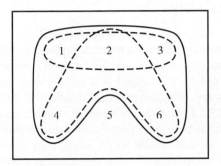

The various events are shown in Figure 13.8. Using the general addition rule,

$$P(\text{even or less than four}) = P(\text{even}) + P(\text{less than four})$$

$$- P(\text{even number less than four})$$

$$= P(\{2, 4, 6\}) + P(\{1, 2, 3\}) - P(\{2\})$$

$$= \tfrac{1}{2} + \tfrac{1}{2} - \tfrac{1}{6} = \tfrac{5}{6}$$

This result also follows from the complement rule:

$$P(\{1, 2, 3, 4, 6\}) = P(\text{not } \{5\}) = 1 - P(\{5\}) = \tfrac{5}{6}$$

Example 13.4   During the assessment of a class of students, 80% passed the examination in mathematics, 85% passed in laboratory work, and 75% passed both. For a student chosen at random from the class, find the probabilities that the student

(a) passed in either mathematics or laboratory work,

(b) passed in mathematics but failed laboratory work,

(c) failed in both.

Solution   Let M and L denote passes in mathematics and laboratory work respectively.
(a) By the general addition rule,

$$P(M \cup L) = P(M) + P(L) - P(M \cap L)$$

$$= 0.8 + 0.85 - 0.75 = 0.9$$

(b) The group of students who passed in mathematics consists of those who passed in both together with those who passed in mathematics but failed in laboratory work, so that

$$P(M \cap \bar{L}) = P(M) - P(M \cap L) = 0.8 - 0.75 = 0.05$$

(c) De Morgan's law (Section 6.2.4, equations (6.7)), together with the result of (a), gives

$$P(\bar{M} \cap \bar{L}) = 1 - P(M \cup L) = 1 - 0.9 = 0.1$$

13.3.4 Conditional probability

Information sometimes arrives in stages, and this happens whenever the outcome of one experiment (or part of an experiment) is relevant to another outcome subsequent to it. For example, the outcome of a seismic survey tells an oil company something about the chances of finding oil if a well is drilled in a certain area, but has no direct causal influence on that discovery. Sometimes a causal influence does exist, as in the case (mentioned in Section 13.2.1) of an opinion poll or vote conducted in a large hall with a microphone passed from person to person. In such circumstances there is strong psychological pressure on individuals to go along with the majority. This is very undesirable in statistical sampling because it can result in a serious bias, so one of the essential features of a sample is **independence**. This requires that future outcomes should *not* depend upon past or present outcomes, but first we need to express the possibility of dependence in general probability terms.

From the start, all probabilities are probabilities of *events* (see Section 13.3.3), so suppose in general that an event A is known to have occurred (representing the existing information). The probabilities of possible future events are now measured relative to the fact that A has occurred. This event must therefore encompass all possibilities compatible with the known information, and can effectively be regarded as a new, revised, sample space in the light of that information. This is the key to understanding the definition and examples that follow.

> The **conditional probability of B given A** is defined as
> $$P(B|A) = \frac{P(A \cap B)}{P(A)}, \quad \text{where } P(A) > 0$$

This represents the new probability of B given that A has occurred, and depends upon the probability of the intersection as shown in Figure 13.9.

Figure 13.9
Venn diagram for conditional probability.

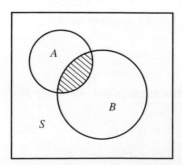

Example 13.5   Someone tosses a die, covers it up and tells you that the number shown is less than four. How does this change the probability that the number is even?

Solution   From the definition,

$$P(\text{even number} \mid \text{less than four}) = \frac{P(\text{even number less than four})}{P(\text{number less than four})}$$

$$= \frac{P(\{2\})}{P(\{1, 2, 3\})} = \frac{\frac{1}{6}}{\frac{1}{2}}$$

$$= \tfrac{1}{3}$$

The information that the number is less than four causes the sample space to shrink to $\{1, 2, 3\}$, and only one entry in this set is even. The outcomes are equally likely, so the probability that the number is even drops from one-half to one-third.

Example 13.6   The probability that a regularly scheduled flight departs on time is $P(D) = 0.83$, the probability that it arrives on time is $P(A) = 0.92$, and the probability that it both departs and arrives on time is $P(A \cap D) = 0.78$. Find the probability that a plane

(a) arrives on time given that it departed on time,

(b) did not depart on time given that it fails to arrive on time.

Solution   (a) This is straightforward from the definition:

$$P(\text{arrives on time} \mid \text{departed on time}) = P(A \mid D)$$

$$= \frac{P(A \cap D)}{P(D)} = 0.94$$

(b) First, using De Morgan's law (Section 6.2.4, equations (6.7)), we have

$$P(\bar{A} \cap \bar{D}) = 1 - P(A \cup D)$$

$$= 1 - P(A) - P(D) + P(A \cap D)$$

by the general addition rule (7). Hence

$$P(\text{did not depart on time} \mid \text{does not arrive on time})$$

$$= P(\bar{D} \mid \bar{A})$$

$$= \frac{P(\bar{A} \cap \bar{D})}{P(\bar{A})}$$

$$= \frac{1 - P(A) - P(D) + P(A \cap D)}{1 - P(A)}$$

$$= 0.375$$

In a sense it is cheating to refer to a conditional 'probability' until it is clear that this quantity actually satisfies the axioms of probability. It is a useful exercise to show that this is the case. Consider the three axioms in turn:

(1)   The event A can be considered as the new sample space, and

$$P(A \mid A) = \frac{P(A)}{P(A)} = 1$$

(2) $P(B \mid A) \geq 0$ because $P(A \cap B) \geq 0$ and $P(A) > 0$

(3)   If $B \cap C = \varnothing$ then $(B \cap A) \cap (C \cap A) = \varnothing$,   and so

$$P[(B \cap A) \cup (C \cap A)] = P(B \cap A) + P(C \cap A)$$

Since $(B \cap A) \cup (C \cap A) = (B \cup C) \cap A$ we have that

$$P(B \cup C \mid A) = \frac{P[(B \cup C) \cap A]}{P(A)} = \frac{P(B \cap A) + P(C \cap A)}{P(A)}$$

$$= P(B \mid A) + P(C \mid A)$$

So conditional probability satisfies the axioms and therefore the various further rules such as the complement rule (4) and the general addition rule (7). Also, because a conditional probability is itself a probability, it is possible to conditionalize again. Thus if the probabilities in the definition are all conditioned upon another event C, we have

$$P(B \mid A \cap C) = \frac{P(A \cap B \mid C)}{P(A \mid C)}$$

Example 13.7

Suppose that on a small tropical island there are only two kinds of day: sunny days and rainy days. The probability that a sunny day is followed by a rainy day is 0.6, and the probability that a rainy day is followed by another rainy day is 0.8. The weather on any day depends upon the previous day's weather but not upon any earlier days. Find the probability that if Thursday is rainy then it will be sunny on Saturday.

Solution

Let T, F and S denote the events that Thursday, Friday and Saturday are sunny respectively (see Figure 13.10). All probabilities must be conditioned upon the assumption that Thursday is rainy. The first step is

$$P(S \mid \overline{T}) = P(S \cap F \mid \overline{T}) + P(S \cap \overline{F} \mid \overline{T})$$

Figure 13.10
Sequences of events
for Example 13.7.

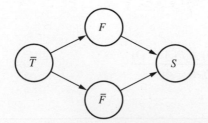

because conditional probabilities obey the addition rule. Also,

$$P(S \cap F \mid \overline{T}) = P(S \mid F \cap \overline{T})P(F \mid \overline{T})$$

from the definition. Now

$$P(F \mid \overline{T}) = P(\text{sunny} \mid \text{rainy})$$

$$= 1 - 0.8 = 0.2$$

because conditional probabilities obey the complement rule. The assumption that the influence of the weather does not extend beyond the previous day implies that

$$P(S \mid F \cap \overline{T}) = P(S \mid F) = P(\text{sunny} \mid \text{sunny})$$

$$= 1 - 0.6 = 0.4$$

(This is known as **independence**, and is described in Section 13.3.5.) Similarly,

$$P(S \cap \overline{F} \mid \overline{T}) = P(S \mid \overline{F} \cap \overline{T})P(\overline{F} \mid \overline{T})$$

$$= P(S \mid \overline{F})P(\overline{F} \mid \overline{T})$$

$$= P(\text{sunny} \mid \text{rainy})P(\text{rainy} \mid \text{rainy})$$

$$= (1 - 0.8)(0.8)$$

Thus

$$P(S \mid \overline{T}) = (0.2)(0.4) + (0.2)(0.8) = 0.24$$

which is the answer required.

We shall not use conditional probabilities very much in this chapter, but the idea of a probability that is conditional upon another event is pervasive. For the moment, however, we must use conditional probability (rather paradoxically) to express the absence of interaction between the events.

13.3.5 Independence

It is possible for the probability of an event B to be raised, lowered or left unchanged by the information that another event A has occurred. For the events shown in Figure 13.11, B_1 is a subset of A and B_2 is disjoint from A, so that

Figure 13.11
Venn diagram illustrating conditional probabilities.

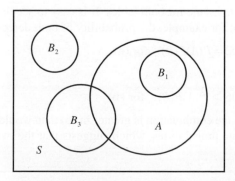

$$P(B_1 \mid A) = \frac{P(B_1)}{P(A)} \geqslant P(B_1) \quad \text{because} \quad A \cap B_1 = B_1 \quad \text{and} \quad P(A) \leqslant 1$$

and

$$P(B_2 \mid A) = 0 < P(B_2) \quad \text{because} \quad A \cap B_2 = \varnothing$$

The probability of B_3 could go either way, or remain unchanged, depending on the probability of the intersection. The situation where the probability is unchanged assumes a special importance.

Events A and B are called **independent** when

$$P(B \mid A) = P(B)$$

In this situation A conveys effectively no information about B. From the definition of conditional probability in Section 13.3.4 it follows that

$$P(A \cap B) = P(A)P(B)$$

The joint probability is the product of the separate probabilities. This shows that independence is symmetric between the two events, so we also have

$$P(A \mid B) = P(A)$$

Example 13.8

Items from a production line can have defects A or B. Some items have both, some just one, but most have neither. Tables (a) and (b) show two alternative sets of joint probabilities:

(a)	B	$\bar{B}$	*Total*
A	0.02	0.08	0.10
$\bar{A}$	0.18	0.72	0.90
Total	0.20	0.80	1.00

(b)	B	$\bar{B}$	*Total*
A	0.06	0.04	0.10
$\bar{A}$	0.14	0.76	0.90
Total	0.20	0.80	1.00

Test for independence in each case.

Solution

The row and column totals shown in the tables are the respective probabilities for the two defects individually, for example

$$P(A \cap B) + P(A \cap \bar{B}) = P(A)$$

and these figures are the same for both tables. It is easy to see that independence holds for (a) but not for (b); for example, the probability of both defects together is

$$P(A \cap B) = 0.02 = P(A)P(B) \quad \text{for (a)}$$

but

$$P(A \cap B) = 0.06 > P(A)P(B) \quad \text{for (b)}$$

The probability of the combination is greater in (b) than would be expected from the product of the separate probabilities, which suggests that the two defects are causally related in some way.

In general, for any number of independent events the probabilities multiply:

$$P(A_1 \cap A_2 \cap \ldots \cap A_n) = P(A_1)P(A_2) \ldots P(A_n)$$

This is called the **product rule** and must be distinguished from the addition rule, which applies (in its basic form, axiom (3)) to exclusive events. Independent events cannot be exclusive unless the probability of at least one of them is zero.

Example 13.9   A card is selected at random from an ordinary pack of 52 playing cards. Find the probabilities that the card drawn is

(a) an ace and a club,   (b) an ace or a club,

(c) an ace and a king,   (d) an ace or a king.

Solution   Let the events that the card is an ace, king and club be denoted by A, K and C respectively.

(a) The events that the card is an ace and that it is a club are independent because there are the same numbers of cards for each suit. These events are not exclusive unless the ace of clubs happens to be missing. Thus

$$P(A \cap C) = P(A)P(C) = (\tfrac{1}{13})(\tfrac{1}{4}) = \tfrac{1}{52}$$

(b) The general addition rule for events (Section 13.3.3) gives

$$P(A \cup C) = P(A) + P(C) - P(A \cap C)$$

$$= \tfrac{1}{13} + \tfrac{1}{4} - \tfrac{1}{52} = \tfrac{4}{13}$$

(c) The events that the card is an ace and that it is a king are mutually exclusive, so

$$P(A \cap K) = 0$$

(d) By the third axiom of probability (Section 13.3.3),

$$P(A \cup K) = \tfrac{1}{13} + \tfrac{1}{13} = \tfrac{2}{13}$$

Example 13.10   If two fair dice are tossed, find the probability of at least one six occurring.

Solution   We shall assume that the throws are causally independent in that the outcome for one die does not relate in any way to the outcome for the other. They will then be statistically independent, and, by the complement and product rules,

$$P(\text{at least one six}) = 1 - P(\text{no six})$$

$$= 1 - P(\text{first die not six})P(\text{second die not six})$$

$$= 1 - (\tfrac{5}{6})^2 = \tfrac{11}{36}$$

Example 13.11   If n people are independently selected, how large does n have to be before there is a better than even chance that at least two of them have the same birthday (not necessarily in the same year and ruling out February 29th)? Assume that all possibilities are equally likely.

Solution   The method of solution to this problem is similar to that for Example 13.10.

P(at least two with the same birthday)

$= 1 - P$(all different birthdays)

$= 1 - P$(2nd different from 1st)P(3rd different from 1st and 2nd)

$\ldots P$(nth different from 1st, $\ldots$, $(n-1)$th)

$$= 1 - \frac{364}{365}\frac{363}{365}\cdots\frac{366-n}{365}$$

$= 0.507$ when $n = 23$

Many people are surprised to find that the answer to Example 13.11 is so small, but this shows that our subjective expectations sometimes have to give way when the rules of probability are properly applied.

In connection with this example, a number of fallacies that 'appear to be most prevalent and injurious to the susceptible gambler' have been identified (R. A. Epstein, *The Theory of Gambling and Statistical Logic*, Academic Press, New York, 1977, p. 393) among which is

> *A tendency to interpret the probability of successive independent events as additive rather than multiplicative. Thus the chance of throwing a given number on a die is considered twice as large with two throws as it is with a single throw.*

The addition and product rules apply in different circumstances and must not be confused.

13.3.6 Exercises

6   If S is the set {bolt, nut, washer, screw, bracket, flange}, and A and B are sets {bracket, nut, flange} and {bolt, bracket} respectively, then what combinations of A and B produce the following sets as outcomes?

(a) {bracket}

(b) {flange, bracket, bolt, nut}

(c) {washer, bolt, screw}

(d) {screw, flange, nut, bolt, washer}

7   Let the sample space S and three events be defined as $S = $ {car, bus, train, bicycle, motorcycle, boat, aeroplane}, $A = $ {bus, train, aeroplane}, $B = $ {train, car, boat}, $C = $ {bicycle}. List the elements of the sets corresponding to the following events:

(a) $\bar{A}$ (b) $A \cap B \cap \bar{C}$ (c) $(\bar{A} \cup B) \cap (\bar{A} \cap C)$

8   If A and B are mutually exclusive events and $P(A) = 0.2$ and $P(B) = 0.5$, find

(a) $P(A \cup B)$ (b) $P(\bar{A})$ (c) $P(\bar{A} \cap B)$

9   From a pack of 52 cards a card is withdrawn at random and not replaced. A second card is then drawn. What is the probability that the first card is an ace and the second card a king?

10   Two ordinary six-faced dice are tossed. Write down the sample space of all possible combinations of values. What is the probability that the two values are the same? What is the probability that they differ by at most 1?

11   The personnel manager of a manufacturing plant claims that among the 400 employees, 312 got a pay rise last year, 248 got increased pension benefits, 173 got both and 43 got neither. Explain why this claim should be questioned.

12 If a card is drawn from a well-shuffled pack of 52 playing cards, what is the probability of drawing

(a) a red king   (b) a 3, 4, 5 or 6

(c) a black card   (d) a red ace or a black queen?

13 In a single throw of two dice, what is the probability of getting

(a) a total of 5,

(b) a total of at most 5,

(c) a total of at least 5?

14 Suppose that you roll a pair of ordinary dice repeatedly until you get either a total of seven or a total of 10. What is the probability that the total then is seven?

15 The 'odds' in favour of an event A are quoted as 'a to b' if and only if $P(A) = a/(a + b)$. The 'odds against' are then 'b to a' (which is the usual way to quote odds in betting situations).

(a) If an insurance company quotes odds of 3 to 1 in favour of an individual 70 years of age surviving another 10 years, what is the corresponding probability?

(b) If the probability of a successful transplant operation is $\frac{1}{8}$, what are the odds against success?

16 Two fair coins are tossed once. Find the conditional probability that both coins show heads given that

(a) the first coin shows a head,

(b) at least one coin shows a head.

17 During the repair of a large number of car engines it was found that part number 100 was changed in 36% and part number 101 in 42% of the cases, and that both parts were changed in 30% of cases. Is the replacement of part 100 connected with that of part 101? Find the probability that in repairing an engine for which part 100 has been changed it will also be necessary to replace part 101.

18 If $P(A) = 0.3$, $P(B) = 0.4$ and $P(B \mid A) = 0.5$, find

(a) $P(A \cap B)$ (b) $P(A \cup B)$ (c) $P(B \mid \bar{A})$

19 Three people work independently at deciphering a message in code. The probabilities that they will decipher it are $\frac{1}{5}, \frac{1}{4}$ and $\frac{1}{3}$. What is the probability that the message will be deciphered?

20 Part of an electric circuit consists of three elements K, L and M in series. Probabilities of failure for elements K and M during operating time t are 0.1 and 0.2 respectively. Element L itself consists of three sub-elements L_1, L_2 and L_3 in parallel, with failure probabilities 0.4, 0.7 and 0.5 respectively, during the same operating time t. Find the probability of failure of the circuit during time t, assuming that all failures of elements are independent.

21 A system can fail (event C) because of two possible causes (events A and B). The probabilities of A, B and $A \cap B$ are known, together with the probabilities of failure given A, given B and given $A \cap B$. Express the following in terms of these known quantities:

(a) $P(A \cup B)$ (b) $P(C \mid A \cap \bar{B})$

(c) $P(C \mid A \cup B)$

22 An advertising agency notes that approximately one in 50 potential buyers of a product sees a given magazine advertisement and one in five sees the corresponding advertisement on television. One in 100 sees both. One in three of those who have seen the advertisement purchase the product, and one in 10 of those who haven't seen it also purchase the product. What is the probability that a randomly selected potential customer will purchase the product?

23 On an infinite chess-board with each side of a square equal to d, a coin of diameter $2r < d$ is thrown at random. Find the probabilities that

(a) the coin falls entirely in the interior of one of the squares,

(b) the coin intersects no more than one side of a square.

13.4 Random variables

13.4.1 Introduction and definition

Now that the foundation of probability theory has been laid, we can begin to consider the data that originates in typical experiments – in particular, numerical data from observations of random variables. It is quite possible for non-numerical outcomes to be of interest, for instance in an experiment where a machine's possible faults might consist of the set {overheated, jammed, misaligned}. Even then, the experiment is likely to be repeated a number of times, and the count for each outcome gives rise to numerical data that can be treated statistically. For the moment, however, let us assume that the outcomes themselves take numerical values.

A **random variable** consists of a sample space of possible numerical values together with a probability over those values.

Random variables vary in their degree of advance predictability. As the following four examples show, the probabilities of the possible values are very dispersed for some random variables, but highly concentrated for others:

(a) *The toss of a die.* No die is perfect, but for this random variable the probabilities of the six values are almost equal.

(b) *Next month's rainfall.* Unless you live in a part of the world that has a very constant climate, the amount of rain that falls in March, say, varies from year to year quite considerably. The probabilities are not quite so dispersed as for the die toss, but there is a high degree of uncertainty.

(c) *A flight delay.* Here there is a high probability of at most a short delay, but a small probability of a very long delay. The probabilities are relatively concentrated.

(d) *The time of tomorrow's sunrise.* Knowing your latitude, longitude, altitude, the date, the direction of sunrise and the height above sea level of the horizon in that direction, you could predict the time very precisely. There would be some small uncertainty because of atmospheric refraction.

The behaviour of a random variable is determined by the profile of its probability distribution. We shall now enlarge upon this for the two common types. The notation convention is to denote a random variable by a capital letter, say X, and an observed value by the corresponding lower-case letter, then x.

13.4.2 Discrete random variables

The distinction between discrete and continuous random variables is inherited from that for sample spaces (Section 13.3.2). First we shall consider the discrete case.

The random variable X, say, has a list of possible values $v_1, v_2, \ldots, v_m$ with probabilities $P(X = v_1), \ldots, P(X = v_m)$ of equalling these values. In other words, each actual value x of X is equal to v_i for some $i = 1, \ldots, m$, and we allow m to be infinite if required. This can be regarded as an idealization of the histogram of data in Section 13.2.2, where m is the number of classes. Typical examples are die tosses, birthdays, and the numbers of defective components in a batch from a production line.

Figure 13.12
Example 13.12: (a)
probability function;
(b) distribution
function.

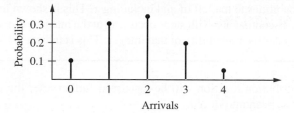

(a)

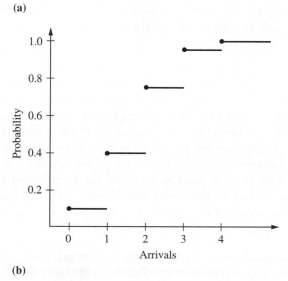

(b)

In general, the behaviour of a discrete random variable can be represented graphically by means of a **probability function**

$$P_X(x) = P(X = x) \quad (-\infty < x < +\infty)$$

and illustrated in Figure 13.12(a) for Example 13.12. Also useful is the **distribution function** $F_X(x)$ defined as

$$F_X(x) = P(X \leqslant x) \quad (-\infty < x < +\infty)$$

and illustrated in Figure 13.12(b). This definition is based on the fact that the set of points in the sample space for which $X \leqslant x$ constitutes an event, and the probability of this event (as a function of x) forms the distribution function. Sometimes this is referred to as the **cumulative distribution function**, because it measures the cumulative probability up to (and including) the value of its argument.

Example 13.12

The number of ships arriving at a container terminal during any one day can be any integer from zero to four, with respective probabilities 0.1, 0.3, 0.35, 0.2, 0.05. Plot the probability and distribution functions.

Solution

The probability function is shown in Figure 13.12(a). The function has zero value except at the five integer points. The value of the distribution function at any point x is the sum of the probabilities to the left of and including x. This is shown in Figure 13.12(b). The function is discontinuous, with steps occurring at the integer points, and the value at each integer includes the probability of that integer. This is indicated by the blob at each step.

The distribution function will be discussed further after the other class of random variables has been introduced.

13.4.3 Continuous random variables

A continuous random variable X can take any value within some interval (v_1, v_2). If this interval is not already infinite, we define the random variable to have zero probability for any value outside it, and hence extend the domain of definition to $(-\infty, +\infty)$. Typical examples are a person's height and weight, component lifetimes, and all measured quantities expressed in units of mass, length, time, temperature, resistance and so on.

In general, the behaviour of a continuous random variable X is described by a probability density function $f_X(x)$ for $-\infty < x < +\infty$ as illustrated in Figure 13.13. As will be explained below, $f_X(x)$ is not the probability that $X = x$: instead, the density function has to be understood in terms of the **distribution function** $F_X(x)$, which measures (as before) the probability that the value of the random variable is less than or equal to the argument x:

$$F_X(x) = P(X \leq x) \quad (-\infty < x < +\infty)$$

In this case, because there are no discrete steps in probability, $F_X(x)$ is continuous and differentiable, and its derivative is called the **probability density function** $f_X(x)$:

$$f_X(x) = \frac{\mathrm{d}}{\mathrm{d}x}[F_X(x)]$$

The significance of the density function is that it indicates for a continuous random variable the concentration of possible observed values along the real axis. This interpretation will be clarified in Section 13.4.4.

Figure 13.13
Typical probability
density function.

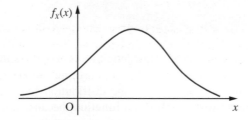

Example 13.13

The lifetime of an electronic component (in thousands of hours) is a continuous random variable with density function

$$f_X(x) = \begin{cases} \frac{1}{2}e^{-x/2} & (x \geq 0) \\ 0 & (x < 0) \end{cases}$$

(This is an example of an **exponential distribution** with parameter $\frac{1}{2}$.) Plot the distribution and density functions.

Figure 13.14
An exponential
distribution (Example
13.13).

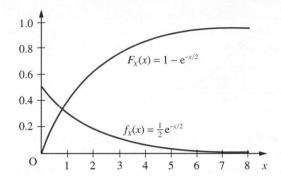

Solution   Integrating the density function gives the distribution function (Figure 13.14):

$$F_X(x) = \int_0^x \tfrac{1}{2} e^{-z/2}\, dz = [-e^{-z/2}]_0^x = 1 - e^{-x/2}$$

for $x \geqslant 0$, and zero for $x < 0$. The variable z is a dummy variable used for integration. The distribution and density functions show that most components have short lifetimes, but a small proportion can survive for much longer.

13.4.4 Properties of density and distribution functions

In order to use the density and distribution functions, we need the following results, which are immediate from the definitions:

(a) $\lim\limits_{x \to -\infty} F_X(x) = 0$ and $\lim\limits_{x \to +\infty} F_X(x) = 1$

Clearly it is impossible for a random variable to have a value less than $-\infty$, and it is certain to have a value less than $+\infty$.

(b)   If $x_1 < x_2$ then $F_X(x_1) \leqslant F_X(x_2)$

Here the event that $X \leqslant x_1$ is a subset of the event that $X \leqslant x_2$, so the probability of the latter must be at least as great as that of the former. From results (a) and (b) it follows that at any point the distribution function is either constant or else increasing, ultimately from its lower limit of zero (at $-\infty$) to its upper limit of one (at $+\infty$).

(c) $P(x_1 < X \leqslant x_2) = F_X(x_2) - F_X(x_1)$

For any random variable the difference between the values of the distribution function at two points is the probability that a value of the random variable will lie between those two points (or is equal to the upper one). For a continuous random variable this is also the area under the density function between those points, by virtue of the relationship between the functions:

$$P(x_1 < X \leqslant x_2) = \int_{x_1}^{x_2} f_X(z)\, dz$$

Figure 13.15
Probability of interval
from density function.

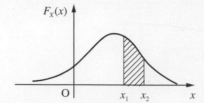

as illustrated in Figure 13.15. This crucial result expresses the significance of the density function and leads to another feature of continuous random variables that should be clearly understood. Setting $x_1 = x_2 = x$, we see that the probability $P(X = x)$ that the random variable has a value exactly equal to x is zero for any x, because the integral is over a domain of length zero. This is in sharp contrast to discrete random variables, which can *only* take certain specific values.

(d) $$\int_{-\infty}^{+\infty} f_X(x)\mathrm{d}x = 1$$

The total area under the density function must be unity because the random variable must have a value somewhere.

Example 13.14

For the distribution of component lifetimes in Example 13.13 find the proportion of components that last longer than 6000 hours.

Solution

Using the distribution function,

$$P(X > 6) = 1 - P(X \leqslant 6) = 1 - F_X(6)$$

$$= 1 - [1 - e^{-6/2}] = e^{-3} \approx 0.05$$

In other words, approximately one in 20 components lasts longer than 6000 hours.

Example 13.15

Two people have agreed to meet in a definite place between six and seven o'clock. Their actual times of arrival are independent and entirely random (no arrival time more likely than any other) within the hour. Find

(a) the density function of the time that the first person arriving has to wait, and

(b) the probability that the meeting will occur if the first person to arrive does not wait for longer than 15 minutes.

Solution

(a) The sample space can be regarded as a unit square as depicted in Figure 13.16(a). Each point represents a pair of arrival times, each measured as part of one hour from six o'clock. Because all arrival times are equally likely for each person and because they arrive independently, all points in the unit square are equally likely. Because the total probability must be one, this implies that the probability of any subset of points is simply equal to the area of that subset. Points along the diagonal lines offset along either axis by a distance w correspond to a waiting time for the first person arriving

Figure 13.16
(a) Sample space
for Example 13.15.
(b) Density and
distribution functions
for waiting time.

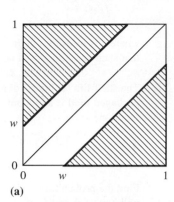

(a)

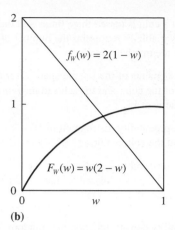

(b)

(W, say) equal to w, because the difference between the arrival times is constant along such lines. The shaded area therefore represents a waiting time greater than w. Putting the two triangles together, we obtain a square of side $1 - w$, so the probability that the waiting time exceeds w is given by

$$P(W > w) = (1 - w)^2$$

The complement of this gives the distribution function of waiting time:

$$F_W(w) = P(W \leqslant w) = 1 - (1 - w)^2 = 2w - w^2$$

and, by differentiation, the density function is

$$f_W(w) = 2(1 - w)$$

both functions being for w between zero and one and illustrated in Figure 13.16(b).

(b) The probability that the meeting will occur is the probability that the waiting time does not exceed 15 minutes:

$$P(W \leqslant \tfrac{1}{4}) = F_W(\tfrac{1}{4}) = \tfrac{7}{16}$$

13.4.5 Exercises

24 Find the distribution of the sum of the numbers when a pair of dice is tossed.

25 At the 18th hole of a golf course the probability that a golfer will score a par four is 0.55, the probability of one under is 0.17, of two under is 0.03, of one over is 0.2 and of two over is 0.05. Plot the (cumulative) distribution function.

26 A difficult assembly process must be undertaken, and the probability of success at each attempt is 0.2. The distribution of the number of independent attempts needed to achieve success is given by the product rule as

$$P(X = k) = (0.2)(0.8)^{k-1} \quad (k = 1, 2, 3, \dots)$$

Plot the distribution function and find the probabilities that the number of attempts will be

(a) less than four,

(b) between three and five.

27　Suppose that a coin is tossed three times and that the random variable W represents the number of heads minus the number of tails.

(a) List the elements of the sample space S for the three tosses of the coin, and to each sample point assign a value w of W.

(b) Find the probability distribution of W, assuming that the coin is fair.

(c) Find the probability distribution of W, assuming that the coin is biased so that a head is twice as likely to occur as a tail.

28　If the probability density function of a random variable X is given by

$$f_X(x) = \begin{cases} c/\sqrt{x} & (0 < x < 4) \\ 0 & \text{(elsewhere)} \end{cases}$$

where c is a constant, find

(a) the value of c,

(b) the distribution function,

(c) $P(X > 1)$.

29　The time interval (X) between successive earthquakes of a certain magnitude has an exponential distribution with density function given by

$$f_X(x) = \begin{cases} \frac{1}{90} e^{-x/90} & \text{if } x \geq 0 \\ 0 & \text{if } x < 0 \end{cases}$$

where x is measured in days. Find the probability that such an interval will not exceed 30 days.

30　The shelf life (in hours) of a certain perishable packaged food is a random variable with density function

$$f_X(x) = \begin{cases} 20\,000(x + 100)^{-3} & (x > 0) \\ 0 & \text{(otherwise)} \end{cases}$$

Find the probabilities that one of these packages will have a shelf life of

(a) at least 200 hours,

(b) at most 100 hours,

(c) between 80 and 120 hours.

31　The wave amplitude X on the sea surface often has the following (Rayleigh) distribution:

$$f_X(x) = \begin{cases} \dfrac{x}{a} \exp\left(\dfrac{-x^2}{2a}\right) & (x > 0) \\ 0 & \text{(otherwise)} \end{cases}$$

where a is a positive constant. Find the distribution function and hence the probability that a wave amplitude will exceed $5.5\,\text{m}$ when $a = 6$.

13.4.6　Measures of location and dispersion

The observable properties of a random variable are determined by its distribution of probabilities (if discrete) or density function (if continuous), but this amount of information is difficult to extract from data. One common approach that is rather simpler is to assume that the random variable is one of a class whose distribution is specified by a formula and which often arises in practice, such as the binomial, Poisson or normal. These distributions will be covered in Section 13.5. Another common approach is to characterize the random variable in terms of two numbers: a measure of **location** ('typical' value) and a measure of **dispersion** ('spread' about that value). In practice, both approaches are used together, with the measures of location and dispersion often providing the parameters for the formula of the distribution.

Mean, median and mode

There are three common meaures of location, the most important of which is the **mean**. For a random variable X this is usually given the symbol μ_X and is defined as

$$\mu_X = \begin{cases} \displaystyle\sum_{k=1}^{m} v_k P(X = v_k) & \text{if } X \text{ is discrete} \\[2ex] \displaystyle\int_{-\infty}^{+\infty} x f_X(x)\mathrm{d}x & \text{if } X \text{ is continuous} \end{cases}$$

This represents a weighted sum of the possible values of X, with weights reflecting their relative likelihood of occurrence, and is effectively the 'centre of gravity' of the distribution.

Another measure of location that is often used is the **median**. For a continuous random variable X this is the point m_X for which

$$P(X \leqslant m_X) = F_X(m_X) = \tfrac{1}{2}$$

In other words, there are equal chances of X being greater than the median or less than the median. For a discrete random variable the median may not be unique, and is any point for which

$$P(X \leqslant m_X) \geqslant \tfrac{1}{2} \quad \text{and} \quad P(X \geqslant m_X) \geqslant \tfrac{1}{2}$$

The median of a distribution does not coincide with the mean unless the distribution has an axis of symmetry, in which case both measures lie on it.

The third measure of location is the **mode**, which is any point for which the probability function $P_X(\text{mode})$ (if discrete) or the density function $f_X(\text{mode})$ (if continuous) is an overall maximum. The mode can therefore be regarded as the most likely value of X to be observed. The mean, median and mode can all differ (see for example Figure 13.17), and can occur in any order.

Figure 13.17
Mode, median and mean for a particular distribution.

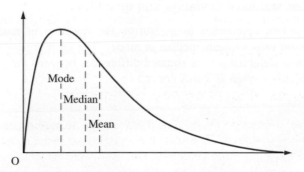

Example 13.16   Find the mean, median and mode for

(a) a simple die toss,

(b) the number of ship arrivals (Example 13.12),

(c) the lifetime distribution (Example 13.13).

Solution   (a) For the toss of a fair die

$$\mu_X = \sum_{k=1}^{6} \tfrac{1}{6}k = 3.5$$

The median is any point in the interval [3, 4], and each possible value is a mode.

(b) For the number of ship arrivals

$$\mu_X = (0)(0.1) + (1)(0.3) + (2)(0.35) + (3)(0.2) + (4)(0.05)$$

$$= 1.8$$

The median is two because $P(X \leqslant 2) = 0.75$ and $P(X \geqslant 2) = 0.6$, both of which exceed one-half. The mode is also equal to two because this is the most likely value.

(c) For the lifetime distribution

$$\mu_X = \int_0^\infty \tfrac{1}{2} x e^{-x/2} \, \mathrm{d}x$$

which we integrate by parts to obtain

$$\mu_X = [-x e^{-x/2}]_0^\infty + \int_0^\infty e^{-x/2} \, \mathrm{d}x = [-2e^{-x/2}]_0^\infty = 2$$

The median is given by

$$F_X(m_X) = 1 - e^{-m_X/2} = \tfrac{1}{2}$$

from which $m_X = 1.386$. The mode, however, is zero because this is the peak of the density function.

Variance, standard deviation and quartiles

There are two approaches to measuring the variation of random variables around their central values (mean, median or mode). The most important such measure is the **variance**, a weighted sum of squared differences between the possible values and the mean, usually written as $\mathrm{Var}(X)$ or σ_X^2:

$$\mathrm{Var}(X) = \sigma_X^2 = \begin{cases} \displaystyle\sum_{k=1}^{m} (v_k - \mu_X)^2 P(X = v_k) & \text{if } X \text{ is discrete} \\[2ex] \displaystyle\int_{-\infty}^{+\infty} (x - \mu_X)^2 f_X(x) \, \mathrm{d}x & \text{if } X \text{ is continuous} \end{cases}$$

This is analogous to 'moment of inertia': it measures how tightly concentrated the possible values are about the mean (centre of gravity). One undesirable feature is that the squaring operation changes the units, so that a random variable measured, say, in volts will have a variance in volts-squared. The remedy is to use the **standard deviation** σ_X, which is defined as the square root of the variance.

The alternative approach to measuring dispersion is to exploit the distribution function $F_X(x)$. Suppose for simplicity that X is a continuous random variable. We have already defined the median by

$$F_X(m_X) = \tfrac{1}{2}$$

The points q_1 and q_3 where

$$F_X(q_1) = \tfrac{1}{4} \quad \text{and} \quad F_X(q_3) = \tfrac{3}{4}$$

are called **quartiles**, and the median can also be described as a quartile q_2. These quartiles divide the range of possible values of X into four successive intervals, for each of which the probability of X falling into the interval is one-quarter. In fact, a finer sub-division into 100 equally likely intervals is also used, the dividing points being called **percentiles**. The first quartile q_1 is then the 25th percentile, and so on. The 10th and 90th percentiles are also known as the first and ninth **deciles**, d_1 and d_9.

The most common measure of dispersion apart from variance (or standard deviation) is the **interquartile range** $q_3 - q_1$. Sometimes the **semi-interquartile range** or **quartile deviation** $\tfrac{1}{2}(q_3 - q_1)$ is quoted instead. The **interdecile range** $d_9 - d_1$ is also used.

Example 13.17

Find the variance and standard deviation for each of the random variables in Example 13.16, and the interquartile range for the lifetime distribution.

Solution

(a) For the toss of a fair die, using $\mu_X = 3.5$,

$$\sigma_X^2 = \left[\sum_{k=1}^{6} \tfrac{1}{6}(k - 3.5)^2 \right] = 2.917$$

from which $\sigma_X = 1.708$.

(b) For the number of ship arrivals, using $\mu_X = 1.8$,

$$\sigma_X^2 = (0 - 1.8)^2(0.1) + (1 - 1.8)^2(0.3) + (2 - 1.8)^2(0.35)$$
$$+ (3 - 1.8)^2(0.2) + (4 - 1.8)^2(0.05) = 1.060$$

from which $\sigma_X = 1.030$.

(c) For the lifetime distribution, using $\mu_X = 2$,

$$\sigma_X^2 = \int_0^\infty \tfrac{1}{2}(x - 2)^2 e^{-x/2}\, dx$$

which we integrate by parts to obtain

$$\sigma_X^2 = -[(x - 2)^2 e^{-x/2}]_0^\infty + \int_0^\infty 2(x - 2)e^{-x/2}\, dx$$

$$= 4 - [4(x - 2)e^{-x/2}]_0^\infty + \int_0^\infty 4e^{-x/2}\, dx = 4$$

from which $\sigma_X = 2$. The quartiles q_1 and q_3 are the solutions of

$$1 - e^{-x/2} = \tfrac{1}{4} \quad \text{and} \quad 1 - e^{-x/2} = \tfrac{3}{4}$$

respectively, from which $q_1 = 0.575$ and $q_3 = 2.773$, and the interquartile range is there-fore $2.773 - 0.575 = 2.198$.

13.4.7 Expected values

The mean and variance are special cases of expected values for a random variable. In general, the **expected value** of a function $h(X)$ of a random variable X is

$$
E[h(X)] = \begin{cases} \displaystyle\sum_{k=1}^{m} h(v_k)P(X = v_k) & \text{if } X \text{ is discrete} \\[2ex] \displaystyle\int_{-\infty}^{+\infty} h(x)f_X(x)\mathrm{d}x & \text{if } X \text{ is continuous} \end{cases}
$$

As before, this is a weighted combination of the possible values. The mean and variance are retrieved by taking $h(X) = X$ and $h(X) = (X - \mu_X)^2$ respectively.

Expected values have many applications. One immediate application is in a useful alternative expression for the variance, obtained by expanding the square. If X is continuous then

$$
\sigma_X^2 = \int_{-\infty}^{+\infty} (x - \mu_X)^2 f_X(x)\mathrm{d}x
$$

$$
= \int_{-\infty}^{+\infty} x^2 f_X(x)\mathrm{d}x - 2\mu_X \int_{-\infty}^{+\infty} x f_X(x)\mathrm{d}x + \mu_X^2 \int_{-\infty}^{+\infty} f_X(x)\mathrm{d}x
$$

$$
= E(X^2) - \mu_X^2
$$

In other words, the variance is the expected value (or mean) of the square minus the square of the mean. The same is true when X is discrete, by a similar proof.

Example 13.18 Find the mean and standard deviation of the waiting time in Example 13.15.

Solution The mean waiting time is

$$
\mu_W = \int_0^1 2(w - w^2)\mathrm{d}w = \tfrac{1}{3}
$$

The mean square is

$$
E(W^2) = \int_0^1 2(w^2 - w^3)\mathrm{d}w = \tfrac{1}{6}
$$

so the standard deviation is

$$
\sigma_W = \sqrt{[\tfrac{1}{6} - (\tfrac{1}{3})^2]} = 0.236
$$

These translate to 20 minutes for the mean and about 14 minutes for the standard deviation.

13.4.8 Independence of random variables

It is possible for two different random variables to be measured for the same object: for example, a person's height and weight. Individually, these random variables have distributions, mean values and variances, which apply to a particular population, but this is not the whole story. It is clear that taller people tend to be heavier than shorter people (although obviously there are exceptions). In this case we say that these variables are **dependent** upon each other (to some degree). The notion of dependence is basically the same as that applying to events, discussed in Section 13.3.4. Furthermore, just as events can be independent (Section 13.3.5), so can random variables. For example, it is plausible that a person's birthday and telephone number are not related in any way, and are therefore independent random variables. Nothing is likely to be learnt about the one from an observation of the other.

For independent events we have the rule that the joint probability is the product of the separate probabilities:

$$P(A \cap B) = P(A)P(B)$$

For independent discrete random variables a similar rule applies:

$$P(X = u_i \cap Y = v_j) = P(X = u_i)P(Y = v_j)$$

where $u_1, u_2, \ldots, u_k$ are the possible values of X and $v_1, v_2, \ldots, v_m$ are the possible values of Y. This effectively specifies a **joint distribution** for the two random variables. If we sum over the possible values of Y then we find

$$\sum_{j=1}^{m} P(X = u_i \cap Y = v_j) = \sum_{j=1}^{m} P(X = u_i)P(Y = v_j)$$

$$= P(X = u_i) \sum_{j=1}^{m} P(Y = v_j)$$

$$= P(X = u_i)$$

Thus the individual probability of one value u_i of X can be obtained from the joint distribution by summing over all values v_j of Y, with X fixed at u_i. In fact, this is true even when the random variables are dependent.

Example 13.19

A new plant at a manufacturing site has to be first installed and then commissioned. The times required for these two stages depend upon different random factors, and can therefore be regarded as independent. Based on past experience, the respective distributions for X (installation time) and Y (commissioning time), both in days, are as follows:

u_i	3	4	5	6
$P(X = u_i)$	0.1	0.4	0.3	0.2

v_j	2	3	4
$P(Y = v_j)$	0.50	0.35	0.15

Find the joint distribution for X and Y, and the probability that the total time will not exceed seven days.

Solution

Because the random variables are independent, the joint distribution is given by the product of the separate distributions:

$$P(X = u_i \cap Y = v_j) = P(X = u_i)P(Y = v_j)$$

with the following result:

Joint probability		u_i 3	4	5	6	Total
	2	0.050	0.200	0.150	0.100	0.50
v_j	3	0.035	0.140	0.105	0.070	0.35
	4	0.015	0.060	0.045	0.030	0.15
Total		0.10	0.40	0.30	0.20	1.00

Note that the row and column totals give the individual distributions for X and Y. The probability that the total time will not exceed seven days is given by the sum of those joint probabilities above the stepped broken line:

$$P(X + Y \leqslant 7) = P(X = 3 \cap Y = 2) + P(X = 3 \cap Y = 3) + P(X = 3 \cap Y = 4)$$

$$+ P(X = 4 \cap Y = 2) + P(X = 4 \cap Y = 3) + P(X = 5 \cap Y = 2)$$

$$= 0.050 + 0.035 + 0.015 + 0.200 + 0.140 + 0.150$$

$$= 0.59$$

13.4.9 Scaling and adding random variables

Example 13.19 has introduced the idea of a sum of random variables, itself a random quantity. The distribution of this quantity can be deduced from the joint distribution; thus in Example 13.19 the probability that the total time (installation plus commissioning) will take exactly seven days is

$$P(X + Y = 7) = P(X = 3 \cap Y = 4) + P(X = 4 \cap Y = 3) + P(X = 5 \cap Y = 2)$$

$$= 0.015 + 0.140 + 0.150 = 0.305$$

A similar calculation can be done for every possible value from the minimum (five days) to the maximum (ten days), and the distribution is then complete.

Example 13.20

Find the distribution of total time for the situation described in Example 13.19, and the expected value of this time.

Solution Proceeding as described above, we obtain the following distribution:

w_k	5	6	7	8	9	10
$P(X + Y = w_k)$	0.050	0.235	0.305	0.265	0.115	0.030

The expected value is then

$$E(X + Y) = \sum_{w_k} w_k P(X + Y = w_k) = 7.25$$

There is, however, an easier way to arrive at the mean of a sum of random variables. The separate means (or expected values) of X and Y are easily found from the values given in Example 13.19:

$$E(X) = \sum_{u_i} u_i P(X = u_i) = 4.60$$

$$E(Y) = \sum_{v_j} v_j P(Y = v_j) = 2.65$$

It has turned out that

$$E(X + Y) = E(X) + E(Y)$$

The mean of the sum of random variables is the sum of the means. That this is a general result is shown as follows:

$$E(X + Y) = \sum_{i=1}^{k} \sum_{j=1}^{m} (u_i + v_j) P(X = u_i \cap Y = v_j)$$

$$= \sum_{i=1}^{k} u_i \left[\sum_{j=1}^{m} P(X = u_i \cap Y = v_j) \right] + \sum_{j=1}^{m} v_j \left[\sum_{i=1}^{k} P(X = u_i \cap Y = v_j) \right]$$

$$= \sum_{i=1}^{k} u_i P(X = u_i) + \sum_{j=1}^{m} v_j P(Y = v_j)$$

$$= E(X) + E(Y)$$

The double summation is over all the possible values of $X + Y$ (which are $u_i + v_j$) times the probability of each combination, and the result in Section 13.4.8 that summing a joint probability over the values of one variable gives the probability of the other variable has been used. Furthermore, because this also holds for dependent variables (as can be seen in *Advanced Modern Engineering Mathematics*), the mean of a sum of random variables is always equal to the sum of the means, whether they are dependent or not.

For the variance of a sum it is not quite so simple. If the mean of X is μ_X and the mean of Y is μ_Y then

$$\text{Var}(X + Y) = E\{[(X + Y) - (\mu_X + \mu_Y)]^2\} = E\{[(X - \mu_X) + (Y - \mu_Y)]^2\}$$

$$= E\{(X - \mu_X)^2 + (Y - \mu_Y)^2 + 2(X - \mu_X)(Y - \mu_Y)\}$$

$$= E\{(X - \mu_X)^2\} + E\{(Y - \mu_Y)^2\} + E\{2(X - \mu_X)(Y - \mu_Y)\}$$

The first two terms on the right-hand side are $\text{Var}(X)$ and $\text{Var}(Y)$ respectively. The third term (which is actually called the **covariance**) is a measure of dependence, and it is shown in *Advanced Modern Engineering Mathematics* that this is always zero for independent variables. Hence if X and Y are independent, the variance of a sum is equal to the sum of the variances:

$$\text{Var}(X + Y) = \text{Var}(X) + \text{Var}(Y)$$

These results for the mean and variance of sums of random variables extend naturally to any number of variables, and apply whether the variables are discrete or continuous.

If we add a constant (c, say) to a random variable X, it follows immediately from the definitions in Section 13.4.6 that the same constant is added to the mean, but the variance does not change. If X is a continuous random variable, with density function $f_X(x)$, say, then

$$E(X + c) = \int_{-\infty}^{+\infty} (x + c)f_X(x)\,\mathrm{d}x = \int_{-\infty}^{+\infty} xf_X(x)\,\mathrm{d}x + c\int_{-\infty}^{+\infty} f_X(x)\,\mathrm{d}x$$

$$= \mu_X + c$$

$$\mathrm{Var}(X + c) = \int_{-\infty}^{+\infty} [(x + c) - (\mu_X + c)]^2 f_X(x)\,\mathrm{d}x$$

$$= \int_{-\infty}^{+\infty} (x - \mu_X)^2 f_X(x)\,\mathrm{d}x = \sigma_X^2$$

If we multiply a random variable X by a constant c, the mean is multiplied by c and the variance by c^2:

$$E(cX) = \int_{-\infty}^{+\infty} cxf_X(x)\,\mathrm{d}x = c\mu_X$$

$$\mathrm{Var}(cX) = \int_{-\infty}^{+\infty} (cx - c\mu_X)^2 f_X(x)\,\mathrm{d}x = c^2 \int_{-\infty}^{+\infty} (x - \mu_X)^2 f_X(x)\,\mathrm{d}x = c^2\sigma_X^2$$

All of these results hold whether X is continuous or discrete.

Example 13.21   If a mean temperature is 58°F, what is the mean temperature in degrees Celsius?

Solution   If T_F and T_C denote temperatures in Fahrenheit and Celsius respectively then

$$T_C = \tfrac{5}{9}(T_F - 32)$$

so

$$E(T_C) = \tfrac{5}{9}[E(T_F) - 32] = \tfrac{5}{9}(58 - 32) = 13.4°\mathrm{C}$$

13.4.10   Measures from sample data

We can now return to the consideration of data, which is the object of the whole exercise. Given that the exact distribution of a random quantity under investigation is usually not known in an experimental context but that the mean and variance at least would be useful characteristics of it, it is reasonable to try to estimate these from the data. Experience shows that quite good estimates of mean and variance can be obtained even from rather small samples, whereas a much larger sample is needed before the histogram gives a good approximation to the whole shape of the true distribution.

Sample average and variance

For a sample $\{X_1, \ldots, X_n\}$ of data, the **sample average** and **sample variance** are defined as

$$\bar{X} = \frac{1}{n} \sum_{i=1}^{n} X_i \quad \text{and} \quad S_X^2 = \frac{1}{n} \sum_{i=1}^{n} (X_i - \bar{X})^2$$

respectively. The **sample standard deviation** is the square root of the sample variance.

The average of the sample, and the average squared deviation from the sample average, are easy to work out from the data, and characterize the data in location and dispersion. It turns out that these approximate the true figures of mean and variance, and the approximations improve as $n \to \infty$ in a sense to be made precise below.

By expanding the square in the formula for the sample variance (as in Section 13.4.7 for the true variance) it is easy to show that an alternative expression (which is useful for hand calculation) is

$$S_X^2 = \overline{X^2} - (\bar{X})^2$$

that is, the average of the square minus the square of the average. When small samples are used in statistics (this is considered in *Advanced Modern Engineering Mathematics*), a different definition of sample variance must be adopted:

$$S_{X,n-1}^2 = \frac{1}{n-1} \sum_{i=1}^{n} (X_i - \bar{X})^2$$

The difference between the two definitions is relatively small. Many scientific calculators provide functions to work out sample average and both forms of sample variance or standard deviation.

Example 13.22 A die was tossed 24 times, producing the following results:

$$4, 6, 2, 4, 2, 1, 5, 1, 3, 1, 3, 4, 5, 4, 3, 1, 6, 5, 6, 3, 1, 2, 4, 6$$

Find the sample average and standard deviation.

Solution The average score over the 24 tosses is

$$\bar{X} = 3.42$$

The average of the squares is 13.667, so the standard deviation is

$$S_X = \sqrt{[13.667 - (3.42)^2]} = 1.73$$

These figures are close to the theoretical values worked out in Examples 13.16 and 13.17.

An issue first raised in Section 1.5 is important here: to how many places of decimals should these results be quoted? The actual average of the data in this example is

3.4166 ..., but the results should be stated with no more significant digits than can be justified statistically. The average might be quoted as 3, 3.4, 3.42, 3.417 and so on, but the appropriate precision depends upon the sample size n.

The sample average itself is a random variable; it has a mean and variance, and it follows from the results in Section 13.4.9 that

$$\text{Var}(\overline{X}) = \text{Var}\left(\frac{X_1 + \ldots + X_n}{n}\right) = \frac{1}{n^2}\text{Var}(X_1 + \ldots + X_n) = \frac{n\sigma_X^2}{n^2} = \frac{\sigma_X^2}{n}$$

since the random variables, X_i, can be reasonably assumed independent here.

The larger the sample size, the smaller the variance of $\overline{X}$ and the greater the precision, but to quantify the precision we also need a value for σ_X. Usually all we have is the estimate S_X, but this is also a random variable and subject to error. It can be shown that in many situations a (rather rough) indication of the accuracy of S_X as an estimate of σ_X is that its relative error (see Section 1.5.3) varies inversely with $\sqrt{(2n)}$:

$$\frac{|S_X - \sigma_X|}{\sigma_X} \approx \frac{1}{\sqrt{(2n)}}$$

Returning to Example 13.22, with $n = 24$, the percentage error in S_X is estimated at 14%, so the error in S_X is likely to be of order 0.2, and the second decimal place has no meaning. The error in $S_X/\sqrt{n}$ is correspondingly of order 0.05 in its value of 0.35. The results of Example 13.22 can therefore be stated more properly as

$$\overline{X} = 3.4 \quad \text{(with likely error of order 0.4)}$$

$$S_X = 1.7 \quad \text{(with likely error of order 0.2)}$$

In practice, these high standards of honesty are not always maintained, and it is very important not to be misled by the spurious precision with which results are often quoted.

Example 13.23

Measured values of resistance (in Ω) for 12 nominally $100\,\Omega$ resistors were as follows:

106, 98, 95, 109, 99, 102, 101, 108, 94, 99, 96, 102

Find the sample average and both forms of sample variance and standard deviation.

Solution

The average of the 12 figures is

$$\overline{X} = 100.75$$

which is slightly high but close to the nominal figure, and much closer to that figure than a 'typical' value from the data. The two results for sample variance and standard deviation are

$$S_X^2 = 22.9, \quad S_X = 4.8$$

and

$$S_{X,n-1}^2 = 25, \quad S_{X,n-1} = 5$$

Despite the small sample size, the difference between the two versions of sample standard deviation is not very large. Furthermore, following the above discussion, an error of order 1.0 is likely in the standard deviation ($\sqrt{\frac{1}{24}}$ is about 20%), so there is no point in distinguishing them, even to the first decimal place. The value of $S_X/\sqrt{n}$ is 1.4, with a likely error of order 0.3, so the average should properly be stated as

$$\overline{X} = 101 \quad \text{(with likely error of order 1.5)}$$

It is worth noting that in Example 13.23 the distribution of the random variable (resistance) is not known, but the sample provides useful information about the mean value and the variability about that value.

As mentioned above, many scientific calculators will work these results out automatically. There are also many statistical packages that run on computers of all sizes, and they will do the same. Alternatively, Figure 13.18 contains a pseudocode listing of an efficient program to compute the sample average and standard deviation of a set of data $X_1, \ldots, X_n$. The algorithm used works as follows.

Let M_k and Q_k respectively represent the average of the first k observations and the sum of squares of deviations of the first k observations about their average:

$$M_k = \frac{1}{k}\sum_{i=1}^{k} X_i \quad \text{and} \quad Q_k = \sum_{i=1}^{k}(X_i - M_k)^2$$

The program exploits the following recursion relations, which are proved in D. Cooke, A. H. Craven and G. M. Clarke, *Statistical Computing in Pascal* (Edward Arnold, London, 1985), pp. 54–5 (© 1985 Edward Arnold Ltd. Reproduced by permission of Edward Arnold (Publishers) Ltd).

Figure 13.18
Pseudocode listing for sample average and variance.

```
{Program to compute the sample average and standard deviation,
x(k) is the array of data,
n is the sample size,
xbar is the sample average,
sx and sxn_1 are the two versions of standard deviation,
Mk and Qk hold running totals,
notation as in Section 13.4.10.}

Mk ← 0
Qk ← 0
for k is 1 to n do
    diff ← x(k) − Mk
    Mk ← ((k − 1)*Mk + x(k))/k
    Qk ← Qk + (1 − 1/k) *diff*diff
endfor
xbar ← Mk
sx ← square_root(Qk/n)
sxn_1 ← square_root(Qk/(n − 1))
```

$$M_k = \frac{1}{k}[(k-1)M_{k-1} + X_k]$$

and

$$Q_k = Q_{k-1} + \left(1 - \frac{1}{k}\right)(X_k - M_{k-1})^2$$

Finally,

$$\overline{X} = M_n, \quad S_X^2 = \frac{Q_n}{n} \quad \text{and} \quad S_{X,n-1}^2 = \frac{Q_n}{n-1}$$

The use of this recurrence method avoids having to make two passes through the data (as required for the original definition of the sample variance), and also avoids the loss of precision involved in subtracting two quantities that often turn out in practice to be large in magnitude and similar in value (as required by the alternative expression).

Sample median and range

The sample average and standard deviation are not the only measures of location and dispersion derived from data. Suppose that the data $\{X_1, \ldots, X_n\}$ are ordered so that

$$X_{(1)} \leq X_{(2)} \leq \ldots \leq X_{(n)}$$

Then a **sample median** that provides an estimate of the true median (Section 13.4.6) can be defined as

$$\text{sample median} = \begin{cases} X_{(k)} & (\text{odd } n = 2k - 1) \\ \frac{1}{2}[X_{(k)} + X_{(k+1)}] & (\text{even } n = 2k) \end{cases}$$

A common measure of dispersion (especially for small samples) is the **sample range** $X_{(n)} - X_{(1)}$, the difference between the largest and smallest elements of the data set, often used in quality control. The ideas of quartiles and percentiles (Section 13.4.6) can also be applied to data, based on the cumulative percentages (Section 13.2.3).

Example 13.24 Find the sample median and range for the data in Examples 13.22 and 13.23.

Solution For the die toss the sorted data is

1, 1, 1, 1, 1, 2, 2, 2, 3, 3, 3, 3, 4, 4, 4, 4, 4, 5, 5, 5, 6, 6, 6, 6

so the sample median is $\frac{1}{2}(3 + 4) = 3.5$, and the sample range is 5.

For the resistors the sorted data is

94, 95, 96, 97, 98, 99, 101, 101, 102, 106, 108, 109

so the sample median is $\frac{1}{2}(99 + 101) = 100$, and the sample range is 15.

13.4.11 Exercises

32 Suppose that the probability distribution for the number of days required to ship a package from London to New York is as follows:

Number of days	2	3	4	5	6	7
Probability	0.05	0.20	0.35	0.25	0.1	0.05

Find the mean of this distribution, and the probability that a particular package arrives in less than five days.

33 The distribution of the daily number of malfunctions of a certain computer is given by the following table:

Number of malfunctions	0	1	2	3	4	5	6
Probability	0.17	0.29	0.27	0.16	0.07	0.03	0.01

Find the mean, the median and the standard deviation of this distribution.

34 Find the average sentence length for the sentences with lengths given in Question 1 in Exercises 13.2.4.

35 The distribution of the number X of independent attempts needed to achieve the first success when the probability of success is 0.2 at each attempt is given by

$$P(X = k) = (0.2)(0.8)^{k-1} \quad (k = 1, 2, 3, \ldots)$$

(see Question 26 in Exercises 13.4.5). Find the mean, the median and the standard deviation for this distribution.

36 You arrive at a railway station knowing only that trains leave for your destination at intervals of one hour. Find the mean and standard deviation of your waiting time.

37 A random variable X has the linear distribution given by

$$f_X(x) = \begin{cases} a - bx & (0 \leqslant x \leqslant \frac{a}{b}) \\ 0 & (\text{otherwise}) \end{cases}$$

where a and b are constants. Show that

(a) $a = \sqrt{2b}$ (b) the median of X is $(\sqrt{2} - 1)/\sqrt{b}$

38 Suppose that the running distance (in thousands of kilometres) that car owners get from a tyre is a random variable with density function

$$f_X(x) = \begin{cases} \frac{1}{30}e^{-x/30} & (x > 0) \\ 0 & (x \leqslant 0) \end{cases}$$

Find

(a) the probability that one of these tyres will last at most 19 000 km,

(b) the mean and standard deviation of X, and

(c) the median and interquartile range of X.

39 If the probability density of the random variable X is

$$f_X(x) = \begin{cases} 30x^2(1 - x)^2 & (0 < x < 1) \\ 0 & (\text{otherwise}) \end{cases}$$

find the probability that X will take a value within two standard deviations of its mean.

40 The distribution of downtime T for breakdowns of a computer system is given by

$$f_T(t) = \begin{cases} a^2 t e^{-at} & (t > 0) \\ 0 & (\text{otherwise}) \end{cases}$$

where a is a positive constant. The cost of downtime derived from the disruption resulting from breakdowns rises exponentially with T:

$$\text{cost factor} = h(T) = e^{bT}$$

Show that the expected cost factor for downtime is $[a/(a - b)]^2$, provided that $a > b$.

41 The mean times for completion of tasks A and B are four and six hours respectively. A particular project involves three tasks of type A and two of type B, all to be performed in succession. What is the expected time for completion of the project? Also, if the standard deviations for A and B are one and two hours respectively, and if all project times are independent, what is the standard deviation of the completion time?

42 An inspection of 12 specimens of material from inside a reactor vessel revealed the following percentages of impurities:

2.3, 1.9, 2.1, 2.8, 2.3, 3.6, 1.4, 1.8, 2.1, 3.2, 2.0, 1.9

Find (a) the sample average and both versions of the sample standard deviation, (b) the sample median and range.

 43 Find the sample average, standard deviation, median and range for the following sample of component lifetimes (in thousands of hours):

5.6, 4.1, 6.0, 5.8, 5.2, 4.3, 6.4, 5.5, 6.0, 5.1, 4.9, 4.2, 4.8, 6.8, 5.6, 5.2, 7.3, 5.4, 4.7, 5.9, 5.0, 6.3, 4.4, 6.0

 44 Find the sample averages and standard deviations for the engine performance data in Figure 13.1.

45 In a problem similar to that in Question 35 the probability of success at the first attempt is 0.2 but the probability of failure at each subsequent attempt (if needed) is half of that for the previous attempt. Find the mean number of attempts needed to achieve the first success.

 46 Find the median and the mode for the Rayleigh distribution

$$f_X(x) = \begin{cases} \dfrac{x}{a}\exp\left(-\dfrac{x^2}{2a}\right) & (x > 0) \\ 0 & (\text{otherwise}) \end{cases}$$

(see Question 31 in Exercises 13.4.5). Also show that the mean is given by

$$\mu_X = \int_0^\infty \exp\left(-\dfrac{x^2}{2a}\right)\mathrm{d}x$$

which can be shown to be $\sqrt{(\tfrac{1}{2}\pi a)}$. Compare these quantities when $a = 6$, and find the interquartile range.

 47 Two people are separately attempting to succeed at a particular task, and each will continue attempting until success is achieved. The probability of success of each attempt for person A is p, and that for person B is q, all attempts being independent. What is the probability that person B will achieve success with no more attempts than person A does?

$$\left(\text{Hint: } \sum_{i=0}^{n-1} x^i = \frac{1 - x^n}{1 - x}\right)$$

 48 Sample values that are several standard deviations away from the sample average are called **outliers**. They are often just measurement or transcription errors, but they can bias a statistical calculation. Which of the following data are more than three sample standard deviations away from the average?

19.4, 18.1, 25.6, 18.2, 20.6, 25.0, 21.8, 15.5, 26.3, 15.8, 18.7, 19.3, 22.3, 20.9, 24.2, 21.4, 23.2, 21.4, 47.1, 23.6, 46.3, 21.2, 27.5, 20.8, 24.7, 25.9, 25.8, 33.4, 30.9, 24.5

13.5 Important practical distributions

A lot of information is required to specify the exact distribution of a random variable, and even more to specify the joint distribution of two or more variables. The mean, variance and covariance are useful measures of the most important properties of random variables, namely location, dispersion and dependence, which can realistically be estimated from data. These measures are of great value in statistics, as can be seen in the companion text *Advanced Modern Engineering Mathematics*. Another short cut is provided by the various classes of distributions that are often used in statistical practice. The user has to supply the values of certain essential parameters, perhaps using estimates of mean and variance to do so, and then the probability distribution is determined by a formula. Experience shows that these classes of distributions (which are idealized in mathematical form) do approximate very well to the actual distributions in many practical situations.

The most important of these classes of distributions are the binomial, Poisson and normal. In this section we cover these, with a particular view towards the statistical applications to follow.

13.5.1 The binomial distribution

Consider first a simple coin-tossing experiment, or any other random situation where only two outcomes are possible. We shall refer to these outcomes as 'success' and 'failure', but any other pair of terms (appropriate to the context) will do. Imagine tossing the coin (or performing the general experiment) n times and counting the number of successes. Clearly the sample space for this random variable Y, say, is $S = \{0, 1, \ldots, n\}$ with values near the middle of the range being more probable than values near the ends. It is this distribution that is sought.

A **Bernoulli trial** is a single observation of a random variable X, say, that can take the values zero or one:

$$P(X = 1) = p \quad \text{and} \quad P(X = 0) = 1 - p$$

for some success probability p.

The mean and variance of X are easily derived:

$$E(X) = 1(p) + 0(1 - p) = p$$

and

$$E(X^2) = 1^2(p) + 0^2(1 - p) = p$$

Hence

$$\sigma_X^2 = p - p^2 = p(1 - p)$$

Now let $\{X_1, \ldots, X_n\}$ denote n independent Bernoulli trials, each with success probability p. The number of successes is

$$Y = X_1 + \ldots + X_n$$

Suppose in general that $Y = k$, where $0 \leqslant k \leqslant n$. Then k of the X_i values are equal to one and $n - k$ are equal to zero. The probability of this occurring is

$$p^k(1 - p)^{n-k}$$

by the product rule (because the separate outcomes are independent).

There are many ways in which the k successes can be distributed among the n trials. For instance, if $n = 5$ and $k = 3$, the result might be $\{1, 1, 0, 1, 0\}$ or $\{0, 1, 1, 1, 0\}$ or $\{1, 0, 0, 1, 1\}$, and so on. As far as we are concerned, these are all equivalent, since we are interested only in the total number of successes and not their particular arrangement among the trials. The number of possible arrangements of the k successes among the n trials is given by the binomial coefficient (see Section 7.7)

$$\binom{n}{k} = \frac{n!}{(n - k)!k!}$$

Each arrangement of successes is exclusive of every other, so the addition rule of probabilities gives us the distribution

$$P(Y = k) = \binom{n}{k}p^k(1 - p)^{n-k} \quad (k = 0, \ldots, n)$$

This is the general form of the **binomial distribution**, with parameters n and p. The mean and variance of the binomial distribution are

$$E(Y) = np \quad \text{and} \quad \text{Var}(Y) = np(1 - p)$$

(this follows from the mean and variance of the Bernoulli random variable and the results on the mean and variance of sums of random variables in Section 13.4.9). Two typical binomial distributions can be seen in Figure 13.19.

Figure 13.19
Binomial distributions:
(a) $n = 12$, $p = 0.2$;
(b) $n = 12$, $p = 0.5$.

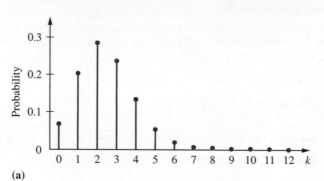

(a)

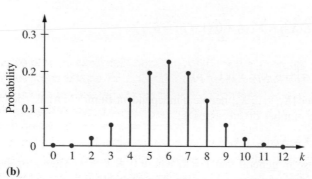

(b)

Example 13.25

A component supplier claims that 95% of its catalogue items are in stock at any time. A particular order for 20 different components is returned with three items missing as being out of stock. Is this likely, given the supplier's claim?

Solution

Each item can be either in stock or out of stock at any time, and the probability of each item being out of stock is 5%. The binomial distribution therefore applies and

$$P(k \text{ out of stock}) = \binom{20}{k} 0.05^k 0.95^{20-k}$$

so

$$
\begin{aligned}
P(3 \text{ or more out of stock}) &= P(3) + P(4) + \ldots + P(20) \\
&= 1 - P(0) - P(1) - P(2) \\
&= 1 - 0.3585 - 0.3774 - 0.1887 \\
&= 0.0755
\end{aligned}
$$

This is unlikely, given the supplier's claim.

There are several points to note about this simple example. The assumed figure of 5% probability of being out of stock is prompted by the supplier's claim, but in reality this will be an average figure, both between components (some may be out of stock more often than others because of supply difficulties) and over time (for the same reason). The independence assumption may not be true – if for instance a consignment of several similar types of components is awaited from a manufacturer and several of these are included in the order.

Most importantly, the probability worked out in the solution is that of three *or more* being out of stock, a result *at least as extreme* as that observed. Any result may have a low probability. What matters here is how far into the 'tail' of the distribution the actual result lies, and this is assessed by the total probability from there to the maximum value of k, which is 20. Note the use of the complement rule to simplify the calculation.

13.5.2 The Poisson distribution

The binomial distribution becomes unwieldy for large values of its parameter n, as illustrated in Examples 13.26 and 13.27. Another discrete distribution that often serves as a useful approximation to the binomial is the following:

$$P(X = k) = \frac{\lambda^k e^{-\lambda}}{k!} \quad (k = 0, 1, 2, \dots)$$

This is the general form of the **Poisson distribution**, with parameter λ. It is shown in *Advanced Modern Engineering Mathematics* that the mean and variance of the Poisson distribution are both equal to λ. These can be derived directly from the definition, but are more easily obtained by using the **moment generating function**, which will also be considered there. Also using this technique, it can be shown that for large n and small p the Poisson distribution approximates the binomial with $\lambda = np$. As a guide, the Poisson approximation can be used if $n \geqslant 25$ and $p \leqslant 0.1$. This is illustrated numerically in Figure 13.20, where binomial and Poisson distributions are compared for $n = 25$, $p = 0.1$ and $\lambda = 2.5$.

Figure 13.20
Binomial and Poisson distributions.

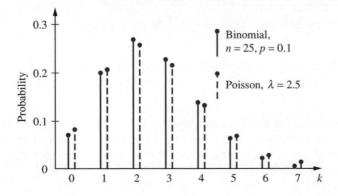

Example 13.26   If 0.04% of cars break down while driving through a certain tunnel, find the probability that at most two break down out of 2000 cars entering the tunnel on a given day.

Solution   The true distribution of breakdowns is binomial:

$$P(k \text{ breakdowns}) = \binom{2000}{k}(0.0004)^k(0.9996)^{2000-k}$$

for which $P(0) = 0.449\,26$, $P(1) = 0.359\,55$ and $P(2) = 0.143\,81$, and

$$P(\text{at most two breakdowns}) = P(0) + P(1) + P(2) = 0.952\,61$$

Because n is large and p small, the Poisson aproximation can also be used, with $\lambda = np = 0.8$, so that

$$P(\text{at most two breakdowns}) \approx e^{-\lambda}(1 + \lambda + \tfrac{1}{2}\lambda^2) = 0.952\,58$$

The Poisson calculation is easier, and the agreement is very good. It would not normally be appropriate to quote such an answer to five significant digits but it is only with such precision that the difference between the two distributions shows up.

Despite its ease of use compared with the binomial distribution some calculations with the Poisson distribution are difficult, especially those involving long summations. The following recurrence formulae are useful both for hand calculation and in computer programs. If X has a Poisson distribution with parameter λ, the successive Poisson probabilities are given by

$$P(X = k) = \frac{\lambda P(X = k - 1)}{k} \qquad (k = 1, 2, \dots)$$

with $P(X = 0) = e^{-\lambda}$. Furthermore, a cumulative property such as

$$P(X \leqslant 3) = e^{-\lambda}\left(1 + \lambda + \frac{\lambda^2}{2!} + \frac{\lambda^3}{3!}\right)$$

can be rewritten in the nested form (see Section 2.4)

$$P(X \leqslant 3) = e^{-\lambda}\left(\left(\left(\frac{\lambda}{3} + 1\right)\frac{\lambda}{2} + 1\right)\lambda + 1\right)$$

This approach can be generalized as follows: let

$$G_n = \frac{\lambda}{n} + 1 \quad \text{and} \quad G_{k-1} = \frac{\lambda}{k-1}G_k + 1 \quad (k = n, n-1, \dots, 2)$$

then

$$P(X \leqslant n) = e^{-\lambda}G_1$$

Example 13.27   A machine produces components that have defect A with probability 0.015 and defect B with probability 0.020, the two defects being independent. If 54 components are packed into a batch, what is the (approximate) probability that the batch contains at least 50 components without defects?

Solution   By the complement and product rules, the probability that a component will have neither defect is

$$P(\bar{A} \cap \bar{B}) = [1 - P(A)][1 - P(B)] = 0.9653$$

so the probability that a component will have at least one defect is 0.0347. If a batch contains at least 50 good components then it contains at most four defective ones, and, from the binomial distribution,

$$P(\text{at most four defective}) = \sum_{k=0}^{4} \binom{54}{k} (0.0347)^k (0.9653)^{54-k}$$

This is rather unwieldy, so we use the Poisson approximation with $\lambda = (54)(0.0347) = 1.874$. The successive values $G_4, \ldots, G_1$ in the recurrence formula above are 1.469, 1.917, 2.797 and 6.241, and hence

$$P(\text{at most four defective}) \approx 6.241\mathrm{e}^{-1.874} = 0.958$$

In other words, about one batch in 24 will contain less than 50 good components.

The binomial and Poisson are discrete distributions, which have the widest application among all discrete random variables. The Poisson distribution is especially useful to engineers because of its importance in statistical quality control. This will be introduced in Section 13.6, but we now turn to the most important of the continuous distributions.

13.5.3 The normal distribution

One class of distributions is awarded the name 'normal' because of the regularity with which random continuous data is found to obey it. This is no coincidence. The central limit theorem (Section 13.5.4) provides an explanation in terms of cumulative independent random parts adding up to a normal whole, a situation that is of great value in statistical inference (considered in *Advanced Modern Engineering Mathematics*). The normal distribution also serves as an approximation to the binomial distribution that complements the Poisson approximation.

The normal distribution has two parameters, which can be shown (see Question 60 in Exercises 13.5.7) to be the mean and standard deviation, so the appropriate symbols μ_X and σ_X are used.

A continuous random variable X has a **normal distribution** with mean μ_X and variance σ_X^2 if

$$f_X(x) = \frac{1}{\sigma_X \sqrt{(2\pi)}} \exp\left[-\frac{1}{2} \left(\frac{x - \mu_X}{\sigma_X} \right)^2 \right] \qquad (-\infty < x < +\infty,\ \sigma_X > 0)$$

The density function is symmetrical about μ_X and has the bell-shaped form shown in Figure 13.21. This distribution is also sometimes referred to by its more traditional name: the **Gaussian distribution**.

The need to declare that a random variable has a normal distribution (with a specified mean and variance) is so common that a special notation exists for the purpose:

$$X \sim N(\mu_X, \sigma_X^2)$$

Calculations involving the normal distribution are complicated by the fact that there is no simple expression for the integral of the density function on an arbitrary interval; in other words, the distribution function $F_X(x)$ does not have a simple explicit form.

Figure 13.21
The normal density
and distribution
functions (for $\mu_X = 0$
and $\sigma_X = 1$).

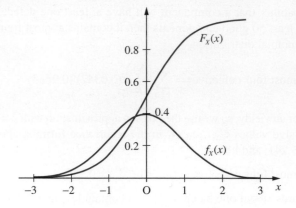

Instead, tables of this function are used. In fact, only a single table is needed: that for the special case of a normal distribution with a mean of zero and a variance of one.

> The **standard normal** cumulative distribution function is
>
> $$\Phi(z) = \frac{1}{\sqrt{(2\pi)}} \int_{-\infty}^{z} e^{-x^2/2} \, dx$$

This function is usually tabulated only for $z \geqslant 0$; for $z < 0$ the symmetry implies that

$$\Phi(-z) = 1 - \Phi(z)$$

A typical table of the standard normal function $\Phi(z)$ is provided in Figure 13.22.

For any random variable X, whether normal or not, subtracting the mean gives a random variable whose mean is zero:

$$E\{X - \mu_X\} = 0$$

The variance is not changed by this subtraction, but then dividing by the standard deviation gives a variable with a variance of one:

$$\mathrm{Var}\left(\frac{X - \mu_X}{\sigma_X}\right) = 1$$

(this follows from the results in Section 13.4.9). It is a property of the normal distribution, not shared by most distributions, that the result of this operation is still normal. It is usual to denote the new random variable by the letter Z:

$$Z = \frac{X - \mu_X}{\sigma_X}$$

This is then a **standard normal** random variable, to which the table applies. Conversely, any normal random variable can be considered to have been obtained from a standard normal random variable by multiplying by the required standard deviation and adding the mean:

$$X = \sigma_X Z + \mu_X$$

It follows that we can use the one table for the standard normal for all calculations involving normal variates.

Figure 13.22
Table of the standard normal cumulative distribution function $\Phi(z)$.

z	.00	.01	.02	.03	.04	.05	.06	.07	.08	.09
.0	.5000	.5040	.5080	.5120	.5160	.5199	.5239	.5279	.5319	.5359
.1	.5398	.5438	.5478	.5517	.5557	.5596	.5636	.5675	.5714	.5753
.2	.5793	.5832	.5871	.5910	.5948	.5987	.6026	.6064	.6103	.6141
.3	.6179	.6217	.6255	.6293	.6331	.6368	.6406	.6443	.6480	.6517
.4	.6554	.6591	.6628	.6664	.6700	.6736	.6772	.6808	.6844	.6879
.5	.6915	.6950	.6985	.7019	.7054	.7088	.7123	.7157	.7190	.7224
.6	.7257	.7291	.7324	.7357	.7389	.7422	.7454	.7486	.7517	.7549
.7	.7580	.7611	.7642	.7673	.7704	.7734	.7764	.7794	.7823	.7852
.8	.7881	.7910	.7939	.7967	.7995	.8023	.8051	.8078	.8106	.8133
.9	.8159	.8186	.8212	.8238	.8264	.8289	.8315	.8340	.8365	.8389
1.0	.8413	.8438	.8461	.8485	.8508	.8531	.8554	.8577	.8599	.8621
1.1	.8643	.8665	.8686	.8708	.8729	.8749	.8770	.8790	.8810	.8830
1.2	.8849	.8869	.8888	.8907	.8925	.8944	.8962	.8980	.8997	.9015
1.3	.9032	.9049	.9066	.9082	.9099	.9115	.9131	.9147	.9162	.9177
1.4	.9192	.9207	.9222	.9236	.9251	.9265	.9279	.9292	.9306	.9319
1.5	.9332	.9345	.9357	.9370	.9382	.9394	.9406	.9418	.9429	.9441
1.6	.9452	.9463	.9474	.9484	.9495	.9505	.9515	.9525	.9535	.9545
1.7	.9554	.9564	.9573	.9582	.9591	.9599	.9608	.9616	.9625	.9633
1.8	.9641	.9649	.9656	.9664	.9671	.9678	.9686	.9693	.9699	.9706
1.9	.9713	.9719	.9726	.9732	.9738	.9744	.9750	.9756	.9761	.9767
2.0	.9772	.9778	.9783	.9788	.9793	.9798	.9803	.9808	.9812	.9817
2.1	.9821	.9826	.9830	.9834	.9838	.9842	.9846	.9850	.9854	.9857
2.2	.9861	.9864	.9868	.9871	.9875	.9878	.9881	.9884	.9887	.9890
2.3	.9893	.9896	.9898	.9901	.9904	.9906	.9909	.9911	.9913	.9916
2.4	.9918	.9920	.9922	.9925	.9927	.9929	.9931	.9932	.9934	.9936
2.5	.9938	.9940	.9941	.9943	.9945	.9946	.9948	.9949	.9951	.9952
2.6	.9953	.9955	.9956	.9957	.9959	.9960	.9961	.9962	.9963	.9964
2.7	.9965	.9966	.9967	.9968	.9969	.9970	.9971	.9972	.9973	.9974
2.8	.9974	.9975	.9976	.9977	.9977	.9978	.9979	.9979	.9980	.9981
2.9	.9981	.9982	.9982	.9983	.9984	.9984	.9985	.9985	.9986	.9986
3.0	.9987	.9987	.9987	.9988	.9988	.9989	.9989	.9989	.9990	.9990
3.1	.9990	.9991	.9991	.9991	.9992	.9992	.9992	.9992	.9993	.9993
3.2	.9993	.9993	.9994	.9994	.9994	.9994	.9994	.9995	.9995	.9995
3.3	.9995	.9995	.9995	.9996	.9996	.9996	.9996	.9996	.9996	.9997
3.4	.9997	.9997	.9997	.9997	.9997	.9997	.9997	.9997	.9997	.9998

z	1.282	1.645	1.960	2.326	2.576	3.090	3.291	3.891	4.417
$\Phi(z)$	.90	.95	.975	.99	.995	.999	.9995	.99995	.999995
$2[1 - \Phi(z)]$	.20	.10	.05	.02	.01	.002	.001	.0001	.00001

Example 13.28   If $X \sim N(4, 4)$, find

(a) $P(X \leqslant 6.7)$

(b) the constant c such that $P(X > c) = 0.1$.

Solution   (a) $P(X \leqslant 6.7) = P\left(\dfrac{X-4}{2} \leqslant \dfrac{6.7-4}{2}\right)$

$$= P(Z \leqslant 1.35) = 0.9115 \quad \text{(from Figure 13.22)}$$

(b) If $P(X > c) = 0.1$ then $P(X \leqslant c) = 0.9$, so that

$$P\left(\dfrac{X-4}{2} \leqslant \dfrac{c-4}{2}\right) = P\left(Z \leqslant \dfrac{c-4}{2}\right) = 0.9$$

from which $\frac{1}{2}(c-4) = 1.282$ (using Figure 13.22); hence $c = 6.564$.

Example 13.28 shows that the standard normal table can be used in either direction: either to find the probability of an interval or to find the interval that gives a particular probability.

Example 13.29   The burning time X of an experimental rocket is a random variable having (approximately) a normal distribution with mean 600 s and standard deviation 25 s. Find the probability that such a rocket will burn for

(a) less than 550 s,   (b) more than 640 s.

Solution   Using the normal table as appropriate,

(a) $P(X < 550) = P\left(\dfrac{X-600}{25} < \dfrac{550-600}{25}\right) = P(Z < -2)$

$$= \Phi(-2) = 1 - \Phi(2) = 0.0228$$

(b) $P(X > 640) = P\left(\dfrac{X-600}{25} > \dfrac{640-600}{25}\right) = P(Z > 1.6)$

$$= 1 - \Phi(1.6) = 0.0548$$

13.5.4   The central limit theorem

The practical methods of statistical inference have foundations in probability theory, and the fundamental assumption underlying many of these methods is that the data has a distribution that is normal. Some statistical methods are **robust** in the sense that they work reliably even under moderate violations of their assumptions, but it is unsatisfactory to rely heavily upon this. If normality of the data were exceptional then this would

Figure 13.23
Continuous signal with
normal distribution.

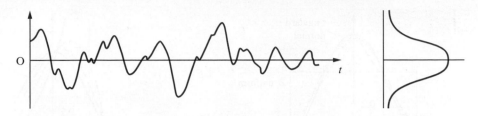

severely limit the scope of those methods that assume it. Fortunately (and as the name implies), the normal distribution arises very frequently in practice; the reason for this will be explained in this section.

Continuous measurements of random phenomena such as noise in electronic circuits or wave elevation on the sea surface give rise to graphs of the form shown in Figure 13.23. If the signal is sampled at regular intervals and a histogram of values built up, it is often found that the histogram closely approximates to a normal density curve. Physically, there are many separate independent random components adding up to produce the measured signal, and it is the total that is normal. There are many sources of noise in an electronic circuit and there are many separate waves on the sea. That the cumulative effect of these, which are often not individually normal, is to produce a total that has that special character is the substance of the following result, which is proved in *Advanced Modern Engineering Mathematics*.

Theorem 13.1   **Central limit theorem**

If $\{X_1, \ldots, X_n\}$ are independent and identically distributed random variables (the distribution being arbitrary), each with mean μ_X and variance σ_X^2, and if

$$W_n = \frac{X_1 + \ldots + X_n}{n} \quad \text{and} \quad Z_n = \frac{X_1 + \ldots + X_n - n\mu_X}{\sigma_X \sqrt{n}}$$

then, as $n \to \infty$, the distributions of W_n and Z_n tend to $W_n \sim N(\mu_X, \sigma_X^2/n)$ and $Z_n \sim N(0, 1)$ respectively.

---

end of theorem

Loosely speaking and with certain exceptions, the sum of independent identically distributed random variables tends to a normal distribution. The following points should be noted.

(a) The standard normal is obtained by subtracting the mean of the total and dividing by the standard deviation.

(b) The distributions converge to the normal in the sense that the cumulative distribution functions converge. This ensures that all observational properties of Z_n will be standard normal for sufficiently large n.

(c) How large n has to be before the normal approximation is good depends upon the underlying population. If the distribution of the variables X_i is symmetric about the mean then convergence to the normal is rapid. Figure 13.24(a) shows the distributions of the uniform random variable X with density function

$$f_X(x) = \tfrac{1}{2}\sqrt{\tfrac{1}{3}} \quad (-\sqrt{3} \leqslant x \leqslant \sqrt{3})$$

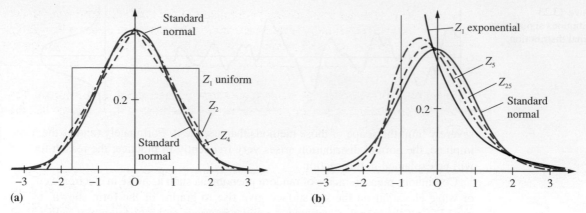

Figure 13.24 Central limit theorem: (a) uniform; (b) exponential.

(which has mean zero and variance one), together with those for Z_2 and Z_4. The normal distribution is also shown. Figure 13.24(b) shows similar results for the exponential random variable X with density function

$$f_X(x) = e^{-(x+1)} \quad (x \geq -1)$$

(which has mean zero and variance one), together with Z_5 and Z_{25}. Convergence is clearly more rapid for the symmetric distribution.

(d) The theorem can be generalized so that the random variables X_i do not need to be identically distributed, which is usually not the case in physical situations.

(e) Even where the data of an experiment is not normally distributed, the central limit theorem implies that the sample average has a normal distribution for large samples. Much valuable statistics exploits this fact.

Example 13.30

In a quality control scheme at a factory, batches of components are accepted or rejected depending on the number of defective items counted in a sample. Rejected batches are inspected and all defective items are replaced with good ones. From the machine reliability statistics it has been calculated that the probabilities of three, four, five, six, and seven defective items in a rejected batch are 0.3, 0.4, 0.2, 0.08 and 0.02 respectively. Fifty rejected batches produced a total of 221 defective items. Does this suggest that the machines are producing more defective items than they should?

Solution

If X represents the number of defective items in a rejected batch, the mean and standard deviation are given by

$$\mu_X = 3(0.3) + 4(0.4) + 5(0.2) + 6(0.08) + 7(0.02) = 4.12$$

$$\sigma_X = \sqrt{[9(0.3) + 16(0.4) + 25(0.2) + 36(0.08) + 49(0.02) - (4.12)^2]} = 0.9928$$

By the central limit theorem, the aggregate count of defectives Y in 50 rejected batches will be approximately normal, and

$$P(Y \geq 221) = P\left(\frac{Y - 50\mu_X}{\sigma_X\sqrt{50}} \geq \frac{221 - 50\mu_X}{\sigma_X\sqrt{50}}\right) = 1 - \Phi(2.137) = 0.0163$$

This probability is rather small, so the performance of the machines must come under suspicion. In fact, a rather more accurate answer to this problem is obtained by making a continuity correction, as explained in Section 13.5.5, but the conclusion is the same.

Example 13.30 is typical of many applications of the central limit theorem. The underlying distribution is certainly not normal, but it is reasonable to assume that the aggregate is approximately normal.

Example 13.31

In Section 1.5.5 it was noted that the maximum error that would occur in the sum of 100 numbers, each of which was rounded to three decimal places, is 0.05. Find the probability of the error in the sum exceeding 0.005 in magnitude, and the expected magnitude of the error.

Solution

We assume that the error in a number rounded to 3dp may be anything between -0.0005 and $+0.0005$, with all values in the range equally likely. In other words, the error in each value is a uniform random variable, X, say, with

$$f_X(x) = \begin{cases} 1000 & (-0.0005 < x < +0.0005) \\ 0 & (\text{otherwise}) \end{cases}$$

from which the mean and variance are given by

$$\mu_X = \int_{-0.0005}^{+0.0005} 1000x\,dx = 0$$

$$E(X^2) = \int_{-0.0005}^{+0.0005} 1000x^2\,dx = 8.333 \times 10^{-8}$$

$$\sigma_X^2 = E(X^2) - \mu_X^2 = 8.333 \times 10^{-8}$$

The error in the sum is a random variable $Y = X_1 + \ldots + X_{100}$. By the central limit theorem, approximately

$$Y \sim N(100\mu_X, 100\sigma_X^2) = N(0, 8.333 \times 10^{-6})$$

so

$$P(Y > 0.005) \approx P\left(Z > \frac{0.005}{0.00289}\right)$$

$$= P(Z > 1.732) = 1 - \Phi(1.732)$$

The error in the sum will exceed 0.005 in magnitude if $Y > 0.005$ or $Y < -0.005$, so, by symmetry,

$$P(|Y| > 0.005) \approx 2[1 - \Phi(1.732)]$$

$$= 2(1 - 0.9584)$$

$$= 0.0832$$

Thus, because the errors tend to cancel each other out, there is only one chance in 12 of the error reaching even $\frac{1}{10}$ of its maximum possible value. Furthermore, the expected value of the error magnitude is

$$E(|Y|) \approx \int_{-\infty}^{+\infty} \frac{|y|}{\sigma_Y \sqrt{(2\pi)}} e^{-y^2/2\sigma_Y^2} dy = 2 \int_0^{\infty} \frac{y}{\sigma_Y \sqrt{(2\pi)}} e^{-y^2/2\sigma_Y^2} dy$$

$$= \frac{1}{\sigma_Y \sqrt{(2\pi)}} \int_0^{\infty} e^{-w/2\sigma_Y^2} dw \quad \text{(by substitution of } w = y^2)$$

$$= \frac{2\sigma_Y^2}{\sigma_Y \sqrt{(2\pi)}} = \sigma_Y \sqrt{\left(\frac{2}{\pi}\right)}$$

With $\sigma_Y^2 = 8.333 \times 10^{-6}$, this gives $E(|Y|) \approx 0.0023$, which is less than $\frac{1}{20}$ of the maximum possible value.

13.5.5 Normal approximation to the binomial

One immediate corollary of the central limit theorem is that the normal distribution can be used to approximate the binomial distribution when n is sufficiently large. This follows from the definition of a binomial random variable as a sum of Bernoulli random variables (see Section 13.5.1). All that has to be done is to choose the parameters of the normal distribution to match the mean np and variance $np(1-p)$. As a rule, the normal approximation can be used when $n \geqslant 25$ and $0.1 \leqslant p \leqslant 0.9$. For values of p outside this range the Poisson approximation can be used.

It may seem surprising that the normal distribution, which is continuous, can be used to approximate a discrete distribution, given the very different character of these two types of random variable. The approximation of a discrete distribution X, say, by a continuous one Y works in the manner indicated in Figure 13.25. The probability that X takes the integer value k is approximated by the area under the density function $f_Y(y)$ between $k - 0.5$ and $k + 0.5$. Similarly, the following integral approximates to the probability that X exceeds k:

$$P(X > k) \approx \int_{k+0.5}^{\infty} f_Y(y) dy$$

Figure 13.25
Continuous
approximation to a
discrete distribution.

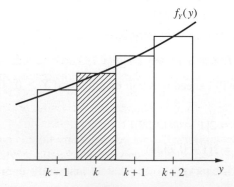

By the same token we would have

$$P(X \geqslant k) \approx \int_{k-0.5}^{\infty} f_Y(y)dy$$

This use of a half-integer shift in the limit of integration is called the **continuity correction** and gives a more accurate result.

Example 13.32   If 70% of airline passengers using a particular route are members of a frequent-flyer club, find the probability that out of a sample of 50 chosen independently, more than 40 will be members of a frequent-flyer club.

Solution   Let X represent the number who are members of a frequent-flyer club. The conditions for a binomial distribution are met, and the mean and variance are 50(0.7) = 35 and 50(0.7)(0.3) = 10.5 respectively. With no continuity correction, we have

$$P(X > 40) = P\left(\frac{X - 35}{\sqrt{(10.5)}} > \frac{40 - 35}{\sqrt{(10.5)}}\right)$$

$$\approx P(Z > 1.543) \quad \text{(where Z is standard normal)}$$

$$= 1 - \Phi(1.543) = 0.061$$

With the continuity correction,

$$P(X > 40) \approx P\left(Z > \frac{40.5 - 35}{\sqrt{(10.5)}}\right)$$

$$= P(Z > 1.697) = 0.045$$

As a percentage, this difference is substantial, so the continuity correction is important.

We now have three special classes of distribution: the binomial, Poisson and normal. The binomial is the most fundamental, and the others provide useful approximations to it in different circumstances, but the Poisson and especially the normal also have very important applications of their own. Increasingly in engineering and in all parts of industry, there are problems arising that involve these and other distributions but which it is not practical to solve without the aid of a computer.

13.5.6   Random variables for simulation

Computer simulations are very widely used in research, design and training. Perhaps the best known is the flight simulator upon which pilots receive much of their training. Simulations are used in research and design wherever a system is too complex for a complete solution to a problem to be obtained theoretically, or where a solution can be obtained but its completeness or accuracy is open to question.

Simulations are **deterministic** if what occurs at any time is completely determined by the state of the system. In contrast, they are **stochastic** if what occurs at any time can be influenced by a chance element that is inherently unpredictable. Stochastic simulations therefore require that random variables (or outcomes) be generated within the program. This may seem a hopeless requirement, considering that computer programs are sequences of deterministic instructions running on deterministic hardware. However, it is possible to generate sequences of numbers that are deterministic and repeatable but that have the appearance of being random. These **pseudo-random numbers** are very useful for simulations, and for other purposes such as the so-called Monte Carlo numerical methods.

Most modern computers contain a software facility for generating pseudo-random numbers with a uniform distribution on the interval $(0, 1)$:

$$f_U(u) = \begin{cases} 1 & (0 < u < 1) \\ 0 & (\text{otherwise}) \end{cases}$$

The successive variables $\{U_1, U_2, \ldots\}$ appear to be uncorrelated, and, although there is some structure in the sequence (and indeed the sequence will eventually repeat itself), it is rare for these deficiencies to cause problems in practice.

Random variables with non-uniform distributions are obtained from the sequence $\{U_1, U_2, \ldots\}$ by applying various transformations. Figure 13.26 contains pseudocode

Figure 13.26
Pseudocode listings for non-uniform random variables.

```
{Bernoulli random variable X, parameter p.}
U ← rnd
if U < p then X ← 1 else X ← 0 endif

{Binomial random variable X, parameters n,p.}
X ← 0
for i is 1 to n do
   U ← rnd
   if U < p then X ← X + 1 endif
endfor

{Exponential random variable X, parameter L, uses log function to base e.}
U ← rnd
X ← − (log(U))/L

{Poisson random variable X, parameter L.}
X ← −1
W ← 1
P0 ← exp(−L)
repeat
   X ← X + 1
   U ← rnd
   W ← W*U
until W < P0

{Normal random variable X, parameters mean, sd.}
T ← 0
for i is 1 to 12 do
   U ← rnd
   T ← T + U
endfor
X ← sd*(T − 6) + mean
```

listings for generating the most common random variables. In each case it is assumed that the system function 'rnd' returns a uniform (0, 1) value, which is stored in the variable U. The variable X contains the required value of the random variable. The binomial is based on the Bernoulli, the Poisson on the exponential, and the normal on the central limit theorem. For a full explanation of how these work see S. J. Yakowitz, *Computational Probability and Simulation* (Addison-Wesley, 1977). Computer packages such as Minitab also often contain facilities for generating random data.

13.5.7 Exercises

49 Eight babies are born in a hospital on a particular day. Find the probability that exactly half of them are boys. (The probability that a baby is a boy is actually slightly greater than one-half, but you can take it as exactly one-half for this exercise.)

50 A town has five fire engines operating independently, each of which spends 94% of the time in its station awaiting a call. Find the probability that at least three fire engines are available when needed.

51 The probability of issuing a drill of high brittleness (a reject) is 0.02. Drills are packed in boxes of 100 each. What is the probability that the number of defective drills is no greater than two?

52 If Z is a random variable having the standard normal distribution, find the probabilities that Z will have a value

(a) greater than 1.14,

(b) less than −0.36,

(c) between −0.46 and −0.09,

(d) between −0.58 and 1.12.

53 Assume that

(a) an aircraft can land safely if at least half of its engines are working,

(b) the probability of an engine failing is 0.1, and

(c) engine failures are independent.

Which is safer, a four-engine plane or a two-engine plane?

54 If on average one in 20 of a certain type of column will fail under a given axial load, what are the probabilities that among 16 such columns, (a) at most two, (b) at least four will fail?

55 A machine makes components, and the probability that a component is defective is p. If components are packed in cartons of 20, what value of p will ensure that 90% of cartons contain at most one defective component?

56 If on average 7% of airline passengers order special meals, find the approximate probability that on a particular flight carrying 85 passengers, eight or more will order special meals.

57 A Geiger counter and a source of radioactive particles are so situated that the probability that a particle emanating from the radioactive source will be registered by the counter is 1/10 000. Assume that during the time of observation, 30 000 particles emanated from the source. What is the probability that the number of particles registered was (a) zero, (b) three, (c) more than five?

58 Assume that in the composition of a book there exists a constant probability 0.0001 that an arbitrary letter will be set incorrectly. After composition, the proofs are read by a proofreader, who discovers 90% of the errors. After the proofreader, the author discovers half of the remaining errors. Find the probability that in a book with 500 000 printing symbols there remain after this no more than six unnoticed errors.

59 Suppose that the actual amount of cement that a filling machine puts into 'six-kilogram' bags is a normal random variable with $\sigma = 0.05$ kg. If only 3% of bags are to contain less than 6 kg, what must be the mean fill of the bags?

60 Prove by making the substitution $u = (x − \mu_X)/\sigma_X$ in the integrals concerned that the mean and variance of the normal distribution are μ_X and σ_X^2 respectively. (*Hint*: for the variance, integrate $u[u \exp(−u^2/2)]$ by parts.)

61 If 23% of all patients with high blood pressure have bad side-effects from a certain kind of medicine, use the normal approximation to the binomial to find the probability that among 120 patients with high blood pressure treated with this medicine, more than 32 will have bad side-effects.

62 In firing at a target, a marksman scores at each shot either 10, 9, 8, 7, or 6, with respective probabilities 0.5, 0.3, 0.1, 0.05, 0.05. If he fires 100 shots, what is the approximate probability that his aggregate score exceeds 940?

63 A fleet car operator has n cars, each of which has probability 8% of being broken down on any particular day. Find the smallest value of n that gives probability 90% that at least 40 cars will be available for use on any one day.

64 The diameter of ball bearings produced by a machine is a random variable having a normal distribution with mean 6.00 mm and standard deviation 0.02 mm. If the diameter tolerance is ±1%, find the proportion of ball bearings produced that are out of tolerance. After several years' use, machine wear has the effect of increasing the standard deviation, although the mean diameter remains constant. The manufacturer decides to replace the machine when 2% of its output is out of tolerance. What is the standard deviation when this happens?

65 A major airline operates 350 flights a day throughout the world. The probability that a flight will be delayed for more than one hour, for any reason, is 0.7%. If more than four flights suffer such delays in any one day, the implications for route organization and crewing become serious. Call such a day a 'flap-day'. Using approximations as appropriate, find the probabilities that

(a) any particular day is a flap-day,

(b) two flap-days (not more) occur in one week,

(c) more than 50 flap-days occur in a year of 365 days.

13.6   Engineering application: quality control

This is a topic of particular relevance to engineers, because the statistical methods of quality control are widely and increasingly used in industry in order to promote the reliability of products. Orders have been won and lost because one manufacturer has implemented quality control in the workplace more than another and the purchaser has used this as a criterion when deciding where to place the order.

Quality control statistics is not particularly difficult, but (as usual) it rests on fundamental results such as the Poisson approximation to the binomial distribution (Section 13.5.2). The methods apply mainly to mass production systems where quality can be measured numerically.

This section will introduce the use of control charts for continuous monitoring of quality, rather than the more traditional batch inspection plans, and control charts are further discussed in *Advanced Modern Engineering Mathematics*.

13.6.1   Attribute control charts

Manufactured items that are elaborate and therefore expensive can each be tested thoroughly before dispatch to consumers, but such item-by-item testing must be ruled out for low-level components on grounds of cost. Some defective items are bound to slip through, and the objective of quality control is to keep the proportion of these within acceptable and agreed limits.

Figure 13.27
Attribute control chart
for Example 13.33.

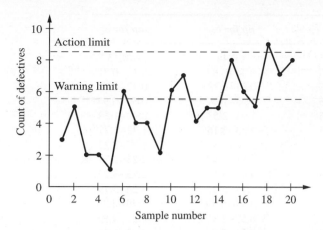

Variations occur in the quality of a product, caused either by variations in the raw material or input or by variations in processing. Quality is monitored by regular testing of samples of output. Assume for now that the test consists in counting the number within the sample which pass or fail according to some performance criterion. A small proportion of defective items in the output is permitted while the process is said to be **in control**. If the actual proportion of defectives rises to an unacceptable level, the process is said to be **out of control**, and the counts of defectives in the samples would be expected to rise. We should like to detect this as soon as possible when it occurs, but without incurring the expense of a large number of false alarms while the process is actually in control.

An essential aid to the quality controller is the **Shewhart control chart**, which is a plot of the successive counts of defective items against sample number. Figure 13.27 is an example of such a chart. Also shown on the chart are two limits on the counts of defectives, corresponding to probabilities of one in 40 and one in 1000 of a sample count falling outside the limit if the process is in control. These are called **warning** and **action limits** respectively, and are denoted by c_W and c_A.

Any sample point falling outside the action limit would normally result in the process being suspended and the problem corrected. Roughly one in 40 sample points will fall outside the warning limit purely by chance, but if this occurs repeatedly or if there is a clear trend upwards in the counts of defectives then action may well be taken before the action limit itself is crossed.

To obtain the warning and action limits, we use the Poisson approximation to the binomial. If the acceptable proportion of defective items is p, usually small, and the sample size is n then for a process in control the defective count C, say, will be a binomial random variable with parameters n and p. Provided that n is not too small, the Poisson approximation can be used (Section 13.5.2):

$$P(C \geqslant c) \approx \sum_{k=c}^{n} \frac{(np)^k e^{-np}}{k!}$$

Equating this to $\frac{1}{40}$ and then to $\frac{1}{1000}$ gives equations that can be solved for the warning limit c_W and the action limit c_A respectively, in terms of the product np. This is the basis of the table shown in Figure 13.28, which enables c_W and c_A to be read directly from the value of np.

Figure 13.28
Shewhart attribute control limits: n is sample size, p is probability of defect, c_W is warning limit and c_A is action limit.

c_W or c_A	np for c_W	np for c_A
1.5	<0.44	<0.13
2.5	0.44–0.87	0.13–0.32
3.5	0.87–1.38	0.32–0.60
4.5	1.38–1.94	0.60–0.94
5.5	1.94–2.53	0.94–1.33
6.5	2.53–3.16	1.33–1.77
7.5	3.16–3.81	1.77–2.23
8.5	3.81–4.48	2.23–2.73
9.5	4.48–5.17	2.73–3.25
10.5	5.17–5.87	3.25–3.79
11.5	5.87–6.59	3.79–4.35
12.5	6.59–7.31	4.35–4.93
13.5	7.31–8.05	4.93–5.52
14.5	8.05–8.80	5.52–6.12
15.5	8.80–9.55	6.12–6.74
16.5	9.55–10.31	6.74–7.37
17.5	10.31–11.08	7.37–8.01
18.5	11.08–11.85	8.01–8.66
19.5	11.85–12.63	8.66–9.31
20.5	12.63–13.42	9.31–9.98
21.5	13.42–14.21	9.98–10.65
22.5	14.21–15.00	10.65–11.33
23.5	15.00–15.80	11.33–12.02
24.5	15.80–16.61	12.02–12.71
25.5	16.61–17.41	12.71–13.41
26.5	17.41–18.23	13.41–14.11
27.5	18.23–19.04	14.11–14.82
28.5	19.04–19.86	14.82–15.53
29.5	19.86–20.68	15.53–16.25
30.5		16.25–16.98
31.5		16.98–17.70
32.5		17.70–18.44
33.5		18.44–19.17
34.5		19.17–19.91
35.5		19.91–20.66

Example 13.33

Regular samples of 50 are taken from a process making electronic components, for which an acceptable proportion of defectives is 5%. Successive counts of defectives in each sample are as follows:

Sample	1	2	3	4	5	6	7	8	9	10	11	12	13	14	15	16	17	18	19	20
Count	3	5	2	2	1	6	4	4	2	6	7	4	5	5	8	6	5	9	7	8

At what point would the decision be taken to stop and correct the process?

Solution The control chart is shown in Figure 13.27. From $np = 2.5$ and Figure 13.28 we have the warning limit $c_W = 5.5$ and the action limit $c_A = 8.5$. The half-integer values are to avoid ambiguity when the count lies on a limit. There are warnings at samples 6, 10, 11, 15 and 16 before the action limit is crossed at sample 18. Strictly, the decision should be taken at that point, but the probability of two consecutive warnings is less than one in 1600 by the product rule of probabilities, which would justify taking action after sample 11.

Example 13.33 shows that the strict practice of waiting for the action limit to be crossed in the Shewhart control chart would be rather conservative. The long sequence of counts that exceed the expected number of defectives would lead to the decision being taken sooner in practice.

13.6.2 United States standard attribute charts

The control chart decribed above, with action and warning limits set by probability of exceedance, is the standard practice in the United Kingdom. In the United States the practice is rather different in that there is usually no warning limit and that action limit (called the **upper control limit**, **UCL**) is set at three standard deviations above the mean. Because the count of defectives is binomial with mean np and variance $np(1 - p)$, this means that

$$UCL = np + 3[np(1 - p)]^{1/2}$$

Example 13.34 Find the UCL and apply it to the data in Example 13.33.

Solution From $n = 50$ and $p = 0.05$ we infer that $UCL = 7.1$, which is between the warning limit c_W and the action limit c_A in Example 13.33. The decision to correct the process would be taken after the 15th sample, the first to exceed the UCL.

Sometimes a **lower limit control**, **LCL**, is defined at three standard deviations below the mean:

$$LCL = np - 3[np(1 - p)]^{1/2}$$

If this is positive, it can be used to test whether the proportion defective in the output is falling significantly below the expected value.

Control charts are also useful for monitoring the output of a manufacturing process where quality depends upon a numerical measure such as dimension, weight or resistance. Charts that are more powerful than the Shewhart charts at detecting variations in the output are also used. These topics are covered in *Advanced Modern Engineering Mathematics*.

13.6.3 Exercises

 66 It is intended that 90% of electronic devices emerging from a machine should pass a simple on-the-spot quality test. The numbers of defectives among samples of 50 taken by successive shifts are as follows:

$$5, 8, 11, 5, 6, 4, 9, 7, 12, 9, 10, 14$$

Find the action and warning limits, and the sample number at which an out-of-control decision is taken. Also find the UCL (United States practice) and the sample number for action.

 67 Thirty-two successive samples of 100 castings each, taken from a production line, contained numbers of defectives as follows:

$$3, 3, 5, 3, 5, 0, 3, 1, 3, 5, 4, 2, 4, 3, 5, 4, 3,$$
$$4, 5, 6, 5, 6, 4, 4, 7, 5, 4, 8, 5, 6, 6, 7$$

If the proportion defective is to be maintained at 0.02, use the Shewhart method (both UK and US standard) to indicate whether this proportion is being maintained, and if not then after how many samples action should be taken.

13.7   Engineering application: clustering of rare events

13.7.1   Introduction

To conclude this chapter, we shall apply some of the probability theory covered so far to an investigation of a serious problem, or rather a family of problems. Failures of engineering systems or structures are rare events, but they have serious consequences. If a number of similar failures occur and a link between them can be found then it may be possible to anticipate and prevent future failures. One aspect of this is the detection of regional variations in the number of failures, which may provide clues as to possible causes.

Problems like this, and their associated difficulties, arise in many fields, and the lessons learned from analysing one can often be applied to others. Some typical examples are as follows:

(a)   near-misses between two aircraft in flight,
(b)   collisions or capsizing of ships at sea,
(c)   accidents involving road vehicles,
(d)   occurrences of environmentally induced diseases such as leukaemia.

These problems can be looked at in various ways, but there is an approach that applies to all of them because of the following common elements:

(a)   a very large number of potential cases,
(b)   a very small proportion of these become actual cases,
(c)   a possible common cause,
(d)   regional variations in the common cause if it exists.

Common causes for (a) and (b) that vary regionally could be dangerous weather conditions or inadequate control over routes taken; for (c) they could be inadequate lighting or hazard warnings, and for (d) they could be proximity to a nuclear installation. For each problem it is important to identify the common cause if it exists – and one clue to its existence is the regional variation. Also, for each problem the main difficulty is the rarity of the cases, and it is this that makes an analysis using probability theory useful.

This case study is expressed in terms of a survey of near-misses in aircraft operations, but the analysis could be applied to any of the above examples. The figures are hypothetical, but the method of analysis is realistic.

13.7.2 Survey of near-misses between aircraft

Suppose that the major airlines cooperate in a survey of near-misses during a period of one year. The region being studied is divided up into 1000 areas in such a way that there are on average 200 flights per year through each area. Suppose that the total number of flights is 200 000 and that the total number of near-misses logged by the pilots is 120. Although a near-miss involves two aircraft, it is recorded as a single incident. At the end of the year the data is examined and two areas in particular stand out. In one area A four incidents occurred in a total of 400 flights, and in another area B two incidents occurred in a total of 150 flights.

The question that it is natural to ask is whether there are any areas, in particular these two, in which the number of near-misses is greater than can be accounted for by chance. If any such area exists, it can be examined to see what makes it special, and this may lead to the discovery of a common cause and appropriate action being taken. To approach this, we shall first assume that the probability that a near-miss will occur is the same for every flight and in every area. This probability is taken to be the total number of near-misses divided by the total number of flights, which gives $p = 6 \times 10^{-4}$.

To assess how unlikely the figures for areas A and B are we need to calculate the probability of the given number *or more* of incidents, as explained in the discussion following Example 13.25. If we assume that the probability p applies independently for every flight then for area A, using the binomial distribution, we have

$$P(4 \text{ or more incidents}) = 1 - \sum_{k=0}^{3} \binom{400}{k} p^k (1 - p)^{400-k}$$

which gives 1.13×10^{-4}. Alternatively, using the Poisson approximation (Section 13.5.2),

$$P(4 \text{ or more incidents}) \approx 1 - \sum_{k=0}^{3} \frac{\lambda^k e^{-\lambda}}{k!} \quad (\text{with } \lambda = 400p = 0.24)$$

$$= 1 - e^{-\lambda} \left(\left(\left(\frac{\lambda}{3} + 1 \right) \frac{\lambda}{2} + 1 \right) \lambda + 1 \right)$$

$$= 1.14 \times 10^{-4}$$

which is close to the exact figure. Similarly, for area B

$$P(2 \text{ or more incidents}) = 1 - (1 - p)^{150} - 150p(1 - p)^{149} = 3.79 \times 10^{-3}$$

$$\approx 1 - e^{-\lambda}(1 + \lambda) \quad (\text{with } \lambda = 150p = 0.09)$$

$$= 3.82 \times 10^{-3}$$

The effectiveness of the Poisson approximation to the binomial is clear from these results.

The four incidents in area A are seen to be much less likely to be due to chance than the two incidents in area B. This is interesting, because common sense would suggest comparing the proportions in the respective areas, which are 1% for A and 1.33% for B. Despite having the lower proportion of incidents, area A provides stronger evidence for a regional anomaly, by more than an order of magnitude.

The extent of anomaly can be judged from the probability that at least one of the 1000 areas in the region will give a result at least as extreme as those observed. After all, with 1000 opportunities for a rare event to occur, the probability that it will occur in at least one of them is significantly enhanced. Using the complement and product rules we have

$$P(\text{at least one event with probability } 1.13 \times 10^{-4} \text{ in 1000 areas})$$

$$= 1 - P \text{ (no such events)}$$

$$= 1 - (1 - 1.13 \times 10^{-4})^{1000} = 0.107$$

Similarly,

$$P(\text{at least one event with probability } 3.79 \times 10^{-3} \text{ in 1000 areas})$$

$$= 1 - (1 - 3.79 \times 10^{-3})^{1000} = 0.978$$

The area B result has a high probability of occurring by chance, somewhere within the 1000 areas. However, there is only one chance in ten that a result as improbable as that in area A would occur anywhere, assuming a constant value of p. It seems that the true probability of a near-miss is higher in that area. Although the number of incidents is small, quite a firm conclusion has been reached.

The most interesting point about this analysis is that the comparison of proportions, which is the most obvious way of judging the results, is so misleading. The reason why it doesn't work is that (as can be seen in *Advanced Modern Engineering Mathematics*) the variance of a sample proportion depends upon the size of the sample, the denominator in that proportion. Dividing the number of incidents by the number of flights in an attempt to normalize the data fails to eliminate the number of flights as a variable, because of its lingering influence on the statistics.

This is as far as the mathematical analysis can proceed. It cannot point to any particular cause without further data. Tracking down the reason for the anomaly can be very difficult in situations like this, but at least the search can be focused on an area. The local weather, the operating procedures and technical support of the flight controllers, and natural sources of interference in the navigational equipment would all be under suspicion.

13.7.3 Exercises

 68 A third area in the near-miss survey recorded five incidents in 800 flights. Should this area also be regarded as unusually risky?

 69 Two adjacent areas recorded two incidents in 250 flights and one incident in 85 flights respectively. Test the combination of the two areas.

13.8   Review exercises (1–13)

1   A continuous random variable X has probability density function given by

$$f_X(x) = \begin{cases} \dfrac{c}{x^4} & \text{for } x \geq 1 \\[2mm] 0 & \text{for } x < 1 \end{cases}$$

where c is constant. Find

(a) the value of the constant c,

(b) the cumulative distribution function of X,

(c) $P(X > 2)$,

(d) the mean of X,

(e) the standard deviation of X.

2   If there are 720 personal computers in an office building and they each break down independently with probability 0.002 per working day, use the Poisson approximation to the binomial distribution to find the probability that more than four of these computers will break down in any one working day.

3   The City Engineer's department installs 10 000 fluorescent lamp bulbs in street lamp standards. The bulbs have an average life of 7000 operating hours with a standard deviation of 400 hours. Assuming that the life of the bulbs, L, is a normal random variable, what number of bulbs might be expected to have failed after 6000 operating hours? If the engineer wishes to adopt a routine replacement policy which ensures that no more than 5% of the bulbs fail before their routine replacement, after how long should the bulbs be replaced?

4   The binomial is a special case of the more general **multinomial distribution**:

$$P(n_1, \ldots, n_k) = \frac{n!}{n_1! \ldots n_k!} (p_1)^{n_1} \ldots (p_k)^{n_k}$$

where $p_1 + \ldots + p_k = 1$ and $n_1 + \ldots + n_k = n$. Each observation of a random variable has k possible outcomes, with probabilities $p_1, \ldots, p_k$, and the observed total numbers of each possible outcome

after n independent observations are made are respectively $n_1, \ldots, n_k$. Suppose that 60% of calls to a telephone banking enquiry service are for account balance requests, 20% are for payment confirmations, 10% are for transfer requests, and 10% are to open new accounts. Find the probability that out of 20 calls to this service there will be 10 balance requests, five payment confirmations, three transfers and two new accounts.

5   A manufacturer has agreed to dispatch small servomechanisms in cartons of 100 to a distributor. The distributor requires that 90% of cartons contain at most one defective servomechanism. Assuming the Poisson approximation to the binomial distribution, write down an equation for the Poisson parameter λ such that the distributor's requirements are just satisfied. Solve by trial and error (approximate solution 0.5), and hence find the required proportion of manufactured servomechanisms that must be satisfactory.

6   Ten thousand numbers are to be added, each rounded to the sixth decimal place. Assuming that the errors arising from rounding the numbers are mutually independent and uniformly distributed on $(-0.5 \times 10^{-6}, +0.5 \times 10^{-6})$, find the limits in which the total error will lie with probability 95%.

7   Suppose that X is a continuous random variable with mean μ_X and variance σ_X^2. By separating the integral in the definition of σ_X^2 into three parts and substituting the respective bounds for $(x - \mu_X)^2$ as follows

$$(x - \mu_X)^2 \geq \begin{cases} (k\sigma_X)^2 & \text{on } (-\infty, \mu_X - k\sigma_X) \\[1mm] 0 & \text{on } (\mu_X - k\sigma_X, \mu_X + k\sigma_X) \\[1mm] (k\sigma_X)^2 & \text{on } (\mu_X + k\sigma_X, +\infty) \end{cases}$$

where k is a constant, prove **Chebyshev's theorem**

$$P(|x - \mu_X| > k\sigma_X) \leq k^{-2}$$

Deduce that for every continuous random variable X the probability is at least $\frac{8}{9}$ that X will take a value within three standard deviations of the mean.

8   The function

$$\Gamma(\alpha) = \int_0^\infty y^{\alpha-1} e^{-y} \, dy \quad (\alpha > 0)$$

is known as the **gamma function**, and the probability density function

$$f_X(x) = \begin{cases} [\Gamma(\alpha)]^{-1} \lambda^\alpha x^{\alpha-1} e^{-\lambda x} & (x > 0) \\ 0 & \text{(otherwise)} \end{cases}$$

defines the **gamma distribution**. Prove that

(a) $\mu_X = \alpha/\lambda$,

(b) $\sigma_X^2 = \alpha/\lambda^2$.

9   If $X_1, \ldots, X_n$ are independent exponentially distributed random variables, each with parameter λ, prove that the random variable whose value is given by the minimum of $\{X_1, \ldots, X_n\}$ also has an exponential distribution, with parameter $n\lambda$. In particular, if a complex piece of machinery consists of six parts, each of which has an exponential distribution of time to failure with mean 2000 hours, and if the machine fails as soon as any of its parts fail, find the probability that the time to failure exceeds 300 hours.

10 Find the expected value of the maximum of four independent exponential random variables, each with parameter λ. In particular, if the time taken for a routine test and service of a jet aircraft engine has an exponential distribution with a mean of three hours, find the mean time to complete a four-engine aircraft if the service times are independent.

11 In the game of craps, two dice are tossed. A total of 7 or 11 wins immediately, a total of 2, 3 or 12 loses. For remaining outcomes, both dice are tossed repeatedly until either a total of 7 appears, which loses, or the original number, which wins. Show that the overall probability of winning is approximately 0.493.

12 A large number N of people are subjected to a blood investigation to test for the presence of an illegal drug. This investigation is carried out by mixing the blood of k persons at a time and testing the mixture. If the result of the analysis is negative then this is sufficient for all k persons. If the result is positive then the blood of each person must be analysed separately, making $k + 1$ analyses in all. Assume that the probability p of a positive result is the same for each person and that the results of the analyses are independent. Find the expected number of analyses, and minimize with respect to k. In particular, find the optimum value of k when $p = 0.01$, and the expected saving compared with a separate analysis for all N people.

13 Error-correcting codes are widely used for data transmission. A message consisting of N binary bits is partitioned into blocks of k bits, and each block is transmitted with some additional parity bits giving a total of n bits per block. The parity bits are used at the receiving end to correct any errors that occur in transmission (bits that get inverted, including the parity bits themselves). Some error-correcting codes can correct only a single error per block; others can correct up to two errors. The number $n - k$ of parity bits is chosen as small as possible to satisfy the relationship:

$$2^{n-k} \geq n + 1 \quad \text{(single-error-correcting code)}$$

or

$$2^{n-k} \geq n^2 + 1 \quad \text{(double-error-correcting code)}$$

(a) Suppose that transmission errors occur independently at an average rate of 1% of bits transmitted. For data blocks k of 4, 8, 16, 32 and 64 bits, find the value of n and the probability of more errors occurring than the code can correct. Do this for single- and double-error-correcting codes.

(b) Find for each type of code the largest block size k that allows a total of $N = 64$ data bits to be transmitted with at least 95% probability of correct overall interpretation at the receiving end. Compare the total numbers of bits transmitted in each case.

AI.1 Trigonometric identities

$$\cos^2 x + \sin^2 x = 1$$

$$1 + \tan^2 x = \sec^2 x$$

$$1 + \cot^2 x = \operatorname{cosec}^2 x$$

$$\sin(x + y) = \sin x \cos y + \cos x \sin y$$

$$\sin(x - y) = \sin x \cos y - \cos x \sin y$$

$$\cos(x + y) = \cos x \cos y - \sin x \sin y$$

$$\cos(x - y) = \cos x \cos y + \sin x \sin y$$

$$\tan(x + y) = \frac{\tan x + \tan y}{1 - \tan x \tan y}$$

$$\tan(x - y) = \frac{\tan x - \tan y}{1 + \tan x \tan y}$$

$$\sin 2x = 2 \sin x \cos x$$

$$\cos 2x = \cos^2 x - \sin^2 x$$

$$= 1 - 2 \sin^2 x$$

$$= 2 \cos^2 x - 1$$

$$\sin x + \sin y = 2 \sin \tfrac{1}{2}(x + y) \cos \tfrac{1}{2}(x - y)$$

$$\sin x - \sin y = 2 \cos \tfrac{1}{2}(x + y) \sin \tfrac{1}{2}(x - y)$$

$$\cos x + \cos y = 2 \cos \tfrac{1}{2}(x + y) \cos \tfrac{1}{2}(x - y)$$

$$\cos x - \cos y = -2 \sin \tfrac{1}{2}(x + y) \sin \tfrac{1}{2}(x - y)$$

$$\sin x \cos y = \tfrac{1}{2}[\sin(x + y) + \sin(x - y)]$$

$$\cos x \sin y = \tfrac{1}{2}[\sin(x + y) - \sin(x - y)]$$

$$\cos x \cos y = \tfrac{1}{2}[\cos(x + y) + \cos(x - y)]$$

$$\sin x \sin y = \tfrac{1}{2}[\cos(x - y) - \cos(x + y)]$$

$$\sin 3x = 3 \sin x - 4 \sin^3 x$$

$$\cos 3x = 4 \cos^3 x - 3 \cos x$$

AI.2 Derivatives and integrals

y	dy/dx	$\int y\,dx$
x^n	nx^{n-1}	$x^{n+1}/(n+1)\quad(n\neq-1)$
$1/x$	$-1/x^2$	$\ln\lvert x\rvert$
$\sin x$	$\cos x$	$-\cos x$
$\cos x$	$-\sin x$	$\sin x$
$\tan x$	$\sec^2 x$	$-\ln\lvert\cos x\rvert$
$\sec x$	$\sec x\tan x$	$\ln\lvert\sec x+\tan x\rvert$
$\cot x$	$-\mathrm{cosec}^2 x$	$\ln\lvert\sin x\rvert$
$\mathrm{cosec}\,x$	$-\mathrm{cosec}\,x\cot x$	$-\ln\lvert\mathrm{cosec}\,x+\cot x\rvert$
$\sin^{-1}x$	$\dfrac{1}{\sqrt{(1-x^2)}}$	$x\sin^{-1}x+\sqrt{(1-x^2)}$
$\cos^{-1}x$	$\dfrac{-1}{\sqrt{(1-x^2)}}$	$x\cos^{-1}x-\sqrt{(1-x^2)}$
$\tan^{-1}x$	$\dfrac{1}{1+x^2}$	$x\tan^{-1}x-\tfrac12\ln(1+x^2)$
$\sec^{-1}x$	$\dfrac{1}{x\sqrt{(x^2-1)}}$	$x\sec^{-1}x-\ln\lvert x+\sqrt{(x^2-1)}\rvert$ $(0<\sec^{-1}x<\tfrac12\pi)$
$\mathrm{cosec}^{-1}x$	$\dfrac{-1}{x\sqrt{(x^2-1)}}$	$(x\,\mathrm{cosec}^{-1}x+\ln\lvert x+\sqrt{(x^2-1)}\rvert$ $(0<\mathrm{cosec}^{-1}x<\tfrac12\pi)$
$\cot^{-1}x$	$\dfrac{-1}{1+x^2}$	$x\cot^{-1}x+\tfrac12\ln(1+x^2)$
e^{ax}	$a\mathrm{e}^{ax}$	e^{ax}/a
a^x	$a^x\ln a$	$a^x/\ln a$
$\sinh x$	$\cosh x$	$\cosh x$
$\cosh x$	$\sinh x$	$\sinh x$
$\tanh x$	$\mathrm{sech}^2 x$	$\ln\cosh x$
$\mathrm{sech}\,x$	$-\mathrm{sech}\,x\tanh x$	$\tan^{-1}(\sinh x)$
$\mathrm{cosech}\,x$	$-\mathrm{cosech}\,x\coth x$	$\ln\lvert\tanh\tfrac12 x\rvert$
$\coth x$	$-\mathrm{cosech}^2 x$	$\ln\lvert\sinh x\rvert$
$\sinh^{-1}x$	$\dfrac{1}{\sqrt{(1+x^2)}}$	$x\sinh^{-1}x-\sqrt{(1+x^2)}$
$\cosh^{-1}x$	$\dfrac{1}{\sqrt{(x^2-1)}}$	$x\cosh^{-1}x-\sqrt{(x^2-1)}$
$\tanh^{-1}x$	$\dfrac{1}{1-x^2}$	$x\tanh^{-1}x+\tfrac12\ln(1-x^2)$ $(\lvert x\rvert<1)$
$\mathrm{sech}^{-1}x$	$\dfrac{-1}{x\sqrt{(1-x^2)}}$	$x\,\mathrm{sech}^{-1}x+\sin^{-1}x$
$\mathrm{cosech}^{-1}x$	$\dfrac{-1}{x\sqrt{(1+x^2)}}$	$x\,\mathrm{cosech}^{-1}x+\sinh^{-1}x$
$\coth^{-1}x$	$\dfrac{1}{1-x^2}$	$x\coth^{-1}x+\tfrac12\ln(x^2-1)$ $(\lvert x\rvert>1)$
$\ln x$	$1/x$	$x\ln x-x$

AI.3   Some useful standard integrals

$f(x)$, $(a > 0)$	$\int f(x)\,\mathrm{d}x$
$\dfrac{1}{a^2 + x^2}$	$\dfrac{1}{a}\tan^{-1}\left(\dfrac{x}{a}\right)$
$\dfrac{1}{\sqrt{(a^2 - x^2)}}$	$\sin^{-1}\left(\dfrac{x}{a}\right)$
$\dfrac{1}{\sqrt{(a^2 + x^2)}}$	$\sinh^{-1}\left(\dfrac{x}{a}\right)$ or $\ln\left[x + \sqrt{(x^2 + a^2)}\right]$
$\dfrac{1}{\sqrt{(x^2 - a^2)}}$	$\cosh^{-1}\left(\dfrac{x}{a}\right)$ or $\ln\left[x + \sqrt{(x^2 - a^2)}\right]$
$\dfrac{1}{a^2 - x^2}$ for $\lvert x \rvert < a$	$\dfrac{1}{2a}\ln\left(\dfrac{a + x}{a - x}\right)$
$\dfrac{1}{x^2 - a^2}$ for $\lvert x \rvert > a$	$\dfrac{1}{2a}\ln\left(\dfrac{x - a}{x + a}\right)$

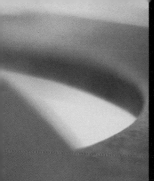

Answers to Exercises

CHAPTER 1

Exercises

1 (a) $1/2$ (b) 2^7 (c) $1/2^{12}$
   (d) 3^2 (e) $1/6$ (f) 2^3

2 (a) $21 + ((4 \times 3) \div 2)$ (b) $(17 - (6^{(2+3)}))$
   (c) $(4 \times (2^3)) - ((7 \div 6) \times 2)$
   (d) $((2 \times 3) - (6 \div 4)) + 3^{(2-5)}$

3 (a) $1393 + 985\sqrt{2}$ (b) $68 + 48\sqrt{2}$
   (c) $1 + \sqrt{2}$ (d) $-1 + \frac{3}{2}\sqrt{2}$

4 (a) $-7 + 5\sqrt{2}$ (b) $-\frac{60}{17} - \frac{41}{17}\sqrt{2}$
   (c) $\frac{5}{11} - \frac{1}{11}\sqrt{3}$ (d) $\frac{28}{11} + \frac{18}{11}\sqrt{5}$

5 $\sqrt{3} + \sqrt{19} > \sqrt{5} + \sqrt{13}$

6 (a) $-2 \leqslant x \leqslant 10$, $[-2, 10]$
   (b) $5 < x < -1$, $(-5, -1)$
   (c) $-3 \leqslant x \leqslant 4$, $[-3, 4]$
   (d) $-24 < x < 0$, $(-24, 0)$

7 (a) $\{x: |x - 4| < 3\}$ (b) $\{x: |x + 3| \leqslant 1\}$
   (c) $\{x: |2x - 43| < 9\}$ (d) $\{x: |8x - 1| \leqslant 5\}$

8 (b) only
   (b), (c) and (d) true

9 (a) $v_a = \frac{1}{2}(v_1 + v_2)$ (b) $v_b = \dfrac{2v_1 v_2}{v_1 + v_2}$

10 $\frac{239}{169}, \frac{577}{408}, \frac{1393}{985}$

11 (a) x^{-1} (b) x^7 (c) x^{-12} (d) x^2
   (e) $1/(2x^4)$ (f) $4x/9$ (g) $x^{5/2} - 2x^{-1/2}$
   (h) $25x^{2/3} + 1/(4x^{2/3}) - 5$ (i) $2 - 1/x$
   (j) $a^{-1}b^{9/2}$ (k) $1/(8b^3 a^{3/2})$

12 (a) $xy(x - y)$ (b) $xyz(x - y + 2z)$
   (c) $(a + b)(x - 2y)$ (d) $(x + 5)(x - 2)$
   (e) $(x + \frac{1}{2}y)(x - \frac{1}{2}y)$ (f) $(9x^2 + y^2)(3x - y)(3x + y)$

13 (a) $(x + 3)/(x + 4)$ (b) $(5 - x)/[(x - 3)(x + 1)]$
   (c) $2/[(x - 2)(x + 12)]$ (d) $3x^2 - 4y^2$

14 $\frac{4}{3}$

15 $7, -2$

16 $s = (m^2 + p^2)t/(m^2 - p^2)$, $m^2 \neq p^2$

17 $t = (u - 1)x^2/(u + 1)$, $u \neq -1$

18 $-1 \pm \sqrt{2}$

21 $10\,\mathrm{m}$

22 (a) $\sqrt{2}$ (b) $(1 + \sqrt{5})/2$

23 (a) $x < 0$ and $x > 5/2$
   (b) $x < 1$ and $x > 2$
   (c) $x < 0$ and $x > 1$
   (d) $x < -4$ and $x > \frac{2}{3}$

24 $-2 < x < 2$

25 (a) $(x + \frac{1}{2})^2 - \frac{49}{4}$ (b) $(x - 1)^2 + 2$
   (c) $\frac{1}{3} - 3(x - \frac{5}{3})^2$ (d) $5 - (x - 2)^2$

27 (a) $A = -\frac{1}{3}, B = \frac{1}{3}$ (b) $A = 8, B = -5$
   (c) $A = \frac{5}{2}, B = -\frac{3}{2}$

28 $A = 2, B = -1, C = 9$

29 (a) 5 (b) 0 (c) -9 (d) 11 (e) 20 (f) -4

30 (a) 120 (b) $\frac{1}{4}$ (c) 35 (d) 10 (e) 84 (f) 70

31 (a) $x^4 - 12x^3 - 54x^2 - 108x + 81$
   (b) $x^3 + \frac{3}{2}x^2 + \frac{3}{4}x + \frac{1}{8}$
   (c) $32x^5 + 240x^4 + 720x^3 + 1080x^2 + 810x + 243$
   (d) $81x^4 + 216x^3 y + 216x^2 y^2 + 96xy^3 + 16y^4$

32 (a) $y = \frac{3}{2}x - 2$ (b) $y = -2x - 1$
   (c) $y = \frac{5}{2}x - \frac{1}{2}$ (d) $y = -\frac{3}{5}x + 3$
   (e) $y = \frac{1}{3}x + \frac{2}{3}$ (f) $y = -3x + 4$

33 $(x - 1)^2 + (y - 2)^2 = 25$

34 $4, (-2, 3)$

35 $x^2 + y^2 + 4x - 6y = 12$

36 $x^2 + y^2 - 6x - 3y + 5 = 0$

37 $x^2 + y^2 = 25^2$

38 $x = \frac{2}{3}, x = -\frac{2}{3}$

39 $(0, 3), (0, -3), \frac{3}{5}, y = \frac{25}{3}, y = -\frac{25}{3}, 10, 8$

40 $(-5, 0), (5, 0), (-4, 0), (4, 0), y = \dfrac{3x}{4}, y = -\dfrac{3x}{4}$

41 $y = -1 \pm \sqrt{(x + 4)}, (y + 1)^2 = x + 4$

43 54.625_{10}

44 $11111111000001_2, 37701_8$
13455_8

45 $11110.100110011001\ldots_2$
$36.463\,146\,31\ldots_8$
Yes

46 (a) $101\,110.100_2$ (b) $10\,101.110\,101\,01_2$

47 (a) 3dp, 6sf   (b) 30dp, 3sf   (c) 0dp, 5sf
   (d) 0dp, 3sf   (e) 0dp, 4sf   (f) 10dp, 3sf

48 The answer claims unjustified accuracy: hypotenuse = 2.236 ± 0.007 m. The angles are also subject to error.

49 (a) Absolute error bound is $\frac{1}{12}$ min, relative error bound is $\frac{1}{420}$
   (b) Absolute error bound is 1.4 min, relative error bound is 0.04
   (c) Absolute error bound is 0.005, relative error bound is $\frac{1}{116}$

50 0.0039, 12.9

51 (a) 3.613 ± 0.0015, relative error bound 0.0004, 3.61
   (b) 2.5351 ± 0.0176, relative error bound 0.007, 2.5
   (c) 22.47 ± 0.015, relative error bound 0.0007, 22.5

52 4.51

53 $10.00 \pm 0.01, \frac{1}{1000}$
$-0.02 \pm 0.01, \frac{1}{2}$
$24.9999 \pm 0.05, \frac{1}{500}$
$0.996\,008 \pm 0.002, \frac{1}{500}$

54

Label	Value	Absolute error bound	Relative error bound
a	3.251	0.0005	
b	3.115	0.0005	
$a - b$	0.136	0.001	0.0074
c	0.112	0.0005	0.0045
$(a - b)/c$	1.2143	0.0145	0.0119
d	9.21	0.005	
$d + (a - b)/c$	10.4243	0.0195	

Result: 10.4

55 $0.7634 \pm 0.000\,72, 0.76$

56 (a) 0.2713 ± 0.0237 (b) 0.2715 ± 0.0072

57 $10^1(0.2709), 10^1(0.2708)$
The second result is more accurate since by adding the small numbers together first their combination is given its proper weight.

58 $10^{-8}(0.6538), 10^{-3}(0.6752)$

59 0.5

1.7 Review exercises

1 (a) $A = \pm QKD/\sqrt{(HD^2 - Q^2K^2)}$
   (b) $-(9 \pm \sqrt{145})/8$

2 (a) $(x - 1)(a - 2)$ (b) $(a - b + c)(a + b - c)$
   (c) $(2k + l - 3m)(2k + l + 3m)$
   (d) $(p - q)(p - 2q)$ (e) $(l + n)(l + m)$

3 (a) 1 cm   (b) 3.812

4 (a) $L = \left(\dfrac{1}{2\pi nC} \pm \sqrt{(Z^2 - R^2)}\right)\Big/(2\pi n)$
   (b) 0.1434; 0.0592; 0.4160
   (L must be positive from practical considerations)

5 (a) $30 - 12\sqrt{6}$ (b) $-53 + 11\sqrt{15}$
   (c) $\frac{1}{23}(14 + 11\sqrt{2})$ (d) $3 + 2\sqrt{2} + 2\sqrt{3} + \sqrt{6}$
   (e) $\frac{1}{2} + \frac{1}{4}\sqrt{2} + \frac{1}{4}\sqrt{6}$

6 5, 6

8 (a) $(-\frac{1}{2}, \frac{3}{2})$ (b) $(-\infty, -5) \cup (-2, 1)$
   (c) $(-3, 1)$ (d) $(-\frac{4}{3}, 0)$

10 $\dfrac{a}{b} > \dfrac{a + c}{b + c} > 1$

11 (b) (i) $1 - \frac{5}{2}x + \frac{5}{2}x^2 - \frac{5}{4}x^3 + \frac{5}{16}x^4 - \frac{1}{32}x^5$
    (ii) $729 - 2916x + 4860x^2 - 4320x^3 + 2160x^4$
      $- 576x^5 + 64x^6$

12 (a) 90.5
   (b) $P_1 = 1, P_2 = 3, P_3 = 5, P_r = (2r - 1)$,
    $\displaystyle\sum_{r=1}^{n} P_r = n^2$

13 (a) $y = 2x + 1$ (b) $y = (x - 7)/3$ (c) $y = 2x - \frac{7}{3}$

14 $(y - 3)^2 + (x - 5)^2 = 25$

15 (a) $(-1, 2), 2$ (b) $(\frac{1}{2}, -\frac{3}{2}), \frac{1}{2}$ (c) $(-\frac{1}{3}, \frac{1}{3}), \sqrt{3}$

16 (i) (a) $(1, 2)$ (b) $(3, 2)$ (c) $x = -1$ (d) $y = 2$
   (ii) (a) $(-2, 1)$ (b) $(-2, -2)$ (c) $y = 4$ (d) $x = -2$

17 $(2, 8), (2, 5), (2, 11), (2, 3), (2, 13), y = -\frac{1}{3}, y = \frac{49}{3}$

18 $\Delta.477\,4\Delta\,774\Delta\ldots_{12}$, where $\Delta_{12} = 10_{10}$

19

	Value	Absolute error bound	Relative error bound
a	7.01	0.005	→ 0.000 7
$\sqrt{a}$	2.647 6	0.000 9	← 0.000 35
b	52.13	0.005	→ 0.000 096
$\sqrt{b}$	7.220 111	0.000 347	← 0.000 048
c	0.010 11	0.000 005	→ 0.000 495
$\sqrt{c}$	0.100 548	0.000 025	← 0.000 25
d	5.631×10^{11}	0.5×10^8	→ 0.000 088 8
$\sqrt{d}$	7.504×10^5	0.33×10^2	← 0.000 044 4

	$\sqrt{a}$	$\sqrt{b}$	$\sqrt{c}$	$\sqrt{d}$
Correctly rounded values	2.65	7.22	0.101	7.504×10^5

20 0.37 ± 0.07

21 1.714   (a) 0.0026, (b) 0.0075

22 6

CHAPTER 2

Exercises

1 $A = 2x(5 + |x|)$

x/m	0	1	2	3	4	5
$Area$/m²	0	12	28	48	72	100

$A(-2) = -28$, area of cutting

2

r/m	0.10	0.15	0.20	0.25	0.30	0.35	0.40
A/m²	3.05	2.12	1.71	1.53	1.47	1.50	1.59

$r^* = 0.32$ according to worked answer: estimated from a graph (not drawn).

4 5 years

5 (a) Increasing on $-1 < x < 0$ and $x > +1$
Decreasing on $x < -1$ and $0 < x < +1$
Maximum at $(0, 0)$, minimum at $(-1, -1)$ and $(1, -1)$
(b) Increasing on $x < 1$, decreasing on $x > 1$, maximum at $(1, -1)$

6 (a) odd   (b) even   (c) neither
(d) neither   (e) odd   (f) even

9 $F(x) = (x - 1)^2$: $f(x)$ shifted by 2 units in positive x direction
$G(x) = (x + 1)^2 - 2$: $f(x)$ shifted by 2 units in negative y direction

11 (a) $\frac{1}{2}(x + 3)$ (b) $\dfrac{4x + 3}{2 - x}$
(c) Restriction of domain to $[0, \infty)$
$\sqrt{(x - 1)}, x \geqslant 1$

15 (a) $3 - 2x$
(b) $\frac{1}{2}x + \frac{5}{2}$
(c) $0.255x + 2.478$ (3dp)

16 (a) 3
(b) -3
(c) $\frac{1}{2}$

17 £$(50 + 0.455x)$, £960, £$(1.20x - 960)$, 800

18 (a) $\frac{2}{3}x^2 + 2x + \frac{1}{3}$
(b) $\frac{2}{5}x^2 - x - \frac{2}{5}$

19 $(x - 2)^2 - 4(x - 2) - 2$

20 (a) irreducible   (b) not irreducible
(c) not irreducible   (d) irreducible

21 (a) minimum at $x = -1$ of 2
(b) minimum at $x = \frac{3}{2}$ of 0
(c) maximum at $x = -\frac{2}{3}$ of $\frac{22}{3}$
(d) maximum at $x = \frac{3}{10}$ of $-\frac{11}{20}$

22 (a) $x < 2$ and $x > 4$
(b) $-\frac{3}{2} < x < 3$

23 $a = 0.311$

24 $m = 0.82, c = 60.9$

25 (a) $(x - 1)(x + 3)(x - 4)$
(b) $(x + 1)(x - 2)(x + 3)$
(c) $(x - 1)(x + 1)(x^2 + 2)$
(d) $x(x - 1)(2x + 3)(x + 2)$
(e) $(x - 1)^2(2x - 1)(x - 2)$
(f) $(x^2 + 9)(x - 2)(x + 2)$

26 2, 7, 139, 527, 524

27 $y = (x - 5)^4 + 15(x - 5)^3 + 80(x - 5)^2 + 165(x - 5) + 81$. Since coefficients are all positive, zeros of y must all lie to the left of $x = 5$, i.e. $x < 5$. Hence the zeros of y lie between $x = 0$ and $x = 5$.

29 $x^3 - 5x^2 + 1$

30 $3x^2 + 22x + 378$

31 (b) $r = 10/(4\pi)^{1/3}$, $h = 20/(4\pi)^{1/3}$

32 $0.096\,\text{m}^3$, $0.1875\,\text{m}^3$

33 $x_0 = 10.94$, width of alley $= 4.92\,\text{m}$

34 (a) $1 + (x+2)/[(x+1)(x-1)]$
  (b) $x^3 - 2x^2 + x + 1 - 3x/(x^2 + x + 1)$

35 (a) $(-5x^2 + x - 2)/[x(x-2)(x^2+1)]$
  (b) $2/[(x-1)^3(x+1)]$
  (c) $(4x^4 - 11x^3 + 10x^2 - 5x + 4)/[(x^2+1)(x-1)^2(x-2)]$

36 (a) $\dfrac{\frac{1}{3}}{x-2} - \dfrac{\frac{1}{3}}{x+1}$

  (b) $\dfrac{1}{x-2} + \dfrac{1}{x+1}$

  (c) $1 + \dfrac{\frac{2}{3}}{x-2} + \dfrac{\frac{1}{3}}{x+1}$

  (d) $\dfrac{\frac{1}{3}}{(x-2)^2} + \dfrac{\frac{2}{9}}{x-2} - \dfrac{\frac{2}{9}}{x+1}$

  (e) $\dfrac{1}{x+1} - \dfrac{x+1}{x^2 + 2x + 2}$

  (f) $\dfrac{-\frac{1}{3}}{x+1} + \dfrac{\frac{1}{12}}{x-2} + \dfrac{\frac{1}{4}}{x+2}$

37 (a) $\dfrac{\frac{1}{3}}{x-4} - \dfrac{\frac{1}{3}}{x-1}$

  (b) $\dfrac{\frac{1}{3}}{x-1} - \dfrac{\frac{1}{3}x + \frac{2}{3}}{x^2 + x + 1}$

  (c) $\dfrac{\frac{5}{9}}{x-2} + \dfrac{\frac{4}{3}}{(x+1)^2} - \dfrac{\frac{5}{9}}{x+1}$

  (d) $1 - \dfrac{3}{x-2} + \dfrac{8}{x-3}$

  (e) $\dfrac{1}{x^2+1} + \dfrac{x-2}{(x^2+1)^2}$

  (f) $\dfrac{x+1}{x^2+4} + \dfrac{2}{x-1} - \dfrac{3}{x+5}$

38 (a) $(\sqrt{2}, \sqrt{2})$, $(-\sqrt{2}, -\sqrt{2})$
  (b) $(\sqrt{2}, \sqrt{2})$, $(-\sqrt{2}, -\sqrt{2})$
  (c) $(-\sqrt{\frac{2}{5}}, -\sqrt{\frac{2}{5}})$, $(\sqrt{\frac{2}{5}}, \sqrt{\frac{2}{5}})$, $(\sqrt{2}, \sqrt{2})$, $(-\sqrt{2}, -\sqrt{2})$
  (d) does not intersect on domain

39 (a) asymptotes: $y = x - 8$, $x = 0$,
    maximum $(\sqrt{15}, -8 + 2\sqrt{15})$,
    minimum $(-\sqrt{15}, -8 - 2\sqrt{15})$
  (b) asymptotes: $y = 1$, $x = 1$ (c) $y = x$, $x = -5$

42 (a) 0.3398, 2.8018, $\frac{3}{2}\pi$
  (b) 1.8235, 4.4597, π
  (c) 2.6779, 5.8195, $\frac{1}{4}\pi$, $\frac{5}{4}\pi$
  (d) $\frac{1}{2}\pi$, $\frac{3}{2}\pi$, $\frac{1}{6}\pi$, $\frac{5}{6}\pi$

43 $\frac{1}{2}\sqrt{3}$, $\sqrt{3}$, $\frac{1}{2}\sqrt{3}$, $\frac{1}{2}$
  $\frac{1}{2}\sqrt{(2-\sqrt{3})}$, $\frac{1}{2}\sqrt{(2+\sqrt{3})}$, $2-\sqrt{3}$
  (a) $\frac{1}{2}\sqrt{3}$ (b) $1/\sqrt{3}$
  (c) $\frac{1}{2}\sqrt{3}$ (d) $\frac{1}{2}\sqrt{(2+\sqrt{3})}$
  (e) $-\frac{1}{2}\sqrt{(2-\sqrt{3})}$ (f) $-(2-\sqrt{3})$

44 (a) $-\sqrt{(1-s^2)}$ (b) $-2s\sqrt{(1-s^2)}$
  (c) $s(3-4s^2)$ (d) $\sqrt{\{\frac{1}{2}[1 + \sqrt{(1-s^2)}]\}}$

46 $x = n\pi$ $(n = \pm 1, \pm 3, \dots)$
  and $x = 0.9273 + 2n\pi$ $(n = 0, \pm 1, \pm 2, \dots)$

47

	a	b	c	d	e	f
$\sin x$	$\frac{1}{2}$	$\pm\frac{1}{2}$	$\pm\sqrt{\frac{1}{2}}$	$\pm\sqrt{\frac{1}{2}}$	$-\frac{1}{2}$	$\pm\frac{1}{2}$
$\cos x$	$\pm\frac{1}{2}\sqrt{3}$	$-\frac{1}{2}\sqrt{3}$	$\pm\sqrt{\frac{1}{2}}$	$\sqrt{\frac{1}{2}}$	$\pm\frac{2}{\sqrt{3}}$	$\pm\frac{1}{2}\sqrt{3}$
$\tan x$	$\pm\sqrt{\frac{1}{3}}$	$\pm\sqrt{\frac{1}{3}}$	-1	± 1	$\pm\sqrt{\frac{1}{3}}$	$\sqrt{\frac{1}{3}}$
$\operatorname{cosec} x$	2	± 2	$\pm\sqrt{2}$	$\pm\sqrt{2}$	-2	± 2
$\sec x$	$\pm 2\sqrt{\frac{1}{3}}$	$-2\sqrt{\frac{1}{3}}$	$\pm\sqrt{2}$	$\sqrt{2}$	$\pm\frac{2}{3}$	$\pm 2\sqrt{\frac{1}{3}}$
$\cot x$	$\pm\sqrt{3}$	$\pm\sqrt{3}$	-1	± 1	$\pm\sqrt{3}$	$\sqrt{3}$

49 (a) $2\sin 2\theta\cos\theta$ (b) $2\sin\frac{3}{2}\theta\sin\frac{1}{2}\theta$
  (c) $2\cos\frac{7}{2}\theta\cos\frac{3}{2}\theta$ (d) $-2\cos\frac{3}{2}\theta\sin\frac{1}{2}\theta$

50 (a) $\frac{1}{2}(\cos 2\theta - \cos 4\theta)$ (b) $\frac{1}{2}(\sin 4\theta + \sin 2\theta)$
  (c) $\frac{1}{2}(\sin 4\theta - \sin 2\theta)$ (d) $\frac{1}{2}(\cos 4\theta + \cos 2\theta)$

51 (a) $2\cos(\theta - \frac{2}{3}\pi)$, $2\sin(\theta - \frac{1}{6}\pi)$
  (b) $\sqrt{2}\cos(\theta - \frac{3}{4}\pi)$, $\sqrt{2}\sin(\theta - \frac{1}{4}\pi)$
  (c) $\sqrt{2}\cos(\theta - \frac{1}{4}\pi)$, $\sqrt{2}\sin(\theta - \frac{7}{4}\pi)$
  (d) $\sqrt{13}\cos(\theta - 0.9828)$, $\sqrt{13}\sin(\theta - 5.6952)$

52 $x = 2n\pi$, $2n\pi \pm \frac{2}{3}\pi$ $(n = 0, \pm 1, \pm 2, \dots)$

53 (a) $\pi/6$ (b) $-\pi/6$ (c) $\pi/3$
  (d) $2\pi/3$ (e) $\pi/3$ (f) $-\pi/3$

56 (a) $(2e+1)e^5$ (b) e^{4x} (c) e^6
  (d) e^9 (e) $e^{x/2}$

58 (a) 3 (b) -2 (c) $-\frac{1}{2}$
  (d) 4 (e) $\frac{1}{2}$ (f) $-\frac{1}{2}$

59 (a) $2\ln x + \ln y$ (b) $\frac{1}{2}\ln x + \frac{1}{2}\ln y$
  (c) $5\ln x - 2\ln y$

60 (a) $\ln 4$ (b) $\ln 3.2$ (c) $\ln 0.75$ (d) $\ln 0.5$

61 (a) $\sqrt{[(1-x)/(1+x)]}$ (b) x^2

63 $\frac{3}{2}\ln(x^2+1) - \frac{1}{3}\ln(x^4+1) - \frac{1}{5}\ln(x^4+4)$

64

	a	*b*	*c*	*d*	*e*	*f*
$\sinh x$	$\pm\frac{3}{4}$	$\frac{8}{15}$	$\frac{7}{24}$	$\pm\frac{12}{5}$	$-\frac{4}{3}$	$\frac{12}{5}$
$\cosh x$	$\frac{5}{4}$	$\frac{17}{15}$	$\frac{25}{24}$	$\frac{13}{5}$	$\frac{5}{3}$	$\frac{13}{5}$
$\tanh x$	$\pm\frac{3}{5}$	$\frac{8}{17}$	$-\frac{7}{25}$	$\pm\frac{12}{13}$	$-\frac{4}{5}$	$\frac{12}{13}$
$\operatorname{cosech} x$	$\pm\frac{4}{3}$	$\frac{15}{8}$	$\pm\frac{24}{7}$	$\pm\frac{5}{12}$	$-\frac{3}{4}$	$\frac{5}{12}$
$\operatorname{sech} x$	$\frac{4}{5}$	$\frac{15}{17}$	$\frac{24}{25}$	$\frac{5}{13}$	$\frac{3}{5}$	$\frac{5}{13}$
$\coth x$	$\pm\frac{5}{3}$	$\pm\frac{17}{8}$	$-\frac{25}{7}$	$\pm\frac{13}{12}$	$-\frac{3}{4}$	$\frac{13}{12}$

65 (a) $\tanh 3x = \dfrac{(3 + \tanh^2 x)\tanh x}{1 + 3\tanh^2 x}$

(b) $\cosh(x + y) = \cosh x \cosh y + \sinh x \sinh y$

(c) $\cos 2x = 1 - 2\sin^2 x$

(d) $\sinh x - \sinh y = 2\sinh\frac{1}{2}(x - y)\cosh\frac{1}{2}(x + y)$

67 (a) 0.7327 (b) 1.3170 (c) 0.5493

68 17.1383 (4dp)

69 1.0074 (4dp)

71 $A = 250, B = -273.26$

72 $\ln(20 \pm 6\sqrt{10}) = 3.6629, 0.02599$

75 (a) Cusp at $x = 0$, maximum at $x = 4$, asymptote $y = -x$

(b) Minimum at $x = 2$, asymptotes $y = \pm\sqrt{x} - 1, x = 1$

78 $\dfrac{ax}{l}H(x) + \dfrac{2a}{l}(l - x)H(x - l) - \dfrac{a}{l}(2l - x)H(x - 2l)$

79 $x[1 - H(x)] - (x - 1)H(x - 1)$

80 $\operatorname{INTPT}(x + \frac{1}{2})$

83 0.9401, 0.005, 0.9425

84 0.04, 0.16, 0.01, 0.00625

85 0.3081, 0.2829, 16.79

86 0.2954, 0.2688, 17.10

87 $f(x) = -\frac{1}{84}x^3 + \frac{85}{84}x$

88

x	3045	3051	3058	3064	3070	3077	3083
y		14.50	14.51	14.52	14.53	14.54	14.55

2.11 Review exercises

1 $h(x) = x - 4 \qquad x \in [0, 200]$
$k(x) = (x^2 - 4)^{1/2} \qquad x \in [-20, -2] \cup [2, 20]$

2 (a) 25.6 cm (b) 2.35 m²

3

Price/£	1.00	1.05	1.10	1.15	1.20	1.25	1.30
Sales/000	8	7	6	5	4	3	2
Revenue/£000	8	7.35	6.60	5.75	4.80	3.75	2.60
Profit/£000	0	0.35	0.60	0.75	0.80	0.75	0.60

4 $g(x) = \begin{cases} 3x + 3 & x \leqslant -3 \\ 2x & -3 < x \leqslant -1 \\ 3x + 1 & -1 < x \leqslant 0 \\ x + 1 & 0 < x \leqslant 1 \\ 3x - 1 & x > 1 \end{cases}$

6 0.37 ± 0.005

8 $(x - 1)^4 + 7(x - 1)^3 + 14(x - 1)^2 + 13(x - 1) + 4$

9 (a) $\dfrac{2}{x - 4} - \dfrac{1}{x - 1}$

(b) $1 - \dfrac{\frac{5}{4}}{x + 1} + \dfrac{\frac{13}{4}}{x - 3}$

(c) $\dfrac{2}{9(x - 1)} + \dfrac{7}{9(x + 2)} - \dfrac{11}{3(x + 2)^2}$

(d) $\dfrac{\frac{1}{13}(5x - 7)}{x^2 - x + 1} + \dfrac{\frac{21}{13}}{x + 3}$

10 (a) $2\cos\frac{3}{2}\theta \sin\frac{1}{2}\theta$

(b) $2\cos\frac{5}{2}\theta \cos\frac{1}{2}\theta$

(c) $-2\sin\dfrac{3\theta}{2}\cos\dfrac{11\theta}{2}$

11 (a) $2\sqrt{5}\sin(\theta - \alpha), \alpha = \tan^{-1}\frac{1}{2}$

(b) $\sqrt{65}\sin(\theta - \alpha), \alpha = -\tan^{-1}8$

(c) $2\sin(\theta + \frac{1}{6}\pi)$

13 2

15 0.0025, 0.300

16 $\sqrt{\frac{1}{3}}$

17 (b) $\frac{1}{2}D\sqrt{3}$

18 1, 2, 0, 1, 3, 7, 5, 5, ...
0.83389, 0.55194

21 $r = 3/(2\sin\theta - \cos\theta)$

CHAPTER 3

Exercises

2 (a) 10 (b) $-3 - j4$
(c) $\frac{1}{25}(47 - j4)$ (d) $-j$
(e) j (f) $5 - j12$
(g) $j\frac{3}{17}$ (h) $-\frac{1}{178}(5 + j8)$

3 (a) $-1 + j, \ 1 - j$
(b) $-2, 1 + j\sqrt{3}, 1 - j\sqrt{3}$

4 $\pm 3 + j2$

5 (a) $3 + j2$ (b) $2 + j3$
   (c) $\frac{1}{13}(2 + j3)$ (d) $2 - j3$

6 (a) $\sqrt{2}, \frac{1}{4}\pi$ (b) $2, -\frac{1}{6}\pi$
   (c) $5, \pi - \tan^{-1}\frac{4}{3}$ (d) $2, -\frac{1}{3}\pi$
   (e) $2, \frac{2}{3}\pi$ (f) $2, -\frac{2}{3}\pi$

7 $w = 5 - j4, z = 2 + j3$

8 $x = \frac{1}{2}, y = -\frac{3}{2}$

9 $2 + j2, \frac{1}{2}$

10 $\frac{1}{5}(7 + j4)$

11 $\frac{1}{130}(451 + j878)$

12 $x = \frac{1}{4}, y = -\frac{3}{4}$

13 $\frac{11}{4} - j\frac{13}{4}$

14 (a) $-\frac{5}{12}\pi, -\frac{11}{12}\pi$ (b) $\frac{1}{2}\sqrt{3} + j\frac{3}{2}$

15 (a) $128, -\frac{1}{3}\pi$ (b) $1024, 0$ (c) $\frac{1}{16}, \frac{2\pi}{3}$

16 (a) $1\angle \pi/2$
   (b) $1\angle 0$
   (c) $1\angle \pi$
   (d) $\sqrt{2}\angle -\pi/4$
   (e) $\sqrt{6}\angle -\pi/4$
   (f) $\sqrt{5}\angle(\pi - \tan^{-1}\frac{1}{2})$
   (g) $\sqrt{13}\angle(\tan^{-1}\frac{2}{3} - \pi)$
   (h) $\sqrt{74}\angle(-\tan^{-1}\frac{5}{7})$
   (i) $5\angle 0$
   (j) $53\angle(\tan^{-1}\frac{28}{45} - \pi)$

18 $\frac{4}{5} + j\frac{7}{5}, \frac{1}{5}\sqrt{65}\angle\tan^{-1}\frac{7}{4}$

19 $x = \dfrac{\tanh u \sec^2 v}{1 + \tanh^2 u \tan^2 v}$

   $y = \dfrac{\tan v \operatorname{sech}^2 u}{1 + \tanh^2 u \tan^2 v}$

   $\dfrac{2\tanh 2}{1 + \tanh^2 2} + j\dfrac{\operatorname{sech}^2 2}{1 + \tanh^2 2}$
   $= 0.9994 + j0.0366$

20 (a) $\frac{1}{2}\cosh 1 - j\frac{1}{2}\sqrt{3}\sinh 1$
   (b) $\cosh\frac{3}{4}$

   (c) $\frac{1}{2}\sinh\dfrac{\pi}{3} + j\dfrac{\sqrt{3}}{2}\cosh\dfrac{\pi}{3}$

   (d) $\dfrac{1}{\sqrt{2}}$

21 (a) $\frac{1}{2}(4n + 1)\pi + j\cosh^{-1}2$
   (b) $\frac{1}{2}(2n + 1)\pi + j(-1)^{n+1}\sinh^{-1}\frac{3}{4}$
   (c) $\frac{1}{2}(4n + 1)\pi + j\cosh^{-1}3$
   (d) $\cosh^{-1}2 + j(2n + 1)\pi$

23 $0.1645 - j0.1214$

24 (a) $-2 + j2, -4$ (b) $-j8, -8 - j8\sqrt{3}$
   (c) $117 + j44, -527 + j336$ (d) $-8, -8 + j8\sqrt{3}$
   (e) $8, -8 + j8\sqrt{3}$ (f) $8, -8 - j8\sqrt{3}$

25 (a) $\frac{1}{8}\cos 4\theta + \frac{1}{2}\cos 2\theta + \frac{3}{8}$
   (b) $\frac{3}{4}\sin\theta - \frac{1}{4}\sin 3\theta$

27 $2^{7/6}\angle(\frac{\pi}{12} + \frac{2}{3}k\pi), k = 0, 1, 2$

28 (a) $2^{1/4}\angle(-\frac{1}{24}\pi + \frac{1}{2}k\pi), k = 0, 1, 2, 3$
   (b) $2\angle(\frac{1}{6}\pi + \frac{2}{3}k\pi), k = 0, 1, 2$
   (c) $18^{-1/3}\angle(\frac{1}{6}\pi - \frac{4}{3}k\pi), k = 0, 1, 2$
   (d) $1\angle(\frac{1}{4}\pi + \frac{1}{2}k\pi), k = 0, 1, 2, 3$
   (e) $4\angle(\frac{1}{3}\pi + \frac{8}{3}k\pi), k = 0, 1, 2$
   (f) $34^{-1/4}\angle(\frac{1}{2}\tan^{-1}\frac{3}{5} - k\pi), k = 0, 1$

29 $1.455 - j0.344, 0.344 + j1.455, -1.455 + j0.344,$
   $-0.344 - j1.455$

30 $2.529 + j2.743, 0.471 + j2.257$

31 $1, -\frac{1}{2} \pm j\frac{1}{2}\sqrt{3}$

   $\cos\dfrac{2k\pi}{n} + j\sin\dfrac{2k\pi}{n}, k = 1, 2, \ldots, n$

   (a) $j2\cot\frac{1}{5}k\pi, k = 1, \ldots, 4$
   (b) $\frac{3}{2}(1 + j\cot\frac{1}{6}k\pi), k = 1, \ldots, 5$

32 $5, 13$

33 (a) $x = 5$, a straight line
   (b) circle centre $(1, 0)$, radius 3
   (c) circle centre $(-\frac{5}{4}, 0)$, radius $\frac{3}{4}$
   (d) half-line, $y = x - 2, x > 2$

34 circle is $|z + 2| = 2$
   line is $\operatorname{Re}((3 + j) + z) = -2$

35 (a) Straight line, $y = 1$
   (b) Circle, centre $(0, 2)$, radius 1
   (c) Circle, centre $(0, \frac{5}{4})$, radius $\frac{3}{4}$
   (d) Circle, centre $(\sqrt{\frac{1}{3}}, 0)$, radius $2\sqrt{\frac{1}{3}}$
   (e) Rectangular hyperbola, $xy = 1$
   (f) Ellipse, foci at $(1, 0), (0, -1)$, through $(0, 0)$
   (g) Hyperbola, foci at $(1, 0), (0, -1)$
   (h) Half-line, $y = x - 2, x > 2$
   (i) Half-line, $y = \sqrt{3}x - \frac{3}{2}\sqrt{3}, x < \frac{3}{2}$
   (j) Circle, centre $(0, 2)$, radius 1

36 (a) $\operatorname{Re}[(3 + j)z] = 2$ (b) $|z + 2| = 2$
   (c) $|z + 1 - j2| = 3$ (d) $\operatorname{Re}(z^2) = 1$

37 (a) Circle, centre $(1, 0)$, radius 2
   (b) Circle, centre $(\frac{1}{2}, 0)$, radius $\frac{3}{2}$
   (c) Circle, centre $(2, 3)$, radius 4
   (d) Half-line, $y = 0, x > 0$
   (e) Circle, centre $(-\frac{13}{8}, 0)$, radius $\frac{15}{8}$
   (f) Semicircle, centre $(\frac{1}{2}, -\frac{1}{2})$, radius $\frac{1}{2}\sqrt{2}$, through $(0, 0)$

38 $x^2 + y^2 - 4x - 2y + 1 = 0$, $|z - 2 - j| = 2$,

$\arg\left(\dfrac{z - j}{z - 4 - j}\right) = \pm\dfrac{\pi}{2}$

39 Part of $x^2 + (y - 1)^2 = 2$

40 $(x - 3)^2 + y^2 = 4$

41 $100 + j100.12$

42 $\frac{8}{3} + j\frac{8}{3}$

3.6 Review exercises

1 $x = \pm\frac{3}{2}$, $y = \pm 2$

2 $\frac{1}{10}(7 + j9)$

3 (a) Circle centre $(-\frac{1}{3}, \frac{4}{3})$, radius $\frac{2}{3}\sqrt{2}$

   (b) $\mathrm{Re}\left(\dfrac{1}{z - 2}\right) = -\dfrac{1}{2}$

5 Centre $(R_2, \frac{1}{2}\omega L)$, radius $\frac{1}{2}\omega L$

6 (a) $32\cos^6\theta - 48\cos^4\theta + 18\cos^2\theta - 1$

12 $419.8 - j238.8$, $0.5928 \times 10^{-3} + j1.0518 \times 10^{-3}$

13 $1 + j3$, $\frac{1}{5}(3 + j11)$, $\frac{1}{5}(7 + j11)$

14 Mod $\frac{25}{13}$, $\arg = -154°17' = -2.6927\,\mathrm{rad}$

15 (a) $0.22 \pm j0.49$ (b) $1.44 + j1.57$ (c) $10.48 + j19.74$
   (d) $0.80 + j0.46$ (e) $1.09 + j0.83$

18 $\theta = \tan^{-1}\left[\dfrac{2R_0 X - 2RX_0}{R^2 + X^2 - R_0^2 - X_0^2}\right]$

19 $4.46 - j2.06$

20 (a) $0.7974 + j0.3685$ (b) $r = 0.8784$
   $\theta = 24°49' = 0.4329\,\mathrm{rad}$,
   1.098

22 (a) $-0.04 + j0.28$ (b) $\pm(0.35 + j0.40)$
   (c) $0.92 + j0.27$ (d) $-1.26 + j1.71$
   (e) $-0.04 + j0.28$

23 $1\angle 18°26'$, $1\angle 108°26'$, $1\angle 198°26'$, $1\angle 288°26'$

24 $2^{1/6}e^{j(1/9 + k/3)\pi}$, $k = 0, \ldots, 5$

26 $-(\omega u + \omega^2 v)$, $-(\omega^2 u + \omega v)$; $\frac{1}{4}r^2 \leqslant -\frac{1}{27}q^3$

27 $1 - j2$, $2\sqrt{5}$

CHAPTER 4

Exercises

1 (a) $(3, 3, 1)$ (b) $(2, 4, \frac{5}{2})$ (c) $(0, 0, 1)$
   (d) $\sqrt{2}$ (e) 3 (f) $\sqrt{3}$
   (g) $(\sqrt{\frac{1}{2}}, \sqrt{\frac{1}{2}}, 0)$ (h) $(\frac{2}{3}, \frac{2}{3}, \frac{1}{3})$

2 $\overrightarrow{PQ} = (4, -5, 11)$, $|\overrightarrow{PQ}| = 9\sqrt{2}$
   direction cosines $4/(9\sqrt{2})$, $-5/(9\sqrt{2})$, $11/(9\sqrt{2})$

3 $\sqrt{134}\,\mathrm{N}$, $(7, 2, 9)/\sqrt{134}$

4 $\alpha = 4$ $\beta = 1$ $\gamma = 2$

5 $\sqrt{21}$ $\sqrt{17}$ $\sqrt{38}$

6 $60°$ or $-60°$ to the positive z axis

7 $\overrightarrow{PQ} = \overrightarrow{QR} = (1, 5, -3)$ and $PQ : QR = 1 : 1$

8 $8\sqrt{2}$ kilometres per hour from the NW

9 distance $= 13/5$, $t = 1/5$

11 $1 - j2$ length $= \sqrt{20}$

12 $\frac{1}{2} - j\frac{1}{2}$

13 $r = (2 - t, 2t, -1 + 4t)$
   $2 - x = \frac{1}{2}y = \frac{1}{4}(1 + z)$, no intersection

15 $\theta = \sin^{-1}\left(\dfrac{W_1^2 + W_2^2 - W_3^2}{2W_1 W_2}\right)$

   $\phi = \sin^{-1}\left(\dfrac{W_2^2 + W_3^2 - W_1^2}{2W_2 W_3}\right)$

16 $F = (-940, 124, -31)\,\mathrm{N}$
   $(7.93\,\mathrm{m}, -1.04\,\mathrm{m}, 0\,\mathrm{m})$, $T = 1342\,\mathrm{N}$

17 $T = 539\,\mathrm{N}$ $S = 389\,\mathrm{N}$

18 (a) 14 (b) 6 (c) $(2, 1, 6)/\sqrt{41}$
   (d) $(12, 0, -6)/\sqrt{5}$ (e) -24 (f) $(12, 4, 8)$

19 (a) $98.0° = 1.711\,\mathrm{rad}$ (b) $64.8° = 1.130\,\mathrm{rad}$
   (c) $\frac{14}{5}$ (d) 3 or -4

20 4 units

21 $\frac{5}{14}$

22 $\sqrt{5}/2$

27 (a) $50\,\mathrm{N\,m}$ (b) $\frac{1}{2}Whpn(n + 1)$

28 $r^2 - (r \cdot \hat{a})^2 = R^2$

29 $\sqrt{3}$, $70.5°$ or $1.23\,\mathrm{rad}$

30 $|X| \leqslant 2.98\,\mathrm{m}$

31 $|\theta| \leqslant 36.9°$ or $0.644\,\mathrm{rad}$

32 (a) $(3, -2, -1)$ (b) $(-1, 1, 0)$ (c) $(5, -4, -2)$
   (d) -1 (e) 1 (f) $(2, 2, 1)$

35 (a) $(8, 1, 6)$, $(4, 1, 3)$ (b) $(-\frac{3}{5}, 0, \frac{4}{5})$ (c) $\frac{5}{2}$

38 $(48, 72, 0)/\sqrt{14}$

39 $(-8, -32, -4)/\sqrt{21}$

40 (a) $(0, 1, -1)$ $(-2, 1, -1)$ $(-2, 1, 0)$
   (b) $(0, 0, 0)$ $(0, 0, -2)$ $(-3, 3, -1)$
   (c) $(-3, 3, -3)$

41 $\pm(-3, 5, 11)/\sqrt{155}$ 0.9968

42 Distance = 1.92

43 $m\omega = eB$

44 15

46 8

48 (a) $(-5, 3, -7)/\sqrt{83}$ (b) $(0, 1, -4)/\sqrt{17}$

49 $\frac{7}{3}(1, 1, 1)$ $\frac{1}{3}(2, -13, 11)$

50 $\begin{vmatrix} u_1 & u_2 & u_3 \\ v_1 & v_2 & v_3 \\ w_1 & w_2 & w_3 \end{vmatrix}$

51 $\alpha = -1/F^2$

52 (a) $(c \cdot a)(b \cdot d) - (c \cdot b)(d \cdot a)$
   (c) $-(a \cdot b)(a \times c)$

53 (a) $(3, 3, 3)$ (b) $(1 + s, 2 + s, 3 + s)$
   (c) $x - 1 = y - 2 = z - 3$

54 $r \cdot (0, -1, 1) = 1$ $-y + z = 1$

56 $(3, 4, 0)$, $43.5° = 0.759$ rad

57 $r \cdot [b \times (c - a)] = a \cdot (b \times c)$

59 $r = (0, -5, 10) + \lambda(1, 2, -3)$

60 $r = (1, 2, 4) + t(1, 1, 2)$ $(-\frac{5}{2}, -\frac{3}{2}, -3)$

61 $79.0° = 1.38$ rad

62 (a) $r \cdot (2, 3, 6) = -28$ (b) 5

63 $r = (1 + 2t, -1 + 4t, 3 - 4t)$, $41.8° = 0.729$ rad

64 $\sqrt{35}$

65 $r \cdot (1, -5, 3) = 28$

67 $r = (-1 + 14t, t, 1 - 8t)$
   $\frac{2}{3}\sqrt{29}$, $r \cdot (-18, 36, -27) = -9$

4.5 Review exercises

1 (a) $\sqrt{93}$ (b) $(17, -3, -10)/\sqrt{398}$
   (c) $85.8° = 1.50$ rad, $47.0° = 0.820$ rad
   (d) $(2, 13, -13)/6$

2 (a) $(3, 4, 5)$ (b) $\sqrt{35}$ (c) $34/3$

3 (a) $(1, 2, 0), (2, 1, 1)$ (b) $\sqrt{5}$
   (c) 1   (d) $112.2° = 1.96$ rad

4 P(2, 4, 4), Q(1, 2, 3)
   $(-2, -4, 0)$

5 $(\frac{2}{3}, \frac{1}{3}, \frac{2}{3})$, $(-\frac{3}{5}, 0, \frac{4}{5})$; $(1, 5, 22)$

7 $(1, 1, 1), (-5, -11, 1)$

8 $E = e(0.550, 0.282, 0.282)$

9 (a) $2x + 3y + 6z + 28 = 0$ (b) 5

10 (a) $2x + 3y - z = 10$; $10/\sqrt{14}$
   (b) $\sqrt{3}/2$

11 (a) -4 (b) 1 or -4

12 (a) 0   (b) $15(1, 1, -2)$

14 $(-90, -36, 12)$, $85.3°$ or 1.49 rad

15 $(11, -12, 5)$; $76.8°$; $(-11, 12, -5)/\sqrt{290}$
   (a) $-11x + 12y - 5z = 8$
   (b) $-11x + 12y - 5z = -4$ (c) $12/\sqrt{290}$

16 $r = (-3, 0, 1) + \lambda(8, -8, -8) + \mu(5, 1, -3)$
   $r \cdot (-1, 2, -1) = -6$

17 $x + 2y - 2z = -1$

18 $x = \dfrac{U}{K}(1 - e^{-KT})$

19 (a) $(0, 0, 1), (1, -1, 0), (0, 1, -1)$
   (b) 1   (c) $3, -3, 2$ (d) $3, 2, 1$

20 $r = (2 - t, 3 - 3t, 2t)$
   (a) $\sqrt{(61/14)}$ (b) $(0, -3, 4)$ (c) $(19, 15, 18)/14$

21 $\alpha = r \cdot a'$ $\beta = r \cdot b'$ $\gamma = r \cdot c'$

22 Taking i along OA and j along OB then
   $F = \omega^2(1.4, 1.65)$ and $OC = (-1.4, -1.65)$ m.

23 $(0, 0, -25)$ N $(-2.5, 2.5, -0.2)$ N m

CHAPTER 5

Exercises

1 (a) not possible   (b) $\begin{bmatrix} 1 \\ 3 \\ 1 \end{bmatrix}$ (c) not possible

   (d) not possible   (e) $\begin{bmatrix} 8 & 9 & 10 \\ 7 & 10 & 9 \end{bmatrix}$

2 $\alpha = 1$ $\beta = -1$ $\gamma = 2$

3 (a) $\lambda = 1$ $\mu = -1$ $\nu = 3$

4 Average $= \begin{bmatrix} 30.5 \\ 27.5 \\ 19.5 \\ 11.5 \\ 11.0 \end{bmatrix}$ Weighted average $= \begin{bmatrix} 32.57 \\ 26.43 \\ 19.14 \\ 11.43 \\ 10.43 \end{bmatrix}$

5 $\begin{bmatrix} 2 & 2 & -1 \\ 2 & 2 & -1 \end{bmatrix}\begin{bmatrix} 0 & 2 \\ 0 & 2 \end{bmatrix}$

$\begin{bmatrix} 1 & 3 \\ 1 & 3 \\ 1 & 1 \end{bmatrix}\begin{bmatrix} 2 & 2 & 2 \\ 2 & 2 & 2 \\ -2 & -2 & -2 \end{bmatrix}$

$\begin{bmatrix} 2 & 2 \\ 2 & 2 \\ -1 & -1 \end{bmatrix}$

6 (a) No, yes, yes, yes, no, no

(b) $\begin{bmatrix} 9 & 7 \\ 7 & 12 \end{bmatrix}\begin{bmatrix} 13 & 18 \\ 8 & 5 \end{bmatrix}\begin{bmatrix} 9 & 5 \\ 18 & 2 \end{bmatrix}$

(c) $\begin{bmatrix} 37 & 33 \\ 26 & 36 \\ 29 & 28 \end{bmatrix}$

7 $x^2 + y^2 + z^2$

$x^2 + 4y^2 + 7z^2 + 5xy + 8xz + 11yz$

$\left.\begin{array}{r} x + 2y + 3z = 2 \\ 3x + 4y + 5z = 3 \\ 5x + 6y + 7z = 4 \end{array}\right\}$

8 (a) $\boldsymbol{A} = \begin{bmatrix} 1 & 1 & 1 \\ 1 & 1 & 1 \end{bmatrix}$ $\boldsymbol{B} = \begin{bmatrix} 1 \\ 1 \\ 1 \end{bmatrix}$

(b) $\boldsymbol{A} = \begin{bmatrix} 1 & 1 & 1 \end{bmatrix}$ $\boldsymbol{B} = \begin{bmatrix} 1 \\ 1 \\ 1 \end{bmatrix}$

(c) $\boldsymbol{A} = \begin{bmatrix} 1 & 0 \\ 0 & 0 \end{bmatrix}$ $\boldsymbol{B} = \begin{bmatrix} 0 & 0 \\ 1 & 1 \end{bmatrix}$

9 $\boldsymbol{A} = \begin{bmatrix} 1 & \frac{5}{2} & \frac{3}{2} \\ \frac{5}{2} & -1 & 2 \\ \frac{3}{2} & 2 & 1 \end{bmatrix} + \begin{bmatrix} 0 & \frac{1}{2} & \frac{1}{2} \\ -\frac{1}{2} & 0 & -2 \\ -\frac{1}{2} & 2 & 0 \end{bmatrix}$

10 $\begin{bmatrix} 41 & 15 & 7 \\ -9 & 63 & -40 \\ -13 & 38 & 41 \end{bmatrix}$

11 £2273.88

12 $\begin{bmatrix} 1 & 3 & 2 \\ 0 & 5 & 2 \\ 2 & -2 & 1 \end{bmatrix}\begin{bmatrix} 3 & 6 & 5 \\ 6 & 7 & 8 \\ -4 & 1 & -3 \end{bmatrix}$

14 $\begin{bmatrix} R_1 & R_2 & 0 & 0 & R_5 & 0 \\ 0 & R_2 & -R_3 & R_4 & 0 & 0 \\ 0 & 0 & 0 & R_4 & -R_5 & R_6 \\ 1 & -1 & -1 & 0 & 0 & 0 \\ 0 & 1 & 0 & -1 & -1 & 0 \\ 0 & 0 & -1 & -1 & 0 & 1 \end{bmatrix}\begin{bmatrix} i_1 \\ i_2 \\ i_3 \\ i_4 \\ i_5 \\ i_6 \end{bmatrix} = \begin{bmatrix} E \\ 0 \\ 0 \\ 0 \\ 0 \\ 0 \end{bmatrix}$

15 $\begin{bmatrix} 799.8 \\ 800 \\ 800.2 \end{bmatrix}\begin{bmatrix} 800 \\ 800 \\ 800 \end{bmatrix}$

16 $h = \frac{1}{3}, k = \frac{2}{3}, l = \frac{1}{3}, m = \frac{1}{6}$

18 $\boldsymbol{A} = \begin{bmatrix} \sqrt{\frac{1}{2}} & -\sqrt{\frac{1}{2}} \\ \sqrt{\frac{1}{2}} & \sqrt{\frac{1}{2}} \end{bmatrix}$

$\begin{bmatrix} a \\ b \end{bmatrix} = \begin{bmatrix} 16 \\ -10 \end{bmatrix} + \boldsymbol{A}\begin{bmatrix} 16 \\ -10 \end{bmatrix}$ and $\boldsymbol{B} = \boldsymbol{A}$

19 $n = 3$

20 Minors $= \begin{matrix} -1 & 0 & 1 \\ -1 & -2 & -1 \\ 2 & -2 & -2 \end{matrix}$

Cofactors $= \begin{matrix} -1 & 0 & 1 \\ 1 & -2 & 1 \\ 2 & 2 & -2 \end{matrix}$

$|\boldsymbol{A}| = 2$

21 (a) -19 (b) 130 (c) -65

22 $\begin{bmatrix} d & -b \\ -c & a \end{bmatrix}$

23 $\begin{bmatrix} 2 & -1 & 0 \\ -4 & 3 & -1 \\ 1 & -1 & 1 \end{bmatrix}$

27 (a) $-1.6569, 9.6569$
(b) $4.6667 \pm j0.62361$
(c) $2, 3 \pm j$

28 (a) -0.1884 (b) 100

30 $x^2(2x + 1)^2(x - 1)^2$

34 Non-singular, singular, non-singular, singular

35 $\begin{bmatrix} -\frac{1}{3} & \frac{2}{3} \\ \frac{2}{3} & -\frac{1}{3} \end{bmatrix}\begin{bmatrix} 1 & 0 & 0 & -1 \\ 0 & 1 & 0 & -1 \\ 0 & 0 & 1 & -1 \\ 0 & 0 & 0 & 1 \end{bmatrix}$

36 $\dfrac{1}{5}\begin{bmatrix} -3 & 2 & 2 \\ 2 & -3 & 2 \\ 2 & 2 & -3 \end{bmatrix}$, $\quad\begin{bmatrix} 0.68 & -0.32 & -0.32 \\ -0.32 & 0.68 & -0.32 \\ -0.32 & -0.32 & 0.68 \end{bmatrix}$

37 $\dfrac{1}{68}\begin{bmatrix} -4 & 4 & 8 \\ 6 & 11 & -12 \\ 36 & -2 & -4 \end{bmatrix}$ $\quad \dfrac{1}{49}\begin{bmatrix} -5 & 22 & 8 \\ 11 & -19 & 2 \\ 13 & -18 & -11 \end{bmatrix}$

38 $A = A^{-1}$, $B^{-1} = \begin{bmatrix} 1 & 0 & 0 & 0 \\ 0 & 0 & 0 & 1 \\ 0 & 1 & 0 & 0 \\ 0 & 0 & 1 & 0 \end{bmatrix}$

$(AB)^{-1} = \begin{bmatrix} 1 & 0 & 0 & 0 \\ 0 & 0 & 0 & 1 \\ 0 & 0 & 1 & 0 \\ 0 & 1 & 0 & 0 \end{bmatrix}$

39 $\dfrac{1}{12}\begin{bmatrix} -7 & -6 & 5 \\ 2 & 0 & 2 \\ 1 & -6 & 1 \end{bmatrix}$

$x = 2,\ y = 1,\ z = 2$

40 $\alpha = 1 \quad x = \lambda \quad y = 5\lambda \quad z = 7\lambda$
$\alpha = -6 \quad x = \mu \quad y = -2\mu \quad z = 0$

41 $-6, -3, -2$

42 (a) $a = 0$ (b) $\dfrac{1}{9}\begin{bmatrix} 1 \\ 7 \\ 6 \end{bmatrix}$ (c) $\begin{bmatrix} \lambda \\ \lambda \\ \lambda \end{bmatrix}$ (d) $\begin{bmatrix} -2 \\ 1 \\ 3 \end{bmatrix}$

43 $\begin{bmatrix} 0.1896 & -0.0604 & -0.0167 & 0.0167 & -0.0021 & -0.0021 \\ -0.0604 & 0.2088 & -0.0551 & -0.0218 & 0.0171 & -0.0021 \\ -0.0167 & -0.0551 & 0.2103 & -0.0564 & -0.0218 & 0.0167 \\ 0.0167 & -0.0218 & -0.0564 & 0.2103 & -0.0551 & -0.0167 \\ -0.0021 & 0.0171 & -0.0218 & -0.0551 & 0.2088 & -0.0604 \\ -0.0021 & -0.0021 & 0.0167 & -0.0167 & -0.0604 & 0.1896 \end{bmatrix}$,

$\begin{bmatrix} 0.1875 \\ -0.0625 \\ 0 \\ 0 \\ -0.0625 \\ 0.1875 \end{bmatrix}$

44 $a = u_1 \quad b = (-u_1 + u_2)/p \quad c = (-u_1 + u_3)/q$
$d = (u_1 - u_2 - u_3 + u_4)/pq$

45 $x = 0.5889u \quad y = 0.4222u \quad z = 0.2222u$

46 $a = -0.4011 \quad b = 1 \quad c = -0.5825$
$f(1) = 1.4345$

47 $y_1 = 1.8936 \quad y_2 = 4.6809 \quad y_3 = 8.1489 \quad y_4 = 10.7660$

48 (a) $\begin{bmatrix} -3 \\ 0 \\ 2 \end{bmatrix}$ (b) $\begin{bmatrix} -3 \\ 2 \\ 4 \end{bmatrix}$ (c) $\begin{bmatrix} -\frac{3}{25} \\ \frac{1}{150} \\ \frac{2}{75} \end{bmatrix}$

49 $x = 1 \quad y = 2 \quad z = 2 \quad t = 3$

50 $x = -0.0833 \quad y = 0.7083 \quad z = 1.9167 \quad t = 2.9583$

51 (a) $\begin{bmatrix} 1.1602 \\ -0.0515 \\ 0.0065 \end{bmatrix}$ (b) $\begin{bmatrix} 2.6844 \\ 0.0234 \\ 2.3569 \end{bmatrix}$ (c) $\begin{bmatrix} -1.3424 \\ 1.2860 \\ 2.4458 \\ 0.5511 \end{bmatrix}$

52 $\begin{bmatrix} -74.17 \\ -25.54 \\ 140.11 \end{bmatrix}$, $\det = 0.002\,725$ $\quad \begin{bmatrix} -142.53 \\ -50.52 \\ 262.77 \end{bmatrix}$, $\det = 0.001\,453$

54 $y_1 = 1.028 \quad y_2 = 4.004 \quad y_3 = 8.993 \quad y_4 = 11.329$

55 $4.5, 8, 10.5, 12, 12.5$

56 $\left|\dfrac{E_0}{Z_3}\right| = \sqrt{\left(\dfrac{L}{C}\right)\dfrac{1}{\alpha}[(2-\alpha^2)^2(1-\tfrac{1}{2}\alpha^2)^2}$
$\qquad\qquad - 2(2-\alpha^2)(1-\tfrac{1}{2}\alpha^2)-(1-\tfrac{1}{2}\alpha^2)^2+1]$
where $\alpha^2 = LC\omega^2$

57 Solution: 1, 2, 2, 3
After 5 iterations: 0.989, 1.99, 1.98, 3.00

58 Solution: -0.083, 0.708, 1.917, 2.958
After 3 iterations: -0.189, 0.634, 1.868, 2.920

59 (a) 0.8 (b) 1.1 (c) no convergence

60 $\begin{bmatrix} 10 \\ 20 \\ -30 \end{bmatrix}$

There is no convergence in 50 iterations, even from a

starting value of $\begin{bmatrix} 10.1 \\ 19.9 \\ -29.9 \end{bmatrix}$ except when $w = 1.8$

61 $I_1 = 0.5172,\ I_2 = 0.4914,\ I_3 = 0.8017$

62 0.1685, 0.3258, 0.5282, 0.7188, 0.9563, 1.0063, 0.8063, 0.6064, 0.4059, 0.2079

63 (a) 2 (b) 3

64 (a) Rank = 2, $(-2 + t, 5 - 2t, t)$
(b) Rank = 2, no solution

65 Rank = 3, rank = 3; $(\mu, -1, 1, -\mu)$

66 (a) $x = -1$, $y = \frac{1}{3}(2 - 4\mu)$, $z = \mu$
  (b) No solution
  (c) $x = \frac{1}{11}(-9 - 45\lambda + 13\mu)$, $y = \frac{1}{11}(5 - 8\lambda + 5\mu)$
    $z = \lambda$, $t = \mu$
  (d) Unique solution $x = -1$, $y = -1$, $z = 1$

68 Rank = 4 implies points not coplanar; rank = 3 implies the points lie on a plane; rank = 2 implies the points lie on a line; rank = 1 implies the four points are identical

70 (a) $\lambda^2 - 4\lambda + 3$, eigenvalues 3, 1
  (b) $\lambda^2 - 3\lambda + 1$, eigenvalues 2.618, 0.382
  (c) $\lambda^3 - 6\lambda^2 + 11\lambda - 6$, eigenvalues 3, 2, 1
  (d) $\lambda^3 - 6\lambda^2 + 9\lambda - 4$, eigenvalues 4, 1, 1
  (e) $\lambda^3 - 12\lambda^2 + 40\lambda - 35$, eigenvalues 7, 3.618, 1.382

71 (a) 2, 0; $\begin{bmatrix} 1 \\ 1 \end{bmatrix}$, $\begin{bmatrix} 1 \\ -1 \end{bmatrix}$ (b) 4, −1; $\begin{bmatrix} 2 \\ 3 \end{bmatrix}$, $\begin{bmatrix} -1 \\ 1 \end{bmatrix}$

  (c) 9, 3, −3; $\begin{bmatrix} -1 \\ 2 \\ 2 \end{bmatrix}$, $\begin{bmatrix} 2 \\ 2 \\ -1 \end{bmatrix}$, $\begin{bmatrix} 2 \\ -1 \\ 2 \end{bmatrix}$

  (d) 3, 2, 1; $\begin{bmatrix} 2 \\ 2 \\ 1 \end{bmatrix}$, $\begin{bmatrix} 1 \\ 1 \\ 0 \end{bmatrix}$, $\begin{bmatrix} 0 \\ 2 \\ -1 \end{bmatrix}$

  (e) 14, 7, −7; $\begin{bmatrix} 2 \\ 6 \\ 3 \end{bmatrix}$, $\begin{bmatrix} 6 \\ -3 \\ 2 \end{bmatrix}$, $\begin{bmatrix} 3 \\ 2 \\ -6 \end{bmatrix}$

  (f) 2, 1, −1; $\begin{bmatrix} 1 \\ -1 \\ -1 \end{bmatrix}$, $\begin{bmatrix} 1 \\ 0 \\ -1 \end{bmatrix}$, $\begin{bmatrix} 1 \\ 2 \\ -7 \end{bmatrix}$

  (g) 5, 3, 1; $\begin{bmatrix} -2 \\ -3 \\ 1 \end{bmatrix}$, $\begin{bmatrix} 1 \\ -1 \\ 0 \end{bmatrix}$, $\begin{bmatrix} 0 \\ -1 \\ 1 \end{bmatrix}$

  (h) 4, 3, 1; $\begin{bmatrix} 2 \\ -1 \\ -1 \end{bmatrix}$, $\begin{bmatrix} 2 \\ -1 \\ 0 \end{bmatrix}$, $\begin{bmatrix} 4 \\ 1 \\ -2 \end{bmatrix}$

72 (a) 5, 1, 1; $\begin{bmatrix} 1 \\ 1 \\ 1 \end{bmatrix}$, $\begin{bmatrix} -2 \\ 1 \\ 0 \end{bmatrix}$, $\begin{bmatrix} -1 \\ 0 \\ 1 \end{bmatrix}$

  (b) 2, 2, −1; $\begin{bmatrix} -1 \\ 1 \\ 0 \end{bmatrix}$, $\begin{bmatrix} 8 \\ 1 \\ 3 \end{bmatrix}$

  (c) 2, 2, 1; $\begin{bmatrix} 3 \\ 1 \\ -2 \end{bmatrix}$, $\begin{bmatrix} 4 \\ 1 \\ -3 \end{bmatrix}$

  (d) 2, 1, 1; $\begin{bmatrix} 2 \\ 1 \\ 2 \end{bmatrix}$, $\begin{bmatrix} 0 \\ -2 \\ 1 \end{bmatrix}$, $\begin{bmatrix} 1 \\ 3 \\ 0 \end{bmatrix}$

73 One eigenvector $\begin{bmatrix} -3 \\ 1 \\ 1 \end{bmatrix}$

74 2, 1, 1; $\begin{bmatrix} -1 \\ 1 \\ 1 \end{bmatrix}$, $\begin{bmatrix} -1 \\ 1 \\ 0 \end{bmatrix}$, $\begin{bmatrix} 1 \\ 0 \\ 1 \end{bmatrix}$

77 3, 2, −6; $\begin{bmatrix} -1 \\ 1 \\ 1 \end{bmatrix}$, $\begin{bmatrix} 0 \\ 1 \\ -1 \end{bmatrix}$, $\begin{bmatrix} 2 \\ 1 \\ 1 \end{bmatrix}$

78 $\begin{bmatrix} 1 \\ -1 \\ 0 \end{bmatrix}$

5.9 Review exercises

1 (a) $\begin{bmatrix} 12 & 18 & -40 \\ 0 & 0 & 8 \\ 0 & 0 & 4 \end{bmatrix}\begin{bmatrix} 12 & 0 & 0 \\ 18 & 0 & 0 \\ -40 & 8 & 4 \end{bmatrix}$

  (b) $\begin{bmatrix} 8 & 9 & -7 \\ 0 & 2 & 0 \\ 0 & 0 & 5 \end{bmatrix}\begin{bmatrix} 12 & 12 & -6 \\ 0 & 4 & 0 \\ 0 & 0 & 16 \end{bmatrix}$

  $\begin{bmatrix} 4 & 14 & -3 \\ 0 & 0 & 5 \\ 0 & 0 & 4 \end{bmatrix}$

2 $\lambda = -1$, $\mu = 2$
  $\lambda = 2$, $\mu = -1$

3 Normal strain = $\begin{bmatrix} 3 \\ -3 \\ 3\sqrt{2} \end{bmatrix}$

  Shear strain = 0

4 $(\alpha - \beta)(\beta - \gamma)(\gamma - \alpha)(\alpha + \beta + \gamma)$

5 $\theta = 1$: $(1 + 2\alpha, -3\alpha, \alpha)$
  $\theta - 2$: $(2\alpha, 1 - 3\alpha, \alpha)$

6 $A^2 = \begin{bmatrix} 1 & 3 & 2 \\ 0 & 5 & 2 \\ 2 & -2 & 1 \end{bmatrix}$ $A^3 = \begin{bmatrix} 3 & 6 & 5 \\ 6 & 7 & 8 \\ -4 & 1 & -3 \end{bmatrix}$

$A^{-1} = \begin{bmatrix} 3 & -2 & -1 \\ 2 & -1 & 0 \\ -4 & 3 & 1 \end{bmatrix}$ $X = \begin{bmatrix} -11 \\ -1 \\ 15 \end{bmatrix}$

7 (a) $P^T = \dfrac{1}{3}\begin{bmatrix} 2 & -2 & -1 \\ 1 & 2 & -2 \\ 2 & 1 & 2 \end{bmatrix}$, the solution $x = P^{-1}b$

exists

(b) $E = \begin{bmatrix} I_x - \dfrac{Q_x^2}{A} & I_{xy} - \dfrac{Q_x Q_y}{A} & 0 \\ I_{xy} - \dfrac{Q_x Q_y}{A} & I_y - \dfrac{Q_y^2}{A} & 0 \\ 0 & 0 & A \end{bmatrix}$

8 (a) $B = \begin{bmatrix} 8 & -10 & 3 \\ 3 & -4 & 1 \\ -1 & 0 & 0 \end{bmatrix}$

(b) $k = 8.2316, k = -1.9316$

9 (a) $\begin{bmatrix} -2 & -3 & -1 \\ -8 & -1 & 29 \\ -6 & 2 & 8 \end{bmatrix}, -22, \dfrac{1}{22}\begin{bmatrix} 2 & 3 & 1 \\ 8 & 1 & -29 \\ 6 & -2 & -8 \end{bmatrix}$,

$z = 2, y = 1, x = 2$

(b) $Y = \begin{bmatrix} 8 & 1 & 9 \\ 18 & 3 & -35 \\ 8 & -4 & -6 \end{bmatrix}$

(c) $Z = \begin{bmatrix} 6 & 11 & 3 \\ 15 & 2 & -59 \\ 16 & -7 & -16 \end{bmatrix}$

10 (a) 3, 0, 2, 1; det = 12

(b) 1, 2, 3, 4

11 1, 2, 3

12 If $c \neq 0$ then rank = 2
if $c = 0$ then rank = 1

13 $a = \begin{bmatrix} 0.0051 \\ 0.9712 \\ -0.3931 \\ -0.0760 \\ 0.0283 \end{bmatrix}$ $f(2) = 0.2200$ $f(3.5) = -0.4228$

16 $E_1 = 4E_2 + 3I_2; I_1 = 3E_2 + \frac{5}{2}I_2$

17 $a = 0.4424, b = -1.5037, c = 1.5023, d = -0.0611$
max at $x = 0.74, f = 0.4065$

18 rank $B = 2, AA^T = I, A^{-1} = A^T$
$x_1 = 2.444, x_2 = -2.556, x_3 = -1.222$

19 (b) $x_1 = 44, x_2 = -48, x_3 = -39, x_4 = 33$

20 (a) $\begin{bmatrix} 0 \\ -1 \\ 2 \end{bmatrix}$ (b) $\begin{bmatrix} -8 \\ -5 \\ \frac{1}{2} \end{bmatrix}$

21 (a) 4, 3, 2; $\begin{bmatrix} 0.5774 \\ 0.5774 \\ -0.5774 \end{bmatrix}, \begin{bmatrix} 0.1961 \\ 0.5883 \\ -0.7845 \end{bmatrix}, \begin{bmatrix} 0 \\ 0.7071 \\ -0.7071 \end{bmatrix}$

(b) 5, 3, -1; $\begin{bmatrix} 0.2033 \\ 0.6505 \\ 0.7318 \end{bmatrix}, \begin{bmatrix} 0.1374 \\ 0.8242 \\ 0.5494 \end{bmatrix}, \begin{bmatrix} -0.4472 \\ -0.8944 \\ 0 \end{bmatrix}$

(c) 9, 6, 3; $\dfrac{1}{3}\begin{bmatrix} -1 \\ 2 \\ 2 \end{bmatrix}, \dfrac{1}{3}\begin{bmatrix} 2 \\ -1 \\ 2 \end{bmatrix}, \dfrac{1}{3}\begin{bmatrix} 2 \\ 2 \\ -1 \end{bmatrix}$

22 $\lambda = 9$ $\alpha = 1$ $\beta = 6$

23 $\begin{bmatrix} 1 \\ 0 \\ 1 \end{bmatrix}, \begin{bmatrix} 0 \\ 1 \\ 0 \end{bmatrix}, \begin{bmatrix} -1 \\ 0 \\ 1 \end{bmatrix}$

24 (a) 3, 1; $\begin{bmatrix} 0.7071 \\ 0.7071 \end{bmatrix}, \begin{bmatrix} 0.7071 \\ -0.7071 \end{bmatrix}$

(b) 0.8794, -1.3473, -2.5321;
$\begin{bmatrix} 0.4491 \\ 0.8440 \\ 0.2931 \end{bmatrix}, \begin{bmatrix} 0.8440 \\ -0.2931 \\ -0.4491 \end{bmatrix}, \begin{bmatrix} 0.2931 \\ -0.4491 \\ 0.8440 \end{bmatrix}$

CHAPTER 6

Exercises

1 $A = \{1, 2, 3, 4, 5, 6, 7, 8, 9\}$
$B = \{-4, 4\}$
$C = \{5, 6, 7, 8, 9, 10\}$
$D = \{4, 8, 12, 16, 20, 24\}$

2 $A \cup B = \{-4, 1, 2, 3, 4, 5, 6, 7, 8, 9\}$
$A \cap B = \{4\}$
$A \cup C = \{n \in \mathbb{N}: 1 \leqslant n \leqslant 10\}$
$A \cap C = \{n \in \mathbb{N}: 5 \leqslant n \leqslant 9\}$
$B \cup D = \{-4, 4, 8, 12, 16, 20, 24\}$
$B \cap D = \{4\}$
$B \cap C = \varnothing$

3 $A \cup B = \{n \in \mathbb{N}: 1 \le n \le 10\}$
$A \cap C = \{1, 5, 9\}$
$A \cap B = \varnothing$
$B \cup C = \{1, 2, 4, 5, 6, 8, 9, 10\}$
$B \cap C = \{4, 8\}$

4

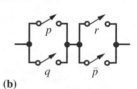

$\bar{A} \cap \bar{B}$ $\bar{A} \cup \bar{B}$ $\overline{A \cap B}$

$\overline{A \cup B}$ $A \cap \bar{B}$

5 (a) $A \cup B = \{1, 2, 3, 4, 5, 6, 7, 8, 9, 10, 12, 14, 16, 18, 20\}$
(b) $A \cap B = \{2, 4, 6, 8, 10\}$
(c) $A \cup C = \{1, 2, 3, 4, 5, 6, 7, 8, 9, 10, 16, 32\}$
(d) $A \cap C = \{2, 4, 8\}$

6 (a) True   (b) False   (c) False

7 (a) $\{n \in \mathbb{N}: 11 \le n \le 32\}$
(b) $\{11, 13, 15, 17, 19, 21, 22, 23, 24, 25, 26, 27, 28, 29, 30, 31, 32\}$
(c) $\bar{A} = \{n \in \mathbb{N}: 11 \le n \le 32\}$
$\bar{B} = \{1, 3, 5, 7, 9, 11, 13, 15, 17, 19, 20, 21, 22, 23, 24, 25, 26, 27, 28, 29, 30, 31, 32\}$
(d) $\overline{A \cap B} = \{1, 3, 5, 7, 9, 11, 12, 13, 14, 15, 16, 17, 18, 19, 20, 21, 22, 23, 24, 25, 26, 27, 28, 29, 30, 31, 32\}$
(e) $\bar{A} \cup \bar{B} = \overline{A \cap B}$ (see (d))

11 (a) $A \cap B$ (b) $\varnothing$ (c) A
(d) U (e) A (f) $A \cup (B \cap C)$
(g) $A \cap (B \cap C)$

14 (a) 5   (b) 25

15 (a) 20   (b) 27

16 4

17 (a) $\bar{C} = \{a, b, d, i\}, \overline{B \cup C} = \{a, b, i\}$
$\bar{B} \cap \bar{C} = \{a, b, i\}, A \cap B \cap D = \varnothing$
$A \cup F = \{a, b, c, f, i\}, D \cup (E \cap F) = \{b, c, e, h, i\}$,
$(D \cup E) \cap F = \{b, c, i\}$
(b) $B \cup C = \{c, d, e, f, g, h\}, C \cup E = \{c, e, f, g, h, i\}$
$D \cup E \cup F = \{b, c, e, f, h, i\}$
(c) $L_1: \{b, c, d, e, f, g, h, i\}$
$L_2: \{b, c, d, e, f, g, h, i\}$
$L_3:$ all elements

18 (a) 1 if $p = 1$, $q = 1$; 0 otherwise
(b) 0 if $p = 0$, $q = 0$; 1 otherwise
(c) 0   (d) 1

19 (a) $p \cdot q + \bar{p} \cdot \bar{q}$ (b) $(p + \bar{p}) \cdot (q + \bar{q})$
(c) $p + q + \bar{p} + \bar{q}$ (d) $p \cdot q + r \cdot s$
(e) $\bar{p} \cdot q \cdot s + \bar{p} \cdot \bar{q} \cdot r \cdot s + p \cdot q \cdot r \cdot s + p \cdot \bar{q} \cdot s + p \cdot q \cdot s$
(f) $p \cdot q \cdot r + p \cdot q \cdot t + p \cdot q \cdot u + p \cdot s \cdot \bar{u} + p \cdot v$

20 $\bar{p} \cdot \bar{q} + r + \bar{s} + \bar{q} \cdot t$

21

p	q	r	$\bar{p}$	$\bar{q}$	$\bar{r}$	$\bar{p}\cdot q\cdot\bar{r}$	$\bar{p}\cdot q\cdot r$	$p\cdot\bar{q}\cdot\bar{r}$	$p\cdot q\cdot r$	f
0	0	0	1	1	1	0	0	0	0	0
0	0	1	1	1	0	0	0	0	0	0
0	1	0	1	0	1	1	0	0	0	1
0	1	1	1	0	0	0	1	0	0	1
1	0	0	0	1	1	0	0	1	0	1
1	0	1	0	1	0	0	0	0	0	0
1	1	0	0	0	1	0	0	0	0	0
1	1	1	0	0	0	0	0	0	1	1

$f = \bar{p} \cdot q \cdot \bar{r} + \bar{p} \cdot q \cdot r + p \cdot \bar{q} \cdot \bar{r} + p \cdot q \cdot r$

22 (a) $p \cdot q$ (b) $\bar{p} \cdot r$
(c) $p \cdot q + \bar{p} \cdot \bar{q}$ (d) $p + q + r$
(e) 1   (f) $q + r$

23 (a) $\bar{p} \cdot q + p \cdot \bar{q} + p \cdot q$
(b) $(p + q) \cdot (p + \bar{q}) + p \cdot (\bar{r} + \bar{q})$
(c) $p \cdot (q + \bar{p}) + (q + r) \cdot \bar{p}$

24

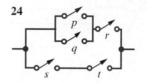

(a)

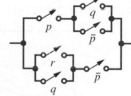

(b)

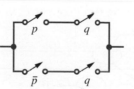

(c)

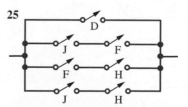

(d)

25

26

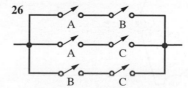

(A, B and C are the panel of three)

27

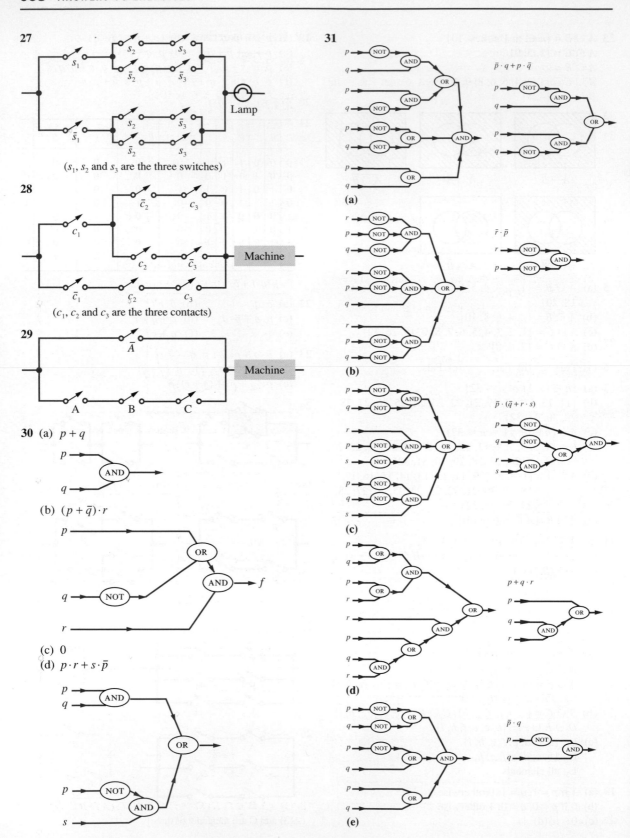

$(s_1, s_2$ and s_3 are the three switches)

28

$(c_1, c_2$ and c_3 are the three contacts)

29

30 (a) $p + q$

(b) $(p + \bar{q}) \cdot r$

(c) 0

(d) $p \cdot r + s \cdot \bar{p}$

31

(a) $\bar{p} \cdot q + p \cdot \bar{q}$

(b) $\bar{r} \cdot \bar{p}$

(c) $\bar{p} \cdot (\bar{q} + r \cdot s)$

(d) $p + q \cdot r$

(e) $\bar{p} \cdot q$

32 (a) Fred is not my brother
(b) 12 is an odd number
(c) There will be no gales next winter
(d) Bridges do not collapse when design loads are exceeded

33 (a) F   (b) T   (c) T   (d) F

34 (a) T   (b) F
(c)–(e) are not propositions
(f) Truth value is not known

35 (a) $A \wedge B$ (b) $A \to C$
(c) $\tilde{A} \to (\tilde{B} \wedge C)$ (d) $\tilde{C} \to B$

36 (a) It is raining and the sun is shining therefore there are clouds in the sky
(b) It is raining therefore there are clouds in the sky and hence the sun is shining
(c) If it is not raining then the sun is shining or there are clouds in the sky
(d) It is not the case that it rains and the sun shines, and there are clouds in the sky

37 (a) $x^2 = y^2 \to x = y$ for positive numbers x and y
(b) $x^2 = y^2 \to x = y$ for $x = 1$ and $y = -1$ (one of many possible answers)

38 (a) $n = 4$ (b) $n = 3$ (c) $n = 7$

39 (a) $B \wedge C \to A$
$\widetilde{B \wedge C} \to \tilde{A}$
(b) If $x^2 + y^2 \geqslant 1$ then $x + y = 1$; if $x^2 + y^2 < 1$ then $x + y \neq 1$
(c) If $3 + 3 = 9$ then $2 + 2 = 4$; if $3 + 3 \neq 9$ then $2 + 2 \neq 4$

40 (a)

A	B	$A \wedge \tilde{A}$
T	F	F
F	T	F

(b)

$\tilde{A}$	$\tilde{B}$	$\tilde{A} \vee \tilde{B}$
F	F	F
F	T	T
T	F	T
T	T	T

(c) and (d)

A	B	C	$A \wedge B$	$A \wedge B \to C$	$\widetilde{A \wedge B \to C}$
F	F	F	F	T	F
F	F	T	F	T	F
F	T	F	F	T	F
F	T	T	F	T	F
T	F	F	F	T	F
T	F	T	F	T	F
T	T	F	T	F	T
T	T	T	T	T	F

47 1, 4, 9, 16

6.7 Review exercises

1 (a) $\overline{A \cup B} = \{8, 9\}$
(b) $C - A = \{3, 7, 8\}$ $\overline{C} \cap \overline{B} = \{6, 9\}$

2 (a) $A \cap B = \{2, 4, 6, 8, 10\}$
(b) $A \cap B \cap C = \{10\}$
(c) $A \cup (B \cap C) = \{1, 2, 3, 4, 5, 6, 7, 8, 9, 10, 11, 12, 14, 20\}$

3 (a) $\overline{A} = \{n \in \mathbb{N}: 11 < n \leqslant 20\}$
(b) $\overline{A} \cup \overline{B} = \{1, 3, 5, 7, 9, 11, 12, 13, 14, 15, 16, 17, 18, 19, 20\}$
(c) $\overline{A \cup B} = \{13, 15, 17, 19\}$
(d) $A \cap (\overline{B \cup C}) = \{1, 3, 5, 7, 9\}$

4 Statement (a) is true

5 (a) $f = A$ $g = U$ (the universal set)
(b)

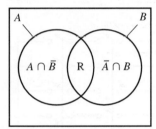

Only equals $A \cup B$ if region R does not exist, that is $A \cap B = \varnothing$

7
(a)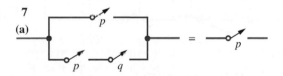

(b)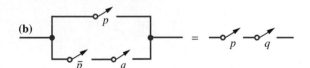

(c)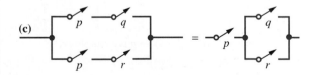

(d)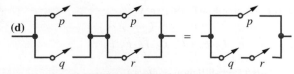

8 (a) $(\overline{A} \cap \overline{B} \cap D) \cap (A \cap C \cap D) \cap (\overline{A \cap B \cap D})$
(b) $(B \cap \overline{C}) \cup (C \cap \overline{B})$

9 (a) $x \cdot y \cdot z \cdot u + \bar{x} \cdot \bar{y} \cdot z \cdot u + x \cdot \bar{u}$

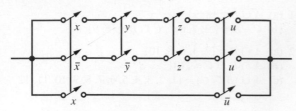

(b) $\bar{x} \cdot \bar{y} + \bar{z} \cdot \bar{u} + x \cdot y \cdot z$

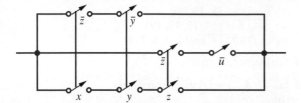

(c) (i) 0   (ii) 0

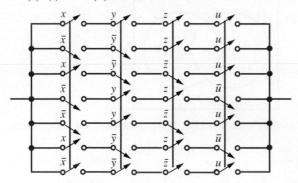

10 (a)

p	q	$p \wedge q$
T	T	T
T	F	F
F	T	F
F	F	F

(b)

p	q	$p \vee q$
T	T	T
T	F	T
F	T	T
F	F	F

(c)

p	q	$p \to q$
T	T	T
T	F	F
F	T	T
F	F	T

     A   B   C

(e)

p	q	$\bar{p}$	$\bar{p} \wedge q$	$\overline{\bar{p} \wedge q}$	$p \vee q$	$\boxed{A} \wedge \boxed{B}$	$\boxed{C} \to p$
T	T	F	F	T	T	T	T
T	F	F	F	T	T	T	T
F	T	T	T	F	T	F	T
F	F	T	F	T	F	F	T

Hence $(p \vee q) \wedge (\bar{p} \wedge q) \to p$ is a tautology.

11 (a) $\bar{q} \cdot p + \bar{p} \cdot q$ (b) $p + q + \bar{r}$
(c) $p \cdot q \cdot r + q \cdot \bar{r} \cdot s$

12 (a) (i) $p \cdot r + q \cdot \bar{r} \cdot s + \bar{q} \cdot r \cdot \bar{s}$
     (ii) $p \cdot \bar{q} \cdot \bar{r} + \bar{p} \cdot \bar{r} \cdot s$

13 $C_1 \cdot \bar{C}_2 \cdot \bar{F}_1 \cdot F_2 \cdot \bar{F}_3 + C_1 \cdot \bar{C}_3 \cdot \bar{F}_1 \cdot \bar{F}_2 \cdot F_3$
$+ C_1 \cdot C_2 \cdot \bar{C}_3 \cdot \bar{F}_1 \cdot \bar{F}_2 \cdot F_3$
where C_i = call button on floor i
and $F_i = 1$ if lift is on floor, 0 otherwise

14 (a) Let the four people be labelled A, B, C, D. The truth table is then as given below:

A	B	C	D	Yes	No	Tie
0	0	0	0	0	1	0
0	0	0	1	0	1	0
0	0	1	0	0	1	0
0	0	1	1	0	0	1
0	1	0	0	0	1	0
0	1	0	1	0	0	1
0	1	1	0	0	0	1
0	1	1	1	1	0	0
1	0	0	0	0	1	0
1	0	0	1	0	0	1
1	0	1	0	0	0	1
1	0	1	1	1	0	0
1	1	0	0	0	0	1
1	1	0	1	1	0	0
1	1	1	0	1	0	0
1	1	1	1	1	0	0

Extracting from this table those inputs that cause a Yes, No or Tie (Y, N or T) we have
(b) $Y = \bar{A} \cdot B \cdot C \cdot D + A \cdot \bar{B} \cdot C \cdot D + A \cdot B \cdot \bar{C} \cdot D$
     $+ A \cdot B \cdot C \cdot \bar{D} + A \cdot B \cdot C \cdot D$
$N = \bar{A} \cdot \bar{B} \cdot \bar{C} \cdot \bar{D} + \bar{A} \cdot \bar{B} \cdot \bar{C} \cdot D$
     $+ \bar{A} \cdot \bar{B} \cdot C \cdot \bar{D} + \bar{A} \cdot B \cdot \bar{C} \cdot \bar{D} + A \cdot \bar{B} \cdot \bar{C} \cdot \bar{D}$
$T = \bar{A} \cdot \bar{B} \cdot C \cdot D + \bar{A} \cdot B \cdot \bar{C} \cdot D + \bar{A} \cdot B \cdot C \cdot \bar{D}$
     $+ A \cdot \bar{B} \cdot \bar{C} \cdot D + A \cdot \bar{B} \cdot C \cdot \bar{D} + A \cdot B \cdot \bar{C} \cdot \bar{D}$
(c) $Y = A \cdot B \cdot D + A \cdot B \cdot C + A \cdot C \cdot D + B \cdot C \cdot D$
$N = \bar{A} \cdot \bar{B} \cdot \bar{C} + \bar{A} \cdot \bar{B} \cdot \bar{D} + \bar{A} \cdot \bar{C} \cdot \bar{D} + \bar{B} \cdot \bar{C} \cdot \bar{D}$
T does not simplify
(d) To modify the circuit we introduce the chairman's vote E. If N denotes No and Y denotes Yes, the new circuit must have the output

$$N_{\text{new}} = (N_{\text{old}} + T) \cdot \bar{E}$$

$$Y_{\text{new}} = (Y_{\text{old}} + T) \cdot E$$

where T = Tie. Hence the modified circuit will be

No (old)

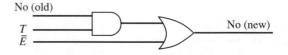

T
$\bar{E}$

No (new)

Yes (old)

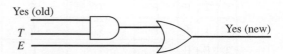

T
E

Yes (new)

A tie is now impossible.

15 $\bar{N} \cdot (\bar{V} + \bar{R} \cdot M)$
i.e. no dope smoking occurs if Neil is absent and *either* Vivian is absent *or* Mike is present *and* Rick is absent.

16 $p \cdot q \cdot \overline{p \cdot r} + q \cdot \overline{r} = q \cdot \overline{r}$

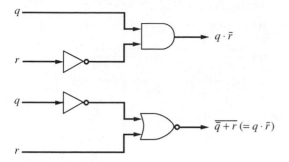

17 (a) F   (b) No   (c) No
    (d) F   (e) F   (f) F
    (g) No   (h) F   (i) T

18 (a) (i)

p	q	$p \rightarrow q$	$q \rightarrow p$	$p \leftrightarrow q$
T	T	T	T	T
T	F	F	T	F
F	T	T	F	F
F	F	T	T	T

(ii)

q	p	$p \rightarrow q$
T	T	T
T	F	F
T	F	F
F	F	T

(b) (i) False   (ii) False

19 $\overline{P}_O \cdot \overline{T} + P_O \cdot \overline{P}_F \cdot \overline{T} + P_O \cdot P_F \cdot T$
    $= \overline{P}_O \cdot \overline{T} + \overline{P}_F \cdot \overline{T} + P_O \cdot P_F \cdot T$ is minimal
    $P_O = 1$ when pressure in oxidizer tank $\geqslant$ required
      minimum
    $P_F - 1$ when pressure in fuel tank $\geqslant$ required minimum
    $T = 1$ when time $\leqslant$ 15 min to lift-off
    $L = 1$ when panel light is on

21 $p \cdot q \cdot r \cdot s$, $p \cdot q \cdot r \cdot \overline{s}$, $p \cdot q \cdot \overline{r} \cdot s$, $p \cdot q \cdot \overline{r} \cdot \overline{s}$, $p \cdot \overline{q} \cdot r \cdot s$,
    $p \cdot \overline{q} \cdot r \cdot \overline{s}$, $p \cdot \overline{q} \cdot \overline{r} \cdot s$, $p \cdot \overline{q} \cdot \overline{r} \cdot \overline{s}$, $\overline{p} \cdot q \cdot r \cdot s$, $\overline{p} \cdot q \cdot r \cdot \overline{s}$,
    $\overline{p} \cdot q \cdot \overline{r} \cdot s$, $\overline{p} \cdot q \cdot \overline{r} \cdot \overline{s}$, $\overline{p} \cdot \overline{q} \cdot r \cdot s$, $\overline{p} \cdot \overline{q} \cdot r \cdot \overline{s}$, $\overline{p} \cdot \overline{q} \cdot \overline{r} \cdot s$,
    $\overline{p} \cdot \overline{q} \cdot \overline{r} \cdot \overline{s}$

	Converse	*Contrapositive*
22 (a)	If I do not go, the train is late	If the train is early, I will go
(b)	If you retire, you will have enough money	If you do not have enough money, you will not retire
(c)	You cannot do it unless I am there	I can do it if you are there
(d)	If I go, so will you	If you do not go nor will I

23 'If you were a member of the other tribe, what would you answer if I asked you if your God was male?' The answer is then definitely false!

CHAPTER 7

Exercises

2 45, 57.5, 63.75, 66.875

3 (a) 40   (b) 16 736   (c) 35

4 $\left(\dfrac{v}{V}\right)^{n+1} \times 100$

5 $\dfrac{1}{n} \displaystyle\sum_{l=1}^{n} (x_l - 10)^2$

6 2.618

7 $\{0, \frac{1}{3}, \frac{16}{25}, \frac{81}{109}, \frac{256}{321}, \frac{625}{751}\}$

8 $\{1, 1.5, 1.4, 1.417, 1.414, 1.414, 1.414\}$

9 1.222

10 5

11 (a) 16, 31   (b) 10, 20, 40, 80

12 £2700k, 11

13 $\dfrac{1 + (n - 2)\sqrt{x}}{1 - x}$

14 3.1365 (4dp)

15 $\frac{2}{5}, \frac{2}{7}, \frac{2}{9}, \frac{2}{11}, \frac{2}{13}, \frac{2}{15}$

16 £66 116, £128 841, after 7.3 years

17 (a) 11 781   (b) 1 205 589   (c) $1 - (\frac{1}{2})^{153}$
     (d) $3^{154} - 1$ (e) 1 217 370   (f) $\frac{153}{154}$

18 9

20 $x = \dfrac{10r}{1 - (1 + \frac{1}{100}r)^{-n}}$,   117.46

21 $2 - \dfrac{2 + n}{2^n}$

22 $\dfrac{a(1 - r^n)}{1 - r} + \dfrac{dr}{(1 - r)^2}[1 + nr^{n-1} - (n - 1)r^n]$

23 £1770, {10 000, 9430, 8792, 8077, 7276, 6379, 5375, 4250, 2989, 1578, −3}

24 $(A + n)/n^2$

25 (a) 0   (b) -3×2^n (c) 0   (d) $75 \times (-2)^n$
     (a) and (c) satisfy recurrence relation

27 (a) $A5^n + B2^n$ (b) $A3^n + B(-2)^n$
     (c) $(\frac{1}{5})^n\left(A\cos\dfrac{n\pi}{2} + B\sin\dfrac{n\pi}{2}\right)$
     (d) $A5^n + Bn5^n$
     (e) $A(-\frac{1}{2})^n + B$

28 (a) $\frac{1}{4} + \frac{13}{12}(5^n) - \frac{1}{12}(2^n)$ (b) $\frac{4}{9} + \frac{14}{9}(-\frac{1}{2})^n + \frac{1}{3}n$

    (c) $-\frac{1}{6}n - \frac{1}{36} + A3^n + B(-2)^n$

29 (b) $(1-n)a^n$ (c) $(3 + (2a^{-10} - 0.3)n)a^n$

30 $T_2 = 2x^2 - 1$, $T_3 = 4x^3 - 3x$, $T_4 = 8x^4 - 8x^2 + 1$

31 (a) $N_t = 2^t$

    (b) $N_t = \left[\left(\dfrac{1 + \sqrt{5}}{2}\right)^{t+1} - \left(\dfrac{1 - \sqrt{5}}{2}\right)^{t+1}\right] \Big/ \sqrt{5}$

32 (a) 0.5, 0.4, 0.3, 0.2353, 0.1923, 0.1632; →0

    (b) 0.4615, 0.4722, 0.4789, 0.4831, 0.4859, 0.4879;
    →0.5

    (c) 2, 2, 1.817, 1.682, 1.585, 1.513; →1

    (d) 1.5, 1.5625, 1.5880, 1.6018, 1.6105, 1.6165;
    →e$^{1/2}$ = 1.6487

    (e) 1.4142, 1.5538, 1.5981, 1.6119, 1.6161, 1.6174;
    $\frac{1}{2}(1 + \sqrt{5})$

    (f) →0, 2, 2.5981, 2.8284, 2.9389, 3.0000; →π

33 (a) 1, 1, $\frac{5}{3}$, 2.5, 3.4, 4.3333; diverges to infinity

    (b) 1, 0, −1, 0, 1, 0; oscillates between 1, 0, −1

    (c) 1, 3, 1, 3, 1, 3; oscillates between 1 and 3

34 (a) 10   (b) 19   (c) 1 000 002

    (d) 25   (e) 18

35 70%

36 2.2, 2.324, 2.418 996, 2.450 262

    Estimate = 2.465 5011

    Limit = 2.465 571

37 (a) convergent   (b) convergent   (c) divergent

    (d) divergent

38 $\frac{9}{4}$

39 (a) $1 - \dfrac{1}{2N+1}$, 1,   (b) $4 - \dfrac{2+N}{2^{N-1}}$, 4

    (c) $\dfrac{1}{4} - \dfrac{1}{2(N+1)(N+2)}$, $\dfrac{1}{4}$

40 (a) divergent   (b) divergent   (c) convergent

42 $\frac{19}{33}$;   (a) $\frac{413}{999}$ (b) $\frac{10}{99}$ (c) 1   (d) $\frac{172\,300}{9999}$

44 1.082 322 1; summation from right allows full account to be taken of the accumulative effect of small terms

46 (a) $|x| < 1$ (b) $x \in \mathbb{R}$ (c) $|x| \leqslant 1$ (d) $|x| < 1$

47 (a) $(-x^2)^r$ $(|x| < 1)$ (b) $\dfrac{x^{2r+1}}{2r+1}$ $(|x| < 1)$

    (c) $(-1)^r(r+1)x^r$ $(|x| < 1)$

    (d) $-\dfrac{(2r-3)!}{2^{2r-2}r!(r-2)!}x^r$ $r > 1\,(|x| < 1)$

(e) $\dfrac{1}{10}\left[2^{r+2} + \dfrac{(-1)^r}{2^r}\right]$ $(|x| < \frac{1}{2})$

(f) $(1+x)x^{4r}$ $(|x| < 1)$

48 (a) $\dfrac{5 \cdot 4}{1 \cdot 2}$ (b) $\dfrac{(-2)(-3)(-4)}{1 \cdot 2 \cdot 3}$

    (c) $\dfrac{(\frac{1}{2})(-\frac{1}{2})(-\frac{3}{2})}{1 \cdot 2 \cdot 3}$ (d) $\dfrac{(-\frac{1}{2})(-\frac{3}{2})(-\frac{5}{2})(-\frac{7}{2})}{1 \cdot 2 \cdot 3 \cdot 4}$

50 $n = 8$,   1.1905,   six multiplications

51 (a) $(1 + 2x^2)^{-1}$ (b) $(1-x)^{-1/2}$

    (c) $1 + \dfrac{1-x}{x}\ln(1-x)$ (d) $\dfrac{x^2}{1-x^2}\ln(1-x^2)$

52 3.1415

53 (a) $\frac{1}{3}$ (b) $-\frac{2}{3}$ (c) 2   (d) 0

54 (a) 3   (b) $\frac{1}{2}$

55 (a) −1   (b) 1   (c) $n-1$ (d) n

58 (a) Undefined at $x = 0$, continuous for $x \neq 0$

    (b) Infinite discontinuity at $x = 2$, continuous for $x \neq 2$

    (c) Finite discontinuity at $x = 0$, continuous for $x \neq 0$

    (d) Finite discontinuities at $x = \pm\sqrt{n}$, $n = 0, 1, 2, \ldots$

59 (a) Upper bound is 7, lower bound 5

    (b) Upper bound is 3, lower bound −1

60 1.75

61 0.830

62 $\alpha = -1.879$, $\beta = 0.347$, $\gamma = 1.532$

63 (a) is convergent, (b) convergent and (c) divergent

    (b), 0.771

64 5.4267, 5.3949, $\varepsilon_1 \simeq 0.05$

65 $\alpha_0 \approx 1.9$, $\alpha_k \approx \frac{1}{2}(2k+1)\pi$

    $\theta_{n+1} = \cos^{-1}(-\mathrm{sech}\,\theta_n)$, $\alpha_0 = 1.8751$

7.12 Review exercises

1 1000, 850, 700, 550, 400, 250, 100
  1000, 681, 464, 316, 215, 147, 100

2 £361, £243, £141, £53 for $r < 23.375$

3 $1 - 0.2(-\frac{1}{2})^r$,   1

5 (a) $A2^n + B3^n$ (b) $(A + Bn)2^n$

    (c) $A2^n + B3^n + \frac{1}{2}4^n$ (d) $A2^n + B3^n + 3^{n-1}n$

8 $200 + 20(-\frac{2}{5})^t$

9 $2 + A\cos n\theta + B\sin n\theta$, $\tan\theta = \dfrac{\sqrt{7}}{3}$

10 $\gamma \approx 0.577\,235$; compare the true value 0.577 216

11 (a) divergent   (b) convergent

    (c) divergent   (d) convergent

12 (a) $\frac{410}{333}$ (b) $\frac{143}{333}$ (c) $\frac{101}{999}$ (d) $\frac{1724}{3333}$

13 (a) convergent   (b) divergent
    (c) convergent   (d) divergent

14 $\dfrac{x(1 + x)}{(1 - x)^3}$

15 $a = 1$, $b = -\frac{1}{3}$, $c = \frac{1}{45}$
    $|x| < 0.2954$
    $\tan 0.29$ is given to 4dp; $\tan 0.295$ has an error of $\frac{1}{2}$ unit in 1dp, but when rounded to 4dp gives an error of 1 unit.

16 $x - \frac{1}{6}x^3$
    $x - \frac{1}{6}x^3 + \frac{3}{40}x^5 - \frac{5}{112}x^7 + \frac{35}{1152}x^9$

17 0.095, 31

19 $4.05\,\text{rad} = 232°$

20 $2.718\,586\,07$
    $2.718\,357\,88$
    $2.718\,281\,81$

21

r	0	1	2	3	4	5	6	7	8
M_r	0	2.64	1.93	2.13	2.06	2.13	1.93	2.64	0

23 $F = \dfrac{W(l - a)^2(l + 2a)}{l^3} - WH(x - a)$

    $M = \dfrac{W(l - a)^2 a}{l^2} - \dfrac{W(l - a)^3(l + 2a)x}{l^3}$
       $+ W(x - a)H(x - a)$

CHAPTER 8

Exercises

1 (a) 0   (b) 1   (c) $2x$

    (d) $3x^2$ (e) $\dfrac{1}{2\sqrt{x}}$ (f) $\dfrac{-1}{(1 + x)^2}$

2 (a) $4x - 5$ (b) -1
    (c) $(1, -15)$, $(\frac{1}{2}m + \frac{3}{2}, \frac{1}{2}m^2 + \frac{1}{2}m - 15)$
    (d) -1 (e) $y = -x - 14$

3 (a) $6x^2 - 6x + 1$ (b) 1
    (c) $(1, 3)$, $(\frac{1}{4}[1 \pm \sqrt{(1 + 8m)}], \frac{1}{4}[12 - 3m \pm m\sqrt{(1 + 8m)}])$
    (d) $m = 1, -1/8$ (e) $y = x + 2$, $y = -0.125x + 3.125$

4 (a) minimum at $x = -1/2$ (b) minimum at $x = 1/3$
    (c) maximum at $x = 3/2$ (d) maximum at $x = 1/2$

5 $3ax^2 + 2bx + c$

6 $\frac{1}{3}\,\text{m}^2\,\text{min}^{-1}$

8 $v(t) = \begin{cases} 1, & 0 \leqslant t < 1 \\ 2t - 1, & 1 \leqslant t \leqslant 2 \\ 3, & 2 \leqslant t < 3 \\ -1, & 3 \leqslant t \leqslant 9 \end{cases}$

9 $3W(2x - l)^2/4l^2$

12 μT

16 $\dfrac{\text{d}x}{\text{d}t} \propto (a - x)(b - x)$

17 (a) $10x - 2$ (b) $12x^2 + 1$ (c) $24x^{23}$
    (d) $45(5x + 3)^8$ (e) $28(4x - 2)^6$
    (f) $-18(1 - 3x)^5$ (g) $3(6x - 1)(3x^2 - x + 1)^2$
    (h) $6(12x^2 - 2)(4x^3 - 2x + 1)^5$
    (i) $-5(4x^3 - 1)(1 + x - x^4)^4$

18 (a) $512(x + 2)^6(3x - 2)^4(9x + 4)$
    (b) $(5x + 1)^2(3 - 2x)^3(37 - 70x)$
    (c) $(\frac{1}{2}x + 2)(x + 3)^3(3x + 11)$
    (d) $12x^3 - 6x^2 - 20x + 11$
    (e) $36x^5 + 20x^3 - 54x^2 + 12x - 15$
    (f) $2(x^2 + x + 1)(x^3 + 2x^2 + 1)^3 \times$
       $(8x^4 + 19x^3 + 16x^2 + 10x + 1)$
    (g) $(x^5 + 2x + 1)^2(2x^2 + 3x - 1)^3 \times$
       $(46x^6 + 57x^5 - 15x^4 + 44x^2 + 58x + 6)$
    (h) $(2x + 1)^2(7 - x)^4(37 - 16x)$
    (i) $(3x + 1)^4(21x^2 + 74x + 19)$

20 $b = 56$, $h = 144$, $w = 90$

21 (a) $1/\sqrt{(1 + 2x)}$ (b) $(3x + 2)/(2\sqrt{x})$
    (c) $(3x + 4)/[2\sqrt{(x + 2)}]$

22 (a) $(2x^2 + 4)/\sqrt{(4 + x^2)}$
    (b) $(9 - 2x^2)/\sqrt{(9 - x^2)}$
    (c) $(2x^2 + 4x + 4)/\sqrt{(x^2 + 2x + 3)}$
    (d) $\frac{2}{3}x^{-1/3} - \frac{1}{4}x^{-3/4}$ (e) $\frac{2x}{3}(x^2 + 1)^{-2/3}$
    (f) $(8x - 3)/[3(2x - 1)^{2/3}]$

23 (a) $-2/(x + 3)^3$ (b) $1/(x + 1)^2$ (c) $1/(x - 2)^2$
    (d) $2(2 - x)/(x^2 - 4x + 1)^2$ (e) $1 - 1/x^2$
    (f) $(6 - x^2)/(x^2 + 5x + 6)^2$ (g) $-1/(x^2 - 1)^{3/2}$
    (h) $2(2x + 1)(2 - 9x - 12x^2)/(3x^2 + 1)^4$

24 $10^3\sqrt{2}$

25 (a) $3\cos(3x - 2)$ (b) $-4\cos^3x\sin x$
    (c) $-6\cos 3x\sin 3x = -3\sin 6x$
    (d) $\frac{5}{2}\cos 5x - \frac{1}{2}\cos x$ (e) $\sin x + x\cos x$
    (f) $-\sin 2x/\sqrt{(2 + \cos 2x)}$ (g) $-a\sin(x + \theta)$
    (h) $4\sec^2 4x$

26 (a) $1/\sqrt{(4 - x^2)}$ (b) $-5/\sqrt{(1 - 25x^2)}$
    (c) $(x\tan^{-1}x + 1)/\sqrt{(1 + x^2)}$ (d) $1/\sqrt{(3 + 2x - x^2)}$
    (e) $3/(1 + 9x^2)$ (f) $1 - \dfrac{x\sin^{-1}x}{\sqrt{(1 - x^2)}}$

27 (a) $2e^{2x}$ (b) $-\frac{1}{2}e^{-x/2}$
    (c) $(2x + 1)\exp(x^2 + x)$ (d) $xe^{5x}(5x + 2)$
    (e) $e^{-x}(1 - 3x)$ (f) $-e^x/(1 + e^x)^2$
    (g) $\frac{1}{2}e^x/\sqrt{(1 + e^x)}$ (h) ae^{ax+b}

28 (a) $2/(2x + 3)$ (b) $2(x + 1)/(x^2 + 2x + 3)$
(c) $1/(x - 2) - 1/(x - 3)$ (d) $(1 - \ln x)/x^2$
(e) $5/[(2x + 1)(1 - 3x)]$ (f) $(2x + 1)/[x(x + 1)]$

29 (a) $3\cosh 3x$ (b) $4\,\text{sech}^2 4x$
(c) $3x^2 \cosh 2x + 2x^3 \sinh 2x$ (d) $\frac{1}{2}\tanh\frac{1}{2}x$
(e) $\cos x \sinh x - \sin x \cosh x$ (f) $-\sinh x/\cosh^2 x$

30 (a) $2/\sqrt{(4 + x^2)}$ (b) $2/\sqrt{(x^2 - 1)}$ (c) $1/(1 - x^2)$
(d) $1 + \dfrac{x\sinh^{-1}x}{\sqrt{(1 + x^2)}}$ (e) $\sqrt{(4 - x^2)}/x$
(f) $[(1 + x^2) - 2x\tanh^{-1}x(1 - x^2)]/[(1 - x^2)(1 + x^2)^2]$

31 $e^{-2\pi a/\omega}$

32 $a = 26,\ b = 39$

33 horizontal side $1/\sqrt{2}$

35 $(1 - t\tan t)/(\tan t + t)$

36 $y = 1 - (\sqrt{2} - 1)x$

37 $y = 4 - 64x$

38 $\frac{3}{4}$

39 (a) $10^x \ln 10$ (b) $-2^{-x}\ln 2$
(c) $(2x^3 + 3x^2 + 6x + 6)(x - 1)^{5/2}[(x + 1)^{1/2}/(x^2 + 2)^2]$

40 $x^2 e^{-2x}((3 - 2x)\sin \pi x + \pi x \cos \pi x)$

41 (a) $(6 + 19x^2 + 12x^4)x/(1 + x^2)^{3/2}$
(b) $(1 - 2x - 2x^2)/(1 + x + x^2)^2$
(c) $-(84x^2 + 6y^2y' + 6xyy'^2)/(1 + 3xy^2)$ where
$y' = -(28x^3 + y^3)/(1 + 3xy^2)$
(d) $(6x - 2y' - 6yy'^2)/(3y^2 + x)$ where
$y' = (1 + y - 3x^2)/(3y^2 - x)$

42 (a) $-(2 + t^2)/(\sin t + t\cos t)^3$
(b) $\frac{1}{8}\,\text{cosec}\,\dfrac{3t}{2}\sec^3\dfrac{t}{2}$

45 (a) $2x - \dfrac{2}{x^3},\quad 2 + \dfrac{6}{x^4}$

46 $\cot\dfrac{\theta}{2},\ -\dfrac{a}{y^2}$

47 $y = x + 1;\ (0, 1),\ (-1, 0)$

48 $\cot\frac{1}{2}\theta$

49 $\left(\dfrac{1 + 2t}{1 + t}\right)^2,\ \dfrac{2(1 + t)^3}{3(1 + 2t)^3}$

50 $-\dfrac{x(2x^2 + 2y^2 - a^2)}{y(2x^2 + 2y^2 + a^2)}$

51 $2, 0$

52 (a) $(\ln x)^{x-1}[1 + (\ln x)\ln\ln x]$ (b) $(2\ln x)x^{\ln x-1}$
(c) $-\frac{5}{3}x(5 - 2x^2)(1 - x^2)^{-1/2}(2x^2 + 3)^{-7/3}$

53 (a) $3^4 e^{3x},\ 3^n e^{3x}$
(b) $-\dfrac{6}{(2 + x)^4},\ (-1)^{n-1}(n - 1)!/(2 + x)^n$
(c) $\dfrac{12}{(1 + x)^5} + \dfrac{12}{(1 - x)^5},$
$\frac{1}{2}n!\left[\dfrac{1}{(1 - x)^{n+1}} + \dfrac{(-1)^n}{(1 + x)^{n+1}}\right]$

54 $a^4 \sin(ax + b)$

57 (a) $(x^2 - 20)\cos x + 10x\sin x$
(b) $(x - 4)e^{-x}$
(c) $216(273x^2 + 39x + 1)(3x + 1)^9$

58 (a) $[(3 - 2x)\ln x + 1]x^2 e^{-2x}$
(b) $[(x - 1)\sin 2x + 2x\cos 2x]\dfrac{e^x}{x^2}$

59 $\frac{1}{12}(145)^{3/2}$

61 $\sqrt{2},\ (1, 2)$

62 $13^{3/2}/16$

63 $1.035, 0.92, 0.88$

65 (a) $5.436 = 2e,\ 8.155 = 3e$
(b) $5.440\ (h = 0.01)$, error depends on h^2
(c) $8.00\ (h = 0.01)$

66 1.5432

68 Brian by $(6\sqrt{3} - 4\sqrt{6})\,\text{s}$

71 -6

75 $F = \begin{cases} 7W/8 - 4Wx/l & 0 < x < l/4 \\ -W/8 & l/4 < x < l \end{cases}$

76 Depth h satisfies $1000\pi h^2(3 - 2h) = 6t$

77 (a) $\frac{1}{7}x^7 + c$ (b) $\frac{1}{3}e^{3x} + c$
(c) $-\frac{1}{5}\cos 5x + c$ (d) $\frac{1}{8}(2x + 1)^4 + c$
(e) $\frac{1}{3}\tan 3x + c$ (f) $2\ln|x| + c$
(g) $-\dfrac{3}{x} + c$ (h) $\frac{1}{2}\sin 2x + c$
(i) $\frac{1}{4}\sec 4x + c$ (j) $\frac{1}{6}(4x - 1)^{3/2} + c$

78 (a) $\frac{9}{5}x^{5/3} + c$ (b) $\frac{2}{3}\sqrt{2}x^{3/2} + c$
(c) $\frac{1}{2}x^4 - \frac{2}{3}x^3 + \ln|x| - 2x + c$
(d) $2e^x + \frac{3}{2}\sin 2x + c$ (e) $\frac{1}{3}x^3 + 3e^x + \dfrac{1}{x} + c$
(f) $\frac{1}{8}(2x + 1)^4 + c$ (g) $-\frac{3}{8}(1 - 2x)^{4/3} + c$
(h) $\frac{8}{7}x^7 + \frac{12}{5}x^5 + 2x^3 + x + c$
(i) $\frac{1}{2}\sin(2x + 1) + c$ (j) $\dfrac{2^x}{\ln 2} + c$
(k) $\frac{1}{16}(4\cos 2x - \cos 8x) + c$
(l) $\frac{1}{24}(\sin 12x + 6\sin 2x) + c$

79 (a) $\frac{1}{5}[\ln|x+1|+4\ln|x-4|]+c$

(b) $\ln|x-2|-\dfrac{2}{(x-2)}+c$

(c) $\ln\left|\dfrac{x}{x+1}\right|+c$ (d) $\ln|x+1|+\dfrac{1}{x+1}+c$

(e) $\frac{1}{2}\ln\left|\dfrac{x-1}{x+1}\right|+c$ (f) $\ln\left|\dfrac{x-1}{x}\right|+\dfrac{1}{x}+c$

(g) $\frac{1}{2}\ln\left|\dfrac{x(x-2)}{(x-1)^2}\right|+c$ (h) $\frac{1}{3}\ln\left|\dfrac{1+2x}{1-x}\right|+c$

(i) $2x+\frac{2}{3}\ln|x-1|-\frac{1}{3}\ln|x^2+2+1|+$
$\dfrac{10}{3\sqrt{3}}\,\tan^{-1}\left(\dfrac{2x+1}{\sqrt{3}}\right)+c$

(j) $2\ln|x|+\ln(x-1)-\tan^{-1}x+c$

(k) $\ln\left|\dfrac{x-1}{x+2}\right|+\dfrac{3}{x+2}+c$

(l) $-2\ln|x-1|+(\frac{1}{2}+\frac{3}{2}\sqrt{5})\ln|x-\frac{1}{2}-\frac{1}{2}\sqrt{5}|$
$-(\frac{3}{2}\sqrt{5}-\frac{1}{2})\ln|x-\frac{1}{2}+\frac{1}{2}\sqrt{5}|+c$

80 (a) $\frac{4}{3}$ (b) $-\frac{1}{156}$ (c) 0 (d) $\frac{1}{2}\pi$
(e) $2^{7/2}/5-\frac{9}{10}$ (f) 1 (g) $\frac{1}{3}\pi$ (h) $\frac{1}{2}\ln 2+\frac{1}{2}$

82 (a) $-\dfrac{1}{x}+c$ (b) $\frac{3}{2}(x+1)^{2/3}+c$

(c) $2x^2-7x-\dfrac{1}{x}+c$ (d) $\sin x-\cos x+c$

(e) $\frac{1}{2}x-\frac{1}{4}\sin 2x+c$ (f) $\frac{1}{2}x+\sin 2x+c$

(g) $\frac{1}{2}x+\frac{1}{4}\sinh 2x+c$ (h) $\frac{1}{5}\cosh(5x+1)+c$

(i) $\frac{1}{24}\ln\left|\dfrac{3+4x}{3-4x}\right|+c$ (j) $\sin^{-1}(x-1)+c$

(k) $\frac{1}{3}\sin^{-1}3x+c$ (l) $\sin^{-1}\frac{1}{2}x+c$

(m) $\sin^{-1}\dfrac{(2x+1)}{\sqrt{5}}+c$ (n) $\sin^{-1}(2x-1)+c$

(o) $\sin^{-1}(x-2)/3+c$ (p) $\frac{1}{2}\tan^{-1}\frac{1}{2}(x+3)+c$

83 (a) $\frac{5}{2}$ (b) $\frac{9}{2}$ (c) 3 (d) $\frac{3}{2}$ (e) $\frac{13}{2}$

84 (a) $-x\cos x+\sin x+c$

(b) $\frac{1}{9}(3x-1)e^{3x}+c$

(c) $\frac{1}{16}x^4(4\ln|x|-1)+c$

(d) $-\frac{1}{13}e^{-2x}(3\cos 3x+2\sin 3x)+c$

(e) $\frac{1}{2}[(x^2+1)\tan^{-1}x]-\frac{1}{2}x+c$

(f) $\frac{1}{4}[(2x\sin 2x+\cos 2x)+c$

85 (a) $x\sin^{-1}x+\sqrt{(1-x^2)}+c$ (b) $x\ln|x|-x+c$
(c) $x\cosh^{-1}x-\sqrt{(x^2-1)}+c$
(d) $x\tan^{-1}x-\frac{1}{2}\ln(x^2+1)+c$

86 (a) $\frac{1}{3}(1+x^2)^{3/2}+c$ (b) $\frac{1}{4}\sin^4 x+c$

(c) $-\dfrac{1}{2(1+x^2)}+c$ (d) $\sqrt{(x^2-1)}+c$

(e) $\ln|x^2+3x+2|+c$ (f) $\ln|\ln|x||+c$

87 $a=\frac{3}{2},\,b=-1$
$\frac{3}{2}\ln(x^2+2x+5)-\frac{1}{2}\tan^{-1}\frac{1}{2}(x+1)+c$

88 (a) $\frac{1}{2}\ln(x^2+4x+5)-\tan^{-1}(x+2)+c$
(b) $-2\sqrt{(5+4x-x^2)}+7\sin^{-1}[(x-2)/3]+c$
(c) $\frac{1}{2}x-\frac{1}{2}\ln|\sin x+\cos x|+c$

89 (a) $\frac{1}{15}(3x^2-2)(1+x^2)^{3/2}+c$
(b) $\frac{1}{24}(6-8\sin^2 x+3\sin^4 x)\sin^4 x+c$

(c) $\dfrac{1}{2}\left(\dfrac{x}{1+x^2}+\tan^{-1}x\right)+c$

(d) $-\sinh^{-1}\left(\dfrac{3}{x}\right)+c$ (e) $-\sqrt{(4-x^2)}+c$

(f) $2\sqrt{x}-6\ln(3+\sqrt{x})+c$

90 (a) $2\sqrt{(1+x)}-2\ln[1+\sqrt{(1+x)}]+c$
(b) $\frac{1}{15}\sin^3 x(5-3\sin^2 x)+c$
(c) $2\sin\sqrt{x}-2\sqrt{x}\cos\sqrt{x}+c$

91 (a) $\ln|\tan\frac{1}{2}x|$

(b) $\ln\left|\dfrac{1+\tan\frac{1}{2}x}{1-\tan\frac{1}{2}x}\right|+c$

(c) $\dfrac{1}{\sqrt{7}}\ln\left|\dfrac{4-\sqrt{7}+3\tan\frac{1}{2}x}{4+\sqrt{7}+3\tan\frac{1}{2}x}\right|+c$

(d) $\frac{1}{13}\ln\left|\dfrac{2+3\tan\frac{1}{2}x}{3-2\tan\frac{1}{2}x}\right|+c$

92 $\ln 4$

93 (a) $\pi-2$ (b) $9\ln 3-\frac{26}{9}$ (c) $\frac{1}{9}(2e^3+1)$

94 (a) $2e(e-1)$ (b) $64/3$ (c) $\frac{1}{18}\pi^2$ (d) $\frac{7}{6912}$

96 $\frac{197}{10}\pi$

97 $2\displaystyle\int_0^{\pi/2}\sqrt{(1+\cos^2 x)}\,dx,\;\frac{1}{2}\pi^2$

99 $\frac{54}{35}\pi,\,\frac{48}{5}\pi$

100 $41.8\,\Omega$

101 (a) $\frac{2}{3},\,(\frac{11}{8},\frac{2}{5})$ (b) $\frac{8}{15}\pi,\,(\frac{21}{16},0)$

103 $20,\,\frac{10}{3},\,130$

104 $(\frac{2}{5},1),\,(\frac{1}{2},0)$

105 $0.6109,\,0.6463,\,0.6549,\,0.6569;\,0.6577$

106 0.1526

107 $5.869\,849$

108 $246\,\text{A}\,\text{h}$

109 76.09

110 1.1114 (4dp)

8.10 Review exercises

1 (a) $(2x + 1)e^{x^2+x}$ (b) $\dfrac{x^2(9 - x)}{(3 - x)^3}$

(c) $5\cos(5x - 1)$

(d) $(\ln\tan x + 2x\operatorname{cosec} 2x)(\tan x)^x$

(e) $\dfrac{1}{\sqrt{(1 - x^2)}}$ (f) $-\tfrac{1}{2}(1 + x)^{-3/2}$

(g) $-\dfrac{1}{1 + x^2}$ (h) $-\dfrac{(2x + 1)}{(x - 1)^2(x + 2)^2}$

(i) $3\cos(3x + 1)$ (j) $x^2(1 + 3\ln x)$

(k) $\dfrac{x^2(9 - x^2)}{(3 - x^2)^2}$ (l) $-\operatorname{sech} 2x$

(m) $\sqrt{\tfrac{1}{2}}\sinh\tfrac{1}{2}x$ (n) $2x\sin 2x + 2(x^2 + 1)\cos 2x$

(o) $\dfrac{4 - x}{(x + 2)^3}$ (p) $\dfrac{e^{\sqrt{x}}}{2\sqrt{x}}$

(q) $2\operatorname{cosec} 2x$ (r) $\dfrac{(8 - 7x)(2x - 1)^{1/2}}{(x + 1)^6}$

(s) $\sin x + x\cos x$ (t) $2xe^{x^2}$ (u) $2^x\ln 2$

(v) $-\dfrac{1}{1 + \sin x}$ (w) $-\dfrac{1}{x\sqrt{(x^2 - 1)}}$

(x) $x^2(3\cos 2x - 2x\sin 2x)$

(y) $\dfrac{3x^2 + 1}{2\sqrt{(x^3 + x + 3)}}$ (z) $-\dfrac{e^{-x}(2 + x)}{(1 + x)^2}$

2 (a) $\tfrac{2}{9}x^{3/2}(3\ln x - 2) + c$

(b) $\ln(x^2 + 2x + 2) + \tan^{-1}(x + 1) + c$

(c) $\ln\tfrac{9}{2}$ (d) $\tfrac{1}{2} + \tfrac{1}{4}\sqrt{\tfrac{1}{3}}\pi$

(e) $\ln\left|\dfrac{(x - 2)^2}{x - 1}\right| + c$ (f) $\tfrac{1}{3}\tan^3 x - \tan x + x + c$

(g) $\tfrac{1}{2} + \tfrac{2}{3}\sqrt{\tfrac{1}{3}}\pi$ (h) $\tfrac{1}{3}\sin^{-1}\tfrac{3}{2}x + c$

(i) $\tfrac{2}{3}\sqrt{(x^3 - 1)} + c$

(j) $\tfrac{1}{2}x^2 - x + 2\ln|x + 1| + c$

(k) $\tfrac{1}{2}\tan^{-1}\tfrac{1}{2}(x + 3) + c$

(l) $2(2 - x)\cos\sqrt{x} + 4\sqrt{x}\sin\sqrt{x} + c$

(m) 1 (n) $\tfrac{1}{4}(\sinh 2 - 2)$

(o) $-\tfrac{1}{30}(1 - 3x)^{10} + c$

(p) $\tfrac{1}{2}\sin x - \tfrac{1}{10}\sin 5x + c$

(q) $x(\ln 2x - 1) + c$ (r) $-e^{-x^2/2} + c$

(s) $\tfrac{1}{2}\cosh^{-1}\tfrac{2}{3}x + c$

(t) $7\pi/2$

(u) $\tfrac{\pi}{8} - \tfrac{1}{4}\ln 2$

(v) $-\tfrac{1}{15}(4 - 3x)^5 + c$

(w) $\tfrac{1}{10}\sin 5x + \tfrac{1}{2}\sin x + c$

(x) $x\sin^{-1}x + \sqrt{(1 - x^2)} + c$

(y) $-e^{-x}(x^2 + 2x + 2) + c$

(z) $2\sqrt{\tfrac{1}{3}}\tan^{-1}[\sqrt{\tfrac{1}{3}}(2x + 1)] + c$

3 $y = 12x - 8, y = \tfrac{1}{12}(49 - x)$

4 $y = \tfrac{1}{5}(4x + 6), y = 1 - x, \tfrac{8}{9}\sqrt{2}$

6 $0.782\,80, \tfrac{1}{4}\pi$, error $= -0.002\,60$

8 (a) $\dfrac{1076}{15}$ (b) $\ln\tfrac{7}{6}$

9 $-\tfrac{1}{4}\sec^4 t, -\tfrac{1}{4}\sec t\operatorname{cosec}^3 t$

10 $-3, 18, 5\sqrt{10}/9$

11 $\tfrac{8}{15}$

12 $\tfrac{3}{4}\pi ab$

13 $\tfrac{3}{2}\pi$

15 $\tfrac{1}{2}(\sinh^{-1}2 + 2\sqrt{5})$

16 $\tfrac{8}{15}$

17 (a) $3\pi a^2$ (b) $4a$

(c) cycloid has cusps at these values   (d) $8a$

18 (a) $\tfrac{13}{15} - \tfrac{1}{4}\pi$ (b) $\tfrac{5}{12} - \tfrac{1}{2}\ln 2$

22 $\pi/2, \tfrac{1}{6}\pi(5^{3/2} - 1)$

25 (a) $\tfrac{54}{35}\pi, \tfrac{48}{5}\pi$ (b) $(\tfrac{\sqrt{3}}{2} - \tfrac{\pi}{6}, \tfrac{5}{6}\pi - \tfrac{5}{2}\sqrt{3})$

27 (a) 5.21×10^6 (b) 7.76×10^6

CHAPTER 9

Exercises

1 (a) Minimum $(1, 0)$, maximum $(\tfrac{2}{3}, \tfrac{1}{27})$, inflection $(\tfrac{5}{6}, \tfrac{1}{54})$

(b) Minimum $(1, 43)$, maximum $(-5, 151)$, inflection $(-2, 97)$

2 (a) Minimum $(-2, -\tfrac{1}{3})$, maximum $(2, -3)$, inflexion $(-(4 - 3\sqrt[3]{4})/(\sqrt[3]{4} - 1), 3(\sqrt[3]{4} + 1)/(\sqrt[3]{4} - 1))$

(b) Maximum $(4, 54, e^{-4})$

(c) Minimum $(0, 0)$, maximum $(2, 4e^{-2})$, inflexions $(2 \pm \sqrt{2}, (4 \pm 2\sqrt{2})e^{-2\mp\sqrt{2}})$

3 $d = 10^3\sqrt{2/\pi}, \quad h = 10^3\sqrt{(2/\pi)}$

4 $d = 8.0$ (1dp), $\quad h = 9.9$

5 $r = l/4$

6 $\tfrac{8}{27}$ (volume of sphere)

7 $(0, 0)$ minimum, $(1/\sqrt{e}, K/(2e))$ maximum.

(Note: $\dfrac{dv}{dx}$ not defined at $x = 0$ but $\dfrac{dv}{dx} \to 0$ as $x \to 0+$)

8 $S = 8\pi a^2$

9 $b \in [80, 82.2]$

10 (a) $x = \tfrac{4}{3}$ (b) $x = 2$

11 £1.25, 7500

12 Distribute wash water equally

13 In year k a volume $(1 - \alpha)/(1 - \alpha^{11-k})$ of standing timber should be felled, α growth factor

14 (a) 1/4   (b) 1/2   (c) 1/4   (d) 2
    (e) 2/3   (f) −4/3   (g) 2   (h) 3/2
    (i) $\pi/4$

15 (b) is convergent, 0.860 334

16 $|f'(x)| > 1$ near $x = \alpha$

17 1.618 034

18 $x_{n+1} = x_n - \frac{1}{10}(x_n^3 - 2x - 1)$

19 (a) $f'(x) < 1$ $(x < 1)$ (b) $f'(x) > 1$ $(x > 1)$

20 1.5, 1.49, 1.48, 1.48, 1.47, 1.47, 1.46, 1.46, 1.45, 1.45;
    $\sqrt{2} = 1.41$

21 $e(1 - \frac{1}{2}x^2 + \frac{1}{6}x^4 - \frac{31}{720}x^6 + \ldots)$

22 $1 + 2x + \frac{1}{2}x^2 + \frac{1}{6}x^3 + \ldots = x + e^x$

23 $y_3 = 1 + 2x + \frac{1}{2}x^2 + \frac{1}{6}x^3 - \frac{1}{24}x^4$
    $y_4 = 1 + 2x + \frac{1}{2}x^2 + \frac{1}{6}x^3 + \frac{1}{24}x^4 - \frac{1}{120}x^5$

24 $x - \frac{1}{3}x^3 + \frac{1}{5}x^5 - \frac{1}{7}x^7 + \ldots$ $(-1 \leqslant x \leqslant 1)$

25 $x + \frac{1}{3}x^3 + \frac{2}{15}x^5 + \frac{17}{315}x^7 + \ldots$
    $\ln \cos x = -\left\{\frac{1}{6}x^2 + \frac{1}{12}x^4 + \frac{1}{45}x^6 + \frac{17}{2520}x^8 + \ldots\right\}$

31 $l > 18$
    % error bound
    $\approx$ (approximation − true value) $\times$ 100/(approximation)
    $\approx \dfrac{(D - d)^4/192^3 + (1.57 - \frac{1}{2}\pi)(d + D)}{2l + 1.57(D + d) + (D - d)^2/4l} \times 100$
    $\approx 1/(\pi l^4)$ for given values

32 (a) $\frac{3}{4}$ (b) $\frac{1}{4}$ (c) $-\frac{3}{2}$ (d) −1   (e) $-\frac{1}{3}$ (f) −1

33 $b_0 = 82.82$, $b_1 = -\frac{5}{24}$, $b_2 = -0.0018$

34 $X = \ln 4 \approx 1.386$

35 0.0006

36 1.175 201 21

37 (a) 1st order   (b) 2nd order   (c) 3rd order

38 −0.1038

39 2.732 051, 4.872 977

40 0.576 368 88

41 $n = 2$, 0.0203

42 (a) 0.643 283   (b) 0.6875

43 0.4627, $h = \frac{1}{256}$

45 $(1, 2t, 3t^2)$, $(0, 2, 6t)$

46 $\dfrac{d\mathbf{r}}{dt} = (1 + 2t^2)\hat{\mathbf{T}}(t)$, where
    $\hat{\mathbf{T}}(t) = \dfrac{1}{1 + 2t^2}\mathbf{i} + \dfrac{2t}{1 + 2t^2}\mathbf{j} + \dfrac{2t^2}{1 + 2t^2}\mathbf{k}$

50 2, 3

51 $\cos y$, $-x \sin y$

52 (a) $3x^2y + 4x + y$, $x^3 + 18y + x$
    (b) $3(x + y^2)^2$, $6y(x + y^2)^2$
    (c) $\dfrac{3x + y}{(3x^2 + y^2 + 2xy)^{1/2}}$, $\dfrac{y + x}{(3x^2 + y^2 + 2xy)^{1/2}}$

53 (a) $e^{xy}(y \cos x - \sin x)$, $e^{xy}x \cos x$
    (b) $\dfrac{y^2 - x^2}{(x^2 + y^2)^2}$, $-\dfrac{2xy}{(x^2 + y^2)^2}$
    (c) $\dfrac{-x^2 - 2xy + 2y^2 + 6}{(x^2 + 2y^2 + 6)^2}$, $\dfrac{x^2 - 4xy - 2y^2 + 6}{(x^2 + 2y^2 + 6)^2}$

54 $-1 + \frac{1}{2}\sqrt{3}$, $-\tan^{-1}2$

55 (a) $-x/z$, $-y/z$ (b) $\dfrac{1 - yz}{xy - 1}$, $\dfrac{1 + xz}{1 - xy}$

57 (a) $2xy + 3yz - 4z^3xy$, $x^2 + 3xz - 2x^2z^3$,
       $3yx - 6z^2x^2y$
    (b) $-ye^{2z}\sin xy$, $-xe^{2z}\sin xy$, $2e^{2z}\cos xy$

60 (a) $\dfrac{2t^3 + 3t - 1}{\sqrt{(t^4 + 3t^2 - 2t + 1)}}$
    (b) $4xt(x^2 - 2t^2)/(2x + 3t)$

61 $2se^x\cos y - 2te^x\sin y$
    $-2te^x\cos y - 2se^x\sin y$

65 (a) 10.5   (b) $19\sqrt{\frac{1}{34}}$

66 $\frac{19}{5}\pi \, \mathrm{cm \, s}^{-1}$

67 $\sqrt{(1 + 4t^2 + 9t^4)}$

68 $-6e^{-2s} + 2e^{-s-t}$, $-6e^{-2t} + 2e^{-s-t}$

70 −3

76 $a = 3$, $b = \frac{3}{2}$

78 0.018 702, 0.02

79 $0.029\,65\,\mathrm{m}^3$, $0.0295\,\mathrm{m}^3$

80 $173 \pm 4\,\mathrm{m}$

81 $-\frac{2}{3}$

82 3%

84 0.5%

85 35% increase

86 (a) $xy^2 + x^2y + x + c$
    (b) $x^2y^2 + y \sin 3x + c$
    (c) Not exact
    (d) $z^3x - 3xy + 4y^3$

87 -1, $y \sin x - x \cos y + \frac{1}{2}(y^2 - 1)$

88 $m = 2$
    $8x^5 + 36x^4y + 62x^3y^2 + 63x^2y^3 + 54xy^4 + 27y^5 + c$

89 (a) (0, 0), maximum; (10, 0), saddle
  (b) (0, 0), maximum   (c) (−1, 3), saddle
  (d) (−1, $\frac{3}{2}$), saddle; (1, $\frac{3}{2}$) minimum
  (e) (0, −1), saddle; (0, 3), saddle; (−1, 1), maximum
  (f) Minimum at ($\frac{1}{2}$, $\frac{1}{3}$); degenerate and stationary sets $x = 0$ and $y = 0$
  (g) (1, 1), minimum

91 Maximum at (0, 0); saddle at ($\frac{1}{3}$, $\frac{1}{3}$)

92 $N = 2000$, $n = 2000$, $P = 250$

93 Minimum at (2/3, 4/3)

94 $a = \dfrac{20(\pi^2 - 16)}{\pi^5}$, $b = \dfrac{12(20 - \pi^2)}{\pi^4}$

95 $x = 2$, $y = 2$

96 Minimum $T = -\frac{1}{4}$ at ($\frac{1}{2}$, 0)
  maximum $T = \frac{9}{4}$ at ($-\frac{1}{2}$, $\pm\frac{1}{2}\sqrt{3}$)

97 $x = \frac{200}{3}$, $\theta = \frac{1}{3}\pi$

98 ($\frac{2}{3}$, $\frac{4}{3}$)

99 ($\frac{1}{3}$, $\frac{1}{2}$, $\frac{1}{6}$)

100 1, 2

101 (−1/3, −2/3, −2/3)

102 −41/4, 7/4

103 $\frac{285}{92}$

9.11 Review exercises

1 Maximum ($\frac{2}{3}$, $\frac{1}{27}$), minimum (1, 0), inflection at $x = \frac{5}{6}$

2 0.2575

5 $\frac{1}{2}$

6 $\frac{1}{8}(\sin 2k - 2k \cos 2k)$

7 2.09

8 Maximum 10.55 when $\theta = 4.42$ and 1.28, minimum 1.45 when $\theta = 2.85$ and 5.99

9 (a) $\dfrac{wL^4}{16EI}$ (b) $\dfrac{L}{2}(1 \pm \sqrt{\frac{1}{3}})$

10 (a) For these series see Section 6.3.5.

11 $L = 100\,\text{m}$, $W = (200/\pi)\,\text{m}$

12 Local minimum (0, 0), local maximum (3, −3) aymptotes $x = 2$, $x = 6$, $y = 1$

13 $\pi/2$

14 (a) $\frac{1}{2}$ (b) $\pi/4$ (c) $\frac{1}{16}$ (d) $\frac{1}{4}$ (e) $\frac{2}{5}$ (f) $\frac{2}{3}$

15 (a) 6, $x = 0$ (b) 3, $x = \frac{3}{2}$ (c) −1, $x = 0$

16 0.785, 0.626, 0.624; 2.62

17 0.4446 cf 0.4425

18 8.155 299, 8.154 959, 8.154 845

19 $4\,\text{m s}^{-1}$, $4\,\text{m s}^{-2}$

20 $(\frac{1}{2}t^2 + \frac{1}{6}t^3 + t)\boldsymbol{i} + (\frac{1}{12}t^4 - t)\boldsymbol{j} + t^2\boldsymbol{k}$

23 −0.21, 0.01

24 0.61%

25 −0.2%

26 −3.33%

27 (b) $2u$

29 (a) 2

32 $f''(z)/(4t\sqrt{t})$

34 $-y/(x^2 + y^2) + \text{const}$

35 Maximum at (0, 0), saddle points at (3, 3), (−3, −3), (1, −1), (−1, 1)

36 Minimum at (0, 0), saddle at ($\frac{1}{2}$, $\frac{3}{2}$)

37 $x = (\frac{2}{3})^{2/3}$, $y = (\frac{2}{3})^{2/3}$, $z = 2^{2/3} \cdot 3^{-1/6}$

38 Saddle at (0, 0) and (0, 4), maximum at (2, 2), minimum at (−2, 2)

39 $x = 0$, $y = \pm 3$ (max); $y = 0$, $x = \pm 3$ (min)

CHAPTER 10

Exercises

1 (a) First-order, dependent variable x, independent variable t, linear, homogeneous, ordinary differential equation
  (b) First-order, dependent variables x and y, independent variable t, linear, homogeneous, partial differential equation
  (c) Second-order, dependent variable x, independent variable t, linear, homogeneous, ordinary differential equation
  (d) Second-order, dependent variables x and y, independent variable t, linear, homogeneous, partial differential equation
  (e) First-order, dependent variable x, independent variable t, nonlinear, ordinary differential equation
  (f) First-order, dependent variables x and y, independent variable t, nonlinear, partial differential equation
  (g) First-order, dependent variable x, independent variable t, linear, nonhomogeneous, ordinary differential equation

(h) Second-order, dependent variable x, independent variable t, linear, nonhomogeneous, ordinary differential equation

2 (a) Second-order nonlinear partial differential equation, dependent variable f, independent variables x and y

(b) Second-order nonlinear ordinary differential equation, dependent variable p, independent variable z

(c) Second-order linear homogeneous partial differential equation, dependent variable h, independent variables x and y

(d) Second-order linear nonhomogeneous ordinary differential equation, dependent variable s, independent variable t

(e) First-order linear nonhomogeneous partial differential equation, dependent variable x, independent variables u and v

(f) Third-order nonlinear ordinary differential equation, dependent variable p, independent variable y

(g) First-order linear nonhomogeneous ordinary differential equation, dependent variable r, independent variable z

(h) First-order nonlinear partial differential equation, dependent variable f, independent variables x, y and z

(i) First-order linear homogeneous ordinary differential equation, dependent variable x, independent variable t

(j) First-order linear nonhomogeneous ordinary differential equation, dependent variable x, independent variable t

(k) Second-order linear nonhomogeneous partial differential equation, dependent variable f, independent variables x and y

(l) Third-order nonlinear ordinary differential equation, dependent variable p, independent variable q

(m) Second-order nonlinear ordinary differential equation, dependent variable x, independent variable y

(n) Second-order linear homogeneous partial differential equation, dependent variable y, independent variables p and t

(o) First-order linear homogeneous ordinary differential equation, dependent variable y, independent variable z

3 (a) $x(t) = \frac{4}{3}t^3 + C$ (b) $x(t) = \frac{1}{20}t^5 - \frac{1}{3}t^3 + Ct + D$

(c) $x(t) = \frac{1}{16}e^{4t} + Ct + D$ (d) $x(t) = Ae^{-6t}$

(e) $x(t) = \ln t + \frac{1}{125}\cos 5t + Ct^2 + Dt + E$

(f) $x(t) = Ae^{2\sqrt{2}t} + Be^{-2\sqrt{2}t}$

4 (a) $x(t) = \frac{2}{3}t^3 + Ct + 2$ (b) $x(t) = -\frac{1}{4}\sin 2t - \frac{t}{\pi} + \frac{5}{2}$

(c) $x(t) = 4t + D$ (d) $x(t) = 2 - t^2$

(e) $x(t) = \frac{1}{2}e^{-2t} + Ct + a - \frac{1}{2}$ (f) $x(t) = C - \cos 2t$

(g) $x(t) = e^{2t}$ (h) $x(t) = \dfrac{e}{e^2 - 1}(e^t - e^{-t})$

5 (a) Under-determined

(b) Fully determined, boundary-value problem

(c) Fully determined, initial-value problem

(d) Under-determined

(e) Fully determined, boundary-value problem

(f) Fully determined, initial-value problem

(g) Under-determined

(h) Under-determined

(i) Fully determined, boundary-value problem

(j) Fully determined, initial-value problem

(k) Fully determined, boundary-value problem

(l) Fully determined, initial-value problem

6 $y(x) = \dfrac{1}{24EI}[w(a - x)^4 - 4R(a - x)^3$
$+ 4a^2(aw - 3R)x - a^3(aw - 4R)]$

At A the boundary condition is $y(a) = 0$ so $R = 3aw/8$
Maximum displacement is $y = 0.005\,42\,wa^4/EI$

11 (a) $x(t) = Ce^{kt}$ (b) $x(t) = Ce^{2t^3}$

(c) $x(t) = Ct^b$ (d) $x(t) = (2a\ln t + C)^{1/2}$

12 (a) $x(t) = (67 - 3\cos t)^{1/3}$ (b) $x(t) = \left(\dfrac{163}{2} - \dfrac{2}{t}\right)^{1/2}$

13 (a) $x(t) = \pm t\sqrt{[2\ln(Ct)]}$ (b) $x(t) = t(\ln Ct^3)^{1/3}$

(c) $x(t) = \dfrac{-t}{\ln Ct}$

14 $x(t) = \pm t(4\ln t + 256)^{1/4}$

15 (a) $x(t) = (t^{1/2} + C)^2$ (b) $x(t) = \cos^{-1}(Ce^{\cos t - t})$

(c) $x(t) = C\exp(\frac{1}{2}e^{t^2})$ (d) $x(t) = (3e^t + C)^{1/3}$

(e) $x(t) = (1 - Ce^{at})^{-1}$ (f) $x(t) = (C - 2\cos t)^{1/2}$

16 (a) $x(t) = -2 \pm (\frac{2}{3}t^3 + 2t)^{1/2}$

(b) $x(t) = \dfrac{4(t - 1)}{4 - t}$ (c) $x(t) = \dfrac{3 + e^{2\sin t}}{3 - e^{2\sin t}}$

(d) $x(t) = -\ln(1 + e^{-a} - e^t)$

(e) $x(t) = [12(t\ln t - t + 1)]^{1/3}$

17 (a) $x(t) = \dfrac{t}{\sqrt{3}}\left(\dfrac{C}{t^3} - 1\right)^{1/2}$ (b) $x(t) = t\cot^{-1}(\ln(1/Ct))$

(c) $x(t) = \dfrac{t}{2}\left[1 \pm \left(\dfrac{C}{t} - 11\right)^{1/2}\right]$ (d) $x(t) = t\sin^{-1}Ct$

(e) $x(t) = t \pm (2t^2 + D)^{1/2}$ (f) $x(t) = -t\ln(-\ln Ct)$

18 (a) $x(t) = \dfrac{2t}{\sqrt{(2 - t^4)}}$ (b) $x(t) = \pm\frac{1}{4}(9t^2 - 32)^{1/2}$

(c) $x(t) = t\ln(\ln\frac{1}{2}t + e^2)$ (d) $x(t) = \pm t[\ln(\ln t^2 + e^4)]^{1/2}$

(e) $x(t) = \dfrac{4t^2}{5 - 4t}$

19 (a) $x(t) = t + 3 \pm (2t + C)^{1/2}$

(b) $x(t) = \frac{1}{2}[\pm(2t + C)^{1/2} - t - 1]$

(c) $x(t) = \frac{1}{2}[\pm(2t + C)^{1/2} - t]$

(d) $x(t) = t - 1 \pm (2t + C)^{1/2}$

(e) $x(t) = Ae^t - 2t - 4$

(f) $x(t) = \frac{1}{2}t - 2 + Ae^t$ (g) $x(t) = -\dfrac{1}{t + C} - 2t$

20 $K = 2/75$, $x(10) = 20/7$, $x(50) = 100/23$, $x \to 5$ as $t \to \infty$

21 $t = \sqrt{(m/Kg)}\tanh^{-1}\frac{1}{2}$

22 $A(t) = \dfrac{1}{\alpha}[1 - (1 + 6\alpha Kt)^{-1/6}]$

23 (a) $x(t) = \pm\sqrt{(C - t^2)}$ (b) $x(t) = \pm\sqrt{(C + t^2)}$

(c) $x(t) = -t \pm \sqrt{(C + 2t^2)}$ (d) $x(t) = t^2 \pm \sqrt{(C + t^4)}$

(e) $\frac{1}{2}x^2 - xt + \frac{1}{2}t^2 - t = C$ (f) $x^2 + xt + t^2 = C$

24 (a) $x(t) = 1 \pm \sqrt{(1 - 2t - t^2)}$

(b) $x(t) = \frac{1}{2}[-t \pm \sqrt{(3t^2 + 4)}]$

(c) $x(t) = \dfrac{2}{t^2}(1 \pm (1 - t^2)^{1/2})$ (d) $x(t) = \dfrac{2 - t}{\cos t}$

25 (a) Not exact   (b) Not exact

(c) $x\sin(x + t) = C$ (d) Not exact

(e) $x + e^{xt} = C$ (f) $(x + \sqrt{t})^2 = C$

(g) Not exact   (h) $t\ln(x + t) = C$

26 (a) $x(t) = \sin^{-1}(1 - t) - t$

(b) $x(t) = [\frac{1}{2}(15 - t)]^{2/3} - 2t$

(c) $x(t) = \pm[t^2 \pm (1 - 4t)^{1/2}]^{1/2}$

(d) $x(t) = 4\exp\left(\dfrac{1}{2} - \dfrac{1}{t}\right) - t$

27 Must have $b = e$; then $ax^2 + 2bxt + ft^2 + C = 0$

28 Must have $h(t) = dg/dt$; then $x = -C/g(t)$

29 Must have $k = -1$; then $x\ln(x + t) + C = 0$

30 Must have $k = 2$; then $x = [\sin^{-1}(C/t^3)]/t$

31 (a) $x(t) = \frac{2}{3} + Ce^{-3t}$

(b) $x(t) = -\frac{1}{4}t - \frac{1}{16} + Ce^{4t}$

(c) $x(t) = -\frac{1}{2}e^{-4t} + Ce^{-2t}$

(d) $x(t) = Ce^{-t^2/2} - 2$

32 (a) $x(t) = -\frac{3}{2} + \frac{7}{2}e^{2t}$

(b) $x(t) = \frac{1}{3}t - \frac{1}{9} + \frac{10}{9}e^{-3t}$

(c) $x(t) = \frac{1}{3}t^3 - 3t\ln t - \frac{3}{2}t$

33 (a) $x(t) = Ce^t - 2t^2 - 5t - 5$

(b) $x(t) = -\frac{1}{4}t^2 - \frac{1}{8} + Ce^{2t^2}$

(c) $x(t) = \left(1 - \dfrac{2}{t^2}\right)\sin t + \dfrac{2}{t}\cos t + \dfrac{C}{t^2}$

(d) $x(t) = \left(\dfrac{1}{t} - \dfrac{3}{t^2} + \dfrac{6}{t^3} - \dfrac{6}{t^4}\right)e^t + \dfrac{C}{t^4}$

(e) $x(t) = \frac{1}{2}\sin 2t \ln(\tan \frac{1}{2}t) + C\sin 2t$

(f) $x(t) = \frac{1}{3}t^2 + Ce^{-2t^3}$ (g) $x(t) = Ce^{-1/t} - 4$

34 (a) $x(t) = \frac{1}{2}(1 - e^{2t^2})$ (b) $x(t) = 2e^{t-1}t^{-t}$

(c) $x(t) = \frac{1}{5}t - \frac{1}{25} + \frac{1}{3}e^{-2t} + \dfrac{18 - 25e^2}{75e^5}e^{-5t}$

(d) $x(t) = 1 + e^{1/t - 1/2}$

(e) $x(t) = \frac{1}{2} + t^2 + \left(1 - \dfrac{2}{t} + \dfrac{2}{t^2}\right)e^t - \dfrac{1}{t^2}(\frac{3}{2} + e)$

(f) $x(t) = U(1 + e^{1 + \cos t})$

35 $T(t) = T_{\text{in}} + Ce^{-AU\alpha t/V}$

36 $Q(t) = \dfrac{2\alpha\rho gh(1 - e^{-pt})}{(2 + 2\alpha\beta + \alpha\gamma_0)} + \dfrac{\alpha\rho ghe^{-pt}}{(1 + \alpha\beta + \alpha\gamma_0)}$

where $p = \dfrac{(2 + 2\alpha\beta + \alpha\gamma_0)A}{2\alpha\rho d}$

37 $X(0.3) = 1.268$

38 $X(0.25) = 2.050\,439$

39 $X(1) = 1.2029$

40 $X(0.5) = 2.1250$

41 $X_a(2) = 2.811\,489$, $X_b(2) = 2.819\,944$

42 $X_a(2) = 1.573\,065$, $X_b(2) = 1.558\,541$

43 $X_a(1.5) = 2.241\,257$, $X_b(1.5) = 2.206\,232$

44 (a) $L = \dfrac{d}{dt} + t^2$ (b) $L = \dfrac{d}{dt} - 6t^2$

(c) $L = \dfrac{d}{dt} - k$

45 (a) independent   (b) dependent

46 (a) $k_1 = 2$, $k_2 = -2$, $k_3 = -1$

(b) $k_1 = 1$, $k_2 = -1$, $k_3 = -1$, $k_4 = 1$

47 (a) $L = \dfrac{d}{dt} - f(t)$ (b) $L = \dfrac{d^3}{dt^3} + \sin t\dfrac{d^2}{dt^2} + 4t^2$

(c) $L = \dfrac{d^2}{dt^2} + \sin t\dfrac{d}{dt} - t - \cos t$

(d) $L = \sin t\dfrac{d}{dt} - \dfrac{\cos t}{t}$ (e) $L = \dfrac{d}{dt} - \dfrac{b}{t}$

(f) $L = \dfrac{d}{dt} - te^{t^2}$ (g) $L = t^2\dfrac{d^2}{dt^2} + (2t - t^2)\dfrac{d}{dt} - t$

(h) $L = t\dfrac{d^2}{dt^2} + 3\dfrac{d}{dt} - t$

48 (a) dependent   (b) independent

(c) independent   (d) independent

(e) dependent   (f) dependent

(g) independent   (h) dependent

(i) dependent   (j) independent

(k) independent

49 (a) $2, -1, 1, 1$ (b) $-3, 3, 2, 1$
(c) $0, 2, -1$ (d) $1, -1, 0, 1$
(e) $0, -1, 0, 1, 6$

50 (a) $x(t) = A + Bt + Ct^2 + Dt^3$ (b) $x(t) = Ae^{pt} + Be^{-pt}$
(c) $x(t) = A\cos pt + B\sin pt + C\cosh pt + D\sinh pt$
(d) $x(t) = A + Be^{-2t}$ (e) $x(t) = A + B\cos 2t + C\sin 2t$
(f) $x(t) = Ae^{-t} + Bte^{-t}$ (g) $x(t) = Ae^t + Bte^t + Ce^{-t}$

51 $LM = \dfrac{1}{t}\dfrac{d^3}{dt^3} - \left(\dfrac{2}{t^2} + 4 + e^t\right)\dfrac{d^2}{dt^2}$

$\qquad + \left(\dfrac{2}{t^3} + \dfrac{4}{t} + 6t + (4t-2)e^t\right)\dfrac{d}{dt}$

$\qquad + (4t - 6t^2 - 1)e^t$

$\qquad ML = \dfrac{1}{t}\dfrac{d^3}{dt^3} - (e^t + 4)\dfrac{d^2}{dt^2} + \left(6t - \dfrac{4}{t} + 4te^t\right)\dfrac{d}{dt}$

$\qquad - 6t^2 e^t + 12$

52 $LM = f_1 f_2 \dfrac{d^2}{dt^2} + \left(f_1\dfrac{df_2}{dt} + f_1 g_2 + f_2 g_1\right)\dfrac{d}{dt}$

$\qquad + f_1\dfrac{dg_2}{dt} + g_1 g_2$

$\qquad ML = f_1 f_2\dfrac{d^2}{dt^2} + \left(f_2\dfrac{df_1}{dt} + f_1 g_2 + f_2 g_1\right)\dfrac{d}{dt}$

$\qquad + f_2\dfrac{dg_1}{dt} + g_1 g_2$

53 (a) $x(t) = Ae^t + Be^{3t/2}$
(b) $x(t) = e^{-t}(A\cos 2t + B\sin 2t)$
(c) $x(t) = Ae^t + Be^{-4t}$
(d) $x(t) = e^{2t}(A\cos 3t + B\sin 3t)$

54 (a) $x(t) = \frac{1}{7}(3e^t - 10e^{-2t/5})$
(b) $x(t) = e^{3t}(2\cos t - 6\sin t)$
(c) $x(t) = \frac{1}{2}(e^{3t} - e^t)$

55 (a) $x(t) = e^{t/4}[A\cos(\frac{1}{4}\sqrt{27}t) + B\sin(\frac{1}{4}\sqrt{27}t)]$
(b) $x(t) = Ae^{(\sqrt{13}-3)t} + Be^{-(\sqrt{13}+3)t}$
(c) $x(t) = e^{-t/2}[A\cos(\frac{1}{2}\sqrt{3}t) + B\sin(\frac{1}{2}\sqrt{3}t)]$
(d) $x(t) = Ae^{4t} + Bte^{4t}$ (e) $Ae^t + Be^{2t/3} + Ce^{-2t/3}$
(f) $x(t) = Ae^{-t} + e^t[B\cos(2\sqrt{2}t) + C\sin(2\sqrt{2}t)]$
(g) $x(t) = A + e^t[B\cos(\sqrt{2}t) + C\sin(\sqrt{2}t)]$

56 $x(t) = e^t(A\cos t + B\sin t + C\cos 2t + D\sin 2t)$

57 (a) $x(t) = e^{t/2}[\cos(\frac{1}{2}\sqrt{5}t) - \sqrt{\frac{1}{5}}\sin(\frac{1}{2}\sqrt{5}t)]$
(b) $x(t) = 2(t - 1)e^{2(t-1)}$
(c) $x(t) = e^{-5t/2}[\cos(\frac{1}{2}\sqrt{7}t) + \sqrt{\frac{1}{7}}\sin(\frac{1}{2}\sqrt{7}t)]$
(d) $x(t) = \frac{1}{6}(7t + 33)e^{-(t+3)/3}$
(e) $x(t) = \frac{7}{2}e^t - 4e^{2t} + \frac{3}{2}e^{3t}$
(f) $x(t) = (2 - 5t + 4t^2)e^{-2(t-1)}$

58 $x(t) = e^{t/2}[A\cos(\frac{1}{2}\sqrt{3}t) + B\sin(\frac{1}{2}\sqrt{3}t) + Ct\cos(\frac{1}{2}\sqrt{3}t)$
$\qquad + Dt\sin(\frac{1}{2}\sqrt{3}t)]$

59 $x(t) = Ae^{4t} + Be^{-t} + Cte^{-t} + Dt^2 e^{-t}$

60 (a) $x(t) = \frac{2}{9} - \frac{1}{3}t + Ae^{-t} + Be^{3t}$
(b) $x(t) = -\frac{1}{5}t^2 + \frac{14}{25}t - \frac{38}{125} + Ae^{(1+\sqrt{6})t} + Be^{(1-\sqrt{6})t}$
(c) $x(t) = -5e^t + Ae^{(\sqrt{5}+1)t/2} + Be^{-(\sqrt{5}-1)t/2}$

61 (a) $x(t) = -\frac{1}{8}\cos 4t + \frac{1}{24}\sin 4t$
$\qquad + e^{3t/2}[A\cos(\frac{1}{2}\sqrt{7}t) + B\sin(\frac{1}{2}\sqrt{7}t)]$
(b) $x(t) = \frac{1}{121}e^{-3t} + Ae^{2t/3} + Bte^{2t/3}$
(c) $x(t) = -\frac{105}{289}\cos 2t + \frac{56}{289}\sin 2t + Ae^{-\left(1-\frac{3}{\sqrt{2}}\right)t}$
$\qquad + Be^{-\left(1+\frac{3}{\sqrt{2}}\right)t}$
(d) $x(t) = \frac{5}{4}t - \frac{33}{16} + e^{-t/2}[A\cos(\frac{1}{2}\sqrt{15}t) + B\sin(\frac{1}{2}\sqrt{15}t)]$
(e) $x(t) = t - 2 + Ae^{-t/4} + Bte^{-t/4}$
(f) $x(t) = -\frac{72}{625}\cos 3t - \frac{21}{625}\sin 3t + Ae^{4t} + Bte^{4t}$
(g) $x(t) = \frac{1}{52}e^{-5t} + e^{2t}[A\cos(\sqrt{3}t) + B\sin(\sqrt{3}t)]$
(h) $x(t) = -t^2 - 6t - 24 + \frac{1}{5}e^{-2t} + Ae^{(\sqrt{7}\sqrt{3}-1)t/2}$
$\qquad + Be^{-(\sqrt{7}\sqrt{3}+1)t/2}$
(i) $x(t) = -\frac{5}{4}te^{-3t} - \frac{4}{65}\cos 2t - \frac{7}{65}\sin 2t + Ae^t + Be^{-3t}$
(j) $x(t) = \frac{1}{16} - \frac{1}{4}t\cos 4t + A\cos 4t + 3\sin 4t$
(k) $x(t) = -\frac{7}{4}t - \frac{3}{4}te^{4t} + A + Be^{4t}$

62 (a) $x(t) = \frac{17}{1460}\cos 2t + \frac{21}{1460}\sin 2t + Ae^t + Be^{-2t}$
$\qquad + Ce^{(2+\sqrt{3})t} + De^{(2-\sqrt{3})t}$
(b) $x(t) = -\frac{1}{12}e^{2t} - \frac{1}{39}te^{-2t} + Ae^t + Be^{-2t} + Ce^{(2+\sqrt{3})t}$
$\qquad + De^{(2-\sqrt{3})t}$
(c) $x(t) = -\frac{1}{2}t^2 - \frac{9}{2}t - \frac{69}{4} - \frac{1}{12}e^{-t} + Ae^t + Be^{-2t}$
$\qquad + Ce^{(2+\sqrt{3})t} + De^{(2-\sqrt{3})t}$

63 (a) $x(t) = \frac{1}{125}\cos t + \frac{11}{250}\sin t - \frac{1}{27}(t + 1)$
$\qquad + (A + Bt + Ct^2)e^{3t}$
(b) $x(t) = -\frac{1}{8}e^t + (A + Bt + Ct^2)e^{3t}$
(c) $x(t) = \frac{1}{6}t^3 e^{3t} - \frac{1}{27}(t + 1) + (A + Bt + Ct^2)e^{3t}$

64 (a) $\omega = 3, \zeta = 1$ (b) $\omega = \sqrt{7}, \zeta = 2\sqrt{\frac{1}{7}}$

65 (a) $a = 1, b = 4$ (b) $p = 1.4, q = 0.25$
(c) $\beta = 2.2, \gamma = 1.21$

66 (a) $\omega = 4p, \zeta = \dfrac{a}{4p}$ (b) $\omega = \dfrac{1}{\sqrt{(2\alpha)}}, \zeta = 7\sqrt{(\frac{1}{2}\alpha)}$
(c) $\omega = 1.78, \zeta = 0.12$ (d) $\omega = 5\eta, \zeta = 4$
(e) $\omega = 0.51, \zeta = 2.48$

67 (a) $\alpha = \pi, \beta = \pi^2$;   (b) $a = 0.4\pi, b = 4\pi^2$
(c) $q = 8, r = 4$ (d) $a = \dfrac{7}{2\pi^2}, b = \dfrac{28}{\pi}$

68 $\Omega_{max} = \omega\sqrt{(1 - 2\zeta^2)}$, only exists if $\zeta^2 < \frac{1}{2}$
$\qquad A(\Omega_{max}) = \dfrac{1}{2\zeta\omega^2\sqrt{(1 - \zeta^2)}}$

69 $2.52\,\mathrm{m\,s^{-1}}$ (approximately 5 knots)

70 $\mu > 621\,\mathrm{N\,m^{-1}\,s}$

71 $73\,\mathrm{pF} > C > 7\,\mathrm{pF}$

72 (a) $\dfrac{dx}{dt} = v,$ $\qquad\qquad x(0) = 1$

$\dfrac{dv}{dt} = 4xt - 6(x^2 - t)v,$ $\quad v(0) = 2$

(b) $\dfrac{dx}{dt} = v,$ $\qquad\qquad x(0) = 0$

$\dfrac{dv}{dt} = \sin v - 4x,$ $\quad v(0) = 0$

73 $X(0.3) = 0.29990$

74 (a) $\dfrac{dx}{dt} = v,$ $\qquad\qquad x(1) = 2$

$\dfrac{dv}{dt} = -4\sqrt{(x^2 - t^2)},$ $\quad v(1) = 0.5$

(b) $\dfrac{dx}{dt} = v,$ $\qquad\qquad x(0) = 1$

$\dfrac{dv}{dt} = w,$ $\qquad\qquad v(0) = 2$

$\dfrac{dw}{dt} = e^{2t} + x^2 t - 6e^t v - tw,$ $\quad w(0) = 0$

(c) $\dfrac{dx}{dt} = v,$ $\qquad\qquad x(1) = 1$

$\dfrac{dv}{dt} = w,$ $\qquad\qquad v(1) = 0$

$\dfrac{dw}{dt} = \sin t - x^2 - tw,$ $\quad w(1) = -2$

(d) $\dfrac{dx}{dt} = v,$ $\qquad\qquad x(2) = 0$

$\dfrac{dv}{dt} = w,$ $\qquad\qquad v(2) = 0$

$\dfrac{dw}{dt} = (x^2 t^2 + tw)^2,$ $\quad w(2) = 2$

(e) $\dfrac{dx}{dt} = v,$ $\qquad\qquad x(0) = 0$

$\dfrac{dv}{dt} = w,$ $\qquad\qquad v(0) = 0$

$\dfrac{dw}{dt} = u,$ $\qquad\qquad w(0) = 4$

$\dfrac{du}{dt} = \ln t - x^2 - xw,$ $\quad u(0) = -3$

(f) $\dfrac{dx}{dt} = v,$ $\qquad\qquad x(0) = a$

$\dfrac{dv}{dt} = w,$ $\qquad\qquad v(0) = 0$

$\dfrac{dw}{dt} = u,$ $\qquad\qquad w(0) = b$

$\dfrac{du}{dt} = t^2 + 4t - 5 + \sqrt{(xt)} - v - (v - 1)u,$ $\quad u(0) = 0$

75 $X(0.65) = -0.83463$

76 $X_{0.01}(0.4) = 0.398022$
$X_{0.005}(0.4) = 0.397919$
step size required is <0.0024
$X_{0.002}(0.4) = 0.397856$

77 s tends to around 6.3%. With double the inflow s tends to about 11.1%.

10.13 Review exercises

1 (a) Second-order nonlinear ordinary differential equation, dependent variable x, independent variable t
(b) Third-order linear homogeneous partial differential equation, dependent variable h, independent variables t and v
(c) First-order nonlinear ordinary differential equation, dependent variable z, independent variable x
(d) Third-order linear nonhomogeneous ordinary differential equation, dependent variable p, independent variable s

2 (a) Under-determined, $x = \frac{1}{6}t^3 + At + 1$
(b) Fully determined, $x(t) = \frac{1}{24}t^4 - \frac{7}{24}t^2 + \frac{1}{4}t$
(c) Over-determined, no solution exists
(d) Fully determined, $x = \frac{1}{16}e^{4t} - \frac{1}{4}te^4 - \frac{1}{16}$

3 $x(t) = \dfrac{C}{\sqrt{(C + e^{-2at})}}$

4 (a) $x(t) = \cos^{-1}(\sin t - 1)$ $\qquad$ **(b)** $x(t) = \ln(\ln t + e^2)$
(c) $x(t) = e^{(t^3 - 8)/3}$ $\qquad$ **(d)** $x(t) = t\cos^{-1}(\cos 1 - \ln t)$
(e) $x(t) = -\frac{1}{2}[t \pm \sqrt{(17t^2 + 16)}]$ $\qquad$ **(f)** $x(t) = t2^t$
(g) $x(t) = t(3 - \ln t)$ $\qquad$ **(h)** $x(t) = t \pm \sqrt{(4 - 6t^2)}$

5 (a) $x(t) = \dfrac{\sqrt{[4 + a(t - 1)]}}{t}$ $\qquad$ **(b)** not exact

(c) $x(t) = \dfrac{\sin^{-1}(\pi - t)}{t}$ $\qquad$ **(d)** not exact

(e) $x(t) = \dfrac{\ln(2 + e^8 - t)}{t}$

6 (a) $x(t) = \frac{9}{4}e^{2t} - \frac{1}{2}t - \frac{1}{4}$ $\quad$ **(b)** $x(t) = \frac{1}{2}(e^{-t} + e^{-t^2})$
(c) $x(t) = \frac{1}{5}(e^{2t} + 9e^{-3t})$ $\quad$ **(d)** $x(t) = 1 + (e - 1)e^{\cos t + 1}$

7 $X_{0.1}(0.4) = 1.125583,\ X_{0.05}(0.4) = 1.142763$
Richardson extrapolation estimates the error as 0.017180, so, to obtain an error less than 5×10^{-3}, a step less than 0.0146 should be used

8 $X_{0.05}(0.25) = 2.003749,\ X_{0.025}(0.25) = 2.004452$
Richardson extrapolation estimates the error as 0.000703, so, to obtain an error less than 5×10^{-4}, a step less than 0.0179 should be used

9 $x(t) = (20 - t) - \dfrac{(20 - t)^3}{400}$

11 $\alpha = 4kT_0^3$

13 $y(t) = y_0 + C\sqrt{(x - x_0)}$

14 Half life is $\ln 2/k$

15 Time to 95% of final value is $\ln(20)L/R$

16 Tyre life is approximately $29\,500$ miles

17 (a) $L = \dfrac{d^2}{dt^2} + \sin t \dfrac{d}{dt} - 9, \quad f(t) = -\cos t$

    (b) $L = \dfrac{d^3}{dt^3} + t\dfrac{d^2}{dt^2} + t(t-4)\dfrac{d}{dt} + 1, \quad f(t) = -e^t$

    (c) $L = \dfrac{d}{dt} - e^{-t}, \quad f(t) = e^t$

    (d) $L = \dfrac{d^2}{dt^2} + 4, \quad f(t) = \cos \Omega t$

    (e) $L = t^2 \dfrac{d^3}{dt^3} - \dfrac{1}{t^2 + 2t + 4}\dfrac{d}{dt}, \quad f(t) = -\ln(t^2 + 4)$

18 (a) $\sin t - \cos t$ (b) 0 (commutative)

    (c) 0 (commutative)   (d) $2\dfrac{d}{dt}$

19 (a) $\dfrac{df}{dt} = \dfrac{dg}{dt}$ (b) $\dfrac{df}{dt} = \dfrac{dg}{dt}$ and $\dfrac{d^2f}{dt^2} = \dfrac{d^2g}{dt^2}$

20 (a) $x(t) = Ae^t + Be^{2t} + \frac{1}{10}\sin t + \frac{3}{10}\cos t$

    (b) $x(t) = Ae^t + Be^{2t} + Ce^{-3t} + \frac{1}{6}t + \frac{7}{36}$

    (c) $x(t) = Ae^t + Be^{2t} + Ce^{-3t} + \frac{1}{5}te^{2t}$

    (d) $x(t) = Ae^{4t} + te^{4t}$

    (e) $x(t) = e^{-3t/2}(A\sin t + B\cos t) + \frac{4}{13}t^2 - \frac{96}{169}t + \frac{736}{2197}$

    (f) $x(t) = e^{-3t/2}(A\sin t + B\cos t) - \frac{16}{75}\cos t + \frac{4}{25}\sin t$

    (g) $x(t) = Ae^{2t} + Be^{4t} + Ce^{-t} + \frac{1}{8}t^2 - \frac{3}{16}t + \frac{13}{64}$

    (h) $x(t) = e^t(A\cos 2t + B\sin 2t) + \frac{1}{8}e^{-t}$

    (i) $x(t) = Ae^{2t} + Be^{4t} + Ce^{-t} - \frac{1}{6}te^{2t} + \frac{1}{6}e^t$

    (j) $x(t) = e^t(A\cos 2t + B\sin 2t) + \frac{1}{5}t + \frac{2}{25} + \frac{1}{4}te^t\sin 2t$

21 (a) $x(t) = \frac{1}{5}(1 - e^{-t}\cos 2t - \frac{1}{2}e^{-t}\sin 2t)$

    (b) $x(t) = -2t + 5 + \frac{7}{2}e^t - \frac{3}{2}e^{-t/3}$

    (c) $x(t) = (12e^{-t} + 30te^{-t} - 12\cos 2t + 16\sin 2t)/25$

    (d) $x(t) = 3e^t - 2 - e^{2t}$

    (e) $x(t) = -\frac{7}{5}e^t + \frac{4}{3}e^{2t} + \frac{1}{15}e^{-4t}$

    (f) $x(t) = \frac{1}{13} + \frac{8}{5}e^{-t} - \dfrac{e^{-2t}}{65}(44\cos 3t - 27\sin 3t)$

22 (a) $\omega = \sqrt{2}, \zeta = \dfrac{7}{2\sqrt{2}}$ (b) $\omega = p^{1/4}, \zeta = \frac{1}{2}p^{3/4}$

    (c) $\omega = \dfrac{\sqrt{q}}{\sqrt{2}}, \zeta = a\sqrt{(2q)}$

    (d) $\omega = \sqrt{(2\alpha)}, \zeta = \dfrac{7}{\sqrt{(2\alpha)}}$

23 (a) $a = 2, b = 4$ (b) $a = 4\pi, b = \pi^2$

    (c) $a = 2, c = 8$ (d) $p = 150, q = 6\sqrt{2}$

24 $x(t) = t - Ce^{-t} + D$

    (a) $x(t) = \frac{1}{12}e^{-2t} + Ce^{4t} + D$

    (b) $x(t) = -\ln(\cos(t + C)) + D$

    (c) $x(t) = Ct^3 + D$

25 (a) $x(t) = \dfrac{t^3}{3k} - \dfrac{t^2}{k^2} + \dfrac{2t}{k^3} + \dfrac{(k^3 - 2)}{k^4}(1 - e^{-kt})$

    (b) $x(t) = \dfrac{U}{k - U}\ln\left(\dfrac{U}{k} + \dfrac{k - U}{k}e^{kt}\right)$

    (c) $x(t) = \frac{2}{15}t^3 + \frac{8}{5}t - \frac{26}{15}$

    (d) $x(t) = -\frac{4}{17}\cos t - \frac{1}{17}\sin t - \frac{4}{17}e^{-4(t-\pi)}$

26 $x(t) = C\tan(\frac{1}{2}Ct + D)$

    (a) $x(t) = Ae^{pt} + B$

    (b) $x(t) = -\ln(t + C) + D$

    (c) $x(t) = \pm\sqrt{(C - \ln(D - t))}$

27 (a) $x(t) = \left[\dfrac{(1 - p)t + 4}{4^p}\right]^{\frac{1}{1-p}}$

    (b) $x(t) = 1$

    (c) $x(t) = (\frac{1}{4} - \frac{2}{3}t)^{-1/2}$

    (d) $x(t) = \frac{1}{2}t^2 + 1$

28 Length of runway is $\dfrac{m}{2(\mu\alpha - \beta)}\ln\left[\dfrac{\mu\alpha - \beta}{T - \mu mg}V_2^2 + l\right]$

Time to take off is $\dfrac{m}{\sqrt{((\mu\alpha - \beta)(T - \mu mg))}}$

$$\arctan\left[\sqrt{\left(\dfrac{\mu\alpha - \beta}{T - \mu mg}\right)}V_2\right]$$

29 $X_{0.025}(2) = 0.847\,035, X_{0.0125}(2) = 0.844\,066$
Richardson extrapolation estimates the error as
$0.002\,969$, so we have $X(2) = 0.84$

32 $R = 2\sqrt{\dfrac{L}{C}}$

CHAPTER 11

Exercises

1 (a) $\dfrac{s}{s^2 - 4}$, $\text{Re}(s) > 2$ (b) $\dfrac{2}{s^3}$, $\text{Re}(s) > 0$

    (c) $\dfrac{3s + 1}{s^2}$, $\text{Re}(s) > 0$ (d) $\dfrac{1}{(s + 1)^2}$, $\text{Re}(s) > -1$

2 (a) 5   (b) -3 (c) 0   (d) 3   (e) 2
    (f) 0   (g) 0   (h) 0   (i) 2   (j) 3

3 (a) $\dfrac{5s - 3}{s^2}$, $\text{Re}(s) > 0$

    (b) $\dfrac{42}{s^4} - \dfrac{6}{s^2 + 9}$, $\text{Re}(s) > 0$

(c) $\dfrac{3s-2}{s^2}+\dfrac{4s}{s^2+4}$, Re(s) > 0

(d) $\dfrac{s}{s^2-9}$, Re(s) > 3

(e) $\dfrac{2}{s^2-4}$, Re(s) > 2

(f) $\dfrac{5}{s+2}+\dfrac{3}{s}-\dfrac{2s}{s^2+4}$, Re(s) > 0

(g) $\dfrac{4}{(s+2)^2}$, Re(s) > −2

(h) $\dfrac{4}{s^2+6s+13}$, Re(s) > −3

(i) $\dfrac{2}{(s+4)^3}$, Re(s) > −4

(j) $\dfrac{36-6s+4s^2-2s^3}{s^4}$, Re(s) > 0

(k) $\dfrac{2s+15}{s^2+9}$, Re(s) > 0

(l) $\dfrac{s^2-4}{(s^2+4)^2}$, Re(s) > 0

(m) $\dfrac{18s^2-54}{(s^2+9)^3}$, Re(s) > 0

(n) $\dfrac{2}{s^3}-\dfrac{3s}{s^2+16}$, Re(s) > 0

(o) $\dfrac{2}{(s+2)^3}+\dfrac{s+1}{s^2+2s+5}+\dfrac{3}{s}$, Re(s) > 0

4 (a) $\frac14(e^{-3t}-e^{-7t})$ (b) $-e^{-t}+2e^{3t}$

(c) $\frac49-\frac13 t-\frac49 e^{-3t}$ (d) $2\cos 2t+3\sin 2t$

(e) $\frac{1}{64}(4t-\sin 4t)$ (f) $e^{-2t}(\cos t+6\sin t)$

(g) $\frac18(1-e^{-2t}\cos 2t+3e^{-2t}\sin 2t)$ (h) $e^t-e^{-t}-2te^{-t}$

(i) $e^{-t}(\cos 2t+3\sin 2t)$ (j) $\frac12 e^t-3e^{2t}+\frac{11}{2}e^{3t}$

(k) $-2e^{-3t}+2\cos(\sqrt{2}t)-\sqrt{\frac12}\sin(\sqrt{2}t)$

(l) $\frac15 e^t-\frac15 e^{-t}(\cos t-3\sin t)$

(m) $e^{-t}(\cos 2t-\sin 2t)$ (n) $\frac12 e^{2t}-2e^{3t}+\frac32 e^{-4t}$

(o) $-e^t+\frac32 e^{2t}-\frac12 e^{-2t}$ (p) $4-\frac92\cos t+\frac12\cos 3t$

(q) $9e^{-2t}-e^{-3t/2}[7\cos(\frac12\sqrt3 t)-\sqrt3\sin(\frac12\sqrt3 t)]$

(r) $\frac19 e^{-t}-\frac1{10}e^{-2t}-\frac1{90}e^{-t}(\cos 3t+3\sin 3t)$

5 (a) $x(t)=e^{-2t}+e^{-3t}$

(b) $x(t)=\frac{35}{78}e^{4t/3}-\frac{3}{26}(\cos 2t+\frac23\sin 2t)$

(c) $x(t)=\frac15(1-e^{-t}\cos 2t-\frac12 e^{-t}\sin 2t)$

(d) $y(t)=\frac1{25}(12e^{-t}+30te^{-t}-12\cos 2t+16\sin 2t)$

(e) $x(t)=-\frac75 e^t+\frac43 e^{2t}+\frac1{15}e^{-4t}$

(f) $x(t)=e^{-2t}(\cos t+\sin t+3)$

(g) $x(t)=\frac{13}{12}e^t-\frac13 e^{-2t}+\frac14 e^{-t}(\cos 2t-3\sin 2t)$

(h) $y(t)=-\frac23+t+\frac23 e^{-t}[\cos(\sqrt2 t)+\sqrt{\frac12}\sin(\sqrt2 t)]$

(i) $x(t)=(\frac18+\frac34 t)e^{-2t}+\frac12 t^2 e^{-2t}+\frac38-\frac12 t+\frac14 t^2$

(j) $x(t)=\frac15-\frac15 e^{-2t/3}(\cos\frac13 t+2\sin\frac13 t)$

(k) $x(t)=te^{-4t}-\frac12\cos 4t$

(l) $y(t)=e^{-t}+2te^{-2t/3}$

(m) $x(t)=\frac54+\frac12 t-e^t+\frac5{12}e^{2t}-\frac23 e^{-t}$

(n) $x(t)=\frac9{20}e^{-t}-\frac7{16}\cos t+\frac{25}{16}\sin t-\frac1{80}\cos 3t$
  $-\frac3{80}\sin 3t$

6 (a) $x(t)=\frac14(\frac{15}{4}e^{3t}-\frac{11}{4}e^t-e^{-2t})$, $y(t)=\frac18(3e^{3t}-e^t)$

(b) $x(t)=5\sin t+5\cos t-e^t-e^{2t}-3$
  $y(t)=2e^t-5\sin t+e^{2t}-3$

(c) $x(t)=3\sin t-2\cos t+e^{-2t}$
  $y(t)=-\frac72\sin t+\frac92\cos t-\frac12 e^{-3t}$

(d) $x(t)=\frac32 e^{t/3}-\frac12 e^t$, $y(t)=-1+\frac12 e^t+\frac32 e^{t/3}$

(e) $x(t)=2e^t+\sin t-2\cos t$
  $y(t)=\cos t-2\sin t-2e^t$

(f) $x(t)=-3+e^t+3e^{-t/3}$
  $y(t)=t-1-\frac12 e^t+\frac32 e^{-t/3}$

(g) $x(t)=2t-e^t+e^{-2t}$, $y(t)=t-\frac72+3e^t+\frac12 e^{-2t}$

(h) $x(t)=3\cos t+\cos(\sqrt3 t)$
  $y(t)=3\cos t-\cos(\sqrt3 t)$

(i) $x(t)=\cos(\sqrt{\frac3{10}}t)+\frac34\cos(\sqrt6 t)$
  $y(t)=\frac54\cos(\sqrt{\frac3{10}}t)-\frac14\cos(\sqrt6 t)$

(j) $x(t)=\frac13 e^t+\frac23\cos 2t+\frac13\sin 2t$
  $y(t)=\frac23 e^t-\frac23\cos 2t-\frac13\sin 2t$

7 $I_1(s)=\dfrac{E_1(50+s)s}{(s^2+10^4)(s+100)^2}$

$I_2(s)=\dfrac{Es^2}{(s^2+10^4)(s+100)^2}$

$i_2(t)=E(-\frac1{200}e^{-100t}+\frac12 te^{-100t}+\frac1{200}\cos 100t)$

9 $i_1(t)=20\sqrt{\frac17}e^{-t/2}\sin(\frac12\sqrt7 t)$

10 $x_1(t)=-\frac32\cos(\sqrt3 t)-\frac7{10}\cos(\sqrt{13}t)$
  $x_2(t)=-\frac12\cos(\sqrt3 t)+\frac32\cos(\sqrt{13}t)$, $\sqrt3$, $\sqrt{13}$

11.5 Review exercises

1 (a) $x(t)=\cos t+\sin t-e^{-2t}(\cos t+3\sin t)$

(b) $x(t)=-3+\frac{13}{7}e^t+\frac{15}{7}e^{-2t/5}$

2 (a) $e^{-t}-\frac12 e^{-2t}-\frac12 e^{-t}(\cos t+\sin t)$

(b) $i(t)=4e^{-t}-3e^{-2t}$
  $+V[e^{-t}-\frac12 e^{-2t}-\frac12 e^{-t}(\cos t+\sin t)]$

3 $x(t)=-t+5\sin t-2\sin 2t$,
  $y(t)=1-2\cos t+\cos 2t$

4 $\frac15(\cos t+2\sin t)$
  $e^{-t}[(x_0-\frac15)\cos t+(x_1+x_0-\frac35)\sin t]$
  $\sqrt{\frac15}$, 63.4° lag

6 (a) (i) $\dfrac{s\cos\phi - \omega\sin\phi}{s^2 + \omega^2}$

(ii) $\dfrac{s\sin\phi + \omega(\cos\phi + \sin\phi)}{s^2 + 2\omega s + \omega^2}$

(b) $\frac{1}{20}(\cos 2t + 2\sin 2t) + \frac{1}{20}e^{-2t}(39\cos 2t + 47\sin 2t)$

7 (a) $e^{-2t}(\cos 3t - 2\sin 3t)$

(b) $y(t) = 2 + 2\sin t - 5e^{-2t}$

8 $x(t) = e^{-8t} + \sin t, \; y(t) = e^{-8t} - \cos t$

9 $q(t) = \frac{1}{500}(5e^{-100t} - 2e^{-200t}) - \frac{1}{500}(3\cos 100t - \sin 100t)$, current leads by approximately $18.5°$

10 $x(t) = \frac{29}{20}e^{-t} + \frac{445}{1212}e^{-t/5} + \frac{1}{3}e^{-2t}$
$\qquad - \frac{1}{505}(76\cos 2t - 48\sin 2t)$

11 (a) $\theta(t) = \frac{1}{100}(4e^{-4t} + 10te^{-4t} - 4\cos 2t + 3\sin 2t)$

(b) $i_1(t) = \frac{1}{7}(e^{4t} + 6e^{-3t}), \; i_2 = \frac{1}{7}(e^{-3t} - e^{4t})$

12 $i(t) = \dfrac{E}{R}[1 - e^{-nt}(\cos nt - \sin nt)]$

13 $i_1(t) = \dfrac{E(4 - 3e^{-Rt/L} - e^{-3Rt/L})}{6R}, \; i_2(t) \to E/3R$

14 $x_1(t) = \frac{1}{3}[\sin t - 2\sin 2t + \sqrt{3}\sin(\sqrt{3}t)]$
$\qquad x_2(t) = \frac{1}{3}[\sin t + \sin 2t - \sqrt{3}\sin(\sqrt{3}t)]$

15 (a) (i) $e^{-t}(\cos 3t + \sin 3t)$ **(ii)** $e^t - e^{2t} + 2te^t$

(b) $y(t) = \frac{1}{2}e^{-t}(8 + 12t + t^3)$

16 (a) $\frac{5}{2}e^{7t}\sin 2t$

(b) $\dfrac{n^2 i}{Ks(s^2 + 2Ks + n^2)}, \; \theta(t) = \frac{i}{K}(1 - e^{-Kt}) - ite^{-Kt}$

17 (a) $v_1 = 250e^{-0.1t}, \; v_2 = (50 + 25t)e^{-0.1t}$

(b) $t = 23.026$

CHAPTER 12

Exercises

1 (a) $f(t) = -\dfrac{1}{4}\pi - \dfrac{2}{\pi}\sum_{n=1}^{\infty}\dfrac{\cos(2n-1)t}{(2n-1)^2}$
$\qquad + \sum_{n=1}^{\infty}\left[\dfrac{3\sin(2n-1)t}{2n-1} - \dfrac{\sin 2nt}{2n}\right]$

(b) $f(t) = \dfrac{1}{4}\pi + \dfrac{2}{\pi}\sum_{n=1}^{\infty}\dfrac{\cos(2n-1)t}{(2n-1)^2} - \sum_{n=1}^{\infty}\dfrac{\sin nt}{n}$

(c) $f(t) = \dfrac{2}{\pi}\sum_{n=1}^{\infty}\dfrac{\sin nt}{n}$

(d) $f(t) = \dfrac{2}{\pi} + \dfrac{4}{\pi}\sum_{n=1}^{\infty}\dfrac{(-1)^{n+1}\cos 2nt}{4n^2 - 1}$

(e) $f(t) = \dfrac{2}{\pi} + \dfrac{4}{\pi}\sum_{n=1}^{\infty}\dfrac{(-1)^{n+1}\cos nt}{4n^2 - 1}$

(f) $f(t) = \dfrac{1}{2}\pi - \dfrac{4}{\pi}\sum_{n=1}^{\infty}\dfrac{\cos(2n-1)t}{(2n-1)^2}$

(g) $f(t) = -\dfrac{4}{\pi}\sum_{n=1}^{\infty}\dfrac{\cos(2n-1)t}{(2n-1)^2} - \sum_{n=1}^{\infty}\dfrac{\sin 2nt}{n}$

(h) $f(t) = \left(\dfrac{1}{2}\pi + \dfrac{1}{\pi}\sinh\pi\right)$
$\qquad + \dfrac{2}{\pi}\sum_{n=1}^{\infty}\left[\dfrac{(-1)^n - 1}{n^2} + \dfrac{(-1)^n\sinh\pi}{n^2 + 1}\right]\cos nt$
$\qquad - \dfrac{2}{\pi}\sum_{n=1}^{\infty}\dfrac{n(-1)^n}{n^2 + 1}\sinh\pi\sin nt$

2 $f(t) = \dfrac{1}{3}\pi^2 + 4\sum_{n=1}^{\infty}\dfrac{\cos nt}{n^2}$
Taking $t = \pi$ gives the required result.

3 $q(t) = Q\left[\dfrac{1}{2} - \dfrac{4}{\pi^2}\sum_{n=1}^{\infty}\dfrac{\cos(2n-1)t}{(2n-1)^2}\right]$

4 $f(t) = \dfrac{5}{\pi} + \dfrac{5}{2}\sin t - \dfrac{10}{\pi}\sum_{n=1}^{\infty}\dfrac{\cos 2nt}{4n^2 - 1}$

5 Taking $t = 0$ and $t = \pi$ gives the required answers.

6 $f(t) = \dfrac{1}{4}\pi - \dfrac{2}{\pi}\sum_{n=1}^{\infty}\dfrac{\cos(4n-2)t}{(2n-1)^2}$
Taking $t = 0$ gives the required series.

7 $f(t) = \dfrac{3}{2} + \dfrac{4}{\pi^2}\sum_{n=1}^{\infty}\dfrac{\cos(2n-1)t}{(2n-1)^2}$
Replacing t by $t - \frac{1}{2}\pi$ gives the following sine series of odd harmonics:
$f\left(t - \dfrac{1}{2}\pi\right) - \dfrac{3}{2} = -\dfrac{4}{\pi^2}\sum_{n=1}^{\infty}\dfrac{(-1)^n\sin(2n-1)t}{(2n-1)^2}$

8 $f(t) = \dfrac{2l}{\pi}\sum_{n=1}^{\infty}\dfrac{(-1)^{n+1}}{n}\sin\dfrac{n\pi t}{l}$

9 $f(t) = \dfrac{2K}{\pi}\sum_{n=1}^{\infty}\dfrac{1}{n}\sin\dfrac{n\pi t}{l}$

10 $f(t) = \dfrac{3}{2} + \dfrac{6}{\pi}\sum_{n=1}^{\infty}\dfrac{1}{(2n-1)}\dfrac{\sin(2n-1)\pi t}{5}$

11 $v(t) = \dfrac{A}{\pi}\left(1 + \dfrac{1}{2}\pi\sin\omega t - 2\sum_{n=1}^{\infty}\dfrac{\cos 2n\omega t}{4n^2 - 1}\right)$

12 $f(t) = \dfrac{1}{3}T^2 + \dfrac{4T^2}{\pi^2}\sum_{n=1}^{\infty}\dfrac{(-1)^n}{n^2}\cos\dfrac{n\pi t}{T}$

13 $e(t) = \dfrac{E}{2}\left(1 - \dfrac{2}{\pi}\sum_{n=1}^{\infty}\dfrac{1}{n}\sin\dfrac{2\pi nt}{T}\right)$

15 $f(t) = -\dfrac{8}{\pi^2} \displaystyle\sum_{n=1}^{\infty} \dfrac{1}{(2n-1)^2} \cos(2n-1)\pi t$

16 (a) $f(t) = \dfrac{2}{3} - \dfrac{1}{\pi^2} \displaystyle\sum_{n=1}^{\infty} \dfrac{1}{n^2} \cos 2n\pi t + \dfrac{1}{\pi} \displaystyle\sum_{n=1}^{\infty} \dfrac{1}{n} \sin 2n\pi t$

(b) $f(t) = \dfrac{1}{\pi} \displaystyle\sum_{n=1}^{\infty} \dfrac{1}{n} \sin 2n\pi t$

$\qquad + \dfrac{2}{\pi} \displaystyle\sum_{n=1}^{\infty} \left[\dfrac{1}{2n-1} + \dfrac{4}{\pi^2(2n-1)^3} \right]$

$\qquad \times \sin(2n-1)\pi t$

(c) $f(t) = \dfrac{2}{3} + \dfrac{4}{\pi^2} \displaystyle\sum_{n=1}^{\infty} \dfrac{(-1)^{n+1}}{n^2} \cos n\pi t$

17 $f(t) = \dfrac{1}{6}\pi^2 - \displaystyle\sum_{n=1}^{\infty} \dfrac{1}{n^2} \cos 2nt$

$f(t) = \dfrac{8}{\pi} \displaystyle\sum_{n=1}^{\infty} \dfrac{1}{(2n-1)^3} \sin(2n-1)t$

18 $f(x) = \dfrac{8a}{\pi^2} \displaystyle\sum_{n=1}^{\infty} \dfrac{(-1)^{n+1}}{(2n-1)^2} \sin \dfrac{(2n-1)\pi x}{l}$

19 $f(x) = \dfrac{2l}{\pi^2} \displaystyle\sum_{n=1}^{\infty} \dfrac{(-1)^{n+1}}{(2n-1)^2} \sin \dfrac{2(2n-1)\pi x}{l}$

20 $f(t) = \dfrac{1}{2}\sin t + \dfrac{4}{\pi} \displaystyle\sum_{n=1}^{\infty} \dfrac{n(-1)^{n+1}}{4n^2-1} \sin 2nt$

21 $f(x) = -\dfrac{1}{2}A - \dfrac{4A}{\pi^2} \displaystyle\sum_{n=1}^{\infty} \dfrac{1}{(2n-1)^2} \cos \dfrac{(2n-1)\pi x}{l}$

22 $T(x) = \dfrac{8KL^2}{\pi^3} \displaystyle\sum_{n=1}^{\infty} \dfrac{1}{(2n-1)^3} \sin \dfrac{(2n-1)\pi x}{L}$

23 $f(t) = \dfrac{1}{2} + \dfrac{1}{2}\cos \pi t + \dfrac{4}{\pi} \displaystyle\sum_{n=1}^{\infty} \dfrac{1}{4n^2-1} \sin 2n\pi t$

$\qquad - \dfrac{2}{\pi} \displaystyle\sum_{n=1}^{\infty} \dfrac{1}{2n-1} \sin(2n-1)\pi t$

26 (c) $1 + 4 \displaystyle\sum_{n=1}^{\infty} \dfrac{(-1)^{n+1}}{n} \sin nt$

12.6 Review exercises

1 $f(t) = \dfrac{1}{6}\pi^2 + \displaystyle\sum_{n=1}^{\infty} \dfrac{2}{n^2}(-1)^n \cos nt$

$\qquad + \displaystyle\sum_{n=1}^{\infty} \left[\dfrac{\pi}{2n-1} - \dfrac{4}{\pi(2n-1)^3} \right] \sin(2n-1)t$

$\qquad - \displaystyle\sum_{n=1}^{\infty} \dfrac{\pi}{2n} \sin 2nt$

Taking $T = \pi$ gives the required sum.

2 $f(t) = \dfrac{1}{9}\pi$

$\qquad + \dfrac{2}{\pi} \displaystyle\sum_{n=1}^{\infty} \dfrac{1}{n^2} \left\{ \cos \dfrac{1}{3}n\pi - \dfrac{1}{3}[2 + (-1)^n] \right\} \cos nt; \ \tfrac{2}{9}\pi$

3 (a) $f(t) = \dfrac{2T}{\pi^2} \displaystyle\sum_{n=1}^{\infty} \dfrac{(-1)^{n+1}}{(2n-1)^2} \sin \dfrac{2(2n-1)\pi t}{T}$

(b) $-\tfrac{1}{4}T$

(c) Taking $t = \tfrac{1}{4}T$ gives $S = \tfrac{1}{8}\pi^2$

4 $y = \dfrac{4P}{\pi\alpha} \displaystyle\sum_{n=1}^{\infty} \dfrac{1}{(2n-1)^2} \sin(2n-1)\alpha \sin(2n-1)x$

6 $f(t) = \dfrac{4}{\pi} \displaystyle\sum_{n=1}^{\infty} \dfrac{(-1)^n \sin(2n-1)t}{(2n-1)^2}$

8 $f(x) = \dfrac{4}{\pi} \displaystyle\sum_{n=1}^{\infty} \dfrac{\cos(2n-1)x}{(2n-1)^2}$

Taking $x = 0$ gives

$\pi^2 = 8 \displaystyle\sum_{n=1}^{\infty} \dfrac{1}{(2n-1)^2}$

9 $f(x) = \displaystyle\sum_{n=1}^{\infty} \dfrac{1}{(2n-1)} \left[1 + \dfrac{2(-1)^{n+1}}{\pi(2n-1)} \right] \sin(2n-1)x$

$\qquad - \displaystyle\sum_{n=1}^{\infty} \dfrac{1}{2n} \sin 2nx$

10 $V = \tfrac{25}{3}(1 - e^{-1.2}) + \displaystyle\sum_{n=1}^{\infty} \dfrac{50(1 - e^{-1.2})}{9 + 25n^2\pi^2}$

$\qquad \times (3 \cos 5n\pi t + 5n\pi \sin 5n\pi t)$

Amplitude of the nth harmonic is

$\dfrac{50(1 - e^{-1.2})}{\sqrt{(9 + 25n^2\pi^2)}} \approx \dfrac{50(1 - e^{-1.2})}{5n\pi} \approx \dfrac{2 \cdot 22}{n}$

13 $f(x) = \dfrac{4}{\pi} \displaystyle\sum_{n=1}^{\infty} \dfrac{(-1)^{n+1}}{(2n-1)^2} \sin(2n-1)x$

$f(x) = \dfrac{1}{4}\pi - \dfrac{2}{\pi} \displaystyle\sum_{n=1}^{\infty} \dfrac{\cos 2(2n-1)x}{(2n-1)^2}$

15 (a) $f(t) = \displaystyle\sum_{n=1}^{\infty} \dfrac{2}{n} \sin nt$

(b) $f(t) = \dfrac{1}{2}\pi + \dfrac{4}{\pi} \displaystyle\sum_{n=1}^{\infty} \dfrac{1}{(2n-1)^2} \cos(2n-1)t$

16 $f(t) = \tfrac{2}{3} + \dfrac{4}{\pi^2} \displaystyle\sum_{n=1}^{\infty} \dfrac{(-1)^{n+1}}{n^2} \cos n\pi t$

17 (a) (i) a constant term and cosine terms with even harmonics

(ii) constant, cosine and sine terms present

(iii) a constant term and sine terms with odd harmonics

(b) $f(t) = \dfrac{\pi^2}{24} + \dfrac{2}{\pi}\left(\dfrac{\pi^2}{4} - 2 \right)\cos t - \dfrac{1}{2}\cos 2t$

$\qquad - \dfrac{2}{\pi}\left(\dfrac{\pi^2}{12} - \dfrac{2}{27} \right)\cos 3t + \dfrac{1}{8}\cos 4t$

18 (a) $f(t) = \dfrac{1}{2}\pi - \dfrac{4}{\pi}\displaystyle\sum_{n=1}^{\infty}\dfrac{1}{(2n-1)^2}\cos(2n-1)t$

(b) $g(t) = \dfrac{4}{\pi}\displaystyle\sum_{n=1}^{\infty}\dfrac{1}{2n-1}\sin(2n-1)t$

19 $g(t) = \dfrac{4}{\pi}\displaystyle\sum_{n=1}^{\infty}\dfrac{1}{2n-1}\sin(2n-1)t$

$f(t) = 1 + g(t)$

CHAPTER 13

Exercises

6 (a) $A \cap B$ (b) $A \cup B$
(c) $S - A$ (d) $S - (A \cap B)$

7 (a) {car, bicycle, motorcycle, boat}
(b) {train}   (c) {car, motorcycle, boat}

8 (a) 0.7   (b) 0.8   (c) 0.5

9 $\dfrac{16}{2652}$

10 $P(\text{same values}) = \dfrac{1}{6}$, $P(\text{differ by at most 1}) = \dfrac{4}{9}$

12 (a) $\dfrac{1}{26}$ (b) $\dfrac{4}{13}$ (c) $\dfrac{1}{2}$ (d) $\dfrac{1}{13}$

13 (a) $\dfrac{1}{9}$ (b) $\dfrac{5}{18}$ (c) $\dfrac{5}{6}$

14 $P(\text{total} = 7 \,|\, 7 \text{ or } 10) = \dfrac{2}{3}$

15 (a) $\dfrac{3}{4}$ (b) 7 to 1

16 (a) $\dfrac{1}{2}$ (b) $\dfrac{1}{3}$

17 $\dfrac{5}{6}$

18 (a) 0.15   (b) 0.55   (c) 0.357

19 0.6

20 0.381

21 (a) $P(A) + P(B) - P(A \cap B)$

(b) $\dfrac{P(C|A)P(A) - P(C|A \cap B)P(A \cap B)}{P(A) - P(A \cap B)}$

(c) $\dfrac{P(C|A)P(A) + P(C|B)P(B) - P(C|A \cap B)P(A \cap B)}{P(A) + P(B) - P(A \cap B)}$

22 0.149

23 (a) $\left(1 - \dfrac{2r}{d}\right)^2$ (b) $1 - \left(\dfrac{2r}{d}\right)^2$

24 $P(2) = \dfrac{1}{36}$ $P(3) = \dfrac{2}{36}, \dots, P(7) = \dfrac{6}{36}$
$P(8) = \dfrac{5}{36}, \dots, P(12) = \dfrac{1}{36}$

26 (a) 0.488   (b) 0.3123

27 (b) $P(-3) = \dfrac{1}{8}$, $P(-1) = \dfrac{3}{8}$
$P(1) = \dfrac{3}{8}$, $P(3) = \dfrac{1}{8}$
(c) $P(-3) = \dfrac{1}{27}$, $P(-1) = \dfrac{6}{27}$
$P(1) = \dfrac{12}{27}$, $P(3) = \dfrac{8}{27}$

28 (a) $\dfrac{1}{4}$

(b) $F_X(x) = \begin{cases} 0 & (x < 0) \\ \dfrac{1}{2}\sqrt{x} & (0 \leqslant x \leqslant 4) \\ 1 & (x > 4) \end{cases}$

(c) $\dfrac{1}{2}$

29 $P(X \leqslant 30) = 0.28$

30 (a) $\dfrac{1}{9}$ (b) $\dfrac{3}{4}$ (c) 0.102

31 $1 - \exp(-x^2/2a)$, 0.0804

32 mean = 4.3, $P(\text{less than 5 days}) = 0.6$

33 mean = 1.8, median = 2,
standard deviation = 1.34

34 Average length = 5.88

35 mean = 5, median = 3
standard deviation = 4.47

36 mean = 30 minutes, standard deviation = 17.3 min

38 (a) 0.47   (b) $\mu_X = 30$, $\sigma_X = 30$
(c) median = 20.8, $q_3 - q_1 = 33.0$

39 0.969

41 24 hours, 3.32 hours

42 (a) $\bar{X} = 2.28$, $S_X = 0.60$, $S_{X,n-1} = 0.63$
(b) sample median = 2.1, range = 2.2

43 $\bar{X} = 5.44$, $S_X = 0.81$, median = 5.45, range = 3.2

44 $\bar{A} = 24.2$, $S_A = 1.76$, $\bar{T} = 16.7$, $S_T = 4.00$
$\bar{B} = 25.4$, $S_B = 1.66$, $\bar{U} = 18.1$, $S_U = 4.93$
where A, T are time, temperature for A, and B, U are
time, temperature for B

45 2.19

46 median = $\sqrt{[2a \ln 2]}$, mode = $\sqrt{a}$
$a = 6$: mean = 3.07, median = 2.88, mode = 2.45,
$q_3 - q_1 = 2.22$

47 $q/(p + q - pq)$

48 47.1 and 46.3

49 $P(\text{4 boys}) = 0.273$

50 0.998

51 0.677

52 (a) 0.1271   (b) 0.3594
(c) 0.1413   (d) 0.5876

53 4 engines

54 (a) 0.957   (b) 0.0071

55 0.027

56 $P(\text{8 or more}) = 0.249$

57 (a) 0.050   (b) 0.224   (c) 0.084

58 0.986

59 6.09

61 0.144

62 0.011

63 46

64 0.3%, 0.0258

65 (a) 0.102   (b) 0.128   (c) 0.011

66 Warning 9.5, action 13.5, sample 12
UCL = 11.4, sample 9

67 UK sample 28, US sample 25

68 P(at least one such area) = 0.133

69 P(at least one such area) = 0.688

13.8 Review exercises

1 (a) 3   (b) $F_X(x) = \begin{cases} 0 & \text{for } x \leqslant 1 \\ 1 - x^{-3} & \text{for } x > 1 \end{cases}$

(c) $\frac{1}{8}$ (d) $\frac{3}{2}$ (e) $\sqrt{3}/2$

2 0.0159

3 60, 6342 hours

4 $P(10, 5, 3, 2) = 0.009$

5 $e^{-\lambda}(1 + \lambda) > 0.9$, proportion = 0.0053

6 $\pm 5.66 \times 10^{-5}$

9 0.407

10 $E(\text{minimum}) = \dfrac{25}{12\lambda}$

$\qquad\qquad = \dfrac{25}{4}$ hours when $\lambda = \dfrac{1}{3}$

12 $E(\text{number of analyses}) = N\,[1 - (1 - p)^k] + \dfrac{N}{k}$

$\qquad\qquad\qquad = 0.196\,N$ when $k = 11$

13 (a) *single*
$k = 4$: $n = 7$, $P(\text{error}) = 0.0020$
$k = 8$: $n = 12$, $P(\text{error}) = 0.0062$
$k = 16$: $n = 21$, $P(\text{error}) = 0.0185$
$k = 32$: $n = 38$, $P(\text{error}) = 0.0555$
$k = 64$: $n = 71$, $P(\text{error}) = 0.1588$
double
$k = 4$: $n = 11$, $P(\text{error}) = 0.0002$
$k = 8$: $n = 17$, $P(\text{error}) = 0.0006$
$k = 16$: $n = 26$, $P(\text{error}) = 0.0022$
$k = 32$: $n = 43$, $P(\text{error}) = 0.0092$
$k = 64$: $n = 77$, $P(\text{error}) = 0.0424$
(b) *single*: $k = 8$, so total 96 bits
double: $k = 64$, so total 77 bits

Index

Italicised page numbers indicate where an entry has been defined in the text

N

O

P

Q

R

T

IMPORTANT: PLEASE READ CAREFULLY

WARNING BY OPENING THE PACKAGE YOU AGREE TO BE BOUND BY THE TERMS OF THE LICENCE AGREEMENT BELOW

SINGLE USER LICENCE AGREEMENT

ONE COPY ONLY

This licence is for a single user copy of the software
PEARSON EDUCATION LIMITED RESERVES THE RIGHT TO TERMINATE THIS LICENCE BY WRITTEN NOTICE AND TO TAKE ACTION TO RECOVER ANY DAMAGES SUFFERED BY PEARSON EDUCATION LIMITED IF YOU BREACH ANY PROVISION OF THIS AGREEMENT

Pearson Education Limited and its licensors own the software
You only own the disk(s) on which the software is supplied.